HTD-Vol. 106

HEAT TRANSFER PHENOMENA IN RADIATION, COMBUSTION, AND FIRES

presented at
THE 1989 NATIONAL HEAT TRANSFER CONFERENCE
PHILADELPHIA, PENNSYLVANIA
AUGUST 6–9, 1989

sponsored by
THE HEAT TRANSFER DIVISION

edited by
R. K. SHAH
GENERAL MOTORS CORPORATION

THE AMERICAN SOCIETY OF MECHANICAL ENGINEERS
United Engineering Center 345 East 47th Street New York, N.Y. 10017

Statement from By-Laws: The Society shall not be responsible for statements or opinions advanced in papers . . . or printed in its publications (7.1.3)

ISBN No. 0-7918-0348-1

PREFACE

RADIATION HEAT TRANSFER

A. M. Smith, The University of Mississippi
T. F. Smith, The University of Iowa

In response to the continued interest in radiation heat transfer research, the K-12 Committee (Committee on Aerospace Heat Transfer) of the ASME Heat Transfer Division is sponsoring a technical session on the subject at the 26th National Heat Transfer Conference in Philadelphia, Pennsylvania, August 6-9, 1989. There are twenty-one papers in this session. Both experimental and theoretical works are included covering various geometries and physical conditions. These papers are on many different topics lying within the following general areas: (1) spectral and grey radiative transfer in absorbing, scattering, diffracting, and emitting media including surface and induced effects; (2) radiation and film boiling heat transfer in energy systems; (3) radiative properties such as reflectance, transmittance, emittance, refractive index, absorption coefficients, and scattering phase function; and (4) the interaction of conduction and/or convection with radiation. The papers provide a representative cross section of current research on radiation heat transfer. We wish to thank the authors for their participation in this session and the many reviewers who assisted us in evaluating the manuscripts.

MULTIDIMENSIONAL RADIATIVE HEAT TRANSFER

A. L. Crosbie, University of Missouri-Rolla
W. A. Fiveland, Babcock and Wilcox

The analysis of multidimensional situations is one of the most challenging problems in radiative heat transfer. The fundamental difficulty is the mathematical complexity associated with carrying out a multidimensional analysis. This difficulty is evident even for the simplest two-dimensional geometry. The papers in this section illustrate the wide range of solution techniques used to solve multidimensional radiative heat transfer problems. The techniques include: the Monte Carlo method, a network analogy, the YIX method, the nodal approximation, the improved differential approximation, and the discrete exchange factor method. Except for one paper on remote sensing of temperature, all the papers deal with radiative transfer in an absorbing and scattering medium. Two of the five scattering papers consider anisotropic scattering; results for a linear scattering phase function are presented in one. Specific results are obtained for three important two-dimensional geometries; a finite cylinder, a truncated cone, and a square enclosure with a partition. Specific results also are obtained for a three-dimensional cubic enclosure. This section was sponsored by ASME Heat Transfer Division's K-8 Committee on Theory and Fundamental Research. The efforts of both authors and reviewers are very much appreciated. We hope that this section devoted to multidimensional radiative heat transfer will provide encouragement for further research in the area.

RADIATION IN ENERGY SYSTEMS

T. W. Tong, Arizona State University
K. Y. Wang, General Electric Company

In response to the need for a forum to address the latest developments in the area of radiation heat transfer in thermal engineering systems, the K-6 Committee of the ASME Heat Transfer Division sponsored a session on Radiation in Energy Systems at the 1989 National Heat Transfer Conference. Six papers were selected for presentation at the session. They typified the diversity of applications involving thermal radiation as a primary mode of heat transfer. The topics addressed by the papers included: seeding a liquid falling film with particles to provide direct absorption of solar energy; radiation effects in MHD generators; determination of particle composition and size distribution in a particle laden plume from infrared data; using sub-micron size fibers to improve the performance of porous radiant burners; development of a model for designing direct-fired batch reheating furnace; and development of a model to predict the heat transfer behavior of steam boilers. The readers should find the papers informative and useful for designing the investigated systems with improved performance. The session organizers would like to thank the authors for contributing their papers to the session and the reviewers for their assistance in evaluating the papers.

FLAME RADIATION PHENOMENA

W. W. Yuen, University of California at Santa Barbara

W. L. Grosshandler, Washington State University

This section is sponsored by the ASME Heat Transfer Division's K-11 Committee on Heat Transfer in Fire and Combustion Systems. The objective is to provide a forum for the presentation and discussion of recent advances in these areas. Three papers are presented. One paper deals with experimental measurements of flame radiation intensity by high-speed spectroscopy. The remaining two papers present analyses relating to the interaction of radiation and conduction in Fused Silica and the effect of turbulence fluctuation on radiative output of high temperature gas.

HEAT AND MASS TRANSFER IN FIRES

N. R. Keltner, Sandia National Laboratories

A. Kulkarni, The Pennsylvania State University

The papers in the session on Heat and Mass Transfer in Fires demonstrate the diverse aspects of fire science. The papers cover topics ranging from fundamental studies to large scale experimental investigations. There are papers on dimensional analysis, numerical modeling, simulation of fire generated flows, the development of techniques for measurements in both the flames and in the fuel, tank car safety, and reducing the effects of the wind on open pool fires. The session developers wish to thank all of the authors for their contributions and all of our colleagues who carefully reviewed the papers.

COMBUSTION FUNDAMENTALS

R. J. Santoro, The Pennsylvania State University

C. K. Law, Princeton University

The session on Combustion Fundamentals emphasizes basic combustion phenomena and measurements. Contributions include studies of tomographic reconstruction procedure for measurement analysis, dense spray ignition studies, and the study of the thermal structure of spray flames. These studies consider the basic mechanisms controlling the phenomena of interest and include both theoretical and modelling aspects of the problem.

COMBUSTION IN MULTIPHASE SYSTEMS

J. N. Chung, Washington State University

R. L. Alpert, Factory Mutual Research Corp.

The objective of this technical session is to provide a forum for the dissemination of new research results concerning multiphase transport phenomena in fire and combustion systems. The papers presented cover a wide range of topics including droplet evaporation at intermediate Reynolds numbers, enhancement of coal combustion by high-intensity acoustic fields, convective thermal ignition process of reactive solid particles, properties of turbulent reacting gas jets, and modelling of combustion in porous radiant burners. The breadth of coverage indicates the current interests in understanding transport processes in many combustion systems. This session is sponsored by the K-11 Committee on Heat Transfer in Fire and Combustion. The session organizers wish to express their sincere appreciation to the individual authors and to the many reviewers who contributed in making this session a worthwhile addition to the symposium literature.

MODELING OF COMBUSTION SYSTEMS

B. Farouk, Drexel University

S. Acharya, Louisiana State University

The physical and chemical processes in combustion systems like furnaces, combustors, kilns, boilers, engines, and other devices are complex. Modeling of such systems have recently provided useful design guidelines and valuable insights to their efficient operation. The papers in this section originated from a call for papers sponsored by ASME K-11 Committee on the modeling of combustion systems. The topics of the papers presented range from modeling fundamental problems associated with combustion systems to the modeling of large coal-fired utility boilers. One of the papers addresses fundamental issues on the modeling of interacting vaporizing fuel sprays. A model for combustion of single coal char particle is proposed in another one. Numerical models for the prediction of both diffusion and premixed turbulent reacting flows are presented in another paper. The two remaining papers are concerned with performance and modeling of combustion devices. Experimental results of burning methane in a laboratory-scale Rijke pulsating combustor is presented in one of them. The remaining paper describes a mathematical model of a boiler developed for use as a diagnostic tool and for optimized corrective design in problem resolution. We thank all the authors for their fine contributions and all those who carefully reviewed and significantly enhanced the quality of the papers.

HEAT AND MASS TRANSFER IN SPRAY SYSTEMS

C. Presser, National Institute of Standards and Technology

A. K. Gupta, University of Maryland

This section contains seven papers which were accepted and presented at the session on Heat and Mass Transfer in Spray Systems during the ASME National Heat Transfer Conference, held on August 6-9, 1989, in Philadelphia, Pennsylvania. The session was sponsored by the ASME K-6 Technical Committee on Heat Transfer in Energy Systems of the Heat Transfer Division. All papers were reviewed in accordance with standards and procedures of the ASME. The papers cited here are listed in the approximate order in which they were presented as posters at the conference. In addition, a workshop entitled, "Challenging Issues in Heat and Mass Transfer for Spray Systems," was arranged to foster author-audience interaction and to identify future research needs.

The purpose of this session was to provide a forum for the dissemination of information pertaining to research currently being carried out in spray flames, and heat and mass transfer in other types of spray systems. Topics that were solicited for, but not restricted to, the session included atomization of liquid and slurry fuels, characterization of sprays and their interaction with the surrounding environment, fuel transport in spray systems with complex aerodynamic patterns, fuel property effects and preparation on combustion processes. Systems of special interest included: gas turbine combustors, furnaces and boilers, internal combustion engines, atomizers for powder and slurry dispersion, aerosols, and spray dryers and reactors. All of the above-mentioned topics have a significant influence upon heat and mass transfer in different types of spray and combustion systems, and are of particular interest to the scientific research community. The papers presented at the session included experimental and numerical studies on several of the aforementioned topics. The session included participants from university, industry, and government, as well as foreign representatives.

The editors wish to express their sincere thanks to those individuals who contributed to this session and the publication of these Proceedings. They are particularly thankful to all of the authors, abstract and manuscript reviewers, workshop participants, and the ASME Headquarters personnel for their complete cooperation throughout the publication process.

HEAT TRANSFER PHENOMENA IN RADIATION, COMBUSTION, AND FIRES

R. K. Shah, Editor

In order to maintain high quality of the papers published by the ASME in the symposium volume or in preprint form, all papers are reviewed by the professionals in the field on a volunteer basis. On behalf of the ASME Heat Transfer Division, Conference Organizers, and Session Organizers, we are grateful to the following reviewers for their time and effort to provide thorough and detailed reviews in a timely manner. This list includes reviewers for papers published or rejected for the technical sessions included in this symposium volume.

Adiga, K. C.
Adorjan, A. S.
Aggarwal, S. K.
Alpert, R. L.
Annamalai, K.
Arpaci, V. S.
Ashgriz, N.
Avedisian, C. T.

Babikian, D. S.
Ballal, D. R.
Bayazitoglu, Y.
Bill, R. G.
Birk, A.
Bobco, R. P.
Brewster, M. Q.
Buckius, R. O.

Cebeci, T.
Cernansky, N. P.
Chan, S. H.
Charagundla, S. R.
Chen, L. D.
Cheung, F. B.
Chung, J. N.

Davis, R.
Debellis, C. L.
Delichatsios, M. A.
Dougherty, R. L.
Drolen, B. L.

Edwards, C. F.
Edwards, D. K.
Ellzey, J. L.
Epstein, M.

Faeth, G. M.
Farag, I. H.
Farouk, B.
Felske, J. D.
Fenton, D. L.
Fernandez-Pello, A. C.
Fischer, S. J.
Fiveland, W. A.

Gill, W.
Greif, R.
Gore, J. P.
Gunaji, M. V.
Gupta, A. K.

Habib, I. S.
Hadjisophocleous, G. V.
Hasegawa, H.
Heskestad, G.
Howell, J. R.

Jamaluddin, A. S.
Joklik, R.

Kallasanath, K.
Kaminski, D. A.
Kashiwagi, T.
Kassemi, M.
Kaviany, M.

Lavine, A.
Law, C. K.
Lee, H. O.
Li, W.
Litkouhi, B.
Litzinger, T. A.
Lloyd, J. R.
Look, D. C.
Love, T. J.

Matthews, L.
McCaffrey, B. J.
Menguc, M. P.
Miller, D. L.
Modest, M. F.

Nakos, J. T.
Naraghi, M. H. M.
Nelson, H. F.

Ofle, D. B.
Ozisik, M. N.

Plumb, O. A.
Presser, C.

Ragland, K.
Ronney, P.

Sahu, R.
Saito, K.
Samuelsen, G. S.
Santoro, R. J.
Sarofim, A. F.
Sathe, S. B.
Schiller, D.
Shumay, R. W.
Sibulkin, M.
Siegel, R.
Slezak, S. E.
Smith, A. M.
Smith, P. J.
Smith, T. F.
Sousa, A. C. M.
Steckler, K. B.
Swathi, P. S.

Takahashi, F.
Talbot, L.
Tewarson, A.
Thynell, S. T.
Tien, C. L.
Tiwari, S. N.
Todd, D. C.
Tong, T. W.
Torrance, K. E.
Turns, B.

Ural, E. A.

Vatistas, G. H.
Venart, J. E. S.
Viskanta, R.

Walsh, P.
Walton, G. N.
Webb, B. W.
Weckman, E. J.
Wilk, R. D.
Williams, F. A.
Williamson, R. B.
Winter, M.
Wong, C. C.
Wood, B. E.
Wortman, J. J.

Yang, K. T.
Yao, S. C.
Yimer, B.
Yucel, A.
Yuen, W. W.

Zachariah, M. R.
Zurlo, J. R.

CONTENTS

RADIATION HEAT TRANSFER

MULTIDIMENSIONAL RADIATIVE HEAT TRANSFER

1989 National Heat Transfer Conference
HTD-Vol. 106, Heat Transfer Phenomena in Radiation, Combustion and Fires

RADIATIVE TRANSFER WITHIN A FINITE PLANE-PARALLEL MEDIUM EXHIBITING FRESNEL REFLECTION AT A BOUNDARY

F. Rokhsaz and R. L. Dougherty
Oklahoma State University
Stillwater, Oklahoma

ABSTRACT

Ambarzumian's method has been applied to obtain the solution for radiative transfer within a homogeneous, one-dimensional, absorbing, isotropically scattering finite medium with a reflective interface. The cases considered imposed collimated incident radiation boundary conditions at the upper or lower boundary. Superposition can be used to obtain the results for other conditions. The formulated equations were presented, along with graphical results from the numerical solution of those equations. Fresnel reflection was assumed to govern interface reflection for the purposes of the numerical calculations. The numerical results include source function, flux, and intensity at the boundaries - both just inside and just outside the medium. The ranges of the variables considered were 1.1-2.0 for refractive index, 0.2-4.0 for optical depth, and 0.1-0.3 for albedo. Results were shown to be accurate to between four and six significant figures, depending upon the parameters selected. Investigation is continuing to widen the range of parameters and extend the method to anisotropic scattering and/or two dimensions.

NOMENCLATURE

A,B - fundamental source functions,
C - auxiliary source function,
e - subscript designating an intensity external to the medium,
E_m - exponential integral (radiative) of order m,
I - intensity,
$I^+(I^-)$-intensity directed in the $+\tau$ $(-\tau)$ direction,
I_0 - magnitude of collimated intensity incident from outside upper boundary,
I_2 - magnitude of collimated intensity incident upon the lower boundary,
K - auxiliary source function,
L - parameter of unit magnitude and positive or negative in sign,
n - ratio of medium refractive index to surroundings refractive index,
q - flux: case 1 (q_1), case 2 (q_2), $A(q_A)$ or $B(q_B)$,
R - reflection function,
S - source function, for case 1 (S_1) or case 2 (S_2),
T - transmission function,
Z_m - modified exponential integral of order m,
δ - Dirac Delta function,
Δ - term representing either source function A or K,
χ - parameter representing one of the source functions A, B, C or K,
Γ - function dependent upon the A and B source functions at both boundaries,
μ - cosine of the polar angle within the medium,
μ_{cr} - cosine of the critical angle for total internal reflection,
μ_1 - cosine of incident polar angle outside the lower boundary,
μ_i - cosine of incident polar angle just within the upper boundary,
μ_e - cosine of exiting polar angle just outside the upper boundary,
μ_o - cosine of incident polar angle outside the upper boundary,
ν - subscript designating frequency dependence,
- dummy variable of integration for μ dependence,
ω - single scattering albedo,
Ω - function dependent upon C and K source functions at both boundaries,
ϕ - azimuthal angle,
Φ - function dependent upon fundamental source functions at both boundaries,
ψ - function dependent upon fundamental source functions at the lower boundary,
ρ - interface reflection coefficient,
τ,τ_o - optical depth, optical thickness of medium,
θ - polar angle,
∇ - term representing either source function B or C.

INTRODUCTION

Refractive index between two materials has a great effect on the radiative properties (intensity, flux, reflection and transmission) in the medium. Therefore, it should be of no surprise to find a considerable amount of research on the effects of refractive index in different types of media, whether they be finite or semi-finite, one- or multi-dimensional.

Shouman and Ozisik have conducted two one-dimensional studies concerning finite, absorbing, emitting and isotropically scattering media. They applied the

F_n method (1) and the source function expansion technique (2) as methods of solution for radiative heat transfer in a two-layer medium having diffusely and specularly reflecting boundaries.

Crosbie has also conducted studies in a finite absorbing, emitting and scattering medium with index of refraction greater than one. In one of his studies (3), he expressed the radiative properties for a medium backed by a diffuse substrate in terms of the properties of a medium backed by a transparent substrate. In another study (4), he used Chandrasekhar's X and Y functions to determine the radiative properties of the medium.

Casti, Kalaba and Ueno (5) have built on Sobolev's work (6) by studying diffuse reflection from a finite medium which absorbs and isotropically scatters radiation, with the bottom surface being a specular reflector. They solved for the diffusely reflected and transmitted radiation fields by integrating a system of ordinary differential equations with known initial conditions.

Houf and Incropera (7) have compared several methods for predicting radiative transfer in a finite, one-dimensional medium with index of refraction other than unity. Each of the methods were examined for their accuracy in predicting the flux and intensity.

There have been other studies (8-12) done on absorbing, emitting and anisotropically scattering finite media using approximate methods. Additional researchers (13-17) have conducted extensive studies on semi-finite media using exact formulations, and on finite non-homogeneous media using approximate techniques.

The purpose of the following work is to present numerical results for the exact formulation of a problem concerning a finite one-dimensional absorbing/isotropically scattering medium for which refraction and reflection at the upper boundary play an important role. Numerical and graphical results are presented for the source function, flux and the intensity--just inside and outside of the boundaries--in the case where the incident radiation is collimated and of a single frequency. Several indices of refraction are investigated along with the effect of incident angle of collimated incident intensity as well as optical thickness.

SUMMARY OF FUNDAMENTAL EQUATIONS

The one-dimensional transport equation for an absorbing/isotropically scattering medium (Fig. 1) is:

$$\mu\, dI_\nu(\tau,\mu,\phi)/d\tau + I_\nu(\tau,\mu,\phi) = S_\nu(\tau,\mu) = (\omega/2)\int_0^{2\pi}\int_{-1}^{1} I_\nu(\tau,\mu',\phi')d\mu'd\phi'. \quad (1)$$

The transport equation can be solved by the standard integrating factor approach to yield equations for $I^+(\tau,\mu)$ and $I^-(\tau,\mu)$, intensities which have components in the positive and negative τ directions, respectively. In the following, the frequency dependence still exists but is not explicitly shown.

The boundary condition at the upper surface of the medium includes both incident intensity and intensity internally reflected at the boundary, while the lower surface is assumed neither to reflect internal radiation nor to refract externally incident radiation:

$$I^+(0,\mu) = I_i^+(\mu)+I^-(0,\mu)\rho(\mu,n), \quad I^-(\tau_o,\mu) = I_i^-(\mu). \quad (2)$$

Two problems (cases 1 and 2) are considered. In the first case (Fig. 1a), collimated radiation is incident upon the upper surface with no radiation incident upon the lower surface:

$$I_i^+(\mu) = I_o\delta(\mu-\mu_i)\delta(\phi-\phi_i)\mu_o/\mu_i, \quad I_i^-(\mu) = 0. \quad (3)$$

The second case (Fig. 1b) interchanges these two incident boundary conditions.

For these cases, it can be shown that the source functions are:

$$S_1(\tau,\mu_o) = (\omega I_o/4\pi)(\mu_o/[1-(1-\mu_o^2)/n^2]^{1/2}) \times [1-\rho(\mu_o,1/n)]B(\tau,\mu_i;n;\tau_o), \quad (4)$$

$$S_2(\tau,\mu_1) = (\omega I_2/4\pi)[\exp(-\tau_o/\mu_1)\rho(\mu_1,n)B(\tau,\mu_1;n;\tau_o) + A(\tau,\mu_1;n;\tau_o)], \quad (5)$$

where subscript 1 on S refers to case 1, subscript 2 on S refers to case 2, and the fundamental source function equations are:

$$B(\tau,\mu;n;\tau_o) = \exp(-\tau/\mu) + (\omega/2)\int_0^{\tau_o} B(t,\mu,n;\tau_o)[E_1(|\tau-t|) + Z_1(\tau+t,n)]dt, \quad (6)$$

$$A(\tau,\mu;n;\tau_o) = \exp[-(\tau_o-\tau)/\mu] + (\omega/2)\int_0^{\tau_o} A(t,\mu,n;\tau_o)[E_1(|\tau-t|) + Z_1(\tau+t,n)]dt, \quad (7)$$

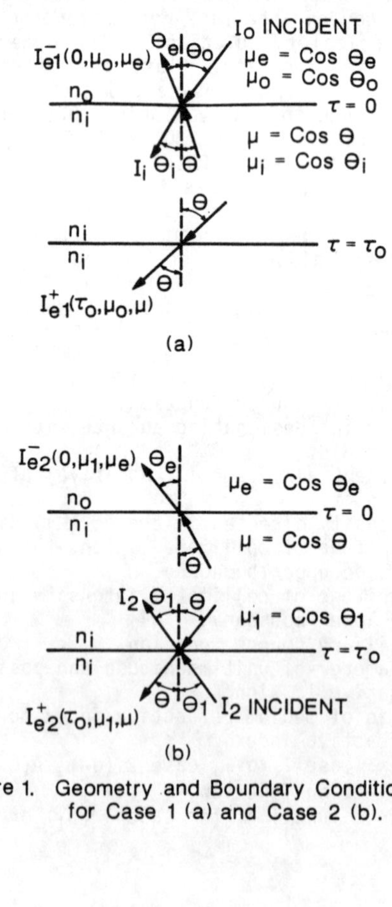

Figure 1. Geometry and Boundary Conditions for Case 1 (a) and Case 2 (b).

with E_m being the standard exponential integral of radiative transfer and Z_m being a modified exponential integral (13):

$$Z_m(\tau,n) = \int_0^1 \exp(-\tau/\mu)\ \rho(\mu,n)\ \mu^{m-2}\ d\mu. \qquad (8)$$

In Eq. (4), μ_0 is related to μ_i by Snell's Law.

The development showing the procedure for arriving at these equations along with the solution procedure are provided in Refs. 18 and 19. The solution procedure involves a modification of Ambarzumian's method to arrive at a set of coupled integral equations for B and A (as well as a pair of auxiliary functions C and K) at the upper and lower boundaries. At the upper boundary, these are:

$$B(0,\mu;n;\tau_0) = 1 + (\omega/2)\int_0^1 R_B(\mu,\mu')[1+\rho(\mu',n)]d\mu'/\mu', \qquad (9)$$

$$A(0,\mu;n;\tau_0) = \exp(-\tau_0/\mu) + (\omega/2)\int_0^1 R_A(\mu,\mu')\ [1+\rho(\mu',n)]d\mu'/\mu'. \qquad (10)$$

The reflection functions of Eqs. (9) and (10) are related to the intensity leaving the upper surface of the medium, and are defined as:

$$R_\chi(\mu,\mu') = \int_0^{\tau_0} \chi(t,\mu;n;\tau_0)\exp(-t/\mu')dt, \qquad (11)$$

where χ represents one of the functions B, A, C or K. These reflection functions can be written in terms of the upper and lower boundary source functions only as:

$$R_B(\mu,\mu') = \mu\mu'[\mu'\ B(0,\mu';n;\tau_0)C(0,\mu;n;\tau_0) - B(0,\mu;n;\tau_0)C(0,\mu';n;\tau_0)]/(\mu'^2-\mu^2) + \mu^2\mu'\ \Phi(\mu,\mu')/(\mu'^2-\mu^2) - \mu\mu'^2\ \Phi(\mu',\mu)/(\mu'^2-\mu^2), \qquad (12)$$

$$R_A(\mu,\mu') = \mu^2\mu'\ \Phi(\mu,\mu')A(\tau_0,\mu;n;\tau_0)/[B(\tau_0,\mu;n;\tau_0)\ x\ (\mu'^2-\mu^2)] + \mu\mu'^2\ \Phi(\mu',\mu)K(\tau_0,\mu;n;\tau_0)/[C(\tau_0,\mu;n;\tau_0)(\mu'^2-\mu^2)] - \mu\mu'[\mu'\ K(0,\mu;n;\tau_0)B(0,\mu';n;\tau_0) + \mu A(0,\mu;n;\tau_0)\ C(0,\mu';n;\tau_0)]/(\mu'^2-\mu^2), \qquad (13)$$

where

$$\Phi(\mu,\mu')=[K(0,\mu';n;\tau_0)B(0,\mu;n;\tau_0)+ A(0,\mu;n;\tau_0)C(0,\mu';n;\tau_0)]/\psi(\mu,\mu'), \qquad (14)$$

and

$$\psi(\mu,\mu') = K(\tau_0,\mu';n;\tau_0)/C(\tau_0,\mu';n;\tau_0) + A(\tau_0,\mu;n;\tau_0)/B(\tau_0,\mu;n;\tau_0). \qquad (15)$$

The defining integral equations for C and K are:

$$C(\tau,\mu;n;\tau_0) = \exp(-\tau/\mu) + (\omega/2)\int_0^{\tau_0} C(t,\mu;n;\tau_0)[E_1(|\tau-t|)-Z_1(\tau+t,n)]dt, \qquad (16)$$

$$K(\tau,\mu;n;\tau_0)= \exp[-(\tau_0-\tau)/\mu] + (\omega/2)\int_0^{\tau_0} K(t,\mu;n;\tau_0)\ [E_1(|\tau-t|)-Z_1(\tau+t,n)]dt. \qquad (17)$$

The equations for C and K which complement Eqs. (9), (10), (12) and (13) are as follows (18,19):

$$C(0,\mu;n;\tau_0) = 1+(\omega/2)\int_0^1 R_C(\mu',\mu)[1-\rho(\mu',n)]d\mu'/\mu', \qquad (18)$$

$$K(0,\mu;n;\tau_0)=\exp(-\tau_0/\mu)+ (\omega/2)\int_0^1 R_K(\mu',\mu)[1-\rho(\mu',n)]\ d\mu'/\mu', \qquad (19)$$

$$R_C(\mu,\mu') = \mu\mu'[\mu'\ B(0,\mu;n;\tau_0)C(0,\mu';n;\tau_0) - \mu B(0,\mu';n;\tau_0)C(0,\mu;n;\tau_0)]/(\mu'^2-\mu^2) + \mu\mu'\ [\mu\ \Phi(\mu,\mu') - \mu'\ \Phi(\mu',\mu)\ ]/(\mu'^2-\mu^2), \qquad (20)$$

$$R_K(\mu,\mu') = [\mu'^2\mu\Phi(\mu,\mu')A(\tau_0,\mu;n;\tau_0)/B(\tau_0,\mu;n;\tau_0) + \mu^2\mu'\ \Phi(\mu',\mu)K(\tau_0,\mu;n;\tau_0)/C(\tau_0,\mu;n;\tau_0)]/(\mu'^2-\mu^2) - \mu\mu'[\mu\ K(0,\mu;n;\tau_0)B(0,\mu';n;\tau_0) + \mu'\ A(0,\mu;n;\tau_0)\ x\ C(0,\mu';n;\tau_0)]/(\mu'^2-\mu^2). \qquad (21)$$

As can be seen from Eqs. (12), (13), (20) and (21), the reflection functions are not only dependent on the fundamental source functions at the upper boundary, but are also dependent on those functions at the lower boundary. Therefore, expressions for the source functions at the lower boundary may be found from Eqs. (6), (7), (16) and (17), employing definitions for the exponential integral and modified exponential integral (Eq. (8)), as:

$$\nabla(\tau_0,\mu;n,\tau_0) = \exp(-\tau_0/\mu)+(\omega/2)[\int_0^1 T_\nabla(\mu',\mu)d\mu'/\mu'+ L\int_0^1 \exp(-\tau_0/\mu')\rho(\mu',n)R_\nabla(\mu',\mu)d\mu'/\mu'], \qquad (22)$$

where ∇ is either source function B (L = +1) or C (L = -1),

$$\Delta(\tau_0,\mu;n;\tau_0) = 1 + (\omega/2)[\int_0^1 T_\Delta(\mu',\mu)\ d\mu'/\mu' + L\int_0^1 \exp(-\tau_0/\mu')\ \rho(\mu',n)\ R_\Delta(\mu',\mu)\ d\mu'/\mu'], \qquad (23)$$

where Δ is either source function A (L = +1) or K (L = -1), and the transmission functions are:

$$T_\chi(\mu',\mu) = \int_0^{\tau_0} \chi(t,\mu;n;\tau_0)\ \exp[-(\tau_0-t)/\mu']\ dt, \qquad (24)$$

where χ is again source function B, A, C or K.

It can be shown (18,19) that these transmission functions are found by solving the following set of equations:

$$4\ \exp(-\tau_0/\mu')/\omega + 2\int_0^1 T_B(\nu,\mu')\ d\nu/\nu +$$

$$\Gamma(\mu;n;\tau_o)[\ A(\tau_o,\mu;n;\tau_o)((\mu - \mu')/\mu\mu')T_B(\mu',\mu) -$$
$$B(\tau_o,\mu;n;\tau_o)\ ((\mu - \mu')/\mu\mu')\ T_A(\mu',\mu)\ ] +$$
$$\Omega(\mu;n;\tau_o)\{K(\tau_o,\mu;n;\tau_o)[T_B(\mu',\mu)/\mu - T_C(\mu',\mu)/\mu'] +$$
$$C(\tau_o,\mu;n;\tau_o)\ [T_K(\mu',\mu)/\mu' + T_A(\mu',\mu)/\mu\ ]\} = 0, \qquad (25)$$

$$4\exp(-\tau_o/\mu)/\omega + 2\int_0^1 T_B(\nu,\mu)\ d\nu/\nu +$$
$$\Gamma(\mu';n;\tau_o)[\ A(\tau_o,\mu';n;\tau_o)((\mu'-\mu)/\mu'\mu)\ T_B(\mu,\mu') -$$
$$B(\tau_o,\mu';n;\tau_o)\ ((\mu'-\mu)/\mu\mu')\ T_A(\mu,\mu')] +$$
$$\Omega(\mu';n;\tau_o)\{K(\tau_o,\mu';n;\tau_o)[T_B(\mu,\mu')/\mu' - T_C(\mu,\mu')/\mu]$$
$$+\ C(\tau_o,\mu';n;\tau_o)[T_K(\mu,\mu')/\mu + T_A(\mu,\mu')/\mu']\} = 0, \qquad (26)$$

$$4\exp(-\tau_o/\mu')/\omega + 2\int_0^1 T_C(\nu,\mu')\ d\nu/\nu +$$
$$\Gamma(\mu;n;\tau_o)\{A(\tau_o,\mu;n;\tau_o)[T_C(\mu',\mu)/\mu - T_B(\mu',\mu)/\mu'] +$$
$$B(\tau_o,\mu;n;\tau_o)[T_A(\mu',\mu)/\mu' + T_K(\mu',\mu)/\mu]\} +$$
$$\Omega(\mu;n;\tau_o)\{K(\tau_o,\mu;n;\tau_o)[T_B(\mu',\mu)/\mu - T_C(\mu',\mu)/\mu'] +$$
$$C(\tau_o,\mu;n;\tau_o)[T_K(\mu',\mu)/\mu' + T_A(\mu',\mu)/\mu]\} = 0, \qquad (27)$$

$$4\exp(-\tau_o/\mu)/\omega + 2\int_0^1 T_C(\nu,\mu)\ d\nu/\nu +$$
$$\Gamma(\mu';n;\tau_o)\{A(\tau_o,\mu';n;\tau_o)[T_C(\mu,\mu')/\mu' - T_B(\mu,\mu')/\mu] +$$
$$B(\tau_o,\mu';n;\tau_o)[T_A(\mu,\mu')/\mu + T_K(\mu,\mu')/\mu']\} +$$
$$\Omega(\mu';n;\tau_o)\{K(\tau_o,\mu;n;\tau_o)[T_B(\mu,\mu')/\mu' - T_C(\mu,\mu')/\mu] +$$
$$C(\tau_o,\mu';n;\tau_o)[T_K(\mu,\mu')/\mu + T_A(\mu,\mu')/\mu']\} = 0, \qquad (28)$$

where

$$\Gamma(\mu;n;\tau_o) = 2/\{\omega[B(0,\mu;n;\tau_o)A(\tau_o,\mu;n;\tau_o) - B(\tau_o,\mu;n;\tau_o)A(0,\mu;n;\tau_o)]\}, \qquad (29)$$

and

$$\Omega(\mu;n;\tau_o) = 2/\{\omega[C(0,\mu;n;\tau_o)K(\tau_o,\mu;n;\tau_o) - C(\tau_o,\mu;n;\tau_o)K(0,\mu;n;\tau_o)]\}. \qquad (30)$$

The solutions of Eqs. (25)-(28) are presented in Refs. 18 and 19. With this final set of equations, the source functions at the upper and lower boundaries, as well as the reflection and transmission functions can now be determined. The intensities and fluxes at the boundaries can also be obtained in terms of the reflection and transmission functions as shown in the following sections.

Intensity

For case 1 (see Fig. 1a) the reflected intensity just outside the upper boundary is (18,19):

$$I^-_{e1}(0,\mu_o,\mu_e) =$$
$$(I_o\omega\mu_o/4\pi\mu\mu_i n^2)[1-\rho(\mu_o,1/n)]\ [1-\rho(\mu,n)]R_B(\mu_i,\mu), \qquad (31)$$

while the transmitted intensity just outside the lower boundary is (18,19):

$$I^+_{e1}(\tau_o,\mu_o,\mu) =$$
$$(I_o\mu_o/\mu_i)[1-\rho(\mu_o,1/n)]\exp(-2\tau_o/\mu_i)\delta(\mu-\mu_i)\delta(\phi-\phi_i) +$$
$$(I_o\omega\mu_o/4\pi\mu\mu_i)[1-\rho(\mu_o,1/n)] \times$$
$$[\exp(-\tau_o/\mu)\rho(\mu,n)R_B(\mu,\mu_i)+T_B(\mu,\mu_i)], \qquad (32)$$

where subscript e1 refers to case 1 external to the medium, and μ_o is related to μ_i by Snell's Law, as are μ_e and μ:

$$\mu_o = [1-(1-\mu_i^2)n^2]^{1/2}, \qquad \mu_e = [1-(1-\mu^2)n^2]^{1/2}. \qquad (33)$$

For case 2 (see Fig. 1b), the reflected intensity just outside the upper boundary is (18,19):

$$I^-_{e2}(0,\mu_1,\mu_e) =$$
$$(I_2\mu/\mu_e)[1-\rho(\mu,n)]\exp(-\tau_o/\mu)\delta(\mu-\mu_1)\delta(\phi-\phi_1) +$$
$$(\omega I_2/4\pi\mu n^2)[1-\rho(\mu,n)] \times$$
$$[\exp(-\tau_o/\mu_1)\rho(\mu_1,n)R_B(\mu,\mu_1) + R_A(\mu,\mu_1)], \qquad (34)$$

while the transmitted intensity just outside the lower boundary is (18,19):

$$I^+_{e2}(\tau_o,\mu_1,\mu) = I_2\exp(-2\tau_o/\mu)\rho(\mu,n)\delta(\mu-\mu_1)\delta(\phi-\phi_1) +$$
$$(\omega I_2/4\pi\mu)\exp(-\tau_o/\mu)\rho(\mu,n) \times$$
$$[\exp(-\tau_o/\mu_1)\rho(\mu_1,n)R_B(\mu,\mu_1)+R_A(\mu,\mu_1)] +$$
$$(\omega I_2/4\pi\mu)[\exp(-\tau_o/\mu_1)\rho(\mu_1,n)T_B(\mu,\mu_1)+T_A(\mu,\mu_1)]. \qquad (35)$$

Flux

For case 1, the fluxes at the upper and lower boundaries can be found from (18,19):

$$q_1(\tau;\mu_o;n) = (\mu_o I_o/\mu_i)[1-\rho(\mu_o,1/n)]q_B(\tau;\mu_i;n), \qquad (36)$$

where τ is replaced by 0 and τ_o for fluxes at the top and bottom, respectively, and where:

$$q_B(0;\mu_i;n) = \mu_i-(\omega/2)\int_0^1 R_B(\mu_i,\mu')[1-\rho(\mu',n)]d\mu', \qquad (37)$$

$$q_B(\tau_o;\mu_i;n) = \mu_i\ \exp(-\tau_o/\mu_i)+(\omega/2)\int_0^1 [T_B(\mu',\mu_i) +$$
$$R_B(\mu_i,\mu')\exp(-\tau_o/\mu')\rho(\mu',n)]d\mu'. \qquad (38)$$

For case 2, the fluxes at the upper and lower boundaries are (18,19):

$$q_2(\tau;\mu_1;n) =$$
$$I_2\ \exp(-\tau_o/\mu_1)\rho(\mu_1,n)q_B(\tau;\mu_1;n) - I_2\ q_A(\tau;\mu_1;n), \qquad (39)$$

where τ is replaced by 0 and τ_o for fluxes at the top and bottom, respectively, and where:

$$q_A(0;\mu_1;n) = \mu_1\ \exp(-\tau_o/\mu_1)$$
$$-\ (\omega/2)\int_0^1 R_A(\mu_1,\mu')[1-\rho(\mu',n)]\ d\mu', \qquad (40)$$

$$q_A(\tau_o;\mu_1;n) = \mu_1+(\omega/2)\int_0^1 [T_A(\mu',\mu_1) + \exp(-\tau_o/\mu')\rho(\mu',n)R_A(\mu_1,\mu')]d\mu'. \quad (41)$$

RESULTS

In the development of the fundamental source function equations, the reflection coefficient, $\rho(\mu,n)$, is a generic term. Therefore, any mathematical expression which satisfies the major restrictions on ρ can be utilized. These restrictions are that it 1) be a real-valued function, 2) have a range of values between zero and unity and 3) be zero for unit refractive index. In our computations, Fresnel's representation for non-polarized intensity has been employed (13).

Computational Approach

Equations (9), (10), (12), (13), (18)-(21), (22), (23) and (25)-(28) have been solved simultaneously by the method of successive approximation. The program is designed to compute the integrals by using Gaussian quadrature points in the μ interval from zero to one. The rapid decrease of Fresnel's coefficient, which occurs as μ increases to become greater than the critical value μ_{cr} makes it necessary to use a dense quadrature around μ_{cr}. Therefore, in order to accommodate the behavior of the Fresnel coefficient, the zero to one interval has been divided into four segments. Two of these intervals are from μ_{cr} to $1.015\mu_{cr}$ and from $1.015\mu_{cr}$ to $1.085\mu_{cr}$. The two remaining sets of quadrature are placed in the two remaining segments of the zero to one interval.

Typical accuracy of these solutions as compared to solving Eq. (6) with quadrature in optical depth [Direct Solution] for one incident angle are shown in Table I. This comparison was created using a quadrature set in the current solution of 25, 32, 25, 25 in the four intervals (in numerical order) between μ of zero and unity, with the Direct Solution used as the basis on which percent error was computed.

Table I
Comparison of Direct Solution to Current Solution for Selected Parameters and Refractive Index of 1.33

ω	τ_o	μ	$B(0,\mu;n;\tau_o)$ (%error)	$B(\tau_o,\mu;n;\tau_o)$ (% error)	$q_B(0;\mu;n)$ (% error)
0.1	0.2	0.01	-0.000039	-0.000080	0.0000049
0.1	0.2	0.10	0.000022	0.0000012	0.0000077
0.1	0.2	0.50	0.000026	-0.000017	0.0000041
0.1	0.2	1.00	0.000058	0.0000069	0.000086
0.1	2.0	1.00	-0.000075	0.000033	0.000011
0.3	3.0	1.00	0.000075	0.0000025	0.00014

Convergence of the successive approximation solution technique is very albedo-sensitive. With the current form of the code, the accuracy decreases to only two or three significant digits when the albedo is increased to 0.5. Therefore, results are presented only for albedos less than or equal to 0.3. Numerical techniques for overcoming this limitation are currently being investigated.

Numerical Results

In this section we will examine graphical results for the source function, flux, reflection and transmission functions, and intensity at the boundaries of the media shown in Figs. 1a and 1b. The effects of optical thickness, albedo and refractive index will be explored.

Figure 2 has been generated to demonstrate the behavior of source function B at the top boundary and source function A at bottom boundary for a refractive index of 1.33. The first observation that we can make is that the source function at the top increases as the optical thickness of the medium increases, similar to the trend for unit refractive index. We can also observe that the increase in the source function is not proportional to the increase in the optical depth. This is evident by comparing the increase in the source function between τ_o=0.2 and τ_o=0.5 with that between τ_o=0.5 and τ_o=1.0.

Note that the B source function for μ_i less than μ_{cr} (≈ 0.66) has no physical significance for case 1 (see Eq. (4)). This is due to the fact that for μ_i less than μ_{cr}, no radiation can have entered the medium. Therefore, the source function cannot physically exist. However, B source function values at low μ_i values are necessary for numerical results, and are physically significant for case 2 (see Eq. 5).

Figure 2 also illustrates the behavior of the A source function at the bottom boundary. We can observe that the A source function at the bottom boundary initially increases as the optical thickness increases. Then it reaches a peak which depends on angle of incidence, and after peaking, decreases with increased optical thickness. Since the A source function represents only a part of the solution for case 2 (see Eq. 5), the reason for this behavior is not as clearly defined as for the B source function, which represents all of the solution for case 1.

However, in examining the equation for A (Eq. (7)), we find that the A source function does not include the contribution to the radiation field which results from the non-attenuated incident radiation being reflected back into the medium by the upper surface. Therefore, considering only the incident radiation, it appears that for a given incident angle, there is an optimum optical thickness where scattering of the incident radiation and reflection of that scattered radiation from the upper surface result in the maximum source function. At that incident angle and optical thickness, an increase in optical thickness appears to reduce the reflected contribution from the upper surface, while a decrease in optical thickness tends to reduce the contribution from scattering of the incident radiation.

Figure 3 presents the A source function at the top boundary and the B source function at the bottom boundary. We observe that as the optical thickness is increased, both source functions decrease. These trends are again similar to those of the case for unit refractive index. Again for μ_i less than μ_{cr}, the values of the B source function at the bottom have no physical significance for case 1, but are physically significant for case 2.

Figure 4 presents the B flux at the top boundary and the A flux at the bottom boundary. The general shapes of the flux curves are similar to those of the source function curves of Fig. 2. Note that the magnitude of the A flux at the lower boundary is larger than the B flux at the upper boundary. As with the source function results, we must keep in mind that the B flux represents the results for case 1, while the A flux represents only part of the case 2 solution. Even so, the difference in the magnitude of the fluxes may be due to the fact that, in both instances, energy is reflected by the upper boundary back into the medium and toward the lower boundary. For the B flux case, that reflected energy may then exit the medium at the lower boundary, resulting in less energy available at the upper surface, where the incident radiation entered. In the A flux case, that reflected energy may also exit

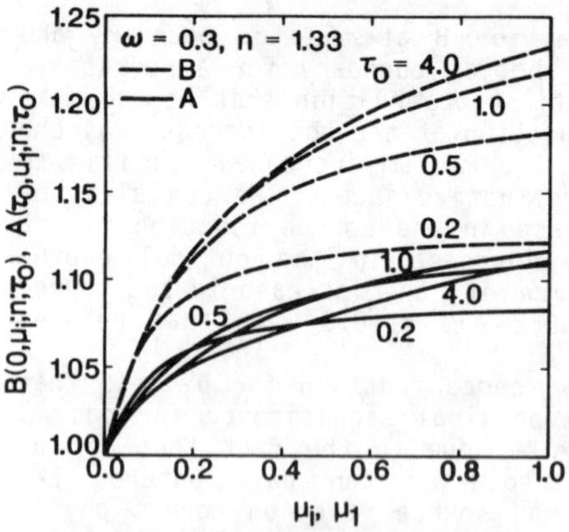

Figure 2. Effect of Optical Thickness on the A and B Fundamental Source Functions.

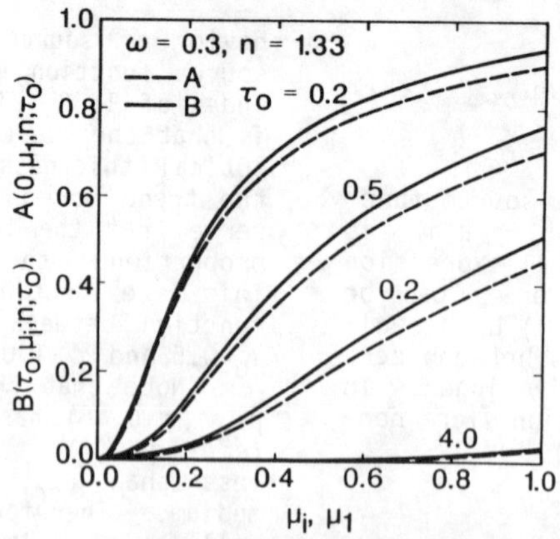

Figure 3. Effect of Optical Thickness on the A and B Fundamental Source Functions.

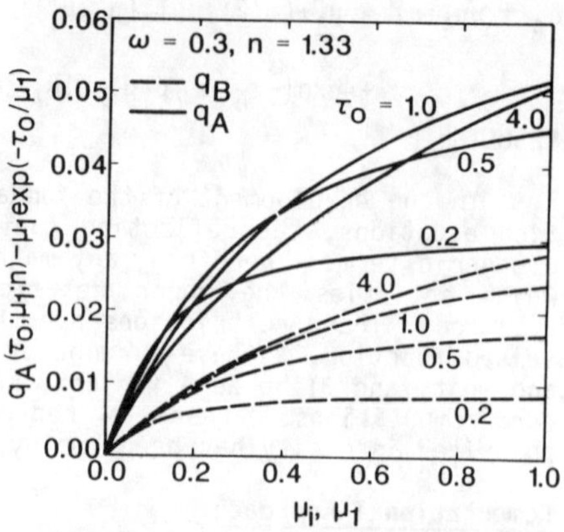

Figure 4. Effect of Optical Thickness on the A and B Fundamental Fluxes.

the medium at the lower boundary, where the incident radiation entered. Thus, it appears that the energy reflected back to the boundary at which the radiation entered the medium is greater if a reflective interface is placed opposite the entry boundary rather than being placed at the entry boundary.

Figure 5 presents the B flux at the upper boundary and the A flux at the lower boundary. Note that these fluxes rise and fall with increasing optical thickness, similar to the trends for unit refractive index.

Figure 6 illustrates the effect of changing the optical thickness on reflection function for incident polar angles corresponding to μ_i values of 0.5 and 1.0 (normal incidence) for case 1. This figure indicates that as the optical thickness increases, reflection within the medium also increases. These are the same trends observed for the source function in Fig. 2. As with the B source function, μ_i less than μ_{cr} is not physically possible for case 1. Therefore, the reflection function for μ_i of 0.5 is not physically meaningful for case 1 (but is possible for case 2).

Figure 7 presents the B-related transmission function for two incident polar angles. For any given angle within the medium (corresponding to μ), the function first rises with increasing optical thickness, peaks, then drops with increasing optical thickness. These trends are very similar to those found for the flux at the lower boundary in Figure 4, due to the scattering/reflective boundary effects discussed with respect to that figure.

The trends of the A reflection function (Figure 8) are very similar to those of the B transmission function of Figure 7. This is due to the fact that, though termed a "reflection function", R_A is given at the boundary opposite to the boundary where the radiation was incident. Therefore the A reflection function is really comparable to transmission. It is interesting to note that the A reflection function is slightly greater than the B transmission function, for the same angles and optical thickness. Although significant at the peak optical thicknesses, the differences become negligible away from the optimum optical thickness.

Due to the fact that increasing albedo increases the source functions at either boundary, the reflection and transmission functions, and the fluxes at either boundary, only two plots are presented to depict the albedo's effect. These are Figs. 9 and 10 for the B fluxes, and for the B reflection and transmission functions. As might be expected, since increasing the albedo increases the scattering and decreases absorption, an increased albedo causes an increase in all of the functions shown.

The refractive index impact on most of the functions of interest is quite small for moderate optical thicknesses and for the albedos presented herein. This is true for the source functions and especially true for the reflection and transmission functions. The greatest effects due to refractive index are shown for the B flux in Fig. 11 and for the A flux in Fig. 12. The B flux at the upper boundary decreases significantly with an increase in refractive index. However, the B flux at the lower boundary increases as refractive index increases. This is probably due to the fact that as the refractive index increases, the critical angle also increases, resulting in more energy being reflected away from the upper boundary and toward the lower

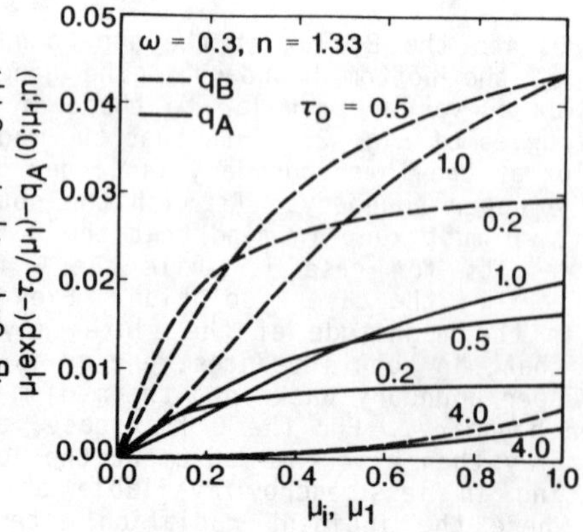

Figure 5. Effect of Optical Thickness on the A and B Fundamental Fluxes.

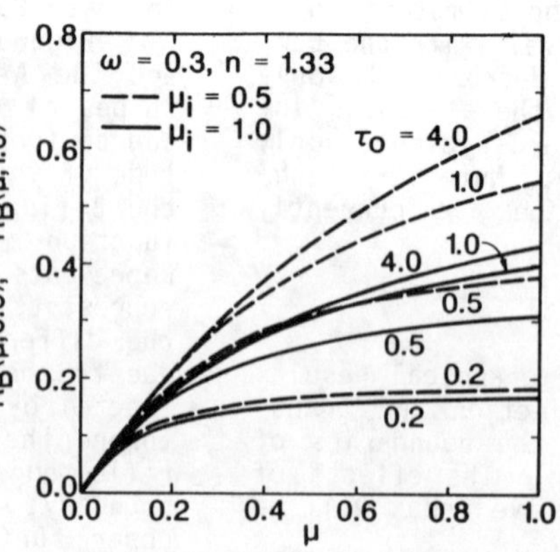

Figure 6. Impact of Optical Thickness on the B Reflection Function.

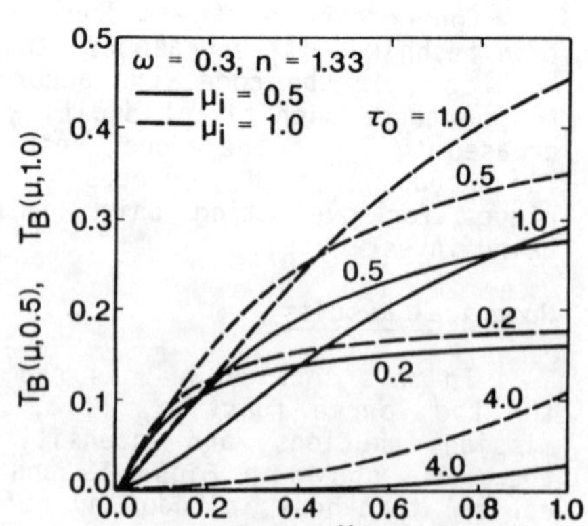

Figure 7. Impact on Optical Thickness on the B Transmission Function.

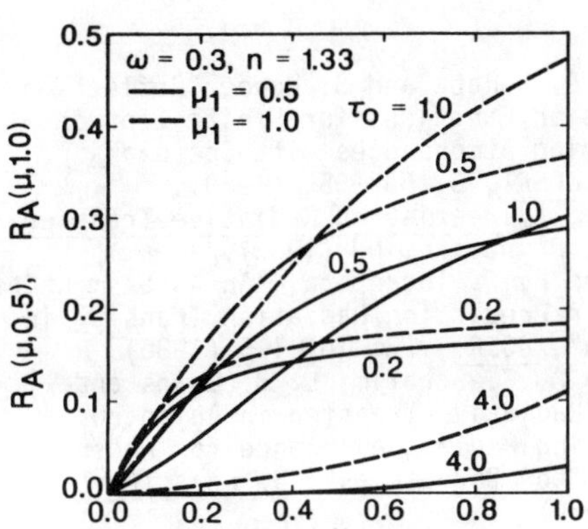

Figure 8. Impact of Optical Thickness on the A Reflection Function.

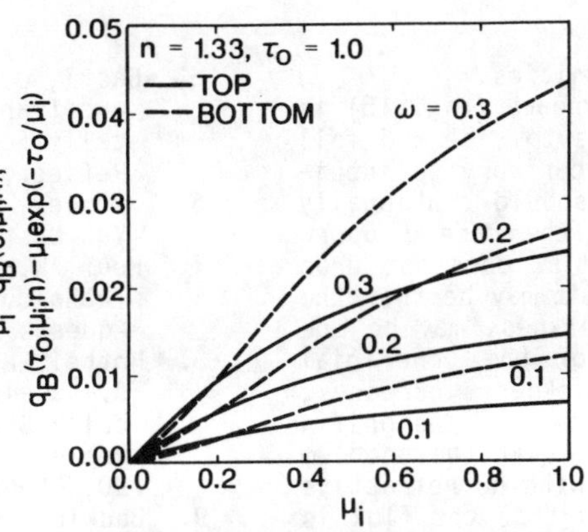

Figure 9. Effect of Albedo on the B Flux at the Boundaries of the Medium.

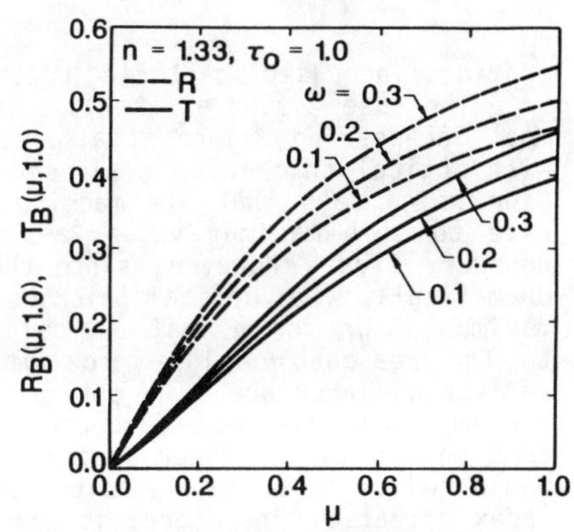

Figure 10. Effect of Albedo on the B Reflection and Transmission Functions.

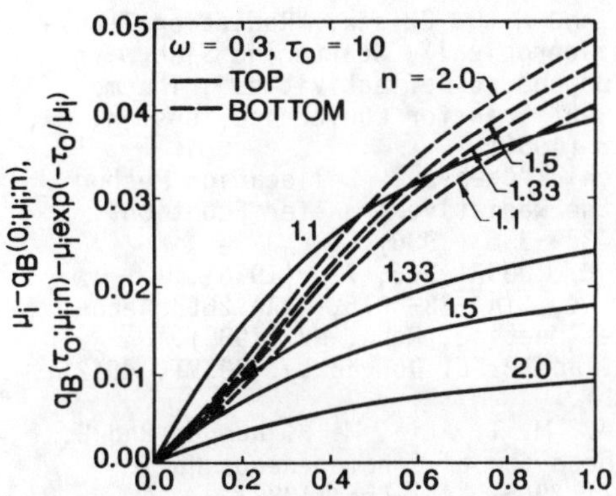

Figure 11. Refractive Index Effects on the B Flux.

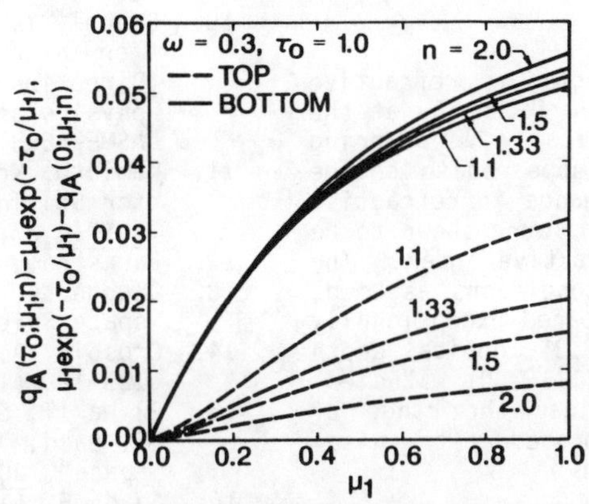

Figure 12. Refractive Index Effects on the A Flux.

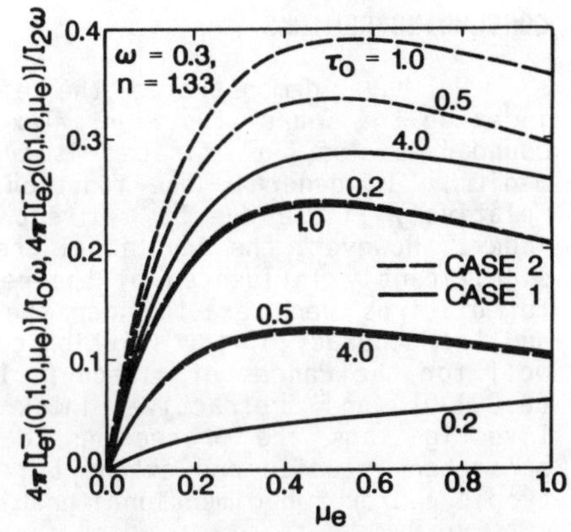

Figure 13. Impact of Optical Thickness on the Case 1 and 2 Reflected Intensities.

boundary. Thus, the upper surface flux decreases while the lower surface flux increases. The same reasoning for the A flux again results in the flux at the upper boundary decreasing with refractive index, while the flux at the lower boundary increases (Fig. 12).

In Figs. 13-16, the functions external to the medium are examined for sensitivity to optical depth. It should be noted however, that in each case shown, the effects due to the nonattenuated portion of the incident radiation are excluded from the plot.

Figure 13 presents the effect of optical depth on the reflected intensities just outside the upper boundary of the medium for both cases. As might be expected from the previous discussion, the case 1 intensity increases with optical depth, since the upper surface is the point of entry for the case 1 radiation. However, since the case 2 incident radiation enters at the lower surface, the "reflected" intensity decreases as the medium thickens.

The transmitted intensities of Fig. 14 show a most interesting trend near the critical angle (corresponding to μ=0.66 for n=1.33). For angles less than the critical angle ($\mu>\mu_{cr}$), the transmitted intensities reduce dramatically. This is obviously due to the upper interface changing from total reflection for angles greater than critical ($\mu<\mu_{cr}$) to very small reflection (~2%) for near normal angles ($\mu\approx 1.0$). Thus, significantly more energy is reflected out of the lower boundary for large angles as opposed to near normal angles. Note that this effect diminishes as optical thickness increases - probably due to the significant number of scattering events occurring between the upper and lower surfaces, tending to de-sensitize or average the lower surface response to upper surface characteristics.

The influence of optical depth on the fluxes at both boundaries for cases 1 and 2 are presented in Figs. (15) and (16). The trend in the case 1 flux at the upper surface (Fig. (15)) is to increase with increasing optical depth. Since more scattering events occur as the depth increases, it would be reasonable to conclude that there would be a greater opportunity for energy to exit the upper surface. But as optical depth increases toward infinite depth, the increased number of scattering events deep within the medium should have a diminishing impact on the upper surface results - as depicted by Fig. 15. On the other hand, the case 1 flux at the lower boundary (Fig. (16) should increase as increased optical thickness causes more scattering events to contribute to energy reaching that surface. Then, as optical thickness continues to increase, the larger number of scattering events will actually hinder energy reaching the lower surface. This rise and fall of the lower boundary case 1 flux is verified by Fig. 16.

The case 2 fluxes of Figs. 15 and 16 present different trends. The upper boundary flux (Fig. 16) rises and falls with optical thickness as does the case 1 boundary flux - for basically the same reasons. However, the optimum optical thickness where peaking occurs is not the largest optical thickness, as for that of the case 1 flux, but depends on the incident angle of radiation. This appears to be due to the trade off between higher angles of incidence increasing the effective optical distance that the incident radiation must travel, resulting in more scattering events; and the contribution of the opposing reflective interface which reflects the energy back into the medium and provides increased scattering opportunities.

The case 2 flux at the lower boundary (Fig. 15) is most interesting, since it also seems to rise and fall with optical thickness. This seems contrary to intuition which says that its magnitude should continually rise to some maximum value, as for the case 1 upper boundary flux. However, since the flux does not drop dramatically with optical thickness, it may be that the maximum shown for a unit optical thickness may be due to the reasonably close proximity of the beneficial reflecting interface. As optical thickness increases, the impact of the interface would become so insignificant on the case 2 lower surface flux that the problem would reduce to a semi-infinite case with no refractive index effects. Therefore, it appears that the flux is boosted somewhat by contributions from the reflective boundary for optimum geometries.

CONCLUDING REMARKS

We have demonstrated the effects of refractive index on the source function, flux and intensity at the boundaries of a finite isotropically scattering medium. In general, the functions appeared to change relatively little due to a direct change in refractive index. However, the functional trends were shown to be significantly influenced by the refractive index. The formulations were exact under the conditions assumed, and the accuracy of the solution compared exceptionally well for the ranges of albedo (0.1-0.3), optical depth (0.2-4.0) and refractive index (1.1-2.0) studied. Investigations are proceeding to widen the range of parameters studied and extend this procedure to anisotropic and/or two-dimensional problems.

REFERENCES

1. Shouman, S. M. and M. N. Ozisik, "Radiative Transfer in an Isotropically Scattering Two-Region Slab with Reflecting Boundaries", JQSRT, **26**, 1-9 (1981).
2. Ozisik, M. N. and S. M. Shouman, "Source Function Expansion Method for Radiative Transfer in a Two-Layer Slab", JQSRT, **24**, 441-449 (1980).
3. Crosbie, A. L., "Emittance of a Finite Scattering Medium with Refractive Index Greater Than Unity", AIAA J., **18**(6), 730-732 (1980).
4. Crosbie, A. L., "Apparent Radiative Properties of an Isotropically Scattering Medium on a Diffuse Substrate", J. Heat Trans., **101**, 68-75 (1979).
5. Casti, J. L., R. Kalaba and S. Ueno, "Reflection and Transmission Functions for Finite Isotropically Scattering Atmospheres with Specular Reflectors", JQSRT, **9**, 537-552 (1969).
6. Sobolev, V. V., A Treatise on Radiative Transfer, Van Nostrand, Princeton, NJ (1963).
7. Houf, W. G. and F. P. Incropera, "An Assessment of Techniques for Predicting Radiation Transfer in Aqueous Media", JQSRT, **23**, 101-115 (1980).
8. Hottel, H. C., A. F. Sarofim, L. B. Evans and I. A. Vasalos, "Radiative Transfer in Anisotropically Scattering Media: Allowance for Fresnel Reflection at the Boundaries", J. Heat Trans., **90**, 56-62 (1968).
9. Buckius, R. O. and M. M. Tseng, "Radiative Heat Transfer in a Planar Medium with Anisotropic Scattering and Directional Boundaries", JQSRT, **20**, 385-402 (1978).
10. Cengel, Y. A. and M. N. Ozisik, "Radiation Transfer in an Anisotropically Scattering Slab with Directional Dependent Reflectivities", Thermophysics and Heat Transfer Conference, Boston, MA, ASME-86-HT-28 (1986).
11. Kamiuto, Kouichi, "Chebyshev Collocation Method for Solving the Radiative Transfer Equation", JQSRT, **35**(4) 329-336 (1986).
12. Barkstrom, B. R., JQSRT, **16**, 725 (1976).
13. Dougherty, R. L., AIAA-88-0076, AIAA 26th Aerospace Sciences Meeting, Reno, NV (1988).
14. Crosbie, A. L. and R. L. Dougherty, JQSRT, **40**(2), 123-128 (1988).
15. El Wakil, S. A., M. T. Atia, M. H. Haggag and E. S. Saad, "Albedo for an Inhomogeneous Half Space", JQSRT, **29**(5), 451-456 (1983).
16. Saad, E. A., A. A. Hendi and S. A. El Wakil, "Radiation Transfer by a Finite-Emitting Inhomogeneous Medium", JQSRT, **34**(1), 65-73 (1985).
17. Lin, J. D. and J. M. Huang, "Radiative Transfer Within a Cylindrical Geometry with Fresnel Reflecting Boundary", AIAA-87-0326, AIAA 25th Aerospace Sciences Meeting, Reno, NV (1987).
18. Rokhaz, F., "Radiative Transfer in a Finite Absorbing/Scattering Medium with Reflective Boundary", M.S., School of Mechanical and Aerospace Engineering, Oklahoma State University (1988).
19. Dougherty, R. L. and Rokhsaz, F., "Formulation of Equations Describing Radiative Transfer within a Finite Medium Having a Reflective Boundary", to be published.

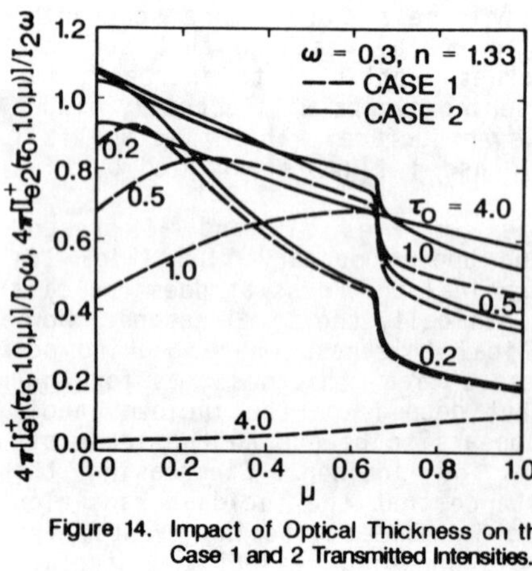

Figure 14. Impact of Optical Thickness on the Case 1 and 2 Transmitted Intensities.

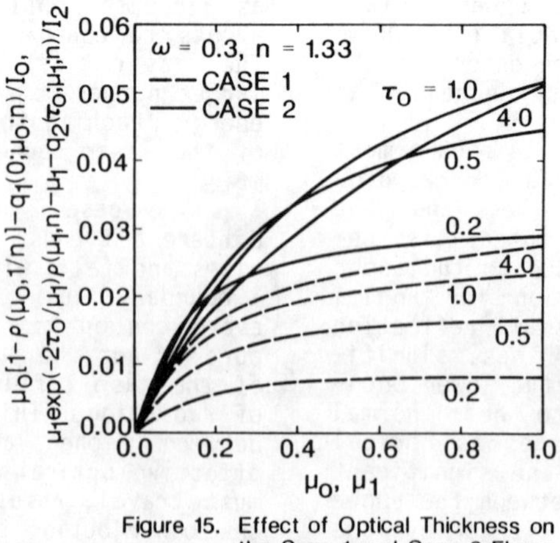

Figure 15. Effect of Optical Thickness on the Case 1 and Case 2 Fluxes.

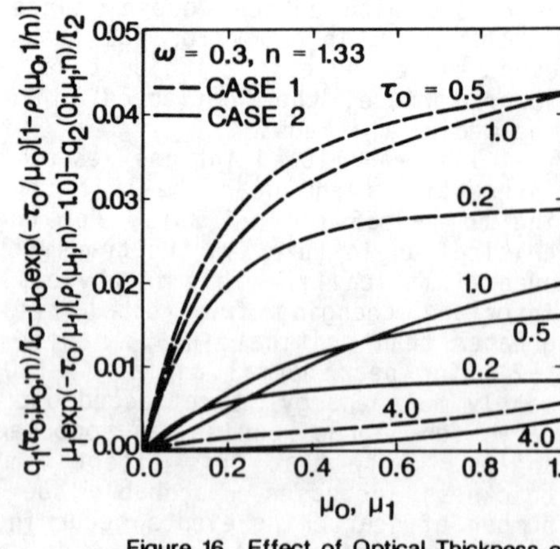

Figure 16. Effect of Optical Thickness on the Case 1 and Case 2 Fluxes.

1989 National Heat Transfer Conference
HTD-Vol. 106, Heat Transfer Phenomena in Radiation, Combustion and Fires

RADIATION TRANSFER IN ABSORBING, EMITTING TWO-DIMENSIONAL MEDIA BOUNDED BY REFLECTING WALLS

S. T. Thynell and W.-Q. Lin
Department of Mechanical Engineering
The Pennsylvania State University
University Park, Pennsylvania

ABSTRACT

Radiation transfer in absorbing, emitting, two-dimensional rectangular media bounded by diffusely reflecting walls is analyzed. The analysis of the considered problem is based on utilizing the integral form of the integro- differential equation of radiative transfer, which is represented by a system of four coupled Fredholm type integral equations. To solve in a highly accurate manner for the unknown functions, power series expansions are employed. A collocation strategy based on the zeroes of the Chebyshev polynomials is subsequently applied to generate a system of linear algebraic equations for the unknown expansion coefficients. A comprehensive parametric study is performed in order to examine the importance of the reflecting walls, as opposed to assuming black bounding walls. Comparison of results and discussion on the accuracy of the well-known P_1 approximation for solving the radiation problem are also presented.

Keywords: Numerical Methods, Radiation

Nomenclature

A_i	expansion coefficients to $q_y^-(b,z)$
b	half the optical width of enclosure
B_i	expansion coefficients to $q_z^-(y,c)$
c	half the optical height of the enclosure
C_i	expansion coefficients to $q_y^+(-b,z)$
D_i	expansion coefficients to $q_z^+(y,-c)$
$F_i(u,v)$ i=1,2,3,4	integral defined in the Appendix
$G(y,z)$	incident radiation
$I(y,z,\theta,\phi)$	radiation intensity
$I_b(T)$	$= n^2 \bar{\sigma} T^4/\pi$, the Planck function
I_{bi}	emission from the i-th wall
$K_n(u,v)$	kernels defined by equation (6a)
$Ki_n(x)$	Bickley function, equation (6b)
n	index of refraction
$q_y(y,z)$	net radiation heat flux in y-direction
$q_y^{\pm}(y,z)$	forward (+) and backward (−) radiation heat flux in y-direction
$q_z(y,z)$	net radiation heat flux in z-direction
$q_z^{\pm}(y,z)$	forward (+) and backward (−) radiation heat flux in z-direction
$\bar{q}$	net heat flux vector
$T(y,z)$	temperature
T_i	wall temperature
$U_{n,i}(u,z)$	integral defined in the Appendix
$V_{n,i}(y,v)$	integral defined in the Appendix
y	optical variable in y-direction
$Y_i^{\pm}(y,v)$	integral defined in the Appendix
z	optical variable in z-direction
$Z_i^{\pm}(u,z)$	integral defined in the Appendix

Greek Symbols

ϵ_i	emissivity of wall surface
θ	polar angle
ϕ	azimuth angle
σ	Stefan-Boltzmann constant
ρ_i	diffuse reflectivity of bounding surface
ξ	$= z/c$
η	$= y/b$
∇	gradient operator

INTRODUCTION

Radiation transfer in multi-dimensional media is of significant interest to workers in numerous fields, including engineering, geophysics and applied physics. An excellent survey on radiative transfer, with particular emphasis on discussing current understanding in order to stimulate future research, has been given by Viskanta (1984). Recently, Howell (1988) reviewed the major events in the development of engineering treatment of radiative transfer in participating media. Relevant to engineering applications, radiative heat transfer is of interest in the design and analysis of energy conversion systems, such as combustors, natural gas and pulverized coal furnaces, engines, prediction of fire spread and safety, the studies of nuclear reactor safety, advanced propulsion devices for outer space, and the interaction of laser beams with semi-transparent media. In these situations, the temperatures are generally high, resulting in a strong contribution from the radiative transfer to the overall transport of energy within the system. Radiative transfer is also of importance in low temperature applications, such as in those involving the analysis of energy transfer through fibrous and porous insulations, cellular and porous materials, photochemical reactors and solar ponds.

To perform analyses of the energy transfer in any of these situations, it is often necessary to consider the effects of convection, conduction and chemical reactions in addition to the radiative transport of energy. Since the coupling between the energy equation and the equation of radiative transfer is highly nonlinear, it is of utmost importance to utilize efficient and accurate solution procedures to the radiation part of the overall problem. Consequently, a great effort has been expended in developing such methods.

Several workers have utilized the integral form of the equation of transfer for studying multi-dimensional radiative transfer, but it appears that these works have been limited to an analysis of cases involving black bounding surfaces. Breig and Crosbie (1975) investigated the effects of cosine varying radiation on the boundary fluxes. Crosbie and Koewing (1979) analyzed the effects of scattering in a planar medium subjected to collimated irradiation. Razzaque et al. (1983) constructed a finite element approach for treating radiative transfer in an absorbing, emitting, two-dimensional medium. Yuen and Ho (1985) employed a power series expansion and collocation strategy for analyzing the effects of heat generation in a two-dimensional medium. Thynell and Özişik (1987a) utilized a power series expansion and the Galerkin method to study radiative transfer in absorbing, emitting and isotropically scattering media. Recently, Yuen and Takara (1988) presented "benchmark" results involving the interaction of conduction and radiation in two-dimensional rectangular media bounded by black walls. They showed that the use of the diffusion or the P_3 approximations, which are highly popular solution approaches, yield highly inaccurate results for both temperatures and temperature gradients. Furthermore, almost all of the previously mentioned works excluded the effects of reflecting bounding surfaces in their analyses, i.e., only black walls were considered. Thus there is a need to continue the development of accurate solution techniques to radiative transport in participating, multi-dimensional media bounded by emitting and reflecting surfaces.

The primary objective of this work is to present a highly accurate solution scheme to radiative transfer in absorbing, emitting two-dimensional media bounded by diffusely reflecting and emitting walls. Effects of the reflecting walls on the radiation fluxes in the y- and z-directions as well as on the incident radiation are investigated extensively. The secondary objective is to compare the results obtained with our approach to those generated by utilizing P_1 approximation.

2 ANALYSIS

2.1 *Mathematical Formulation*

We consider radiative heat transfer in absorbing, emitting, two-dimensional rectangular media bounded by diffusely emitting and reflecting opaque walls. The radiative properties are assumed gray and spatially homogeneous, but the analysis also is applicable if it is applied for a specific wavelength. The medium has spatially distributed and varying energy sources of intensity $I_b[T(y,z)]$, where $T(y,z)$ is the temperature. I_b is the Planck function, which for the gray analysis becomes $I_b = n^2\sigma T^4/\pi$. A schematic of the physical model and coordinates is illustrated in Fig. 1. The mathematical description to this problem is given by (Özişik, 1973)

$$\left[\sin\theta\sin\phi\frac{\partial}{\partial y} + \cos\theta\frac{\partial}{\partial z} + 1\right] I(y,z,\theta,\phi) = I_b[T(y,z)],$$
$$\text{in } |y| < b, \quad |z| < c, \quad 0 \le \theta \le \pi, \quad -\pi \le \phi \le \pi, \tag{1}$$

and the boundary conditions are given by

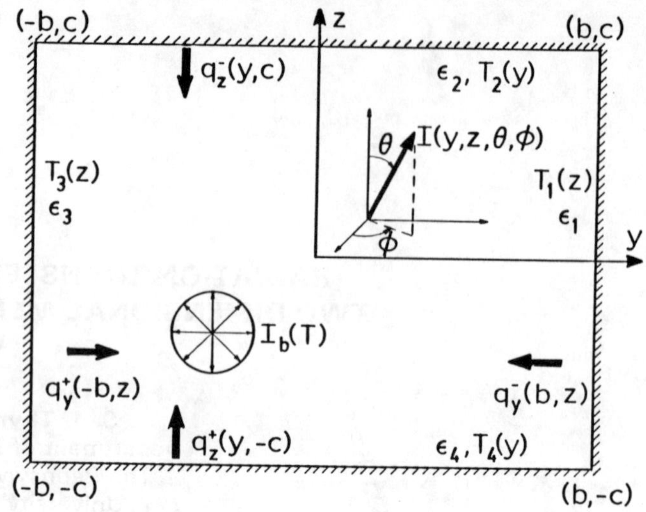

Fig. 1. Schematic of physical model and coordinates.

$$I(b,z,\theta,\phi) = \epsilon_1 I_{b1}(z) + \frac{\rho_1}{\pi} q_y^+(b,z),$$
$$-c < z < c, \quad 0 \le \theta \le \pi, \quad -\pi \le \phi \le 0, \tag{2a}$$
$$I(y,c,\theta,\phi) = \epsilon_2 I_{b2}(y) + \frac{\rho_2}{\pi} q_z^+(y,c),$$
$$-b < y < b, \quad \tfrac{\pi}{2} \le \theta \le \pi, \quad -\pi \le \phi \le \pi, \tag{2b}$$
$$I(-b,z,\theta,\phi) = \epsilon_3 I_{b3}(z) + \frac{\rho_3}{\pi} q_y^-(-b,z),$$
$$-c < z < c, \quad 0 \le \theta \le \pi, \quad 0 \le \phi \le \pi, \tag{2c}$$
$$I(y,-c,\theta,\phi) = \epsilon_4 I_{b4}(y) + \frac{\rho_4}{\pi} q_z^-(y,-c),$$
$$-b < y < b, \quad 0 \le \theta \le \tfrac{\pi}{2}, \quad -\pi \le \phi \le \pi. \tag{2d}$$

Here, $I(y,z,\phi,\theta)$ is the radiation intensity, y and z are the optical variables, θ and ϕ are the polar and azimuth angles, respectively, $\epsilon_i = 1 - \rho_i$, $i = 1,2,3,4$ the emissivities of the opaque surfaces from the inside, and $I_b[T(y,z)]$ is the spatially distributed energy source due to the temperature $T(y,z)$ of the medium, and for simplicity we define at the wall $I_{bi} = I_b(T_i)$, e.g., $I_{b1}(z) = I_b[T_1(z)]$. The forward and backward radiation heat fluxes in the y- and z-directions, respectively, are defined by

$$q_y^+(y,z) = 2\int_{\theta=0}^{\pi}\int_{\phi=0}^{\frac{\pi}{2}} I(y,z,\theta,\phi)\sin^2\theta\sin\phi\, d\phi\, d\theta, \tag{3a}$$

$$q_z^+(y,z) = 2\int_{\theta=0}^{\frac{\pi}{2}}\int_{\phi=-\frac{\pi}{2}}^{\frac{\pi}{2}} I(y,z,\theta,\phi)\cos\theta\sin\theta\, d\phi\, d\theta, \tag{3b}$$

$$q_y^-(y,z) = 2\int_{\theta=0}^{\pi}\int_{\phi=0}^{\frac{\pi}{2}} I(y,z,\theta,\pi-\phi)\sin^2\theta\sin\phi\, d\phi\, d\theta, \tag{3c}$$

$$q_z^-(y,z) = 2\int_{\theta=0}^{\frac{\pi}{2}}\int_{\phi=-\frac{\pi}{2}}^{\frac{\pi}{2}} I(y,z,\pi-\theta,\phi)\cos\theta\sin\theta\, d\phi\, d\theta. \tag{3d}$$

The net radiation flux in y- and z-directions are now specified as

$$q_y(y,z) = q_y^+(y,z) - q_y^-(y,z), \tag{4a}$$
$$q_z(y,z) = q_z^+(y,z) - q_z^-(y,z). \tag{4b}$$

Equation (1) subject to boundary conditions given by equations (2) represents a singular integro-differential equation due to the possibility of different surface temperatures and emissivities, as well as the effects of the corners of the enclosure. As such, equation (1) is difficult to solve accurately, but the integro-differential equation is transformable into a corresponding integral equation of the Fredholm type, in which all such singularities are being smoothed out. In the present case, however, it is necessary to construct four such integral equations. The reason for this is that the incident flux on one of the surfaces depends on the radiation flux leaving from all elemental surface areas of the other three surfaces. By following the analysis of either Smith (1964) or Lin and Tsai (1988), it is readily shown that the dependent variables are the outgoing fluxes at the walls (the radiosities), resulting in the following system of coupled integral equations:

$$\begin{aligned} q_y^+(y,z) = & F_1(y,z) + \frac{1}{\pi}\int_{-c}^{c} \frac{(b+y)}{(z-z')} K_3(b+y; z-z')\, q_y^+(-b,z')\, dz' \\ & + \frac{1}{\pi}\int_{-b}^{y} [K_3(y-y'; c-z) q_z^-(y',c) \\ & + K_3(y-y'; c+z)\, q_z^+(y',-c)]\, dy', \end{aligned} \quad (5a)$$

$$\begin{aligned} q_z^+(y,z) = & F_2(y,z) + \frac{1}{\pi}\int_{-b}^{b} \frac{(c+z)}{(y-y')} K_3(y-y'; c+z)\, q_z^+(y',-c)\, dy' \\ & + \frac{1}{\pi}\int_{-c}^{z} [K_3(b-y; z-z')\, q_y^-(b,z') \\ & + K_3(b+y; z-z')\, q_y^+(-b,z')]\, dz', \end{aligned} \quad (5b)$$

$$\begin{aligned} q_y^-(y,z) = & F_3(y,z) + \frac{1}{\pi}\int_{-c}^{c} \frac{(b-y)}{(z-z')} K_3(b-y; z-z')\, q_y^-(b,z')\, dz' \\ & + \frac{1}{\pi}\int_{y}^{b} [K_3(y'-y; c-z)\, q_z^-(y',c) \\ & + K_3(y'-y; c+z)\, q_z^+(y',-c)]\, dy', \end{aligned} \quad (5c)$$

$$\begin{aligned} q_z^-(y,z) = & F_4(y,z) + \frac{1}{\pi}\int_{-b}^{b} \frac{(c-z)}{(y-y')} K_3(y-y'; c-z)\, q_z^-(y',c)\, dy' \\ & + \frac{1}{\pi}\int_{z}^{c} [K_3(b-y; z-z')\, q_y^-(b,z') \\ & + K_3(b+y; z-z')\, q_y^+(-b,z')]\, dz', \end{aligned} \quad (5d)$$

where the functions $F_i(y,z)$, $i = 1,2,3,4$ are defined in the Appendix. If isotropically scattering particulate matter is present within the medium, then the blackbody emission term $I_b[T(y,z)]$ would be replaced with $(1-\omega)I_b[T(y,z)] + \frac{\omega}{4\pi}G(y,z)$ in the representation of the functions $F_i(y,z)$, $i = 1,2,3,4$; ω denotes the single scattering albedo. The kernels are given by

$$K_n(u;v) = 2\,u\,v\,Ki_n[(u^2+v^2)^{1/2}]/(u^2+v^2)^{n/2}, \quad (6a)$$

where $Ki_n(x)$ are the Bickley functions (Bickley and Naylor, 1935; Yuen and Wong, 1983), which is an integral of the form

$$Ki_n(x) = \int_0^{\pi/2} \exp(-x/\cos\theta)\cos^{n-1}\theta\, d\theta. \quad (6b)$$

2.2 *Solution Method*

To solve for the outgoing fluxes, $q_y^{\pm}(\mp b, z)$ and $q_z^{\pm}(y, \mp c)$, it is necessary to construct a simultaneous solution of equations (5). First, we integrate the boundary conditions, equations (2), over the appropriate hemispherical solid angles to obtain

$$q_y^-(b,z) = \epsilon_1 \pi I_{b1}(z) + \rho_1 q_y^+(b,z), \qquad -c < z < c, \quad (7a)$$

$$q_z^-(y,c) = \epsilon_2 \pi I_{b2}(y) + \rho_2 q_z^+(y,c), \qquad -b < y < b, \quad (7b)$$

$$q_y^+(-b,z) = \epsilon_3 \pi I_{b3}(z) + \rho_3 q_y^-(-b,z), \qquad -c < z < c, \quad (7c)$$

$$q_z^+(y,-c) = \epsilon_4 \pi I_{b4}(y) + \rho_4 q_z^-(y,-c), \qquad -b < y < b. \quad (7d)$$

Next, we evaluate equation (5a) at $y = b$, equation (5b) at $z = c$, equation (5c) at $y = -b$, and equation (5d) at $z = -c$, then eliminate the respective fluxes into the walls by utilizing equations (7). This results in the following set of integral equations for the four unknown radiosities:

$$\begin{aligned} q_y^-(b,z) = & \epsilon_1 \pi I_{b1}(z) + \rho_1 F_1(b,z) \\ & + \frac{\rho_1}{\pi}\int_{-c}^{c} \frac{2b}{(z-z')} K_3(2b; z-z')\, q_y^+(-b,z')\, dz' \\ & + \frac{\rho_1}{\pi}\int_{-b}^{b} [K_3(b-y'; c-z)\, q_z^-(y',c) \\ & + K_3(b-y'; c+z)\, q_z^+(y',-c)]\, dy', \end{aligned} \quad (8a)$$

$$\begin{aligned} q_z^-(y,c) = & \epsilon_2 \pi I_{b2}(y) + \rho_2 F_2(y,c) \\ & + \frac{\rho_2}{\pi}\int_{-b}^{b} \frac{2c}{(y-y')} K_3(y-y'; 2c)\, q_z^+(y',-c)\, dy' \\ & + \frac{\rho_2}{\pi}\int_{-c}^{c} [K_3(b-y; c-z')\, q_y^-(b,z') \\ & + K_3(b+y; c-z')\, q_y^+(-b,z')]\, dz', \end{aligned} \quad (8b)$$

$$\begin{aligned} q_y^+(-b,z) = & \epsilon_3 \pi I_{b3}(z) + \rho_3 F_3(-b,z) \\ & + \frac{\rho_3}{\pi}\int_{-c}^{c} \frac{2b}{(z-z')} K_3(2b; z-z')\, q_y^-(b,z')\, dz' \\ & + \frac{\rho_3}{\pi}\int_{-b}^{b} [K_3(b+y'; c-z)\, q_z^-(y',c) \\ & + K_3(b+y'; c+z)\, q_z^+(y',-c)]\, dy', \end{aligned} \quad (8c)$$

$$\begin{aligned} q_z^+(y,-c) = & \epsilon_4 \pi I_{b4}(y) + \rho_4 F_4(y,-c) \\ & + \frac{\rho_4}{\pi}\int_{-b}^{b} \frac{2c}{(y-y')} K_3(y-y'; 2c)\, q_z^-(y',c)\, dy' \\ & + \frac{\rho_4}{\pi}\int_{-c}^{c} [K_3(b-y; c+z')\, q_y^-(b,z') \\ & + K_3(b+y; c+z')\, q_y^+(-b,z')]\, dz'. \end{aligned} \quad (8d)$$

To approximate the solution to the radiosities, we assume power series expansions of the forms

$$q_y^-(b,z) = \sum_{i=0}^{I} A_i z^i, \qquad |z| \le c, \quad (9a)$$

$$q_z^-(y,c) = \sum_{i=0}^{I} B_i y^i, \qquad |y| \le b, \quad (9b)$$

$$q_y^+(-b,z) = \sum_{i=0}^{I} C_i z^i, \qquad |z| \le c, \quad (9c)$$

$$q_z^+(y,-c) = \sum_{i=0}^{I} D_i y^i, \qquad |y| \le b. \quad (9d)$$

Inspection of equations (9) reveals that the series expansions are terminated at the same value ($i = I$); however, it is not necessary to assume such truncation and in this work it is used for simplicity. To generate a system of linear algebraic equations for the unknown expansion coefficients, equations (9) are substituted into equations (8) and then a collocation method based on the zeroes of the Chebyshev polynomials is utilized (Thynell and Özişik, 1987b). The resulting system of linear algebraic equations is given by

$$\sum_{i=0}^{I}[\pi A_i z_j^i - \rho_1 B_i Y_i^-(b;c-z_j) - \rho_1 C_i U_{3,i}(2b;z_j) - \rho_1 D_i Y_i^-(b;c+z_j)] = \epsilon_1 \pi I_{b1}(z_j) + \rho_1 F_1(b,z_j),$$

$$\sum_{i=0}^{I}[-\rho_2 A_i Z_i^-(b-y_j;c) + \pi B_i y_j^i - \rho_2 C_i Z_i^-(b+y_j;c) - \rho_2 D_i V_{3,i}(y_j;2c)] = \epsilon_2 \pi I_{b2}(y_j) + \rho_2 F_2(y_j,c),$$

$$\sum_{i=0}^{I}[-\rho_3 A_i U_{3,i}(2b;z_j) - \rho_3 B_i Y_i^+(-b;c-z_j) + \pi C_i z_j^i - \rho_3 D_i Y_i^+(-b;c+z_j)] = \epsilon_3 \pi I_{b3}(z_j) + \rho_3 F_3(-b,z_j),$$

$$\sum_{i=0}^{I}[-\rho_4 A_i Z_i^+(b-y_j;-c) - \rho_4 B_i V_{3,i}(y_j;2c) - \rho_4 C_i Z_i^+(b+y_j;-c) + \pi D_i y_j^i] = \epsilon_4 \pi I_{b4}(y_j) + \rho_4 F_4(y_j,-c), \qquad j=0,1,\ldots,I. \tag{10}$$

Definition of the functions $U_{n,i}(u;z)$, $V_{n,i}(y;v)$, $Y_i^{\pm}(y;v)$, and $Z_i^{\pm}(u;z)$ is presented in the Appendix. The collocation points, y_j and z_j, $j = 0,1,\ldots,I$, are computed from

$$\frac{y_j}{b} = \frac{z_j}{c} = \cos\left(\frac{2j+1}{2(I+1)}\pi\right), \qquad j=0,1,\ldots,I. \tag{11}$$

Once a solution has been obtained for the unknown expansion coefficients, the forward and backward radiation fluxes in the y- and z-directions, respectively, are computed from

$$q_y^+(y,z) = F_1(y,z) + \frac{1}{\pi}\sum_{i=0}^{I}[B_i Y_i^-(y;c-z) + C_i U_{3,i}(b+y;z) + D_i Y_i^-(y;c+z)], \tag{12a}$$

$$q_z^+(y,z) = F_2(y,z) + \frac{1}{\pi}\sum_{i=0}^{I}[A_i Z_i^-(b-y;z) + C_i Z_i^-(b+y;z) + D_i V_{3,i}(y;c+z)], \tag{12b}$$

$$q_y^-(y,z) = F_3(y,z) + \frac{1}{\pi}\sum_{i=0}^{I}[A_i U_{3,i}(b-y;z) + B_i Y_i^+(y;c-z) + D_i Y_i^+(y;c+z)], \tag{12c}$$

$$q_z^-(y,z) = F_4(y,z) + \frac{1}{\pi}\sum_{i=0}^{I}[A_i Z_i^+(b-y;z) + B_i V_{3,i}(y;c-z) + C_i Z_i^+(b+y;z)]. \tag{12d}$$

The gradient of the net radiation heat flux is related to the temperature and incident radiation by

$$\nabla \bullet \vec{q}(y,z) = 4\pi I_b[T(y,z)] - G(y,z), \tag{13}$$

where the incident radiation is evaluated from

$$\pi G(y,z) = \sum_{i=0}^{I}[A_i U_{2,i}(b-y;z) + B_i U_{2,i}(b+y;z) + C_i V_{2,i}(y;c-z) + D_i V_{2,i}(y;c+z)] + \int_{z'=-c}^{z'=c}\int_{y'=-b}^{y'=b} \pi I_b[T(y',z')]\frac{K_1(y-y';z-z')}{(y-y')(z-z')}\,dy'\,dz'. \tag{14}$$

3 RESULTS AND DISCUSSION

To illustrate the application of the previously described method of solution, we consider an enclosure containing a medium whose temperature is uniform, $I_b[T(y,z)] = 1$, and whose bounding surfaces are nonemitting (cold), $I_{bi} = 0$, $i = 1,2,3,4$. Because of these assumptions, we note that the radiosities must be symmetric functions, permitting us to exclude the odd order terms in the expansions given by equations (9). We also note that the emissivities of the opposing bounding surfaces are identical, resulting in $q_y^-(b,z) = q_y^+(-b,z)$ and $q_z^-(y,c) = q_z^+(y,-c)$. Thus the numerical effort has been simplified significantly. For this physical situation, we will discuss the accuracy of the solution procedure, the effects of reflecting bounding surfaces as opposed to assuming black surfaces, and compare the results to the P_1 approximation.

In Tables 1 and 2, we show the convergence of the incident radiation along the centerline $y = 0$ for the case of a uniform source of unit strength within the medium and cold bounding surfaces with $\epsilon_i = 0.99$ and $\epsilon_i = 0.5$, i =1, 2, 3, 4, respectively. Examination of these tables indicate that rapid convergence is achieved for all the optical dimensions considered. To obtain results of two-digit accuracy (within 1 per cent of highest order expansion), only one or at most two terms are needed in the expansions given by equations (9). That is, two or four unknown expansion coefficients need to be solved for in the system of linear algebraic equations. The high degree of accuracy is, of course, attributable to the smoothness of the radiosities as well as the fact that medium temperature is uniform.

Table 1. Results of the incident radiation along the centerline for a unit source within the medium and nonemitting boundaries with $b = 0.5$ and ϵ_i=0.99, i= 1,2,3,4, $\xi = z/c$.

I	$G(0,z)$ for $c = 0.05$					
	ξ=−1.0	ξ=−0.8	ξ=−0.6	ξ=−0.4	ξ=−0.2	ξ=0.0
2	1.5600	1.7572	1.8634	1.9296	1.9667	1.9787
4	1.5601	1.7572	1.8635	1.9297	1.9668	1.9787
6	1.5601	1.7572	1.8635	1.9297	1.9667	1.9787
8	1.5601	1.7572	1.8635	1.9297	1.9667	1.9787
I	$G(0,z)$ for $c = 0.5$					
	ξ=−1.0	ξ=−0.8	ξ=−0.6	ξ=−0.4	ξ=−0.2	ξ=0.0
2	3.9599	5.4241	6.0722	6.4394	6.6327	6.6933
4	3.9600	5.4243	6.0724	6.4397	6.6329	6.6935
6	3.9600	5.4243	6.0724	6.4397	6.6329	6.6935
8	3.9600	5.4243	6.0724	6.4397	6.6329	6.6935
I	$G(0,z)$ for $c = 2.5$					
	ξ=−1.0	ξ=−0.8	ξ=−0.6	ξ=−0.4	ξ=−0.2	ξ=0.0
2	4.2886	7.5921	8.1648	8.3495	8.4168	8.4342
4	4.2878	7.5919	8.1656	8.3503	8.4172	8.4345
6	4.2878	7.5923	8.1658	8.3502	8.4171	8.4344
8	4.2878	7.5924	8.1657	8.3502	8.4171	8.4344

Table 2. Results of the incident radiation along the centerline for a unit source within the medium and nonemitting boundaries with $b = 0.5$ and ϵ_i=0.5, i= 1,2,3,4, $\xi = z/c$.

I	$G(0,z)$ for $c = 0.05$					
	ξ=−1.0	ξ=−0.8	ξ=−0.6	ξ=−0.4	ξ=−0.2	ξ=0.0
2	3.0856	3.2555	3.3461	3.4026	3.4343	3.4445
4	3.0899	3.2604	3.3514	3.4080	3.4398	3.4501
6	3.0888	3.2590	3.3501	3.4068	3.4386	3.4489
8	3.0895	3.2598	3.3508	3.4075	3.4394	3.4497
I	$G(0,z)$ for $c = 0.5$					
	ξ=−1.0	ξ=−0.8	ξ=−0.6	ξ=−0.4	ξ=−0.2	ξ=0.0
2	7.1763	8.0496	8.4384	8.6637	8.7848	8.8232
4	7.1816	8.0609	8.4506	8.6757	8.7965	8.8348
6	7.1823	8.0608	8.4508	8.6761	8.7969	8.8352
8	7.1824	8.0614	8.4511	8.6763	8.7971	8.8354
I	$G(0,z)$ for $c = 2.5$					
	ξ=−1.0	ξ=−0.8	ξ=−0.6	ξ=−0.4	ξ=−0.2	ξ=0.0
2	7.6766	9.5919	9.9796	10.128	10.191	10.210
4	7.6432	9.5802	10.001	10.153	10.206	10.218
6	7.6460	9.5933	10.009	10.150	10.202	10.215
8	7.6490	9.5965	10.006	10.149	10.203	10.216

In Figs. 2-4, we illustrate the profiles of the net radiant heat fluxes $q_y(b,\xi)$ and $q_z(\eta,c)$ for the different values of the emissivities $\epsilon_1 = \epsilon_3 = 0.5$, 0.99, and $\epsilon_2 = \epsilon_4 = 0.5$, 0.99, and for the optical dimensions $(b,c) = (0.05, 0.5)$, (0.05, 2.5), and (0.5, 2.5), respectively. Examination of these figures reveals that the net radiant heat flux along the top wall ($z = c$), $q_z(\eta,c)$, is very sensitive to the emissivity of the four bounding surfaces. The maximum net flux in the z-direction is obtained for the case $\epsilon_1 = \epsilon_3 = 0.50$ and $\epsilon_2 = \epsilon_4 = 0.99$, because a significant amount of the radiant energy emitted by the medium is reflected by the side walls onto and absorbed by the top wall. Thus, the minimum net radiant flux is readily deduced to occur when the top wall is more reflecting and the vertical surfaces are nearly black ($\epsilon_1 = \epsilon_3 = 0.99$ and $\epsilon_2 = \epsilon_4 = 0.50$). We also note in these figures that even for large aspect ratios (c/b=10 in Fig. 2, and to a smaller extent c/b=50 in Fig. 3) that the assumption of uniform net radiant heat fluxes at the bounding surfaces may introduce an appreciable amount of errors into the analysis. Thus a one-dimensional analysis (a plane-parallel medium approximation) should be used cautiously and after consideration to the results shown in Figs. 2-4. In Fig. 4, however, the effect of the emissivity of the adjacent and opposing bounding surfaces on the net heat fluxes at the walls diminishes due to the increase in the optical dimensions of the enclosure. That is, the direct radiant communication between the surfaces, in particular the between the opposing ones, is weakened significantly. Hence, the top group of four lines corresponds to strong absorption ($\epsilon_2 = \epsilon_4 = 0.99$), whereas the bottom group of four lines corresponds to appreciable reflection ($\epsilon_2 = \epsilon_4 = 0.5$). In addition, the net heat flux in the y-direction, $q_y(b,\xi)$, is in all cases only weakly affected by the surface emissivities of top and bottom surfaces due to the large values of the aspect ratios considered in these figures.

In Figs. 5-7, we show the results of the net radiant heat flux $q_y(b,\xi)$ for the various bounding surface emissivities ϵ_1=ϵ_3=0.5, 0.99, and ϵ_2=ϵ_4=0.5, 0.99, using both the "exact" solution methodology described in this paper as well as the P_1 approximation for a square enclosure whose dimensions $b = c = 0.05$, 0.5 and 2.5, respectively. Inspection of these figures reveals that the P_1 approximation is, as expected, inaccurate for the smaller optical dimensions ($b = c$ =0.05

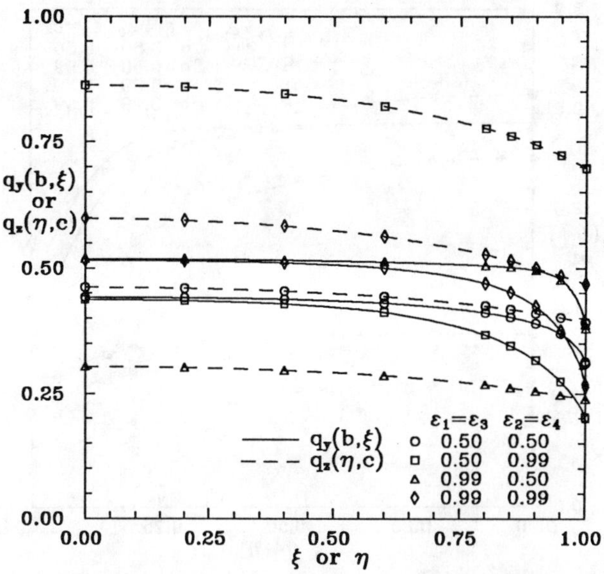

Fig. 2. Effects of cold, diffusely reflecting bounding surfaces on the net radiative heat fluxes $q_y(b,\xi)$ and $q_z(\eta,c)$ along the walls at $y = b$ and $z = c$, respectively, for a rectangular enclosure whose medium is at uniform temperature with $I_b[T(y,z)] = 1$; b =0.05, c =0.5, $\eta = y/b$, and $\xi = z/c$.

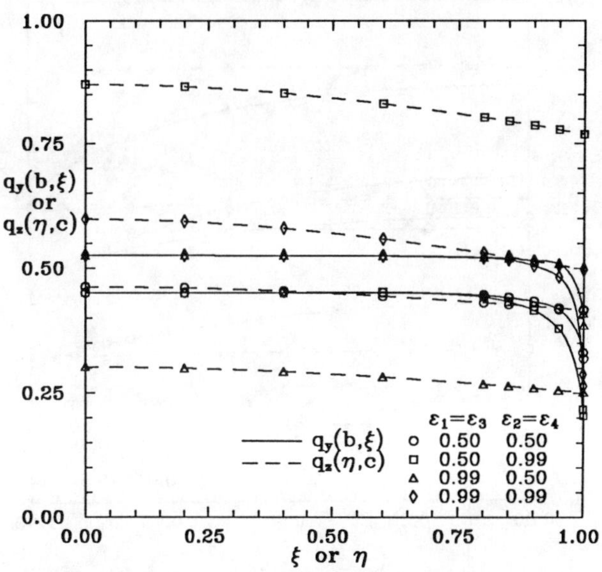

Fig. 3. Effects of cold, diffusely reflecting bounding surfaces on the net radiative heat fluxes $q_y(b,\xi)$ and $q_z(\eta,c)$ along the walls at $y = b$ and $z = c$, respectively, for a rectangular enclosure whose medium is at uniform temperature with $I_b[T(y,z)] = 1$; b =0.05, c =2.5, $\eta = y/b$, and $\xi = z/c$.

and 0.5). The difference between the "exact" solution and the P_1 approximation is in some cases more than 40%. It is observed that the difference is especially large in those cases in which the bounding emissivities are different. Although the medium emits the radiant energy isotropically, the radiation intensity becomes highly discontinuous due to the different bounding emissivities. The P_1 approximation, with its poor representation of the intensity, is not capable of handling such discontinuities very accurately. The approach presented in this

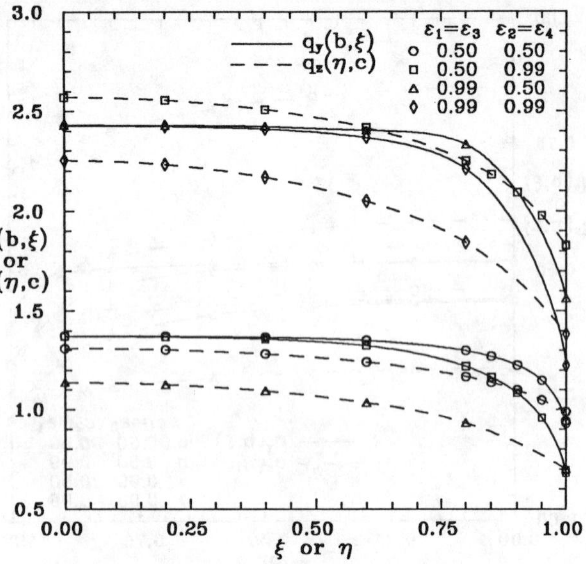

Fig. 4. Effects of cold, diffusely reflecting bounding surfaces on the net radiative heat fluxes $q_y(b,\xi)$ and $q_z(\eta,c)$ along the walls at $y = b$ and $z = c$, respectively, for a rectangular enclosure whose medium is at uniform temperature with $I_b[T(y,z)] = 1$; b =0.5, c =2.5, $\eta = y/b$, and $\xi = z/c$.

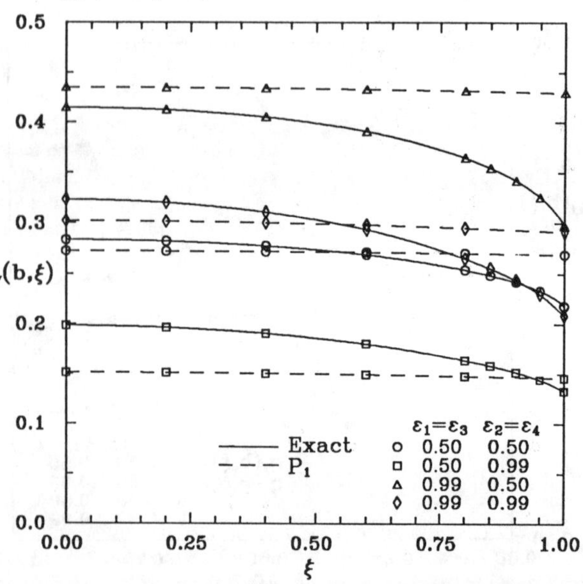

Fig. 5. Results of the net radiative heat flux $q_y(b,\xi)$ along the wall at $y = b$ obtained from the "exact" solution and the P_1 approximation for a square enclosure whose medium is at uniform temperature with $I_b[T(y,z)] = 1$; $b = c$ =0.05 and $\xi = z/c$.

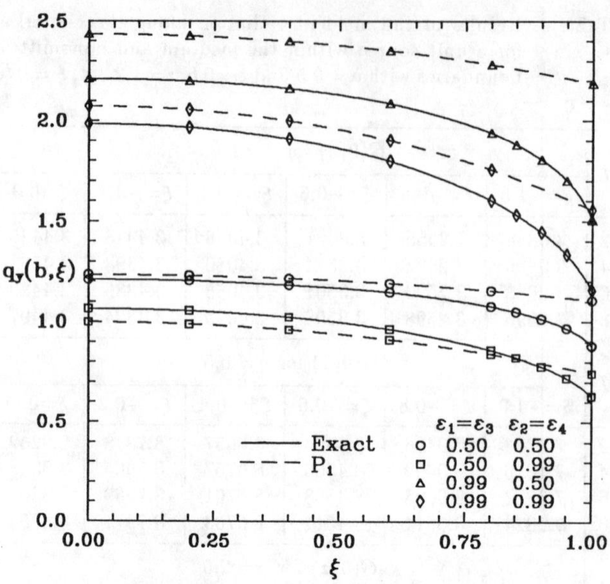

Fig. 6. Results of the net radiative heat flux $q_y(b,\xi)$ along the wall at $y = b$ obtained from the "exact" solution and the P_1 approximation for a square enclosure whose medium is at uniform temperature with $I_b[T(y,z)] = 1$; $b = c$ =0.5 and $\xi = z/c$.

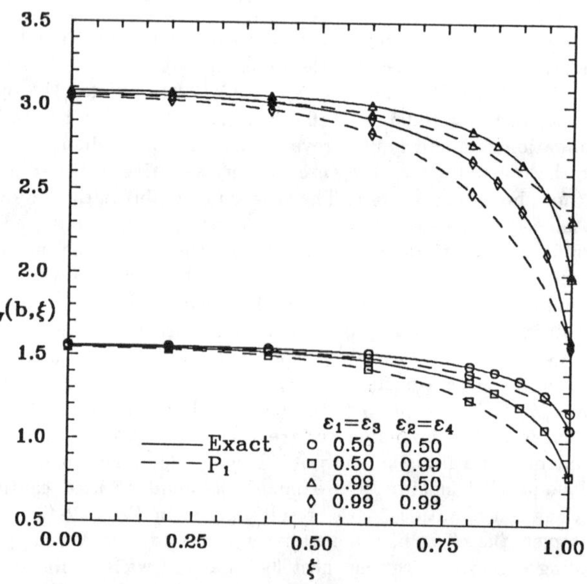

Fig. 7. Results of the net radiative heat flux $q_y(b,\xi)$ along the wall at $y = b$ obtained from the "exact" solution and the P_1 approximation for a square enclosure whose medium is at uniform temperature with $I_b[T(y,z)] = 1$; $b = c$ =2.5 and $\xi = z/c$.

work does not have such difficulty with the discontinuous intensity, at least not directly. The reason for this is that the method is based on seeking a solution to solid angle integrated functions, which become much smoother and are much more readily approximated by polynomial expansions. However, as the optical dimensions of the enclosure increases ($b = c$ =2.5), the difference between the "exact" solution and the P_1 approximation is almost negligible near the center of the surface ($y = 0$, $z = c$), but a significant difference is still observed near the corners of the enclosure. For such large optical dimensions, the intensity field near the top center of the enclosure is largely due to the isotropic emission by the medium; hence the observed satisfactory agreement between the "exact" solution and the P_1 approximation.

In Figs. 8-10, we illustrate the incident radiation $G(0,\xi)$ along the centerline $y = 0$ for the various bounding surface emissivities $\epsilon_1 = \epsilon_3 =$ 0.5, 0.99, and $\epsilon_2 = \epsilon_4 =$ 0.5, 0.99, using both the previously described "exact" solution method and the P_1 approximation for a square enclosure whose dimensions are $b = c =$ 0.05, 0.5 and 2.5, respectively. Examination of these figures reveals that the incident radiation in the interior of the enclosure is underestimated by the P_1 approximation

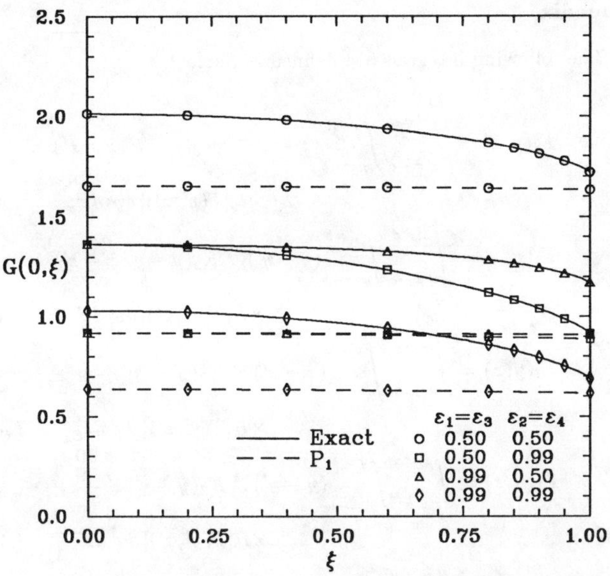

Fig. 8. Results of the incident radiation $G(0,\xi)$ along the centerline $y = 0$ obtained from the "exact" solution and the P_1 approximation for a square enclosure whose medium is at uniform temperature with $I_b[T(y,z)] = 1$; $b = c = 0.05$ and $\xi = z/c$.

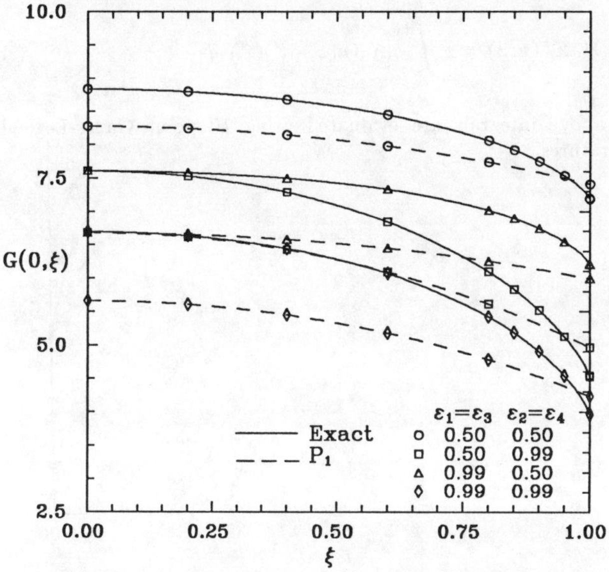

Fig. 9. Results of the incident radiation $G(0,\xi)$ along the centerline $y = 0$ obtained from the "exact" solution and the P_1 approximation for a square enclosure whose medium is at uniform temperature with $I_b[T(y,z)] = 1$; $b = c = 0.5$ and $\xi = z/c$.

by more than 40% in those cases involving the smaller optical dimensions ($b = c = 0.05$ and 0.5). It should be noted that it is of utmost importance to compute the incident radiation within the medium as accurate as possible. The reason for this is that the overall conservation of energy yields expressions involving the gradient of the net radiant heat, which may be written as $\nabla \bullet \overline{q} = 4n^2\sigma T^4 - G$. The P_1 approximation is inaccurate in its prediction of both the incident radiation in the interior as well as the net flux along the bounding walls in the case of smaller optical dimensions. It is also of interest to note that the results shown in Figs. 8 and 9 indicate that the P_1 approximation

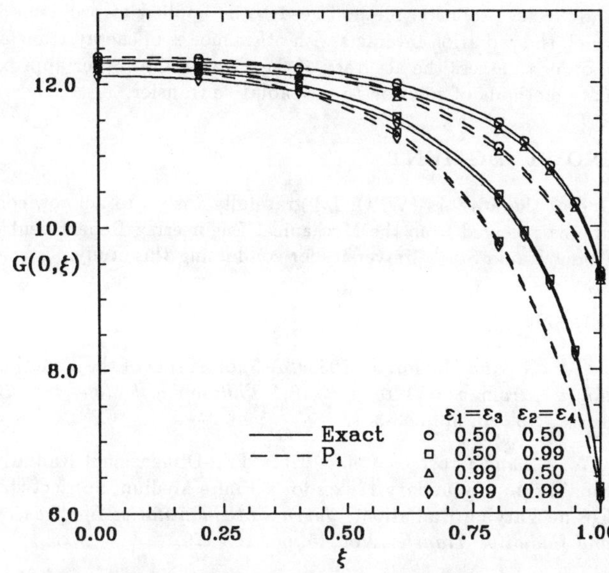

Fig. 10. Results of the incident radiation $G(0,\xi)$ along the centerline $y = 0$ obtained from the "exact" solution and the P_1 approximation for a square enclosure whose medium is at uniform temperature with $I_b[T(y,z)] = 1$; $b = c = 2.5$ and $\xi = z/c$.

is inaccurate for both the nearly black and the nonblack bounding surfaces; that is, the corners of the enclosure have a significant effect on the incident radiation as the surfaces become nearly black. The highly diffusely reflecting surfaces have the effect of smoothing out the intensity distribution, resulting in better agreement between the "exact" solution and the P_1 approximation in the calculation of the incident radiation. For the larger optical dimensions ($b = c = 2.5$), shown in Fig. 10, the accuracy of the P_1 approximation improves considerably in the interior of the medium due to a more isotropic angular distribution of the radiation intensity; near the bounding surfaces, an appreciable difference exists between the "exact" and the P_1 approximation.

SUMMARY AND CONCLUSIONS

A method of solution for analyzing radiative transfer in absorbing, emitting, two-dimensional rectangular media bounded by diffusely emitting and diffusely reflecting surfaces has been presented. The method is based on using the integral form of the equation of transfer, which is represented by four coupled integral equations of the Fredholm type for the four radiosities. The radiosities are represented in terms of power series, whose expansion coefficients are solved for by employing standard methods. Based on the presented results for an absorbing, emitting medium at uniform temperature and bounded by cold surfaces, the following conclusions have been reached.

1. Only two or at most three terms may be needed in the expansions to obtain results to within one percent accuracy. This implies that an 8×8 or at most 12×12 coefficient matrix may need to be inverted in the analysis of problems in which the radiative transfer interacts with other modes of energy transfer.

2. The required integrals need to be evaluated only for different optical dimensions, thus reducing the computing times significantly if parametric studies need to be performed.

3. The P_1 approximation yields results for both the net heat flux at the bounding surfaces and the incident radiation within the medium that for small or intermediate optical dimensions are significantly in error compared to the "exact" solution.

4. The "exact" solution method need to be applied to problems in which the radiation interacts with other modes of energy transfer, in order to assess the accuracy of the P_1 as well as other approximate methods of solution to the radiative transfer.

ACKNOWLEDGMENT

One of the authors (W.-Q. L.) gratefully wishes to acknowledge the support received from the Mechanical Engineering Department of The Pennsylvania State University for conducting this study.

REFERENCES

Bickley, W. G., and Naylor, J., 1935, "A Short Table of the Functions $Ki_n(x)$, from $n = 1$ to $n = 16$," *Philosophical Magazine*, 7th Series, No. 20, pp. 343-347.

Breig, W. F., and Crosbie, A. L., 1975, "Two-Dimensional Radiative Equilibrium: Boundary Fluxes for a Finite Medium Subjected to Cosine Varying Radiation," *Journal of Quantitative Spectroscopy and Radiative Transfer*, Vol. 15, pp. 163-179.

Crosbie, A. L., and Koewing, J. W., 1979, "Two-Dimensional Radiative Transfer in a Finite Scattering Planar Medium," *Journal of Quantitative Spectroscopy and Radiative Transfer*, Vol. 21, pp. 573-595.

Howell, J. R., 1988, "Thermal Radiation in Participating Media: The Past, The Present and Some Possible Futures," *Journal of Heat Transfer*, Vol. 110, No. 4(B), pp. 1220-1229.

Lin, J.-D., and Tsai, J.-H., 1988, "Exact Formulation of Anisotropic Scattering in an Arbitrary Enclosure," *Journal of Quantitative Spectroscopy and Radiative Transfer*, Vol. 39, pp. 299-308.

Özişik, M. N., 1973, *Radiative Transfer*, Wiley, New York.

Razzaque, M. M., Klein, D. E., and Howell, J. R., 1983, "Finite Element Solution to Radiative Transfer in a Two-Dimensional Rectangular Enclosure with Gray Participating Media," *Journal of Heat Transfer*, Vol. 105, pp. 933-936.

Smith, M. G., 1964, "The Transport Equation with Plane Symmetry and Isotropic Scattering," *Proceedings of the Cambridge Philosophical Society*, Vol. 60, pp. 909-921.

Thynell, S. T., and Özişik, M. N., 1987a, "Radiation Transfer in Isotropically Scattering, Rectangular Enclosures," *Journal of Thermophysics and Heat Transfer*, Vol. 1, No. 1, pp. 69-76.

Thynell, S. T., and Özişik, M. N., 1987b, "Radiation Transfer in Absorbing, Emitting, Isotropically Scattering, Homogeneous Cylindrical Media," *Journal of Quantitative Spectroscopy and Radiative Transfer*, Vol. 38, pp. 413-426.

Viskanta, R., 1984, "Radiative Heat Transfer," *Fortschritte der Verfahrenstechnik*, Vol. 22, pp. 51-81.

Yuen, W. W. and Takara, E. E., 1988, "Analysis of Combined Conductive-Radiative Heat Transfer in a Two-Dimensional Rectangular Enclosure With a Gray Medium," *Journal of Heat Transfer*, Vol. 110, pp. 468-474.

Yuen, W. W., and Ho, C. F., 1985, "Analysis of Two-Dimensional Radiative Heat Transfer in a Gray Medium With Internal Heat Generation," *International Journal of Heat and Mass Transfer*, Vol. 28, No. 1, pp. 17-23.

Yuen, W. W., and Wong, L. W., 1983, "Numerical Computation of an Important Integral Function in Two-Dimensional Radiative Transfer," *Journal of Quantitative Spectroscopy and Radiative Transfer*, Vol. 29, pp. 145-149.

Appendix

The following integrals are defined in Sect. 2:

$$F_1(y,z) = \int_{z'=-c}^{z'=c}\int_{y'=-b}^{y'=y} (z-z')^{-1} K_2(y-y';z-z') \times I_b[T(y',z')]\,dy'\,dz', \quad (A1)$$

$$F_2(y,z) = \int_{z'=-c}^{z'=z}\int_{y'=-b}^{y'=b} (y-y')^{-1} K_2(y-y';z-z') \times I_b[T(y',z')]\,dy'\,dz', \quad (A2)$$

$$F_3(y,z) = \int_{z'=-c}^{z'=c}\int_{y'=y}^{y'=b} (z-z')^{-1} K_2(y'-y;z-z') \times I_b[T(y',z')]\,dy'\,dz', \quad (A3)$$

$$F_4(y,z) = \int_{z'=z}^{z'=c}\int_{y'=-b}^{y'=b} (y-y')^{-1} K_2(y-y';z'-z) \times I_b[T(y',z')]\,dy'\,dz', \quad (A4)$$

$$U_{n,i}(u;z) = \int_{-c}^{c} \frac{u^{n-2}}{(z-z')} K_n(u;z-z')(z')^i\,dz', \quad (A5)$$

$$V_{n,i}(y;v) = \int_{-b}^{b} \frac{v^{n-2}}{(y-y')} K_n(y-y';v)(y')^i\,dy', \quad (A6)$$

$$Y_i^{\pm}(y;v) = \pm\int_{y}^{\pm b} K_3(y-y';v)(y')^i\,dy', \quad (A7)$$

$$Z_i^{\pm}(u;z) = \pm\int_{z}^{\pm c} K_3(u;z-z')(z')^i\,dz'. \quad (A8)$$

The above integrals are evaluated using 10 point Gauss-Legendre quadratures.

1989 National Heat Transfer Conference
HTD-Vol. 106, Heat Transfer Phenomena in Radiation, Combustion and Fires

RADIATIVE HEAT TRANSFER IN NON-RECTANGULAR ENCLOSURES

M. H. N. Naraghi
Department of Mechanical Engineering
Manhattan College
Riverdale, New York

ABSTRACT

A method for analysis of radiative heat transfer in non-rectangular enclosures with absorbing-emitting and scattering media is presented. In this method, an enclosure is subdivided into a number of rectangular and right triangular elements and the discrete exchange factor (DEF) method is used to discretize each element. Different ways of discretization of rectangular and triangular pieces are discussed and numerical results for triangular enclosures with various surface and gas conditions are presented. The difficulties of this method in analysis of regions in the vicinity of acute angles are also discussed.

NOMENCLATURE

$\overline{dgg}$	direct exchange factor between gases
$\overline{DGG}$	total exchange factor between gases
$\overline{dgs}$	direct exchange factor between gas and surface
$\overline{DGS}$	total exchange factor between gas and surface
$\overline{dsg}$	direct exchange factor between surface and gas
$\overline{DSG}$	total exchange factor between surface and gas
$\overline{dss}$	direct exchange factor between surfaces
$\overline{DSS}$	total exchange factor between surfaces
E	emissive power
E_{ref}	reference emissive power
I	$[\delta_{i,j}]$, identity matrix
K_a	absorption coefficient
K_s	scattering coefficient
K_t	$= K_a + K_t$, extinction coefficient
L_x & L_y	dimensions in x and y directions, respectively
N_g	number of discrete points in gas
N_s	number of discrete points on surface
$\mathbf{r}_j$	position vector
q_s''	surface heat flux
q_g'''	gas heat source
w_i	numerical integration weight factor
W	$[w_i\delta_{i,j}]$
x & y	Cartesian coordinates
X & Y	dimensionless Cartesian coordinates

Greek Symbols

α	absorptivity
β	corner angle
ϵ	emissivity
θ	dimensionless temperature
ρ	reflectivity
$\tau(\mathbf{r})$	transmittance along path $\mathbf{r}$
ω_0	K_s/K_t, albedo for scattering

Subscripts

g	designates gas
s	designates surface

INTRODUCTION

The one-dimensional analysis of radiative heat transfer, although helpful to the understanding of the physical phenomena involved in the problem and the influence of different parameters on the solution, is not applicable to the real problem which is usually multidimensional. One of the multidimensional problems which has received considerable attention in recent years is radiative transfer in rectangular enclosures. Crosbie and Schrenker (1984) contributed exact solutions for rectangular enclosures with non-scattering (absorbing) medium. Approximate methods have been reported by Modest (1975), Shah (1979), Ratzel and Howell (1982) and Iyer and Mengüç (1988). Finite element and discrete-ordinate methods have been used by Razzaque et al. (1981) and Fiveland (1984), respectively, to analyze radiative transfer in rectangular enclosures. Yuen and coworkers (1984, 1985) used the point allocation method to analyze radiative equilibrium in rectangular enclosures. Most recently, Naraghi and Kassemi (1988) used the discrete exchange factor (DEF) method to analyze radiative transfer in rectangular enclosures.

All of the aforementioned works deal with rectangular geometries where boundary conditions for both analytical and numerical solutions can be easily expressed mathematically. In most practical problems, however, enclosures have irregular (non-rectangular) configurations where exact and approximate analytical solutions cannot be obtained. Furthermore, most numerical methods may have difficulties at the boundaries of these enclosures. To date, the most versatile methods for analysis of radiative transfer in enclosures with complex geometries are the Monte Carlo and the zonal methods. The Monte Carlo method was first applied by Howell and Perlmutter (1964) to a one-dimensional radiative transfer problem and later used by Taniguchi (1969) to analyze a three-dimesional enclosure. The application of Monte Carlo method to problems with non-rectangular geometries are reported by Howell et al. (1965, 1966). Although the Monte Carlo method is suitable for analysis of enclosures with complex configurations, its statistical error makes it undesirable when accurate results are needed. The zonal method has been widely used for radiative analysis of most practical problems. This method was developed by Hottel and coworkers (1958, 1967) and later Noble (1975) and Naraghi and Chung (1985) used completely different approaches to obtain explicit formulations for total exchange areas in terms of direct exchange areas. The major difficulty in using the zonal method is the evaluation of direct exchange areas. To avoid this difficulty, Larsen and Howell (1985) proposed an alternative scheme in which the exchange factors are determined experimentally. In both methods the zones are isothermal; hence, the enclosure has to be subdivided into very fine zones if the temperature gradient within the enclosure is large.

Most recently, Naraghi et al. (1988) introduced the idea of the continuous exchange factor (CEF) method which is based on the radiative exchange between differential gas volumes and surface areas. The integral equations in this method relate the continuous total and direct exchange factors. A numerical integration method is used to discretize the integral equations and derive equations for the discrete exchange factor (DEF) method. In the DEF method, radiative exchange between differential surface areas and gas volumes located at nodal points is considered; hence, fewer integrations are needed for the evaluation of the direct exchange factors as compared to the zonal method. It has also been shown (Naraghi and Kassemi, 1988) that it provides accurate results for all range of optical thicknesses even when a small number of nodes is used (e.g., 3×3 or 5×5). Nevertheless, it shares the shortcoming of the other techniques when it is applied to irregularly shaped (non-rectangular) enclosures.

The intent of this work is to extend the application of the DEF method to non-rectangular enclosures with isotropically scattering media. Any non-rectangular enclosure can be subdivided into a number of rectangular and triangular elements. By piecewise integration of the governing equations, a solution can be obtained for the entire enclosure.

FORMULATION

Consider the two-dimensinal enclosure shown in Fig. 1. To apply the exchange factor method (Naraghi et al. 1988), the enclosure is subdivided into a number of rectangular and triangular elements (see Fig. 2). The elements next to the surfaces can, in general, be approximated by right triangles where the surface of the enclosure approximately coincide to the hypotenuse of these triangles. In this paper, attention is focused on two-dimensional configurations. Nevertheless, the same technique can be applied to three-dimensional enclosures which can be broken into rectangular parrallelpiped and pyramid elements.

Let m and n be numbers of gas (volume) and surface (area) pieces, respectively. The governing equations of the CEF method become

$$\overline{DSS}(\mathbf{r}_i,\mathbf{r}_j) = p_{ss}(\mathbf{r}_i,\mathbf{r}_j)\alpha + \sum_{k=1}^{n}\int_{S_k} p_{ss}(\mathbf{r}_i,\mathbf{r}_k)\rho\overline{DSS}(\mathbf{r}_k,\mathbf{r}_j) \quad (1)$$

$$\overline{DSG}(\mathbf{r}_i,\mathbf{r}_j) =$$
$$[p_{sg}(\mathbf{r}_i,\mathbf{r}_j) + \sum_{k=1}^{n}\int_{S_k}\overline{DSS}(\mathbf{r}_i,\mathbf{r}_k)(\rho/\alpha)p_{sg}(\mathbf{r}_k,\mathbf{r}_j)](1-\omega_0) \quad (2)$$

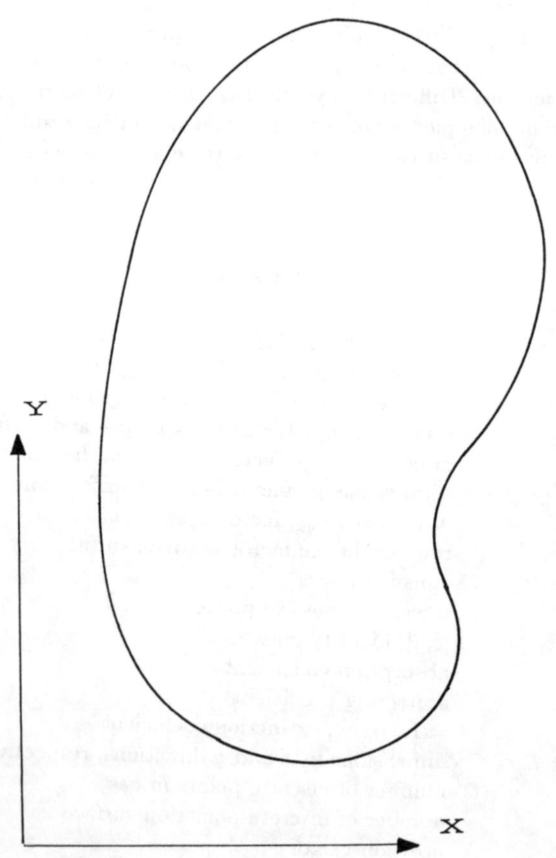

Figure 1: Schematic of a non-rectangular enclosure.

$$\overline{DGS}(\mathbf{r}_i,\mathbf{r}_j) = p_{gs}(\mathbf{r}_i,\mathbf{r}_j)\alpha + \sum_{k=1}^{n}\int_{S_k} p_{gs}(\mathbf{r}_i,\mathbf{r}_k)\rho\overline{DSS}(\mathbf{r}_k,\mathbf{r}_j) \quad (3)$$

and

$$\overline{DGG}(\mathbf{r}_i,\mathbf{r}_j) = [p_{gg}(\mathbf{r}_i,\mathbf{r}_j) + \sum_{k=1}^{n}\int_{S_k} p_{gs}(\mathbf{r}_i,\mathbf{r}_k)\rho p_{sg}(\mathbf{r}_k,\mathbf{r}_j)$$
$$+\sum_{k=1}^{n}\sum_{l=1}^{n}\int_{S_k}\int_{S_l} p_{gs}(\mathbf{r}_i,\mathbf{r}_k)\rho\overline{DSS}(\mathbf{r}_k,\mathbf{r}_l)(\rho/\alpha)p_{sg}(\mathbf{r}_l,\mathbf{r}_j)](1-\omega_0) \quad (4)$$

where

$$p_{gg}(\mathbf{r}_i,\mathbf{r}_j) = \overline{dgg}(\mathbf{r}_i,\mathbf{r}_j) + \sum_{k=1}^{m}\int_{V_k}\overline{dgg}(\mathbf{r}_i,\mathbf{r}_k)\omega_0 p_{gg}(\mathbf{r}_k,\mathbf{r}_j) \quad (5)$$

$$\begin{aligned} p_{ss}(\mathbf{r}_i,\mathbf{r}_j) &= \overline{dss}(\mathbf{r}_i,\mathbf{r}_j) + \sum_{k=1}^{m}\int_{V_k}\overline{dsg}(\mathbf{r}_i,\mathbf{r}_k)\omega_0\overline{dgs}(\mathbf{r}_k,\mathbf{r}_j) \\ &+ \sum_{k=1}^{m}\sum_{l=1}^{m}\int_{V_k}\int_{V_l}\overline{dsg}(\mathbf{r}_i,\mathbf{r}_k)\omega_0 p_{gg}(\mathbf{r}_k,\mathbf{r}_l)\omega_0\overline{dgs}(\mathbf{r}_l,\mathbf{r}_j) \end{aligned} \quad (6)$$

$$p_{gs}(\mathbf{r}_i,\mathbf{r}_j) = \overline{dgs}(\mathbf{r}_i,\mathbf{r}_j) + \sum_{k=1}^{m}\int_{V_k}\overline{dgg}(\mathbf{r}_i,\mathbf{r}_k)\omega_0 p_{gs}(\mathbf{r}_k,\mathbf{r}_j) \quad (7)$$

$$p_{sg}(\mathbf{r}_i,\mathbf{r}_j) = \overline{dsg}(\mathbf{r}_i,\mathbf{r}_j) + \sum_{k=1}^{m}\int_{V_k} p_{sg}(\mathbf{r}_i,\mathbf{r}_k)\omega_0\overline{dgg}(\mathbf{r}_k,\mathbf{r}_j) \quad (8)$$

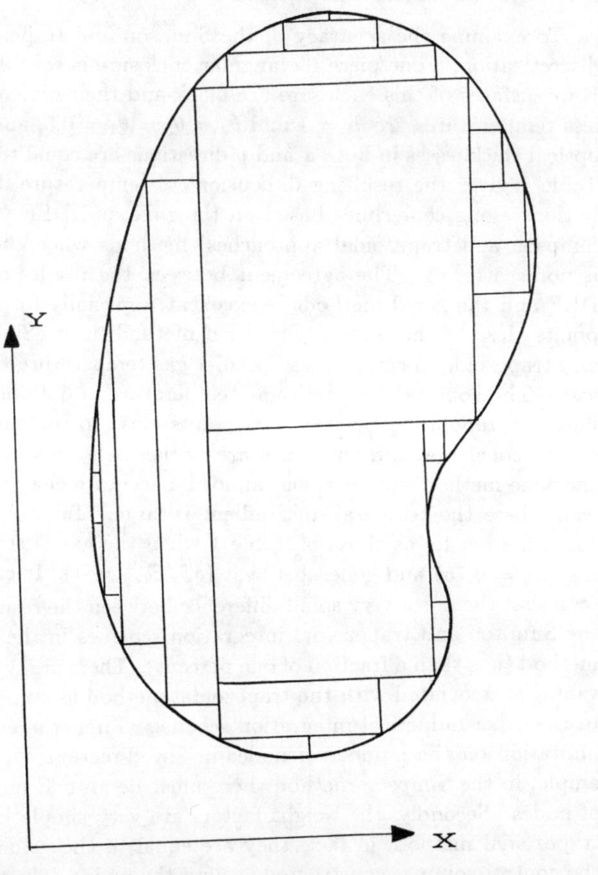

Figure 2: The non-rectangular enclosure divided into rectangular and triangular elements.

S_k is the area of surface element k and V_k is the volume element k. Note that the above equations relate continuous total exchange factors ($\overline{DSS}(\mathbf{r}_i,\mathbf{r}_j)$, $\overline{DSG}(\mathbf{r}_i,\mathbf{r}_j)$, $\overline{DGS}(\mathbf{r}_i,\mathbf{r}_j)$ and $\overline{DGG}(\mathbf{r}_i,\mathbf{r}_j)$) and direct exchange factors ($\overline{dss}(\mathbf{r}_i,\mathbf{r}_j)$, $\overline{dsg}(\mathbf{r}_i,\mathbf{r}_j)$, $\overline{dgs}(\mathbf{r}_i,\mathbf{r}_j)$ and $\overline{dgg}(\mathbf{r}_i,\mathbf{r}_j)$). The definition and equations for calculating direct exchange factors are given by Naraghi and Kassemi (1988) for two-dimensional enclosures and Naraghi and Litkouhi (1989) for three-dimensional enclosures.

Evaluating the continuous exchange factors in close form using equations (1)-(8) is not tractable. Therefore, a numerical integration method is used to discretize the above equations and radiative exchange between a number of discrete nodes is considered. The positions of these nodes are determined by the ordinates of the numerical quadrature chosen for the discretization of CEF equations. The following explicit matrix formulations for the discrete total exchange factor matrices in terms of the discrete direct exchange factor matrices, surface and gas properties are obtained:

$$\begin{aligned} \overline{\mathbf{DSS}} &= [I - \{\overline{\mathbf{dss}} + \overline{\mathbf{dsg}}\omega_0 W_g[I - \overline{\mathbf{dgg}}\omega_0 W_g]^{-1}\overline{\mathbf{dgs}}\}\rho W_s]^{-1}. \\ &\{\overline{\mathbf{dss}} + \overline{\mathbf{dsg}}\omega_0 W_g[I - \overline{\mathbf{dgg}}\omega_0 W_g]^{-1}\overline{\mathbf{dgs}}\}\alpha \end{aligned} \quad (9)$$

$$\begin{aligned} \overline{\mathbf{DSG}} &= [I - \{\overline{\mathbf{dss}} + \overline{\mathbf{dsg}}\omega_0 W_g[I - \overline{\mathbf{dgg}}\omega_0 W_g]^{-1}\overline{\mathbf{dgs}}\}\rho W_s]^{-1}. \\ &\overline{\mathbf{dsg}}[I - \omega_0 W_g\overline{\mathbf{dgg}}]^{-1}(1-\omega_0) \end{aligned} \quad (10)$$

$$\begin{aligned} \overline{\mathbf{DGS}} &= [I - \overline{\mathbf{dgg}}\omega_0 W_g]^{-1}\overline{\mathbf{dgs}}. \\ &[I - \rho W_s\{\overline{\mathbf{dss}} + \overline{\mathbf{dsg}}\omega_0 W_g[I - \overline{\mathbf{dgg}}\omega_0 W_g]^{-1}\overline{\mathbf{dgs}}\}]^{-1}\alpha \end{aligned} \quad (11)$$

and

$$\begin{aligned} \overline{\mathbf{DGG}} &= [I - \overline{\mathbf{dgg}}\omega_0 W_g]^{-1}\overline{\mathbf{dgg}}(1-\omega_0) + [I - \overline{\mathbf{dgg}}\omega_0 W_g]^{-1}. \\ &\overline{\mathbf{dgs}}W_s[I - \{\overline{\mathbf{dss}} + \overline{\mathbf{dsg}}\omega_0 W_g[I - \overline{\mathbf{dgg}}\omega_0 W_g]^{-1}\overline{\mathbf{dgs}}\}\rho W_s]^{-1}\overline{\mathbf{dsg}}. \\ &[I - \omega_0 W_g\overline{\mathbf{dgg}}]^{-1}(1-\omega_0) \end{aligned} \quad (12)$$

where $\overline{\mathbf{DSS}}$, $\overline{\mathbf{DSG}}$, $\overline{\mathbf{DGS}}$ and $\overline{\mathbf{DGG}}$ are partitioned matrices of total exchange factors between differential surfaces and volumes at the discrete points determined by the ordinates of the numerical quadrature method, and $\overline{\mathbf{dss}}$, $\overline{\mathbf{dsg}}$, $\overline{\mathbf{dgs}}$ and $\overline{\mathbf{dgg}}$ are partitioned matrices of direct exchange factors. For example, $\overline{\mathbf{DSS}}$ is given by

$$\overline{\mathbf{DSS}} = \left[\begin{array}{c|c|c|c} \overline{\mathbf{DSS}}(1,1) & \overline{\mathbf{DSS}}(1,2) & .... & \overline{\mathbf{DSS}}(1,m) \\ \hline \overline{\mathbf{DSS}}(2,1) & \overline{\mathbf{DSS}}(2,2) & .... & \overline{\mathbf{DSS}}(2,m) \\ \hline \vdots & \vdots & \vdots & \vdots \\ \hline \overline{\mathbf{DSS}}(m,1) & \overline{\mathbf{DSS}}(m,2) & .... & \overline{\mathbf{DSS}}(m,m) \end{array}\right] \quad (13)$$

where $\overline{\mathbf{DSS}}(i,j) = \left[\overline{DS_kS_l}(i,j)\right]$ is a partition of matrix $\overline{\mathbf{DSS}}$ and its elements are total exchange factors between nodes on surface elements i and j. W_g and W_s are diagonal matrices of the numerical integration weight factors. In the previous works (Naraghi et al., 1988 and Naraghi and Kassemi, 1988), the Gaussian quadrature method was used in the DEF method. In the Gaussian quadrature method, the positions of the nodal points are determined by the roots of an orthogonal polynomial and they are unevenly distributed. Other numerical integration methods, i.e., Simpson and trapezoidal, are more suitable for complex geometries, especially when finite difference compatible grids are employed.

Once the total exchange factors between nodal points are evaluated, then discretized form of nodal balance of energy equation can be used to relate nodal heat and emissive powers. These equations are given by

$$\left\{ \begin{array}{c} \mathbf{q_s}'' \\ -- \\ \mathbf{q_g}''' \end{array} \right\} = \left[\mathbf{I} - \overline{\mathbf{DZZ}}^T \mathbf{W}\right] \left\{ \begin{array}{c} \mathbf{E_s} \\ -- \\ \mathbf{E_g} \end{array} \right\} \tag{14}$$

where

$$\overline{\mathbf{DZZ}} = \left[\begin{array}{ccc} \overline{\mathbf{DSS}} & | & \overline{\mathbf{DSG}} \\ --- & - & --- \\ \overline{\mathbf{DGS}} & | & \overline{\mathbf{DGG}} \end{array} \right]$$

$$\mathbf{W} = \left[\begin{array}{ccc} \mathbf{W}_s & | & 0 \\ --- & - & --- \\ 0 & | & \mathbf{W}_g \end{array} \right]$$

$\mathbf{q_s}''$ and $\mathbf{q_g}'''$ are vectors of surface nodal heat flux and gas nodal heat source, respectively, and $\mathbf{E_s}$ and $\mathbf{E_g}$ vectors of nodal emissive powers.

After subdividing the enclosure into rectangular and triangular elements, each element has to be discretized according to the ordinates of the numerical integration method chosen. The following sections contain detailed descriptions of the procedures for discretization of the rectangular and triangular elements.

Rectangular Elements

The discretization of rectangular elements is identical to those of rectangular enclosures which are described in the previous work (Naraghi and Kassemi, 1988) using the Gaussian quadrature discretization. It has been shown that the Gaussian quadrature method provides accurate results even when a small number of nodes is implemented. In the Gaussian quadrature method, however, the nodal points are unevenly spaced, making them incompatible with the finite difference nodal points. In this work, the other numerical integration methods, namely, the Simpson and trapezoidal methods will be used for discretization of rectangular pieces. Fig. 3 shows positions of an evenly spaced set of nodal points for a typical rectangular piece. The distance between two neighboring nodes are h_x and h_y in the x and y directions, respectively. Boundaries of this piece can be a part of the surface of the enclosure or another gas piece. It should be noted that on the surface boundaries the surface and gas nodes are coincident

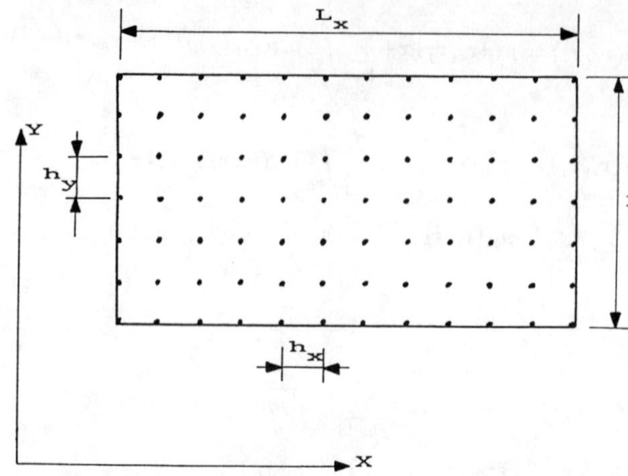

Figure 3: A rectangular element (enclosure) with equally spaced nodes.

and at the corners, surface nodes from two neighboring surfaces overlap, causing a singularity condition at these nodes. A complete procedure for treatment of singularities, i.e., calculation of direct exchange factors between coincident nodes for three-dimensional rectangular enclosures is presented by Naraghi and Litkouhi (1989). The same method can be used here for a two-dimensional enclosure.

To examine the accuracy of the Simpson and trapezoidal discretization, a one piece rectangular enclosure is considered. Four surfaces of this enclosure are black and their dimensionless temperatures are $\theta_1 = 1$ and $\theta_2 = \theta_3 = \theta_4 = 0.5$, and the optical thicknesses in both x and y directions are equal to one. Table 1 gives the resulting dimensionless temperature distributions along centerlines based on the zone and DEF (using Simpson and trapezoidal approaches) methods when the gas is non-scattering. The agreement between the results of the DEF and the zonal methods is excellent, especially for those points close to the center. The DEF method, in its Simpson and trapezoidal form, can also predict gas temperature at the wall. The zone method wall gas temperature is obtained by linear extrapolation between two points close to the surface. In the zonal method, the zones are isothermal, consequently the zone method's predictions cannot be accurate close to the wall where the temperature gradient is large. Table 2 gives the same results as those of Table 1 when the gas is scattering ($\omega_0 = 0.75$) and generates heat ($q_g'''/E_{ref} = 4$). It can be seen that there is a very small difference between the results of the Simpson and trapezoidal integration schemes in the DEF method (less than a fraction of one percent). There are two advantages associated with the trapezoidal method as compared to the other numerical integration schemes. First, there is no limitation on the number of nodes in any direction. For example, in the Simpson method there must be an odd number of nodes. Secondly, the weight factors are very simple in the trapezoidal method, in fact, they are equal to the volume of the control volumes constructed around the nodes. These features of the trapezoidal integration make it a very attractive method for complex configurations.

Table 1: A comparison between the results of centerline temperature distribution based on the zone and DEF (Simpson and trapezoidal) methods for a square enclosure with black surfaces when $\theta_1 = 1.$, $\theta_2 = \theta_3 = \theta_4 = 0.5, LK_t = 1., \omega_0 = 0.$ and $q_g''' = 0$.

	$\theta(0.5, y)$		
	Zone Method	DEF Method	
	Larsen, 1983	Simpson	Trapezoidal
0.	.8956*	.9081	.9055
.1	.8611	.8641	.8619
.2	.8274	.8278	.8277
.3	.7957	.7957	.7958
.4	.7659	.7659	.7659
.5	.7382	.7381	.7381
.6	.7125	.7124	.7124
.7	.6882	.6882	.6882
.8	.6649	.6649	.6649
.9	.6415	.6401	.6414
1.	.6175*	.6082	.6102
	$\theta(x, 0.5)$		
0.	.6707*	.6561	.6588
.1	.6949	.6933	.6950
.2	.7147	.7149	.7150
.3	.7280	.7282	.7282
.4	.7356	.7357	.7357
.5	.7382	.7381	.7381
.6	.7356	.7357	.7357
.7	.7280	.7282	.7282
.8	.7147	.7149	.7150
.9	.6949	.6933	.6950
1.	.6707*	.6561	.6588

Table 2: A comparison between the results of centerline temperature distribution based on the zone and DEF (Simpson and trapezoidal) methods for a square enclosure with black surfaces when $\theta_1 = 1.$, $\theta_2 = \theta_3 = \theta_4 = 0.5, LK_t = 1., \omega_0 = 0.$ and $q_g'''/E_{ref} = 4$.

	$\theta(0.5, y)$		
	Zone Method	DEF Method	
	Larsen, 1983	Simpson	Trapezoidal
0.	1.512*	1.503	1.505
.1	1.516	1.516	1.517
.2	1.520	1.520	1.520
.3	1.520	1.520	1.520
.4	1.518	1.519	1.519
.5	1.516	1.516	1.517
.6	1.513	1.513	1.513
.7	1.507	1.507	1.508
.8	1.500	1.500	1.500
.9	1.489	1.487	1.489
1.	1.473*	1.461	1.464
	$\theta(x, 0.5)$		
0.	1.479*	1.465	1.468
.1	1.494	1.492	1.494
.2	1.505	1.505	1.505
.3	1.512	1.511	1.512
.4	1.515	1.515	1.515
.5	1.516	1.516	1.517
.6	1.515	1.515	1.515
.7	1.512	1.511	1.512
.8	1.505	1.505	1.505
.9	1.494	1.492	1.494
1.	1.479*	1.465	1.468

Triangular Elements

Fig. 4 depicts a typical triangular element. The optical thicknesses along x, y and the hypotenuse of the triangle are K_tL_x, K_tL_y and $K_t\sqrt{L_x^2 + L_y^2}$, respectively. The easiest way to discretize the triangular pieces is to break sides $\overline{AB}$ and $\overline{AC}$ into an identical number of nodes ($n_x = n_y = n$). Under this condition the grid sizes in x and y directions become $h_x = L_x/(n-1)$ and $h_y = L_y/(n-1)$, respectively. This will ensure a uniform distribution of nodal points over the entire triangular piece. It is found that the uniform grid is trouble free for triangles with $L_x \approx L_y$ (aspect ratio$\approx$ 1). However, for other aspect ratios it is difficult to evaluate the exchange factors for nodal points at corners with acute angles. It should be noted that the conservation equation is used in the DEF method to evaluate exchange factors between coincident surface and gas nodes and the gas node to itself (Naraghi and Litkouhi, 1989). This approach yields negative direct exchange factors for nodes located at corners with angles less than 45 degrees. Often, simply ignoring these corner nodes results in little or no effect on the temperature and heat values of the other nodes. However, if necessary, the problems due to the corner nodes

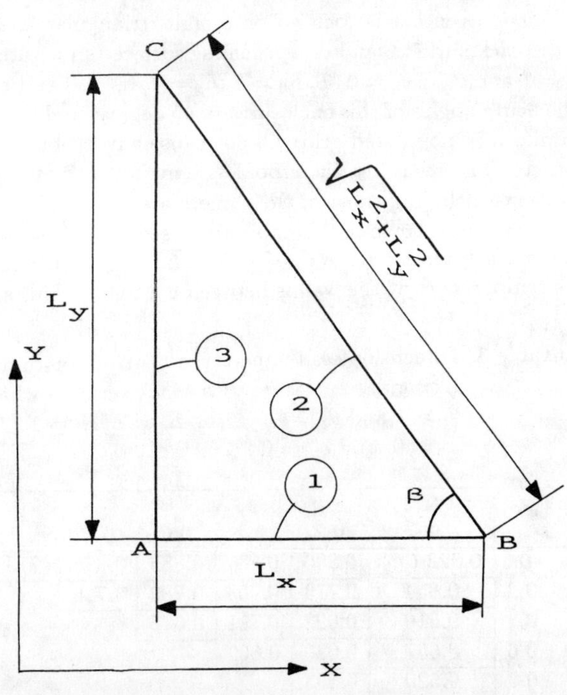

Figure 4: A typical triangular element.

*Based on extrapolation

can be overcome by generating a finer nodal mesh in the corner region.

The triangular enclosures with acute angles are the most difficult enclosures to analyze using the DEF method. Apart from the aforementioned difficulties in predicting the exchange factors between coincident surface and gas nodes and gas node to itself at the corners, there is also another singularity which is caused by coincident surface nodes at corners and edges. A complete discussion of the singularity at the coincident corner nodes is presented elsewhere and their values for right corners are given (Naraghi and Litkouhi, 1989). For corners with other angles, like that of corner B in Fig. 4, the exchange factor between coincident nodes i and j from surfaces 1 and 2 is given by

$$w_i\overline{ds_is_j} = \frac{1}{2}(1 - \sin\beta) \qquad (15)$$

Once all pieces are discretized and the partitioned matrices of direct exchange factors ($\overline{\mathbf{dss}}$, $\overline{\mathbf{dsg}}$, $\overline{\mathbf{dgs}}$, and $\overline{\mathbf{dgg}}$) is evaluated, then equations (9)-(12) can be used to evaluate total exchange factors between nodal points of different pieces and equations (14) and (15) can be used to relate nodal temperature (or emissive power) and heats. It should be noted that if the line connecting two nodes intersect the surface of the enclosure, the direct exchange factor between those nodes is zero.

RESULTS AND DISCUSSION

As discussed in the previous section the triangular enclosures are the most difficult enclosures to analyze, due to their acute angles. In this section, the results of surface heat flux and gas temperatures for two-dimensional triangular enclosures with different surface and gas properties will be presented.

First attention is focused on a right triangular enclosure with black surfaces and dimensionless surface temperatures of $\theta_1 = 1$ and $\theta_2 = \theta_3 = 0$, $L_xK_t = L_yK_t = 1$, $\omega_0 = 0$ and $q_g = 0$. The acute angles of this enclosure are 45 degrees and the use of a uniformly populated grid will not cause any problem at the corners. The resulting dimensionless temperatures at different positions including those of the corners are presented in Table 3. The results shown in this table are based on an 11×11 grid. The surface temperature at $Y = 0$ ($\overline{AB}$) is $\theta_1 = 1$, while the gas temperature at the wall is between 0.8 to 0.9. This sudden change of temperature is due to the radiative slip condition at the wall.

Table 3: Dimensionless temperature distribution in a triangular enclosure with black surfaces when $\theta_1 = 1$ and $\theta_2 = \theta_3 = 0$, $L_xK_t = L_yK_t = 1$, $\omega_0 = 0.$ and $q_g''' = 0$.

		X					
		0.	0.2	0.4	0.6	0.8	1.0
	0.	0.822 (A)	0.890	0.885	0.883	0.878	.779 (B)
	0.2	0.654	0.773	0.795	0.789	0.726	
Y	0.4	0.616	0.692	0.711	0.676		
	0.6	0.567	0.620	0.609			
	0.8	0.520	0.533				
	1.0	0.413 (C)					

Table 4: Dimensionless temperature distribution in a triangular enclosure with black surfaces when $\theta_1 = 1$ and $\theta_2 = \theta_3 = 0$, $L_xK_t = L_yK_t = 1$, $\omega_0 = 0.$ and $q_g'''/E_{ref} = 4$.

		X					
		0.	0.2	0.4	0.6	0.8	1.0
	0.	0.964 (A)	1.088	1.098	1.092	1.087	0.950 (B)
	0.2	1.088	1.124	1.127	1.114	1.056	
Y	0.4	1.098	1.127	1.125	1.071		
	0.6	1.092	1.114	1.071			
	0.8	1.087	1.056				
	1.0	0.950 (C)					

To examine the effect of gas heat generation, the above enclosure when $q_g'''/E_{ref} = 4$ and all surfaces are cold is analyzed. The resulting gas temperature distribution is given in Table 4 and wall heat flux is presented in Fig. 5. As can be seen from the temperature distribution, the lowest gas temperatures correspond to those nodes at corners with smaller angles. The heat flux from surfaces $\overline{AB}$ and $\overline{AC}$ are identical due to the symmetry of the configuration. Furthermore, the maximum heat flux on surface $\overline{BC}$ is at its center, while on surface $\overline{AB}$ the maximum heat flux is shifted slightly toward point A. Next, consideration is given to the same enclosure when the dimensionless temperatures are $\theta_1 = 1$, $\theta_2 = 0$ and $\theta_3 = 0$ and the gas generates heat $q_g'''/E_{ref} = 4$, with a scattering albedo of $\omega_0 = 0.5$. The resulting gas temperature distribution is shown in Fig. 6. The temperature distribution along lines of constant Y deviate from the normal trend close

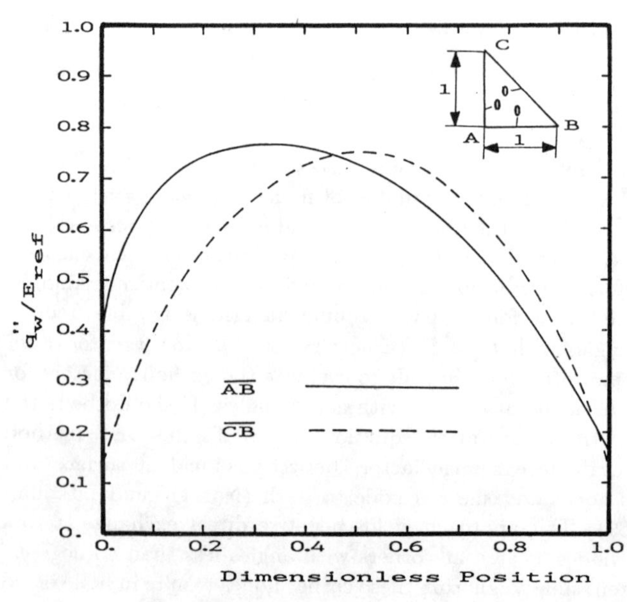

Figure 5: Dimensionless wall heat flux for a triangular enclosure when when all surfaces are black and cold and $L_xK_t = L_yK_t = 1.$, $q_g/E_{ref} = 4$ and $\omega_0 = 0.$.

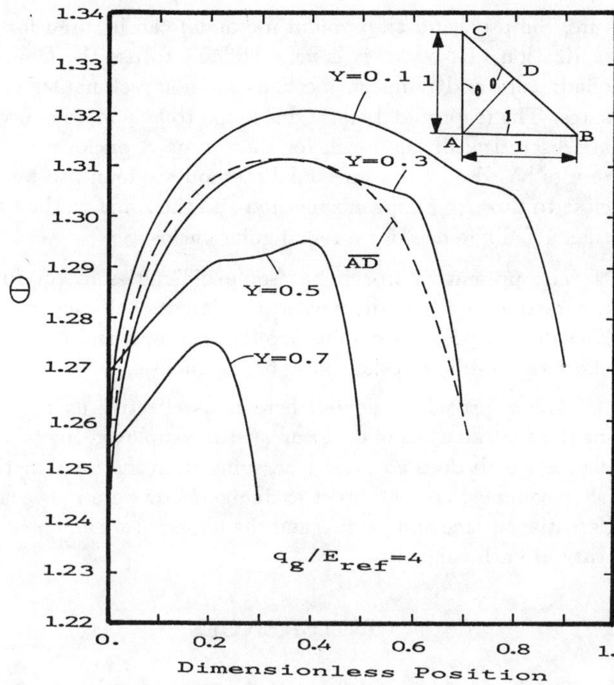

Figure 6: Dimensionless gas temperature distribution for a triangular enclosure when the surface are black and $\theta_1 = 1$, $\theta_2 = \theta_3 = 0$ $L_xK_t = L_yK_t = 1.$, $q_g/E_{ref} = 4$, and $\omega_0 = 0.5$.

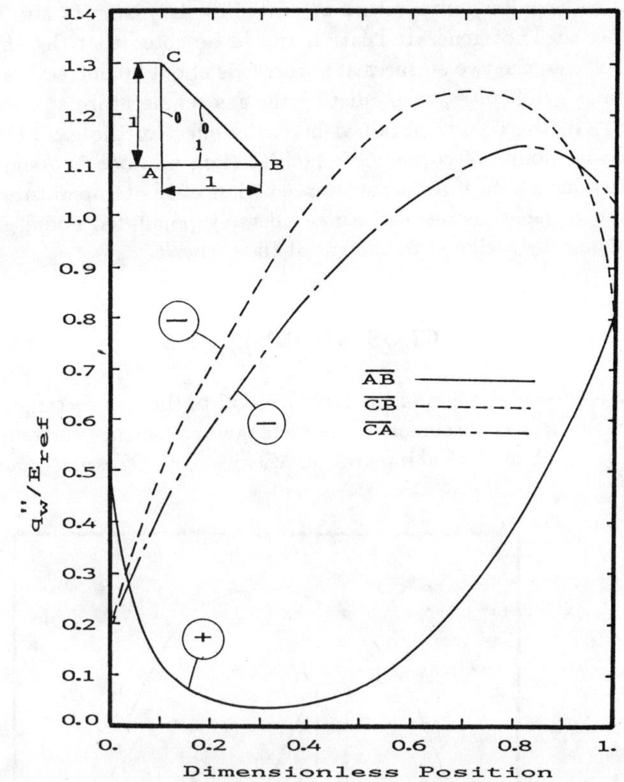

Figure 7: Dimensionless wall heat flux for a triangular enclosure when the surface are black and $\theta_1 = 1$, $\theta_2 = \theta_3 = 0$, $L_xK_t = L_yK_t = 1.$, $q_g/E_{ref} = 4$, and $\omega_0 = 0.5$.

to the surface $\overline{BC}$. This deviation might be due to the angle of 45 degrees between lines of constant Y and surface $\overline{BC}$, and it cannot be noticed from the temperature distribution along line $\overline{AD}$ which is perpendicular to side $\overline{BC}$. Fig. 7 shows the heat fluxes across different surfaces of the same enclosure. The heat flux crossing surface $\overline{AB}$ is positive and those of surfaces $\overline{BC}$ and $\overline{CA}$ are negative, i.e., heat gain to the enclosure from side $\overline{AB}$ and heat loss from $\overline{BC}$ and $\overline{AC}$. The lowest heat loss from the enclosure is at point C which is the farthest point from the hot surface $\overline{AB}$. The maximum heat losses, however, are close to points A and B on sides $\overline{CA}$ and $\overline{CB}$, respectively. The heat gain from side $\overline{AB}$ are largest close to corners A and B (especially corner B) and its lowest value is about in the middle of $\overline{AB}$ slightly shifted toward corner A. The heat gain from side $\overline{AB}$ is due to the gas heat generation (q_g'''). Indeed, if the gas heat generation increases, the minimum heat gain becomes negative (i.e., some heat loss from side $\overline{AB}$).

Finally, to examine the effect of the aspect ratio, consideration is given to an enclosure with dimensionless side lengths of $L_xK_t = 0.5$ and $L_yK_t = 1$. The surfaces of the enclosure are black and cold and the gas is non-scattering and generates heat ($q_g'''/E_{ref} = 4$). The resulting gas heat flux distribution for different surfaces of the enclosure are shown in Fig. 8. The heat fluxes of surfaces $\overline{AB}$ and $\overline{AC}$ are nearly the same close to corner A, but deviate significantly away from A. The heat flux is larger close to corner B than that of corner C since angle B is larger than angle C. Fig. 9 shows the resulting gas temperature distribution for the same enclosure when the

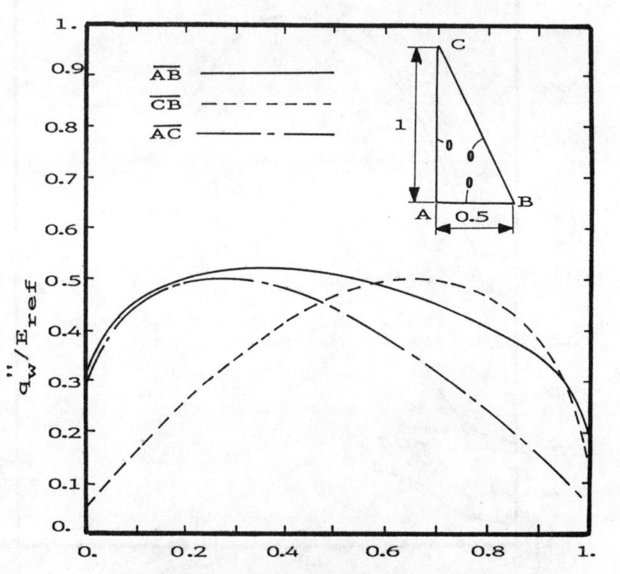

Figure 8: Dimensionless wall heat flux for a triangular enclosure when the surface are black and cold and $L_xK_t = 0.5$, $L_yK_t = 1.$, $q_g/E_{ref} = 4$, and $\omega_0 = 0$.

dimensionless temperature of surface 1 ($\overline{AB}$) is unity and the gas does not generate heat. It should be noted that the angle between the two surfaces at corner C is about 26 degrees. This causes difficulties in predicting the gas temperature at corner C. In the results obtained here, the effect of the gas at the nodal points of corner C is ignored (i.e., no node is assigned to corner C). If an accurate prediction of gas temperature at the corner C is needed, a more densely populated nodal grid should be assigned to the gas at the corner C.

CLOSING REMARKS

The application of the DEF method to the non-rectangular enclosures is discussed. For rectangular elements (or enclosures), all numerical integration schemes (e.g., Gaussian quadrature, Simpson and trapezoidal methods) can be used for discretization. However, it is very difficult to use the Gaussian quadrature and Simpson methods for non-rectangular enclosures. The trapezoidal method is found to be a more appropriate discretization approach for these type of enclosures. The results based on the trapezoidal method are found to be very close to those of Simpson's method and the zonal method (less than 1% difference) for a rectangular enclosure.

The present approach has some difficulties in producing accurate results at corners with acute angles between surfaces. This difficulty is only confined to those corners and it does not effect the results of other parts of the enclosure.

The approach presented here is as effective as the zonal method for analysis of enclosures with complex geometric configuration. It does not, however, suffer from the zonal methods shortcomings, i.e., its direct exchange factors are between differential surface and/or gas, and the temperature and heat may vary at each element.

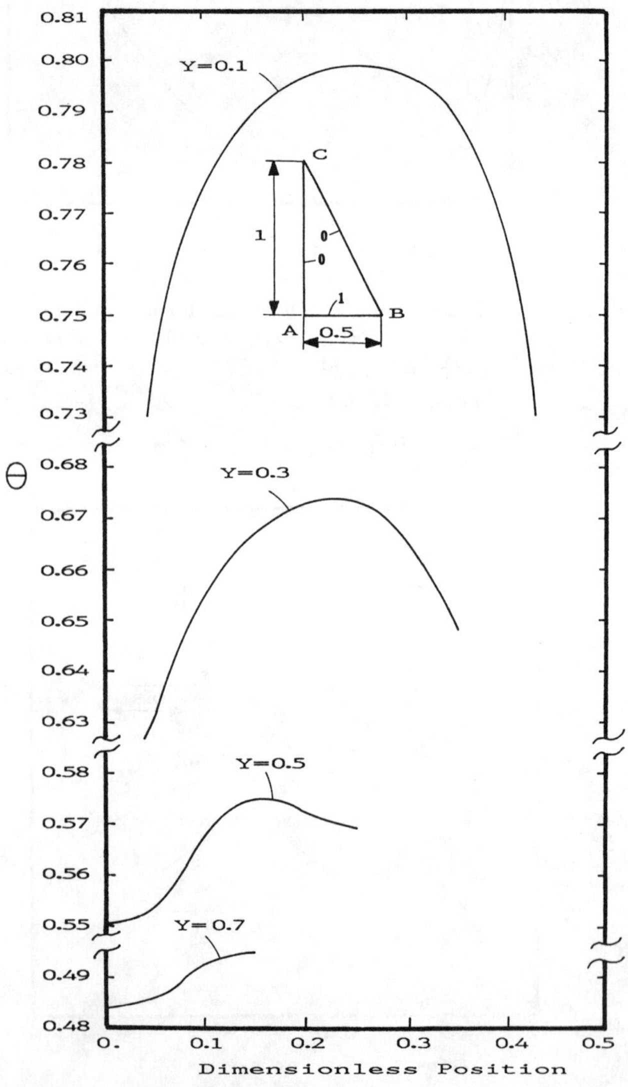

Figure 9: Dimensionless gas temperature distribution for a triangular enclosure when the surface are black and $\theta_1 = 1$, $\theta_2 = \theta_3 = 0$ $L_x K_t = 0.5$, $L_y K_t = 1.$, $q_g = 0$, and $\omega_0 = 0$.

REFERENCES

Crosbie, A. L., and Schrenker, R. G., 1984, *Radiative Transfer in a Two-Dimensional Rectangular Medium Exposed to Diffuse Radiation,* J. Quant. Spectroc. and Radiat. Transfer, Vol. 31, No. 4, pp. 339-372.

Fiveland, W. A., 1984, *Discrete-Ordinate Solution of the Radiative Transport Equation for Rectangular Enclosures,* Journal of Heat Transfer, Trans. ASME, Vol. 106, pp. 699-706.

Hottel, H. C. and Cohen, E. S., 1958, *Radiant Heat Exchange in a Gas-Filled Enclosure: Allowance of Non-Uniformity of Gas Temperature,* AIChE Journal, Vol. 4, No. 1, pp. 3-14.

Hottel, H. C. and Sarofim, A. F., 1967, *Radiative Transfer,* McGraw-Hill, New York.

Howell, J. R., and Perlmutter, M., 1964, *Monte Carlo Solution of Thermal Transfer through Radiant Media between Gray Walls,* Journal of Heat Transfer, Trans. ASME, Vol. 86, pp. 116-122.

Howell, J. R., Strite, M. K. and Renkel, H., 1965, *Heat Transfer Analysis of Rocket nozzle Using Very High-Temperature Propellants,* AIAA Journal, Vol. 3, No. 4, pp. 669-673.

Howell, J. R. and Strite, M. K., 1966, *Heat Transfer in Rocket Nozzles Using High-Temperature Hydrogen Propellant With Real Property Variations,* Journal of Spacecraft Rockets, Vol. 3, No. 7, pp. 1063-1068.

Iyer, R. K. and Mengüç, M. P., 1988, *Quadruple Spherical Harmonics Approximation for Two-Dimensional Rectangular Enclosures,* ASME Proceedings of the 1988 National Heat Transfer Conference, Editor H. R. Jacobs, Vol. 1, pp. 251-258.

Larsen, M. E., 1983, , *The Exchange Factor Method: An Alternative Zonal Formulation for Analysis of Radiating Enclosures Containing Participating Media,* Ph.D. Dissertation, The University of Texas at Austin.

Larsen, M. E. and Howell, J. R., 1985, *The Exchange Factor Method: An Alternative Basis for Zonal analysis of Radiating Enclosures,* Journal of Heat Transfer, Trans. ASME, Vol. 107, No. 4, pp. 936-942.

Modest, M. F., 1975, *Radiative Equilibrium in a Rectangular Enclosure Bounded by Gray Walls,* J. Quant. Spectroc. Radiat. Transfer, Vol. 15, No. 6, pp. 445-461.

Naraghi, M. H. N. and Chung, B. T. F., 1985, *A Unified Matrix Formulation for the Method, A Stochastic Approach,* International Journal of Heat and Mass Transfer, Vol. 28, pp. 245-251.

Naraghi, M. H. N., Chung, B. T. F. and Litkouhi, B., 1988, *A Continuous Exchange Factor Method for Radiative Exchange in Enclosures with Participating Media,* Journal of Heat Transfer, Trans. ASME, Vol. 110, pp. 456-462.

Naraghi, M. H. N. and Kassemi, M., 1988, *Radiative Transfer in Rectangular Enclosures: A Discretized Exchange Factor Solution,* ASME Proceedings of the 1988 National Heat Transfer Conference, Houston, Texas, July 24-27, Editor H. R. Jacobs, Vol. 1. pp. 259-267.

Naraghi, M. H. N. and Litkouhi, B., 1989, *Discrete Exchange Factor Solution of Radiative Heat Transfer in Three-Dimensional Enclosures,* to be presented at the MULTIDIMENSIONAL RADIATIVE TRANSFER session of the 26th National Heat Transfer Conference.

Noble, J. J., 1975, *The Zone Method: Explicit Matrix Relations for Total Exchange Areas,* International Journal of Heat and Mass Transfer, Vol. 18, pp. 261-269.

Ratzel, A. C. and Howell, J. R., 1982, *Two-Dimensional Radiation in Absorbing-Emitting-Scattering Media Using the P-N Approximation,* ASME paper 82-HT-19, 1982.

Razzaque, M. M., 1981, *Finite Element Analysis of Combined Heat Transfer Including Radiation in Gray Radiating Media,* Ph.D. dissertation, The University of Texas at Austin.

Shah, N. G., 1979, *New Method of Computation of Radiation Heat Transfer in Combustion Chamber,* Ph.D. Thesis, Imperial College of Science and Technology.

Taniguchi, H, 1969, *The Radiative Heat Transfer in a Three Dimensional System Calculated by Monte Carlo Method,* Bull. JSME, Vol. 12, pp. 67-78.

Yuen, W. W. and Wong, L. W., 1984, *Analysis of Radiative Equilibrium in a Rectangular Enclosure With Gray Medium,* Journal of Heat Transfer, Trans. ASME, Vol. 106, pp. 433-440.

Yuen, W. W. and Ho, C. F., 1985, *Analysis of Two Dimensional Radiative Heat Transfer in a Gray Medium with Internal Heat Generation,* International Journal of Heat and mass transfer, Vol. 28, No. 1, pp. 17-23.

1989 National Heat Transfer Conference
HTD-Vol. 106, Heat Transfer Phenomena in Radiation, Combustion and Fires

A NONGRAY P-N APPROXIMATION FOR RADIATIVE TRANSFER

D. A. Kaminski and J. P. Moder
Department of Mechanical Engineering, Aeronautical Engineering, and Mechanics
Rensselaer Polytechnic Institute
Troy, New York

Abstract

A simple technique for modeling radiative heat transfer in a participating medium with an arbitrarily-varying spectral absorption coefficient is described. The method is applied to both the P-1 and P-3 differential approximations. An analytical solution for the heat flux in an isothermal nongray planar medium is obtained. This analysis imposes no restrictions on the variation of the spectral absorption coefficient with wavelength. For a nongray planar medium in radiative equilibrium, an iterative numerical technique is developed. The numerical method is applied to the case of a two-level absorption coefficient, for which previous exact and approximate results are available. Comparisons of the results of this approach with those of previous investigators is very good. The method is easily extended to multi-dimensional geometries.

Nomenclature

a	absorption coefficient
a_λ	spectral absorption coefficient
c	half-spacing between plates (see Figure 2)
d	spacing between plates $d=2c$
e_{bj}	blackbody emissive power of band j
e_{bwj}	blackbody emissive power of band j evaluated at the wall temperature
$e_{b\lambda}$	blackbody spectral emissive power
i_{kj}	kth moment of intensity of band j, $i_{kj} = i_{k\lambda j}\ \Delta\lambda_j$
$i_{k\lambda}$	kth moment of spectral intensity of radiation, $k=0,1,....,N$
q_R	radiative heat flux
S	net rate of heat production by processes other than radiation
T_1	wall temperature at $x=0$
T_2	wall temperature at $x=d$
x	spatial coordinate, $x=0$ at the bottom plate
z	spatial coordinate, $z=0$ half way between plates
α	band height in Fig. 3
α_1,α_2	constants in equ. (29)
$\Delta\lambda_j$	width of band j
ε_λ	spectral emissivity at wall
θ	dimensionless temperature, $\theta = T/T_1$
θ_1	dimensionless wall temperature at $x=0$
θ_2	dimensionless wall temperature at $x=d$
λ_c	cutoff wavelength, see Fig. 3
$\hat{\lambda}$	dimensionless cutoff wavelength $\hat{\lambda} = hc/\lambda_c kT_1$
τ_0	optical thickness, $\tau_0 = ad$ or $\max(a_1,a_2,a_3,...,a_n)d$
τ_j	optical depth, $\tau_j = a_j\ x$

Subscripts

j	jth band
k	kth moment of intensity
w	wall values

Introduction

The analysis of radiative heat transfer in a medium which absorbs, emits, and scatters radiation presents a difficult mathematical challenge. An exact analysis requires the solution of the equation of radiative transfer, which is an integro-differential equation in the radiation intensity. Exact solutions are generally confined to simple geometries and boundary conditions.

There is a wide range of available approximate and

numerical techniques for describing radiation in participating media. One of the simplest of these is the P-1 differential approximation (Jeans, 1917, and Krook, 1955). In the P-1 approximation, the radiation intensity at a point is expanded in a series of spherical harmonics. The series is truncated, and the resulting simplified radiation transport equation takes the form of a differential equation.

The accuracy of the differential approximation can be improved by retaining additional terms in the series of spherical harmonics which describes the radiation intensity. This method, called the P-3 approximation, generally yields results that are more accurate than the P-1 approximation, when compared to the exact solution (Bayazitoglu and Higenyi, 1979). The improved accuracy is accompanied by an increase in complexity.

Although the P-N approximations are receiving widespread use in gray participating media, as in the work of Ratzel and Howell (1983), Menguc and Viskanta (1986), and Kaminski (1988), very little work has been done in non-gray media. An early paper by Kung and Sibulkin (1969) uses the P-1 method in a medium with a two-level absorption coefficient. More recently, Modest (1979) has presented a box model for the P-1 approximation. His method has the advantage of being no more difficult to solve than the gray case, but all the bands must have the same value of the absorption coefficient. Yuen and Rasky (1981) have presented an elegant analytical technique for a nongray case. Yucel and Bayazitoglu(1983) present extensive results for a medium in radiative equilibrium with a two-level absorption coefficient, for both nongray P-1 and nongray P-3 cases.

The work presented here differs from that of previous investigators in basic approach. The method is easy to apply in a variety of situations and is very general. Results are in good agreement with those of previous investigators.

Analysis

To illustrate the technique for modeling nongray radiative transfer in a participating medium, a simple geometry is chosen. The physical system consists of two infinite, parallel plates. Between the plates is a nongray medium which absorbs and emits radiation, but does not scatter. Neglecting scattering does not represent a major limitation of the method. It is not difficult to include isotropic scattering or linear anisotropic scattering in the formulation, as described by Modest and Azad (1980). Scattering is ignored here because it is not central to the issue at hand and is easy to add later.

There is no need for a method which is useful only in one-dimensional plane-parallel geometries. The P-N methods have proven to be useful in one-dimensional cylindrical and spherical geometries, as illustrated by Bayazitoglu and Higenyi (1979) and Azad and Modest (1981), as well as in many multidimensional geometries, as described by Ratzel and Howell (1983), Menguc and Viskanta (1986), and Kaminski (1988). The plane parallel geometry is chosen here for simplicity and for ease of comparison with known nongray solutions, but the method is in no way restricted to one-dimensional or plane-parallel geometries.

Analysis for Isothermal Medium

Two different cases are explored in this study. In the first case, the bounding walls are black and at a constant temperature and the medium is isothermal at a different temperature. For the P-1 method, the governing equations in spectral form are (Yucel and Bayazitoglu, (1983)):

$$\frac{di_{1\lambda}}{dz} = a_\lambda(4e_{b\lambda} - i_{0\lambda}) \qquad (1)$$

$$\frac{di_{0\lambda}}{dz} = -3\, a_\lambda\, i_{1\lambda} \qquad (2)$$

The absorption coefficient, a_λ, can vary arbitrarily with wavelength. The spectrum is now divided into a finite number of bands as shown in Fig. 1. Both the band width and the band height can vary with wavelength. Equations (1) and (2) are now integrated over a single band, j, with the result:

$$\frac{di_{1j}}{dz} = a_j\,(4e_{bj} - i_{0j}) \qquad (3)$$

$$\frac{di_{0j}}{dz} = -3\, a_j\, i_{1j} \qquad (4)$$

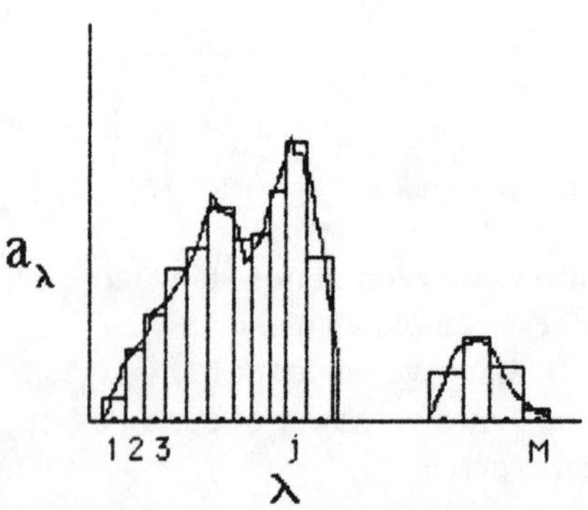

Figure 1 Variation of absorption coefficient with wavelength

where

$$e_{bi}(T) = \int_{\lambda_j}^{\lambda_{j+1}} e_{b\lambda}(T)\, d\lambda \tag{5a}$$

$$i_{ki}(T) = \int_{\lambda_j}^{\lambda_{j+1}} i_{ki}(T)\, d\lambda \qquad k=0,1 \tag{5b}$$

The integral of the Planck function in equ. (5a) cannot be expressed analytically; however, Wiebelt (1966) has presented useful curve fits in the form:

$$\frac{1}{\sigma}\int_0^{\lambda T} \frac{e_{b\lambda}(\lambda)}{T^5}\, d(\lambda T) = \frac{15}{\pi^4} \sum_{m=1,2,\dots} \frac{e^{-mv}}{m^4}\{[(mv+3)mv+6]mv+6\} \qquad v \geq 2 \tag{6a}$$

$$\frac{1}{\sigma}\int_0^{\lambda T} \frac{e_{b\lambda}(\lambda)}{T^5}\, d(\lambda T) = 1 - \frac{15}{\pi^4} v^3\left(\frac{1}{3} - \frac{v}{8} + \frac{v^2}{60} - \frac{v^4}{5040} + \frac{v^6}{272{,}160} - \frac{v^8}{13{,}305{,}600}\right) \qquad v<2 \tag{6b}$$

where $v = C_2/\lambda T$ and $C_2 = 14{,}388\ \mu m\ K$. For the coordinate system shown in Fig. 2, and black bounding surfaces, the boundary conditions are:

$$2\, i_{1i} = 4\, e_{bwi} - i_{0i} \qquad \text{at } z=-c \tag{7}$$

$$2\, i_{1i} = i_{0i} - 4\, e_{bwi} \qquad \text{at } z=c \tag{8}$$

Equations (3) and (4) with boundary conditions (7) and (8) are easily solved to give:

$$\frac{i_{0i} - 4\, e_{bi}}{4\,(e_{bwi} - e_{bi})} = \frac{e^{\sqrt{3} a_i z} + e^{-\sqrt{3} a_i z}}{e^{\sqrt{3} a_i c}\left(1 + \frac{2\sqrt{3}}{3}\right) + e^{-\sqrt{3} a_i c}\left(1 - \frac{2\sqrt{3}}{3}\right)} \tag{9}$$

The first moment of the intensity, $i_{1\lambda}$, physically represents the radiative flux in the z-direction. The radiative flux integrated over wavelength is therefore:

$$q_R = \int_0^{\infty} i_{1\lambda}\, d\lambda \tag{10}$$

The divergence of the radiative flux equals the net rate of heat production in the medium:

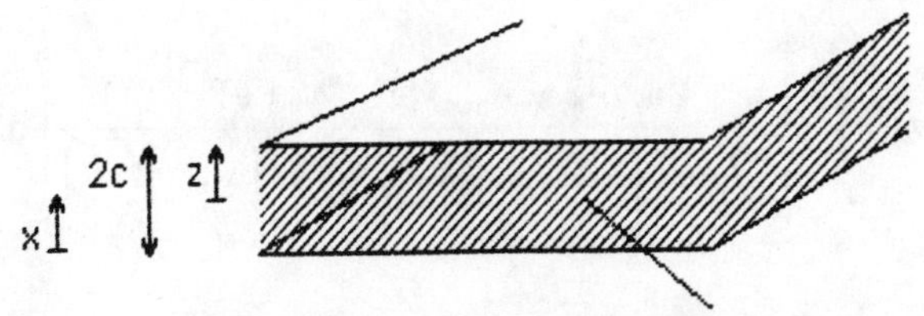

Nongray medium at constant temperature

Figure 2 Physical system

$$\frac{dq_R}{dz} = S(z) = \int_0^{\infty} \frac{di_{1\lambda}}{dz}\, d\lambda \tag{11}$$

Combining equations (11) and (1) gives:

$$S(z) = \int_0^{\infty} a_\lambda\,(4\, e_{b\lambda} - i_{0\lambda})\, d\lambda \tag{12}$$

If equ. (12) is applied to a medium which has been represented by M bands, as shown in Fig.1, the net rate of heat production is:

$$S(z) = \sum_{i=1}^{M} a_i\,(4\, e_{bi} - i_{0i}) \tag{13}$$

Equ. (9) can now be substituted into equ. (13) to give:

$$S(z) = \sum_{i=1}^{M} \left\{ \frac{4\, a_i\,(e_{bi} - e_{bwi})\,[e^{\sqrt{3} a_i z} + e^{-\sqrt{3} a_i z}]}{e^{\sqrt{3} a_i c}\left(1 + \frac{2\sqrt{3}}{3}\right) + e^{-\sqrt{3} a_i c}\left(1 - \frac{2\sqrt{3}}{3}\right)} \right\} \tag{14}$$

Equ. (14) can easily be evaluated numerically for any arbitrary spectral variation in the absorption coefficient. This same technique can be applied in many other situations. If the temperature field is specified in the medium and an analytical solution is known, it will be possible to derive a simple expression analogous to equ. (14) which describes an arbitrary nongray case.

Equ. (14) is in a form useful for numerical computation. Note that an exact form can be written by taking the limit of equ. (14) as the number of bands approaches infinity, to give:

$$S(z) = \int_0^{\infty} \left[\frac{4\, a_\lambda (e_{b\lambda} - e_{bw\lambda}) \left[e^{\sqrt{3} a_\lambda z} + e^{-\sqrt{3} a_\lambda z} \right]}{e^{\sqrt{3} a_\lambda c} \left(1 + \frac{2\sqrt{3}}{3} \right) + e^{-\sqrt{3} a_\lambda c} \left(1 - \frac{2\sqrt{3}}{3} \right)} \right] d\lambda \quad (15)$$

The case of an isothermal gas between parallel plates with arbitrarily varying absorption coefficient has now been solved. If greater accuracy is required, the P-3 approximation can be used instead of the P-1 approximation. In the derivation of the P-N approximations, the radiation intensity is expanded in a series of spherical harmonics (Jeans (1917)).

$$i'(\bar{r}, \theta, \phi) = \sum_{k=0}^{N} \sum_{m=-k}^{k} A_k^m(\bar{r})\, Y_k^m(\theta, \phi) \quad (16)$$

where $A_k^m(\bar{r})$ are the functions to be determined and $Y_k^m(\theta, \phi)$ are the spherical harmonics. The P-1 approximation is obtained by letting N=1 (resulting in 4 terms) and the P-3 approximation is obtained by letting N=3 (resulting in 16 terms). As a result, the P-3 method is generally more accurate than the P-1 method. The price for the increased accuracy is a more complicated set of equations. The P-3 equations, in spectral form, are (Yucel and Bayazitoglu, (1983)):

$$\frac{di_{0\lambda}}{dz} = a_\lambda \left(-10\, i_{1\lambda} + \frac{35\, i_{3\lambda}}{3} \right) \quad (17)$$

$$\frac{di_{1\lambda}}{dz} = a_\lambda (4 e_{b\lambda} - i_{0\lambda}) \quad (18)$$

$$\frac{di_{2\lambda}}{dz} = -\, a_\lambda\, i_{1\lambda} \quad (19)$$

$$\frac{di_{3\lambda}}{dz} = a_\lambda \left(\frac{4\, e_{b\lambda}}{3} - i_{2\lambda} \right) \quad (20)$$

The boundary conditions, of the type derived by Marshak (1947), for these equations are:

$$3\, \varepsilon_\lambda i_{0\lambda} \pm 16\, (2 - \varepsilon_\lambda)\, i_{1\lambda} + 15\, \varepsilon_\lambda i_{2\lambda} = 32\, \varepsilon_\lambda e_{b\lambda} \quad (21)$$

$$(3\varepsilon_\lambda - 5)\, i_{0\lambda} \pm 16\, (1 - \varepsilon_\lambda)\, i_{1\lambda} + 15\, (1 + \varepsilon_\lambda)\, i_{2\lambda} \pm 32 i_{3\lambda} = 32\, \varepsilon_\lambda e_{b\lambda} \quad (22)$$

where the plus sign is applied at the lower boundary and the minus sign is applied at the upper boundary.

To apply the P-3 equations to the case of an arbitrarily-varying absorbtion coefficient, equs. (17) through (22) are now integrated over a single band, j, to give:

$$\frac{di_{0j}}{dz} = a_j \left(-10\, i_{1j} + \frac{35\, i_{3j}}{3} \right) \quad (23)$$

$$\frac{di_{1j}}{dz} = a_j (4 e_{bj} - i_{0j}) \quad (24)$$

$$\frac{di_{2j}}{dz} = -a_j\, i_{1j} \quad (25)$$

$$\frac{di_{3j}}{dz} = a_j \left(\frac{4\, e_{bj}}{3} - i_{2j} \right) \quad (26)$$

$$3\, \varepsilon_j\, i_{0j} \pm 16\, (2 - \varepsilon_j)\, i_{1j} + 15 \varepsilon_j i_{2j} = 32\, \varepsilon_j\, e_{bj} \quad (27)$$

$$(3\varepsilon_j - 5)\, i_{0j} \pm 16\, (1 - \varepsilon_j)\, i_{1j} + 15\, (1 + \varepsilon_j)\, i_{2j} \pm 32 i_{3j} = 32\, \varepsilon_j e_{bj} \quad (28)$$

Equs. (23) through (26) can be combined into a single fourth order differential equation, which is:

$$\frac{d^4 i_{0j}}{d\tau_j^4} - 10\, \frac{d^2 i_{0j}}{d\tau_j^2} + \frac{35}{3}\, i_{0j} = \frac{140}{3}\, e_{bj} \quad (29)$$

where τ_j is the optical thickness given by $\tau_j = a_j\,(z + d/2)$. For the case of an isothermal medium, the solution to equ. (29) is:

$$i_{0j}(\tau_j) = A_{1j} \sinh(\alpha_1 \tau_j) + A_{2j} \sinh(\alpha_2 \tau_j) + A_{3j} \cosh(\alpha_1 \tau_j) + A_{4j} \cosh(\alpha_2 \tau_j) + 4 e_{bj} \quad (30)$$

where $\alpha_1 = 2.941$ and $\alpha_2 = 1.161$. Algebraic forms for the four constants A_{1j}, A_{2j}, A_{3j}, A_{4j}, were obtained by use of the symbolic manipulation program MACSYMA (Rand (1984)). Since the algebraic equations occupied about forty pages using 12 pt. type, they are not reproduced here. A more fruitful approach is to employ MACSYMA to calculate the values of $A_{1j}, A_{2j}, A_{3j}, A_{4j}$ for a particular case. As an illustration, consider the case of a single band with absorbtion coefficient $a = 2.5\ cm^{-1}$ centered at 2.75 μm and with bandwidth 0.5 μm. The participating medium is held at 800 K and both walls are black at 500 K. The walls are separated by 0.4 cm. For this case the values of $A_{1j}, A_{2j}, A_{3j}, A_{4j}$ are given in Table 1. The resulting values of i_0 for this case are listed in Table 2. Values of i_0 derived by the P-1 method are also given in this table for comparison.

Once the values of $A_{1j}, A_{2j}, A_{3j}, A_{4j}$ are computed for all bands of interest, the net rate of heat production can be computed by equ. (13). In this way, the P-3 method can be used to estimate the rate of heat production for an arbitrary nongray absorption coefficient.

A1 =	0.1924014
A2 =	0.0818620
A3 =	- 0.2138496
A4 =	- 0.1564874

Table 1 Values of constants in equ. (29) for single band example in W/cm^2

z/d	i_0 (W/cm^2) P-1	i_0 (W/cm^2) P-3
0.0	0.314983	0.317882
0.1	0.354807	0.374447
0.2	0.384604	0.414570
0.3	0.405270	0.441270
0.4	0.417425	0.456539
0.5	0.421437	0.461506
0.6	0.417425	0.456539
0.7	0.405270	0.441270
0.8	0.384604	0.414570
0.9	0.354807	0.374447
1.0	0.314983	0.317882

Table 2 Values of i_0 for single band example

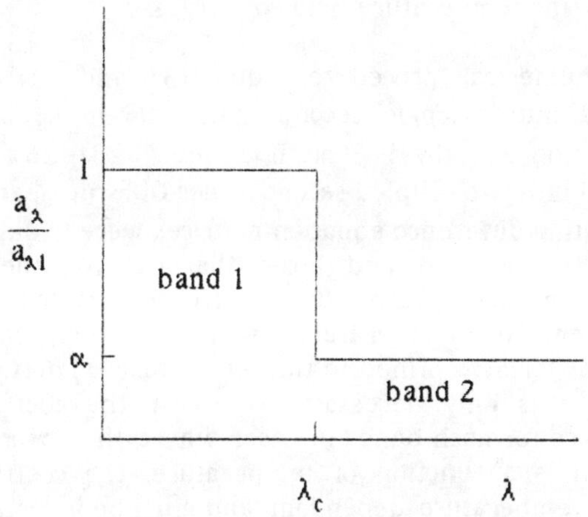

Figure 3 Spectral absorption coefficient for two-band model

Analysis for a Medium in Radiative Equilibrium

In the case described above, the temperature of the medium was known. In the next case to be considered, the medium is in radiative equilibrium with temperature distribution to be determined. In such cases, spectral calculations are less straightforward than they are when the temperature distribution is known. For the radiative equilibrium calculation, the bounding plates are at two different temperatures. Unfortunately, there is no easy analytical form for the nongray solution in the case of an arbitrarily-varying spectral absorption coefficient. A numerical solution is undertaken for a particular variation of the absorption coefficient. The model chosen is a two-level absorption coefficient, as shown in Fig. 3. Note that the method is in no way limited to this simple variation of the absorption coefficient and that the arbitrarily-varying absorption coefficient of Fig. 1 could just as easily have been analysed.

The steps in the numerical procedure are as follows:

1. A temperature distribution between the plates is assumed.

2. The value of i_{0j} is computed by the finite difference solution of either equs. (3) and (4) (for P-1 method) or equs. (23)-(26) (for P-3 method) for each band j.

3. With the values of i_{0j} from step *2*, the roots of equ. (13) are sought. Note that the value of e_{bj} in equ. (13) depends on temperature. At each nodal point in the finite difference grid, the temperature is found for which S(z)=0.

4. Steps 2 and 3 are performed alternately until the temperature field converges.

In the numerical procedure, equs. (3) and (4) were combined into a single second-order ODE in i_{0j} before discretization. For the P-3 method, equs. (23)-(26) were combined into two coupled second-order ODEs in i_{0j} and i_{2j}. The resulting difference equation matrices were tridiagonal for the P-1 method and box-tridiagonal for the P-3 method. In each case, the coefficient matrix was independent of temperature, with all temperature-dependent terms confined to the source term matrix. As a result, it was only necessary to invert the coefficient matrix once for each band. In cases where the absorption coefficient is a function of temperature, the coefficient matrix is temperature-dependent, and must be inverted on each iteration.

In formulating the difference equations, central differences were used throughout. The root finding routine used in step 3 was simple interval halving, although a more refined method, such as the secant method, could have been used.

The results were compared with those of Yucel and Bayazitoglu (1983), who investigated the P-1 and P-3 cases, and with those of Crosbie and Viskanta (1970), who presented the exact solution. A comparison of the present P-1 results with the calculations of these two earlier investigators is given in Table 3. Similar comparisons for the P-3 case are shown in Table 4. For the most part, the present results are in good agreement with those of Yucel and Bayazitoglu (1983). The present results are compared with the exact solution graphically in Figures 4 and 5.

The results are somewhat sensitive to various parameters in the numerical solution. These parameters include the number of grid points, the convergence criterion for the temperature field iteration, and the convergence criterion for the root finding procedure. In the case of the P-1 calculations, grid-independence was achieved with 121 grid points and a convergence criterion of $3.3333 \cdot 10^{-6}$. The results for the P-1 case are quite close to those of Yucel and Bayazitoglu, as shown in Table 3. In the case of the P-3 solution, the number of grid points necessary to achieve full grid-independence exceeded the capabilities of the available compiler. To find a grid-independent and parameter-independent solution, a

x/d		$\alpha = 0$	$\alpha = 0.1$	$\alpha = 0.4$	$\alpha = 0.5$	$\alpha = 0.6$	$\alpha = 0.8$	$\alpha = 1$
0	Exact	0.9483	0.9426	0.9364	0.9359	0.9359	0.9365	0.9377
	Y&B	0.9377	0.9315	0.9235	0.9228	0.9227	0.9234	0.9250
	p-1	0.9377	0.9315	0.9236	0.9228	0.9227	0.9235	0.9251
0.2	Exact	0.9195	0.9125	0.9043	0.9034	0.9029	0.9027	0.9031
	Y&B	0.9156	0.9090	0.8996	0.8985	0.8979	0.8978	0.8985
	p-1	0.9156	0.9089	0.8997	0.8985	0.8979	0.8979	0.8986
0.5	Exact	0.8781	0.8710	0.8604	0.8585	0.8571	0.8550	0.8537
	Y&B	0.8781	0.8715	0.8606	0.8586	0.8570	0.8550	0.8537
	p-1	0.8780	0.8714	0.8607	0.8586	0.8571	0.8551	0.8538
0.8	Exact	0.8272	0.8114	0.8091	0.8059	0.8030	0.7981	0.7939
	Y&B	0.8329	0.8275	0.8160	0.8129	0.8101	0.8050	0.8005
	p-1	0.8329	0.8274	0.8161	0.8129	0.8101	0.8051	0.8007
1	Exact	0.7744	0.7712	0.7558	0.7554	0.7500	0.7414	0.7334
	Y&B	0.7962	0.7925	0.7815	0.7776	0.7737	0.7658	0.7581
	p-1	0.7962	0.7923	0.7816	0.7776	0.7737	0.7659	0.7583
Q_R	Exact	0.6703	0.6427	0.5846	0.5702	0.5576	0.5362	0.5188
	Y&B	0.6811	0.6585	0.6037	0.5891	0.5761	0.5538	0.5357
	p-1	0.6805	0.6594	0.6031	0.5892	0.5759	0.5536	0.5355

Table 3. Comparison of P-1 results with those of Yucel and Bayazitoglu (1983), and Crosbie and Viskanta (1970). $\bar{\lambda}=3$, $\tau_0 = 1$, $\theta_2 = 0.5$

x/d		α = 0	α = 0.1	α = 0.4	α = 0.5	α = 0.6	α = 0.8	α = 1
0	Exact	0.9483	0.9426	0.9364	0.9359	0.9359	0.9365	0.9377
	Y&B	0.9452	0.9389	0.9319	0.9315	0.9316	0.9325	0.9340
	p-3	0.9453	0.9395	0.9327	0.9326	0.9327	0.9334	0.9347
0.2	Exact	0.9195	0.9125	0.9043	0.9034	0.9029	0.9027	0.9031
	Y&B	0.9200	0.9129	0.9043	0.9033	0.9030	0.9031	0.9036
	p-3	0.9203	0.9133	0.9053	0.9045	0.9043	0.9043	0.9049
0.5	Exact	0.8781	0.8710	0.8604	0.8585	0.8571	0.8550	0.8537
	Y&B	0.8781	0.8709	0.8600	0.8581	0.8567	0.8548	0.8537
	p-3	0.8783	0.8714	0.8613	0.8595	0.8582	0.8563	0.8554
0.8	Exact	0.8272	0.8114	0.8091	0.8059	0.8030	0.7981	0.7939
	Y&B	0.8264	0.8207	0.8086	0.8054	0.8024	0.7973	0.7930
	p-3	0.8265	0.8214	0.8101	0.8071	0.8041	0.7992	0.7952
1	Exact	0.7744	0.7712	0.7558	0.7554	0.7500	0.7414	0.7334
	Y&B	0.7812	0.7779	0.7663	0.7621	0.7577	0.7492	0.7411
	p-3	0.7813	0.7783	0.7677	0.7636	0.7597	0.7514	0.7435
Q_R	Exact	0.6703	0.6427	0.5846	0.5702	0.5576	0.5362	0.5188
	Y&B	0.6717	0.6464	0.5882	0.5734	0.5604	0.5386	0.5210
	p-3	0.6716	0.6480	0.5935	0.5796	0.5675	0.5466	0.5304

Table 4. Comparison of P-3 results with those of Yucel and Bayazitoglu (1983), and Crosbie and Viskanta (1970). $\bar{\lambda}=3$, $\tau_0 = 1$, $\theta_2 = 0.5$

refinement procedure was implemented. Each P-3 data point was generated for three different sets of increasingly refined parameters. The final results presented in Table 4 are an extrapolation to zero grid size from the 3 data points generated for each case. The differences between the present results and the earlier results are possibly due to errors in extrapolation. The errors are greatest in the calculation of the heat flux. Since this parameter is calculated as the derivative of the radiation intensity, numerical errors are more prone to occur in the heat flux than elsewhere.

In all cases investigated, convergence was rapid. The P-3 method converged more rapidly than the P-1 method. It was not necessary to use underrelaxation on the temperature field to avoid divergence although this technique could be used in future applications if necessary.

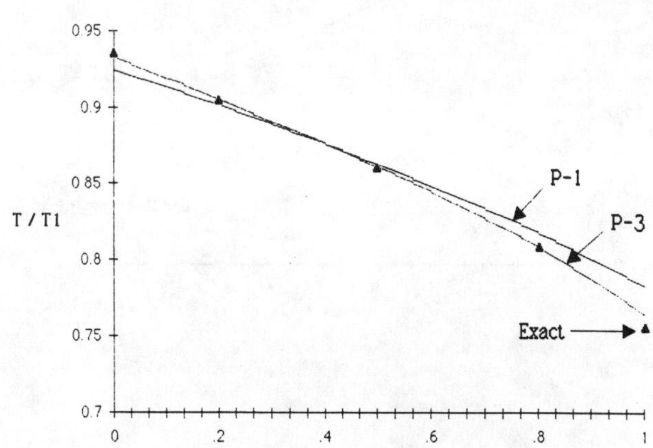

Figure 4. Comparison of P-1 and P-3 solutions with exact solution. (α = 0.4)

Conclusions

The technique presented here has several advantages. It is easy to understand, it can be applied for any spectral variation of the absorption coefficient, and it can be used in multi-dimensional cases. The method is particularly straightforward when the temperature field in the medium is known. In applications in which radiation is coupled to other modes of heat transfer, the radiation equations are often solved iteratively with the energy equation. In that case, an assumed temperature field is passed to the radiation calculation and a computed radiation source term is passed back to the energy equation. In an iterative scheme like this, it would be easy to take nongray effects into account with very little modification of existing codes. In cases where radiation source terms are specified and temperatures are unknown, the spectral calculations are more involved. Nevertheless, the extra coding needed to take nongray effects into account is minimal.

Previous investigators have applied more complicated techniques to non-gray analysis than the simple, straightforward technique described here. As a result, very few investigators studying coupled-mode analyses choose to use non-gray radiating media in their formulation. For gases, the gray assumption is often a very poor approximation. At this time, when the speed and cost of numerical computation have so sharply declined, it is no longer necessary to avoid non-gray formulations in many cases. In this work, we have shown that a straightforward finite-difference approach can produce results comparable to more sophisticated techniques.

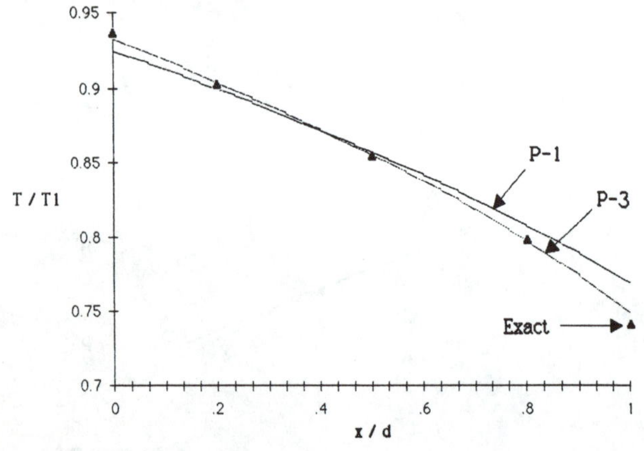

Figure 5. Comparison of P-1 and P-3 solutions with exact solution. ($\alpha = 0.8$)

References

Azad F. H., and M. F. Modest, "Evaluation of the Radiative Heat Flux in Absorbing, Emitting and Linear-Anisotropically Scattering Cylindrical Media" J. of Heat Transfer, Vol. 103, pp. 350-356 (1981)

Bayazitoglu Y. and J. Higenyi, "Higher-Order Differential Equations of Radiative Transfer: P_3 Approximation", AIAA Journal, Vol. 17 No.4 pp. 424-431 (1979)

Crosbie, A. L., and R. Viskanta, "Rectangular Model for Nongray Radiative Transfer," AIAA Journal, Vol. 8, pp. 2055-2057 (1970)

Jeans, J. H., "The Equations of Radiative Transfer of Energy", Monthly Notices of Royal Astronomical Society, Vol. 78, pp. 28-36 (1917)

Kaminski, D.A., "Radiative Heat Transfer From a Gray, Absorbing-Emitting Medium in a Conical Enclosure" ASME Proc. of 1988 National Heat Transfer Conference, HTD-96 Vol. 1 pp. 201-207 (1988)

Krook, M., "On the Solution of Equations of Transfer,I", The Astrophysical Journal, Vol. 122, pp. 488-497 (1955)

Kung, H.C. and M. Sibulkin, "Radiative Transfer in a Nongray Gas Between Parallel Walls", J. Quant. Spectrosc. Radiat. Transfer, Vol. 9, pp. 1447- 1461 (1969)

Marshak, R. E., "Note on the Spherical Harmonics Method as Applied to the Milne Problem for a Sphere", Physical Review, Vol. 71, pp. 433-446 (1947)

Menguc, M. P. , and R. Viskanta, "Radiative Transfer in Axisymmetric, Finite Cylindrical Enclosures", Journal of Heat Transfer, Vol. 108, pp. 271-276 (1986)

Modest, M. F., "A Simple Differential Approximation for Radiative Transfer in Nongray Gases," Journal of Heat Transfer, Vol. 101, pp. 735-736 (1979)

Modest, M. F., and F. H. Azad, "The Differential Approximation for Radiative Transfer in an Emitting, Absorbing and Anisotropically Scattering Medium", J. Quant. Spectrosc. Radiat. Transfer, Vol. 23, pp. 117-120 (1980)

Rand, R.H. "Computer Algebra in Applied Mathematics: An Introduction to MACSYMA," Pitman Publishing Inc. (1984)

Ratzel, A. C., and J. R. Howell, "Two-Dimensional Radiation in Absorbing-Emitting Media Using the P-N Approximation" ASME J. of Heat Transfer, Vol. 105, pp. 333-340 (1983)

Wiebelt, John A., *Engineering Radiation Heat Transfer*, Holt, Rinehart and Winston, Inc. (1966)

Yucel, A. and Y. Bayazitoglu, "P-N Approximation for Radiative Heat Transfer in a Nongray Medium", AIAA Journal, Vol. 21, No. 8, pp. 1196-1203 (1983)

Yuen, W.W. and D. J. Rasky, "Application of the P-1 Approximation to Radiative Transfer in a Non-Gray Medium", Journal of Heat Transfer, Vol. 103, pp. 182-184 (1981)

1989 National Heat Transfer Conference
HTD-Vol. 106, Heat Transfer Phenomena in Radiation, Combustion and Fires

RADIATIVE TRANSFER IN PARTITIONED ENCLOSURES

A. Yücel
Nuclear Science Center
Louisiana State University
Baton Rouge, Louisiana

ABSTRACT

Radiative heat transfer through absorbing, emitting and isotropically scattering media in complex enclosures is studied numerically. Specifically, a two-dimensional square enclosure partially divided by vertical baffles are considered. The discrete ordinates method is used to obtain solutions to the equation of radiative transfer. The changes in the temperature distributions and the heat transfer rates due to the presence of partitions are analyzed. Results are presented for various optical thickness values. The role of heat conduction across the baffle and the effects of baffle geometry and location are discussed.

NOMENCLATURE

d	Baffle width
h	Baffle height
I	Dimensionless radiation intensity, $i/4\hat{\sigma}T_o^4$
k_B	Thermal conductivity of baffle
k_m	Thermal conductivity of medium
L	Width and height of enclosure
N_B	Baffle conduction-radiation paremeter, $k_B(\kappa+\sigma)/4\hat{\sigma}T_o^3$
N_m	Medium conduction-radiation parameter, $k_m(\kappa+\sigma)/4\hat{\sigma}T_o^3$
Q_C	Dimensionless conductive heat flux, $q_C/4\hat{\sigma}T_o^4$
Q_R	Dimensionless radiative heat flux, $q_R/4\hat{\sigma}T_o^4$
Q_b	Bottom wall heat transfer, Eqs. (A3)
Q_c	Cold wall heat transfer, Eq. (A2)
Q_h	Hot wall heat transfer, Eq. (A1)
Q_t	Top wall heat transfer, Eqs. (A4)
T	Temperature
T_c,T_h	Cold and hot wall temperatures
T_o	Reference temperature
X,Y	Dimensionless coordinates, x/L and y/L
X_c	Baffle centerline position
κ	Absorption coefficient
θ	Dimensionless temperature, T/T_o
μ,ξ	Direction cosines (ordinates)
σ	Scattering coefficient
$\hat{\sigma}$	Stefan-Boltzmann constant
τ	Optical thickness, $(\kappa+\sigma)L$
$\boldsymbol{\Omega}$	Direction vector, $\mu\,\mathbf{i} + \xi\,\mathbf{j}$
ω	Single scattering albedo, $\sigma/(\kappa+\sigma)$
∇	Dimensionless gradient operator, $\mathbf{i}\,(\partial/\partial X) + \mathbf{j}\,(\partial/\partial Y)$

INTRODUCTION

In diverse applications such as modeling of room fires, nuclear reactor safety analysis, and furnace or solar collector design, prediction of heat transfer in complex enclosures may be required. Heat transfer processes in such applications are often complicated by the presence of internal partitions. Relevant work in these areas have mainly focused on natural convection in partitioned enclosures (Yang and Lloyd, 1985; Zimmerman and Acharya, 1987). At elevated temperatures, thermal radiation can play a major role in energy transport across the enclosure. While there has been a strong interest in multi-dimensional radiative heat transfer problems in recent years, these studies have considered plain rectangular or cylindrical enclosures (Viskanta, 1984; Howell, 1988). The understanding of radiative transfer through participating media in partitioned enclosures is limited and requires further study.

This paper presents a parametric study of radiative transfer in a two-dimensional square enclosure, partitioned by vertical baffles protruding from the top and bottom walls. The problem under consideration is of particular importance in the study of smoke and fire spread in building fires (Larson and Viskanta, 1976; Yang, 1986; Hoffmann and Markatos, 1988): the present model can be readily extended to coupled flow analysis to study the influence of radiative transfer on the temperature and flow fields in such applications. Previously, Chang et al (1983) have studied the radiation-natural convection interactions in a partitioned square enclosure, but employed a cumbersome radial flux method based on a polar description for radiation exchange. In this study, the discrete ordinates method is used to calculate the radiation field, which is coupled with the temperature calculation for the partitions.

The method is readily adaptable to multi-zone enclosures and compatible with many of the algorithms currently being used for flow and temperature calculations. Numerical solutions are generated for various baffle configurations and optical thickness values. The effects of baffle properties on the temperature field and wall heat transfer rates are discussed.

ANALYSIS

The physical model considered is shown in Figure 1. The vertical baffles are aligned and have a finite thermal conductivity. The gray absorbing, emitting and isotropically scattering medium is in local thermodynamic equilibrium. All physical properties are assumed to be constant. The constant radiative and thermal property assumptions are invoked primarily for numerical convenience; they are often used to reduce the number of independent parameters in generalized analyses (Larson and Viskanta, 1976).

Governing Equations

With the above assumptions, the radiative transfer equation governing the steady state radiation intensity I is given in dimensionless form by

$$\mu\,(\partial I/\partial X) + \xi\,(\partial I/\partial Y) + \tau I = \tau\,[(1-\omega)\,\theta^4 + \omega \int_{4\pi} I d\Omega]/4\pi \quad (1)$$

where τ is the optical thickness, ω is the scattering albedo, and θ is the temperature of the medium. In the absence of convection and heat generation, the conservation of energy equation governing the temperature distribution in a radiating and conducting medium is given by

$$N_m \nabla^2 \theta - \tau \nabla \cdot \mathbf{Q_R} = 0 \quad (2)$$

where N_m is the conduction-radiation parameter for the medium. While the above formulation is generalized to allow for simultaneous heat transfer by conduction and radiation, this study is restricted to pure radiative transfer through the medium, i.e. $N_m = 0$. In the baffle regions, which are opaque to radiation, the energy equation reduces to

$$N_B \nabla^2 \theta = 0 \quad (3)$$

Integrating Eq. (1) over the whole solid angle yields a relation for the divergence of the radiative heat flux vector

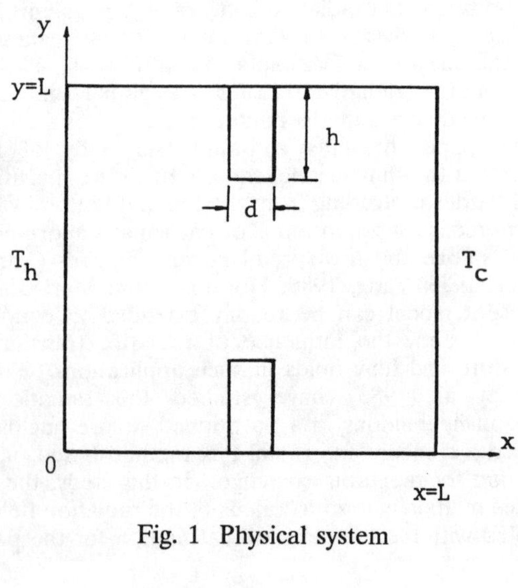

Fig. 1 Physical system

$$\nabla \cdot \mathbf{Q_R} = (1-\omega)\,\tau\,[\,\theta^4 - \int_{4\pi} I\, d\Omega] \quad (4)$$

which is substituted in the energy equation. The radiative and conductive heat flux vectors are calculated by

$$\mathbf{Q_R} = \int_{4\pi} (\mu\,\mathbf{i} + \xi\,\mathbf{j})\, I\, d\Omega \quad (5a)$$

$$\mathbf{Q_C} = -\,Pl\,\nabla\theta \quad (5b)$$

Boundary conditions

The side walls of the enclosure are isothermal and maintained at T_h (hot wall) and T_c (cold wall). Temperature along the end walls varies linearly between these two temperatures. For simplicity, all surfaces are assumed to be black. Thus, the intensity of radiation leaving a bounding surface is simply the blackbody intensity of radiation.

For the baffles, the energy balance at the baffle-medium interface requires

$$(-N_m \nabla\theta + \tau\,\mathbf{Q_R}) \cdot \mathbf{n} = -N_B \nabla\theta \cdot \mathbf{n} \quad (6)$$

where **n** denotes the unit vector normal to the surface interface and N_B is representative of the baffle thermal conductivity.

Solution Procedure

Equations (1)-(4) with the associated boundary conditions provide a complete mathematical formulation of the problem. The two-dimensional conduction equation is solved by a control-volume based finite difference procedure described by Patankar (1980). The solutions to the equation of radiative transfer are obtained by the discrete ordinates method. In the discrete ordinates formulation, the solid angle is discretized by selecting an ordered set of directions with predetermined weights. Quantities involving integrals over the solid angle are evaluated by sums over the discrete directions with weights as the weighing functions. The spatial discretization introduced to obtain a finite difference form of the radiative transfer equation employs the same control volumes (cells) as conduction solution method. The discrete ordinate transport equation corresponding to Eq. (1) is written as

$$\mu_m\,(I_{i+1} - I_i)/\Delta X_i + \xi_m\,(I_{j+1} - I_j)/\Delta Y_j + \tau I = S \quad (7)$$

In this notation, I is the center intensity for the spatial cell bounded by X_i, X_{i+1} and Y_j, Y_{j+1}; μ and ξ are the cosines of the respective angles the direction of travel $\mathbf{\Omega}$ makes with the X and Y directions. Equation (7) represents the radiative energy balance for the control volume shown in Fig. 2. The first two terms on the left hand side of the above equation represent the loss of radiant energy from the cell volume

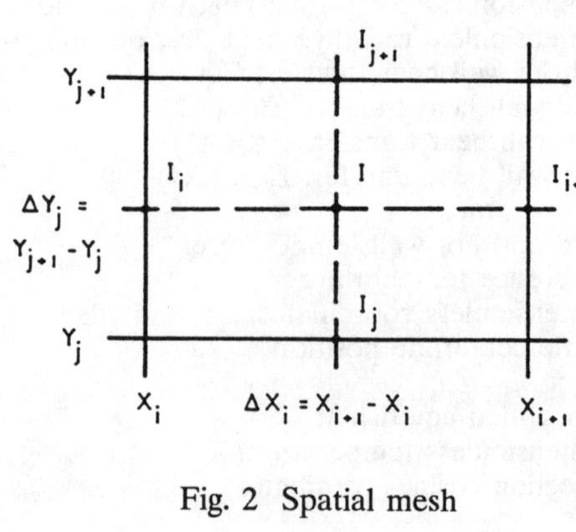

Fig. 2 Spatial mesh

across the spatial boundaries, while the third term is the combined loss by absorption and scattering. The source term on the right hand side includes gains due to inscatter from other directions and due to emission. The cell center and cell boundary intensities in Eq. (7) are related to each other by a selected interpolation scheme. Two most commonly used cell interpolation relations are the linear (diamond difference) and the step schemes which are mathematically expressed as

$$I = a\, I_{i+1} + (1\text{-}a)\, I_i = a\, I_{j+1} + (1\text{-}a)\, I_j \qquad (8)$$

with a = 0.5 for the linear model and a = 1 for the step model. With the introduction of the above interpolation relations into Eq. (7), the resulting discrete ordinates equations can be solved by proceeding in the direction of photon travel on the spatial mesh, i.e. sweeping away from the boundary into the mesh. Depending on the particular quadrant for a given direction, the sweep is initiated at the respective corner of the enclosure and proceeds towards the opposite corner. For the current problem, this is implemented by traversing horizontally towards the opposing side wall for each vertical Y-level. When the horizontal sweep for a given Y-level is completed, a new horizontal sweep is initiated at the next (up or down) level. Horizontal sweeps that intersect the bottom or top baffle are terminated at the baffle surface and restarted at the opposite side of the baffle. The discrete ordinates methodology is discussed in detail by Lewis and Miller (1984).

The coupled conduction and radiative transfer equations are solved through an iterative procedure which is outlined below:

1) An initial temperature distribution is assumed.

2) The radiative transfer equation is solved for the given temperature field. The incident radiant energy term in Eq. (4) and the boundary radiative heat flux terms in Eq. (6) are determined.

3) Eq. (6) is used to update the baffle surface temperature required for the radiation calculation.

4) Eqs. (2) and (3) are solved to obtain a new temperature distribution.

5) A new iteration is completed by repeating Steps 2-4. The procedure is continued until pointwise convergence on temperature etc. is obtained.

RESULTS AND DISCUSSION

In this section, isotherm contour plots, and heat transfer rates and variations along the bounding walls are presented. As stated before, conduction in the medium is neglected (N_m = 0). Thus the medium is in radiative equilibrium and the temperature field is independent of the scattering albedo ω; however, we assume ω = 0 for simplicity. With these specifications, the energy equation for the medium reduces to

$$\theta^4 - \int_{4\pi} I\, d\Omega = 0 \qquad (9)$$

The ratio of side wall temperatures is fixed at $T_c/T_h = 0.5$ in this study ($\theta_h = 1$ and $\theta_c = 0.5$ with $T_o = T_h$). Temperature contours are plotted with isotherm values ranging from 0.95 to 0.65 with 0.05 decrements. It should be noted that the sharp turns in the isotherms that are particularly obvious at the baffle-medium interfaces are not realistic: they are artificially created by the contour plotting software which does not recognize the temperature discontinuity that exists at the surfaces for a purely radiating medium. These sharp bends in the isotherms are not totally without merit, however, as they serve to illustrate the magnitude of the temperature drops at the walls.

Figure 3a shows the temperature contours for an unpartitioned enclosure. The temperature variation across the enclosure is flatter for an optically thin medium with sharper temperature discontinuities at the walls. The medium temperature increases near the hot wall and decreases near the cold wall as the optical thickness increases. Radiative transfer through an optically thicker medium resembles a local diffusion process; the temperature distribution assumes a more conduction-like (linear) variation between the hot and cold walls.

Figure 3b shows the isotherms for an enclosure divided by centrally located baffles with d/L = 0.1 and h/L = 0.25. In general, the high temperature region to the left of the baffles remains largely unaffected by the presence of baffles. This is also true for temperatures in the mid-portion of the right compartment which receives the hot wall radiation through the aperture. At the top and bottom regions of the right compartment, temperatures fall significantly since the baffles shield the cold wall from the hot wall radiation and emit radiation at lower surface temperatures. Thus the isotherms are clamped at the top and bottom resulting in a "bulge" in the middle for the right compartment. As before, medium temperatures rise near the hot wall and fall near the cold wall as optical thickness increases. Consequently, the bulge in the cold region isotherms is more pronounced for an optically thin medium.

The baffle thermal conductivity can have a strong influence on the temperature distribution in the enclosure. For higher conductivity (higher N_B), the stronger conduction heat transfer leads to more uniform baffle temperatures that are close to the end wall temperature at the base of the baffle. For lower thermal conductivity (lower N_B), the heat transfer into the baffle on the left surface (radiation received from the hot wall) and the heat transfer out of the baffle on the right surface (radiation emitted to the cold wall) are reduced significantly, approaching adiabatic conditions on the baffle surfaces. Thus the temperature distribution at the top and bottom portions of the left compartment becomes more uniform as temperature increases near the baffle. Similarly, the corresponding temperature distribution on the right compartment is also more uniform but with reduced temperatures near the baffle. Because of increased temperatures on the left and decreased temperatures on the right, large temperature gradients develop across the baffle. While the high temperatures on the left and the low temperatures on the right penetrate further towards the baffle, the central portion temperatures are relatively unchanged. As a result, the high temperature region isotherms on the left curve backwards, while the characteristic bulge in the cold region isotherms on the right becomes more pronounced.

The heat flux variations along the hot and cold walls for unpartitioned and partitioned enclosures are shown in Figure 4. As expected, heat transfer rates decrease when baffles are present. Heat fluxes for a partitioned enclosure are more peaked near the midplane due to streaming of radiation through the aperture. This behavior is accentuated for an optically thin medium since the radiative interactions between the side walls and the facing baffle surfaces, and between the side walls themselves across the middle, are stronger and become more distinct. This is illustrated by the kinks in the curves at the baffle tip levels. On the other hand, as the optical thickness increases, heat fluxes decrease and become smoother as a result of increased local attenuation (decreased penetration),. The heat fluxes also decrease with decreasing

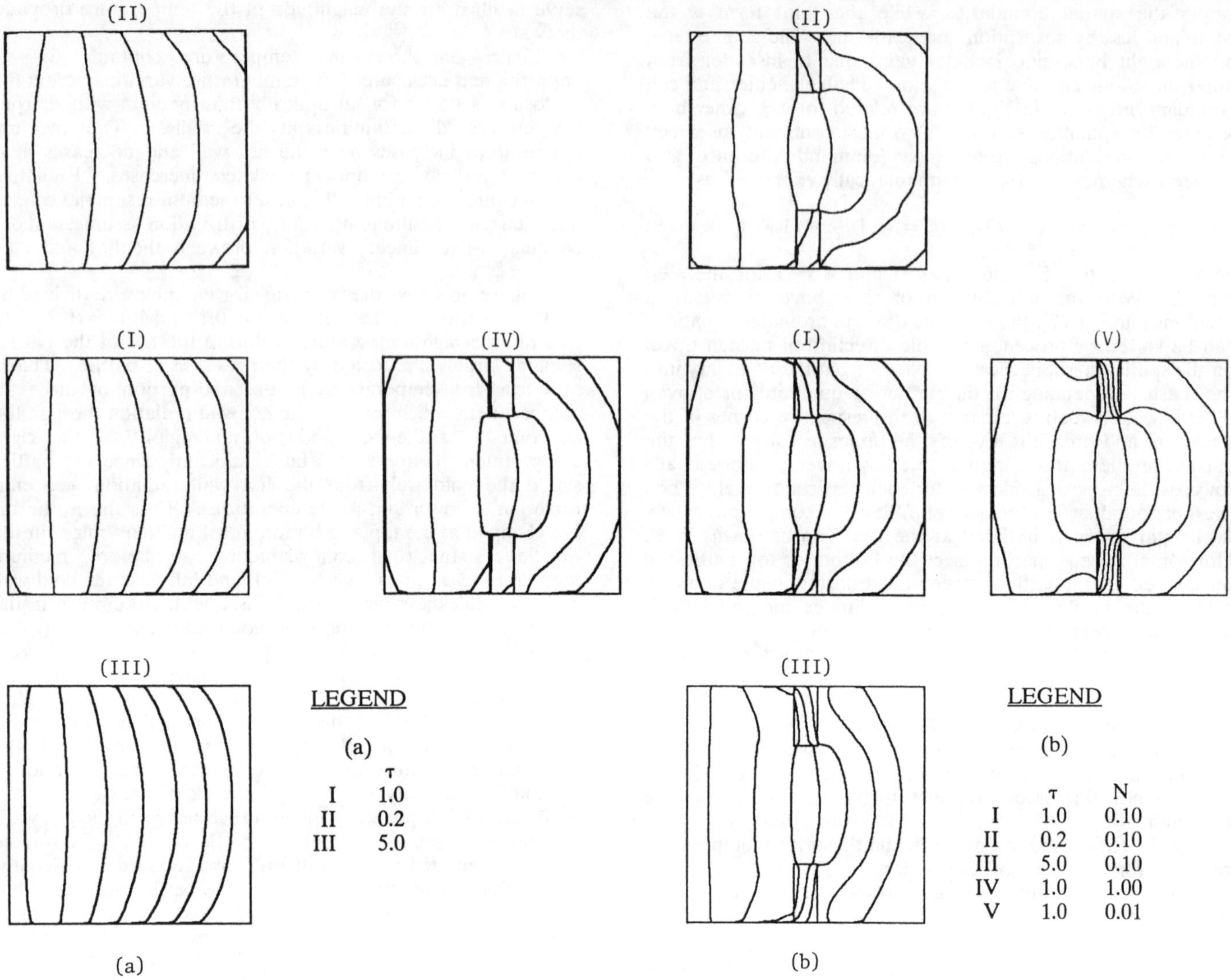

Fig. 3 Isotherms for a) unpartitioned enclosure, b) partitioned enclosure with $d/L=0.1$, $h/L=0.25$ and $X_c=0.5$ ($N_B=0.422N$)

baffle thermal conductivity due to reduced interaction between the hot wall and left baffle surface, and between the cold wall and the right baffle surface.

The heat flux variations along the bottom wall are shown in Figure 5. The heat flux variations along the top wall are identical but in the opposite direction due to the symmetry of the problem. Recall that the bottom wall temperature T_b varies linearly between the side walls. The corresponding emissive power decreases nonlinearly as T_b^4. Near the hot wall, heat is transferred into the enclosure through the bottom surface since the wall emissive power is initially close to the hot wall emissive power. This is also borne out by the concavity in the isotherms close to the hot wall. As the bottom wall emissive power falls off rapidly moving away from the hot wall, the direction of heat flow is reversed and heat is transferred out of the enclosure. For a partitioned enclosure, the net flow out of the bottom wall is higher on the left (hotter) compartment side and lower on the right (colder) compartment side. For clarity of figure, the conductive heat flux variations along the base of the baffle are not plotted.

The total (integrated) heat transfer results along the side and end walls are tabulated in Table 1. The heat transferred across the enclosure is lower for a partitioned enclosure. Heat transfer out of the end walls by conduction at the base of the baffle can be significant at low optical thickness values.

The effects of baffle geometry and location on the isotherms are illustrated in Figure 6. When the baffles are short, the temperature field is largely unperturbed except in the vicinity of the stub-like baffles. Longer baffles promote isolation of the high temperature region from the cold temperature region except near the small gap region where a backward bulge develops in the isotherms. The streaming of radiation through the reduced aperture is reflected in the cold region isotherms. When the baffle thickness is doubled, temperatures increase in the left compartment and decrease in the right compartment, accompanied by the backward curving of the mid-portion isotherms. On the other hand, reducing the baffle thickness by half has relatively little effect on the temperature distribution. Higher temperatures prevail over most of the enclosure when the baffles are placed closer to the cold wall, while lower temperatures persist overall when the baffles are aligned closer to the hot wall.

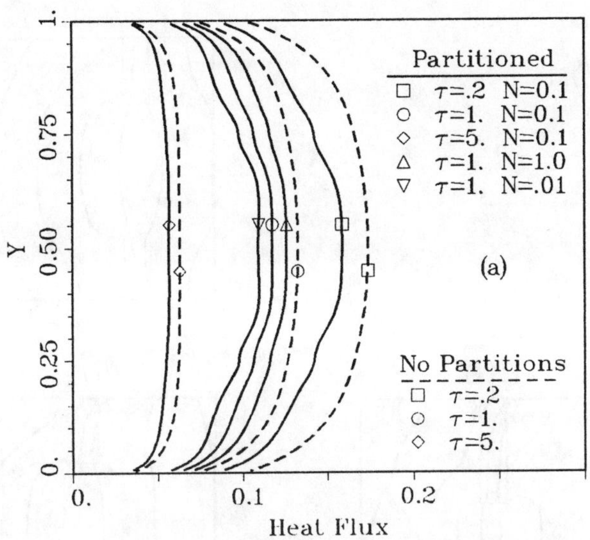

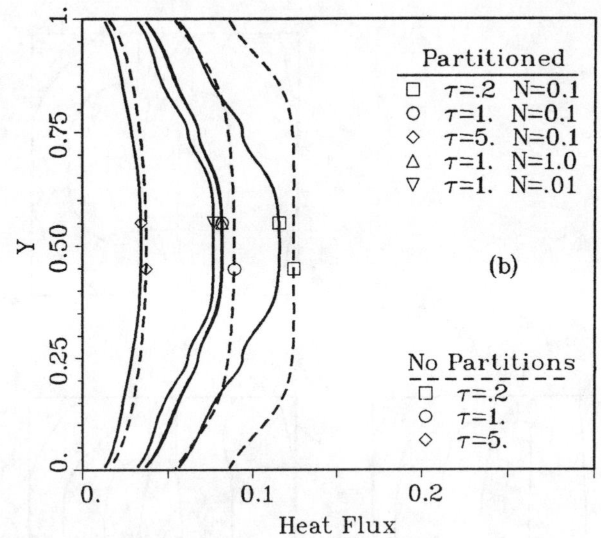

Fig. 4 Side wall heat flux variations for an unpartitioned enclosure and a partitioned enclosure with d/L=0.1, h/L=0.25 and X_c=0.5 (N_B=0.422N)
a) Hot wall, b) Cold wall

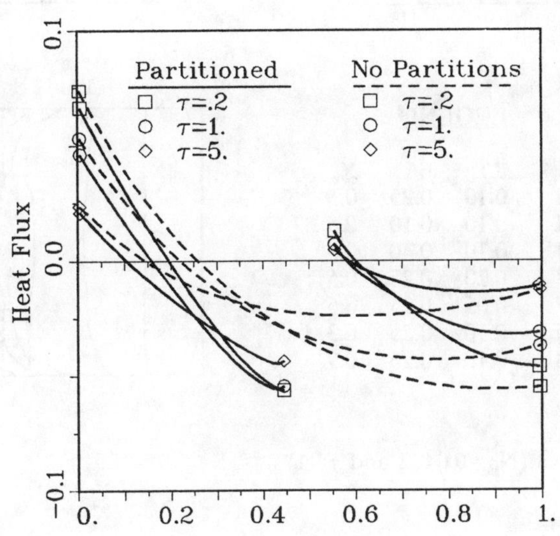

Fig. 5 Bottom wall heat flux variations for an unpartitioned enclosure and a partitioned enclosure with d/L=0.1, h/L=0.25 and X_c=0.5 (N_B=0.0422)

TABLE 1

Total heat transfer for the side and end walls

a) Unpartitioned

τ	Hot Wall Rad	Cold Wall Rad	End Wall Total	End Wall Rad	End Wall Cond
0.2	0.159	0.116	0.021	0.021	-
1.0	0.121	0.081	0.020	0.020	-
5.0	0.060	0.032	0.014	0.014	-

b) Partitioned (N_B=0.0422; d/L=0.1, h/L=0.25, X_c=0.5)

τ	Hot Wall Rad	Cold Wall Rad	End Wall Total	End Wall Rad	End Wall Cond
0.2	0.138	0.092	0.023	0.014	0.010
1.0	0.103	0.066	0.019	0.016	0.003
5.0	0.053	0.028	0.013	0.012	0.001

The total side and end wall heat transfer results corresponding to these configurations are presented in Table 2. As expected, the side wall heat transfer increases as the baffle height decreases or the baffle thickness increases. In general the end wall heat transfer is less influenced by the baffle height or thickness. Heat transfer rates on all surfaces decrease when baffles are located closer to the hot wall and increase when baffles are placed closer to the cold wall.

Computational Aspects

The discrete ordinate solutions were obtained using a fully symmetric S8 quadrature set (40 directions) which satisfies the half-range first moments of intensity (Fiveland, 1988). While S4 quadrature (12 directions) is adequate for many two-dimensional applications (Fiveland, 1985; Yucel et al, 1988), it could not accurately represent radiation streaming through the aperture between the baffles in the problem considered. Experiments also showed that the linear interpolation scheme led to unrealistic spatial oscillations in the radiation intensity field because of negative flux fix-ups. Thus the step method was used in this study. Step interpolation is a lower order accurate method compared to the linear scheme for a given mesh size. However, positive intensities are guaranteed and therefore spatial oscillations are not encountered in the step method. A 42x42 mesh was found to be sufficient for accuracy, which was packed near the boundaries and inside the baffles for partitioned enclosures. Experiments with denser meshes (75x75) or higher order quadratures (S12 quadrature

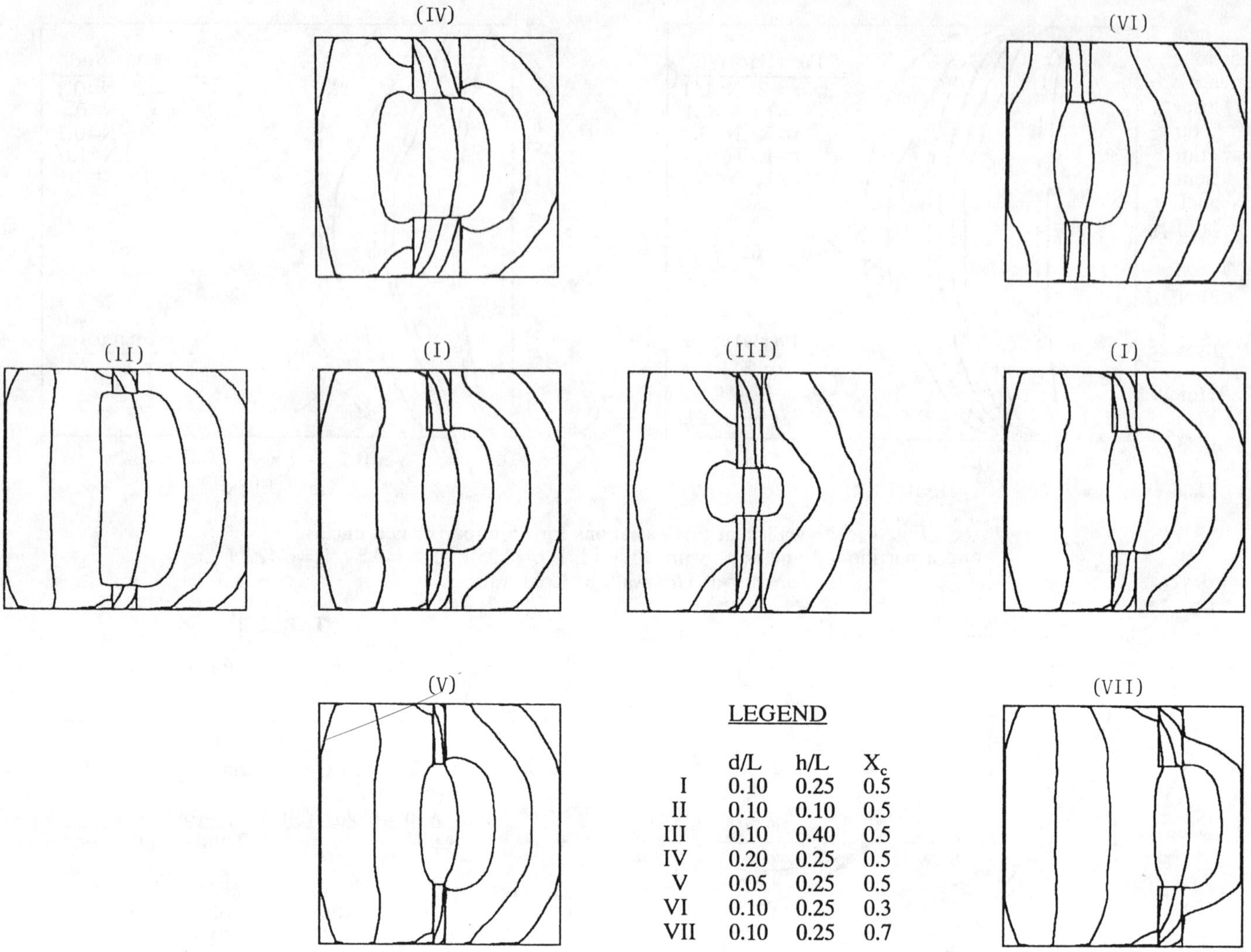

Fig. 6 Isotherms for partitioned enclosures (N_B=0.0422 and τ=1)

TABLE 2

Total heat transfer for the side and end walls for partitioned enclosures with N_B = 0.0422 and τ = 1.

			Hot Wall	Cold Wall	End Wall		
d/L	h/L	X_c	Rad	Rad	Total	Rad	Cond
0.10	0.25	0.5	0.103	0.066	0.019	0.016	0.003
0.10	0.10	0.5	0.116	0.078	0.019	0.016	0.003
0.10	0.40	0.5	0.088	0.050	0.019	0.016	0.003
0.20	0.25	0.5	0.098	0.063	0.018	0.013	0.005
0.05	0.25	0.5	0.106	0.068	0.019	0.017	0.002
0.10	0.25	0.3	0.096	0.066	0.015	0.014	0.001
0.10	0.25	0.7	0.110	0.064	0.023	0.018	0.005

set with 84 directions) yielded only minor differences in the results. Furthermore, the overall energy balance for the enclosure requires

$$R \equiv Q_h - Q_c + Q_b - Q_t \qquad (10)$$

where Q_h, Q_c, Q_b and Q_t represent the total heat transfer on the hot, cold, bottom and top walls respectively (See Appendix for detailed definitions). In all cases considered, R differed from zero in the fourth digit or higher following the decimal point. The number of iterations for convergence varied between 50 to 100 iterations. On the average, an iteration required about 0.6 CPU seconds for a nonoptimized scalar run on the LSU IBM 3090-600E mainframe computer.

CONCLUSION

A numerical study of radiative transfer through a participating medium in a square enclosure partially divided by vertical baffles is conducted. The temperature distribution and the heat transfer variations are altered by the presence of partitions. The baffle conductivity, geometry and location can have a strong influence on the heat transfer processes in the enclosure.

Numerical solutions to the radiative transfer equation were generated by discrete ordinates method. The method is efficient and suitable for coupled flow and heat transfer calculations. Work in progress involves combined conduction and radiation heat transfer in partitioned enclosures.

APPENDIX

Expressions for Total Heat Transferred at the Enclosure Walls

1. Hot Wall

$$Q_h = \int_{Y=0}^{1} [Q_R \cdot i]_{X=0}\, dY \quad \text{(A1)}$$

2. Cold Wall

$$Q_c = \int_{Y=0}^{1} [(Q_R \cdot i]_{X=1}\, dY \quad \text{(A2)}$$

3. Bottom Wall

a) Unpartitioned Case

$$Q_b = \int_{X=0}^{1} [(Q_R \cdot j]_{Y=0}\, dX \quad \text{(A3a)}$$

b) Partitioned Case ($a = X_c - d/2L$, $b = X_c + d/2L$)

$$Q_b = \int_{X=0}^{a} [Q_R \cdot j]_{Y=0}\, dX + \int_{X=b}^{1} [Q_R \cdot j]_{Y=0}\, dX + \int_{X=a}^{b} [Q_C \cdot j]_{Y=0}\, dX \quad \text{(A3b)}$$

4. Top Wall

a) Unpartitioned Case

$$Q_t = \int_{X=0}^{1} [(Q_R \cdot j]_{Y=1}\, dX \quad \text{(A4a)}$$

b) Partitioned Case ($a = X_c - d/2L$, $b = X_c + d/2L$)

$$Q_t = \int_{X=0}^{a} [Q_R \cdot j]_{Y=1}\, dX + \int_{X=b}^{1} [Q_R \cdot j]_{Y=1}\, dX + \int_{X=a}^{b} [Q_C \cdot j]_{Y=1}\, dX \quad \text{(A4b)}$$

REFERENCES

Chang, L., Yang, K. T. and Lloyd, J. R., 1983, "Radiation-Natural Convection Interactions in Two-Dimensional Complex Enclosures," *J. Heat Transfer*, Vol. 105, pp. 89-95.

Fiveland, W. A., 1985, "Discrete Ordinates Solutions of the Radiative Transport Equation for Rectangular Enclosures," *J. Heat Transfer*, Vol. 106, pp. 669-706.

Fiveland, W. A., 1988, "Three-Dimensional Radiative Heat-Transfer Solutions by the Discrete-Ordinates Method," *J. Thermophysics and Heat Transfer*, Vol. 2, pp. 309-316.

Hoffmann, N. and Markatos, N. C., 1988, "Thermal Radiation Effects on Fires in Enclosures," *Applied Mathematical Modeling*, Vol.12, pp. 129-140.

Howell, J. R., 1988, "Thermal Radiation in Participating Media: The Past, the Present and Some Possible Futures," *J. Heat Transfer*, Vol. 110, pp. 1220-1229.

Larson, D. W. and Viskanta, R., 1976, "Combined Laminar Free Convection and Radiation in a Rectangular Enclosure," *J. Fluid Mechanics*, Vol. 78, pp. 65-85.

Lewis, E. E. and Miller, W. F., Jr., 1984, *Computational Methods of Neutron Transport*, Wiley, New York.

Patankar, S. V., 1980, *Numerical Heat Transfer and Fluid Flow*, Hemisphere Publishing Co., New York.

Viskanta, R., 1984, "Radiative Heat Transfer," *Fortschritte der Verfahrenstechnik*, Vol. 22, pp. 51-81.

Yang, K. T., 1986, "Numerical Modeling of Natural-Convection Radiation Interactions in Enclosures," *Heat Transfer 1986*, C. L. Tien et al, ed., Hemisphere, Washington, D.C., Vol. 1, pp. 303-329.

Yang, K. T. and Lloyd, J. R., 1985, "Turbulent Buoyant Flow in Vented Simple and Complex Enclosures," *Natural Convection: Fundamentals and Applications*, S. Kakac et al, ed., Hemisphere, Washington, D.C., pp. 303-329.

Yücel, A., Acharya, S. and Williams, M. L., 1989, "Natural Convection and Radiation in a Square Enclosure," *Numerical Heat Transfer, Part A: Applications*, Vol.15, pp. 261-278.

Zimmerman, E. and Acharya, S., 1987, "Free Convection Heat Transfer in a Partially Divided Vertical Enclosure with Conducting End Walls," *Int. J. Heat Mass Transfer*, Vol. 30, pp. 319-331.

1989 National Heat Transfer Conference
HTD-Vol. 106, Heat Transfer Phenomena in Radiation, Combustion and Fires

THREE-DIMENSIONAL SPECTRAL RADIATIVE HEAT TRANSFER SOLUTIONS BY THE DISCRETE-ORDINATES METHOD

W. A. Fiveland and A. S. Jamaluddin
The Babcock & Wilcox Company
Research and Development Division
Alliance, Ohio

ABSTRACT

Radiative heat transfer in three-dimensional enclosures with a participating medium was predicted using the discrete-ordinates method formulated for spectral variations in the radiative properties. The spectral distribution of radiation was subdivided into a finite number of bands within which the properties were assumed uniform. The method uses a control volume formulation of the discrete-ordinates equations for each band. Solutions were found for each band and summed to obtain the total contribution. The solution strategy and stability are discussed in the paper. Results have been obtained for the S_4 method, a 24-flux approximation. Solutions are benchmarked with several methods. The favorable results indicate that the spectral discrete-ordinates method offers an alternative that is easily adaptable and usable in existing control volume transport codes.

NOMENCLATURE

A	north-south areas, m^2
a_n	coefficients of a Legendre series
B	east-west areas, m^2
C	front-back areas, m^2
$e_{b\lambda}$	Planck's blackbody function
G	incident energy, $\int_{4\pi} I d\Omega$, W/m^2
I	radiant intensity, $W/(m^2 \cdot Sr)$
L	enclosure dimension, m
L_m	mean beam length, m
n	unit normal
N	number of terms in summation
q	heat flux, W/m^2
r	position vector, m
S	source term, W/m^3
T	temperature, °K
V_p	volume of pth control volume, m^3
w_m	weight function in a direction −m (fractional area of a unit sphere)
x	coordinate, m
y	coordinate, m
z	coordinate, m
$\bar{x}$	fractional distance, x/L_x
$\bar{y}$	fractional distance, y/L_y
$\bar{z}$	fractional distance, z/L_z
α	finite-difference weighting factor ($.5 \leq \alpha \leq 1.0$)
β	extinction coefficient, $\sigma + \kappa$, m^{-1}
ε	surface emittance
κ	absorption coefficient, m^{-1}
λ	wavelength, μm
μ, ξ, η	ordinates $\mu = \cos\theta$, $\xi = \sin\theta \sin\phi$, $\eta = \sin\theta \cos\phi$
η	waveband, cm^{-1}
Ω	outgoing direction of radiation
Φ	phase function
ρ	surface reflectance
σ	scattering coefficient, m^{-1}
$\bar{\sigma}$	Boltzmann's constant, 5.669×10^{-8} $W/(m^2 \cdot K^4)$
Θ	angle between incoming and outgoing intensity
∇	gradient vector
Δ	fractional value

Superscripts

′	(prime) incoming value
m′	incoming direction

Subscripts

b	blackbody
g	gas
m	outgoing ordinate direction
m′	incoming ordinate direction
N,S,E W,F,B	node points around p

n,s,e w,f,b	control volume faces surrounding node point p
p	control volume center
r	radiative
x	coordinate direction
y	coordinate direction
z	coordinate direction

<u>Boldface</u>

Vectorial quantity

INTRODUCTION

In many high-temperature devices, including combustion, radiant heat transfer is the dominant mode of energy transfer. Accurate solutions of the thermal performance of these processes are strongly linked to the solution of the radiation heat transfer. An extensive survey of multidimensional radiative models has been reported elsewhere by Viskanta (1984) and Viskanta and Menguc (1987). Existing methods (Howell, 1968, Lockwood and Shah, 1981, Menguc and Viskanta, 1985, and Fiveland, 1988) of solving multidimensional problems are based on radiatively gray property assumptions. This paper describes a spectral discrete-ordinates method applicable for absorption, scattering, and re-emission of radiant energy in two- and three-dimensional Cartesian geometries. The method is an efficient, accurate, and stable computational tool for analysis of radiative heat transfer.

Radiation properties are often very nongray. Mixtures of absorbing gases such as water vapor and carbon dioxide and absorbing and scattering particles like coal, char, and fly ash, need to be evaluated on a spectral basis. In this paper, the nongray distribution of radiation is subdivided into a finite number of wavebands, within which the properties are constant. The discrete-ordinates equations for two- or three-dimensional geometries are derived. The picket-fence method is used (Viskanta and Menguc, 1987), where each band is solved separately and total surface fluxes and incident energy are found by adding the contributions of the individual bands. The theoretical treatment is presented, and a criterion for stability is given. The paper presents several examples to demonstrate the method and benchmark it against the numerically exact zone method. Predictions indicate that the spectral formulation is a good alternative for performing radiative calculations.

ANALYSIS

Consider the radiative transfer equation for the enclosure shown in Figure 1. The balance of monochromatic energy passing in a specified direction ($\mathbf{\Omega}$) through a differential volume in an emitting-absorbing and scattering medium can be written as follows:

$$(\mathbf{\Omega}\cdot\nabla)I(\mathbf{r},\mathbf{\Omega}) = -(\kappa + \sigma)I(\mathbf{r},\mathbf{\Omega}) + \kappa I_b(\mathbf{r}) + \frac{\sigma}{4\pi}\int_{\mathbf{\Omega}' = 4\pi} I(\mathbf{r},\mathbf{\Omega}')\,\Phi(\mathbf{\Omega}' \to \mathbf{\Omega})\,d\Omega' \quad (1)$$

where $\Phi(\mathbf{\Omega}' \to \mathbf{\Omega})$ is the phase function of energy transfer from the incoming direction ($\mathbf{\Omega}'$) to the outgoing direction ($\mathbf{\Omega}$) [see Figure 1]; $I(\mathbf{r},\mathbf{\Omega})$ is the radiation intensity, which is a function of position and direction; $I_b(\mathbf{r})$ is the intensity of a blackbody radiation at the temperature of the medium; and κ and σ are the absorption and scattering coefficients of the medium, respectively. The expression on the left-hand side represents the gradient of the intensity in the specified direction ($\mathbf{\Omega}$). The three terms on the right represent the changes in intensity due to absorption and out-scattering, emission and in-scattering, respectively.

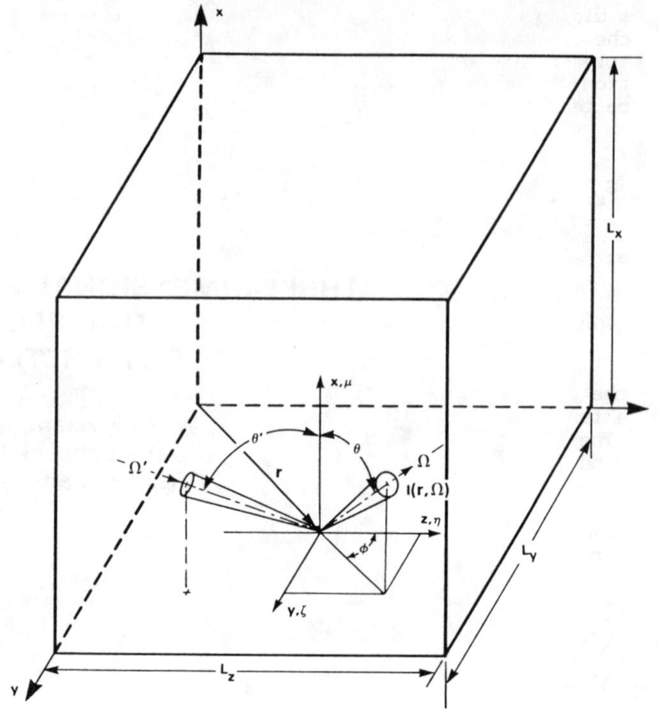

Figure 1. Coordinate system

If the surface bounding the medium emits and reflects radiant energy, the radiative boundary condition for Equation 1 is given by:

$$I(\mathbf{r},\mathbf{\Omega}) = \varepsilon I_b(\mathbf{r}) + \frac{\rho}{\pi}\int_{\mathbf{n}\cdot\mathbf{\Omega}'<0} |\mathbf{n}\cdot\mathbf{\Omega}|\,I(\mathbf{r},\mathbf{\Omega}')\,d\Omega' \quad (2)$$

where $I(\mathbf{r},\mathbf{\Omega})$ is the intensity of radiant energy leaving a surface at a boundary location, ε and ρ are the diffuse surface emissivity and surface reflectivity, respectively, and $\mathbf{n}$ is the unit normal vector at the boundary location.

Real systems exhibit a distribution of radiant energy that is nonuniform. Surface conditions in glass furnaces and combustors are typically nongray. Mixtures of absorbing gases and absorbing and scattering particles need to be evaluated on a spectral basis. However, practical considerations prohibit a wavelength-by-wavelength analysis. In this paper, the range of wavelengths is subdivided into a finite number of bands of width ($\Delta\lambda$) within which the radiation properties are assumed constant (radiatively gray). Equations 1 and 2 are written for each band, and calculations are performed. Once the band calculations are complete, the total net heat flux is calculated.

The radiative transfer equation for each band is approximated using the discrete-ordinates method (Fiveland, 1988). For the multidimensional case, the transfer equation is also solved for a number of ordinate directions, and the integrals are replaced by a quadrature summed over the ordinate directions. For each band, the equation can be written:

$$\mu_m \frac{\partial I^m}{\partial x} + \xi_m \frac{\partial I^m}{\partial y} + \eta_m \frac{\partial I^m}{\partial z} = -\beta I^m + \kappa I_b + \frac{\sigma}{4\pi} S_m \quad (3)$$

For a discrete direction Ω_m, the values μ_m, ξ_m, and η_m are the direction cosines. The parameters β, κ, and σ, represent, respectively, the band extinction, absorption, and scattering coefficients. S_m is the source term for incoming radiation:

$$S_m = \sum_{m'=1}^{N_{m'}} w_{m'} \; \Phi(m',m) \; I^{m'} \tag{4}$$

where:

$$\Phi(m',m) = \sum_{n=0}^{N} (2n+1) \; a_n P_n \; (\mu_m \mu_{m'} + \xi_m \xi_{m'} + \eta_m \eta_{m'}) \tag{5}$$

and the a_n represent Legendre coefficients for the phase function in each wave band.

The finite difference approximation for each waveband is (see Equations 23 of Fiveland, 1988)

$$I_p^m = \frac{\mu_m AI_w^m + \xi_m BI_s^m + \eta_m CI_b + \alpha(S_1 + S_2) \; V_p}{\mu_m A + \xi_m B + \eta_m C + \alpha \; \beta \; V_p} \tag{6}$$

where:

$A = \Delta y \; \Delta z$

$B = \Delta x \; \Delta z$

$C = \Delta x \; \Delta y$

$S_1 = \kappa \; \Delta I_b$

$S_2 = \frac{\sigma}{4\pi} \sum_{m'}^{N_{m'}} w_{m'} \; \Phi_{m'm} \; I_p^{m'}$

The face intensities are related to the cell center intensities as follows:

$$I_p^m = \alpha I_n^m + (1 - \alpha) I_s^m \tag{7}$$

$$I_p^m = \alpha I_e^m + (1 - \alpha) I_w^m \tag{8}$$

$$I_p^m = \alpha I_f^m + (1 - \alpha) I_b^m \tag{9}$$

The parameter α varies between 0.5 and 1; if $\alpha = 0.5$, Equations 7 - 9 represent the second-order diamond-difference scheme proposed by Carlson and Lathrop (1968) and if $\alpha = 1.0$, a first-order difference method results. The solution of Equation 6 follows the same procedure as in Fiveland (1988) and uses the ordinates listed there. The fractional energy emitted in each band is written as:

$$\Delta I_b = \Delta F \; \frac{\bar{\sigma} T^4}{\pi} \tag{10}$$

where the quantity, ΔF, is:

$$\Delta F = F(0 \to \lambda_2 T) - F(0 \to \lambda_1 T) \tag{11}$$

and

$$F(0 \to \lambda_1 T) \equiv \frac{\int_0^{\lambda_1} e_{b\lambda} \, d_\lambda}{\int_0^{\infty} e_{b\lambda} \, d_\lambda} \tag{12}$$

Chang and Rhee (1984) have presented a function for the fractional emission in each waveband:

$$F(0 \to \lambda T) = \frac{15}{\pi^4} \sum_{n=1}^{\infty} \frac{e^{-nv}}{n} \left(v^3 + \frac{3v^2}{n} + \frac{6v}{n^2} + \frac{6}{n^3}\right) \tag{13}$$

where $v \equiv \frac{C_2}{\lambda T}$ and $C_2 = 14388 \; \mu m^\circ K$.

The total radiant flux from a surface is the sum of the net fluxes computed for each band:

$$q_{net} = \sum_{k=1}^{N_B} \pi \left\{ \varepsilon_k \; \Delta I_{b,k}(r) - \frac{\varepsilon_k}{\pi} \sum_{m'=1}^{N_{m'}} w_{m'} \; \mu_{m'} \; I_k^{m'} \right\} \tag{14}$$

The total incident energy is the sum of the incident energies in each of the bands:

$$G = \sum_{k=1}^{N_B} \sum_{m=1}^{N_m} w_m \; I_k^m \tag{15}$$

Solution Stability

When Equation 6 is used with Equations 7 - 9 to predict face intensity, the extrapolated face intensity may be negative, which is physically unrealistic. For each band, the conditions to maintain positive intensities are (Fiveland, 1988):

$$\Delta x \le \frac{|\mu|}{\beta(1-\alpha)} \; \phi, \quad \Delta y \le \frac{|\xi|}{\beta(1-\alpha)} \; \phi, \quad \Delta z \le \frac{|\eta|}{\beta(1-\alpha)} \; \phi \tag{16}$$

where:

$$\phi = \frac{\alpha^3 + (1-\alpha)^2 \; (2 - 5\alpha)}{\alpha}$$

The algorithm for each waveband is similar and is started with a base value of α of 0.5. If negative intensities are encountered as band-by-band calculations proceed, the calculation is stopped and the value of α is increased by 10%. This increase in α is repeated until only positive intensities are computed or $\alpha = 1.0$ is reached. The first-order accurate method ($\alpha = 1$) produces positive intensities and, therefore, stable solutions regardless of the ordinate set chosen or the local radiative properties.

RESULTS

This paper focuses on presentation of S_4 solutions, a 24-flux approximation, which was shown to be computationally efficient for the amount of numerical overhead (Fiveland, 1988). All solutions were judged converged based on a preset error of 0.01% in the incident energy.

Case 1

A unit cubical enclosure studied by Fiveland (1988) is shown in Figure 1. This test case was set up to demonstrate that the spectral method with a number of bands all having the same properties reproduces the gray calculation. The walls of the enclosure are considered radiatively black ($\varepsilon = 1$), with unit emissive powers on the planes at $\bar{x} = 1$, $\bar{y} = 0$, and $\bar{z} = 0$ (emitting walls), and zero emissive powers on the receiving walls ($\bar{x} = 0$, $\bar{y} = 1$ and $\bar{z} = 1$). The medium is in radiative equilibrium ($\nabla \cdot q = 0$). The total incident energy in a computational cell was found from a summation of the band incident energies using Equation 15. Temperature was determined from the relation

$$- \nabla \cdot q = \kappa(G - 4\sigma \; T^4) \tag{17}$$

Solutions were found with an extinction coefficient of $\kappa = 1.0\ m^{-1}$ and a computational grid of 125 control volumes (5x5x5).

Predictions were obtained for a gray solution (1 band) and a spectral solution (10 bands) with all bands having identical gas absorption coefficients. Figure 2 shows the predicted wall heat flux along the wall for the emitting and receiving walls. Consider the net wall heat flux on the emitting wall at $\bar{x} = 1$. In Figure 2, the wall heat flux is either $q(\bar{y} = 1/2)$ versus distance, $\bar{z}$, or $q(\bar{z} = 1/2)$ versus distance, $\bar{y}$, since a line of symmetry exists on the plane $\phi = 45°$ in Figure 1. The surface heat flux for the gray and spectral discrete-ordinates solutions coincided, showing that the numerical scheme was coded properly.

Case 2

Edwards and Nelson (1962) and Seigel and Howell (1972) predicted radiative transfer between parallel, nongray plates spaced 25.4 mm (1"). The plates were made of tungsten, and maintained at temperatures of 1111°K (2000°R) and 556°K (1000°R), respectively. Pure CO_2 at 1.013 MPa (10 atm) pressure and 556°K (1000°R) is between the plates. Using the tabulated data in Seigel and Howell (1972), the absorption coefficients used in the spectral calculations (see Table 1) were obtained from the equation

$$\kappa_g = -\frac{1}{L_m} \ln(1 - \varepsilon_g) \tag{18}$$

Table 1. Spectral Data for Case 2

η Band cm^{-1}	Δλ μm	ε_w	κ_g m^{-1}
∞ → 3750	0 → 2.67	.73	0
3750 → 3573	2.67 → 2.798	.69	∞
3573 → 2430	2.798 → 4.115	.61	0
2430 → 2221	4.115 → 4.502	.65	∞
2221 → 1141	4.502 → 8.764	.45	0
1141 → 1013	8.764 → 9.871	.46	1.743
1013 → 849	9.871 → 11.78	.37	1.348
849 → 779	11.78 → 12.83	.32	0
779 → 555	12.83 → 18.01	.26	49.05
555 → 0	18.01 → ∞	.37	0

The 1-dimensional geometry was modeled using the 3-D model with a ratio of $L_y:L_x$ and $L_y:L_z$ of 1:24 for a 4x4x4 computational grid. Under the assumed conditions, nearly 100% of energy leaving the 1111°K plate was received by the 556°K plate. The four end walls are modeled as reflecting surfaces with an emittance approaching zero ($\varepsilon = .0001$).

The predicted local and average heat fluxes at the receiving surface are shown in Figure 3. The fluxes obtained from the network method used by Edwards and Nelson and the exact solution presented by Seigel and Howell are also shown in the figure. Table 2 shows the numerical values of the average fluxes.

Table 2. Predicted Average Heat Flux on the Receiving Surface

This study	29,991	w/m^2
Edwards and Nelson (1962)	29,562	w/m^2
Seigel and Howell (1972)	30,417	w/m^2

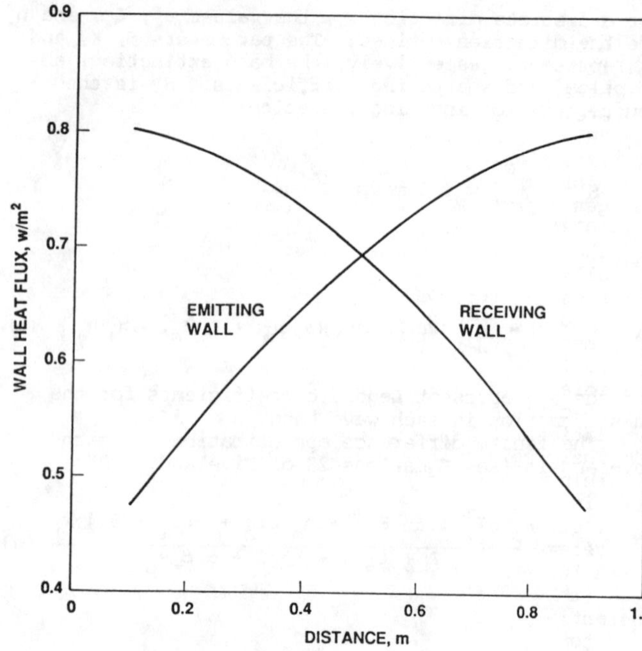

Figure 2. Predicted wall heat flux profiles at half elevation using gray and spectral calculations (Case 1)

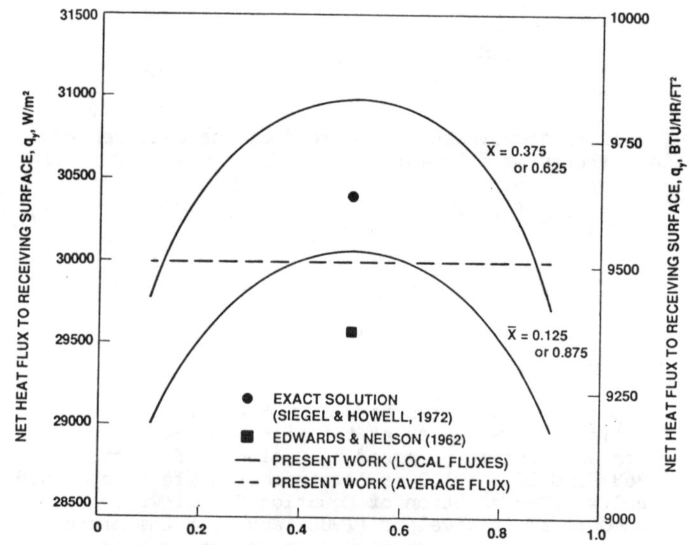

Figure 3. Spectral heat flux to the receiving surface (Case 2)

Most of the radiative transfer is in the transparent regions between the CO_2 absorption bands. In the bands $1141 \leq \eta \leq 2221$, $2430 \leq \eta \leq 3573$, and $3750 \leq \eta \leq \infty$, for example, the radiative transfer is 20.3%, 33.6%, and 42.6%, respectively, of the total. Only 3.5% of the radiative transfer occurs in the other bands.

The profiles of heat flux shown in Figure 3 are not uniform due to the end effects caused by modeling the one-dimensional geometry as three-dimensional. The variation in heat flux (<3%, see Figure 3) can be reduced by increasing $L_x:L_y$ and $L_z:L_y$ at the expense of additional computational time. As expected, the fluxes are somewhat higher for the two center-rows of the grid.

Case 3

A cubical enclosure with side dimension 0.964 m (√10 ft) was modeled to compare the spectral discrete-ordinates and zone methods. The walls of the enclosure are maintained at a temperature of 1089°K (1960°R) and an emittance of $\varepsilon = 0.6$. The isothermal gas mixture at a temperature of 1367°K (2460°R) consists of 10% water vapor, 10% carbon dioxide, and 80% nitrogen (by volume) maintained at a total pressure of 0.1013 MPa (1 atm).

The predicted radiant transfer between the gas and walls is shown in Figure 4 for the discrete-ordinates and the zone methods (Hottel and Sarofim, 1967). Discrete-ordinates and zone calculations were performed by evaluating the spectral absorption using Edwards (1976) wide-band models. Surface and gas properties used in the analyses are shown in Table 3. A 5x5x5 computational grid was used in performing the calculations.

The predicted average heat flux, q_y, at the wall is within 0.7% for either method. The curved profiles shown in the discrete-ordinates solution are caused by corner effects. Identical results were obtained for the average heat fluxes q_x and q_z.

A 10x10x10 grid was also used with similar results (the average heat flux agreed to within 0.1%). Substantially longer computer time (190.6 s as compared to 28.9 s for 5x5x5 grids on a SUN 4.0/260 workstation) was required for the larger grid.

Case 4

This case demonstrates the importance of performing spectral calculations. The enclosure is the same as Case 3. Calculations are performed using a 5x5x5 computational grid for several examples: gray walls and gray gas; gray walls and spectral gas; spectral walls and gray gas; and spectral walls and spectral gas. The measured normal spectral wall emittance (DeBellis, et al., 1989) shown in Figure 5 was used to model the spectral walls and to determine a gray wall emittance of 0.681 for the calculations with gray walls. The gas mixture in Case 3 (Table 3) was used to model the spectral gases and was used to compute a gray gas emissivity of 0.778 for calculations with a gray gas.

Table 3 - Spectral Data for Case 3

Wavelength (μm) Min	Max	Gas Transmissivity	Surface Emissivity	Absorption Coefficient (m^{-1})
0.100	1.357	1.0000	0.600	0.0003
1.357	1.402	0.9000	0.600	0.1873
1.402	1.812	1.0000	0.600	0.0003
1.812	1.918	0.9000	0.600	0.1873
1.918	1.928	0.8100	0.600	0.3747
1.928	1.930	0.9000	0.600	0.1873
1.930	2.429	1.0000	0.600	0.0003
2.429	2.574	0.7278	0.600	0.5650
2.574	2.911	0.5368	0.600	1.1063
2.911	2.938	0.7278	0.600	0.5650
2.938	4.149	1.0000	0.600	0.0003
4.149	4.654	0.1972	0.600	2.8871
4.654	5.425	1.0000	0.600	0.0003
5.425	7.372	0.5937	0.600	0.9272
7.372	9.299	1.0000	0.600	0.0003
9.299	9.572	0.9000	0.600	0.1873
9.572	10.254	1.0000	0.600	0.0003
10.254	10.585	0.9000	0.600	0.1873
10.585	12.514	1.0000	0.600	0.0003
12.514	12.886	0.2640	0.600	2.3681
12.886	17.922	0.0829	0.600	4.4281
17.922	24.390	0.2640	0.600	2.3681

Figure 6 exhibits the predicted wall heat flux, q_y versus $\bar{z}$ at $\bar{x} = 0.5$, for the examples. For this application, the figure shows that the gas properties significantly affect the wall heat flux; the top two curves are predicted heat flux profiles for the gray gas and the lower curves are predicted profiles for the spectral gas. There is more than a 30% reduction in predicted heat flux for the examples with spectral gas. The wall emittance has a lesser effect; wall heat flux is reduced less than 4% when calculations simulate spectral walls instead of gray walls.

CONCLUSIONS

A spectral discrete-ordinates method was formulated to demonstrate the importance of spectral calculations in nongray applications. The discrete-ordinates method is found to be accurate in performing spectral radiation calculations. The calculations emphasize the need to collect spectral property data in furnaces and show that consideration for nongray radiation may substantially improve furnace heat transfer predictions.

ACKNOWLEDGMENT

The authors want to thank the Gas Research Institute (GRI) for permitting use of the spectral emittance results shown in Figure 5; these were collected under GRI Contract 5087-232-1579.

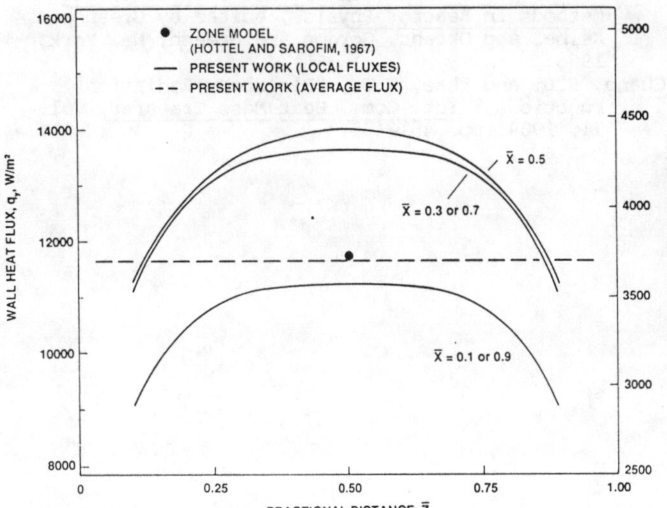

Figure 4. Spectral heat flux on the walls (Case 3)

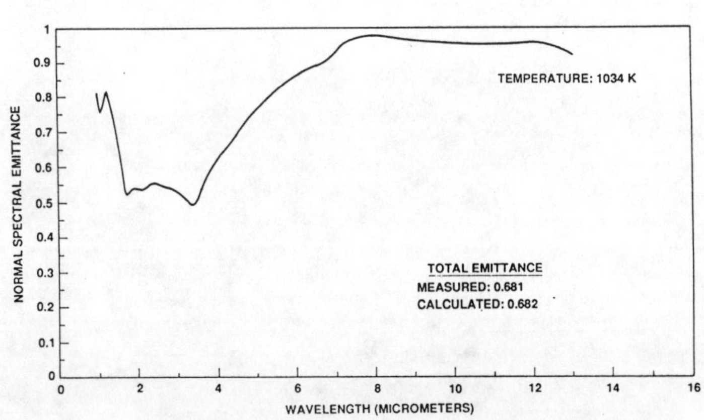

Figure 5. Spectral emittance of a refractory material (DeBellis, et al., 1989)

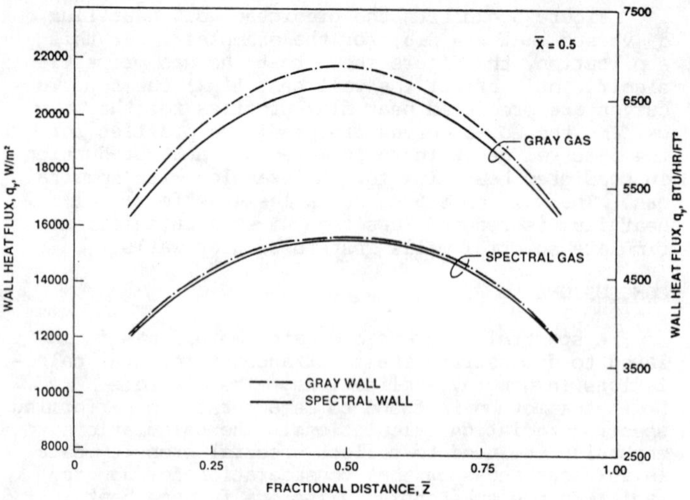

Figure 6. Effect of spectral calculations on the predicted wall heat flux (Case 4)

REFERENCES

Carlson, B.G. and Lathrop, K.D., Transport Theory -- The Method of Discrete Ordinates in Computing Methods in Reactor Physics, edited by Greenspan, Kelber and Okrent, Gordon and Breach, New York, 1968.

Chang, S.L. and Rhee, K.T., "Blackbody Radiation Functions," Int. Comm. Heat Mass Transfer, Vol. 11, 1984, pp. 451-455.

DeBellis, C.L., Fiveland, W.A. and Bailey, R.T., "Effect of Refractory Emittance in Industrial Furnaces," Annual Report, Gas Research Institute, Contract No. 5087-232-1579, January 1989.

Edwards, D. K., "Molecular Gas Band Radiation," Advances in Heat Transfer, Vol. 12, 1976, pp. 115-193.

Edwards, D. and Nelson, K., "Rapid Calculation of Radiant Energy Transfer Between Nongray Walls and Isothermal H_2O and CO_2 Gas," Journal of Heat Transfer, Vol. 84, No. 4, 1962, pp. 273-278.

Fiveland, W.A., Journal of Thermophysics and Heat Transfer, Vol. 2, No. 4, 1988, pp. 309-316.

Hottel, H. and Sarofim, P., Radiative Transfer, McGraw-Hill, 1967.

Howell, J.R., "Application of Monte Carlo to Heat Transfer Problems," Advances in Heat Transfer, Vol. 5, edited by T.F. Irvine, Jr. and J.P. Hartnett, Academic Press, New York, 1968, pp. 1-54.

Lockwood, F.C., and Shah, N.G., "New Radiation Solution Method for Incorporation in General Combustion Prediction Procedures," Proceedings of Eighteenth Symposium (International) on Combustion, The Combustion Institute, Pittsburgh, 1981, pp. 1405-1414.

Menguc, M. and Viskanta, R., "Radiative Transfer in Three-Dimensional Rectangular Enclosures," Journal of Quantum Spectroscopy and Radiative Transfer, Vol. 33, 1985, pp. 533-549.

Seigel, R. and Howell, J., Thermal Radiation Heat Transfer, McGraw-Hill, 1972, p. 586.

Viskanta, R., "Radiative Heat Transfer," Fortschritte der Verfahrenstechnik, Vol. 22, 1984, pp. 51-81.

Viskanta, R. and Menguc, M., "Radiation Heat Transfer in Combustion Systems," Progress in Energy and Combustion Science, Vol. 13, 1987, pp. 97-160.

1989 National Heat Transfer Conference
HTD-Vol. 106, Heat Transfer Phenomena in Radiation, Combustion and Fires

A FINITE-VOLUME METHOD FOR PREDICTING RADIANT HEAT TRANSFER IN ENCLOSURES WITH PARTICIPATING MEDIA

G. D. Raithby and E. H. Chui
Department of Mechanical Engineering
University of Waterloo
Waterloo, Ontario, Canada

ABSTRACT

A new "finite-volume" method is proposed to predict radiant heat transfer in enclosures with participating media. The method can conceptually be applied with the same non-orthogonal computational grids used to compute fluid flow and convective heat transfer. A fairly general version of the method is derived, and details are illustrated by applying it to several simple benchmark problems. Test results indicate that good accuracy is obtained on coarse computational grids, and that solution errors diminish rapidly as the grid is refined.

NOMENCLATURE

A_S	surface area of control volume, m^2
$A_{S,f}$	surface area of panel f of control volume, m^2
B_P	source term in algebraic equation, e.g. (24)
D	depth of 1D plane layer, m
$\vec{e}_j$	unit vector in the j direction
I	radiation intensity, W/m^2Sr
$\overline{I}$	scattered intensity, W/m^2Sr, see (2)
I_a	average intensity, (13)
I_b	blackbody intensity, $\sigma T^4/\pi$
I_B^ℓ	discrete intensity at the bottom surface within ω^ℓ
I_P^ℓ	discrete intensity at node P within ω^ℓ
I_S	intensity at surface of control volume
I_T^ℓ	discrete intensity at the top surface within ω^ℓ
K	absorption coefficient, m^{-1}
L_x, L_y	dimensions of rectangular cavity in the x and y directions, m
m	index denoting the ω^ℓ with boundaries at θ_+^m and θ_-^m. See Figures 3B and 7.
n	index denoting the ω^ℓ with boundaries at ϕ_+^n and ϕ_-^n. See Figure 7.
$\vec{n}$	unit vector normal to a control volume surface
N^ℓ	total number of discrete solid angles
N_x, N_y, N_z	number of control volumes in the x, y, and z-directions respectively
N^θ, N^ϕ	number of discrete angles in the θ and ϕ directions
q	heat flux, W/m^2
Q	total radiant heat transfer rate, W
Q_f^ℓ	radiant heat transfer across surface panel f
$\vec{r}$	position in space
R	$(KI_b + \sigma^s\overline{I})/\kappa$, W/m^2Sr
$\overline{R}, \tilde{R}$	see (21)
s	distance along a beam in the $\vec{s}$ direction, m
$\vec{s}$	unit vector in the direction for which I is to be calculated
$\vec{s}^\ell$	unit vector in the centre of discrete solid angle ω^ℓ
S	distance from the interpolation point to the integration point, see Figures 2B and 7
$\overline{S}$	average value of S over the solid angle ω^ℓ
T	temperature, K
$\forall$	volume, m^3
x, y, z	Cartesian coordinates, m
x_1, x_2, x_3	Cartesian coordinates, $x_1 = x, x_2 = y, x_3 = z, m$
x^*	$(\kappa + K)D$, used for one-dimensional problem
Greek	
ϵ	surface emissivity
θ	polar angle, measured from the z-axis
κ	extinction coefficient, $K + \sigma^s$, m^{-1}
ρ	surface reflectivity
σ	Stefan-Boltzman constant, 5.729×10^{-8} W/m^2K^4
σ^s	scattering coefficient, m^{-1}
ϕ	azimuthal angle in the $x-y$ plane, measured from x axis, Figure 7
Φ	phase function for scattering from direction $\vec{s}'$ to direction $\vec{s}$
ω	solid angle
ω^ℓ	solid angle of the discrete bundle centered at $\vec{s}^\ell$

Subscript

B	bottom surface
P, N, E	labels for nodes within control volumes
f	at the integration point f
uf	at the interpolation point, Figure 2A
T	top surface
j	coordinate index, $j = 1, 2, 3$

Superscript

ℓ	in the solid angle ω^ℓ
$*$	nondimensional

INTRODUCTION

Fluid flow and convective heat transfer can now be predicted with reasonable economy and accuracy in very complex geometries. Most methods use finite difference, finite volume or finite element methods to reduce the continuous equations of motion to a set of coupled algebraic equations that can be solved by computer. For many situations, e.g. flows with combustion, it would be desirable to add a solution for radiant exchange to the code that solves the fluid flow and internal energy equations; it would be convenient if the radiation model were based on the same philosophy and perhaps even used the same computational grid, as the fluid flow solver. Among the solution methods that are currently available, such as the many variations on Monte Carlo, Zonal, Ray Tracing, Flux and Discrete Ordinates methods, none is fully satisfactory for this purpose except in the limit when the diffusion approximation applies (optically dense medium). Since excellent recent reviews of the state-of-the-art are provided by Viskanta and Mengüç (1984,1987), the competing methods for predicting radiative heat exchange will not be reviewed here.

According to the finite volume method, a solution domain is subdivided by a computational mesh that defines nodes for each variable and a control volume surrounding each node. To compute flows in complex domains, the flexibility allowed by a non-orthogonal boundary-fitted mesh is desirable. The present paper describes a new method for predicting radiative heat transfer that also uses such a grid. Besides using the same mesh as the flow solver, the method can account for absorption, emission and scattering in the medium, permits spectral dependence through a broad-band model (Viskanta and Mengüç, 1987), allows radiation properties to depend on other variables (like local pressure, concentration and temperature), and will permit desired reflectivity properties of the wall to be incorporated. The method is also fully conservative in the sense that the global conservation balance for each discrete intensity component, as well as for the radiative flux, is exactly satisfied by the discrete equations.

The model has, until now, only been exercised for simple cases. These include heat transfer in rectangular enclosures, with diffuse gray isothermal walls, that contain gray emitting-absorbing and/or isotropically scattering media. To limit the model description to these cases would, however, make it difficult to appreciate the general applicability of the model. On the other hand, to provide the model in its full generality would result in notational complexity that would encumber the presentation. As a compromise, generality is restricted in those areas where removing the restriction is obvious and simple (e.g. gray medium and diffuse wall). The complexity of a nonisotropically scattering medium and a non-orthogonal grid is retained in the model development, even though these are not used in the test problems.

THE FINITE VOLUME APPROXIMATION

The Volume-Integrated Equation for Intensity

Referring to the definition sketch, Figure 1A, $\vec{r}$ denotes a location in space, $d\omega$ is a differential solid angle that is centered on the unit vector $\vec{s}$, and s is distance along a particular pencil of radiation that lies in the $\vec{s}$ direction. Over the path length ds in the direction of the unit vector $\vec{s}$, the rate of increase of total intensity $I(\vec{r}, \vec{s})$ is (Siegel, 1981)

$$\frac{dI}{ds} = -\,(K + \sigma^s)I + (KI_b + \sigma^s \overline{I}) \tag{1}$$

where the first term on the right side is attenuation through absorption and outscattering and the second term is augmentation due to emission and inscattering. Because of the gray-medium approximation, the absorption coefficient K and the scattering coefficient σ^s are appropriate wave-length averaged quantities. The inscattering term is given by

$$\overline{I}(\vec{r}, \vec{s}) = \frac{1}{4\pi} \int_{4\pi} I(\vec{r}, \vec{s}')\, \Phi\,(\vec{s}, \vec{s}')\, d\omega' \tag{2}$$

where $I(\vec{r}, \vec{s}')$ is the intensity at $\vec{r}$ within the solid angle $d\omega'$ that lies in the direction of the unit vector $\vec{s}'$.

Integration of (1) over any specified volume yields the desired integral equation. Choosing the hexahedron in Figure 1B for purposes of illustration, the volume is subdivided into pencils along $\vec{s}$ of cross-sectional area dA^n. Multiplication of (1) by the differential volume $d\forall = dA^n ds$, carrying out the integration, and multiplying of the resulting equation by $d\omega$, yields

$$\int_{A_S} I_S(\vec{s} \cdot \vec{n}) dA_S d\omega = \int_{\forall} [-\,(K + \sigma^s)I + KI_b + \sigma^s \overline{I}] d\forall\, d\omega \tag{3}$$

where A_S and $\forall$ are the surface area and volume of the hexahedron, $\vec{n}$ is the unit surface normal, and I_S is the intensity at the surface. This equation demands that the net radiant energy within $d\omega$ that leaves through all surfaces of the volume is exactly balanced by the excess, within the volume, of inscattering and emission over outscattering and absorption.

Discrete Equation, Level 1 Approximation

According to the finite volume method, space within the interior of the domain of interest is subdivided into discrete non-overlapping volumes, and a single "node" is located centrally within each volume. The control volume in Figure 1B is redrawn in Figure 2, showing its central node P, and nodes E and S for the neighbouring control volumes to the east and south.

Since direction is also an independent variable, it is consistent with the finite volume method to subdivide direction into N^ℓ discrete, non-overlapping, solid angles of size ω^ℓ, $\ell = 1, 2, 3, \ldots$ which sum to 4π. The number of angles and their size distribution is, like the discretization of space, at the discretion of the analyst.

The objective of the finite volume method is to find I_P^ℓ, the intensity at each node P associated with each discrete bundle ω^ℓ. An algebraic equation for I_P^ℓ is obtained by introducing approximations into (3).

Choosing the volume in (3) to be the control volume associated with node P in Figure 2, with surface area $A_{S,P}$, and integrating the equation over the discrete solid angle ω^ℓ, (3) becomes

$$\int_{\omega^\ell}\int_{A_{S,P}} I_S(\vec{s}\cdot\vec{n})dA_S d\omega$$

$$= \int_{\omega^\ell}\int_{\forall_P} [-(K+\sigma^s)I + KI_b + \sigma^s\overline{I}]d\forall\, d\omega \qquad (4)$$

The simplest approximation of the right side of (4) is

$$\int_{\omega^\ell}\int_{\forall_P} [-(K+\sigma^s)I + KI_b + \sigma^s\overline{I}]d\forall\, d\omega$$

$$\approx [-(K_P + \sigma_P^s)\, I_P^\ell + K_P\, I_{bP}^\ell + \sigma_P^s\, \overline{I}_P^\ell]\forall_P\, \omega^\ell \qquad (5)$$

In carrying out the volume integrations, all variables have been assumed constant over $\forall_P$ and ω^ℓ. A more accurate representation, involving, for example, some linear distribution between nodes, is possible but is both more complex and may have serious negative computational effects (like a strong coupling between intensities in different directions that is not consistent with (1)).

As the first step to approximating the left side of (4), the surface of the control volume is subdivided into N_f surface panels of area $A_{S,f}$, a few of which are shown in Figure 2A. Denoting the radiant transfer through panel f and within ω^ℓ by Q_f^ℓ, the left side of (4) is just the sum of Q_f^ℓ, $f = 1,2,\cdots,N_f$. The simplest approximation for Q_f^ℓ is

$$Q_f^\ell = A_{S,f}\int_{\omega^\ell} I_f(\vec{s}\cdot\vec{n}_f)d\omega \qquad (6)$$

where the intensity and unit normal to the surface have been approximated by their values at the integration point "f", located "centrally" on the panel.

The simplest Level 1 approximation of (3) is therefore

$$\sum_{f=1}^{N_f} Q_f^\ell = \sum_{f=1}^{N_f} A_{S,f}\int_{\omega^\ell} I_f(\vec{s}\cdot\vec{n}_f)d\omega$$

$$= [-(K_P + \sigma_P^s)I_P^\ell + K_P I_{b,P} + \sigma_P^s\,\overline{I}_P^\ell]\forall_P\,\omega^\ell (7)$$

Discrete Equation, Level 2 Approximation

To close the set of equations, relations are needed between the panel integration-point values I_f and nodal point values I_P, I_E etc. Since several different closure relations, all starting from (7), are possible, these are defined as the Level 2 approximations.

In the present study, I_f was found by tracing back along the path taken by the ray in reaching f (i.e. along the -$\vec{s}$ direction) until a location was reached at which the intensity can be reasonably found by interpolation between nodes. The location of the interpolation point will be specified later when the computation of grids are generated. For now, it is sufficient to denote its location by uf as shown in Figure 2 and the intensity (known by the interpolation) by I_{uf}. To estimate the value of I_f from I_{uf} for the given $\vec{s}$ direction, rewrite (1) as

$$\frac{dI}{ds} + \kappa I = \kappa R \qquad (8A)$$

where

$$\kappa = K + \sigma^s \quad \text{and} \quad R = (KI_b + \sigma^s\overline{I})/\kappa \qquad (8B)$$

Let R be approximated by the first two terms of a Taylor series expansion about point f, and treat κ as constant at κ_f. Carrying out the integration of (8A) over distance S from uf to f, shown in Figure 2B, the intensity in direction $\vec{s}$ at integration point f is

$$I_f = I_{uf}e^{-\kappa_f S} + R_f(1-e^{-\kappa_f S}) - \frac{(\partial R/\partial s)_f}{\kappa_f}(1-e^{-\kappa_f S}(1+\kappa_f S)) \qquad (9)$$

$(\partial R/\partial s)_f$, the directional derivative in the $\vec{s}$ direction, is related to the gradient of R, $\vec{\nabla} R$, by

$$\left(\frac{\partial R}{\partial s}\right)_f = (\vec{\nabla} R)_f\cdot\vec{s}$$

Introducing (9) into (6), and carrying out the integration over ω^ℓ, gives the following estimate of the radiant heat transfer within ω^ℓ that crosses surface panel f

$$Q_f^\ell \approx A_{S,f}\, I_{uf}^\ell\, e^{-\kappa_f\overline{S}}\, N_f^\ell + A_{S,f}R_f^\ell(1-e^{-\kappa_f\overline{S}})N_f^\ell$$
$$- \frac{A_{S,f}}{\kappa_f}(1-e^{-\kappa_f\overline{S}}(1+\kappa_f\overline{S}))\sum_{j=1}^{3}\left(\frac{\partial R^\ell}{\partial x_j}\right)_f D_{jf}^\ell \qquad (10)$$

where R^ℓ is the value of R for the bundle ω^ℓ, where the gradient of R has been arbitrarily written in Cartesian coordinates x_j, defined by the unit vectors $\vec{e}_j$, as

$$\vec{\nabla} R^\ell = \sum_{j=1}^{3}\left(\frac{\partial R^\ell}{\partial x_j}\right)\vec{e}_j$$

and where N_f^ℓ and D_{jf}^ℓ are integrals that depend only on the panel orientation for a given ω^ℓ

$$N_f^\ell = \int_{\omega^\ell}(\vec{s}\cdot\vec{n}_f)d\omega \qquad D_{jf}^\ell = \int_{\omega^\ell}(\vec{e}_j\cdot\vec{s})(\vec{s}\cdot\vec{n}_f)\,d\omega \quad (11)$$

In the integration over ω^ℓ in (6) that led to (10), I_{uf} was treated as constant, and S was held constant at $\overline{S}$. Their values are evaluated along the ray $\vec{s}^{\,\ell}$ that lies in the centre of the solid angle ω^ℓ. To include the effect of varying S and I_{uf} over ω^ℓ would dramatically increase complexity. Furthermore, the practice of using constant values is consistent with the other approximations embodied in the discrete equation representation of the continuous equations.

Before continuing, it is of interest to demonstrate that the approximation of the heat flux given by (10) is reasonable. For concreteness, consider the case where the surface panel lies in a z-plane, with surface normal $\vec{e}_3$. For the optically thin limit ($\kappa_f \to 0$), the radiant flux becomes

$$Q_f^\ell \to A_{S,f}\, I_{uf}^\ell\int_{\omega^\ell}(\vec{s}\cdot\vec{e}_3)d\omega$$

The intensity in this limit is propagated correctly (i.e. without attenuation) from the interpolation point uf to the integration point on the panel.

In the optically thick limit ($\kappa_f \to \infty$),

$$Q_f^\ell \to A_{S,f}\, R_f^\ell\int_{\omega^\ell}(\vec{s}\cdot\vec{e}_3)d\omega$$
$$- \frac{A_{S,f}}{\kappa_f}\sum_{j=1}^{3}\left(\frac{\partial R^\ell}{\partial x_j}\right)_f\int_{\omega^\ell}(\vec{e}_j\cdot\vec{s})(\vec{s}\cdot\vec{e}_3)\,d\omega$$

Summing over ℓ to obtain the net radiant heat transfer crossing the panel, and for isotropic scattering (so that R becomes independent of angle ℓ)

$$Q_f = \sum_{\ell} Q_f^\ell = A_{S,f}\left(-\frac{4\pi}{3\kappa_f}\right)\left(\frac{\partial R}{\partial x_3}\right)_f$$

This is the correct diffusion approximation for the optically thick limit (Siegel, 1981). Thus the expression for Q_f^ℓ given by (10) has retained sufficient generality to capture both the optically thin and thick limits. Note that, had the third term in (9) been omitted (which is the usual solution (Siegel, 1981) to 8A for constant κ and R), the radiation heat transfer across the panel would not have recovered the diffusion approximation.

Returning to the development of the discrete equation for I_P^ℓ, (10) is substituted into (7). The only remaining issues are how to interpolate I_{uf}^ℓ from nodal-point values like I_P^ℓ, and how to find R_f^ℓ. As already mentioned, the interpolation will be illustrated when the method is implemented on a specific mesh. The value of R_f^ℓ is obtained by a straightforward linear interpolation using values of R^ℓ, such as R_P^ℓ, that are calculated and stored at the nodes.

To find R_P^ℓ at the nodes requires an evaluation of both its components (see (8B)) $\overline{I}_P^\ell$ and $I_{b,P}$. $\overline{I}_P^\ell$ is obtained from the following discrete approximation to (2)

$$\overline{I}_P^\ell = \frac{1}{4\pi}\sum_{\ell'} I_P^{\ell'}\,\Phi(\ell',\ell)\omega^{\ell'} \qquad (12)$$

where $\Phi(\ell',\ell)$ is the phase function for scattering from the solid angle $\omega^{\ell'}$ to the solid angle ω^ℓ. For isotropic scattering, $\Phi = 1$ and (12) becomes

$$\overline{I}_P^\ell = \frac{1}{4\pi}\sum_{\ell'} I_P^{\ell'}\,\omega^{\ell'} = I_{aP} \qquad (13A)$$

where I_{aP} is the discrete approximation at node P of the average intensity I_a:

$$I_a(\vec{r}) = \frac{1}{4\pi}\int_{4\pi} I(\vec{r},\vec{s}\,')\,d\omega' \qquad (13B)$$

The value of the second component of R_P, i.e. $I_{bP} = \sigma T_P^4/\pi$, is found from the thermal energy balance for the P-control volume. The increase in internal energy due to radiation crossing the boundaries of the control volume in Figure 2A is obtained by replacing the integration limit ω^ℓ in (4) by 4π.

$$-\int_{A_{S,P}}\int_{4\pi} I(\vec{s}\cdot\vec{n})d\omega dA_S = 4\pi\int_{\forall_P} K(I_a - I_b)d\forall \qquad (14A)$$

$$\approx 4\pi\; K_P(I_{a,P} - I_{b,P})\forall_P \qquad (14B)$$

The scattering contribution drops out of this energy balance for both isotropic and nonisotropic scattering. The temperature is established, in general, by a complex interaction (Siegel, 1981) of this term with advection and diffusion of energy, and with work done by the pressure and other stresses. For the simplest case of radiative equilibrium, the fluid temperature adjusts to make the net surface heat transfer in (14A) zero, so that

$$I_{b,P} = I_{a,P} \qquad (15)$$

For the special case of isotropic scattering and radiative equilibrium, combining (8B), (13) and (15) results in

$$R_P^\ell = R_P = I_{a,P} = I_{b,P} \qquad (16)$$

Summary and Discussion

This section has provided the details of a general finite volume method, except for the interpolation to obtain I_{uf}^ℓ and the application of boundary conditions. These details, and the performance of the method, are illustrated in the following sections by application to some simple problems for which benchmark solutions are available.

ONE-DIMENSIONAL RADIANT HEAT TRANSFER IN A PLANE MEDIUM

Problem Description

Very precise solutions (Heaslet, 1965) (see (Viskanta and Mengüç, 1987) for other references) have been obtained for radiant heat transfer between two isothermal diffuse, flat plates, at temperatures T_B and T_T and emissivities ϵ_B and ϵ_T (see Figure 3), that enclose a gray isotropically scattering, or absorbing-emitting medium with constant scattering and absorption coefficients, and that is in radiative equilibrium. The finite volume method is first applied to this problem.

Discrete Equations

The natural control volumes for this problem are shown in Figure 3A. The depth Δz of each control volume is arbitrary and nodes are located midway between the control volume faces, except for the first and last nodes that are located on the plates. Integration points, like n and s that are denoted by $\times$, lie on the volume faces. On the walls, the integration points b and t lie in the medium distance δ from the walls, where $\delta \to 0$.

For this problem, the radiation intensity depends on z and θ (Figure 3B). The discrete solid angle ω^ℓ is defined as the annular cone between θ_-^ℓ and θ_+^ℓ. The interpolation point uf that is upstream of the integration point n is taken to lie on the plane of node P, as shown in the figure. From the assumption that the intensity varies spatially only with z, I_{uf}^ℓ in (10) can be replaced by I_P^ℓ for $\vec{s}\cdot\vec{e}_3 > 0$ and by I_N^ℓ for $\vec{s}\cdot\vec{e}_3 < 0$. For $\vec{s}\cdot\vec{e}_3 > 0$, the radiant heat transfer within ω^ℓ that travels through the control volume surface n that lies north of node P is, according to Eq. 10 and Fig. 3B,

$$Q_n^\ell = F_n^\ell I_P^\ell + G_n^\ell R_n - H_n^\ell\left(\frac{\partial R}{\partial z}\right)_n \qquad (17)$$

where the angular dependence of R has been dropped due to the isotropic scattering approximation. In this equation

$$F_n^\ell = A_S e^{-\kappa S^\ell} N^\ell \qquad G_n^\ell = A_S(1 - e^{-\kappa S^\ell})N^\ell \qquad (18A)$$

$$H_n^\ell = \frac{A_S}{\kappa}\,(1 - e^{-\kappa S^\ell}(1 + \kappa S^\ell))D_3^\ell \qquad (18B)$$

$S^\ell = (z_n - z_P)/\cos\theta^\ell$ is defined here as the distance S in Figure 3B for the angle $\theta = \theta^\ell = (\theta_-^\ell + \theta_+^\ell)/2$. From (11) for this geometry

$$N^\ell = \int_{\omega^\ell}(\vec{s}\cdot\vec{e}_3)d\omega = -\pi[\cos^2\theta_+^\ell - \cos^2\theta_-^\ell] \qquad (18C)$$

$$D_3^\ell = \int_{\omega^\ell}(\vec{s}\cdot\vec{e}_3)^2 d\omega = -\frac{2\pi}{3}[\cos^3\theta_+^\ell - \cos^3\theta_-^\ell] \qquad (18D)$$

and the solid angle is

$$\omega^\ell = \int_{\theta_-^\ell}^{\theta_+^\ell} 2\pi \sin\theta d\theta = -2\pi[\cos\theta_+^\ell - \cos\theta_-^\ell] \qquad (18E)$$

Substitution of (17) for Q_n^ℓ, and a similar equation for the heat transfer through the south face Q_s^ℓ, into (7), and collection of all terms involving R into a "source term", b_P^ℓ, results in the following equation for I_P^ℓ (for $\vec{s}^\ell \cdot \vec{e}_3 > 0$)

$$a_P^\ell I_P^\ell = a_S^\ell I_S^\ell + a_N^\ell I_N^\ell + b_P^\ell \qquad (19A)$$

where

$$a_P^\ell = F_n^\ell + \kappa \forall_P \omega^\ell \qquad a_S^\ell = F_s^\ell \qquad a_N^\ell = 0 \qquad (19B)$$

$$b_P^\ell = [G_s^\ell R_s - H_s^\ell \left(\frac{\partial R}{\partial z}\right)]_s - [G_n^\ell R_n - H_n^\ell \left(\frac{\partial R}{\partial z}\right)]_n + \kappa R_P \forall_P \omega^\ell \qquad (19C)$$

The values of R and $\frac{\partial R}{\partial z}$ at an integration point (i.e. at a face) are found by linear interpolation of the nodal point values or, at the top and bottom plates, by linear extrapolation from interior nodal values.

For a diffuse top surface, at temperature T_T, the boundary condition for the surface intensity I_T^ℓ for radiation into the medium, $\vec{s}^\ell \cdot \vec{e}_3 < 0$, is

$$(1-\epsilon_T) \sum_{\vec{s}^{\ell'} \cdot \vec{e}_3 > 0} Q_t^{\ell'} + \epsilon_T A_S \sigma T_T^4 = \sum_{\vec{s}^{\ell} \cdot \vec{e}_3 < 0} Q_T^\ell = A_S \pi I_T^\ell \qquad (20)$$

The terms on the left hand side of (20) represent reflected and emitted radiation from the T-surface, which consitute the total radiation into the medium. The last equality defines the intensity through the gray-surface approximation. A similar equation at the bottom surface relates the intensities into the medium ($\vec{s}^\ell \cdot \vec{e}_3 > 0$) to the bottom surface temperature and to intensities incident on the bottom.

Solution Method

For fixed nodal values of R and upward intensities at the bottom surface, I_P^ℓ values are found for a given ℓ by application of (19A) to each interior node, starting at the bottom and sweeping upward. This is repeated for each ℓ for which $\vec{s}^\ell \cdot \vec{e}_3 > 0$. Equation (17) provides the incident fluxes onto the top surface, from which (20) gives the downward intensities. Equations similar to (19A), but with a_S^ℓ rather than a_N^ℓ equal to zero, are then solved for the downward intensities and incident fluxes on the lower surface. The lower surface boundary condition yields the upward boundary intensities to complete one cycle.

If the nodal values of R are fixed, one cycle yields the desired result if $\epsilon_B = \epsilon_T = 1.0$. Several cycles will be required for reflective boundaries.

The determination of R is, of course, part of the solution procedure. The simplest "explicit" method is to obtain the intensities for an assumed R in the manner just described, to substitute these into (13A) and thus obtain new R-values from (16). This cycle is repeated until convergence is achieved. Convergence was deemed satisfactory when the change, between successive updates of R, in net heat flux at both surfaces was less than 0.001%.

Results

When a fine grid (see later) was used, the non-dimensional heat transfer, $q^* = q/\sigma(T_B^4 - T_T^4)$, obtained using the finite volume method agreed to within 1 in the 4th significant figure with the predictions reported by Heaslet and Warming (1965). The gas temperature, and temperature jump at the boundary, were also predicted correctly.

Even for a coarse grid (see Figure 3) of 4 uniform control volumes in the z-direction and with polar angle θ divided into 8 equal intervals, the finite volume method predicts the correct heat transfer to within about 1%, for a wide range of optical depths, as shown in Table 1. The CPU time required was less than 0.5 s on a MicroVax II computer.

Having established that the finite volume method does yield the correct results as the grid is refined, it is of major interest to establish how quickly the accuracy improves with grid refinement.

κ^*	$q^* = q/\sigma(T_B^4 - T_T^4)$ coarse grid finite volume results	exact results	% error	CPU sec.
0.0	1.0000	1.0000	0.0	0.1
0.1	0.9125	0.9157	0.3	0.2
0.5	0.6964	0.7040	1.1	0.3
1.0	0.5469	0.5532	1.1	0.4
2.0	0.3865	0.3900	0.9	0.4

Table 1: Comparison of finite volume results for uniform grid of 4 control volumes and 8 angles with "exact" results (Heaslet and Warming, 1965). $\kappa^* = \kappa D$

For $\kappa^* = \kappa D$ of 1.0, the heat transfer was obtained using $N_z = 64$ uniform control volumes and $N^\ell = N^\theta = 96$ equal angle intervals. To study error reduction with N_z, q^* was re-calculated on a sequence of grids of size $(N_z \times 96)$, $N_z = 2, 4, \cdots, 64$. The results in Figure 4 indicate the q^* is obtained to within 0.1% with $N_z = 2$, and for $N_z \geq 4$ the error becomes proportional to $(N_z)^{-2}$ (i.e. quadratic convergence). To study error reduction with angular refinement, q^* was calculated for grids of $(64 \times N^\theta)$, $N^\theta = 2, 4, 6, \cdots, 96$. The error decreases from about 10% for $N^\theta = 2$, and for $N^\theta \geq 4$, and decreases roughly as $(N^\theta)^{-2}$.

Another issue of major interest is the number of updates of R required to obtain convergence. For black surfaces, and for a change of q^* at the surfaces of less than 0.001% between R updates, the number of iterations required by the explicit method is plotted against $\kappa^* = \kappa D$ in Figure 5. As $\kappa^* \to 0$, the correct solution is obtained without iteration because the intensities become independent of R. The number rises sharply with κ^* but convergence is satisfactory for $\kappa^* \lesssim 4.0$, which includes the range of interest for many practical applications. The number of iterations increases sharply with κ^* for $\kappa^* > 4.0$. A similar behaviour, of increasingly difficult convergence with increasing κ^*, has been encountered in the use of the discrete ordinates method (Viskanta and Mengüç, 1987). The next section addresses this issue.

Implicit Solution for Strongly Participating Medium

For applications that involve strongly participating media ($\kappa^* \geq 4$ for a plane layer), the explicit update of R becomes unsatisfactory. This section describes a method that enhances convergence for large κ^*. The method is, for simplicity, presented in the context of the present 1D problem, but the derivation is kept sufficiently general to include non-isotropic scattering.

To re-introduce non-isotropic scattering, R in (19C) is replaced by R^ℓ, and the nodal values of R are split as follows

$$R_P^\ell = \frac{K_P I_{b,P} + \sigma_P^s I_{a,P}}{\kappa_P} + \frac{\sigma^s}{\kappa_P}(\overline{I}_P^\ell - I_{a,P}) = \overline{R}_P + \tilde{R}_P^\ell \quad (21)$$

For isotropic scattering, $\tilde{R}= 0$. Expressing the integration point values of R_f^ℓ and $\left(\frac{\partial R^\ell}{\partial z}\right)_f$, $f = s$ and n, in (19C) in terms of nodal point values using linear interpolation, substituting the resultant b_P^ℓ expression into (19A), and dividing by a_P^ℓ, yields an equation for I_P^ℓ of the form

$$I_P^\ell = \frac{a_S^\ell}{a_P^\ell} I_S^\ell + \frac{a_N^\ell}{a_P^\ell} I_N^\ell - A_P^\ell \overline{R}_P + A_S^\ell \overline{R}_S + A_N^\ell \overline{R}_N + B_P^\ell \quad (22)$$

where terms involving $\tilde{R}^\ell$ have been grouped into B_P^ℓ. Multiplication of (22) by ω^ℓ, summation over all ω^ℓ, and division of the resulting equation by 4π yields an equation of the form

$$I_{a,P} = -\hat{A}_P \overline{R}_P + A_S \overline{R}_S + A_N \overline{R}_N + B_P \quad (23)$$

For the special case of isotropic scattering and radiative equilibrium, $\overline{R}_P = I_{a,P}$ (from (16)) so that (23) becomes

$$A_P I_{a,P} = A_S I_{a,S} + A_N I_{a,N} + B_P \quad (24)$$

which is an implicit equation for $I_{a,P} = I_{b,P}$ that is simply solved by application of the Thomas (or tridiagonal) algorithm. This replaces the "explicit" update of R described in the previous section. It is also easily shown that, as $\kappa^* \to \infty$, (24) becomes the discrete representation of the diffusion approximation of radiation heat transfer. For large κ^*, therefore, the first solution for $I_{a,P}$ will yield nearly the correct $I_{a,P} = R_P$ values, which, used in the b_P^ℓ term in (19A), will yield nearly converged values of I_P^ℓ, so that very few cycles will lead to the converged result.

Figure 5 shows the number of cycles required by this implicit method to obtain surface heat transfers that changed less than 0.001% between R-update cycles. The maximum number of cycles is required for $\kappa^* \approx 20$. For $\kappa^* \to 0$ and $\kappa^* \to \infty$, the converged results are obtained in one solution cycle.

Although the method has not yet been applied to problems involving non-isotropic scattering and radiative nonequilibrium, it is interesting to speculate how the implicit method could be used to accelerate convergence. Splitting $\overline{R}$ in (23) into its I_a and I_b components results in an equation of the form

$$A_P^{aa} I_{a,P} = A_S^{aa} I_{a,S} + A_N^{aa} I_{a,N} - A_P^{ab} I_{b,P} + A_N^{ab} I_{b,N} + A_S^{ab} I_{b,S} + B_P^a \quad (25)$$

where the first superscript on the coefficients refers to the equation and the second to the variable it multiplies (either a for the variable I_a or b for the variable I_b).

The thermal energy equation has diffusion and convection terms that are linear in T, and a source term given by (14B) that contains I_a and $I_b = \sigma T^4/\pi$. If radiation dominates the internal energy balance, it is appropriate to linearize the discrete energy equation in the form

$$A_P^{bb} I_{b,P} = A_S^{bb} I_{b,S} + A_N^{bb} I_{b,N} + A_P^{ba} I_{a,P} + B_P^b \quad (26)$$

Equations (25) and (26) can be efficiently solved for $I_{a,P}$ and $I_{b,P}$ (e.g. use a (2×2) block tridiagonal solver). If radiation only weakly affects the medium temperature, it would be more appropriate to linearize the equations such that T instead of I_b appears as the dependent variable in (25) and (26). In either case, iteration is required to converge to the solution of the equations in which some terms involve T and others T^4.

TWO-DIMENSIONAL EXCHANGE IN RECTANGULAR ENCLOSURES

Problem Description

The finite volume method will now be applied to a more complex class of problems that involve two-dimensional (2D) radiative heat transfer in rectangular enclosures with isothermal, diffusely reflecting, gray walls. The enclosed medium is also assumed to be gray, and to scatter isotropically. Figure 6A shows a typical enclosure. The objective is to predict radiative heat transfer at the boundaries, and the medium temperature under the assumption of radiative equilibrium. To test the finite volume method, several benchmark problems will be solved. To simplify notation in the development that follows, the (x_1, x_2, x_3)-coordinates will be used interchangeably with (x, y, z).

Discretization

Space is divided into control volumes of unit depth in the z-direction and with boundaries defined by the solid lines in Figure 6A. Nodes, also called grid points, are located centrally in each interior volume as well as on the boundary. Each control volume face is divided into two panels with an integration point, such as $1, 2, 3$ in Figure 6B, located at the centre of each panel.

The angular orientation of intensity is defined by a polar angle θ, measured from the $x_3 = z$ axis, and the azimuthal angle ϕ, that lies in the (x, y) plane measured from the x-axis. The solid angle ω^ℓ is defined by the range $\theta_-^m \le \theta \le \theta_+^m$, $\phi_-^n \le \phi \le \phi_+^n$. All results shown are for N^θ equal increments of θ in the range $0 \le \theta \le \pi/2$ and for N^ϕ equal increments in ϕ in the range $0 \le \phi \le 2\pi$, where, for simplicity, N^ϕ was always chosen to be a multiple of 4.

To apply the general equations of section 2 to the grid system just described, some details must be decided. The central ray within ω^ℓ, i.e. the one along $\vec{s}^\ell$, is defined by (θ^m, ϕ^n) where each angle lies in the middle of its range (e.g. $\phi^n = (\phi_+^n + \phi_-^n)/2$). The path length $\overline{S}^\ell$ is taken as the distance from the integration point to the point where the $\vec{s}_\ell$ that passes through the integration point intersects the closest member of the set of surfaces that join the grid points. For example, in Figure 6B, for integration point $f = 1$

$$\overline{S}^\ell = S_{xy}^n / \sin\theta^m$$

where S_{xy}^n is the projection of $\overline{S}^\ell$ onto the $x - y$ plane.

Intensities at u_1, u_2, etc. in Figure 6B must be related to the nodal intensities by interpolation functions. This step requires great care. I_{u1}^ℓ could, for example, be found by linear interpolation between I_N^ℓ and I_P^ℓ, but this would cause I_N^ℓ to appear in the I_P^ℓ equation. This is undesirable because the equation for I_P^ℓ can no longer be solved by marching, and because such dependence of "downstream" intensities is inconsistent with the transportive properties of the differential equation for intensity. A better alternative is to find I_{u1} by extrapolation using I_P^ℓ and I_S^ℓ, but this practice applied at each integration point, results in a computational molecule that connects I_P^ℓ with as many as 6 upstream nodal intensities. While this is fairly easily accommodated, a simpler approach was used for the present application. When (for $\vec{s}^{\ell}$ upward and to the right) an interpolation point lies on a solid line in Figure 6B, linear interpolation between nodes was used. When it lies on a broken line, the interpolation value was taken as the value at the nearest "upstream" node on the line. According to this scheme I^{ℓ}_{u1} and I^{ℓ}_{u4} would be both approximated by I^{ℓ}_{P} for $(\vec{s}^{\ell} \cdot \vec{e}_j) \geq 0,\ j = 1, 2$. I^{ℓ}_{u2} and I^{ℓ}_{u3} would both be found by a linear interpolation involving I^{ℓ}_{P} and I^{ℓ}_{W}.

If linear interpolation or extrapolation were used to find all interpolation-point values, the solution error should decrease quadratically with mesh refinement, as for the 1D problem, but it is well known (from experience in approximating fluid advection) that "wiggles" in the solution can occur. The interpolation practice adopted here is less likely to admit wiggles, but it is expected that errors due to spatial discretization will decrease with grid refinement at a rate which is between a linear and quadratic. These are features that will be examined when the benchmark problems are solved.

Solution Procedure

For intensities in those $\vec{s}^{\ell}$ directions that are upward to the right, the discrete equation for I^{ℓ}_{P} has the form

$$a^{\ell}_{P} I^{\ell}_{P} = a^{\ell}_{W} I^{\ell}_{W} + a^{\ell}_{S} I^{\ell}_{S} + a^{\ell}_{SW} I^{\ell}_{SW} + b^{\ell}_{P} \tag{27}$$

Given the boundary values of I^{ℓ} along the left and bottom surfaces in Figure 6A, nodal values of I^{ℓ}_{P} are found by direct substitution into (27), sweeping the domain in the positive x and y-directions. With new incident intensities on the right boundary known, boundary conditions are applied to update the intensities leaving this surface, and an equation similar to (27) establishes nodal values for $\vec{s}^{\ell}$ in the range $(\vec{s} \cdot \vec{e}_1) < 0$ and $(\vec{s} \cdot \vec{e}_2) > 0$. Repeating this for all four sectors leads to a new I^{ℓ}_{P} solution, for a given R_P distribution. R_P is then explicitly updated by calculating $I_{aP} = R_P$ from Eq. (13A).

For simplicity, the coefficients of (27) were recalculated each time the set of intensities associated with a particular ω^{ℓ} was updated. The run times for the problems solved could be dramatically reduced by storing coefficients or even by using simple, computationally-inexpensive approximations to the exponential terms. The intensities were stored in a four-dimensional matrix, also for simplicity. For isotropic scattering, it is only necessary to store the complete I^{ℓ}_{P} values at boundary nodes so that four-dimensional storage is not required.

Comparison With 1D Solutions

The 2D solution was applied first to the 1D problem described in the last section, for a range of surface emissivities. As the grid was refined, the 2D model converged to the same result as the 1D model.

Problem 1: Black Enclosure With Emitting Medium

The first benchmark problem is the prediction of surface heat transfer in a square enclosure ($L_x = L_y = L$) with cold walls (at 0 K) which contains a gas at uniform temperature, T_g. The gas does not scatter ($\sigma^s = 0$) but does emit and absorb ($K \neq 0$). With the gas temperature known, the value of R in (10) is prescribed.

The local heat flux calculated to fall on the lower wall is plotted in Figures 7 and 8 for the cases of a strongly emitting medium ($KL = 10$) and weakly emitting medium ($KL = 0.1$) respectively. The "exact" solution was obtained by analytically integrating the radiation transfer equation for a given angle to the point of interest on the wall from the "upstream" wall, and then numerically integrating over all angles to find the local heat transfer. The present finite volume model could be made to agree with these exact results to any desired tolerance by refining the spatial and angular grids.

The S_4 discrete ordinates solution by Fiveland (1984) to this problem, using a (20×20) uniform spatial grid and with twelve quadrature points for the angular integration are shown on Figure 7 for $KL = 10$. Truelove (1987) was able to improve Fiveland's discrete ordinates solution by changing the quadrature points. Comparisons are made here with Fiveland's solution, because the quadrature he used was not tuned to this particular class of problem. The present finite volume solutions for 20 uniform control volumes in each direction, $(N_x \times N_y) = (20 \times 20)$, and with a uniform angular discretization of $(N^{\theta} \times N^{\phi}) = (1 \times 4)$ and (2×8) are also shown in Figure 7, and they give reasonable agreement with the exact solution. Reducing the spatial grid to (8×8) but using an expanding mesh which doubles the size of each successive control volume moving from the corners to the centerlines ($F_x = F_y = 2.0$ in Figure 7), and using a uniform (1×8) angular discretization, results in excellent agreement with the exact solution. The solution time for $(20 \times 20) \times (2 \times 8)$ problem was $15s$ on a MicroVax II computer.

Similar results appear in Figure 8 for the $KL = 0.1$ case. The finite volume results obtained using all the coarse grids shown in Figures 7 and 8 predict a total heat transfer on the bottom which agrees with the exact values to within 0.8%.

The results for this simple problem illustrate that the finite volume method converges to the exact solution as the angular and spatial grids are refined, and that predictions of good accuracy can be obtained on quite coarse grids.

Problem 2: 2D Heat Transfer in an Enclosure With One Hot Wall

The second 2D problem involves radiative exchange in rectangular enclosures with the left, top and right wall temperatures set to T_c and the bottom wall set a higher temperature T_h. The medium scatters isotropically and is assumed to be in radiative equilibrium. The objective is to predict the heat transfer that leaves the bottom surface as well as the temperature of the medium. Careful solutions to this problem have been reported by Crosbie and Schrenker (1984) for a range of cavity aspect ratios and scattering coefficients. Comparisons with finite volume predictions are presented here for 3 cases.

The "exact" (Crosbie and Schrenker, 1984) dimensionless local heat transfer q^* leaving the hot wall and temperature distribution along a vertical line in the centre of a square enclosure are plotted as solid lines in Figure 9 for a medium with $\sigma^s L = 1.0$. Application of the finite volume method with a very fine mesh $(N_x \times N_y) \times (N^{\theta} \times N^{\phi}) = (31 \times 31) \times (5 \times 24)$ is shown to give almost exact agreement (local heat fluxes agree to well within 0.1%). Use of a coarse $(8 \times 8) \times (2 \times 12)$ grid, such as would be of more practical interest, also yields excellent accuracy, with an average heat transfer $\overline{q}^*$ that is only 0.1% too high. To obtain a solution on this grid, in which the absolute value of the heat transfer on all walls changed by less than 0.05% between R-update cycles, required 8 cycles and $32s$ on a MicroVax II computer.

As previously discussed, it is important to know how quickly the solution error decreases with grid refinement. To perform this evaluation, the error in the local heat transfer at the centre of the hot wall ($x/L = 0.5$) was examined. Retaining (32×32) uniform control volumes, fixing N^ϕ at 24, and varying N^θ, Figure 10 shows that the error diminishes quadratically with N^θ. Similarly, for N^θ fixed at 5, increasing N^ϕ results in a quadratic decrease in error. For a fine angular grid (5×24) and with refinement of a uniform spatial grid, the error is seen in Figure 10 to decrease somewhere between linear and quadratic with increasing $N_x = N_y$. These results are in agreement with the expectations that were previously discussed.

For an enclosure with an aspect ratio of $L_x/L_y = 5$, the "exact solution" of Crosbie and Schrenker (1984) and a fine-grid $(61 \times 25) \times (7 \times 32)$ finite volume solution are plotted in Figure 11 for $\sigma^s L_y = 1.0$. The finite volume predictions of heat flux agree with the exact values to within 0.2%, and the temperature is also closely predicted. Figure 11 also shows that a much coarser grid $(10 \times 5) \times (3 \times 20)$, where the control volumes widths in the x-direction expand by a factor of 1.5 moving from the outer walls towards the centre ($F_x = 1.5$), yields virtually identical results, with an average heat transfer ($\overline{q}^*$) just 0.2% below the fine grid prediction. The prediction on the $(10 \times 5) \times (3 \times 20)$ grid required 5 R-update cycles and 17.6s to achieve a total absolute value of heat transfer on the boundaries that changed less than 0.05% between cycles.

The last test problem described in this paper has been specifically chosen to demonstrate that, as mentioned earlier, profile wiggles can occur under adverse conditions. Results for a cavity with $\sigma^s L_y = 1.0$ and $L_y/L_x = 10$ are compared to the predictions of Crosbie and Schrenker in Figure 13. Calculations on a fine $(31 \times 61) \times (5 \times 24)$ grid, with uniform control volumes in the x-direction and with volumes in the y-direction that increase by an expansion factor of $F_y = 1.05$, yielded local heat transfer predictions that are in excellent agreement with Crosbie and Schrenker (Figure 12). The predicted centreline temperature near the bottom wall (right side of Figure 12) is also very accurate but, when the aspect ratio of the control volumes becomes large in the centre to top of the cavity, the predicted temperatures oscillate (or wiggle) slightly about the profile of Crosbie and Schrenker. This "wiggle" is a consequence of the assumed linear profile of I^ℓ between nodes in Figure 6B, and is analogous to the wiggles that are often tolerated when a linear profile between nodes is used in the approximation of the advective transport terms in fluid flows. Such wiggles can be overcome (Roache, 1972) at the expense of accuracy by adopting a simple "upwind" approximation, or by using a more elaborate discretization.

Figure 12 also shows predictions using a coarse $(6 \times 10) \times (2 \times 12)$ mesh that has uniform spacing in the x-direction and expands rapidly ($F_y = 1.5$) in the y-direction away from the heated wall. This grid yields good temperature predictions, and heat transfer rates that are within 0.3% of exact values.

SUMMARY AND DISCUSSION

A new finite volume method has been proposed for the prediction of radiative heat transfer that can be implemented on non-orthogonal grids of the type employed to predict complex fluid flows. The theory is advanced for an absorbing-emitting and scattering gray medium in an enclosure with gray diffuse walls, but extensions to a broad bandwith model and to include walls with other reflective properties are straightforward. For strongly participating media, the method captures the diffusion approximation to the radiant heat transfer equations. The method ensures that the intensity in a given direction, as well as the net radiative heat transfer, satisfy the global conservation constraints. The approximations of the solid-angle integrals also prevent any occurrence of the "ray effects" encountered in discrete ordinate methods (Viskanta, 1987). The method also appears to be the first which shares the philosophy of, and can use the same non-orthogonal boundary fitted mesh as, modern methods for computing fluid flows <u>and</u> which also converges with grid refinement to the exact solution of the radiation transport equation.

The authors are indebted to anonymous reviewers for pointing out that it may be undesirable for radiation grid to be congruent with the fluid flow grid since the need for grid refinement may occur in different regions. Such situations are already encountered when the internal energy equation is solved over extensive regions of the flow in which the temperature is nearly constant, but in these regions a fine grid is needed to compute the fluid flow. It is common practice in such cases to use the same grid for both the fluid flow and heat transfer for the sake of simplicity. Similarly, for fluid flow and radiation different grids could certainly be used, but at the expense of complexity. Even if different grids were used, it is still desirable to use a philosophically consistent method.

The method has been demonstrated for a 1D problem, and several simple 2D problems, for which benchmark solutions and other approximate solutions are available. The method is shown to yield good accuracy on coarse grids, and errors due to the discrete representations of the continuous equations reduce rapidly with grid refinement.

ACKNOWLEDGMENTS

This work was supported by a contract, negotiated through the Unsolicited Proposal Program, from the Combustion and Carbonization Research Laboratory of the Energy Resources Laboratory, CANMET, which is part of the Energy Mines and Resources Department of the Canadian Federal Government. The authors gratefully acknowledge the support and assistance of the Scientific Advisor, Mr. Pat Hughes. Our thanks also to Professor K.G.T. Hollands of the Mechanical Engineering Department at the University of Waterloo for his encouragement, criticism and enthusiasm.

LIST OF REFERENCES

Crosbie, A.L., and Schrenker, R.G., 1984, "Radiative Transfer in a Two-Dimensional Rectangular Medium Exposed to Diffuse Radiation", *J. Quant. Spectrosc. Radiat. Transfer*, Vol. 31, pp. 339-372.

Fiveland, W.A., 1984, "Discrete-Ordinates Solutions of the Radiation Transport Equation for Rectangular Enclosures", *J. Heat Transfer*, Vol. 106, Nov., pp. 699-706.

Heaslet, M.A. and Warming, R.F., 1965, "Radiative Transport and Wall Temperature Slip in Absorbing Planar Medium", *Int. J. Heat Mass Transfer*, Vol. 8, pp. 979-994.

Roache, P.J., 1972, *Computational Fluid Dynamics*, Hermosa Publ., Albuquerque, N.M.

Siegel, R. and Howell, J.R., 1981, *Thermal Radiation Heat Transfer*, Hemisphere Publishing Corp.

Truelove, J.S., 1987, "Discrete Ordinate Solutions of the Radiation Transport Equation", *J. Heat Transfer*, Vol. 109, Nov., pp. 1048-1051.

Viskanta, R., 1984, "Radiative Heat Transfer", *Fortschrift der Verfahrenstechnik*, Vol. 22A, pp. 51-81.

Viskanta, R. and Mengüç, M.P., 1987, "Radiation Heat Transfer in Combustion Systems", *Prog. Energy Combust. Sci.*, Vol. 13, pp. 97-160.

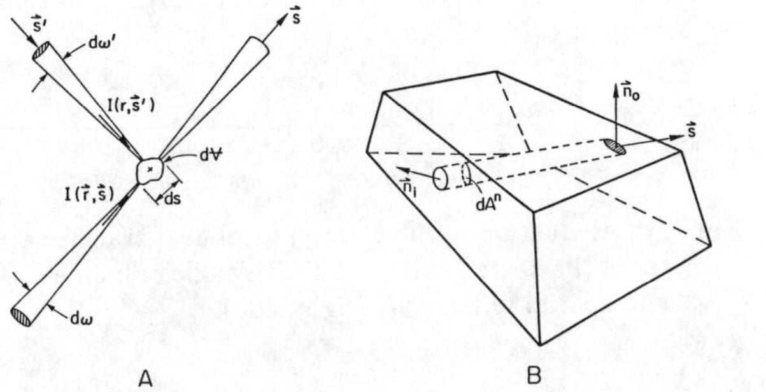

Fig. 1. Part A shows notation used in the model. The control volume, for which the integral equation for intensity is derived, appears in B.

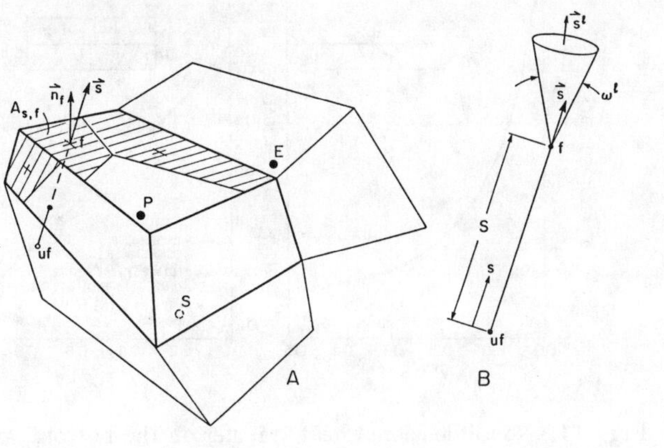

Fig. 2. Discrete surface panels on the discrete volume centered at P, and neighbour control volumes. The intensity I_f^ℓ, from which the flux through $A_{s,f}$ is found, is expressed as an "upstream" value I_{uf}^ℓ found by interpolation from nodal values, and a change between uf and f.

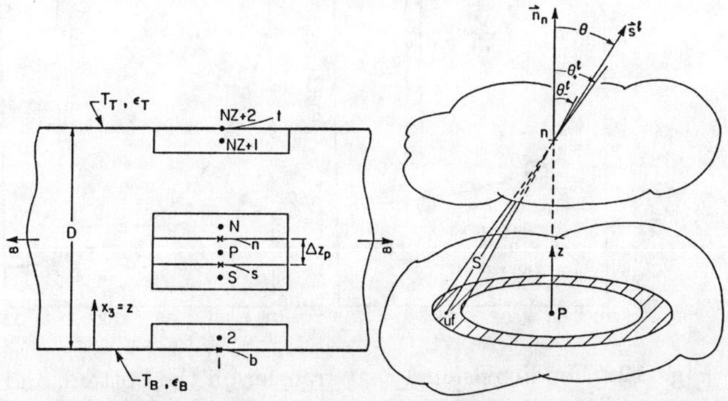

Fig. 3. Geometry and control volumes for radiation in a plane participating medium contained between isothermal gray walls.

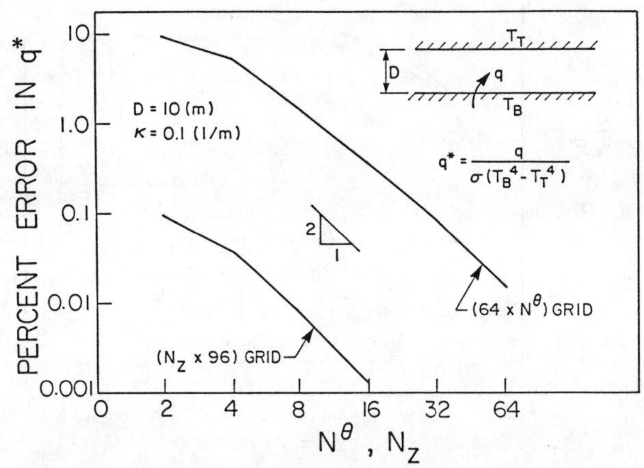

Fig. 4. Results of computational experiments to determine the rate at which the error in heat transfer from the surface decreases with grid refinement. The slope of 2 : 1 on the figure denotes quadratic convergence.

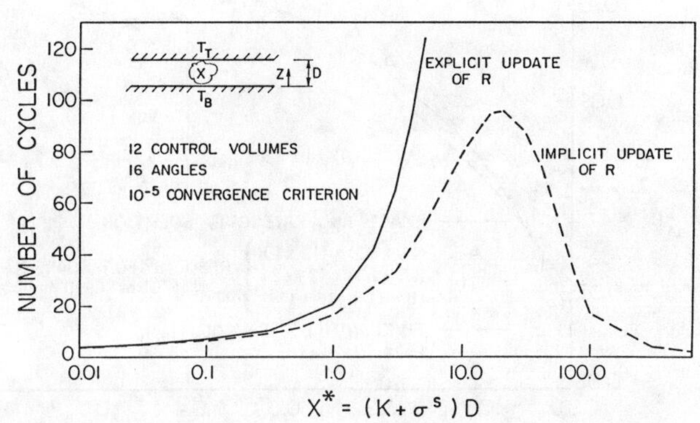

Fig. 5. The number of R-update cycles required to obtain a wall heat transfer that changed less than 0.001% between cycles.

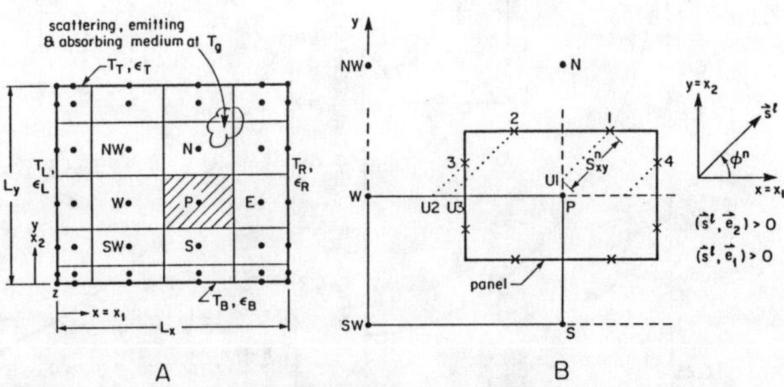

Fig. 6. Schematic of a rectangular cavity with isothermal walls with a Cartesian computation grid. Part B shows details related to determination of the integration point intensities for those intensities that are directed upward and to the right.

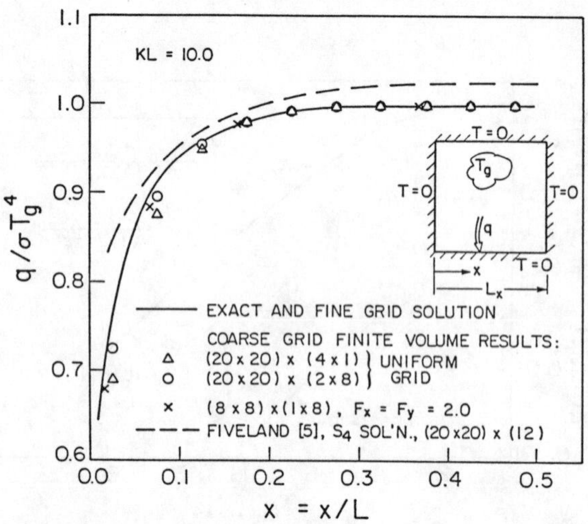

Fig. 7. Nondimensional heat transfer on the wall of a square cavity containing an emitting gas at uniform temperature T_g and with zero wall temperature for an optically thick case ($KL = 10.0$).

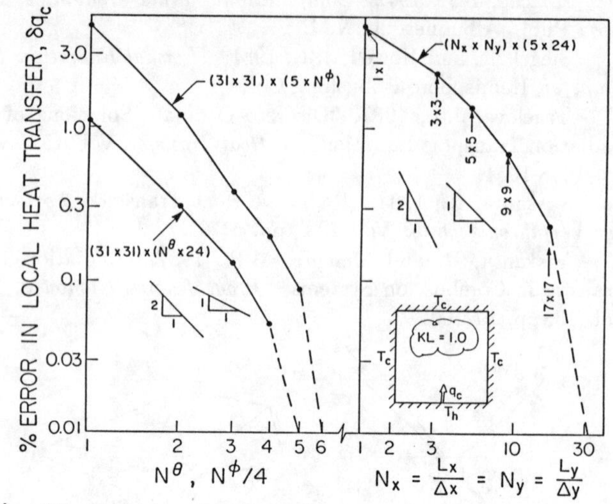

Fig. 10. Reduction in the error in local heat transfer at the center of the heated cavity wall with angle refinement (left side) and spatial refinement (right side).

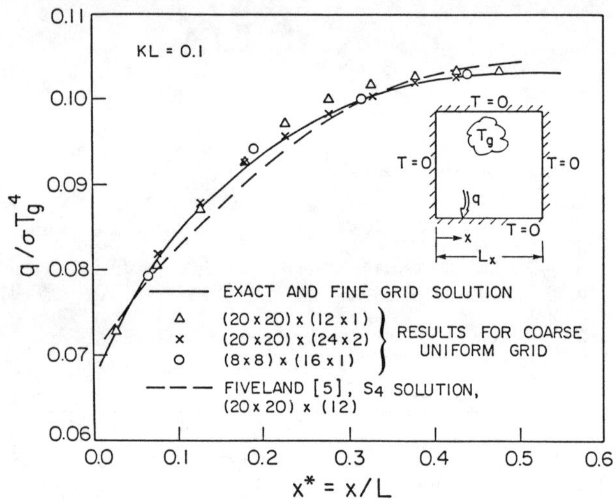

Fig. 8. Nondimensional heat transfer on the wall of a square cavity containing an emitting gas at uniform temperature T_g and with zero wall temperature for an optically thin case ($KL = 0.1$).

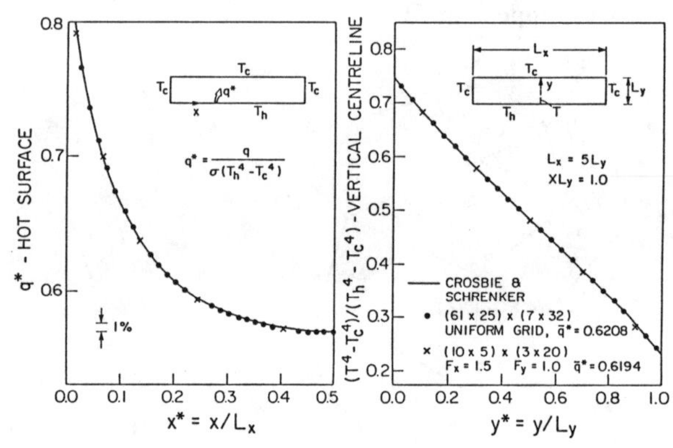

Fig. 11. Nondimensional heat transfer on the bottom, and centerline temperature distribution, in a $5:1$ aspect ratio cavity enclosing a scattering gas with $\sigma^s L_y = 1.0$.

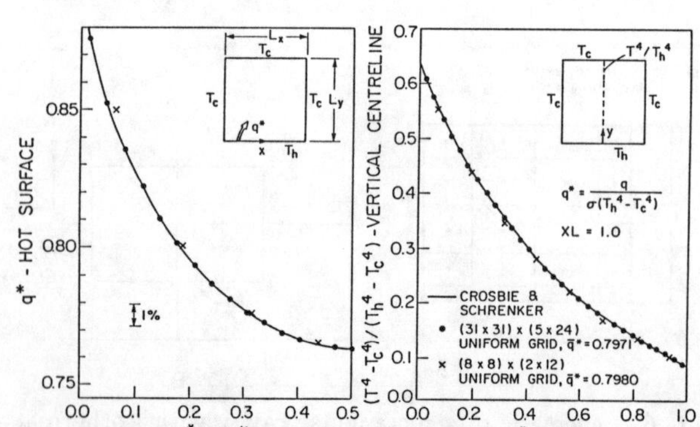

Fig. 9. Nondimensional heat transfer on the bottom, and centerline temperature distribution, in a square cavity with one heated wall enclosing a scattering gas with $\sigma^s L = 1.0$.

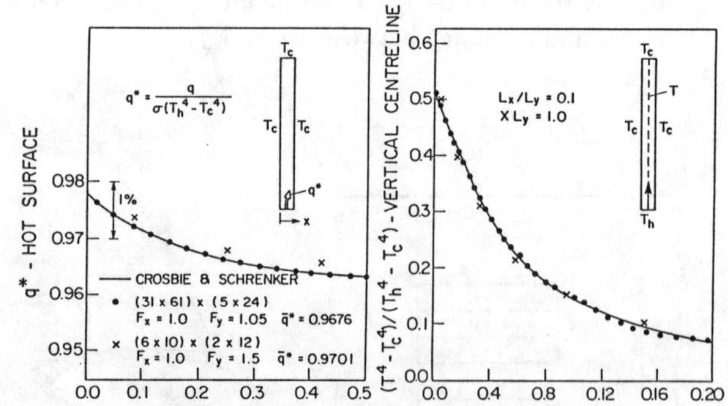

Fig. 12. Nondimensional heat transfer on the bottom, and centerline temperature distribution, in a $1:10$ aspect ratio cavity enclosing a scattering gas with $\kappa L_y = 1.0$.

1989 National Heat Transfer Conference
HTD-Vol. 106, Heat Transfer Phenomena in Radiation, Combustion and Fires

DIFFRACTION EFFECTS ON RADIATIVE HEAT TRANSFER THROUGH PERFORATED SHIELDS

C. A. Rivera-Carpio and C. L. Tien
Department of Mechanical Engineering
University of California
Berkeley, California

ABSTRACT

This work examines the effects of diffraction on the radiative heat transfer through perforated shields. A simple radiative heat transfer model is presented to account for diffraction effects. The model uses transmission coefficients through one and more than one perforated shield. Approximate transmission coefficients may be obtained from an analysis of short-wave electromagnetic diffraction, which provides the leading-order correction to geometrical optics predictions. Results are presented for the case of shields with regularly spaced slits. Diffraction effects on the overall radiative heat transfer depend on the emissivity and on the fraction of open area of the shields, the size of the openings compared to wavelength, and the spacing between shields. For highly reflective shields, diffraction effects are predicted for openings as large as 100 times the characteristic wavelength of thermal radiation. Diffraction increases the radiative heat transfer in multilayer insulation systems with perforated shields by as much as 30 percent.

NOMENCLATURE

a, b, c, d, e, f	= thermal resistance, equation (16)
B	= radiosity
B	= half-width of slit
C	= half-spacing between slits
D	= spacing between shields, diffraction coefficient
D	= diffraction matrix coefficient
e	= unit vector
E	= electric field vector
F	= diffraction function, equation (B.2)
f	= diffraction pattern
H	= magnetic field vector
i	= $\sqrt{-1}$
k	= wavenumber $= 2\pi/\lambda$
K	= modified Fresnel integral
N	= number of perforations per unit shield area
n	= number of shields
Q	= radiative heat flux
r	= thermal resistance, equations (17)-(18)
r	= radial distance
r	= position vector
s	= radial distance in edge coordinate system
s	= position vector in edge coordinate system
S	= Poynting vector
t	= transmission coefficient for apertures
T	= transmission coefficient for shields, temperature
u	= field component
x, y	= Cartesian coordinates
α	= angle of incidence
ε	= emissivity
λ	= wavelength
μ	= permeability
ω	= angular frequency
Φ, ϕ	= radiant power
ψ	= polar angle in edge coordinate system
τ	= fraction of open area $= B/C$
θ	= polar angle
ν	= wave frequency; argument of modified Fresnel integral
χ	= transmission index

Subscripts

d	= doubly diffracted
e	= electric
h	= magnetic
i	= resistance identification

j,k	=	edge identification
m	=	multiply diffracted
r	=	radiative
o	=	reference, origin
s	=	singly diffracted
t	=	transmitted
I	=	one shield
II	=	two shields

Superscripts

c	=	conventional
d	=	diffracted
i	=	incident
r	=	reflected

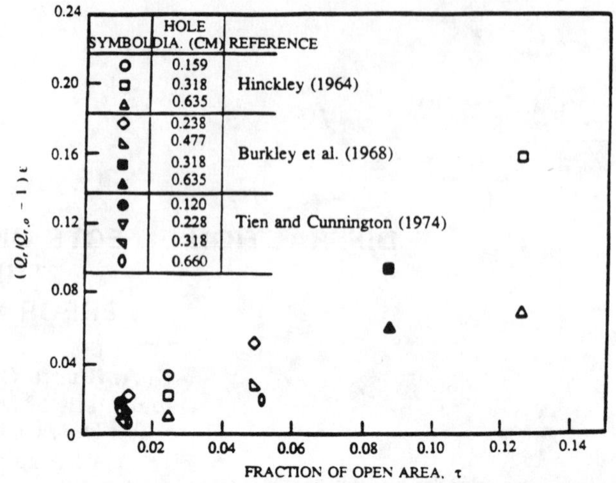

Fig. 1 Experimental data; perforation size effect on radiative heat flux ratio in multilayer insulations.

INTRODUCTION

Conventional radiative heat transfer analyses assume that diffraction effects are negligible. However, diffraction effects may be significant under certain conditions and, therefore, must be properly taken into account. The derivation of engineering models which include diffraction effects on radiative heat transfer is not only a fundamental problem but also one of practical significance.

This study examines the effects of diffraction on the transfer of thermal radiation. In particular, diffraction through apertures is considered, which is of concern in engineering systems consisting of perforated shields. Furthermore, the case under study will allow a qualitative comparison between theoretical predictions and experimental data reported for multilayer insulations. Perforated shields are often used in these systems to facilitate evacuation.

Experimental data on the radiative heat flux in multilayer insulation systems with perforated shields (Hinckley, 1964; Burkley et al., 1968; Tien and Cunnington, 1974) show that the size of the perforations has a definite effect on the radiative heat transfer and that this effect may be significant. The perforation size effect on the radiative transfer is shown in Fig. 1, as reported by Tien and Cunnington (1974). The vertical axis in Fig. 1 is the ratio of the additional radiative heat transfer that arises because of the perforations, $Q_r - Q_{r,o}$, to that corresponding to non-perforated multilayer systems, $Q_{r,o}$. The heat flux ratio is multiplied by the emissivity of the radiation shields, ε, to account for variations in this parameter among experiments ($\varepsilon \sim 0.027 - 0.04$). Contrary to what might be expected, the data shows that for a given fraction of open area, the radiative heat flux increases as the size of the perforations decreases.

Several analytical studies have been conducted to predict the radiative heat transfer in perforated multilayer insulations (e.g., Tien and Cunnington, 1974; Khrustalev et al., 1984). In these studies, surface radiant exchange is modeled based on geometrical optics concepts, either through view factors or fraction of open area. Potential causes to explain the perforation size effect were investigated by Tien and Cunnington (1974). These models cannot account for the perforation size effect on the radiative heat transfer. To date this problem has remain unsolved. It should be noted that in these systems the size of the openings is often between one to two orders of magnitude larger than wavelength. This suggests that the methods of geometrical optics might be questionable; geometrical optics methods ignore wavelength effects.

The relationship between geometrical optics and the more fundamental description of electromagnetic fields provided by Maxwell's equations is well established (Born and Wolf, 1980; Kline and Kay, 1965). Geometrical optics is the "zeroth" approximation in which the wave nature of thermal radiation is essentially ignored. Geometrical optics is an approximation valid when the wavelength is *infinitesimally* small in comparison with the size of the scattering object or with distances over which changes occur in the spatial radiant energy distribution. Often, geometrical optics provides an adequate description of the propagation of radiant energy in engineering systems. However, care must be exercised when applying this approximation to systems where these assumptions may break down.

Fundamental to the study of radiative transfer through perforated shields is the recognition of the physical process of electromagnetic wave diffraction, and the realization that conventional radiative transfer models based on geometrical optics may be unrealistic. According to geometrical optics, rays travel in straight lines and therefore opaque objects cast sharp shadows. However experimental investigations reveal that even a thin, well-defined edge, in the presence of a localized radiating source, does not produce a sharp shadow; rather, there is a significant penetration of radiant energy into the region called the *geometrical shadow*. This effect is one aspect of the physical wave phenomenon called *diffraction*. Diffraction is attributed to the fact that the wavelength is not zero. Geometrical optics fails to account for diffraction effects, such as the field in the shadow regions.

The effect of diffraction on the radiative heat transfer through perforated shields can be easily understood by considering Young's classical double-slit experiment. By allowing a monochromatic beam of light to pass through a slit in an otherwise opaque screen S_1, and interposing a second screen S_2 containing two slits, as shown in Fig. 2, Young achieved, because of diffraction, an interference pattern on the final screen S_3. If the portion of the incident wave front encountering the small area of

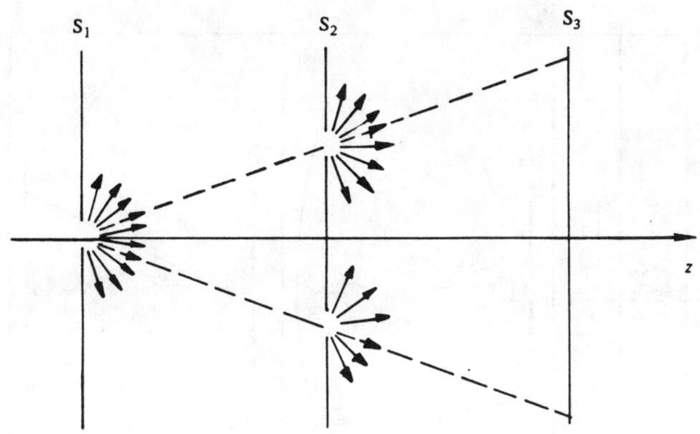

Fig. 2 Diffraction effects on transmission.

the slit on S_1 merely travels straight through along the x axis, as is predicted according to geometrical optics, it would never reach the second set of slits nor would it give rise to the radiant energy distribution observed on screen S_3. For radiant energy to be transmitted through this system, light emerging from the first slit and from the double-slit must spread in all directions, or *diffract*. Diffraction gives rise to transmission of radiant energy through systems consisting of more than one non-overlapping perforated shield. This is a fundamental diffraction effect; transmission of radiant energy through more than one perforated shield is not considered in conventional radiative transfer analyses of these systems. A fundamental investigation of radiative heat transfer through perforated shields must, therefore, be based on an analysis of transmission of radiant energy.

The transmission of radiant energy through a system of perforated shields is analytically investigated using an approximate technique known as the Geometric Theory of Diffraction, GTD. It was conceived by J. B. Keller (1957a, 1957b, 1962) and has developed continuously since then, and it is now an established analytical technique for the prediction of short-wavelength electromagnetic diffraction phenomena (Jones, 1979; James, 1980). Basically, GTD is a diffraction ray theory that provides the leading order correction to geometrical optics. It describes the diffracted field in terms of ray concepts. The concept of diffracted rays was developed by Keller from the asymptotic evaluation of the known exact solution to scattering from special problems. These rays are produced by incident rays which hit edges, corners, or vertices of boundary surfaces. The theory provides a systematic means for determining the diffracted field; it assigns a field value to each point on a ray and postulates that the total field at a point is the sum of the field values on all rays through that point. Although the theory is an approximation strictly valid when the diffracting body is large in comparison with the wavelength, inclusion of higher order diffraction effects can extend this technique, in some cases, to structures having a dimension less than the wavelength (see e.g., Yu and Rudduck, 1967). GTD is a simple and successful technique for determining the diffracted field, which would otherwise be very cumbersome and less accurate, if at all possible, to calculate by other methods (e.g. physical optics).

The effect of diffraction on the radiative heat transfer, in systems consisting of several perforated shields, is investigated using a radiation network model. The model accounts for diffraction effects through transmission coefficients through more than one perforated shield. These transmission coefficients are derived using GTD. The radiation network is solved for an increasing number of perforated shields and results are obtained for conditions typically found in multilayer insulation systems. The effects of the emissivity of the shields and the spacing between shields are also investigated.

ANALYSIS

Transmission Analysis. This section investigates the transmission of radiant energy through a system consisting of one and two non-overlapping perforated shields. In order to get an appraisal of radiant transmission through such systems, transmission of an incident plane wave through a system of staggered perforated shields, with regularly spaced slits, is studied. Only waves incident normal to the edges of the apertures are considered.

Transmission through one shield

Consider a plane monochromatic wave incident on an opaque shield with regularly spaced slits of width $2B$, as shown in Fig. 3a. The shield is on the plane $x = 0$ and the incident wave propagates parallel to the x,y plane with its direction of propagation making an angle α with the x axis. The transmission coefficient for one perforated opaque shield, T_I, is defined as the fraction of the incident radiant energy that is transmitted through it. According to geometrical optics, the transmission coefficient for an opaque shield with openings is equal to the fraction of open area, τ. More generally, the transmission coefficient through one perforated shield, T_I, may be written in the form

$$T_I = \chi_I \ \tau \tag{1}$$

where χ_I is a *transmission index* for a perforated shield, and is equal to one according to geometrical optics. For a given aperture shape, χ_I is, in general, a function of the size of the openings (in terms of wavelength), the spacing between apertures, and the state of polarization and direction of propagation of the incident wave.

The transmission coefficient for an *opaque* shield with N perforations per unit area may be determined from

$$T_I = \frac{N \phi_{t,I}}{\Phi_o} \tag{2}$$

where $\phi_{t,I}$ is the radiant energy per unit time transmitted through a given aperture, and Φ_o is the radiant power incident normal to a shield of unit area. The latter Φ_o is equal to the intensity of the incident field multiplied by the projected area normal to the direction of incidence, of a shield of unit area.

When the apertures are several wavelengths apart, interaction effects between apertures may be neglected. In this case, $\phi_{t,I}$ is given by

$$\phi_{t,I} = t_I \ \phi_o \tag{3}$$

where t_I is the transmission coefficient for an isolated aperture, and ϕ_o is the radiant power incident on an aperture; ϕ_o is equal to the intensity of the incident field multiplied by the projection of the aperture normal to the direction of propagation. Combining equations (2) and (3) yields

$$T_I = t_I \ \tau \tag{4}$$

From equations (1) and (4) it follows that, when interaction effects between apertures are negligible, the transmission index for a shield with apertures of equal shape, χ_I, is equal to the transmission coefficient for a single aperture, t_I.

Transmission coefficients for single apertures obtained using GTD have been found to be in very good agreement with exact solutions where they exist (Keller, 1962). It will prove useful at this stage to briefly review the general concepts for determining the transmission coefficient for a single aperture, and at the same time introduce the major features of GTD.

The parameter t_I is most conveniently determined by applying the forward scattering theorem (Born and Wolf, 1980), which allows calculating the transmission coefficient from the amplitude of the transmitted far field, $\mathbf{E}(r,\theta)$, in the direction of the incident field, i.e., $\theta = \alpha$.

Let the incident electric field vector $\mathbf{E}^i$ at position $\mathbf{r}$ be given by

$$\mathbf{E}^i(\mathbf{r}) = \mathbf{e}_o \exp(i[\omega t - k(\mathbf{e}^i.\mathbf{r})]) \qquad (5)$$
$$= \mathbf{e}_o \exp(i[\omega t - kr\cos(\theta - \alpha)])$$

where $k = 2\pi/\lambda$ is the wavenumber, λ wavelength, $\omega = 2\pi v$ angular frequency, v wave frequency, t time, and $\mathbf{e}_o$ and $\mathbf{e}^i$ are unit amplitude electric vector and unit propagating vector, respectively. The resultant electric field vector $\mathbf{E}(\mathbf{r})$ in the far zone region, $|\mathbf{r}| >> 2B$ and $-\pi/2 < \theta < \pi/2$, is equal to the sum of electric fields on edge-diffracted rays through $\mathbf{r}$ (Menendez and Lee, 1977). If only singly diffracted rays are considered, as is shown in Fig. 3c, $\mathbf{E}(\mathbf{r})$ is given by

$$\mathbf{E}(\mathbf{r}) = \sum_{j=1}^{j=2} \mathbf{E}_j^d(\mathbf{s}_j) \ , \quad \mathbf{s}_j = \mathbf{r} - \mathbf{r}_{o,j}, \quad j = 1,2 \qquad (6)$$

where $\mathbf{E}_j^d$ is the electric vector on the diffracted ray associated with the edge N_j, which is located at $\mathbf{r}_{o,j}$, and $\mathbf{s}_j$ is the vector distance from the edge N_j.

The edge-diffracted-ray electric vector may be expressed in the form (James, 1980)

$$\mathbf{E}_j^d(\mathbf{s}_j) = \frac{\exp(-iks_j)}{s_j^{1/2}} \mathbf{D}_j(\mathbf{e}^i, \mathbf{e}_{s_j}) \mathbf{E}^i(\mathbf{r}_{o,j}) \qquad (7)$$

where s_j and $\mathbf{e}_{s_j}$ are the radial distance from the edge and the corresponding unit vector, respectively. $\mathbf{D}(\mathbf{e}^i, \mathbf{e}_{s_j})$ is a diffraction matrix with its elements proportional to $k^{-1/2}$. It represents a transformation function that maps the incident field onto the edge-diffracted field. $\mathbf{D}$ depends on the directions of the incident and diffracted rays, the local nature of the diffracting object at the point of diffraction, and the propagation constant $k = 2\pi/\lambda$.

From equations (5)-(7) and the forward scattering theorem, the transmission index for a perforated shield, χ_I, can be readily determined. The following results are obtained (see appendix A):

For electric polarization,

$$\chi_{I,e} \sim t_{I,e}(\alpha) = 1 - \frac{1}{2\pi^{1/2}(kB)^{3/2}\cos\alpha} \times \qquad (8)$$
$$\left[\frac{\cos[2kB(1-\sin\alpha) - \pi/4)]}{1-\sin\alpha} + \frac{\cos[2kB(1+\sin\alpha) - \pi/4]}{1+\sin\alpha}\right]$$

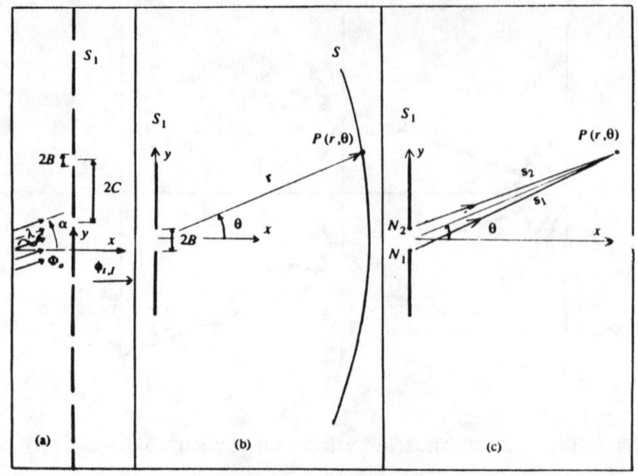

Fig. 3 Transmission through one perforated shield: a) physical system, b) geometry for transmission through an aperture on S_1, c) description of electromagnetic field in terms of diffracted rays.

and for magnetic polarization,

$$\chi_{I,h} \sim t_{I,h}(\alpha) = 1 - \frac{1}{32\pi^{1/2}(kB)^{5/2}\cos\alpha} \times \qquad (9)$$
$$\left[\frac{\sin^2(\pi/2+\alpha/2)}{\cos^4(\pi/4+\alpha/2)}\sin[2kB(1-\sin\alpha) - \pi/4] + \right.$$
$$\left. \frac{\sin^2(\pi/4-\alpha/2)}{\cos^4(\pi/4-\alpha/2)}\sin[2kB(1+\sin\alpha) - \pi/4]\right]$$

The correction to the geometrical transmission index is of the order of $(kB)^{-3/2}$, for electric polarization, and of order $(kB)^{-5/2}$, for magnetic polarization.

Transmission through two shields

Figure 4a shows the physical system under consideration. It consists of two identical perforated shields, parallel to each other, arranged so that the apertures do not overlap. The transmission coefficient for this two-shield system may be expressed in the form

$$T_{II} = \chi_{II}\tau^2 \qquad (10)$$

where χ_{II} is the *transmission index* for two perforated shields. For a given aperture shape, χ_{II} is, in general, a function of the fractional open area, the size and shape of the openings, the spacing between shields, and the direction and polarization of the incident field. According to geometrical optics, χ_{II} is zero.

The transmission coefficient through two shields, T_{II}, may be determined from

$$T_{II} = \frac{N\phi_{t,II}}{\Phi_o} \qquad (11)$$

where $\phi_{t,II}$ is the transmitted power through an aperture on the second shield S_2 (Fig. 4a), N number of apertures per unit area, and Φ_o incident power normal to a unit area on S_1.

When interaction effects due to the proximity of apertures in a shield are negligible, the transmitted power through one aperture on S_2, $\phi_{t,II}$, is independent of the number of apertures on S_2. In this case $\phi_{t,II}$ may be determined using the model shown in Fig. 4b.

Denoting by S(r) the Poynting vector at location **r** (Fig. 4b), the power transmitted through an aperture on shield S_2 is

$$\phi_{t,II} = \int_{-\pi/2}^{\pi/2} \mathbf{S}(\mathbf{r}) \cdot \mathbf{e}_r r d\theta \qquad (12)$$

The determination of S(r) using GTD requires that all diffracted rays passing through **r** be accounted. In this study a first-order solution is considered. Also the case investigated corresponds to two parallel shields several wavelengths apart. Since the amplitude of the diffracted field decreases with the path-length, diffracted rays that are reflected several times between shields and then transmitted through S_2 may then be neglected. The far field at **r** is then approximately given by the sum of the fields on doubly edge-diffracted rays through **r**. A doubly diffracted ray arises when a singly diffracted ray hits a second edge. Two of these doubly diffracted rays are shown in Fig. 4c. For the case of incoherent radiation, S is given by

$$\mathbf{S} = \sum_j \sum_k \mathbf{S}_{j,k}(\mathbf{s}_{j,k}\ ,\ \mathbf{s}_k)$$

$$\mathbf{s}_k = \mathbf{r} - \mathbf{r}_{o,k},\quad k = 1, 2 \qquad (13)$$

$$\mathbf{s}_{j,k} = \mathbf{r}_{o,k} - \mathbf{r}_{o,j},\quad j = ..., -2, -1, 1, 2, ...,\quad k = 1, 2$$

where $\mathbf{S}_{j,k}$ is the Poynting vector of the doubly diffracted ray due to the edges M_j and N_k, which are located at $\mathbf{r}_{o,j}$ and $\mathbf{r}_{o,k}$, respectively.

The Poynting vector of a doubly diffracted field is given by

$$\mathbf{S}_{j,k} = \frac{1}{2}\mathrm{Re}\left[\mathbf{E}_{j,k} \cdot conj(\mathbf{H}_{j,k})\right] \qquad (14)$$

$$\mathbf{H}_{j,k} = \frac{\omega}{k\,\mu_o}\mathbf{e}_{s_{j,k}} \cdot \mathbf{E}_{j,k} \qquad (15)$$

where $\mathbf{E}_{j,k}$, $\mathbf{H}_{j,k}$, and $\mathbf{e}_{s_{j,k}}$ are the electric field vector, magnetic field vector, and unit propagating vector, respectively, of the doubly diffracted field from the edges located at $\mathbf{r}_{o,j}$ and $\mathbf{r}_{o,k}$. In equation (15), *conj* denotes complex conjugate. Explicit expressions for $\mathbf{S}_{j,k}$ are given in appendix B.

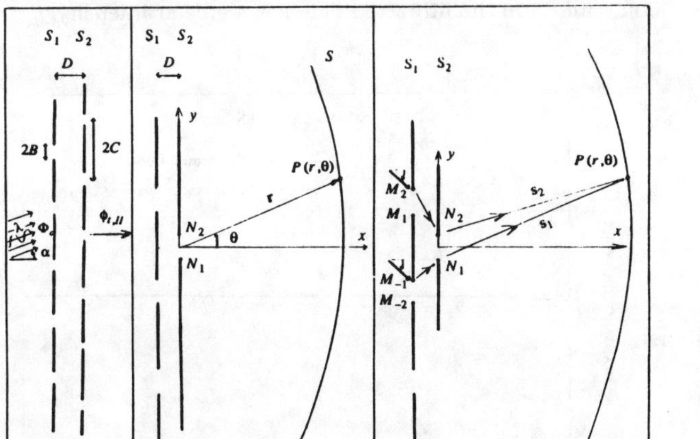

Fig. 4 Transmission through two perforated shields: a) physical system, b) geometry for transmission through an aperture on S_2, c) description of the electromagnetic field in terms of diffracted rays.

Radiative Transfer Analysis. Radiative transfer through perforated shields has been modeled by several investigators (Tien and Cunnington, 1974; Khrustalev et. al., 1984). In contrast to these models, this section presents an engineering radiative transfer model that attempts to account for diffraction effects. A special feature of the model is its reduction to conventional radiative transfer analyses in the special case when the wavelength of thermal radiation is much smaller than the size of the perforations. The proposed model attempts to account for diffraction and to predict when diffraction effects become important.

The physical system under consideration consists of two constant-temperature black surfaces with several perforated shields in between, as shown in Fig. 5a. Radiative transfer through the perforated shields is modeled using a radiation network approach. The radiation network method incorporates, in a simple and direct way, results from radiative transmission analyses.

In investigations concerning wave propagation through perforated shields, it was noted that radiant energy may be transmitted through more than one perforated shield. Transmission coefficients for one and two non-overlapping perforated shields have been presented in the previous section. Here, these results are used in a radiative transfer network to two degrees of approximation. These approximations are 1) radiant energy transmitted through a perforated shield is completely intercepted by the next consecutive shield; and 2) radiant energy transmitted through the first perforated shield and partially transmitted through the second one is completely intercepted by the third shield.

The network approach to radiative heat transfer analyses may then be classified according to the number of radiant exchange interactions that each radiating shield has. Radiation network analyses based on the first approximation assume that each radiating shield (i.e. both surfaces) interchanges radiant energy with four surfaces, two on each side of a shield. This approximation is exactly equivalent to that which is based on geometrical optics, provided that the transmission coefficient T_I is equal to the fractional open area τ.

Radiative transfer models based on the second approximation differ from the above in that they do account for diffraction effects to a first order approximation. Models based on the second of these approximations assume that each radiating surface is radiatively coupled to six surfaces. Radiative heat transfer models that account for diffraction to higher levels of approximation are clearly possible; however, preliminary calculations suggest that the six-interaction model is a good approximation.

Four-interaction model

The radiation network in the four-interaction model is shown in Fig. 5b. Following the convention shown in Fig. 5a, B_i and B_i' represent the radiosities of the left-hand and right-hand surfaces of shield i, respectively. The network resistances a, b, c, d, and e are

$$a = \frac{1}{(1-\tau)},\quad \frac{b}{2} = \frac{(1-\varepsilon)}{\varepsilon(1-\tau)},\quad c = \frac{1}{\chi_I \tau(1-\tau)}, \qquad (16)$$

$$d = \frac{1}{(1-\tau)^2},\quad e = \frac{1}{\chi_I \tau}$$

where ε is the shield emissivity, τ the fraction of open area, and χ_I the *transmission index* for one shield.

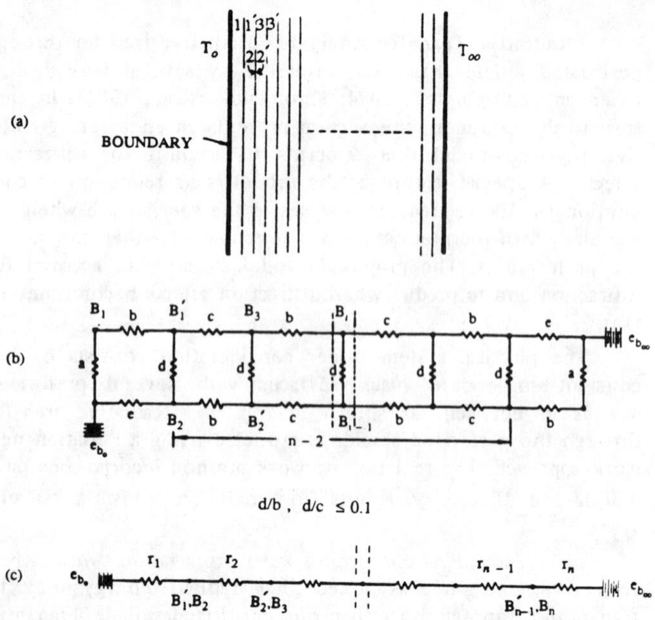

Fig. 5 Four-interaction model: a) physical system, b) radiation network, c) approximate radiation network.

For small values of ε and τ ($\varepsilon < 0.2$ and $\tau < 0.1$), $d/b < 0.15$ and $d/c < 0.15$. The four-interaction network, for a system consisting of n shields, may then be approximated by the network shown in Fig. 5c, where the new resistances r_1, r_i ($i = 2, ..., n-1$), r_n are

$$r_1 = r_n = \frac{(a+b)e}{a+b+e} \; ; \; r_i = \frac{b+c}{bc} \; , \; i = 2, ..., n-1 \; , \qquad (17)$$

Six-interaction model

The radiation network in the six-interaction model, for small values of ε and τ, is shown in Fig. 6. Diffraction effects on the radiative heat transfer are clearly illustrated by comparison of Fig.6 with Fig. 5c. In Fig. 6 the resistances r_1', r_i' ($i = 2, ..., n-1$), r_n', f' are

$$r_1' = r_n' = \frac{(a+b)e'}{a+b+e'} \; ; \; r_i' = \frac{bc'}{bc'} \; , \; i = 2, ..., n-1 \; ; \qquad (18)$$

$$f' = \frac{1}{\chi_{II}\,\tau^2(1-\tau)}$$

where $c' = 1/\tau(\chi_I - \chi_{II}\tau)(1-\tau)$ and $e' = 1/\tau(\chi_I - \chi_{II}\tau)$. The parameters χ_I and χ_{II} are the *transmission indexes* for one and two shields, respectively.

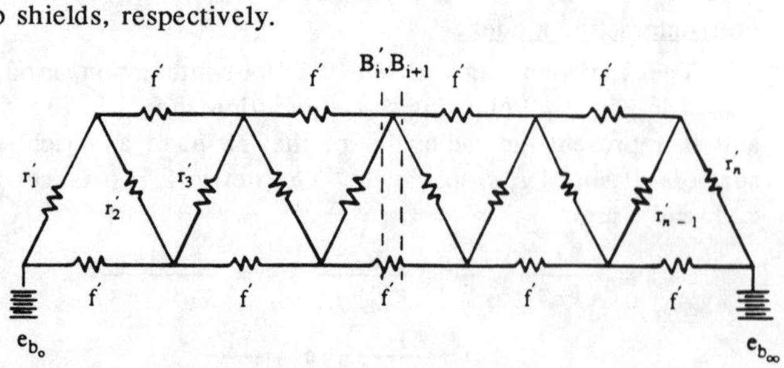

Fig. 6 The approximate radiation network of the six-interaction model for the physical system shown in Fig. 5a.

RESULTS AND DISCUSSION

Transmission Results. Results for χ_I and χ_{II} are presented for the case of normal incidence.

Figure 7 shows the transmission index for one perforated shield, χ_I, as a function of the size of the slit, B/λ. For shields with slits larger than the wavelength, the transmission index is between 0.97 and 1.0 and monotonically approaches unity as the size of the perforation increases. It then follows that for an *isolated* shield, higher-order corrections to geometrical optics results are small and may be neglected for $B/\lambda > 1$. In this case the transmission coefficient T_I is equal to the fraction of open area.

The transmission coefficient for two shields with non-overlapping perforations, T_{II}, is shown in Fig. 8 as a function of the open area, τ, and the size of the perforations, B/λ, for a spacing between shields equal to 10 wavelengths. Transmission increases as the open area increases and, for a given open area, it increases further as the size of the slits decreases.

Figure 9 shows the transmission index for two shields, χ_{II}, as a function of the fraction of open area, τ, and the size of the slit, B/λ, for $D/\lambda = 10$. The numerical value of the transmission index χ_{II}, introduced in equation (10), describes how well the transmission coefficient for a two-shield system is approximated by the product of transmission coefficients for isolated shields. Figure 9 shows that this approximation will often underestimate the actual transmission. For example, for $D/\lambda = 10$, transmissions through a two-shield system as much as seven times that obtained by multiplying individual transmissions are predicted. The error involved in this approximation increases significantly as the size of the openings and the open area decreases.

The effect of spacing between shields on the transmission index, χ_{II}, is shown in Fig. 10 for $B/\lambda = 5$ and $B/\lambda = 25$. When closely spaced shields are moved apart, the transmission through this system increases because apertures on the second shields will be seen at angles closer to the forward direction of propagation, and there the diffracted field has a greater intensity.

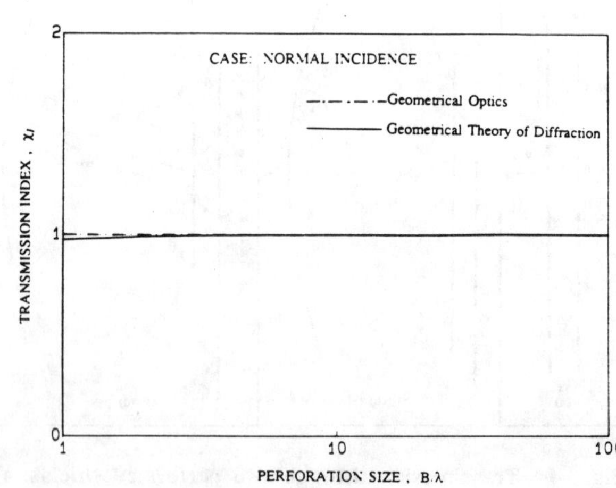

Fig. 7 Transmission index for one perforated shield as a function of perforation size.

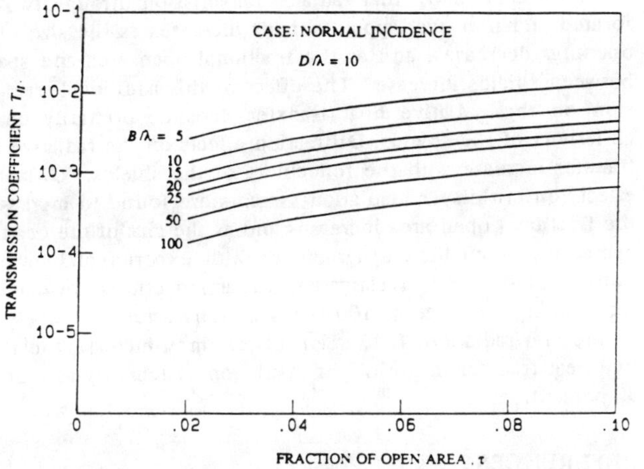

Fig. 8 Transmission coefficient for two perforated shields as a function of open area and perforation size.

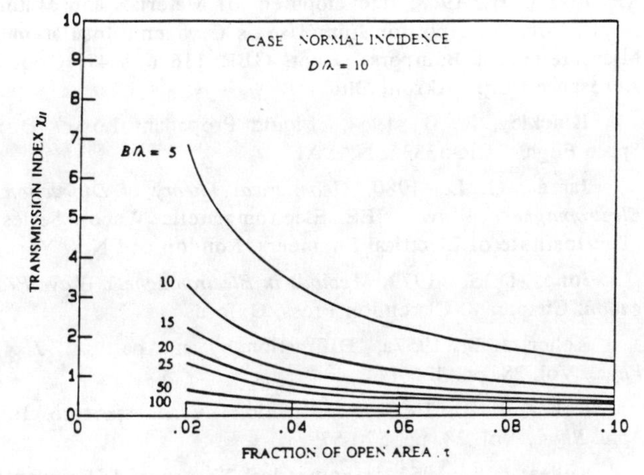

Fig. 9 Transmission index for two perforated shields as a function of open area and perforation size.

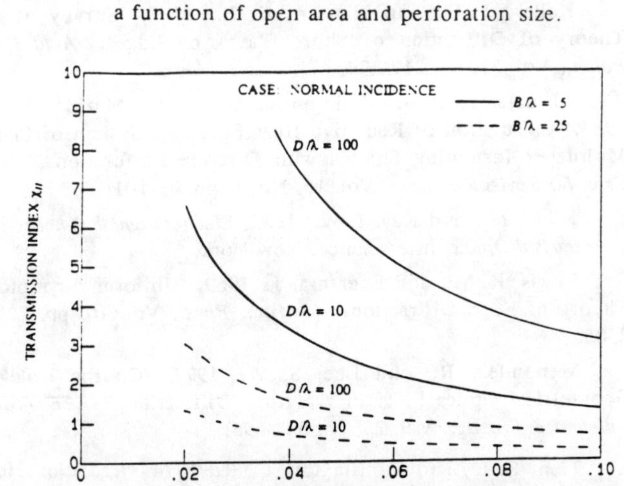

Fig. 10 Transmission index for two perforated shields; effect of spacing between shields.

Radiative Heat Transfer Results. Since transmission through perforated shields is wavelength dependent and radiative heat transfer covers a wide range of wavelengths, diffraction effects on the radiative heat transfer must be implemented using some sort spectral scheme. The simplest scheme is one that uses effective transmission properties for the entire spectral range. In this section results for the radiative heat flux are presented using this approach. It is assumed that effective transmission coefficients may be determined using transmission coefficients for normal incidence at a characteristic wavelength. Transmission coefficients for normal incidence are reasonable estimates for the hemispherical transmission coefficients. A characteristic wavelength may be obtained, e.g., using Wien's law for a mean temperature, or by averaging two or more wavelengths, where each one is obtained using Wien's law at specific temperatures. The final determination of the characteristic wavelength requires a comparison of different effective transmission models against experimental data, which is outside the scope of the present investigation.

Figure 11 shows the radiative heat flux in a system consisting of several perforated shields for conditions typically found in multilayer insulation systems ($\varepsilon = 0.05$, $D/\lambda = 10$). The vertical axis is the ratio of the overall radiative heat flux, Q_r, to that corresponding to non-perforated multilayer systems, $Q_{r,o}$. This figure shows that for a given open area, the radiative heat flux increases as the size of the perforations decreases and that this size effect increases as the fraction of open area increases, in qualitative agreement with the experimental data shown in Fig. 1. For the case under investigation, diffraction effects on the radiative heat transfer are observed for width of slits as large as 100 times the wavelength, and may increase the radiative heat transfer in multilayer insulation systems by as much as 30 percent, depending upon the open area and the size of the perforations.

Figure 12 shows the effect of emissivity on the radiative heat flux. In this figure, the vertical axis is the ratio of the radiative heat flux as predicted by the six-interaction model, Q_r^d, to that predicted by conventional models which neglect diffraction, Q_r^c. Diffraction effects on the radiative heat transfer increase as the emissivity decreases because in these cases the transfer of

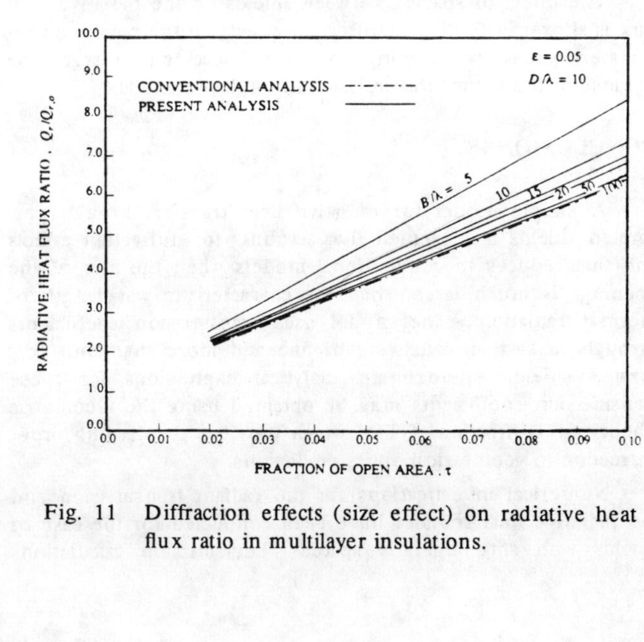

Fig. 11 Diffraction effects (size effect) on radiative heat flux ratio in multilayer insulations.

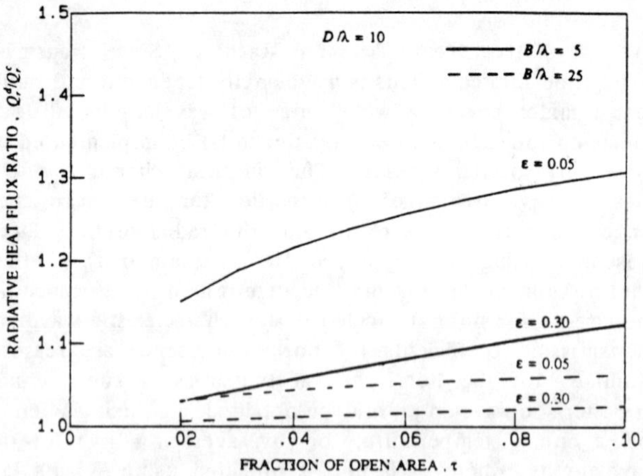

Fig. 12 Diffraction effects on radiative heat flux; effect of emissivity of shields.

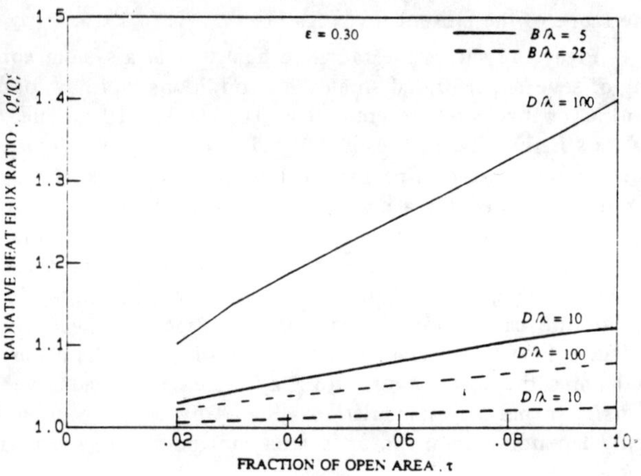

Fig. 13 Diffraction effects on radiative heat flux; effect of spacing between shields.

thermal radiation through two shields, which is due to diffraction effects, becomes comparable to that between two consecutive shields.

The effect of spacing between shields on the radiative heat flux is shown in Fig. 13. Diffraction effects on the radiative heat transfer increase when going from closed spacing to a larger one because of the higher transmission through two shields.

CONCLUSIONS

A simple model for radiative heat transfer through perforated shields is presented that accounts for diffraction effects and that reduces to conventional models when the size of the openings is much larger than the characteristic wavelength of thermal radiation. The model uses transmission coefficients through a system consisting of one and more than one perforated shield. Approximate analytical expressions for these transmission coefficients may be obtained using the Geometric Theory of Diffraction, GTD, which provides the leading order correction to geometrical optics predictions.

Numerical investigations for the radiant transmission and the radiative heat transfer have been conducted for the case of shields with slits regularly spaced. Transmission calculations based on GTD show that radiant transmission through two perforated and non-overlapping shields increases as the size of the openings decreases, and as the fractional open area and spacing between shields increase. The effect of this additional transmission on the radiative heat transfer depends primarily on the emissivity of the shields. Diffraction effects on the radiative heat transfer increase with the reflectivity of the shields. Diffraction effects on multilayer insulation systems are found to increase as the fractional open area increases and as the size of the openings decreases, in qualitative agreement with experimental observations. For the case investigated, diffraction effects are expected for openings as large as 100 times the characteristic wavelength of thermal radiation. Diffraction effects may increase the radiative heat transfer in multilayer insulation systems by as much as 30 percent.

REFERENCES

Born, M. and Wolf, E., 1980, *Principles of Optics*, sixth edition, Pergamon Press, New York.

Burkley, R. A., Rollins, S. R., Apisa, J., Shriver, C., and Kariotis, A. H., 1968, "Development of Materials and Materials Application Concepts for Joint Use as Cryogenic Insulation and Micrometeoroid Bumpers," Rept. GER 11676 S/47, Goodyear Aerospace Corp., Akron, Ohio.

Hinckley, R. B., 1964, "Liquid Propellant Losses During Space Flight," CR-53336, NASA.

James, G. L., 1980, *Geometrical Theory of Diffraction for Electromagnetic Waves*, IEE Electromagnetic Waves Series 1, (The Institute of Electrical Engineers, London and New York).

Jones, D. S. , 1979, *Methods in Electromagnetic Wave Propagation*, Chapter 8, Clarendon Press, Oxford.

Keller, J. B., 1957a, "Diffraction by an Aperture," *J. Appl. Phys.*, Vol. 28, pp. 426-444.

Keller, J. B., 1957b, "Diffraction by an Aperture. II," *J. Appl. Phys.*, Vol. 28, pp. 570-579.

Keller, J. B., 1962, "Geometrical Theory of Diffraction," *J. Opt. Soc. Am.*, Vol. 52, pp. 116-130.

Keller, J. B., and Hansen, E. B., 1965, "Survey of the Theory of Diffraction of Short Waves by Edges," *Acta Phys. Polon.*, Vol. 27, pp. 217-234.

Khrustalev, B. A., Filimonov, S. S., and Mazilin, I. M., 1984, "Calculation of Radiative Heat Flux in a System of Planar Multilayer Insulating Shields with Discrete Perforations," *Heat Transfer-Soviet Research*, Vol. 16, No. 1, pp. 93-101.

Kline, M., and Kay, I. W., 1965, *Electromagnetic Theory and Geometrical Optics*, Interscience, New York.

Lewis, R. M., and Boersma, J., 1969, "Uniform Asymptotic Theory of Edge Diffraction," *J. Math. Phys.*, Vol. 10, pp. 2291-2305.

Menendez, R., and Lee, S. W., 1977, "On the Role of Geometrical Optics Field in Aperture Diffraction," *IEEE Trans. Antenna & Propag.* Vol. 25, pp. 688-695.

Tien, C. L., and Cunnington, G. R., 1974, "Radiation Heat Transfer in Multilayer Insulation Having Perforated Shields," *Progress in Astronautics and Aeronautics* , Vol. 35, pp. 65-74.

Yu, J. S., and Rudduck, R. C., 1967, "On Higher-Order Diffraction Concepts Applied to a Conducting Strip," *IEEE Trans. in Antenna & Propag.*, Vol. 15, pp. 662-668.

APPENDIX A

Diffracted Field for One-Shield Model

Consider a plane monochromatic wave incident on a slit of width $2B$, in a thin and perfectly conducting shield. Only incident waves propagating normal to the edges of the slit are considered.

Choose the plane of the shield to be at $x = 0$, and the center of the slit to be at the origin of the x,y coordinate system (Fig. 3b). Let the incident field u^i ($u^i = E_z$ for electric polarization, and $u^i = H_z$ for magnetic polarization) be given by

$$u^i = \exp\left[i[\omega t - k(x\cos\alpha + y\sin\alpha)]\right] \quad (A.1)$$

where $k = 2\pi/\lambda$ is the wavenumber, $\omega = 2\pi\nu$ angular frequency, ν wave frequency, t time, and α angle between the direction of propagation of the incident wave and the x-axis.

Two singly diffracted rays pass through each point $P(r,\theta)$, (Fig. 3c), so equation (6) becomes

$$u^d(P) = u^i(N_1)D(s_1,\alpha,\psi_1)\frac{\exp(-iks_1)}{s_1^{1/2}} + \quad (A.2)$$

$$u^i(N_2)D(s_2,-\alpha,-\psi_2)\frac{\exp(-iks_2)}{s_2^{1/2}}$$

where $u^i(N_1)$ and $u^i(N_2)$ are the incident fields at the edges N_1 and N_2 respectively, and ψ_1 and ψ_2 the angle that each diffracted ray makes with the x-axis. The diffraction coefficient D is given by (James, 1980),

$$D_e(s,\alpha,\psi) = -s^{1/2}[\varepsilon^i K(v^i) - \varepsilon^r K(v^r)] \quad \text{(elect.)(A.3a)}$$

$$D_h(s,\alpha,\psi) = -s^{1/2}[\varepsilon^i K(v^i) + \varepsilon^r K(v^r)] \quad \text{(mag.)}$$

$$\varepsilon^{i,r} = sgn\left[\cos\left(\frac{(\psi+\pi/2) \mp (\alpha+3\pi/2)}{2}\right)\right] \quad (A.3b)$$

$$v^{i,r} = \sqrt{2ks}\left|\cos\left(\frac{(\psi+\pi/2) \mp (\alpha+3\pi/2)}{2}\right)\right| \quad (A.3c)$$

where K in equation (A.3a) is the modified Fresnel integral, and sgn in equation (A.3b) is the sign function ($sgn(x) = 1$ for $x > 0$, otherwise $sgn(x) = -1$). In equations (A.3b) and (A.3c), the negative (positive) sign corresponds to the superscript i (r).

In the far zone, $ks_1 >> 1$ and $ks_2 >> 1$, and K may be approximated by its asymptotic expression

$$K(x) = \sqrt{\frac{i}{\pi}}\exp(ix^2)\int_x^{\infty} \exp(-it^2)\,dt \sim \frac{1}{2x\sqrt{i\pi}} \text{ ; large } x \quad (A.4)$$

and D becomes

$$D_e(s,\alpha,\psi) \sim -\frac{\exp(-i\pi/4)}{2\sqrt{2\pi k}}\left[\csc\left(\frac{\psi-\alpha}{2}\right)+\sec\left(\frac{\psi+\alpha}{2}\right)\right] \quad (A.5)$$

$$D_h(s,\alpha,\psi) \sim -\frac{\exp(-i\pi/4)}{2\sqrt{2\pi k}}\left[\csc\left(\frac{\psi-\alpha}{2}\right)-\sec\left(\frac{\psi+\alpha}{2}\right)\right]$$

- Electric Polarization

From equations (A.2)-(A.5) the field at point P, in the far zone, is given by

$$u^d(P) = -\frac{\exp(-ik[-B\sin\alpha + s_1] - i\pi/4)}{2(2\pi ks_1)^{1/2}} \times \quad (A.6)$$

$$\left[\csc\left(\frac{\psi_1 - \alpha}{2}\right) + \sec\left(\frac{\psi_1 + \alpha}{2}\right)\right]$$

$$-\frac{\exp(-ik[B\sin\alpha + s_2] - i\pi/4)}{2(2\pi ks_2)^{1/2}} \times$$

$$\left[\csc\left(\frac{-\psi_2 + \alpha}{2}\right) + \sec\left(\frac{-\psi_2 - \alpha}{2}\right)\right]$$

Also $s_1 = r + B\sin\theta$, $s_2 = r - B\sin\theta$, $\psi_1 = \psi_2 = \theta$. Equation (A.6) becomes

$$u^d(r,\theta) = \left(\frac{k}{2\pi r}\right)^{1/2}\exp(-ikr - i\pi/4)f_s(\theta) \quad (A.7)$$

where $f_s(\theta)$ is the diffraction pattern of the slit due to *singly* diffracted rays. From equation (A.6)

$$f_s(\theta) = i\frac{\sin[kB(\sin\theta - \sin\alpha)]}{k\sin\left(\frac{\theta-\alpha}{2}\right)} - \frac{\cos[kB(\sin\theta - \sin\alpha)]}{k\cos\left(\frac{\theta+\alpha}{2}\right)} \quad (A.8)$$

According to the forward scattering theorem, the transmission cross section of the slit (per unit length) is equal to the imaginary part of the diffraction pattern $f_s(\theta)$ in the forward direction, $\theta = \alpha$. The transmission coefficient, t_1, defined as the ratio of the transmission cross section to the geometric cross section normal to the direction of propagation, is

$$t_{1,e}(\alpha) = 1 \quad (A.9)$$

This result is identical to that obtained from geometrical optics.

If doubly diffracted rays are also considered, the diffraction pattern becomes

$$f_m(\theta) = f_s(\theta) + f_d(\theta) \quad (A.10)$$

where f_d is given by

$$f_d(\theta) = -\frac{1}{k(\pi kB)^{1/2}}\left[\frac{\exp(-i[k2B(1-\sin\theta)+\pi/4])}{1-\sin\theta} + \frac{\exp(-i[k2B(1+\sin\theta)+\pi/4])}{1+\sin\theta}\right] \quad (A.11)$$

Applying the forward scattering theorem yields for the transmission coefficient t_1

$$t_{1,e} = 1 - \frac{1}{2\pi^{1/2}(kB)^{3/2}\cos\alpha}\left[\frac{\cos[2kB(1-\sin\alpha)-\pi/4]}{1-\sin\alpha} + \frac{\cos[2kB(1+\sin\alpha)-\pi/4]}{1+\sin\alpha}\right] \quad (A.12)$$

From equations (A.9) and (A.12) it follows that the correction to the geometrical optics result is of the order $(kB)^{-3/2}$. In the special case of normal incidence, $\alpha = 0$, equation (A.12) becomes

$$t_{1,e}(\alpha=0) = 1 - \frac{\cos(2kB - \pi/4)}{\pi^{1/2}(kB)^{3/2}} \quad (A.13)$$

- Magnetic Polarization

When singly diffracted rays are considered, the transmission cross section for a slit is equal to the its projected geometric cross section. When doubly diffracted rays are also included the transmission cross section is given by (Keller, 1962),

$$t_{1,m}(\alpha) = 1 - \frac{1}{32\pi^{1/2}(kB^{5/2})\cos\alpha} \times \quad \text{(A.14)}$$

$$\left[\frac{\sin^2(\pi/2+\alpha/2)}{\cos^4(\pi/4+\alpha/2)}\sin[2kB(1-\sin\alpha)-\pi/4] + \frac{\sin^2(\pi/4-\alpha/2)}{\cos^4(\pi/4-\alpha/2)}\sin[2kB(1+\sin\alpha)-\pi/4]\right]$$

The correction to geometrical optics is of the order $(kB)^{-5/2}$. For normal incidence, $\alpha = 0$, and equation A.14 becomes

$$t_{1,m}(\alpha{=}0) = 1 - \frac{\sin(2kB-\pi/4)}{8\pi^{1/2}(kB)^{5/2}} \quad \text{(A.15)}$$

APPENDIX B

Diffracted Field for Two-Shield Model

This appendix presents analytical expressions for calculating $S_{j,k}$. The transmission coefficient, T_{II}, can then be obtained from equations (11)-(13).

Consider a plane monochromatic wave incident upon the two-shield structure shown in Fig. 4b, consisting of two parallel thin-conducting shields with slits. Only incident waves normal to the edges are considered.

Let the screen S_2 to be at $x = 0$, and the center of the slit, on S_2, to be at the origin of the x,y coordinate system. Let also the incident field u^i ($u^i = E_z$ for electric polarization, and $u^i = H_z$ for magnetic polarization) be given by

$$u^i = \exp\left\{i[\omega t - k(x\cos\alpha + y\sin\alpha)]\right\} \quad \text{(B.1)}$$

where $K = 2\pi/\lambda$ is the wavenumber, $\omega = 2\pi\nu$ angular frequency, ν wave frequency, t time, and α angle between the direction of propagation of the incident and the x-axis.

Doubly diffracted fields arriving at point $P(r,\theta)$ are shown in Fig 5c. Higher order diffracted rays are not included because of their complexity. Let $u_{j,k}$ by the z-component of the doubly diffracted field associated with edges M_j and N_k. To determine $u_{j,k}$ repeated use is made of expressions similar to those used in equation (A.2). The calculations are simplified by introducing the function $F(s,\alpha,\phi)$ defined by

$$F(s,\alpha,\psi) = D(s,\alpha,\psi)\frac{\exp(-iks)}{s^{1/2}} \quad \text{(B.2)}$$

where the diffraction coefficient D is given by equation (A.3). For example, the field at N_2, along the singly diffracted ray emerging from the edge M_2 in direction to N_2, is $u^i(M_2)$ $F(s_{2,2}, -\alpha, \psi_{2,2})$, where $s_{2,2}$ is the path-length of the diffracted ray and $\psi_{2,2}$ the the angle (absolute value) between the diffracted ray and the x-axis. Then the fields $u_{j,k}$ on each doubly diffracted ray arriving at P are given by

$$u_{2,2} = u^i(M_2)\,F(s_{2,2},-\alpha,\psi_{2,2})\,F(s_2,\psi_{2,2},-\psi_2)$$

$$u_{1,2} = u^i(M_1)\,F(s_{1,2},\alpha,-\psi_{1,2})\,F(s_2,\psi_{1,2},-\psi_2)$$

$$u_{-1,2} = u^i(M_{-1})\,F(s_{-1,2},-\alpha,-\psi_{-1,2})\,F(s_2,-\psi_{-1,2},-\psi_2)$$

$$u_{-2,2} = u^i(M_{-2})\,F(s_{-2,2},\alpha,\psi_{-2,2})\,F(s_2,-\psi_{-2,2},-\psi_2)$$

(B.3)

$$u_{2,1} = u^i(M_2)\,F(s_{2,1},-\alpha,\psi_{2,1})\,F(s_1,-\psi_{2,1},\psi_1)$$

$$u_{1,1} = u^i(M_1)\,F(s_{1,1},\alpha,-\psi_{1,1})\,F(s_1,-\psi_{1,1},\psi_1)$$

$$u_{-1,1} = u^i(M_{-1})\,F(s_{-1,1},-\alpha,-\psi_{-1,1})\,F(s_1,\psi_{-1,1},\psi_1)$$

$$u_{-2,1} = u^i(M_{-2})\,F(s_{-2,1},\alpha,\psi_{-2,1})\,F(s_1,\psi_{-2,1},\psi_1)$$

where $\psi_{j,k}$ is the angle (absolute value) between the diffracted ray leaving M_j and arriving at N_k and the x-axis, and ψ_k the angle between the diffracted ray arriving at the observation point from N_k and the x-axis.

In the far zone, $ks_1 \gg 1$ and $ks_2 \gg 1$, and the asymptotic expression for F in equation (B.3), with s_1 and s_2 as path argument, applies. Also $s_1 = r + B\sin\theta$, $s_2 = r - B\sin\theta$, $\psi_1 = \theta$, and $\psi_2 = \theta$.

The Poynting vector $\mathbf{S}_{j,k}(r,\theta)$, for an incident wave of intensity I^i, is given by

$$\mathbf{S}_{j,k}(r,\theta) = I^i\,|u_{j,k}|^2\mathbf{e}_r \quad \text{(B.4)}$$

1989 National Heat Transfer Conference
HTD-Vol. 106, Heat Transfer Phenomena in Radiation, Combustion and Fires

COBRA/TF ANALYSIS OF SVEA SPRAY COOLING EXPERIMENTS

J. S. Chiou and L. E. Hochreiter
Westinghouse Electric Corporation
Pittsburgh, Pennsylvania

ABSTRACT

This paper presents the results of several COBRA/TF (WCOBRA/TRAC) calculations which modeled the SVEA spray cooling BWR rod bundle experiments. A subchannel radiation model which was originally developed for COBRA/TF was modified such that more accurate rod-to-rod, radiation could be calculated. The analysis indicated that the majority of the heat transfer in the spray tests was by rod-to-rod and rod-to-surface radiation, and that the spray flow provided some additional dispersed flow film boiling heat transfer. The spray tests were highly non-equilibrium with vapor superheats in the bundle center that were only 150°C below that of the heater rod surface. The agreements between the COBRA/TF calculations and the test data were excellent for the uniform spray tests.

The non-uniform spray tests were modeled with coarse noding to see the possible crossflow effects between minibundles due to the non-uniformity of the spray. It was observed that the crossflow through the minibundle gap areas was significant at the top of the bundle due to the more rapid quenching of adjacent channels. The detailed non-uniform spray prediction yielded higher temperatures than the data. It is believed that this is due to the early calculated dryout of the water-cross since the crossflow effects were not modeled. The non uniform SVEA calculations do show that the minibundles are both hydraulically coupled as well as thermally coupled which reduces the sensitivity to the individual bundle spray flow.

NOMENCLATURE

a	=	absorptance
A	=	area
B	=	radiosity
C_{ij}	=	geometry configuration factor defined in Equation (7)
d	=	hydraulic diameter
d_d	=	liquid droplet
D_i	=	the radiant energy arriving at surface i
E	=	black body radiant energy
F_{ij}	=	view factor from surface i to surface j
h	=	convective heat transfer coefficient
k_v	=	vapor conductivity
L	=	axial characteristic length
N_ℓ'''	=	droplet number density
Q	=	heat flux
T	=	temperature
ε	=	emissivity
ρ	=	density
μ	=	dynamic viscosity
β	=	expansion coefficient
σ	=	Stefan-Boltzmann constant
σ_{ij}	=	scattering coefficient from surface i to surface j
τ	=	transmittance
Ω	=	total radiant energy from fluid

INTRODUCTION

Core heat transfer during the post CHF and reflood phases of a Loss-of-Coolant Accident (LOCA) transient is characterized by the single phase vapor or dispersed phase film boiling heat transfer regimes in the unflooded portions of the core. Heat transfer to the vapor and droplets is poor, resulting in high temperatures at the surface of the fuel rods. As the cladding temperature increases, heat transfer due to thermal radiation accounts for an increasing portion of the total heat flux from the fuel rods particularly for Boiling Water Reactors (BWR) since cold surfaces such as channel walls are available as additional radiation heat sinks.

Detailed rod-to-rod radiation computer analysis program have been written and used to calculate the radiation heat flux in rod bundles (Evans, 1970). These models solve a planar problem in which there is no axial dependency, and the fluid conditions are input such as the sink temperature and the convection or film boiling heat transfer coefficient. There is no thermal coupling between the convective wall heat flux and the radiation heat flux in the calculation. None of these models can calculate the induced flow patterns within a steam filled bundle due to temperature differences between the high temperature fuel rods and the cool bundle channel wall, as well as the feedback effect of the convection or film boiling heat transfer on the radiation heat transfer. Also many of these programs treat the entire surface of the rod as one temperature and do not permit a gradient across the fuel rod (Slifer and Hench, 1971). As a result, the radiation heat transfer will be over-estimated in the central region of the bundle and under-estimated at the edge of the bundle facing the colder wall.

State-of-the-art best estimate computational models, such as BWR TRAC (Taylor et al., 1984), have coupled the radiation and convective heat flux calculations together in a three-dimensional calculational scheme. In this model a lumped-system approximation of the net radiation heat exchange is given. The model uses an enclosure of discrete, diffuse, gray surfaces and calculates the incident and reflective radiation components. Each fuel rod is one surface at a calculated surface temperature and the channel wall is a given separate calculated wall temperature. The radiation fluxes are coupled to the total wall heat flux and to a three-dimensional fluid calculation such that there is thermal feedback between the wall temperatures, total wall heat flux, and any thermal induced flow in the system. Normally the smallest computational cell in BWR TRAC is several BWR fuel assemblies.

The COBRA/TF (Paik et al., 1985) (WCOBRA/TRAC) subchannel radiation model combines elements of both the above rod-to-rod radiation calculational schemes along with the complete thermal coupling to the three-dimensional fluid calculation within the BWR rod bundle. In the COBRA/TF case the smallest computational cell can be a fraction of a rod bundle subchannel. The COBRA/TF subchannel radiation model permits the use of individual surfaces on each fuel rod such that three-dimensional azmuthal temperature profiles within a single fuel rod can be modeled. Also COBRA/TF solves the fuel rod bundle three-dimensional flow field simultaneously with the solution of the rod surface temperatures. In this fashion, the radiation and convective fluxes at a surface are not independent but are thermally coupled to produce a more accurate fuel rod temperature response which results in a more accurate total rod bundle heat transfer prediction.

As part of a joint ASEA ATOM, Swedish State Power Board, Swedish State Inspectrate, and Westinghouse Electric Corporation program; Westinghouse was directed to analyze a series of spray heat transfer experiments, (SVEA), using the COBRA/TF vessel model of WCOBRA/TRAC. The objective of the analysis was to help explain the results of the SVEA experiments and the spray cooling performance of the SVEA bundle in terms of the basic thermal-hydraulic mechanisms.

WCOBRA/TRAC MODEL

The COBRA/TF computer code provides a two-fluid, three-field representation of two-phase flow. Each field is treated in three dimensions and is compressible. Continuous vapor, continuous liquid and entrained liquid drop are the three fields. The conservation equations for each of the three fields and for heat transfer, from and within the solid structures in contact with the fluid, are solved using a semi-implicit, finite-difference, numerical technique on an Eulerian mesh. The COBRA/TF vessel model features extremely flexible noding for both the hydrodynamic mesh and the heat transfer solution. This flexibility provides the capability to model the wide variety of geometries encountered in vertical components of nuclear reactor primary systems. The flexible noding scheme allows representation of single hot subchannel with its unique hardware and power characteristics while permitting coarser noding for the remainder of the assembly and core. In order to analyze in detail the radiation heat transfer in a boiling water reactor fuel assembly, a subchannel radiation model was added to COBRA/TF.

The following major assumptions have been made in developing the subchannel radiation model:

1) All solid structures in the enclosure are divided into finite isothermal surfaces.

2) Radiation in the axial direction is neglected.

3) All surfaces are treated as gray bodies.

4) All surfaces are assumed to be infinite in length so that Hottel's cross string method (Siegel and Howell, 1988) can be used to generate geometric configuration factors.

5) Subchannels and their corresponding surfaces can be lumped together to give an averaged response if desired.

The subchannel radiation model consists of an enclosure that includes a fluid channel and the four 1/4 rods that face the fluid channel as shown in Figure 1. The radiant heat flux between these four quarter rods, and between the rods and the fluid contained within the channel are calculated in detail. The fluid within the channel is assumed homogeneous with respect to temperature, density and droplet distribution.

Each radiation channel is treated as an enclosure with M surfaces, where M is equal to the total number of isothermal solid surfaces plus the number of spaces (gaps) between solid surfaces that permit radiant heat exchanges between the radiation channel and fluid and structures that surround the radiation channel. These M surfaces completely bound the radiation channel, as shown in Figure 1. The surfaces labeled $\overline{T}_1$, $\overline{T}_2$, $\overline{T}_3$ and $\overline{T}_4$ represent the four one-quarter rod surfaces that are located at the four corners of the radiation channel. Q_5, Q_6, Q_7 and Q_8 represent the radiant heat flux between the radiation channel and surrounding fluid channels. T_5, T_6, T_7 and T_8 are the "average" temperatures of the structures within each of the surrounding fluid channels. These average temperatures are obtained by taking the area and view factor weighted average of the solid surface temperature and the gap temperature to the 4th power.

That is;

$$\overline{T}_j^{\,4} = \frac{\sum_{i=1}^{7} A_{ji}\, T_{ji}^{4}\, F_{ji}}{\sum_{i=1}^{7} A_{ji}\, F_{ji}}\,, \quad j = 5, 6, 7, 8 \tag{1}$$

Where j represents one of the channels which surrounds the radiation channel, and the summation i is over the seven surfaces (4 solid surfaces and 3 gaps) in each surrounding channel and the gap temperature is defined as the weighted average of the solid surface temperatures outside that particular gap. For example, the temperature of gap 56 is expressed by:

$$\overline{\overline{T}}_{56}^{\,4} = \frac{\sum_{i=1}^{4} A_{56i}\, T_{56i}^{4}\, F_{56i}}{\sum_{i=1}^{4} A_{56i}\, F_{56i}} \tag{2}$$

In this fashion, the gap temperature reflects the adjacent sub channel surface temperatures and permits one subchannel to "see" further into the rod array, by using a gap temperature which reflect the rod temperatures in that subchannel.

The average emissivity for the surrounding channel is calculated in a similar way except that a unity value is used for gaps in the average.

The heat flux between the radiation channel and the surrounding fluid channels are calculated using temperatures and emissivities averaged in this way for both channels. The radiant heat flux through the gap separating the two channels is calculated by considering the subchannels on either side of the gap as cavities. The heat flux across the gap is given by:

$$\frac{Q_j}{A_{gap\ j}} = \frac{\sigma\ \bar{T}_j^4 - \sigma\ \bar{T}_o^4}{1 + (\frac{1}{\bar{\epsilon}_j} - 1)\frac{A_{gap\ j}}{A_{tot,j}} + (\frac{1}{\bar{\epsilon}_o} - 1)\frac{A_{gap\ j}}{A_{tot,o}}} \tag{3}$$

where:

$$A_{tot,j} = \sum_{i=1}^{7} A_{ji} \quad , \quad j = 5,\ 6,\ 7,\ 8$$

$$A_{tot,o} = \sum_{i=1}^{8} A_{oi} \quad , \quad i \neq j \tag{4}$$

The view factors, beam lengths and surface areas for six radiation channel geometry types have been coded to simplify the generation of input data for the radiation model (Chiou et al., 1988).

The detailed derivation of the system of radiation equations on a subchannel basis is given in Chiou et al. (1988). The final radiosity equation for each of the surfaces (and gaps between the rods) within the radiation channel forms a set of M equations. These equations are solved for the radiosity of the surfaces. The old time surface, vapor, and droplet temperatures are used from the previous time step. The vapor and droplet (if any) flows are assumed to be optically thin mediums for radiation penetration and the typical droplet size is assumed much larger than the radiation wave length such that the radiation can be absorbed and emitted as in a scattering regime. The emissivities for vapor and droplet can then be expressed by:

$$E_V = 1 - \exp\ (-A_V L_m) \tag{5}$$

$$E_L = 1 - \exp\ (-A_L L_m) \tag{6}$$

where the radiation mean beam length (L_M) was set to be 0.9 D_R, the vapor absorption coefficient is obtained from Abu-Romia and Tien's data (1967), and the droplet absorption coefficient by Hottel and Sarofim (1967) is adopted as:

$$A_v = P[(\frac{1000}{T})^2 - 0.03\ (\frac{1000}{T})^4] \tag{7}$$

$$A_\ell = 0.74\ \frac{\pi}{4}\ d_d^2\ N_\ell''' \tag{8}$$

where N_ℓ''' is the droplet number density.

Once the radiosity for each surface has been obtained, the solid surface heat flux is obtained from:

$$\frac{Q_i}{A_i} = \frac{\varepsilon_i}{1-\varepsilon_i}\ [E_i - B_i] \tag{9}$$

Heat that is not transmitted or scattered to another surface of the enclosure is absorbed by the coolant. Therefore, the radiant heat flux to the vapor and liquid droplets is given by:

$$Q_g = \frac{a_g}{a} \sum_{i=1}^{M} A_i\ (B_i - E_g)\ \left(1 - \sum_{j=1}^{M} F_{ij}\ \tau_{ij} - \sum_{j=1}^{M} F_{ij}\ \sigma_{ij} \right) \tag{10}$$

$$Q_d = \frac{a_d}{a} \sum_{i=1}^{M} A_i\ (B_i - E_d)\ \left(1 - \sum_{j=1}^{M} F_{ij}\ \tau_{ij} - \sum_{j=1}^{M} F_{ij}\ \sigma_{ij} \right) \tag{11}$$

These heat fluxes to the solid surfaces and to the fluid are added as explicit source terms to the respective energy equations such that the correct amount of evaporation and vapor superheat are calculated.

SVEA UNIFORM SPRAY TEST ANALYSIS

Figure 2a shows the test section of SVEA spray cooling experiments (Eklund et al., 1987) performed in ASEA-ATOM laboratories. The 8 x 8 bundle is separated into four minibundles by the water-cross, which is the unique design of SVEA/QUAD+ fuel assemblies as shown in Figure 2b. The heater rods are powered by electric heating elements with a step cosine power shape. The axial power factor and the heater rod geometries and material are shown in Figure 3. For a loss of cooling accident (LOCA) experiment, the subcooled water will be injected into water-cross as well as into the regular fuel rod channel and the bypass channel. The channel walls are quenched shortly after the initiation of the cooling spray. With the existence of water-cross, the view factors from the fuel rods to the cold boundaries increase and enhance the radiation heat transfer from the fuel rods to the boundary walls.

SVEA spray Test No. 44 was simulated using the COBRA/TF code with the subchannel radiation model as given in Chiou et al. (1988). The test conditions are given in Table 1. The test results indicated that the boundary walls of the bypass channel and the water-cross channel were wetted immediately after the spray began, and no dryout was found during the entire test period. From the heat transfer point of view, either the bypass wall or the water-cross wall will provide the same T_{sat} boundary to the inside of the minibundle, therefore, the minibundle is assumed symmetric to both x and y axes, and only 1/8 of a minibundle is modeled in this analysis. The channel connection and the axial description of the COBRA/TF model are shown in Figure 4 and the detailed subchannel model of the 1/8 rod bundle is depicted in Figure 5.

TABLE 1
SVEA TEST CONDITIONS

	SVEA Test No. 44	SVEA Test No. 37
Pressure	2 Bar	2 Bar
Power	400 kW, ANS 79 + 2σ decay	400 kW, ANS 79 + 2σ decay
Spray	40 g/s per minibundle uniform	60,45,45,10 g/s
Venting Type	Combine	Combine
Flashing	Yes	Yes
Bypass Spray	130 g/s	130 g/s
Water-cross Spray	10 g/s	10 g/s

Two calculations were performed for this simulation. Both calculations used 0.6 for heater rod emissivity, but one calculation had 0.6 for wall (dry) emissivity and the other used 0.96 for wall (wet) emissivity. The wet wall emissivity value simulates the liquid film on the wall from the spray flow.

The comparison of the calculated and measured rod temperatures for both cases at two elevations are presented in Figure 6 for central rods, Figure 7 for side rods, and Figure 8 for corner rods. Level 4 is at the bundle midplane at the peak power elevation while level 7 is near the top of the bundle at 3 meters in elevation from the bottom of the heated length. The calculation with $\varepsilon_w = 0.96$ has better agreement at the mid-elevation especially at the side and corner rods, see Figures 7 and 8. However, the difference between both calculations is small and both have a fairly good overall agreement with data.

The calculated vapor temperatures, surface temperatures, the ratios of radiation heat flux, and the mean heat transfer coefficient for these elevations are shown in Figure 9 for $\varepsilon_w = 0.96$ case. The mean saturated heat transfer coefficient is defined as

$$\bar{h}_s = \frac{\bar{q}''_t - \bar{q}''_{rod}}{\bar{T}_r - T_{sat}} \qquad (12)$$

where $\bar{q}''_t$, $\bar{q}''_{rod}$, and $\bar{T}_r$ are the rod average total heat flux, radiation heat flux, and rod temperature weighted by surface fraction of a rod. The calculated h_s is from 1.58 to 4.94 W/M^2-°C (.28 to .87 Btu/Hr-ft^2-°F) for the $\varepsilon_w = 0.96$ case and from 1.93 to 8.8 W/M^2-°C (0.34 to 1.55 Btu/Hr-ft^2-°F) for the $\varepsilon_w = 0.6$ case. The much higher convective film boiling heat transfer coefficients are calculated at level 7, since the convective heat flux is calculated by weighting the transition boiling heat flux with the film boiling heat flux. This is because the effect of top down quench front begins to affect the upper portion (about 10 ft) of the heater rod at t = 300 seconds. The local volumetric fractions for each field at each elevation indicate that the liquid film will exist in the boundary channels (channels 7, 8, and 9) for $\varepsilon_w = 0.6$, but not the case of $\varepsilon_w = 0.96$. This is probable because the liquid film will be evaporated as a result of the higher radiation heat flux from rod to the wall when ε_w is changed from 0.6 to 0.96. For the $\varepsilon_w = 0.96$ case, although the dryout phenomena was shown in the internal boundary channels, the wall boundary remains close to T_{sat} because the liquid film exists in the bypass channel, on the other side of the wall.

As the quench front advances down from the top, the fraction of the heat flux due to radiation decreases and the convective film boiling fraction correspondingly increases. The vapor superheat temperature also decreases as seen in Figure 9. At the bundle midplane, level 4, the heat transfer is always in a dispersed flow with no liquid film on the walls and a small amount of entrained flow. The calculations show a counterflow limiting effect at the top of the bundle in which liquid, as a film, runs down and is entrained at the quench front into droplets. The droplets can fall in some subchannels or be swept upward in other subchannels. The flow remains highly dispersed with significant non equilibrium present. The steam/water mixture in a given subchannel provides a limited heat sink and the majority (greater than 90%) of the total heat transfer is rod-to-rod, or rod-to-surface radiation.

SVEA NON UNIFORM SPRAY TEST ANALYSIS

In addition to a uniform SVEA spray test, a non-uniform spray test was modeled with COBRA/TF. The objective of this calculation was to assess the non-uniform spray effects between minibundles on the resulting heater rod temperatures and heat transfer. Two different models were used in the COBRA/TF analysis so assess the non-uniform spray effects as well as to determine the relative proportions of convective (dispersed flow film boiling) and radiation heat fluxes.

A coarse model, COBRA/TF model was used to examine the minibundle-to-minibundle flow while a detailed subchannel model of the starved minibundle was used to obtain local heat transfer information. The first model used is referred to as the coarse noded model. In this model one-half SVEA bundle symmetry is modeled and each minibundle is modeled as a single channel along with the water-cross and the bypass region. The cross-section of the SVEA bundle which is modeled is shown in Figure 10 with the COBRA/TF model of the SVEA bundle. The three minibundles are channels 4, 5, and 6, and these are interconnected with gaps at each elevation that will allow flow to move from one minibundle to another depending on the pressure differences. The gap between the minibundles at the water-cross and channel boundary is modeled as a 1 mm (0.040 inches) longitudial gap with a loss coefficient of 5.0 as given from Idel'chik (1966). The spray flows into the minibundles are non-uniform with channel 4 receiving 60 gms/sec, channel 5 receiving 45 gms/sec and channel 6 receiving only 10 gms/sec such that channel 6 is the starved minibundle. All channels have a CCFL region where the heater rod extension and the fuel assembly upper tie plate is modeled (channels 9, 10, 11) and is connected to a common upper plenum (similar to Figure 4). The heater rods are modeled in each channel as a single rod to represent the averaged behavior of the 16 rods in a minibundle.

One of the purposes of the coarse noded SVEA calculation was to investigate the magnitude of the leakage or crossflow between the minibundles when a non-uniform spray condition existed. To determine the effect of the crossflows (liquid, vapor, and entrained liquid) through the gaps on the axial flows, two types of comparisons were made. First the total crossflow (liquid, vapor, and entrained flow) was normalized on the total axial flow at the same elevations. When calculating the totals of the axial flow, the absolute values of the liquid and entrained flow were used since in countercurrent flow the liquid and vapor have opposite signs. As time progresses, the crossflow into the top of bundle 6 (the starved minibundle) increases, with a slight increase in the flow out of bundle 6 at lower elevations. At 300 seconds, there is a significant crossflow into the starved channel while more flow is pushed out at lower elevations as shown in Figure 11. This trend also continues at later times.

Since the spray flows are non-uniform, the heat transfer rates, and hence the steam generation rates in the different minibundles, are also non-uniform. With the higher spray flows in the other two minibundles, a slightly larger fraction of the total surface heat flux is convective film boiling such that more vapor is generated.

The largest source of vapor generation is the top down quenching effect of the heater rods due to the falling film moving down the rods from the unheated rod region. At the top of the channel, the vapor flow significantly increases due to the heat release to the fluid from the heater rod quenching. Since the flows are non-uniform, the quench front progression in the starved minibundle tends to lag the other two bundles. The flow in a given minibundle will crossflow into adjacent bundles depending on the heat release and corresponding vapor generation at the top of the bundle due to the heater rod quenching. The generated steam flow plus the axial upflow of the steam, in a given minibundle, tends to equilibrate with the adjacent bundles. However, the vapor velocities at the top of the bundle and tie plate region are still higher for the starved minibundle.

The entrained liquid flow that crossflows into the starved minibundle is usually larger than the 10 gms/sec which was sprayed into the starved channel. This liquid falling down through the bundle helps to cool the rods. However, since the main convection flow is steam cooling, the steam flow out of the bundle will act to decrease the cooling below the top quench position.

The coarse channel analysis with COBRA/TF does show that crossflow exists in the minibundles when non-uniform spray conditions exist and promotes improved cooling of the starved bundle and in effect, helps to compensate for the reduced flow in that minibundle. Since it was not clear from the coarse-noded calculation how much of a benefit the crossflow into and out of the minibundle would be a COBRA/TF model was set up simulating only the starved minibundle with no crossflow. This should be a conservative analysis since the minibundles are hydraulically and thermally coupled.

The starved minibundle analysis with COBRA/TF used a one-half minibundle as shown in Figure 12. The COBRA/TF model for this bundle is also depicted in this figure and is similar to the COBRA/TF model used for the uniform spray test except that the water-cross boundary had to be modeled separately from the channel since the water-cross could dry out during the non uniform spray experiments and the calculation.

Several COBRA/TF calculations were made to best represent this experiment. In modeling this type of subchannel experiment, the COBRA/TF code must choose whether a subchannel is in a hot wall or wetted wall regime to properly model the liquid film flow, and liquid deentrainment. The side subchannels in the SVEA bundle present a challenge to the code since there is both heated and wetted perimeters in the same fluid subchannel. The code then assumes that a hot wall regime exists and the resulting void fraction in the side channels increases. It was felt that this higher void fraction was inappropriate for channels which had a wetted wall, so the code was forced to select the wet wall regime for the side channels. This results in generally lower void fraction in the channels adjacent to the bypass channel.

A second problem that was encountered was the dryout of the water-cross. The COBRA/TF model assumes symmetry along the center of the water gap separating the two minibundles. Initial COBRA/TF calculations indicated a dryout of the water-cross, at about the same time as the data, but then the calculated water-cross wall temperature quickly increased to the values approaching the heater rod temperatures. In the experiment, while one side of the water-cross wall would dry out in the starved minibundle, the opposite wall, which had more flow, remained at T_{sat} and acts as a heat sink for the water-cross wall which had a dryout. This effect was modeled by including a radiation term from the dryout wall to the opposite wall which was assumed to be at T_{sat}. This effect represents the thermal coupling of the minibundles in the SVEA design.

Comparison of the predicted heater rod temperatures to the transient heater rod data are shown in Figures 13 and 16 for selected rods in the COBRA/TF model. Emissivity values of 0.6 for the heater rods and 0.96 for the cross and channel were used. The COBRA/TF rod temperatures were calculated by area weighting the individual heater rod surface temperatures to obtain an average rod temperature. In general, the agreement is quite reasonable until later in time at which the predicted temperatures exceed the measured values. At this time, the water-cross has been calculated to dryout and the radiation to the other side of the water-cross, across the water-cross gap, also results in drying out that surface. In the test, the other side of the water-cross did not dryout. Also at 361 seconds after power on, the bypass channel is also calculated to dryout. Figures 17 and 18 show cross-section maps of the COBRA/TF SVEA model and show the individual surface temperatures, subchannel vapor temperature, and the fraction of the total heat flux, which is radiation, at 300 seconds. As these figures indicate, the flow in the bundle is highly nonequilibrium, with vapor superheats approximately that of rod surfaces. The radiation component of the total heat flux is still over 90% for most surfaces. The calculated void fractions in the bundle are quite high, and in some cases, cooling is by steam only. The spray flow penetrates down subchannels 8, 9, 10 into the bundle and countercurrent flow exists in these subchannels. Cocurrent upflow exists in nearly all the other subchannels. The downward liquid film flow exists at the top of all the channels, until the top quench front on the heater rods is reached, then this flow becomes entrained. The calculations indicate that at the same elevations, those subchannels with countercurrent flow (with the entrained spray water) have a lower vapor temperature than that in the cocurrent upflow channels.

As this calculation progressed in time, beyond 300 seconds, the water-cross wall, across from the starved minibundle water-cross wall, dried out, as well as, the bypass channel wall. This resulted in calculated rod temperatures which exceeded the data.

It is believed that this may be due, in part, to the higher wall emissivity of 0.96, which is more typical of wall with a liquid film. The higher emissivity increases the radiation heat flux to the wall and helps dry it out. As the wall drys out, the emissivity should decrease to a value of 0.6, however, in these calculations the wall emissivity was kept constant, thus the walls heated up more quickly, and had less of a chance to re-rewet as time progressed. The test data indicated that dry walls would eventually rewet, the code calculations did not show this. The higher wall temperatures propagated into the rod bundle increasing the rod surface temperatures such that the code predictions and the data tended to diverge.

The COBRA/TF model for these tests was felt to be conservative since it only partially included the thermal coupling with the other subchannels. What these comparisons indicate is that there are compensating factors, such as the radiation heat transfer across the water-cross and gap crossflows, which help reduce the effects of the core spray flow non-uniformities.

CONCLUSION

The COBRA/TF code was used to analyze the SVEA spray experiments, and to gain insight into the heat transfer mechanisms responsible for spray cooling heat transfer in SVEA bundles. The flow regime in these experiments was highly dispersed, cocurrent, and countercurrent non-equilibrium two-phase flow. The void fractions were near unity and the majority of the heat transfer (greater than 90%) was due to rod-to-rod, and rod-to-surface radiation. This supports the previous conclusion that the primary heat transfer benefit of the SVEA design is the increased surface radiation due to the water-cross. The effects of the spray maldistribution were second order since the heat transfer is radiation dominated. However, it is apparent that there is flow communication between bundles that aids a starved minibundle, as well as thermal communication via radiation across the water-cross gap which also aids the starved bundle. Convective heat transfer coefficients, based on the saturated fluid temperatures, of 5.7 to 17. W/M^2-HR-°C (1-to-3 Btu/hr-ft^2 °F), were calculated using the COBRA/TF calculated convective film boiling heat flux, and are consistant with those obtained from experimental data.

ACKNOWLEDGEMENTS

SVEA spray cooling test was performed by ASEA-ATOM as a joint project cosponsored by the Swedish Nuclear Power Inspectorate (SKi), the Swedish State Power Board (SSPB), and ASEA-ATOM (AA). The authors would like to express their gratitude to the above parties for their permission to use the test data. Special thanks are also extended to Mr. Hjalmar Wijkstrom of AA for his pertinent technical suggestions during the course of this study and Ms. Denise Henderson for her help in assembling this paper.

REFERENCES

Abu-Romia, M. M. and Tien, C. L., "Appropriate Mean Absorption Coefficients for Infrared Radiation of Gases", J. of Heat Transfer (ASME), pp. 321-327, 1967.

Chiou, J. S., Hochreiter, L. E., and Thurgood, M. J., "Radiation Heat Transfer in BWR Rod Bundles", Presented at 25th National Heat Transfer Conference, Houston, Texas, August, 1988.

Eklund, R., Backwall, M-O, and Wijkstrom, H., "SVEA Spray Cooling Test", RP 87-04/SVEA 87-09, March, 1987.

Evans, D. R., "MOXY Rod-to-Rod Radiation Code", Quarterly Technical Report, Nuclear Safety Program Division, Idaho Nuclear Corporation, INEL-1320, August 1970.

Hottel, H. C. and Sarofim, A. F., Radiative Transfer, McGraw Hill Book, 1967.

Idel'chik, I. E., Handbook of Hydraulic Resistance, AEC-tr-6630, Translated from Russian by A Barouch, 1966, Available from the U.S. Department of Commerce.

Paik, C. Y., Hochreiter, L. E., Kelly, J. M., and Kohrt, R. J., "Analysis of FLECHT-SEASET 163-Rod Blocked Bundle Data Using COBRA/TF", NRC/EPRI/Westinghouse Report No. 15, October 1985.

Siegel, R. and Howell, J. R., Thermal Radiation Heat Transfer, McGraw-Hill Book, 1972.

Slifer, B. C. and Hench, J. E., "Loss-of-Coolant Accident and Emergency Core Cooling Models for General Electric Boiling Water", NEDO-10329 (April 1971).

Taylor, D. D. et al., "TRAC-BD1/MOD 1: An advanced Best Estimate Computer Program for Boiling Water Reactor Transient Analysis", Vol. 1: Model Description, NUREG/CR-3633, April 1984.

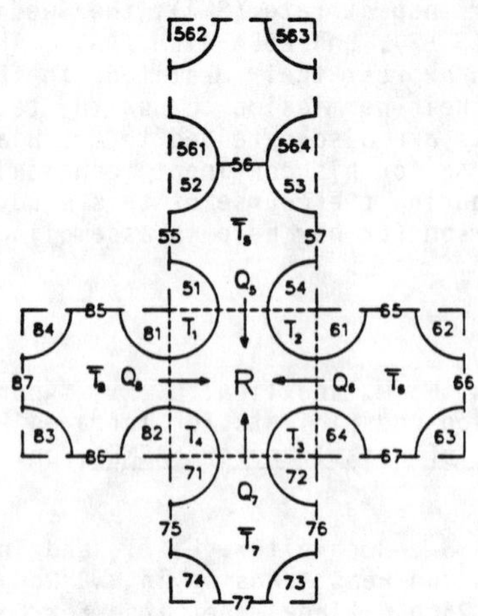

SUBCHANNEL 0 IS THE FLUID CHANNEL

Figure 1. Typical Subchannel Enclosure

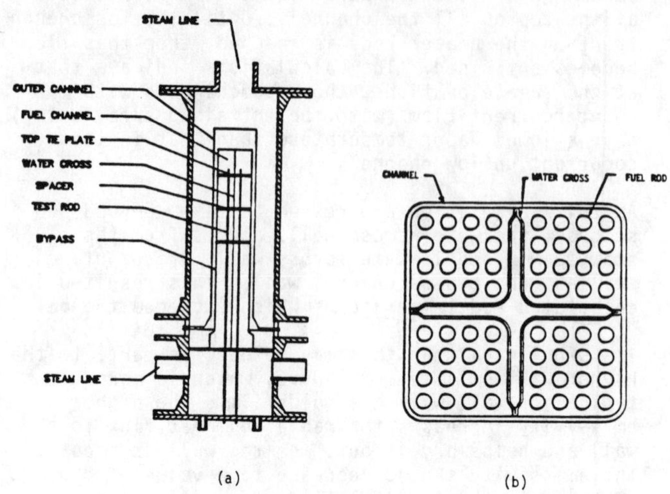

Figure 2. **The Test Section of SVEA Spray Cooling Experiments**

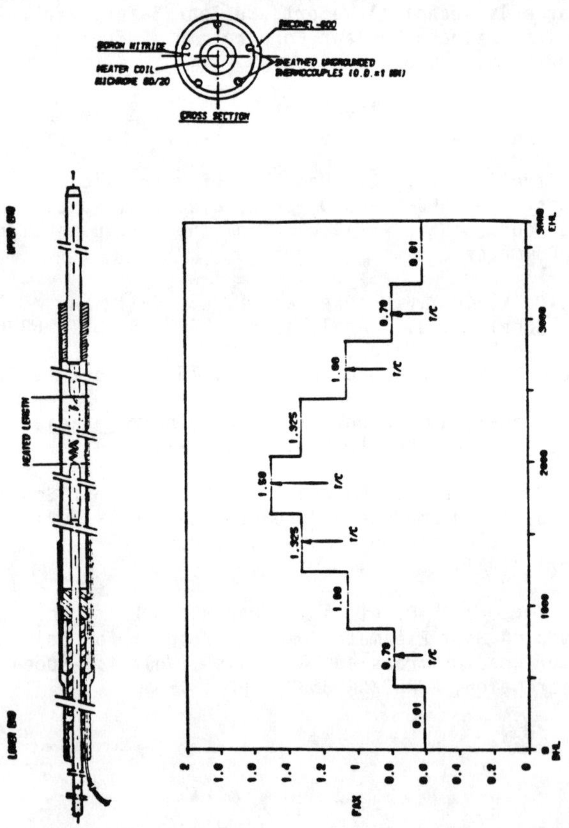

Figure 3. **Heater Rod Geometry and the Power Shape of Heated Length**

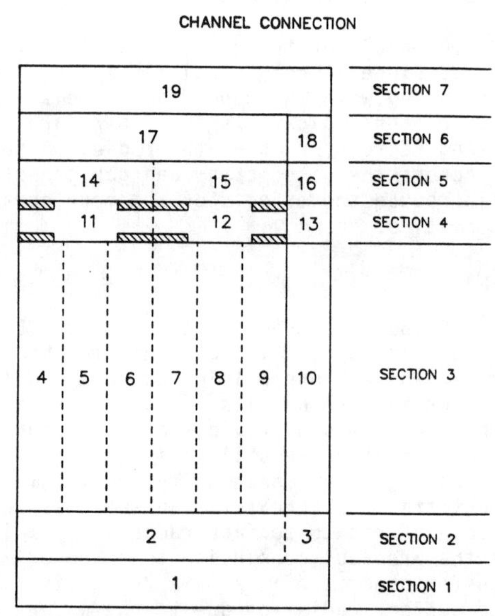

Figure 4. COBRA-TF Model for SVEA Uniform Spray Test, Channel Connection

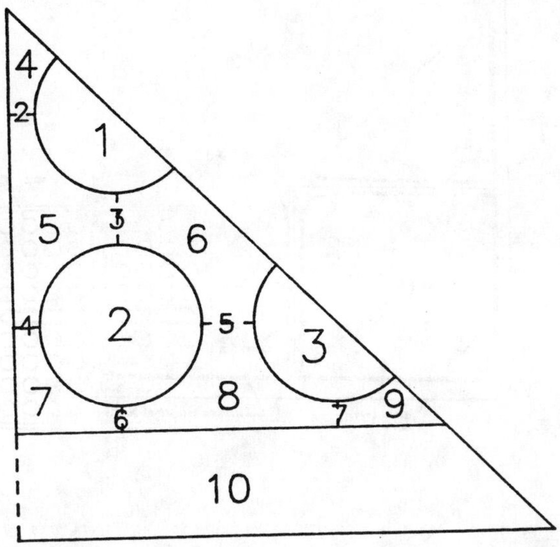

Figure 5. COBRA-TF Model for SVEA Uniform Spray Test, 1/8 of a Minibundle

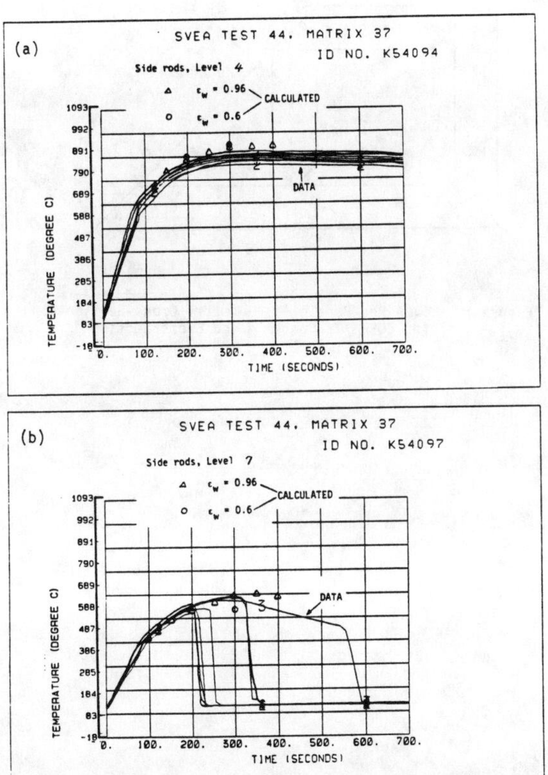

Figure 7. Comparison of Side Rod Temperature at (a) Level 4, (b) Level 7 for Uniform Spray, Test 44

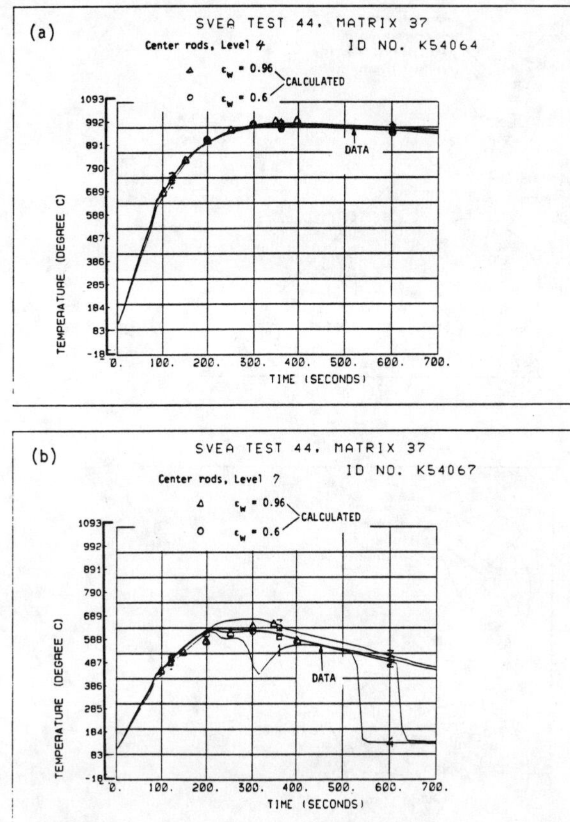

Figure 6. Comparison of Center Rod Temperature (a) Level 4 (b) Level 7 for Uniform Spray, Test 44

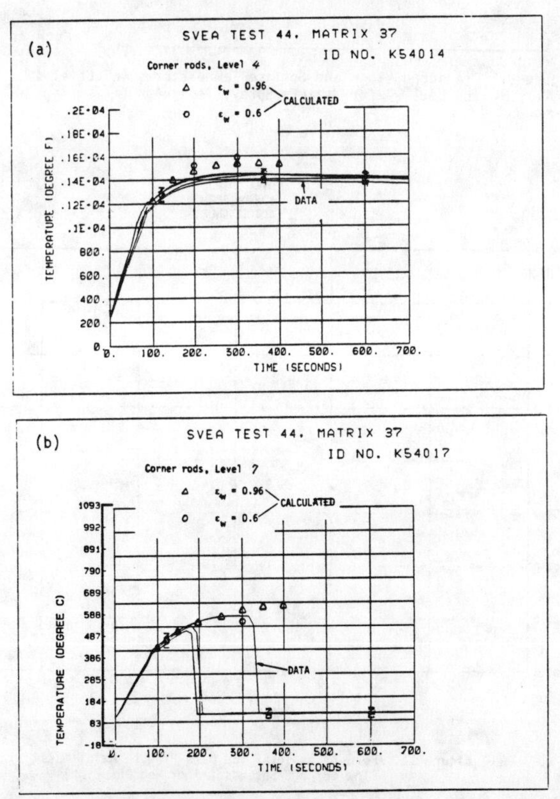

Figure 8. Comparison of Corner Rod Temperature at (a) Level 4, (b) Level 7 for Uniform Spray, Test 44

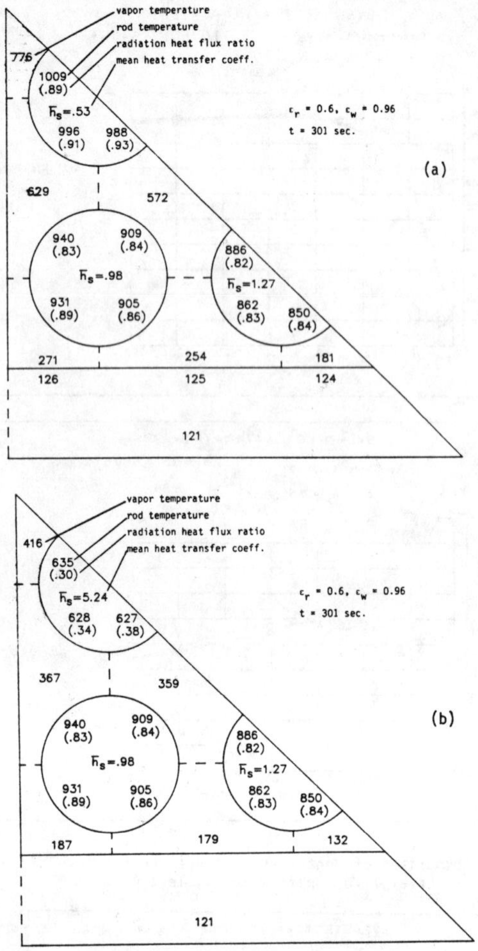

Figure 9. Predicted Vapor and Surface Temperatures (a) Level 4, (b) Level 7 for Uniform Spray, Test 44

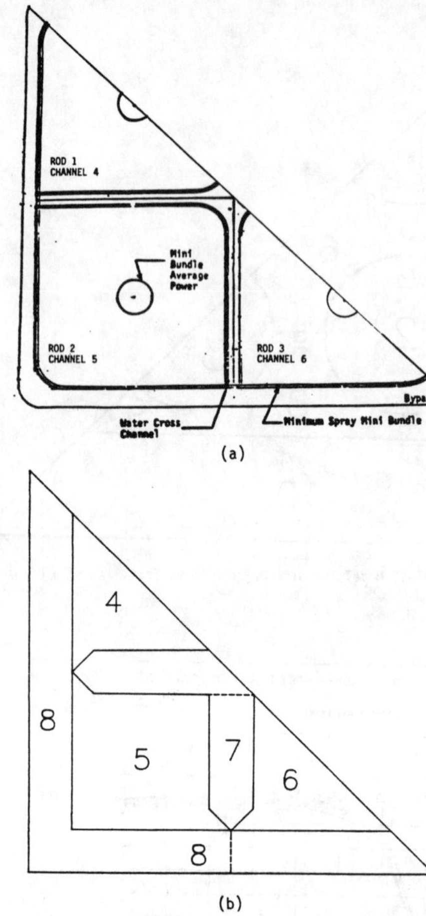

Figure 10. SVEA Rod Bundle (a) Geometry Cross-Section (b) COBRA/TF Coarse Noded Cross-Section

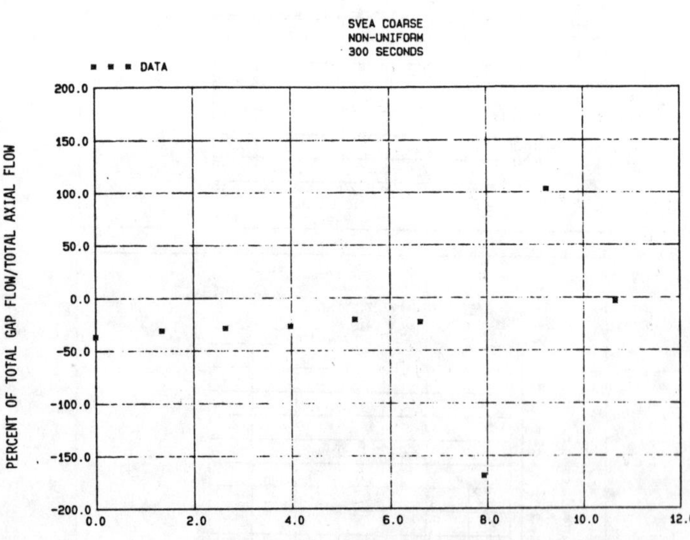

Figure 11. Percent of Total Gap Mass Flow to Axial Mass Flow for 300 Seconds (After Rod Power On)

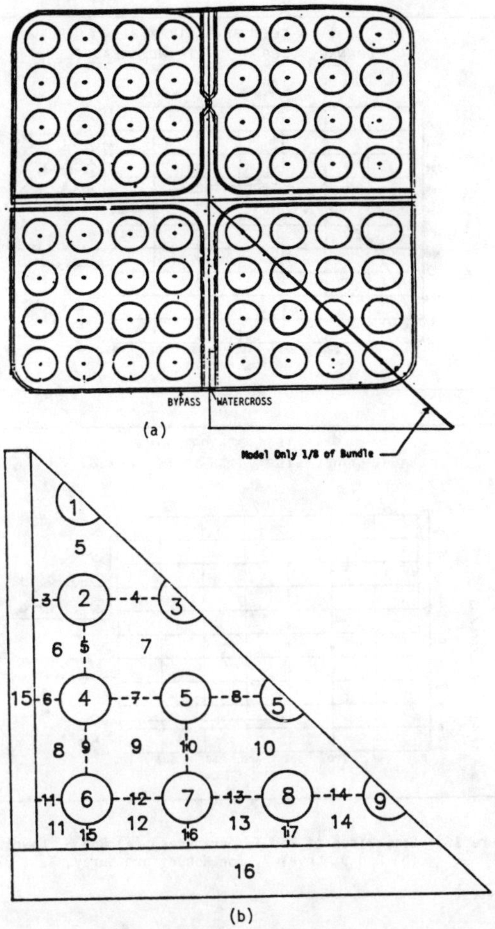

Figure 12. COBRA-TF Detailed Model for (a) SVEA Nonuniform Spray Test and (b) Test #37 (Specific)

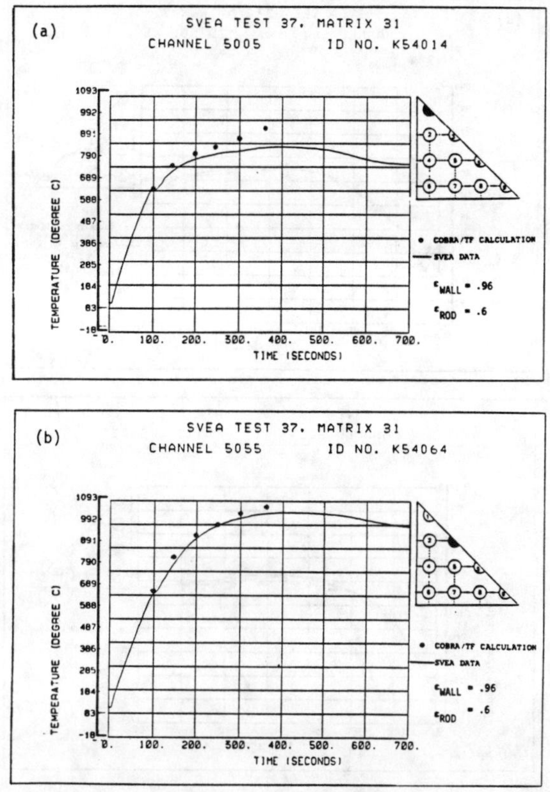

Figure 13. Comparison of Rod Temperatures (a) Rod 1, Level 4 (b) Rod 3, Level 4 for Nonuniform Spray, Test 37

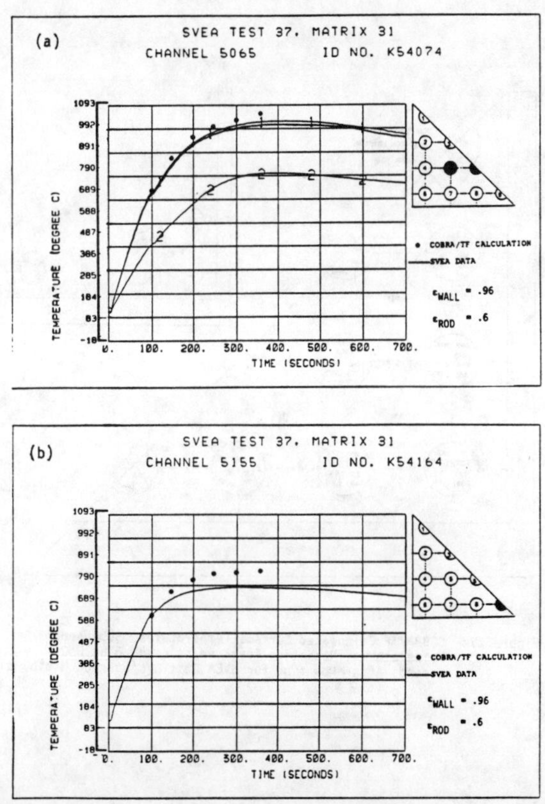

Figure 14. Comparison of Rod Temperatures (a) Rod 5, Level 4 (b) Rod 9, Level 4, for Nonuniform Spray, Test 37

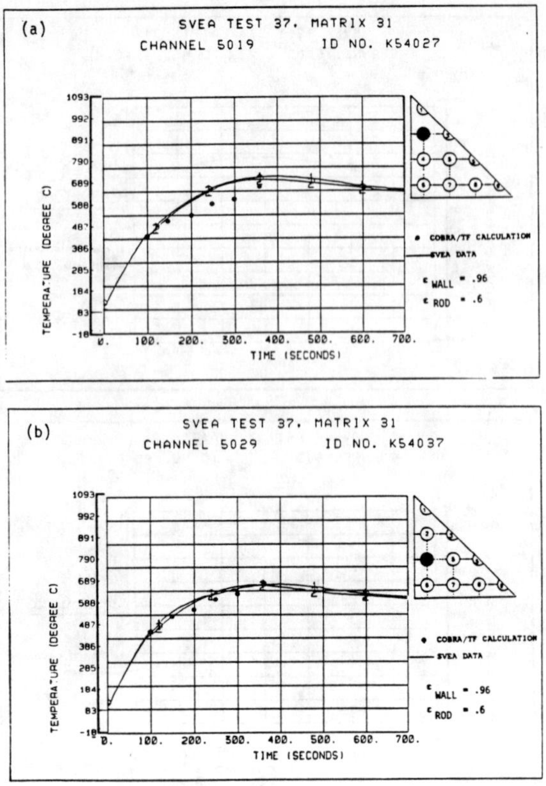

Figure 15. Comparison of Rod Temperatures (a) Rod 2, Level 7 (b) Rod 4, Level 7 for Nonuniform Spray, Test 37

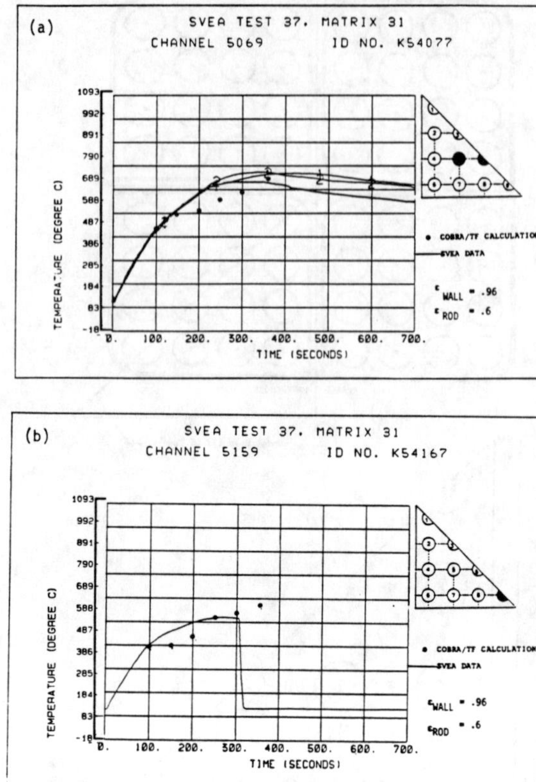

Figure 16. Comparison of Rod Temperatures, (a) Rod 5, Level 7 (b) Rod 9, Level 7, for Nonuniform Spray, Test 37

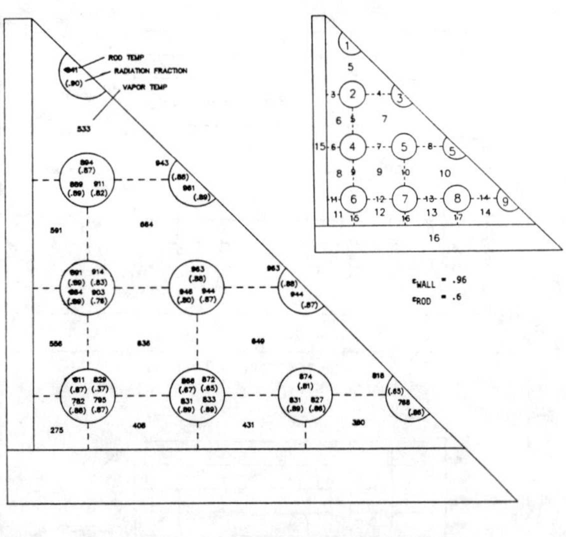

Figure 17. COBRA/TF Calculated Surface Temperatures, Subchannel Temperatures, Radiation Fraction at Level 3 at 300 Seconds After Power on for SVEA Test 37 (Starved Minibundle)

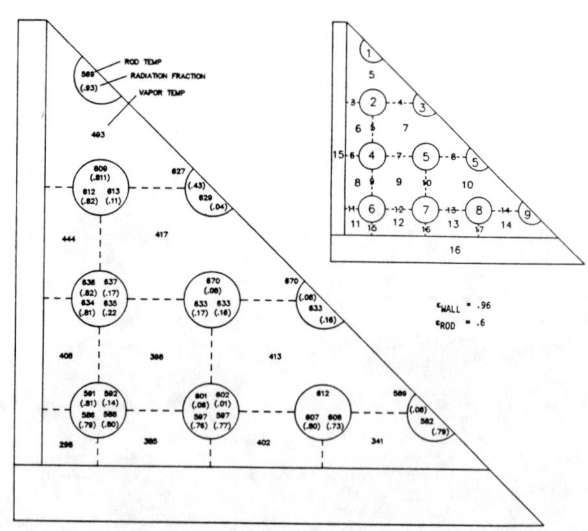

Figure 18. COBRA/TF Calculated Surface Temperatures, Subchannel Temperatures, Radiation Fraction at Level 7 at 300 Seconds After Power on for SVEA Test 37 (Starved Minibundle)

1989 National Heat Transfer Conference
HTD-Vol. 106, Heat Transfer Phenomena in Radiation, Combustion and Fires

LAMINAR NATURAL CONVECTION HEAT TRANSFER ALONG A NEEDLE WITH THERMAL RADIATION

G. H. Lee and K. H. Kim
Department of Engineering Science and Mechanics
The University of Tennessee
Knoxville, Tennessee

ABSTRACT

The purpose of this study is to investigate radiative heat transfer effect on laminar natural convection along a thin needle. In particular, the emphasis is placed on the skin friction factor and heat transfer rate as well as the velocity and temperature profiles along the transformed spatial variables. The medium is assumed to be optically thick, hence the diffusion approximation is used to represent the linearized radiative heat flux term. The study is initiated by first applying the general similarity via the one-parameter group theory on the boundary layer equations in cylindrical coordinates, and then requiring the condition of constant conformally invariant transformation properties. Numerical results are presented for various values of radiative heat transfer and Prandtl number.

NOMENCLATURE

c	= absorptivity
c_p	= specific heat
f	= stream function in similarity transformation
f'	= similarity velocity function
f"	= transformed skin friction
g	= similarity temperature function
g'	= transformed heat transfer rate
G_r	= Grashof number
$\overline{h}$	= average heat transfer coefficient
k	= thermal conductivity
L	= characteristic length
N	= conduction-to-radiation ratio
$\overline{N}_u$	= Nusselt number
P	= pressure
P_r	= Prandtl number
r	= spatial variable in r-direction
$\overline{r}$	= dimensionless spatial variable in r-direction
T	= temperature
u_r	= velocity in r-direction
$\overline{u}_r$	= dimensionless velocity in r-direction
u_z	= velocity in z-direction
$\overline{u}_z$	= dimensionless velocity in z-direction
z	= spatial variable in z-direction
$\overline{z}$	= dimensionless spatial variable in z-direction
α	= constant
β	= thermal expansion coefficient
η	= independent similarity variable
η_o	= needle shape parameter
θ	= dimensionless temperature
λ	= thermal radiation wave length
μ	= dynamic viscosity
ρ	= density
σ	= Stefan-Boltzman constant
τ	= shear stress
ψ	= stream function

1 Introduction

Over the past two decades, considerable attention has been devoted to predicting the flow and heat transfer problems where the flow is two-dimensional (e.g., vertical plates and cylinders, tilted plates, and horizontal circular cylinders) or axisymmetric bodies (e.g., spheres) [1,2,3,4,5]. Particular interest is directed toward the prediction of skin friction and heat transfer rate at the wall as well as temperature and velocity profiles along the isothermal thin needle. In this study a parabolic needle shape is assumed to represent a very small diameter cylinder compared to its length such as wires and sensores of hot wire.

Cebeci and Na [2] solved an isothermal needle problem using similarity technique. Narain [1] extended his study to include a constant heat flux at the wall. Raithby et. al. [4] reported experimental results for a needle problem considering curvature effect. Narain [3] solved slender cylinder problem by using finite difference scheme based on non-similarity concept.

An isothermal needle of high temperature is assumed to be immersed in a large fluid reservoir. The Boussinesq approximation is used; in this the fluid density is assumed constant except for the generation of buoyancy forces. Other fluid properties are taken as constant. In this study the skin friction and heat transfer rate along the needle, as well as the temperature and velocity profiles in the fluid are investigated.

Another aspect of this study is to investigate the influence of thermal radiation. The effect of thermal radiation on heat transfer of the laminar boundary layer flow in an absorbing and emitting medium has been studied by several investigators. Viskanta and Grosh [6] solved the laminar flow problem over a wedge with radiative heat transfer effect, and Kim, et. al. [7] solved the same problem for the case of local similarity approach. Novotny and Yang [8] and Goulard [9] have considered optically thick approximation for boundary layer flow over a flat plate. Kim and Ozisik [10] solved heat transfer problem of an absorbing, emitting, isotropically scattering stagnation point boundary layer flow by treating the radiation part of the problem exactly with the normal-mode-expansion technique. In this study the Rosseland approximation is employed to represent heat flux due to thermal radiation.

The needle problem of physical dimensions is transformed into a non-dimensional scaled form. The non-dimensional equations are transformed into ordinary differential equations through the application of one-parameter group theory described in [11], and the resulting differential equations are solved numerically.

2 Analysis

The scaled, dimensionless, and steady boundary layer equations in cylinderical coordinates for the geometry shown in Fig. 1 are

Continuity

$$\frac{\partial \overline{v}_r}{\partial \overline{r}} + \frac{1}{\overline{r}}\overline{v}_r + \frac{\partial \overline{v}_z}{\partial \overline{z}}, \tag{1}$$

Momentum

$$\overline{v}_r\frac{\partial \overline{v}_z}{\partial \overline{r}} + \overline{v}_z\frac{\partial \overline{v}_z}{\partial \overline{z}} = \frac{1}{\overline{r}}\frac{\partial \overline{v}_z}{\partial \overline{r}} + \frac{\partial^2 \overline{v}_z}{\partial \overline{r}^2} + \theta, \tag{2}$$

Energy

$$\overline{v}_r\frac{\partial \theta}{\partial \overline{r}} + \overline{v}_z\frac{\partial \theta}{\partial \overline{z}} = \frac{1}{Pr}(1+\frac{4}{3N})(\frac{1}{\overline{r}}\frac{\partial \theta}{\partial \overline{r}} + \frac{\partial^2 \theta}{\partial \overline{r}^2}). \tag{3}$$

The boundary conditions for the surface and for the outer edge of the boundary layer are

$$\begin{array}{llll} \overline{r} = \overline{r}_o(\overline{z}), & \overline{v}_r = 0, & \overline{v}_z = 0, & \theta = 1, \\ \overline{r} \longrightarrow \infty, & & \overline{v}_z = 0, & \theta = 0. \end{array} \tag{4}$$

Where $\overline{z}, \overline{r}, \overline{v}_z$, and $\overline{v}_r$ are the scale variables defined by

$$\begin{array}{lll} \overline{r} = \frac{r}{L}(Re)^{1/2}, & \overline{z} = \frac{z}{L}\frac{w^2}{\beta g(T_o - T_\infty)L}, & \\ \overline{v}_r = \frac{v_r}{w}(Re)^{1/2}, & \overline{v}_z = \frac{v_z}{w}\frac{w^2}{\beta g(T_o - T_\infty)L}, & \\ Re = \frac{Lw}{\nu}, & Pr = \frac{\nu}{\alpha}, & \theta = \frac{T - T_\infty}{T_o - T_\infty}, \end{array} \tag{5}$$

and the subscripts "∞" and "o" denote free stream and surface conditions, respectively. The conduction-to-radiation ratio, N is defined as

$$N = \frac{k\beta}{4\lambda^2\sigma T_\infty^3} \tag{6}$$

The nondimensionalized equations become a highly complex set of partial differential equations that can not be easily solved numerically.

Defining dimensionless stream function as

$$\overline{v}_r = \frac{1}{\overline{r}}\frac{\partial \psi}{\partial \overline{r}}, \tag{7}$$

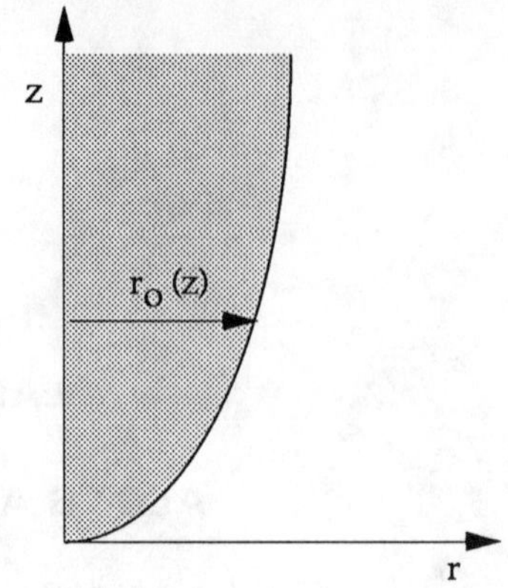

Figure 1: Needle geometry of the problem.

then, the continuity equation is identically satisfied, and the momentum and energy equation are transformed to

$$\frac{1}{\overline{r}^2}\frac{\partial \psi}{\partial \overline{z}}\frac{\partial^2 \psi}{\partial \overline{r}^2} + \frac{1}{\overline{r}^3}\frac{\partial \psi}{\partial \overline{z}}\frac{\partial \psi}{\partial \overline{r}} + \frac{1}{\overline{r}^2}\frac{\partial \psi}{\partial \overline{r}}\frac{\partial^2 \psi}{\partial \overline{z}\partial \overline{r}} = \frac{1}{\overline{r}}\frac{\partial^3 \psi}{\partial \overline{r}^3} - \frac{1}{\overline{r}^2}\frac{\partial^2 \psi}{\partial \overline{r}^2} + \frac{1}{\overline{r}^3}\frac{\partial \psi}{\partial \overline{r}} - \theta, \tag{8}$$

$$-\frac{1}{\overline{r}}\frac{\partial \psi}{\partial \overline{z}} + \frac{1}{\overline{r}}\frac{\partial \psi}{\partial \overline{r}}\frac{\partial \theta}{\partial \overline{z}} = \frac{1}{P_r}(1+\frac{4}{3N})(\frac{\partial^2 \theta}{\partial \overline{r}^2} + \frac{1}{\overline{r}}\frac{\partial \theta}{\partial \overline{z}}). \tag{9}$$

The boundary conditions (4) then become

$$\begin{array}{lll} \overline{r} = \overline{r}(\overline{z}), & \frac{\partial \psi}{\partial \overline{z}} = 0, & \theta = 1, \\ \overline{r} \longrightarrow \infty, & \frac{1}{\overline{r}}\frac{\partial \psi}{\partial \overline{z}} = 0, & \theta = 0. \end{array} \tag{10}$$

Applying the general similarity via the one-parameter group theory by Birkhoff [12], and requiring the condition of constant conformally invariant transformation properties, the following transformation variables are obtained.

$$\eta = (\frac{\overline{r}}{\overline{z}^\alpha})^2, \qquad f(\eta) = \frac{\psi}{\overline{z}}, \qquad g(\eta) = \frac{\theta}{\overline{z}^{(1-4\alpha)}} \tag{11}$$

where

$$\alpha = -\frac{\alpha_2}{\alpha_1}, \qquad 1 - 4\alpha = \frac{\alpha_4}{\alpha_1}. \tag{12}$$

With these transformations, the boundary conditions (10) become

$$\overline{r}(\overline{z}) = \eta^{\frac{1}{2}}\overline{z}^\alpha, \qquad f = f' = 0, \qquad \overline{z}^{1-4\alpha}g = 1, \tag{13}$$

$$\eta \longrightarrow \infty, \qquad f'(\infty) = 0, \qquad g = 0 \tag{14}$$

Examination of the transformed boundary conditions shows that the boundary conditions can be transformed only if $\alpha = \frac{1}{4}$; then the equation of the needle surface is given by

$$\frac{\overline{r}^2}{\overline{z}^{\frac{1}{2}}} = \eta_o = constant. \tag{15}$$

Hence, by using the transformation given by the Eq. (11) and by taking $\alpha = \frac{1}{4}$, the momentum and energy equation for laminar free convection with thermal radiation can be transformed into the following ordinary differential equations:

$$8\eta f''' + 8f'' + 4ff'' - 2f'^2 + g = 0, \tag{16}$$

$$\eta g'' + [\frac{1}{2}Pr\frac{1}{1+\frac{4}{3N}}f + 1]g' = 0, \qquad (17)$$

with the boundary conditions

$$at \quad \eta = \eta_o, \quad f(\eta_o) = 0, \quad f'(\eta_o) = 0, \quad g(\eta_o) = 1$$
$$at \quad \eta \longrightarrow \infty, \quad f'(\eta) = 0, \quad g(\eta) = 0. \qquad (18)$$

The set of Eqs. (16) and (17) with the boundary conditions (18) constitutes similarity equations for natural free convection problem along a needle.

2.1 Skin friction and heat flux

Defining shear stress τ_o on the needle as

$$\tau_o = \mu(\frac{\partial v_z}{\partial r})_o. \qquad (19)$$

Applying the transformation variables, Eq. (19) can be written as

$$\frac{\tau_o(Re)^2}{\frac{1}{2}\rho w^2(Gr)^{\frac{3}{4}}} = \frac{32}{5}(\eta_o)^{\frac{1}{2}}f''(\eta_o), \qquad (20)$$

where Gr is Grashop number defined as

$$Gr = \frac{g\beta(T_o - T_\infty)L^3}{\nu^2}. \qquad (21)$$

The shear parameter, $f''(\eta_o)$ indicates the skin friction on the surface of the needle.

The heat flux at the surface is defined as

$$q_o = -k(\frac{\partial T}{\partial r})_o. \qquad (22)$$

In terms of similarity variables, Eq. (22) can be written as

$$q_o = -2k\frac{(T_o - T_\infty)}{L}(Re\eta_o)^{\frac{1}{2}}\frac{g'(\eta)}{z^{\frac{1}{2}}}. \qquad (23)$$

The average heat transfer at the surface becomes

$$\overline{q}_o = \frac{1}{\overline{z}_L}\int_0^{\overline{z}_L} L_o q_o(z)dz. \qquad (24)$$

The integration of Eq. (24) results in

$$\overline{q}_o = -\frac{8}{3}k\frac{(T - T_\infty)}{L}(\eta_o)^{\frac{1}{2}}g'(\eta)(Gr)^{\frac{1}{4}} \qquad (25)$$

This average heat transfer rate can be expressed in terms of an average Nusselt number, $\overline{N}_u$ defined as

$$\overline{N}_u = \frac{\overline{h}L}{k} \qquad (26)$$

where $\overline{h}$ is the average heat transfer coefficient.

From Eq. (23), the following expression for $\overline{N}_u$ can be written as

$$\frac{\overline{N}_u}{(Gr)^{\frac{1}{4}}} = -\frac{8}{3}(\eta_o)^{\frac{1}{2}}g'(\eta_o). \qquad (27)$$

Hence, the slope of g indicates the heat transfer rate at the sruface.

3 Result and Discussion

The set of similarity equations (16) and (17) with the boundary conditions (18) are integrated by the fourth-order Runge-Kutta method with Gill coefficient by employing a modified version of the scheme developed by Nachtsheim and Swigert [13] for the successive iterative guesses on the initial conditions of the momentum and energy equations. Numerical results for similar dimensionless skin friction, heat transfer rate at the wall, and velocity and temperature distributions are presented in terms of conduction-to-radiation ratio, N.

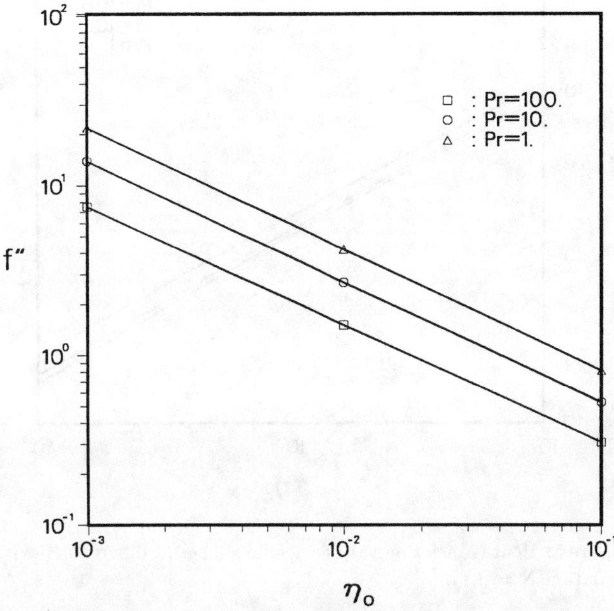

Figure 2: Skin friction on various needle shapes at different P_r without radiation.

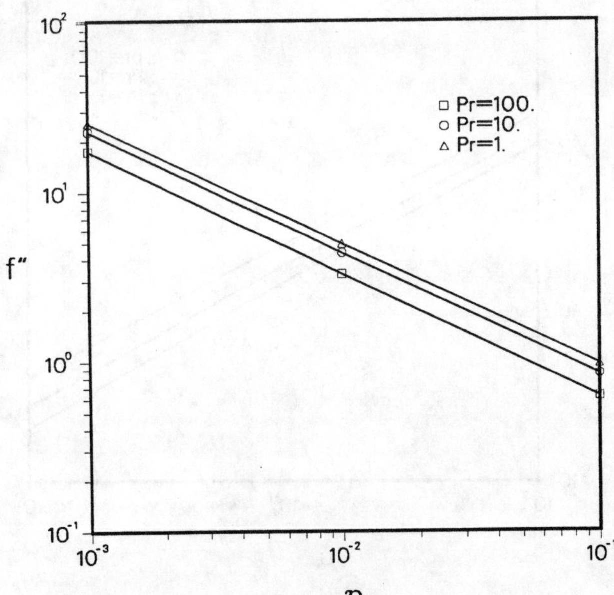

Figure 3: Skin friction on various needle shapes at different P_r with radiation(N = 0.1).

The results for the skin friction with and without thermal radiation effect are presented in Figs. (2) and (3), respectively. In Fig. (2) the effect of the Prandtl number and the needle shape is evident . Increasing the Prandtl number and the size of needle shape tends to decrease the friction factor in the absence of thermal radiation. However, the friction factors are increased substantially in the presence of thermal radiation as shown in Fig. (3).

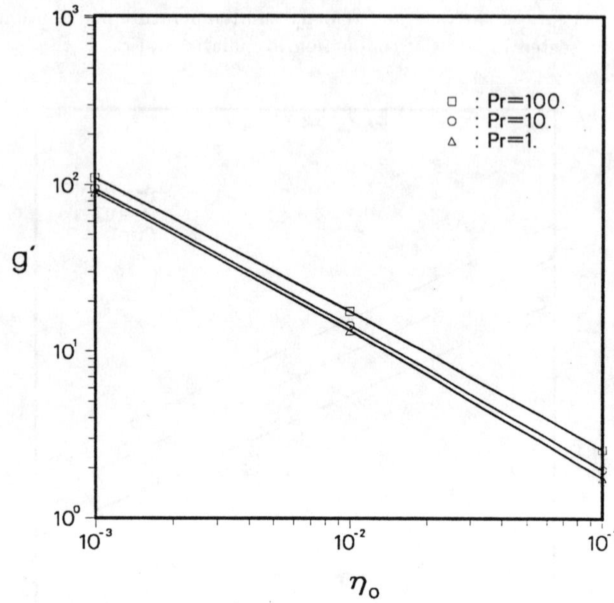

Figure 5: Heat transfer on various needle shapes at different P_r with-radiation(N = 0.1).

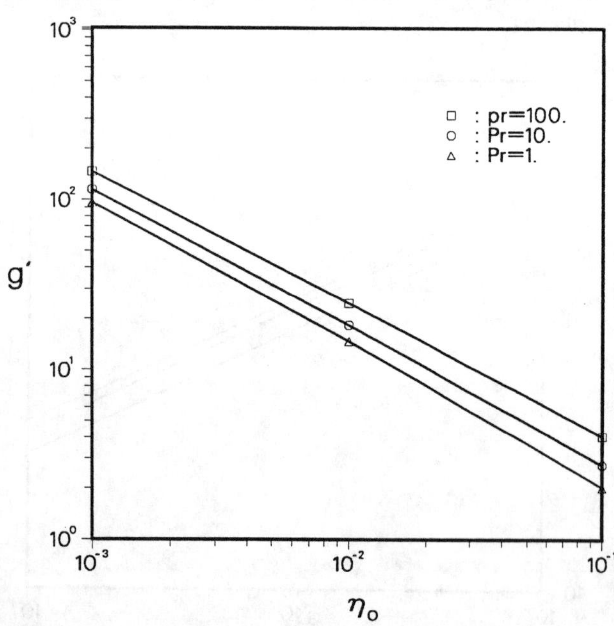

Figure 4: Heat transfer on various needle shapes at different P_r without radiation.

Figs. (4) and (5) show the heat transfer rates as a function of the Prandtl number, needle shape, and thermal radiation. It is observed in both Figs. (4) and (5) that the increase of the Prandtl number results in the increase of heat transfer rate, contrary to the case of friction factor. It is interesting to note that the heat transfer rates decrease in the presence of thermal radiation.

The effect of thermal participation and the Prandtl number on the velocity and temperature distribution is shown in Figs. (6) and (7), respectively. The radiation participation is seen to increase both the thermal boundary layer and the temperature gradient at the wall of the slender needle.

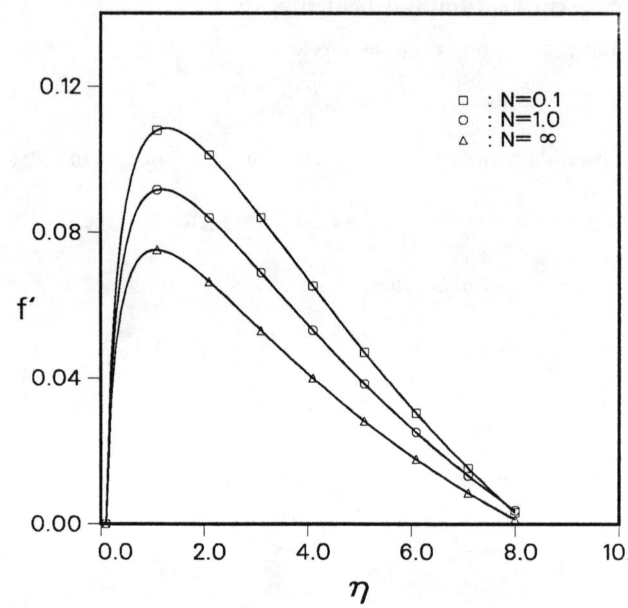

Figure 6: Similarity velocity profile at η_o = 0.1 and P_r = 10.

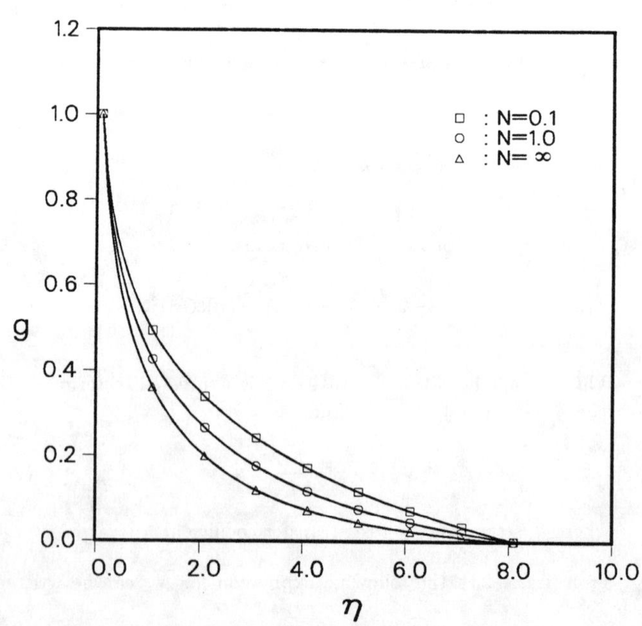

Figure 7: Similarity temperature profile at η_o = 0.1 and P_r = 10.

4 Conclusion

In summary, several conclusions can be made regarding the effect of the needle shape, Prandtl number, and thermal radiation on a laminar free convection along an isothermal slender needle. First, as the needle shape becomes more slender, the skin friction factor and the heat transfer rate increase regardless the presence of thermal radiation. Second, the Prandtl number affects both the skin friction and heat transfer rate in such a manner that increasing the Prandtl number decreases the skin friction and increases the heat transfer rate. Third, increasing the thermal radiation has the effect of decreasing the heat transfer. Finally, the thermal and momentum boundary layers increase in the presence of thermal radiation. This is partly due to the fact that the Rosseland approximation enhances energy diffusion and momentum transport in the fluid medium. Present studies can be extended for a pair of needles which may provide valuable information for a bundle of hot wires used in various measuring devices.

References

[1] Narain, J. P., and Uberoi, M. S.,"*Laminar Free Convection from Vertical Thin Needles*," The Physics of Fluids, Vol. 15, No. 5, pp. 928-929, 1972.

[2] Cebeci, T. and Na, T. S., "*Laminar Free-Convection Heat Transfer from a Needle*," The Physics of Fluids, Vol. 12, No. 3, pp. 463-467. 1969.

[3] Narain, J. p.,"*Free and Forced Convective Heat Transformation of Slender Cylinders*," Letters in Heat and Mass Transfer, Vol. 3, pp. 21-30, 1976.

[4] Raithby, G. D., et. al.,"*Free Convection Heat Transfer from Spheroids*," Journal of Heat Transfer, pp. 452-458, August, 1976.

[5] Raithby, G. D. and Hollands, K. G. T.,"*Free Convection Heat Transfer from Vertical Needles*," Journal of Heat Transfer, pp. 522-523, August, 1976.

[6] Viskanta, R. and Grosh, R. J.,"*Boundary Layer in Thermal Radiation Absorbing and Emitting Media*," International Journal of Heat and Mass Transfer, Vol. 5, pp. 795-806, 1962.

[7] Kim, K. H., Ozisik, M. N., and Mulligan, J. C.,"*A Non-Similar Solution of Heat Transfer in External Non-Newtonian Flow with Thermal Radiation*," AIAA Paper No. 73-116, 1973.

[8] Novotny, J. L. and Yang, K. T.,"*The Interaction of Thermal Radiation in Optically Thick Boundary Layers*," ASME Paper No. 67 HT-9, 1967.

[9] Goulard, R., "*The Transition from Blackbody to Rosseland Formulations in Optically Thick Flows*," International Journal of Heat and Mass Transfer, Vol. 7, pp. 1145-1146, 1964.

[10] Kim, K. H. and Ozisik, M. N.,"*Application of Normal-Mode-Expansion Technique to Absorbing, Emitting, and Scattering Stagnation Point Boundary Layer Flow*," Development in Theoretical and Applied Mechanics, Proceedings of the Tenth Southeastern Conference on Theoretical and Applied Mechanics, Vol. 10, pp. 681-701, 1980.

[11] Hansen, A. G.,*Similarity Analysis of Boundary Value Problems in Engineering*,Prentice Hall, Englewood Cliffs, NJ, 1964

[12] Birkhoff, G., *Hydrodynamics*, Princeton University Press, Princeton, N. J., Chapters 4 and 5, 1960.

[13] Nachtsheum, P. R., and Swigert, P.,"*Satisfaction of Asymptotic Boundary Condition in Numerical Solution of Systems of Nonlinear Equations of Boundary Layer Type*," NASA Technical Note D3004, National Aeronautics and Space Administration, 1965.

1989 National Heat Transfer Conference
HTD-Vol. 106, Heat Transfer Phenomena in Radiation, Combustion and Fires

COMBINED RADIATION AND NATURAL CONVECTION IN A PARTICIPATING MEDIUM BETWEEN HORIZONTAL CONCENTRIC CYLINDERS

Z. Tan and J. R. Howell
Department of Mechanical Engineering
University of Texas at Austin
Austin, Texas

Combined radiation and laminar natural convection in a participating medium between horizontal concentric cylinders is investigated numerically in this paper. The YIX method is used to discretize the exact radiation transport equations of integral form, and finite differences are applied to the Navier-Stokes equations. The nonlinear SOR iteration scheme is applied to solve the resulting nonlinear algebraic equations. Radiation is found to have significant influence on the flow and temperature fields, as well as on heat transfer. At high Rayleigh numbers, radiation can decrease the stability of the flow.

NOMENCLATURE

(All quantities are dimensionless.)

e_g	=	radiative emissive power of medium
e_s	=	radiative emissive power of enclosure
$\mathbf{n}$	=	inward unit normal vector at $\mathbf{r}$
Pr	=	Prandtl number
q_c	=	convection heat flux on S
$\mathbf{q}_r$	=	radiation flux vector
q_s	=	radiation heat flux on S
Q, Q_{comb}	=	scaled total heat flux
Qr	=	scaled radiation flux
$\mathbf{r}$	=	$(x,y) = (r,\theta)$, position vector
$R(\omega_k)$	=	distance from $\mathbf{r}$ to boundary along ω_k
r_a	=	radius of inner cylinder
Ra	=	Rayleigh number, $\rho g\beta(r_b-r_a)^3(T_b-T_a)/\mu\alpha$
r_b	=	radius of outer cylinder
Rc	=	radiation-conduction parameter, $r_a\sigma T_c^3/\lambda$
S	=	boundary of Ω
S_n	=	Bickley's function of nth order
T	=	temperature
T_h, T_c	=	temperatures on hot and cold walls
u,v	=	velocity in r and θ components
α	=	thermal diffusivity
ε	=	emissivity
λ	=	thermal conductivity
σ	=	Stefan-Boltzmann constant
τ	=	optical depth based on r_a
ψ	=	stream function
ω	=	vortex
ω_k	=	discrete angular coordinates in Eq.(4)
ω_o	=	albedo for scattering
Ω	=	medium

Subscript

a	=	inner cylinder
b	=	outer cylinder

INTRODUCTION

Heat transfer by combined radiation and natural convection is important in applications such as shut down of pebble-bed nuclear reactors and in analysis of building insulation. In spite of the high nonlinearity and complexity of these problems, several numerical solutions are available for different geometries. Larson (1981) studied a fire spreading process in enclosure. Chang, Yang and Lloyd (1983) investigated the combined radiation and natural convection in two-dimensional partially divided rectangular enclosures by using a simple and fast approximation for radiation transport. Yücel and Bayazitoglu (1980) and Trevisan and Bejan (1986) investigated the onset of free convection in a fluid layer with a free surface as driven by an external radiation source, and Arpaci and Bayazitoglu (1973) studied the radiation effects on thermal stability of a participating medium in a vertical slot. The natural convection induced by irradiation in an absorbing medium was studied by Webb and Viskanta (1986) by both experimental and numerical techniques. A one-dimensional radiation model was used in their work. Tan and Howell (1989b) studied the coupled radiation and natural convection in a square medium by using the full radiation transport equations. Much recent work is reviewed by Yang (1986).

No literature on combined radiation and natural convection between concentric horizontal cylinders has been found by the present authors, but much work on pure convection between concentric cylinders exists. Extensive reviews of both numerical and experimental results for this problem can be found in Kuehn and Goldstein (1976a, 1976b), Shih (1987), and Yang and Yang (1988).

In this paper, we present a numerical study of the problem of combined radiation and natural convection in a participating medium between two concentric cylinders. A newly developed method, the YIX method (Tan and Howell, 1989a) which solves the exact two-dimensional radiation tranport equations of integral form, is used for the radiation equations and is coupled with finite difference schemes for the fluid flow and energy equations. Influences of the Rayleigh number, and the radiation-conduction parameter on heat transfer, flow, and temperature distribution are investigated and discussed. Results for combined radiation and conduction with different optical properties are also presented.

FORMULATION

Consider two concentric cylinders a and b, with radii $r_a=1$ and $r_b=2.6$, respectively. The cylinder walls are gray and diffuse and kept at different constant temperatures $T_a=2$ and $T_b=1$. An emitting, absorbing and scattering Boussinesq fluid is placed in the annulus. For simplicity, all physical properties are assumed to be constant in space.

In this paper, all variables and parameters have been nondimensionalized.

The governing Navier-Stokes equations in stream function-vortex form are

$$\nabla^2\psi = \omega \tag{1a}$$

$$\frac{1}{Pr}\left[u\frac{\partial\omega}{\partial r} + \frac{v}{r}\frac{\partial\omega}{\partial\theta}\right] = \nabla^2\omega - Ra\left[\sin\theta\,\frac{\partial T}{\partial r} + \frac{\cos\theta}{r}\frac{\partial T}{\partial\theta}\right] \tag{1b}$$

$$u\frac{\partial T}{\partial r} + \frac{v}{r}\frac{\partial T}{\partial\theta} = \nabla^2 T + Rc\nabla\cdot\mathbf{q}_r \tag{1c}$$

in $1 < r < r_b/r_a,\ 0 < \theta < \pi$, where

$$\nabla^2 := \frac{\partial^2}{\partial r^2} + \frac{1}{r}\frac{\partial}{\partial r} + \frac{1}{r^2}\frac{\partial^2}{\partial\theta^2}\,, \quad u := \frac{1}{r}\frac{\partial\psi}{\partial\theta}, \quad \text{and } v := -\frac{\partial\psi}{\partial r}$$

θ is measured clockwise from the positive vertical line (y axis), Pr is the Prandtl number, Ra the Rayleigh number based on the gap width, r_b-r_a, Rc the radiation-conduction parameter defined in this case by

$$Rc = \frac{r_a\sigma T_a^3}{\lambda}$$

$\mathbf{q}_r$ is the radiation energy flux vector, which relates Eqs.(1) with the radiation transport equations of integral form (Tan, 1989a)

$$4e_g(\mathbf{r}) - \frac{1}{(1-\omega_0)\tau}\nabla\cdot\mathbf{q}_r(\mathbf{r}) =$$

$$\int_\Omega [e_g(\mathbf{r}') - \frac{\omega_0}{4(1-\omega_0)\tau}\nabla\cdot\mathbf{q}_r(\mathbf{r}')]\,\frac{S_1(\tau|\mathbf{r}-\mathbf{r}'|)}{|\mathbf{r}-\mathbf{r}'|}\,dA(\mathbf{r}')$$

$$+ \int_S [e_s(\mathbf{r}') - \frac{1-\varepsilon}{\varepsilon}q_s(\mathbf{r}')]\,\frac{S_2(\tau|\mathbf{r}-\mathbf{r}'|)}{|\mathbf{r}-\mathbf{r}'|}\cos(\mathbf{r}-\mathbf{r}',\mathbf{n}')\,dl(\mathbf{r}')$$

$$\mathbf{r}\in\Omega \tag{2a}$$

$$e_s(\mathbf{r}) - \frac{1}{\varepsilon}q_s(\mathbf{r}) =$$

$$\int_\Omega [e_g(\mathbf{r}') - \frac{\omega_0}{4(1-\omega_0)\tau}\nabla\cdot\mathbf{q}_r(\mathbf{r}')]\,\frac{S_2(\tau|\mathbf{r}-\mathbf{r}'|)}{|\mathbf{r}-\mathbf{r}'|}\cos(\mathbf{r}'-\mathbf{r},\mathbf{n})\,dA(\mathbf{r}')$$

$$+ \int_S [e_s(\mathbf{r}') - \frac{1-\varepsilon}{\varepsilon}q_s(\mathbf{r}')]\,\frac{S_3(\tau|\mathbf{r}-\mathbf{r}'|)}{|\mathbf{r}-\mathbf{r}'|}\cos(\mathbf{r}'-\mathbf{r},\mathbf{n})\cos(\mathbf{r}-\mathbf{r}',\mathbf{n}')\,dl(\mathbf{r}')$$

$$\mathbf{r}\in S \tag{2b}$$

where $\Delta x_j \equiv x_j - x_j'$ (j=1,2), and S_i is Bickley's function defined by

$$S_i(t) = \frac{2}{\pi}\int_0^{\pi/2} e^{-t/\cos\theta}\cos^{i-1}\theta\,d\theta$$

τ is the optical thickness defined by the product of the extinction coefficient of the medium and the gap width, $\Omega\equiv\{1\le r\le r_b/r_a,\ 0\le\theta\le\pi\}$ is the region of the medium, ω_0 the albedo for single scattering, and S the boundary of Ω. To apply the YIX method, it is convenient to use Cartesian coordinates for the radiation equations (Eqs.(2)).

The boundary conditions are:

$$T=2,\ \psi=u=v=0, \qquad r=r_b,\ 0\le\theta\le\pi \tag{3a}$$

$$T=1,\ \psi=u=v=0, \qquad r=r_a,\ 0\le\theta\le\pi \tag{3b}$$

$$\frac{\partial T}{\partial\theta}=0, \qquad \theta=0,\pi,\ r_a\le r\le r_b \tag{3c}$$

To save time and storage only the right half of the system is used in our computations due to the symmetry of the system.

In this work, we take Pr=0.71, and unit optical thickness is assumed unless declared.

NUMERICAL METHOD

The medium domain is divided into 21×31 elements by equally distributed constant r- and θ-lines, and each boundary wall is relevantly divided into 31 equal segments. The nodes are located at the centers of each element or segment.

Equations (1) are discretized by centered, second order finite difference schemes. The second order finite difference scheme is also applied to all boundary conditions.

To discretize the radiation transport equations, a new developed method, called the YIX method, which has been successfully applied to various one- and two-dimensional problems, is applied. The YIX method solves the exact radiation transport equations (say, as a special case, Eqs.(2)) written in an equivalent distance-angular integral form. The essence of the method is to construct integral formulae for the distance integral and store the resulting abscissas, so singularities are removed and numerous evaluations of the time-consuming kernel functions (e.g., Bickley's function in the two-dimensional case) are avoided. For the angular integration, equally spaced integration points are chosen. For example, assuming $\nabla\cdot\mathbf{q}_r(\mathbf{r})=0$ for a 2-D problem with constant properties and without scattering and surface reflection, the approximation for equation (2a) is

$$4e_g(\mathbf{r}) \approx \frac{1}{M}\sum_{k=1}^{M}\{\lambda[\tfrac{1}{2}e_g(\mathbf{r}) + \sum_{i=1}^{n-1}e_g(\mathbf{r}+\omega_k t_i)] + [P_{n+1} + (R(\omega_k) - t_n)Q_{n+1}]e_g(\mathbf{r}+\omega_k t_n)\}$$

$$+ \frac{1}{M^*}\sum_{k=1}^{M^*}\cos(-\omega_k,\mathbf{n}')S_2(\tau R(\omega_k))e_s(\mathbf{r}+\omega_k R(\omega_k)) \tag{4}$$

where M and M* are the number of integration points ω_k for the angular integration, $R(\omega_k)$ the distance from $\mathbf{r}$ to boundary along direction ω_k, and λ, t_i, P_i and Q_i the precalculated constants which eliminate the evaluation of S_1 for space integration. The number of

Table 1 Comparison between the results of Pandey and Cogley (1983) and present work for the 1-D radiation transfer problem in a black concentric cylinder enclosure with T_b=0, T_a=1, r_b/r_a=10 and τ=2/9.

$(r-r_a)/(r_b-r_a)$	present NR=27	present NR=57	Pandey and Cogley
0.1	.3070	.3105	.315
0.3	.1759	.1807	.185
0.5	.1154	.1188	.122
0.7	.0781	.0776	.080
0.9	.0474	.0462	.047

distance integration points, n, along ω_k is determined by $t_n \leq R(\omega_k) < t_{n+1}$.

Since the construction of the numerical formulae is with respect to the coordinate of optical length, therefore, the method applies with a little extension for systems with inhomogeneous media and both optically thin and optically thick regions. It can also deal with complex geometries, which is demonstrated in this work. More detailed description and discussion of the accuracy of this method can be found in Tan and Howell(1989a).

To solve the global nonlinear algebraic equations, the nonlinear SOR (Successive Over Relaxiation) iterating scheme (Chen, 1982; Tan,1989b) is applied.

For convenience, the total heat flux on the wall, $q_c + Rc\,q_s$, is scaled by 1+Rc:

$$Q \equiv Q_{comb} \equiv \frac{q_c + Rc\,q_s}{1+Rc}$$

so that it varies between q_c and q_s. Correspondingly, we define

$$Q_r \equiv \frac{Rc}{1+Rc} q_s$$

The convergence criterion used is

$$\frac{\max |\psi^n - \psi^{n-1}|}{\max |\psi^n|} + \max |T^n - T^{n-1}| < 10^{-5}$$

where the superscript n denotes the nth iteration.

All computations were performed on a CDC 170/750 computer. Although no divergence was experienced using the nonlinear SOR technique, the CPU time depends on the initial guesses and relaxation parameters chosen. In general, 200–500 seconds is needed for convergence in the case Ra=10^4 for all Rc.

RESULTS AND DISCUSSIONS

The computer program was designed to solve the combined heat transfer problem in absorbing, emitting and isotropically media with reflecting boundaries. Extension to anisotropic scattering can also be done easily. However, if scattering and wall reflection are included, the parameters are many. Moreover, since the behavior of the influences of those factors on radiation transport is quite well known and we are mostly interested in how radiation and natural convection are coupled, our study is basically on the influence of the radiation-conduction parameter (Rc) for unit optical thickness (τ=1), non-scattering (ω_0=0) and non-reflecting (ε=1) cases at various Rayleigh number, Ra. In the simpler case Ra=0, i.e., a radiation-conduction system, the problem is one-dimensional, so including τ, ω_0 and ε would not greatly complicate the analysis.

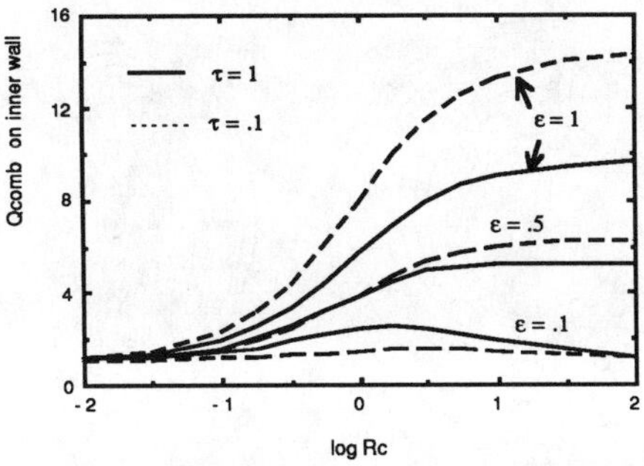

Fig. 1a Combined radiation and conduction between concentric cylinders. Influence of wall emissivity and optical thickness on combined flux.

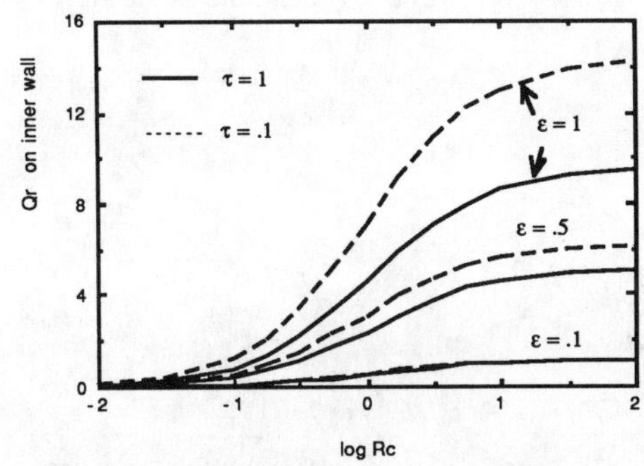

Fig. 1b Combined radiation and conduction between concentric cylinders. Influence of wall emissivity and optical thickness on radiative flux.

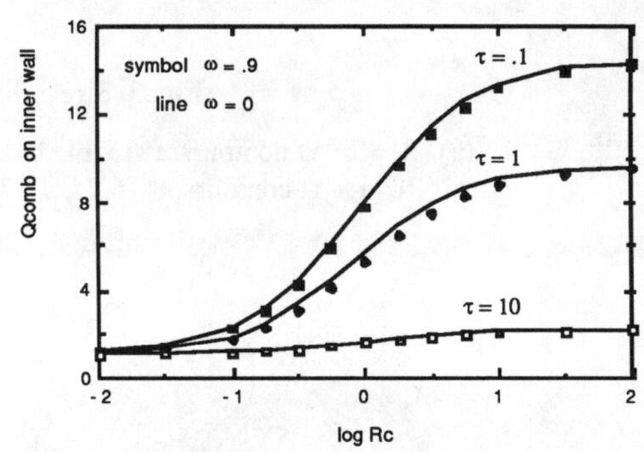

Fig. 1c Combined radiation and conduction between concentric cylinders. Influence of scattering albedo and optical thickness on combined flux.

1. Comparison With Other Work for Pure Radiation or Convection Problems

To verify the accuracy, our computations are compared with other available solutions in simple cases.

Table 1 gives the comparison between our results with the solution of Pandey and Cogley (1983) for pure radiation in a black concentric cylinder enclosure with T_b=0, T_a=1, r_b/r_a=10 and τ=2/9. For numerical comparison, the number of elements in the r-direction is taken to be Nr=27 and Nr=57. Fair agreement is found. Discrepancies among the data may be because: i) Pandey and Cogley solved an

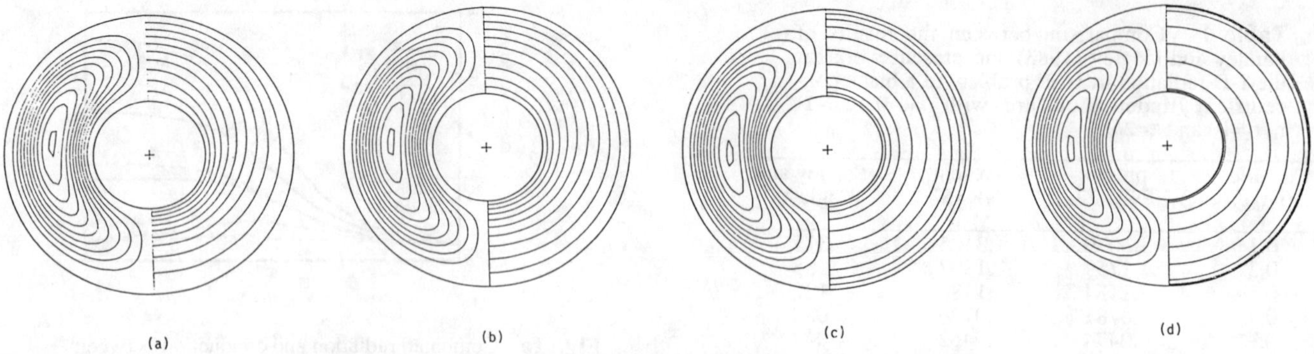

Fig. 2 Streamlines and temperature contours when Ra=10^3.

(a) Rc=0. ψ contours at 2.341, 2.129(–.2365)0. **(b)** Rc=.1. ψ contours at 2.258, 2.052(–.2281)0.
(c) Rc=1. ψ contours at 1.811, 1.647(–.1830)0. **(d)** Rc=∞. ψ contours at 1.310, 1.191(–.1323)0.

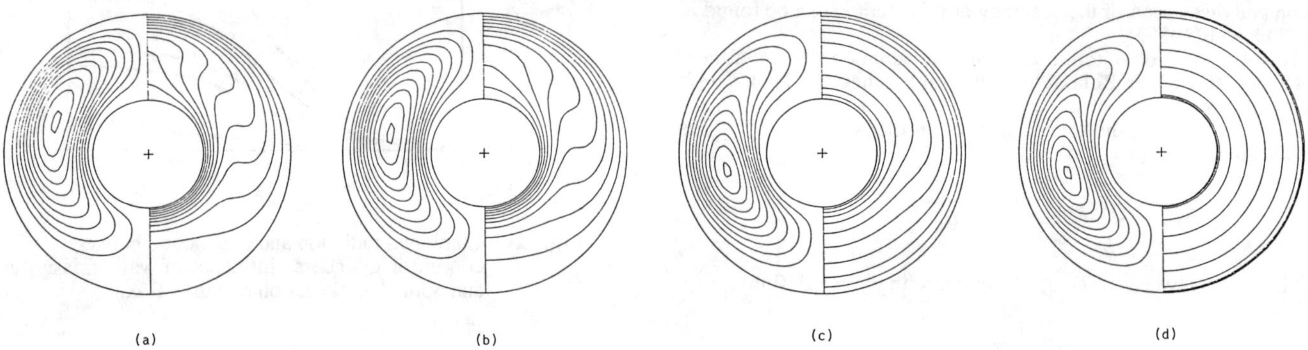

Fig. 3 Streamlines and temperature contours when Ra=10^4.

(a) Rc=0. ψ contours at 13.14, 11.94(–1.327)0. **(b)** Rc=.1. ψ contours at 13.98, 12.70(–1.412)0.
(c) Rc=1. ψ contours at 16.25, 14.78(–1.642)0. **(d)** Rc=∞. ψ contours at 13.89, 12.63(–1.403)0.

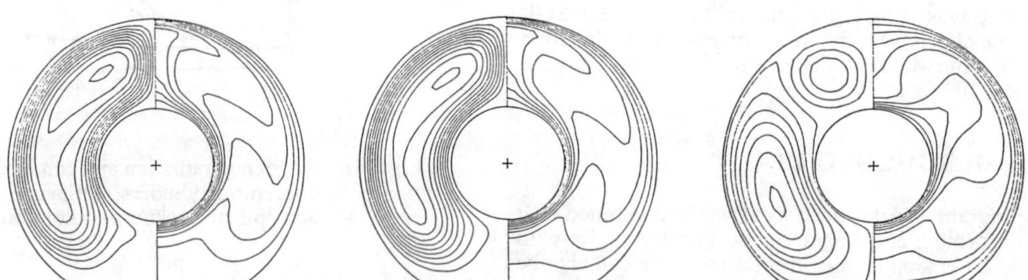

Fig. 4 Streamlines and temperature contours when Ra=10^5.

(a) Rc=0. ψ contours at 31.05, 28.23(–3.136)0. **(b)** Rc=.1. ψ contours at 33.15, 30.13(–3.348)0.
(c) Rc=1. ψ contours at 48.88, 44.43(–4.937)0.

approximate transport formulation while we solve an exact one; ii) the radius radio r_b/r_a=10 is quite large, so that the error of numerical integration used in the YIX method for the Nr=27 case is significant. Many more integration points are used in the Nr=57 case for the YIX method shown in the table, so that solution should be quite accurate. Our results are also in very good agreement with the Monte Carlo data of Perlmutter and Howell (1964) when r_b/r_a=2 (not shown here). When T_b=0, T_a=1 and r_b/r_a=2.6 and with the Nr (=21) and the integration points used in the formal computations, our solution gives q_s=9.551. The error is 0.6% in comparison with the results using Nr=51 and many more integration points.

The equivalent conductivities of our solutions are compared with the results of Kuehn and Goldstein (1976) for pure natural convection. At low Ra ($\leq 10^3$), the agreement is excellent. At high Ra (10^5), our solutions give q_s=3.661, about 6.6% higher than the result of Yang and Yang (1988) who used more grid points (48×40).

A coarse grid system (11×21) was used to check the accuracy of the formal computations. When Rc=1 and for Ra=10^4 and 10^5, the difference in max ψ using the coarse and normal grid systems is about 7% and 12%, respectively, while the difference in both Q_{comb} and Q_r is less than 3%.

2. Combined Radiation and Conduction

Figure 1a and 1b show the influences of wall emissivity and optical thickness on scaled combined and radiative heat fluxes, Q_{comb} and Q_r, respectively, on the inner cylinder, as a function of Rc ($0.01 \leq Rc \leq 100$), when ω_0=0 (no scattering). When ε=1, radiation is found to be of about the same importance to heat transfer as convection would be at about Rc=0.1. As ε decreases, Q_r should also decrease according to Eqs.(2), so radiation becomes less important; that is, Q_r/Q_{comb} decreases. A decrease in ε also reduces the combined heat fluxes.

As τ decreases, the radiation flux Q_r simply increases due to the reduction of the "radiant resistance" of the medium. On the other hand, the influence of τ on the combined heat flux is very different. When τ decreases, if the wall is a blackbody (ε=1), then Q_{comb} increases; however, if the wall is highly reflective (ε=0.1), then Q_{comb} decreases with decreasing τ for all $0<Rc<\infty$. At intermediate ε(=0.5), the influence of decreasing τ is to decrease Q_{comb} if Rc is small and to increase it if Rc is large.

Studies are also performed on the influence of scattering albedo on scaled combined heat flux on the inner cylinder when ε=1. The effects are found insignificant for all optical thicknesses. Figure 1c shows the results when ω_0=0.9. It is seen that the influence is very small (the maximum change is at Rc=0.56 for τ=1 and is less than 7%).

3. Combined Radiation and Natural Convection

Computations were performed at Ra=10^3, 10^4 and 10^5. As mentioned above, we let $\tau=\varepsilon=1$ and ω_0=0 for these cases.

a. Influence of Rc on flow and temperature patterns. Figures 2–4 show the streamlines and temperature contours at different Rayleigh numbers and with (a) Rc=0 (pure convection), (b) Rc=0.1, (c) Rc=1, and (d) Rc=∞ (convection induced by pure radiation), respectively. As Rc increases, the temperature profiles at all Rayleigh numbers are less distorted and more uniform in the θ-direction. When Rc=∞, heat convection and diffusion terms are zero by Eqs.(1) and (2), and heat transfer and temperature profiles are independent of the Rayleigh number and determined by one-dimensional radiation only, so there is a temperature slip on the walls. The flow pattern, however, changes when Ra varies (cf. Eqs.(1) and (2)).

The influence of Rc on max $|\psi|$ (and, therefore, the total circulating mass) at different Ra is interesting. At Ra=10^3, the max $|\psi|$ keeps decreasing monotonically from 2.365 to 1.323 as Rc increases from 0 to ∞. On the other hand, at Ra=10^5, the max $|\psi|$ is found increasing monotonically from 31.36 to 49.37 as Rc increases from 0 to 1, and no convergent solution could be obtained when $Rc \geq 10$. At the intermediate Rayleigh number 10^4, the max $|\psi|$ is found to increase from 13.27 to 16.59 in $0<Rc<1.78$, then it decreases to 14.03 at Rc=∞ (Fig. 5).

At all Rayleigh numbers, the position of the core of the main circulating flow moves downward when Rc increases. At Ra=10^3 and

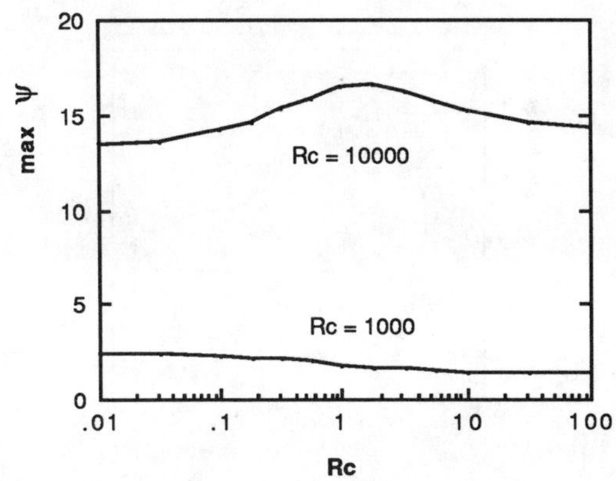

Fig. 5 Influence of radiation on max ψ when Ra=10^3 and 10^4.

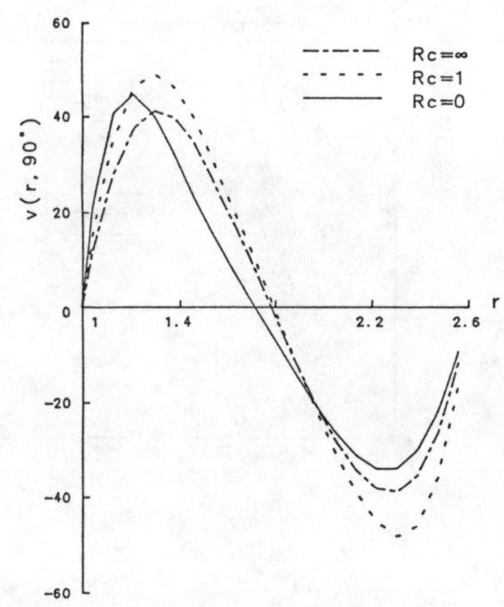

Fig. 6 Influence of radiation on θ-velocity at θ=90°, when Ra=10^4.

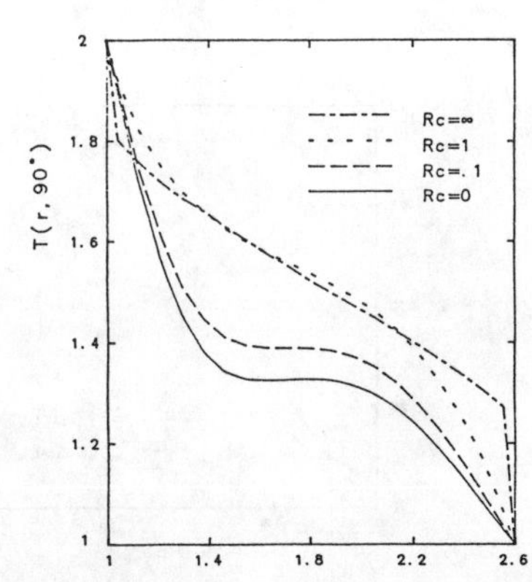

Fig. 7 Influence of radiation on temperature distribution at θ=90°, when Ra=10^4.

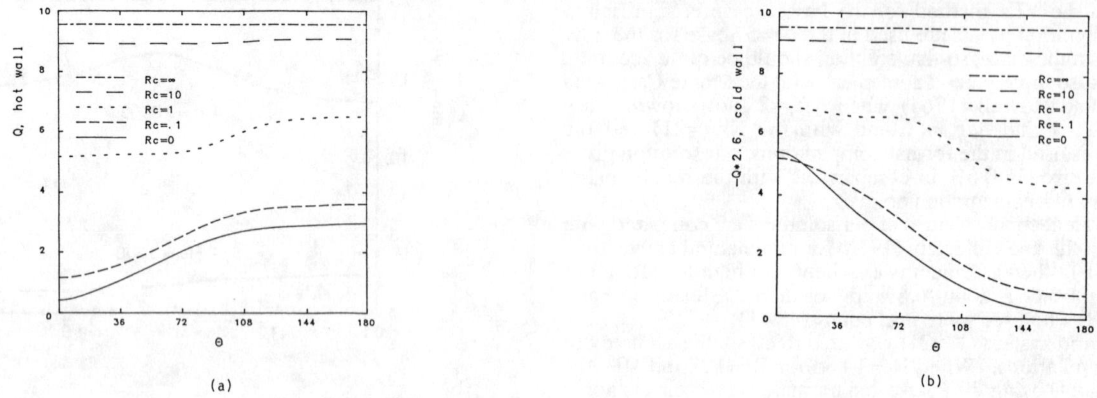

Fig. 8 Influence of radiation on combined heat fluxes. Ra=10^4. **(a)** hot wall; **(b)** cold wall.

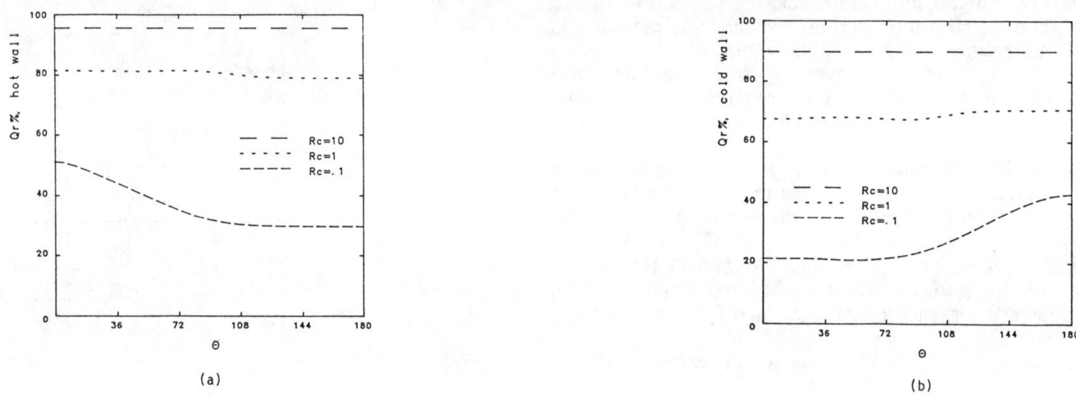

Fig. 9 Influence of radiation on percentage radiation flux distributions. Ra=10^4.

(a) hot wall; **(b)** cold wall

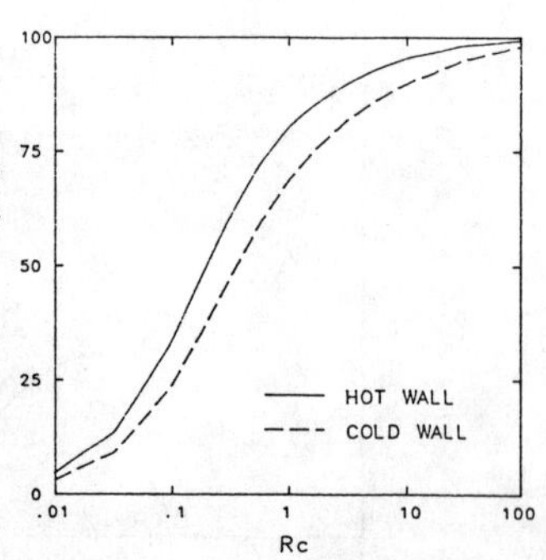

Fig. 10 Influence of radiation on percentage total radiation flux on walls. Ra=10^4.

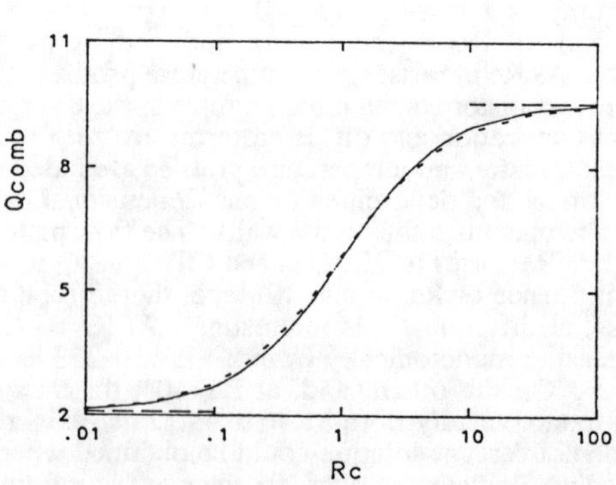

Fig. 11 Influence of radiation on scaled combined heat flux. The short dashed line are results calculated by addition approximation. The long dashed lines denote pure convection and radiation limits. Ra=10^4.

10^4, only one core is found for all Rc. Especially in the case of Ra=10^3, as radiation becomes more important (i.e., Rc increases), the core which is initially above the horizontal symmetry line (θ=90°) is pushed downward until it reaches the horizontal line at Rc=∞; correspondingly, the flow field becomes more symmetric to the horizontal line (Figs. 2a–2d). At Ra=10^5, the core is initially single and at a very high position (~θ=30°) in the pure convection (Rc=0) case. When Rc increases to 1, a secondary core appears and the main stream core moves below the horizontal line (~θ=120°). Combined with the results of above paragraph, we may conclude that the stability of flow can be improved by radiation at low Rayleigh numbers and is degraded by radiation at high Rayleigh numbers. This is different from the stability of a radiant fluid between horizontal parallel plates heated below, which is always found to be improved as radiation becomes important (Arpaci and Gozum, 1973). This phenomenon can be explained as follows. As mentioned above, the temperature in the annulus tends to be one-dimensional with radius when Rc increases, so the isotherms near the upper part of the inner wall become more horizontal and more dense than in the non-radiation case. Since the inner wall is hot, this results in the same condition at this local region as for Rayleigh-Benard flow. Note that when Rc=0, the core is always above the horizon which suggests that the lower part of the annulus has a tendency toward becoming unstable. Therefore, the presence of radiation, which pushes the core downward, can reduce or remove the tendency toward instability in the lower part but increase the possibility of instability in the upper part. At lower Rayleigh numbers, those tendencies compromise so well that the fluid becomes more stable; and an onset of instability cannot occur because the viscous force is very large. At higher Rayleigh numbers, however, the viscous force on the upper part cannot sustain the gravity force, and instability occurs.

Figure 6 shows the influence of Rc on the θ–velocity distribution at θ=90° when Ra=10^4. A peak and a valley is found at about r=1.3 and 2.3 for each line. The highest peak and lowest valley is found when Rc=1. This corresponds to the fact that the total circulating mass is the maximum near Rc=1 (at Rc=1.78, as mentioned above). It can be seen that the thickness of the boundary layer on the inner cylinder increases as Rc increases.

Corresponding to Fig.6, Fig.7 shows the influence of Rc on the temperature profile at θ=90° when Ra=10^4. The bulk temperature is found to be increased as radiation becomes more important. At small Rc, there is a region around r=1.8 where the temperature gradient is almost zero (see also Figs. 3a and 3b).

b. Influence of Rc on heat transfer. We shall discuss only the case Ra=10^4. As expected, the scaled combined heat flux on both the hot (inner) cylinder and the cold (outer) cylinder, Q, increases and becomes more uniform when Rc increases (Figs. 8a, 8b and 11). Figures 9a, 9b and 10 depict the percentage radiative heat flux on both walls. Radiation is found important even at Rc=0.1. On the hot wall, the radiation flux is larger than on the cold wall, due to the fact that the radiation flux is proportional to the difference of the fourth power of temperatures of the surface and the average surroundings. At the lower part of the inner cylinder and the upper part of the outer cylinder, the radiation flux is relatively small because convection in these regions is strong. When Rc>10, the convective flux is quite small. Convection is negligible if Rc>100 while radiation is negligible if Rc<0.01.

It is found that the combined heat transfer can be well approximated by simple addition (Fig. 11) in the black wall case. This formula also works at other Rayleigh numbers as long as the cylinder walls are black. If the walls are reflective, the approximation works poorly, as found in the radiation-conduction case (Rc=0) where the maximum error of the addition approximation is about 60 percent when ε=0.1 and τ =1.

CONCLUSIONS

In this paper, combined radiation and natural convection in a participating medium between concentric cylinders is studied by solving numerically the exact formulations. The following conclusions are obtained:

1. The numerical method used is very efficient. The results are accurate in comparison with limiting cases in the literature. To get better results, more grid points or higher order approximations may be needed.
2. The presence of radiation has great influence on the patterns of flow and temperature as well as heat transfer. At low Rayleigh numbers, the total circulating mass of the flow is increased by the presence of radiation. At high Rayleigh numbers, radiation decreases the circulating mass.
3. In the black wall case, the simple addition formula gives an excellent approximation to the combined heat flux.

ACKNOWLEDGEMENTS

This work is supported by the Texas Advanced Research Program, Grant Number 4587, "Innovative Methods for Analyzing Complex Systems Involving Radiative Transfer."

REFERENCES

Arpaci, V.S., and Bayazitoglu, Y., 1973, "Thermal Stability of Radiating Fluids: Asymmetric Slot Problem," *Phy. Fluids,* **16**, 589-593.

Arpaci, V.S., and Gozum, D., 1973, "Thermal Stability of Radiating Fluids: The Benard Problem," *Phy. Fluids,* **16**, 581-588.

Chang, L.C., Yang, K.T., and Lloyd, J.R., 1983, "Radiation-Natural Convection Interaction in Two-Dimensional Complex Enclosures," *ASME J. of Heat Transfer*, **105**, 89-95.

Chen, Haigen, 1982, "Zonal Model for Reheating Furnaces With Obstructing Walls," (in Chinese) Master's Thesis, Northeast University of Technology, Shenyang.

Kuehn, T.H., and Goldstein, R.J., 1976, "An Experimental and Theoretical Study of Natural Convection in the Annulus Between Horizontal Concentric Cylinders," *J. Fluid Mech.*, **76**, 695-715.

Kuehn, T.H., and Goldstein, R.J., 1976, "Correlating Equations for Natural Convection Heat Transfer Between Horizontal Circular Cylinders," *Int. J. Heat Mass Transfer,* **19**, 1127-1134.

Larson, D.W., 1981, "Enclosed Radiation and Turbulent Natural convection Induced by a Fire," in *Numerical Methods in Heat Transfer*, Eds. R.W. Lewis, K. Morgan, and O.C. Zienkiewicz, John Wiley and Sons.

Pandey, D.K., and Cogley, A.C., 1983, "An Integral Solution Procedure for Radiative Transfer in Concentric Cylinder Media," ASME Paper, 83-WA/HT-78.

Perlmutter, M., and Howell, J.R., 1964, "Radiant Transfer Through a Gray Gas Between Concentric Cylinders Using Monte Carlo,"*ASME J. of Heat Transfer*, **86**, 169-179.

Shih, T.-M., 1987, "A Literature Survey on Numerical Heat Transfer (1984-1985)," *Num. Heat Transfer,* **11**, 1-29.

Tan, Zhiqiang, 1989a, "Radiative Heat Transfer in Multidimensional Emitting, Absorbing and Anisotropic Scattering Media -- Mathematical Formulation and Numerical Method", *ASME J. of Heat Transfer,* **111**, 141-147.

Tan, Zhiqiang, 1989b, "Combined Radiative and Conductive Heat Transfer in Two-Dimensional Emitting, Absorbing and Anisotropic Scattering Square Media", accepted for publication in *Int. Comm. Heat Transfer*.

Tan, Zhiqiang, and Howell, J.R., 1989a, "A New Numerical Method for Radiation Heat Transfer in Participating Media," to be submitted.

Tan, Zhiqiang, and Howell, J.R., 1989b, "Combined Radiation and Natural Convection in a Two-Dimensional Participating Square Medium," to be submitted.

Trevisan, O. V. and Bejan, Adrian, 1986, "Convection Driven by the Nonuniform Absorption of Thermal Radiation at the Free Surface of a Stagnant Pool," *Num. Heat Trans.*, **10,** 483-506.

Webb, B.W., and Viskanta, R., 1987, "Radiation-Induced Buoyancy-Driven Flow in Rectangular Enclosures: Experiment and Analysis," *ASME J. of Heat Transfer*, **109**, 427-433.

Yang, H.Q., and Yang, K.T., 1988, "Natural Convection Suppression in Horizontal Annuli by Azimuthal Baffles," *Int. J. Heat Mass Transfer*, **31**, 2123-2135.

Yang, K.T., 1986, "Numerical Modelling of Natural Convection-Radiation Interactions in Enclosures," *Heat transfer*, The Eighth Inter-national Heat Transfer Conference, Vol.1, pp.131-140, Hemisphere, Washington, D.C.

Yücel, Adnan, and Bayazitoglu, Yildiz, 1980, "The Onset of Convection in Free Surface Layers with Radiative Heat Generation," *Lett. Heat Mass Trans.*, **7**, 391-396.

1989 National Heat Transfer Conference
HTD-Vol. 106, Heat Transfer Phenomena in Radiation, Combustion and Fires

TWO MAPPING TECHNIQUES FOR CALCULATING RADIATIVE HEAT TRANSFER WITH SCATTERING*

W. C. A. Gu and K. K. Murata
Sandia National Laboratories
Albuquerque, New Mexico

ABSTRACT

The problem of radiative heat transfer through a gray, emitting, absorbing, and scattering medium with uniform optical properties is reduced to one without scattering through two techniques. One uses scaling laws, and the other uses a self-consistent effective gas temperature. The scaling laws are derived via the P_1 approximation to the radiative transfer equation and can be applied to multidimensional problems with nonisothermal media. The effective temperature method is presently restricted to isotropic scattering and isothermal media. Both methods are evaluated in the current study as a function of scattering albedo, wall emissivity, and optical thickness for two different geometries, and two sets of wall and gas temperatures. The effects of scattering anisotropy are also assessed for the P_1 method. The numerical results show that for these cases the scaling method is reasonably accurate for optically thick media with a scattering albedo less than 0.8, and the effective temperature technique is reasonably accurate for optically thin media for all albedos.

NOMENCLATURE

D = diffusion coefficient, m
I = radiation angular intensity, W/m^2-sr
$\hat{n}_w$ = outward unit vector normal to a surface element
P = scattering phase function
q = radiant heat flux, W/m^2
r = radial coordinate, m
s = surface area, m^2
S = total surface area, m^2
T = temperature, K
v = volume, m^3
V = total volume, m^3

Greek Symbols

ϵ = emissivity
θ_o = scattering angle
μ = angular coordinate, see Fig. 1
$\bar{\mu}_o$ = average cosine of the scattering angle
ξ = optical depth
ρ = reflectivity
σ = attenuation coefficient due to radiation/media interactions, m^{-1}
$\hat{\sigma}$ = Stefan-Boltzmann constant, W/m^2-K^4
ϕ = scalar flux of thermal radiation, W/m^2
$\tilde{\phi}$ = volume-averaged scalar flux, W/m^2
ω = scattering albedo
$\hat{\Omega}$ = solid angle

Subscripts

a = absorption
e = extinction
eff = net efflux of thermal radiation
m = medium property
s = scattering
t = total length
w = wall property
1 = boundary index
2 = boundary index

Superscripts

---

*This work is supported by the United States Nuclear Regulatory Commission under FIN# A1198 and was performed at Sandia National Laboratories, which is operated for the United States Department of Energy under contract number DE-AC04-76DP00789

' = angular coordinate of the scattering direction, or the effective temperature

1. INTRODUCTION

Radiative heat transfer is one of the dominant modes of heat transfer in many engineering applications involving high temperatures. For radiative transfer through purely gaseous media, the effects of scattering are negligible in most heat transfer calculations due to the fact that the interactions between the thermal radiation and gas molecules or atoms occur in the Rayleigh limit. In this limit scattering is negligible compared to the absorption and transmission of radiation. However, in the presence of suspended solid or liquid aerosol particles, scattering often must be considered.

By nature, the equation of radiative transfer with scattering is an integro-differential equation whose solution is quite difficult. Although advanced numerical solution techniques are available for solving the general equation such as the P_n method (e.g., Bell and Glasstone, 1970), the discrete ordinates or the S_n method (e.g., Bell and Glasstone, 1970), and the Monte Carlo method (Meyer, 1956), such detailed numerical solutions are not warranted in many practical applications for a number of reasons. First, the solution of the radiative transfer equation is quite sensitive to the optical properties of the transport medium and such optical properties for materials of interest are seldom available in the literature, (e.g., the complex indices of refraction for most dielectric materials are only reported for visible light at room temperature.) Even if the indices of refraction are known, the shape, chemical structure, and size distribution of the aerosols may not be known accurately. Assuming this information were available, in some applications one cannot afford computationally to take the resulting wavelength dependence of the optical properties into account with high accuracy. Second, the results of radiant energy exchange are of little use in many cases without considering simultaneously the thermal-hydraulic response of the system to the radiation. The solution of the coupled problem can be a formidable task.

The above considerations indicate that methods for solving the scattering problem to moderate accuracy may be quite useful if they are efficient. The present paper proposes and gives the results of a preliminary study for two such methods. These methods have been developed for potential use with CONTAIN, a code that can analyze coupled aerosol, fission product, and thermal-hydraulic behavior in severe nuclear reactor accidents (Murata et al., 1989). The radiative transport problem is solved every timestep in this code, and consequently there is a premium on computational efficiency.

Methods with moderate accuracy similar to one of the present methods have been proposed before. The equation of radiative transfer reduces to a differential equation when scattering is absent so that the effort required to solve the problem is reduced considerably. Although scattering itself often cannot be neglected in the presence of aerosols, one can take an unconventional approach to the transport problem with scattering by first mapping it onto a nonscattering one, and then seeking the solution of the mapped problem. Such approaches have been reported in the literature (Goswami and Vachon, 1977; Lee and Buckius, 1982, 1983, and 1986). The basis of these earlier approaches is either the P_1 approximation to the transport equation or the two-flux method.

In the present study, two complementary mapping methods have been developed for potential application to the multidimensional problem, one using the P_1 approximation and the other using an effective medium temperature. The P_1 mapping is derived by requiring the P_1 solutions of the original and the mapped problems to be identical and can be applied to multidimensional problems with nonisothermal media. While similar to the P_1 approach used in Lee and Buckius (1983), the present approach has a number of advantages. First, when the equation for the scalar flux is scaled, as in the present approach, rather than that for the heat flux, it becomes apparent that nonisothermal media can be treated. (The difficulty is that the temperature gradient term in the heat flux equation introduces a second dimensionless group.) Second, when gray wall boundary conditions are used, a factor containing the wall emissivity can be combined with a second dimensionless group introduced by the boundary conditions. Because there are two parameters to scale in the present treatment (the optical depth and the wall emissivity), the scaling difficulties found by Lee and Buckius to arise from the presence of two dimensionless groups are absent, and the present method is exact within the P_1 approximation.

It is well known that the P_1 approximation is poor when the optical thickness of the system decreases (Duderstadt and Hamilton, 1976). Unfortunately, the proposed P_1 scaling laws transform the original problem with a high scattering albedo into an optically thin problem. Therefore, it is expected that the current P_1 mapping method will fail in two circumstances: (1) for optically thin systems, and (2) for high scattering albedos. (It should be noted that in contrast the P_1 approximation itself generally improves with scattering albedo at fixed optical depth.)

A second mapping method, an effective temperature method, has been developed for optically thin systems. In the effective temperature method, the emission and the scattering terms of the radiative transfer equation are combined into an effective emission term characterized by an effective emission temperature. Generally, this effective emission temperature is unknown and must be obtained by iteration.

Both the P_1 mapping technique and the effective temperature method have been assessed through a number of calculations for isothermal media comparing the mapped solutions with those obtained by a reference discrete ordinates method. While the generality of these results remains to be demonstrated, the results for the cases investigated indicate that the P_1 mapping method will result in less than 5 percent error in the net efflux of radiation (defined in Eq. (19)) for an optical depth greater than 4 and a scattering albedo less than 0.8, while the effective temperature method will result in less than 5 percent error for an optical depth less than 4. Furthermore, the error in both mapping methods is improved considerably when the transport medium is surrounded by reflective (nonblack) boundaries.

2. THEORY

The P_1 approximation to the radiative transfer equation can be derived by taking the zeroth and the first moments of the radiative transfer equation and assuming that the angular intensity is at most linearly anisotropic (see, for example, Duderstadt and Hamilton, 1976). The present scaling method uses the P_1 scalar flux equation, i.e.,

$$-D\nabla^2\phi(\vec{r})+\sigma_a\phi(\vec{r})=4\sigma_a\hat{\sigma}T^4(\vec{r}) \quad . \tag{1}$$

It should be noted that Eq. (1) applies only to a gray medium with uniform optical properties. The boundary condition for a diffusely reflecting wall in the P_1 approximation can be written as

$$D\vec{\nabla}\phi(\vec{r}_w)\cdot\hat{n}_w=\frac{\epsilon_w}{2(1+\rho_w)}\phi(\vec{r}_w)-\frac{2\epsilon_w}{1+\rho_w}\hat{\sigma}T_w^4 \quad , \tag{2}$$

where D is the diffusion coefficient defined as

$$D=[3(\sigma_e-\bar{\mu}_o\sigma_s)]^{-1} \quad , \tag{3}$$

and $\bar{\mu}_o$ is the first moment of the scattering phase function.

In the first mapping method, the original problem with scattering is solved by assuming that it is equivalent to a nonscattering one. The P_1 equation and the boundary condition for the mapped problem without scattering can be written as

$$-\bar{D}\nabla^2\bar{\phi}(\vec{r})+\bar{\sigma}_a\bar{\phi}(\vec{r})=4\bar{\sigma}_a\hat{\sigma}T^4(\vec{r}) \quad , \tag{4}$$

and

$$\bar{D}\vec{\nabla}\bar{\phi}(\vec{r}_w)\cdot\hat{n}_w=\frac{\bar{\epsilon}_w}{2(1+\bar{\rho}_w)}\bar{\phi}(\vec{r}_w)-\frac{2\bar{\epsilon}_w}{1+\bar{\rho}_w}\hat{\sigma}T_w^4 \quad , \tag{5}$$

where

$$\bar{D}=(3\bar{\sigma}_a)^{-1} \quad , \tag{6}$$

and the bar indicates the mapped quantities.

The solutions for ϕ in the original and mapped problems are identical, if and only if, the following relations are satisfied:

$$\frac{\sigma_a}{D}=\frac{\bar{\sigma}_a}{\bar{D}} \quad , \tag{7}$$

and

$$\frac{\epsilon_w}{D(1+\rho_w)}=\frac{\bar{\epsilon}_w}{\bar{D}(1+\bar{\rho}_w)} \quad . \tag{8}$$

Therefore after rearrangement, the scaling laws given by Eqs. (7) and (8) can be written as:

$$\bar{\sigma}_a=\sigma_e\sqrt{(1-\omega)(1-\bar{\mu}_o\omega)} \quad , \tag{9}$$

and

$$\bar{\rho}_w=2\left[1+\sqrt{\frac{1-\bar{\mu}_o\omega}{1-\omega}}\frac{\epsilon_w}{1+\rho_w}\right]^{-1}-1 \quad , \tag{10}$$

where ω is the single scattering albedo σ_s/σ_a. It should be noted that $\bar{\rho}_w$ may be negative, in which case the corresponding wall emissivity is greater than one. (The solutions for ϕ in this case are not unphysical because they correspond to the original problem with a properly bounded emissivity.) Eq. (9) is equivalent to the square-root scaling law derived by Lee and Buckius (1982, 1983).

Finally, since $\vec{q}(\vec{r})$ and $\phi(\vec{r})$ are related in the P_1 approximation according to

$$\vec{q}(\vec{r})=-D\vec{\nabla}\phi(\vec{r}) \quad , \tag{11}$$

the radiant energy flux of the original problem is given by

$$[\vec{q}(\vec{r})]_{\text{original}}=\sqrt{\frac{1-\omega}{1-\bar{\mu}_o\omega}}[\vec{q}(\vec{r})]_{\text{mapped}} \quad . \tag{12}$$

The second mapping method, or the effective temperature method, applies to problems with isotropic scattering. It is assumed here for simplicity that the medium temperature and effective temperature are constant. (This assumption can be relaxed through a generalization of the equations below.) For an isotropic scattering medium, the scattering integral in the radiative transfer equation, Eq. (A-1), reduces to $\sigma_s\phi(\vec{r})/4\pi$, and the right-hand side of Eq. (A-1) becomes independent of angle. Furthermore, if the optical depth is small and the spatial variation of the scattered flux is small, it is possible to represent the total radiation source (emission and scattering) with an effective radiation source characterized by an effective emission temperature T′:

$$\frac{\sigma_a}{\pi}\hat{\sigma}T^4+\frac{\sigma_s}{4\pi}\phi(\vec{r})=\frac{\sigma_e}{\pi}\hat{\sigma}(T')^4 \quad . \tag{13}$$

The radiative transfer equation for the mapped problem thus becomes a nonscattering one:

$$\hat{\Omega}\cdot\vec{\nabla}I(\vec{r},\hat{\Omega})+\sigma_e I(\vec{r},\hat{\Omega})=\frac{\sigma_e}{\pi}\hat{\sigma}(T')^4 \quad . \tag{14}$$

Generally, the effective temperature is unknown and must be found by iteration. It is defined implicitly by the volume averages of the radiation flux equation and Eq. 13. The radiation flux equation is obtained by integrating Eq. (14) over all solid angles, which results in

$$\vec{\nabla}\cdot\vec{q}(\vec{r})+\sigma_e\phi(\vec{r})=4\sigma_e\hat{\sigma}(T')^4 \quad . \tag{15}$$

From the divergence theorem, the volume-averaged scalar flux is found to be

$$\tilde{\phi}=4\hat{\sigma}(T')^4-\frac{1}{\sigma_e V}\oint_S\vec{q}(\vec{r})\cdot d\vec{s} \quad . \tag{16}$$

The volume average of Eq. (13) gives

$$(T')^4=(1-\omega)T^4+\frac{\omega}{4\hat{\sigma}}\tilde{\phi} \quad . \tag{17}$$

Therefore, one may write

$$T^4-(T')^4=\frac{\omega}{4\hat{\sigma}\sigma_e V(1-\omega)}\oint_S\vec{q}(\vec{r})\cdot d\vec{s} \quad . \tag{18}$$

Equation (18) defines T' implicitly and is solved iteratively for T'.

In the next section, the two mapping methods will be assessed against more exact numerical methods.

3. EVALUATION

To examine the validity of both the P_1 mapping technique and the effective temperature method, one-dimensional planar and spherical geometries are studied quantitatively. Figure 1 summarizes the coordinate systems used in the following calculations.

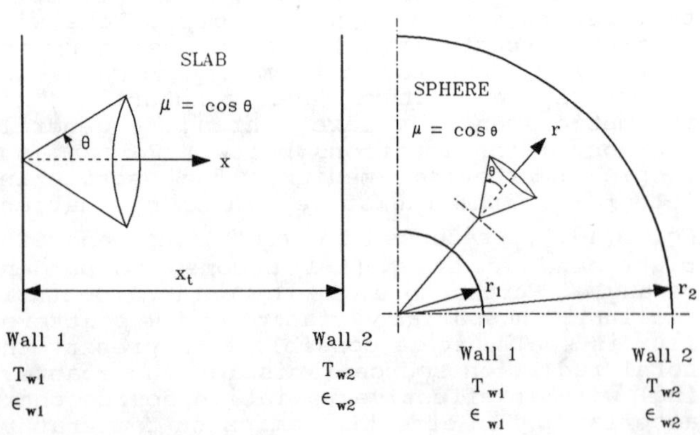

Fig. 1. Graphic representation of the coordinate systems used in the numerical calculations.

In the following assessment of the mapping methods, the original problems with scattering and the mapped problems without scattering were all solved by an S_{32} discrete ordinates method except for the mapped problems in planar geometry. For the latter cases the well-known analytic solution was used. The discrete ordinates solutions were obtained using 32 angular directions, 201 spatial nodes, and a convergence criterion of 10^{-6} for the fractional error in angular intensities. (The P_1-mapped problem is not solved here in the P_1 approximation because in the intended application, it will not be solved in the P_1 approximation but by essentially exact net enclosure methods.)

Finally, it is assumed in the assessment that one is interested in the net radiant heat fluxes at walls, or equivalently, the volume-averaged scalar flux, $\tilde{\phi}$, which is proportional to the total absorption of radiant energy by the medium. The net radiant heat fluxes at the boundaries, or efflux ϕ_{eff}, and the volume-averaged scalar flux are related by the radiation flux equation according to

$$\phi_{eff}=\frac{1}{\sigma_a V}\oint_S\vec{q}(\vec{r})\cdot d\vec{s}=4\hat{\sigma}T_m^4-\tilde{\phi} \quad . \tag{19}$$

In view of the large number of input parameters, the physical dimensions were fixed in the calculations. They are: x_t = 1 m, r_1 = 1 m, and r_2 = 2 m. Two sets of boundary temperatures were studied: (1) $T_{w1} = T_{w2} = T_w$, and T_w/T_m = 1.5; and (2) T_{w2}/T_{w1} = 1.5, and T_{w2}/T_m = 1.2. The combinations of geometries and temperatures considered are given by the three cases listed in Table 1. Each case was evaluated for the two mapping methods as a function of wall emissivity, albedo, and optical depth. The P_1 mapping method was also evaluated for two different scattering phase functions. The first,

$$P_1(\theta_o)=1 \quad , \tag{20}$$

represents isotropic scattering and the second,

$$P_2(\theta_o)=\frac{8}{3\pi}(\sin\theta_o-\theta_o\cos\theta_o) \quad , \tag{21}$$

represents diffuse scattering from a large sphere (Siegel and Howell, 1972). Hereafter, these two phase functions will be referred to as the isotropic scattering and anisotropic scattering phase functions, respectively.

The percentage error in the net radiation efflux is presented in Figures 2 through 8 for various wall emissivities. The percentage error of the mapping methods is defined by

$$\left|\frac{(\phi_{eff})_{mapping}-(\phi_{eff})_{Sn}}{(\phi_{eff})_{Sn}}\right|\times 100 \quad . \tag{22}$$

Table 1. Cases considered.

	Geometry	T_m	T_{w1}	T_{w2}
Case I	Planar	T_m	$1.5T_m$	$1.5T_m$
Case II	Planar	T_m	$0.8T_m$	$1.2T_m$
Case III	Spherical	T_m	$1.5T_m$	$1.5T_m$

The coordinate ξ_t in these figures is equal to $\sigma_e x_t$ and $\sigma_e(r_2-r_1)$ in planar and spherical geometry, respectively. Error contour plots are shown for the P_1 mapping technique in Figures 2 through 5, and for the effective temperature method in Figures 6 through 8.

There are several conclusions that can be drawn from the numerical results:

1. As shown in Figures 2 through 5, the error in the efflux in the P_1 mapping method increases as the scattering albedo increases. This can be attributed to the fact that the mapped system as shown in Eq. (9) becomes optically thinner as scattering albedo increases, and the P_1 approximation is poor for an optically thin system. The errors are not a strong function of the cases and the two scattering phase functions studied. However, it remains to be seen whether the P_1 mapping method is suitable for a more irregularly shaped phase function, as found in general from Mie theory (van de Hulst, 1981). As shown in Figures 2 through 5, an error in the efflux of less than 5 percent can be expected using the P_1 mapping technique for an optical depth greater than about four, and a scattering albedo less than about 0.8.

2. As shown in Figures 6 through 8, the efflux in the effective temperature method is accurate for small to moderate values of optical thickness. For an optical thickness less than about four, the effective temperature method is very accurate for the entire range of the scattering albedos. Note that this conclusion is drawn for the efflux for the cases considered; other quantities and cases are presently under investigation.

3. The accuracy of both mapping methods studies improves considerably as the wall emissivity decreases. For a gray wall, part of the thermal radiation that reaches the boundary is reflected back into the system. As a result, radiant energy becomes more evenly distributed than it would with black walls, and consequently both methods become more suitable.

4. In Figures 2 through 5 for the P_1 mapping technique, the errors are observed to decrease sharply as the optical depth becomes extremely small. This behavior is surprizing since it is known that the P_1 approximation fails for a transparent medium. However, it can be shown that the P_1 approximation predicts the correct transparent medium wall fluxes in the limit as ξ_t approaches zero. Therefore, both the P_1 approximation and the P_1 mapping technique predict the correct behavior in this limit.

4. CONCLUSIONS

Two mapping methods based on the P_1 approximation and an effective temperature have been developed for radiative heat transfer problems with scattering. In the P_1 mapping technique, a radiative transfer problem with scattering medium is transformed into a non-scattering one by scaling the absorption coefficient and the wall emissivities within the P_1 approximation. The scaling laws thus obtained are not restricted to the one-dimensional planar geometries used in previous scaling approaches and may in principle be applied to nonisothermal problems as well. The effective temperature method applies to optically thin systems with isotropic scattering and isothermal media, provided the spatial variation of the scattered flux is small.

Both mapping methods have been examined in a number of cases involving an isothermal gas and both planar and curved geometries. Note, however, that although a large number of calculations were required to generate the results in Figures 2-8, the cases presented in this paper are not comprehensive. For example, it would be desirable to run cases with a hot gas radiating to cold walls and with a relatively small hot surface radiating to cold walls that have a much larger area. In the latter case, it is not clear that the net efflux is the appropriate quantity to use in evaluating the accuracy of the mapping methods. The errors in the transmitted flux and gas temperature under equilibrium conditions are also of interest. These other cases and quantities are presently under investigation.

It is concluded that for the cases considered the current P_1 mapping technique gives a reasonably accurate net efflux for an optical depth greater than about four and a scattering albedo less than about 0.8, and the effective temperature method gives a reasonably accurate one for an optical depth less than about four.

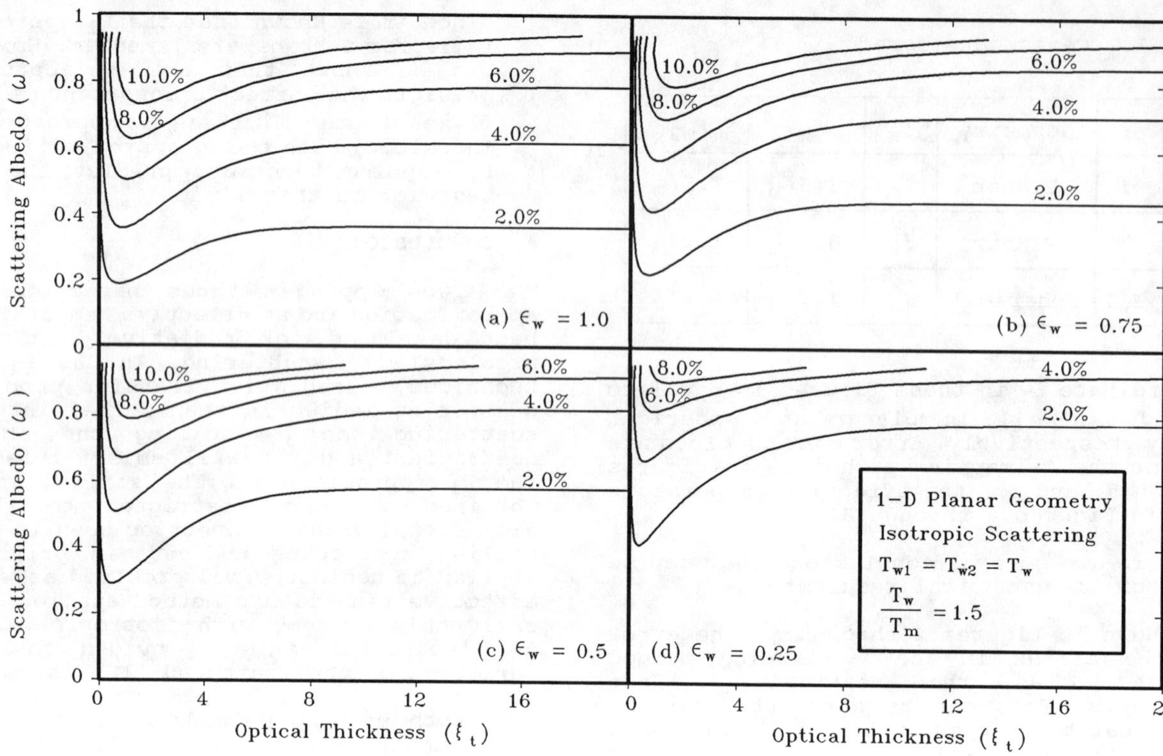

Fig. 2. Contours of constant percentage error in the net efflux, Eq. (22), predicted by the P_1 mapping technique for Case I of Table 1 and isotropic scattering.

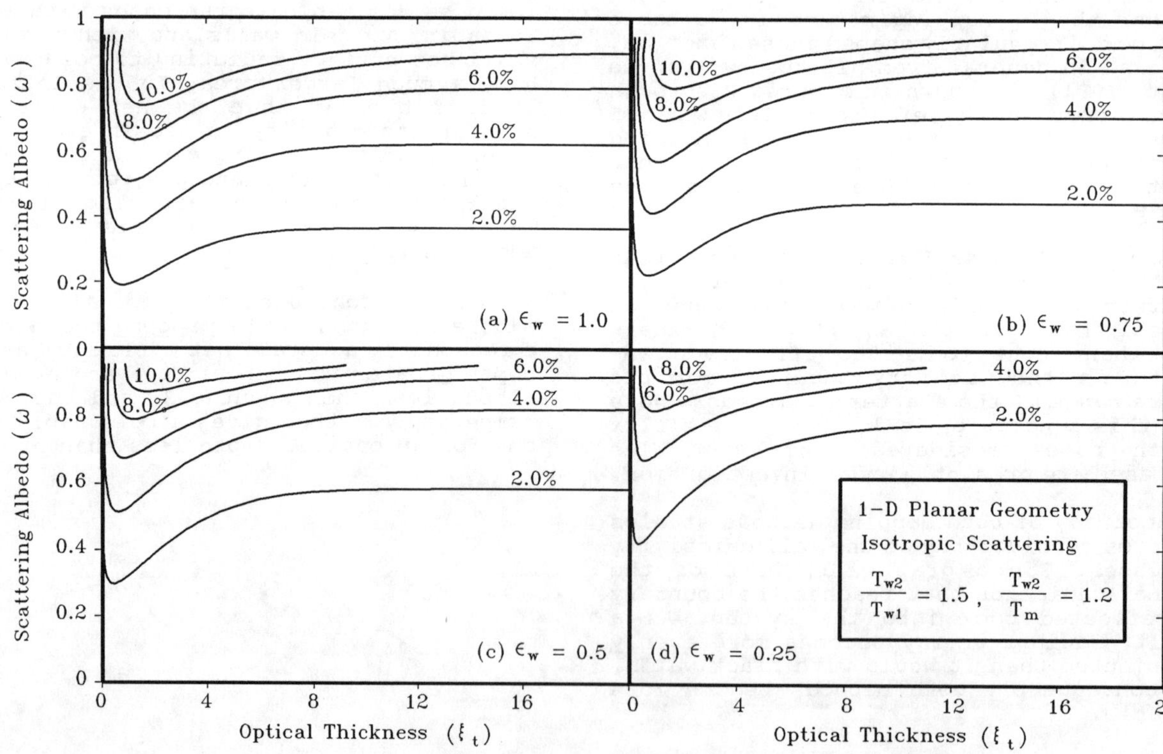

Fig. 3. Contours of constant percentage error in the net efflux, Eq. (22), predicted by the P_1 mapping technique for Case II of Table 1 and isotropic scattering.

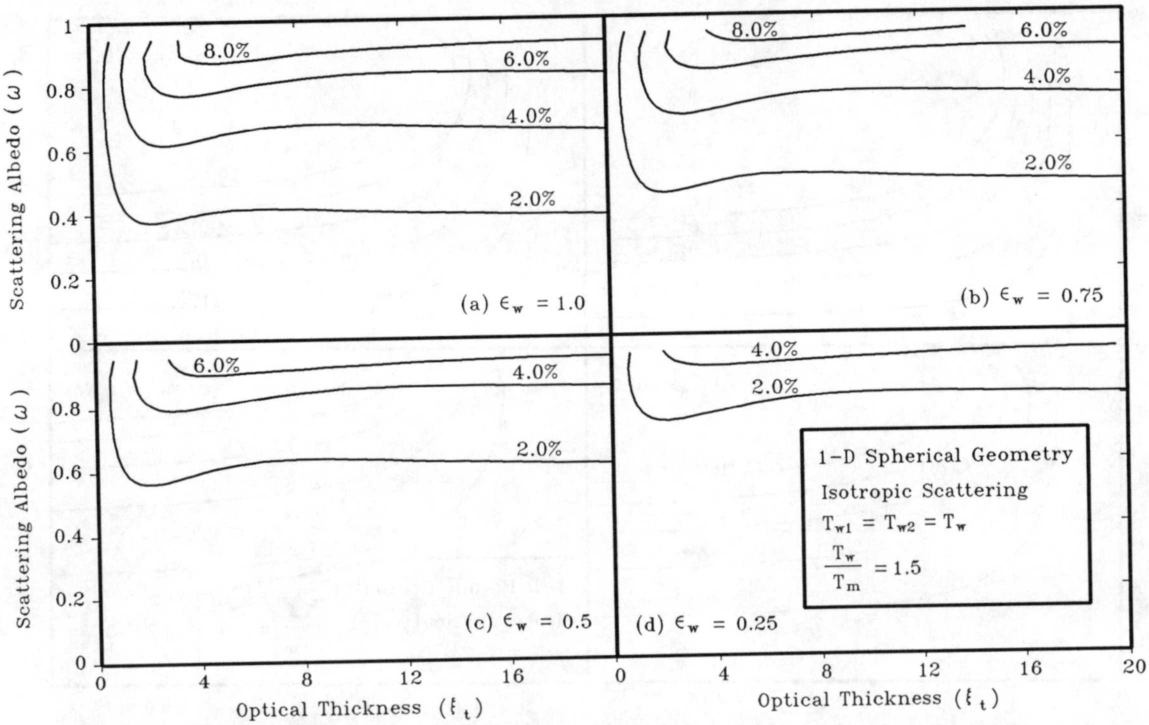

Fig. 4. Contours of constant percentage error in the net efflux, Eq. (22), predicted by the P_1 mapping technique for Case III of Table 1 and isotropic scattering.

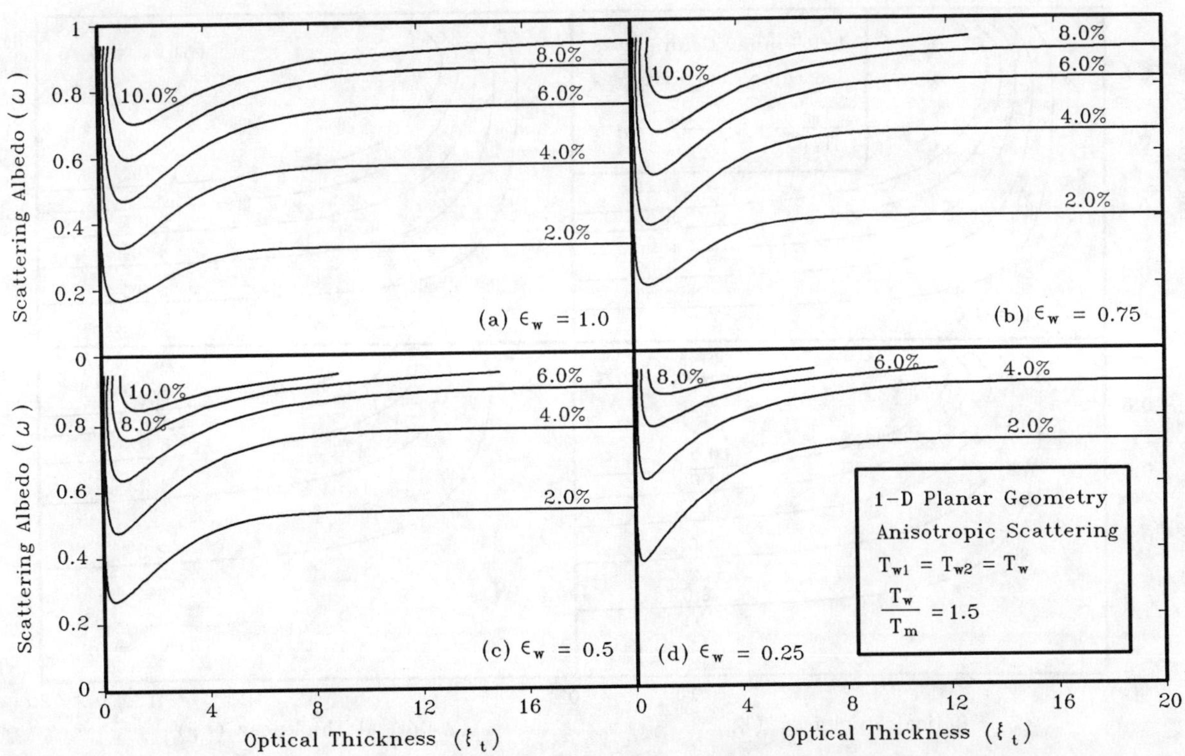

Fig. 5. Contours of constant percentage error in the net efflux, Eq. (22), predicted by the P_1 mapping technique for case I of Table 1 and anisotropic scattering.

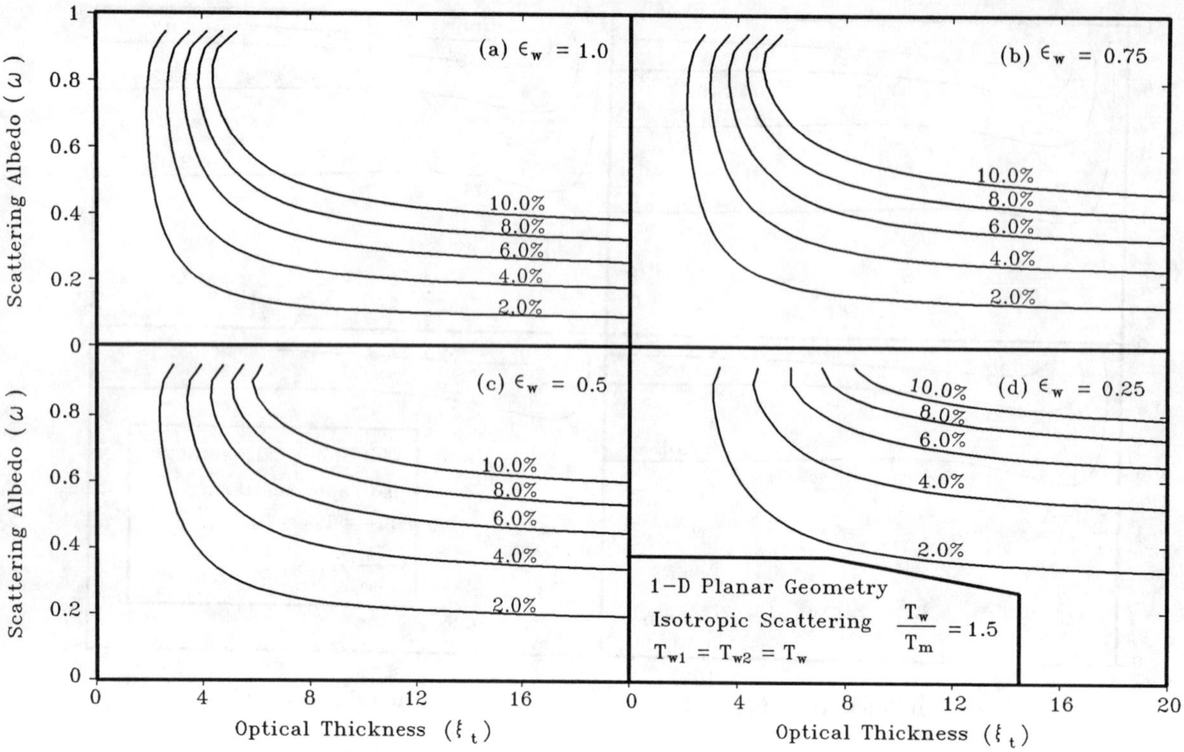

Fig. 6. Contours of constant percentage error in the net efflux, Eq. (22), predicted by the effective temperature technique for Case I of Table 1.

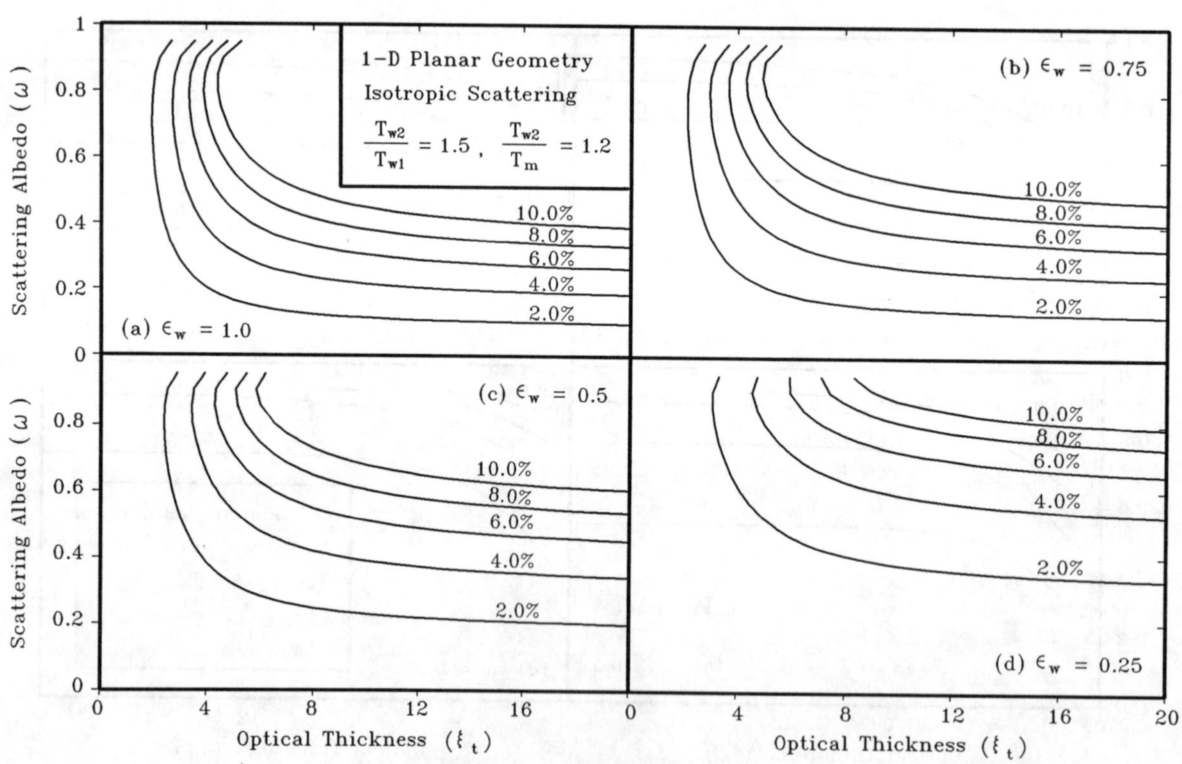

Fig. 7. Contours of constant percentage error in the net efflux, Eq. (22), predicted by the effective temperature technique for Case II of Table 1.

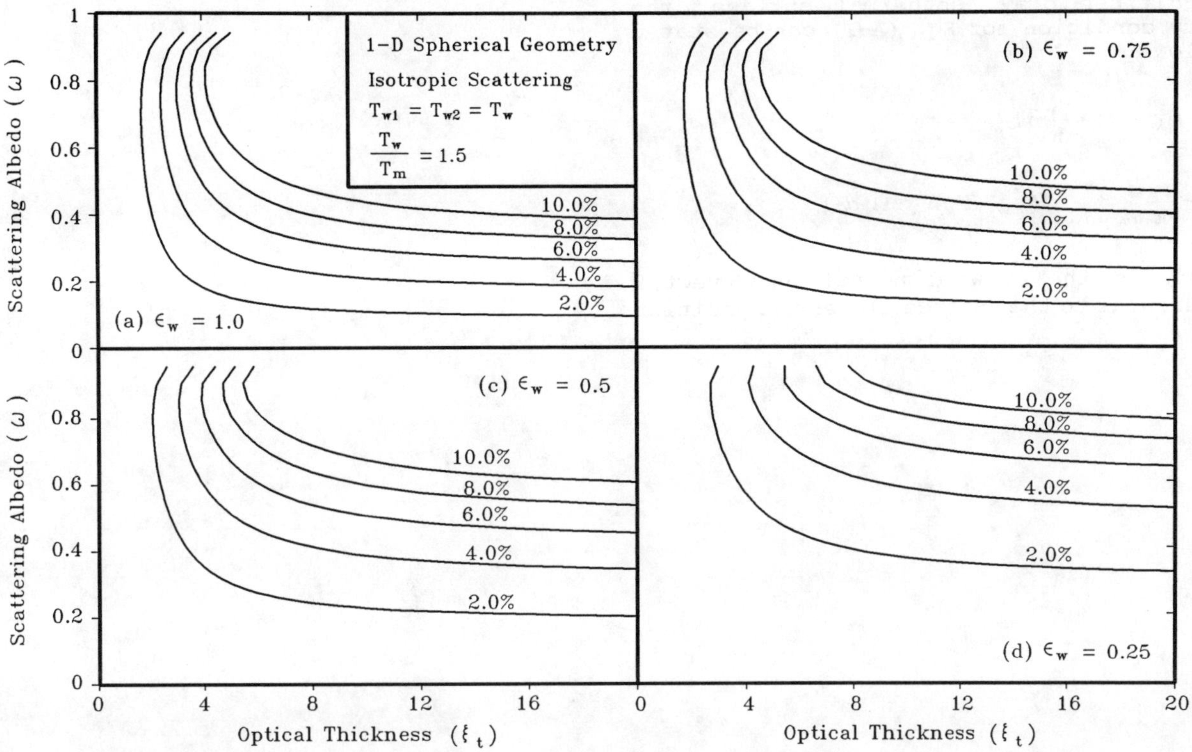

Fig. 8. Contours of constant percentage error in the net efflux, Eq. (22), predicted by the effective temperature technique for Case III of Table 1.

REFERENCES

Bell, G. I., and Glasstone, S., 1970, Nuclear Reactor Theory, Van Nostrand, Princeton, NJ.

Duderstadt, J. J., and Hamilton, L. J., 1976, Nuclear Reactor Analysis, John Wiley & Sons, Inc., New York.

Goswami, D. Y., and Vachon, R. I., 1977, "Radiative Heat Transfer Analysis Using an Effective Absorptivity for Absorption, Emission and Scattering," International Journal of Heat and Mass Transfer, Vol. 20, pp. 1233-1239.

Lee, H., and Buckius, R. O., 1982, "Scaling Anisotropic Scattering in Radiation Heat Transfer for a Planar Medium," Journal of Heat Transfer, Vol. 104, pp. 68-75.

Lee, H., and Buckius, R. O., 1983, "Reducing Scattering to Nonscattering Problems in Radiation Heat Transfer," International Journal of Heat and Mass Transfer, Vol. 26, No. 7, pp. 1055-1062.

Lee, H., and Buckius, R. O., 1986, "Combined Mode Heat Transfer Analysis Utilizing Radiation Scaling," Journal of Heat Transfer, Vol. 108, pp. 626-632.

Meyer, H. A. (ed.), 1956, Symposium on Monte Carlo Methods, John Wiley & Sons, Inc., New York.

Murata, K. K., et al., 1989, "User's Manual for CONTAIN 1.1, A Computer Code for Severe Nuclear Reactor Accident Containment Analysis," (Draft), SAND87-2309, Sandia National Laboratories, Albuquerque, NM.

Siegel, R., and Howell, J. R., 1972, Thermal Radiation Heat Transfer, McGraw-Hill Book Co., New York.

van de Hulst, H. C., 1981, Light Scattering by Small Particles, Dover Publications, Inc., New York.

APPENDIX A. THE P_1 APPROXIMATION

For a gray medium with uniform optical properties, the radiative transfer equation is given by

$$\hat{\Omega}\cdot\vec{\nabla}I(\vec{r},\hat{\Omega}) + \sigma_e I(\vec{r},\hat{\Omega}) = \frac{\sigma_a}{\pi}\hat{\sigma}T^4(\vec{r}) + \frac{\sigma_s}{4\pi}\int_{4\pi} P(\hat{\Omega}\cdot\hat{\Omega}')I(\vec{r},\hat{\Omega}')d\hat{\Omega}' \quad . \qquad (A-1)$$

For a diffuse gray isothermal surface, the boundary condition for Eq. (A-1) can be stated as follows:

$$I(\vec{r}_w, \hat{\Omega}\cdot\hat{n}_w > 0) = \frac{\epsilon_w}{\pi}\hat{\sigma}T_w^4 - \frac{\rho_w}{\pi}\int_{\hat{n}_w\cdot\hat{\Omega}'<0} I(\vec{r}_w, \hat{\Omega}')\hat{n}_w\cdot\hat{\Omega}' d\hat{\Omega}' \quad , \qquad \text{(A-2)}$$

where $\hat{n}_w$ is the outward normal unit vector perpendicular to the surface element at position $\vec{r}_w$.

1989 National Heat Transfer Conference
HTD-Vol. 106, Heat Transfer Phenomena in Radiation, Combustion and Fires

GRAY GAS RADIATION EFFECT ON THE ENHANCEMENT OF RADIATIVE HEAT TRANSFER BY POROUS LAYER

Z.-G. Du and S.-Y. Ko
Institute of Engineering Thermophysics
Chinese Academy of Sciences
Beijing, People's Republic of China

Abstract

The effect of gas radiation on the Effective Energy Conversion Method(EECM) proposed by R. Echigo has been studied by using the two phase radiative transfer equations proposed earlier by the present authors. The study results indicated that the EECM will be seriously restricted when radiative gases are involved. Since the solid porous layers in EECM are usually very thin and the upstream gas temperatures are higher than that of the porous layer, the gas radiation will still be very important even if gas radiative absorption coefficient α_g is much smaller than that of the solid porous material α_p. These restrictions may eliminate the advantage of the EECM with non-radiative gases.

Nomenclature

- A_{21} apparent surface area of an individual porous particle
- B_p blackbody intensity
- B_1, B_2 diffusive radiation boundary condition on the upstream and downstream side of the porous layer
- B_u, B_d radiation sources in the upstream and downstream of the porous layer
- c specific heat capacity
- F ratio of radiative heat transfer to gas enthalpy
- g gas phase
- G quantity in Eq.(12)
- h_p heat transfer coefficient between gas and porous matrix
- H dimensionaless total radiative flux
- I_g, I_p specific intensity of the gas phase and the solid porous phase
- L_u, L_d equivalent thickness of gases upstream and downstream of the porous layer
- n_p number density of porous particles
- p porous phase
- q_{gr}, q_{pr} radiative heat flux in gas or porous phase
- q_r total radiative heat flux ($=q_{gr}+q_{pr}$)
- Q quantity in Eq.(41)
- R ratio of gas-porous matrix convection heat transfer to radiative heat transfer
- R_t equivalent thermal resistance
- T_g, T_p temperature of gas and porous material
- T_e temperature of the surroundings
- u_g gas flow velocity
- x coordinate
- x^* dimensionless coordinate
- α_g, α_p gray absorption coefficients of gas and porous material
- γ ratio of gas radiation to porous radiation($=\alpha_g/\kappa_p$)
- δ thickness of porous layer
- η dimensionaless G in Eq.(35)
- θ_g, θ_p dimensionaless temperatures of gas and porous matrix
- κ total extinction coefficient of the gas and porous material
- κ_p extinction coefficient of porous material
- ρ mass density of gas
- σ_p gray scattering coefficient of porous material
- τ optical thickness of porous layer (including gas radiation)
- τ_p optical thickness due to porous radiative properties
- ω_s scattering abledo
- Ω direction

* present address: Ecole Polytechnique de Montreal, Case Postale 6079, succ."A", Montreal, Quebec H3C 3A7

1 Introduction

Enhancement of heat transfer by radiation has received increasing attention in such high temperature applications as the design of high performance heat exchangers, the combustion of low calorific gases and so forth. The common feature of the radiation enhancement is to augment the radiative heat transfer by adding various solid components of relative high radiative ability to the high temperature gas flow. These radiative solid components may be fine solid particles [Hasengawa,etc,1986], solid plates [Mori,1980, 1986,Hasengawa, etc,1974] and porous materials [Echigo,1982]. All these methods have been reported as having noticeable advantages over those without the added radiative solid components.

Among these proposed radiation enhancement methods, a promising one is the Effective Energy Conversion Method (EECM) suggested by Echigo [1982]. In this method, a thin porous layer is installed downstream of a high temperature gas flow. Owing to the strong radiative ability and high convective heat transfer characteristics of the porous materials, the enthalpy of the high temperature gas can be effectively transferred to the porous matrix. Then, a considerable amount of this transferred energy is recovered by the upstream radiation of the porous layer. As a result, an extreme high temperature drop is produced within the porous layer (about 300 K/cm), which corresponds to a high equivalent thermal resistance. Also, a uniform radiation field is formed upstream of the porous layer. Since 1982, Echigo and his associates [1982,1986a,b,c] have made substantial contributions to this method. The main feature of their works is to develop new types of high performance heat exchangers, to improve the heating equipment and to combust low carlorific gases by using the most important characteristic of the method, namely, the negative radiative flux against the flow direction from the porous layer. It is this characteristics characteristic that leads leads to the radiative energy recovery ability of the EECM, which has been studied extensively [Echigo,1982; Wang and Tien,1984;Ko, Du and Xu,1986]. These studies emphasized the influences of the optical thickness, of the scattering of the porous layer and of the formation of the uniform high temperature field in the upstream gas flow. As the typical experiment and application by Echigo were performed through electrical heating (or combustion of city gas or clean oil), the produced high temperature gases can be considered as non-radiating. Thus all these previous studies were carried out under the condition that the flowing high temperature gases possessed no radiative ability. Therefore, no information has been provided on the gas radiation effect. Nevertheless, the gas radiation will be of practical importance because in most industrial applications the high temperature gases involved are mostly radiative gases, such as gases containing water vapor or carbon dioxide. So the application and extension to such radiative gases have naturally become a true challenge for the EECM. Therefore, the primary objective of this research is to ascertain if the enhancement effect of the EECM still persists when radiative gases are involved.

2 Conceptual Consideration

A conceptual consideration about the effect of gas radiation on the energy recovery ability would be helpful, since it can provide us with a qualitative understanding of the physical phenomena thus avoid possible confusion in the mathematical formulation. As shown in Fig.(1) a thin porous layer is installed inside a flow channel.The radiative property of such a thin porous layer can be described by its absorption coefficient α_p. When a high temperature gas flowing into the porous layer is non-radiative, the porous layer will receive the enthalpy of the gas by convection only. Since the non-radiative gases present no resistance to radiation , the porous layer will radiate the received energy towards its upstream and downstream without any resistance. As a result, there is a net negative radiative heat flux produced on the upstream side of the porous layer, and this leads to a considerable amount of recovered radiative energy from the porous layer. At the same time, an extremely high temperature drop is produced within the thin porous layer.

Now, let us consider a more general case with radiative flowing gas. This flowing gas possesses radiative ability described by its absorption coefficient α_g. In this case, as the porous layer radiates energy towards its upstream and downstream, the flowing gas will exert resistance to this radiative energy transfer process , because the gas will also radiate energy to the porous layer. If α_g is conceptually assumed to be large enough and α_p is not significant, the flowing system will behave as only one medium with one radiative property α_g . That is , the porous layer will only attribute a negligible quantity to the energy balance of the flowing gas. Thus there will be no high temperature drop presented and also, the energy recovery ability of the porous layer will be greatly diminished. Instinctively, the porous layer can recover energy in a non-radiative gas flow because it is "foreign" matter added into the flowing system. The "foreign" characteristics of it lies in its solid radiative ability which will qualitatively influence the energy balance of the flowing gas. However,if the gas possesses radiative ability as well, the porous layer will not be " foreign " matter in the system and its influence on the energy balance of the flowing gas becomes quantitative effect. With increasing gas radiation ability, this quantitative influence will be decreased or even become negligible when gas radiation predominates. Therefore, the radiative energy recovery ability of the porous layer will be diminished. In industrial applications, although the radiative ability of gases is not comparable with that of solid porous materials, the effect of gas radiation can still be important due to the following:

a) a gas possessing radiative ability changes the "foreign" characteristics of the porous layer qualitatively;

b) the porous layer is usually made very thin to avoid large flow resistance, while the upstream gas layer is usually thicker and at higher temperatures. Thus, the quantity of the gas radiation from such a thick radiative gas layer could be comparable with, or even larger than that of the porous layer.

3 Problem Formulation

The physical model for studying the effect of gas radiation on the energy recovery ability of the EECM is shown in Fig.(2). Since the thickness δ of the porous layer is rather small, the temperature and radiation field within the layer are assumed to be one dimensional. The gas velocity u_g is constant. There may exist radiative sources in the open spaces upstream of the porous layer to simulate the practical situation where thin wired meshes may be installed as in furnaces and high temperature heat exchangers [Echigo,1982,1986a,b,c].

Following Echigo [1982] and Wang and Tien [1984], the porous layer can be considered as having homogeneous composition. For a system involving radiative gases, the energy balance equation for the flowing gas within the porous layer will become:

$$\rho_g c_g u_g (dT_g/dx)+dq_{gr}/dx+h_p n_p A_p (T_g-T_p)=0 \qquad (1)$$

where T_g and T_p are the temperatures of gas and porous material respectively, h_p is the heat transfer coefficient between the gas and porous material and n_p

is the number density of particles in the porous layer. The A_p is the apparent surface area of an individual particle. The term q_{gr} represents the radiative heat flux of flowing gas. Obviously, the presence of q_{gr} will make the energy balance of the flowing gas very different from that of non-radiating gases systems.

For the porous materials, the energy balance requires:

$$h_p n_p A_p (T_g - T_p) = dq_{pr}/dx \tag{2}$$

where q_{pr} is the radiative heat flux of the porous material. Using Eq.(2), Eq.(1) can be rewritten as:

$$\rho_g c_g u_g (dT_g/dx) + d(q_{gr} + q_{pr})/dx = 0 \tag{3}$$

4 Determination of Radiative Heat Fluxes q_{gr} and q_{pr}

It can be noted that for the present analysis there are two radiative heat fluxes q_{gr} and q_{pr} in the porous layer to describe the radiation fields produced by gas radiation and solid porous material radiation, respectively. These two radiative heat fluxes should be determined by the reduced two phase radiative transfer equations proposed earlier[Du, Ko, 1987, 1988] as follows:

Form I

$$\Omega\nabla I = \alpha_g B(T_g) + \alpha_p B(T_p) - (\alpha_g + \alpha_p + \sigma_p) I + (1/4\pi)\sigma_p \int I(\Omega) d\Omega \tag{4}$$

Form II

$$\Omega\nabla I_g = \alpha_g [B(T_g) - I_g] - \alpha_p I_p \tag{5}$$

$$\Omega\nabla I_p = \alpha_p [B(T_p) - I_p] - \alpha_p I_g - \sigma_p) I + (1/4\pi)\sigma_p \int I(\Omega) d\Omega \tag{6}$$

where, the gray approximation for radiation and isotropic scattering for the porous material (which means ψ=1) are postulated. The gas scattering has been neglected since it is usually negligible. Now, by using Eqs.(4),(5),(6), we can start to determine the radiative heat fluxes q_{gr}, q_{pr} and q_r.

For one dimensional radiative transfer, it is customary to divide the radiation intensity $I(x,\cos\theta)$ into forward and backward parts as:

$$I(x,\cos\theta) = \begin{cases} I^+(x,\cos\theta) & \cos\theta > 0 \\ I^-(x,\cos\theta) & \cos\theta < 0 \end{cases} \tag{7}$$

where θ is the angle between the positive direction and the direction of radiation propagation. Hence Eq.(4) becomes:

$$\cos\theta (dI^+/dx) + \kappa I^+ = [\alpha_g B(T_g) + \alpha_p B(T_p)] + (1/4\pi)\sigma_p G(x) \tag{8}$$

$$\cos\theta (dI^-/dx) + \kappa I^- = [\alpha_g B(T_g) + \alpha_p B(T_p)] + (1/4\pi)\sigma_p G(x) \tag{9}$$

where

$$\kappa = \alpha_g + \alpha_p \tag{10}$$
$$\kappa_p = \alpha_p + \sigma_p \tag{11}$$

The diffusive radiative boundary conditions for Eqs.(8),(9) are:

$$I^+(x=0,\ \cos\theta) = B_1/\pi \tag{13}$$
$$I^-(x=\delta,\ \cos\theta) = B_2/\pi \tag{14}$$

It is easy to obtain the solution of Eqs.(8) and (9) with boundary conditions (13), (14) and then determine the total radiative heat flux q_r as:

$$\begin{aligned} q_r(r) &= \int I(r,\Omega)\Omega d\Omega \\ &= 2B_1E_3(\tau) - 2B_2E_3(\tau_0 - \tau) + (2/\kappa)\int\{[\alpha_g\sigma T_g{}^4 + \alpha_p\sigma T_p{}^4 \\ &+ \sigma_p G(t)/4]E_2(\tau - t)\}dt + (2/\kappa)\int\{[\alpha_g\sigma T_g{}^4 + \alpha_p\sigma T_p{}^4 \\ &+ \sigma_p G(t)/4]E_2(\theta - t)\}dt \end{aligned} \tag{15}$$

with

$$G(\tau) = 2B_1E_2(\tau) + 2B_2E_2(\tau_0 - \tau) + \\ +(2/\kappa)\int\{[\alpha_g\sigma T_g{}^4 + \alpha_p\sigma T_p{}^4 + \sigma_p G(t)/4]E_1(|\tau - t|)\}dt \tag{16}$$

where τ and τ_0 are the optical thicknesses defined by:

$$\tau = \kappa x \tag{17}$$
$$\tau_0 = \kappa\delta \tag{18}$$

and $E_1(x)$, $E_2(x)$ and $E_3(x)$ are the well-known exponential integral functions:

$$E_n(x) = \int_0^1 \mu \exp(-x/\mu) d\mu \tag{19}$$

To determine q_{gr} or q_{pr}, one should use the two phase radiative transfer equation (Form II), i.e., Eqs.(5) and (6). From Eq.(5) one has:

$$I = -(1/\alpha_g)\Omega\nabla I_g + B(T_g) \tag{20}$$

Substituting it into Eq.(6) one obtains:

$$\Omega\nabla I_p = \alpha_p B(T_p) - \kappa_p B(T_g) - (\kappa_p/\alpha_g)\Omega\nabla I_g + (\sigma_p/4\pi)\int I(\Omega) d\Omega \tag{21}$$

In order to integrate Eq.(21), the following equations are used:

$$\int \Omega I_g(r,\Omega) d\Omega = \nabla q_{gr}(r) \tag{22}$$

$$\int \Omega I_p(r,\Omega) d\Omega = \nabla q_{pr}(r) \tag{23}$$

With Eqs.(22),(23), integrating Eq.(21) over 4π solid angle and for one dimensional case, one obtains:

$$dq_{pr}/dx = (\kappa_p/\alpha_g) dq_{gr}/dx + 4\alpha_p\sigma T_p{}^4 - 4\kappa_p\sigma T_g{}^4 - \sigma_p G(x) \tag{24}$$

and

$$(1 + \kappa_p/\alpha_g) dq_{pr}/dx = (\kappa_p/\alpha_g)(dq_{gr}/dx + dq_{pr}/dx) \\ + 4\alpha_p\sigma T_p{}^4 - 4\kappa_p\sigma T_g{}^4 - \sigma G(x) \tag{25}$$

Since $q_r = q_{gr} + q_{pr}$, Eq.(25) can be written as:

$$dq_{pr}/dx = (\kappa_p/\kappa) dq_r/dx + 4(\alpha_g\alpha_p/\kappa)\sigma T_p{}^4 \\ - 4(\alpha_p\kappa_p/\kappa)\sigma T_g{}^4 + (\alpha_p\sigma_p/\kappa)G(x) \tag{26}$$

5 Radiation Boundary Condition

Until now, the radiation heat fluxes q_r and q_{pr} in the system energy balance equations (2) and (3) have been determined by Eqs.(15) and Eq.(26) from the two phase radiative transfer equations (4), (5) and (6). The radiation boundary conditions B_1 and B_2 presented in Eqs.(15) and (26) remain to be determined.

From Eqs.(13),(14), we have

$$B_1=\pi I_+(x=0,\cos\theta) \quad (27)$$

$$B_2=\pi I_-(x=\delta,\cos\theta) \quad (28)$$

So the question is reduced to the determination of the forward specific intensity on the upstream side of the porous layer $I^+(x=0)$ and the backward specific intensity on the downstream side of the porous layer $I^-(x=\delta)$. This means that to determine B_1 and B_2 one should know the radiation fields upstream and downstream of the porous layer which are controlled by the upstream and downstream gas temperatures and radiation fields. In the present study, the upstream gas temperature is assumed to be uniform at T_{gin}. By doing so, one could avoid the multidimensional radiative transfer problem in the upstream radiative gas without losing generality. The same consideration applies to the downstream gas radiation. Therefore, if the gas in the open space upstream and downstream of the porous layer have the equivalent thicknesses L_u and L_d, one can obtain the difussive radiation boundary conditions B_1 and B_2 as follows:

$$B_1=2B_uE_3(\tau_u)+2\sigma T_{gin}^4[0.5-E_3(\tau_u)] \quad (29)$$

$$B_2=-2B_dE_3(\tau_d)-2\sigma T_d^4[0.5-E_3(\tau_d)] \quad (30)$$

where B_u and B_d are the radiation sources in the upstream and downstream gases. τ_u and τ_d are defined by:

$$\tau_u = \alpha_g L_u \quad (31)$$

$$\tau_d = \alpha_g L_d \quad (32)$$

T_d is the equivalent downstream gas temperature for downstream gas radiation and could be taken as:

$$T_d=(T_{gout}+T_e)/2 \quad (33)$$

where T_{gout} is the gas temperature on the downstream side of the porous layer and T_e the surrounding temperature. In fact, because the gas temperature upstream of the porous layer should always be higher than T_{gin}, B_1 determined by Eq.(29) will give somewhat conservative estimate for the gas radiation effect on the energy recovery ability. The same is true for B_2 determined by Eq.(30). The detailed derivation of Eqs.(29),(30) is given in the Appendix.

6 Nondimensional Equations and Numerical Solution

All the control equations for studying the gas radiation effect have been derived as Eqs.(2), (3), (15), and (26). The boundary conditions are Eqs.(29), (30) and the entrance condition of the flowing gas temperature is:

$$T_g(x=0)=T_{gin} \quad (34)$$

For parametric studies, the control equations and boundary conditions are nondimensionalized by introducing the following dimensionaless variables :

$$\left.\begin{aligned} x^* &= x/\delta \\ \theta_g &= T_g/T_{gin} \\ \theta_p &= T_p/T_{gin} \\ \theta_e &= T_e/T_{gin} \\ H(x^*)&=q_r(x^*)/(\sigma T_{gin}^4) \\ \eta(x^*)&=G(x^*)/(\sigma T_{gin}^4) \\ \mathcal{B}_1 &= B_1/(\sigma T_{gin}^4) \\ \mathcal{B}_2 &= B_2/(\sigma T_{gin}^4) \end{aligned}\right\} \quad (35)$$

and parameters:

$$\left.\begin{aligned} \gamma &= \alpha_g/\kappa_p \\ \omega_s &= \sigma_p/\kappa_p \\ F &= \sigma T_{gin}^3/(\rho_g c_g u_g) \\ R &= h_p n_p A_p/(\kappa_p \sigma T_{gin}^3) \end{aligned}\right\} \quad (36)$$

where F and R represent ratios of radiative heat transfer to gas enthalpy and porous matrix -gas convective heat transfer to radiative heat transfer, ω_s is the scattering albedo of porous material. The parameter γ describes the relative importance of gas radiation to the porous radiation.

Thus, the control equations reduce to:

$$d\theta_g/dx^*+FdH/dx^* = 0 \quad (37)$$

$$R(\theta_g-\theta_p)=(1/\tau_0)dH/dx^*+4[\gamma/(1+\gamma)]$$
$$[(1-\omega_s)\theta_p-\theta_g^4+\omega_s\eta] \quad (38)$$

$$H(x^*)=2\mathcal{B}_1E_3(\tau_0x^*)-2\mathcal{B}_2E_3[\tau_0(1-x^*)]$$
$$+2\tau_0\int_0^{x^*}Q(y)E_2[\tau_0(x^*-y)]dy-2\tau_0\int_{x^*}^1 Q(y)E_2[\tau_0(y-x^*)]dy \quad (39)$$

$$\eta(x^*)=2\mathcal{B}_1E_2(\tau_0x^*)+2\mathcal{B}_2E_2[\tau_0(1-x^*)]$$
$$+\ 2\tau_0\int_0^1 Q(y)E_1[\tau_0|(x^*-y)|]dy \quad (40)$$

$$Q(y)=[\gamma/(1+\gamma)][\theta_g^4+(1-\omega_s)\theta_p^4+\omega_s\eta] \quad (41)$$

$$\mathcal{B}_1=2\mathcal{B}_uE_3(\tau_u)+2\theta_{4gin}[0.5-E_3(\tau_u)] \quad (42)$$

$$\mathcal{B}_2=2\mathcal{B}_dE_3(\tau_d)+2[(\theta_{gout}+\theta_e)/2]^4\ [0.5-E_3(\tau_d)] \quad (43)$$

$$\theta_g(x=0) = 1. \quad (44)$$

The equations (37) to (41) together with boundary conditions (42) to (44) constitute the system of nonlinear integro-differential equations. The complexity of the problem arises from the simultaneous existence of convection which is a " local " phenomenon and radiation which is an " action at a distance " phenomenon. To solve the set of equations, it is clear that the analytical solution is not feasible and that a numerical solution becomes inevitable. In this study, an iterative solution method is employed. The values of $\theta_g(x^*)$ and $\theta_p(x^*)$ are first set to determine the dimensionaless radiative heat flux $H(x^*)$ from Eqs.(39)

to (41). Then the new values of $\theta_g(x^*)$ and $\theta_p(x^*)$ are calculated from Eqs.(37) and (38) respectively. Repeated computations are performed until all the calculated $\theta_g(x^*)$ satisfy certain imposed convergence criteria. In the numerical scheme, a variable grid method for x is designed to account for the steep changes of $\theta_g(x^*)$ and $\theta_p(x^*)$. In computation, some suitable weight factors are needed in order to find corrected new $\theta_g(x^*)$ and $\theta_p(x^*)$ for the next iterative step. All numerical computation was performed on Computer IBM 4300.

7 Results and Discussion

In the forgoing analysis, the effect of gas radiation on the effective energy conversion method is obviously dependent on the parameter γ which is the measure of relative importance of gas radiation to porous radiation. So the numerical solutions were generated for the variety of combinations of the ratio γ and other involved parameters.

The effect of gas radiation on the temperature distributions and on the radiation field within the porous layer is shown in Figs.(3), (4), (5) and (6). The parameters R, κ_p are chosen based on the ones reported by Echigo and F is chosen within the range of interest for practical high temperature gases, such as carbon dioxide and water vapor at 1000 K . Because the absorption coefficient of gases α_g is usually small compared with that of solid porous materials as indicated by Sparrow (1978), the gas radiation ratio γ varies from 0.00 to 0.10. The curves of γ = 0.00 in all figures correspond to the systems with nonradiating gases studied by Echigo (1982) and Wang and Tien (1984). When considering the systems with radiative gases, that is, when $\gamma > 0$, the large temperature drop within the porous layer will be reduced. This fact is reflected by the increase of the outlet gas temperature from porous layer $\theta_{gout}[= \theta_g(x=1)]$ with the increase of γ . At the same time, the radiative fluxes $H(\tau_0)$ on the downstream side of the porous layer will increase. This means more energy (received by porous layer through convection) will be radiated towards the downstream. From Figs.(5) and (6) it is found that the negative values of the radiative heat fluxes on the upstream side of the porous layer will become positive when γ is larger than certain values, i.e., when gas radiation ability is high. This phenomenon is of fundamental importance to the method when gas radiation is involved and shall be discussed in more detail by combining with Fig.(7).

Shown in Fig.(7) are the radiative heat fluxes H(0) and $H(\tau_0)$ on the upstream and downstream sides of the porous layer as a function of the ratio γ at different porous optical thicknesses τ_p $(=\kappa_p\delta)$. As indicated by Echigo, the negative value of radiative flux means the radiation propagation is in the upstream direction of the porous layer. It is this negative radiative heat flux that contributes to the energy recovery ability of the effective energy conversion method. When the flowing gas is nonradiating and no radiation source exists in the upstream gas, the radiative heat flux on the upstream side of the porous layer will always be negative since the porous layer is the only matter with radiative ability in the system. Thus it can radiate energy toward its upstream and downstream without any resistance. However, when the flowing gas possesses radiative ability, the situation becomes different. In this case, the upstream gas will also radiate energy to the porous layer as discussed in Section 1. This exerts resistance to the porous upstream radiation and results in a decreasing radiative flux on the upstream side of the porous layer, as shown in Fig.(7). If the gas radiation is strong enough, for instance, when τ_p = 4 , α_g = 10 m^{-1}, κ_p = 200 m^{-1} and γ=10/200=0.05 (these values correspond to typical values in practical applications) there will be no negative radiative flux H(0) on the upstream side. That is to say, there will be no net radiative energy recovered by the porous layer. In this case, the energy received by porous layer due to convection from gas enthalpy is mainly radiated to the downstream. This is reflected by the increase of downstream side radiative flux $H(\tau_0)$ with an increase of the gas radiation ratio γ .

From Fig.(7) one can find that the value of γ on which the upstream radiative flux H(0) becomes positive depends on the optical thickness of the porous material , i.e., the radiative ability of the porous layer. If there were no porous layer, or the radiative ability of the porous layer were rather poor as to make τ_p approach zero, the radiative flux on the " imagined " upstream side would be zero (if nonradiating gas involved) or positive (if radiating gas involved), and there would be no energy recovery phenomenon. The other limit is when τ_p approaches infinity. In this case, the effect of gas radiation may not be pronounced. In industrial applications, the radiative extinction coefficient and the thickness of the porous layer are always within certain finite values, for instance, δ = 1 or 2 cm, τ_p finite.

Shown in Fig.(8) are the ratios of upstream side radiative heat flux with gas radiation $[H(0)]_\gamma$ to flux without gas radiation $[H(0)]_{\gamma=0}$ based on the results of Fig.(7). From this figure one can clearly see that the energy recovery ability decreases rather sharply with the increase of gas radiation. Correspondingly, Fig.(9) shows the influence of the ratio γ on the outlet gas temperatures at different porous optical thicknesses τ_p. As expected, the outlet gas temperature will increase with increasing gas radiation. The effect of upstream radiation source B_1 has also been shown in these figures.

The scattering effect of the porous materials has also been studied in the present work. Unless the porous material is strong backward-scattering, which is rarely found in practice, the scattering of porous media will generally decrease the energy recovery ability. This is shown in Fig.(10); the results for γ=0.00 in this figure correspond to those by Wang and Tien [1984]. This decrease can be explained as follows: larger scattering albedo means poor emission ability, which is the main factor of porous upstream radiation. When the gas radiation ability is taken into account, the influence of ω_s is found to decrease with the increase of the gas radiation ratio γ . This phenomenon is also observed in Fig.(11) where the increase of the outlet gas temperature is slowed down by an increasing gas radiation ratio .

The upstream side radiative heat flux H(0) and the outlet gas temperature θ_{gout} as functions of parameters F and R are shown in Fig.(12) to Fig.(15) for different values of the ratio γ . The results in these figures reveal that the gas radiation will reduce the influence of F and R in general. This is because the most important characteristics of the EECM has been reduced by gas radiation. This greatly reduces the temperature drop within the porous layer and thus decrease the influences of F, R and other parameters involved.

Another interesting aspect of the EECM is that the high temperature drop within the porous layer characterizes a very high equivalent thermal resistance to the flow system. This high thermal resistance makes the porous layer behave as a high insulation material. We can define the equivalent thermal resistance as

$$R_t = \Delta T/Q_L \qquad (45)$$

where

$$\Delta T = T_{gin} - T_{gout} \qquad (46)$$

Q_L is the total energy out of the porous layer towards the downstream gas:

$$Q_L=\rho_g c_g u_g T_{gout}+q_{pr}(x=\delta)+q_{gr}(x=\delta) \qquad (47)$$

where the first term on the right-hand of Eq.(47) is the gas enthalpy flowing out of the porous layer; q_{pr} and q_{gr} represent the radiative energy out of the porous layer by gas radiation and solid porous radiation.

Let us consider the case with gases possessing no radiative ability. When there is no porous layer installed in the flowing system, $q_{pr} = q_{gr} = 0$, $T_{gin} = T_{gout}$, so that $\Delta T = 0$. Therefore, in this case there is no equivalent thermal resistance existing in the system. However, if a porous layer is installed in the flowing system, there will be a very high temperature drop ΔT produced within it, and $\rho_g c_g u_g T_{gout}$ will decrease greatly. Since q_{pr} is roughly proportional to the fourth power of T_{gout}, q_{pr} will not offset the decrease of $\rho_g c_g u_g T_{gout}$ and Q_L will decrease greatly. As a result, a very high equivalent thermal resistance R_t is produced in the flowing system. This case corresponds to the analyses of Echigo[1982] and Wang and Tien[1984].

When the gases possess radiative ability, the results will be very different. As indicated in Fig.(8), when the gas radiation is involved, the outlet gas temperature T_{gout} will increase greatly, which results in the corresponding decrease of the temperature drop within the porous layer. At the same time, the total energy loss Q_L from the porous layer will increase because both $\rho_g c_g u_g T_{gout}$ and $q_{pr}(x=\delta)$ will increase with increasing T_{gout}. In addition, there will be a new radiative loss $q_{gr}(x=\delta)$ due to the gas radiation. That is, with increasing gas radiation, Q_L will increase while ΔT will decrease. This will obviously reduce the equivalent thermal resistance produced by the porous layer.

R_t in Eq.(45) can be rewritten as:

$$R_t=(\theta_{gin}-\theta_{gout})/(\theta_{gout}+H(\tau_0)/F) \qquad (48)$$

The influence of the gas radiation ratio γ on R_t is plotted in Fig.(16). The results clearly demonstrate the above analyses: From Fig.(16) it can be seen that for $\tau_p = 5$, and $\gamma = 0$, the porous layer equivalent thermal resistance will reach $R_t = 1.2 \times 10^{-2}$ m^2°C/W. For the case with radiative gases, and $\gamma = 0.05$, R_t will decrease to about 0.6×10^{-2} m^{2}C/W. That is , the equivalent thermal resistance R_t will be reduced rather quickly with increasing gas radiation.

8 CONCLUDING REMARKS

The effect of gas radiation on the Effective Energy Conversion Method EECM has been investigated by utilizing the two phase radiative transfer equations. As expected by conceptual considerations, the results indicated that the EECM will encounter severe restrictions when extending it to the systems with radiative gases. The negative radiative heat flux H(0) in the upstream direction from the porous layer will be greatly decreased by gas radiation. As the gas radiation ratio increases to a certain value, H(0) will become positive, which means no radiative energy being recovered by the porous layer in this case. At the same time, the temperature drop within the porous layer is reduced greatly. The high equivalent thermal resistance produced by the porous layer is also decreased correspondingly if the radiative gases are present.

Acknowledgements

the authors are grateful to professor E.Bilgen for editing this paper. The work was sponsored by Chinese Academy of Sciences, P.R.China. This support is grateful acknowledged.

References

Du, Zhan-Gong and Ko, Shao-Yen, "Two Phase Radiative Transfer Equations for Two Phase Radiative Media at Different Phase Temperature Distributions", Presented at the 7th National Conference on Heat and Mass Transfer,China.1987

Echigo, R., "Effective Energy Conversion Method Between Gas Enthalpy and Thermal Radiation and Application to Industrial Furnaces", Proceedings of the 7th International Heat Transfer Conference, Munchen, pp.361-366, 1982

Echigo, R.,et al., "Heat Transfer Augmentation in High Temperature Heat Exchangers", High Temperature Equipment, Edited by A. E. Sheindlin, Hemisphere Publishing Corporation, 1986a

Echigo, R.,et al., "Analytical and Experimental Studies on Radiation Propagation in Porous Media with Internal Heat Generation", Proceedings of the 8th International Heat Transfer Conference, San Francisco, pp. 827-832, 1986b

Hasegawa, S., Echigo, R., et al., "Augmentation on Heat Transfer by Thermal Radiation Shielding Plate Placed in a Duct", Proc. 5th Int. Heat Transfer Conf., Tokyo, Vol. I, pp. 108-112, 1974

Hasegawa, S., Echigo, R., Shimizu, A., "Convective and Radiative Heat Transfer of Flowing Gaseous-Solid Suspensions", Review paper, Handbook of Heat and Mass Transfer, Vol. 1, Ed., Cheremisinoff, N. P., pp. 523-557, 1986

Ichimiya, K., Kobayashi, K., Echigo, R., "Fundamental Study of Temperature Distribution of a High Temperature Steel Slab (1st report)", Trans JSME B, pp.904-909, 1986c

Ko, Shao-Yen, Du, Zhan-Gong, Xu, Jing-Zhong, "Formation of High Temperature Field upstream of Porous Layer in Duct", J. Engineering Thermophysics, Vol.9, No.1, 1988

Ko, Shao-Yen, Du, Zhan-Gong and Xu, Jing-Zhong, "Two Phase Radiative Transfer Equations and Its Application in High Temperature Porous Flow", Presented at the 2nd Int. Symposium on Heat Transfer,Beijing,September, 1988

Mori, Y., Yamada, Y., et al., "Radiation Effects on the Performance of Radiative Gas Heat Exchangers at High Temperatures", Int. J. Heat Mass Transfer, Vol. 23, pp. 1079-1089, 1980

Mori, Y., "High Temperature Heat Exchangers", Proc.

of the 8th Int. Heat Transfer Conf., San Francisco, Vol.1, pp. 259-268, 1986

Siegel. R., Howell, John R., Thermal Radiation Heat Transfer, Second Edition, Hemisphere Publication Corporation, 1981

Sparrow, E. M.,et al., Radiation Heat Transfer , Augmented Edition, McGraw-Hill, Fig.7-5, 1978

Wang, K. Y., and Tien, C. L., "Thermal Insulation in Flow Systems", Trans. ASME Journal of Heat Transfer, Vol.106, pp. 453-459, 1984

Appendix

Consider a radiation system consisting of media g and p with interface I-I as shown in Fig.(A1). To determine the interface radiation condition B for the medium p, one should first find the radiative transfer equations for the medium g and p as follows:

for medium g:

$$I_g^+(\tau_g,\mu)=I_g^+(0,\mu)\exp(-\tau_g/\mu)+(1/\pi)\int\sigma T_g^4\exp[-(\tau_g-t)/\mu]dt/\mu \qquad (A1)$$

$$I_g^-(\tau_g,\mu)=I_g^-(\tau_u,\mu)\exp[-(\tau_u-\tau_g)/\mu]+(1/\pi)\int\sigma T_g^4\exp[-(\tau_g-t)/\mu]dt/\mu \qquad (A2)$$

for medium p:

$$I_p^+(\tau_p,\mu)=I_p^+(0,\mu)\exp(-\tau_p/\mu)+(1/\pi)\int\sigma T_p^4\exp[-(\tau_p-t)/\mu]dt/\mu \qquad (A3)$$

$$I_p^-(\tau_p,\mu)=I_p^-(\tau_{p0},\mu)\exp[-(\tau_{p0}-\tau_p)/\mu]+(1/\pi)\int\sigma T_p^4\exp[-(\tau_p-t)/\mu]dt/\mu \qquad (A4)$$

where I_g^+, I_g^- and I_p^+, I_p^- are the forward and backward specific intensities as defined the same way as in Eq.(7), μ is the simplified form of $\cos\theta$. The scattering of the medium p is neglected here because it has no influence on the determination of the interface radiation condition B_1 with medium g (corresponding to gas medium).

Obviously, the interface condition at I-I is:

$$I_p^+(0,\mu) = I_g^+(\tau_u,\mu) \qquad (A5)$$

$$I_g^-(\tau_u,\mu) = I_p^-(0,\mu) \qquad (A6)$$

As the diffusive condition B_1 is defined by:

$$B_1 = \pi I_p^+(0,\mu) \qquad (A7)$$

Only Eq.(A5) is used to determine B_1. Multiply Eq.(A5) by μ then integrate over 4π solid angle, combining with Eq.(A1) one has :

$$2\pi\int I_p^+(0,\mu)\mu d\mu=2\pi\int I_g(\tau_u,\mu)\mu d\mu \qquad (A8)$$

Note that one can assume a diffusive radiation boundary for the medium g as well as:

$$B_u = \pi I_g^+(0,\mu) \qquad (A9)$$

Combining Eqs.(A7), (A8), (A9) and (A1) one can find :

$$B_1 = 2B_u\int_0^1\exp(-\tau_u/\mu)\mu d\mu + 2\int_0^{\tau_u}\sigma T_g^4\int_0^1\exp[-(\tau_u-t)/\mu]d\mu dt \qquad (A10)$$

Recall the definition of exponential integral function (19), Eq.(A10) becomes:

$$B_1=2B_uE_3(\tau_u)+2\int_0^{\tau_u}\sigma T_g^4E_2(\tau_u-t)dt \qquad (A11)$$

When the temperature field T_g within the medium g is assumed to be uniform at T_{gin}, one obtains Eq.(28). Eq.(29) for B_2 can be obtained in the same way.

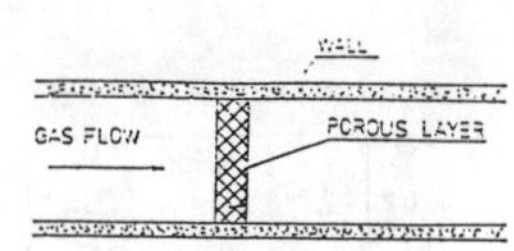

FIGURE 1 FLOW CHANNEL WITH POROUS LAYER

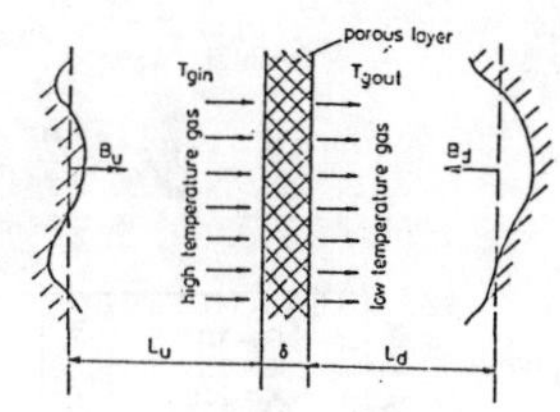

FIGURE 2 PHYSICAL MODEL

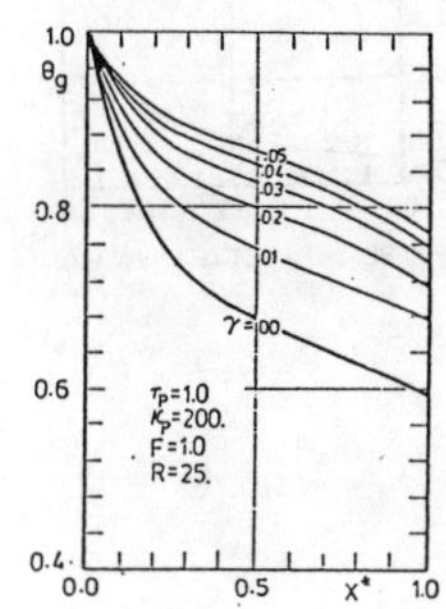

FIGURE 3 TEMPERATURE DISTRIBUTION WITHIN THE POROUS LAYER [CASE 1]

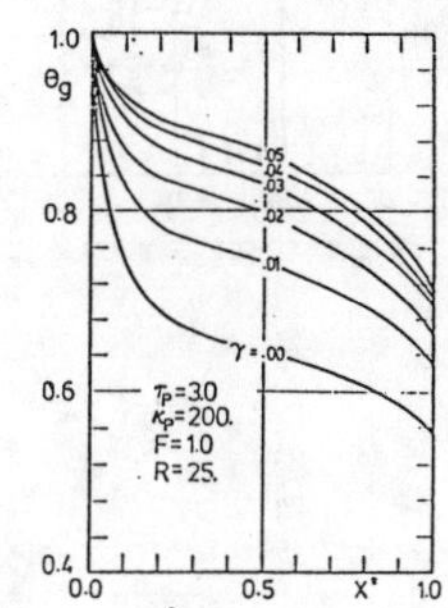

FIGURE 4 TEMPERATURE DISTRIBUTION WITHIN THE POROUS LAYER [CASE 2]

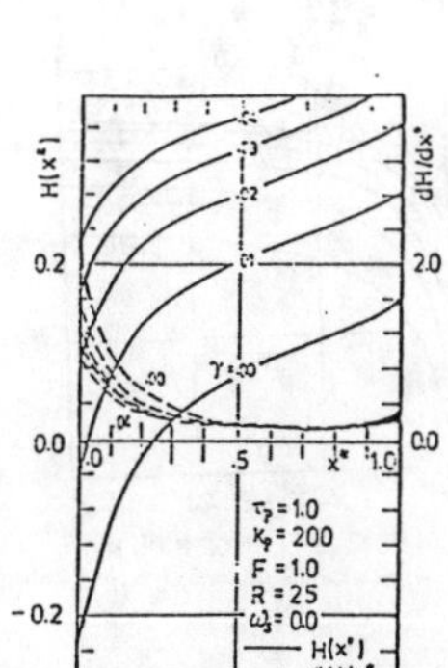

FIGURE 5 RADIATION FIELD WITHIN THE POROUS LAYER [CASE 1]

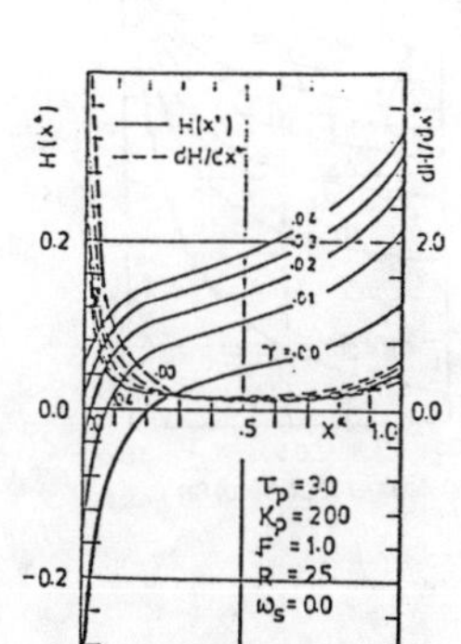

FIGURE 6 RADIATION FIELD WITHIN THE POROUS LAYER [CASE 2]

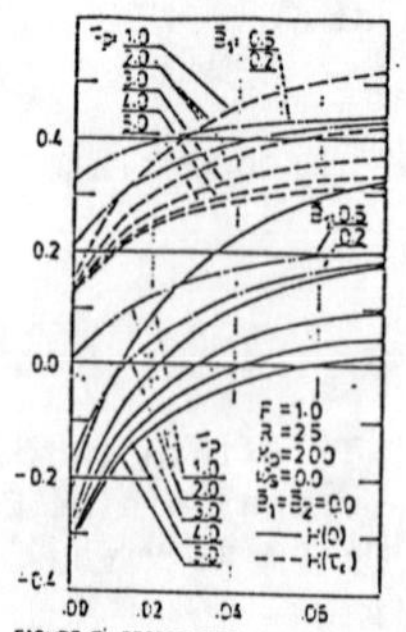

FIGURE 7 EFFECT OF γ ON $H(0)$, $H(\tau_1)$

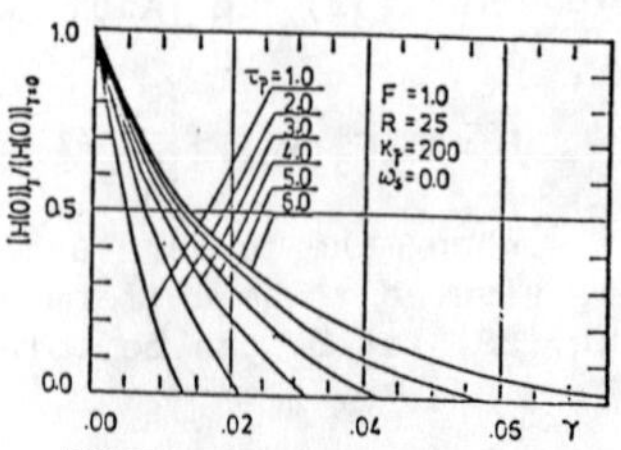

FIGURE 8 EFFECT OF γ ON $[H(0)]_\gamma/[H(0)]_{\gamma=0}$

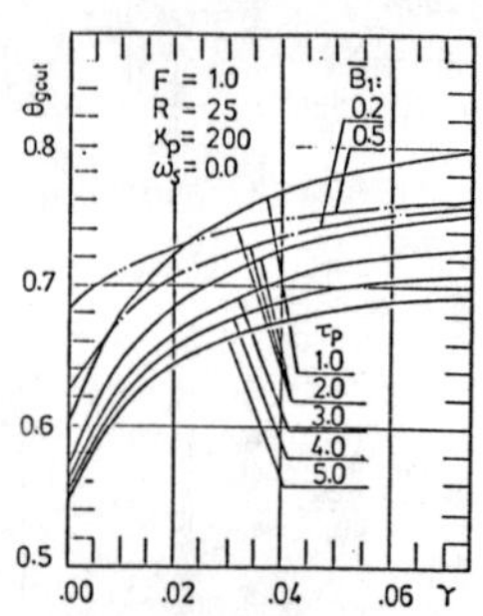

FIGURE 9 EFFECT OF γ ON θ_{gout}

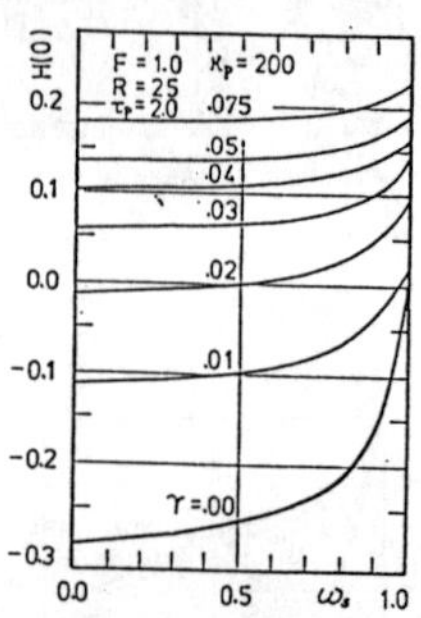

FIGURE 10 EFFECT OF ω_s ON $H(0)$

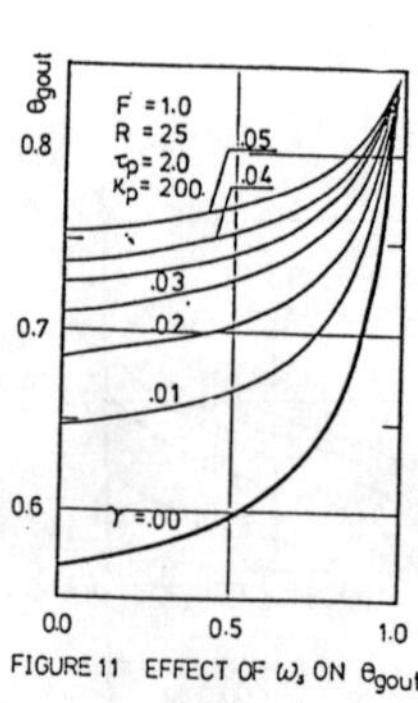

FIGURE 11 EFFECT OF ω_s ON θ_{gout}

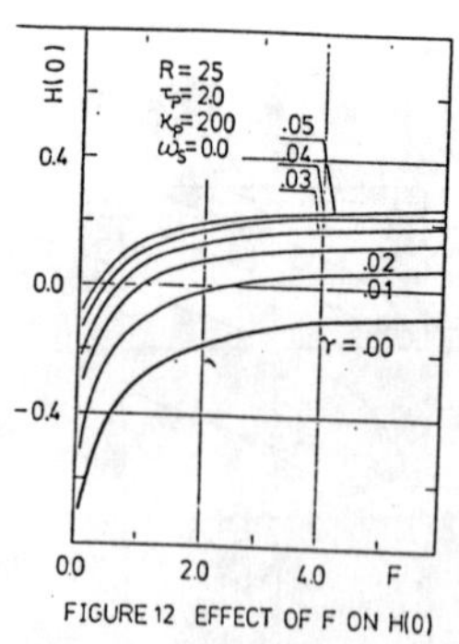

FIGURE 12 EFFECT OF F ON $H(0)$

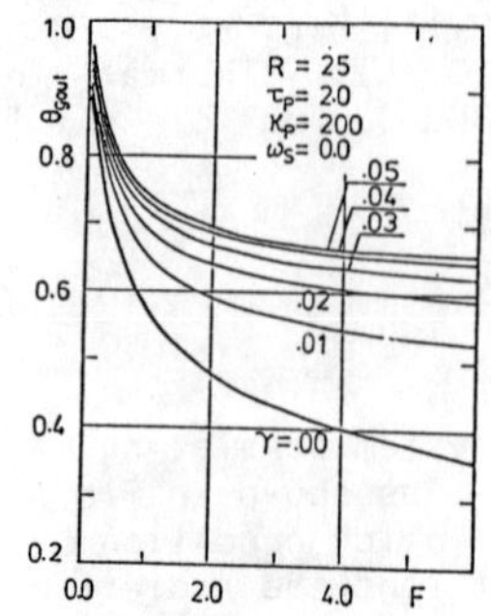

FIGURE 13 EFFECT OF F ON θ_{gout}

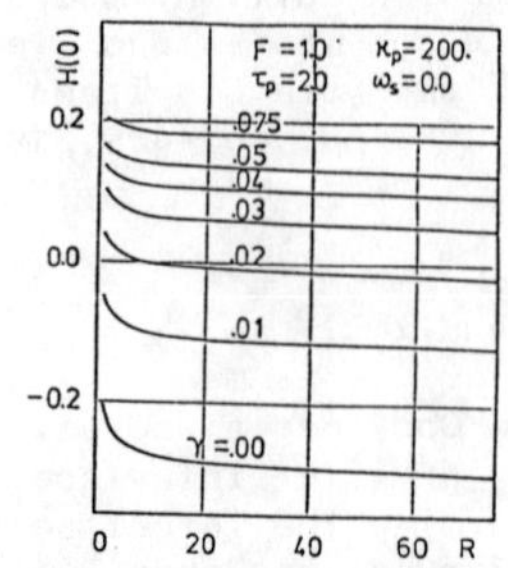

FIGURE 14 EFFECT OF R ON $H(0)$

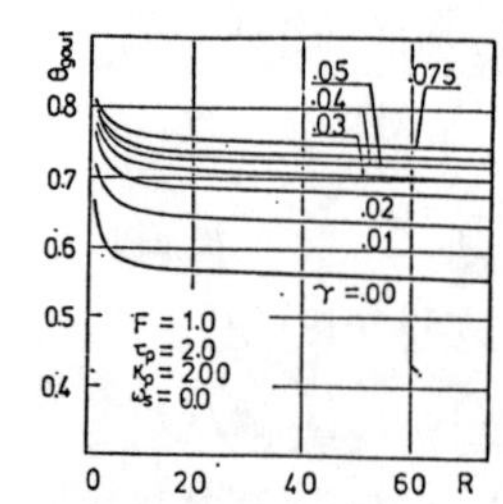

FIGURE 15 EFFECT OF R ON θ_{gout}

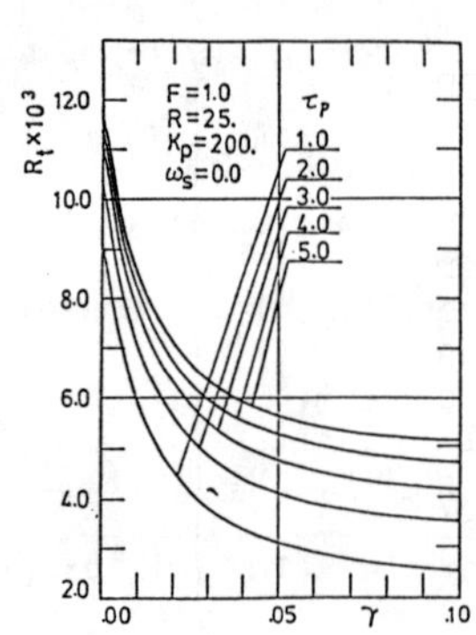

FIGURE 16 EFFECT OF γ ON R_t

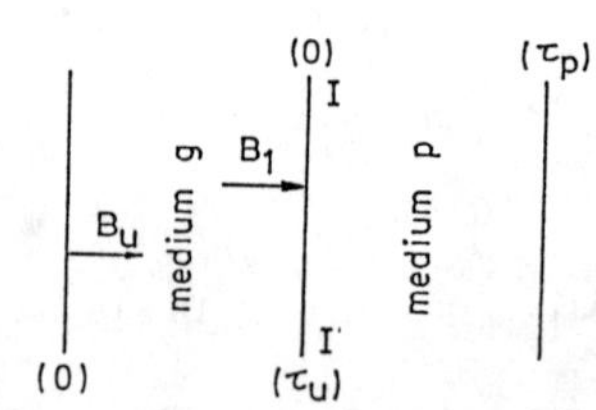

FIGURE A1 THE PHYSICAL MODEL
FOR THE DETERMINATION OF B_1, B_2

1989 National Heat Transfer Conference
HTD-Vol. 106, Heat Transfer Phenomena in Radiation, Combustion and Fires

HEAT TRANSFER IN THE NON-REACTING ZONE OF A CEMENT ROTARY KILN

P. S. Ghoshdastidar and V. K. Anandan Unni
Department of Mechanical Engineering
Indian Institute of Technology
Kanpur, India

ABSTRACT

A new approach is used to numerically model the heat transfer processes occurring in a rotary kiln employed for drying and preheating of wet solids. The model includes radiation exchange among hot gas, refractory wall and the solid surface, transient conduction in the wall and mass and energy balances of the hot gas and the solids. Numerical results ard obtained for the non-reacting zone of a cement rotary kiln. The length predicted by this model is found to be shorter than that of the said kiln. However, the axial solid and gas temperatures vs. percent kiln length match very well with those of the actual kiln. There is very little circumferential variation of refractory wall temperature. The radial wall temperature distribution is linear.

NOMENCLATURE

A	Area
a_{rf}	Thermal diffusivity of the refractory
C_{p_s}	Specific heat of the solid
C_{p_g}	Specific heat of the gas
C_{p_l}	Specific heat of water (liquid)
C_v	Specific heat of the water-vapour
D	Diameter of the kiln
E_b	Black body emission per unit area
F_{ij} or F_{gj}	Shape factor between surface i and j or between gas and surface j
G	Irradiation per unit area
h_r	Radiation heat transfer coefficient from outer refractory wall to ambient, as defined in eq. (5)
h_c	Convection heat transfer coefficient from the outer refractory wall to ambient
h_{cs}	Heat transfer coefficient from inner refractory wall to the bulk solid
$\dot{m}$	Mass flow rate
L	Latent heat of vaporisation of water
ℓ	Length of the kiln
N	Number of surface elements on the wall and the solid
R	Radius of the kiln
T	Temperature (K)
t	Time
U	Circumferential velocity of the kiln wall
x	Coordinate in the radial direction
y	Coordinate in the circumferential direction
Δz	Size of each axial segment of the kiln

Greek letters

α	Fill-angle
γ_g	Absorptance of the gas
ε	Emissivity
σ	Stefan-Boltzmann constant, 5.67×10^{-8} $W/m^2\text{-}K^4$
θ	As defined in Fig. 2(a)
τ_g	Transmittivity of the gas

Subscripts

black	Black bodies
g	Gas
gray	Gray bodies
i	Element number of the wall or the solid
j	Element number of the wall or the solid
l	Liquid
r	Radiation
rf	Refractory
s	Solid
sat	Saturation
v	Vapour
z	Axial coordinate or in the axial direction
z+Δz	At a distance z+Δz from the kiln inlet
∞	Ambient

Special Symbol

Σ	Summation

INTRODUCTION

A rotary kiln employed for drying and preheating of wet solids consists of a refractory lined cylindrical shell mounted at a slight incline from the horizontal plane. The kiln is very slowly rotated about its longitudinal axis. The wet solids are fed into the upper end of the cylinder and during the process, are dried and heated by the countercurrent flow of the hot gas. Finally, it is transferred to the lower end where it reaches the desired temperature and is discharged. Fig. 1 shows the schematic diagram of the kiln.

In the present study, the kiln can be considered to consist of three sections. In the first section, the wet solids are heated to the boiling point of the entrained liquid. In the second region, the liquid evaporates at constant temperature until the feed is completely dry. In the third region, the solids are heated to some specified temperature and then are discharged from the kiln. The influence of mass transfer has not been considered in this work.

The efficiency of the rotary kiln heater is better because of improved heat transfer to the solids. The rotation of the kiln enhances the contact of hot gas with the solids which assists in the drying and heating. In addition, heat is transferred from the refractory wall to the solids in the region of contact.

Fig. 2(a) shows the nomenclature of the kiln. The angle, α , is called the fill-angle which shows the region of volume containing in solid. Fig. 2(b) shows the heat transfer processes occurring in the kiln.

Although rotary kilns are used in many chemical and metallurgical industries, the amount of informations that have been published regarding the heat transfer analysis and design of these units are very small. The heat transfer in rotary kilns for making cement was investigated by Sass (1967). A heat transfer model was developed for the axial temperature distribution of the gas and solid in the kiln. Circumferential and radial heat transfer and temperature variations in the kiln wall were neglected. Simultaneous mass transfer and chemical reaction were also not considered. Manitius et al. (1974) developed a similar model for a rotary kiln used to produce aluminium oxide. Heat generated by the exothermic chemical reactions was included in the energy balance. Manson and Unger (1979) described in the heat transfer processes in a rotary kiln incinerator but did not perform a detailed analysis. Recently Ghoshdastidar et al. (1985) developed a detailed heat transfer model for the rotary kiln waste incinerator. This model included radiation between the flame and refractory surface, transient conduction in the wall and chemical reaction in the waste material. Numerical results were obtained for burning of Polymethyl-methacrylate (Plexiglas). Circumferential and radial refractory wall temperature variations, axial variation of solid waste temperature and mass flow rate, for various fill angles, were presented. The predicted length of the kiln needed to completely burn the waste was also discussed.

In the present study, a heat transfer model is developed which includes radiation exchange among hot gas, the refractory wall and the solid surface. Radiation heat transfer is modeled by dividing wall and the solid surface into surface elements. Since the refractory wall is alternately heated and cooled during each revolution, unsteady heat conduction in the wall is modeled by a finite-difference technique. The continuity and energy equations are solved for the hot gas and the solid assuming only an axial temperature variation. The present model is in line with Ghoshdastidar et al. (1985).

A computer program is developed to obtain a numerical solution to the heat transfer equations. Results are presented for a particular rotary kiln, in this case, the non-reacting zone of the cement rotary kiln of Sass (1967). The radial and circumferential temperature variations in the refractory wall are presented. The axial gas and solid temperature variations and the predicted length of the kiln needed to completely dry and preheat the solids are compared with the results predicted by Sass (1967) and also with the measured values. The deviations which are due to uncertainties in some of the input date are discussed.

PROBLEM FORMULATION

Thermal Radiation Among Hot Gas, Refractory Wall and Solid Surface

Heat is exchanged among the hot gas, the inner wall of the kiln and the solid surface by radiation and convection. In this study, the convection has been neglected since radiation is expected to be the predominant transfer mode as the temperature of the gas is very high. Particle radiation has also not been considered. Since the refractory surface temperature varies circumferentially and axially, the wall is divided into surface elements as shown in Fig.3. Each of the refractory elements, that are not covered by solids, exchange radiation with the gas, solid and other refractory surface elements. Each axial segment of the refractory surface is divided into fifteen surface elements of equal size. The solid surface is divided into five surface elements.

The surfaces are assumed diffuse and gray and therefore the shape factors can be used to determine the radiation exchange between the elements. It is assumed that the surface elements exchange radiation only with surfaces of the same axial segment. That is, axial radiation heat transfer is neglected. Since the ratio of length and diameter (ℓ/D) of the rotary kiln varies from 2 to 10 (1985), the segment lengths (Δz in Fig. 3) are sufficiently long so that most of the radiation exchange occurs between the surface elements in the same axial segment.

The radiation heat transfer is calculated by using the theory (Holman, 1981) of a gray enclosure containing a gas which emits, absorbs and transmits radiation. Thus, if the gas volume is enclosed by N black surfaces at different temperatures, the net energy gain for a particular surface can be expressed as,

$$q_j = G_j A_j - E_{bj} A_j \quad (1)$$

where G_j and E_{bj} are irradiation and black body emission per unit area of surface j respectively. Also, Irradiation of surface j = Irradiation from the gas + Irradiation from all other surfaces that are transmitted through the gas.

$$\text{i.e. } G_j A_j = A_g F_{gj} \varepsilon_g(T_g) E_{bg} + \sum_{i=1}^{N} A_i F_{ij} \tau_g(T_i) E_{bi} \quad (2)$$

where $\varepsilon_g(T_g)$ and $\tau_g(T_i)$ mean that emissivity of the gas is evaluated at T_g and transmissivity of the gas is evaluated at T_i respectively. $\tau_g(T_i) = 1 - \gamma_g(T_i)$, where $\gamma_g(T_i)$ is the gas absorptance for radiation from the black surface, i, at T_i and is a function of both T_i and T_g.

Heat exchange in gray enclosure can now be calculated as given by Hottel (Holman,1981)

$$q_{gray} = q_{black} \cdot \frac{\varepsilon+1}{2} \quad \text{for } \varepsilon > 0.8$$

where ε is the emissivity of the enclosure walls. In this problem emissivities of refractory and solid are taken as 0.9 and 0.95 respectively.

For the preceding analysis, shape factors are calculated as shown in Appendix A. For the purpose of calculating emissivity and other characteristics of the gas, the composition of the gas has been taken from Weber (1963). The reason for this selection was the non-availability of the gas composition data for the cement kiln mentioned in Sass (1967). The details of gas emissivity calculations are given in Appendix B.

Substituting equation (2) in eq. (1), yields an expression for the heat gain of each surface, q_j, which is calculated.

$$q_j = A_g F_{gj} \varepsilon_g(T_g) E_{bg} + \sum_{i=1}^{N} A_i F_{ij} \tau_g(T_i) E_{bi} - E_{bj} A_j \quad 2(a)$$

The temperature of solid surface elements are taken to be uniform in each axial segment.

Conduction Heat Transfer in the Refractory Wall

Since wall elements of the kiln are alternately heated and cooled during each revolution by the hot gas and the solid respectively, quasi-steady heat conduction is present. Only heat conduction in the radial direction of the wall is taken into account, assuming negligible conduction in the circumferential and axial direction, as the circumference and length of the kiln are much greater than the thickness of the kiln. For a coordinate system that is fixed to the rotary kiln, the temperature varies with time and consequently an energy storage term is needed in the energy equation. But, for a coordinate system that is stationary, the temperature remains constant and hence the heat storage term is not required. However, the kiln wall now has a velocity with respect to the coordinate system and hence the convective term has to be introduced. Because the nonrotating coordinate system is more conveniently combined with the radiation boundary condition discussed in the previous section, this coordinate system is chosen for xhe analysis.

The coordinate system used to model the wall heat conduction is shown in Fig. 4. Because the refractory thickness is small compared to the diameter, Cartesian coordinates are used. The x and y coordinates are in the radial and circumferential directions, respectively. The energy equation for the wall then becomes,

$$U \frac{\partial T}{\partial y} = a_{rf} \frac{\partial^2 T}{\partial x^2} \quad (3)$$

where U is the circumferential velocity of the wall and a_{rf} is the thermal diffusivity of the refractory wall. The left hand term of eq. (3) accounts for the energy convected by the refractory wall and the right hand term is the radial heat conduction.

The boundary condition for the inner surface of the refractory is dependdnt upon whether the surface element is exposed to the hot gas or covered with the solid. Elements exposed to gas are heated by radiation and the heat input is q_j, which is obtained from eq. 2(a). Surface elements covered by solids exchange heat with the solids primarily by surface contact. Helmrich and Schugerl (1980) investigated the heat transfer between the kiln and the solids over a range of particle diameters. They expressed heat transfer rates as product of heat transfer coefficient and the difference between the wall and bulk particle temperature. The same approach is used in the present study. Therefore,

$$q_j = h_{cs} A_{j_i} (T_{j_i} - T_s) \quad (4)$$

where q_j is the heat transfer rate, h_{cs} is the contact heat transfer coefficient between the refractory wall and the solid, A_{j_i} is the inner surface element area, T_{j_i} is the inner refractory wall surface temperature and T_s is the solid temperature. Since, the heat transfer coefficient, h_{cs}, is not available for the non-reacting part of the cement kiln in the present study, a value for the reacting zone of the kiln as given by Helmrich and Schugerl (1980) has been used. The idea is to use the value (h_{cs}) obtained for a certain portion of the kiln rather than using a wild guess in the absence of the information for the part of the kiln under present investigation. The value of h_{cs} taken here is 64 W/m^2-K.

The boundary condition for the outer surface of the refractory wall is the heat loss by convection and radiation. Suryanarayana et al. (1986) have investigated the convection and radiation losses from a rotary kiln for various wind velocities, orientation of the kiln, rotational speed of the kiln etc. In the present investigation, convection heat transfer coefficient, h_c, is taken as 10 W/m^2-K considering wind is blowing over the kiln and using the graphs of Suryanarayana et al. (1986) for the inclination (3 degrees) and the rotational speed of the present kiln. The radiation heat transfer coefficient is calculated from:

$$h_r = \epsilon_{rf} (T_{j_o} + T_\infty)(T_{j_o}^2 + T_\infty^2) \qquad (5)$$

where T_{j_o} and T_∞ are the outer wall temperature and ambient temperature respectively.

The combined heat loss from the outer surface is written as:

$$g_j = (h_c + h_r)A_{j_o}(T_{j_o} - T_\infty) \qquad (6)$$

where A_{j_o} is the area of the outer surface element.

Eq. (3) is solved numerically by first transforming it into finite-difference form. In the region not covered by the solid, circumferential segmentation (y-direction) is chosen so that elements are aligned with the radiation surface elements (of Fig.3). Five additional elements are taken for the region covered by solid. The refractory wall is sectioned into ten radial elements separated by equal distance. In solving eq. (3), a dummy storage term has been used. This is because, the boundary conditions for eq. (3) are temperature dependent and the temperatures are not known at the start.

An explicit scheme has been used to march ahead in time. Since the initial temperature distribution of the refractory wall is not known, an arbitrary temperature is assumed at all grid points in the refractory wall. The solution converges when there is no further change in the temperature at each grid point as $t \to \infty$. This temperature distribution represents the steady state temperature distribution of the wall.

Energy Transport in the Solid

As mentioned earlier, the kiln can be considered to consist of three sections. In the first section wet solids are heated to the boiling point of the entrained liquid, in the second section liquid evaporates at constant temperature and in the third section dry solids are heated to the required temperature.

In the first and third sections, the physical processes are similar except that in the former the solids are wet while solids are dry in the latter. Fig. 5 shows the energy balancd on an element of solid contained in an axial segment either in the first or in the third section of the kiln. It can be seen that,

$$\dot{m}_{s,z} C_{p_s} T_{s,z} + q_1 + q_2 = \dot{m}_{s,z+\Delta z} C_{p_s} T_{s,z+\Delta z} \qquad (7)$$

But since no evaporation takes place,

$$\dot{m}_{s,z} = \dot{m}_{s,z+\Delta z} \qquad (8)$$

Then, using eq. (8), eq. (7) can be written as,

$$T_{s,z+\Delta z} = T_{s,z} + \frac{(q_1 + q_2)}{\dot{m}_{s,z} C_{p_s}} \qquad (9)$$

where q_1 is the net radiation heat transfer on the solid surface and q_2 is the total heat transfer from the inner refractory wall (covered by the solids) to the solids.

Since there are five elements on the surface of the solid, total radiation heat transfer q_1 is,

$$q_1 = \sum_{j=1}^{5} q_j \qquad (9a)$$

Similarly, the heat transfer, q_2, from the refractory inner wall is given by

$$q_2 = \sum_{j=1}^{5} h_{cs} A_{j_i} (T_{j_i} - T_s) \qquad (9b)$$

In the second section of the kiln, the evaporation of the entrained liquid takes place and hence the solid temperature remains constant at the boiling point of the entrained liquid.

Figure 6 shows the mass balance of the solids in the second section. The mass balance equation can be written as,

$$\dot{m}_{s,z+\Delta z} = \dot{m}_{s,z} - \dot{m}_v \qquad (10)$$

where $\dot{m}_v$ is the rate of evaporation in that segment.

Figure 7 shows the energy balance of the wet solids in the second section. The energy balance equation can be written as,

$$\dot{m}_{s,z} C_{p_s} T_{s,z} + q_1 + q_2 = \dot{m}_v (L + C_{p_l} T_{s,z}) + \dot{m}_{s,z+\Delta z} C_{p_s} T_{s,z} \qquad (11)$$

From eq. (10) and eq. (11), $\dot{m}_v$ is obtained as,

$$\dot{m}_v = \frac{q_1 + q_2}{L} \qquad (12)$$

where L is the latent heat of vaporisation of the entrained liquid.

Energy Balance in the Gas

Figure 8 shows the energy balance of hot gas in the first and third section. It is to be noted that here the gas flow is countercurrent to that of solids.

The energy balance equation is written as,

$$\dot{m}_{g,z+\Delta z} C_{p_g} T_{g,z+\Delta z} = \dot{m}_{g,z} C_{p_g} T_{g,z} + q_r \qquad (13)$$

where q_r is the total heat lost from the hot gas by radiation. But, the heat lost by the gas is gained by wall surface and solid surface. So,

$$q_r = \sum_{j=1}^{20} q_j \qquad (14)$$

where q_j is given by eq. 2(a).

Since, $\dot{m}_{g,z} = \dot{m}_{g,z+\Delta z}$ in the first and third segment, eq. (13) can be written as

$$T_{g,z+\Delta z} = T_{g,z} + \frac{q_r}{\dot{m}_{g,z} \cdot C_{p_g}} \quad (15)$$

In the second region, the gas is gaining water-vapor as the solid is evaporating in that region. Fig. 9 shows the mass balance of the gas in the second section.

The mass balance equation can be written as,

$$\dot{m}_{g,z+\Delta z} = \dot{m}_{g,z} - \dot{m}_v \quad (16)$$

Also, in the second section, superheating of the vapor takes place. Fig. 10 shows the energy balance of the gas in the second section. If C_v is the specific heat of the vapor, the energy equation can be written as,

$$\dot{m}_{g,z+\Delta z} C_{p_g} T_{g,z+\Delta z} = \dot{m}_{g,z} C_{p_g} T_{g,z} + q_r + \dot{m}_v C_v (T_{g,z} - T_{sat}) \quad (17)$$

From Eq. (17),

$$T_{g,z+\Delta z} = \frac{\dot{m}_{g,z}}{(\dot{m}_{g,z} - \dot{m}_v)} T_{g,z} + \frac{q_r + \dot{m}_v C_v (T_{g,z} - T_{sat})}{(\dot{m}_{g,z} - \dot{m}_v) C_{p_g}} \quad (18)$$

Thus, the temperature of the gas in the next axial segment can be calculated.

It should be noted that the temperature of the gas and the solid in each axial segment have been taken as uniform in this analysis.

METHOD OF SOLUTION

A computer program was developed to obtain numerical results for this rotary kiln problem. The computations had been carried out in the DEC 1090 computer system. For solving the heat transfer problem, the kiln is divided into segments of equal length Δz, in the axial direction. In this problem, Δz is taken as 1m. Input data required for the computer program are given in Table-I. The solution is initiated at the inlet of the kiln and proceeds to the exit. In the first section (in which solids are being preheated to the boiling point of the entrained liquid), the temperature of the solid and the gas at the end of the axial segments are calculated using eq. (9) and eq. (15) respectively. The computations for the first section stop at the end of that axial segment for which the temperature of the solid is equal to the boiling point of the entrained liquid. Therefore, the sum of the axial segments upto that point is the length of the first section (L_1).

In the second section, the rate of evaporation ($\dot{m}_v$) and the temperature of the gas in the axial segments are calculated using eq. (12) and eq. (18), respectively. The temperature of the solid remains constant in this section. The end of the second section is indicated at the point where the cumulative $\dot{m}_v$ is equal to the predetermined total amount of water-vapour to be evaporated per sec. Therefore, the sum of the lengths of the axial segments, starting from the segment where evaporation begins to that segment where rest of the water evaporates, gives the length of the second section (L_2).

In the third section, the solid is completely dry and the heating of the solid to the desired temperature is achieved. Here, similar to first section, eq. (9) and eq. (15) are used to calculate the temperatures of the solid and the gas respectively in the axial segments. Once the temperature of the dry solid reaches the required value, the computation stops and hence the length of the third region (L_3) is calculated.

The total length (ℓ) of the kiln required to dry the wet solids and preheat it to a desired temperature = $L_1+L_2+L_3$.

Table I : Input Data to the Program

I. Rotary Kiln	
A. Diameter	: 1.95 m
B. Refractory	
1. Thickness	: 0.3 m
2. Thermal conductivity	1.3274 W/m-K
3. Specific heat	: 692.556 J/kg-K
4. Density	: 3000 kg/m^3
5. Emissivity	: 0.9
C. Rotational speed	: 1.9 rpm
II. Solid	
A. Inlet temperature	: 366.66 K
B. Mass flow rate (dry)	: 2.996 kg/sec
C. Specific heat	: 1122.54 J/kg-K
D. Water in solid feed	: 18% (by weight)
E. Emissivity	: 0.95
III. Gas	
A. Outlet temperature	: 935.77 K
B. Specific heat	: 1214.7 J/kg-K
C. Mass flow rate	: 5.9075 kg/sec
IV. Water	
A. Latent heat of vaporisation	: 2.26x10^6 J/kg
B. Specific heat (vapour)	: 2083.2 J/kg-K
V. Convection Heat Transfer Coefficient	
A. Between inner refractory surface and the solids	: 64 W/m^2-K
B. Between outer refractory surface and the surrounding	: 10 W/m^2-K
VI. Fill-Angle	: 90°

RESULTS AND DISCUSSION

The accuracy of the kiln simulation has been tested by comparing the present computer output with that produced by the model of Sass (1967) as well as with the actual kiln operating data obtained from Sass (1967). The cement kiln was actually 56.4 meters long, but as the present simulation neglects chemical reaction, only the 27.3 meters portion of the kiln in which no chemical reaction occurs is used for comparison with the computer results.

Figure 11 shows a comparison of temperature-profiles of the solid and the gas as a function of percent kiln length with those predicted by Sass's (1967) model as well as with actual measured data. Although some of the predicted temperatures are slightly low as compared to the other two, it is apparent that the simulation has accurately predicted the gas and solid temperature profile in the kiln, the fraction of the kiln used for water evaporation and the fraction used for heating the solids.

However, the kiln length predicted by the present model turns out to be 20 meters as compared with 25.5 meters predicted by Sass (1967) and 27.3 meters of the actual kiln. This deviation may be attributed to uncertainties in some of the input data used in this study. They are as follows:

(i) Sass (1967) did not mention anything regarding the composition of the hot gas. The gas composition is required for the gas emissivity calculations. So, the present authors have used the gas composition data of Weber (1963) for a typical cement kiln. However, the gas used in the actual kiln might not be having the same composition.

(ii) The fill-angle (90°) used here has been indirectly calculated from the data given by Sass (1967). The density of the solids used in this calculation is taken as that of cement. In the absence of reliable information regarding the reactant mixture density, this is the best that can be done. Also, the fill-angle calculated by using the product (cement) density matches very closely with the fill-angle (90°) measured from the diagram of the kiln given in Sass (1967). However, Sass (1967) has not explicitly indicated the actual fill-angle. Also, the assumption that the fill-angle remains constant throughout the length of the kiln is not strictly correct, as during drying the water evaporates and hence the level of solids drop.

(iii) Actual heat transfer coefficient from the inner wall of the kiln to the solid (h_{cs}) is not known. Only an estimate has been used.

(v) Actual ambient temperature outside the kiln is not stated in Sass's (1967) paper. Also, the actual heat transfer coefficient from the outside of the kiln to the surroundings is not known. Here also, an estimate has been used.

(v) The actual speed of rotation of the kiln also is not mentioned in Sass's (1967) work as he did not use it in his model. Here, also, an estimate has been used.

(vi) No measurement of the solid-exit temperature or the gas-inlet temperatures are available. The values used in Sass (1967) are based on estimates made by operating personnel.

The authors could change the data mentioned in (i) through (vi) in the preceding column. But the major constraint was the computer time. For each axial segment of the kiln, it took approximately 5000 iterations to obtain converged solution for the steady state temperature distribution in the kiln wall. Since axial segment length (Δz) was taken as 1 m and the kiln length predicted by the present model was 20 m, twenty such computations were required. Hence, the total computer time taken was enormously large (approximately 45 minutes). That's why, playing with the data was ruled out.

Figure 12 shows the circumferential temperature variation of the inner kiln wall in the first axial segment of the kiln. The minimum temperature occurs at a point ($\theta \approx 90^\circ$) where the wall emerges from the solid. After the wall comes out of the solid, it experiences a rise in temperature because it is then exposed to radiation from the hot gas. The temperature rise continues until it is again covered by the solid ($\theta = 360^\circ$ or 0°). However, it is clear from the graph that the difference between the minimum temperature (742K) and the maximum temperature (748 K) is very small. This lack of appreciable circumferential temperature variation may be due to the assumption of the uniform or averaged temperature of the gas in each axial segment. However, it may be noted here that Sass (1967) neglected circumferential temperature variation in the kiln wall and still got good results. So, the finding of the present model does tally with Sass's assumption.

Figure 13 shows the radial temperature distribution of the kiln wall in the first axial segment of the kiln for $\theta = 0^\circ$. It is clear that the temperature profile is linear.

CONCLUSIONS

A new approach has been used to model heat transfer phenomena in the non-reacting zone of a cement rotary kiln. The present model is more rigorous than the one developed by Sass (1967). Basically, this model can be used to design an actual rotary kiln employed for drying and preheating of wet solids in many industries.

Although the length of the kiln predicted by the present model shows a smaller value than that of the actual kiln, the axial gas and solid temperature distribution vs. percent length of the kiln matches well with the actual kiln profiles. The assumed reasons why the predicted kiln length doesn't match are the uncertainties in some of the input data. This can be overcome by checking the results of the theoretical model with a companion laboratory scale experimental set-up. For example, the heat transfer coefficient, h_{cs}, from the inner refractory wall to the solid, should be taken from the date of an actual experimental work.

Lastly, future efforts should be made to develop the model of an entire cement kiln which includes the reacting zone as well.

REFERENCES

1. Ghoshdastidar, P.S., Rhodes, C.A., and Orloff, D.I., 1985, "Heat transfer in a rotary kiln during incineration of solid waste," 23rd National ASME Heat Transfer Conference, Denver, Colorado, August 4-7, ASME Paper No. 85-HT-86.
2. Helmrich, H. and Schugerl, K., 1980, "Rotary Kiln reactors in Chemical Industry," Ger. Chem. Engg., Vol. 3, pp. 194-202.
3. Holman, J.P., 1981, Heat Transfer, pp. 341-348, 5th International Student Edition, McGraw-Hill.
4. Jacob, M., 1957, Heat Transfer, Vol. II, pp. 19-21, John Wiley & Sons, Inc., New York.
5. Manitius, A., Kureyuz, E., and Kawecki, W., 1974, "Mathematical model of the Aluminium oxide rotary kiln," Ind. Eng. Chem. Process Design and Development, Vol. 13, No. 2, pp. 132-142.
6. Manson, L., and Unger, S., 1979, "Hazardous material incineration design criteria", EPA Report No. 600/2-79-198, October.
7. Sass, A., 1967, "Simulation of the heat transfer phenomena in a rotary kiln," Ind. Eng. Chem. Process Design and Development, Vol. 6, No. 4, pp.532-535.
8. Suryanarayana, N.V., Lyon, J.E., and Kim, N.K., 1986, "Heat shield for high temperature kiln," Ind. Eng. Chem. Process Design and Development, Vol. 25, pp. 843-849.
9. Weber, P., 1963, Heat Transfer in Rotary Kilns, Bauverlag Gmbh, Wiesbaden-Berlin.

APPENDIX A

Calculation of Shape Factors

The shape factors between the twenty refractory surface and solid surface elements are determined assuming the surfaces as infinite parallel strips. The equation for shape factors for infinite parallel strips was derived by Hottel (1957). Surfaces (1) and (2) in Fig. A-1 are strips perpendicular to the XY plane and is measured from the normal to the surface (1). The shape factor F_{1-2} is given by,

$$F_{1-2} = \frac{\sin \phi' - \sin \phi''}{2} \quad \text{(A-1)}$$

A-1 Shape Factors between the Elements of the Kiln Wall

Equation (A-1) is written in terms of the circumferential location of the two surfaces. For notations shown in Fig. A-2, the shape factor between surface (1) and (2) is given by

$$F_{1-2} = \frac{\Delta\Theta \cdot \cos\beta}{4} \quad \text{(A-2)}$$

where $\Delta\Theta$ is the arc angle of surface (2) in radian. The angle β is defined as

$$\beta = \frac{\pi - |(\Theta_2 - \Theta_1)|}{2} \quad \text{(A-3)}$$

where Θ_1 and Θ_2 are the circumferential positions of surface elements (1) and (2) respectively with respect to the reference axis. For convenience, the reference axis is chosen perpendicular to the solid surface. Because of the absolute bracket in eq. (A-3), the shape factors F_{1-2} and F_{2-1} are equal for the same $\Delta\Theta$ as required by the reciprocity relations of shape factors.

A-2 Shape Factors between Elements of Kiln Wall and Solid Surface

In Fig. A-3, surface (1) is a solid surface element and surface (2) is a refractory wall element. Surface (1) is located at a distance 'b' from the reference axis and Θ_2 is the circumferential angle for surface (2). Angle 'α' is the fill-angle of the solid material. Additional terms which are shown in Fig. A-4 are needed to determine the shape factor.

The shape Factor F_{1-2} is given by,

$$F_{1-2} = \frac{D \cos\Psi_1 \cos\Psi_2 \ \Delta\Theta}{4e} \quad \text{(A-4)}$$

where D is the inside diameter of the kiln. The angle Ψ_1 and Ψ_2 are shown in Fig. A-3 and are respectively the angles between the normals to surface (1) and (2) and the ray extending between the surfaces. The distance 'e' is the length of this ray. Expressions for Ψ_1, Ψ_2 and e in terms of parameters shown in Fig. A-4 were derived by Ghoshdastidar et al. (1985) and are as follows:

$$\Psi_3 = \tan^{-1} (2b/D \cos\alpha/2) \quad \text{(A-5a)}$$

$$c = (D \cos \alpha/2)/2 \cos \Psi_3 \quad \text{(A-5b)}$$

$$e = [c^2+D^2/4 - c D \cos (2\pi - \Theta_2+\Psi_3)]^{1/2} \quad \text{(A-5c)}$$

$$\Psi_2 = \sin^{-1} [c \sin (2\pi - \Theta_2 + \Psi_3)/e] \quad \text{(A-5d)}$$

$$\Psi_1 = \Theta_2 - \Psi_2 - \pi \quad \text{(A-5e)}$$

APPENDIX B

Calculation of Gas Emissivity

For the calculation of the emissivity of the hot gas at different temperatures, the composition of the gas is required. Since, the gas composition for this kiln was not known, the composition (by volume) of hot gas for a typical cement rotary was taken from Weber (1963) and is as follows:

CO_2 - 15.897%
O_2 - 1.418%
N_2 - 41.784%
H_2O - 40.9%

Out of the above gases, only CO_2 and H_2O (vapour) are the ones which emit, absorb and transmit radiation. For the calculation of emissivity, the Mean beam length theory is used.

According to Holman (1981), the mean beam length, Le can be approximated as,

$$Le = 3.6 \left(\frac{V}{A}\right) \qquad (B\text{-}1)$$

where V is the total volume of the gas and A is the total surface area. For the fill-angle of $\alpha = 90^{\circ}$ in the present work,

$$Le = 3.6\left[\frac{\frac{3}{4}\pi R^2 + \frac{1}{2}R^2}{\frac{3}{2}\pi R + \sqrt{2}\,R}\right]$$

i.e.

$$Le = 1.6783\ R$$

where R is the inside radius of the kiln.

The pressure inside the kiln is assumed to be 1 atmosphere. Then, the partial pressures of CO_2 and H_2O (Vapour) are 0.1589 atm. and 0.409 atm. respectively. Radius of the kiln is 3.25 ft.

Using the above values and the method given in (Holman, 1981), the emissivities tabulated in Table B-1 are calculated:

Table B-1 : Emissivities of Hot Gas for Different Temperature Ranges

Temperature range °K From	To	Emissivity
273.0	347.0	0.6304
347.0	486.0	0.5524
486.0	625.0	0.5358
625.0	764.0	0.5408
764.0	903.0	0.5210
903.0	1042.0	0.5028
1042.0	1181.0	0.4680
1181.0	1320.0	0.4170
1320.0	1459.0	0.3838
1459.0	1598.0	0.3506
1598.0	1737.0	0.3202

FIGURES FOR THE MAIN BODY

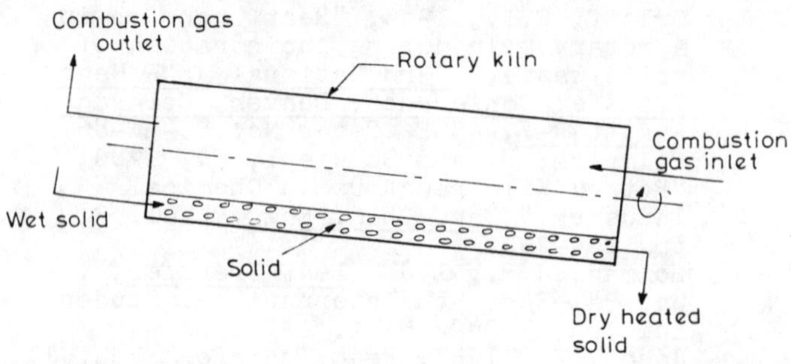

Fig.1 Schematic diagram of a rotary kiln

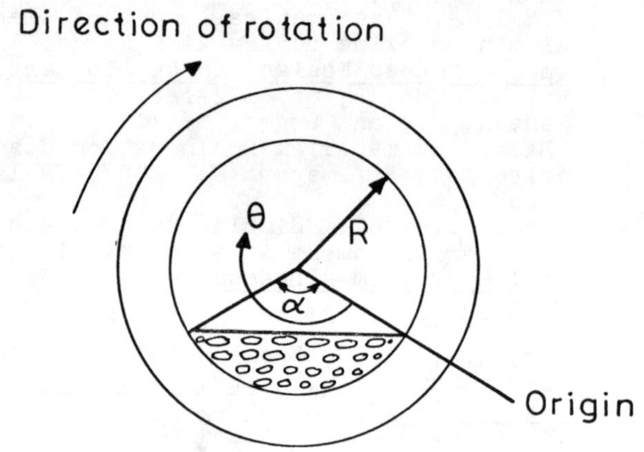

Fig.2(a) Nomenclature of the kiln

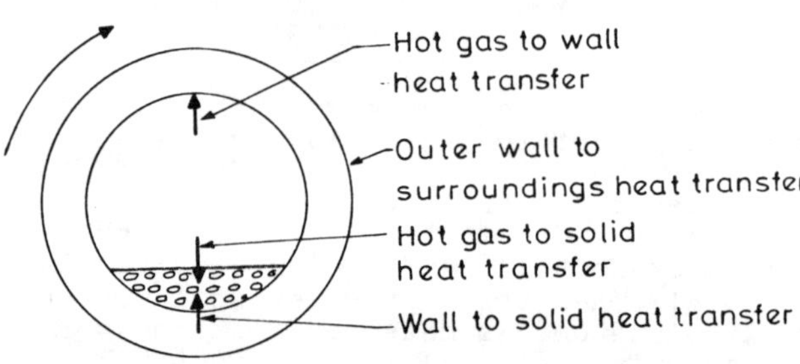

Fig.2(b) Schematic diagram showing heat transfer processes in a rotary kiln

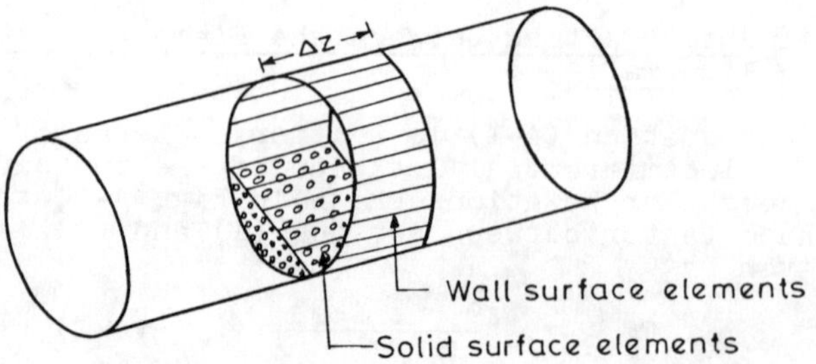

Fig.3 Section of the kiln showing surface elements of the wall and the solid

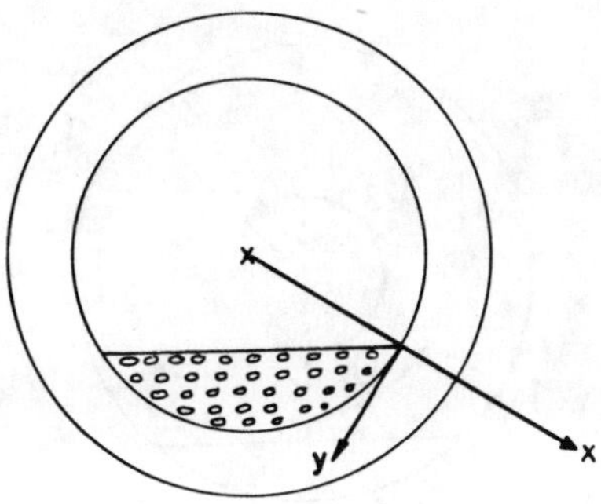

Fig. 4 Coordinate system for refractory wall conduction

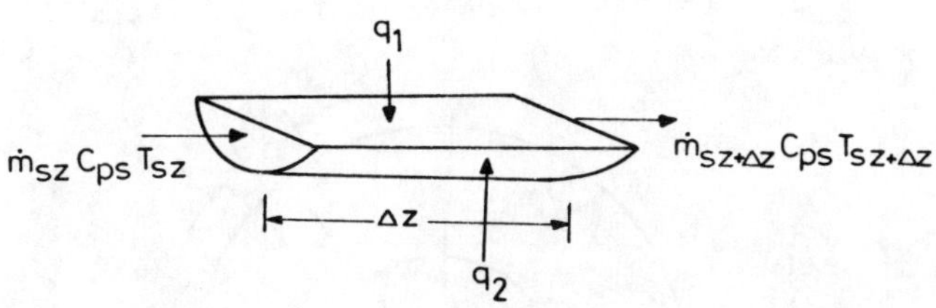

Fig.5 Energy balance of solid element contained in an axial segment in the 1st or 3rd section of the kiln.

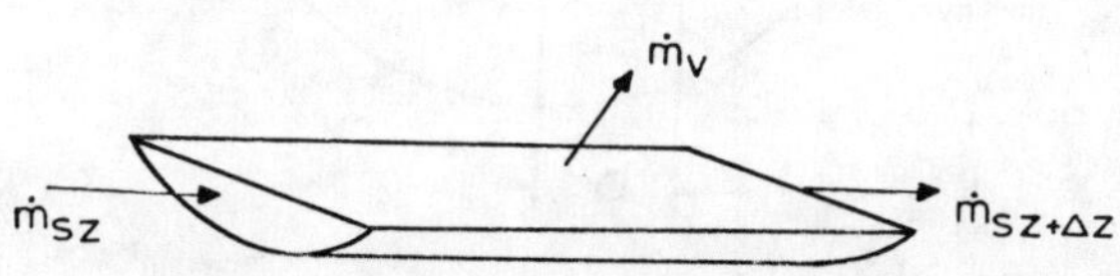

Fig.6 Mass balance of solid element in the second section of the kiln

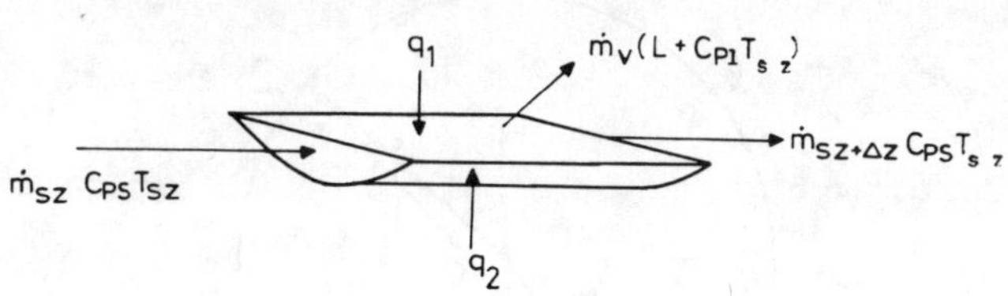

Fig.7 Energy balance of solid element in the second section of the kiln

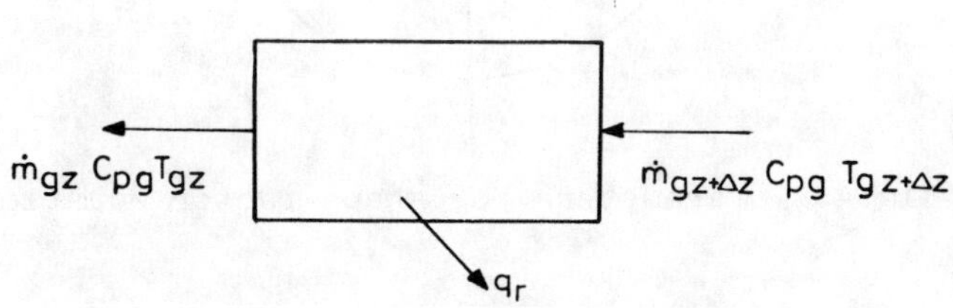

Fig.8 Energy balance of gas in the 1st and 3rd sections

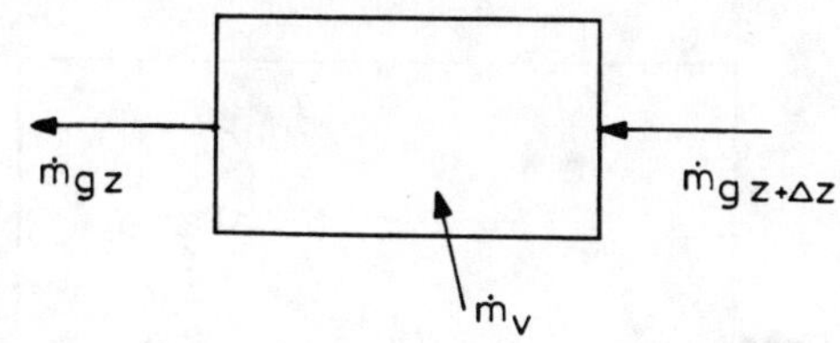

Fig.9 Mass balance of gas in the second section

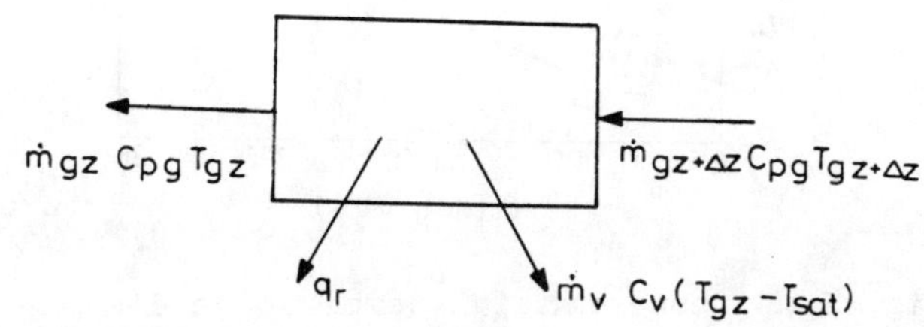

Fig.10 Energy balance of gas in the second

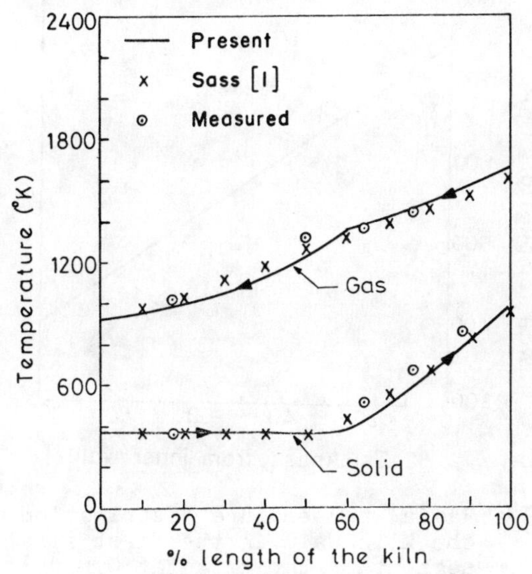

Fig.11 Axial temperature distribution of solid and gas

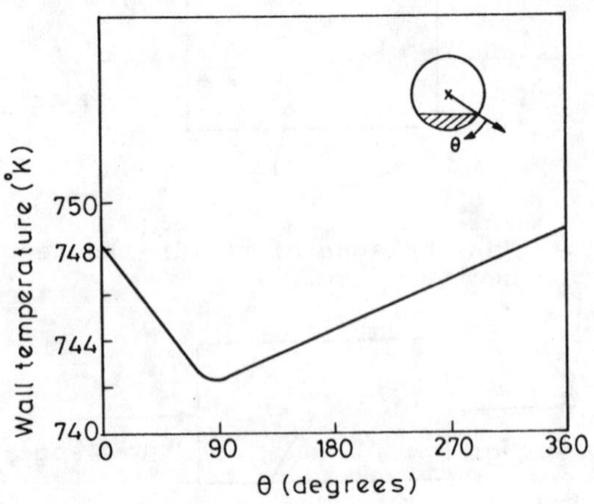

Fig.12 Circumferential temperature distribution in the inner kiln wall in the first axial segment of the kiln

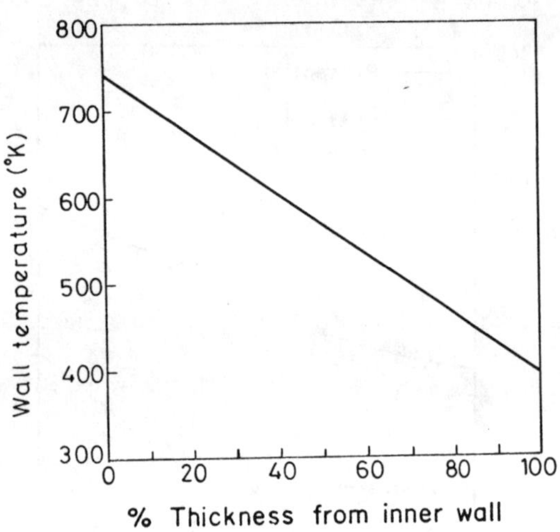

Fig.13 Radial temperature distribution in the kiln wall in the first axial segment of the kiln (at θ = 0°)

FIGURES FOR APPENDIX A

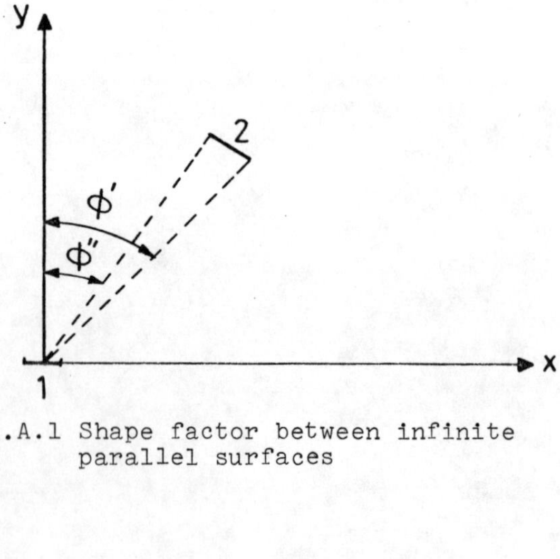

Fig.A.1 Shape factor between infinite parallel surfaces

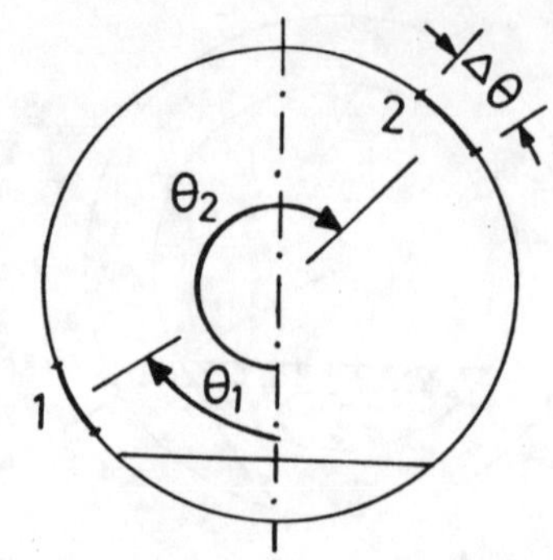

Fig.A.2 Shape factor nomenclature when both surfaces are on the refractory wall

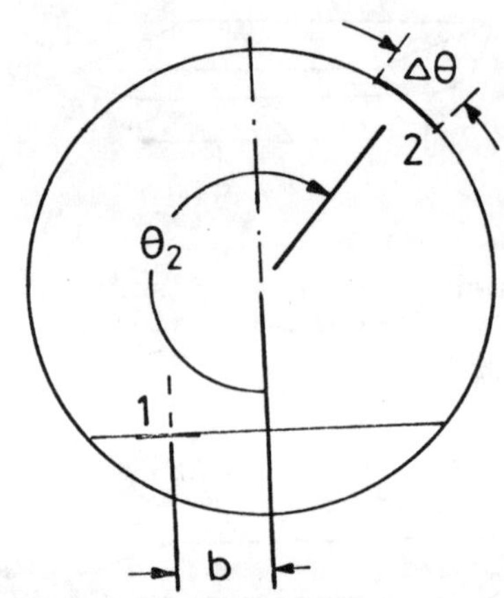

Fig.A.3 Shape factor nomenclature when one surface is on refractory and the other is on solid

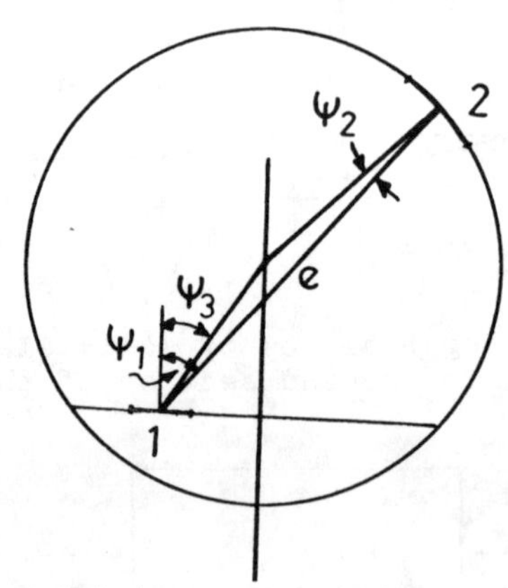

Fig.A.4 Parameters for shape factor equation

1989 National Heat Transfer Conference
HTD-Vol. 106, Heat Transfer Phenomena in Radiation, Combustion and Fires

THE OPTICAL PROPERTIES OF MOLTEN COAL SLAG

J. L. Ebert and S. A. Self
Stanford University
Mechanical Engineering Department
Stanford, California

ABSTRACT

The optical properties of flyash are needed to predict radiative heat transfer in coal combustors. Since flyash chiefly consists of homogeneous spheres of slag, Mie theory can be used to determine the absorption and scattering characteristics of dispersions of flyash using optical properties of bulk slags. In this paper a method for measuring the complex refractive index, $m = n + ik$, of molten slag (1600° C) with known composition is presented. Results of transmittance measurements show that the absorption index, k, can increase by a factor of 2.5 as the temperature is increased from 20°C to 1600° C. The current measurements are consistent with other recently available data at lower temperatures.

NOMENCLATURE

h	depth below slag surface
I_{col}	collected spectral power
I_o	incident spectral power
I_1	collected power for mirror at depth h
I_2	collected power for mirror at depth $h + \Delta h$
m	$= n + ik$ complex refractive index
R_{Pt}	spectral reflectance of platinum mirror
T	temperature
α_λ	spectral absorption coefficient
Δh	change in submerged mirror depth
λ	wavelength
Φ_λ	spectral scattering phase function
σ_λ	spectral scattering coefficient

INTRODUCTION

Flyash suspended in the gaseous products of coal combustors and gasifiers strongly influences radiation heat transfer (Boothroyd and Jones, 1984, Ebert and Self, 1988, Gupta, et al.,1983, Goodwin and Mitchner, in press, Mengüc and Viskanta, 1988, Sarofim, 1978, and Wall et al.,1979). Particle emission enhances heat transfer, while scattering tends to reduce it. To account for the effects of the flyash a knowledge of the optical properties of the suspension is needed, *i.e.* its contribution to the spectral absorption coefficient α_λ, the scattering coefficient σ_λ, and the phase function Φ_λ of the medium.

Flyash consists primarily of vitreous, spherical particles of mixed metal oxides, having a broad size distribution and a composition which, while reasonably homogeneous for individual particles, is quite heterogeneous from particle to particle. The spectral absorption and scattering cross–sections and the phase function for an individual, homogeneous, isotropic sphere can be calculated from Mie theory when the diameter, wavelength, and complex refractive index, $m \equiv n + ik$, are given. The latter is a material property which depends on composition, wavelength, and temperature. Given, for a particular ash, the particle loading and the size and refractive index distributions, the properties of the suspension can be calculated. The size and composition distributions can be determined by microanalytical techniques, but information on the dependence of the complex refractive index $m(C, \lambda, T)$ on composition, wavelength and temperature has not been available until recently (Goodwin and Mitchner, 1989, and Goodwin, 1986).

It is impractical to determine m from measurements on individual particles and it is difficult to extract reliable values of m from measurements of extinction, emission, and scattering on a heterogeneous polydispersion. At best one might extract an effective average value of m for a particular ash. Consequently, we have adopted the approach of measuring $m(C, \lambda, T)$ on bulk samples of synthetic slags of known composition, taking advantage of well established techniques for bulk samples. Such measurements on thin, optically–polished wafers of (solid) slag for $T \leq 900°C$ have been reported recently (Goodwin and Mitchner, 1989, and Goodwin, 1986) and reveal that $m(\lambda)$ is very spectrally dependent (non-gray). The current work, reported here, extends such measurements to 1600°C, when the slag is above the liquidous temperature.

BACKGROUND

Flyash typically contains many constituents which do not play an important role in determining its optical properties, either because they are present in minor or trace amounts or because they do not provide infrared absorption mechanisms. Therefore, a base composition of silica, alumina, and calcia, the three major constituents of most bulk ashes, was chosen to represent typical flyash. The composition of the base slag (SA00) is shown in Table 1. In addition to these major constituents, iron oxide in the flyash effects the absorption index, k, owing to its strong absorption bands in the wavelength range 1-4 μm. Although many individual flyash particles have compositions which vary considerably from this simple composition, the optical properties of most flyash particles are reasonably well represented by it.

Goodwin's (1986) measurements of the refractive index, m, were made on optically polished, solid wafers and were limited to temperatures below about 1200 K. They showed that the effect of temperature on the absorption index, k, can be significant. In the wavelength range from 1-4 μm the absorption is due to the Fe^{2+} valence state of iron in the slag. In this range the temperature dependence of k is stronger when the ferrous/ferric ratio (Fe^{2+}/Fe^{3+}) is lower, which probably indicates that the it is the absorption due to intervalence charge transfer which is mostly effected by temperature. When the slag is produced in a reducing environment, or when the iron content is low (below 1%), the effect of temperature is smaller and little or no change in k was observed with changing temperature.

The strongest temperature effect is in the wavelength range beyond 4 μm , where the absorption is due primarily to the O–Si–O vibrational absorption band centered near 9 μm . In this range, multi-phonon absorption, much like combination and overtone bands in gases, accounts for a wide band of strong absorption. The temperature dependence of this multi-phonon absorption mechanism is relatively strong. For the wavelength range between 4 μm and 8 μm Goodwin found that k could increased by a factor of nearly two for the temperature range from 292 K to 1197 K. No measurements above 8 μm were reported.

EXPERIMENTAL APPARATUS AND PROCEDURES

An iron-bearing synthetic slag (SA05) was produced by mixing powdered oxides with proportions given in Table 1. The powder mixture was added to an alumina crucible and placed in an electric furnace. The temperature was raised to 1600°C at a rate of 10°C per minute. The mixture was held at that temperature for approximately 12 hours, and then cooled at rate a of 10°C per minute. The resultant slag appeared homogeneous with a dark brown, almost black, color. The slag was glassy (amorphous) apart from a region near the alumina crucible where crystals, rich in alumina and low in calcia and iron, grew outward from the boundary. The crystals were thin and long, separated by regions of amorphous material. Microprobe analysis and backscatter electron images showed that the bulk of the slag was reasonably homogeneous.

Table 1. Composition of Synthetic Slags

Oxide	Base (SA00)	SA05
SiO_2	58.0	55.1
Al_2O_3	29.0	27.6
CaO	13.0	12.4
Fe_2O_3	0.0	4.9

A technique for measuring the transmittance and reflectance of molten slag was devised and the apparatus constructed. Figure 1 shows a schematic of the apparatus for measuring the transmittance of thin slag layers at 1600° C. The broadband radiation source is a Nernst Glower, which is operated at an effective blackbody temperature of 1700°C. The ceramic filament in the Nernst Glower is imaged by a concave spherical mirror (M1), with a focal length of 0.5 meters, onto an aperture (All mirrors, except those inside the furnace, are gold front surface mirrors). The glower image is then chopped at 670 Hz so that the detector signal can be measured synchronously using a lock–in amplifier. This eliminates unmodulated background radiation. The concave mirror M3 images the glower filament onto a platinum mirror located in an alumina crucible inside the furnace. The image is reflected from the platinum mirror and collected by concave mirror M5, which reforms the glower filament image on the entrance slit of the monochromator. A longpass filter is placed in front of the entrance slit to the monochromator to remove higher order diffraction. The monochromator is an Ebert type grating monochromator and descriminates selected spectral ranges which are detected by the HgCdTe infrared detector. The monochromator slits are opened to $2mm$ to allow maximum power to reach the detector.

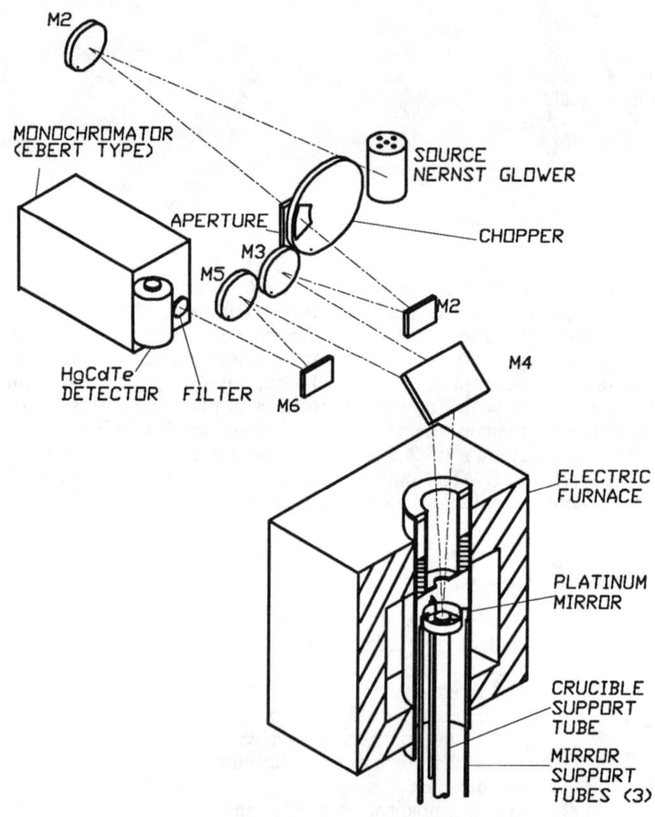

Figure 1. Schematic of the apparatus for determining the infrared optical properties of molten slag. Mirrors M1,M3, and M5 are concave gold and mirrors M2, M4, and M6 are plane gold. The source beam is chopped to allow synchronous detection of the signal from the infrared detector.

The transmittance of the slag is measured by a "*submerged reflector method*" which is illustrated in Figure 2. In this technique a platinum mirror is submerged below the surface of a crucible filled with molten slag. The platinum mirror is suspended by three platinum rods, which are supported on three alumina tubes. A small alumina rod is attached to the top of each alumina tube. A "V–shaped" groove is cut into the top of each rod to keep the platinum mirror from moving laterally. The three tubes are mounted on vertical translation stages which allow adjustment of the angular orientation of the platinum mirror relative to the incident beam.

The alumina crucible is mounted on a larger central tube which can be translated vertically with micrometer precision. The depth of the platinum mirror in the molten slag is adjusted by translating the central tube while leaving the platinum mirror support tubes fixed.

The platinum mirror is tilted slightly from the plane of the molten slag surface to allow the first surface reflection and the multiple internal reflections to be rejected from the collection optics. The intensity of the collected beam is

$$I_{col} = I_0 R_{Pt} exp(\frac{-4\pi k(2h)}{\lambda}) \tag{1}$$

where I_0 is the incident beam intensity, k is the absorption index (imaginary part of the refractive index), λ is the wavelength, and h is the depth of the mirror below the slag surface. The factor R_{Pt} is the reflectance of the platinum mirror and is not generally known. However we have measured it at 1600° $C$and found it to be slightly above 0.55 at $\lambda \sim 1\ \mu m$, increasing to a nearly constant value of about 0.9 at longer wavelengths ($\lambda > 9\ \mu m$). At room temperature, platinum has similar spectral characteristics but shifted to slightly higher values of reflectance.

The transmittance of the thin slag layer is determined by measuring two signals, one at mirror depth h, and one at mirror depth $h+\Delta h$. The ratio can be taken to eliminate R_{Pt} to find k in terms of known quantities. If I_1 is the signal at h and I_2 is the signal at $h+\Delta h$, then

$$k = \frac{-\lambda}{8\pi\Delta h} ln(I_2/I_1). \tag{2}$$

The change in depth Δh is measured accurately using a dial indicator attached to the central crucible support tube.

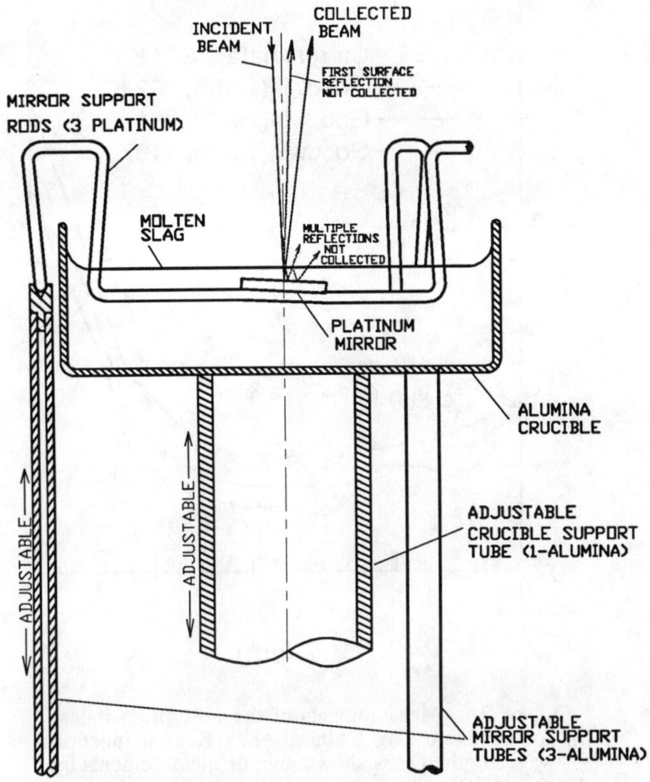

Figure 2. Schematic of the submerged reflector apparatus. The platinum mirror can be tilted independently of the slag crucible so that the first surface reflection from the molten slag surface and multiple internal reflections are rejected from the collection optics. The thin slag layer above the platinum mirror can be varied in thickness by vertical translation of the slag crucible.

RESULTS

Results of the submerged reflector measurements for $k(\lambda)$ are presented in Figure 3. In these experiments, approximately $30ml$ of slag SA05 was placed in an alumina crucible and brought to 1600° Cin the furnace. The platinum mirror was aligned and the crucible of slag was raised until the platinum mirror was submerged. The monochromator dial was moved to the appropriate wavelength setting and the detector signal I_1 was measured. The depth of the platinum mirror was changed by moving the slag crucible Δh and the second signal I_2 was measured. Then the monochromator was tuned to a new wavelength and the procedure repeated.

The depth of the platinum mirror ranged from approximately 100 μm to 1000 μm as the slag crucible was raised and lowered from one measurement to the next. Generally, a Δh of 200 $\mu m$was used for most wavelengths although the decrease in signal due to large k values at longer wavelengths required Δh =100 μm .

Noise becomes a significant factor at the longer wavelengths where SA05 becomes strongly absorbing. However most of the scatter seen in this data is attributable to the beam steering caused by density gradients in the hot column of air above the platinum mirror. The beam steering causes the image on the monochromator entrance slit to move noticeably. It is believed that much of this noise can be overcome by using a source of larger area so that motion of the image on the entrance slit will not significantly alter the radiant power entering the monochromator.

The measurements shown if Figure 3 compare well with earlier measurements of Goodwin (1986) at lower temperatures. The absorption index in the range from 1 $\mu m$to 4 $\mu m$is due primarily to the iron content of the slag. The increase in k beginning at 4 $\mu m$is due to the shoulder of a strong silica vibrational absorption band which is centered near 9 μm . Goodwin (1986) reports a correlation of k with the Fe^{2+} valence state of iron with k increasing as Fe^{2+} concentration increases. This slag was produced in an oxidizing environment (air) as was Goodwin's slag SA05. Mössbauer spectra of Goodwin's slag SA05 indicated that approximately 60% of the available iron is in the Fe^{2+} valence state. Therefore, it is likely that this slag also has the same Fe^{2+}/Fe ratios. Slags produced in a reducing environment are expected to have generally higher Fe^{2+}/Fe ratios (Goodwin, 1986).

Figure 4 shows the change in k with temperature at several different wavelengths. The results are essentially in agreement with the trends indicated by the lower temperature measurements of Goodwin (1986). For this slag, SA05, the imaginary refractive index increases with increasing temperature. The largest increase shown in Figure 4 is for a wavelength of 4 μm, where k increases from 1.95×10^{-4} at room temperature to nearly 4.9×10^{-4} at 1600° C. An important result of these measurements is the apparent continuity in k from the solid state at lower temperatures to the liquid state at 1600° C. This is to be expected for a vitreous material for which there is no phase change between the solid and liquid states.

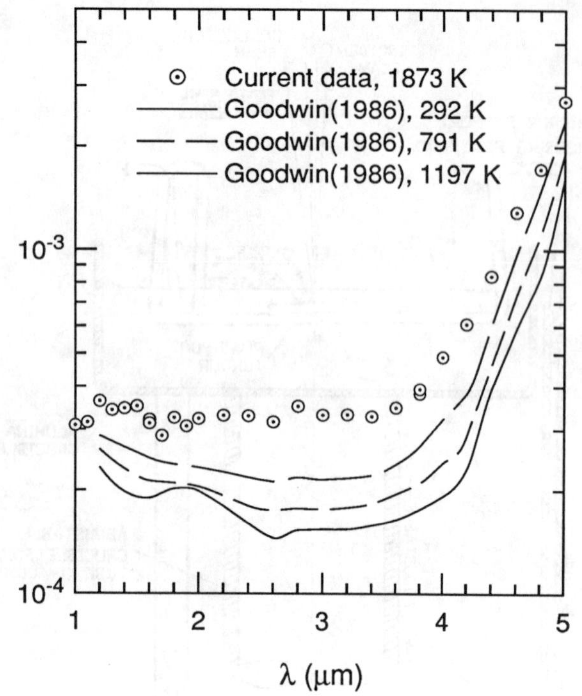

Figure 3. Measurements of the absorption index, k, for synthetic slag SA05 at 1873 K as a function of wavelength. Lines show previous measurements by Goodwin[10] at lower temperatures.

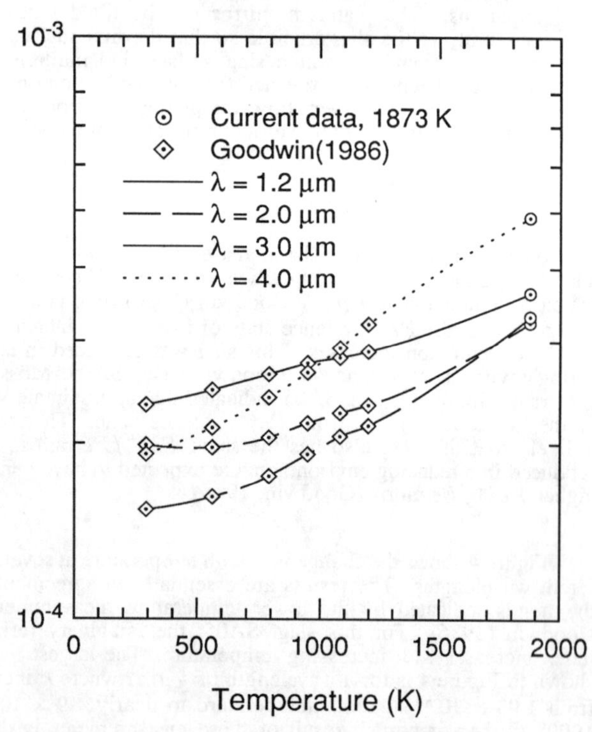

Figure 4. The absorption index, k, for synthetic slag SA05, as a function of temperature.

CONCLUSIONS

An experimental method for measuring the optical properties of molten slag has been presented. The "submerged reflector method" was used to measure the transmittance of a synthetic slag at 1600° Cin the wavelength range 1-5 μm. The effect of temperature was illustrated by comparing this data to data from Goodwin (1986) and the results were found to be in general agreement. For the results presented, k increases by as much as a factor of 2.5 at $\lambda \sim 4\ \mu m$ as the temperature is increased from $20°C$ to $1600°\ C$. This increase is somewhat less for shorter wavelengths.

Experiments are currently underway to obtain reflectance data which will provide the real refractive index, n, and k over a the entire wavelength range from 1-12 μm. Also, experiments to determine n and k for various slag compositions are being made. Results of these experiments will be reported as they become available.

ACKNOWLEDGEMENTS

This work was supported by NSF (Grant No. CBT-8516363) and by DOE (Contract No. DE-AC22-87PC 79963).

REFERENCES

Boothroyd, S. A., and Jones, A. R., 1984, "Radiative Transfer Scattering Data Relevant to Flyash", *Journal of Physics D: Applied Physics*, Vol. 17, pp. 1107 – 1114.

Ebert, J.L. and Self, S. A.,1988, "Radiation Heat Transfer in a Dispersion of Flyash" *Proceedings of the Engineering Foundation Conference, Mineral Matter and Ash Deposition from Coal*, Santa Barbara, CA, February, in press.

Goodwin, D.G., 1986, "Infrared Optical Constants of Coal Slags", PhD Thesis, HTGL Topical Report No. T-255, Stanford University, July.

Goodwin, D.G., and Mitchner, M., 1989, "The Infrared Optical Constants of Coal Slags: Dependence on Chemical Composition", *AIAA Journal of Thermophysics and Heat Transfer*, Vol. 3, No. 1, pp. 53-60.

Goodwin, D.G. and Mitchner, M., in press, "Flyash Radiative Properties and Effects on Radiative Heat Transfer in Coal-Fired Systems," *International Journal of Heat and Mass Transfer*.

Gupta, R. P., Wall, T. F., and Truelove, J. S., 1983, "Radiative Scatter by Fly Ash in Pulverized–Coal–Fired Furnaces: Application of the Monte Carlo Method to Anisotropic Scatter", *International Journal of Heat and Mass Transfer*, Vol. 26, pp. 1649 – 1660.

Mengüc, M. P. and Viskanta, R. 1988 ,"Effect of Fly-ash Particles on the Spectral and Total Radiation Blockage", *Combustion Science and Technology*, Vol. 60, pp. 97–115

Sarofim, A. F., 1978, "Flame Emissivities: Alternate Fuels in Combustion and Chemical Kinetics", C. T. Bowman and J. Birkeland (Eds.), *Progress in Aeronautics and Astronautics*, Vol. 62, AIAA, New York.

Wall, T. F., Lowe, A., Wibberley, L. J., and Stewart, I. McC., 1979 "Mineral Matter in Coal and the Thermal Performance of Large Boilers", *Progress in Energy and Combustion Science*,Vol.5, pp. 1–29.

1989 National Heat Transfer Conference
HTD-Vol. 106, Heat Transfer Phenomena in Radiation, Combustion and Fires

THE PLANCK MEAN AND ROSSELAND MEAN ABSORPTION COEFFICIENTS FOR PARTICLE LADEN MEDIA

A. Selamet and V. S. Arpaci
Department of Mechanical Engineering
and Applied Mechanics
The University of Michigan
Ann Arbor, Michigan

ABSTRACT

A recent study by Selamet and Arpaci (1989) on particle laden media extends by the Penndorf expansion the Planck and Rosseland mean absorption coefficients based on the Rayleigh limit of the Lorenz-Mie theory. The Penndorf-based Planck and Rosseland mean absorption coefficients relative to those of Rayleigh-based have been expressed as

$$\kappa^P_{P,R}/\kappa^R_{P,R} = 1 + \Pi_{P,R}\,(M_1,\ldots,\, N_1,\ldots,\, \pi DT/C_2)\ ,$$

where superscripts P and R denote Penndorf and Rayleigh; subscripts P and R, the Planck mean and Rosseland mean absorption coefficients, respectively; Π, the Penndorf correction depending on M's and N's and the dimensionless number $\pi DT/C_2$; M's and N's being the explicit functions of refractive and absorptive indices of particles; D, the particle diameter; T, the temperature; and C_2, the second radiation constant.

The present study extends Selamet and Arpaci by investigating the variation of the ratio of absorption coefficient against $\pi DT/C_2$ as a function of absorptive and refractive indices (m=2–i, m=1.5–0.5i), and approximates the rather involved form of the foregoing ratio by a simple empirical relation,

$$\kappa^P_{P,R}/\kappa^R_{P,R} = 1 + C\left(\pi DT/C_2\right)^p\ ,$$

where C and p are the constants fitted by nonlinear least squares damped Newton-Raphson method. Also, a numerical integration carried out for the Planck mean absorption coefficient based on the Lorenz-Mie theory provides, depending on the refractive indices, $\pi DT/C_2 \cong 0.8-1.0$ as the bound of Penndorf-based Planck mean absorption coefficient.

NOMENCLATURE

a, b	expansion coefficients defined by equations (A1) and (A2)
c	$= 2.9979\times10^8$ [m/s], speed of light
C	cross section [m^2]
C	coefficient in fit expressions
C_1	$= 3.74\times10^{-16}$ [W.m^2], first radiation constant
C_2	$= 1.4388\times10^{-2}$ [m.K], second radiation constant
D	diameter [m]
$E_{b\lambda}$	spectral emissive power of blackbody [W/m^3]
f_v	soot volume fraction [m^3 particle/m^3]
h	$= 6.6262\times10^{-34}$ [J.s], Planck's constant
i	complex unit
$\mathcal{I}$	"the imaginary part of"
k	absorptive index
k	$= 1.3806\times10^{-23}$ [J/K], Boltzmann's constant
ℓ	summation index
m	complex refractive index
n	refractive index
M_j, N_j	functions of n and k defined following equations (6) and (7), and by equations (17) and (22)
N	particle number density [number of particles/m^3]
P	normalized size distribution function [m^{-1}]

Q	efficiency factor
$\mathcal{R}$	"the real part of"
s	argument of Riemann zeta and Gamma function
T	temperature [K]
x	variable defined by equation (12)

Greek symbols

α	size parameter defined by equation (4d)
β	$= m\alpha$
Γ	Gamma function
ζ	Riemann zeta function
κ	volumetric spectral coefficient $[m^{-1}]$
κ_P	Planck mean absorption coefficient $[m^{-1}]$
κ_R	Rosseland mean absorption coefficient $[m^{-1}]$
λ	wavelength [m]
ν	frequency $[s^{-1}]$
ξ	Riccati-Bessel functions
Ξ	function defined by equation (14)
Π	Penndorf correction
ψ	Riccati-Bessel functions

Subscripts

a	absorption
b	blackbody
e	extinction
λ	spectral
P	related to Planck mean absorption coefficient
R	related to Rosseland mean absorption coefficient
s	scattering

Superscripts

E	Exact
p	Exponent in fit expression
P	Penndorf
R	Rayleigh
'	differentiation with respect to the argument

1. INTRODUCTION

Thermal radiation from particulate laden media often plays an appreciable role in industrial combustion applications (Hottel and Sarofim 1967; Thring and Lowes 1972; Tien and Lee 1982; Viskanta and Mengüç 1987), in fires (de Ris 1979), as well as in atmospherical and extraterrestrial problems (van de Hulst 1957; Liou 1980; Bohren and Huffman 1983). Examples are fluidized beds, oil and gas-fired furnaces, radiant burners, solid propellant rockets, gas turbine combustors, internal combustion engines, natural fires, clouds, fog, atmospheric and interstellar dust. The radiation in these applications depends on the spectral (volume or mass) coefficients which can be obtained from the classical Lorenz-Mie (LM) theory for isotropic, homogeneous spherical particles (see, for example, van de Hulst 1957; Kerker 1969; Deirmendjian 1969; Born and Wolf 1986; Jones 1979; Bayvel and Jones, 1981; Bohren and Huffman 1983). The Rayleigh approximation of the LM-theory is widely utilized when the particle diameter is small relative to the wavelength of radiation (see, for example, Hottel and Sarofim 1967; Penner and Olfe 1968; Dalzell and Sarofim 1969; Siegel and Howell 1981; Thring and Lowes 1972; Felske and Tien 1973, 1977; Buckius and Tien 1977; Hubbard and Tien 1978; Sarofim and Hottel 1978; Tien and Lee 1982; Bard and Pagni 1981; Santoro *et al.* 1983; and Beier *et al.* 1984). This approximation is usually assumed to be accurate up to the size parameter $\alpha = \pi D/\lambda \cong 0.3$, D being the particle diameter, λ the wavelength of radiation. Kerker *et al.* (1978), Bayvel and Jones (1981), and in light of Penndorf's studies (1960, 1962a) Ku and Felske (1984) and recently Selamet and Arpaci (1989) have contributed to improvements on this criterion.

Selamet and Arpaci (1989) also developed expressions for Planck and Rosseland mean coefficients based on the Penndorf extension and demonstrated the improvement over Rayleigh-based expressions.[1] Several studies have been conducted to derive these coefficients. Felske and Tien (1977) developed analytical expressions for mean coefficients in particulate laden media by utilizing Rayleigh limit of LM-theory, followed by the work of Hubbard and Tien (1978) who considered further the wavelength dependence of refractive index and used the Rayleigh limit. Brosmer and Tien (1985) calculated the Planck mean coefficient for acetylene. Various other researchers also investigated the Planck and Rosseland mean absorption coefficients in a wide variety of problems: see, for example, Blanco *et al.* (1983) who also used Rayleigh limit for Planck mean absorption of amorphous Carbon and graphite grains, as well as Henning *et al.* (1983) on application to circumstellar dust grains, Mengüç and Viskanta (1985) on coal polydispersions, Pollack *et al.* (1985) on dust grains in nebulae, Ganapol and Pomraning (1987) on the use of mean coefficients in non-equilibrium radiative transfer equation (RTE),

[1] Note the need of these mean coefficients for the well-known thin gas and thick gas approximations, respectively. A large body of literature employing these limits is available: see, for example, Lord and Arpaci (1970), Sparrow and Cess (1978), Selamet and Arpaci (1988) for thin gas, and Arpaci (1968), Arpaci and Larsen (1969), and Viskanta and Anderson (1975) for thick gas.

Tsakiris and Eidmann (1987) on hot plasmas, and Voshchinnikov and Il'in (1982) on non-spherical grains. The foregoing studies are the motivation for the present work on Planck and Rosseland mean coefficients. Following a brief review of a recent study on the mean coefficients based on the Penndorf expansion (Selamet and Arpaci 1989), the first objective is to investigate these coefficients further depending on refractive indices, and to replace them with useful approximate expressions. The second objective is to carry out a numerical integration for the Planck mean absorption coefficient based on the LM-theory and to examine limit expressions against the result of this integration.

The study consists of six sections: following this introduction, Section 2 reviews the derivation of efficiency factors for LM-theory, and the Rayleigh and Penndorf limits of these factors. Section 3 applies these approximations to Planck and Rosseland mean absorption coefficients and develops improved analytical expressions which are fit by simplified formulae in Section 4 by employing nonlinear least squares damped Newton-Raphson method. Section 5 introduces exact numerical integration for Planck mean absorption coefficient from LM-theory and Section 6 concludes the study.

2. PENNDORF EXPANSION

Under the assumptions of single and independent scattering, the spectral coefficients of RTE for a polydisperse media (see, for example, Buckius and Hwang 1980; Mengüç and Viskanta 1985) are

$$\kappa_{a,s}(m,N,\lambda) = \int_0^\infty Q_{a,s}(m,D,\lambda)\,\frac{\pi D^2}{4}\,N\,P(D)\,d(D)\ , \quad (1)$$

and

$$\kappa_e(m,N,\lambda) = \kappa_a(m,N,\lambda) + \kappa_s(m,N,\lambda)\ , \quad (2)$$

where subscripts a, s, and e denote absorption, scattering, and extinction, respectively, $m = n - ik$ is the usual complex refractive index of particles with respect to surrounding medium, i the complex unit, N the particle number density, λ the wavelength, Q the efficiency factor, D the diameter, and P(D) the normalized particle size distribution function which satisfies

$$\int_0^\infty P(D)d(D) = 1\ .$$

For monodisperse medium, equations (1) and (2) are reduced, in terms of the definition of cross section $C = Q\,\frac{\pi D^2}{4}$, to

$$\kappa_{a,s,e} = Q_{a,s,e}\,\frac{\pi D^2}{4}\,N \equiv C_{a,s,e}\,N \quad (3)$$

The LM-theory derived for isotropic, homogeneous spheres gives for the extinction and scattering efficiencies (refer to, for example, van de Hulst),

$$Q_e(m,\alpha) = \frac{2}{\alpha^2}\sum_{\ell=1}^{\infty}(2\ell+1)\,\mathcal{R}\left(a_\ell+b_\ell\right)\ , \quad (4a)$$

$$Q_s(m,\alpha) = \frac{2}{\alpha^2}\sum_{\ell=1}^{\infty}(2\ell+1)\left(|a_\ell|^2+|b_\ell|^2\right)\ , \quad (4b)$$

and for the absorption efficiency,

$$Q_a(m,\alpha) = Q_e(m,\alpha) - Q_s(m,\alpha)\ , \quad (4c)$$

where

$$\alpha = \frac{\pi D}{\lambda} \quad (4d)$$

is the size parameter and $\mathcal{R}$ indicates "the real part of", | | "the absolute value of", a_ℓ and b_ℓ are the expansion coefficients (see, Appendix A). Equations (3) and (4) coupled with information on N allow the computation of spectral coefficients.

In the Rayleigh limit where $\alpha \ll 1$, the efficiency factors are known to be

$$Q_a^R(m,\alpha) = -4\alpha\mathcal{I}\left(\frac{m^2-1}{m^2+2}\right)\ ,\ Q_s^R(m,\alpha) = \frac{8}{3}\alpha^4\left|\frac{m^2-1}{m^2+2}\right|^2 , \quad (5a,b)$$

and

$$Q_e^R(m,\alpha) = Q_a^R(m,\alpha) + Q_s^R(m,\alpha)\ , \quad (5c)$$

where the superscript R is for **R**ayleigh; $\mathcal{I}$, "the imaginary part of". Insertion of $m=n\text{-}ik$ into equations (5a,b) yields explicit expressions for efficiency factors in terms of α, n and k. These are

$$Q_a^R = 12\left(\frac{N_1}{M_1}\right)\alpha\ ,\ Q_s^R = \frac{8}{3}\left(1 - 3\,\frac{M_2}{M_1}\right)\alpha^4\ , \quad (6a,b)$$

and

$$Q_e^R = Q_a^R + Q_s^R\ , \quad (6c)$$

where

$$M_1 = N_1^2 + (2 + N_2)^2\ ,\ M_2 = 1 + 2N_2\ ,$$

$$N_1 = 2nk\ ,\ N_2 = n^2 - k^2\ .$$

The foregoing results have been extended by

Penndorf (1960, 1962a, 1962b) to larger particles by a series expansion. For scattering and extinction efficiency factors, the Penndorf expansion yields, after some rearrangement,

$$Q_s^P = Q_s^R \left[1 + 2 \frac{\alpha^2}{M_1} \left(\frac{3}{5} M_3 - 2 N_1 \alpha \right) \right] , \qquad (7a)$$

$$Q_e^P = Q_a^R + 2\alpha^3 \left[N_1 \left(\frac{1}{15} + \frac{5}{3} \frac{1}{M_4} + \frac{6}{5} \frac{M_5}{M_1^2} \right) + \frac{4}{3} \frac{M_6}{M_1^2} \alpha \right] , \qquad (7b)$$

and

$$Q_a^P = Q_e^P - Q_s^P , \qquad (7c)$$

where superscript P is for **P**enndorf, and the remaining M's are defined to be

$$M_3 = N_3 - 4 \ , \ M_4 = 4 N_1^2 + (3 + 2N_2)^2 ,$$

$$M_5 = 4(N_2 - 5) + 7N_3 , M_6 = (N_2 + N_3 - 2)^2 - 9 N_1^2 ,$$

and

$$N_3 = (n^2 + k^2)^2 = N_1^2 + N_2^2 .$$

The next section proceeds to a brief review of modified mean coefficients recently developed by Selamet and Arpaci (1989).

3. MODIFIED PLANCK AND ROSSELAND MEAN ABSORPTION COEFFICIENTS

The LM-theory and its Rayleigh limit have been extensively used in combustion problems involving RHT and/or optical diagnostics (OD). The majority of OD studies on combustion attempt to extract information about the properties of combustion generated particles such as soot formed by burning of hydrocarbon (HC) fuels.[2] Depending on the chemical composition and HC content, the refractive and absorptive indices of soot are experimentally found to fall into $1.5 \leq n \leq 2.5$ and $0.5 \leq k \leq 1.5$ in the visible range (Janzen 1979). Although n and k depend on the wavelength (Stull and Plass 1960; Foster and Howarth 1968; Lee and coworkers 1981, 1982; and Charalampopoulos and Felske 1987), this dependence remains well within the spread of the experimental data in the visible range. An inspection of heat transfer and combustion literature (Dalzell and Sarofim 1969; Janzen 1979; Pluchino *et al.* 1980; Lee and Tien 1981; Tien and Lee 1982; Drolen and Tien 1987; D'Alessio and coworkers 1981, 1982; Driscoll *et al.* 1979; and Flower 1983) reveal three widely used complex refractive indices

$$m = 1.5 - 0.5\,i \ , \ \text{(Dalzell and Sarofim} - 1969)$$
$$m = 2.0 - 1.0\,i \ , \ \text{(Janzen} - 1979)$$
$$m \cong 1.75 - 0.75\,i \ , \ \text{(Pluchino et al.} - 1980)$$

which are utilized in the remainder of the present study.

In a particulate laden medium exhibiting continuous radiation, spectrally integrated quantities such as the Planck mean and the Rosseland mean absorption coefficients (see, for example, Sparrow and Cess 1978),

$$\kappa_P = \frac{\int_0^\infty \kappa_{a\lambda} E_{b\lambda} \, d\lambda}{\int_0^\infty E_{b\lambda} \, d\lambda} \qquad (8)$$

and

$$\kappa_R = \frac{\displaystyle\int_0^\infty \frac{\partial E_{b\lambda}}{\partial T} d\lambda}{\displaystyle\int_0^\infty \left(\frac{1}{\kappa_{a\lambda}} \right) \frac{\partial E_{b\lambda}}{\partial T} d\lambda} , \qquad (9)$$

are needed. Here

$$E_{b\lambda} = \frac{C_1}{\lambda^5 \left(e^{C_2/\lambda T} - 1 \right)} \qquad (10)$$

is the spectral Planck function for blackbody emissive power; C_1 the first radiation constant; $C_2 = hc/k$ the second radiation constant, h being Planck's constant, c the speed of light, and *k* Boltzmann's constant. Inserting

$$N = \frac{f_v}{\pi D^3/6}$$

into equation (3) and combining with equation (7) for Q_a yields the following expression for the spectral absorption coefficient based on the Penndorf expansion:

$$\kappa_{a\lambda}^P = \left(18\,\pi \frac{N_1}{M_1} f_v \right) \frac{1}{\lambda}$$

$$+ \left[N_1 \left(\frac{1}{5} + \frac{5}{M_4} + \frac{18}{5} \frac{M_5}{M_1^2} \right) \pi^3 D^2 f_v \right] \frac{1}{\lambda^3} \qquad (11)$$

$$+ \left[4 \left(\frac{M_6}{M_1^2} - 1 + 3 \frac{M_2}{M_1} \right) \pi^4 D^3 f_v \right] \frac{1}{\lambda^4} ,$$

[2] For details of soot structure, refer to Selamet and Arpaci (1989).

where the first term represents the Rayleigh contribution to be denoted by $\kappa_{a\lambda}^{R}$. To proceed with the development of the Penndorf-based Planck mean absorption coefficient, substitute equations (10) and (11) into equation (8) and conveniently express integration in the latter in terms of

$$x = \frac{h\nu}{kT} = \frac{C_2}{\lambda T} \qquad \left(\lambda = \frac{C_2}{T}\frac{1}{x}, \quad d\lambda = -\frac{C_2}{T}\frac{dx}{x^2} \right), \tag{12}$$

ν being the frequency, T the temperature , to obtain

$$\begin{aligned} \kappa_P^P = & \; 18 \frac{\Xi(5)}{\Xi(4)} \frac{N_1}{M_1} f_v \left(\frac{\pi T}{C_2} \right) \\ & + \frac{\Xi(7)}{\Xi(4)} N_1 \left(\frac{1}{5} + \frac{5}{M_4} + \frac{18}{5} \frac{M_5}{M_1^2} \right) f_v D^2 \left(\frac{\pi T}{C_2} \right)^3 \\ & + 4 \frac{\Xi(8)}{\Xi(4)} \left(\frac{M_6}{M_1^2} - 1 + 3 \frac{M_2}{M_1} \right) f_v D^3 \left(\frac{\pi T}{C_2} \right)^4 , \end{aligned} \tag{13}$$

where $\Xi(s)$ is defined to be

$$\Xi(s) = \zeta(s)\,\Gamma(s) = \int_0^\infty \frac{x^{s-1}}{e^x - 1} dx , \tag{14}$$

ζ being the Riemann zeta function (Abramowitz and Stegun 1972), and Γ the Gamma function; and others have dimensions C_2 [m.K], T [K], D [m], and κ_P^P [m^{-1}]. Equation (13) may be rearranged by dividing both sides with the first term on the right hand side

$$\kappa_P^R = \frac{\Xi(5)}{\Xi(4)} \left(18 \frac{N_1}{M_1} \right) f_v \left(\frac{\pi T}{C_2} \right) \tag{15}$$

as

$$\frac{\kappa_P^P}{\kappa_P^R} = 1 + \frac{\Xi(7)}{\Xi(5)} M_I \left(\frac{\pi D T}{C_2} \right)^2 + \frac{\Xi(8)}{\Xi(5)} M_{II} \left(\frac{\pi D T}{C_2} \right)^3 , \tag{16}$$

where κ_P^R is the Planck mean absorption coefficient based on the Rayleigh limit, and M_I and M_{II} are defined to be

$$M_I = \frac{M_1}{18} \left(\frac{1}{5} + \frac{5}{M_4} + \frac{18}{5} \frac{M_5}{M_1^2} \right) , \tag{17a}$$

$$M_{II} = \frac{2}{9} \frac{M_1}{N_1} \left(\frac{M_6}{M_1^2} - 1 + 3 \frac{M_2}{M_1} \right) . \tag{17b}$$

Evaluation of Ξ function in terms of Riemann-zeta and Gamma functions gives for equation (15)

$$\kappa_P^R = 3.8322 \left(18 \frac{N_1}{M_1} \right) f_v \left(\frac{\pi T}{C_2} \right) , \tag{18}$$

and for equation (16),

$$\frac{\kappa_P^P}{\kappa_P^R} = 1 + 29.173\, M_I \left(\frac{\pi DT}{C_2} \right)^2 + 203.347\, M_{II} \left(\frac{\pi DT}{C_2} \right)^3 , \tag{19a}$$

or,

$$\kappa_P^P / \kappa_P^R = 1 + \Pi_P^P (\underbrace{M_1, \ldots, N_1, \ldots,}_{\text{EM}-\text{waves}} \; \underbrace{\pi DT/C_2}_{\text{Quanta}}) , \tag{19b}$$

where κ_P^R given by equation (18) is the Rayleigh-based Planck mean absorption coefficient originally derived by Felske and Tien (1977), and Π_P is the Penndorf correction to this expression. This correction depends on M's and N's (given in terms of n and k) and the dimensionless $\pi DT/C_2$ (combining the effects of diameter and temperature). The former shows the mean effect of the interaction of electromagnetic waves with spherical particles, while the physical significance of the latter is clear from quantum mechanics which assumes that the radiation-matter interaction is mostly between photons and vibrational modes of matter (Phillips and Arpaci 1972, 1975, Selamet and Arpaci 1989).

A similar, but somewhat more involved development for Rosseland mean absorption coefficient (see, Selamet and Arpaci 1989, for details) yields, in terms of Rayleigh-based Rosseland mean

$$\kappa_R^R = \frac{4}{3} \frac{\Xi(4)}{\Xi(3)} \left(18 \frac{N_1}{M_1} \right) f_v \left(\frac{\pi T}{C_2} \right) , \tag{20}$$

the ratio as

$$\frac{\kappa_R^P}{\kappa_R^R} = \frac{1}{1 + \frac{5}{3} \frac{\Xi(5)}{\Xi(3)} M_{III} \left(\frac{\pi D T}{C_2} \right)^2 + 2 \frac{\Xi(6)}{\Xi(3)} M_{IV} \left(\frac{\pi D T}{C_2} \right)^3} , \tag{21}$$

where M_{III} and M_{IV} are defined to be

$$\begin{aligned} M_{III} & \cong -10.9375 \left[1 - \frac{1 + 0.7086 M_I + 0.576 M_{II}}{(1 + 0.16 M_I + 0.064 M_{II})(1 + 0.64 M_I + 0.512 M_{II})} \right] , \\ M_{IV} & \cong 11.7188 \left[1 - \frac{1 + 0.8 M_I + 0.6613 M_{II}}{(1 + 0.16 M_I + 0.064 M_{II})(1 + 0.64 M_I + 0.512 M_{II})} \right] . \end{aligned} \tag{22}$$

Evaluation of Ξ function gives, for equation (20),

$$\kappa_R^R = 3.6016 \left(18 \frac{N_1}{M_1}\right) f_v \left(\frac{\pi T}{C_2}\right), \tag{23}$$

and, for equation (21),

$$\frac{\kappa_R^P}{\kappa_R^R} = \frac{1}{1 + 17.253\, M_{III}\left(\frac{\pi DT}{C_2}\right)^2 + 101.560\, M_{IV}\left(\frac{\pi DT}{C_2}\right)^3}, \tag{24a}$$

or, introducing Π_R^P,

$$\kappa_R^P/\kappa_R^R = 1 + \Pi_R^P(\underbrace{M_1,\ldots, N_1,\ldots,}_{\text{EM-waves}} \underbrace{\pi DT/C_2}_{\text{Quanta}}). \tag{24b}$$

Here, κ_R^R is the Rayleigh-based Rosseland mean absorption coefficient originally derived by Felske and Tien (1977), Π_R is the Penndorf correction to this expression.

Consider now illustrative examples in terms of $m = 2 - i$ and $m = 1.5 - 0.5i$. The results of calculations for κ_P^P and κ_R^P respectively from equations (19a) and (24a) are depicted in Fig. 1 against $\pi DT/C_2$. The interpretation of these results for $m = 2 - i$ is given by Selamet and Arpaci (1989). The present study investigates the effect of m variation on the Penndorf extended mean coefficients. Figure 1 shows, as m changes from $m = 2 - i$ to $1.5 - 0.5i$, the decreasing effect of Penndorf extension. Figures 2 and 3 demonstrate the explicit effect of diameter and temperature. The smooth variation of ratios for different refractive indices observed in Fig. 1 suggests that equations (19a) and (24a) be replaced with simple empirical relations which are developed next.

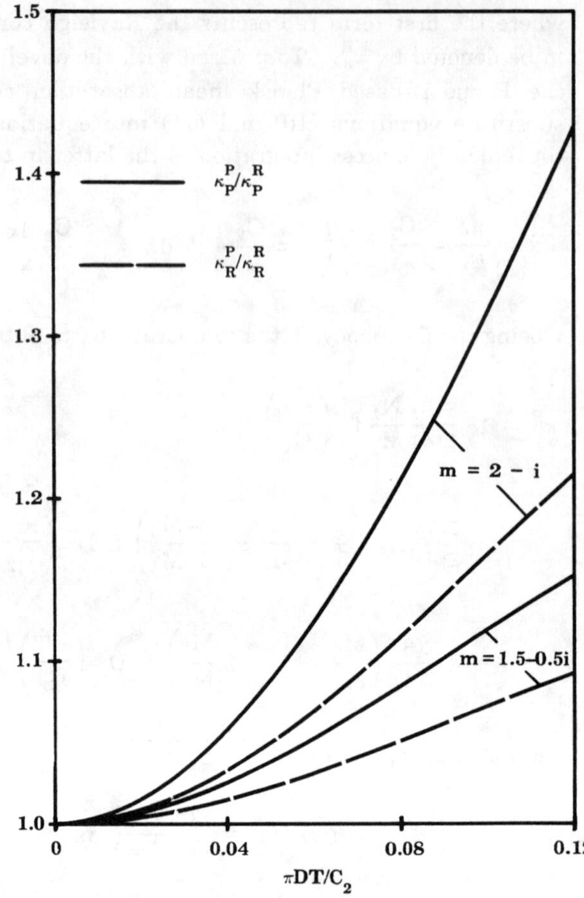

Figure 1. The Penndorf correction to the Rayleigh based Planck and Rosseland mean coefficients as a function of $\pi DT/C_2$, for typical values of m

4. APPROXIMATE MEAN COEFFICIENTS

Let the Planck and Rosseland coefficients given by equations (19a) and (24a) be approximated by

$$\kappa_{P,R}^P/\kappa_{P,R}^R = 1 + C\left(\pi DT/C_2\right)^p, \tag{25}$$

where C and p are constants to be fitted by nonlinear least squares damped Newton-Raphson method. Table 1 and 2 summarize fit parameters for Planck and Rosseland mean absorption coefficients, respectively. The last column in these tables represents the maximum percent error in equation (25) over the range $0 \le \pi DT/C_2 \le 0.12$.

A computational study of the exact mean coefficients is useful for bounds of the approximate coefficients and is carried out in the next section.

Table 1. Fit parameters for the ratio of Planck mean absorption coefficients

m	C	p	Max. Error
1.5–0.5i	4.3575	1.5699	0.32%
1.75–0.75i	9.3935	1.6735	0.32%
2.0–1.0i	19.3635	1.7927	0.27%

Table 2. Fit parameters for the ratio of Rosseland mean absorption coefficients

m	C	p	Max. Error
1.5–0.5i	2.8023	1.5966	0.20%
1.75–0.75i	5.0433	1.6381	0.29%
2.0–1.0i	7.2473	1.6491	0.40%

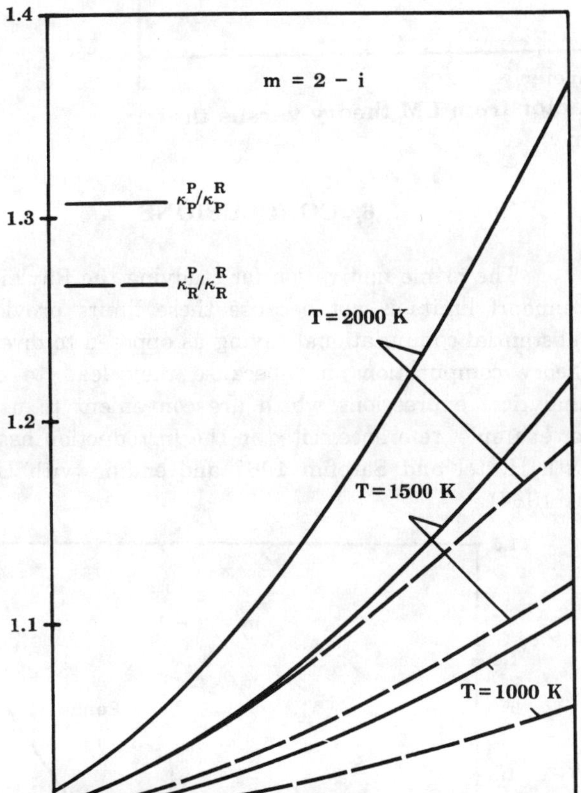

Figure 2. The Penndorf correction to the Rayleigh based Planck and Rosseland mean coefficients as explicit function of diameter and temperature for m=2-i

5. EXACT MEAN COEFFICIENTS

Reconsider the Planck mean absorption coefficient defined by equation (8). Section 3, employing the Rayleigh limit and Penndorf extension for the efficiency factor, and hence for the spectral absorption coefficient, provided convenient analytical expressions (equations 18, 19a). Here, these expressions are examined against an accurate numerical integration which employs the absorption efficiency factor directly from LM−theory.

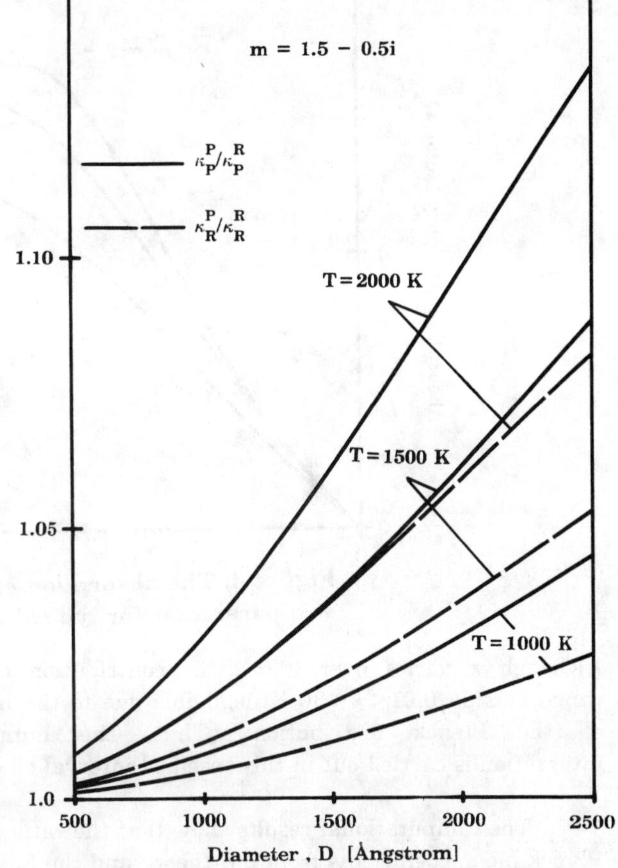

Figure 3. The Penndorf correction to the Rayleigh based Planck and Rosseland mean coefficients as explicit function of diameter and temperature for m=1.5–0.5i

The numerator of equation (8) can be rewritten, for monodisperse media, as

$$\int_{\lambda=0}^{\infty} \kappa_{a\lambda}^{E} E_{b\lambda}\, d\lambda = N\frac{\pi D^2}{4} C_1 \int_{\lambda=0}^{\infty} Q_a(m,D,\lambda)\frac{d\lambda}{\lambda^5\left(e^{C_2/\lambda T}-1\right)}, \quad (26)$$

where superscript E denotes the numerical integration based on the exact LM-code for computation of Q_a, to be briefly explained at the end of this section. The variation of Q_a for various values of m is shown in Fig. 4 (Selamet 1989). Transforming the integration in equation (26) into the x domain (recall equation 12) and combining the result with equations (8) and (18) yields

$$\frac{\kappa_P^E}{\kappa_P^R} = 0.0033486\, \frac{M_1}{N_1}\left(\frac{\pi DT}{C_2}\right)^{-1} \int_{x=0}^{\infty} Q_a(m,D,x)\frac{x^3\, dx}{(e^x-1)}. \quad (27)$$

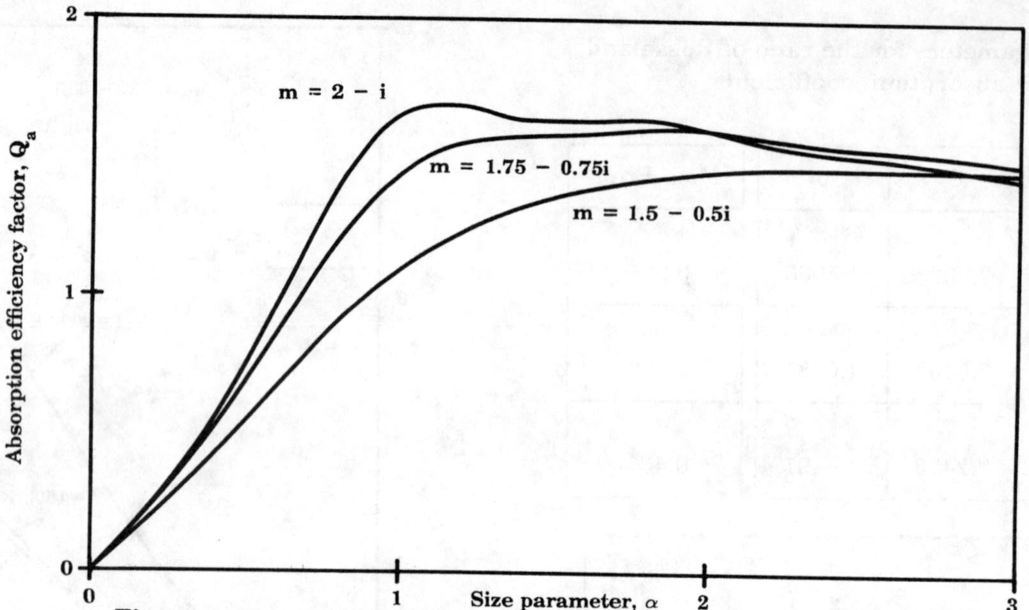

Figure 4. The absorption efficiency factor from LM theory versus the size parameter for typical values of m

Although x varies over $(0,\infty)$, the contribution of the range outside $0.01<x<30$ is negligible due to the nature of the Planck distribution. Thus, the numerical integration is carried out in this specified interval of x.

The computational results show that the ratio of the Planck mean coefficients from LM-theory and the Rayleigh limit is also a function of the dimensionless number $\pi DT/C_2$, as expected. This allows a direct comparison between limit expressions and computational results. One such comparison depicted in Fig. 5 for $m = 2 - i$, illustrates the accuracy of the Penndorf-based Planck mean coefficient up to about $\pi DT/C_2 \cong 0.8-1.0$. The deviation in Fig. 5 for large $\pi DT/C_2$ turns out to be less for $m=1.75\text{–}0.75i$ and $m=1.5\text{–}0.5i$. Here, the figures for the last two refractive indices are not included because of space considerations.

The exact LM calculations of this section for equations (4a-c) require the evaluation of a_ℓ and b_ℓ from equations (A1) and (A2) which can be reduced to a combination of half integer order Bessel functions with real and complex arguments. Several techniques are available in the literature (see Selamet 1989 or Selamet and Arpaci 1989, for details and references). Of particular interest is the one improved by Wiscombe (1979, 1980) who developed efficient recurrence formulas for angular functions and generated an empirical criterion for allowing up-recurrence in the computation of half order Bessel functions. The present study, following the Wiscombe criterion, carries out computations with a fast up-recurrence scheme are (see, also, Selamet 1989). A similar study on the Rosseland coefficient is being carried out and will be reported later.

6. CONCLUSIONS

The prime motivation for studying the Rayleigh and Penndorf limits is not because these limits provide any substantial computational saving as opposed to direct LM-theory computations but because they lead to explicit analytical expressions which are convenient to use (see, for example, references cited in the Introduction beginning with Hottel and Sarofim 1967 and ending with Beier *et al.* 1984).

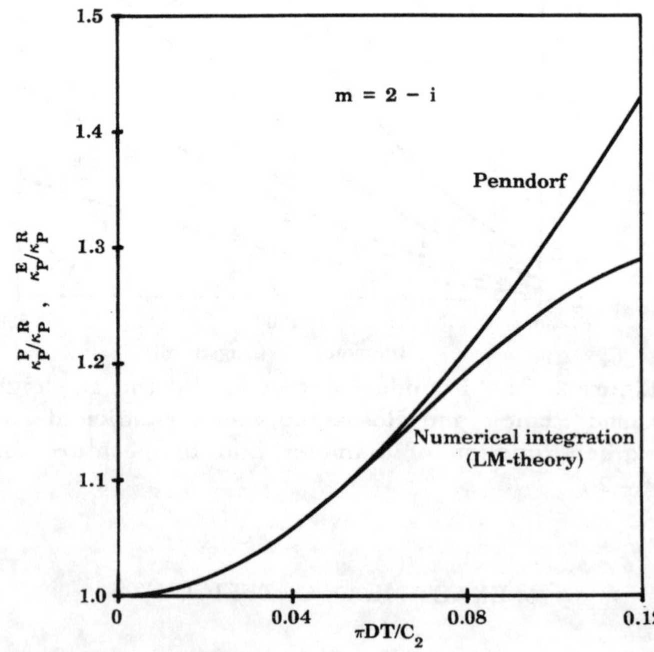

Figure 5. Comparison of Penndorf-based and Lorenz-Mie-based Planck mean absorption coefficient

The Penndorf correction is shown to cover approximately the spectral range of continuous radiation from soot particles. For thin clouds, this correction is applied to the Planck mean absorption coefficient. Similar development is performed for thick clouds leading to an expression for the Rosseland mean absorption coefficient based on the Penndorf correction. The refractive index dependency of Penndorf-based mean coefficients is investigated. Approximate fit expressions to analytical forms of improved Planck and Rosseland mean coefficients are developed. The exact numerical integrations employing LM-theory are carried out. A comparison of integration results with Penndorf-extended Planck coefficient confirms the importance and validity of the Penndorf extension. Similar computations for the Rosseland mean are being carried out and will be reported later.

REFERENCES

Abramowitz, M. and **Stegun, I.A.** 1972, *Handbook of Mathematical functions* (10th Print), NBS-Applied Mathematics Series 55, pp. 807, 811. US Govt. Printing Office, Washington.

Arpaci, V.S. 1968, Effect of thermal radiation on the laminar free convection from a heated vertical plate, *Int. J. Heat Mass Transfer* **11**, 871–881.

Arpaci, V.S. and **Larsen, P.S.** 1969, A thick gas model near boundaries, *AIAA J.* **7**, 602–606.

Bard, S. and **Pagni, P.J.** 1981, Carbon particulate in small pool fire flames, *J. Heat Transfer* **103**, 357–362.

Bayvel, L.P. and **Jones, A.R.** 1981, *Electromagnetic Scattering and its Applications*. Applied Science, New Jersey.

Beier, R.A., Pagni, P.J. and **Okoh, C.I.** 1984, Soot and radiation in combusting boundary layers, *Comb. Sci. and Tech.* **39**, 235–262.

Blanco, A., Falcicchia, G., and **Merico, F.** 1983, Planck mean absorption efficiency for amorphous Carbon and Graphite grains, *Astrophysics and Space Science* **89**, 163–168.

Bohren, C.F. and **Huffman, D.R.** 1983, *Absorption and Scattering of Light by Small Particles*. Wiley, New York.

Born, M. and **Wolf, E.** 1986, *Principles of Optics* (6th Edn). Pergamon Press, Oxford.

Brosmer, M.A. and **Tien, C.L.** 1985, Thermal radiation properties of Acetylene, *J. Heat Transfer* **107**, 943–948.

Buckius, R.O. and **Tien, C.L.** 1977, Infrared flame radiation, *Int. J. Heat Mass Transfer* **20**, 93–106.

Buckius, R.O. and **Hwang, D.C.** 1980, Radiation properties for polydispersions: Application to coal, *J. Heat Transfer* **102**, 99–103.

Charalampopoulos, T.T. and **Felske, J.D.** 1987, Refractive indices of soot particles deduced from in-situ laser light scattering measurements, *Combust. Flame* **68**, 283–294.

Dalzell, W.H. and **A. F. Sarofim** 1969, Optical constants of soot and their application to heat flux calculations, *J. Heat Transfer* **91**, 100 – 104.

D'Alessio, A. 1981, Laser light scattering and fluorescence diagnostics of rich flames produced by gaseous and liquid fuel, in *Particulate Carbon: Formation During Combustion* (Edited by D. C. Siegla and G. W. Smith), pp. 207–259. Plenum Press, New York.

Deirmendjian, D. 1969, *Electromagnetic Scattering on Spherical Polydispersions*. American Elsevier Publ. Co., New York.

de Ris, J. 1979, "Fire radiation – A review," *Seventeenth Symposium (International) on Combustion*, pp. 1003–1016. The Combustion Institute, Pittsburgh.

Driscoll, J.F., Mann, D.M. and **McGregor, W.K.** 1979, Submicron particle size measurements in an Acetylene-Oxygen flame, *Comb. Sci. and Tech.* **20**, 41–47.

Drolen, B.L. and **Tien, C.L.** 1987, Absorption and scattering of agglomerated soot particulate, *J. Quant. Spectrosc. Radiat. Transfer* **37**, 433–448.

Felske, J.D. and **Tien, C.L.** 1973, Calculation of the emissivity of luminous flames, *Comb. Sci. and Tech.* **7**, 25–31.

Felske, J.D. and **Tien, C.L.** 1977, The use of the Milne-Eddington absorption coefficient for radiative heat transfer in combustion systems, *J. Heat Transfer* **99**, 458–465.

Flower, W.L. 1983, Optical measurements of soot formation in premixed flames, *Comb. Sci. and Tech.* **33**, 17–33.

Foster, P.J. and **Howarth, C.R.** 1968, Optical constants of carbon and coals in the infrared, *Carbon* **6**, 719–729.

Ganapol, B.D. and **Pomraning, G.C.** 1987, On the use of the Rosseland and Planck mean absorption coefficients in the non-equilibrium radiative transfer equation, *J. Quant. Spectrosc. Radiat. Transfer* **37**, 297–309.

Henning, T., **Gürtler, J.**, and **Dorschner, J.** 1983, Observationally-based infrared efficiencies and Planck means for circumstellar dust grains, *Astrophysics and Space Science* **94**, 333–349.

Hottel, H.C. and **Sarofim, A.F.** 1967, *Radiative Transfer*, Chap.12. McGraw-Hill, New York.

Hubbard, G.L. and **Tien, C.L.** 1978, Infrared mean absorption coefficients of luminous flames and smoke, *J. Heat Transfer* **100**, 235–239.

Janzen, J. 1979, The refractive index of colloidal carbon, *J. Colloid Interface Sci.* **69**, 436–447.

Jones, A.R. 1979, Scattering of electromagnetic radiation in particulate laden fluids, *Prog. Energy Combust. Sci.* **5**, 73–96.

Kerker, M. 1969, *The Scattering of Light and Other Electromagnetic Radiation.* Academic Press, New York.

Kerker, M., **Scheiner, P.**, and **Cooke, D.D.** 1978, The range of validity of the Rayleigh and Thomson limits for Lorenz-Mie scattering, *J. Optical Society of America* **68**, 135–137.

Ku, J.C. and **Felske, J.D.** 1984, The range of validity of the Rayleigh limit for computing Mie scattering and extinction efficiencies, *J. Quant. Spectrosc. Radiat. Tr.* **31**, 569–574.

Lee, S.C. and **Tien, C.L.** 1981, Optical constants of soot in hydrocarbon flames, *Eighteenth Symposium (International) on Combustion*, pp. 1159–1166. The Combustion Institute, Pittsburgh.

Lee, S.C., **Yu, Q.Z.**, and **Tien, C.L.** 1982, Radiation properties of soot from diffusion flames, *J. Quant. Spectrosc. Radiat. Transfer* **27**, 387–396.

Liou, K.N. 1980, *An Introduction to Atmospheric Radiation.* Academic Press, New York.

Lord, H.A. and **Arpaci, V.S.** 1970, Effect of nongray thermal radiation on laminar forced convection over a heated horizontal plate, *Int. J. Heat Mass Transfer* **13**, 1737–1751.

Mengüç M.P. and **Viskanta, R.** 1985, On the radiative properties of polydispersions: A simplified approach, *Combust. Sci. and Tech.* **44**, 143–159.

Menna, P. and **D'Alessio, A.** 1982, Light scattering and extinction coefficients for soot forming flames in the wavelength range 0.2μ-0.6μ, *19th Symposium (International) on Combustion*, pp. 1421–1428. The Combustion Institute, Pittsburgh.

Penndorf, R.B. 1960, Scattering coefficients for absorbing and nonabsorbing aerosols, Tech. Report RAD-TR-60-27; AFCRL-TN-60-667, Avco Corp., Wilmington, Mass.

Penndorf, R.B. 1962a, Scattering and Extinction coefficients for small absorbing and nonabsorbing aerosols, *J. Opt. Soc. Am.* **52**, 896–904.

Penndorf, R.B. 1962b, Scattering and extinction coefficients for small spherical aerosols, *J. Atmos. Sci.* **19**, 193.

Penner, S.S. and **Olfe, D.B.** 1968, *Radiation and Reentry*, Chap.4. Academic Press, New York.

Phillips, W.F. and **Arpaci, V.S.** 1972, Diatomic gas-thermal radiation interaction: a model equation for the internal fluid, *J. Plasma Physics* **7**, 235–246.

Phillips, W.F. and **Arpaci, V.S.** 1975, Monatomic plasma-thermal radiation interaction: a weakly-ionized kinetic model, *J. Plasma Physics* **13**, 523–537.

Pluchino, B., **Goldberg, S.S.**, **Dowling, J.M.**, and **Randall, C.M.** 1980, Refractive index measurements of single micron-sized carbon particles, *Appl. Optics* **19**, 3370–3372.

Pollack, J.B., **McKay, C.P.**, and **Christofferson, B.M.** 1985, A calculation of the Rosseland mean opacity of dust grains in Primordial Solar system Nebulae, *Icarus* **64**, 471–492.

Santoro, R.J., **Semerjian, H.G.**, and **Dobbins, R.A.** 1983, Soot particle measurements in diffusion flames, *Combust. Flame* **51**, 203–218.

Sarofim, A.F. and **Hottel, H.C.** 1978, Radiative transfer in combustion chambers: Influence of alternative fuels, *Proceedings of the 6th International Heat Transfer Conference*, Vol. **6**, pp. 199–217. Toronto, Canada.

Selamet, A. and **Arpaci, V.S.** 1988, Entropy production in radiation-affected boundary layers, *AIAA Thermophysics, Plasmadynamics and Lasers Conference* Paper No AIAA-88-2640, San Antonio, Texas.

Selamet, A. 1989, Radiation affected laminar flame propagation, Ph.D. thesis, University of Michigan, Ann Arbor, Michigan.

Selamet, A. and **Arpaci, V.S.** 1989, Rayleigh Limit – Penndorf extension, to appear in *Int. J. Heat Mass Transfer*.

Siegel, R. and **Howell, J.R.** 1981, *Thermal Radiation Heat Transfer* (2nd Edn), Chap.17. Hemisphere Publ. Co., McGraw-Hill, Washington.

Sparrow, E.M. and **Cess, R.D.** 1978, *Radiation Heat Transfer*. Hemisphere Publ. Co., Washington, McGraw-Hill, New York.

Stull, V.R. and **Plass, G.N.** 1960, Emissivity of dispersed carbon particles, *J. Opt. Soc. Am.* **50**, 121–129.

Thring, M.W. and **Lowes, T.M.** 1972, Luminous radiation from industrial flames, *Combust. Sci. and Tech.* **5**, 251–256.

Tien, C.L. and **Lee, S.C.** 1982, Flame Radiation, *Prog. Energy Combust. Sci.* **8**, 41–59.

Tsakiris, G.D. and **Eidmann, K.** 1987, An approximate method for calculating Planck and Rosseland mean opacities in hot, dense plasmas, *J. Quant. Spectrosc. Radiat. Transfer* **38**, 353–368.

van de Hulst, H.C. 1957, *Light Scattering by Small Particles*. Wiley, New York.

Viskanta, R. and **Anderson, E.E.** 1975, Heat transfer in semitransparent solids, in *Adv. Heat Transfer* (Edited by J.P. Hartnett and T.F. Irvine), Vol. 11, pp. 318–441.

Viskanta, R. and **Mengüç, M.P.** 1987, Radiation heat transfer in combustion systems, *Prog. Energy Combust. Sci.* **13**, 97–160.

Voshchinnikov, N.V. and **Il'in V.B.** 1982, Planck mean radiation-pressure cross sections for nonspherical grains. I, *Astrophysics* **18**, 353–360.

Wiscombe, W.J. 1979, Mie scattering calculations; Advances in technique and fast, vector-speed computer codes, Tech. Note NCAR/TN-140+STR. National Center for Atmospheric Research, Boulder, Colorado.

Wiscombe, W.J. 1980, Improved Mie scattering algorithms, *Appl. Optics* **19**, 1505–1509.

APPENDIX A

The expansion coefficients from LM-theory are given (see, for example, van De Hulst 1957, p.123; Kerker 1969, p.45) as

$$a_\ell = a_\ell(\alpha,m) = \frac{\psi_\ell(\alpha)\,[\psi_\ell'(\beta)/\psi_\ell(\beta)] - m\,\psi_\ell'(\alpha)}{\xi_\ell(\alpha)\,[\psi_\ell'(\beta)/\psi_\ell(\beta)] - m\,\xi_\ell'(\alpha)}\,, \tag{A1}$$

$$b_\ell = b_\ell(\alpha,m) = \frac{m\,\psi_\ell(\alpha)\,[\psi_\ell'(\beta)/\psi_\ell(\beta)] - \psi_\ell'(\alpha)}{m\,\xi_\ell(\alpha)\,[\psi_\ell'(\beta)/\psi_\ell(\beta)] - \xi_\ell'(\alpha)}\,, \tag{A2}$$

where $\beta = m\alpha$, and ψ, ξ being the Riccati-Bessel functions, and the prime denoting the differentiation with respect to the argument. These coefficients are obtained, after some manipulations and simplifications, in the following form (Selamet 1989)

$$a_\ell = \left[1 + i(-1)^\ell \frac{T_{13} - mT_{34}}{T_1 - mT_2}\right]^{-1}, \quad b_\ell = \left[1 + i(-1)^\ell \frac{mT_{13} - T_{34}}{mT_1 - T_2}\right]^{-1} \tag{A.3a,b}$$

where

$$T_1 = -\frac{\ell}{\beta} + \frac{J_{\ell-\frac{1}{2}}(\beta)}{J_{\ell+\frac{1}{2}}(\beta)}\,, \quad T_2 = -\frac{\ell}{\alpha} + \frac{J_{\ell-\frac{1}{2}}(\alpha)}{J_{\ell+\frac{1}{2}}(\alpha)}\,, \tag{A.4a,b}$$

$$T_3 = \frac{J_{-\ell-\frac{1}{2}}(\alpha)}{J_{\ell+\frac{1}{2}}(\alpha)}\,, \quad T_4 = \frac{J_{-\ell-\frac{3}{2}}(\alpha)}{J_{\ell+\frac{1}{2}}(\alpha)}\,, \tag{A.4c,d}$$

and

$$T_{13} = T_1T_3\,, \quad T_{34} = \left(\frac{\ell+1}{\alpha}T_3 + T_4\right). \tag{A.4e}$$

Here, T_1 and T_{13} are complex terms, while the remaining ones are real. These functions can readily be evaluated from the recursion formula

$$J_{\ell+\frac{1}{2}}(z) = \frac{2\ell-1}{z} J_{\ell-\frac{1}{2}}(z) - J_{\ell-\frac{3}{2}}(z)\,, \tag{A.5a}$$

where the first few functions of lowest order are known explicitly as

$$J_{+\frac{1}{2}}(z) = \left(\frac{2}{\pi z}\right)^{1/2} \sin z\,, \quad J_{-\frac{1}{2}}(z) = \left(\frac{2}{\pi z}\right)^{1/2} \cos z\,. \tag{A.5b,c}$$

1989 National Heat Transfer Conference
HTD-Vol. 106, Heat Transfer Phenomena in Radiation, Combustion and Fires

TWO-DIMENSIONAL SCATTERING FACILITY WITH MULTIPLE POSITION DATA ACQUISITION CAPACITY

D. C. Look, Jr.
Thermal Radiative Transfer Group
Mechanical, Aerospace and Mechanics Department
University of Missouri-Rolla
Rolla, Missouri

ABSTRACT

A newly developed, computer controlled experimental facility to measure two-dimensional multiple scattering of light at three different positions with respect to the scattering volume is discussed. Data acquired at all three positions are presented. In particular, the results of several measurements of the light scattered radially are favorably but qualitatively compared to the existing theory.

NOMENCLATURE

- c = effective scattering coefficient = $6\ C_{sca}/(\pi d^3)$
- C_{sca} = scattering cross section
- d = particle diameter
- g = asymmetry factor
- I_i = magnitude of the normally incident intensity at the surface of the scattering medium
- I_{exp} = measured intensity leaving the medium
- L = liquid carrier (water) depth
- n_m = refractive index of liquid carrier (water)
- N = number density of scattering center
- P_i = input power to the medium
- r = radial distance from the center of the laser beam
- r_o = effective laser beam radius
- V = voltage
- x = particle size parameters = $n_m\ \pi d/\lambda_o$
- z = distance measured from the top surface downward
- η = particle volume concentration = $N\pi d^3/6$
- κ = absorption coefficient of scattering medium (latex particles)
- λ_o = wavelength of light in air
- τ_o = optical depth to the bottom (or thickness) of the scattering medium in the direction of the incident light.
- τ_r = radial optical thickness
- τ_R = value of τ_r when r equals tank radius
- τ_z = optical thickness from top surface

INTRODUCTION

The scattering of thermal radiation in the backscatter, radial, or transmitted directions may be viewed as individually important in many situations. The study of each direction of scatter may be used in various applications ranging from underwater visibility and meteorology to medicine and areas of national defense.

The study of backscattered light is receiving attention recently. Pal and Carswell (1985) are attempting to identify scattering particle size and multiple scattering effects from the polarization anisotropy of lidar scattering from clouds. General formulations describing multiple backscattering of polarized light incident on a slab of spherical particles has been given by Cheung and Ishimaru (1982). Their results are applicable to incident linearly or circularly polarized light. Fainman et al. (1987) have used backscattered light from rough surfaced conductors to separate first and second order effects. Nelson and Look (1988) have used two-dimensional backscatter to investigate substrate effects. Their effort was to demonstrate a relationship between physical variables defining regions in which substrate effects are evident.

Transmitted radiation has been used as a diagnostic tool for years. For example, Cheung and Ishimaru (1982) also produced a formulation describing the effects of distributed spherical particles on the transmission of incident linearly and circularly polarized light. The dependence of reconstructed holograms on the relative orientation of polarization has been presented by Kostuk and Sincerbox (1988).

A major force driving the study of radial scatter is the characterization of missile or rocket plumes (Nelson, 1984, a and b). Identification of the plume signature is dependent on the radial scattering from particles in the plume (e.g., Al_2O_3 and/or C). Radial scattering has been used as a tool to study combustion processes (Johnston et al., 1986). Comparison of theoretical calculations and experimental data are presented by Nelson and Satish (1988). The theory was limited to a single scattering approximation to obtain fairly good agreement.

The purpose of this paper is to present the features of a newly developed experimental facility, present some characteristic data acquired with this facility, and discuss in more detail some interesting characteristics of radial scattering.

2-D SCATTERING FACILITY

A new experimental facility has been constructed which allows for accurate and precise two dimensional scattering data to be acquired at three positions outside of a cylindrical scattering volume: top, side and bottom. That is, this facility has the capability for measurements of backscattered, radially scattered and transmitted (forward scattered) light. In addition to the multiple position data acquisition capability, the following five areas have been addressed in the new facility: 1) the low light level of the detected radiation, 2) the need for a small acceptance angle and aperture, 3) the need for accurate positioning of the detector probe, 4) the need to control incident polarization, and 5) the need for an automated data-acquisition system with a real time display of the data.

Facility Components

The present facility will allow for the acquisition of scattering data at three different positions with respect to the scattering volume. Installed as part of this facility are the following components:

Vibration Insulated Table (VIT): A high quality VIT, which is used as a base for the whole experiment, very nearly eliminates vibration problems introduced in the second floor Thermal Radiation Laboratory. The table top is of optical quality with a damping core. The vibrations transmitted from the floor of the laboratory to the scattering center carrier (typically water) are of such a small magnitude as to not effect the data acquired even with the smallest aperture acceptance angle detection probes. That is, the top surface water ripple is so small that stable readings can be made. Measurements of the vibration characteristics of the VIT are presented in Table 1. The numbers in the sketch in Table 1 indicate the positions of the measurements 1) on the floor, under the VIT, 2) on the VIT, rigidly mounted, and 3) on the VIT but with the 4-post isolation system installed. The 4-post isolation system was tuned to the VIT weight and produced a dramatic reduction in vibration amplitude.

Argon Ion Laser: A three watt argon ion continuous-wave laser tuned to 0.5145 μm is the light source. This laser provides enough power for accurate measurements of scattered light at all positions with respect to the scattering volume. Its output is a 2 mm diameter, single line, vertically polarized, TEM_{00} beam. Figure 1 is a copy of the actual laser output variation with time (as read by a photodiode). Note that approximately 1 hour and 40 minutes is required for long term stability and that the short term stability is within approximately 3 percent at all times.

Linear Positioning Device (LPD): The LPD has been mounted on the VIT and geared to facilitate the measurements of the back-scattered, radial-scattered, and transmitted light. This device allows for extremely accurate and precise positioning of the detector probes which is especially critical in regions near the laser beam where there are large variations in the scattered light with position. The device presently mounted on the VIT is a computer controlled, stepping motor actuated, 8 inch maximum travel platform. The flatness of travel (vertical variation) measured by the manufacturer is at most 3.05 μm while the straightness of travel perpendicular to the displacement axis, horizontally, is at most 6.62 μm.

TABLE 1: Vibration Characteristics of the VIT.

Position Number	AMPLITUDE (mils) Floor(Table)(Pad)Iso Sup.			FREQUENCY (CPM) Floor(Table)(Pad)		
	Floor	(VIT)	4-post	Floor	(VIT)	4-post
1	0.020	0.075	—	1300	1500	______
2	0.020	0.075	0.003	1300	1500	______
3	0.020	0.070	0.003	1300	1500	______
4	0.021	0.060	0.003	1300	1500	______
5	0.021	0.055	0.002	1300	1500	______
6	0.021	0.048	0.002	1300	1500	______
7	0.021	0.040	0.002	1300	1500	______
8	0.021	0.040	0.002	1300	1500	______
9	0.021	0.038	0.002	1300	1500	______
10	0.020	0.040	—	1300	1600	______
11	0.020	0.040	0.003	1300	1600	______
12	0.019	0.035	0.003	1300	1600	______
13	0.019	0.032	0.003	1400	1600	______
14	0.016	0.032	0.004	1400	1600	______
15	0.016	0.030	—	1400	1600	______
16	0.015	0.030	0.003	1400	1500	______
17	0.016	0.038	0.003	1400	1500	______
18	0.016	0.042	0.002	1500	1500	______
19	0.014	0.054	0.002	1400	1500	______
20	0.014	0.065	0.002	1400	1500	______
21	0.013	0.075	0.002	1400	1500	______
22	0.013	0.080	0.003	1300	1500	______
23	0.013	0.085	0.003	1300	1500	______
24	0.013	0.095	—	1300	1500	______
25	0.014	0.090	0.003	1300	1500	______
26	0.015	0.085	0.003	1300	1500	______
27	0.017	0.080	0.004	1300	1500	______
28	0.019	0.078	0.003	1300	1500	______
29	—	0.060	0.002	—	1500	______
30	—	0.042	0.002	—	1500	______
31	—	0.035	0.002	—	1500	______

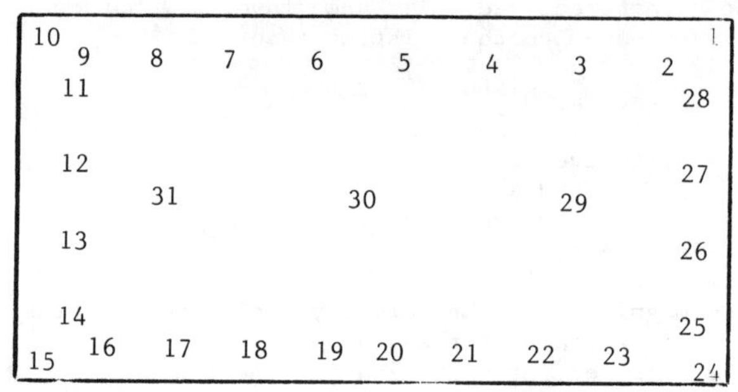

VIT

Cylindrical Glass Tank: A high quality 24-1/2 cm diameter, 24-1/2 cm deep glass tank is used to contain the scattering medium. The glass is of the highest optical quality and essentially imperfection free to reduce inadvertent scattering by the container.

Lock-In Amplifier and Chopper: A state-of-the-art A-C detection-display system is used to measure the signals from the three detector probes via three fiber-optic bundles. The device presently in use is a single phase lock-in with voltage ranges from 0.01 μV to 0.5V and coupled to a computer through an IEEE-488 interface. The chopper is a 50% duty cycle, continuously variable frequency device.

Three Position Linear Actuator (TPLA): The TPLA is a linear translation actuated device which serves as an optical equivalent to a single pole-triple throw electrical switch (Look, 1988). This device was built so that three fiberoptic bundles may be moved such that only one of them is positioned in front of the optical sensor at a time with essentially no "cross-talk".

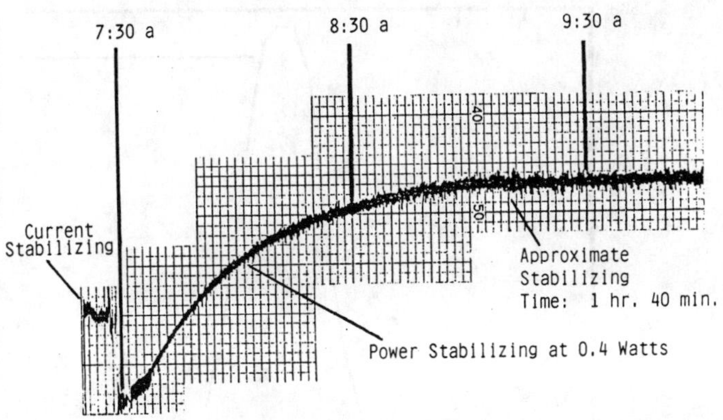

Fig. 1. Long Term and Short Term Laser Stability
Laser Current: 18 amp
Power: 0.4 watts
Recorder Full Scale: 5 volts
Time Seal: 6 cm/hr

Thus, the light from the other two bundles cannot be seen by the optical sensor. The positioning of the fiberoptic bundle terminations is prompted by the computer management system.

Photomultiplier Tube (PMT): A very stable, cooled PMT serves as the optical sensor. This sensor is a so-called blue-sensitive tube, selected because it optimally matched the output of the blue argon ion laser line (0.5145 μm).

Computer Management System (CMS): An IBM-AT based CMS has been developed. The CMS is used to

a) acquire experimental data after receiving appropriate input parameters which dictate the number of positions to take data, scattering particle characteristics, and so on;
b) receive three sets of data pairs: $(r,V)_{top}$, $(z,V)_{side}$, and $(r,V)_{bottom}$ as per instructions cited in (a);
c) manipulate the experimental data to the desired form; and
d) produce a hardcopy printout of all data, raw and manipulated, a variety of graphical displays, a floppy disk record of the raw and manipulated data, and appropriate parameters.

Discussion of Facility

Figure 2 is a schematic of the overall setup. The argon ion laser light enters the large beaker from the top, is scattered by the particles immersed within the fluid and is detected at three positions outside of the tank. All of the indicated components are mounted on the 4 by 8 foot VIT system. In fact, the argon ion laser is mounted on a large, elevated A frame. Also mounted on this frame are several carriages upon which are placed various components - the chopper, a reflecting prism, and other light conditioning equipment. At the table top level, additional equipment has been mounted. Most important of these devices is the LPD. This device is manipulated by a controller. Mounted on this device is a series of sprockets and chains which moves the top, side and bottom detector optics to preselected positions. Restrictive probes are on the ends of three multi-fiber optics bundles in order to reduce and control the acceptance angle. These probes have been characterized to have acceptance angles of less than one degree. The fiber optic bundles transport the detected light to the TPLA. The switch selects light from only one fiber bundle at a time to enter the photomultiplier. The photomultiplier is housed in a thermoelectrically cooled housing which greatly lowers the noise level. The photomultiplier is attached directly to a high quality lock-in amplifier-voltmeter. Figure 3 is a photograph of the entire system.

Scattering Particle Characteristics: The uniformly sized particles used as scattering centers are all made from polystyrene latex with a specific gravity of approximately 1.05. The particle volume concentrations (volume of particles per unit volume of scattering mixture) are varied by carefully mixing a small volume of particles in the large volume of water. Depending on the volume needed, the requisite volume of particles is determined either by measured volumes of suspended particles or by measured weights of the particle suspension. This appears to be accurate for dispensing "uniform" volumes. If a non-uniform distribution of particle sizes is required (e.g., latex paint) the same procedure is used to dispense the particles.

Uniform distributions of the scattering centers in the carrier (e.g., latex particles in water) are obtained by stirring the medium (scattering centers and carrier) vigorously. It then takes a couple of minutes for the medium to become stationary. The data acquisition process may then proceed. The stability of the medium has been tested by letting the medium set for up to 24 hours and then retaking the data. Negligible changes in the data were observed for small particles (less than 2 μm diameter particles). Thus no noticeable or detectable irreversible settling or coagulating occurs in periods up to 24 hours.

The particles were obtained from Duke Company and are sized at specific diameters with very small variances (see Table 2). Although these data were provided by Duke, transmission measurements through known concentrations of particles have been acquired as a check and the results obtained agree with theory. Further, the relative angular variation of scattered intensity has been measured using a Brice Phoenix spectrophotometer and showed good agreement with Mie theory.

A relative error analysis has been made using typical values of voltages, lengthes, concentrations, and so on. The resulting estimated average relative error is approximately ± 1/4 percent for the optical thicknesses and ± 1.5 percent for the non-dimensional intensities.

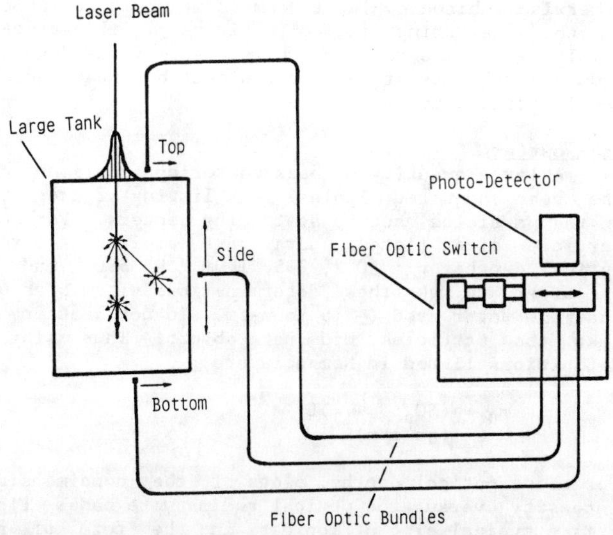

Fig. 2. Schematic of Experimental Set-up

Fig. 3. Photograph of Experimental Set-up
1. Laser, 2. Chopper, 3. Prism, 4. Positioner sprockets, 5. Controller, 6. Lockin, 7. Printer, 8. Tank, 9. Switch, 10. Power meter, 11. IBM-AT, 12. VIT on Four-Post isolation

TABLE 2: Properties of Particles used as Scattering Centers for $\lambda = 0.5145\mu$

d(μm)	Std. Dev.(μm)	x	g	c (1/cm)
0.123	0.0049	1.003	0.1774	4327.1
0.246	0.0043	2.006	0.6636	15292.6
0.494	0.0049	4.028	0.8566	37472.5
0.72	0.0053	5.87	0.9000	49200.0
1.24	0.013	10.11	0.9250	44631.9

g and c were calculated from Mie theory

EXPERIMENTAL MEASUREMENTS

Transmission

The transmittance measurement capability of this facility was used to characterize the reflected but otherwise unhindered laser beam as it would be incident on the scattering medium. Figure 4 illustrates a sample set of the raw data while Figure 5 was developed from several sets of data to indicate how near Gaussian the incident beam is.

Backscatter

A large quantity of backscattering experiment data has been acquired. Table 2 is a listing of properties of the particles used as scattering centers. For each particle size listed, data were acquired at six (6) optical depths, τ_o: 0.2, 0.5, 1.0, 2.0, 5.0, and 10. In order to make these determinations, it was assumed that the water used ($\simeq$ 18 Mohm-cm) did not scatter and that the particles did not absorb. Thus using the definitions listed in Nomenclature,

$$\tau_o = (NC_{sca} + \kappa)L = (\eta c + \kappa)L. \qquad (1)$$

For these optical depths, plots of the nondimensional intensity versus the optical radius were made. Figure 6 is a typical plot of top data in the form of non-dimensional intensity versus the optical radius. The point to note in the plot is the consistency of the data demonstrated by the fact that the curves fall on

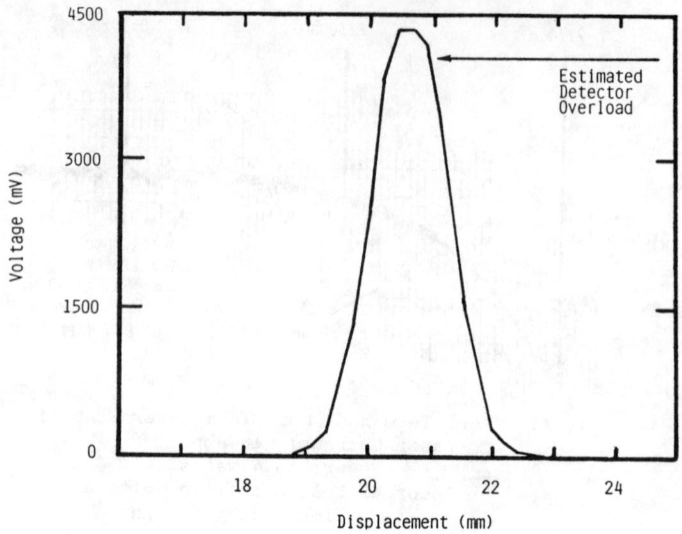

Fig. 4. Laser Beam Profile (Raw Data)

top of each other throughout most of the acquisition regardless of the particle concentration (τ_o). The non-dimensional intensity and the optical radius are defined as follows:

$$\left(\frac{r}{r_o}\right)^2 \frac{I_{exp}}{I_i} \propto r^2\ (V/P_i) \qquad (2)$$

and $\tau_r = (\eta c + \kappa)r. \qquad (3)$

Figure 7 is the same type of information as is presented in Figure 6 but for the various sized particles for an optical depth of 10. In both Figures 6 and 7, the optical thickness is actual, not effective.

Radial Scattering

Figure 8 presents the unusual shape of the detected radial scattering outside of the cylindrical glass tank as a function of optical depth. The axes are defined as

$$\frac{I_{exp}}{I_i} \propto \frac{V}{P_i} \qquad (4)$$

and $\tau_z = (\eta c + \kappa)z. \qquad (5)$

The interesting point to note from these figures is that the received power first increases as the detector probe proceeds down from the upper surface, peaks, then decreases. The optical depth for this data is $\tau_o = 10$, while maximum radial optical value (to the edge of the tank) is $\tau_R \simeq 3.75$. Figure 9 presents further information: the peak occurs at different optical depths depending upon the diameter of the scattering particle. Table 3 is a listing of estimated optical depths of the peak dependent upon the particle size.

Finally, Figure 10 presents the variation of the radial scattering with the optical depth (i.e., the scattering particle concentration). Note that as the optical thickness increases the shape of the radial scatter intensity varies differently with the optical depth. For a relatively thin scattering medium, the slope of the non-dimensional intensity increases with depth. As the thickness increases the "humped shape" takes over, peaking at larger τ_z as τ_o increases.

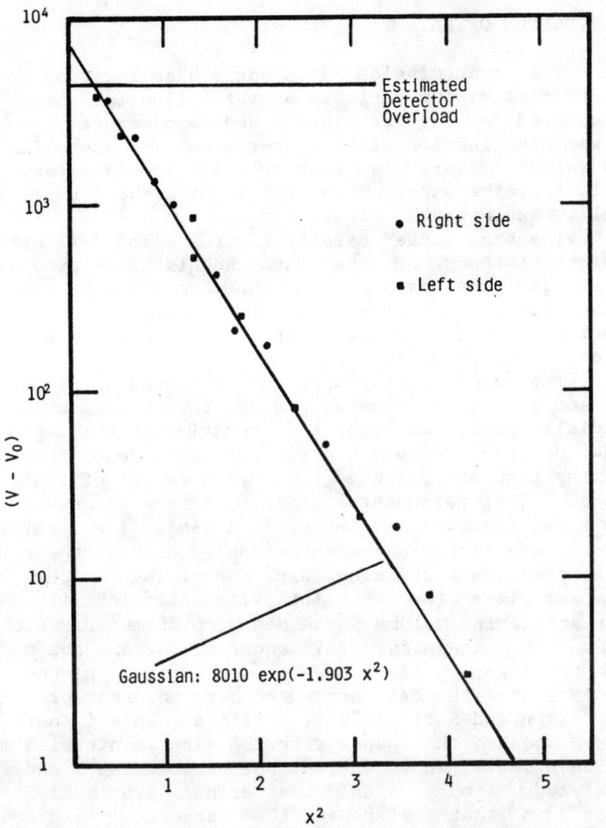

Fig. 5. Laser Beam Profile Compared to Gaussian Curve (V_0 is dark level voltage)

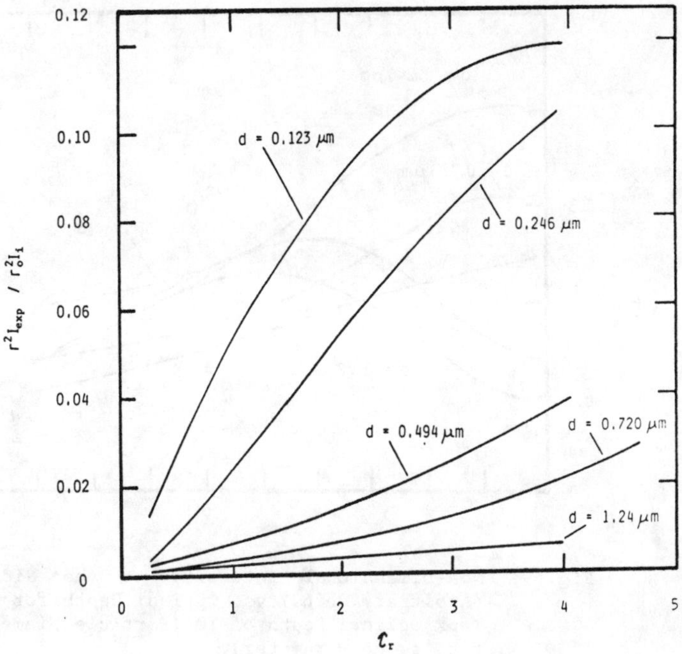

Fig. 7. Non-Dimensional Intensity at the Top (Arbitrary Units) vs. Radial Optical Thickness for a Tank Optical Depth of 10 (Particle Diameter is the Parameter)

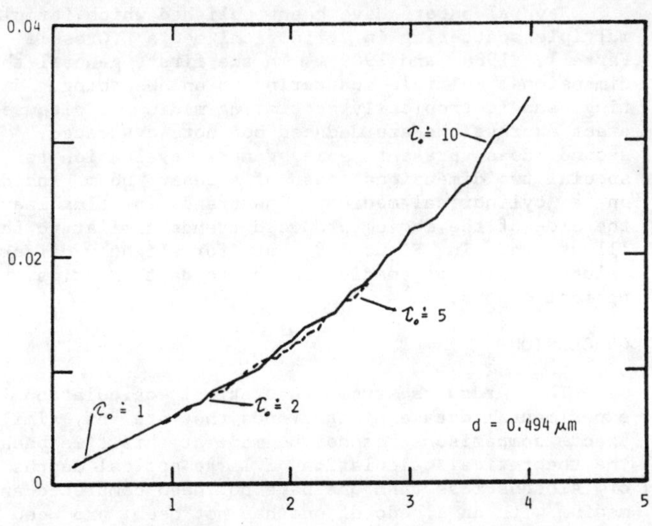

Fig. 6. Non-Dimensional Intensity at the Top (Arbitrary Units) vs. Radial Optical Thickness for 0.494 μm Diameter Particles (Tank Optical Depth is the Parameter)-raw data.

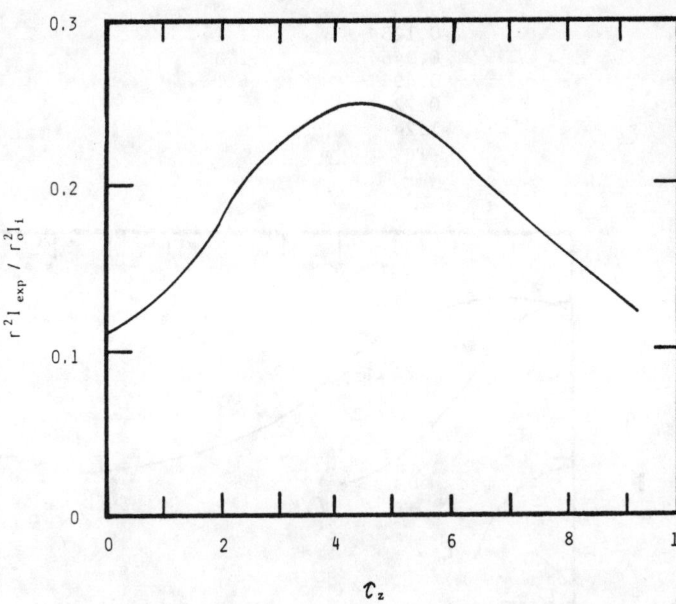

Fig. 8. Non-Dimensional Intensity at the Side (Arbitrary Units) vs. Optical Depth For 0.494 μm Diameter Particles and a Tank Optical Depth of 10.

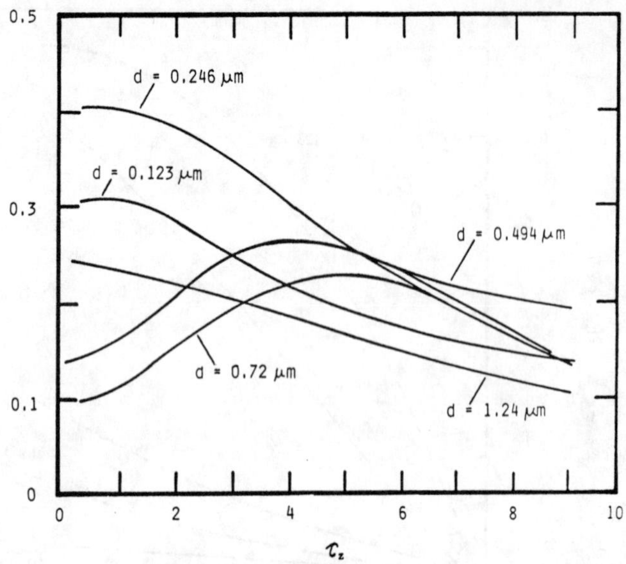

Fig. 9. Non-Dimensional Intensity at the Side (Arbitrary Units) vs. Optical Depth for a Tank Optical Depth of 10 (Particle Diameter is the Parameter).

TABLE 3: Estimated Optical Depth Where The Lateral Scattering is Maximum per Particle Diameter in the case of $\tau_o = 10$.

d(μm)	τ_z($\simeq$ peak)
0.123	1.1
0.246	1.8
0.494	4.1
0.72	4.1
1.24	0.4

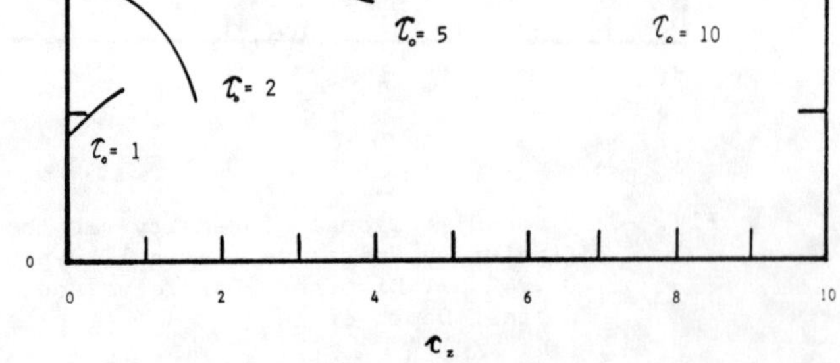

Fig. 10. Non-Dimensional Intensity at the Side (Arbitrary Units) vs. Optical Depth For 0.246 μm Diameter Particles (Tank Optical Depth is the Parameter).

DISCUSSION OF DATA

The transmission data for a beam incident on the scattering medium enclosed by the cylindrical tank are presented as Figures 4 and 5 and demonstrates that the power distribution of the laser beam is approximately Gaussian. There is scatter about the Gaussian line; this is to be expected considering the short term stability exhibited in Figure 1.

The backscatter data of Figures 6 and 7 illustrate the consistency of the data acquisition procedure. That is, the data exhibit the same shapes and variations with scattering particle diameter and radial optical thickness as some earlier publications indicate (Look et al., 1981).

The most exciting data are presented as Figures 8, 9 and 10. These figures present the variations in the radially scattered light as a function of the depth of the detector or probe positioned outside of the scattering tank but directed toward the center of the laser beam. The parameters of interest are the scattering particle diameter and the optical depth (i.e., physical depth and particle concentration). Each of these figures presents a striking and surprising shape which characterizes the light intensity variation with depth for scattering volume whose depth to diameter ratio is less than but approximately equal to one and for moderate to large optical depths. At first glance, it appears that the data presented here and that presented in Nelson and Satish, 1988, conflict. This is not the case because that paper reported experiments with much thinner media, included isotropic scattering, and was analyzed with a single scattering approximation. In fact, in that reference the largest disagreement between that experimental data and the single scattering approximation occurs for the largest optical depth. Thus the single scattering approximation cannot be successfully used for comparison with the data presented here.

Several papers have been published which formulate multiple scattering in cylindrical media (Crosbie and Farrell, 1984 and 1989). In the first, general three dimensional multiple scattering in an absorbing, emitting, and isotropically scattering medium is discussed; exact expressions are deduced but not evaluated. The second does present some general evaluation for the special two dimensional case of a laser beam incident on a cylindrical medium. The trends for flux leaving the side of the medium produced trends similar to those illustrated in Figure 8 but for slightly different values of albedos, medium radius to depth ratios, and optical depths.

CONCLUSION

The results from theoretical calculation and experimental data exhibit trends that are very similar. Exact comparisons cannot be made at this time because the theoretical calculation of large optical depths and of albedos less than one have not been completed and a medium with an albedo of one has not been produced in the laboratory.

ACKNOWLEDGEMENTS

The author acknowledges the partial support of the National Science Foundation through grants NSF-CBT-8501099 and NSF-CBT-8705292 for this research. Further, discussions with Drs. A. L. Crosbie and H. F. Nelson were very helpful.

REFERENCES

Carswell, A. I., and Pal, S. R., 1985, "Polarization Anisotropy in Lidar Multiple Scattering From Atmospheric Clouds," Applied Optics, Vol. 24, pp. 3464-3471 Also Applied Optics, 1980, Vol. 24, pp. 4123-4125.

Cheung, R. L-T., and Ishimaru, A., 1982, "Transmission, Backscattering, and Depolarization of Waves in Randomly Distributed Spherical Particles," Applied Optics, Vol. 21, No. 20, pp. 3792-3798.

Crosbie, A. L., and Farrell, J. B., 1984, "Exact Formulation of Multiple Scattering in a Three-Dimensional Cylindrical Geometry," Journal of Quantitative Spectroscopy and Radiative Heat Transfer, Vol. 31, No. 5, pp. 397-416.

Crosbie, A. L., and Farrell, J. B., 1989, "Multiple Scattering in a Two-Dimensional Finite Cylindrical Medium Exposed to a Laser Beam," Journal of Quantitative Spectroscopy and Radiative Heat Transfer, To be published.

Fainman, Y., Shamer, J., Peri, D., and Brunfeld, A., 1987, "Polarization Properties of Speckle Patterns Scattered From Rough Conductors," Journal of the Optical Society of America, A, Vol. 4, No. 12, pp. 2205-2212.

Johnston, S. C., Dibble, R. W., Schefer, R. W., Ashurst, W. T., and Kollmann, W., 1986, "Laser Measurements and Stochastic Simulations of Turbulent Reacting Flows," AIAA Journal, Vol. 24, pp. 918-937.

Kostuk, R. K., and Sincerbox, G. T., 1988, "Polarization Sensitivity of Noise Recorded in Silver Halide Volume Holograms," Applied Optics, Vol. 27, No. 14, pp. 2993-2998.

Look, Jr., D. C., 1988, "Optical Single-Pole Triple-Throw Switch," Optical Engineering, Vol. 27, No. 3, pp. 253-255.

Look, Jr., D. C., Nelson, H. F., and Crosbie, A. L., 1981, "Anisotropic Two-Dimensional Scattering: Comparison of Experiment and Theory," Journal of Heat Transfer, Vol. 103, No. 1, pp. 127-134.

Nelson, H. F., 1984a, "Influence Particulates on Infrared Emission from a Cylindrical Cloud of Particles," Journal of Spacecrafts and Rockets, Vol. 21, pp. 425-432.

Nelson, H. F., 1984b, "Influence of Scattering on Infrared Signatures of Rocket Plumes," Journal of Spacecrafts and Rockets, Vol. 21, pp. 508-510.

Nelson, H. F., and Look, Jr., D. C., 1988, "Substrate Effects on Two-Dimensional Laser Beam Backscattering," Journal of Thermophysics and Heat Transfer, Vol. 2, No. 3, pp. 203-208.

Nelson, H. F., and Satish, B. V., 1988, "Radial Scattering of a Laser Beam in Anisotropic Scattering Media," Journal of Thermophysics and Heat Transfer, Vol. 2, No. 2, pp. 104-109.

1989 National Heat Transfer Conference
HTD-Vol. 106, Heat Transfer Phenomena in Radiation, Combustion and Fires

AEROSOL THERMOPHORESIS WITH CONDUCTION AND RADIATION

J. Hannon, Y. Yener and J. W. Cipolla, Jr.
Department of Mechanical Engineering
Northeastern University
Boston, Massachusetts

ABSTRACT

The effect of heat conduction and thermal radiation on thermophoresis of an absorbing, emitting and isotropically scattering gray aerosol immersed in a radiatively non-participating background gas between two parallel boundaries is studied under steady-state conditions. The boundaries are considered to be opaque, gray, diffusely emitting and diffusely reflecting, and are maintained at different but uniform temperatures. A simplified model is proposed, and an iterative solution scheme is developed. The radiative transfer equation is solved approximately using the Galerkin method. This is used together with the solutions of the particle concentration and heat conduction equations iteratively to determine the temperature and particle concentration distributions in the mixture. The effects of various parameters of the problem on the concentration and temperature profiles are determined numerically and presented in graphical forms.

NOMENCLATURE

C	aerosol particle concentration
C_o	average aerosol particle concentration
D	Brownian diffusion coefficient
$G = \left(\frac{d\theta}{d\eta}\right)_{\eta=0}$	
k	thermal conductivity
K	thermophoretic coefficient
L	distance between boundaries
N_{CR}	conduction-to-radiation parameter
Q^r	dimensionless radiative heat flux
$Sc = \frac{\nu}{D}$	Schmidt number
T	temperature
T_o, T_L	boundary temperatures
$\underline{V}_T$	thermophoretic velocity
x	transverse coordinate

Greek Symbols

ϵ_o, ϵ_L	emissivities of the boundaries
$\eta = \frac{x}{L}$	dimensionless transverse coordinate
$\phi = \frac{C}{C_o}$	dimensionless particle concentration
$\gamma = K \cdot Sc$	
μ	direction cosine
ν	kinematic viscosity
κ	absorption coefficient
ρ_o, ρ_L	reflectivities of the boundaries
σ	scattering coefficient
$\bar{\sigma}$	Stefan-Boltzmann constant
τ	optical variable
τ_o	optical thickness
$\theta = \frac{T}{T_o}$	dimensionless temperature
$\theta_L = \frac{T_L}{T_o}$	
ω	single-scattering albedo

INTRODUCTION

Temperature gradients in an aerosol-gas mixture can affect the spatial distribution of the aerosol particles in the background gas through the effect known as *thermophoresis*. This non-hydrodynamic effect occurs whenever the aerosol particle diameter is comparable to the mean-free-path of the background gas, and is caused by the disparity in momentum transfer to the particles from collisions with gas molecules that originate in regions of the gas that differ in temperature. As a consequence, the particles experience a net force in the direction of decreasing temperature which, when balanced by viscous or molecular drag, gives rise to a steady particle velocity of the form

$$\underset{\sim}{V}_T = -K \frac{\nu}{T} \nabla T \tag{1}$$

where ν is the kinematic viscosity of the gas. In addition, K is the so-called thermophoretic coefficient which is a dimensionless quantity and depends on the gas-to-particle ratio of thermal conductivities and on the ratio of the mean-free-path to the particle radius (*i.e.*, the Knudsen number).

This coupling between temperature gradients and the spatial distribution of the aerosol particles, particularly when the aerosol itself is partially responsible for the non-uniform temperature distribution in the gas, has not only intrinsic interest from a theoretical viewpoint, but is also important for various fields of application, including material processing techniques that involve particle deposition from high temperature gas streams, (Morse and Cipolla, 1984), the combustion of ultrafine powders, and the formation and motion of soot in flames (Dobbins and Megaridis, 1986).

The interaction of thermal radiation and thermophoresis of an absorbing, emitting, isotropically scattering gray aerosol contained in a non-participating background gas has recently been studied under the condition of radiative equilibrium (Tse *et al.*, 1987). It has been shown that thermophoresis can have a dramatic effect on the spatial distribution of aerosol and on the temperature in the medium. A study of steady and unsteady behavior of an absorbing aerosol irradiated by an external laser has been given by Cipolla and Morse (1987). In this work, however, emission and scattering by the aerosol particles have been neglected.

The objective of the current work is to study, under steady-state conditions, the interaction of heat conduction, thermal radiation and thermophoresis of an absorbing, emitting, isotropically scattering gray aerosol contained in a radiatively non-participating background gas between two parallel boundaries. The boundaries are considered to be opaque, gray, diffusely emitting and diffusely reflecting, and are maintained at different but uniform temperatures. It is assumed that convection effects may be neglected and that the aerosol is monodisperse and in thermal equilibrium with the non-participating background gas. It is also assumed that aerosol does not deposit onto the walls. As it was discussed by Tse *et al.* (1987), this later assumption is at variance with the conventional assertion that all aerosol particles impinging on a solid surface adhere to it due to the strongly attractive van der Walls forces operative at small separations (Friedlander, 1977). However, we retain this condition because, in the steady-state case, the one-dimensional aerosol continuity equation has an exact quadrature to be used in conjunction with the solutions of the heat conduction and radiative transfer equations.

The absorption, emission and scattering coefficients of the aerosol are considered to be proportional to the local concentration of the particles. On the other hand, the concentration is determined, at least partially, by the temperature gradients in the gas. Therefore, the interaction between the temperature and concentration distributions is strongly non-linear. In the following sections, first a formulation of this non-linear problem is given, and then a simplified model based on the physics of the problem is introduced. An iterative calculation scheme is developed for numerical evaluations. Finally, results showing the effects of various parameters of the problem on the concentration and temperature profiles are given in graphical forms.

FORMULATION

Consider an absorbing, emitting, isotropically scattering and gray aerosol immersed in a radiatively non-participating background gas between two large parallel surfaces located in the planes x = 0 and x = L. The bounding surfaces are considered to be opaque and gray-diffuse, and are maintained at different but uniform temperatures. Let the uniform surface temperatures be T_0 at x = 0 and T_L at x = L, respectively.

Under the influence of thermophoresis, the net aerosol flux at any x in the mixture will have two components: a conventional Brownian diffusive flux and the temperature dependent thermophoretic flux. With the assumption that aerosol does not deposit onto the bounding surfaces (*i.e.*, perfectly reflecting boundaries), in a one-dimensional system the net aerosol flux at any x must vanish under steady-state conditions. Thus,

$$-D \frac{dC}{dx} - CK \frac{\nu}{T} \frac{dT}{dx} = 0 \tag{2}$$

where C(x) and T(x) are the aerosol particle concentration and temperature distributions in the mixture, respectively, and D is the Brownian diffusion coefficient.

The solution to equation (2) can be written in dimensionless form as

$$\phi(\eta) = \frac{\theta^{-\gamma}(\eta)}{\int_0^1 \theta^{-\gamma}(\eta) d\eta} \tag{3}$$

where the dimensionless quantities are defined as

$$\phi(\eta) = \frac{C(\eta)}{C_o}, \quad \theta(\eta) = \frac{T(\eta)}{T_o}, \quad \text{and } \eta = \frac{x}{L} \tag{4a,b,c}$$

Here, C_o represents the average aerosol concentration in the mixture. In addition, the quantity γ in equation (3) is defined as $\gamma = K \cdot Sc$, where $Sc = \nu/D$.

Equation (3) gives the aerosol concentration distribution in terms of temperature distribution. The temperature distribution, on the other hand, is obtained from the solution of the energy equation. In the absence of convective effects, the one-dimensional steady-state energy equation applicable to the present problem, can be written, in dimensionless form, as

$$\frac{d^2\theta}{d\eta^2} - \frac{\tau_o}{N_{CR}}\frac{dQ^r}{d\eta} = 0 \qquad (5)$$

subject to the boundary conditions

$$\theta(0) = 1 \text{ and } \theta(1) = \theta_L \qquad (6a,b)$$

where $\theta_L = T_L/T_o$. In equation (5), τ_o is the optical thickness of the mixture and is given by

$$\tau_o = \int_o^L (\kappa+\sigma)dx \qquad (7)$$

where $\kappa(x)$ and $\sigma(x)$ are the absorption and scattering coefficients, which are proportional to the aerosol concentration. Thus, if we let

$$\kappa = \kappa_o\cdot\phi \text{ and } \sigma = \sigma_o\cdot\phi \qquad (8a,b)$$

then the optical thickness reduces to

$$\tau_o = (\kappa_o+\sigma_o)L \qquad (9)$$

In addition, the quantity N_{CR} in equation (5) is the so-called conduction-to-radiation parameter and is defined by

$$N_{CR} = \frac{k(\kappa_o+\sigma_o)}{4\bar{\sigma}T_o^3} \qquad (10)$$

where k is the thermal conductivity of the background gas, which is considered to be constant, and $\bar{\sigma}$ is the Stefan-Boltzmann constant.

The dimensionless radiative heat flux $Q^r(\eta)$ in equation (5) is related to the dimensionless radiation intensity $\psi(\eta,\mu)$ in the mixture by

$$Q^r(\eta) = \frac{1}{2}\int_{-1}^{1}\psi(\eta,\mu)\mu d\mu \qquad (11)$$

where μ is the direction cosine (as measured from the positive x-axis) of the propogating radiation.

For a one-dimensional, plane-parallel, absorbing, emitting and isotropically scattering medium, the dimensionless radiation intensity satisfies the equation of radiative transfer given in the form

$$\mu\frac{\partial\psi}{\partial\tau} + \psi(\tau,\mu) = (1-\omega)\theta^4(\tau) + \frac{\omega}{2}\int_{-1}^{1}\psi(\tau,m')d\mu' \qquad (12)$$

where ω is the scattering albedo given by

$$\omega = \frac{\sigma_o}{\sigma_o+\kappa_o} \qquad (13)$$

and τ is the optical variable defined as

$$\frac{\tau(\eta)}{\tau_o} = \int_o^\eta \phi(\eta')d\eta' \qquad (14)$$

The boundary conditions for equation (12) at $\tau = 0$ and $\tau = \tau_o$ can be written as

$$\psi(0,\mu) = \epsilon_o + 2\rho_o\int_o^1\psi(0,-\mu')\mu'd\mu', \quad \mu>0 \qquad (15a)$$

$$\psi(\tau_o,-\mu) = \epsilon_L\theta_L^4 + 2\rho_L\int_o^1\psi(\tau_o,\mu')\mu'd\mu', \quad \mu>0 \qquad (15b)$$

where ϵ_o and ϵ_L are the emissivities, and ρ_o and ρ_L are the reflectivities of the boundaries at $\tau = 0$ and $\tau = \tau_o$, respectively.

FORMAL SOLUTIONS

If the derivative of the dimensionless radiative heat flux $dQ^r/d\eta$ is considered known, the solution of the energy equation (5) subject to the boundary conditions (6a,b) can be obtained by direct integration as

$$\theta(\eta) = 1 + (\theta_L-1)\eta + \frac{\tau_o}{N_{CR}}\left(\int_{\eta'=o}^{\eta}Q^r(\eta')d\eta' - \eta\int_{\eta'=o}^{1}Q^r(\eta')d\eta'\right) \qquad (16)$$

The temperature distribution $\theta(\eta)$ can be obtained from equation (16) provided that the distribution of the radiative heat flux $Q^r(\eta)$ is known in the medium. The aerosol particle distribution $\phi(\eta)$ can then readily be determined from equation (3). The radiative heat flux $Q^r(\eta)$, on the other hand, is related to the intensity of radiation $\psi(\eta,\mu)$ in the medium as indicated by equation (11). The solution of the radiative transfer equation (12), subject to the conditions (15a,b), in turn, gives the radiation intensity $\psi(\tau,\mu)$, provided that the temperature distribution $\theta(\tau)$ is known in the medium. Equation (14) is the relation that relates the physical variable η to the optical variable τ, which is dependent upon the particle concentration distribution $\phi(\eta)$.

An approach to evaluate the above results numerically would, therefore, be based upon an iterative scheme as follows: First assume a temperature distribution $\theta(\eta)$ in the medium, use it to determine the particle concentration distribution $\phi(\eta)$ from equation (3), and then obtain the optical variable τ as a function of the physical variable η from equation (14). Next, convert $\theta(\eta)$ to $\theta(\tau)$, and then solve the radiation part of the problem with this assumed temperature distribution to obtain $Q^r(\tau)$. Finally, convert $Q^r(\tau)$ to $Q^r(\eta)$, and the obtain the temperature distribtution $\theta(\eta)$ from equation (16). Compare the calculated temperature distribution with the assumed one, and repeat this procedure until the difference between the calculated and the assumed temperature profiles become less than a prescribed accuracy at all points of the region. However, when $\gamma >> 1$ and $\theta_L > 1$, evaluation of $\phi(\eta)$ from equation (3) causes numerical difficulties as $\theta^{-\gamma}(\eta)$ decays very rapidly near the cold boundary. To circumvent this difficulty in numerical evaluations, we propose a simplified method based on the physics of the problem as described in the section below.

A SIMPLIFIED MODEL

Let η^* be defined such that $\theta^{-\gamma}(\eta) \approx 0$ when $\eta^* < \eta < 1$, where $\eta^* << 1$. Moreover, the values of $\theta(\eta)$ in the region $0 < \eta < \eta^*$ are close to unity, that is, $\theta(\eta)-1 << 1$ in $0 < \eta < \eta^*$. It is, therefore, possible to introduce the following approximation:

$$\theta(\eta) \approx e^{G\cdot\eta} \quad \text{in} \quad 0 < \eta < \eta^* \qquad (17)$$

where

$$G = \left(\frac{d\theta}{d\eta}\right)_{\eta=o}$$

$$= (\theta_L - 1) + \frac{\tau_o}{N_{CR}} \left[Q^r(0) - \int_o^1 Q^r(\eta) d\eta \right] \tag{18}$$

With the approximation (17), the relation (14) for the optical variable $\tau(\eta)$ reduces to

$$\frac{\tau(\eta)}{\tau_o} \approx 1 - e^{-\gamma G \eta} \tag{19}$$

which yields

$$\eta = -\frac{1}{\gamma G} \ln \left[1 - \frac{\tau}{\tau_o}\right], \quad 0 \leq \tau < \tau_o \tag{20}$$

and

$$\theta(\tau) = \left[1 - \frac{\tau}{\tau_o}\right]^{-1/\gamma} \tag{21}$$

Moreover, the aerosol concentration from equation (3) reduces to

$$\phi(\eta) = \gamma \, G \, \theta^{-\gamma}(\eta) \tag{22}$$

It is obvious from equation (19) that $\tau(\eta^*) \approx \tau_o$. This indicates that the region $\eta^* < \eta < 1$ is virtually free of aerosol particles and, therefore, transparent to radiation between the aerosol laden region $0 < \eta < \eta^*$ and the boundary at $\eta = 1$. The temperature distribution in this particle-free region is determined by conduction only. On the other hand, all the aerosol particles in the mixture will have accumulated in the region $0 < \eta < \eta^*$ next to the cold boundary. In this cold region, the temperature and particle distributions are governed by the highly non-linear interaction between heat conduction, thermal radiation and thermophoretic motion.

With the approximations introduced in this section, the solution of the interaction problem is now reduced to that of iterative calculations for G. Before we start the calculations, we first solve the radiation part of the problem with the temperature distribution $\theta(\tau)$ as described by equation (21) for given values of γ, τ_o, θ_L, ω, and the radiative properties of the surfaces, ϵ_o, ϵ_L, ρ_o, and ρ_L, by using the Galerkin method as developed for radiative transfer problems by Ozisik and Yener (1982), and obtain $Q^r(\tau)$. To start the calculations, we first assume a value for G and use it to convert $Q^r(\tau)$ to $Q^r(\eta)$ by making use of equation (20). Next, we calculate G from equation (18) and compare it with the assumed value. We repeat this procedure until G converges to a value within a prescribed accuracy. Once we obtain a value for G, we re-evaluate $Q^r(\eta)$ and use it in equation (16) to obtain $\theta(\eta)$. Finally, we determine the aerosol concentration distribution from equation (22).

RESULTS AND DISCUSSIONS

The principal parameters that affect the temperature and aerosol concentration distributions include $\gamma = K \cdot Sc$; the conduction-to-radiation parameter, N_{CR}; the single-scattering-albedo, ω; the optical thickness, τ_o; the ratio of boundary temperatures, θ_L; and the emissivities and reflectivities of the boundary surfaces, ϵ_o, ϵ_L, ρ_o, and ρ_L.

Figures 1 and 2 show the effect of γ on the concentration and temperature profiles. Figure 1 indicates that for a given value of γ, the concentration is higher near the cold boundary and decreases toward the hot boundary. Furthermore, the concentration levels are higher near the cold boundary for larger values of γ. This is reasonable because γ measures the relative importance of thermophoresis compared to Brownian diffusion in the mixture. For large γ, thermophoresis becomes more important causing a drift of particles toward the cold boundary in spite of the very large concentration gradients that exist there. In addition, figure 2 clearly shows the presence of a particle-free and, therefore, conduction-dominated region away from the cold boundary, as well as a very small region near the cold boundary where radiation alters the temperature profiles. It is obvious from figure 2 that the thickness of the particle-laden zone near the cold boundary decreases as γ increases. Moreover, as the thickness of this very thin zone gets smaller, temperatures decrease near the cold boundary. This is because more particles attain temperatures close to the cold boundary temperature. This is also reasonable, because in the limit as $\gamma \to \infty$, all of the particles would have accumulated at the cold boundary and would have the temperature of the cold boundary, and the temperature profile in the whole region would be governed by conduction only.

Figures 3 and 4 show the effect of conduction-to-radiation parameter N_{CR} on the concentration and temperature profiles. The conduction-to-radiation parameter measures the relative importance of conduction to radiation in the mixture. For a large value of N_{CR}, conduction dominates the problem and vice versa. Since the concentration and temperature profiles interact because of the radiation absorbed, emitted and scattered by the aerosol particles, when radiation is more dominant, that is, when N_{CR} is small, more accumulation of the particles near the cold boundary takes place as these two figures indicate.

The effect of the optical thickness τ_o of the mixture on the concentration and temperature profiles are shown in figures 5 and 6. The medium becomes radiatively more thicker as the optical thickness τ_o increases. Hence, the effect of increasing τ_o is qualitatively the same as the effect of decreasing N_{CR}. Figure 7 demonstrates the effect of the single-scattering-albedo ω on the particle distribution. It is obvious that when absorption is significant (*i.e.*, when ω is small) more particles accumulate near the cold boundary as radiation becomes more effective. Figure 8 illustrates the effect of various combinations of the reflectivities of the boundaries on the particle concentration profiles.

It is obvious that whenever radiation is important, as in the case of high-temperature ultra-fine aerosols, thermophoretic motions can significantly alter the temperature and concentration profiles of the particles. This problem is also being investigated currently under steady-state conditions. Investigations of this problem including convection effects and the use of proper aerosol boundary conditions seem warranted.

ACKNOWLEDGEMENTS

The partial support of NSF through Grant CBT-8603634 is gratefully acknowledged.

REFERENCES

Cipolla, J.W. Jr., and Morse, T.F., 1987, "Thermophoresis in an Absorbing Aerosol," *Journal of Aerosol Science*, Vol. 18, No, 3, pp. 245-260.

Dobbins, R.A., and Megaridis, C.M., 1986, "Soot Morphological Characteristics by means of Thermophoretic Sampling," (*preprint*).

Friedlander, S.K., 1977, *Smoke, Dust, and Haze: Fundamentals of Aerosol Behavior*, Wiley-Interscience, p. 44.

Morse, T.F. and Cipolla, J.W. Jr., 1984, "Laser Modification of Thermophoretic Deposition," *Journal of Colloid and Interface Science*, Vol, 97, pp. 137-148.

Ozisik, M.N., and Yener, Y., 1982, "The Galerkin Method for Solving Radiation Transfer in Plane-Parallel Participating Media," *Journal of Heat Transfer*, Vol. 104, pp. 351-354.

Tse, T.S., Yener, Y., and Cipolla, J.W. Jr., 1987, "Aerosol Thermophoresis and Radiative Transfer," *Fundamentals and Applications of Radiation Heat Transfer*, eds. A.M. Smith and T.F. Smith, HTD-Vol. 72, pp. 59-65, ASME, 1987.

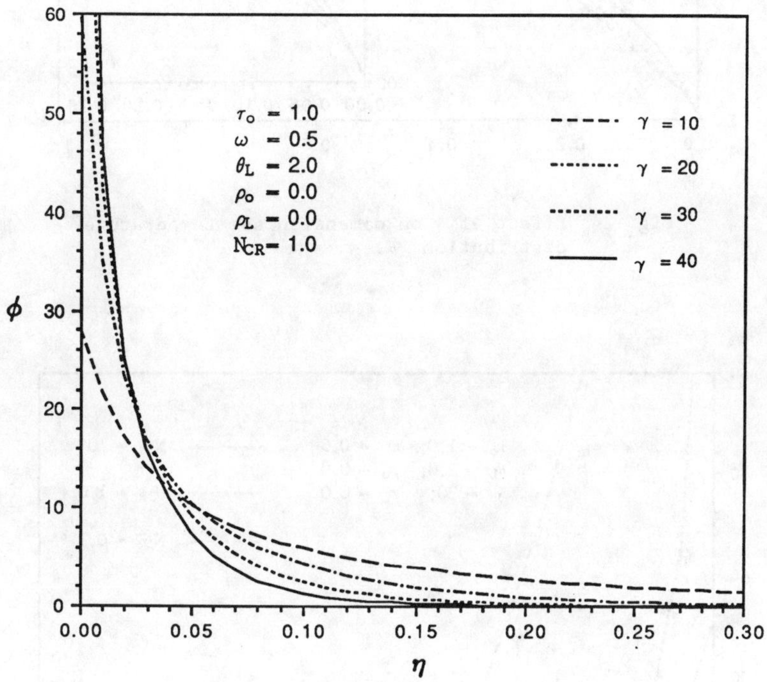

Fig. 1 Effect of γ on dimensionless concentration distribution ϕ.

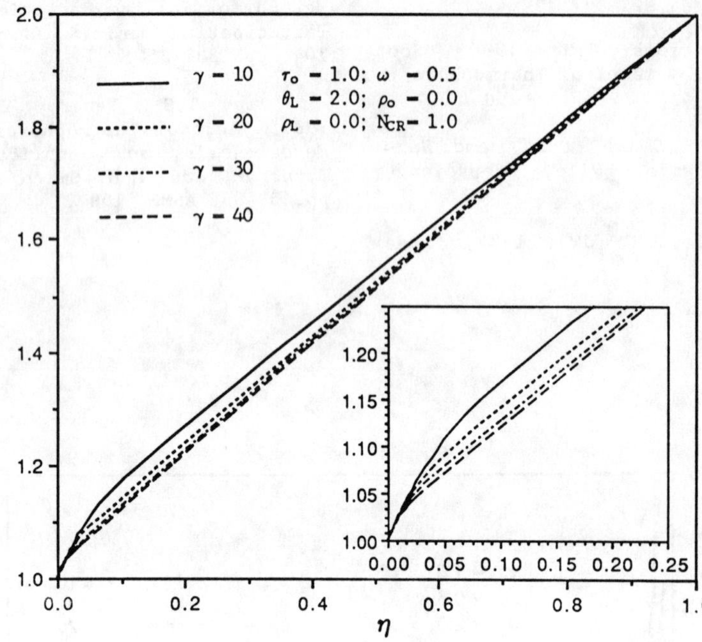

Fig. 2 Effect of γ on dimensionless temperature distribution θ.

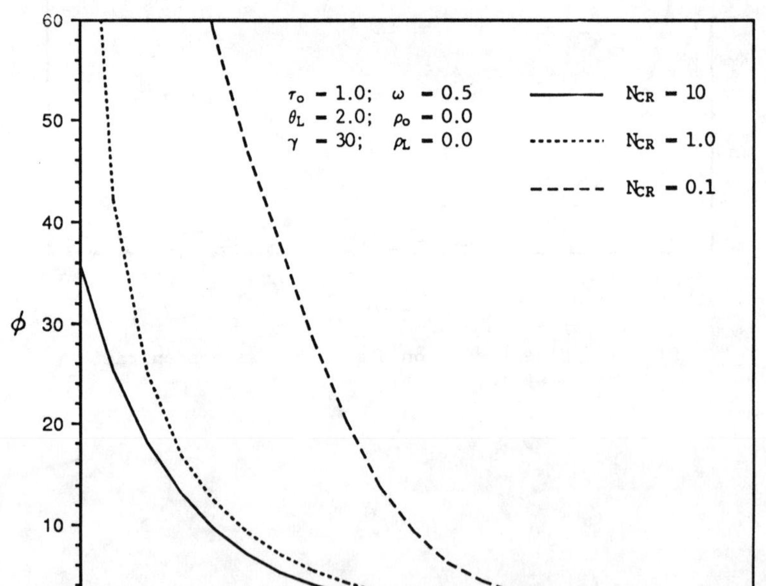

Fig. 3 Effect of N_{CR} on dimensionless concentration distribution ϕ.

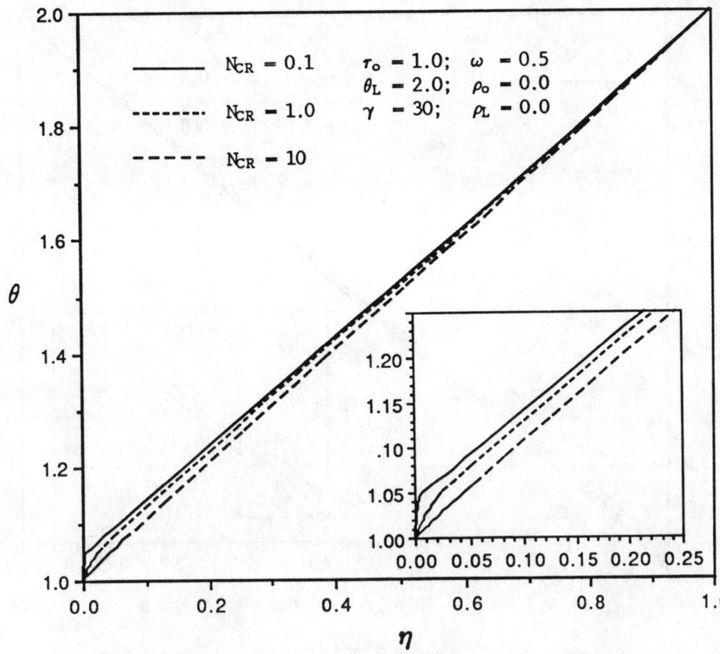

Fig. 4 Effect of N_{CR} on dimensionless temperature distribution θ.

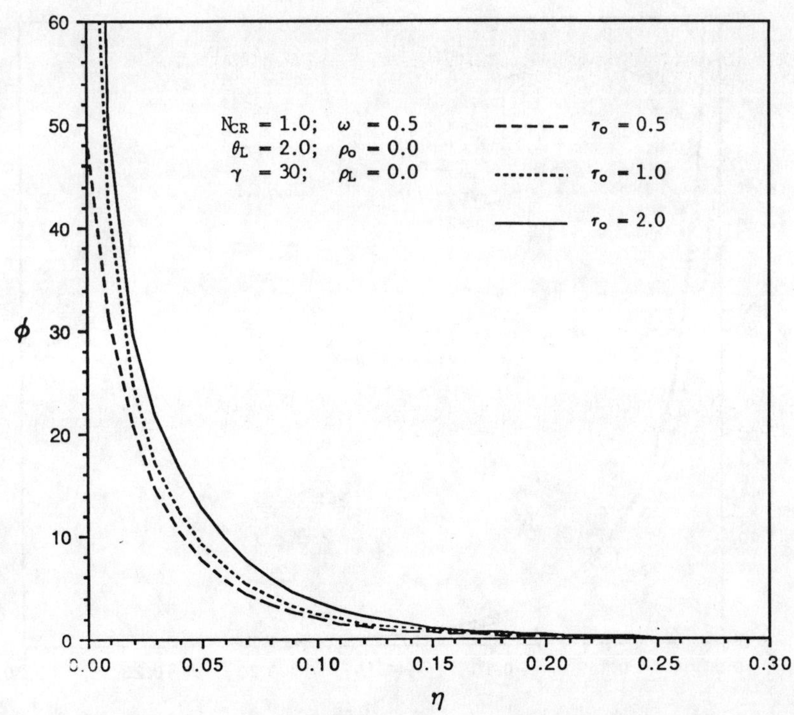

Fig. 5 Effect of τ_o on dimensionless concentration distribution ϕ.

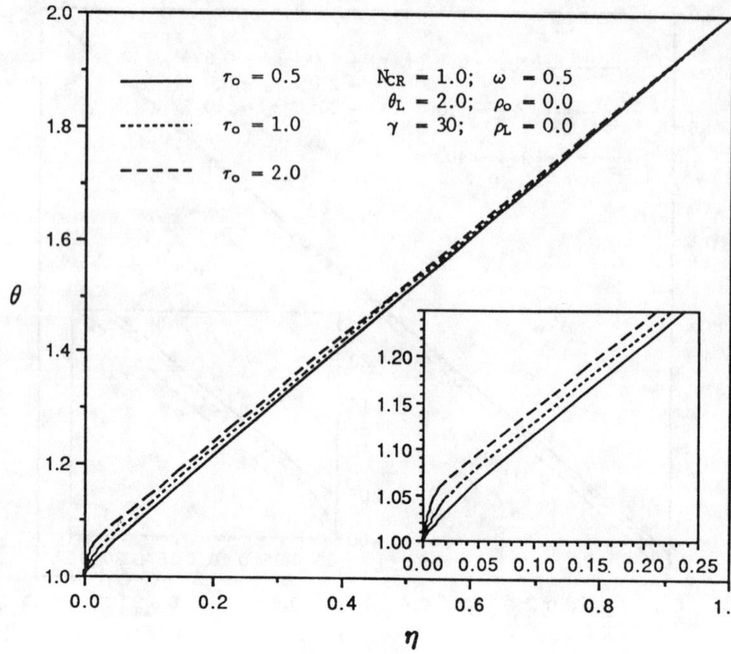

Fig. 6 Effect of τ_o on dimensionless temperature distribution θ.

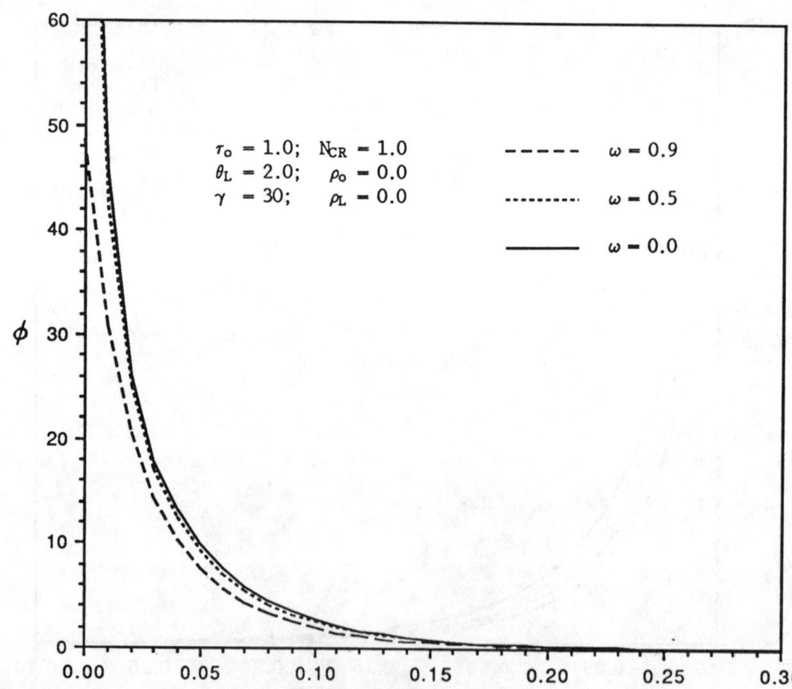

Fig. 7 Effect of ω on dimensionless concentration distribution ϕ.

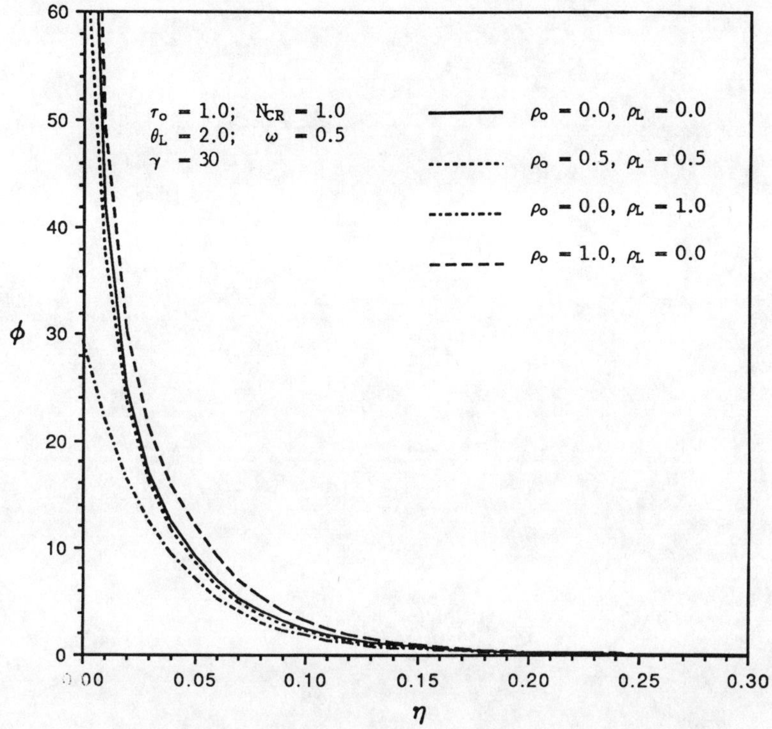

Fig. 8 Effect of reflectivities of the boundaries on dimensionless concentration disrtibution ϕ.

1989 National Heat Transfer Conference
HTD-Vol. 106, Heat Transfer Phenomena in Radiation, Combustion and Fires

THE EFFECTS OF RADIATION ON PARTICLE DEPOSITION IN MCVD: THE OPTICALLY THIN CASE

L. P. Paz and J. W. Cipolla, Jr.
Department of Mechanical Engineering
Northeastern University
Boston, Massachusetts

T. F. Morse
Laboratory for Lightwave Technology
Brown University
Providence, Rhode Island

ABSTRACT

The effect of radiation from the aerosol particles generated in the MCVD process on the temperature and concentration fields, as well as on the deposition efficiency, is considered in the limit where the aerosol can be modelled as an optically thin medium. The flow considered is the two dimensional, steady, laminar flow of a constant property gas-aerosol mixture through a tube with variable wall temperature irradiated by a laser beam. The radiation from the aerosol is modelled as a heat source term in the energy equation and diffusion effects are shown to be negligible in the concentration equation. It is found that the inclusion of radiation from the aerosol particles reduces in general the temperature gradients in the flow field, thereby lowering the thermophoretic effect responsible for the aerosol deposition on the inner wall of the tube. As a consequence the aerosol deposition efficiency is lower than what would be predicted if spontaneous emission were neglected.

INTRODUCTION

The Modified Chemical Vapor Deposition Process (MCVD) is widely used in the fabrication of optical fiber preforms (Simpkins *et al.*, 1979; Walker *et al.*, 1980; Jablonowski *et al.*, 1986; Wood *et al.*, 1987). In this process reactant gases flow through a rotating glass tube and are brought to reaction temperature by an external torch, which traverses slowly in the same direction as the gases. The gaseous products of the reaction rapidly nucleate into an ultrafine aerosol that is deposited onto the tube inner wall by the action of thermophoresis. As the torch traverses, the particulate deposit is consolidated into a thin vitreous, pore-free layer by a sintering mechanism. After several layers have deposited the tube is collapsed into a solid rod and then drawn into a thin fiber in a special drawing tower. For a more complete description of the process refer to MacChesney *et al.*, 1974. The MCVD process has been analysed quite extensively and the governing heat convection and aerosol concentration equations have been solved by several researchers under various conditions. The particulate deposition mechanism has been established as thermophoresis (Simpkins *et al.*, 1979; Walker *et al.*, 1980; Weinberg *et al.*, 1982a, b) and more recently the effect of laser radiation on the aerosol deposition has been studied by Morse *et al.*, 1985, Wang *et al.*, 1985, Ravikumar, 1985, and Cipolla *et al.*, 1987, and shown to be an effective mechanism to enhance the aerosol deposition efficiency in the MCVD process. In these studies it was also shown that the laser radiation absorbed by the aerosol is effectively conducted into the background gas so that the temperature of the gas–aerosol mixture can be obtained from a single energy equation with a heat source to model the radiation effects associated with the aerosol absorption of laser radiation. These latter studies also demonstrate that the aerosol can reach rather high temperatures in the presence of laser radiation, so that radiation from the aerosol should also be considered in the calculation of the temperature field. The study of this effect is then the main purpose of this work. In what follows we distinguish between the radiation absorbed by the aerosol from the monochromatic, collimated laser beam (the laser radiation) and the nearly diffuse radiation field associated with emission and absorption at the aerosol temperature (aerosol radiation).

The system considered is the steady, laminar, constant property flow of an isothermal gas mixture in a horizontal cylindrical tube of radius R. The tube is axially irradiated with a 10.6 μm CO_2 laser and heated externally with an oxy-hydrogen torch, which can raise the wall temperature in the heating section as high as 2000 K. As the gas mixture flows through the heating section silicon tetrachloride and oxygen react to produce silicon dioxide and chlorine gas. This chemical reaction $SiCl_4+O_2 \longrightarrow SiO_2+2Cl_2$ is very temperature dependent and is known to proceed to completion if the reaction temperature is at least 1300 $^\circ C$ (Walker *et al.*, 1980). The silicon dioxide thus formed rapidly coagulates into an aerosol of particles with diameters that range between 0.1 and 0.3 μm (Morse *et al.*, 1984). In the current high MCVD flow rates, the concentration of the aerosol particles so generated can be as high as $10^{12}\ cm^{-3}$(Kim and Pratsinis, 1988), which in turn leads to high aerosol absorption coefficient.

The system of equations that governs the temperature and concentration fields was solved numerically with the laser radiation and the aerosol radiation incorporated in the energy equation as a source term that couples the energy and aerosol concentration equations in a non linear fashion.

The effects of Brownian diffusion are also examined and shown to be negligible. The temperature and concentration fields are thus obtained by solving the energy and aerosol concentration equations without a diffusion term in the concentration equation. In the solution of the concentration equation a particle mass conservation scheme is introduced and shown to yield the same results as when the full aerosol continuity equation, including the effect of diffusion, was used. This technique substantially reduces the computation time for high Schmidt number aerosols, which is the case in the MCVD and was previously used by Ravikumar (1985).

The results obtained showed that the temperature and concentration fields are affected substantially by the inclusion of the radiation term in the energy equation, namely the radial temperature gradients within the tube are always reduced, which in turn, reduces the thermophoretic motions of the aerosol and consequently the deposition efficiency of the process.

RADIATIVE PROPERTIES OF THE AEROSOL

As discussed in the Introduction, the aerosol particles are expected to have diameters ranging from 0.1 to 0.3 μm. For the typical temperature ranges encountered in the MCVD process (500 K to 2000 K) the peak of the blackbody hemispherical spectral emissive power occurs at wavelengths varying from 1.5 μm to about 5 μm. It follows then that the spectral absorption coefficient of the aerosol particles can be obtained from Mie's theory in the limit where the wavelength of the incident radiation is much larger than the diameter of the particle. In this limit, known as the Rayleigh limit, the spectral absorption coefficient is given by Van de Hulst (1981) as,

$$\kappa_\lambda = c\frac{\pi D^3}{\lambda} Im\left(\frac{m^2-1}{m^2+2}\right)$$

where c is the aerosol concentration in particles per m^3, D is the particle diameter in m, λ is the wavelength of the incident radiation in m and m is the complex index of refraction of the particles. A plot of the spectral absorption coefficient of SiO_2 as a function of wavelength is shown in Fig 1. It can be seen that there are essentially four important absorption bands

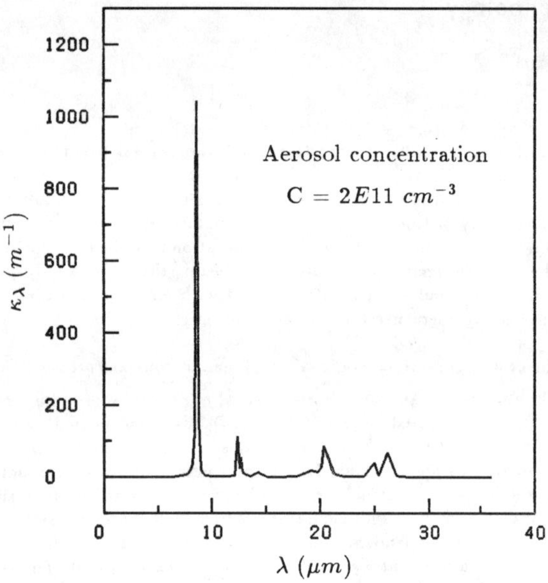

Fig. 1 — Spectral absorption coefficient of SiO_2 as a function of wavelength

with an extremely high peak in the 8.4 to 8.8 μm range. The complex index of refraction of SiO_2 in these computations was obtained from Peterson *et al.* (1969). For purposes of emission and assuming that Kirchoff's law is valid, the Planck mean absorption coefficient can be defined as (Özisik, 1973)

$$\kappa_P = \frac{\int_0^\infty \kappa_\lambda I_{\lambda b}(T)d\lambda}{I_b(T)}$$

κ_P is a function of temperature and is expected to decrease as the temperature increases due to the shift of the peaks of the spectral blackbody emissive flux to the lower wavelengths where the spectral absorption coefficient essentially vanishes. Its order of magnitude, however, depends directly on the aerosol concentration which, with the present high flow rates can be expected to be as high as $10^{12}\ cm^{-3}$ (Kim and Pratsinis 1988), giving mean absorption coefficients of the order of 4 m^{-1}. A plot of the Planck mean absorption coefficient as a function of temperature with an aerosol concentration of $2\times10^{11} cm^{-3}$ is shown in Fig 2.

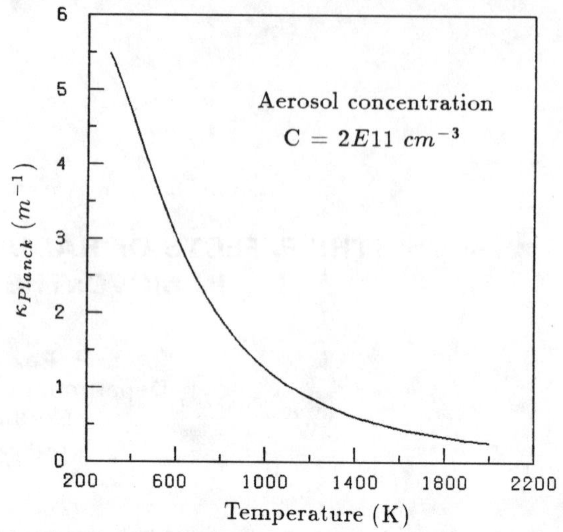

Fig. 2 — Planck mean absorption coefficient of SiO_2 as a function of temperature.

GOVERNING EQUATIONS

The flow is modelled as steady, laminar, incompressible, constant property flow of an aerosol-gas mixture initially at T_{in} and c_{in} through a horizontal circular tube of radius R. The flow is assumed to be hydrodynamically fully developed, therefore its velocity is given by the well known Poiseuille equation $u(r) = 2\bar{U}\left[1-\left(\frac{r}{R}\right)^2\right]$ (see Fig 3). The tube wall temperature closely approximates the distribution obtained experimentally by Walker *et al.*(1980) and is reproduced in Fig 4. It is understood that the constant property assumption may be somewhat severe in light of the large temperature variations that are expected to occur, but since the main focus of this study is on the effects of radiation from the aerosol particles, it is believed that the results obtained will be applicable, at least qualitatively, to the variable property case. The governing equations will then be the energy equation and the aerosol concentration equation which are given by

$$\rho c_p u(r)\frac{\partial T}{\partial z} = k\nabla^2 T - \nabla\cdot\mathbf{q_r} \tag{1}$$

$$u(r)\frac{\partial c}{\partial z} = \nabla\cdot(D\nabla c - c\mathbf{V_T}) \tag{2}$$

The initial and boundary conditions are

$$T(0,r) = T_{in} \qquad c(0,r) = c_{in}$$

$$c(z,R) = 0 \qquad T(z,R) = T_w(z)$$

$$\frac{\partial T}{\partial r}(z,0) = \frac{\partial c}{\partial r}(z,0) = 0 \tag{3}$$

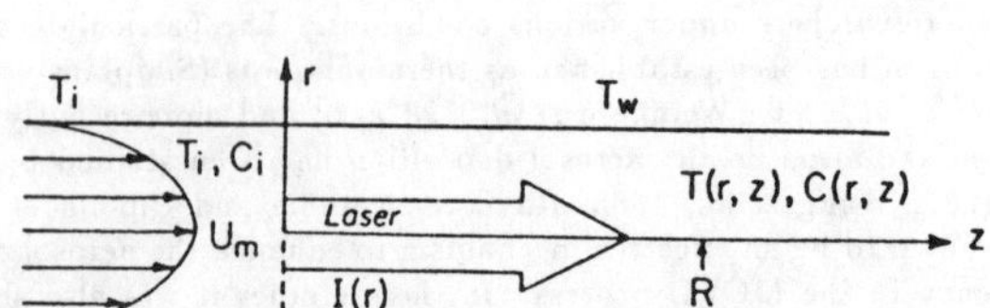

Fig. 3 — Geometry of the flow.

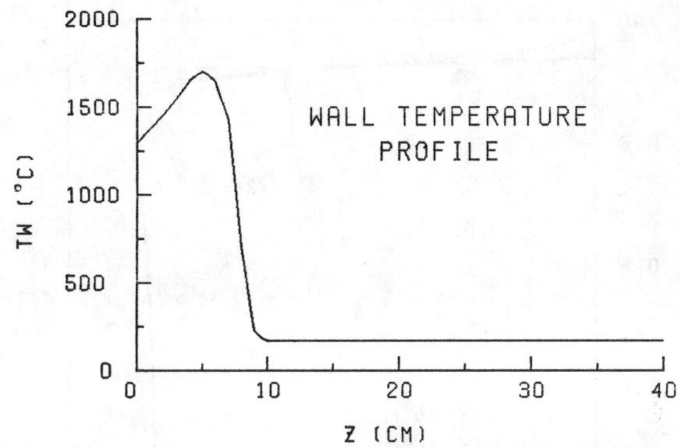

Figure 4 — Tube wall temperature profile

In these equations ρ, c_p, $\bar{U}$ and K are the background gas density, specific heat, mean velocity and thermal conductivity. In the concentration equation D represents the Brownian diffusion coefficient of the aerosol particles and $\mathbf{V_T}$ is the thermophoretic velocity given by

$$\mathbf{V_T} = -K_T \frac{\nu}{T} \nabla T$$

where ν is the kinematic viscosity of the gas, T the local gas temperature and K_T the thermophoretic coefficient which is proportional to the ratio of the gas and particle thermal conductivities and the Knudsen number, defined as the ratio of the gas mean free path and the particle radius. An empirical expression for K_T is given by Talbot *et al.*(1980). Its value is of order 1 and for the particular gas aerosol combination encountered in the MCVD process it is found to be about 0.5 (Morse, Cipolla, 1984). In this work, however, the value of K_T is taken to be 0.9 to account partially for the neglect of the variation of the gas properties with temperature. The radial and axial variables are normalized with respect to the tube radius and the product of the tube radius and the Peclet number, respectively, and the temperature and concentration normalized with respect to the inlet temperature and concentration. Defining $\theta = \frac{T}{T_{in}}$, $\phi = \frac{c}{c_{in}}$ and introducing these definitions in the energy and concentration equations one obtains

$$(1 - r^2)\frac{\partial \theta}{\partial z} = \frac{1}{r}\frac{\partial}{\partial r}(r\frac{\partial \theta}{\partial r}) + S \tag{4}$$

where S is a source term defined as

$$S = -\frac{R^2}{KT_{in}} \nabla \cdot \mathbf{q_r} \tag{5}$$

$$(1 - r^2)\frac{\partial \phi}{\partial z} = \frac{Pr}{Sc}\frac{1}{r}\frac{\partial}{\partial r}(r\frac{\partial \phi}{\partial r}) + K_T Pr \frac{1}{r}\frac{\partial}{\partial r}(r\frac{\phi}{\theta}\frac{\partial \theta}{\partial r}) \tag{6}$$

where Sc is the Schmidt number defined as $Sc = \frac{\nu}{D}$. The boundary and initial conditions are

$$\theta(z,1) = \theta_w(z) \qquad \phi(z,1) = 0$$

$$\frac{\partial \theta}{\partial r}(z,0) = \frac{\partial \phi}{\partial r}(z,0) = 0 \tag{7}$$

$$\theta(0,r) = \phi(0,r) = 1$$

SOURCE TERM EVALUATION

The source term that appears in the energy equation is directly proportional to the divergence of the radiative heat flux which, neglecting scattering and assuming gray behavior, is given by (Özisik, 1973)

$$\nabla \cdot \mathbf{q_r} = \kappa_P 4\sigma T^4 - \int_0^\infty \kappa_\lambda \int_{4\pi} I d\Omega \, d\lambda \tag{8}$$

The two contributions to the intensity are the incident laser radiation and the radiation from the aerosol particles, which are assumed to be spectrally separated. This is reasonable because the temperatures within the tube are expected to be fairly high and the aerosol will be emitting mostly at wavelengths considerably smaller than the wavelength of the incident laser radiation. Therefore the spectral distribution of the radiation intensity will have high values at the low wavelengths and it will decay exponentially to zero at higher wavelengths with the exception of $10.6\mu m$ where it will have a sharp peak due to the laser radiation. At this wavelength the contribution to the intensity due to emission from the aerosol is expected to be essentially zero. A similar approach is given by Modest and Tabanfar (1983). Equation 8 then becomes

$$\nabla \cdot \mathbf{q_r} = \kappa_P \phi[4\sigma T^4 - \int_{4\pi} I d\Omega] - \kappa_{laser}\phi I_0(r) exp\Big(-RPe\int_0^z \kappa_{laser}\phi d\zeta\Big) \tag{9}$$

where κ_{laser} is the absorption coefficient at the laser wavelength and κ_P is the Planck mean absorption coefficient, each evaluated at the initial aerosol concentration. For a Gaussian laser of beam width β the radial dependence of the initial ($z = 0$) total angle-integrated radiation intensity is given by

$$I_0(r) = I_0 \exp(-\beta^2 r^2) \tag{10}$$

where I_0 is the peak laser intensity related to the laser power intersected by the tube by

$$P_{tube} = \frac{I_0 \pi R^2}{\beta^2}[1 - e^{-\beta^2}]$$

The intensity in the second term of Eq.(9) is that due to aerosol radiation and may be approximated in the optically thin limit by

$$I = I_{wall} = \frac{\sigma T_w^4}{\pi}$$

The optical thickness of the aerosol is $\tau_0 = \kappa_P R$, which for a tube radius of 1 cm, will make the aerosol an optically thin medium if $\kappa_P << 100\ m^{-1}$. The source term takes then the form

$$S = S_0 F(r,\beta)\phi \exp\Big(-\kappa_{laser} RPe \int_0^z \phi d\zeta\Big) + N\phi(\theta_w^4 - \theta^4) \tag{11}$$

where

$$F(r,\beta) = \frac{\beta^2}{(1 - e^{-\beta^2})} e^{-\beta^2 r^2} \tag{12}$$

and

$$S_0 = \frac{\kappa_{laser} P_{tube}}{\pi K T_{in}} \qquad N = \frac{4\sigma\kappa_P R^2 T_{in}^3}{K}$$

RESULTS AND DISCUSSION

The energy and aerosol concentration equations derived in the previous section along with the boundary and initial conditions form a system of second order partial differential equations, which are non-linearly coupled through the source term in the energy equation. This will require then that the two equations be solved iteratively. Furthermore the non-dimensional temperature also appears in this term raised to the fourth power, which is not a desirable feature in the numerical solution of the energy equation. In order to treat this nonlinearity a linearization scheme suggested by Patankar (1980) is adopted here. It basically consists in expanding the term involving θ^4 in a Taylor series about its value in the previous iteration. Thus

$$\theta^4 = \theta_{old}^4 + 4\theta_{old}^3(\theta - \theta_{old})$$

where higher order terms have been neglected. The source term is then expressed as

$$S = S_0 F(r,\beta)\phi \exp\Big(-\kappa_{laser} RPe \int_0^z \phi d\zeta\Big) + N\phi(\theta_w^4 - 4\theta\theta_{old}^3 + 3\theta_{old}^4) \tag{13}$$

which still couples both equations through its dependence on the aerosol concentration. The iterative scheme adopted to solve for the temperature and aerosol concentration fields will now be described. At every axial step the energy equation is solved for the temperature field with a source term initially based on the converged concentration field from the previous axial step. The concentration equation is then solved for an updated concentration field, which is again used in the energy equation to generate a new temperature field. This procedure is continued until convergence is obtained. The criterion for convergence is a maximum allowed difference of 10^{-4} in both the normalized temperature and aerosol concentration between sucessive iterations.

Both equations are parabolic, therefore they lend themselves to an implicit numerical solution. The energy equation is readily solved by the Cranck-Nicholson method and the concentration equation by a fully implicit finite differencing scheme. It was found that a fully implicit scheme provides a more satisfactory treatment of the initial jump between the initial concentration $\phi(0,r) = 1$ and the wall concentration $\phi(z,1) = 0$. The finite differencing scheme adopted allowed for variable radial and axial steps, which proved to be very convenient when the diffusion term was considered in the concentration equation. When this term was neglected, constant radial and axial steps were usually adopted in the numerical solution. The tube radius was divided in 50 equally spaced intervals when diffusion was neglected and several hundred unequally spaced intervals when diffusion was included. The axial steps were taken between .0005 and .001 non dimensional units (z/RPe) depending on the magnitude of the laser heating and the initial aerosol concentration.

EFFECTS OF DIFFUSION AND THE AEROSOL MASS CONSERVATION SCHEME

In previous studies of the deposition process in MCVD (Morse, *et al.*, 1985 ; Cipolla *et al.*, 1987) it was argued that, for a cold wall, the effects of Brownian diffusion would only be felt in a very thin layer adjacent to the tube wall since the Schmidt number was of the order of 10^5. The diffusion term in the aerosol concentration equation was then neglected transforming it in a first order equation which was then easier to solve. In this case, though, a wall boundary condition on the aerosol concentration cannot be imposed, so the aerosol concentration at the wall has to be obtained from the final solution of the system of equations. This procedure is acceptable for particle deposition on cold walls since the themophoretic particle flux at the outer edge of the thin diffusion layer is equal to the diffusive flux at the wall. For a hot wall, however, the aerosol moves away from the wall and therefore the lack of a wall boundary condition can lead to unrealistic aerosol concentrations within the tube. Furthermore there will be a region close to the tube wall from which the aerosol will be evacuated and so the correct aerosol concentrations in that region will be zero in the limit of high Schmidt number. The exact location of the boundary of this region can be obtained by requiring that the aerosol mass flow rate remains constant in regions where there is no deposition. Numerically this mass conservation scheme is implemented by requiring that at every axial location the aerosol mass flow rate given by

$$\dot{m} = \int_{r=0}^{r=r_c} m_p c_{in} \phi(r) u(r) 2\pi r dr,$$

where m_p is the mass of a single aerosol particle and c_{in} the initial aerosol concentration, be equal to its value in the previous axial step. At each iteration r_c is allowed to vary between 0 and 1 until the value of the above integral equals the one obtained in the previous step. The aerosol concentration is then set to zero for values greater than r_c and the iterations are continued.

In order to test the validity of the mass conservation scheme a simpler problem was solved with neither laser nor aerosol radiation and without the diffusion term in the concentration equation. The background gas mixture entrance temperature and the aerosol concentration are constant and the wall temperature is given by Fig 4. In the absence of diffusion the mass conservation scheme outlined above was implemented and the aerosol concentration profiles were compared with the results obtained with diffusion. A sample of the results obtained is shown in Figures 5 and 6. In both figures several different profiles are shown in the case where diffusion was considered so that the effect of the Schmidt number in the aerosol concentration could be examined too. It can be seen that as the Schmdit number is larger than 10^4 the concentration profiles are essentially indistinguishable; furthermore they coincide with the profiles obtained without the diffusion term. So it can be concluded that Brownian diffusion is negligible in the high Schmidt number MCVD process and furthermore, in the region under the torch, where the tube wall is hotter than the aerosol-gas mixture, the correct aerosol concentrations can be found by solving the no-diffusion aerosol concentration equation coupled with the the mass conservation scheme just

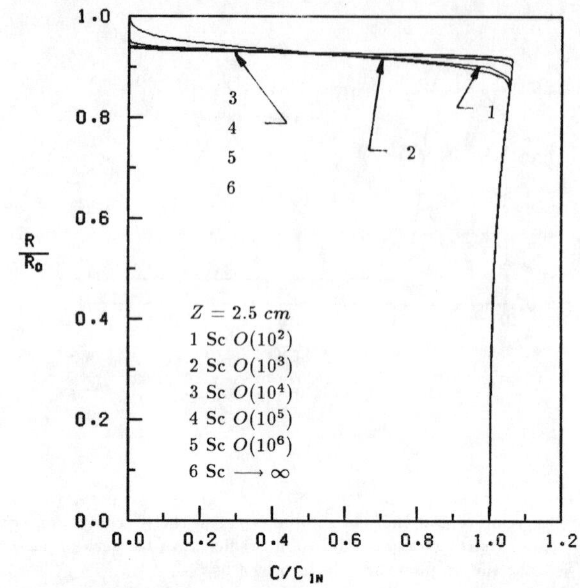

Fig. 5 — Effect of the Schmidt number on the aerosol concentration under the torch (Z = 2.5 cm).

developed. The desirability of this approach stems from the fact that much coarser grids can be used in the numerical solution, since the thin diffusion boundary layer and the steep gradients that develop in the flow need not be resolved if diffusion is neglected. The computational time saved in each run was found to be enormous since the inclusion of diffusion required several hundred nodes in the radial direction whereas the no-diffusion problem only required a maximum of fifty. Typical running times were 5–10 CPU minutes without diffusion as opposed to more than three hours with diffusion on a VAX 11/785. As a final note it should also be pointed out that the locus of points specifying the boundary of the particle-free zone under the torch was found to coincide with the trajectory followed by a particle that enters the heating area of the tube adjacent to the wall.

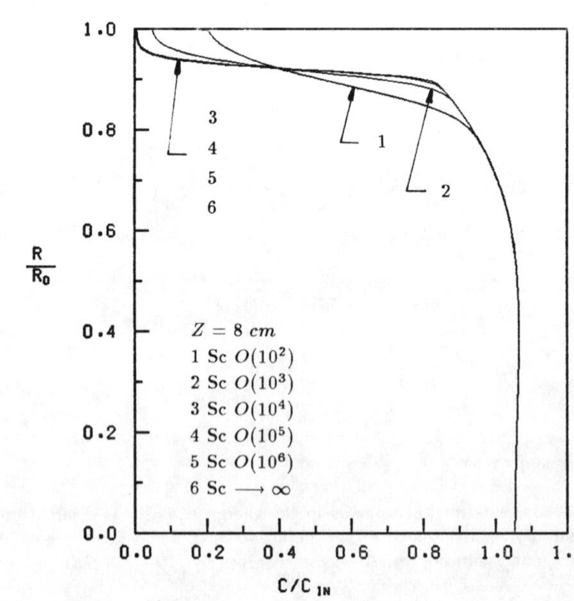

Fig. 6 — Effect of the Schmidt number on the aerosol concentration downstream of the torch (Z = 8 cm).

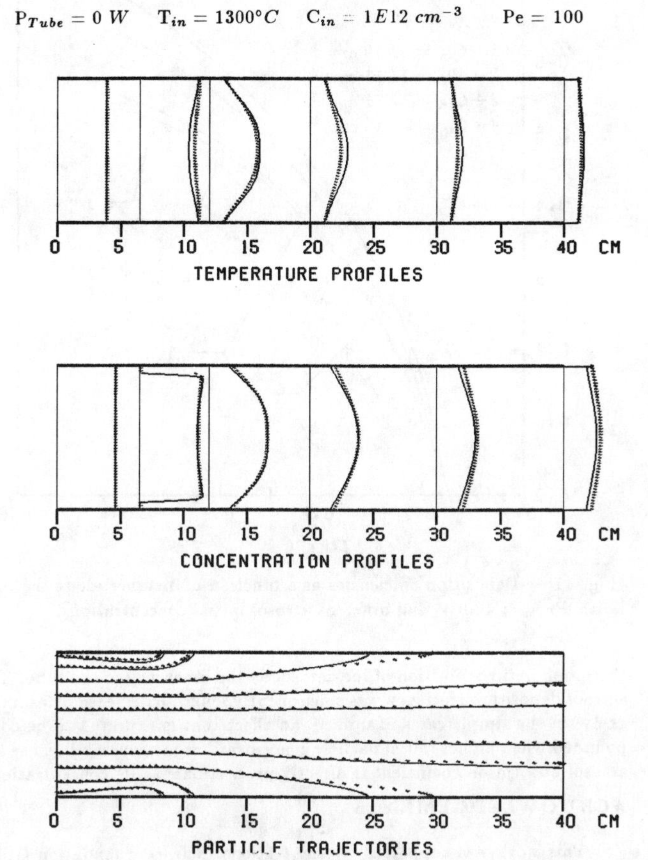

Fig. 7 — Temperature and concentration profiles, and particle trajectories with(· ·) and without(——) aerosol radiation.

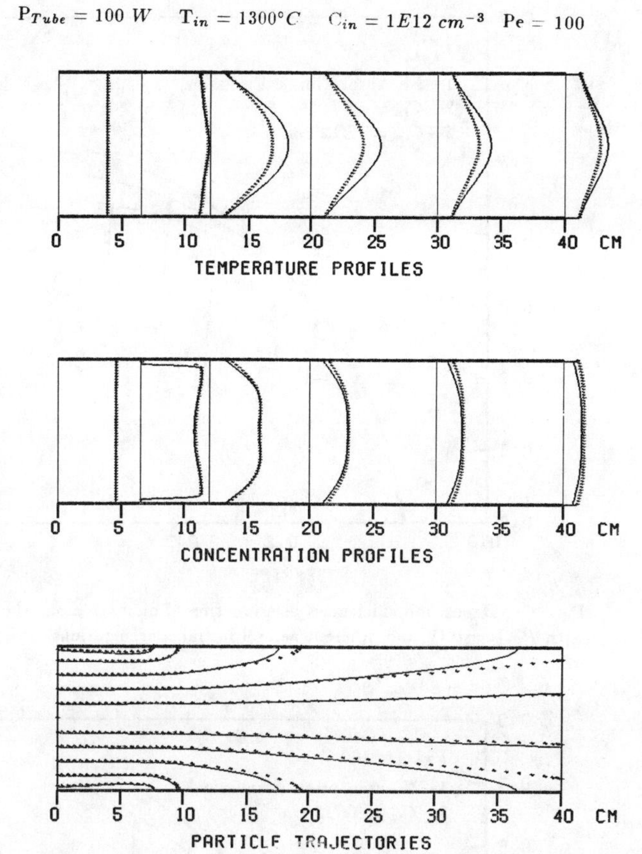

Fig. 8 — Temperature and concentration profiles, and particle trajectories with(· ·) and without(——) aerosol radiation.

RADIATION FROM THE AEROSOL

In this section the effect of aerosol radiation resulting from the high temperatures common in MCVD is examined, when, from a radiative point of view the aerosol particles are modelled as an optically thin medium. The system of equations to be solved is then the energy and the no-diffusion aerosol concentration equations.

In the problem solved the initial aerosol-gas temperature was taken to be the reaction temperature of 1300 $°C$, the wall temperature was given by Fig 4, the tube radius was 1 cm, the Peclet number 100 and the background gas was taken to be oxygen only, with constant properties evaluated at 1100 $°C$. The initial aerosol concentration was varied between $5 \times 10^{10}\ cm^{-3}$ and $10^{12}\ cm^{-3}$ so that the initial Planck mean absorption coefficient of the aerosol evaluated at 1100 ° C varied between 0.2 and 4.0 m^{-1}. A sample of some of the results obtained is shown in Figures 7 through 11.

It can be seen that the energy radiated from the aerosol has a detrimental effect on the radial temperature gradients that are generated within the tube. In regions under the torch, where the wall is hotter than the gas-aerosol mixture, the aerosol absorbs radiation from the wall and acts as a heat source and therefore the aerosol-gas temperatures there are higher than they would be without the inclusion of aerosol radiation (fig 7). When the tube is irradiated with laser energy the absorption of laser radiation dominates the temperature distribution such that the inclusion of aerosol radiation loses importance specially at high laser powers (fig. 8). Downstream of the torch where the wall is colder than the aerosol-gas mixture, the aerosol emits radiation to the wall and acts as a heat sink thereby reducing the aerosol-gas mixture temperature specially at high laser powers.

In essence it can be seen that the inclusion aerosol radiation in the analysis results in temperatures that are everywhere closer to the wall temperature in the deposition zone. It follows then, that the radial temperature gradients within the tube become less steep thereby decreasing the effect of thermophoresis in the motion of the aerosol. The particle trajectories shown in figures 7 and 8 clearly illustrate this effect, there it can be seen that with or without laser radiation a particle entering the tube at a particular radial location will deposit on the wall sooner if the radiation from the aerosol is neglected. The aerosol concentrations will then be slightly higher downstream of the torch if radiation from the aerosol is considered, which means that fewer particles were deposited. Also when aerosol radiation is considered, the boundary of the aerosol–free region moves closer to the tube wall under the torch and downstream of the torch the aerosol does not move as rapidly to the wall as when te aerosol radiation is neglected. The deposition efficiency, defined as the fraction of the incoming aerosol that deposits on the tube wall up to a certain distance along the tube, is then expected to be reduced when radiation from the aerosol is considered. This effect is illustrated in figure 9 where the deposition efficiency is plotted as a function of the distance along the tube with and without aerosol radiation being considered. The reductions are as high as 7% in the short distances used in the computations (40 cm) and are expected to be higher in longer tubes. In the presence of laser radiation the same type of behaviour was encountered although not as severe. In fig 10 the deposition efficiency is shown to decrease by about 4% when radiation from the aerosol is considered and the laser power is 100 W. Finally is fig. 11 it is shown that the effect of increasing the initial aerosol concentration, thereby increasing the absorption of laser radiation increases the deposition efficiency even when the radiation from the aerosol is considered, a result that should be expected from previous studies (Morse and Cipolla, 1984).

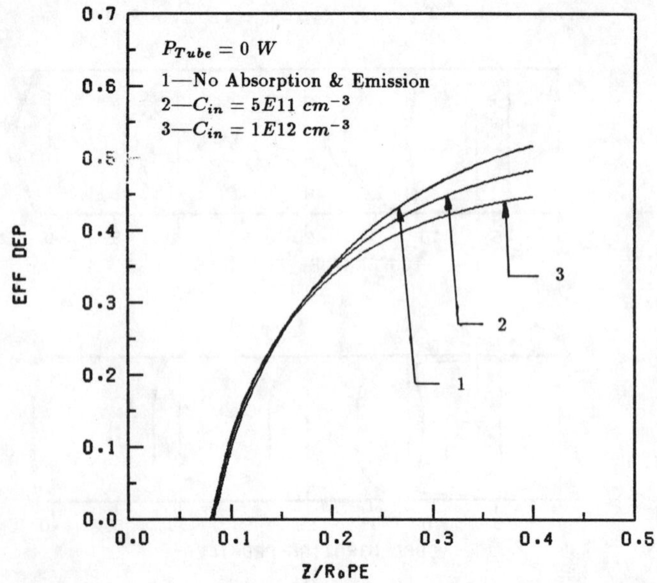

Fig. 9 — Deposition efficiencies as a function of distance along the tube with $P_{Tube} = 0\ W$ and different aerosol initial concentrations.

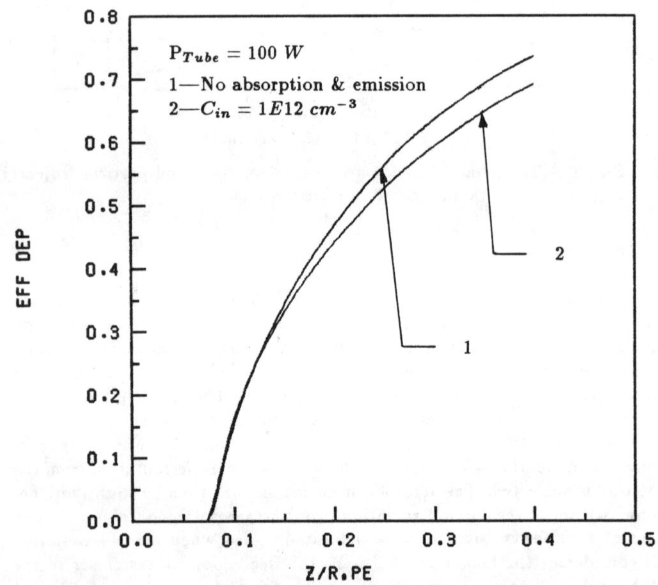

Fig. 10 — Deposition efficiencies as a function of distance along the tube with $P_{Tube} = 100\ W$ and different aerosol initial concentrations.

CONCLUSIONS

A study was made of the influence of radiation from the aerosol in typical MCVD conditions. The two dimensional, laminar constant property system of equations generated was solved numerically by an iterative scheme, and, when diffusion was neglected from the particle concentration equation, a mass conservation scheme was developed to obtain the correct solution of the concentration equation in regions under the torch, where the aerosol particles move inwards towards the center of the tube. Furthermore, it was found that the inclusion of radiation from the aerosol resulted in a decrease of the radial temperature gradients which are responsible for the

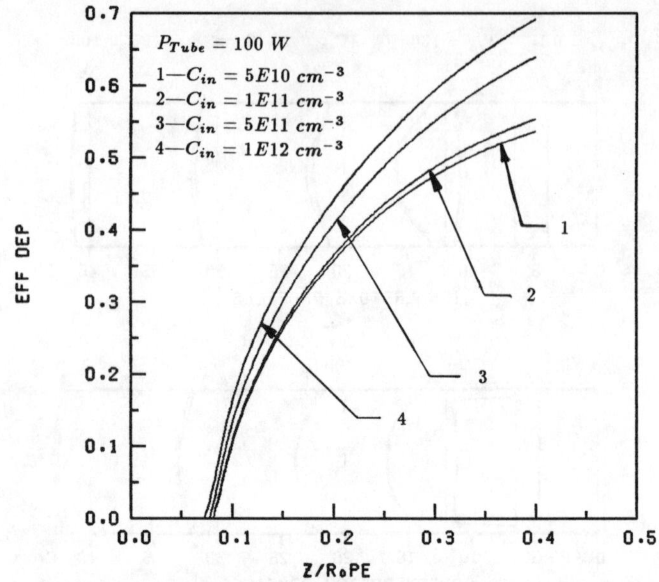

Fig. 11 — Deposition efficiencies as a function of distance along the tube with $P_{Tube} = 100\ W$ and different aerosol initial concentrations.

thermophoretic deposition of the aerosol on the inner wall of the tube. The aerosol deposition efficiency was consequently found to decrease by as much as 6% in the simple cases examined, an effect that is expected to be compounded when higher initial particle concentrations are generated, since the aerosol absorption coefficient is directly proportional to its concentration.

ACKNOWLEDGEMENTS

This work was supported by the National Science Foundation Grant no. CBT-8603634. The computations were performed in the computation center of Northeastern University.

REFERENCES

Cipolla, J. W. Jr, Morse, T. F., (1987) "Thermophoresis in an absorbing aerosol", *J. Aerosol Science*, Vol 18, No 3, pp245–260.

Choi, M., Baum, H. R., Greif, R., (1987) "The heat transfer problem for the modified chemical vapor deposition process", *J. Heat Transfer*, Vol 109, pp642–646.

Jablonowski, D. P., (1986) "Fiber manufacture at AT&T with the MCVD process", *Journal of Lightwave Technology*, Vol LT-4, no 8, pp1016–1019.

Kim, Kyo-Seon, and Sotiris E. Pratsinis (1988) "Manufacture of Optical Waveguide Preforms by Modified Chemical Vapor Deposition" *AIChE Journal* Vol 34 pp912–921.

MacChesney, J. B., O'Connor, P. B., Presby, H. M., 1974b, "A new technique for preparation of low-loss and graded index optical fibers", *Proc. IEEE*, Vol 62, pp1278–1279.

Modest, M. F. and Tabanfar, S., (1983) "A Multi-dimensional Differential Approximation for Absorbing/Emitting and Anisotropically Scattering Media with Collimated Radiation" *J. Quant. Spectr. and Rad. Transfer* Vol 29, No 4, pp 339–351.

Morse, T. F., Cipolla, J. W. Jr., (1984) "Laser modification of thermophoretic deposition", *J. Colloid & Interface Science*, Vol 97, No 1, pp137–148.

Morse, T. F., Wang, C. Y., and Cipolla, J. W. Jr., (1985) "Laser induced thermophoresis and particulate deposition efficiency", *J. Heat Transfer*, Vol 107, pp155–160.

Özisik, N., (1973) *Radiative Transfer*, 1st Edition, p34 & p255, Werbel & Peck, NY.

Patankar, S. V., (1980) *Numerical Heat Transfer and Fluid Flow*, p48–49, Hemisphere Publishing Corporation, NY.

Peterson, J. T., Weinman, J. A., (1969) "Optical properties of quartz dust particles at infrared wavelengths", *J. of Geophysical Research*, Vol 74, No 28, pp6947–6952.

Ravikumar, S., (1985) "Thermal Radiation and Thermophoresis in aerosols", PhD thesis, Northeastern University, Boston, Ma.

Simpkins, P. G., Greenberg-Kosinski, S., MacChesney, J. B., (1979) "Thermophoresis: The mass transfer mechanism in modified chemical vapor deposition", *J. Applied Physics*, 50(9), pp5676–5681.

Talbot, L., Cheng, R. K., Schefer, R. W., Willis, D. R., (1980) *J. Fluid Mechanics*, Vol 101, p. 737.

Van de Hulst, H. C., (1981) *Light scattering by small particles*, p70, Dover NY.

Walker, K. L., Geyling, F. T., Nagel, S. R., (1980) "Thermophoretic deposition of small particles in the Modified Chemical Vapor Deposition (MCVD) Process", *J. American Ceramic Society*, Vol 63, No 9–10, pp552–558.

Weinberg, M., Subramanian, R. S., 1982a, "Thermophoretic deposition in a tube with variable wall temperature", *J. of Colloid & Interface Science*, Vol 87, No 2, pp579–580.

Weinberg, M., 1982b, "Thermophoretic Deposition Efficiency in the Modified Chemical Vapor Deposition Process", *J. of the American Ceramic Society*, Vol 65 pp81–87.

Wood, D. L., Walker, K. L., MacChesney, J. B., Simpson, J. R. and Csencsits, R., (1987) "Germanium chemistry in the MCVD process for optical fiber fabrication", *Journal of Lightwave Technology*, vol LT-5, no 2, pp277–285.

1989 National Heat Transfer Conference
HTD-Vol. 106, Heat Transfer Phenomena in Radiation, Combustion and Fires

CLOSED FORM SOLUTION OF THE GENERAL THREE-DIMENSIONAL RADIATION CONFIGURATION FACTOR PROBLEM WITH MICROCOMPUTER SOLUTION

D. Alciatore and S. Lipp
The University of Texas at Austin
Austin, Texas

W. S. Janna
Memphis State University
Memphis, Tennessee

A closed form based solution to the general problem of finding the radiation configuration factor between a differential element and a three dimensional solid object was developed. The method applies to any 3-D object which can be represented as a series of connected flat-faceted surfaces. The approach was based on insight from the Nusselt Unit Sphere Method. The developed algorithm was implemented in a microcomputer program which solves the problem and provides graphical output.

INTRODUCTION

Solving for the radiation configuration factor between a differential element and a finite area is essential in solving more general radiation problems. Various approaches have been taken in the past to solve this problem. Many authors have approached the problem by directly applying the integral expressions for the configuration factor to generate closed form solutions. The most notable authors taking this approach are Moon (1936), de Bastos (1961), and Sparrow (1963) for their application of Stokes' Theorem involving contour integration. Howell (1982) has also made a tremendous contribution in compiling many specific closed form solutions. The only obvious drawback with this traditional closed form approach is that it is limited to problems with well-defined geometry. The integral expressions become extremely cumbersome even for simple geometry.

A second approach to the problem is to apply the Unit Sphere Method developed by Nusselt (1928). Nusselt introduced solid angle geometry to simplify the problem and to provide a graphical interpretation. This graphical interpretation can be applied to determine the configuration factors experimentally. Many mechanical and photographic apparatus have been designed to implement this geometric solution. Most notable are Hamilton's Mechanical Linkage (Hamilton and Morgan, 1952), Eckert's shadow casting photographic method (Eckert and Drake, 1959), and Farrel's Scintilascope (Farrel, 1976). The advantage of these methods is that they can be applied to any arbitrarily complex object provided a model can be built. The obvious disadvantage in these methods is that they require special equipment and physical models.

Nusselt's Unit Sphere Method has also been applied in the form of direct graphical procedures. Hooper and Juhasz (1952) were probably the first to present a useful graphical procedure. Alciatore et al. (1988) have also presented an engineering graphics solution. These graphical methods are valuable in that they are intuitive, straightforward, and require very little equipment or skill. The major drawback is that they are very time-consuming and require that drawings be constructed.

A third approach is a microcomputer solution. Alciatore et al. (1988) have presented one such application which solves for the configuration factor between an elemental area and a defined 3-D contour. The computer solution is advantageous due to its speed and accuracy, but the contour is often difficult to define for certain geometries, such as arbitrary solid objects. The goals of this paper are to extend the computer solution to general 3-D objects and to develop a closed form solution which finds the configuration factor between a differential element and a 3-D contour directly.

MICROCOMPUTER IMPLEMENTATION

The first requirement of the computer algorithm was to provide a means for describing a general 3-D faceted closed surface. Two input routines were chosen for this purpose. One routine reads a standard wireframe database which contains vertex, edge, and surface information. This data can be read from a pre-existing file or entered at the keyboard. The second input routine allows the user to define the profile of a general surface of revolution. The program can take this profile and generate the complete wireframe information for the solid. This expedites entering geometry for axially symmetric bodies such as cylinders, cones, spheres, etc. Both input routines result in a faceted surface description which is illustrated in Figure 1. In this description, the one place where care must be taken is in the ordering of the facet edges. This ordering establishes the direction of the outward normal, which then determines whether or not the facet is visible

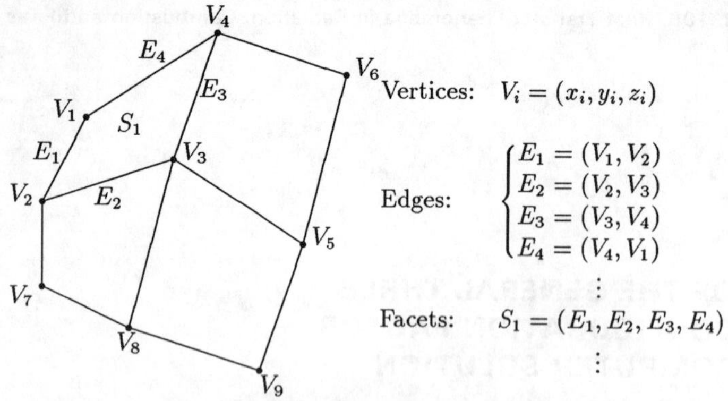

Fig. 1. Faceted Surface Description

to the differential element.

Once the geometry is defined, the next step is to determine the contour which describes the border between the radiated (visible) and non-radiated (non-visible) facets. Once this contour is defined, the closed form solution developed in the next section can be applied directly. The result is the radiation configuration factor between a differential element and the defined 3-D object.

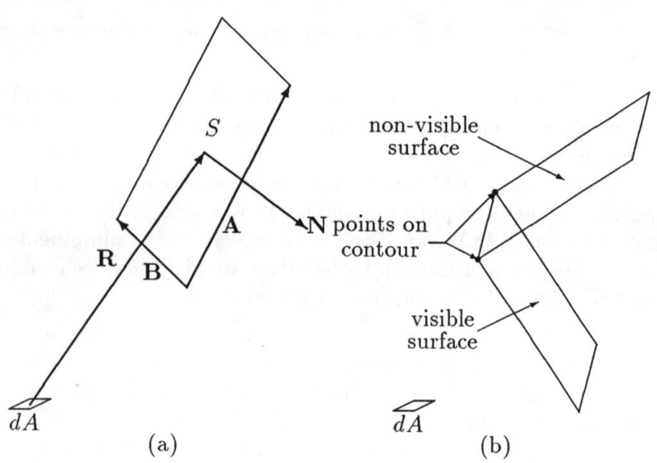

Fig. 2. Determining Surface Visibility and Object Contour

The contour is found using the common computer graphics technique called Back Plane Culling (BPC). BPC is a hidden surface removal algorithm which can be applied to wireframe data. The algorithm allows you to determine which facets are seen (radiated to) from the differential element. BPC is generally restricted to convex polyhedra although it is tolerant of low degrees of concavity. Figure 2(a) illustrates the vectors used in determining facet visibility. The two adjacent edges defined by vectors **A** and **B** define the facet outward normal **N**. Given that the line of sight from dA to facet S is defined by **R**, the following test can be used to determine visibility:

If $\mathbf{R} \cdot \mathbf{N} < 0$, then the facet is visible.

If $\mathbf{R} \cdot \mathbf{N} \geq 0$, then the facet is non-visible.

Once the visibility test has been performed for each facet, the locus of points defining the contour can be determined as illustrated in Figure 2(b). The ordering of these points is important and can be determined from the wireframe edge connectivity information. Having defined the contour for the object, the configuration factor may be calculated directly.

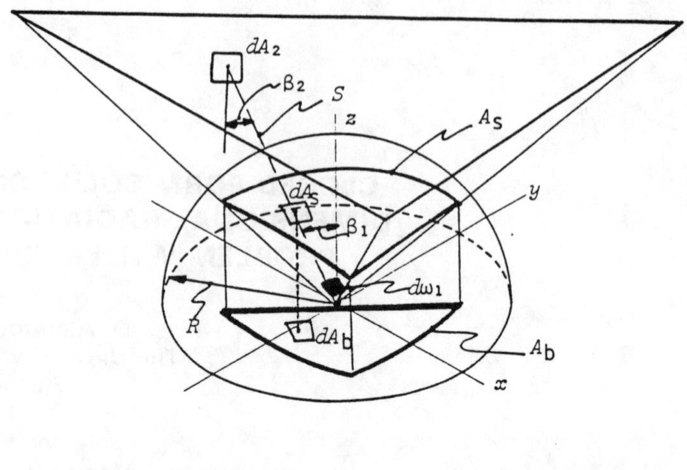

Fig. 3. Nusselt's Unit Sphere Method

CLOSED FORM SOLUTION

The definition of the configuration factor (Siegel and Howell, 1972) between a differential element dA_1 and a finite area A_2 is:

$$F_{dA_1-A_2} = \iint_{A_2} \frac{\cos\beta_1 \cos\beta_2}{\pi S^2}\, dA_2 = \frac{1}{\pi}\iint_{A_2} \cos\beta_1\, d\omega_1\,, \tag{1}$$

where S is the length from dA_1 to dA_2, β_1 is the angle between the normal to dA_1 and S, β_2 is the angle between the normal to dA_2 and S, and $d\omega_1$ is the solid angle defined by:

$$d\omega_1 = \cos\beta_2 \frac{dA_2}{S^2}\,.$$

(Refer to Figure 3 for an illustration of terms). This solid angle geometry was introduced by Nusselt (1928) to reinterpret the configuration factor as:

$$F_{dA_1-A_2} = \frac{1}{\pi R^2}\iint_{A_2} \cos\beta_1\, dA_s\,,$$

where A_s is the area A_2 projected onto the hemisphere of radius R centered at dA_1. This projection onto the hemisphere is accomplished along the lines S. Thus,

$$F_{dA_1-A2} = \frac{A_b}{\pi R^2} = \frac{\textit{projection of } A_s \textit{ onto the base of the hemisphere}}{\textit{area of a circle with the radius of the hemisphere}}\,. \tag{2}$$

Determination of an analytic solution to the configuration factor between an elemental area and a closed contour using Nusselt's Unit Sphere Method is simply a matter of analytic geometry. With the N vertices of the contour given by

$$\mathbf{P}_i = (X_i, Y_i, Z_i) \qquad i = 1, \ldots, N \qquad \mathbf{P}_{N+1} = \mathbf{P}_1\,, \tag{3}$$

the following vectors may be defined:

$$\mathbf{P}_{i(i+1)} = (X_{i+1} - X_i, Y_{i+1} - Y_i, Z_{i+1} - Z_i) = \mathbf{P}_{i+1} - \mathbf{P}_i\,. \tag{4}$$

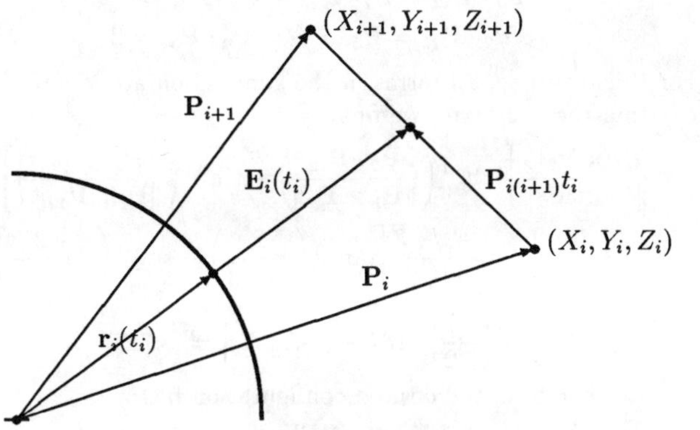

Fig. 4. Vectors Used in Edge Projection

These vectors describe the edges of the contour. To project these vectors onto the sphere of radius R, the direction cosines from the origin to the locus of points defining the edges of the contour must be found. The direction cosines are just the coordinates of the point divided by the distance from the origin. The edges of the contour may be parameterized as:

$$\mathbf{E}_i(t_i) = \mathbf{P}_i + \mathbf{P}_{i(i+1)}t_i \qquad 0 \le t_i \le 1 \,.$$

Thus the distance from the origin to points along the edges of the contour may be written as $\|\mathbf{E}_i(t_i)\|$, where $\|\cdot\|$ is the scalar magnitude of the vector quantity. This magnitude is:

$$\begin{aligned}\|\mathbf{E}_i(t_i)\| &= \left(\mathbf{E}_i(t_i)\cdot\mathbf{E}_i(t_i)\right)^{\frac{1}{2}} \\ &= \left[\mathbf{P}_i\cdot\mathbf{P}_i + 2\mathbf{P}_i\cdot\mathbf{P}_{i(i+1)}t_i + \mathbf{P}_{i(i+1)}\cdot\mathbf{P}_{i(i+1)}t_i^2\right]^{\frac{1}{2}} \,.\end{aligned}$$

The projection of the ith segment onto the sphere is defined by:

$$\mathbf{r}_i(t_i) = \frac{R}{\|\mathbf{E}_i(t_i)\|}\mathbf{E}_i(t_i) = (x_i(t_i), y_i(t_i), z_i(t_i)) \,,$$

where $\mathbf{r}_i(t_i)$ describe the sphere point coordinates. Projecting these coordinates onto the $X-Y$ plane just means eliminating the z_i coordinate. The vector geometry is illustrated in Figure 4.

The area enclosed by these parametric curves will be determined using an application of Green's theorem. With $A = \iint_A dA = \frac{1}{2}\oint_C x\,dy - y\,dx$, x_i, y_i, dx_i, and dy_i, with $i = 1, \ldots, N$, are the terms necessary to determine the configuration factor. In general,

$$d\zeta_i = \frac{d\zeta_i}{dt_i}\,dt_i \qquad \text{for } \zeta = x,\ y, \text{ or } z \,.$$

Hence by differentiating each parametric curve with respect to t_i and then eliminating the Z component, the integration over the projected contour becomes a matter of summing the line integral contributions from individual segments. First note that:

$$\frac{d\mathbf{E}_i(t_i)}{dt_i} = \mathbf{P}_{i(i+1)} \qquad \text{and} \qquad \frac{d}{dt_i}\frac{1}{\|\mathbf{E}_i(t_i)\|} = -\frac{\mathbf{P}_{i(i+1)}\cdot\mathbf{E}_i(t_i)}{[\mathbf{E}_i(t_i)\cdot\mathbf{E}_i(t_i)]^{\frac{3}{2}}}$$

Thus the integrand can be rewritten as:

$$x_i\,dy_i - y_i\,dx_i = \left(x_i\frac{dy_i}{dt_i} - y_i\frac{dx_i}{dt_i}\right)dt_i \,,$$

or, equivalently, the line integrand is equal to the Z component of the vector $\mathbf{r}_i(t_i) \times \frac{d\mathbf{r}_i(t_i)}{dt_i}$ multiplied by dt_i. The cross product is given by:

$$\begin{aligned}&\mathbf{r}_i(t_i) \times \frac{d\mathbf{r}_i(t_i)}{dt_i} \\ &= \frac{R}{\|\mathbf{E}_i(t_i)\|}\mathbf{E}_i(t_i) \times \left[\frac{R}{\|\mathbf{E}_i(t_i)\|}\mathbf{P}_{i(i+1)} - \frac{R\mathbf{P}_{i(i+1)}\cdot\mathbf{E}_i(t_i)}{\|\mathbf{E}_i(t_i)\|^3}\mathbf{E}_i(t_i)\right] \\ &= \frac{R^2}{\|\mathbf{E}_i(t_i)\|^2}\mathbf{E}_i(t_i) \times \mathbf{P}_{i(i+1)} \\ &= \frac{R^2}{\|\mathbf{E}_i(t_i)\|^2}\mathbf{P}_i \times \mathbf{P}_{i(i+1)} \,.\end{aligned}$$

The cross product terms in the previous equations do not depend on t_i, so they may be taken out of the integral. All that needs be integrated is $\frac{1}{\|\mathbf{E}_i(t_i)\|^2}$. The integration is over $0 \le t_i \le 1$. Therefore, Equation 2 may be written as:

$$\begin{aligned}&\int_0^1 \frac{dt_i}{\|\mathbf{E}_i(t_i)\|^2} \\ &= \frac{1}{\|\mathbf{P}_{i(i+1)}\|^2}\int_0^1 \frac{dt_i}{t_i^2 + 2\frac{\|\mathbf{P}_i\|}{\|\mathbf{P}_{i(i+1)}\|}\cos\beta_i t_i + \left(\frac{\|\mathbf{P}_i\|}{\|\mathbf{P}_{i(i+1)}\|}\right)^2} \\ &= \frac{1}{\|\mathbf{P}_{i(i+1)}\|^2}\int_0^1 \frac{dt_i}{\left(t_i + \frac{\|\mathbf{P}_i\|}{\|\mathbf{P}_{i(i+1)}\|}\cos\beta_i\right)^2 + \left(\frac{\|\mathbf{P}_i\|}{\|\mathbf{P}_{i(i+1)}\|}\sin\beta_i\right)^2} \\ &= \frac{1}{\|\mathbf{P}_i\|\|\mathbf{P}_{i(i+1)}\|\sin\beta_i}\int_{\frac{\pi}{2}-\beta_i}^{\tan^{-1}\left(\frac{\|\mathbf{P}_{i(i+1)}\|+\|\mathbf{P}_i\|\cos\beta_i}{\|\mathbf{P}_i\|\sin\beta_i}\right)} d\theta \\ &= \frac{1}{\|\mathbf{P}_i\|\|\mathbf{P}_{i(i+1)}\|\sin\beta_i}\left[\tan^{-1}\left(\frac{\|\mathbf{P}_{i(i+1)}\|+\|\mathbf{P}_i\|\cos\beta_i}{\|\mathbf{P}_i\|\sin\beta_i}\right) + \beta_i - \frac{\pi}{2}\right] \,,\end{aligned}$$

where β_i is the angle between $\mathbf{P}_i$ and $\mathbf{P}_{i(i+1)}$.

At this point, some simplifications are in order. First,

$$\begin{aligned}\sin\beta_i &= \frac{\|\mathbf{P}_i \times \mathbf{P}_{i(i+1)}\|}{\|\mathbf{P}_i\|\|\mathbf{P}_{i(i+1)}\|} \\ \cos\beta_i &= \frac{\mathbf{P}_i\cdot\mathbf{P}_{i(i+1)}}{\|\mathbf{P}_i\|\|\mathbf{P}_{i(i+1)}\|} \,.\end{aligned}$$

This greatly simplifies the argument of $\tan^{-1}$:

$$\begin{aligned}\frac{\|\mathbf{P}_{i(i+1)}\| + \|\mathbf{P}_i\|\cos\beta_i}{\|\mathbf{P}_i\|\sin\beta_i} &= \frac{(\mathbf{P}_{i(i+1)} + \mathbf{P}_i)\cdot\mathbf{P}_{i(i+1)}}{\|\mathbf{P}_i \times \mathbf{P}_{i(i+1)}\|} \\ &= \frac{\mathbf{P}_{i+1}\cdot\mathbf{P}_{i(i+1)}}{\|\mathbf{P}_i \times \mathbf{P}_{i+1}\|} \,.\end{aligned}$$

Also,

$$\begin{aligned}\frac{\mathbf{P}_i \times \mathbf{P}_{i(i+1)}\big|_Z}{\|\mathbf{P}_i\|\|\mathbf{P}_{i(i+1)}\|\sin\beta_i} &= \frac{\mathbf{P}_i \times \mathbf{P}_{i(i+1)}\big|_Z}{\|\mathbf{P}_i \times \mathbf{P}_{i(i+1)}\|} \\ &= \frac{\mathbf{P}_i \times \mathbf{P}_{i+1}\big|_Z}{\|\mathbf{P}_i \times \mathbf{P}_{i+1}\|} \,,\end{aligned}$$

where $|_Z$ indicates the Z-component of the vector. Thus the configuration factor between the differential element dA and the closed contour C (in closed form) is:

$$\begin{aligned}F_{dA-C} = \frac{1}{2\pi}\Bigg|\sum_{i=1}^{N} &\frac{\mathbf{P}_i \times \mathbf{P}_{i+1}|_Z}{\|\mathbf{P}_i \times \mathbf{P}_{i+1}\|} \\ &\cdot\left[\tan^{-1}\left(\frac{\mathbf{P}_{i+1}\cdot\mathbf{P}_{i(i+1)}}{\|\mathbf{P}_i \times \mathbf{P}_{i+1}\|}\right) - \sin^{-1}\left(\frac{\mathbf{P}_i\cdot\mathbf{P}_{i(i+1)}}{\|\mathbf{P}_i\|\|\mathbf{P}_{i(i+1)}\|}\right)\right]\Bigg| \,,\end{aligned} \qquad (5)$$

where the vectors involved are defined by Equations 3 and 4. The absolute value is used to allow the contour to be traversed in either a positive or negative direction.

EXAMPLE

Figure 5 contains the microcomputer solution's graphic output for various orientations of a tetrahedron. The figure illustrates the ability of the program to determine which facets contribute to the configuration factor. This ability allows for the desired outer contour to be defined, thus allowing application of the closed form solution (Equation 5). To illustrate the use of the closed form solution, case iv from Figure 5 will be presented as an example.

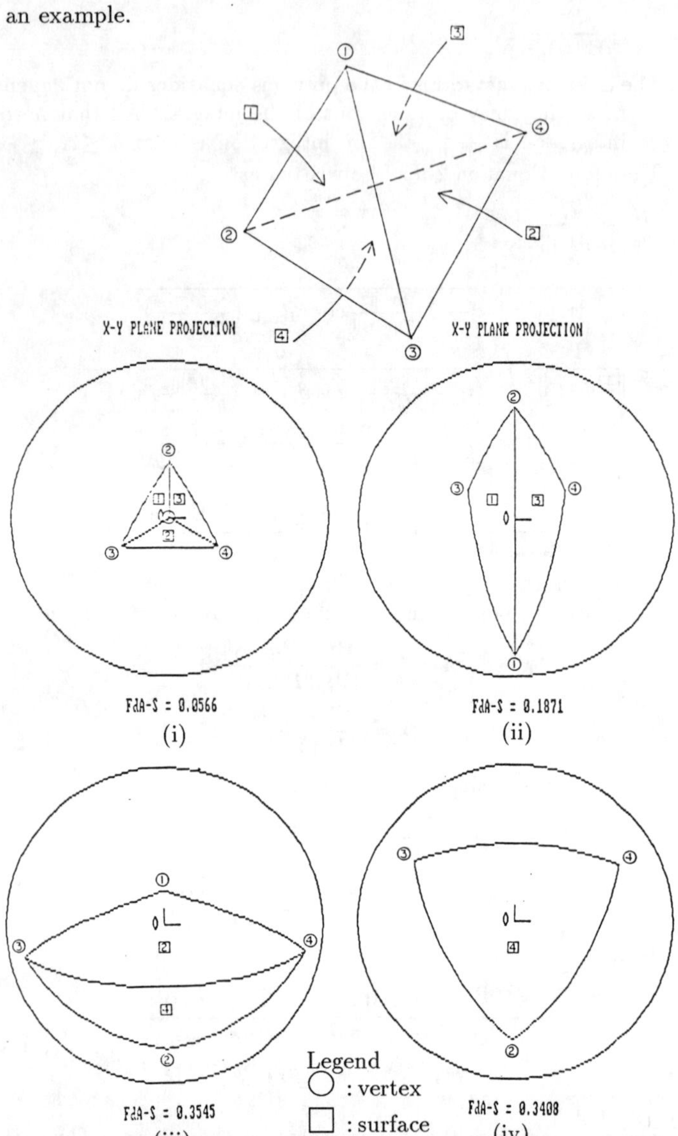

Fig. 5. Microcomputer Solution for a Tetrahedron

The contour vertices for case iv are defined by:

$$\mathbf{P}_1 = \mathbf{P}_4 = (0, -2/\sqrt{3}, 1) \quad \mathbf{P}_2 = (-1, 1/\sqrt{3}, 1) \quad \mathbf{P}_3 = (1, 1/\sqrt{3}, 1)$$

$$\|\mathbf{P}_1\| = \|\mathbf{P}_2\| = \|\mathbf{P}_3\| = \|\mathbf{P}_4\| = \sqrt{7/3}$$

The terms necessary for the use of Equation 5 are:

$$\mathbf{P}_1 \times \mathbf{P}_2 = (-\sqrt{3}, -1, -2/\sqrt{3}) \qquad \mathbf{P}_2 \times \mathbf{P}_3 = (0, 2, -2/\sqrt{3})$$

$$\mathbf{P}_3 \times \mathbf{P}_4 = (\sqrt{3}, -1, -2/\sqrt{3})$$

$$\|\mathbf{P}_1 \times \mathbf{P}_2\| = \|\mathbf{P}_2 \times \mathbf{P}_3\| = \|\mathbf{P}_3 \times \mathbf{P}_4\| = 4/\sqrt{3}$$

$$\mathbf{P}_{12} = (-1, \sqrt{3}, 0) \qquad \mathbf{P}_{23} = (2, 0, 0) \qquad \mathbf{P}_{34} = (-1, -\sqrt{3}, 0)$$

$$\|\mathbf{P}_{12}\| = \|\mathbf{P}_{23}\| = \|\mathbf{P}_{34}\| = 2$$

$$\mathbf{P}_1 \cdot \mathbf{P}_{12} = \mathbf{P}_2 \cdot \mathbf{P}_{23} = \mathbf{P}_3 \cdot \mathbf{P}_{34} = -2$$

$$\mathbf{P}_2 \cdot \mathbf{P}_{12} = \mathbf{P}_3 \cdot \mathbf{P}_{23} = \mathbf{P}_4 \cdot \mathbf{P}_{34} = 2$$

For this example, all terms in the summation are identical. Calculating the i=1 term, we find:

$$\frac{\mathbf{P}_1 \times \mathbf{P}_2|_Z}{\|\mathbf{P}_1 \times \mathbf{P}_2\|}\left[\tan^{-1}\left(\frac{\mathbf{P}_2 \cdot \mathbf{P}_{12}}{\|\mathbf{P}_1 \times \mathbf{P}_2\|}\right) - \sin^{-1}\left(\frac{\mathbf{P}_1 \cdot \mathbf{P}_{12}}{\|\mathbf{P}_1\|\|\mathbf{P}_{12}\|}\right)\right]$$

$$= \frac{-2/\sqrt{3}}{4/\sqrt{3}}\left[\tan^{-1}\left(\frac{2}{4/\sqrt{3}}\right) - \sin^{-1}\left(\frac{-2}{\sqrt{7/3}\,2}\right)\right]$$

$$= -\frac{1}{2}\left[.7137 - (-.7137)\right] = -.7137\ .$$

Therefore, the desired configuration factor is:

$$F_{dA-C} = \frac{1}{2\pi}\left|\sum\right| = \frac{3}{2\pi}(.7137) = .3408\ .$$

This value can be verified with a closed form solution for a parallel polygon centered above a differential element (Howell, 1982, p. 31):

$$F_{dA-C} = \frac{nH}{\pi\sqrt{1+H^2}} \tan^{-1}\sqrt{\frac{R^2 - H^2}{1+H^2}}, \qquad H = \frac{h}{\ell},\ R = \frac{r}{\ell}$$

where n is the number of vertices, r is the radius from the center to a vertex, h is the perpendicular distance between the center and a side, and ℓ is the height above the differential element. For the triangle above, $n = 3$, $r = \frac{2}{\sqrt{3}}$, $h = \frac{1}{\sqrt{3}}$, $l = 1$, and

$$F_{dA-C} = \frac{3}{2\pi}\tan^{-1}\left(\frac{\sqrt{3}}{2}\right) = .3408$$

which agrees identically with Equation 5.

CONCLUSIONS

A closed form solution was developed for the problem of finding the radiation configuration factor between an elemental area and a general 3-D contour. The contour referred to is defined as a closed, segmented curve which could be the perimeter of a surface, or the visible outer contour of a 3-D object. The method is therefore very general.

A microcomputer program was written to apply the closed form solution to the case of a general 3-D object. The algorithm requires that the solid object be represented in wireframe form (vertex, edge, and surface information). The program's output consists of the desired configuration factor, and graphical output relating to Nusselt's interpretation of the problem.

As with most fundamental work with radiation configuration factors, the results apply only to diffuse radiation traveling through a non-absorbing medium.

REFERENCES

Alciatore, D., Shamburger, E. S., and Janna, W. S., 1988, "Microcomputer Solution to the Mathematical Equivalent of the Graphical Method for Finding Radiation Shape Factors," *The Engineering Design Graphics Journal*, Vol. 52, No. 1.

de Bastos, R., 1961, "Computation of Radiation Configuration Factors by Contour Integration," M.S. thesis, Oklahoma State University, Stillwater.

Eckert, E. R. G. and Drake, R. M., 1959, *Heat and Mass Transfer*, Mc-Graw-Hill Book Co., Inc., New York.

Farrel, R., 1976, "Determination of Configuration Factors of Irregular Shape," *Journal of Heat Transfer*, Vol. 98, No. 2, pp. 311-313.

Hamilton, D. C. and Morgan, W. R., 1952, *Radiant Interchange Configuration Factors*, NACA TN 2836.

Hooper, F. C. and Juhasz, E. S., 1952, "Graphical Evaluation of Radiation Interchange Factor," ASME Paper 52-F-19, Presented at the ASME Fall Meeting in Chicago.

Howell, J. R., 1982, *A Catalog of Radiation Configuration Factors*, McGraw-Hill Book Co., Inc., New York.

Moon, P., 1936, *Scientific Basis of Illumination Engineering*, McGraw-Hill Book Co., Inc., New York. Republished by Dover Publications Inc., New York, 1961.

Nusselt, W., 1928, "Graphische Bestimmung des Winkelverhaltnisses bei der Wärmestrahlung," VDI Z., Vol. 72, p. 673.

Siegel, R., and Howell, J. R., *Thermal Radiation Heat Transfer*, McGraw-Hill Book Co., 1972.

Sparrow, E. M., 1963, "A New and Simpler Formulation for Radiative Angle Factors," *Journal of Heat Transfer*, Vol. 85, Series C, No. 2.

1989 National Heat Transfer Conference
HTD-Vol. 106, Heat Transfer Phenomena in Radiation, Combustion and Fires

THE INTERACTION OF SPECULAR RADIATION AND THERMAL FINITE ELEMENT CHARACTERISTICS

A. F. Emery and M. Lobo
Department of Mechanical Engineering
University of Washington
Seattle, Washington

ABSTRACT

Thermal analyses of complex structures exposed to high radiant fluxes have traditionally been based on control volume or finite difference approaches. Recently, the widespread use of finite element methods in thermal stress problems has resulted in the introduction of radiative exchange capabilities into finite element codes. Unfortunately, typical isoparametric elements may give rise to serious errors under certain conditions. This paper investigates the effects of radiative fluxes on finite element temperature predictions and isolates the parameters which control the extent of the error in the predicted solution. Variations of the standard isoparametric element are examined and the qualities of an optimal element are presented. In the presence of specular radiation the range of applicability of current finite element codes is even more limited due to the possibilty of locally intense radiative flux. In addition, the problem of computing the radiation is compounded. The best available specular method, the Monte Carlo, is too slow to be used economically. In this paper we present an accelerator, based on tree searching and oct-cell constructions, which increases the speed of Monte Carlo method to the point where it is a viable approach.

NOMENCLATURE

Roman

h	Heat transfer coefficient, $W/(m^2 -^\circ K)$
h_e	Effective surface transfer coefficient, $W/(m^2 -^\circ K)$
K	Number of degrees of freedom per element
k	Thermal conductivity, $W/(m-K)$
L	Length of bar or plate, m
l_e	Element length, m
m	Ratio $hL^2/k\delta$
N	Number of elements
N_i	Basis function
q	Incident heat flux, W/m^2
q_{conv}	Convective incident heat flux W/m^2
q_{rad}	Radiative incident heat flux W/m^2
q_{tot}	Total incident heat flux W/m^2
R	Region/ Area, m^2
∂R	Boundary of a region
s	Ratio $\sqrt{h/k\delta}$
T	Temperature, $^\circ K$
T_∞	Ambient temperature, $^\circ K$
t	Time, sec
x, y	Space coordinate, m

Greek

$\aleph_i$	Basis functions for lagrangian variables
δ	Bar width or plate thickness, m
ϵ	Emissivity
ϕ_i	Nodeless lagrangian variables
σ	Boltzman's constant, $W/(m^2 -^\circ K^4)$
ξ	Natural coordinate (local to element)

1. INTRODUCTION

Thermal and particularly thermal-mechanical analyses of space structures are frequently carried out using finite element methods. Most of these programs based upon the FEM employ a library of basic elements, viz., triangular and quadrilateral two-dimensional elements, tetrahedral and brick three-dimensional elements, and special elements for one-dimensional conduction (Marlowe et. al., 1988). Beckett and Chu (1973) used the finite element method to solve the transient heat conduction problem of a plane slab with non-linear radiative-convective boundary conditions. An implicit iteration technique was used to converge to the solution at each time step. Lee and Jackson (1976) developed a finite element formulation to solve combined radiative-conductive problems in two dimensions. Their formulation included the capability to handle both specular and diffuse surfaces, but was restricted to triangular elements. Although some libraries contain higher order elements, most use linear iso-parametric ones and depend upon high element densities to achieve satisfactory accuracy of solution. Because convective and radiative heating can occur only on the exposed faces of the elements, it is necessary to identify such surfaces; this means carrying along extra information about each face of each element. Since the number of exposed faces is usually small in relation to the total number, this imposes undesirable increases in required memory and logic. To mitigate these effects and to provide for more flexibility, it is common to define special surface elements which represent surface heat transfer loadings and whose nodes are coincident with the nodes which define a face of an underlying three dimensional element as shown in Figure 1. It is also common to expand the capabilities of these special elements by giving them a thickness and permitting conduction in the lateral (x,y) directions. In this way, these elements can be used to model thin or shell like structures.

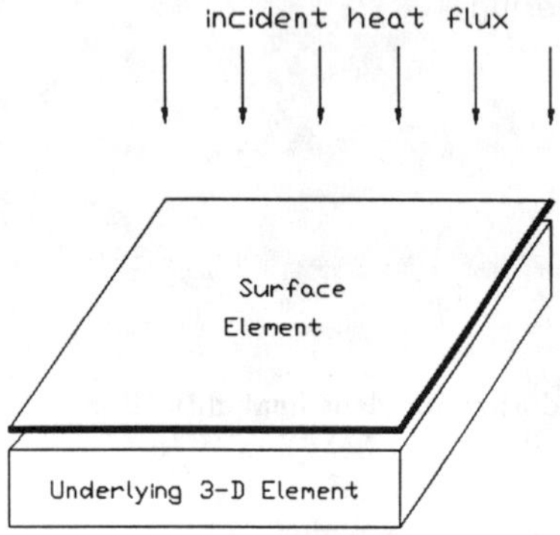

Figure 1. Surface Element Overlayed on a 3-D Brick Element

In most cases these elements perform well. However in the presence of high convective fluxes and particularly high radiative fluxes, anomalous results have been noted (Robinson, 1988). These anomalies have been attributed to the inability of the linear elements to model adequately the high thermal gradients associated with the high surface fluxes. Because the usual cure of increasing the element density results in prohibitive computing costs, it has been suggested that higher order surface elements might be overlayed on the usual linear three-dimensional elements. In addition to the problem of merging such dissimilar elements, they do not seem to have produced the desired cure.

The following sections describe the characteristics of such one and two-dimensional higher order elements and discuss the conditions under which they can be used. It is found that there is an explicit limit to the size of these elements if realistic results are to be produced. Enforcement of this limit increases the number of surface elements and results in severe limitations on the usual radiative load calculations. The final section describes one method of accelerating the radiative calculations to improve the situation.

2. CHARACTERISTICS OF THE PROBLEM

Consider the problem illustrated in Figure 2. It shows a thin plate which is heated on its top surface by a combined convective and radiative heat flux. The heat incident on the top surface is conducted in the x and y directions to the two edges which are held at prescribed temperatures.

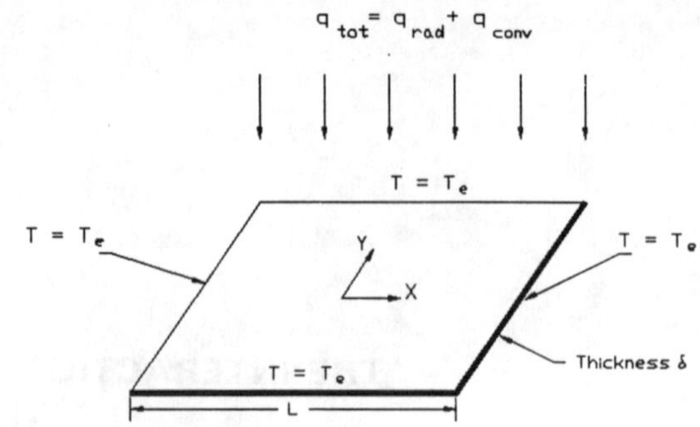

Figure 2. Schematic of the problem

An analytical solution to the problem is available if the incident heat flux is purely convective with a constant heat transfer coefficient. Figure 3a shows a characteristic temperature profile when the heat transfer coefficient is high relative to the thermal conductivity. The temperature is essentially constant over the central portion of the plate, but drops sharply at the two ends. Thus the heat fluxes are extremely high in the vicinity of the ends and small in the central region.

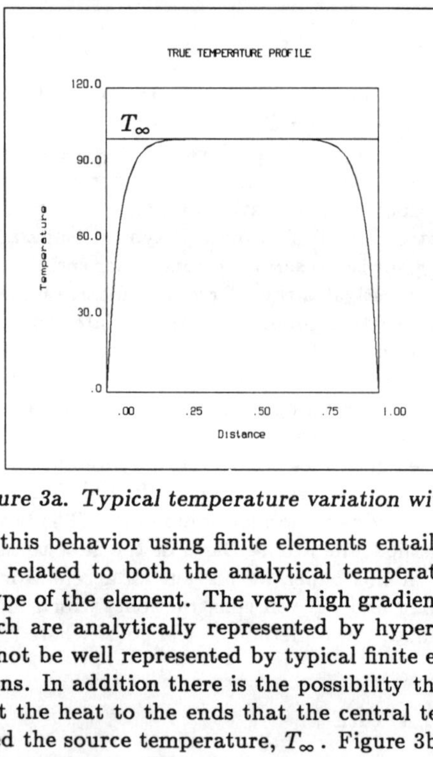

Figure 3a. Typical temperature variation with X

Modeling this behavior using finite elements entails problems which are related to both the analytical temperature profile and the type of the element. The very high gradients near the ends, which are analytically represented by hyperbolic functions, cannot be well represented by typical finite element basis functions. In addition there is the possibility that in order to conduct the heat to the ends that the central temperature may exceed the source temperature, T_∞. Figure 3b shows the results computed using linear isoparametric elements.

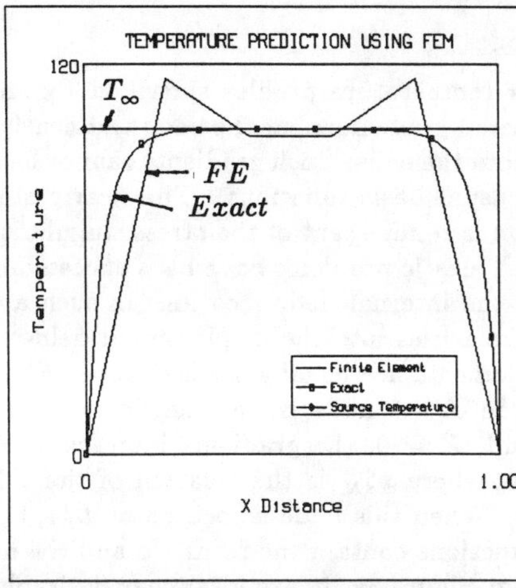

Figure 3b. Finite Element simulation of the problem of Figure 3a

The question is how to eliminate this serious error. Several approaches have been tried: higher order elements, shifted nodes to introduce singular like behavior of the heat flux; Lagrangian degrees of freedom. We have found that higher order elements work under restrictive conditions, but in general do not eliminate the problem of overprediction depicted in Figure 3b. Shifted nodes, which introduce singular basis functions to the element, improve the results, but the degree of shifting is limited. Lagrangian degrees of freedom can materially improve the computations. Unfortunately, the region of applicability of the different methods depends upon the effective surface convection, the thermal conductivity and the dimensions of the element. These characteristics are described in the following section for different types of surface heating.

If the radiation is specular, the possibility of locally intense radiative flux is increased. Elements which work satisfactorily with diffuse radiation frequently fail in the presence of specular radiation. In addition to the problem of computing the temperature, we have the problem of efficiently computing the radiation. A recent survey paper by Emery (1988a), gave a number of examples of the difficulty of computing diffuse radiation exchange factors. Even the best of methods, when applied to specular radiation, falls far short of the desired performance.

Currently, the best specular method is the Monte Carlo, but it is too slow to be considered seriously. Consequently, an accelerator has to be developed. Different possibilities for accelerators exist, each with a unique mode of operation. However, the underlying principle behind all accelerators is a subdivision of the entire domain into smaller subdomains. This procedure is called clipping. A more detailed description of various clipping operations is given by Emery (1988b). In Section 5 an accelerator which performs clipping by dividing the entire domain into cubical cells arranged as a tree structure is described. This procedure is referred to as the oct-cell method in this paper and is based on the octree representation of surfaces or objects in three dimensional space (Wyvill et al., 1986).

3. ANALYSIS OF THERMAL FINITE ELEMENT CHARACTERISTICS

Consider a region R with a bounding surface ∂R. The equation of conservation of energy is given by

$$k\nabla^2 T + Q(T, T_\infty) + q = 0 \tag{1}$$

where Q represents a temperature dependent heat source and q is a constant heat source. The boundary condition on ∂R is assumed to be a Dirichlet condition

$$T = g \quad on \quad \partial R$$

For a surface receiving radiation and convection from a source at T_∞

$$Q(T, T_\infty) = \sigma\epsilon(T_\infty^4 - T^4) + h(T_\infty - T) \tag{2}$$

A finite element simulation of (1) with Q given by (2) is difficult because of the presence of the non-linear radiation term. In order to develop a useful variational formulation equation (2) was linearized by defining an effective surface coefficient

$$h_e = \sigma\epsilon(T^2 + T_\infty^2)(T + T_\infty) + h \tag{3}$$

Using this effective surface coefficient in a variational formulation, the integral to be minimised is given by

$$I(T) = \int\int_R \frac{k}{2}\Big[\Big(\frac{\partial T}{\partial x}\Big)^2 + \Big(\frac{\partial T}{\partial y}\Big)^2 + \frac{h_e}{k}(T^2 - TT_\infty) - \frac{2qT}{k}\Big] dxdy \tag{4}$$

Since h_e is now a function of both the source and surface temperatures, an iterative procedure is required to converge to the final solution.

In order to develop an understanding of the nature of the problem, one-dimensional flow was considered. The heat incident on the top surface of the plate is conducted away towards the ends, which are maintained at T_e

To investigate the influences of various parameters on the finite element solutions, three different approaches were followed

i) Use higher order elements
ii) Use lagrangian variables
iii) Permit shifting of the interior nodes.

To make a comprehensive study of each approach, the following forms of incident heat were considered:

a) a prescribed spatially constant heat flux q
b) convective heat from a source at T_∞ and
c) radiative heat from a source at T_∞.

Each case will be described separately.

3.1 Higher Order Basis functions

Using a traditional finite element approach, the temperature profile in each element is expressed in terms of ξ by

$$X(\xi) = \sum_{k=1}^{K} X_k N_k(\xi), \qquad T(\xi) = \sum_{k=1}^{K} T_k N_k(\xi) \qquad K \le 5$$
$$-1 \le \xi \le 1 \tag{5}$$

where $N_k(\xi)$ are the isoparametric basis functions which satisfy

$$N_k(\xi_i) = \begin{cases} 0, & i \neq k; \\ 1, & i = k \end{cases}$$

and X_k and T_k are the locations and temperatures of the temperatures of the kth node. Generally K is limited to 3, but we used $K \leq 5$ with equally spaced nodes to study their features. Because all nodal temperatures except for $k = 1$ and K are internal to the element, their associated band widths are small and the increments of computational cost are significantly less than if an equal number of nodes were used with simpler elements.

When the true temperature profile is piecewise quadratic, finite element predictions using $K \geq 3$ are excellent; even two-noded linear elements give good results. This merely serves to confirm the belief that simple temperature distributions can be very accurately simulated by finite elements, whose basis functions possess the same character. Other distributions are usually treated by using more elements so that each element need model only a limited range of the temperature profile in much the same way a cubic spline is used. Unfortunately, if the temperature profile possesses narrow zones of high variation, i.e. large values of $\frac{d^nT}{dx^n}$, $n > K$, this approach is only a crude approximation.

3.2 Lagrangian Variable Approach

An alternative to the use of many simple or few complex isoparametric elements, where higher order basis functions are directly associated with an increased number of nodes per element, Lagrangian (nodeless) variables ϕ_i may be introduced to model more complex temperature distributions. The temperature distribution in the element is expressed as:

$$T(\xi) = \sum_{i=1}^{K} T_i N_i(\xi) + \sum_{i=1}^{J} \aleph_i(\xi)\phi_i \tag{6a}$$

where

$$\aleph_i(\xi) = 0 \tag{6b}$$

on the element edges; thus these Lagrangian degrees of freedom affect the inter-element distribution without affecting the intra-element continuity of temperature. However, they do affect the inter-element fluxes. As an example the two node element may be augmented by using the basis functions ($K = 2, J \leq 3$)

$$N_1(\xi) = \frac{1-\xi}{2} \qquad N_2(\xi) = \frac{1+\xi}{2}$$

$$\aleph_1(\xi) = 1 - \xi^2 \qquad \aleph_2(\xi) = \xi(1-\xi^2)$$

$$\aleph_3(\xi) = 1 - \xi^4 \tag{6c}$$

The specific form of these extra functions is quite arbitrary, subject only to Equation 6b, and the degree of satisfaction is dependent upon their success in matching the exact temperature profile. Some care must be exercised in their choice. Because these additional functions are internal to a single element, they contribute to the calculated temperature gradient, but this gradient is not continuous between elements. Thus it is possible to introduce significant discontinuities in inter-element gradients which degrade the overall accuracy of the solution.

3.3 Shifted Nodes

The high flux temperature profiles shown in Figure 3a have very high thermal gradients near the ends and nearly zero gradients away from the ends. Such gradients cannot be simulated by using the usual basis functions. The nearly singular nature of the flux is reminiscent of the stress singularities near a crack tip. Such elastic problems have been successfully treated by modifying the internal node locations in such a way as to introduce singularities into the displacement fields (Barsoum, 1976); such elements are usually termed Super Elements. As the center node of a 3 node isoparametric element is shifted toward the end, $X = 0$, the gradients increase in the region, $0 \leq x \leq x_{SN}$, where x_{SN} is the location of the shifted mid-element node. When this node is located at $L/4$, the temperature basis functions contain the term $\sqrt{x}$ and the heat flux is singular. From Figure 3a, it would appear desirable to introduce a profile of the form $T = x^{1/n}$, with $n \approx 7 - 10$, giving the best results. Unfortunately shifting the mid- node to the left of $L/4$ gives a non-unique mapping. Thus we are limited to a square root singularity in the heat flux.

4. *EXAMPLE CASES*

Let us examine the behavior of the different improvements with respect to the three types of heat fluxes.

4.1 Constant Flux

When the heat flux is constant, the temperature profile is given by

$$T(x) = -\frac{qx^2}{2k} + \left(\frac{qL}{2k} + \frac{T_2 - T_1}{L}\right)x + T_1 \tag{7}$$

Because the heat flux is well behaved, the shifted node element is not needed. Both the higher order elements and the Lagrangian variables perform very well. No real difference can be detected between any of the Lagrangian elements as long as $\aleph_1$ is included. Although the profile is quadratic, even the linear element, $K = 2$, gives very good results as long as the number of elements is 5 or greater.

4.2 Convective Flux

When the flux is convective, with a constant heat transfer coefficient, the temperature is given by

$$\frac{T(x) - T_\infty}{T_1 - T_\infty} = \frac{1}{sinh(sL)}\left[\frac{T_2 - T_\infty}{T_1 - T_\infty} sinh(sx) + sinh(s(L-x))\right]$$
$$s^2 = \frac{h}{k\delta} \tag{8}$$

Linear elements give the poor results shown in Figure 3b. Higher order isoparametric elements gave results of the form illustrated in Figures 4. In general sharp spikes are produced at the node adjacent to the end nodes. These diminish slightly as the distance from the end is increased, but more oscillations are introduced near the center of the element. If the source temperature is higher than the specified boundary conditions, the finite element solution tends to approach the true solution from below when the highest order basis function is even, and from above when it is odd. Regardless of the nature of the basis functions, all higher order elements predict temperatures in excess of the source temperature at some point within the region.

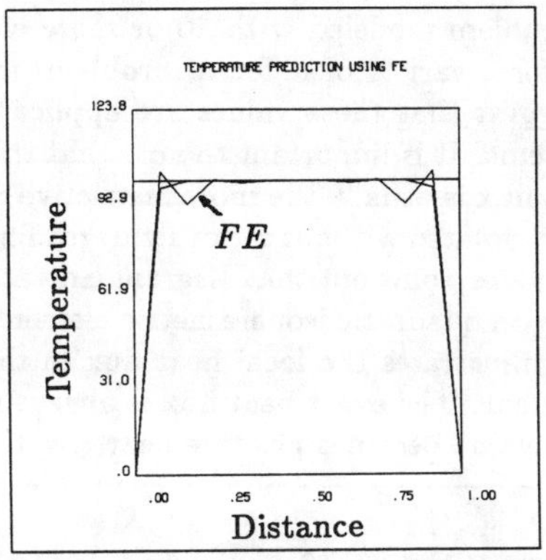

Figure 4a. Temperature profile (six 3-noded elements)

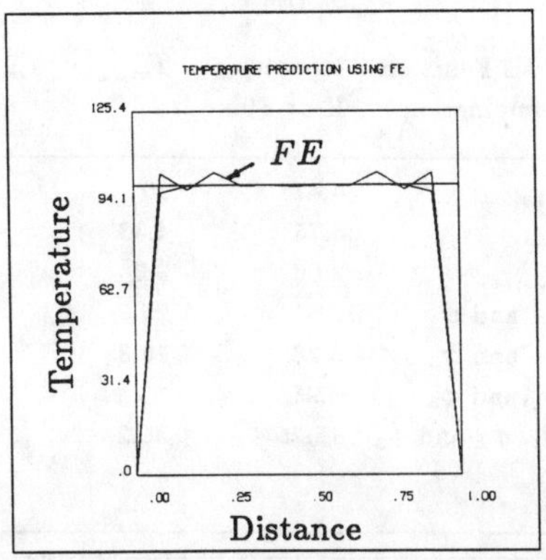

Figure 4b. Temperature profile (four 4-noded elements)

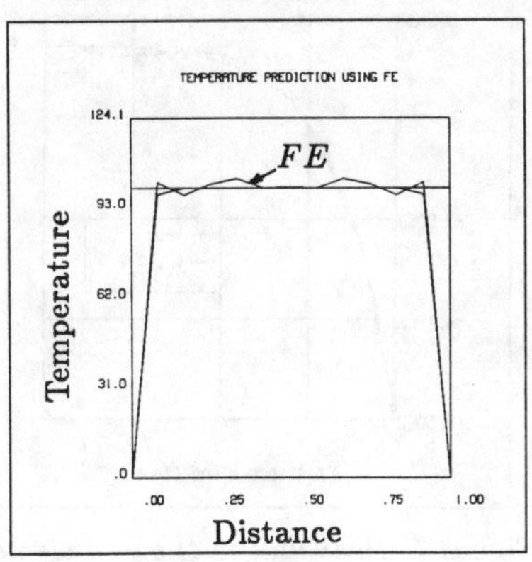

Figure 4c. Temperature profile (three 5-noded elements)

Lagrangian elements tend to be more conservative. Elements whose Lagrangian basis functions are even again approach from below and those whose are odd approach from above. The odd functions show an overshoot in the nodal temperatures, and then only near the ends. However, evaluation of the temperature profile within the element displays results very similar to those achieved with the higher order elements.

Figure 5 shows the results obtained with the shifted mid-node element. As shown, the agreement improves as the shifting is increased, but because the shifting is limited, the agreement is never as satisfactory as desired. In particular, as the node is shifted the gradients at $x = 0$ are overestimated, but if the gradients are based on nodal values only they are underestimated.

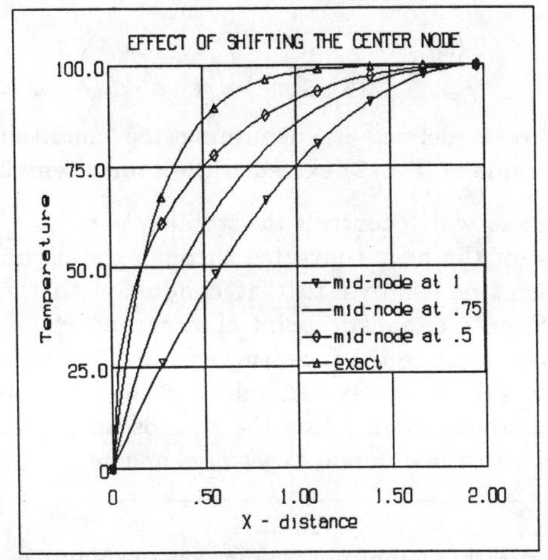

Figure 5. Effect of shifting the center node of a quadratic element

All of the methods significantly underestimate the flux at the ends and overestimate it in the center. In all cases, increasing the number of elements, i.e., decreasing their size, reduced the error, but the reduction was not directly proportional to the size of the element, rather to the size raised to a power less than one.

4.3 Radiative Flux

The only real difference between the convective and radiative flux cases is the variation of the effective heat transfer coefficient over the surface of the region. All methods predicted qualitatively comparable temperature profiles and there was no substantial difference between the convective and radiative results except that the errors, particularly near the prescribed boundary temperatures, were larger. For a boundary temperature of $0° C$ and a source temperature of $3000° C$, the errors were of the order of $500° C$, an unacceptable amount. However, the comparable performance of the constant heat transfer coefficient for convection and the variable coefficient for radiation suggest that information garnered from the convective case can be applied to the radiative case.

5. STABILITY RESTRICTIONS

The examples of the preceeding section clearly indicate that the predicted temperatures will often exceed the source temperature. This phenomenon is in violation of thermodynamics and is therefore unacceptable. We define an element to be "unstable" when the predicted temperatures are physically impossible. The problem is to determine a criterion which will ensure that the element does not become unstable. Following the approach used to define "stability" of explicit finite difference schemes, we investigated the simple problem of convective heating of a one-dimensional bar of length L, thickness δ, thermal conductivity k, heat transfer coefficient h, temperature at the ends of the bar of zero, and source temperature T_∞. The bar was represented by two elements whose basis functions were either linear, quadratic, or quartic. In addition, the bar was also represented by a finite volume differencing scheme using a linear or quadratic temperature distribution. The criterion for stability was that no nodal temperature could exceed the source temperature. For the quadratic element, the temperature solution was given by

$$\frac{T(x)}{T_\infty} = \frac{5m^2}{2(5+2m^2)}\left(2\frac{x}{l} - \left(\frac{x}{l}\right)^2\right) \tag{9}$$

and stability is defined by determing the conditions under which the value of T(L/2) exceeded the source temperature.

The parameter which controls the stability is $m^2 = \frac{hl^2}{k\delta}$, which is the ratio of the heat convected through the lateral surface of an element of length l to that conducted to the ends of the bar. From a simplistic point of view, $m^2 < 1$ for stability. However, because the true temperature distribution varies along the length, m can exceed unity, without the solution becoming unstable. Table 1 lists the bounds on the value of m associated with the different types of elements

TABLE 1
Analytically Computed Bounds on the Value of m

Element Type	m
i) Linear finite volume, FD(1)	$\sqrt{2}$
ii) Linear Finite element, FE(1)	$\sqrt{6}$
iii) Quadratic finite volume, FD(2)	$\sqrt{6}$
iv) Quadratic finite element, FE(2)	$\sqrt{10}$
v) Quadratic + quartic Lagrangian element, FE(4)	$\sqrt{12.2}$

Figure 6 illustrates the effect of changing the value of m upon the maximum predicted temperature as compared to the exact value. It is clear from the tabulated values, that the Lagrangian elements are much to be preferred in that they can support a much longer element than can the simpler approaches.

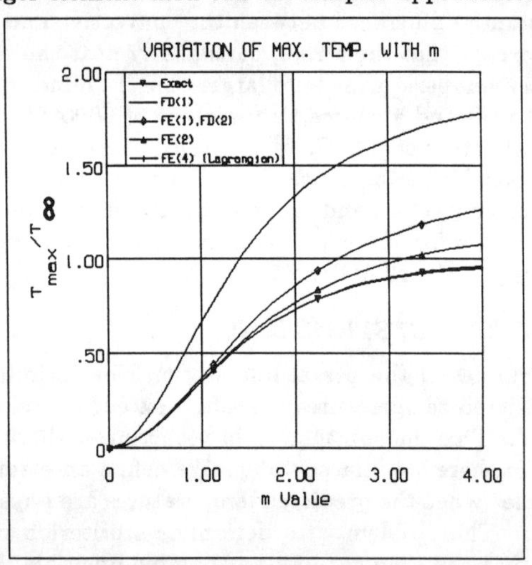

Figure 6. Maximum bar temperature as a function of m

The values listed in Table 1 ensure that no nodal value will exceed the source temperature. For higher order and shifted mid-node elements, the maximum nodal temperature represents the maximum temperature of the element. However in the Lagrangian element, the distribution is such that local maxima may exceed the nodal temperatures. In addition, the complex basis functions often yield heat fluxes which are of opposite sign to the average heat flux. Table 2 lists the maximum values of m to ensure a monotonic heat flux and stable elements for the bar problem modeled with 10 or more elements. Numerical tests for a variety of different problems and boundary conditions suggest that these values are applicable to a wide range of problems. It is important to note that the monotonicity of the heat flux is usually the more restrictive condition but can be slightly violated without seriously degrading the results. The tabular values point out that Lagrangian variables should only be used with quadratic isoparametric elements (linear and ϕ_1). Figure 7 illustrates the local heat flux in the element at the end of the bar. The exact heat flux is always negative, but the numerical value becomes positive near $\xi = 1$.

TABLE 2. Values of m^2 for various combinations of the ϕ_i

Basis Function Combination	Monotonic Heat Flux	$T_{max} < T_\infty$
ϕ_1	4.47	10
ϕ_2	5.75	4.33
ϕ_3	2.06	9.3
ϕ_1 and ϕ_2	21	25.9
ϕ_2 and ϕ_3	3.28	13.8
ϕ_1 and ϕ_3	3.95	12.2
ϕ_1, ϕ_2 and ϕ_3	15.25	30.2

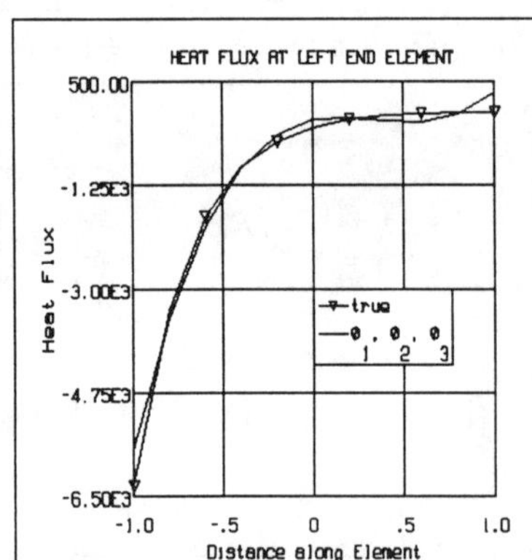

Figure 7. Heat flux in the end element

Although this criterion is simple to apply in principle, it poses severe restrictions on practical problems. Given a value of h, one can size the mesh to ensure stability. However, in a radiation problem, where surface temperatures are to be solved for, the value of h_e is not known. Thus one must construct the mesh based upon the maximum value of h_e, and this will often lead to unacceptably fine meshes.

6. TWO DIMENSIONAL PROBLEMS

The values listed in Table 1 apply only to One-Dimensional problems. Because of the more complex element basis functions, comparable results can only be obtained for two dimensional problems by solving a set of non-linear equations. Instead, a number of two-dimensional problems were examined. Figure 8 shows temperature contours which are typical of an "unstable" two- dimensional element. It was found that when the value of m^2 based upon the length of the element in the direction of the major heat flux satisfied the values listed in Tables 1 and 2, the element was stable.

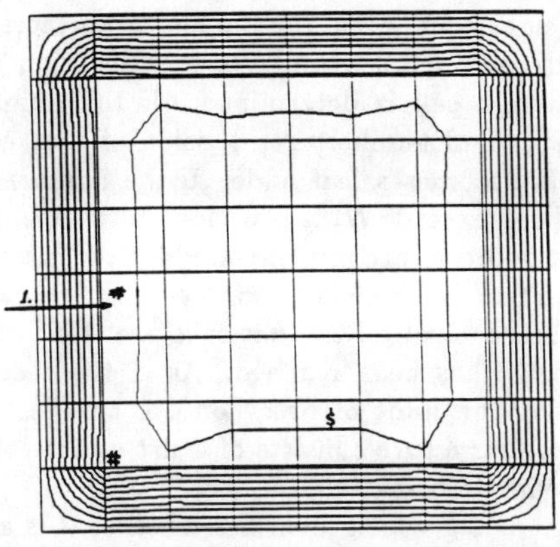

Figure 8. 2-D Contours

7. *SPECULAR CALCULATIONS*

Specular radiation is characterized by sharp changes in radiation intensity. An approximate representation of the effects of specular radiation can be obtained by subjecting a bar to a constant flux along part of its length. Thus a part of an element may be exposed to substantial irradiation while the rest is unheated. Within the context of finite elements, the source term associated with such heating is given by

$$\int_{R_i} q\left[\sum_k T_k N_k(\xi)\right] d\xi \qquad (10)$$

where R_i denotes the area which is irradiated. Because the basis functions span the entire element, even if a node is not irradiated, there will be a contribution to its source term. This is in contrast to the usual finite volume or difference approach in which the source term will be zero. Figure 9 compares the resulting temperatures calculated for a bar which is exposed to a constant radiative flux over $0.333 < X < 1.0$ and which convects and radiates to a source temperature of $0°C$ and whose ends are $x = 0, x = L$ are insulated. Two different element sizes were tested. It is clear that accurate predictions require that the edge of the radiative flux be aligned with an element edge or that the element size be small enough to ensure that the flux is correctly estimated otherwise the profiles are unacceptably smeared. The smearing of the irradiated flux over an entire element has the effect of reducing the average intensity leading to underpredictions of the temperature. For realistic radiation fields, this implies that either a large number of surface elements or a problem specific mesh density will be necessary to model the irradiation. Such fine discretization will then lead to lengthy radiation calculations since these are proportional to the number of surface elements raised to a power greater than unity.

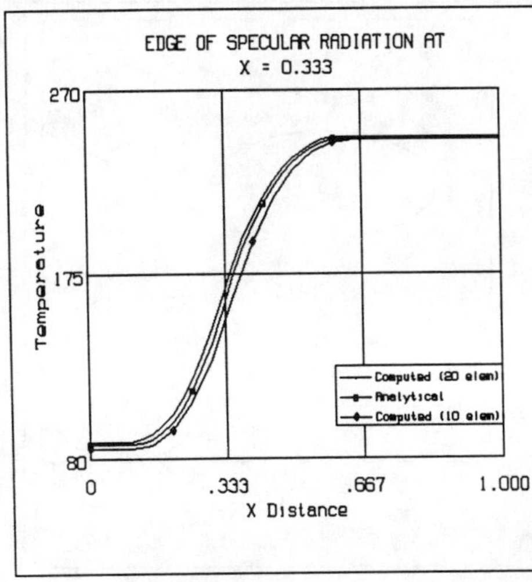

Figure 9. Finite Element and Finite Difference simulations of specular radiation

8. *A DESCRIPTION OF THE OCT-CELL METHOD*

A prerequisite for the evaluation of the radiation exchange between two surfaces is the viewfactor between them. As structures become increasingly complex, even the computation of diffuse radiation viewfactors becomes a formidable task. Unfortunately, most codes which can efficiently compute diffuse radiation viewfactors cannot be used for specular radiation. The Monte Carlo method, which is the best available method for specular radiation is too slow to be used productively. Various techniques have been developed to increase the speed of the Monte Carlo method in order to improve its use in numerical simulations of radiation problems. One of these approaches, the oct-cell method, is described here.

The oct-cell method is based on the octree representation of surfaces or objects in three dimensional space. In an octree representation, information about a collection of objects or surfaces in three dimensions is stored as a tree. The word octree is used because each branch of the tree has eight sub-branches. As a simplified illustration of the octree principle consider the two dimensional objects shown in Figure 10a. The objects are first enclosed by the smallest possible square that contains all of them. This square is called the master cell. The master cell is now divided into four equal squares which are the children of the master cell, Figure 10b. Each child cell is now examined to see if it is (i) empty, (ii) full or (iii) partially full. If the cell is empty or full, further division of that cell is terminated and the cell is flagged to identify whether it is full or empty. The cell is now called a leaf node. If the cell is only partially full it is divided further into four equal cells and the procedure followed for the first division is repeated, Figure 10b.

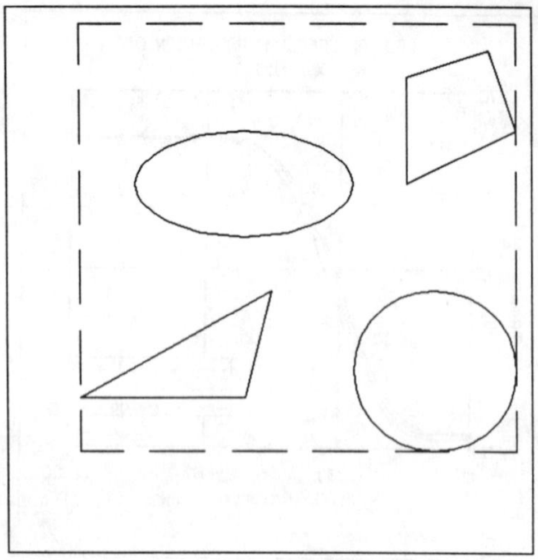

Figure 10a. Objects enclosed by Master Cell

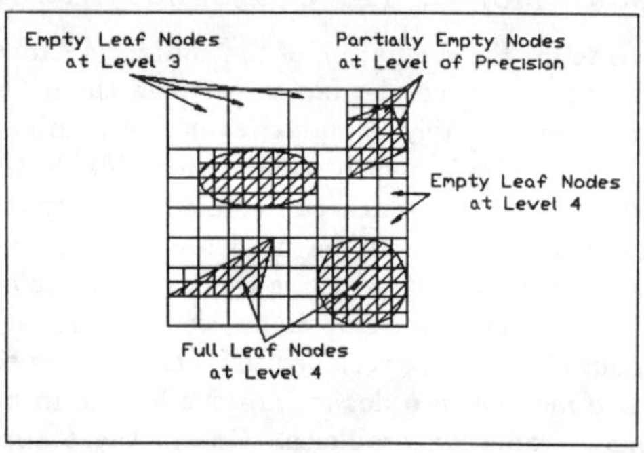

Figure 10b. Octree Division of Master Cell

This process of recursive cell division is continued until all cells are leaf nodes or until the level of precision has been reached. The level of precision is a pre-specified minimum cell size. Figure 10c illustrates how this process can be represented as a tree structure. Note that the cells at the level of precision are also leaves of the tree and are therefore called leaf nodes too. The generalization of this process to three dimensions results in an octree structure. Each cell now has to be divided into eight equal cells (corresponding to the eight octants in 3-D).

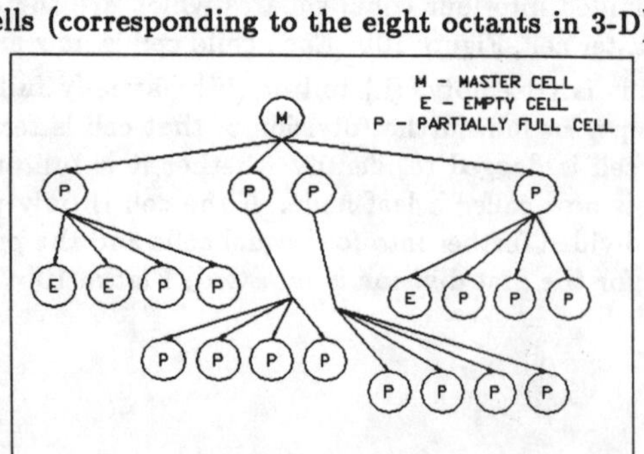

Figure 10c. Tree Representation of Division in Figures 10a,b

8.1 Creation of the Oct-cell Structure

The oct-cell structure is generated exactly like the octree. The difference between the octree and the oct-cell structure lies in their respective functions. An octree is used to represent surfaces or objects in three dimensional space. The oct-cell structure is itself an octree. However, the information stored in each oct-cell is different. After the master cell is divided into eight equal cells, the number of surfaces contained , either wholly or partially, and the identification of each surface in each child cell is determined. If the number is less than a pre-specified number, N, division of that cell is discontinued and it becomes a leaf node. If the number of surfaces in the child cell exceeds N, it is divided further into eight equal cells, and the procedure implemented for the master cell is repeated. This recursive division, exactly as in the case of an octree, is continued until all the oct-cells are leaf nodes or the limit of precision has been reached. As in the case of the octree the cells at the limit of precision are also leaf nodes. The final oct-cell structure consists of a list of oct-cells which store the following information:

1. A flag which identifies whether it is a leaf node.
2. Pointers to its children if it is not one of the above and a list of the surfaces it contains if it is.

8.2 Application to Monte Carlo Simulation

In Monte Carlo ray tracing rays, originating from a random point on a surface are checked to see if they are intercepted by any surface. In a traditional Monte Carlo simulation of a problem involving obstructors every single surface in the domain has to be checked. If the ray is intercepted by a surface, the distance between the point of origination of the ray and the point at which it intersects the surface has to be computed. The ray is then credited to the surface which has the shortest interception distance. For problems such as the ones shown later in Figures 11a-d the cost of such a procedure is exorbitant.

The division of the domain into oct-cells alleviates this problem by performing a crude sorting of the surfaces based on their distances from the point of origination of the ray. The leaf node in which the ray originates is first determined and only the surfaces contained in that particular oct-cell are checked to see if they intercept the ray. If the ray intersects one or more than one surface the shortest interception distance is computed and the ray is credited to that surface. If no intersection is found the ray is traced to the next leaf node and the surfaces in that oct-cell are checked for intersection. Once again the shortest intersection distance among all intersected surfaces in the current oct-cell is determined and the ray is credited to that surface. If no intersection is found the ray is traced to the next leaf node. Thus the ray is traced through the leaf nodes which constitute the master cell until it is intercepted by some surface or exits from the domain.

The objective of reducing the execution time is achieved because a much smaller number of surfaces needs to be processed. All surfaces not contained in the oct-cells through which the ray passes are eliminated. However a price must be paid in that the ray must be traced through the leaf nodes. But the oct-cells are generated so that their bounding planes are normal to the axes. This simplifies the ray tracing procedure and minimises the time required for ray tracing. Consequently the time for ray tracing is only a small fraction of the total execution time for problems with many obstructors.

9. RESULTS OF USING THE OCT-CELL ACCELERATOR

Testing of the application of the oct-cell code to both Monte Carlo ray tracing and Clipping was performed.

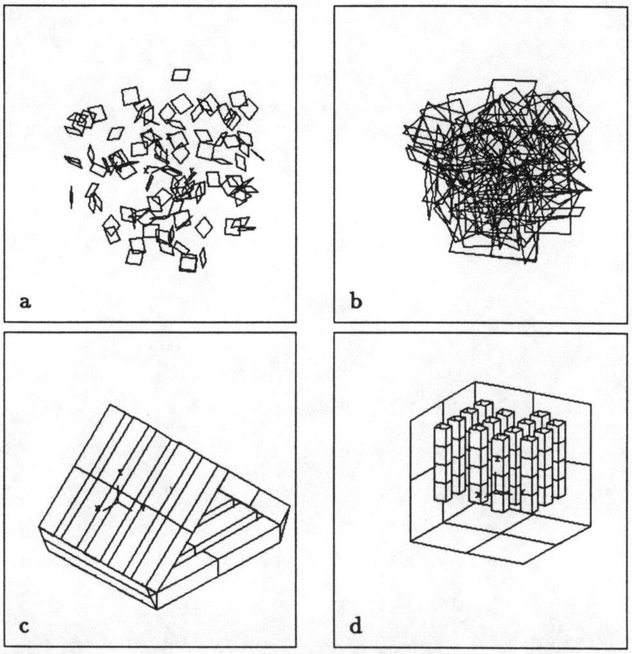

Figures 11. Test Cases

The Oct-cell creation and ray tracing codes were incorporated into RADSIM, a Monte Carlo routine supplied by R.C. Corlett (1988). Four different test cases were run. The four test cases are shown in Figures 11a-d. Figures 11a and 11b consist of 100 square surfaces randomly oriented and with their centroids randomly located in a cube with dimensions 100x100x100 units. In Figure 11a the squares measure 10 units on a side and in Figure 11b they are 40 units on a side. Figure 11c is a truss structure which is divided into 226 rectangular surfaces for analysis purposes. Figure 11d represents a furnace with 248 surfaces. The test cases were run with about 60 interactive users on a VAX 8700. Results of the test cases are shown in Table 3.

TABLE 3

Comparison of Execution Times for Monte Carlo with and without Oct-cell Division Using 10000 Rays

Structure	Execution Time Using Oct-cells (t_1) (sec)	Execution Time Using Standard Monte Carlo (t_2) (sec)	Ratio (t_2/t_1)
Fig 11a	10.53	27.16	2.58
Fig 11b	9.69	27.92	2.88
Fig 11c	21.60	59.39	2.75
Fig 11d	19.17	60.50	3.16

10. CONCLUSIONS

Special two dimensional surface elements provide a computationally useful technique for incorporating convective and radiative loads into finite element thermal analyzers. However care must be taken to ensure that the parameters $\frac{hl^2}{k\delta}$ not exceed critical values if physically realistic temperatures are to be computed. The same limitations are applicable to 3-D elements which are subject to high lateral conduction fluxes (In this case δ represents the effective thickness through which lateral conduction takes place). Based upon this criterion, high incident fluxes or heat transfer coefficients may require that the element be smaller in lateral size. Such small sizes in conjunction with the need to maintain reasonable element aspect ratios may lead to undesirable mesh density, if uniform meshes are used. Work still remains to be done to determine the optimal connection of these surface elements to the underlying 3-D elements. The high element density required to treat high resolution fluxes may lead to excessive computational times when determining specular fluxes. The Oct-cell method described can lead to significant cost savings and should be implemented in general purpose Monte Carlo radiation load programs.

References

1. Marlowe, M.B., Moore, R.A., Whetstone, W.D., "SPAR Thermal Analysis Processors Reference Manual, System Level 16", CR-159162, 1979, NASA.
2. Beckett, R.E., Chu, S.-C., "Finite Element Method Applied to Heat Conduction in Solids with Nonlinear Boundary Conditions," Journal of Heat Transfer, Vol. 95, Feb. 1973, pp 126-129.
3. Jackson, C.E. Jr., Lee, H.P., "Finite Element Solution for Radiative-Conductive Analyses with Mixed Diffuse-Specular Surfaces," AIAA Progress in Astronautics and Aeronautics: Radiative Transfer and Thermal Control, Editor A.M. Smith, Vol 49., pp 25-46, 1976.
4. Robinson, J., Private Communication, 1988.
5. Emery, A.F., et al., "A Comparative Study of Methods for Computing the Diffuse Radiation Viewfactors for Complex Structures", AIAA/ ASME/ ASCE/ ASHRAE 29th SDM Conference, April 18-20, 1988, Williamsburg, Virginia.
6. Emery, A.F., "VIEWC Users Manual," Version 5.6.6, November 1988, Seattle, Wash.
7. Wyvill, A., Kanii, T., Shirai, Y, "Space Division for Ray Tracing in CSG," IEEE CG & A, Vol 6., pp 28-34, 1986
8. Barsoum, R.S., "Application of Quadratic Isoparametric Finite Elements in Linear Fracture Mechanics," International Journal for Num. Methods in Engineering, Vol. 10, 1976, pp 25-37.
9. Corlett, R.C., "Draft: A User's Manual for RADSIM-PC," April 1988, University of Washington, Seattle, Wash.

1989 National Heat Transfer Conference
HTD-Vol. 106, Heat Transfer Phenomena in Radiation, Combustion and Fires

TWO-DIMENSIONAL INVERSE RADIATION HEAT TRANSFER ANALYSIS USING MONTE CARLO TECHNIQUES

W. J. Wu
University of Michigan-Dearborn
Dearborn, Michigan

G. P. Mulholland
New Mexico State University
Las Cruces, New Mexico

ABSTRACT

The use of Monte Carlo techniques for recovering temperature distributions in two-dimensional geometries where remote spectral emission data is available is investigated. The Monte Carlo simulation consists of tracing all spectral energy bundles emitted from the physical system for an assumed temperature distribution. The optimization procedure is then used to vary the assumed temperature distribution, expressed by Legendre polynomials, in order that the polynomial which best matches the measured spectral emission data can be determined. The technique is verified by using calculated emission values with added random errors and comparing the recovered temperature distributions with known values.

NOMENCLATURE

A	surface area
C_j	Legendre polynolmial coefficient
e	emissive power
F	configuration factor
$\mathbf{f}$	detected spectral emission function vector
fi	detected spectral emission function
fm,fn	orthogonal functions
$\mathbf{g(T)}$	$=\|\mathbf{f(T)}-\mathbf{q}\|$
i	radiation intensity
Pi	Legendre polynomial
Q	energy rate, energy per unit time
$\mathbf{q}$	radiative flux vector; detected spectral emission vector
q	energy flux, energy per unit area and time
$\mathbf{T}$	temperature vector
T	temperature
$\mathcal{T}$	temperature parameter vector
x,y,z	coordinates in rectangular system
ϵ	emissivity; error vector
θ	polar angle measured from normal of surface
ϕ	circumferential angle
ω	solid angle

SUBSCRIPTS

b	blackbody
i	component i; incoming
λ	wavelength dependent
$\Delta\lambda$	value integrated over wavelength interval $\Delta\lambda$

1. INTRODUCTION

Since the 1950's, there has been a growing interest in the study of inverse radiation heat transfer. A principal reason for this interest is the many applications in engineering, applied physics, meteorology, and other fields in which the temperature distribution in harsh environments is desired. Instead of a direct measurement, an inverse radiation heat transfer approach referred to as thermal remote sensing is used. This technique remotely measures the spectral emissions from the region of interest and recovers the temperature distribution from these remotely sensed data.

Kaplan (1959) has pioneered the theory and application of inversion methods used in thermal remote sensing. He points out that the structure of the atmosphere, including the three-dimensional distribution of temperature and water vapor, can be obtained from the spectral variation of its thermal radiation as viewed from a reconnaissance aircraft or earth satellite. In the inversion process, the selection of spectral intervals to be used must be carefully planned in order that the remote measurements be capable of unambiguous interpretation. The vertical thermal structure was determined by Kaplan (1959). As pointed out by Hommert and Viskanta (1978), the advantages of thermal remote sensing are:

1. it is noninterfering since it does not require probes;
2. it is passive in that there is no need for an external source of electromagnetic radiation;
3. the measurements can be made from remote locations provided that the effect, if any, of the intervening medium is accounted for; and
4. relatively simple instrumentation is required for emission measurements.

The limitations of thermal remote sensing result from detector sensitivity, spectral resolution, noise level, and measurement errors. Also, the mathematics of inversion are inherently unstable and errors can be amplified (Viskanta and Anderson, 1975). Chupp (1973) has reviewed the theory and applications of inversion techniques. He notes that inversion techniques have been developed following primarily three methods: linear, nonlinear, and iterative. The iterative method has been used extensively. In this approach, an initial guess of the temperature distribution is made and the equation of transfer is used to calculate the emerging radiation. The calculated emissions are then compared to the measured ones to determine the next guess for the temperature profile. Some articles (Dunn, 1981, 1983) have been reported which use an inverse Monte Carlo technique. However, it seems that little work has been done applying the Monte Carlo technique in the field of inverse radiation heat transfer for solving multidimensional problems.

The primary objective of this research is to develop and evaluate the inverse problem using Monte Carlo techniques applicable for multidimensional geometries. It is recognized that Monte Carlo solutions on conventional serial type computers usually have large CPU requirements. However, with the emergence of parallel type computers many complicated radiation heat transfer problems may be solved in less time using Monte Carlo on parallel machines than by using conventional techniques on serial machines. In this paper, the general two-dimensional inverse radiation heat transfer problem is solved using Monte Carlo techniques with the ultimate goal to use this procedure on a parallel computer.

2. ANALYSIS

2.1 PHYSICAL MODEL

The two-dimensional problem considered is illustrated in Figure 1. The region of interest is a rectangular surface which has a significant temperature gradient in the x and y directions. The emissivity of the surface may vary with temperature and/or wavelength and is assumed to be known. The spectral emissions from the surface can be measured by detectors located at various positions above the surface. Utilizing the detected spectral emission data, our main objective is to estimate the temperature distribution on the surface by applying the Monte Carlo technique.

In order to verify the Monte Carlo technique and to generate spectral emission values, solutions for the temperature distribution on the rectangular surface have been calculated separately using a finite-difference analysis. These emission values are then treated as data that would be obtained by detectors located a suitable distance from the surface. In the remainder of this paper, the term "data" will be used to denote these numerically obtained values, referred to as exact data, and values obtained by adding a numerically added random error to the exact data. Three cases of the problem are treated to test the performance of the technique. In Case 1 a smooth temperature profile on a gray surface is considered. Case 2 estimates the temperature distribution on a surface whose emissivity varies with temperature. In Case 3, we deal with a transient temperature profile which has a peak at the center.

2.2 SPECTRAL EMISSION DATA

The spectral emissions which would be measured by a detector are evaluated analytically for this study. To calculate the spectral energy rate leaving the rectangular surface and incident upon the detector, we consider the differential elements dA_1, dA_2, ..., dA_n of the rectangular surface to be isothermal at temperatures T_1, T_2, ..., T_n respectively. The entrance aperture of the detector has a finite area A_0 and is assumed isothermal at low temperature T_0. The elements dA_i and dA_0, as shown in Figure 2, have their normals at angles θ_i and θ_0 to the line of length s joining them.

The spectral energy rate reaching dA_0 from dA_i is

$$dQ_{\lambda,di-d0} = \epsilon_{\lambda,i}(\lambda,\theta,\phi,T_i) i_{\lambda b,i} cos\theta_i dA_i dw_i, \qquad (1)$$

where $\epsilon_{\lambda,i}(\lambda,\theta,\phi,T_i)$ is the directional spectral emissivity, $i_{\lambda b,i}$ is the blackbody spectral intensity, dw_i is related to the blackbody spectral hemispherical emissive power by

$$i_{\lambda b} = \frac{e_{\lambda b}}{\pi} \qquad (2)$$

and the solid angle dw_i is related to dA_0 and s by the relation

$$dw_i = \frac{cos\theta_0 dA_0}{s^2} \qquad (3)$$

Equation (1) can then be written as

$$dQ_{\lambda,di-d0} = \epsilon_{\lambda,i} e_{\lambda b,i} d\lambda dA_i \left(\frac{cos\theta_i cos\theta_0}{\pi s^2}\right) dA_0. \qquad (4)$$

The spectral energy rate leaving each differential surface element dA_i and incident upon the entrance aperture of the detector is then found by integrating over A_0

$$dQ_{\lambda,di-0} = \epsilon_{\lambda,i} e_{\lambda b,i} d\lambda dA_i F_{di-0}, \qquad (5)$$

where F_{di-0} is the configuration factor for plane element dA_i to circular disk A_0. The configuration factor is given by (Siegel and Howell, 1981)

$$F_{di-0} = \frac{1}{2}\left(1 - \frac{1+H^2-R^2}{\sqrt{Z^2-4R^2}}\right). \qquad (6)$$

where $H = h/a$, $R = r/a$, and $Z = 1 + H^2 + R^2$.

Substitution of Equation (6) into Equation (5) yields

$$dQ_{\lambda,di-0} = \epsilon_{\lambda,i} e_{\lambda b,i} d\lambda dA_i \frac{1}{2}\left(1 - \frac{1+H^2-R^2}{\sqrt{Z^2-4R^2}}\right). \qquad (7)$$

The spectral flux for a small wavelength interval $\Delta\lambda$ measured by the detector is

$$q\Delta\lambda = \frac{1}{A_0}\int_{\lambda}^{\lambda+\Delta\lambda}\int_{Ai} \epsilon_{\lambda,i} e_{\lambda,i} \frac{1}{2}\left(1 - \frac{1+H^2-R^2}{\sqrt{Z^2-4R^2}}\right) dA_i d\lambda. \qquad (8)$$

To determine exact spectral emission data which would be measured by a detector, the rectangular surface temperatures must be known. These temperature distributions are obtained by solving the direct problem. The distributions are then used to evaluate the blackbody spectral hemispherical emissive power, $e_{\lambda b,i}$; this value is then used to determine the spectral flux from Equation (8). The spectral flux is then treated as "data", Equation (9). The solution technique to be discussed uses the spectral flux "data" (either exact or with random errors) to predict the temperature distribution (inverse problem). This is a common procedure used to evaluate the validity of inverse-type solution techniques.

2.3 METHOD OF INVERSION

To express the spectral emissions measured by the detector, a system of implicit equations can be written as

$$\begin{aligned} f_1(T_1, T_2, \cdots, T_n) &= q_1 \\ f_2(T_1, T_2, \cdots, T_n) &= q_2 \\ &\vdots \\ f_m(T_1, T_2, ..., T_n) &= q_m, \end{aligned} \qquad (9)$$

where

$m \equiv$ number of wavebands,

$n \equiv$ number of elements,

$q_i \equiv$ spectral emission detected in waveband i,

and

$T_j \equiv$ temperature of the element j.

In vector form, Equation (9) can be written as

$$\mathbf{f}(\mathcal{T}) = \mathbf{q}. \tag{10}$$

An n-vector $\mathcal{T}$ satisfying the system of m equations must be obtained. For estimating $\mathcal{T}$ in Equation (10), an adaptive random search algorithm for the global optimization problems was proposed by Baba and Sawaragi (1977) who also proved its convergence.

A disadvantage of this algorithm is that it may need a lot of effort in order for the global minimum to be reached. To increase the speed of convergence, a modified version of Baba and Sawaragi's algorithm was used. The algorithm is as follows:

1. Generate a sequence of N random vectors $\mathbf{T}_{01}, \mathbf{T}_{02}, \cdots, \mathbf{T}_{0N}$, uniformly distributed over D.
2. Calculate $\mathbf{g}(\mathbf{T}_{0k}) = \| \mathbf{f}(\mathbf{T}_{ok})\text{-q} \|$, k=1,···,N.
3. Find a random vector $\mathbf{T}_k$ by

$$min(\mathbf{g}(\mathbf{T}_{01}), \mathbf{g}(\mathbf{T}_{02}), \cdots, \mathbf{g}(\mathbf{T}_{0N})).$$

4. $k \leftarrow 1$.
5. Generate $\mathbf{t}_k$ from $\mathrm{N}(0,\sum)$.
6. If $\mathbf{T}_k + \mathbf{t}_k \notin D$, set $\mathbf{T}_{k+1} = \mathbf{T}_k$ and go to step 5.
7. Calculate $\mathrm{g}(\mathbf{T}_k + \mathbf{t}_k)$.
8. If $\mathrm{g}\ (\mathbf{T}_k + \mathrm{t}_k)\ \langle\ \mathrm{g}(\mathbf{T}_k)$, $\mathbf{T}_{k+1} = \mathbf{T}_k + \mathbf{t}_k$; otherwise, $\mathbf{T}_{k+1} = \mathbf{T}_k$.
9. $k \leftarrow k + 1$
10. Go to step 5.

2.4 LEGENDRE POLYNOMIAL APPROXIMATION

The temperature distribution can be approximated by using various types of functional relations. Legendre polynomials are chosen because they equally weight the interval of interest and are relatively easy to evaluate. In general, the two-dimensional temperature distribution can be expressed in terms of Legendre polynomials which are orthogonal on the interval (-1, 1)

$$\int_{y=-1}^{y=1} \int_{x=-1}^{x=1} f_m(x,y) f_n(x,y) dx\ dy = 0 \quad \text{for } m \neq n. \tag{11}$$

We then have Legendre polynomials for the two-dimensional temperature distribution of the form

$$\begin{aligned} T(x,y) = {} & C_0[1 + C_{01}P_2(y) + C_{03}P_3(y) + \cdots] \\ & + C_1P_1(x)[1 + C_{11}P_1(y) + C_{12}P_2(y) + C_{13}P_3(y) + \cdots] \\ & + C_2(x)[1 + C_{21}P_1(y) + C_{22}P_2(y) + C_{23}P_3(y) + \cdots] \\ & + C_3P_3(x)[1 + C_{31}P_1(y) + C_{32}P_2(y) + C_{33}P_3(y) + \cdots] \\ & + \cdots \end{aligned} \tag{12}$$

where $\mathrm{P}_i(x)$ and $\mathrm{P}_j(y)$ are Legendre ploynomials in the x and y directions, respectively. Equation (12) is the Legendre polynomial form employed in the Monte Carlo inversion to recover the temperature profile for Cases 1 and 2. In Case 3, because of the roughness of the temperature profile, the surface is divided into smaller regions. Legendre polynomials are then used to approximate the temperature profile for each region.

The Monte Carlo simulation for the radiation field is applied in the inversion procedure, and the Monte Carlo inversion techniques are used to find the approximate temperature distribution. In the polynomial approach, the $C's$ are iterated upon until the best fit of the emission data is obtained. A linear temperature fit (n=2) is first employed and then succesivly higher order fits are attempted with the previous results used to start the solution. The previously determined coefficients have normal distributions with smaller standard deviations for random search. In this way, the added higher order polynomial coefficient will be able to converge more efficiently. However, using smaller standard deviations for the lower order terms may decrease the probability of arriving at a global minimum. There are many ways for determining the "best" model for the temperature distribution. One approach is to calculate the coefficients (parameters) and the residual sum of squares for n=2,3,4,···. Then an F test is used to select the Legendre polynomial coefficients to be used in the temperature approximation.

3 RESULTS

3.1 EFFECTS ON THE MONTE CARLO SIMULATION

In the application of the inverse Monte Carlo technique to our two-dimensional problem, two factors must be taken into consideration. They are the number of energy bundles emitted from the surface of interest and the size of each surface element dA_i. If one increases the number of energy bundles, the root-mean square error will be reduced. However, the computer time required to calulate the spectral emission will be increased. Similarly, as one makes the element size smaller better agreement with exact values can be expected: but, reducing element size increases the memory space required by the computer. The choice of element size and the number of energy bundles used depends on the complexity of the problem and the desired accuracy. With careful planning and practice, one can choose realistic values for both parameters.

3.2 MONTE CARLO INVERSION RESULTS

By using calculated spectral emission data, the Monte Carlo method is employed to recover the temperature distribution for three cases. The inversion results are given in the following sections.

Case 1, Gray Surface Results

In this case we consider a stainless steel block 10.16 cm X 10.16 cm X 1.905 cm, Figure 3. The block is assumed to be at steady-state, have diffuse surfaces, and have a thermal conductivity of 24 W/m K. All surfaces except the upper surface are well insulated and a heat source is applied at the lower surface.

The upper surface is assumed to be coated with a thin film having emissivity ϵ=0.95 and is exposed to air at room temperature. The value of the average convection heat transfer coefficient on the upper surface is assumed to be 15 W/m^2K. A finite-difference solution for the temperature profile on the upper surface is shown in Figure 4. To simulate an actual experiment, the spectral energy emerging from the gray surface which would be measured by four equidistant detectors located 10 cm above the surface is used as input "data". This information is obtained by means of Equation (8) and the appropriate configuration factor for each detector.

Exact Emission Results. To examine the inverse Monte Carlo technique for solving the gray surface case, exact emission data is first used to estimate the temperature distribution on the rectangular surface. Then exact emission data with random errors added will be used to make the estimation.

The recovered temperature profile using exact emission data is shown in Figure 5. Figures 6, 7, and 8 compare the temperature distribution with the predicted one at three locations in the y direction, namely y=5.08 cm, y=2.54 cm, and y=1.27 cm. The differences between predicted and recovered temperature distributions are show in Figure 9 for illustrative purposes only. The deterioration of the recovered temperature distribution close to the boundaries is common for inverse-type solution techniques. It arises from the decreasing amount of information available close to a boundary. In the interior, information is available in a large region surrounding a point; this information is utilized to make the prediction. However, close to a boundary, one has information only on one side of the point and consequently predictions in these regions suffer accordingly. This particular problem has been addressed by many authors, Beck (1977).

The standard deviation of the recovered temperature distribution is 3.87 K. This is 0.41% of the maximum temperature in this case. The maximum deviation between predicted and recovered temperatures is found to be 20.60 K (2.20% of the maximum temperature). The results demonstrate that the temperature distribution can be estimated fairly accurately using a reasonable number of spectral emission measurements.

Random Data Error Results. To simulate experimental data, random errors are added to exact emission values. Data with 5% and 10% random errors are used to examine how Monte Carlo inversion results are affected by the "data" (spectral emitted energy) errors for the gray surface case.

Figure 10 gives results for the effect of 5% data error on the recovered temperature distribution. The effect of larger data error (10%) on the recovered temperature profile is shown in Figure 11. Note that 5% random data error increases the standard deviation from 3.87 K (0.41% of the maximum temperature) to 5.71 K (0.61% of the maximum temperature). However, 10% random data error increases the standard deviation to 6.86 K (0.73% of the maximum temperature). The maximum differences between predicted and recovered temperature profiles are -28.45 K (-3.04% of the maximum temperature) and -28.74 K (-3.07% of the maximum temperature), respectively, for 5% and 10% data errors. This is certainly within acceptable limits for this amount of error.

A typical computer run (IBM-4381D) took approximately 2.5 hours, required 2 Mbytes of memory, used 2500 elements and had 1000 spectral energy bundles emitted from each element. The "data" were obtained by calculating the spectral emitted energy, Equation 8, in 10 wavelength bands between 3 μm and 3.45 μm; these wavelengths were the center of each band with bandwidths equal to 0.05 μm.

Nonconstant Surface Emissivity

In the previous case, the emissivity of the surface was assumed constant. However, emissivity may change with temperature and/or wavelength. Case 2 considers emissivity to be a variable. In this case, the stainless steel block is exposed to the same conditions as Case 1 except that the emissivity of the upper surface is assumed to vary with temperature according to the relation.

$$\epsilon = 0.07[1 + 0.0004(T - 1000)]. \tag{15}$$

As before, the numerical solution for the temperature profile is calculated (direct problem) and the spectral emission data with or without data error are then determined, Equation (8), and used to recover the temperature profile on the surface. The comments made concerning computer usage and calculation of "data" for Case 1 also apply to this case.

Exact Emission Results. When the exact emission data are used to recover the temperature distribution on the surface, the error in the recovered temperatures is due to computional errors or the inexactness of the Monte Carlo technique.

A detailed comparison for the recovered temperature distribution against the predicted one at y=5.08 cm is given in Figure 12. The standard deviation of the recovered temperature profile is found to be 4.97 K which is 0.52% of the maximum temperature on the nongray rectangular surface. The results demonstrate the utility of the inverse Monte Carlo technique for solving the two-dimensional problem in which the variations of optical properties are taken into consideration.

Random Data Error Results. Spectral emission "data" with 5% and 10% random errors added to the exact emission "data" were also used to obtain the temperature profile on the surface.

Figure 13 shows results for the effect of 5% data error on the recovered temperature distribution. The effect of 10% data error is shown in Figure 14. Note that 5% random data error increases the standard deviation from 4.97 K to 5.35 K (0.56% of the maximum temperature) and 10% random data error gives a standard deviation of 8.33 K (0.56% of the maximum temperature). For 5% data error, the maximum difference between predicted and recovered temperature profiles is 23.42 K (2.47% of the maximum temperature) while the maximum difference increases to as much as -37.99 K (4.01% of the maximum temperature) for 10% data error. However, the deviations are all within 5% of the maximum temperature.

Pseudo-Transient Results

So far, we have treated two cases in which temperature profiles are smooth. Now, we turn our attention to a transient problem which has a peak in the temperature profile on the surface. A tungsten plate, as illustrated in Figure 15, is initially at room temperature. Suddenly the lower surface of the plate is heated by a beam of incident heat flux of 100 W/cm^2 uniformly spread over a small centrally located area. A finite-difference analysis has been performed to calculate the transient temperature distribution. The resulting profile at time equal to 60 seconds is shown in Figure 16. Proceeding as before, we use this temperature distribution to determine spectral emission "data", Equation (8), at time equal to 60 seconds; 10 wavelength bands ($\Delta\lambda = 0.1\mu m$) between 2 μm and 2.9 μm were used to obtain the spectral information. This "data" is then used to recover the temperature distribution. For other values of time, the procedure is repeated. Typical computer runs required similar amounts of time as Cases 1 and 2.

Using exact emission data, we obtain the recovered temperature profile. A detailed comparison for the recovered temperature distribution against the predictions along the line x = y is given in Figure 17. The recovered temperatures for the transient case have a standard deviation of 202.14 K which is 13.84% of the maximum temperature. We note that using Legendre polynomial approximations for the whole surface does not successfully recover the temperature profile for this particular case.

In order to obtain a better temperature estimation, the entire surface was divided into smaller regions and the spectral emissions from each division was "measured". Legendre polynomials were then used to approximate the temperature distributions for each division. The surface was divided into four equal square regions. Using exact emission data and Legendre polynomials to estimate the temperatures for each division, the recovered temperature profile along the line x=y is illustrated in Figure 18. The recovered temperatures have a standard deviation of 247.87 K (16.97% of the maximum temperature) for the four-region approximation. It seems that dividing the surface into four divisions does not improve the recovered temperatures. The rectangular surface is further divided into 16 equal square regions. As before, exact emission data and Legendre polynomials are used to approximate the temperature profile for each region. A comparison of recovered and predicted temperatures along the line x=y is given in Figure 19. For the 16-region approximation, the recovered temperatures have a standard deviation of 55.28 K which is 3.78% of the maximum temperature. The maximum temperature difference between recovered and predicted values is -207.53 K, which is -14.20% of the maximum temperature. The recovered temperature distribution is significantly improved by using the 16-region approximation.

For situations where the temperature is a rapidly changing function of position, the Legendre polynomial approximation is not able to adequately estimate the temperature profile for the entire rectangular surface. The Monte Carlo inversion results indicate that by using this approximation for successively smaller regions, the temperature distribution can be recovered; however, this procedure does increase the computer time required for a solution. It appears that temperature approximations other than polynomials, for example a linear combination of sine functions or a gaussian profile, are better suited for this type of temperature profile. The physical situation should dictate the type of approximation to be implemented.

SUMMARY

The usefulness of the Monte Carlo technique for solving inverse radiation heat transfer problems has been demonstrated. Basically, the problem is to recover internal temperature distributions when one only has remotely measured spectral data. It has been shown that the Legendre polynomial approximation is adequate for temperature distributions which are not rapidly varying functions; for those situations, a distribution which is similar to the expected profile should be chosen.

An advantage Monte Carlo techniques have is the ease which complicated geometries and boundary conditions can be accomodated. This was not exploited in this paper since our purpose was to demonstrate the feasibility of this approach for inverse problems. This was accomplished by using simple geometries and comparing known solutions with solutions obtained using this technique. Future work will include complicated geometries.

The Monte Carlo technique for solving either direct or inverse radiation heat transfer problems is best suited for parallel type machines. Use of serial machines requires large amounts of computer time and are impractical. Future work for this project includes testing the code on a parallel machine and developing approximations which will enable us to handle severe thermal environments.

REFERENCES

Baba, N. T. S., and Y. Sawaragi, "A Modified Convergence Theorem for a Random Optimiation Algorithm," *Information Sciences,* vol. 13, 1977.

Beck, J. V., and K. J. Arnold, *Parameter Estimation in Engineering and Science,* John Wiley & Sons, Inc., New York, 1977.

Chupp, R. E., "Development and Experimental Evaluation of a Technique to Determine the Temperature Distribution in Semitransparent Solids from Remotely Sensed Spectral Emission Data., " Ph.D. Thesis, Purdue University, Lafayette, Indiana, 1973.

Dunn, W. L., "Inverse Monte Carlo Analysis," *Journal of Computational Physics,* vol. 41, pp. 154-166, 1981.

Dunn, W. L., "Inverse Monte Carlo Solutions for Radiative Transfer in Inhomogeneous Media," *Journal of Quantitative Spectroscopy and Radiative Transfer,* vol. 29, no. 1, pp. 19-26, 1983.

Hommert, P. J., and R. Viskanta, "Hight Temperature Gas Diagnostics by Spectral Remote Sensing," *International Journal of Heat and Mass Transfer,* vol. 21, pp. 769-781, 1978.

Kaplan, L. D., "Inference of Atmospheric Structure from Remote Radiation Measurements," *Journal of the Optical Society of America,* vol. 49, pp. 1004-1007, 1959.

Siegel, R., and J. R. Howell, *Thermal Radiation Heat Transfer,* 2nd ed., McGraw-Hill Book Company, New York, 1981.

Viskanta, R., and E. E. Anderson, "Heat Transfer in Semitransparent Solids," *Advances in Heat Transfer,* vol. 11, pp. 317-441, 1975.

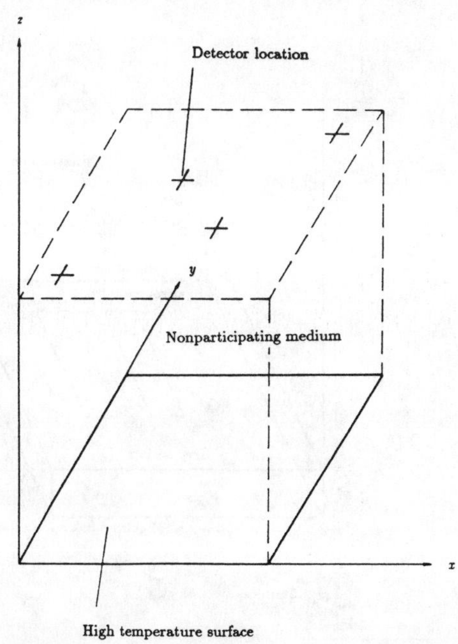

Figure 1 Two-dimensional problem.

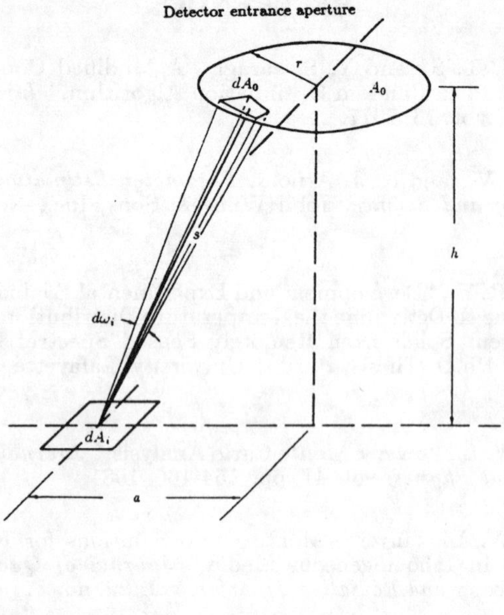

Figure 2 Energy rate incident on the entrance aperture of the detector from a differential surface element.

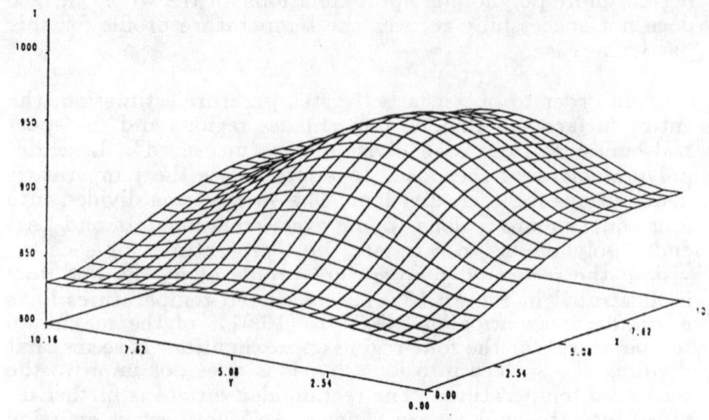

Figure 4 Predicted temperature distribution, Case 1.

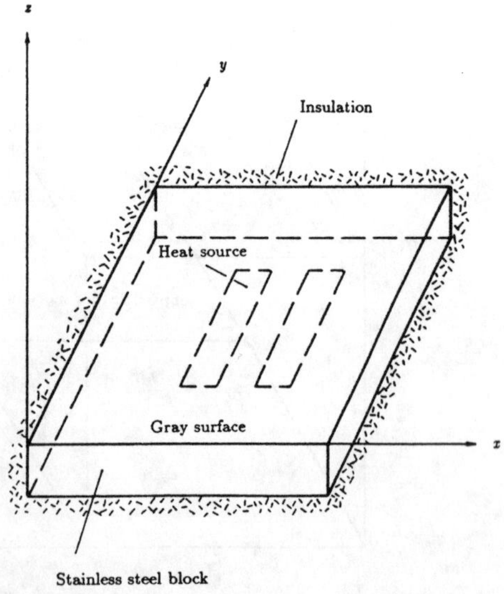

Figure 3 Two-dimensional problem, Case 1.

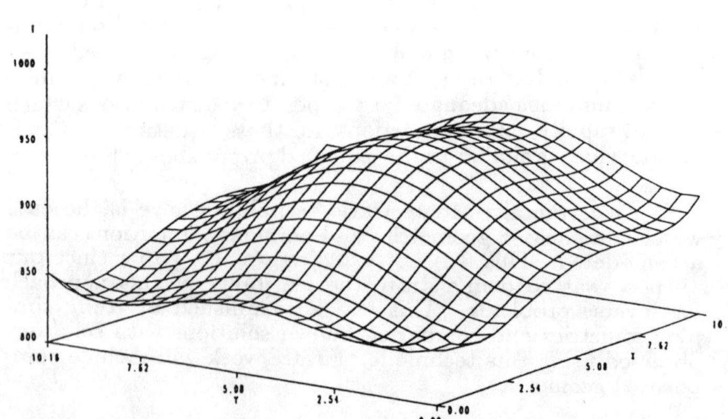

Figure 5 Recovered temperature distribution, Case 1, exact emission data.

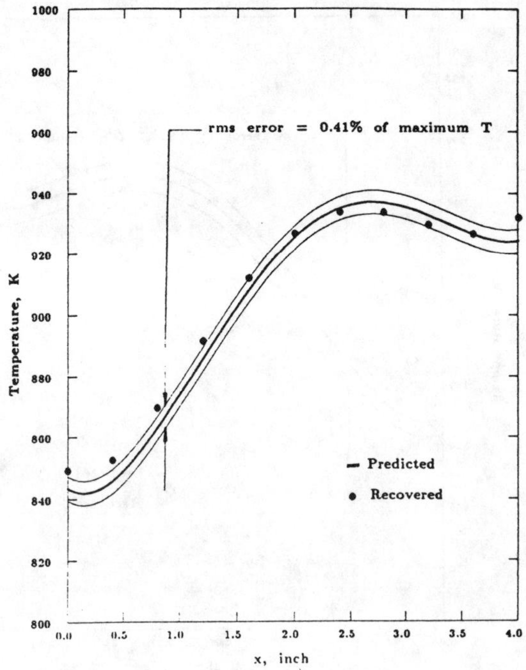

Figure 6 Comparison of predicted and recovered temperature distribution, Case 1, exact emission data, $y = 5.08$ cm.

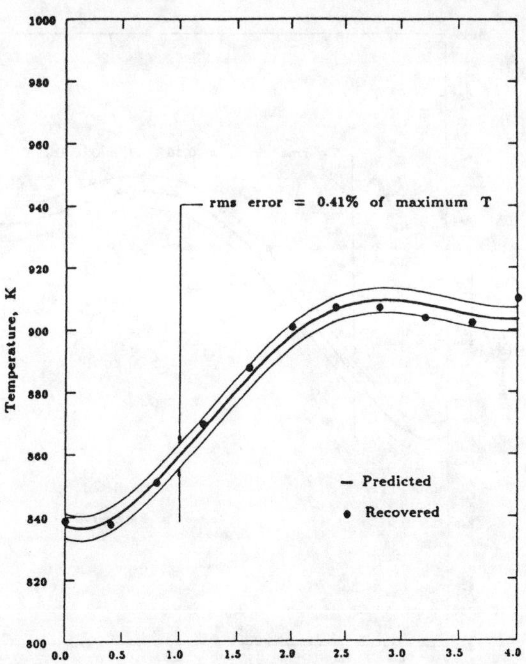

Figure 7 Comparison of predicted and recovered temperature distributions, Case 1, exact emission data, $y = 2.54$ cm.

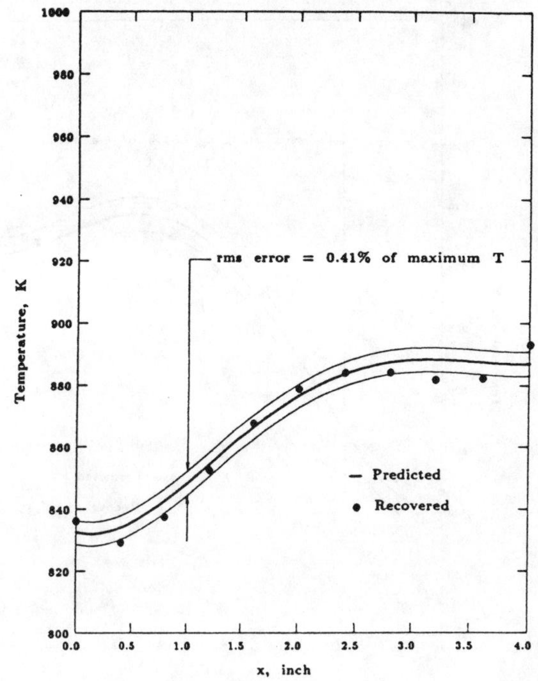

Figure 8 Comparison of predicted and recovered temperature distributions, Case 1, exact emission data, $y = 1.27$ cm.

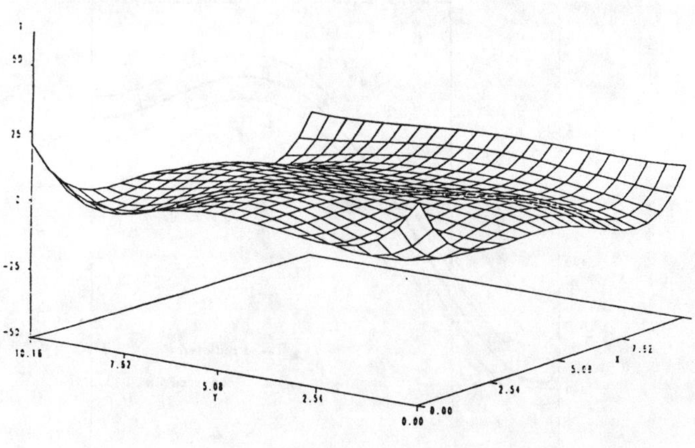

Figure 9 Differences between predicted and recovered temperature distribtions, Case 1, exact emission data.

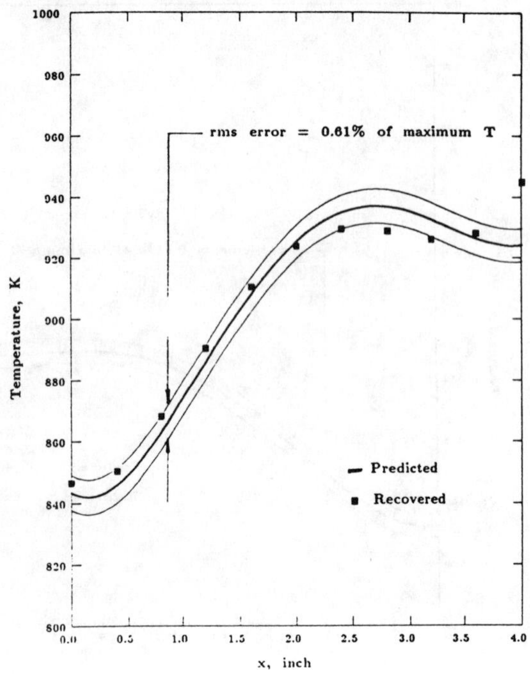

Figure 10 Comparison of predicted and recovered temperature distributions, Case 1, 5% data error, $y = 5.08$ cm.

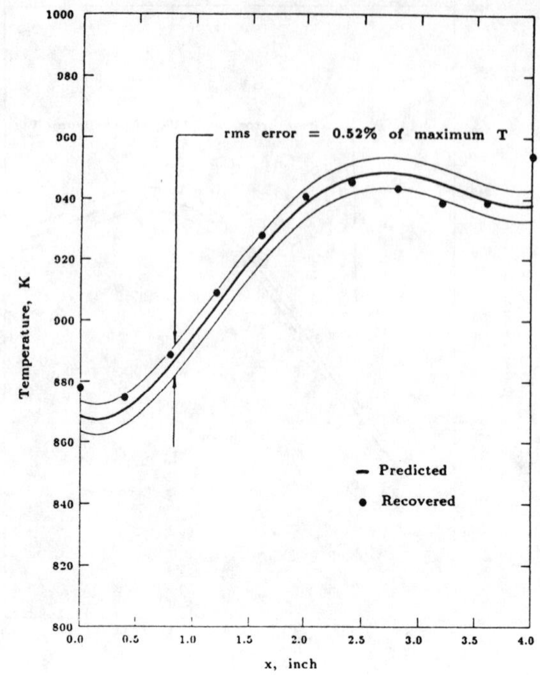

Figure 12 Comparison of predicted and recovered temperature distribtions, Case 2, exact emission data, $y = 5.08$ cm.

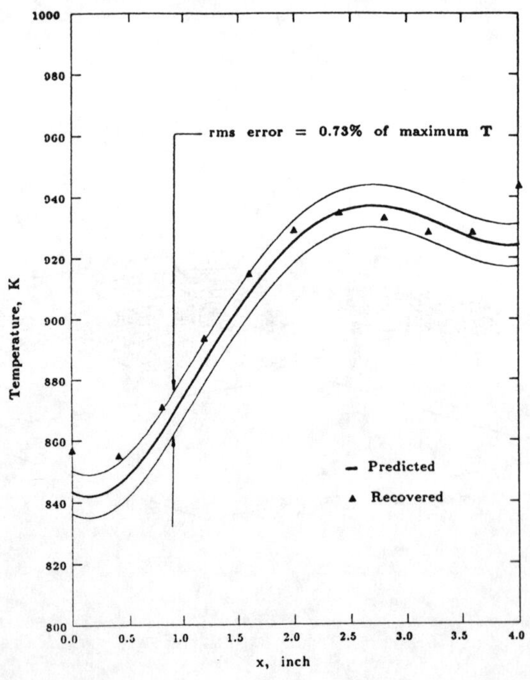

Figure 11 Comparison of predicted and recovered temperature distributions, Case 1, 10% data error, $y = 5.08$ cm.

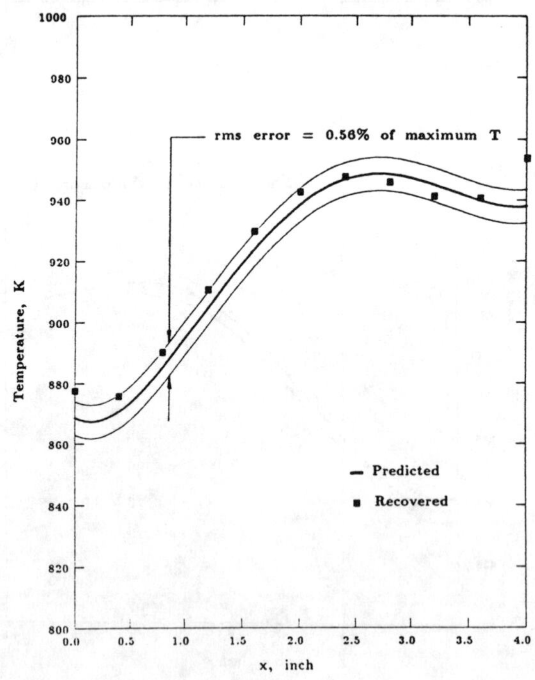

Figure 13 Comparison of predicted and recovered temperature distributions, Case 2, 5% data error, $y = 5.08$ cm.

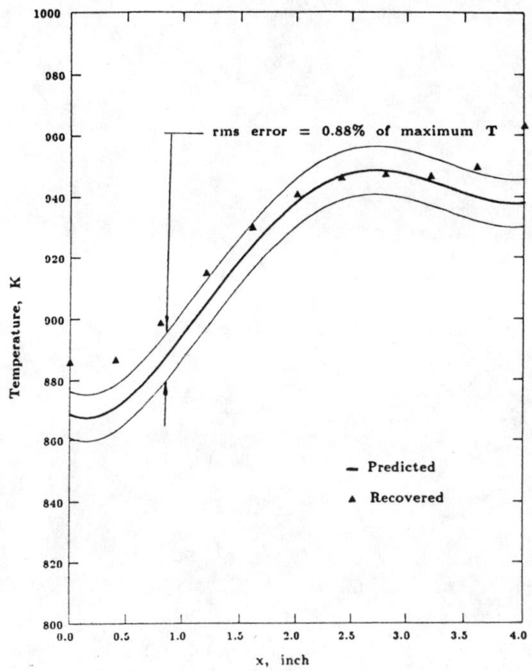

Figure 14 Comparison of predicted and recovered temperature distributions, Case 2, 10% data error, $y = 5.08$ cm.

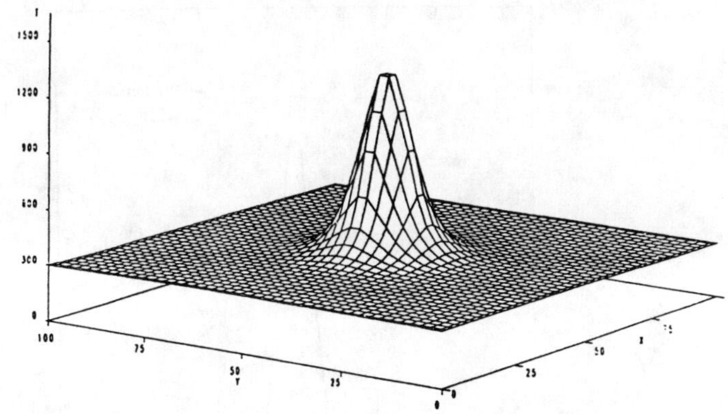

Figure 16 Predicted transient temperature distribution, Case 3.

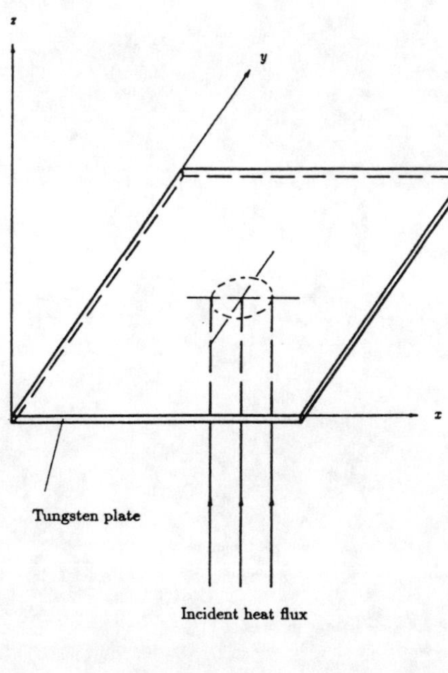

Figure 15 Two-dimensional problem, Case 3.

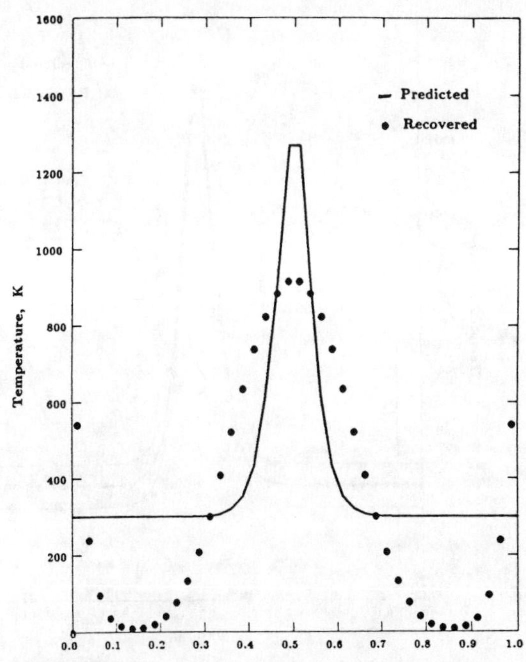

Figure 17 Comparison of predicted and recovered transient temperature distributions, Case 3, exact emission data, $x = y$.

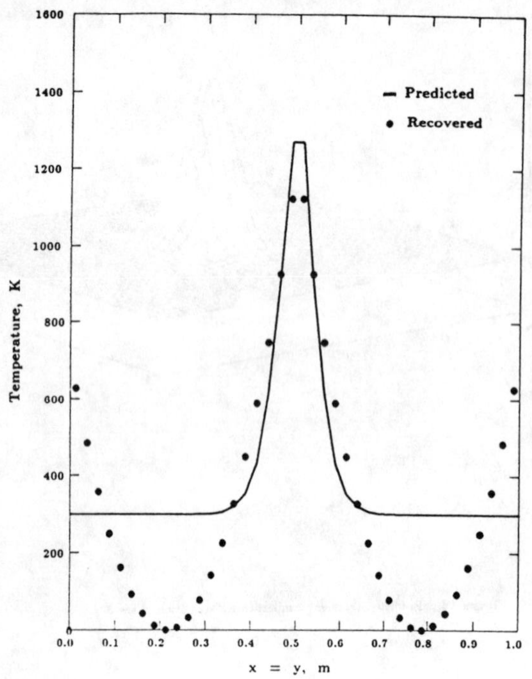

Figure 18 Comparison of predicted and recovered transient temperature distributions, Case 3, exact emission data, four-region approximation, $x = y$.

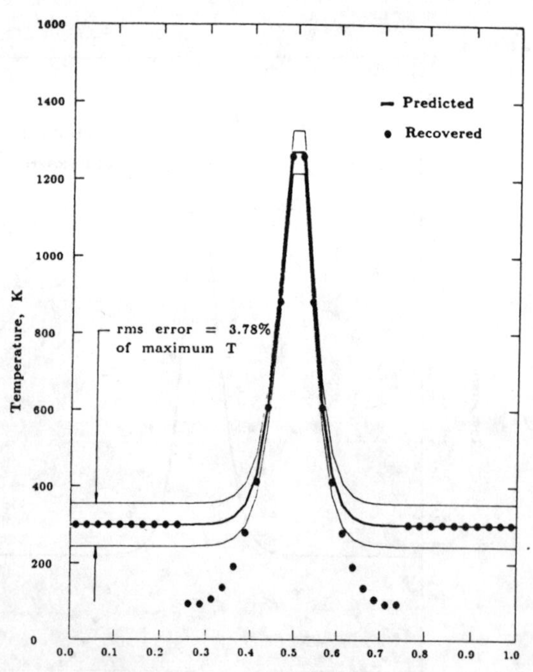

Figure 19 Comparison of predicted and recovered transient temperature distributions, Case 3, exact emission data, 16-region approximation, $x = y$.

1989 National Heat Transfer Conference
HTD-Vol. 106, Heat Transfer Phenomena in Radiation, Combustion and Fires

DEVELOPMENT OF A NETWORK ANALOGY AND EVALUATION OF MEAN BEAM LENGTHS FOR MULTI-DIMENSIONAL ABSORBING/ISOTROPICALLY SCATTERING MEDIA

W. W. Yuen
Department of Mechanical and Environmental Engineering
University of California at Santa Barbara
Santa Barbara, California

ABSTRACT

Based on Hottel's zonal formulation, a network analogy is developed for the analysis of radiative transfer in general multi-dimensional absorbing/isotropically scattering media. Applying the analogy to the analysis of an isothermal medium and assuming that the incoming and outgoing flux density is homogeneous within the medium, the effect of scattering on the evaluation of mean beam lengths is illustrated. Two concepts of mean beam length, an absorption mean beam length (AMBL) and an extinction mean beam length (EMBL), are introduced and shown to be important for the analysis of radiative transfer in practical systems. Both mean beam lengths differ significantly from the conventional mean beam length in systems of moderate and large optical thicknesses. Relations between AMBL and EMBL and their limiting behavior are developed analytically. Numerical results for a sphere radiating to its surface and an infinite parallel slab radiating to one of its surface are presented to demonstrate quantitatively the mathematical behavior of the two mean beam lengths.

1. INTRODUCTION

Radiative transfer in an absorbing/scattering medium is an important aspect in many practical engineering problems such as the modeling of energy transport in combustion chambers and furnaces. Mathematically, an exact solution to the radiative equation of transfer, coupled with the fluid flow and convective heat transfer occurring inside a furnace, can be extremely complicated and therefore unsuitable for practical engineering applications. A great deal of research efforts in radiative transfer in the past fifty years have thus been directed toward the development of approximation methods. Sarofim (1986) and Viskanta and Menguc (1987) have given excellent reviews of the various available techniques.

One of the most important concepts utilized in practical radiative transfer calculations, particularly in applications for non-gray media, is mean beam length (MBL). Introduced originally by Hottel (1927, 1967), MBL of a gas volume radiating to a boundary is defined as the radius of an "equivalent" hemisphere which produces a flux to the center of its base equal to the average flux radiated to the area of interest by the actual volume of gas. Since spectral absorption data for all common gaseous species are generally measured in one-dimensional line-of-sight experimental systems, the MBL concept provides an important theoretical link through which the existing one-dimensional spectral absorption data can be applied to radiative transfer calculations in complex multi-dimensional systems.

In recent years, scattering has been recognized to be important in many particle-laden combustion systems such as fires and pulverized coal furnaces (DeRis, 1978, Wessel, 1985, Menguc and Viskanta, 1986). While it is well known that scattering can have significant effect on the total radiative heat transfer, its effect on MBL has not yet been established. To the best of the author's knowledge, an approximate expression for the scattering mean beam length in the limit of an optically-thin weakly-scattering medium (Cartigny, 1986) appears to be the only reported work addressing this important issue. Indeed, the lack of an accurate definition for MBL in an absorbing/scattering medium has led to large uncertainty on the result of many existing furnace calculations (Menguc and Viskanta, 1986, Viskanta and Menguc, 1987).

In this work, a theoretical formulation of MBL's in an absorbing/scattering medium is derived. In section 2, a network analogy for the calculation of radiative transfer in an emitting, absorbing and isotropically scattering medium is developed. By applying the analogy to an isothermal medium and assuming that the incoming and outgoing flux density is homogeneous, two concepts of MBL, an "absorption" mean beam length (AMBL) and an "extinction" mean beam length (EMBL), are introduced in section 3. Mathematically, AMBL is defined as the radius of a purely absorbing hemisphere with the same absorption coefficient as the medium under consideration which produces a flux at the center of its base equal to the average heat flux radiated to the area of interest. The concept of AMBL is important in estimating the emission from non-gray gaseous species in the presence of scattering particles. EMBL, on the other hand, is defined as the radius of an absorbing/scattering hemisphere with the same extinction coefficient and scattering albedo as the medium under consideration which produces the equivalent heat flux. The concept of EMBL is important in the scaling of scattering/absorbing media. Based on evaluation of AMBL for an absorbing/scattering hemispherical medium radiating to its base, a universal relation between AMBL and EMBL is also presented in section 3. In section 4, exact mathematical expressions for the two MBL's are generated for two enclosures, a sphere and an infinite parallel slab. Numerical results are presented to illustrate parametrically the importance of scattering on the evaluation of MBL's. Finally, a conclusion of the present work and some comments on the future direction of research in this general area are presented in section 5.

2. NETWORK ANALOGY FOR AN ABSORBING/SCATTERING MEDIUM

The analogy between electrical network and radiative transfer in an enclosure with non-participating medium was first developed by Oppenheim (1956). Bevans and Dunkle (1960) extended the idea to include the effect of an isothermal absorbing medium. The extension, however, is quite complicated mathematically and applicable only for a homogeneous isothermal absorbing medium. The practical application for the Bevans and Dunkle's formulation is thus limited. Utilizing results generated by the two-flux model, Tong and Tien (1980) noted that the heat transfer in a planar absorbing/scattering medium can be expressed by a network representation. But they did not consider extension to general multi-dimensional systems. To the best of the author's knowledge, formulation of the network analogy for a general non-isothermal multi-dimensional enclosure with an absorbing/scattering medium has not appeared in the literature.

For an isotropically scattering and absorbing medium, the set of radiative exchange relations generated by the Hottel's zonal method (Hottel and Sarofim, 1967) is a natural basis for the development of the network analogy. Specifically, in an enclosure with N isothermal surface zones and M gas zones, the energy balance on each zone can be written as

$$A_i H_i = \sum_{j=1}^{N} \overline{s_i s_j}\, W_j + \sum_{j=1}^{M} \overline{s_i g_j}\, W_{g,j} \tag{1}$$

$$4KV_i H_{g,i} = \sum_{j=1}^{N} \overline{g_i s_j}\, W_j + \sum_{j=1}^{M} \overline{g_i g_j}\, W_{g,j} \tag{2}$$

In the above expressions, H_i and W_i are the irradiation and radiosity (emission plus reflection) of surface A_i while $4H_{g,i}$ and $4KW_{g,i}$ can be interpreted as the incoming and outgoing (emission plus scattering) flux density of a volume element V_i. K is the extinction coefficient. $\overline{s_i s_j}$, $\overline{s_i g_j}$ and $\overline{g_i g_j}$ are the direct exchange factor between A_i and A_j, A_i and V_j and V_i and V_j respectively. The mathematical expression for these exchange factors are given by Hottel and Sarofim (1967) and will not be repeated here. In general, they satisfy the following reciprocity and closure relations

$$\overline{s_i s_j} = \overline{s_j s_i} \tag{3a}$$

$$\overline{s_i g_j} = \overline{g_j s_i} \tag{3b}$$

$$\overline{g_i g_j} = \overline{g_j g_i} \tag{3c}$$

$$\sum_{i=1}^{N} \overline{s_i s_j} + \sum_{i=1}^{M} \overline{g_i s_j} = A_j \tag{4a}$$

$$\sum_{i=1}^{M} \overline{g_i g_j} + \sum_{i=1}^{N} \overline{s_i g_j} = 4KV_j \tag{4b}$$

The parameters W_i, H_i, $W_{g,i}$ and $H_{g,i}$ are related to the surface emissive power E_i and the gaseous emissive power $E_{g,i}$ by

$$W_i = \varepsilon_i E_i + (1 - \varepsilon_i) H_i \tag{5a}$$

$$W_{g,i} = (1 - \omega_0) E_{g,i} + \omega_0 H_{g,i} \tag{5b}$$

Q_i and $Q_{g,i}$, the net heat transfer from surface A_i and volume V_i respectively, can be written as

$$Q_i = A_i (W_i - H_i) = \frac{A_i \varepsilon_i}{1 - \varepsilon_i}(E_i - W_i) \tag{6a}$$

$$Q_{g,i} = 4V_i (W_{g,i} - H_{g,i}) = 4KV_i \left(\frac{1 - \omega_0}{\omega_0}\right)(E_{g,i} - W_{g,i}) \tag{6b}$$

where ε_i is the surface emissivity of A_i and ω_0 is the scattering albedo of the medium.

To develop the necessary equation for the network analogy, the first half of equations (6a) and (6b), together with the reciprocity and closure relations, can be readily combined with equations (1) and (2) to yield

$$Q_i = \sum_{j=1}^{N} \overline{s_i s_j}\,(W_i - W_j) + \sum_{j=1}^{M} \overline{s_i g_j}\,(W_i - W_{g,j}) \tag{7a}$$

and

$$Q_{g,i} = \sum_{j=1}^{N} \overline{g_i s_j}\,(W_{g,i} - W_j) + \sum_{j=1}^{M} \overline{g_i g_j}\,(W_{g,i} - W_{g,j}) \tag{7b}$$

Based on equations (6a), (6b), (7a) and (7b), the analogy between electrical network and radiative heat transfer in an absorbing/scattering medium is quite apparent. Specifically, they are

E_i = internal potential of surface A_i
W_i = external potential of surface A_i
$\dfrac{1-\varepsilon_i}{A_i \varepsilon_i}$ = internal resistance at surface A_i

$E_{g,i}$ = internal potential of volume V_i
$W_{g,i}$ = external potential of volume V_i
$\dfrac{\omega_0}{4KV_i(1-\omega_0)}$ = internal resistance at volume V_i

$1/\overline{s_i s_j}$ = resistance between surfaces A_i and A_j
$1/\overline{s_i g_j}$ = resistance between surfaces A_i and V_j
$1/\overline{g_i g_j}$ = resistance between surfaces V_i and V_j

Schematic representations of the network analogy for a surface element A_i and a volume element V_i are shown in Fig. 1a and 1b respectively. Note in contrast to Bevans and Dunkle (1960), the present network analogy does not required the restrictive assumption of an isothermal gas volume. It can be applied to any enclosure containing an absorbing/isotropically scattering medium.

3. FORMULATION OF MEAN BEAM LENGTHS

In the formulation of mean beam lengths, the problem of interest is to determine the radiative heat flux incident on an area A from a homogeneous isothermal gas volume (with volume V_g, emissive power E_g, extinction coefficient K and scattering albedo ω_0). The area A, in general, is a part of the total boundary of the enclosure. Let A_c be the remaining area of the enclosure and assuming that H_g and W_g are uniform in the gas volume, the network analogy is applicable with an equivalent network as shown in Fig. 2. The corresponding network equations can be readily solved to yield the following heat flux expression at surface A

$$q = \frac{(E_g \overline{sg}/A)(1-\omega_0)}{1 - \frac{\omega_0 \overline{gg}}{4KV_g}} \tag{8}$$

3a. Concept of Absorption Mean Beam Length (AMBL)

The absorption mean beam length (AMBL), L_{ab}, is defined as the radius of a hemispherical volume of purely absorbing gas (with the same temperature and absorption coefficient as that of the actual mixture) which produces a flux to the center of its base equal to the actual heat flux. Mathematically, L_{ab} is given by the relation

$$1 - e^{-(1-\omega_0)KL_{ab}} = \frac{(1-e^{-KL_e})(1-\omega_0)}{1 - \frac{\omega_0 \overline{gg}}{4KV_g}} \tag{9}$$

where the exchange factor, $\overline{sg}$, is written in term of the conventional mean beam length, L_e as

$$\frac{\overline{sg}}{A} = 1 - e^{-KL_e} \tag{10}$$

The concept of AMBL is useful when it is necessary to estimate the effect of scattering particles on the emission of a second species (e.g. the evaluation of the strength of a gaseous absorption band in the presence of scattering particles in furnace calculation). Note that the mean beam length expression considered by Cartigny (1986) is the absorption mean beam length.

3b. Concept of Extinction Mean Beam Length (EMBL)

The extinction mean beam length (EMBL), L_{ex}, is defined as the radius of a hemispherical volume of scattering and absorbing gas (with the same temperature, extinction coefficient and scattering albedo as that of the actual mixture) which produces a flux to the center of its base equal to the actual heat flux. This concept is important for the scaling of a mixture in which the effect of absorption and scattering are not readily separable (e.g. the assessment and comparison of optical thicknesses for two geometrically dissimilar sooty flames). Unlike L_e and L_{ab}, closed form expression for L_{ex} is not available for an arbitrary enclosure since the heat flux to the center of the base of an absorbing/scattering hemisphere cannot be expressed in closed form. A universal relation between L_{ab} and L_{ex}, however, can be readily generated based on the expression of L_{ab} for a hemispherical volume of gas radiating to the center of its base. Since L_{ex} for a hemisphere radiating to its base is simply its radius, L_{ex} and L_{ab} for an arbitrary enclosure are related by

$$1 - e^{-(1-\omega_0)KL_{ab}} = \frac{(1-e^{-KL_{ex}})(1-\omega_0)}{1 - \omega_0\left(\frac{\overline{gg}}{4KV_g}\right)_{hemisphere}} \tag{11}$$

where $(\overline{gg}/4KV_g)_{hemisphere}$ is the self-exchange factor per unit volume for a hemispherical volume of gas with optical radius KL_{ex}. Specifically, the evaluation of EMBL for an arbitrary enclosure requireds first the evaluation of AMBL based on equation (9) and then the evaluation of EMBL based on equation (11).

The evaluation of $(\overline{gg}/4KV_g)_{hemisphere}$, which has not appeared in the literature, is presented in Appendix A. Numerical values for KL_{ab} for different values of KL_{ex} and ω_0 are presented in Fig. 3. It can be readily observed that, in general, $L_{ab} \le L_{ex}$. In a hemispherical geometry, the primary effect of scattering is to reduce heat transfer. A larger absorbing/scattering hemisphere is required to generate the same heat flux to the center of its base as that of a purely absorbing hemisphere.

3c. Limiting Behavior of L_{ex} and L_{ab}

In the optically thin limit ($KL_{ex} \to 0$) and independent of ω_0, equations (9) and (11) are reduced to

$$L_e = L_{ex} = L_{ab} \tag{12}$$

Based on numerical data presented in Fig. 3, L_{ex} and L_{ab} are essentially identical for $KL_{ex} \le 1.0$.

Physically, the three mean beam lengths are also expected to be identical in the limit of no scattering ($\omega_0 = 0$). Numerical data, however, show that the convergence of L_{ab} to L_{ex} is quite slow in the optically thick limit. In Fig. 4, the relation between L_{ab} and L_{ex} in the weak-scattering limit ($\omega_0 \le 0.1$) are presented. Note that even for a "small" scattering albedo (say, $\omega_0 = 0.01$), the difference between L_{ex} and L_{ab} can be quite substantial in systems with moderate and large optical thicknesses. Equation (12), in the more restrictive optically-thin weakly-scattering limit ($KL_{ab} \to 0, \omega_0 \to 0$), is also deduced by Cartigny (1986).

In the strongly scattering limit ($\omega_0 \to 1.0$), equation (9) can be simplified to yield

$$KL_{ab} = \frac{\overline{sg}/A}{1 - \frac{\overline{gg}}{4KV_g}} \tag{13}$$

and the relation between L_{ab} and L_{ex} becomes

$$KL_{ab} = \frac{(1-e^{-KL_{ex}})}{1 - \omega_0\left(\frac{\overline{gg}}{4KV_g}\right)_{hemisphere}} \tag{14}$$

Note that AMBL and EMBL are both non-zero even in the pure scattering limit ($\omega_0 = 1.0$). Similar to the conventional MBL L_e, L_{ab} and L_{ex} are fundamental geometric parameters of the enclosure.

If A is the total surface of the enclosure (i.e. $A_c = 0$ in Fig. 2), equation (13) can be further simplified (using the reciprocity and closure relations) to yield

$$L_{ab} = \frac{4V_g}{A} \tag{13a}$$

which is identical to the optically thin limiting expression for L_e. Physically, this result is not surprising because from the pure absorption consideration, a pure scattering medium is optically thin. It is important to emphasize that equation (13a) holds for all strongly scattering media independent of optical thickness. L_{ex}, in the pure scattering limit, must still be evaluated using equation (14).

4. NUMERICAL EXAMPLES

To illustrate quantitatively the effect of scattering on AMBL and EMBL, numerical values are tabulated for two specific geometries, a sphere and an infinite parallel slab. These two cases are selected because many practical furnaces can be approximated by these geometries and closed form expressions for the necessary exchange factors $\overline{sg}/A$ and $\overline{gg}/(4KV_g)$ are available from standard references (Hottel and Sarofim, 1967).

For an absorbing-scattering spherical medium radiating to its surface, L_{ab} is given by

$$1 - e^{-(1-\omega_0)KL_{ab}} = \frac{\frac{\overline{sg}}{A}(1-\omega_0)}{1-\omega_0\left(1-\frac{3}{4KR}\frac{\overline{sg}}{A}\right)} \tag{14}$$

where

$$\frac{\overline{sg}}{A} = 1 - \frac{1}{2(KR)^2}\left[1-(2KR+1)e^{-2KR}\right] \tag{15}$$

with R being the radius of the sphere. Numerical values of AMBL and EMBL for different KR and ω_0 are presented in Fig. 5a and 5b. While L_{ex}, in general, increases monotonically with ω_0, L_{ab} first decreases and then increases with increasing ω_0. Physically, the slight decrease in L_{ab} with ω_0 at small albedo can be attributed to the slight enhancement of heat transfer by scattering in an optically thick medium. As ω_0 increases, the effect of extinction by scattering becomes dominant and L_{ab} increases. A comparison between Fig. 5a and 5b also shows that, in general, L_{ex} is much greater than L_{ab}, particularly in the limit of large albedo.

For a parallel-slab of absorbing-scattering medium radiating to one of its two bounding surfaces, L_{ab} is given by

$$1 - e^{-(1-\omega_0)KL_{ab}} = \frac{\frac{\overline{sg}}{A}(1-\omega_0)}{1-\omega_0\left(1-\frac{1}{2KD}\frac{\overline{sg}}{A}\right)} \tag{16}$$

where

$$\frac{\overline{sg}}{A} = 1 - 2E_3(KD) \tag{17}$$

with D being the thickness of the slab and $E_3(x)$ the exponential integral function (Hottel and Sarofim, 1967). Numerical results of L_{ex} and L_{ab} for a slab are shown in Fig. 6a and 6b. Their qualitative behavior is similar to that for a sphere presented in Fig. 5a and 5b.

5. CONCLUSIONS

A network analogy is developed for the analysis of radiative transfer in an absorbing and isotropically scattering medium. Based on a network analysis for an isothermal medium and assuming that the incoming and outgoing flux density are homogeneous, the traditional concept of mean beam length is extended to include the effect of scattering. Two concepts of mean beam length (an absorption mean beam length, AMBL and an extinction mean beam length, EMBL) are shown to be useful in characterizing the radiative transfer in a scattering medium. Based on an analysis of a hemispherical absorbing/scattering medium, a universal relation between AMBL and EMBL is developed. Numerical values for AMBL and EMBL for two enclosures are presented to show that the two mean beam lengths differ significantly from each other and also from the conventional mean beam length in systems with moderate or large optical thickness. The use of the conventional definition of mean beam length in general absorbing/scattering media can thus lead to significant error, except in the optically thin limit.

The general behavior of AMBL and EMBL are illustrated by numerical results generated by two specific systems, a spherical medium radiating to its surface and an infinite parallel slab radiating to one of its surfaces. For a fixed extinction coefficient and physical dimension, EMBL is shown to increase monotonically with scattering albedo while AMBL first decreases and then increases with scattering albedo.

Since an isothermal absorbing/scattering medium can, in general, have non-uniform incoming and outgoing flux density, expressions for AMBL and EMBL developed in the present work are approximate. Additional exact numerical calculations must also be carried out to develop a more precise relation between the two mean beam lengths and their mathematical behavior. Systems with different geometrical shapes must be analyzed in order to generate quantitative relations which are useful for practical applications. These efforts are currently under way and results will be presented in future publications.

6. REFERENCES

Bevans J. T. and Dunkle, R. V., 1960, "Radiative Exchange Within an Enclosure," *ASME Journal of Heat Transfer,* Vol. 82, pp. 1-19.

Cartigny, J. D., 1986, "Mean Beam Length for A Scattering Medium," *Heat Transfer - 1986,* C. L. Tien, V. P. Carey and J. K. Farrell (Eds.), Vol. 2, pp. 769-772.

DeRis, J., 1978, "Fire Radiation - A Review," *The Seventeenth Symposium (International) on Combustion,* The Combustion Institute, Pittsburgh, pp. 1003-1016.

Hottel, H. C., 1927, *Transaction American Institute of Chemical Engineers,* Vol. 19, p. 173.

Hottel, H. C. and Sarofim, A. F., 1967, *Radiative Transfer,* McGraw Hill, Inc.

Menguc, M. P. and Viskanta, R., 1986, "An Assessment of Spectral Radiative Transfer Predictions for a Pulverized Coal-Fired Furnace," *Heat Transfer - 1986,* C. L. Tien, V. P. Carey and J. K. Farrell (Eds.), Vol. 2, pp. 815-820.

Oppenheim, A. K., 1956, "Radiation Analysis by the Network Method," *Trans. ASME,* Vol. 78, pp. 725-735.

Sarofim, A. F., 1986, "Radiative Heat Transfer in Combustion: Friend or Foe," *The Twenty-First Symposium (International) on Combustion,* The Combustion Institute, Pittsburgh, pp. 1-23.

Tong, T. W. and Tien, C. L., 1980, "Resistance-Network Representation of Radiative Heat Transfer with Particulate Scattering," *Journal of Quantitative Spectroscopy and Radiative Transfer,* Vol. 24, pp. 491-503.

Viskanta, R. and Menguc, M. P., 1987, "Radiation Heat Transfer in Combustion Systems," *Progress in Energy and Combustion Science,* Vol. 13, pp. 97-160.

Wessel, R. A., 1985, "Effective Radiative Properties for Particle-Gas Mixtures," *Heat Transfer in Fire and Combustion Systems,* C. K. Law, Y. Jaluria, W. W. Yuen and K. Miyasaka (Eds.), pp. 239-249.

7. APPENDIX A

The evaluation of $\overline{gg}/4KV_g$ for a hemispherical volume of gas is presented in this section. Let A_1 and A_2 be the hemispherical surface and the base surface respectively, equations (4a) and (4b) can be readily utilized to yield

$$\frac{\overline{gg}}{4KV_g} = 1 - \frac{\overline{gs_1}}{4KV_g} - \frac{\overline{gs_2}}{4KV_g} \qquad \text{(A1)}$$

where the two surface-gas exchange factors can be written as

$$\frac{\overline{s_1g}}{A_1} = 1 - \frac{\overline{s_1s_1}}{A_1} - \frac{\overline{s_1s_2}}{A_1} \qquad \text{(A2)}$$

and

$$\frac{\overline{s_2g}}{A_2} = 1 - \frac{\overline{s_2s_1}}{A_2} \qquad \text{(A3)}$$

The evaluation of $\overline{gg}/4KV_g$ thus requires the evaluation of the two surface exchange factors $\overline{s_1s_1}$ and $\overline{s_1s_2}$.

The relevant geometry and coordinate system utilized in the evaluation of $\overline{s_1s_2}$ is shown in Fig. A1. For the differential areas dA_1 and dA_2 as shown, the differential exchange factor $\overline{ds_1ds_2}$ is given by

$$\overline{ds_1ds_2} = dA_1dA_2\frac{\cos\theta\cos\alpha\, e^{-Kd}}{\pi d^2} \qquad \text{(A4)}$$

where d is the distance between dA_1 and dA_2 and θ and α are as defined in Fig. A1.

Mathematically, it can be readily shown that a direct numerical integration based on a cartesian coordinate is ineffective in the evaluation of $\overline{s_1s_1}$ and $\overline{s_1s_2}$ because the integrand becomes singular as $d \to 0$. To overcome this difficulty, equation (A4) is rewritten in terms of a differential solid angle $d\omega_2$ as

$$\overline{ds_1ds_2} = dA_1d\omega_2\frac{\cos\theta\, e^{-Kd}}{\pi} \qquad \text{(A5)}$$

where $d\omega_2$ is given by

$$d\omega_2 = \frac{dA_2\cos\alpha}{d^2} = \sin\theta\, d\theta\, d\phi \qquad \text{(A6)}$$

with ϕ being the azimutual angle measured in the surface of hemisphere A_2' as shown in Fig. A1. In terms of the angles β, θ and ϕ, d is given by

$$d = \frac{R\cos\beta}{\cos\theta\cos\beta - \sin\beta\sin\theta\cos\phi} \qquad \text{(A7)}$$

The integration over dA_2 becomes

$$\overline{ds_1s_2} = \frac{dA_1}{\pi}F(\beta) \qquad \text{(A8)}$$

with

$$F(\beta) = \int_0^{2\pi}\int_0^{\theta_1} e^{-Kd}\cos\theta\sin\theta\, d\theta\, d\phi \qquad \text{(A9)}$$

$$+ 2\int_{\phi_m(\beta,\theta)}^{\pi}\int_{\theta_1}^{\theta_2} e^{-Kd}\cos\theta\sin\theta\, d\theta\, d\phi$$

where

$$\theta_1 = \frac{1}{2}\left(\frac{\pi}{2} - \beta\right) \qquad \text{(A10a)}$$

$$\theta_2 = \frac{1}{2}\left(\frac{\pi}{2} + \beta\right) \qquad \text{(A10b)}$$

and

$$\phi_m(\beta,\theta) = \cos^{-1}(\cos 2\theta\, cot\, \beta) \qquad \theta \le \frac{\pi}{4} \qquad \text{(A11)}$$

$$= \pi + \cos^{-1}(\cos 2\theta\, cot\, \beta) \qquad \theta > \frac{\pi}{4}$$

Carrying out the integration over A_1, one obtains

$$\frac{\overline{s_1s_2}}{R^2} = 2\int_0^{\frac{\pi}{2}} F(\beta)\sin\beta\, d\beta \qquad \text{(A12)}$$

In a similar manner, the exchange factor $\overline{s_1s_1}$ can be expressed as

$$\frac{\overline{s_1s_1}}{R^2} = 4\int_0^{\frac{\pi}{2}}\sin\beta\, d\beta\int_{\theta_1}^{\theta_2}\phi_m(\beta,\theta)\, e^{-2KR\cos\theta}\cos\theta\sin\theta\, d\theta \qquad \text{(A13)}$$

$$+ \frac{\pi}{(KR)^2}\left(1 - \int_0^{\frac{\pi}{2}}\left[2KR\left(\frac{1-\sin\beta}{2}\right)^{1/2} + 1\right] e^{-2KR\left(\frac{1-\sin\beta}{2}\right)^{1/2}}\sin\beta\, d\beta\right)$$

Based on equations (A1), (A2), (A3), and (A8) through (A13), numerical values of $\overline{gg}/4KV_g$ are tabulated for various KR and presented in Table A1. These data, together with equation (11), generate the universal relation between the two mean beam lengths, L_{ab} and L_{ex}, as shown in Fig. 3.

NOMENCLATURE

A_i = ith surface area in an enclosure
D = thickness of the parallel slab system
E_i = blackbody emissive power of surface A_i
$E_{g,i}$ = blackbody emissive power of volume $V_{g,i}$
$E_3(x)$ = exponential integral function
$F(\beta)$ = function defined in equation (A9)
$\overline{g_ig_j}$ = volume-volume direct exchange factor between $V_{g,i}$ and $V_{g,j}$
$\overline{g_is_j}$ = volume-gas direct exchange factor between $V_{g,i}$ and A_j
H_i = irradiation into surface A_i
$4H_{g,i}$ = incoming flux density into volume $V_{g,i}$
K = extinction coefficient
L_{ab} = absorption mean beam length
L_e = conventional mean beam length (defined for a pure absorption medium)
L_{ex} = extinction mean beam length
q = heat flux
Q_i = total heat transfer from surface A_i
$Q_{g,i}$ = total heat transfer from volume $V_{g,i}$
R = radius
$\underline{D}$ = thickness of the parallel slab system
$\overline{s_is_j}$ = surface-surface direct exchange factor between A_i and A_j
$\overline{s_ig_j}$ = surface-gas direct exchange factor between A_i and V_j
$V_{g,i}$ = ith gas volume in an enclosure
W_i = radiosity from surface A_i
$4KW_{g,i}$ = outgoing flux density from volume $V_{g,i}$
α = angle defined in Fig. A1
β = angle defined in Fig. A1
ε_i = emissivity of surface A_i

ω_0 = scattering albedo
θ = angle defined in Fig. A1
θ_1 = angle defined by equation (A10a)
θ_2 = angle defined by equation (A10a)
ϕ = azimutual angle used in equations (A6), (A7) and (A9)
ϕ_m = angle defined by equation (A11)

KR	$\overline{s_1 s_1}/A_1$	$\overline{s_1 s_2}/A_1$	$\overline{gg}/4KV_g$
0.00	0.5000e+00	0.5000e+00	0.0000e+00
0.01	0.4944e+00	0.4962e+00	0.7716e-02
0.05	0.4727e+00	0.4813e+00	0.2858e-01
0.10	0.4471e+00	0.4633e+00	0.5266e-01
0.20	0.4006e+00	0.4301e+00	0.1029e+00
0.30	0.3597e+00	0.3999e+00	0.1490e+00
0.40	0.3241e+00	0.3724e+00	0.1917e+00
0.50	0.2920e+00	0.3474e+00	0.2301e+00
0.60	0.2641e+00	0.3245e+00	0.2665e+00
0.70	0.2392e+00	0.3037e+00	0.2998e+00
0.80	0.2171e+00	0.2845e+00	0.3308e+00
0.90	0.1976e+00	0.2670e+00	0.3597e+00
1.00	0.1802e+00	0.2510e+00	0.3867e+00
2.00	0.8044e-01	0.1458e+00	0.5770e+00
5.00	0.1740e-01	0.5283e-01	0.7935e+00
10.00	0.4681e-02	0.2382e-01	0.8914e+00

Table A1:
Numerical Values for the Various Exchange Factors for a Hemispherical Gas Volume

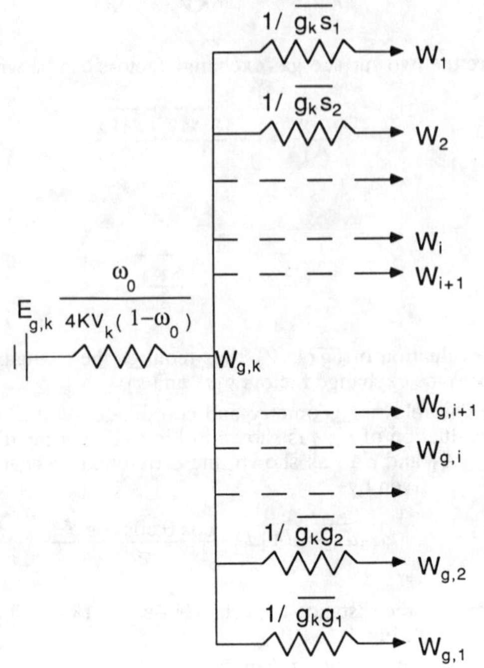

Fig. 1b: Network Representation for a Volume Element V_k.

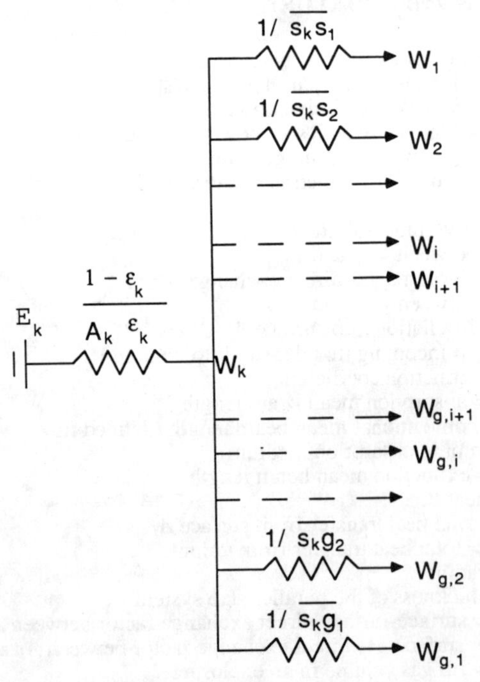

Fig. 1a: Network Representation for a Surface Element A_k.

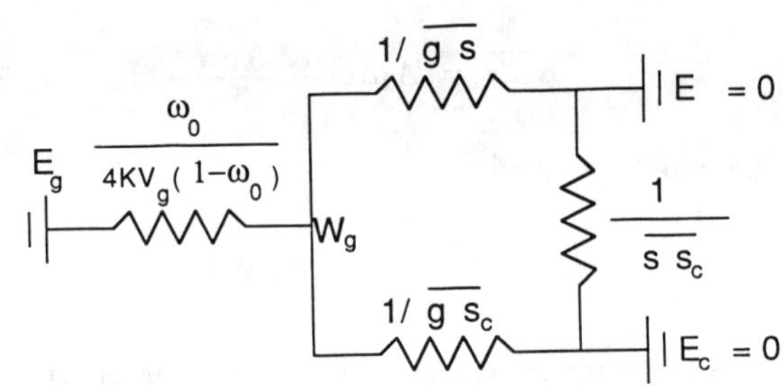

Fig. 2: Network Representation for the Mean Beam Length Calculation.

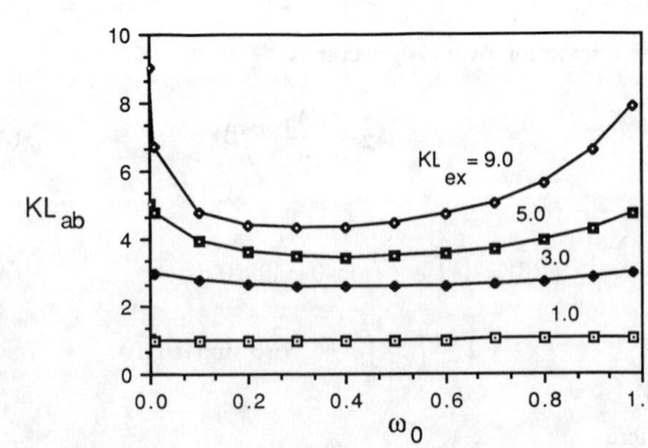

Fig. 3: Universal Relation between KL_{ab}, KL_{ex} and ω_0.

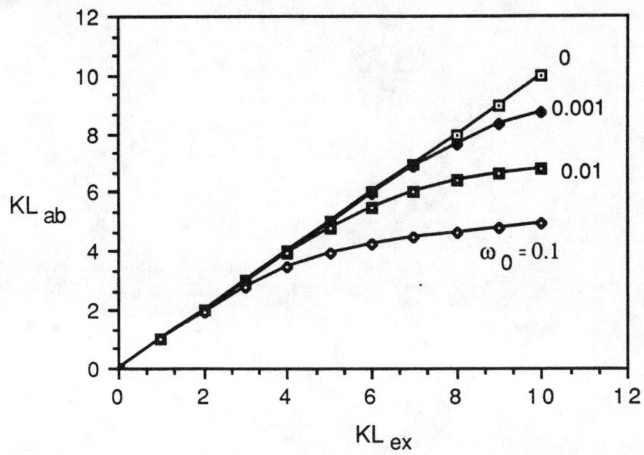

Fig. 4: Relation between KL_{ab} and KL_{ex} in the Weak Scattering Limit.

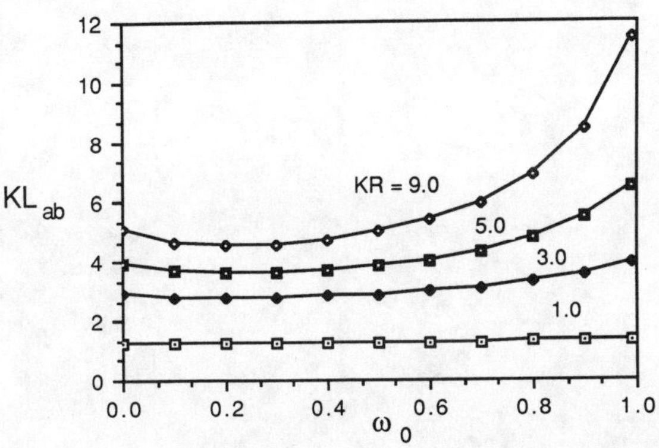

Fig. 5a: Absorption Mean Beam Length for a Sphere Radiating to its Surface for Different Optical Radius (*KR*) and Albedo (ω_0)

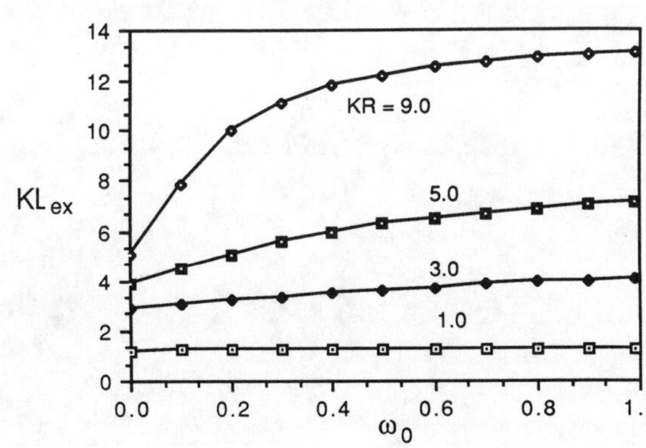

Fig. 5b: Extinction Mean Beam Length for a Sphere Radiating to its Surface for Different Optical Radius (*KR*) and Albedo (ω_0)

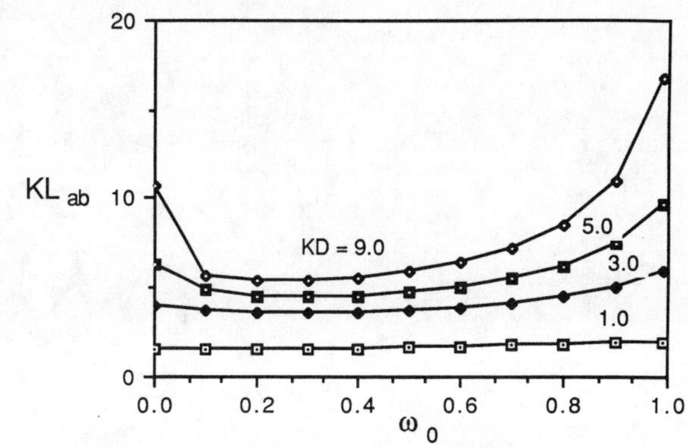

Fig. 6a: Absorption Mean Beam Length for an Infinite Parallel Slab Radiating to One of its Surfaces for Different Optical Slab Thickness (*KD*) and Albedo (ω_0)

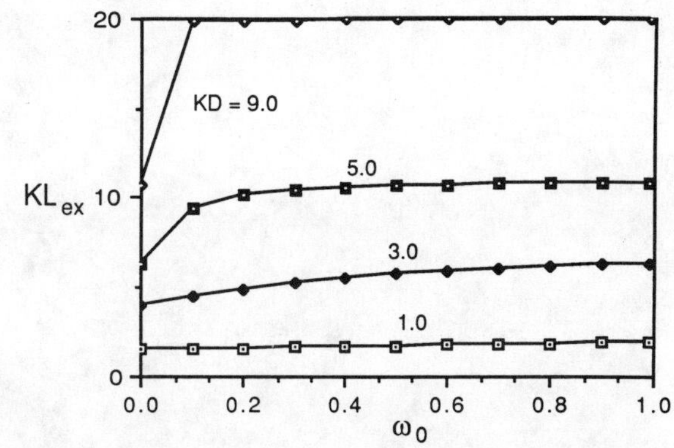

Fig. 6b: Extinction Mean Beam Length for an Infinite Parallel Slab Radiating to One of its Surfaces for Different Optical Slab Thickness (*KD*) and Albedo (ω_0)

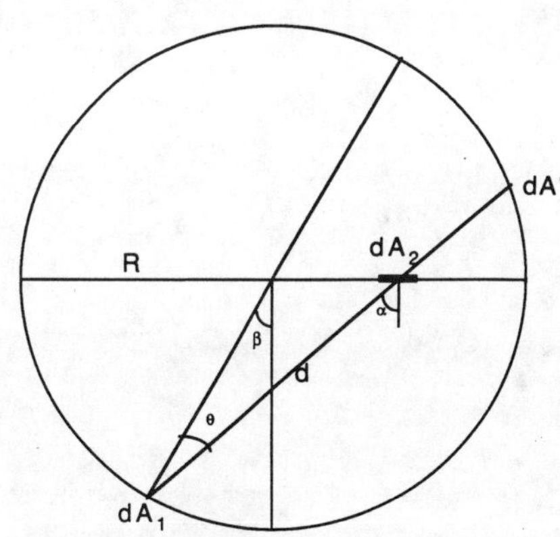

Fig. A1: Geometry and Coordinate System for the Evaluation of $\left(\frac{\overline{gg}}{4KV_g}\right)_{hemisphere}$

1989 National Heat Transfer Conference
HTD-Vol. 106, Heat Transfer Phenomena in Radiation, Combustion and Fires

RADIATION HEAT TRANSFER IN A PARTIALLY DIVIDED SQUARE ENCLOSURE WITH A PARTICIPATING MEDIUM

Z. Tan and J. R. Howell
Department of Mechanical Engineering
The University of Texas at Austin
Austin, Texas

This paper investigates radiation heat transfer in a two-dimensional emitting, absorbing and linearly anisotropically scattering square medium containing an obstructing partition. A newly developed numerical method, the YIX method, is applied to solve the exact radiation transport equations of integral form. Influences of the length and emissivity of an insulating or conducting partial partition, the emissivity of the enclosure, the scattering parameters, and the optical length on heat transfer and temperature are studied. It is found that the partition length and type have significant influence on heat transfer and temperature distributions, and anisotropic scattering in the medium cannot be neglected. Comparison with results by another method shows that the accuracy of the present computations is very satisfactory.

NOMENCLATURE

(All quantities are dimensionless.)

e_g	=	radiative blackbody emissive power of medium
e_s	=	radiative blackbody emissive power of enclosure
$\mathbf{n}$	=	inward unit normal vector at $\mathbf{r}$
q_r	=	radiation flux vector
q_s	=	heat flux on S
$\mathbf{r}$	=	(x,y), position vector
$R(\omega_k)$	=	distance from $\mathbf{r}$ to boundary along ω_k
S	=	boundary of Ω
S_n	=	Bickley's function of nth order
w_1, w_2	=	radiation flux in x and y directions, respectively
β	=	coefficient for linear anisotropic scattering
$\delta(\mathbf{r},\mathbf{r}')$	=	switch function for obstruction defined by Eq.(1d)
ε	=	emissivity
τ	=	optical depth
ω	=	albedo for scattering
ω_k	=	discrete angular coordinates in Eq.(3)
Ω	=	medium

INTRODUCTION

It is well known that the presence of obstructing partitions has a significant influence on radiation heat transfer. Many industrial furnace designs include internal partitions. Solution of one-dimensional problems in non-participating media (e.g., radiation shields) is available in most heat transfer textbooks. Only a few solutions of multi-dimensional problems in participating media, however, are available. Using the zonal method, Chen (1982) studied the influence of obstructing walls on the heating process in reheating furnaces. Since a coarse grid system was used, Chen's results have low accuracy. Chang, Yang and Lloyd (1983) studied combined radiation and convection heat transfer in a partially divided square medium using a radiation model of low-order accuracy. Detailed investigation of the influence of dividing walls on radiation in participating media is still needed.

Theoretically, any numerical technique that solves the exact radiation transport equation of integro-differential or integral form, e.g., the Monte Carlo method (Howell, 1968), the zonal method (Hottel and Cohen, 1958), the finite element method (Chung and Kim, 1984; Razzaque, Howell and Klein, 1984), the product integration method (Tan, 1989a; Tan, 1989b), or the discrete ordinates methods (Fiveland, 1984), can be used after some modification, to solve such problems. However, many methods are inefficient in dealing with complex geometries such as systems with obstructing walls. In this paper, a newly developed method, the YIX method which has been applied to various kinds of problems (Tan and Howell, 1989), is used. Radiation heat transfer in a two-dimensional emitting, absorbing and linearly anisotropic scattering square medium with an insulated or conducting partition is studied. The influences of various parameters on the heat flux and emissive power on the boundary walls are discussed. Some of the present results are compared with those given by another technique.

FORMULATION

Consider a gray, emitting, absorbing and linearly anisotropically scattering square medium, bounded by a gray diffusely reflecting enclosure. The upper and lower walls are insulated, while the left and the right walls are kept at constant nondimensional emissive powers e_s=0.5 and e_s=−0.5, respectively. An infinitesimally thin obstructing partition is placed in the system with the position shown in Fig.1. The partition may be perfectly insulated, or perfectly conducting. For simplicity, all physical properties are assumed to be constant in space, except in the perfectly conducting case where the emissivity of the partition can be a constant different from the emissivity of the enclosure surfaces. It is also assumed that there is no heat source or sink in the system.

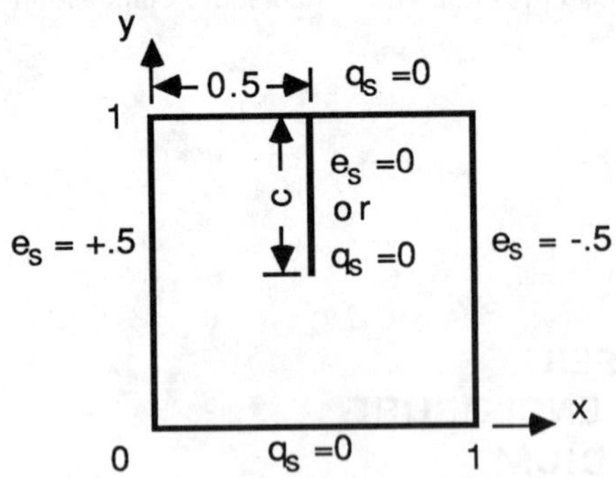

Fig. 1. Geometry and coordinates of a two-dimensional radiation problem.

In nondimensional form, the integral equations for the present two-dimensional radiation transport are (Tan, 1989a):

$$4e_g(\mathbf{r}) - \frac{1}{(1-\omega)\tau}\nabla\cdot\mathbf{q}_r(\mathbf{r}) = \int_\Omega [e_g(\mathbf{r}') - \frac{\omega}{4(1-\omega)\tau}\nabla\cdot\mathbf{q}_r(\mathbf{r}')]\frac{S_1(\tau|\mathbf{r}-\mathbf{r}'|)}{|\mathbf{r}-\mathbf{r}'|}\delta(\mathbf{r},\mathbf{r}')dA(\mathbf{r}')$$

$$+\frac{\omega\beta}{4}\int_\Omega \frac{\Delta x_1 w_1(\mathbf{r}') + \Delta x_2 w_2(\mathbf{r}')}{|\mathbf{r}-\mathbf{r}'|}\frac{S_2(\tau|\mathbf{r}-\mathbf{r}'|)}{|\mathbf{r}-\mathbf{r}'|}\delta(\mathbf{r},\mathbf{r}')dA(\mathbf{r}')$$

$$+\int_S [e_s(\mathbf{r}') - \frac{1-\varepsilon}{\varepsilon}q_s(\mathbf{r}')]\frac{S_2(\tau|\mathbf{r}-\mathbf{r}'|)}{|\mathbf{r}-\mathbf{r}'|}\cos(\mathbf{r}-\mathbf{r}',\mathbf{n}')\delta(\mathbf{r},\mathbf{r}')dl(\mathbf{r}') \qquad \mathbf{r}\in\Omega \qquad (1a)$$

$$w_j(\mathbf{r}) = \int_\Omega [e_g(\mathbf{r}') - \frac{\omega}{4(1-\omega)\tau}\nabla\cdot\mathbf{q}_r(\mathbf{r}')]\frac{\Delta x_j}{|\mathbf{r}-\mathbf{r}'|}\frac{S_2(\tau|\mathbf{r}-\mathbf{r}'|)}{|\mathbf{r}-\mathbf{r}'|}\delta(\mathbf{r},\mathbf{r}')dA(\mathbf{r}')$$

$$+\frac{\omega\beta}{4}\int_\Omega \frac{\Delta x_1 w_1(\mathbf{r}') + \Delta x_2 w_2(\mathbf{r}')}{|\mathbf{r}-\mathbf{r}'|}\frac{\Delta x_j}{|\mathbf{r}-\mathbf{r}'|}\frac{S_3(\tau|\mathbf{r}-\mathbf{r}'|)}{|\mathbf{r}-\mathbf{r}'|}\delta(\mathbf{r},\mathbf{r}')dA(\mathbf{r}')$$

$$+\int_S [e_s(\mathbf{r}') - \frac{1-\varepsilon}{\varepsilon}q_s(\mathbf{r}')]\frac{\Delta x_j}{|\mathbf{r}-\mathbf{r}'|}\frac{S_3(\tau|\mathbf{r}-\mathbf{r}'|)}{|\mathbf{r}-\mathbf{r}'|}\cos(\mathbf{r}-\mathbf{r}',\mathbf{n}')\delta(\mathbf{r},\mathbf{r}')dl(\mathbf{r}') \qquad \mathbf{r}\in\Omega,\ j=1,2 \qquad (1b)$$

$$e_s(\mathbf{r}) - \frac{1}{\varepsilon}q_s(\mathbf{r}) = \int_\Omega [e_g(\mathbf{r}') - \frac{\omega}{4(1-\omega)\tau}\nabla\cdot\mathbf{q}_r(\mathbf{r}')]\frac{S_2(\tau|\mathbf{r}-\mathbf{r}'|)}{|\mathbf{r}-\mathbf{r}'|}\cos(\mathbf{r}'-\mathbf{r},\mathbf{n})\delta(\mathbf{r},\mathbf{r}')dA(\mathbf{r}')$$

$$+\frac{\omega\beta}{4}\int_\Omega \frac{\Delta x_1 w_1(\mathbf{r}') + \Delta x_2 w_2(\mathbf{r}')}{|\mathbf{r}-\mathbf{r}'|}\frac{S_3(\tau|\mathbf{r}-\mathbf{r}'|)}{|\mathbf{r}-\mathbf{r}'|}\cos(\mathbf{r}'-\mathbf{r},\mathbf{n})\delta(\mathbf{r},\mathbf{r}')dA(\mathbf{r}')$$

$$+\int_S [e_s(\mathbf{r}') - \frac{1-\varepsilon}{\varepsilon}q_s(\mathbf{r}')]\frac{S_3(\tau|\mathbf{r}-\mathbf{r}'|)}{|\mathbf{r}-\mathbf{r}'|}\cos(\mathbf{r}'-\mathbf{r},\mathbf{n})\cos(\mathbf{r}-\mathbf{r}',\mathbf{n}')\delta(\mathbf{r},\mathbf{r}')dl(\mathbf{r}') \qquad \mathbf{r}\in S \qquad (1c)$$

where e_g and e_s are nondimensional blackbody emissive powers at a space point and a surface point respectively, q_r is the radiation flux vector, q_s the radiation heat flux on a surface, w_i the radiation flux in the *i* th direction, $\Delta x_j \equiv x_j - x_j'$ (j=1,2), and S_i is Bickley's function defined by

$$S_i(t) \equiv \frac{2}{\pi}\int_0^{\pi/2} e^{-t/\cos\theta}\cos^{i-1}\theta\, d\theta$$

and $\delta(\mathbf{r},\mathbf{r}')$ is a switch function for obstructions defined by

$$\delta(\mathbf{r},\mathbf{r}') \equiv \begin{cases} 1 & \text{if } \mathbf{r} \text{ and } \mathbf{r}' \text{ can see each other} \\ 0 & \text{if } \mathbf{r} \text{ and } \mathbf{r}' \text{ cannot see each other} \end{cases} \qquad (1d)$$

$x_1\equiv$, $x_2\equiv y$, $\mathbf{r}\equiv(x,y)$, Ω is the medium ($-.5\le x\le .5$, $-.5\le y\le .5$), and S denotes all surfaces including the partition. In the present problem, $\nabla\cdot\mathbf{q}_r=0$ because there are no internal heat sources or sinks and no other modes of energy transfer.

The surface boundary conditions needed to solve Eqs.(1) are:

$$e_s(0,y) = 0.5, \quad e_s(1,y) = -0.5, \qquad 0<y<1 \qquad (2a)$$

$$q_s(x,0) = q_s(x,1) = 0, \qquad 0<x<1 \qquad (2b)$$

$$q_s(.5\pm\delta,y) = 0, \qquad c<y<1 \quad \text{in insulated case} \qquad (2c)$$

$$e_s(.5\pm\delta,y) = 0, \qquad c<y<1 \quad \text{in conducting case} \qquad (2d)$$

where 2δ is the thickness of the partition, and $\delta \ll 1$. Equation (2d) applies in the conducting case because of the symmetry of the problem around x=0.5.

NUMERICAL METHOD

To solve Eqs.(1) for the present system, the square medium is divided into 12×12 equal square elements, and the four boundary walls are divided into 4×12 equal segments. The number of segments on the partition depends on its length. For convenience, five values of the partition length, c, are considered: 0, 0.25, 0.5, 0.75 and 1; hence the corresponding numbers of segments on the two partition surfaces (each side) are 0, 2×3, 2×6, 2×9, and 2×12.

In this paper, the YIX method, which has been applied successfully to various problems, is used to solve the radiation problem to obtain the emissive power and/or heat flux on those discrete points.

The YIX method solves the exact radiation transport equations (say, as a special case, Eqs.(1)) written in an equivalent distance-angular integral form. The essence of the method is to construct integral formulae for the distance integral and store the resulting abscissas, so singularities are removed and numerous evaluations of the time-consuming kernel functions (e.g., Bickley's function in the two-dimensional case) are avoided. For the angular integration, equally spaced integration points are chosen. For example, assuming $\nabla\cdot\mathbf{q}_r(\mathbf{r})=0$ for a 2-D problem with constant properties and without scattering and surface reflection, the approximation for Eq.(1a) is

$$4e_g(\mathbf{r}) \approx \frac{1}{M}\sum_{k=1}^{M}\{\lambda[\tfrac{1}{2}e_g(\mathbf{r}) + \sum_{i=1}^{n-1}e_g(\mathbf{r}+\omega_k t_i)] + [P_{n+1} + (R(\omega_k) - t_n)Q_{n+1}]e_g(\mathbf{r}+\omega_k t_n)\}$$

$$+\frac{1}{M^*}\sum_{k=1}^{M^*}\cos(-\omega_k,\mathbf{n}')S_2(\tau R(\omega_k))e_s(\mathbf{r}+\omega_k R(\omega_k)) \qquad (3)$$

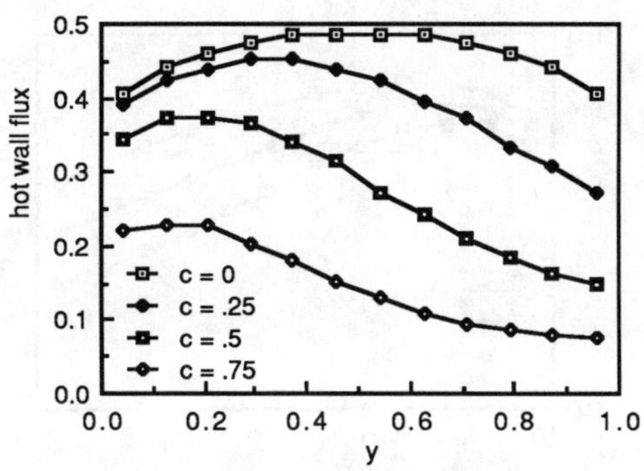

Fig. 2a. Influence of partition length on hot wall flux, when ε=1, τ=1, and ωβ=0. Insulated partition case.

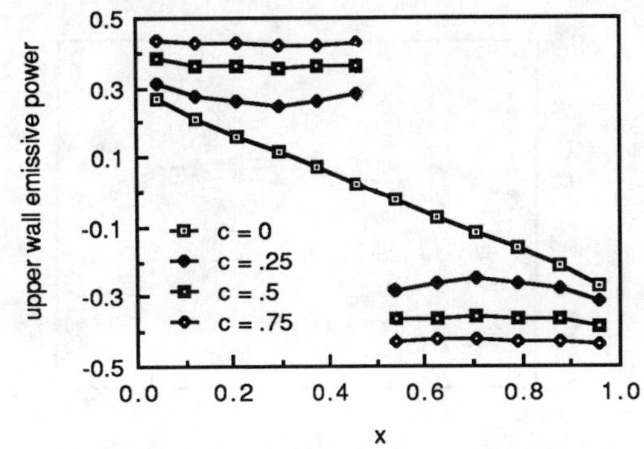

Fig. 2b. Influence of partition length on upper wall emissive power, when ε=1, τ=1, and ωβ=0. Insulated partition case.

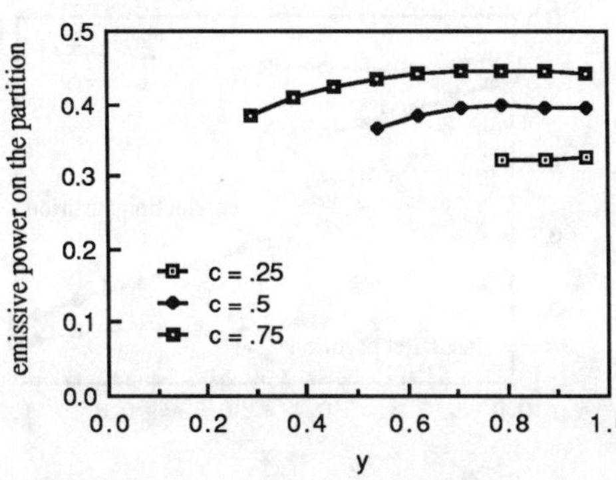

Fig. 2c. Influence of partition length on emissive power on partition, when ε=1, τ=1, and ωβ=0. Insulated partition case.

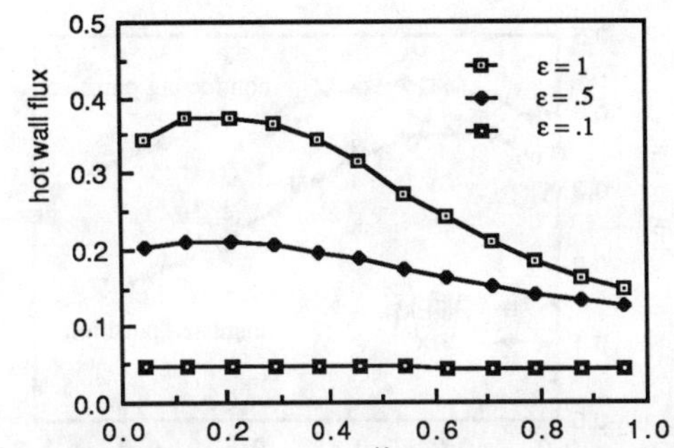

Fig. 3. Influence of emissivity on hot wall flux, when c=0.5, τ=1, and ωβ=0. Insulated partition case.

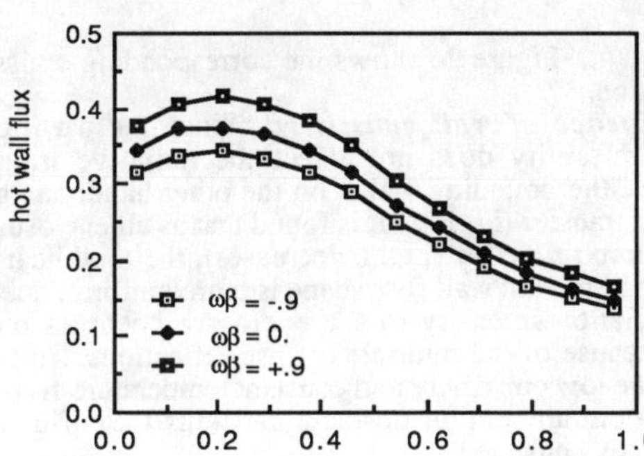

Fig. 4. Influence of anisotropic scattering on hot wall flux, when c=0.5, ε=1, and τ=1. Insulated partition case.

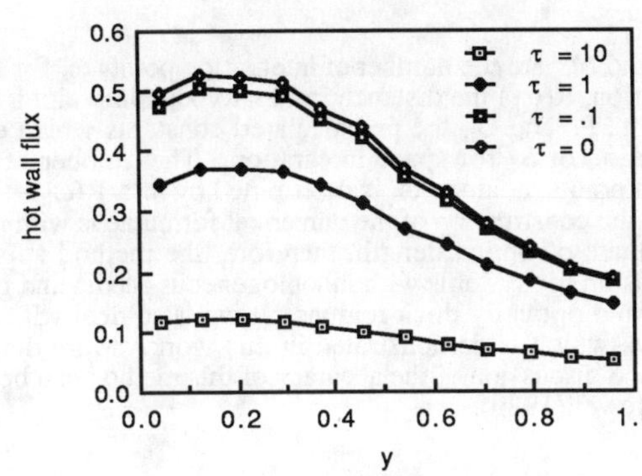

Fig. 5. Influence of optical thickness on hot wall flux, when c=0.5, ε=1, and ωβ=0. Insulated partition case.

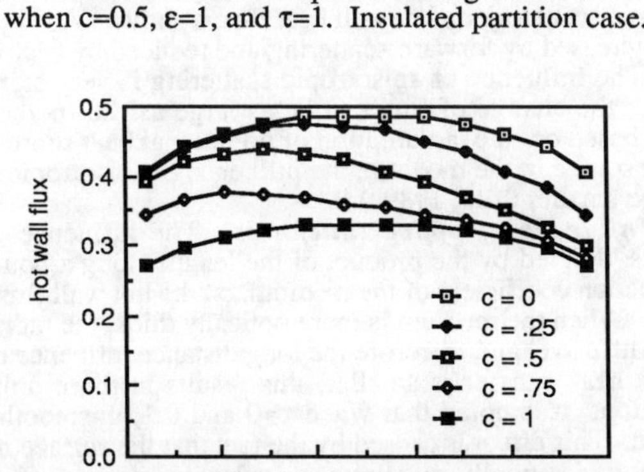

Fig. 6a. Influence of partition length on hot wall flux, when ε=1, τ=1, and ωβ=0. Conducting partition case.

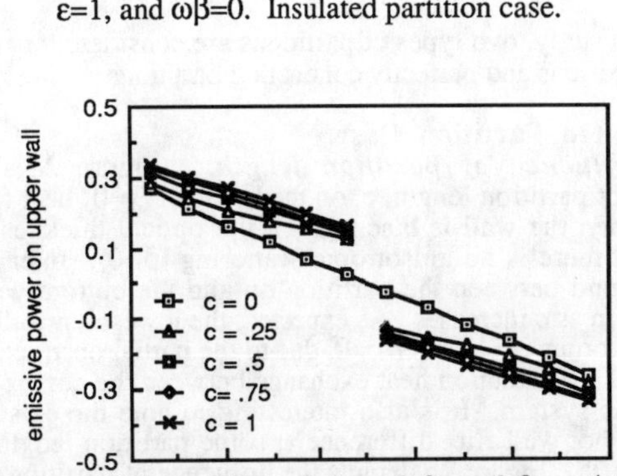

Fig. 6b. Influence of partition length on emissive power on upper wall, when ε=1, τ=1, and ωβ=0. Conducting partition case.

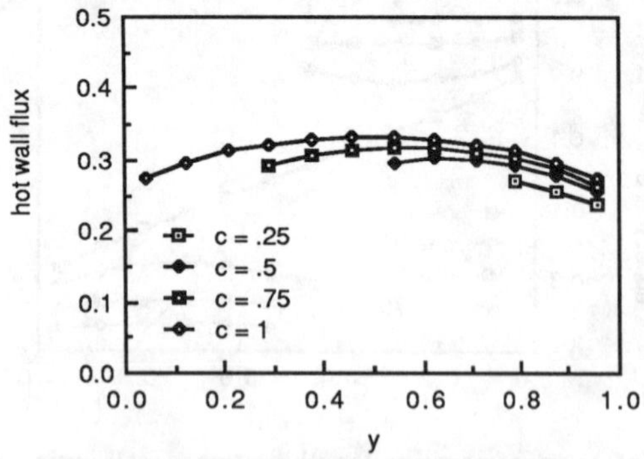

Fig. 6c. Influence of partition length on hot wall flux, when ε=1, τ=1, and ωβ=0. Conducting partition case.

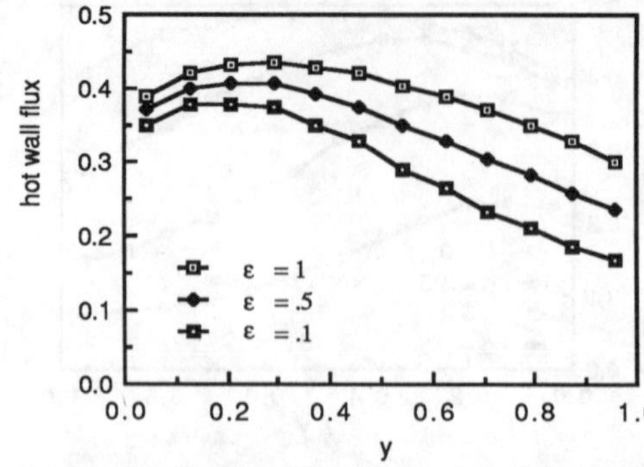

Fig. 7. Influence of partition emissivity on hot wall flux, when ε=1, τ=1, and ωβ=0. Conducting partition case.

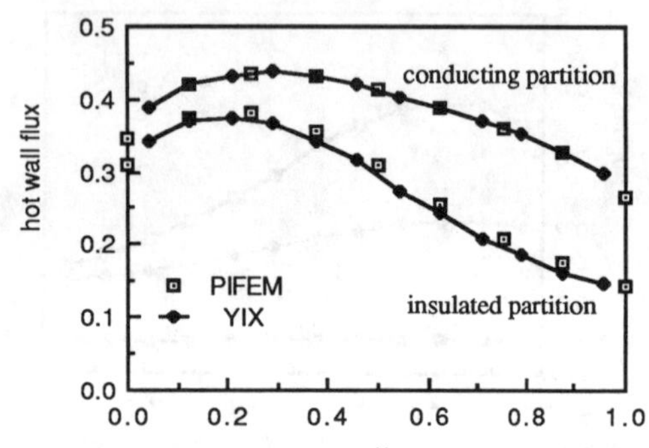

Fig. 8a. Comparative hot wall flux using YIX and PIFE methods, when c=0.5, ε=1, τ=1, and ωβ=0.

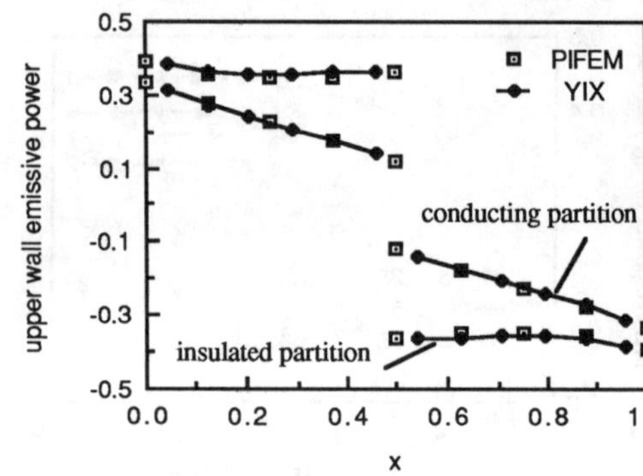

Fig. 8b. Comparative upper wall emissive power using YIX and PIFE methods, when c=0.5, ε=1, τ=1, and ωβ=0.

where M and M* are the number of integration points ω_k for the angular integration, $R(\omega_k)$ the distance from **r** to boundary along direction ω_k, and λ, t_i, P_i and Q_i the precalculated constants which eliminate the evaluation of S_1 for space integration. The number of distance integration points, n, along ω_k is determined by $t_n \leq R(\omega_k) < t_{n+1}$.

Since the construction of the numerical formulae is with respect to the coordinate of optical length, therefore, the method applies with little extension for systems with inhomogeneous media and both optically thin and optically thick regions. It can also deal with complex geometries, which is demonstrated in this work. More detailed description and discussion of the accuracy of this method can be found in Tan and Howell(1989).

RESULTS AND DISCUSSIONS

In this study, two types of partitions are considered: perfectly insulated partitions and perfectly conducting partitions.

1. Insulated Partition Case.

a. Influence of partition length. Figure 2a shows the influence of partition length, c, on the hot wall (x=0) heat flux distribution, when the wall is black (ε=1), the optical thickness is unity (τ=1), and there is no anisotropic scattering (β=0). In each case, a peak is found between the partition tip and the bottom wall, and it moves down as c increases. As expected, the lowest hot wall flux is at the upper corner (y=1) when c≠0, due to the partial obstruction by the partition of the radiation heat exchange between the corner and other parts of the system. It is also interesting to note the existence of a maximum hot wall flux difference at some partition length between 0.25 and 0.75. Figure 2b depicts the influence of partition length on the upper wall emissive power under the same conditions as in Fig.2a. As long as a partition is present, the temperature on the upper wall is discontinuous at the partition base point (x=0.5). It is noted that when c≥0.25, the upper wall emissive power on either the left or right half is nearly constant. Figure 2c shows the corresponding emissive power on the partition.

b. Influence of wall emissivity. Since the partition is insulated, its emissivity does not affect the radiative transfer. The emissivity of the bounding walls, on the other hand, has great influence on heat transfer (Fig.3). It is found that as all enclosure surfaces are made more reflective (i.e., ε decreases), the total heat transfer is decreased, and the hot wall flux shape is more uniform, due to the fact that the radiation intensity in the enclosure becomes more nearly isotropic because of the multiple diffuse reflections. In fact, reflections from the low emissivity and constant temperature surfaces enable radiation to circumvent an obstacle and thus the influences of the obstruction are weakened.

c. Influence of anisotropic scattering. In Fig.4 the effect of strong forward (ωβ=+0.9) and backward (ωβ=−0.9) linear anisotropic scattering on hot wall heat flux is shown. As expected, the flux is increased by forward scattering and reduced by backward scattering. The influence of anisotropic scattering is very significant in this case; the change of flux can be as large as 20 percent. This result is based on the assumption of no internal heat source. If there is a heat source in the medium, the influence of anisotropic scattering should be smaller (Tan, 1989a).

d. Influence of optical thickness. The influence of optical thickness, defined by the product of the length along a boundary and the extinction coefficient of the medium, on the hot wall flux is shown in Fig.5. When the medium is more optically thick, the radiation field is more diffusive, and therefore the long distance influence of the partition on heat transfer is smaller; this results in more uniform flux distribution. It is noted that when τ=0 and 0.1, unsmooth lines are predicted. This effect is caused by the fact that the surface integration calculation via equally distributed angular integration points may be inaccurate if the surface emissive power or flux is discontinuous along the boundary. Discussion and possible remedies for this form of inaccuracy are found in Tan and Howell(1989).

2. Conducting Partition Case.

In this case, because the emissive power in the y-direction is continuous across the centerline (x=0.5) in the presence of the partition, we expect the influence of the conducting partition to be smaller than for an insulated one.

a. Influence of partition length. In Figs.6a-6c the influences of partition length on hot wall heat flux, upper wall emissive power, and heat flux on the partition are shown. Similar trends in the hot wall flux as affected by the partition are found as for the insulated partition case; however, the flux level is higher. The upper wall emissive power is less affected in this case (Fig.6b). In fact, as $c>0.25$, the upper wall emissive power distribution is almost unchanged as c varies, although a jump in emissive power at the partition is still present. The change of the heat flux on the partition as c varies is also quite small.

b. Influence of partition emissivity. Because in this case the partition is not insulated, the emissivity of the partition does affect the radiation heat transfer process. Figure 7 shows the variation of hot wall flux as the emissivity of the partition changes while all other surfaces remain black. The influences are significant; and it is noted that as ε decreases, the hot wall flux becomes less uniform. This is because when the partition is more reflective, less energy is exchanged between the partition and other parts of the system, so the hot wall flux at points nearer the partition is smaller. At the limit $\varepsilon=0$, $q_s=0$ by Eqs.(1) and the system is equivalent to that with an insulated partition.

3. Numerical Accuracy of the Computations

It is difficult to give an exact error bound for the present computations since no exact solution is available in the literature. In all cases, the overall total dimensionless energy is balanced to the maximum error of 1.4×10^{-4}. However, as one can prove, this error only reflects the numerical integration accuracy, not the discretization error. Some information can be obtained by comparing present calculations with results using other methods.

The results in the case $c=0.5$, $\varepsilon=1$, $\tau=1$ and $\omega\beta=1$ for both insulated and conducting partitions using the YIX method and the Product Integration Finite Element Method (PIFEM) (Tan, 1989b) are compared and shown in Figs. 8a and 8b. Excellent agreement is found for the upper wall emissive power for both cases. The agreement in hot wall flux is also excellent in the conducting partition case. In the insulated partition case, a small discrepancy between the results is found (the maximum difference is 0.016, or less than five percent, at y=0.4).

CONCLUSIONS

The YIX method was applied to study the radiation transfer in a two-dimensional emitting, absorbing and anisotropically scattering square medium in the presence of an obstructing partition. Influences of some parameters are investigated and discussed. The following conclusion are obtained.

1) The YIX method can be applied to problems in complex geometries. Comparisons with the product integration finite element method (PIFEM) show that very satisfactory results can be obtained by the YIX method.

2) The presence of a partition can significantly reduce radiative heat transfer. When the medium is partially divided, the hot wall flux distribution is more uneven than with no partition or in the completely divided case. The influences of optical thickness, emissivities and anisotropic scattering parameters are also important to the behavior of partially divided systems.

3) The obstructing effect is greater for an insulated partition than for a conducting partition.

ACKNOWLEDGEMENTS

This work is supported by the Texas Advanced Research Program, Grant Number 4587, "Innovative Methods for Analyzing Complex Systems Involving Radiative Transfer."

REFERENCES

Chang, L.C., Yang, K.T., and Lloyd, J.R., 1983, "Radiation-Natural Convection Interaction in Two-Dimensional Complex Enclosures," *ASME J. of Heat Transfer*, **105**, 89-95.

Chen, Haigen, 1982, "Zonal Mathematical Model for Reheating Furnaces With Obstructing Walls," (in Chinese), Master's Thesis, Northeast University of Technology, Shenyang.

Chung, T.J., and Kim, J.Y., 1984, "Two-Dimensional, Combined-Mode Heat Transfer by Conduction, Convection and Radiation in Emitting, Absorbing and Scattering Media – Solution by Finite Elements," *ASME J. of Heat Transfer*, **106**, 448-452.

Fiveland, W. A., 1984, "Discrete Ordinates Solutions of the Radiative Transport Equation for Rectangular Enclosures," *J. Heat Transfer*, **106**, 699-706.

Hottel, H.C., and Cohen, E.S., 1958, "Radiant Heat Transfer in a Gas-Filled Enclosure: Allowance for Non-Uniformity of Gas Temperature," *AIChE J.*, **4**, 3-14.

Howell, J. R., 1968, "Application of Monte Carlo to Heat Transfer Problems," in *Advances in Heat Transfer*, **5**, J. P. Hartnett and T. Irvine, eds., Academic Press, New York.

Razzaque, M.M., Howell, J.R, and Klein, D.E., 1984, "Coupled Radiative and Conductive Heat Transfer in a Two-Dimensional Rectangular Enclosure With Gray Participating Media Using Finite Elements," *ASME J. of Heat Transfer*, **106**, 613-619.

Tan, Zhiqiang, 1989a, "Radiative Heat Transfer in Multidimensional Emitting, Absorbing and Anisotropic Scattering Media -- Mathematical Formulation and Numerical Method", *ASME J. of Heat Transfer,* **111**, 141-147.

Tan, Zhiqiang, 1989b, "Product-Integration Finite Element Method for Two-Dimensional Radiation Heat Transfer in Participating media," to be submitted.

Tan, Zhiqiang, and Howell, J.R., 1989, "A New Numerical Method for Radiation Heat Transfer in Participating Media," to be submitted.

1989 National Heat Transfer Conference
HTD-Vol. 106, Heat Transfer Phenomena in Radiation, Combustion and Fires

RADIATIVE HEAT TRANSFER FROM AN ABSORBING AND ISOTROPICALLY SCATTERING MEDIUM CONTAINED WITHIN A TRUNCATED CONE

J. D. Lin and C. S. Wang
Department of Mechanical Engineering
National Chiao Tung University
Hsinchu, Taiwan

ABSTRACT

Radiative heat transport from an isotropically scattering medium with truncated conical geometry and spatially varying radiative properties is formulated. Solutions for the radiation flux distribution are obtained by an iterative method and the nodal approximation technique. The radiant flux distribution is presented at the cone base as well as along the axial direction at a given distance from the center line of the cone. The results show that the effects of cone angle on the heat flux at the exterior base region are significant. Comparison of the results with existing data for cylindrical enclosures is also presented and shows good agreement.

NOMENCLATURE

I	intensity
I_b	black body intensity
K	absorption coefficient
L	length of cone
l	distance along the beam of radiation
$\hat{m}$	unit vector normal to lateral surface
$\hat{n}$	unit vector normal to base
$\hat{r}$	position vector
R_o,D	radius and diameter of cone base at Z=0
q	radiant flux
S	radiation source function
T	temperature
V	volume
Z,R,θ	cylindrical coordinates
σ	scattering coefficient
α	cone angle
β	extinction coefficient
Γ	boundary surface
ν	frequency
Ω	solid angle

superscript

*	refer to a point outside the cone
"	refer to a point within the cone
',s	refer to a boundary point

subscript

ν	spectral

INTRODUCTION

Thermal radiation from a participating medium in a conical geometry plays an important role in many engineering applications. For instance, radiation from a solid propellant rocket plume to aft-end equipment and structure is one of factors to be considered in order to design insulation and shielding. Identification as well as detection of a rocket are other factors to be considered. Other applications include heat transfer in MHD generators and other cone-shaped equipment which involve high temperature gases and particulate suspensions. Investigations of radiative heat transfer in a truncated-cone chamber can be found in the literature[Gorshkova, 1986; Kaminski, 1988]. Gorshkova(1986) applied the zonal method to solve for the radiation in a truncated-conical chamber. The medium is treated as isotropically scattering with uniform and constant radiation properties. Kaminski(1988) employed both a Monte-Carlo technique and a P-1 differential approximation to investigate radiative heat transport from a participating, but non-scattering medium also with uniform and constant radiation properties in a truncated, conical enclosure. However, in most of above-mentioned applications, the medium is not isothermal and materials properties are not uniform. The present formulation is motivated by a desire to include the spatially varying radiation properties in prediction of radiation from a truncated conical geometry.

In terms of plume radiation, several studies have been performed for similar geometries. Bobco(1966 and 1969; and Edwards, 1969) presented a model with opaque conical frustum to describe irradiation of a differential area which is simulated as the aft-end of a rocket. In his model, plume radiation is obtained in terms of the cone's radiosity which varies along the axial distance from the cone apex. Edwards and the others(1982, 1985) also used the conical frustum geometry and formulated the radiosity of plume. There are also some investigations in which the model is considered as an isothermal semi-infinite cylinder with a homogeneous particle distribution(Tien and Abu-Romia, 1964; Stockham and Love, 1969; Pearce, 1978; and Nelson, 1984).

In this study, radiative heat transport within a truncated cone-shaped medium is investigated. The medium is assumed to be absorbing, emitting and isotropically scattering. Spatially varying radiation properties are included. Integral formulation of radiative heat transfer is presented for such a medium. Applications of such a formulation may extend to conical frustum geometry for prediction of plume radiation. The formulation may also be applied to predict soot radiation in a combustion system shaped like a truncated cone. The iterative solutions of radiant flux are obtained using the nodal-approximation technique. Comparison of the results with existing data for a cylindrical enclosure is presented and shows good agreement. The numerical results of radiant flux distribution also show that the effects of cone angle on the radiant flux at the exterior base region are significant.

MATHEMATICAL FORMULATION

We consider a participating medium in a truncated conical enclosure(See Fig. 1). The medium is assumed to be absorbing, emitting and isotropically scattering with spatially varying properties. The analysis is simplified by assuming all the boundary walls to be black and all except the cone base(Z=0) kept at zero absolute temperature. This assumption may not be realistic for thermal engineering equipment, it is specified to reduce the complexity. Nevertheless, the present model may be realistic when it is applied to plume radiation problems.

If local thermodynamic equilibrium is established, the integral form of radiative transfer equation is obtained as[Lin, 1987]

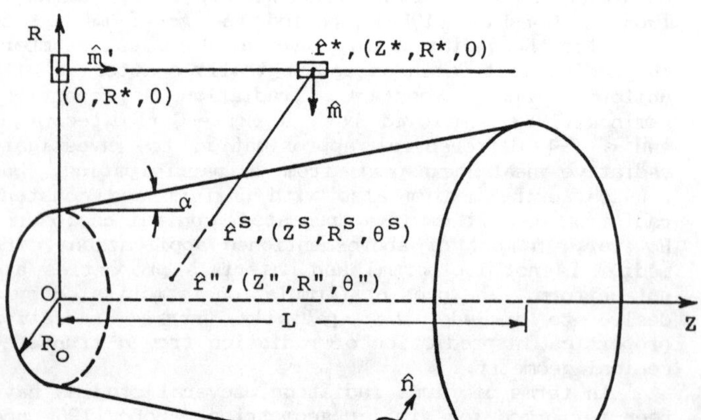

Fig. 1 Diagram of the physical system

$$S_\nu(\hat{r})=K_\nu(\hat{r})I_{\nu b}(T)+\frac{\sigma_\nu(\hat{r})}{4\pi}\int_{4\pi} I\,d\Omega$$

$$=K_\nu(\hat{r})I_{\nu b}(T)+\frac{\sigma_\nu(\hat{r})}{4\pi}\iiint_V \frac{S_\nu(\hat{r}'')}{|\hat{r}-\hat{r}''|^2}$$

$$\times\exp\left[-\int_{\hat{r}''}^{\hat{r}}\beta_\nu(\hat{r}''')dl'\right]dV'' \qquad (1)$$

$$+\frac{\sigma_\nu(\hat{r})}{4\pi}\iint_\Gamma \frac{I_\nu(\hat{r}')[\hat{n}\cdot(\hat{r}-\hat{r}')]}{|\hat{r}-\hat{r}'|^3}\exp\left[-\int_{\hat{r}'}^{\hat{r}}\beta_\nu(\hat{r}''')dl'\right]dA'.$$

This equation reduces in cylindrical coordinates to

$$S_\nu(Z,R)=K_\nu(Z,R)I_{\nu b}[T(Z,R)]$$

$$+\frac{\sigma_\nu(Z,R)}{4\pi}\int_0^L\int_0^{Z''\tan\alpha}\int_0^{2\pi} S_\nu(Z'',R'')$$

$$\times G(Z,R,Z'',R'',\theta'')H(Z,R,Z'',R'',\theta'')d\theta''dR''dZ''$$

$$+\frac{\sigma_\nu(Z,R)}{2\pi}\int_0^{R_o}\int_0^{\pi}\frac{I_\nu(0,R'')ZR''}{L^3(Z,R,0,R'',\theta'')}$$

$$\times\exp\{-L(Z,R,0,R'',\theta'')\int_0^1\beta_\nu[Z'(t),R'''(t)]dt\}d\theta''dR'', \qquad (2)$$

where $G(Z,R,Z'',R'',\theta'')=\exp\{-L(Z,R,Z'',R'',\theta'')$

$$\times\int_0^1\beta_\nu[Z'''(t),R'''(t)]dt\}, \qquad (3a)$$

$$H(Z,R,Z'',R'',\theta'')=R''/[L(Z,R,Z'',R'',\theta'')]^2, \qquad (3b)$$

$$L(Z,R,Z'',R'',\theta'')=[(Z-Z'')^2+R^2+R''^2-2RR''\cos\theta'']^{1/2}, \qquad (3c)$$

$$Z'''(t)=(1-t)Z+tZ'', \qquad (3d)$$

$$R'''(t)=[(1-t)^2R^2+t^2R''^2+2t(1-t)RR''\cos\theta'']^{1/2}, \qquad (3e)$$

and $Z'(t)=(1-t)Z.$ (3f)

Net radiant heat flux at point $\hat{r}$ in the direction of $\hat{m}$ can be expressed as

$$q_\nu(\hat{r})=\int_{4\pi} I_\nu(\hat{r},\Omega')[\hat{m}\cdot(\hat{r}'-\hat{r})]d\Omega' \qquad (4a)$$

$$=\iiint_V \frac{S_\nu(\hat{r}'')}{|\hat{r}-\hat{r}''|^3}\exp\left[-\int_{\hat{r}''}^{\hat{r}}\beta_\nu(\hat{r}''')dl'\right][\hat{m}\cdot(\hat{r}''-\hat{r})]dV''$$

$$+\iint_\Gamma \frac{I_\nu(\hat{r}')}{|\hat{r}-\hat{r}'|^4}\exp\left[-\int_{\hat{r}'}^{\hat{r}}\beta_\nu(\hat{r}''')dl'\right]$$

$$\times[\hat{m}\cdot(\hat{r}'-\hat{r})][\hat{n}\cdot(\hat{r}'-\hat{r})]dA'. \qquad (4b)$$

When the point $\hat{r}$ is located at the cone base(Z=0) but outside the cone, the expression for net radiant flux becomes

$$q_\nu(\hat{r})=\iiint_V \frac{S_\nu(\hat{r}'')}{|\hat{r}-\hat{r}''|^3}\exp[-\int_{\hat{r}''}^{\hat{r}^s}\beta_\nu(\hat{r}''')dl'][\hat{m}.(\hat{r}''-\hat{r})]dV'' \qquad (5)$$

where $\hat{r}^s$ is the intersection point of the cone lateral surface and the line segment $\overline{rr''}$ between points $\hat{r}$ and $\hat{r}''$. Equation (5) with $\hat{m}$ in the direction along the center line of the cone(cf. Fig.1) in cylindrical coordinates is written as

$$q_\nu(0,R^*)=\int_0^L\int_0^{Z''\tan\alpha}\int_0^{2\pi} S_\nu(Z'',R'')$$

$$\times G_f(0,R^*,Z'',R'',\theta'')H_f(0,R^*,Z'',R'',\theta'')d\theta''dR''dZ'', \qquad (6a)$$

where $G_f(Z^*,R^*,Z'',R'',\theta'')=\exp\{-L_f(Z^*,R^*,Z'',R'',\theta'')$

$$\times\int_0^1\beta_\nu[Z^V(t),R^V(t)]dt\},$$

$$H_f(Z^*,R^*,Z'',R'',\theta'')=Z''R''/[L(Z^*,R^*,Z'',R'',\theta'')]^3,$$

L_f is the distance from the point $\hat{r}^s$, which is the intersection point of $\overline{r^*r''}$ and cone lateral surface, to point $\hat{r}''$. Point $[Z^V(t), R^V(t)]$ is on the line $\overline{r^sr''}$.

$$L_f=[(Z^s-Z'')^2+R^{s2}+R''^2-2R^sR''\cos(\theta''-\theta^s)]^{1/2},$$

$$Z^V(t)=(1-t)Z^s+tZ'',$$

and $R^V(t)=[(1-t)^2R^{s2}+t^2R''^2+2t(1-t)R^sR''\cos(\theta''-\theta^s)]^{1/2}$.

And the intersection point is represented as

$$Z^s=(1-t^s)Z^*+t^sZ'',$$

$$R^s=[(1-t^s)^2R^{*2}+t^{s2}R''^2+2t^s(1-t^s)R^*R''\cos\theta'']^{1/2}.$$

and $\theta^s=\sin^{-1}(t^sR''\sin\theta''/R^s)$

with $t^s=(-B-\sqrt{B^2-4AC})/(2A)$,

$A=R^{*2}+R''^2-2R^*R''\cos\theta''-(Z''-Z^*)^2\tan^2\alpha$,

$B=-2R^{*2}+2R^*R''\cos\theta''-2D\tan\alpha(Z''-Z^*)-2Z^*(Z''-Z^*)\tan^2\alpha$,

$C=R^{*2}-(Z^*\tan\alpha+D)^2$,

and D is the diameter of cone base.

The distribution along the axial direction of irradiance on a differential area with the normal $\hat{m}$ perpendicular to the cone center line is given at a distance from the center line (cf. Fig.1) by

$$q_\nu(Z^*,R^*)=\int_0^L\int_0^{Z''\tan\alpha}\int_0^{2\pi} S_\nu(Z'',R'')G_f(Z^*,R^*,Z'',R'',\theta'')$$

$$\times H_f(Z^*,R^*,Z'',R'',\theta'')\frac{R^*-R''\cos\theta''}{Z''}d\theta''dR''dZ'' \qquad (6b)$$

for $R^*\geq R_o+L\tan\alpha$ and the cone base kept at zero absolute temperature.

SOLUTION METHODOLOGY

It is difficult to obtain the analytic solution of source function governed by Eq. (1). An iterative method is used in the present study to get solution of source function at certain points called nodes[Lin, 1988]. The distribution of the source function between two adjacent nodes along both the radial and axial directions is assumed to be linear. This allows the equation for the source function to be rewritten as

$$S_\nu(Z,R)=K_\nu(Z,R)I_{\nu b}[T(Z,R)]+\frac{\sigma_\nu(Z,R)}{4\pi}\sum_{i=1}^{N}\sum_{j=1}^{M}\int_{Z_i}^{Z_i+\Delta Z_i}$$

$$\times\int_{R_{i,j}+(R_{i+1,j}-R_{i,j})Z''}^{R_{i,j+1}+(R_{i+1,j+1}-R_{i,j+1})Z''}\int_0^{2\pi}S_{ap}(Z'',R'')$$

$$\times G(Z,R,Z'',R'',\theta'')H(Z,R,Z'',R'',\theta'')d\theta''dR''dZ''$$

$$+\frac{\sigma_\nu(Z,R)}{2\pi}\int_0^{R_o}\int_0^{\pi}\frac{I_\nu(0,R'')ZR''}{L^3(Z,R,0,R'',\theta'')}\exp\{-L(Z,R,0,R'',\theta'')$$

$$\times\int_0^1\beta_\nu[Z'(t),R'''(t)]dt\}d\theta''dR'',$$

with $S_{ap}(Z'',R'')$ defined as

$$S_{ap}=(1-\bar{x})(1-\bar{y})S_{i,j}+(1-\bar{x})\bar{y}S_{i,j+1}+\bar{x}(1-\bar{y})S_{i+1,j}$$

$$+\bar{x}\bar{y}S_{i+1,j+1}\ ,$$

$$0\leq\bar{x}\leq1,\ 0\leq\bar{y}\leq1,$$

where Z" and R" are nondimensionalized by the sides of trapezoid to give $\bar{x}$ and $\bar{y}$(See Fig.2) and

$$Z''=Z_i+\Delta Z_i\bar{x},$$

$$R''=(1-\bar{x})(R_{i,j}+\bar{y}\Delta R_{i,j})+\bar{x}(R_{i+1,j}+\bar{y}\Delta R_{i+1,j}).$$

The subscripts i and j denote the quantities at the node (i, j). $\Delta Z_i=Z_{i+1}-Z_i$ and $\Delta R_{i,j}=R_{i,j+1}-R_{i,j}$. The integrals in above equation can be approximated using Gaussian quadrature integral formulas. The only unknowns $S_{i,j}$ at the nodes are then determined using an iterative method in the following form

$$S_{i,j}^{n+1}=K_{i,j}I_b(T_{i,j})+\frac{\sigma_{i,j}}{4\pi}\sum_{i=1}^{N}\sum_{j=1}^{M}\int_{Z_i}^{Z_i+\Delta Z_i}$$

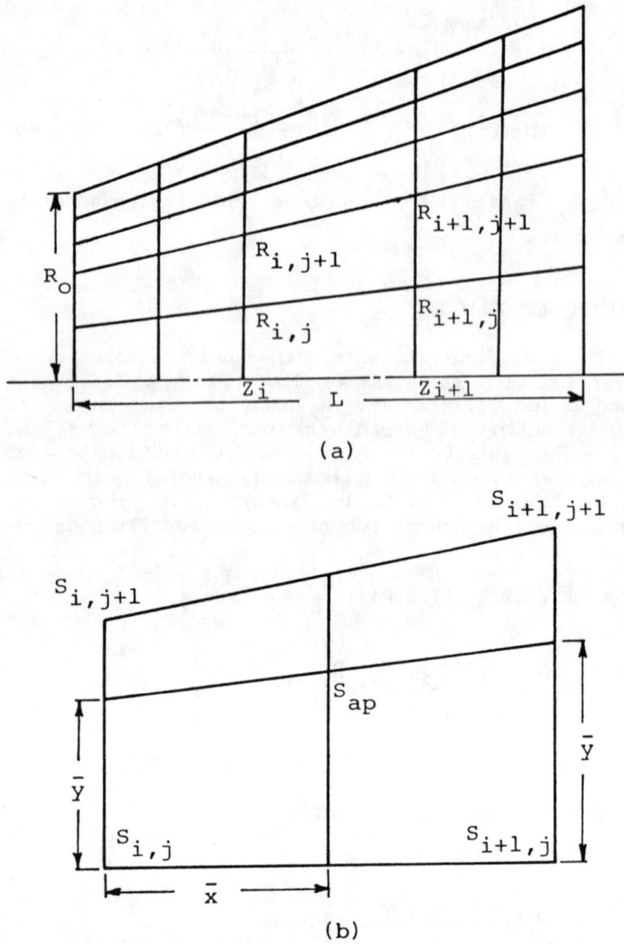

Fig. 2 (a) Node diagram, (b) Approximate source function in an element

$$\times\int_{R_{i,j}+(R_{i+1,j}-R_{i,j})Z''}^{R_{i,j+1}+(R_{i+1,j+1}-R_{i,j+1})Z''}\int_0^{2\pi} S_{ap}^n$$

$$\times G(Z,R,Z'',R'',\theta'')H(Z,R,Z'',R'',\theta'')d\theta''dR''dZ''$$

$$+\frac{\sigma_{i,j}}{2\pi}\int_0^{R_o}\int_0^{\pi}\frac{I_\nu(0,R'')Z_iR''}{L^3(Z_i,R_j,0,R'',\theta'')}\exp\{-L(Z_i,R_j,0,R'',\theta'')$$

$$\times\int_0^1\beta_\nu[Z'(t)_{\text{at } Z=Z_i},R''(t)_{\text{at } R=R_j}]dt\}d\theta''dR'',$$

where the superscript (n+1) indicates that this is (n+1)st iteration. The net radiant flux, Eqs. (6), is then obtained as

$$q(0,R^*)=\sum_{i=1}^{N}\sum_{j=1}^{M}\int_{Z_i}^{Z_i+\Delta Z_i}$$

$$\times\int_{R_{i,j}+(R_{i+1,j}-R_{i,j})Z''}^{R_{i,j+1}+(R_{i+1,j+1}-R_{i,j+1})Z''}\int_0^{2\pi} S_{ap}$$

$$\times G_f(0,R^*,Z'',R'',\theta'')H_f(0,R^*,Z'',R'',\theta'')d\theta''dR''dZ''$$

and the irradiance distribution along the axial direction is obtained as

$$q(Z^*,R^*)=\sum_{i=1}^{N}\sum_{j=1}^{M}\int_{Z_i}^{Z_i+\Delta Z_i}$$

$$\times\int_{R_{i,j}+(R_{i+1,j}-R_{i,j})Z''}^{R_{i,j+1}+(R_{i+1,j+1}-R_{i,j+1})Z''}\int_0^{2\pi} S_{ap}$$

$$\times G_f(Z^*,R^*,Z'',R'',\theta'')H_f(Z^*,R^*,Z'',R'',\theta'')$$

$$\times\frac{R^*-R''\cos\theta''}{Z''}d\theta''dR''dZ''.$$

NUMERICAL RESULTS

In the following numerical results, the truncated cone is divided into 5 non-uniform subdomains in the radial direction and into at least 120 subdomains axially whose amount depends on the length of the cone. Integrals encountered in calculations of the nodal parameters $S_{i,j}$ are performed with 6-point Gaussian quadrature. The iterations are performed until the difference between the source functions obtained in two

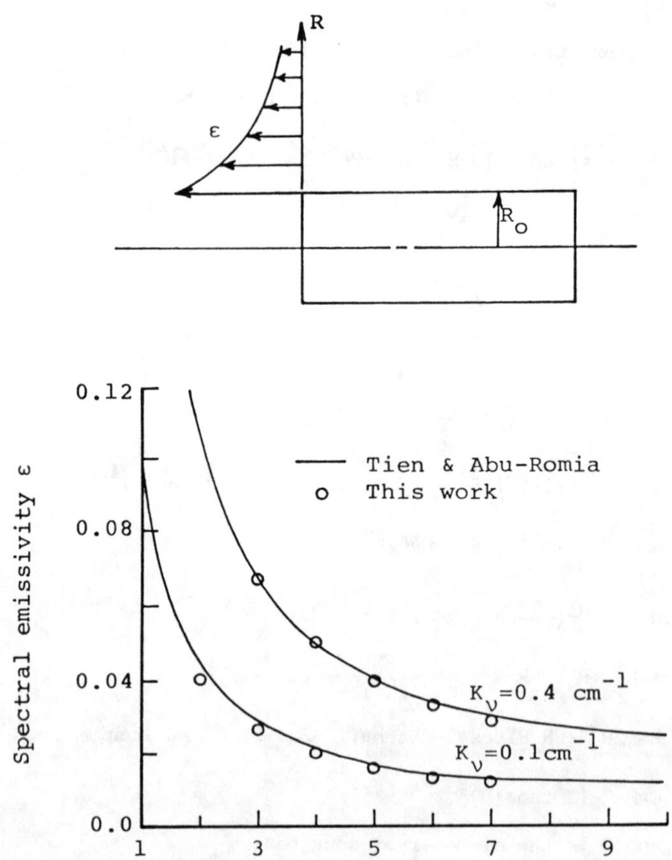

Fig. 3 Spectral apparent emissivity in base plane of a semi-infinite cylindrical gas body

successive iterations is less than one half percent at all nodes.

The truncated cone geometry reduces to a finite or semi-infinite cylindrical geometry when the cone angle is zero. Comparisons of the the results for such cases with existing data are shown in Figs. 3 and 4. In these comparisons, the medium is assumed to be isothermal with uniform and constant properties. Figure 3 shows the results of apparent emissivity obtained in this study compared to Tien and Abu-Romia for a semi-infinite cylindrical, nonscatter-ing medium of radius 1 cm with the cone base at zero absolute temperature. The apparent emissivity is defined as the ratio of the radiative energy flux on the base plane to that of a black body at the temperature of the medium(or of the cone base as is in Fig. 4). Results show an excellent agreement. Figure 4 shows a comparison of the same for a cylindrical, purely and isotropically scattering medium of finite length. The cone base surface is kept at a constant and uniform temperature which acts as the radiation source to simulate the plume radiation from solid propellant rocket motors. The medium is also assumed to have uniform and constant radiation properties. Results also shows a good agreement.

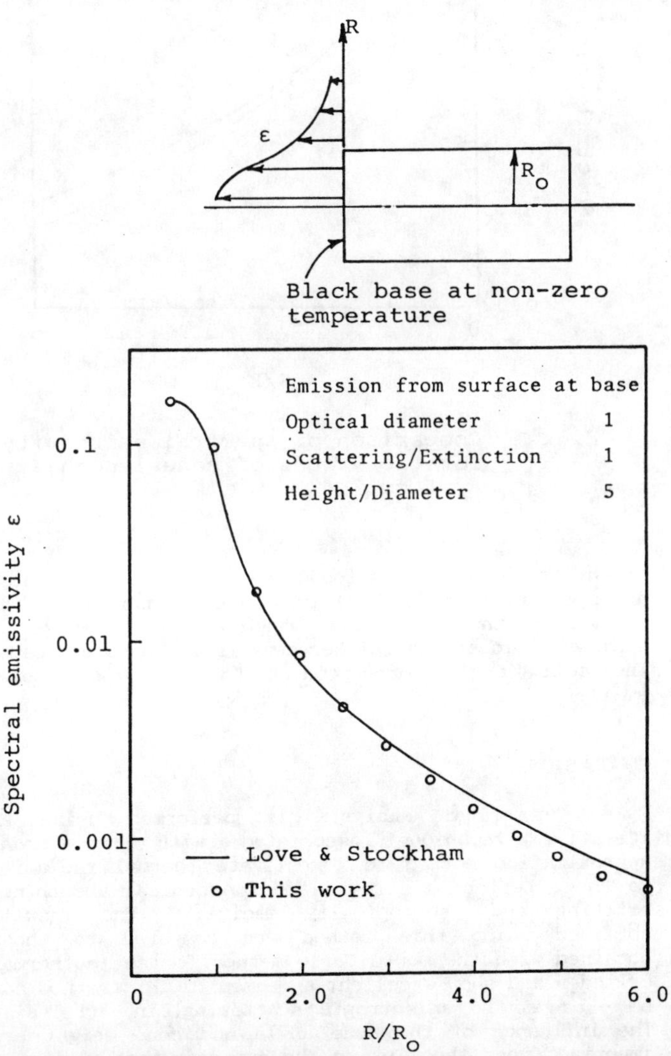

Fig. 4 Spectral emissivity versus radial location of an isothermal cylindrical cloud

In the following, we present some numerical results for media enclosed in truncated cone enclosure to investigate the effects of cone angle and cone height on the radiative heat flux distribution. The irradiance is calculated for the cone base kept at zero absolute temperature. Figure 5 shows the effect of cone angle on the apparent emissivity from an isothermal nonscattering gas medium with the absorption coefficient distribution $K_{\nu}=K_{\nu o}[R_o/(Z\tan\alpha+R_o)]^2$ and base radius of 1 cm, where $K_{\nu o}$ is the absorption coefficient on the cone base. This assumption is specified to account for the change of the medium concentration along the axial direction. It is seen that the effect of cone angle is significant, especially when the absorption coefficient is large. The effects of cone angle are also shown in the following figures for scattering media.

Figures 6 through 8 concern the case of an isothermal conical geometry with the base radius of 1 cm and isotropically scattering. Such a case can be

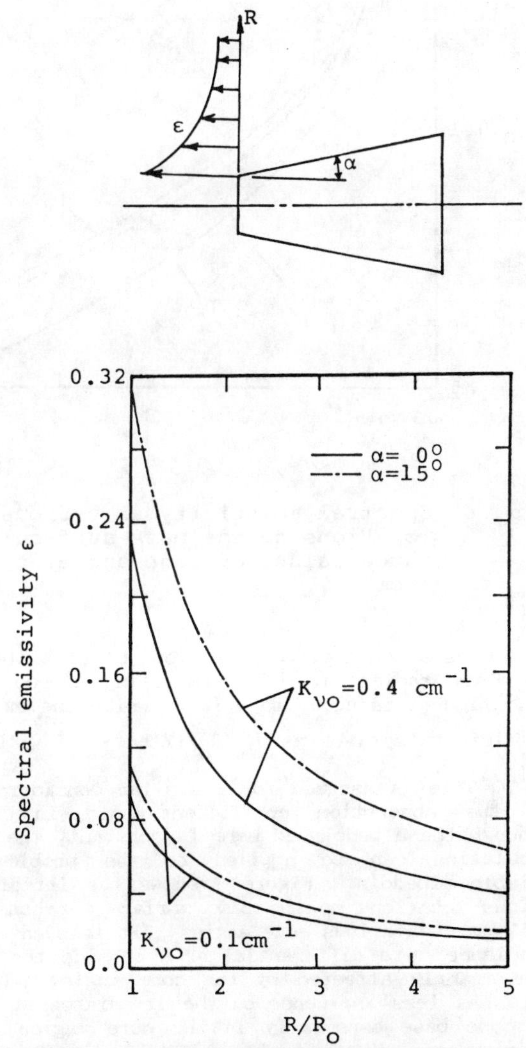

Fig. 5 Spectral apparent emissivity in base plane of a semi-infinite non-scattering conical body for two values of α and R_o=1 cm

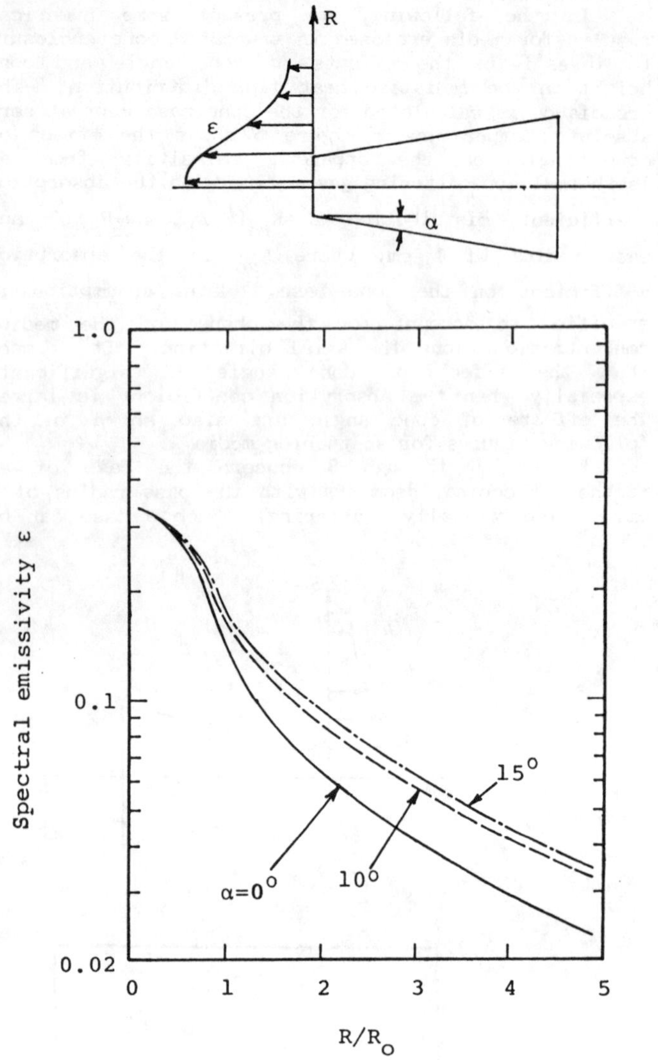

Fig. 6 Spectral emissivity at various locations on the base surface for three values of cone angle at R_o= 1 cm

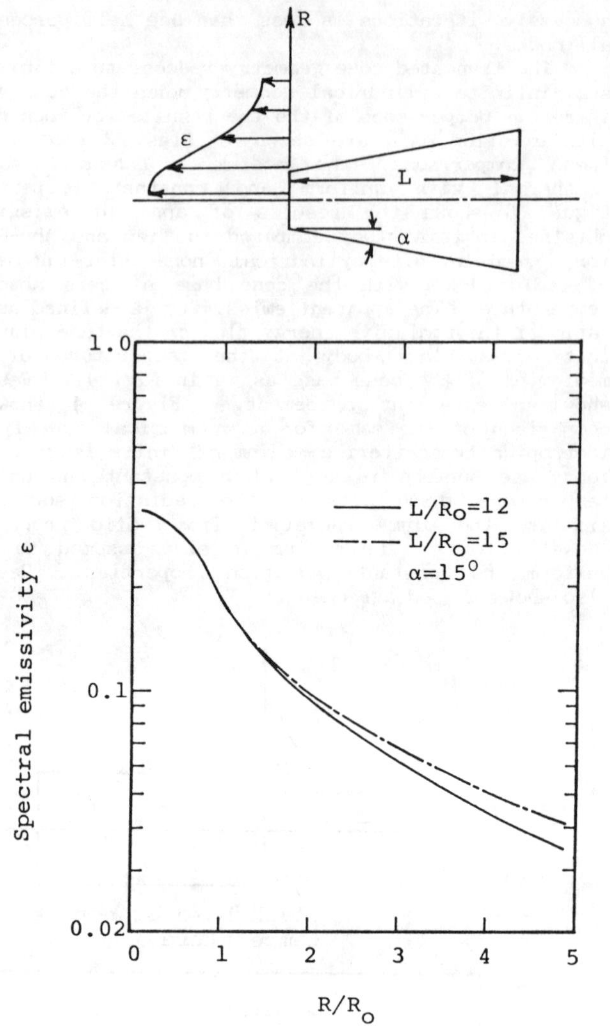

Fig. 7 Comparison of spectral emissivity for two values of cone length at R_o=1 cm

treated as a model of conical cloud of particles as in the plume radiation. The scattering albedo used in these figures is equal to 0.722, while the extinction coefficient is given as $\beta_{\nu o}[R_o/(Z\tan\alpha+R)]^2$ with $\beta_{\nu o}$=11 cm^{-1}. This is assumed for a similar consideration as for the absorption coefficient used in Fig. 5. Although the albedo used here is constant, the present formulation can be applied to the problems with variable albedo's. Figure 6 shows the irradiance at various locations on the base surface for cone height of 12 cm and various cone angles. It is seen that the irradiance on a differential area outside the cone is significantly affected by the cone angle. The cone angle has less influence on the irradiance at area in the cone base especially in the core region for the present case. Figure 7 shows the effect of cone height for isothermal conical geometry with cone angle of 15 degree. It is shown that the influence of cone height is important only in the region far away from the cone. The effect of cone angle on the irradiance distribution along the axial direction is shown in Fig. 8, for cone height of 15 cm. It is found that the cone angle plays an important role in axial irradiance distribution. It is noticed that the axial irradiance distribution is smaller at locations further away from the cone because the solid angle corresponding to the cone becomes smaller.

CONCLUSION

A numerical analysis is performed using an iterative technique associated with the nodal approximation method to investigate thermal radiation to the exterior base region from a conical absorbing, emitting and isotropically medium. The results obtained using this method are compared to those obtained with the Monte Carlo method for an isothermal cylinder and show excellent agreement. This method can be extended to anisotropic scattering[Lin and Tsai]. The influence of the cone angle and cone height are important when the flux on the exterior base region is considered. The medium density(i.e., radiation properties) also affects the heat flux on an area exterior to the base. The mathematical model presented

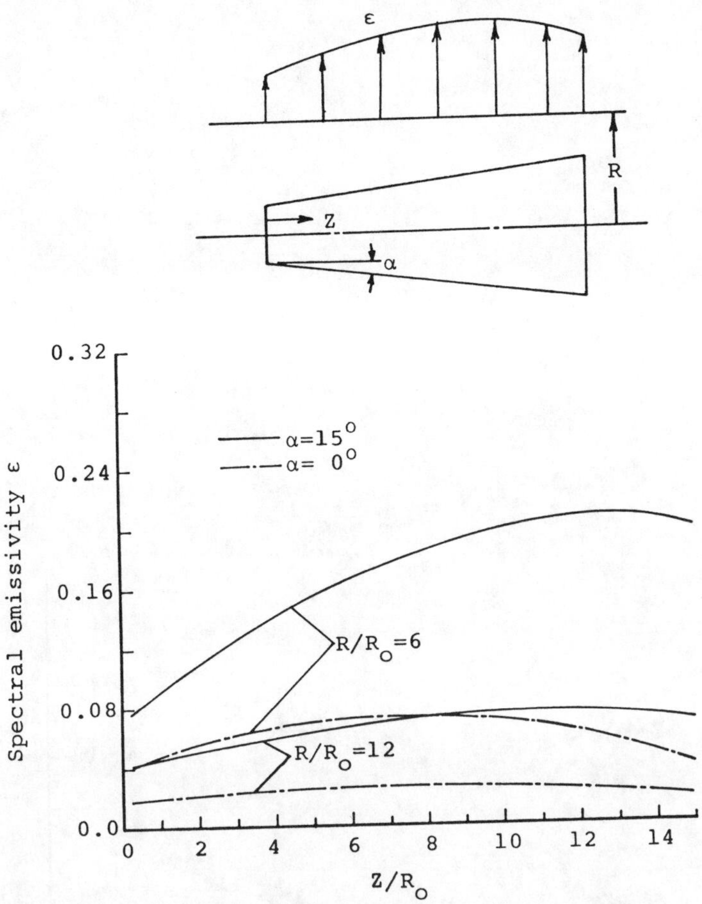

Fig. 8 Spectral emissivity at various axial locations for various values of cone angle and R_o=1 cm

here can be applied to calculate the radiative heat transfer in a combustion chamber shaped like a truncated cone in which both gases and particulates act as radiation heat sources.

REFERENCES

Bobco, R.P., 1966, "Radiation from Conical Surfaces with Nonuniform Radiosity," AIAA Journal, Vol. 4, pp. 544-546.

Bobco, R.P., 1969, "Reply by Author to D.K. Edwards on Radiation from Conical Surfaces with Nonuniform Radiosity," AIAA Journal, Vol. 7, p.1659.

Edwards, D.K., 1969, "Comment on Radiation from Conical Surfaces with Nonuniform Radiosity," AIAA Journal, Vol. 7, pp.1656-1658.

Edwards, D.K. and Bobco, R.P., 1982, "Effect of Particle Size Distribution on the Radiosity of Solid Propellent Rocket Motor Plumes," Progress in Astronautics and Aeronautics, Vol. 83, pp.169-188.

Edwards, D.K., Sakurai, Y. and Babikian, D.S., 1985, "Examination of an Alternative to the Bobco Plume Radiative Model," AIAA Paper, 85-1096.

Gorshkova, S.N., 1986, "Radiative Heat Transfer in a Truncated-Cone Chamber in Which the Heat Source Lies within the Gas Space," Heat Transfer-Soviet Research, Vol. 18, pp.39-43.

Kaminski, D.A., "Radiative Heat Transfer from a Gray, Absorbing-Emitting Medium in a Conical Enclosure," ASME Proceedings of the 1988 National Heat Transfer Conference, HTD-Vol. 96, pp.201-207.

Lin, J.D., 1987, "Exact Expressions for Radiative Transfer in an Arbitrary Geometry Exposed to Radiation," J. Quant. Spectrosc. Radiat. Transfer, Vol. 37, pp.591-601.

Lin, J.D., 1988, "Radiative Transfer within an Arbitrary Isotropically Scattering Medium Enclosed by Diffuse Surfaces," AIAA J. Thermophysics and Heat Transfer, Vol. 2, No. 1, pp.68-74.

Lin, J.D. and Tsai, J.H., "Radiation-Conduction Interaction in a Planar Anisotropically Scattering Medium with Flux Boundary," to appear in the Numerical Heat Transfer journal.

Nelson, H.F., 1984, "Influence of Particulates on Infrared Emission from Tactical Rocket Exhausts," J. Spacecraft and Rockets, Vol. 21, pp.425-432.

Pearce, B.E., 1978, "Radiative Heat Transfer within a Solid-Propellant Rocket Motor," J. Spacecraft and Rockets, Vol. 15, pp.125-128.

Stockham, L.W. and Love, T.J., 1969, "Radiative Heat Transfer from a Cylindrical Cloud of Particles," AIAA Journal, Vol. 6, pp.1935-1940.

Tien, C.L. and Abu-Romia, M.M., 1964, "A Method of Calculation Rocket Plume Radiation to the Base Region," J. Spacecraft and Rockets, Vol. 1, pp.433-435.

1989 National Heat Transfer Conference
HTD-Vol. 106, Heat Transfer Phenomena in Radiation, Combustion and Fires

THE IMPROVED DIFFERENTIAL APPROXIMATION FOR RADIATIVE TRANSFER IN GENERAL THREE-DIMENSIONAL MEDIA

M. F. Modest
Department of Mechanical Engineering
The Pennsylvania State University
University Park, Pennsylvania

Abstract

The improved differential approximation first proposed by Modest for non-scattering media at radiative equilibrium is extended to the most general case of a three-dimensional, absorbing/emitting and anisotropically scattering medium bounded by non-black walls. The model reduces to the correct optically-thin and optically-thick limits for all situations. Some one- and two-dimensional solutions are presented, comparing the present method with exact solutions obtained from Monte Carlo calculations, as well as with approximate solutions obtained from (i) the *P-1* or *ordinary differential approximation* and (ii) the so-called *modified differential approximation* first proposed by Olfe and extended by Modest. Unlike the *P-1* approximation the present method and the modified differential approximation are seen to give results of excellent accuracy for all optical conditions, at the price of a very moderate increase in computational effort. While comparable in accuracy, the improved differential approximation generally is computationally considerably more efficient.

Nomenclature

A	surface area,
a_1	anisotropy factor in scattering phase function,
E_n	exponential integral functions,
G	incident intensity, equation (6),
I, I_b	(blackbody) radiative intensity,
J	surface radiosity, equation (14),
L	length of example cylinder,
$\hat{\mathbf{n}}$	unit surface normal,
Q	total radiative flux,
$q,\mathbf{q}$	radiative flux per unit area, equation (7),
R	radius of example cylinder,
r	radial distance,
S	radiative source function, equation (2),
s	distance along direction $\hat{\mathbf{s}}$,
$\hat{\mathbf{s}}$	unit direction vector,
SV	volume integral in equations (14) and (17)
β	radiative extinction coefficient,
ϵ	surface emissivity,
θ	angle between surface normal and direction vector,
τ	optical coordinate, equation (4),
Φ	scattering phase function,
ψ	azimuthal angle,
ω	solid angle,
ω_s	single scattering albedo.

SUBSCRIPTS

r	radial component,
z	axial component.

SUPERSCRIPTS

$*$	evaluated with ordinary differential approximation

1 Introduction

Exact calculations of radiative heat flux or its divergence require the solution to very complicated integral equations, in particular if the radiatively participating medium scatters radiation. Large numbers of approximate methods have been proposed, such as 2-flux and 6-flux methods, kernel approximations, *P-N* approximations, etc. (see, for example, standard textbooks such as the ones by Siegel and Howell (1981) and Özişik (1973)). All these methods are either tailored to very specific situations (geometry, black walls, radiative equilibrium and/or non-scattering media, etc.), or they become very cumbersome and/or inaccurate in more general situations. One particularly popular approximate method is the so-called *differential approximation* or *P-1 approximation*. This method reduces the calculation of radiative fluxes and temperature fields to the solution of a single linear elliptic partial differential equation, which may be applied to any three-dimensional geometry, with anisotropic scattering as well as boundary reflection (see, for example, Modest and Azad (1980)). Unfortunately, the differential approximation is accurate only in optically thick media, and may become very inaccurate in optically thin situations, in particular in two- and three-dimensional geometries. This is due to the fact that, in this method, the directional variation of the intensity is expanded into a Legendre polynomial series truncated after only two terms (P_0 and P_1), i.e., good accuracy can be expected only in situations with near-isotropic intensity fields. To increase the accuracy, a number of improvements have been proposed. Olfe (1967,1968,1970,1973) and Glatt and Olfe (1973) separated wall emission (which may have strong directional variations) from medium emission (which is isotropic) in what they called the *modified differential approximation*. While very accurate, their model was limited to non-scattering media of simple geometry bounded by black walls. Modest (1974,1975) and Modest and Stevens (1978) combined the *P-1 approximation* with an optically-thin correction term in what they called the *improved differential approximation*. Even more accurate than Olfe's model, their method was applicable to multi-dimensional situations with boundary reflection; however, the method was limited to radiative equilibrium and to non-scattering media. Other investigators [Bayazitoglu and Higenyi (1979), Ratzel and Howell (1983), Mengüc and Viskanta (1986)] used a higher order differential approximation, the *P-3 approximation*. This method is applicable to general situations but --- while more accurate than the ordinary differential approximation --- the method is considerably more involved and may still be very inaccurate in optically thin situations.

Recently, Modest (1989) demonstrated that Olfe's *modified differential approximation* (or even a higher-order *modified P-N approximation*) may be formulated for the general three-dimensional case with an absorbing/emitting, anisotropically scattering medium bounded by non-black walls. The method was shown to be extremely accurate for all optical situations, at the cost of a moderate increase in computational effort.

It is the purpose of the present paper to show that also the *improved differential approximation* may be extended to the most general case. This model also reduces to the correct optically-thin and optically-thick limits for all situations, resulting in similar accuracy as the *modified differential approximation*. Some one- and two-dimensional solutions are presented, comparing the present method with exact solutions obtained from Monte Carlo calculations, as well as with approximate solutions obtained from the *P-1* or ordinary differential approximation and the *modified differential approximation* [Modest (1989)]. While of similar accuracy as the *modified differential approximation* the present method is seen to be computationally more efficient for many practical applications and, therefore, may be considered to be an alternative to that method.

2 Analysis

Consider a three-dimensional enclosure bounded by opaque, diffusely reflecting walls. The equation of radiative transfer for an absorbing, emitting and scattering medium may be written as [Siegel and Howell (1981)]

$$\frac{dI}{d\tau}(\boldsymbol{r},\hat{\boldsymbol{s}}) = S(\boldsymbol{r},\hat{\boldsymbol{s}}) - I(\boldsymbol{r},\hat{\boldsymbol{s}}), \tag{1}$$

where I is radiative intensity, τ is optical distance along the ray traveling into a direction $\hat{\boldsymbol{s}}$ as indicated in Fig. 1, and S is the radiative source term given by

$$S(\boldsymbol{r},\hat{\boldsymbol{s}}) = (1-\omega_s)I_b(\boldsymbol{r}) + \frac{\omega_s}{4\pi}\int_{4\pi} I(\boldsymbol{r},\hat{\boldsymbol{s}}')\Phi(\hat{\boldsymbol{s}}\cdot\hat{\boldsymbol{s}}')\,d\omega'. \tag{2}$$

Here ω_s is the single scattering albedo, I_b is the black body intensity (Planck function), Φ is the scattering phase function and ω denotes solid angle. Equation (2) states that the radiative source inside the medium consists of contributions due to emission and to scattering into direction $\hat{\boldsymbol{s}}$ from all other directions. The above equations govern the total radiative intensity for a gray medium, or the spectral intensity if the medium is non-gray.

Equation (1) may be solved along a pencil of rays (cf. Fig. 1)

$$I(\boldsymbol{r},\hat{\boldsymbol{s}}) = \frac{J_w}{\pi}(\boldsymbol{r}')e^{-\tau} + \int_0^\tau S(\boldsymbol{r}'+s'\hat{\boldsymbol{s}},\hat{\boldsymbol{s}})e^{-(\tau-\tau')}d\tau', \tag{3}$$

where

$$\tau = \int_0^s \beta ds' \tag{4}$$

is the optical distance between that point on the wall from where the beam emanates and the point under consideration (see Fig. 1), β is the radiative extinction coefficient, and J_w is the radiosity at the wall. Here it has been assumed that the walls are purely diffuse reflectors, i.e., the outgoing intensity at the wall is J_w/π.

Similar to Olfe (1967) as generalized by Modest (1989), we will now assume that the intensity at any point consists of two components: one that may be traced back to coming from the wall (by emission or reflection of incoming radiation), and the remainder which may be traced back to the radiative source term (i.e., radiative intensity released into a given direction by emission and scattering). The subtle difference between the present model and the *modified differential approximation* (Modest (1989)) is that here J_w includes the reflected portion of the incoming radiation due to the radiative source within the medium.

Unlike the intensity emanating from the enclosure wall, which may display very irregular behavior (especially in optically thin situations), intensity emanating from inside the medium gener-

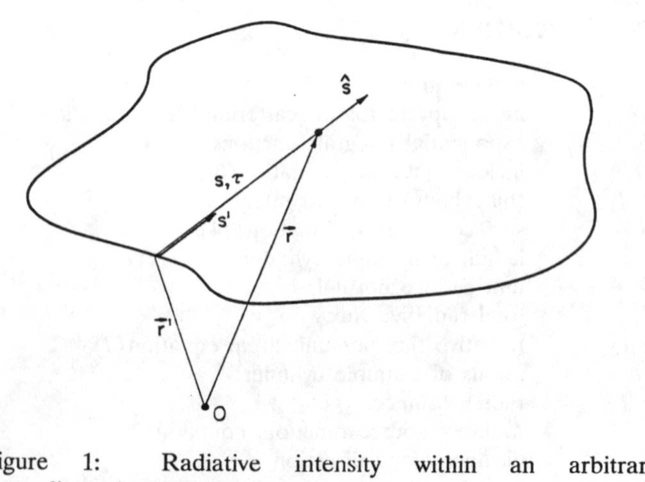

Figure 1: Radiative intensity within an arbitrary three-dimensional enclosure

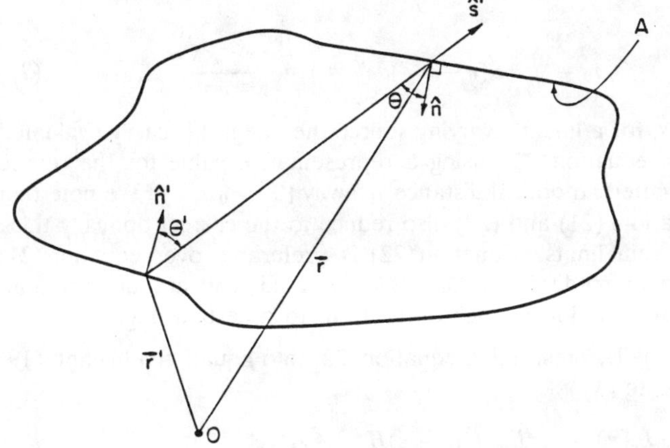

Figure 2: Coordinates for the determination of wall-related radiosity

ally varies very slowly with direction. This is due to the fact that emission and isotropic scattering result in an isotropic radiation source. Only for highly anisotropic scattering may the radiation source --- and, therefore, at least locally also the intensity --- display irregular directional behavior. Therefore, in the present discussion we will assume that the radiative source is adequately represented by the *P-1* or *differential approximation*, in which the intensity is expressed as a series of spherical harmonics, which is truncated after the first harmonic (extension to higher order *P-N approximations* is trivial but tedious). With this assumption the medium intensity may be written as [Modest (1976)]

$$I^* \simeq \frac{1}{4\pi}(G^* + 3\mathbf{q}^* \cdot \hat{\mathbf{s}}), \tag{5}$$

where G^* and $\mathbf{q}^*$ are incident intensity and flux, respectively, defined by

$$G^*(\boldsymbol{r}) = \int_{4\pi} I^*(\boldsymbol{r}, \hat{\mathbf{s}})\, d\omega, \tag{6}$$

$$\mathbf{q}^*(\boldsymbol{r}) = \int_{4\pi} I^*(\boldsymbol{r}, \hat{\mathbf{s}})\hat{\mathbf{s}}\, d\omega, \tag{7}$$

and the asterisk indicates approximate values obtained from the *P-1 approximation*. Such smooth directional behavior can only occur for mildly anisotropic scattering, i.e., expanding the scattering phase function into a series of spherical harmonics this series must also be truncated after the first harmonic, leading to a *linear-anisotropic* phase function,

$$\Phi(\hat{\mathbf{s}} \cdot \hat{\mathbf{s}}') = 1 + a_1 \hat{\mathbf{s}} \cdot \hat{\mathbf{s}}'. \tag{8}$$

Substituting equation (5) and equation (8) into equation (1) leads then to the *differential approximation* for anisotropically scattering media [e.g., Modest and Azad (1980)]

$$\frac{1}{\beta}\nabla \cdot \mathbf{q}^* = (1 - \omega_s)(4\pi I_b - G^*), \tag{9}$$

$$\frac{1}{\beta}\nabla G^* = -3(1 - \frac{a_1\omega_s}{3})\mathbf{q}^*, \tag{10}$$

together with its boundary condition

$$2\left(\frac{2}{\epsilon} - 1\right)\mathbf{q}^* \cdot \hat{\mathbf{n}} = 4\pi I_{bw} - G^*. \tag{11}$$

Once this system of equations has been solved (for a given temperature field, or for the case of radiative equilibrium), the radiative source for linear-anisotropic scattering follows from equations (2) and (8) as

$$S^*(\boldsymbol{r}, \hat{\mathbf{s}}) = (1 - \omega_s)I_b(\boldsymbol{r}) + \frac{\omega_s}{4\pi}[G^*(\boldsymbol{r}) + a_1\mathbf{q}(\boldsymbol{r}) \cdot \hat{\mathbf{s}}]. \tag{12}$$

Substituting equation (12) into equation (3) now gives an *improved* expression for the radiative intensity (as compared to equation (5)),

$$I(\boldsymbol{r}, \hat{\mathbf{s}}) \simeq \frac{J_w}{\pi}(\boldsymbol{r}')e^{-\tau} + \int_0^\tau S^*(\boldsymbol{r}' + s'\hat{\mathbf{s}}, \hat{\mathbf{s}})e^{-(\tau - \tau')}d\tau'. \tag{13}$$

We note that equation (13) reduces to the correct optically thin limit, $I(\boldsymbol{r}, \hat{\mathbf{s}}) \to J_w(\boldsymbol{r}')/\pi$ as well as the correct optically thick limit, since the *P-1 approximation* predicts the correct radiative source (but not necessarily the correct intensity or radiative flux, e.g., near a discontinuity in emissive power).

The radiosity may be eliminated from equation (13) by invoking the definition of the radiosity as the sum of emission plus reflected irradiation, leading to (*cf.* Fig. 2)

$$\begin{aligned} J_w(\boldsymbol{r}) &= \epsilon\pi I_{bw} + (1 - \epsilon)H(\boldsymbol{r}) \\ &= \epsilon\pi I_{bw} + (1 - \epsilon)\left[\int_A J_w(\boldsymbol{r}')e^{-\tau}\frac{cos\theta cos\theta'}{\pi s^2}dA \right. \\ &\quad \left. + \int_V \beta S^*(\boldsymbol{r}' + s'\hat{\mathbf{s}}', \hat{\mathbf{s}}')e^{-(\tau-\tau')}\frac{cos\theta}{s^2}dV\right], \end{aligned} \tag{14}$$

where

$$H(\boldsymbol{r}) = \int_{\hat{\mathbf{s}}' \cdot \hat{\mathbf{n}} < 0} I(\boldsymbol{r}, \hat{\mathbf{s}}')|\hat{\mathbf{s}}' \cdot \hat{\mathbf{n}}|d\omega' \tag{15}$$

is the irradiation onto the surface. Equation (14) is similar to the standard integral equation for the radiosity in an enclosure without a participating medium, except for the attenuation factor $e^{-\tau}$ and the volume integral over the radiative source. Thus the radiosity field may be calculated by standard methods such as breaking up the enclosure surface into N small subsurfaces of constant radiosity. Assuming that the attenuation term may be approximated by the value between node centers, this leads to

$$J_i = \epsilon_i\pi I_{bi} + (1 - \epsilon_i)\left[\sum_{j=1}^{N} J_j e^{-\tau_{ij}}F_{i-j} + SV_i\right], \qquad i = 1, N, \tag{16}$$

where the F_{i-j} are the configuration factors between the subsurfaces and the SV_i are the volume integrals indicated in equation (14).

If the heat flux at the enclosure wall is of importance, it may now be readily calculated from

$$(\mathbf{q} \cdot \hat{\mathbf{n}})_i = \frac{\epsilon_i}{1 - \epsilon_i}(\pi I_{bwi} - J_i) \tag{17}$$

for non-black walls or

$$(\mathbf{q} \cdot \hat{\mathbf{n}})_i = \pi I_{bwi} - H_i = \pi I_{bwi} - \sum_{j-1}^{N} J_j e^{-\tau_{ij}}F_{i-j} - SV_i, \tag{18}$$

for black walls. Inside the medium the divergence of the flux (if the temperature field is given) or the temperature field (for radiative equilibrium) are generally desired. Since equation (9) is exact (i.e., does not only hold for the *P-1 approximation*), the incident intensity needs to be calculated. From equations (6) and (13)

$$G(\boldsymbol{r}) = \frac{1}{\pi}\int_{4\pi} J_w(\boldsymbol{r}')e^{-\tau}d\omega + \int_{4\pi}\int_0^{\tau} S^*(\boldsymbol{r}_w + s'\hat{\mathbf{s}},\hat{\mathbf{s}})e^{-(\tau-\tau')}d\tau' e^{-(\tau-\tau')}d\tau' d\omega$$
$$= \frac{1}{\pi}\int_A J_w(\boldsymbol{r}')e^{-\tau}\frac{\cos\theta' dA}{s^2} + \int_V \beta\, S^*(\boldsymbol{r}_w + s'\hat{\mathbf{s}},\hat{\mathbf{s}})\, e^{-(\tau-\tau')}d\tau'\, e^{-(\tau-\tau')}\frac{dV}{s^2}. \quad (19)$$

Similarly, from equations (6) and (7) it follows that the radiative flux $\mathbf{q}(\boldsymbol{r})$ may be evaluated from the same equation with an additional unit vector $\hat{\mathbf{s}}$ in each integral.

In summary, for a solution by the *improved differential approximation* first the solution to the standard *P-1 approximation*, equations (9) through (11), must be found. This is followed by the solution of equation (14) (for an enclosure with non-black walls), which requires the solution to one surface and one volume integral for each surface node. If internal flux or incident intensity are also desired, an additional surface and volume integral must be evaluated for each internal point. This numerical complexity should be compared with that of the *modified differential approximation* [Modest (1989)], with which it competes. In the latter method two surface integrals must be calculated for each internal and surface nodes. Thus, the *modified differential approximation* appears to be more efficient if incident intensity for all internal nodes is required, while the *improved differential approximation* will be more efficient if only surface fluxes are needed. On the other hand, temperature-dependent radiative properties, which lead to an iterative solution process for the *modified differential approximation*, cause no additional difficulty with the present method. A direct computer time comparison will be given with the discussion of illustrative examples.

Simplified Model

In order to make the present method more efficient it would be very desirable to reduce the numerical complexity of equations (14) and (19). This may be achieved by further approximating the integral in equation (13). To this effect we will assume that the radiative source varies linearly from the point under consideration, $\boldsymbol{r}$, towards the point at the wall, $\boldsymbol{r}'$ (i.e. into the $-\hat{\mathbf{s}}$ direction), or

$$S^*(\boldsymbol{r}' + s'\hat{\mathbf{s}},\hat{\mathbf{s}}) \simeq S^*(\boldsymbol{r},\hat{\mathbf{s}}) - (\tau-\tau')\frac{dS^*}{d\tau'}(\boldsymbol{r},\hat{\mathbf{s}}). \quad (20)$$

Substituting this into equation (13) then leads to:

$$\int_0^{\tau} S^*(\boldsymbol{r}' + s'\hat{\mathbf{s}},\hat{\mathbf{s}})\, e^{-(\tau-\tau')}d\tau' = S^*(\boldsymbol{r},\hat{\mathbf{s}})(1-e^{-\tau}) - \frac{dS^*}{d\tau'}(\boldsymbol{r},\hat{\mathbf{s}})[1-(1+\tau)e^{-\tau}]. \quad (21)$$

To eliminate $dS^*/d\tau'$ from equation (21), we use the assumed linearity of the source function in reverse, i.e., we set

$$\int_0^{\tau} S^*(\boldsymbol{r}' + s'\hat{\mathbf{s}},\hat{\mathbf{s}})e^{-(\tau-\tau')}d\tau' \simeq S^*(\boldsymbol{r} - s_0\hat{\mathbf{s}},\hat{\mathbf{s}})(1-e^{-\tau}), \quad (22)$$

where the geometrical distance s_0 is related to the optical distance

$$\tau_0 = \int_0^{s_0} \beta(\boldsymbol{r} - s''\hat{\mathbf{s}})ds'' = 1 - \frac{\tau e^{-\tau}}{1-e^{-\tau}}. \quad (23)$$

Thus, for a linearly varying source the integral is easily evaluated from equation (22), using a representative value for the source evaluated an optical distance τ_0 away ($0 < \tau_0 < 1$). We note that equations (21) and (22) also reduce to the correct optically thick and thin limits. Equation (22) is preferable over equation (21) since at moderate optical thicknesses, it can account for non-linear variation of the source term to a certain degree.

Finally, substituting equation (22) into equations (14) and (19) leads to

$$J_w(\boldsymbol{r}) = \epsilon\pi I_{bw} + (1-\epsilon)H = \epsilon\pi I_{bw} + (1-\epsilon)\int_A \left[\frac{J_w}{\pi}(\boldsymbol{r}')e^{-\tau}\cos\theta' + S^*(\boldsymbol{r} - s_0\hat{\mathbf{s}}',\hat{\mathbf{s}}')(1-e^{-\tau})\right]\frac{\cos\theta}{s^2}dA, \quad (24)$$

and

$$G(\boldsymbol{r}) = \int_A \left[\frac{J_w}{\pi}(\boldsymbol{r}')e^{-\tau}\cos\theta' + S^*(\boldsymbol{r} - s_0\hat{\mathbf{s}}',\hat{\mathbf{s}}')(1-e^{-\tau})\right]\frac{dA}{s^2}. \quad (25)$$

Again, the heat flux $\mathbf{q}(\boldsymbol{r})$ may be evaluated similarly by including the unit vector $\hat{\mathbf{s}}$ in the integral. Thus, using this simplified version of the *improved differential approximation*, only surface integrals need to be evaluated. But, unlike for the *modified differential approximation*, these surface integrals need to be evaluated only at those locations where a flux or incident intensity is desired.

3 Illustrative Examples

One-Dimensional Slab at Radiative Equilibrium

We will first consider the very simple case of a one-dimensional gray absorbing/emitting and isotropically scattering slab at radiative equilibrium between two isothermal black walls, since this example is a standard by which the accuracy of many models has been assessed. It follows immediately from equation (16) that $J_1 = \sigma T_1^4$ and $J_2 = \sigma T_2^4$, where T_1 and T_2 are the temperatures of the two walls. From equation (9) it follows that $4\pi I_b = G^*$ and, therefore, $S^* = G^*/4\pi$. Since $\nabla \cdot \mathbf{q}^* = 0$ for radiative equilibrium, it follows that S^* varies linearly between the plates. Thus the full [equation (19)] as well as the simplified *improved differential approximation* [equation (25)] give identical results. The local flux is readily calculated as

$$\frac{q(\tau)}{\sigma(T_1^4 - T_2^4)} = \frac{1 + E_3(\tau) + E_3(\tau_0-\tau) - \frac{3}{2}E_4(\tau) - \frac{3}{2}E_4(\tau-\tau_0)}{1 + \frac{3\tau_0}{4}}, \quad (26)$$

where the E_n are exponential integrals and τ_0 is the optical distance between the two plates. At the walls equation (26) is identical to the solution given by Olfe (1967). Figure 3 shows that the ordinary differential approximation has a maximum error of $\approx 3.3\%$ as compared to $\approx 1.3\%$ for the *improved differential approximation* (both at $\tau_0 = 1$). This, however, in no

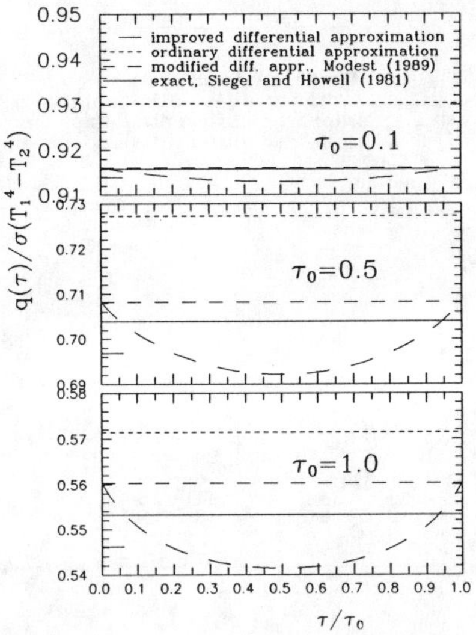

Figure 3: Normalized flux for a one-dimensional slab at radiative equilibrium, bounded by isothermal black plates; influence of optical depth on radiative flux obtained by various methods.

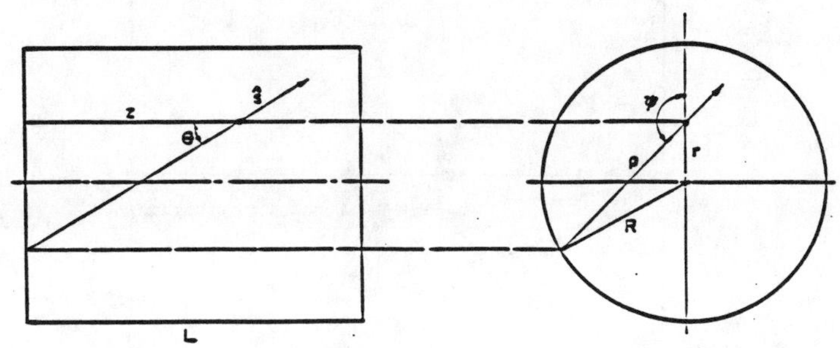

Figure 4: Coordinate system for two-dimensional cylinder.

way demonstrates the power of the present method since this problem is one of the very few where the ordinary differential approximation actually reduces to the correct optically thin limit. However, equation (26) is not constant across the gap as radiative equilibrium would dictate, but varies slightly between the walls (where the flux is slightly overpredicted) and the centerline (where the flux is slightly underpredicted). Figure 3 shows this situation for a number of otpical thicknesses, comparing the exact values with those obtained from the *ordinary differential approximation*, the *modified differential approximation*, and the present method. While the first two approximate methods slightly overpredict the flux, the average from the present method (indicated by short horizontal lines at the left of the figure) underpredict the flux a small amount.

Two-Dimensional Cold Cylinder With Isotropic Scattering

As a second example, shown in Fig. 4, we will consider a cold gray cylindrical medium of radius R and length L, which absorbs and isotropically scatters radiation, bounded radially and at one of the flat faces by cold black walls. The second face is isothermal at Temperature T and has an emissivity of ϵ. The choice of this example has been prompted by the fact that the author has available to him a *Monte Carlo* code for a two-dimensional cylinder with isotropic scattering and given temperature distributions within the medium and along the walls, making comparison of the present method with "exact" results possible. To make the *Monte Carlo* calculations numerically efficient and the results more readily displayable the medium and two of the walls are assumed cold, and the faces and sidewalls are treated as a single nodes (resulting in total or average flux evaluations).

For this case it is advantageous to eliminate the flux from equations (9) and (11), resulting in

$$\frac{1}{\tau_r}\frac{\partial}{\partial \tau_r}\left(\tau_r\frac{\partial G^*}{\partial \tau_r}\right) + \frac{\partial^2 G^*}{\partial \tau_z^2} - 3(1-\omega_s)G^* = 0, \tag{27}$$

subject to the boundary conditions:

$$\begin{aligned} \tau_z = 0: &\quad -\frac{2}{3}\left(\frac{2}{\epsilon}-1\right)\frac{\partial G^*}{\partial \tau_z} + G^* = 4\sigma T^4, \\ \tau_z = \tau_L: &\quad \frac{2}{3}\frac{\partial G^*}{\partial \tau_z} + G^* = 0, \\ \tau_r = \tau_R: &\quad \frac{2}{3}\frac{\partial G^*}{\partial \tau_r} + G^* = 0. \end{aligned} \tag{28}$$

Here the τ_r and τ_z are optical coordinates in the radial and axial directions respectively. While this system may be solved by separation-of-variables, resulting in a generalized Fourier series in Bessel functions, we prefer here a finite-difference solution, using successive over-relaxation (SOR), with the over-relaxation parameter determined by the optimizing routine of Hagemann and Young (1981). Typical CPU times for the *ordinary differential approximation* are 8 sec on a VAX 11/780 (for $L/R = 2, \tau_R = 1, \omega_s = 0.5$ and a 16×32 nodal system).

For comparison of the *improved differential approximation* with *Monte Carlo* results and other approximate methods, we will evaluate the average flux at the two faces of the cylinder, i.e., at $\tau_z = 0$ and $\tau_z = \tau_L$. As a first step the radiosity at all three surfaces must be determined. Since the periphery and the face at $\tau_z = \tau_L$ are cold and black, the radiosities are zero there. The radiosity at $\tau_z = 0$ follows from equation (14) as

$$J_{w0}(\tau_r) = \epsilon\sigma T^4 + (1-\epsilon)H_0(\tau_r), \tag{29}$$

where -- for the full model -- the irradiation is found from equations (12) and (16) as

$$\begin{aligned} H_0(\tau_r) = \frac{\omega_s}{2\pi}\int_0^{\pi/2}\int_0^{\pi}\int_0^{\tau} & G^*\left(\tau_r' = \sqrt{\tau''\sin\theta(\tau''\sin\theta + 2\tau_r\cos\psi) + \tau_r^2}, \tau_z'' = \tau''\cos\theta\right) \\ & \times e^{-\tau''}\,d\tau''\,d\psi\,\sin\theta\,\cos\theta\,d\theta, \end{aligned} \tag{30}$$

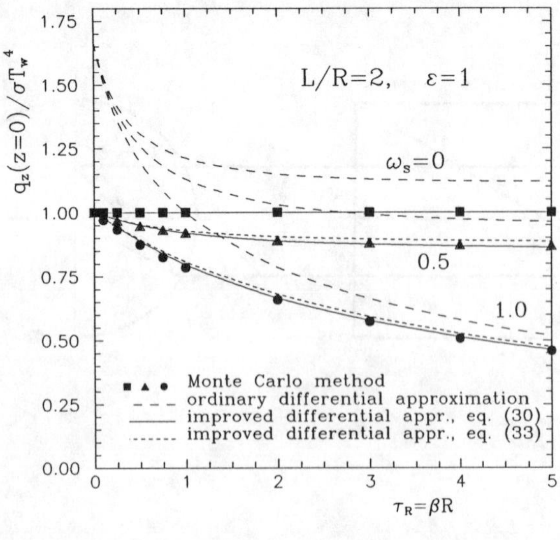

Figure 5: Normalized flux for cold cylinder enclosed by cold, black liner, one cold face plate and one hot face plate; influence of optical depth and scattering albedo on average flux at hot face plate.

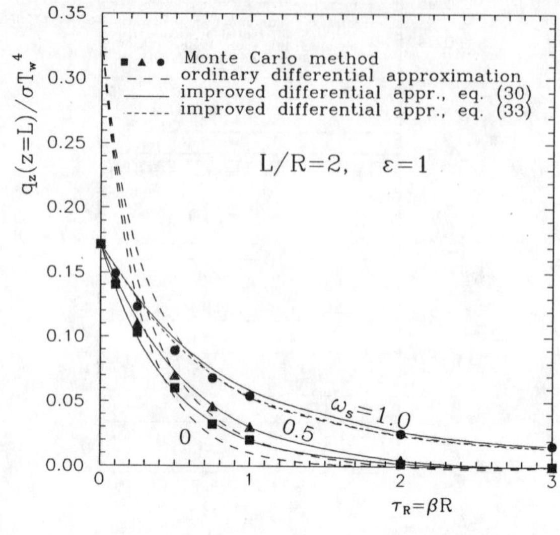

Figure 6: Normalized flux for cold cylinder enclosed by cold, black liner, one cold face plate and one hot face plate; influence of optical depth and scattering albedo on average flux at cold face plate.

with

$$\tau = \begin{cases} \left[\sqrt{\tau_R^2 - \tau_r^2 \sin\psi} - \tau_r \cos\psi\right] / \sin\theta, & \psi < \psi_{RL}, \\ \tau_L / \cos\theta, & \psi > \psi_{RL}, \end{cases} \tag{31}$$

and

$$\psi_{RL} = \cos^{-1}\left\{\frac{\tau_R^2 - \tau_r^2 - \tau_L^2 \tan^2\theta}{2\tau_L\tau_R \tan\theta}\right\}. \tag{32}$$

On the other hand, for the simplified model the irradiation follows from equations (22) and (23) as

$$H_0(\tau_r) \simeq \frac{\omega_s}{2\pi} \int_0^{\pi/2} \int_0^{\pi} G^*\left(\tau_r' = \sqrt{\tau_0 \sin\theta(\tau_0 \sin\theta + 2\tau_r \cos\psi) + \tau_r^2}, \tau_z' = \tau_0 \cos\theta\right) \times (1 - e^{-\tau}) d\psi \sin\theta \cos\theta d\theta, \tag{33}$$

with τ again from equation (31). Here we have again chosen the solid-angle version of equation (16) to avoid the $1/s^2$ singularity in the integral.

The irradiation onto the cold face consists of two terms, i.e. the contribution from the medium plus that from the hot face. The contribution from the medium, $H_{Lm}(\tau_r)$, is found by evaluating G^* in equations (30) or (32) at $\tau_L - \tau_z$ instead of at τ_z. The hot-surface contribution is readily calculated from equation (14) as

$$H_{Lw}(\tau_r) = \frac{2}{\pi} \int_0^{\tan^{-1}[(\tau_R+\tau_r)/\tau_L]} \int_{\psi^*}^{\pi} J_{wo}\left(\tau_r' = \sqrt{\tau_L \tan\theta(\tau_L \tan\theta + 2\tau_r \cos\psi) + \tau_r^2}\right) e^{-\tau_L/\cos\theta} d\psi \sin\theta \cos\theta d\theta \tag{34}$$

Since the wall is cold and black, its flux follows as

$$\frac{q_z}{\sigma T^4}(\tau_r, \tau_L) = H_{Lm}(\tau_r) + H_{Lw}(\tau_r). \tag{35}$$

Some sample results are shown in Figs. 5 through 8.

Figure 5 shows the total normalized flux from the hot surface as function of optical thickness (based on radius) $\tau_R = \beta R$ and scattering albedo ω_s, i.e., the radiative flux integrated over the entire hot surface. In this figure the hot wall is assumed to be black ($\epsilon = 1$), and the length-to-radius ratio is kept constant at $L/R = 2$. It is observed that the *(full) improved differential approximation,* equation (30), follows the Monte Carlo results very closely for all optical thicknesses and scattering albedos, while the ordinary *P-1-approximation* displays serious errors for optically thin situations. Indeed, it is simple to show that for the ordinary differential approximation, as $\tau_R \to 0$,

$$\left[\frac{q_z(r, z=0)}{\sigma T^4}\right]_{ODA} = \left[\frac{Q_z(z=0)}{\sigma T^4 \pi R^2}\right]_{ODA} = \frac{\epsilon}{2-\epsilon}\left(2 - \frac{\epsilon}{1 + (2-\epsilon)\frac{L}{R}}\right), \tag{36}$$

$$\left[\frac{q_z(r, z=L)}{\sigma T^4}\right]_{ODA} = \left[\frac{Q_z(z=L)}{\sigma T^4 \pi R^2}\right]_{ODA} = \frac{\epsilon}{1 + (2-\epsilon)\frac{L}{R}}, \tag{37}$$

while the exact result would be

$$\left[\frac{Q_z(z=0)}{\sigma T^4 \pi R^2}\right]_{IDA} = \left[\frac{Q_z(z=0)}{\sigma T^4 \pi R^2}\right]_{\text{exact}} = \epsilon, \tag{38}$$

$$\left[\frac{Q_z(z=L)}{\sigma T^4 \pi R^2}\right]_{IDA} = \left[\frac{Q_z(z=L)}{\sigma T^4 \pi R^2}\right]_{\text{exact}} = \epsilon\left[1 + \frac{1}{2}\left(\frac{L}{R}\right)^2 - \frac{L}{R}\sqrt{1 + \frac{1}{4}\left(\frac{L}{R}\right)^2}\right], \tag{39}$$

where the latter equation is simply the emissivity multiplied by the configuration factor between parallel disks. Thus it is

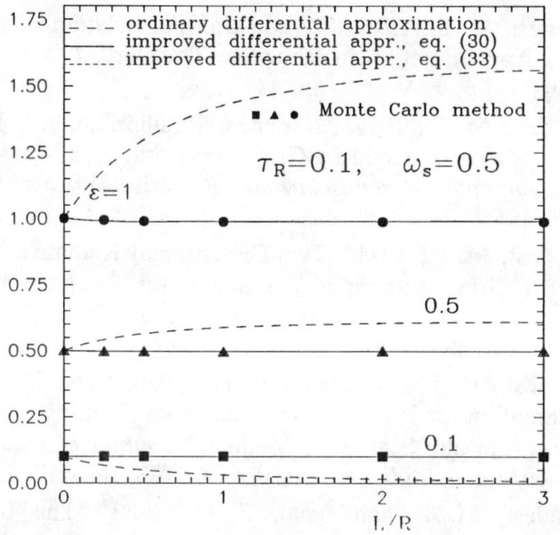

Figure 7: Normalized flux for cold cylinder enclosed by cold, black liner, one cold face plate and one hot face plate; influence of aspect ratio and surface emissivity on average flux at hot face plate.

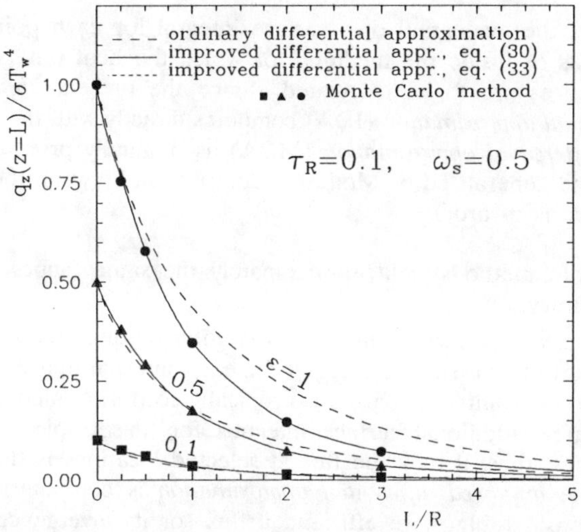

Figure 8: Normalized flux for cold cylinder enclosed by cold, black liner, one cold face plate and one hot face plate; influence of aspect ratio and surface emissivity on average flux at cold face plate.

seen that the ordinary *P-1-approximation* gives errors of up to 100% (for large aspect ratios combined with high emissivities), while the *improved differential approximation* always goes to the correct optically thin limit and is, indeed, accurate for all optical thicknesses and all scattering albedos. Also included in the figure are results from the simplified version of the *improved differential approximation*, equation (32). While not as accurate as the full model, the accuracy is nonetheless excellent. It is interesting to note that both versions of the *improved differential approximation* approach the correct optically thick limit, while the *P-1 approximation* does not: in the vicintiy of a discontinuity in temperature no medium can be "optically thick" and the *P-1 approximation* produces a minimum error of approximately 13%.

Figure 6 shows the identical case for the opposite (cold) face plate, where the *P-1-approximation* shows the same lack of accuracy, while both versions of the *improved differential approximation* again give good results, with only slight errors for intermediate optical depths, particularly for large scattering albedos (the case of $\omega_s = 0$ is not shown to avoid crowding of the figure).

Figures 7 and 8 show the influence of aspect ratio L/R and of hot-surface emissivity ϵ on the average flux at the two faces, all for a small optical thickness of $\tau_R = 0.1$ (where the *P-1-approximation* is often inaccurate) and for a scattering albedo of $\omega_s = 0.5$. For $L/R = 0$ one has the optically transparent one-dimensional case and, as expected, even the *P-1-approximation* gives the correct answer, while for growing aspect ratio the predicted flux at the hot face and the error increase rapidly until they reach an asymptotic value (at $L/R \approx 4$). Heat fluxes at the hot face plate predicted by Monte Carlo and by the *improved differential approximation* remain approximately independent of aspect ratio; with increasing L/R the flux first decreases slightly (never more than 1%) before going back to the original value. Differences between the full and simplified present model are hardly discernable in the figures, both doing very well under all conditions. The *P-1-approximation* does fairly well at the cold plate, except for large emissivities, for which the predicted decrease in flux is much too small. Typical CPU times for the *improved differential approximation* on a VAX 11/780 are 4:14 min for the full model, and 30 sec for the simplified model (again for the case of $L/R = 2, \tau_R = 1, \omega_s = 0.5$ and a 16×32 nodal system). This should be compared with 3:06 min for the *modified differential approximation* (1989).

Due to the limitations of the available Monte Carlo code only isotropic scattering was considered in the examples. However, due to its linear-anisotropic description of the intensity field, the *P-1 approximation* is known to perform equally well for linear-anisotropic scattering, e.g., Modest and Azad (1980). While highly anisotropic scattering cannot be treated by the *P-1* or, therefore, the present approximation, it has been well established that highly anisotropic scattering can always be replaced by an equivalent linear-anisotropic phase function for *heat transfer* calculations, e.g., Modest and Azad (1980). Thus it is reasonable to expect that the present method will do equally well for any arbitrary scattering situation.

4 Conclusions

The *improved differential approximation* first proposed by Modest for non-scattering media at radiative equilibrium has been extended to the most general case of a three-dimensional, absorbing/emitting and anisotropically scattering medium bounded by non-black walls. The present model reduces to the correct optically-thin and optically-thick limits for all situations. Comparison with exact solutions obtained from Monte Carlo calculations demonstrates that, unlike the standard *P-1* approximation, the present method gives results of excellent accuracy for all optical conditions.

The results indicate that a simplified version of the present method also gives results of excellent accuracy under all conditions, making the full approximation unnecessary. This method requires the evaluation of a surface integral for each point, at the walls or inside the medium, for which the heat flux (or its divergence) are to be calculated. Since the present *improved differential approximation* (IDA) competes directly with the *modified differential approximation* (MDA) as originally prsented by Olfe and generalized by Modest, a comparison between the two methods is in order:

- both methods yield approximately the same (superb) accuracy;
- the *modified differential approximation* requires the evaluation of one surface integral for each internal and external nodal point for non- or isotropically scattering media, and three additional surface integrals for anisotropic scatterings; thus, if only the flux at selected locations is desired, the *improved differential approximation* is computationally considerably more efficient, if flux (or its divergence) are needed at every nodal point, the *modified differential approximation* tends to become equally efficient for isotropic scattering, but remains less efficient for anisotropic scattering;
- since the *improved differential approximation* is based on *P-1 approximation* results, no iterations are necessary for temperature-dependent properties (as is the case for the *modified differential approximation*);

Both methods give excellent results, with a moderate increase in computer time as compared to the *modified differential approximation*, and it will depend on the problem at hand to determine which method is preferable under what conditions.

References

Bayazitoglu, Y. and Higenyi, J., 1979, "The Higher-Order Differential Equations of Radiative Transfer: P--3 Approximation", *AIAA Journal*, Vol. 17, pp. 424--431.

Glatt, L. and Olfe, D. B., 1973, "Radiative Equilibrium of a Gray Medium in a Rectangular Enclosure", *Journal of Quantitative Spectroscopy and Radiative Transfer*, Vol. 13, pp. 881--895.

Hageman, Louis A. and Young, D.M., 1981, *Applied Iterative Methods*, Academic Press.

Mengüc, M. P. and Viskanta, R., 1986, "Radiative Transfer in Axisymmetric, Finite Cylindrical Enclosures", ASME JOURNAL OF HEAT TRANSFER, Vol. 108, pp. 271--276.

Modest, M. F., 1989, "The Modified Differential Approximation for Radiative Transfer in General Three-Dimensional Media", *Journal of Thermophysics and Heat Transfer*, Vol. 3, No. 2, Paper No. AIAA 89--0600, 1989 Aerospace Sciences Conference.

Modest, M. F., 1976, "Photon-Gas Formulation of the Differential Approximation in Radiative Transfer", *Letters in Heat and Mass Transfer*, Vol. 3, pp. 111--116.

Modest, M. F., 1975, "Radiative Equilibrium in a Rectangular Enclosure Bounded by Gray Non-Isothermal Walls", *Journal of Quantitative Spectroscopy and Radiative Transfer*, Vol. 15, pp. 445--461.

Modest, M. F., 1974, "Two-Dimensional Radiative Equilibrium of a Gray Medium in a Plane Layer Bounded by Gray Non-Isothermal Walls", ASME JOURNAL OF HEAT TRANSFER, Vol. C96, pp. 483--488.

Modest, M. F. and Azad, F. H., 1980, "The Differential Approximation for Radiative Transfer in an Emitting, Absorbing and Anistropically Scattering Medium", *Journal of Quantitative Spectroscopy and Radiative Transfer*, Vol. 23, pp. 117--120.

Modest, M. F. and Azad, F. H., 1980, "The Influence and Treatment of Mie-Anisotropic Scattering in Radiative Heat Transfer", ASME JOURNAL OF HEAT TRANSFER, Vol. 102, pp. 92--98.

Modest, M. F. and Stevens, D., 1978, "Two Dimensional Radiative Equilibrium of a Gray Medium Between Concentric Cylinders", *Journal of Quantitative Spectroscopy and Radiative Transfer*, Vol. 19, pp. 353--365.

Olfe, D. B., 1968, "Application of a Modified Differential Approximation to Radiative Transfer in a Gray Medium Between Concentric Sphere and Cylinders", *Journal of Quantitative Spectroscopy and Radiative Transfer*, Vol. 8, pp. 899--907.

Olfe, D. B., 1967, "A Modification of the Differential Approximation for Radiative Transfer", *AIAA Journal*, Vol. 5, No. 4, pp. 638--643.

Olfe, D. B., 1970, "Radiative Equilibrium of a Gray Medium Bounded by Nonisothermal Walls", *Progress in Astronautics and Aeronautics*, Vol. 23, pp. 295--317.

Olfe, D. B., 1973, "Radiative Equilibrium of a Gray Medium in a Rectangular Enclosure", *Journal of Quantitative Spectroscopy and Radiative Transfer*, Vol. 13, pp. 881--895.

Özişik, M. N., 1973, *Radiative Transfer and Interaction With Conduction and Convection*, John Wiley and Sons, New York.

Ratzel, A. C. and Howell, J. R., 1983, "Two-Dimensional Radiation in Absorbing-Emitting-Scattering Media Using the P-N Approximation", ASME JOURNAL OF HEAT TRANSFER, pp. 333--340.

Siegel, R. and Howell, J. R., 1981, *Thermal Radiation Heat Transfer*, Hemisphere Publ. Corp., New York.

1989 National Heat Transfer Conference
HTD-Vol. 106, Heat Transfer Phenomena in Radiation, Combustion and Fires

DISCRETE EXCHANGE FACTOR SOLUTION OF RADIATIVE HEAT TRANSFER IN THREE-DIMENSIONAL ENCLOSURES

M. H. N. Naraghi and B. Litkouhi
Department of Mechanical Engineering
Manhattan College
Riverdale, New York

ABSTRACT

The discrete exchange factor method is used to analyze radiative heat transfer in three-dimensional rectangular enclosures. Three numerical integration approaches, Gaussian quadrature, Simpson and trapezoidal are used for discretization of the governing equations of the continuous exchange factor method. The accuracy of the results based on these approaches is examined and the relative difference between the results of different numerical methods is found to be within 5%. The treatment of singular nodes such as those at corner points and common edges is discussed. Comparisons are made between the results of the present approach and those of the zonal method, exact solution, discrete-ordinate method and P-3 approximation.

NOMENCLATURE

$\overline{dgg}(\mathbf{r}_i, \mathbf{r}_j)$	direct exchange factor between gases
$\mathbf{dgg}$	matrix of direct exchange factors between discrete gas nodes
$\overline{DGG}(\mathbf{r}_i, \mathbf{r}_j)$	total exchange factor between gases
$\overline{\mathbf{DGG}}$	matrix of total exchange factors between discrete gas nodes
$\overline{dgs}(\mathbf{r}_i, \mathbf{r}_j)$	direct exchange factor between gas and surface
$\mathbf{dgs}$	matrix of direct exchange factors between discrete gas and surface nodes
$\overline{DGS}(\mathbf{r}_i, \mathbf{r}_j)$	total exchange factor between gas and surface
$\overline{\mathbf{DGS}}$	matrix of total exchange factors between discrete gas and surface nodes
$\overline{dsg}(\mathbf{r}_i, \mathbf{r}_j)$	direct exchange factor between surface and gas
$\mathbf{dsg}$	matrix of direct exchange factors between discrete surface and gas nodes
$\overline{DSG}(\mathbf{r}_i, \mathbf{r}_j)$	total exchange factor between surface and gas
$\overline{\mathbf{DSG}}$	matrix of total exchange factors between discrete surface and gas nodes
$\overline{dss}(\mathbf{r}_i, \mathbf{r}_j)$	direct exchange factor between surfaces
$\mathbf{dss}$	matrix of direct exchange factors between discrete surface nodes
$\overline{DSS}(\mathbf{r}_i, \mathbf{r}_j)$	total exchange factor between surfaces
$\overline{\mathbf{DSS}}$	matrix of total exchange factors between discrete surface nodes
E	emissive power
E_{ref}	reference emissive power
h	distance between two neighboring nodes
I	$[\delta_{i,j}]$, identity matrix
K_a	absorption coefficient
K_s	scattering coefficient
K_t	extinction coefficient
L_x, L_y, L_z	dimensions of rectangular enclosure in x, y and z directions
N_g	number of gas nodes
N_s	number of surface nodes
q''_s	surface heat flux
q'''_g	gas heat source
$\mathbf{r}_j$	position vector
w_i	numerical quadrature weight factor
W	$[w_i\delta_{i,j}]$

Greek Symbols

α	absorptivity
ϵ	emissivity
θ_i	angle between normal to dA_i and connecting line between position $\mathbf{r}_i$ and $\mathbf{r}_j$
ρ	reflectivity
$\tau(\mathbf{r})$	transmittance along path $\mathbf{r}$
τ_0	optical thickness
ω_0	K_s/K_t, albedo for scattering

Subscripts

g designates gas
s designates surface

INTRODUCTION

The radiative heat transfer is a significant mode of energy transfer in high temperature devices such as combustion chambers and furnaces. Consequently, considerable efforts were focused on developing models for analyzing the radiative transfer over the past 30 years. In general, the solution methods of the radiative transfer problem fall into three categories: exact, approximate and numerical.

The exact solutions of multidimensional enclosures are reported by Crosbie and Schrenker (1982). These solutions are limited to specific conditions and usually require numerical integration to obtain the final results. Approximate solutions were obtained using the P-3 approximation (Mengüç and Viskanta, 1985). The P-N approximation usually provides results with reasonable accuracy at higher orders, e.g., P-5. For three-dimensional problems, P-5 approximation involves laborious derivations. Because of the difficulties with the exact and approximate solutions, numerical models became very attractive tools of analyzing the radiative transfer problem. The direct discretization of the radiative transfer equation can be obtained using the discrete-ordinate method. Most recently, Fiveland (1988) used the discrete-ordinate method to analyze three-dimensional rectangular enclosures and has shown a good agreement with the benchmark solutions. The most versatile numerical methods are the Monte Carlo method (Howell, 1968) and the Zonal method (Hottel and Sarofim, 1963 and 1967). Taniguchi (1969) used the Monte Carlo method to analyze a three-dimensional enclosure. Although the Monte Carlo method is suitable for the general radiative transfer problem, its statistical error can become very large. Larsen and Howell (1983, 1985) used the zonal and an alternative zonal method (exchange factor method) to analyze three-dimensional rectangular enclosures. In the zonal method and its alternate form, the zones are isothermal; hence, these methods can become inaccurate when the temperature gradient is large. Furthermore, the evaluation of direct exchange factors (areas) in the zonal method involves multiple integrations and its alternate form requires an experimental setup.

Most recently, Naraghi et al. (1988) developed the continuous exchange factor (CEF) method in which the radiative exchange between differential gas volumes and surface areas are evaluated. The integral equations in this method relate the continuous total and direct exchange factors. In the discrete exchange factor (DEF) method, the integral equations of the continuous exchange factor method are discretized using a numerical integration scheme, namely Gaussian quadrature, and explicit matrix formulations relating discrete total and direct exchange factors are obtained. It has been shown that the DEF method, in its Gaussian quadrature form, provides accurate results for one-dimensional (Naraghi et al., 1988) and two-dimensional (Naraghi and Kassemi, 1988) radiative transfer problems even when a small number of nodes is used. In these works, other numerical integration methods such as Simpson's rule and the trapezoidal method were not implemented due to two reasons: 1) Inferior accuracy compared to Gaussian quadrature; 2) Singularities due to coincident nodes at the surface-gas interface and corners.

The intent of this paper is to extend the application of the DEF method to three-dimensional enclosures, employing different numerical integration schemes. Comparison is made between the results of the DEF method and those of the benchmark solutions. The Gaussian quadrature method provides the most accurate results. However, since the nodal points are unevenly placed in this method it is not desirable when finite difference compatible nodes are needed. Nodal point compatibility is especially desirable when combined modes of heat transfer, conduction-radiation and convection-radiation are analyzed. Hence, the results based on Simpson and trapezoidal discretizations are presented and their accuracy are examined. The procedure for calculating exchange factors between singular nodes at common edges and corners is also discussed.

ANALYSIS

Consider the three dimensional rectangular gray enclosure shown in Fig. 1. The six boundaries (surfaces 1, 2, 3, 4, 5 and 6) are assumed to be diffuse opaque surfaces and the medium consists of an absorbing-emitting and isotropically scattering gas. The dimensions of the medium in the x, y and z directions are L_x, L_y and L_z, respectively. The direct exchange factor between two differential zones $\overline{dzz}(\mathbf{r}_i, \mathbf{r}_j)$ is the fraction of energy that is emitted from a differential surface or gas at $\mathbf{r}_i$ and reaches another differential surface or gas at $\mathbf{r}_j$ by direct radiation. The direct exchange factors between differential surfaces and/or gases are given by the following equations (Hottel and Sarofim, 1967):

$$\overline{dss}(\mathbf{r}_i, \mathbf{r}_j) = \frac{cos\theta_i cos\theta_j \tau(\mathbf{r}_i - \mathbf{r}_j) dA_j}{\pi |\mathbf{r}_i - \mathbf{r}_j|^2} \tag{1}$$

$$\overline{dsg}(\mathbf{r}_i, \mathbf{r}_j) = \frac{K_t cos\theta_i \tau(\mathbf{r}_i - \mathbf{r}_j) dV_j}{\pi |\mathbf{r}_i - \mathbf{r}_j|^2} \tag{2}$$

$$\overline{dgs}(\mathbf{r}_i, \mathbf{r}_j) = \frac{cos\theta_j \tau(\mathbf{r}_i - \mathbf{r}_j) dA_j}{4\pi |\mathbf{r}_i - \mathbf{r}_j|^2} \tag{3}$$

and

$$\overline{dgg}(\mathbf{r}_i, \mathbf{r}_j) = \frac{K_t \tau(\mathbf{r}_i - \mathbf{r}_j) dV_j}{4\pi |\mathbf{r}_i - \mathbf{r}_j|^2} \tag{4}$$

where $\tau(\mathbf{r}_i - \mathbf{r}_j) = e^{-\int_{\mathbf{r}_i}^{\mathbf{r}_j} K_t dr}$ and $K_t = K_a + K_s$. For a homogeneous gas $\tau(\mathbf{r}_i - \mathbf{r}_j) = e^{-K_t |\mathbf{r}_i - \mathbf{r}_j|}$.

The radiative energy emitted from a differential surface or volume will reach another differential surface or volume within the enclosure, hence the direct exchange factors satisfy the following conservation of energy equations

$$\int_S \overline{dss}(\mathbf{r}_i, \mathbf{r}_j) + \int_V \overline{dsg}(\mathbf{r}_i, \mathbf{r}_j) = 1 \tag{5}$$

and

$$\int_S \overline{dgs}(\mathbf{r}_i, \mathbf{r}_j) + \int_V \overline{dgg}(\mathbf{r}_i, \mathbf{r}_j) = 1 \tag{6}$$

The total exchange factor between two differential zones $\overline{DZZ}(\mathbf{r}_i, \mathbf{r}_j)$ is the fraction of energy that is emitted from a

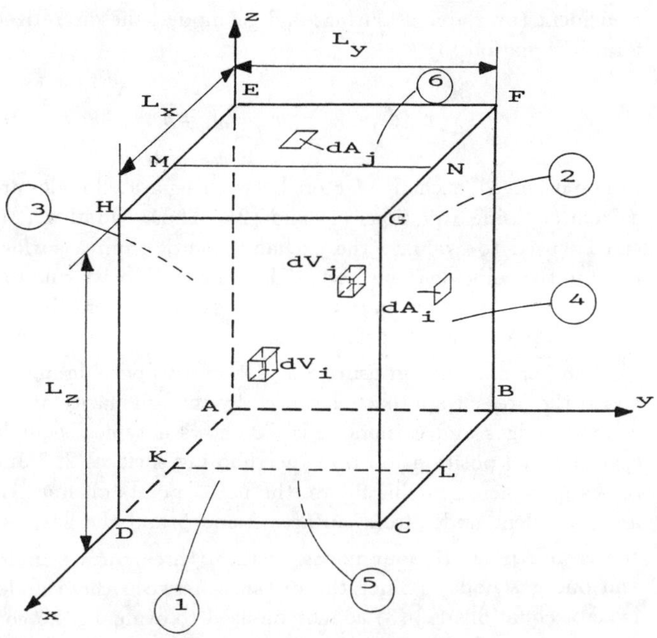

Figure 1: Schematic of a three-dimensional rectangular enclosure

differential surface or gas at $\mathbf{r}_i$ and reaches another differential surface or gas at $\mathbf{r}_j$ by direct radiation and multiple reflection and scattering from the surfaces and gas, respectively. As was the case for the direct exchange factors, there are also four total exchange factors: $\overline{DSS}(\mathbf{r}_i, \mathbf{r}_j)$, $\overline{DSG}(\mathbf{r}_i, \mathbf{r}_j)$, $\overline{DGS}(\mathbf{r}_i, \mathbf{r}_j)$ and $\overline{DGG}(\mathbf{r}_i, \mathbf{r}_j)$ and they satisfy the following conservation of energy conditions:

$$\int_S \overline{DSS}(\mathbf{r}_i, \mathbf{r}_j) + \int_V \overline{DSG}(\mathbf{r}_i, \mathbf{r}_j) = 1 \qquad (7)$$

and

$$\int_S \overline{DGS}(\mathbf{r}_i, \mathbf{r}_j) + \int_V \overline{DGG}(\mathbf{r}_i, \mathbf{r}_j) = 1 \qquad (8)$$

A number of integral equations relating the continuous total exchange factors to the continuous direct exchange factors are derived (Naraghi et al. 1988). These integral equations are:

$$\overline{DSS}(\mathbf{r}_i, \mathbf{r}_j) = p_{ss}(\mathbf{r}_i, \mathbf{r}_j)\alpha + \int_S p_{ss}(\mathbf{r}_i, \mathbf{r}_k)\rho \overline{DSS}(\mathbf{r}_k, \mathbf{r}_j) \qquad (9)$$

$$\overline{DSG}(\mathbf{r}_i, \mathbf{r}_j) =$$
$$[p_{sg}(\mathbf{r}_i, \mathbf{r}_j) + \int_S \overline{DSS}(\mathbf{r}_i, \mathbf{r}_k)(\rho/\alpha)p_{sg}(\mathbf{r}_k, \mathbf{r}_j)](1-\omega_0) \qquad (10)$$

$$\overline{DGS}(\mathbf{r}_i, \mathbf{r}_j) = p_{gs}(\mathbf{r}_i, \mathbf{r}_j)\alpha + \int_S p_{gs}(\mathbf{r}_i, \mathbf{r}_k)\rho \overline{DSS}(\mathbf{r}_k, \mathbf{r}_j) \qquad (11)$$

and

$$\overline{DGG}(\mathbf{r}_i, \mathbf{r}_j) = [p_{gg}(\mathbf{r}_i, \mathbf{r}_j) + \int_S p_{gs}(\mathbf{r}_i, \mathbf{r}_k)\rho p_{sg}(\mathbf{r}_k, \mathbf{r}_j)$$
$$+ \int_S \int_S p_{gs}(\mathbf{r}_i, \mathbf{r}_k)\rho \overline{DSS}(\mathbf{r}_k, \mathbf{r}_l)(\rho/\alpha)p_{sg}(\mathbf{r}_l, \mathbf{r}_j)](1-\omega_0) \qquad (12)$$

where $p_{ss}(\mathbf{r}_i, \mathbf{r}_j)$, $p_{sg}(\mathbf{r}_i, \mathbf{r}_j)$, $p_{gs}(\mathbf{r}_i, \mathbf{r}_j)$ and $p_{gg}(\mathbf{r}_i, \mathbf{r}_j)$ are the probabilities that radiation reaches a differential gas volume or surface at $\mathbf{r}_j$ having been emitted from a differential gas volume or surface at $\mathbf{r}_i$, allowing multiple scattering but excluding wall reflection. These probabilities can be evaluated from the following equations:

$$p_{gg}(\mathbf{r}_i, \mathbf{r}_j) = \overline{dgg}(\mathbf{r}_i, \mathbf{r}_j) + \int_V \overline{dgg}(\mathbf{r}_i, \mathbf{r}_k)\omega_0 p_{gg}(\mathbf{r}_k, \mathbf{r}_j) \qquad (13)$$

$$p_{ss}(\mathbf{r}_i, \mathbf{r}_j) = \overline{dss}(\mathbf{r}_i, \mathbf{r}_j) + \int_V \overline{dsg}(\mathbf{r}_i, \mathbf{r}_k)\omega_0 \overline{dgs}(\mathbf{r}_k, \mathbf{r}_j)$$
$$+ \int_V \int_V dsg(\mathbf{r}_i, \mathbf{r}_k)\omega_0 p_{gg}(\mathbf{r}_k, \mathbf{r}_l)\omega_0 \overline{dgs}(\mathbf{r}_l, \mathbf{r}_j) \qquad (14)$$

$$p_{gs}(\mathbf{r}_i, \mathbf{r}_j) = \overline{dgs}(\mathbf{r}_i, \mathbf{r}_j) + \int_V \overline{dgg}(\mathbf{r}_i, \mathbf{r}_k)\omega_0 p_{gs}(\mathbf{r}_k, \mathbf{r}_j) \qquad (15)$$

and

$$p_{sg}(\mathbf{r}_i, \mathbf{r}_j) = \overline{dsg}(\mathbf{r}_i, \mathbf{r}_j) + \int_V p_{sg}(\mathbf{r}_i, \mathbf{r}_k)\omega_0 \overline{dgg}(\mathbf{r}_k, \mathbf{r}_j) \qquad (16)$$

Once the continuous total exchange factors are determined, the surface heat flux and gas heat source can be related to emissive powers of the medium and the surface via the following equations:

$$q_s''(\mathbf{r}_i) = E_s(\mathbf{r}_i) - \int_S E_s(\mathbf{r}_j)\overline{DSS}(\mathbf{r}_j, \mathbf{r}_i) - \int_V E_g(\mathbf{r}_j)\overline{DGS}(\mathbf{r}_j, \mathbf{r}_i) \qquad (17)$$

and

$$q_g'''(\mathbf{r}_i) = E_g(\mathbf{r}_i) - \int_S E_s(\mathbf{r}_j)\overline{DSG}(\mathbf{r}_j, \mathbf{r}_i) - \int_V E_g(\mathbf{r}_j)\overline{DGG}(\mathbf{r}_j, \mathbf{r}_i) \qquad (18)$$

The discrete exchange factor (DEF) method is obtained by discretizing the above integral equations using a numerical integration technique. In the DEF method radiative transfer between nodal points are evaluated. Naraghi et al. (1988) derived a set of explicit matrix formulations to relate matrices of nodal direct and total exchange factors. These matrices are given by:

$$\overline{\mathbf{DSS}} = [I - \{\overline{\mathbf{dss}} + \overline{\mathbf{dsg}}\omega_0 W_g[I - \overline{\mathbf{dgg}}\omega_0 W_g]^{-1}\overline{\mathbf{dgs}}\}\rho W_s]^{-1}.$$
$$\{\overline{\mathbf{dss}} + \overline{\mathbf{dsg}}\omega_0 W_g[I - \overline{\mathbf{dgg}}\omega_0 W_g]^{-1}\overline{\mathbf{dgs}}\}\alpha \qquad (19)$$

$$\overline{\mathbf{DSG}} = [I - \{\overline{\mathbf{dss}} + \overline{\mathbf{dsg}}\omega_0 W_g[I - \overline{\mathbf{dgg}}\omega_0 W_g]^{-1}\overline{\mathbf{dgs}}\}\rho W_s]^{-1}.$$
$$\overline{\mathbf{dsg}}[I - \omega_0 W_g \overline{\mathbf{dgg}}]^{-1}(1-\omega_0) \qquad (20)$$

$$\overline{\mathbf{DGS}} = [I - \overline{\mathbf{dgg}}\omega_0 W_g]^{-1}\overline{\mathbf{dgs}}.$$
$$[I - \rho W_s\{\overline{\mathbf{dss}} + \overline{\mathbf{dsg}}\omega_0 W_g[I - \overline{\mathbf{dgg}}\omega_0 W_g]^{-1}\overline{\mathbf{dgs}}\}]^{-1}\alpha \qquad (21)$$

and

$$\overline{\mathbf{DGG}} = [I - \overline{\mathbf{dgg}}\omega_0 W_g]^{-1}\overline{\mathbf{dgg}}(1-\omega_0) + [I - \overline{\mathbf{dgg}}\omega_0 W_g]^{-1}.$$
$$\overline{\mathbf{dgs}} W_s[I - \{\overline{\mathbf{dss}} + \overline{\mathbf{dsg}}\omega_0 W_g[I - \overline{\mathbf{dgg}}\omega_0 W_g]^{-1}\overline{\mathbf{dgs}}\}\rho W_s]^{-1}.$$
$$\overline{\mathbf{dsg}}[I - \omega_0 W_g \overline{\mathbf{dgg}}]^{-1}(1-\omega_0) \qquad (22)$$

where $\overline{\mathbf{DSS}} = [\overline{D S_i S_j}]$, $\overline{\mathbf{DSG}} = [\overline{D S_i G_j}]$, $\overline{\mathbf{DGS}} = [\overline{D G_i S_j}]$ and $\overline{\mathbf{DGG}} = [\overline{D G_i G_j}]$ are matrices of total exchange factors between nodal points, $\overline{\mathbf{dss}} = [\overline{d s_i s_j}]$, $\overline{\mathbf{dsg}} = [\overline{d s_i g_j}]$, $\overline{\mathbf{dgs}} = [\overline{d g_i s_j}]$

and $\overline{\mathbf{dgg}} = [\overline{dg_ig_j}]$ are matrices of direct exchange factors and $W_g = [w_{g_i}\delta_{ij}]$ and $W_s = [w_{s_i}\delta_{ij}]$ are diagonal matrices of the numerical integration weight factors. The position of nodal points are based on the ordinates of the numerical integration method chosen for discretization of the CEF equations. If the Gaussian quadrature (GQ) method is selected, the nodal points are unevenly placed while in the Simpson and trapezoidal methods, the nodes are equally spaced. For a three-dimensional enclosure, the weight factors for the surface and gas nodes are the product of two and three single integral weight factors, respectively, i.e., for surface node (i, j), $w_s = w_iw_j$ and for the gas nodes (i, j, k), $w_g = w_iw_jw_k$. These weight factors for a two-dimensional system become $w_s = w_i$ and $w_g = w_iw_j$, and for a one-dimensional system $w_s = 1$ and $w_g = w_i$. The Gaussian quadrature weight factors and the position of nodal points for single integration can be obtained from the Gaussian quadrature tables given by Stroud and Secrest (1966). These weight factors for Simpson and trapezoidal methods, respectively, are: $\{h/3, 4h/3, 2h/3, 4h/3, ..., 2h/3, 4h/3, h/3\}$ and $\{h/2, h, h, h, ..., h, h, h/2\}$, where h is the distance between two neighboring nodes. The discrete total exchange factors calculated from equations (19)-(22) satisfy the discretized form of equations (7) and (8), which are given by

$$\sum_{j=1}^{N_s} w_{s_j}\overline{DS_iS_j} + \sum_{j=1}^{N_g} w_{g_j}\overline{DS_iG_j} = 1 \tag{23}$$

and

$$\sum_{j=1}^{N_s} w_{s_j}\overline{DG_iS_j} + \sum_{j=1}^{N_g} w_{g_j}\overline{DG_iG_j} = 1 \tag{24}$$

Once the discrete total exchange factors are determined, then the discretized form of equations (17) and (18) can be used to relate nodal heat and emissive powers. These equations are given by

$$q''_{s_i} = E_{s_i} - \sum_{j=1}^{N_s} w_{s_j}E_{s_j}\overline{DS_jS_i} - \sum_{j=1}^{N_g} w_{g_j}E_{g_j}\overline{DG_jS_i} \tag{25}$$

and

$$q'''_{g_i} = E_{g_i} - \sum_{j=1}^{N_s} w_{s_j}E_{s_j}\overline{DS_jG_i} - \sum_{j=1}^{N_g} w_{g_j}E_{g_j}\overline{DG_jG_i} \tag{26}$$

where $E_{s_j} = \epsilon_j\sigma T_j^4$ and $E_{g_j} = 4K_t(1-\omega_0)\sigma T_j^4$ are surface and gas nodal emissive powers, respectively.

It should be noted that the governing equations of the CEF and DEF methods, given by equations (5)-(26) for one-, two-, and three-dimensional systems are the same. The multidimensional effect only appears on the direct exchange factors, given by equations (1)-(4), and the numerical integration weight factors, w_s and w_g, in equations (19)-(26).

The use of the Gaussian quadrature in the DEF method has two advantages. First, it provides more accurate results than that of other numerical approaches; second, in the Gaussian quadrature method, none of the gas nodes are located on the surface boundaries. In the other numerical integration methods such as Simpson and trapezoidal, the first gas node is right on the surface making the distance between the surface and the gas node zero, causing, equations (2) and (3) to become singular. To evaluate the exchange factor between two coincident (overlapped) surface and gas nodes, the discretized form of equation (5)

$$\sum_{j=1}^{N_s} w_{s_j}\overline{ds_is_j} + \sum_{j=1}^{N_g} w_{g_j}\overline{ds_ig_j} = 1 \tag{27}$$

can be used. All exchange factors between separated nodes are calculated using equations (1) and (2). Thus, equation (27) can be used to evaluate the exchange factor from a surface node to the coincident gas node. Then the reciprocity rule can be implemented to evaluate the exchange factor from the gas node to the coincident surface node.

Another difficulty in using Simpson and trapezoidal methods in the present approach is caused by the singularity at the common edges and corners. Fig. 2 shows a typical equally spaced nodal positions on three neighboring surfaces 2, 3 and 5. As is depicted in this figure, the nodal points on line $\overline{AB}$ are coincident nodes from surfaces 2 and 5 and the gas. At the corner point A, four nodes overlap (three surface nodes and one gas node). Since the distance between these nodes is zero, equations (1)-(3) cannot be used to evaluate the corresponding exchange factors. Consider two coincident nodes i and j located along the edge line $\overline{AB}$ on surfaces 5 and 2, respectively. It can be shown that half of the radiative energy emitted from a differential area located at node i will reach surface 2. Furthermore, the exchange factors between node i and the nodes on surface 2 other than node j are zero ($\theta_2 = \pi/2$); hence, one can conclude that half of all the radiative energy emitted from the differential area associated with node i will reach node j; that is

$$w_{s_i}\overline{ds_is_j} = 0.5$$

Similarly, it can be shown that the exchange factor between corner nodes k and l located on surfaces 5 and 3, respectively, is given by

$$w_{s_k}\overline{ds_ks_l} = 3/8$$

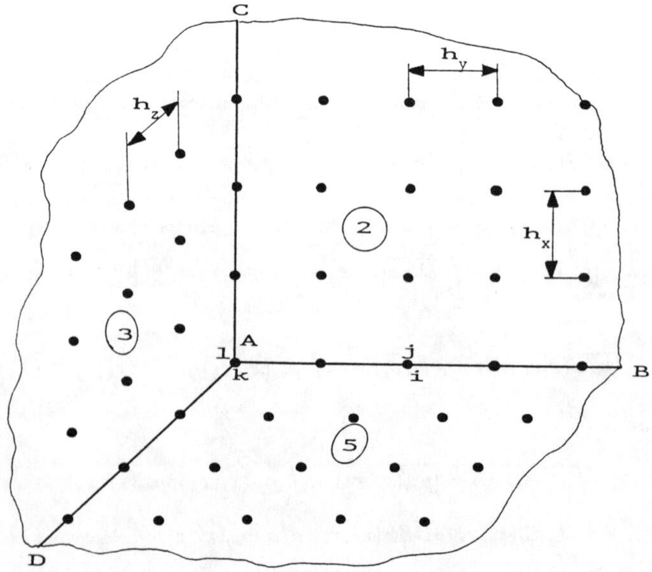

Figure 2: Position of equally spaced nodes on three neighboring surfaces (common edges and a corner)

In calculating the direct exchange factor from a gas node to itself, the distance between the nodes is zero and consequently, equation (4) becomes singular. The direct exchange factors from gas nodes must satisfy equation (7), which in discretized form is given by

$$\sum_{j=1}^{N_s} w_{s_j} \overline{dg_i s_j} + \sum_{j=1}^{N_g} w_{g_j} \overline{dg_i g_j} = 1 \tag{28}$$

All exchange factors in equation (28) are calculated using equations 3 and 4 except the exchange from a gas node to itself. Thus, equation (28) may be used to evaluate the direct exchange factor from a gas node to itself.

RESULTS AND DISCUSSION

The DEF method is used to study radiative transfer in the three-dimensional rectangular enclosure shown in Fig. 1. In order to establish the accuracy of the solution methodology, the results of the present approach based on the three numerical quadrature methods, Gaussian quadrature (GQ), Simpson and trapezoidal are compared to the existing results, primarily to those generated by the zonal method and the exact solutions. Effects of different variables and properties of the enclosure on the accuracy of the results of the present approach are also examined. In the following comparisons the number of nodes in the DEF method are identical to those of the zone method.

Fig. 3 shows the predictions of wall centerline heat flux and gas centerline emissive powers based on the zonal (Larsen, 1983), discrete-ordinate (Fiveland, 1988, first-order, $\alpha = 1$ and diamond-difference, $\alpha = 0.5$) and DEF methods for a black cubical enclosure for the case where dimensionless emissive power of surfaces 1, 4 and 5 is unity and other surfaces are cold and the gas is non-scattering with $K_t L_x = 1$ and $q_g''' = 0.$. The prediction of the DEF method for the emissive power at the center of the enclosure $E(.5,.5,.5)$, in its three forms, satisfy the exact solution criteria presented by Crosbie and Schrenker (1982), i.e.

$$E_{center} = \frac{1}{6}(E_1 + E_2 + E_3 + E_4 + E_5 + E_6) \tag{29}$$

As can be seen from this figure, for the other points, the agreement between the results of all approaches is excellent. As expected, the best prediction of the DEF method is based on the Gaussian quadrature (GQ) discretization and the accuracy is reduced with the Simpson and trapezoidal approaches; however, the latter approaches are capable of predicting the gas emissive power (or temperature) for particles sitting on the surface and heat fluxes along the edges of the enclosure. Fig. 3 also shows the predictions of wall heat flux along the edge lines $\overline{AB}$ and $\overline{AD}$ of surface 5 based on the Simpson approach. The gas particle emissive powers sitting on the center of hot and cold surfaces are 0.791 and 0.209, respectively.

Fig. 4 shows the resulting wall heat flux and emissive power for the same enclosure when all surfaces are cold and the gas

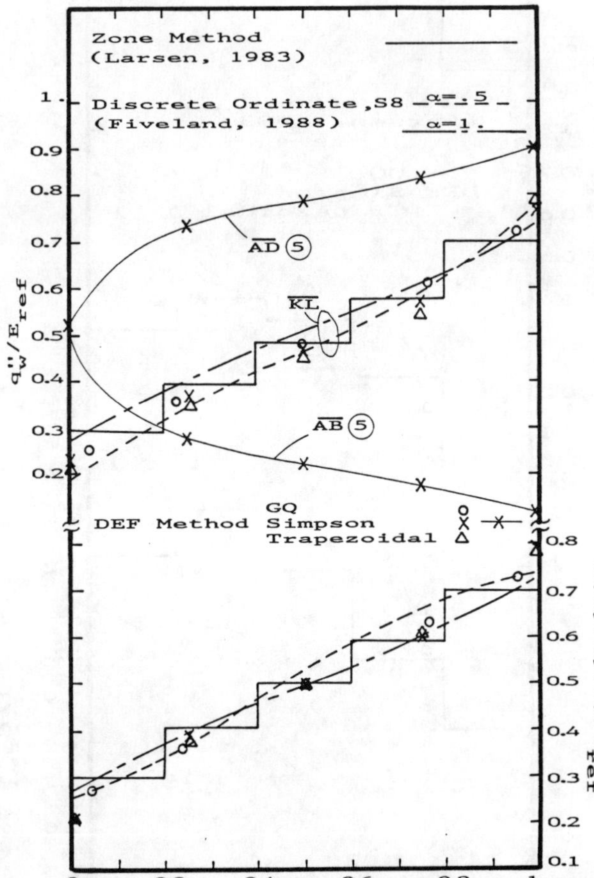

Figure 3: Dimensionless emissive power along centerline for a cubical enclosure with black walls when $K_t = 1$, $q_g''' = 0.$, $\omega_0 = 0.$, $E_1/E_{ref} = E_4/E_{ref} = E_5/E_{ref} = 1$ and $E_1 = E_2 = E_6 = 0.$

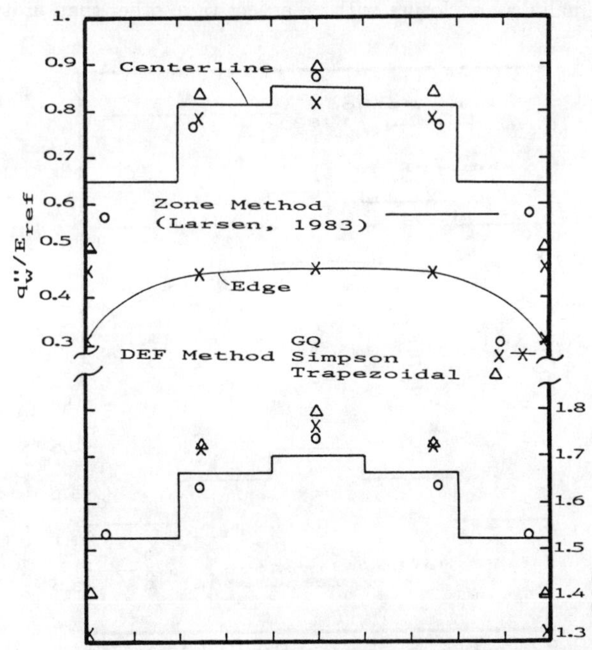

Figure 4: Dimensionless wall centerline heat flux and gas centerline emissive power for a cubical enclosure with black cold walls when $K_t = 1.$, $q_g'''/E_{ref} = 4$ and $\omega_0 = 0.$.

generates heat ($q_g'''/E_{ref} = 4$) based on the zone and the DEF methods. In all these cases, it is found that the maximum relative differences between the Gaussian quadrature results and those of Simpson and trapezoidal are within 5%. To examine the effect of surface emissivity, the same enclosure with surface emissivity of 0.3 is considered and the resulting wall heat flux and gas centerline emissive power are shown in Fig. 5.

Consideration is also given to a black cubical enclosure with gas characteristics of $q_g'''/E_{ref} = 4$, $K_tL_x = 1$ and $\omega_0 = 0.5$ and dimensionless emissive powers of surfaces 2, 4 and 5 are unity and surfaces 1, 3 and 6 are cold. The resulting wall centerline heat flux and gas emissive power are shown in Fig. 6 and the wall heat fluxes along the edges on different surfaces are given in Fig. 7. When the conditions on two neighboring surfaces are the same, the heat flux remains the same across the edge; however, when the surface conditions are different, there will be a sudden change of surface heat flux across the common edges. For example, the heat flux along $\overline{AB}$ on surfaces 5 and 2 are the same, while along $\overline{AD}$ the heat flux abruptly changes from surfaces 5 to 3.

To examine the accuracy of the DEF method in solving problems with different extinction coefficients, two cases with optical thicknesses of K_tL_x =2 and 0.5 are considered. The resulting predictions of the present approach and those of the zonal method are given in Figs. 8 and 9. As can be seen from these figures, the agreement between different solutions is excellent and the DEF results for the center gas emissive power match the exact solution criteria presented by Crosbie and Schrenker (1982). It should be noted that the results of the DEF method presented in Figs. 3 through 9 are based on a 5×5×5 grid.

Finally, an enclosure with an aspect ratio other than unity is considered. The specifications of this enclosure are: $L_x = L_y = 2m$, $L_z = 4m$, $q_g = 5KW/m^3$, $\epsilon_1 = \epsilon_2 = \epsilon_3 = \epsilon_4 = \epsilon_6 = 0.7$, $\epsilon_5 = 0.85$, $T_1 = T_2 = T_3 = T_4 = 900K$, $T_5 = 400K$, $T_6 = 1200K$, $K_t = 0.5m^{-1}$ and $\omega_0 = 0.$. The same enclosure was analyzed by Mengüç and Viskanta (1985) using the zonal method and P-3 approximation and Fiveland (1988) using the discrete-ordinate method. Fig. 10 shows the resulting gas temperature predictions at three axial positions based on the zonal method, P-3 approximation, discrete-ordinate S8 and DEF (Simpson) methods. The results of the DEF method are based on a 7×7×11 grid. The predictions of the DEF method consistently agree with those of the zonal method, while there are noticeable differences between the predictions of the P-3 approximation, the discrete-ordinate method and those of the zonal method especially close to the hot surface. The DEF method predictions of gas temperature based on the Simpson and trapezoidal approaches are so close that they cannot be

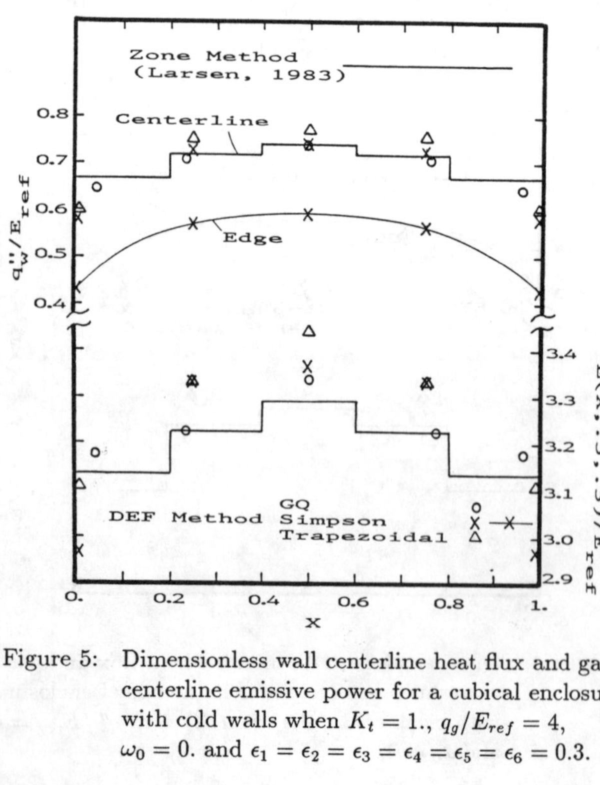

Figure 5: Dimensionless wall centerline heat flux and gas centerline emissive power for a cubical enclosure with cold walls when $K_t = 1.$, $q_g/E_{ref} = 4$, $\omega_0 = 0.$ and $\epsilon_1 = \epsilon_2 = \epsilon_3 = \epsilon_4 = \epsilon_5 = \epsilon_6 = 0.3$.

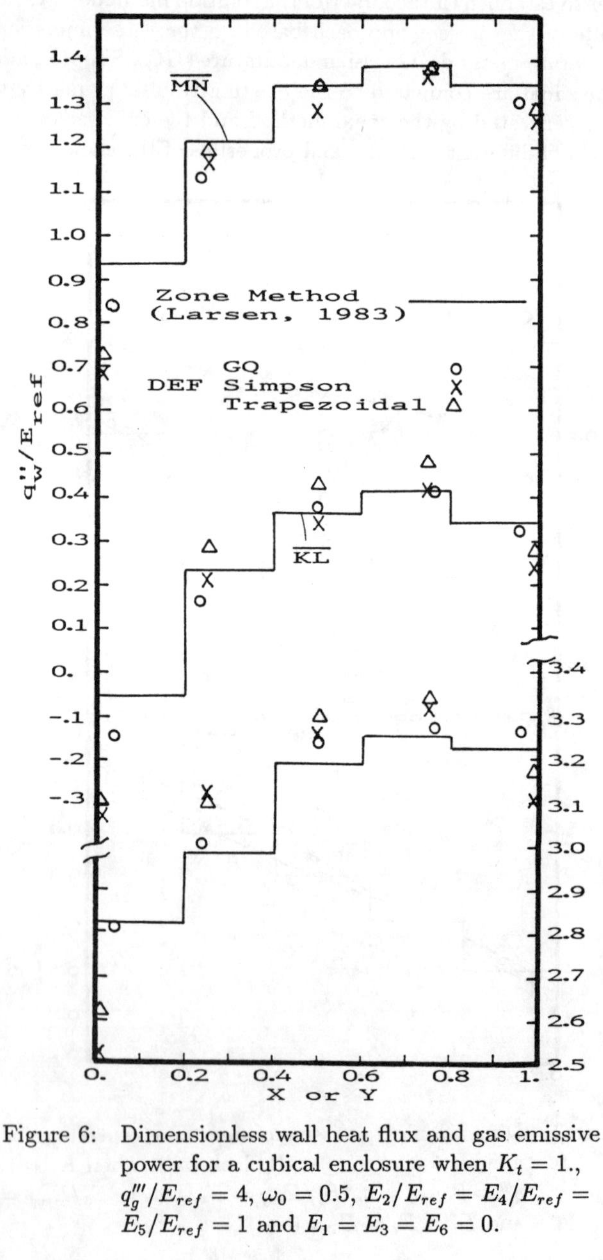

Figure 6: Dimensionless wall heat flux and gas emissive power for a cubical enclosure when $K_t = 1.$, $q_g'''/E_{ref} = 4$, $\omega_0 = 0.5$, $E_2/E_{ref} = E_4/E_{ref} = E_5/E_{ref} = 1$ and $E_1 = E_3 = E_6 = 0$.

distinguished from the graph of Fig. 10. To better observe the differences, they are presented in Table 1. As can be seen from Fig. 10, the temperature of gas at the side wall predicted by the DEF method appears considerably different from those of other methods. Indeed, the gas temperature gradient close to the surface is large and the predictions of the DEF method seem to be quite reasonable. It should be noted that in the pure radiation, the gas temperature at the surface may be different from that of surface temperature due to the temperature slip condition.

CONCLUDING REMARKS

The DEF method is used to analyze radiative transfer in three-dimensional rectangular gray enclosures with isotropically scattering medium. The results for a variety of conditions using three numerical integration methods: Gaussian quadrature, Simpson and trapezoidal are presented. The Gaussian quadrature method provides the most accurate results. However, when finite-difference compatible nodes are needed for combined mode heat transfer problems, the Simpson method is more appropriate. It should be noted that in the trapezoidal method, the weight factors are equal to the volume and the area of the control volume constructed around the gas and surface nodes, respectively. Thus, in an enclosure with an irregular configuration, nodal points can be located arbitrarily. Then, by assigning weight factors equal to the volumes of the control volumes of various nodes, results with the accuracy close to that of the trapezoidal method can be obtained.

The DEF method in its Simpson and trapezoidal forms can predict gas temperature (or emissive power) at the surface and heat fluxes along the edge lines and the corner points. Since the radiative exchange between nodal points is considered in the DEF method, no integration is needed to evaluate direct exchange factors. However, for nonhomogeneous media, at least one integration is required to evaluate the gas transmissivity. If the gas nonhomogeneity is due to temperature dependent properties, the present approach becomes very difficult to implement. Under this condition, the symmetry between every pair of nodes will disappear and an iterative procedure must

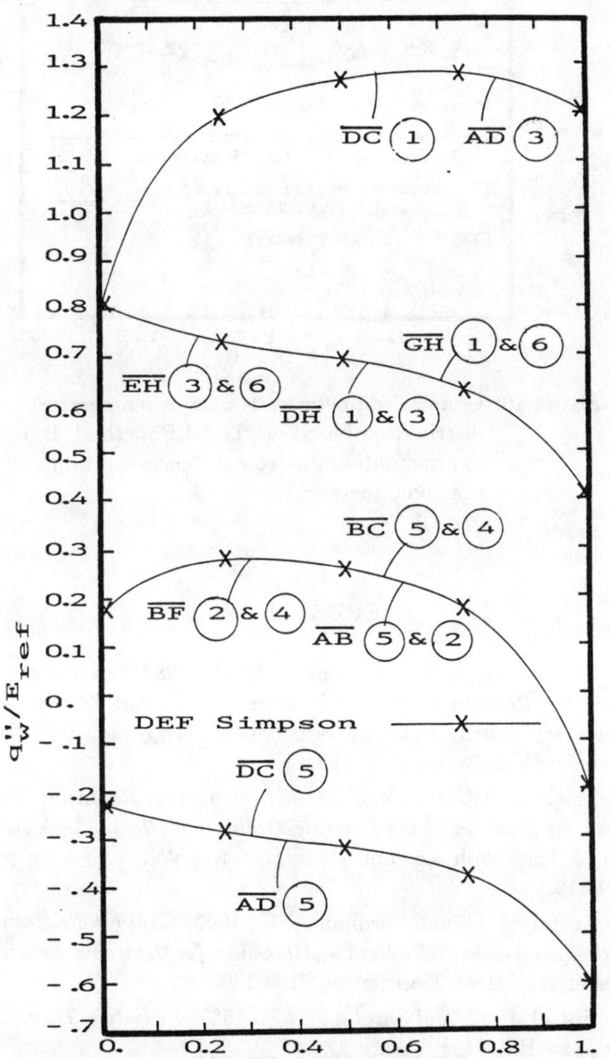

Figure 7: Dimensionless wall heat flux along edges of a cubical enclosure when $K_t = 1.$, $q_g/E_{ref} = 4$, $\omega_0 = 0.5$, $E_2/E_{ref} = E_4/E_{ref} = E_5/E_{ref} = 1$ and $E_1 = E_3 = E_6 = 0$.

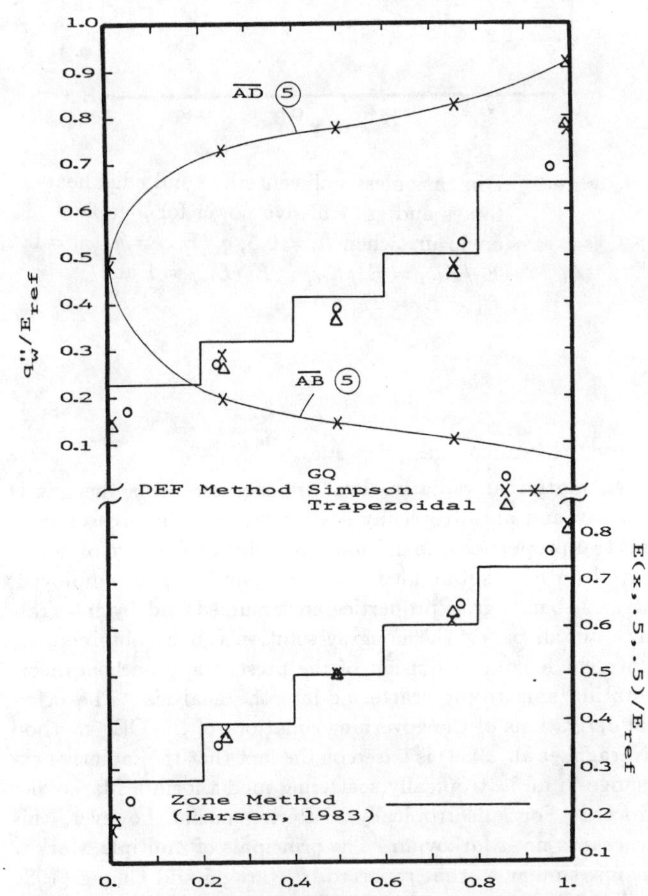

Figure 8: Dimensionless wall centerline and edge heat fluxes and gas emissive power for a cubical enclosure when $K_t = 2.$, $q_g/E_{ref} = 4$, $\omega_0 = 0.$ $E_2/E_{ref} = E_4/E_{ref} = E_5/E_{ref} = 1$ and $E_1 = E_3 = E_6 = 0$.

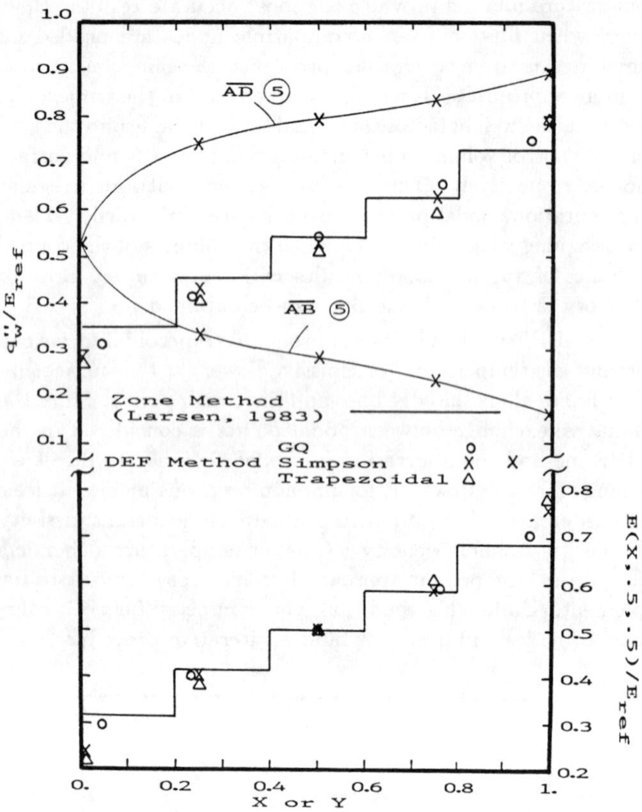

Figure 9: Dimensionless wall centerline and edge heat fluxes and gas emissive power for a cubical enclosure when $K_t = 0.5$, $q_g/E_{ref} = 4$, $\omega_0 = 0$. $E_2/E_{ref} = E_4/E_{ref} = E_5/E_{ref} = 1$ and $E_1 = E_3 = E_6 = 0$.

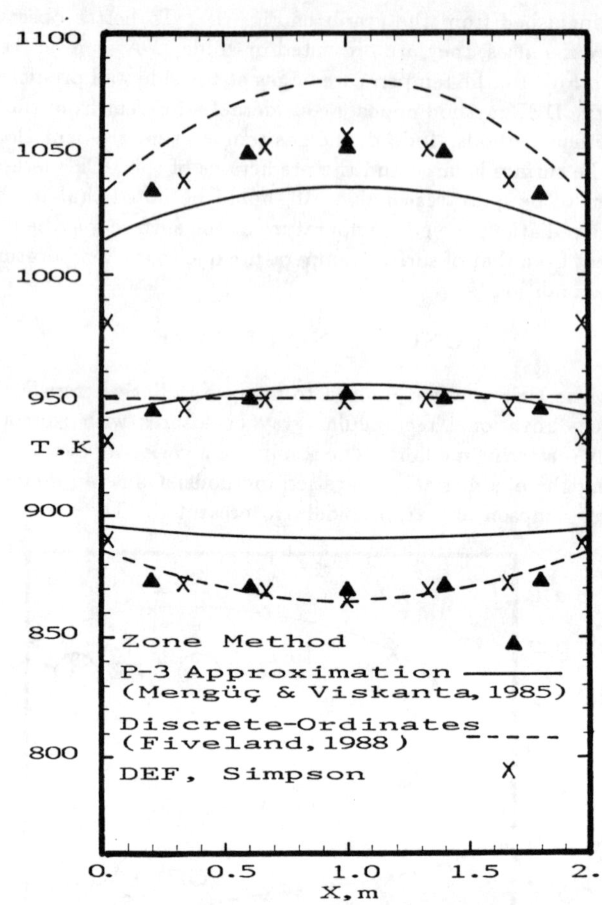

Figure 10: Comparison between results of temperature distribution based on the DEF method, P-3 approximation, discrete-ordinate solution and the zonal method.

be used to evaluate gas properties.

In most real radiative heat transfer problems the gas is nongray and anisotropically scattering. For those cases with spectral properties, a band model in which the range of active wave lengths is subdivided into finite bands can be employed. At each band, gray properties are assumed and by integrating over all bands the nongray solution can be obtained. A more challenging extension of the present approach is incorporating anisotropic scattering into the analysis. The original derivations of the governing equations of the CEF method (Naraghi et al. 1988) is based on the fact that the radiative exchange in an isotropically scattering media form a Markovian process. For anisotropically scattering media, however, this process is not Markovian. The principals of multiple Markov chains, similar to that reported by Naraghi and Chung (1984 and 1986) may be used to formulate the anisotropically scattering problem. Efforts are presently in progress to incorporate the anisotropic scattering into the present formulation.

REFERENCES

Crosbie, A. L. and Schrenker, R. G., 1982 *Exact Expressions for Radiative Transfer in three-dimensional Rectangular Geometry,* Journal of Quantitative Spectroscopy and Radiative Transfer, Vol. 28, pp. 507-526.

Fiveland, W. A., 1988, *Three-Dimensional Radiative Heat Transfer Solution by the Discrete-Ordinate Method,* AIAA Journal of Thermophysics and Heat Transfer, Vol. 2, No. 4, pp. 309-316.

Hottel, H. C. and Sarofim, A. F., 1963, *Gaseous Radiation with Temperature Gradient - Allowance for Isotropic Scatter,* Research in Heat Transfer, pp. 139-150.

Hottel, H. C. and Sarofim, A. F., 1967, *Radiative Transfer,* McGraw-Hill, New York.

Howell, J. R., 1968, *Application of Monte Carlo to Heat Transfer Problems,* Advances in Heat Transfer, Edited by T. F. Irvine Jr., and J. P. Harnett, Academic Press, Vol. 5, pp. 1-54.

Table 1: Comparison between nodal temperatures (°K) of Simpson and trapezoidal versions of DEF method

	$z = 3.6m$		$z = 2.0m$		$z = 0.4m$	
x	Simpson	Trapezoidal	Simpson	Trapezoidal	Simpson	Trapezoidal
0	980	986	932	934	890	890
1/3	1040	1042	944	946	874	876
2/3	1054	1055	948	949	870	870
1	1060	1059	949	950	866	868

Larsen, M. E., 1983, *The Exchange Factor Method: An Alternative Zonal Formulation for Analysis of Radiating Enclosures Containing Participating Media,* Ph.D. Dissertation, The University of Texas at Austin.

Larsen, M. E. and Howell, J. R., 1985, *The Exchange Factor Method: An Alternative Basis for Zonal analysis of Radiating Enclosures,* Journal of Heat Transfer, Trans. ASME, Vol. 107, No. 4, pp. 936-942.

Mengüç, M. and Viskanta, R., 1985, *Radiative Transfer in Three-Dimensional Rectangular Enclosures,* J. Quant. Spectroc. and Radiat. Transfer, Vol. 33, pp. 533-549.

Naraghi, M. H. N. and Chung, B. T. F., 1984, *A Stochastic Approach for Radiation Exchange in Enclosures with Non-Participating Medium,* Journal of Heat Transfer, Trans. ASME, Vol. 106, pp. 690-698.

Naraghi, M. H. N. and Chung, B. T. F., 1986, *A Stochastic Approach for Radiative Exchange in Enclosures with Directional-Bidirectional Properties,* Journal of Heat Transfer, Trans. ASME, Vol. 108, No. 2, pp. 264-270.

Naraghi, M. H. N., Chung, B. T. F. and Litkouhi, B., 1988, *A Continuous Exchange Factor Method for Radiative Exchange in Enclosures with Participating Media,* Journal of Heat Transfer, Trans. ASME, Vol. 110, pp. 456-462.

Naraghi, M. H. N. and Kassemi, M., 1988, *Radiative Transfer in Rectangular Enclosures: A Discretized Exchange Factor Solution,* ASME Proceedings of the 1988 National Heat Transfer Conference, Houston, Texas, July 24-27, Editor H. R. Jacobs, Vol. 1. pp. 259-267, also Journal of Heat Transfer, Trans. ASME (in press).

Stroud, A. H. and Secrest, D., 1966, *Gaussian Quadrature Formulas,* Prentice Hall.

Taniguchi, H, 1969, *The Radiative Heat Transfer in a Three Dimensional System Calculated by Monte Carlo Method,* Bull. JSME, Vol. 12, pp. 67-78.

1989 National Heat Transfer Conference
HTD-Vol. 106, Heat Transfer Phenomena in Radiation, Combustion and Fires

ANALYSIS OF COMBINED RADIATION AND CONVECTION IN A PARTICULATE LADEN FALLING FILM

S. Kumar and C. L. Tien
Department of Mechanical Engineering
University of California
Berkeley, California

Abstract

This paper presents a simple but realistic model of the interaction of radiative flux with convective transport in an absorbing, emitting, and scattering falling film. The analysis accounts for the spectral variations of the radiative intensities, emissions, reflections, refractions, scattering, and radiative properties. Such a model is required to design Direct Absorption Solar Receivers and to numerically evaluate their performance. Previous analyses in the literature have either neglected the flow field or have not considered scattering in the film and have assumed the properties to be spectrally constant. The present model is easy to understand and to numerically implement on the computer.

Nomenclature

a_n	coefficients in the expansion of phase function.
C	function defined in equation (26)
C_p	specific heat at constant pressure
D	diameter of the particles
f	incident intensity distribution
f_v	particle solid volume fraction
f_w	particle weight fraction
g	incident integrated-intensity distribution
h	convective heat transfer coefficient at film surface
i	$\sqrt{-1}$
I	intensity
J	integrated intensity
k	thermal conductivity
L	length of the one-dimensional medium
m	complex refractive index, $n+i\kappa$
$\dot{m}$	mass flow per unit width
M	half of the order of quadrature
n	index of refraction
N	degree of scattering anisotropy
P_n^k	Legendre function
q	heat flux
Q	radiative efficiency
T	temperature
u	velocity along x-direction
U	bulk mean velocity
w	weights for numerical quadrature
x,y,z	Cartesian coordinates
α	size parameter, $\pi D/\lambda$; bulk absorptivity of metal
ε	emissivity
ζ	integrated scattering phase function
θ	polar angle
κ	index of absorption
λ	wavelength of incident radiation
μ	direction cosine with respect to y-direction, $\cos\theta$; coefficient of viscosity
ρ	reflectivity; density
σ	radiative coefficients, $\sigma_e=\sigma_a+\sigma_s$
ϕ	azimuthal angle
Φ	scattering phase function

Subscripts

a	absorption
b	blackbody
e	extinction
in	inlet
L	length of medium
r	radiative
s	scattering
w	wall
λ	spectral
0	coordinate origin
∞	ambient

Superscripts

d	diffuse
s	specular

Constants

C_1	$5.9544\times10^{-17}\ W\cdot m^2$

C_2	$1.4388\times10^4\ \mu m\cdot K$
σ	$5.6696\times10^{-8}\ W/(m^2\cdot K^4)$

1. Introduction

Analysis of combined radiative and convective transport in a falling film is of fundamental importance in the study of combined mode heat transfer applications where neither mode can be neglected. One such system, based on absorption of concentrated solar radiative flux by a falling film, is the Direct Absorption Solar Receiver (Bohn *et al.*, 1986). The advantages of the direct absorption concept is that there are no intervening surfaces between the working fluid and the incident energy, thereby avoiding the associated temperature drop, improving thermal efficiency, and reducing installation costs (Anderson *et al.*, 1986).

Previous studies of gravity driven films subjected to radiation boundary conditions were confined to semi-transparent fluids of constant properties flowing over an opaque surface where scattering of radiation was neglected (Webb and Viskanta, 1985; Bohn and Wang, 1988). Radiative transfer in constant property falling particle layers was also analyzed (Houf and Grief, 1985), but the flow field and energy equations were not considered. The present analysis takes into account the spectral variation of the radiative properties of the fluid and particles, incident solar radiation, reflections, refractions, and emissions in the system, and the anisotropic scattering by particles. The incident flux is assumed to be directionally anisotropic, for example, concentrated solar flux contained in a conical envelope due to incidence from focusing mirrors.

This paper presents a realistic model of combined radiation-convection that is easy to formulate and numerically implement, and which does not neglect any important physical aspect of the system. The analytical/numerical formulation is used to examine the particle-laden liquid Direct Absorption Solar Receiver (DAR) developed by the Solar Energy Research Institute (SERI). Parametric solutions are obtained numerically for a given set of operating conditions.

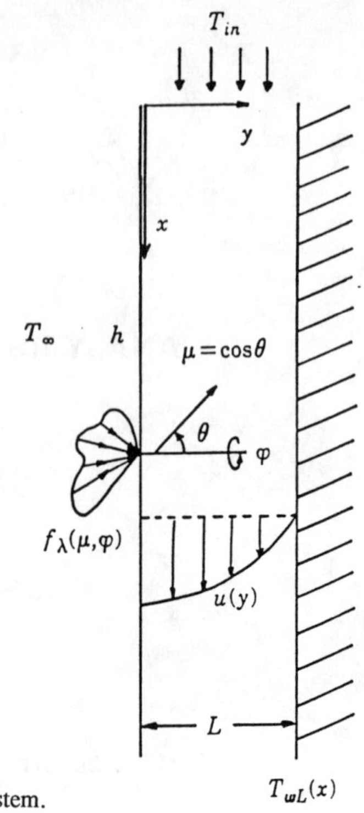

Fig. 1. The Coordinate System.

2. Theoretical Developments

Under consideration is a falling film on a wall that is exposed to concentrated solar radiation at its free surface. Small particles in the film are added to improve the radiation absorption characteristics. The solar flux is absorbed due to the radiative participation of the absorbing/scattering/emitting particle-laden film and is convected away.

2.1 Radiative Transfer

The radiation intensity field in the falling film is two-dimensional because the temperature field in the film, and therefore its associated thermal emission, is two-dimensional in nature. However, since the thickness of the film is much less than its height and the resulting temperature and intensity gradients are steeper across the film than along it, the radiative transfer in the film is modeled locally as a quasi-one-dimensional medium of length L. The variation of radiative intensity along the film is neglected in comparison with the variation across it. This one-dimensional assumption is valid for the films being considered here by scaling arguments. The curvature of the falling film is also neglected because the thickness of the film is much less that the radius of curvature. With this consideration the spectral equation of transfer within the film is written as (see figure 1)

$$\mu\frac{\partial I_\lambda(x,y,\mu,\phi)}{\partial y} = \sigma_{a\lambda}I_{b\lambda}(T(x,y)) - \sigma_{e\lambda}I_\lambda(x,y,\mu,\phi) + \frac{\sigma_{s\lambda}}{4\pi}\int_{\mu'=-1}^{1}\int_{\phi'=0}^{2\pi} I_\lambda(x,y,\mu',\phi')\Phi(\mu',\phi'\rightarrow\mu,\phi)\,d\mu' d\phi' \quad , \quad (1)$$

where I is the intensity (energy per unit projected area per unit steradian centered around (μ,ϕ) per unit wavelength interval $d\lambda$ around λ), ϕ is the azimuthal angle, $\mu = \cos\theta$, θ is the polar angle measured from the y-direction, T is the temperature, and σ are the radiative transfer coefficients, the subscripts a, e, and s referring to absorption, extinction, and scattering respectively, where extinction is the sum of absorption and scattering. The subscripts b and λ refer to black-body and spectral values respectively, and $I_{b\lambda}$ in equation (1) is given as (Siegel and Howell, 1981)

$$I_{b\lambda}d\lambda = I_{b\lambda(film)}d\lambda_{film} = \frac{n_{film}^2 2C_1}{\lambda^5(\exp(C_2/\lambda T)-1)}d\lambda \quad , \quad (2a)$$

$$I_{b\lambda}(T) = \frac{n_{film}^2 2C_1}{\lambda^5(\exp(C_2/\lambda T)-1)} \quad . \quad (2b)$$

Here n_{film} is the index of refraction of the medium (film) into which the black-body emission is being evaluated. In the above equation $I_{b\lambda}$ is the intensity per unit wavelength interval $d\lambda$ in vacuum/air, not in the medium. This is because it is desirable to keep track of the radiative energy in terms of the incident wavelengths that have been measured outside the film (i.e., in air). The wavelength of a given bundle of radiative energy varies as it propagates along the medium since it is a function of the material properties which vary with temperature. The constants C_1 and C_2 are equal to $0.59544\times10^{-16}\ W\cdot m^2$ and $14388\ \mu m\cdot K$. Assuming the particles in the film to be spherical, the scattering phase function in the equation of transfer is expressed as (see Kumar and Felske, 1986)

$$\Phi(\mu',\phi'\rightarrow\mu,\phi) = \sum_{n=0}^{N}\sum_{m=0}^{\infty}(2-\delta_{0m})a_n^m P_n^m(\mu)P_n^m(\mu')\cos m(\phi-\phi') \quad . \quad (3)$$

Here N is the degree of anisotropy of scattering, P_n^m are the Legendre functions, and δ is the Kronecker delta function. The radiative coefficients are related to the individual particle properties by the following relations

$$\sigma_i = \frac{3f_v Q_i(\alpha,m)}{2D} \quad , \quad i = a,e,s \quad , \quad (4)$$

where D is the diameter of the monodispersed particles, $\alpha = \pi D/\lambda_{film}$ is the size parameter, $\lambda_{film} = \lambda/n_{film}$ is the wavelength in the film, $m = n + i\kappa$ is the normalized refractive index of the particles ($= m_{particle}/n_{film}$), n is the index of refraction, κ is the index of refraction, f_v is the solid volume fraction, and Q is the corresponding radiative efficiency of the particle as computed by Mie theory (Kerker, 1961; Dave 1968; Bohren and Huffman, 1983).

The general radiative boundary conditions for a one-dimensional medium are given by

$$I_\lambda(x,0,+\mu,\phi) = f_\lambda(\mu,\phi) + \varepsilon_{0\lambda} I_{b\lambda}(T_{w0}(x)) + \rho^s_{0\lambda}(-\mu) I_\lambda(x,0,-\mu,\phi)$$
$$+ \frac{1}{\pi}\int_{\mu'=0}^{1}\int_{\phi'=0}^{2\pi} \rho^d_{0\lambda}(-\mu') I_\lambda(x,0,-\mu',\phi')\mu' d\mu' d\phi' \ , \quad \mu>0 \ , \quad (5a)$$

$$I_\lambda(x,L,-\mu,\phi) = \varepsilon_{L\lambda} I_{b\lambda}(T_{wL}(x)) + \rho^s_{L\lambda}(+\mu) I_\lambda(x,L,+\mu,\phi)$$
$$+ \frac{1}{\pi}\int_{\mu'=0}^{1}\int_{\phi'=0}^{2\pi} \rho^d_{L\lambda}(+\mu') I_\lambda(x,L,+\mu',\phi')\mu' d\mu' d\phi' \ , \quad \mu>0 \ , \quad (5b)$$

where ε and ρ are the emissivity and reflectivities of the boundaries and f_λ is the function specifying the form of the incident radiation after it has penetrated the film and refraction has been accounted for. The subscripts 0 and L specify the values at $y=0$ and $y=L$ respectively and w denotes the boundary or the wall. Superscripts s and d indicate specular and diffuse respectively. The diffuse reflectivity is the direction-hemispherical reflectivity (Siegel and Howell, 1981) which is a measure of the energy from the given direction being reflected diffusely into the entire hemisphere. The critical-angle reflection effects are incorporated in the above formulation by letting the direction-hemispherical diffuse reflectivity be set to unity for polar angles of incidence greater than the critical angle. The black-body intensities $I_{b\lambda}$ are evaluated from equation (2).

For the particular physical system at hand, the liquid DAR, the $y=0$ surface is exposed to incoming solar flux and the $y=L$ surface is an opaque wall. This simplifies the boundary conditions to ($\mu>0$)

$$I_\lambda(x,L,-\mu,\phi) = \varepsilon_{L\lambda} I_{b\lambda}(T_{wL}(x)) + \int_{\mu'=0}^{1}\int_{\phi'=0}^{2\pi} \frac{\rho^d_{L\lambda}(+\mu')}{\pi} I_\lambda(x,L,+\mu',\phi')\mu' d\mu' d\phi'$$

$$I_\lambda(x,0,+\mu,\phi) = f_\lambda(\mu,\phi) + \int_{\mu'=0}^{1}\int_{\phi'=0}^{2\pi} \frac{\rho^d_{0\lambda}(-\mu')}{\pi} I_\lambda(x,0,-\mu',\phi')\mu' d\mu' d\phi' \ . \quad (6)$$

The specular components of reflection are neglected. This is justified since the back wall is expected to be rendered rough by the continued corrosive contact with hot molten salts. The free surface is also assumed diffuse because of small ripples and waviness at the surface. However, the specular component can be easily introduced numerically since it is easier to numerically account for the specular component than the diffuse which involves evaluating integrals.

The azimuthal angle ϕ does not appear explicitly in the equation of transfer, equation (1), and the ϕ dependence of I is introduced via the boundary conditions, equations (5) and (6). The level of complexity of the equation of transfer and the boundary conditions is higher due to the additional integration over ϕ that has to be performed. A reduction in the number of independent variables in the functional dependence of the dependent variable is greatly desired, especially from the viewpoint of reducing the numerical effort. To simplify these equations it was proposed (Kumar and Felske, 1986) to expand the intensity as a Fourier series in ϕ. This method permits the intensities to be exactly evaluated but requires the evaluation of a corresponding Fourier series for the boundary function $f_\lambda(\mu,\phi)$, which is usually complicated. Since the goal of the analysis is to predict the heat fluxes and temperatures in the film and the individual intensities are not of interest, an "integrated intensity" J is defined as

$$J_\lambda(x,y,\mu) = \int_{\phi=0}^{2\pi} I_\lambda(x,y,\mu,\phi)d\phi \quad (7)$$

which simplifies the equation of transfer to

$$\mu\frac{\partial J_\lambda(x,y,\mu)}{\partial y} = 2\pi\sigma_{a\lambda} I_{b\lambda}(T(x,y)) - \sigma_{e\lambda} J_\lambda(x,y,\mu)$$
$$+ \frac{\sigma_{s\lambda}}{2}\int_{\mu'=-1}^{1} J_\lambda(x,y,\mu')\zeta(\mu'\to\mu)d\mu' \ , \quad (8)$$

where ζ is obtained by integrating equation (3) over all ϕ:

$$\zeta(\mu'\to\mu) = \sum_{n=0}^{N} a_n P_n(\mu) P_n(\mu') \ , \quad a_0 = 1 \ . \quad (9)$$

The corresponding boundary conditions on J are ($\mu>0$)

$$J_\lambda(x,L,-\mu) = 2\pi\varepsilon_{L\lambda} I_{b\lambda}(T_{wL}(x)) + \frac{1}{\pi}\int_{\mu'=0}^{1} \rho^d_{L\lambda}(+\mu') J_\lambda(x,L,+\mu')\mu' d\mu' \quad (10a)$$

$$J_\lambda(x,0,+\mu) = g_\lambda(\mu) + \frac{1}{\pi}\int_{\mu'=0}^{1} \rho^d_{0\lambda}(-\mu') J_\lambda(x,0,-\mu')\mu' d\mu' \ , \quad (10b)$$

where

$$g_\lambda(\mu) = \int_{\phi=0}^{2\pi} f_\lambda(\mu,\phi)d\phi \ . \quad (11)$$

The introduction of J has simplified the equations considerably by eliminating the integrals involving ϕ in the equation of transfer and boundary conditions, thereby reducing the number of dependent variables by one. Even though the individual intensities cannot be found, the radiative heat flux and its divergence are easily computed as

$$q_r(x,y) = \int_\lambda \int_{\phi=0}^{2\pi}\int_{\mu=-1}^{1} I_\lambda(x,y,\mu,\phi)\mu d\mu d\phi d\lambda \ ,$$
$$= \int_\lambda \int_{\mu=-1}^{1} J_\lambda(x,y,\mu)\mu d\mu d\lambda \ , \quad (12)$$

$$\frac{\partial q_r(x,y)}{\partial y} = 4\pi\int_\lambda \sigma_{a\lambda} I_{b\lambda}(T(x,y))d\lambda - \int_\lambda \sigma_{a\lambda}\int_{\mu=-1}^{1} J_\lambda(x,y,\mu)d\mu d\lambda \ . \quad (13)$$

2.2 Energy Equation

The radiative energy in the falling film is in equilibrium with the other modes of energy transport. The boundary conditions and the temperature term strongly link the equation of transfer to the energy equation. The energy equation is

$$\rho C_p u(y)\frac{\partial T(x,y)}{\partial x} = k\frac{\partial^2 T(x,y)}{\partial y^2} - \frac{\partial q_r(x,y)}{\partial y} \ , \quad (14)$$

where ρ is the film density, C_p the specific heat, k the coefficient of thermal conductivity, and u is the velocity in the x-direction. The film properties are functions of temperature. A fully developed velocity profile is assumed in the film which can be treated as laminar (Bohn and Wang, 1988) and is given by

$$u(y) = \frac{3\dot m}{2\rho L}\left[1 - \frac{y^2}{L^2}\right] = \frac{3}{2}U\left[1 - \frac{y^2}{L^2}\right] \ , \quad (15)$$

where $\dot m$ is the mass flow rate per unit length in the z-direction and U is the bulk-mean velocity.

The boundary conditions for the temperature field are given by

$$x=0 \ , \ 0<y<L \qquad T(0,y) = T_{in} \ ,$$
$$y=0 \ , \ x>0 \qquad h(T_\infty - T(x,0)) = -k\partial T(x,y)/\partial y|_{y=0} \ ,$$
$$y=L \ , \ x>0 \qquad q_r(L) - k\partial T(x,y)/\partial y|_{y=L} = 0 \ , \quad (16)$$

where h is the convective heat transfer at the film surface ($y = 0$) and T_∞ is the ambient temperature. The $y = L$ surface is insulated and the convective and radiative energies are balanced. The free surface of the film is exposed to the ambient and loses heat by convection.

3. Numerical Algorithm

3.1 Radiative Transfer

The equation of transfer, equation (8), along with its corresponding boundary conditions, equation (10), are solved by using the Differential-Discrete-Ordinate Method (Kumar *et al.*, 1988). The first step is to replace the integral over μ' by a quadrature. If the μ_i's are the quadrature points between -1 and $+1$ corresponding to a $2M$-order quadrature, and w_i's are the corresponding weights, the equation of transfer is reduced to the following system of ordinary differential equations after assuming that $\partial J/\partial x \approx 0$ locally:

$$\mu_i \frac{dJ_i(x,y)}{dy} = 2\pi\sigma_a I_b(T(x,y)) - \sigma_e J_i(x,y) + \frac{\sigma_s}{2}\sum_{\substack{j=-M \\ j\neq 0}}^{M} w_j J_j(x,y)\zeta(\mu_j \rightarrow \mu_i)$$

$$i = -M,..,M \ , \ i \neq 0 \ , \qquad (17)$$

where $J_i(x,y) = J(x,y,\mu_i)$. Here the subscript λ has been suppressed from J, σ, and I_b for clarity. The order $2M$ of the quadrature contains an even number of points to avoid the value $\mu = 0$ which has no distinct boundary condition at the two edges of the vertical plane. For the sake of simplicity $i = -M,..,-1$ corresponds to the negative μ and $i = 1,..,M$ to the positive. Also the μ's are so ordered that $\mu_{-j} = -\mu_j$ and $w_{-j} = w_j$. The above equation constitutes a $2M^{th}$-order system of ordinary differential equations at each x-location. It is a two-point boundary-value problem where $2M$ boundary conditions are specified by the following:

$$J_{-i}(x,L) = 2\pi\varepsilon_L I_b(T_{wL}(x)) + \frac{1}{\pi}\sum_{j=1}^{M}\rho_L^d(+\mu_j)w_j J_{+j}(x,L)\mu_j \ , \ i = 1,..,M \ ,$$

$$J_{+i}(x,0) = g_i + \frac{1}{\pi}\sum_{j=1}^{M}\rho_0^d(-\mu_j)w_j J_{-j}(x,0)\mu_j \ , \ i = 1,..,M \ . \qquad (18)$$

Here, too, the subscript λ has been omitted from the variables J, ε, I_b, and ρ for clarity. Equation (2) is used to evaluate the I_b terms in equations (17) and (18a). The divergence of the radiative flux in the medium is expressed by the quadrature as

$$\frac{\partial q_r(x,y)}{\partial y} = \int_\lambda \sigma_{a\lambda}\left[4\pi I_{b\lambda}(T(x,y)) - \sum_{\substack{j=-M \\ j\neq 0}}^{M} w_j J_{j\lambda}(x,y)\right]d\lambda \ , \qquad (19)$$

and the radiative flux at the $y=L$ boundary as

$$q_r(x,L) = \int_\lambda\left[-\pi\varepsilon_{L\lambda}I_{b\lambda}(T_{wL}(x)) + \left[1-\frac{\rho_{L\lambda}^d}{2\pi}\right]\sum_{j=1}^{M}w_j J_{j\lambda}(x,L)\mu_j\right]d\lambda \ . \qquad (20)$$

3.2 Energy Equation

The energy equation is solved by the method of finite differences. The divergence of the radiative flux is treated as a volumetric heating source. This is due to the following considerations. Unlike momentum energy whose direction of flow is clearly defined, the radiative energy flows in both directions along the same axis. It is therefore difficult to ascribe a nodal identifier to the radiative flux at the control volume interfaces, and this makes it awkward for using it in the finite difference scheme. On the other hand, the divergence of the radiative flux is easily determined from the equation of transfer and, according to the energy equation, can be unambiguously treated as a volumetric heating source. After incorporating the boundary conditions the corresponding equations are listed as below:

$$\left[\rho C_p u(y_j) + \frac{2\Delta x}{\Delta y^2}k\right]T(x_{i+1},y_j) - \frac{\Delta x}{\Delta y^2}k\left[T(x_{i+1},y_{j+1}) + T(x_{i+1},y_{j-1})\right]$$

$$= \rho C_p u(y_j)T(x_i,y_j) - \Delta x\frac{\partial q_r(x_{i+1},y_j)}{\partial y} \ , \quad 0<y_j<L \ , \qquad (21a)$$

$$\left[\frac{1}{2}\rho C_p u(y_0) + \frac{\Delta x}{\Delta y^2}k + \frac{\Delta x}{\Delta y}h\right]T(x_{i+1},y_0) - \frac{\Delta x}{\Delta y^2}kT(x_{i+1},y_1)$$

$$= \frac{1}{2}\rho C_p u(y_0)T(x_i,y_0) + \frac{\Delta x}{\Delta y}hT_\infty - \frac{\Delta x}{2}\frac{\partial q_r(x_{i+1},y_0)}{\partial y} \ , \quad y_0=0 \ , \qquad (21b)$$

$$\frac{\Delta x}{\Delta y^2}k\left[T(x_{i+1},y_L) - T(x_{i+1},y_{L-1})\right]$$

$$= \frac{\Delta x}{\Delta y}q_r(x_{i+1},y_L) - \frac{\Delta x}{2}\frac{\partial q_r(x_{i+1},y_L)}{\partial y} \ , \quad y_L=L \ . \qquad (21c)$$

The properties k, ρ, C_p are functions of temperature and are evaluated at (x_{i+1},y_j) for equation (21a), at (x_{i+1},y_0) for equation (21b), and at (x_{i+1},y_L) for equation (21c). The above equations yield an implicit system of finite difference equations for evaluating the temperatures at the nodes. These implicit equations are inherently stable for all combinations of Δx and Δy. The expressions for the divergence of the radiative flux and the radiative flux at the boundaries is obtained from equations (19) and (20). The above equations can be used to compute the temperature field by marching forward in the x-direction. Since the radiative flux and its divergence at x_{i+1} are not known before the temperatures at this x-location are computed, equations (21) cannot be directly used. Approximate forms of the above equations are then used which are obtained by replacing the value of the radiative flux and its divergence at x_{i+1} by the values obtained at the previous node x_i. The properties k, ρ, C_p are evaluated at (x_i,y_j) for equation (21a), at (x_i,y_0) for equation (21b), and at (x_i,y_L) for equation (21c). The above forms a tridiagonal linear algebraic system of equations. Equations (21) and its approximate forms can be used together in an iterative scheme.

3.3 Solution Algorithm

The radiative transfer coefficients σ_a, σ_e, σ_s, and the coefficients of the scattering phase function ζ are obtained from Mie theory (Dave, 1968; Bohren and Huffman, 1983) via equations (4) and (9). The $2M^{th}$-order system of ordinary differential equations specified by selecting a Gaussian quadrature for equations (17) and (18) are solved by using the subroutine DVCPR of the commercially available IMSL* software library. Equations (21) for evaluating the temperatures at subsequent x-locations are solved by the linear equation solver subroutine LEQT2B of the same software library.

4. Applications

Numerical solutions of the liquid DAR having the following specifications are obtained:

- Film: nitrate salt with optical constants and properties given by the following: $n_{film} = 1.462 - 1.4\times10^{-4}\times T$, where T is temperature in °C, $\kappa_{film} = 0$. Here n and κ are the indices of refraction and absorption respectively. The density and viscosity of the film are presented in Table 1. The specific heat and coefficient of thermal conductivity are given as $C_p = 2.3874\times10^3 - 1.1785\times T \ J/(kg\cdot K)$ and $k = 0.39 + 1.29\times10^{-4}\times T \ W/(m\cdot K)$, where T is the temperature in K.
- Particles: cobalt oxide (Co_3O_4), optical constants presented in Table 2. The density of the cobalt oxide is 6070 kg/m^3 (i.e., a specific gravity of 6.07).

* IMSL, 2500 ParkWest Tower One, 2500 CityWest Blvd, Houston, TX 77042

Table 1. Molten Salt Density and Viscosity.			
$T\ (K)$	$\rho\ (kg/m^3)$†	$T\ (K)$	$\mu\times10^3\ (kg/m\cdot s)$
625	1870	570	3.038
630	1866	580	2.879
635	1863	590	2.728
640	1860	600	2.585
645	1857	610	2.450
650	1854	620	2.323
655	1850	630	2.205
660	1847	640	2.094
665	1844	650	1.992
670	1840	660	1.899
675	1837	670	1.812
680	1834	680	1.735
685	1830	690	1.665
690	1827	700	1.604
695	1823	710	1.551
700	1820	720	1.506
705	1816		
710	1813		
715	1809		
720	1806		

† represented accurately by the linear relationship:

$$\rho = 1870 - \frac{64}{95}(T - 625) \ .$$

Table 2. Cobalt Oxide Optical Constants.†		
$\lambda(\mu m)$	n	κ
1.2	2.6	0.15
1.15	2.7	0.15
1.1	2.75	0.2
1.05	2.85	0.15
1.0	2.9	0.1
0.95	3.0	0.2
0.9	3.1	0.35
0.85	3.2	0.4
0.8	3.2	0.73
0.75	3.0	0.93
0.725	2.9	0.9
0.7	2.8	0.8
0.65	2.7	0.7
0.6	2.7	0.8
0.55	2.7	0.9
0.5	2.6	0.97
0.45	2.4	1.02
0.4	2.2	1.05
0.35	2.1	1.05

†(from Tanaka, 1979), λ is the wavelength.

- Metallic substrate: oxidized stainless steel, $\alpha_L = \varepsilon_L = 0.9 \neq f(\lambda) \neq f(T)$ where α and ε denote the absorptivity and surface emissivity of the metal.

The operating conditions are given by the following:

- Incoming radiation: solar spectrum radiation, isotropic within a cone whose axis of symmetry forms a 40° angle with the normal to the film and has 37° half angle. The flux on a plane perpendicular to the cone axis is 1380 kW/m^2.
- Film: parabolic velocity profile assumed with a thickness Reynolds number $4\rho UL/\mu$ equal to 70,000. Inlet temperature T_{in} of the film is 288°C.
- Other boundary conditions: metallic substrate insulated on the side not in contact with the salt film, as specified by equation (16).
- Coefficient of convective loss h at the film surface is 10 $W/(m^2{\cdot}K)$. The ambient temperature T_∞ is 27°C.

Parametric solutions are obtained by varying the particle diameter, the particle loading, and the film thickness.

The independence of the surface emissivity from wavelength and temperature simplifies the boundary condition at $y = L$ considerably, yielding

$$q_r(x,L) = -\pi\varepsilon_L I_b(T_{wL}(x)) + \left[1 - \frac{\rho_L^d}{2\pi}\right] \int_\lambda \sum_{j=1}^{M} w_j J_{j\lambda}(x,L)\mu_j \, d\lambda \ , \qquad (22)$$

where $\varepsilon_L = \alpha_L = 0.9$, $\rho_L^d = 1 - \alpha_L = 0.1$, and

$$I_b(T_{wL}) = \int_\lambda I_{b\lambda}(T_{wL}) d\lambda = \frac{n_{film}^2 \sigma T_{wL}^4}{\pi} \ . \qquad (23)$$

The value of σ is $5.6696\times10^{-8} W/(m^2{\cdot}K^4)$. Before computing $f_\lambda(\mu,\phi)$ it is necessary to evaluate the reflectivity from the film surface. These are obtained from the expressions for the directional-hemispherical reflections for dielectric films (Siegel and Howell, 1981):

$$\rho_0^d(\theta_0) = \frac{1}{2}\frac{\sin^2(\theta_0-\theta)}{\sin^2(\theta_0+\theta)}\left[1 + \frac{\cos^2(\theta_0+\theta)}{\cos^2(\theta_0-\theta)}\right] \ , \qquad (24)$$

where θ_0 is the polar angle on the air side of the film and is related to the polar angle θ ($=\cos^{-1}\mu$) in the film by the refraction relationship $\sin\theta_0 = n_{film}\sin\theta$. The solar radiative flux is incident isotropically in a cone of 37° half angle and inclined at 40° from the film normal. The intensity outside the film in this cone is proportional to $I_{b\lambda}(T_{sun})$ for $3°\leq\theta_0\leq77°$ and $-\phi_c\leq\phi\leq\phi_c$ where $I_{b\lambda}$ is the intensity in vacuum/air (equation (2) without n_{film}^2) corresponding to the temperature T_{sun} of the sun ($\approx 5780K$),

$$\phi_c(\theta_0) = \cos^{-1}\left[\frac{\cos37° - \cos\theta_0\cos40°}{\sin40°\sin\theta_0}\right] \ . \qquad (25)$$

With this consideration, the functional form of f_λ in the film, after accounting for the refraction and external reflection effects, is given by

$$f_\lambda(\mu,\phi) = CI_{b\lambda}(T_{sun})(1-\rho_0^d(\theta_0)) \ , \ 3°\leq\theta_0\leq77° \ , \ -\phi_c(\theta_0)\leq\phi\leq+\phi_c(\theta_0) \qquad (26a)$$

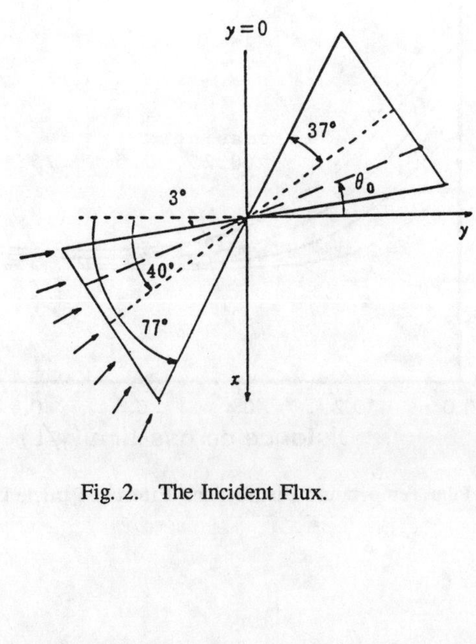

Fig. 2. The Incident Flux.

where

$$\theta_0(\mu) = \sin^{-1}\left[n_{film}\sqrt{1-\mu^2}\right] \;, \qquad C = \frac{1}{\sigma T_{sun}^4}\frac{1380\times10^3}{(1-\cos^2 37°)} \;. \tag{26b}$$

The expressions for $I_{b\lambda}$ are available from equation (2) after replacing the n_{film}^2 term by unity. The azimuthal angles ϕ inside and outside the film are the same. The integrated value g_λ is obtained as

$$g_\lambda(\mu) = 2\phi_c(\theta_0)CI_{b\lambda}(T_{sun})(1-\rho_0^d(\theta_0)) \;, \quad 0<\mu\leq1 \;. \tag{27}$$

The reflectivities for internal reflection at the y=0 surface of the film are also obtained via equation (24) which is modified as

$$\rho_0^d(-\mu) = \frac{1}{2}\frac{\sin^2(\theta-\theta_0)}{\sin^2(\theta+\theta_0)}\left[1+\frac{\cos^2(\theta+\theta_0)}{\cos^2(\theta-\theta_0)}\right] \;, \quad 0<\mu\leq1 \;, \quad 0\leq\theta_0\leq90° \tag{28a}$$

$$\rho_0^d(-\mu) = 1 \;, \quad 0<\mu\leq1 \;, \quad n_{film}\sqrt{1-\mu^2} > 1 \;. \tag{28b}$$

Equation (28b) indicates that total internal reflection is achieved if the angle of incidence within the film is greater than the critical angle for total internal reflection.

5. Numerical Results

Before presenting the results a few numerical difficulties are addressed. Firstly the convergence of the radiative transfer equation is very sensitive to the initial guess provided. A scheme for creating good initial guesses is adopted as follows. A non-scattering absorbing/emitting medium with no reflections at the boundaries is analytically solved and provided as a guess for scattering/absorbing/emitting medium computations. The output from this is used as a guess for the case when reflection at the $y=L$ surface is added, and subsequently its solution is used as initial guess for the complete radiative transport problem as described in the previous sections. Secondly, finer mesh sizes are required near the boundaries. This is needed to accurately capture the exponential decay of the intensities with distance.

Finally, higher quadrature is required for cases where the scattering is highly forward. The reason for this is evident by examining the right-hand side of equation (17). The slope $\partial J_i/\partial y$ at $y=0$ should always be negative since J_i has to decay; the loss by extinction minus the in-

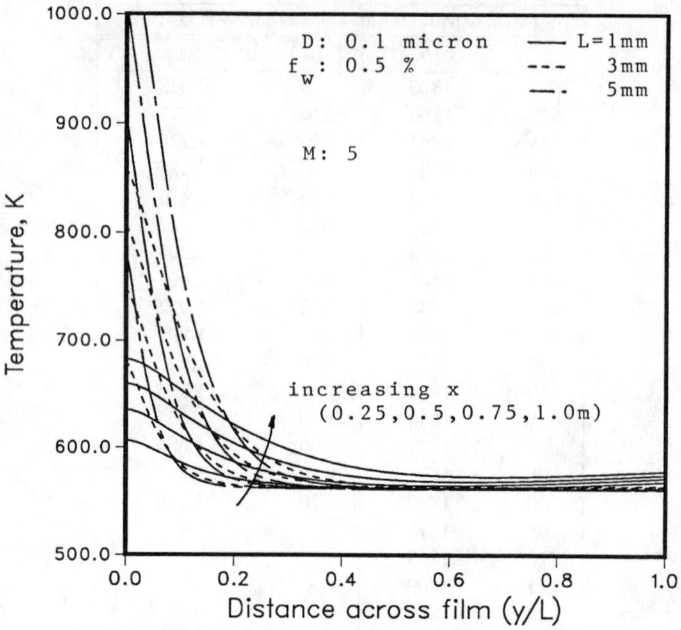

Fig. 3b. Film Temperature Distribution. Effect of Film Width.

scattering of the highly concentrated incident solar flux cannot be offset by the insignificant gain due to emission. But if M is small and $\zeta(\mu_i\rightarrow\mu_i)$ is large due to high forward scattering, $\sigma_s w_i \zeta(\mu_i\rightarrow\mu_i)$ will be numerically much greater than $2\sigma_e$, rendering the right side of equation (17) positive. This leads to spurious results that oscillate wildly instead of decaying with distance. To avoid these oscillations it is necessary to reduce the value of $\sigma_s w_i \zeta(\mu_i\rightarrow\mu_i)$ by selecting smaller-valued weights w_i, which is achieved by increasing the quadrature. One method of recognizing whether the numerical method is converging to reasonable values is to check the sign of J_λ during the computations. A non-physical negative value is an indication that the order of the quadrature

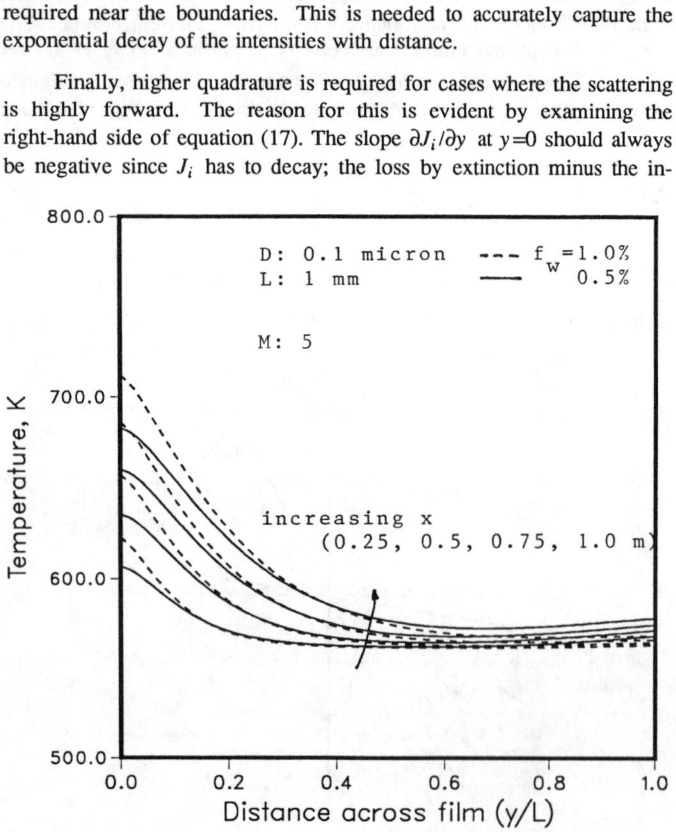

Fig. 3a. Film Temperature Distribution. Effect of Particle Loading.

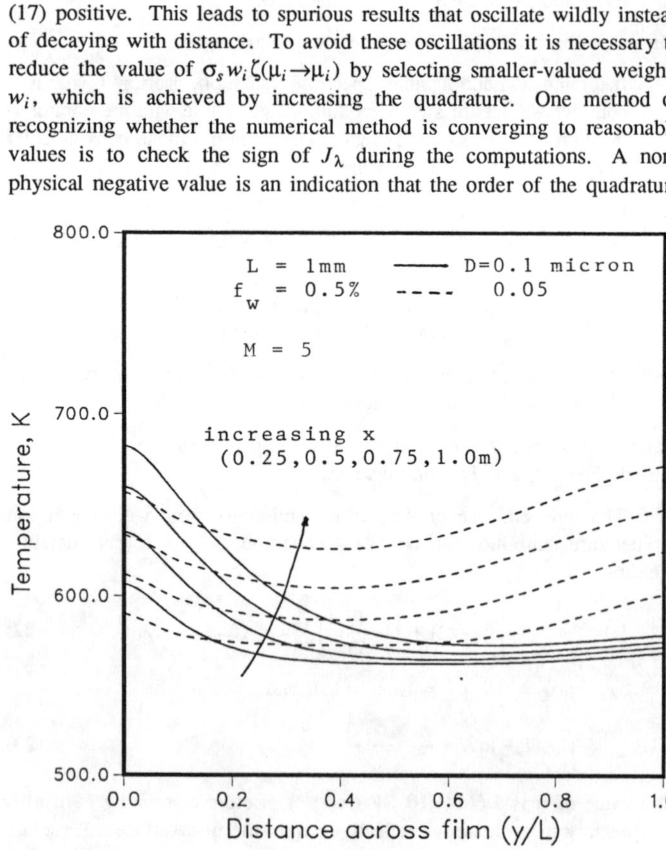

Fig. 3c. Film Temperature Distribution. Effect of Particle Diameter.

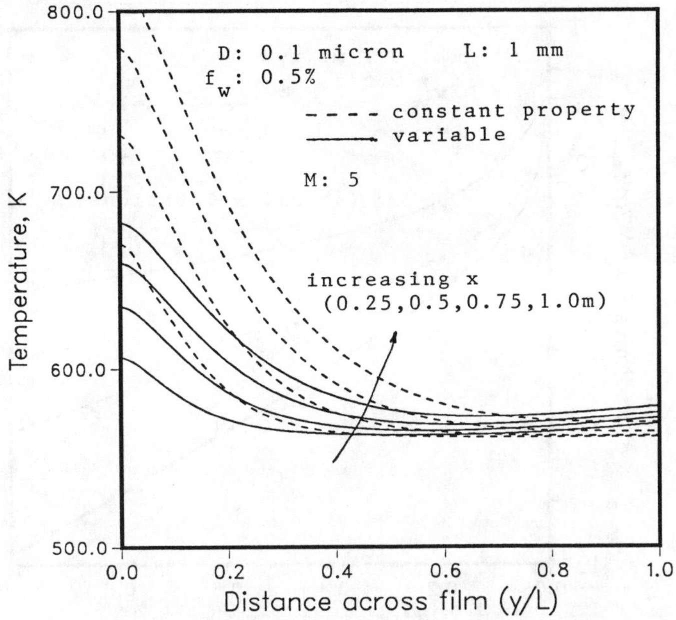

Fig. 3d. Film Temperature Distribution. Effect of Properties.

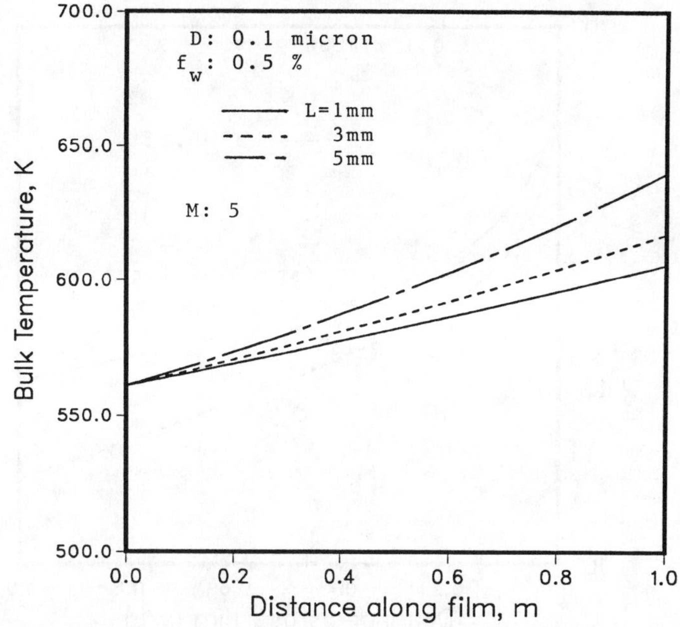

Fig. 4. Bulk Temperature Distribution along the Film.

M should be increased. Similarly negative values of the temperatures or non-monotonic temperature distributions are a symptom of insufficient mesh points and coarse grids.

The computational time increases with higher quadratures and finer mesh points. Due to the spectral nature of the radiative transfer calculations, the time taken to numerically analyze a 1 m long film for the given set of operating parameters is large, taking upto a few hours on a VAX-750 operating on the UNIX software.

Some illustrative numerical results are presented in figures 3, 4, and 5. The spatial increments Δy and Δx are $L/40$ and $0.1m$ respectively, and the wavelength increment $\Delta\lambda$ is $0.05 \mu m$. The temperature distribution within the film is sketched in figure 3 and the variation of bulk temperature along the film in figure 4. The temperature level in the film increases with increasing the film thickness or the particle volume fraction. Increasing the width L of the film has the effect of reducing the velocity since the Reynolds number is kept constant. This increases the residence time of the fluid/particle mixture and thus causes the temperatures to rise. By increasing the particle loading the absorptivity of the film increases and higher amounts of radiation are absorbed before reaching the back wall, increasing the temperature levels. Decreasing the diameter reduces the absorption within the film and hence greater amounts of direct solar radiation reach the back wall. The insulated wall absorbs this radiation and heats the fluid adjacent to the wall. If the thermophysical properties of the film are kept constant and the medium is considered to be gray with constant radiative properties corresponding to a wavelength of 0.5 μm, higher temperatures are predicted in the film. The properties are $\rho = 1840\ kg/m^3$ and $\mu = 2.0\times10^{-3}\ kg/m\cdot s$. For similar reasons the radiative flux decreases more rapidly if the particle loading is larger or the width of the film is increased, see figures 6.

6. Summary

This paper outlines a simple analytical framework for analyzing combined radiation and convection that incorporates all the important physical features such as the spectral nature of the radiation and radiation properties, reflections and refractions at the interfaces, total internal reflection, anisotropic scattering, thermal emission in the film and by the walls, and the non-isotropic and complicated shape of the incident solar radiation. The resulting model is very easy to understand and to program on a computer.

Acknowledgement.

The support from Solar Energy Research Institute through Contract IM-8-18077-1 is acknowledged.

References

Anderson, J.V., Short, W., Wendelin, T.J., and Weaver, N., 1986, "Direct Absorption Receiver Assessment," *SERI/TR-253-3080*, Solar Energy Research Institute, Golden, Colorado 80401.

Bohn, M.S., Green, H.J., Yeagle, G., Siebarth, J., Asbell, O.D., and Brown, T.J., 1986, "Direct Absorption Receiver Experiments and Concept Feasibility," *SERI/TR-252-2884*, Solar Energy Research Institute, Golden, Colorado 80401.

Bohn, M.S., and Wang, K.Y., 1988, "Experiments and Analysis on the Molten Salt Direct Absorption Receiver Concept," *ASME Journal of Solar Energy Engineering*, Vol. 110, pp. 45-51.

Bohren, C.F., and Huffman, D.R., 1983, *Absorption and Scattering of Light by Small Particles*, John Wiley and Sons, New York.

Dave, J.V., 1968, "Subroutines for Computing the Parameters of the Electromagnetic Radiation Scattered by a Sphere," Report 320-3237, IBM Palo Alto Scientific Center.

Houf, W.G., and Grief, R., 1985, "Radiative Transfer in a Solar Absorbing Particle Laden Flow," presented at *ASME/AIChE 23rd National Heat Transfer Conference (Denver)*, August 4-7.

Kerker, M., 1961, *The Scattering of Light and Other Electromagnetic Radiation*, Academic Press, New York.

Kumar, S., and Felske, J.D., 1986, "Radiative Transport in a Planar Medium Exposed to Azimuthally Unsymmetric Incident Radiation," *Journal of Quantitative Spectroscopy and Radiative Transfer*, Vol. 35,

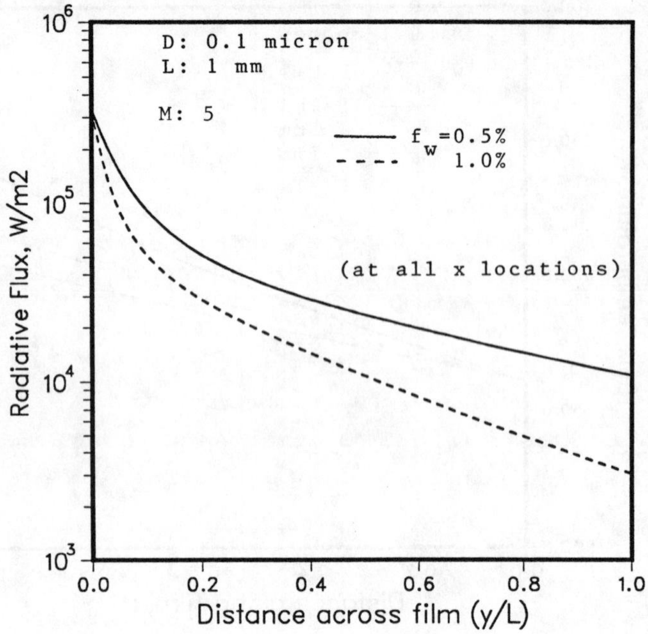

Fig. 5a. Radiative Flux Across the Film. Effect of Particle Loading.

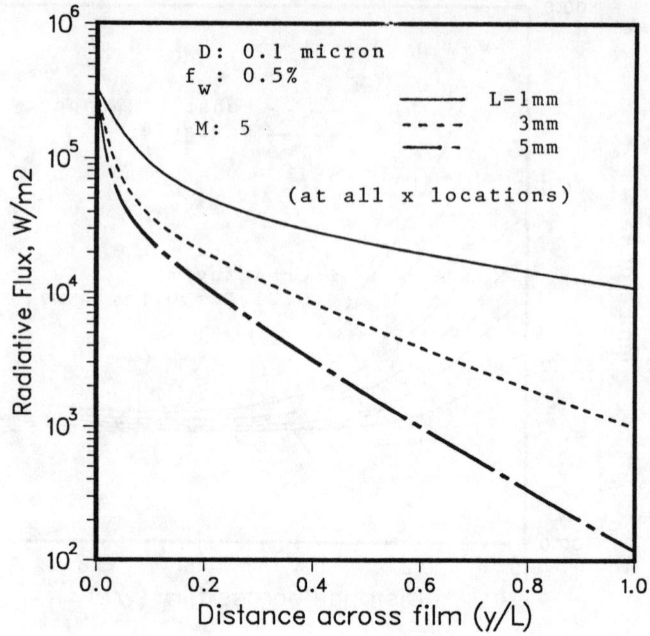

Fig. 5b. Radiative Flux Across the Film. Effect of Film Width.

pp. 187-212.

Kumar, S., Majumdar, A., and Tien, C.L., 1988, "The Differential-Discrete-Ordinate Method for Solving the General Equation of Radiative Transfer," presented at *ASME/AIChE 25th National Heat Transfer Conference (Houston)*, July 24-27, HTD Vol-96, pp. 179-185.

Siegel, R., and Howell, J.R., 1981, *Thermal Radiation Heat Transfer* (2nd edition), Hemisphere Publishers, New York.

Tanaka, T., 1979, "Optical Constants of Polycrystalline 3d Transition Metal Oxides in the Wavelength Region 350 to 1200 nm," *Japanese Journal of Applied Physics*, Vol. 18, pp. 1043-1047.

Webb, B.W., and Viskanta, R., 1985, "Analysis of Heat Transfer and Solar Radiation Absorption in an Irradiated Thin, Falling Molten Salt Film," *ASME Journal of Solar Energy Engineering*, Vol. 107, pp. 113-119.

1989 National Heat Transfer Conference
HTD-Vol. 106, Heat Transfer Phenomena in Radiation, Combustion and Fires

RADIATION HEAT TRANSFER IN AN OPEN CYCLE, ULTRA-HIGH TEMPERATURE HYDROGEN-FUELED MHD GENERATOR

J. P. Moder, D. A. Kaminski, and L. N. Myrabo
Department of Mechanical Engineering, Aeronautical Engineering and Mechanics
Rensselaer Polytechnic Institute
Troy, New York

ABSTRACT

A coupled gas dynamics/radiative heat transfer analysis of an open cycle magnetohydrodynamic generator, using high temperature (~ 15,000 K) unseeded hydrogen as the working fluid, is presented. The radiation heat transfer is modeled using a non-gray, two-dimensional P-1 approximation, which permits an arbitrary variation of the spectral absorption coefficient with frequency. The gas dynamics is modeled using a set of quasi-1-D equations which account for friction and convective, plus radiative, heat transfer, as well as the interaction between the ionized gas and applied magnetic field. The radiative heat transfer is coupled to the gas dynamics through the energy equation, and through the temperature and pressure dependent absorption coefficient. The local gas temperature, pressure, Mach number, and radiative wall heat flux, as well as the total electric work extracted and net radiative heat lost from the hydrogen gas, are presented for a range of wall boundary conditions.

NOMENCLATURE

ENGLISH SYMBOLS

A	Cross-sectional area of the MHD generator, $A = D_z D_y$
a_ω	Absorption coefficient
a_m	Arithmetic mean over a particular band
a_P	Planck mean over a particular band
a_R	Rosseland mean over a particular band
a^*	Equal to a_P for optically dense or a_R for optically thin gas
$\langle a \rangle^{(0)}$	Mean absorption coefficient appearing in band-averaged P-1 equations
$\langle a \rangle^{(1)}$	Mean absorption coefficient appearing in band-averaged P-1 equations
B	Applied magnetic field strength
dB/B	Parameter characterizing distortion of B
C_P	Temperature derivative of h at constant pressure
C_T	Pressure derivative of h at constant temperature
D_e	Exit height of MHD generator duct
D_i	Entrance height of MHD generator duct
D_P	Temperature derivative of ρ at constant pressure
D_T	Pressure derivative of ρ at constant temperature
D_y	Constant width of MHD generator duct
D_z	Local height of MHD generator duct
e_b	Black body emissive power
E_y	Electric field due to external load
G	Non-uniformity factor for electrical conductivity
h	Enthalpy per unit mass
H_o	Stagnation enthalpy per unit mass
$H_{o,ch}$	Stagnation enthalpy per unit mass of hydrogen entering heating chamber
H_R	Extraction ratio — measure of generator efficiency
h_P	Planck's constant
I_o	Solid angle integrated intensity
J_y	Local Faraday current density (assumed positive in negative y-direction)
K	Loading parameter for MHD generator
L	Axial length of MHD generator
L_c	Axial length of conical duct
l_{mb}	Mean beam length
M	Local Mach number
$\dot{m}$	Mass flow rate of hydrogen through MHD generator
N_b	Number of photon energy bands
$\hat{n}_s$	Unit vector normal to surface s
$\mathcal{O}$	Optical thickness
P	Local static pressure
$\wp$	Perimeter of MHD generator duct, $\wp = 2(D_y + D_z)$

Q_N	Energy required to heat hydrogen to entrance temperature of MHD generator
$\vec{q}$	Heat flux vector
q_w	Heat flux into walls of MHD generator
Q_w	Local heat loss per unit mass to walls of generator
$Q_{w,tot}$	Total heat loss per unit mass to walls of generator
r_i	Radius of conical duct at entrance ($x = 0$)
r_e	Radius of conical duct at exit ($x = L_c$)
S_{cone}	Total surface area of gas in conical duct
S_{gen}	Total surface area of gas in MHD generator duct
S_M	Momentum loss or sink term
S_E	Energy loss or sink term
T	Local static temperature
T_s	Temperature of surface s
T_w	Temperature of generator (& heating chamber) walls
u	Local velocity component in x-direction
$\mathcal{V}(\Delta x)$	Volume of gas over length Δx
w	Local electrical work per unit mass produced by MHD generator
w_{tot}	Total electrical work per unit mass produced by MHD generator
x,y,z	rectangular coordinates
r, x	cylindrical coordinates

GREEK SYMBOLS

ϵ_s	Emissivity of surface s
ϵ_w	Emissivity of the gray-diffuse generator walls
μ_o	Permeability of free space
ω	Photon energy, $\omega = h_P\nu$
$\omega_e\tau_e$	Hall parameter, the product of the electron cyclotron frequency times the electron mean collision time
ν	Frequency of radiation
ρ	Local mass density
σ	Scalar electrical conductivity
σ_{en}	Electrical conductivity due to electron-neutral collisions
σ_{ei}	Electrical conductivity due to electron-ion collisions
σ_r	Stefan-Boltzmann constant
τ_w	Shear stress in gas at walls of MHD generator

SUBSCRIPTS

e	Value at exit of MHD generator, property of electrons
$conv$	Convective heat transfer quantity
i	Value at entrance of MHD generator
j	Quantity integrated over photon energy band j
ω	Photon energy (or frequency) dependent quantity
rad	Radiative heat transfer quantity
s	Property of surface s
w	Property of wall of MHD generator

SUPERSCRIPTS

$\overline{f} = 1/A \iint f\, dA$	Value of any function f averaged over cross-sectional area of MHD generator

OTHER

$max\{\ \}$	Maximum value within MHD generator
$\langle\ \rangle_y$	Integration over y in MHD generator
$\langle\ \rangle_z$	Integration over z in MHD generator

INTRODUCTION

The calculation of radiative heat transfer in high temperature, nonisothermal gases is important in many applications, including rocket nozzles, combustors, and MHD generators. The strong dependence of the absorption coefficient on frequency in most gases demands the use of a nongray radiation model for a reasonable description of the radiative transport. Additionally, the radiative heat transfer and dynamics of the gas flow are coupled through the energy equation, and the state dependent absorption coefficient. Finally, these devices require multidimensional analysis.

The exact analysis of the radiative transport in a nongray, nonisothermal, multidimensional gas coupled with the gas dynamics requires the solution of an integro-differential energy equation — integrated with respect to solid angle and frequency — coupled with the differential conservation equations for mass and momentum. Due to the mathematical complexity of these equations, approximate and numerical techniques are required to describe the gas dynamics and radiative heat transfer. Computational methods for the numerical simulation of MHD gas flows have received considerable attention and several powerful 3-D numerical codes have been developed. Also, a wide range of approximate techniques for describing radiation in participating media exist; however, methods applicable to nongray, nonisothermal, multidimensional gas flows are lacking.

One of the more popular approximate techniques for analyzing radiative heat transfer is the P-N approximation or the spherical harmonics method (Jeans, 1917 and Krook, 1955), in which the directional dependence of the radiative intensity is represented by a series of spherical harmonics. Truncating this series after $(N+1)^2$ terms and substituting into a set of N moments of the radiative equation of transfer leads to a coupled set of elliptic linear partial differential equations (PDE's). The accuracy, as well as the mathematical complexity, of this method increases as more terms are retained in the series. It should be noted that this method may be applied to the spectral as well as the total (i.e., integrated over all frequencies) radiative intensity. Thus, one may generate nongray, as well as gray, P-N approximations. Finally, the inclusion of scattering offers no particular difficulty, as illustrated by Modest and Azad (1980) for the cases of isotropic and linear anisotropic scattering.

Although the P-N approximations are receiving widespread use in multidimensional, gray participating media, as in the work of Ratzel and Howell (1983) and Mengüc and Viskanta (1986), very little work has been done in nongray media. An early paper by Kung and Sibulkin (1969) employs the P-1 method in a medium with a two-level absorption coefficient. More recently, Modest (1979) has presented a box model for the P-1 approximation. His method has the advantage of being no more difficult to solve than the gray case, but requires the absorption coefficient to have the same value over all bands. Yuen and Rasky (1981) have presented an elegant analytical technique for a nongray case. Yucel and Bayazitoglu (1983) present extensive results for a medium in radiative equilibrium with a two-level absorption coefficient, for both the nongray P-1 and P-3 approximations. In all of these nongray studies, only the one-dimensional parallel plate medium is considered.

Three dimensional gas dynamics codes for the analysis of MHD generator flows have been developed by Demetriades et al. (1983), Doss and Ahluwalia (1982), and Liu et al. (1986) which investigate the coupled gas dynamics/electrodynamics problem. These codes employ the partially parabolized Navier-Stokes equation and the turbulent kinetic energy equation, along with some form of the generalized Ohm's law and the Maxwell relations for the current density and electric field. An important phenomena accounted for in these 3-D models is the vorticity generated by the MHD body forces, which directs a secondary flow from cathodes to anodes along the centerline of the electrode walls in Faraday MHD generators. Analyses have also been performed for the coupled gas dynamics/radiation problem. Biberman et al. (1979) considered the coupled analysis of an open cycle, natural gas fueled MHD generator using quasi-one-dimensional gas dynamics equations and the Curtis-Godson approximation (Edwards and Weiner, 1966) to model the radiative transport. Smith et al. (1981) have considered the coupled problem using the parallel plate solution for a constant absorption coefficient. Ahluwalia and Im (1981) have investigated the coupled problem employing a one-dimensional P-1 approximation which included both absorption and particle scattering.

This study is concerned with the coupled gas dynamics/radiative heat transfer analysis of a particular Faraday MHD generator envisioned for both space-based and propulsive applications. This MHD generator has a diverging duct with a rectangular cross-section. The gas dynamics is described by a set of quasi-1-D differential equations which account for friction and convective, plus radiative, heat transfer, as well as the interaction between the ionized gas and the applied magnetic field. The radiative heat transfer in the nonisothermal, essentially nonscattering, hydrogen gas is described using a nongray P-1 approximation. The frequency dependence of the absorption coefficient is approximated by a finite number of photon energy bands (or, equivalently, frequency bands), N_b, across which the absorption coefficient is constant. There is no restriction on the width of any band or the value of the absorption coefficient across any band. A set of N_b band-averaged equations is generated by integrating the nongray P-1 equations over each band.

In the radiation analysis, the actual MHD generator duct geometry is approximated by a diverging conical duct of constant divergence angle. The magnitude of the error introduced by this approximation is expected to be on the same order as that introduced into the gas dynamics model by the quasi-1-D MHD equations. The gas flow and boundary conditions are assumed to be axisymmetric, making the radiation problem two-dimensional. Since the purpose is to solve the coupled gas dynamics/radiative heat transfer problem, the development of a 3-D radiation model would not significantly improve the accuracy of the analysis unless a 3-D gas dynamics model was also developed; such a task was beyond the scope of this study. Axial distributions of gas temperature, pressure, Mach number, and radiative wall heat flux are presented for several wall emissivities and wall temperatures. The total radiative heat transfer to the generator walls is found to be very sensitive to the wall emissivity, as might be expected.

THE MHD GENERATOR

The geometry of the MHD generator is shown in Fig. 1. The cross section of the generator duct is rectangular, with a constant width D_y and a variable height D_z. The core flow u_c (i.e., the flow along the duct centerline) is in the x-direction, the applied magnetic field B is in the z-direction, and the induced Faraday current density J_y, resulting from the flow of the ionized gas perpendicular to the applied field B, is in the negative y-direction. The two walls perpendicular to the current flow consist of electrode and insulator segments. Since the direction of J_y represents the flow of ions, the electrons flow in the opposite direction; thus, the near wall electrodes ($y = -D_y/2$) function as cathodes and the far side electrodes ($y = D_y/2$) will function as anodes. An electrical load is connected across each anode-cathode pair forming a closed circuit through which the induced current can flow. In general, an axial or Hall current density J_x will also be present, causing a smaller current density J_y and thus a loss in generator performance.

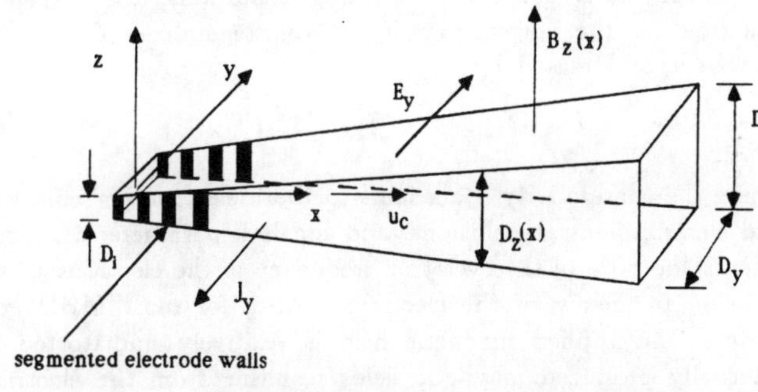

Figure 1: MHD Generator Geometry.

The purpose of these ultra high performance MHD generators is to provide a lightweight electrical power source which produces large amounts (on the order of megawatts) of electrical power efficiently. A good measure of efficiency for these high temperature, open cycle MHD generators is the extraction ratio, H_R. We define H_R as the ratio of electric work extracted (per unit mass of hydrogen) to energy per unit mass, Q_N, required to heat the hydrogen gas to the entrance temperature of the generator, $\overline{T}_i$.

$$H_R = w_{\text{tot}}/Q_N \tag{1}$$

The hydrogen fuel is stored as liquid para hydrogen and used to regeneratively cool the walls of the MHD generator. Before entering the MHD generator, the hydrogen is further heated to the MHD generator entrance temperature of 20,000 K in what will be referred to as the heating chamber. Several methods exist for heating the hydrogen, including a Laser Supported Plasma (Jeng and Keefer, 1987 and 1986, and Keefer et al., 1985) across the entrance of the generator or radiation from the fissioning plasma of a gas core nuclear reactor (Rosa and Myrabo, 1984). The maximum possible temperature of the hydrogen gas entering the heating chamber is the temperature of the chamber walls, T_w. Ignoring any irreversibilities (losses) in heating the hydrogen from T_w to $\overline{T}_i$, the *minimum* value for the required energy Q_N is given by

$$Q_N = H_{oi} - H_{o,ch} \tag{2}$$

where H_{oi} and $H_{o,ch}$ are the stagnation enthalpies per unit mass of the hydrogen at the generator entrance and the heating chamber entrance, respectively.

GAS DYNAMICS COMPUTATIONAL MODEL

Governing Equations

The high pressures which exist throughout the generator allow the partially ionized hydrogen gas to be considered in local thermodynamic equilibrium (LTE); so, although the gas consists of several species, its physical properties may be described by a single temperature and pressure. Also, for the conditions encountered in the MHD generator, the gas as a whole may be treated as a continuum and electrically neutral, except for a small sheath layer near the walls. Thus, the flow may be adequately modeled by a single set of global continuum equations consisting of the conservation of mass, linear momentum, and energy, and a transport equation for the current density (usually referred to as the generalized Ohm's law in one of its simplified forms).

For MHD generator operating conditions, the transport equation for the current density or the generalized Ohm's law may be approximated by

$$\vec{J} = \sigma(\vec{E} + \vec{V} \times \vec{B}) - \frac{\omega_e \tau_e}{B}(\vec{J} \times \vec{B}) \tag{3}$$

where the conductivity σ accounts for both electron-ion collisions and electron-neutral collisions, and the Hall parameter $\omega_e \tau_e$ represents the ratio of the cyclotron frequency of the electrons to the collision frequency of the electrons. Also, for most MHD generators, the applied magnetic field is relatively undistorted by internally produced magnetic fields resulting from the electrical currents within the gas. The Maxwell equation for the magnetic induction, $\vec{B}$, consistent with the assumptions of zero net charge density (electrical neutrality) and a relative permeability of μ_o is given by

$$\vec{\nabla} \times \vec{B} = \mu_0 \vec{J} \tag{4}$$

An order of magnitude approximation of this equation leads to a parameter describing the distortion of the applied magnetic field given by

$$\frac{dB}{B} = \frac{\mu_0 J_y D_z/2}{B} \tag{5}$$

where J_y and $D_z/2$ are characteristic values for the current density perpendicular to the applied magnetic field B and the length over which the magnetic field acts, and dB is a measure of the distortion or change in the magnetic field due to the internal currents J_y. It is assumed in this analysis that for dB/B less than 0.1, the distortion of the applied magnetic field is negligible. With this assumption, the magnetic field distribution becomes a known quantity and the Maxwell relation given by Eqn. 4 is not required.

The gas flow was assumed to be steady and would almost certainly be a fully developed turbulent flow in any real application, especially for the high velocities typical of the generators considered here. To simplify the gas dynamics analysis, the governing equations were integrated or averaged over the cross-sectional area of the generator duct. The applied magnetic field B is a function of the axial coordinate x only (an axially decreasing, cross-sectionally uniform field is used) and has only a z-component, as indicated previously. The shear stress and heat flux were assumed uniform around the perimeter of the generator duct at a given x location, and the pressure was assumed to be a function of x only. The electrode walls were assumed to be infinitely segmented so that the cross-sectionally averaged Hall current $\overline{J}_x = 0$, although an average Hall electric field $\overline{E}_x$ would be present. Finally, making some assumptions consistent the velocity and temperature profile shapes in a turbulent flow (Myrabo et al., 1987), the integration of the equations over the cross-sectional area of the generator yielded the following set of quasi-one-dimensional MHD equations:

$$\text{Continuity} \quad \overline{\rho}\,\overline{u} A = \dot{m} \quad \text{or} \quad \frac{1}{\overline{\rho}}\frac{d\overline{\rho}}{dx} + \frac{1}{\overline{u}}\frac{d\overline{u}}{dx} + \frac{1}{A}\frac{dA}{dx} = 0 \tag{6}$$

$$\text{Momentum} \quad \overline{\rho}\,\overline{u}\frac{d\overline{u}}{dx} + \frac{dP}{dx} = -\left[\tau_w \frac{\wp}{A} + \overline{J}_y B\right] = -S_M \tag{7}$$

$$\text{Energy} \quad \overline{\rho}\,\overline{u}\frac{d}{dx}\left(\overline{h} + \frac{1}{2}\overline{u}^2\right) =$$

$$-\left[\overline{J}_y \overline{E}_y + (q_w)_{conv}\frac{\wp}{A} + \frac{1}{A(x)}\iint_{A(x)}\left(\vec{\nabla}\cdot\vec{q}_{rad}\right)dA\right] = -S_E \tag{8}$$

$$\text{Ohm's Law} \quad \overline{J}_y = \frac{\overline{\sigma}}{G}\overline{u}B\,(1-K) \tag{9}$$

where

$$G = \overline{\sigma}\left\langle \frac{1}{\langle\sigma\rangle_z}\right\rangle_y + (\omega_e\tau_e)^2\left[\overline{\sigma}\left\langle\frac{1}{\langle\sigma\rangle_z}\right\rangle_y - 1\right] \tag{10}$$

$$K = \frac{\overline{E}_y}{\overline{u}B} \tag{11}$$

and

$$\overline{\rho} \approx \rho(\overline{T}, P), \quad \overline{h} \approx h(\overline{T}, P), \quad \overline{\sigma} \approx \sigma(\overline{T}, P)$$

In the above equations, all quantities represent Reynolds averaged mean turbulent quantities, except the externally applied magnetic field, B. The loading parameter, K, is determined by the external electrical load at a given axial location and the non-uniformity factor G represents the influence of the Hall current and of the variation in electrical conductivity from the centerline to the wall on the effective conductivity in the Ohm's law. All quantities in the above equations are positive, except the radiative volume loss term, $\left(\vec{\nabla}\cdot\vec{q}_{rad}\right)$, which may be positive or negative.

The differentials of the enthalpy and density can be expressed as

$$dh = \left(\frac{\partial h}{\partial T}\right)_P dT + \left(\frac{\partial h}{\partial P}\right)_T dP = C_P\,dT + C_T\,dP$$

$$d\rho = \left(\frac{\partial \rho}{\partial T}\right)_P dT + \left(\frac{\partial \rho}{\partial P}\right)_T dP = D_P\,dT + D_T\,dP$$

Expressions for the local wall shear stress τ_w and the convective heat flux flowing into the wall, $(q_w)_{conv}$, were obtained by approximating the flow at any point near each wall of the generator as a steady, compressible, turbulent flat plate boundary layer flow, with the boundary layers merged along the duct centerline. The skin friction coefficient derived from the inner variable method of Christoph and White (1974) was used to calculate the wall shear stress, and the Reynolds analogy for compressible turbulent boundary layer flows (White, 1974) was used to calculate the convective wall heat flux. The viscosity and (scalar) thermal conductivity of hydrogen were approximated by curve fits of data obtained by Yos (1963).

For the temperature and pressure range of the MHD generators considered in this study, the hydrogen gas was essentially completely dissociated, except near the cold walls, where no ionization occurred. For the calculation of τ_w and $(q_w)_{conv}$, the composition and certain properties of hydrogen at the wall temperature T_w were obtained by applying the law of mass action to the dissociation reaction

$$H_2 \rightleftharpoons 2H$$

only. The calculation of composition and gas properties at the higher average gas temperature, $\overline{T}$, assumed complete dissociation and considered only the reaction

$$H \rightleftharpoons H^+ + e^-$$

Using Boltzmann statistics, relatively simple analytic expressions were developed for the thermodynamic properties D_P, D_T, C_P, C_T, h and ρ. These results were compared with data generated by Patch (1971), who considered spin-equilibrated hydrogen in chemical equilibrium in the Debye-Huckel approximation and accounted for the species H^-, H_2^+, and H_3^+. The maximum percent difference was 12% for values of ρ, h, C_P and the electron number density.

Finally, the scalar electrical conductivity σ appearing in the generalized Ohm's Law is defined by

$$\frac{1}{\sigma} = \frac{1}{\sigma_{en}} + \frac{1}{\sigma_{ei}}$$

This relation may be interpretted as the summation of two types of resistance to electron motion in the gas — short range collisions with neutrals and long range collisions with ions. The calculation of σ_{en} was based on the mean free length concept accounting for momentum transfer between electrons and hydrogen atoms. The conductivity σ_{ei} was calculated from the Fokker-Planck formula for the conductivity of a fully ionized gas.

Numerical Solution

In order to carry out the iterative solution of the coupled gas dynamics/radiation heat transfer problem, the cross-sectional area distribution of the MHD generator duct, $A(x)$, was fixed. Since the actual MHD generator geometry is approximated by a conical duct of constant divergence angle for the radiative heat transfer analysis, the closest similarity between the actual MHD generator geometry and the conical duct will occur for a linear height distribution $D_z(x)$. For a specified area distribution, equations 6-11 can be reduced to two coupled, first order, nonlinear differential equations:

$$\frac{dT}{dx} = \frac{\overline{p}\,(1-\overline{p}C_T)\,\overline{u}^3\,dA/dx - A\,(1-D_T\overline{u}^2)\,S_E}{\dot{m}C_P\,(1-M^2)} - \frac{(AD_T\overline{u}^3 - \dot{m}C_T)\,S_M}{\dot{m}C_P\,(1-M^2)} = f(\overline{T},P,x) \quad (12)$$

$$\frac{dP}{dx} = \frac{\overline{p}^2\overline{u}^2C_P\,dA/dx - \overline{u}AD_P\,S_E - A\,(\overline{p}C_P - D_P\overline{u}^2)\,S_M}{\overline{p}AC_P\,(1-M^2)} = g(\overline{T},P,x) \quad (13)$$

The solution was obtained using fourth order Runge-Kutta to march down the duct. At each step, the electrical work extracted per unit mass, Δw, and the heat loss to the walls per unit mass, ΔQ_w, were calculated according to

$$\Delta Q_w = (\Delta Q_w)_{conv} + (\Delta Q_w)_{rad} = \frac{1}{\dot{m}} \int_x^{x+\Delta x} \left[(q_w)_{conv} + (q_w)_{rad}\right] \wp(x)\,dx \quad (14)$$

$$\Delta w = H_o(x) - H_o(x+\Delta x) - \Delta\,(Q_w)_{conv} - \frac{1}{\dot{m}} \iiint_{\mathcal{V}(\Delta x)} \left(\vec{\nabla}\cdot\vec{q}_{rad}\right) d\mathcal{V} \quad (15)$$

The initial conditions consisted of the entrance temperature, pressure, magnetic field, and the mass flow rate. The loading parameter, wall temperature and emissivity, and the cross-sectional area distribution $A(x)$ completed the problem specification. The parameters K, T and ϵ_w were assumed constant for the entire generator. The parameters dB/B and $\omega_e\tau_e$ were also calculated at each step through the generator.

The above method of solution is restricted to gas flows which are either subsonic or supersonic throughout the MHD generator, due to the singularity at the sonic point. If the flow passes through a Mach number of unity, the above set of differential equations represent a shooting or eigenvalue problem, since only certain initial conditions will allow smooth passage through the sonic point. The gas dynamics computational model has not been modified to handle this problem at present; therefore, MHD generators transitioning from subsonic to supersonic flow are not considered in this study.

RADIATIVE HEAT TRANSFER COMPUTATIONAL MODEL

Governing Equations

For the P-1 approximation, the spectral equations governing the radiative transport are given by

$$\vec{\nabla}\cdot\vec{q}_\omega = 4a_\omega e_{b\omega} - a_\omega I_{o\omega} \quad (16)$$

$$\vec{\nabla} I_{o\omega} = -\,3a_\omega \vec{q}_\omega \quad (17)$$

Integrating the above equations over a photon energy band j, the following band-averaged P-1 equations are obtained.

$$\vec{\nabla}\cdot\vec{q}_j = 4a_{Pj}e_{bj} - \langle a\rangle_j^{(0)} I_{oj} \quad (18)$$

$$\vec{\nabla} I_{oj} = -\,3\langle a\rangle_j^{(1)}\vec{q}_j \quad (19)$$

Three mean absorption coefficients are generated by integrating the spectral equations over a photon energy band:

$$a_{Pj} = \frac{\int_{\Delta\omega_j} a_\omega e_{b\omega}\,d\omega}{\int_{\Delta\omega_j} e_{b\omega}\,d\omega} \quad (20)$$

$$\langle a\rangle_j^{(0)} = \frac{\int_{\Delta\omega_j} a_\omega I_{o\omega}\,d\omega}{\int_{\Delta\omega_j} I_{o\omega}\,d\omega} \quad (21)$$

$$\langle a\rangle_j^{(1)} = \frac{\int_{\Delta\omega_j} a_\omega \vec{q}_\omega\,d\omega}{\int_{\Delta\omega_j} \vec{q}_\omega\,d\omega} \quad (22)$$

The mean absorption coefficient a_{P_j} is the familiar Planck mean for a particular band j and is a physical property of the gas. The mean absorption coefficients defined by Eqns. 21 and 22 depend upon the solution of a particular problem and must be approximated in order to make the band-averaged P-1 equations useful. Finkelman and Chien (1968) have shown that the asymptotic values for these mean absorption coefficients as the optical thickness approaches infinity are given by

$$\langle a\rangle_j^{(0)} \sim a_{Pj} \quad \text{and} \quad \langle a\rangle_j^{(1)} \sim a_{Rj}$$

where a_{R_j} is the Rosseland mean over a particular band defined by

$$a_{Rj} = \frac{\displaystyle\int_{\Delta\omega_j} \frac{\partial(e_{b\omega})}{\partial e_b}\, d\omega}{\displaystyle\int_{\Delta\omega_j} \frac{1}{a_\omega}\frac{\partial(e_{b\omega})}{\partial e_b}\, d\omega} \tag{23}$$

In the optically thin limit, we expect to find

$$\langle a\rangle_j^{(0)} \sim a_{Pj} \quad \text{and} \quad \langle a\rangle_j^{(1)} \sim a_{Pj}$$

No studies have been made concerning the behavior of the above mean absorption coefficients in the optically intermediate range, that is, the optical thickness range between optically dense and optically thin. In this study, the above mean absorption coefficients will be approximated by their optically thin or dense limits across a particular band, depending on the optical thickness of the gas across that particular band.

The spectral P-1 boundary condition at any point along a diffuse surface is given by

$$\left(\vec{q}_{rad,\omega}\cdot\hat{n}_s\right)\Big|_s = \frac{1}{2}\frac{\epsilon_{s\omega}}{2-\epsilon_{s\omega}}\left[4e_{b\omega}(T_s) - I_{o\omega}|_s\right] \tag{24}$$

For a nongray surface, the band-averaged boundary condition will generate two means for the parameter $\epsilon_{s\omega}/(2-\epsilon_{s\omega})$ similar to the absorption means a_P and $\langle a\rangle^{(0)}$. In this case, one mean will be a physical property of the surface and the other mean will depend on the solution of a specific problem. In the case of a diffuse, semi-gray surface, in which the value of the spectral wall emissivity $\epsilon_{s\omega}$ is constant over each band j, the band-averaged P-1 boundary condition becomes

$$\left(\vec{q}_{rad,j}\cdot\hat{n}_s\right)\Big|_s = \frac{1}{2}\frac{\epsilon_{s,j}}{2-\epsilon_{s,j}}\left[4e_{bj}(T_s) - I_{oj}|_s\right] \tag{25}$$

In order to simplify the radiation analysis, the actual MHD generator duct geometry was approximated by a diverging conical duct of constant divergence angle, with gray-diffuse, uniform temperature walls. The geometry and cylindrical coordinate system for the conical duct are depicted in Fig. 2. Assuming an axisymmetric distribution of gas properties and axisymmetric wall boundary conditions, the radiation problem becomes two-dimensional.

The conical duct shape is completely determined by the axial length L_c, and the entrance and exit radii, r_i and r_e. The parameters r_i, r_e and L_c were selected such that the radiative transport in the conical duct modeled as closely as possible that in the actual MHD generator duct. The following three conditions were used to determine the conical duct dimensions for a given set of actual MHD generator dimensions:

1. $L_c = L$ (equivalent axial length)
2. $S_{cone} = S_{gen}$ (equivalent gas surface area)
3. $\pi r_i^2 = D_y D_i$ (equivalent entrance cross-sectional area)

The first condition is mainly one of computational convenience, allowing equivalent axial step sizes in the radiation and gas dynamics computational models. Condition (2) is probably the most important condition for radiative transport similarity between the two geometries. Condition (3) was enforced since the maximum gas temperature occurs at the entrance of the generator and steadily decreases thereafter; thus, a net radiative heat flux from the entrance plane into the cooler gas within the generator will exist, the magnitude of which is proportional to the entrance cross-sectional area.

If the radiative heat $q_{rad,j}$ is eliminated between Eqns. 18 and 19 and the result is expanded in axisymmetric cylindrical coordinates, the band-averaged P-1 equation for the integrated intensity becomes

$$\frac{1}{r}\frac{\partial}{\partial r}\left(\frac{r}{a_j^*}\frac{\partial I_{oj}}{\partial r}\right) + \frac{\partial}{\partial x}\left(\frac{1}{a_j^*}\frac{\partial I_{oj}}{\partial x}\right) = -3a_{Pj}\left(4e_{bj} - I_{oj}\right) \tag{26}$$

where the mean absorption coefficient a_j^* is defined by

$$a_j^* = \left\{\begin{array}{ll} a_{Pj}\,; & \text{Gas is Optically Thin over Band } j \\ a_{Rj}\,; & \text{Gas is Optically Dense over Band } j \end{array}\right\} \tag{27}$$

If the radiative heat flux is eliminated between Eqns. 19 and 25, the band-averaged P-1 boundary condition for I_{oj} at a diffuse, semi-gray surface becomes

$$\left(\frac{\vec{\nabla} I_{oj}}{3a_j^*}\cdot\hat{n}_s\right)\Big|_s = -\frac{1}{2}\frac{\epsilon_{s,j}}{2-\epsilon_{s,j}}\left[4e_{bj}(T_s) - I_{oj}|_s\right] \tag{28}$$

Numerical Solution

For each band j, Eqn. 26 was solved numerically for the band-averaged integrated intensity, I_{oj}, assuming the temperature and density distributions were known throughout the gas. Note that the temperature-density dependent mean absorption coefficients a_{Pj} and a_j^* may vary throughout the gas. The entrance and exit planes of the duct were approximated as black bodies at uniform temperatures $\overline{T}_i$ and $\overline{T}_e$, respectively. The walls of the conical duct were treated as gray-diffuse surfaces with constant wall temperature T_w. The black body emissive power for each band j, $e_{bj}(T)$, was calculated using the curve fits recommended

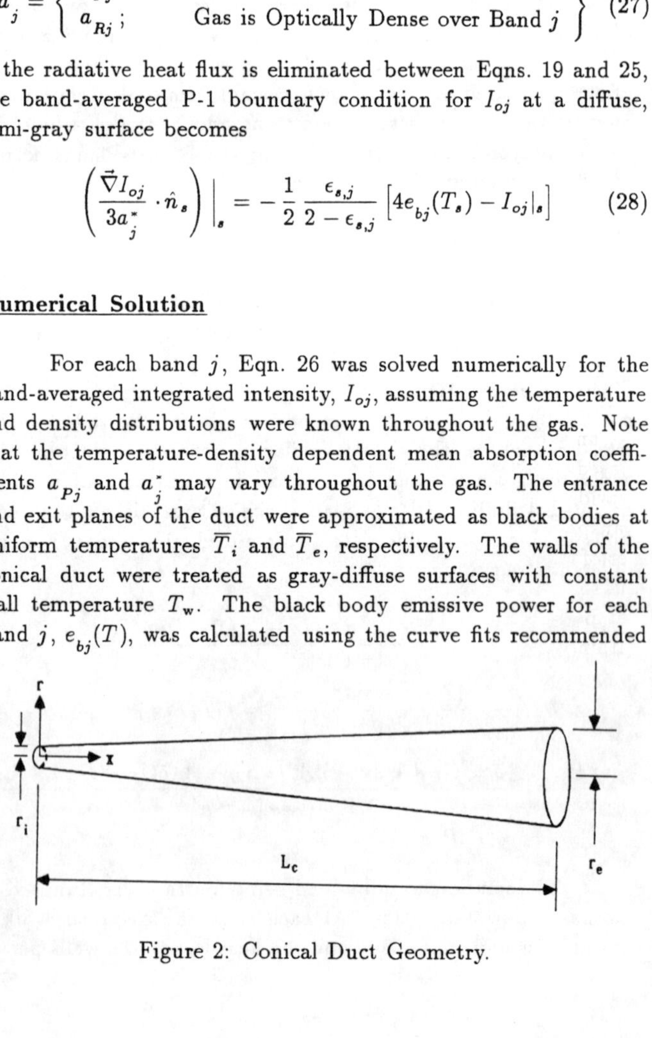

Figure 2: Conical Duct Geometry.

by Wiebelt (1966).

The computational grid for numerical simulation consisted of many non-overlapping control volumes. As indicated in Fig. 3, the majority of the control volumes were circular rings of rectangular cross-section. Along the sloping surface of the cone, rings of triangular cross-section were used to fit the shape of the gas boundary. Control volumes were uniformally spaced along the x axis and, as illustrated in Fig. 3, were not completely uniform along the r axis, but were uniformally spaced in the regions $0 \leq r \leq r_i$ and $r_i \leq r \leq r_e$. A set of discretized algebraic equations for I_{oj} was derived by integrating Eqn. 26 over each control volume. Equation 28 was employed for control volume surfaces along the boundary of the gas. The local radiative heat flux into the duct wall from a triangular control volume was computed using Eqn. 25, summed over all N_b bands.

$$(q_w)_{rad} = \sum_{j=1}^{N_b} \left(\vec{q}_{rad,j} \cdot \hat{n}_s\right)\Big|_s \tag{29}$$

Also, applying Eqn. 18 at x locations corresponding to grid points, the following two quantities were calculated for use in the gas dynamics computational model:

$$\frac{1}{A(x)} \iint_{A(x)} \left(\vec{\nabla} \cdot \vec{q}_{rad}\right) dA \tag{30}$$

$$\frac{1}{\dot{m}} \iiint_{\mathcal{V}(\Delta x)} \left(\vec{\nabla} \cdot \vec{q}_{rad}\right) d\mathcal{V} \tag{31}$$

The above integrals reduce to summations of nodes along the r axis since, for the numerical simulation, the volume loss term $\vec{\nabla} \cdot \vec{q}_{rad}$ is constant over each control volume.

The P-1 solution for an isothermal gas in a conical duct, with cold black walls, and cold black entrance and exit planes, was compared with a Monte-Carlo solution for a range of duct geometries and optical thicknesses. The Monte-Carlo method (Haji-Sheikh and Sparrow, 1969) is a statistical technique in which a large number of energy bundles are followed as they transverse the participating medium. Within the limits of accuracy imposed by any statistical technique, the Monte-Carlo method provides an exact solution to the radiative transport. For conical duct dimensions similar to those used to approximate the MHD generator in this study, the maximum percent difference of 7% between the P-1 and the Monte-Carlo solution occurred at an optical thickness (based on mean beam length) of 5. Thus, the P-1 approximation provides a very good engineering approximation of the radiative transport in this conical geometry for an isothermal gas. The number of control volumes was varied to test the grid independence of the solution. Although an extrapolation to zero grid spacing was not performed, the isothermal P-1 solution was estimated to be within 5% of the zero grid spacing solution for approximately 500 control volumes.

Data for the spectral variation of the absorption coefficient of hydrogen gas at 400 temperature-density states was generated at Los Alamos National Laboratory by Magee (1988) using the code LEDCOP. The data was generated for a temperature grid of 0.7 to 1.8 eV in steps of 0.05 eV, a density grid of 10^{-8} to 10^{-4} g/cm^3 in steps of 1,2,4,6,8 per decade (i.e., 1×10^{-8}, 2×10^{-8}, $\ldots 8 \times 10^{-5}$, 1×10^{-4}), and a photon energy grid of 0.0035 eV to 50 eV in steps of 0.0035 eV for $\omega \leq 14$ eV (region with significant line radiation), followed by steps of 0.1 eV out to 50 eV (con-

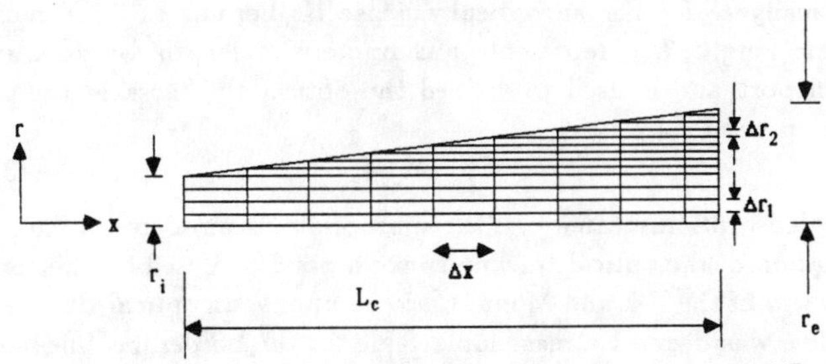

Figure 3: Computational Grid for P-1 Solution in Conical Duct.

tinuum radiation only). Radiation intensities (of the hydrogen gas and generator walls) are insignificant in the regions $\omega > 50$ eV and $\omega < 0.0035$ eV for the temperature and density ranges considered here.

The solution of the band-averaged P-1 equations requires that the spectral behavior of the absorption coefficient be approximated by a finite number of bands N_b. Figure 4 shows the spectral variation of the mass absorption coefficient, a_ω/ρ, at a temperature and density representative of the average state in the MHD generator (for the case of no radiation). An 8-band model was decided on as the minimum number of bands required to adequately model the spectral variation — a minimum number of bands was chosen since computation time for convergence of the iterative solution was not known a priori. The band intervals and the value of the arithmetic mean of the absorption coefficient across each band j,

$$a_{mj} = \frac{\int_{\Delta\omega_j} a_\omega \, d\omega}{\int_{\Delta\omega_j} d\omega} \tag{32}$$

are also displayed in Fig. 4.

As discussed earlier in the formulation of the band-averaged P-1 equations, this study treats each band j as optically dense or

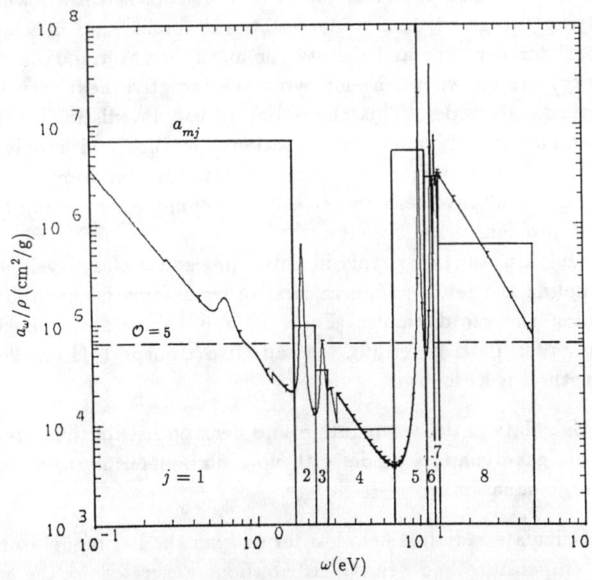

Figure 4: Absorption Coefficient vs. Photon Energy. $T = 1.35$ eV (15,666.75 K) and $\rho = 1 \times 10^{-5}$ g/cm^3. (Data was produced by the code LEDCOP at Los Alamos National Laboratory.)

optically thin. For an optically dense isothermal gas, the mean beam length, l_{mb}, represents a characteristic length for radiative transport and is used to defined the optical thickness of the gas over a band j by

$$\mathcal{O}_j = a_{m_j} l_{mb} \tag{33}$$

The absorption mean a_{m_j} is the appropriate mean to use in the calculation of the optical thickness over a band j. Based on the comparison of the P-1 and Monte-Carlo solutions, an optical thickness of five was deemed a reasonable value for the borderline thickness between optically dense and optically thin. The following criterion was used to designate each band j as optically dense or thin:

$$\begin{aligned} \mathcal{O}_j &< 5 \implies \text{Treat as Optically Thin} \\ \mathcal{O}_j &> 5 \implies \text{Treat as Optically Dense} \end{aligned}$$

Using the above criterion (with a_{m_j} evaluated at the temperature-density state indicated in Fig. 4 and l_{mb} = 7.7 cm for the actual MHD generator duct considered here) all bands except 3 and 4 were treated as optically dense. Bands 5–7 are very dense, while bands 2 and 3 are essentially on the borderline between dense and thin.

ITERATIVE SOLUTION

The governing PDE's for the P-N approximations are elliptic; however, for many applications the governing PDE's for the gas dynamics are parabolic (i.e., subsonic flow using the partially parabolized Navier-Stokes equations) or hyperbolic (supersonic flow), at least over most of the flow. Typically, the description of the gas dynamics is accomplished using a finite difference or finite element numerical method. Although the P-N approximation is well-suited for this solution environment, it should be noted that the direct coupling of the these additional *elliptic* PDE's to those governing the gas dynamics will result in an overall problem which is *elliptic*. In cases in which the coupling is not too strong, it may be possible to achieve the coupled solution using some iterative procedure in which the local radiative heat flux is known when employing the gas dynamics computational model, and in which the local temperature and density (or any other thermodynamic property) are known when employing the radiative heat transfer computational model. This allows one to use an existing tested and familiar gas dynamics computational method, with only a small modification. From a numerical methods standpoint, this may often be a convenient and expedient technique for solving the coupled problem.

In this study, a simple iterative procedure was developed for coupling the set of nonlinear, first order differential equations governing the gas dynamics (Eqns. 12 and 13) with the linear, second order PDE governing the radiative transport (Eqn. 26). The method is as follows:

1. Calculate initial temperature and density distributions from the gas dynamics model with no radiation terms in the energy equation.
2. Calculate radiative heat loss terms from the P-1 model using temperature and density distributions generated by the gas dynamics model.
3. Calculate temperature and density distributions from the gas dynamics model using radiative heat loss terms generated by the P-1 model.
4. Repeat steps 2 and 3 until the change in the temperature and density distributions satisfies the convergence criterion.

In step 3, it was necessary to employ a relaxation method such that, at any x location during the gas dynamics solution, the percent change in the energy sink term, S_E, from one iteration to the next did not exceed ±50%. The convergence criterion imposed on the temperature and density distributions required that the error in both the temperature and the density, at each axial node, fall below a tolerance of 10^{-6}, from one iteration to the next. The error for each distribution was represented by both the standard deviation (std),

$$\text{std} = \frac{\sqrt{\sum_{i=1}^{N} (Y_i - Z_i)^2}}{\frac{1}{N}\sum_{i=1}^{N} Y_i}$$

and the maximum nodal error (mne),

$$\text{mne} = max\left\{\left|\frac{Y_i - Z_i}{Y_i}\right|\right\}_{i=1 \text{ to } N}$$

where Y_i is the value of $\overline{T}(x)$ or $\overline{\rho}(x)$ at node i for the current iteration and Z_i is the value of $\overline{T}(x)$ or $\overline{\rho}(x)$ at node i from the previous iteration.

The sensitivity of the solution to changes in the wall emissivity and wall temperature was investigated. A total of 21 cases was considered, using wall emissivities of 0.01, 0.03, 0.05, 0.07, 0.10, 0.20 and 0.30, and wall temperatures of 1500, 2000 and 2500 K. The same MHD generator design (i.e., initial conditions, generator parameters, and specified area distribution $A(x)$), except for wall temperature, was used as the starting point for all iterative solutions. The characteristics of the T_w = 2500 K design are described in Table 1. The temperature, pressure, and Mach number distributions are displayed in Figs. 5–7, respectively, and correspond to the **No Radiation** line. The starting designs

Entrance Conditions and Generator Parameters					
$\overline{T}_i$	= 20,000 K	B_i	= 3.10 Tesla	D_y	= 0.20 m
P_i	= 45.00 bars	K	= 0.98	D_i	= 2.00 cm
$\dot{m}$	= 0.50 kg/s	L	= 1.00 m	D_i	= 8.00 cm

Resulting Design			
H_R	= 29.57 %	$max\{\omega_e\tau_e\}$	= 0.0974
w_{tot}	= 356.28 MJ/kg	P_e	= 5.814 bars
$Q_{w,tot}$	= 19.33 MJ/kg	$\overline{T}_e$	= 15,117 K
Q_N	= 1.205 GJ/kg	B_e	= 2.714 Tesla
$max\{dB/B\}$	= 0.0206	M_i	= 0.371
L/D_y	= 5.000	M_e	= 0.598

Table 1: Starting Point MHD Generator Design.

for the other two wall temperatures differ only slightly from the T_w = 2500 K design.

The conical duct corresponding to the above MHD generator geometry has the following dimensions:

$$r_i = 0.03568 \text{ m}, \quad r_e = 0.11495\ m, \quad L_c = 1.0\ m$$

These dimensions were determined using the three conditions for r_i, r_e and L_c discussed earlier in the Radiative Computational Model section.

Convergent solutions were obtained for the twenty-one cases considered. Axial distributions of the local gas temperature, pressure, Mach number and radiative wall heat flux are shown in Figs. 5–8 for $T_w = 2500$ K, at various wall emissivities. Figure 9 displays the total radiative heat loss to the walls and total electrical work extracted for various wall emissivities at $T_w = 2500$ K.

DISCUSSION OF RESULTS

The maximum number of iterations required for convergence occurred for the highest wall emissivity, $\epsilon_w = 0.3$, while the lowest number of iterations required occurred for $\epsilon_w = 0.05$ to 0.1. This was true for all wall temperatures considered. The number of iterations ranged between 10 to 23, the number required increasing steadily for higher wall emissivities, except between $\epsilon = 0.01$ and $\epsilon = 0.05$. Here, a slight decrease in the number of iterations was observed for all three wall temperatures; this is probably a relaxation dependent phenomenon.

The trends illustrated in Figs. 5–9 indicate that the coupled solution to the walls is very sensitive to the wall emissivity. As might be expected, gas temperatures decrease and radiative heat fluxes increase as the wall emissivity is increased. Comparison of generator exit quantities at a particular wall emissivity for the three different wall temperatures established that solutions were not sensitive to wall temperature. This is not surprising since $T_w \ll \overline{T}$ throughout the generator for wall temperatures considered in this study — the walls are essentially cold in all three cases. The maximum percent difference in exit quantities occurred for the lowest wall emissivity, $\epsilon_w = 0.01$, and decreased for higher wall emissivities. The exit pressure was the most sensitive to changes in wall temperature, having a maximum percent difference of 9% between $T_w = 1500$ K and $T_w = 2500$ K for $\epsilon_w = 0.01$.

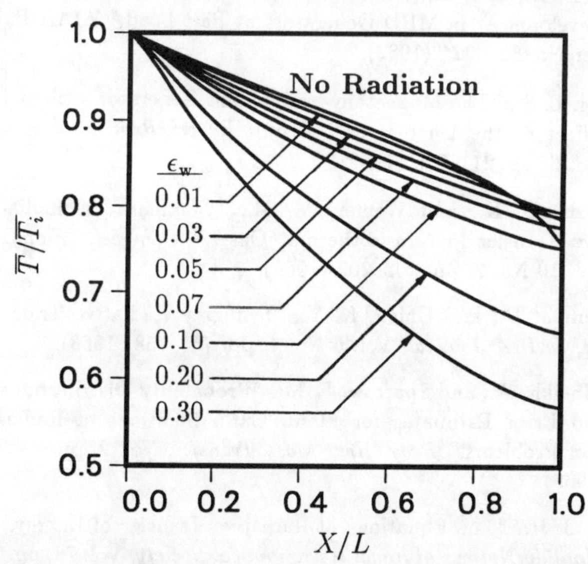

Figure 5: Non-dimensional Temperature Distribution in MHD Generator at Several Wall Emissivities and $T_w = 2500$ K.

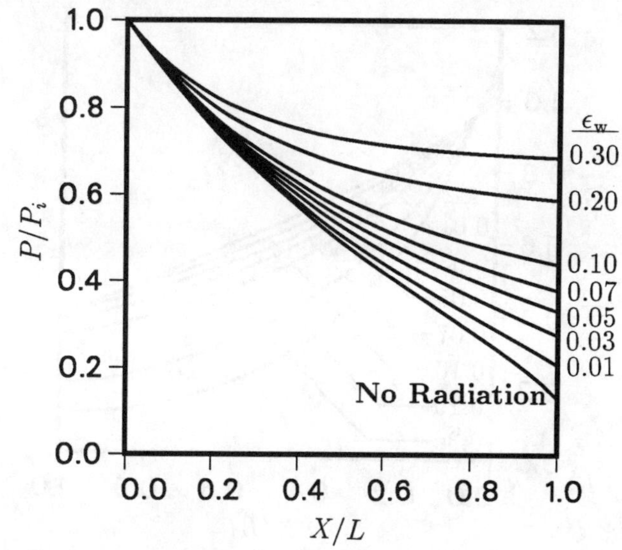

Figure 6: Non-dimensional Pressure Distribution in MHD Generator at Several Wall Emissivities and $T_w = 2500$ K.

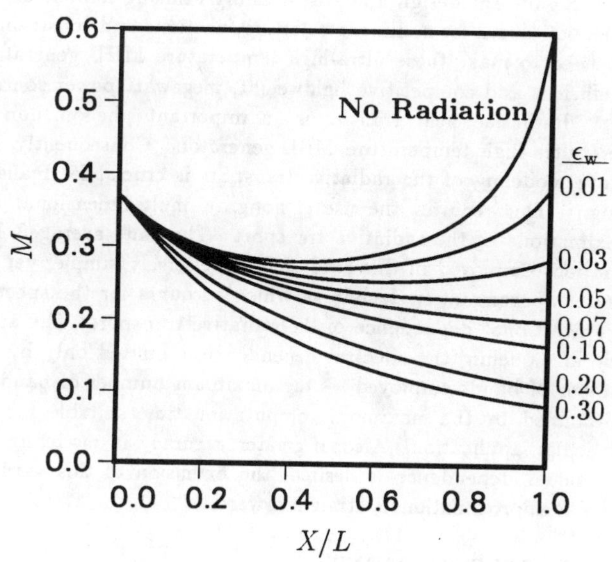

Figure 7: Mach Number Distribution in MHD Generator at Several Wall Emissivities and $T_w = 2500$ K.

CONCLUSIONS

The purpose of this study was to develop a method for solving the coupled gas dynamics/radiative heat transfer problem and to assess the importance of radiative heat transfer in these ultra-high temperature MHD generators. This has been accomplished for wall emissivities not greater than 0.3 using a simple iterative procedure, and the magnitude and sensitivity of radiative heat losses has been established for various wall boundary conditions. The MHD generators presented here are not optimal designs; that is, the extraction ratios are too small and for all but the $\epsilon_w = 0.01$ cases, the heat loss into the generator walls is too

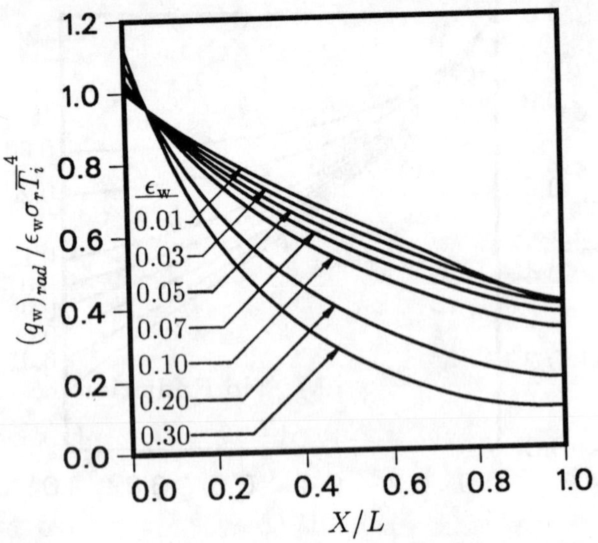

Figure 8: Non-dimensional Radiative Wall Heat Flux Distribution in MHD Generator at Several Wall Emissivities and T_w = 2500 K.

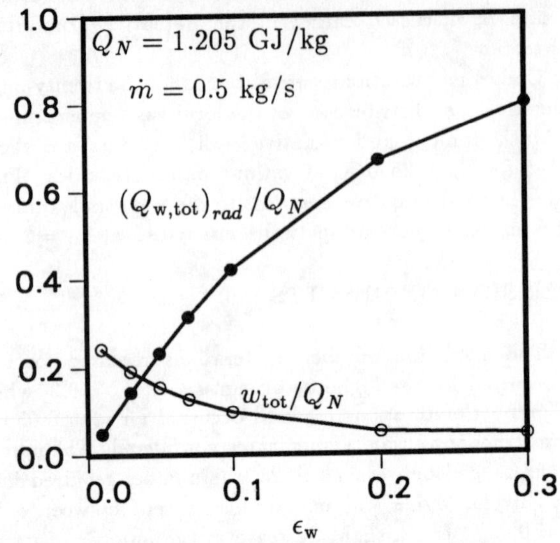

Figure 9: Non-dimensional Total Radiative Wall Heat Loss and Total Electrical Work Extracted in MHD Generator at Several Wall Emissivities and T_w = 2500 K.

large. Significant design analysis, possibly utilizing numerical optimization codes for nonlinear objective functions and constraints, is needed to make these ultra-high temperature MHD generators an efficient and competitive lightweight, megawatt power source.

Radiative heat transfer is an important phenomenon in these ultra-high temperature MHD generators. Consequently, accurate modeling of the radiative transport is crucial for realistic design. This requires the use of nongray, multidimensional approximations of the radiative transport. The band-averaged P-1 equations employed in this study offer a relatively simple, yet accurate, approximation technique which accounts for the spectral and directional dependence of the radiative transport. The accuracy in modeling the spectral dependence is limited only by the number of bands employed — the maximum number of bands is determined by the maximum computation time suitable for the particular application. Also, if greater accuracy in modeling the directional dependence is desired, the extension of this work to the P-3 approximation is straightforward.

ACKNOWLEDGEMENT

This research was partially supported by NASA Lewis Research Center (Cleveland, OH) under contract NAG 3-916, entitled Futuristic Propulsion Systems, R. W. Luidens, program manager.

REFERENCES

Ahluwalia, R. K., and Im, K. H.,"Combined Conduction, Convection, Gas Radiation and Particles Radiation in MHD Diffusers," *J. Heat Mass Transfer*, Vol. 24 No. 8, pp. 1421–1430 (1981).

Azad, F. H., and Modest, M. F., "Evaluation of the Radiative Heat Flux in Absorbing, Emitting and Linear-Anisotropic Scattering Cylindrical Media," *J. of Heat Transfer*, Vol. 103, pp. 350–356 (1981).

Biberman, L. M., Zheleznyak, M. B., Zatelepin, V. N., Lyubimov, G. A., Mein, S. A., and Mnatsakanyan, A. Kh., "Heat Transfer in the Channel of a High-Power MHD Generator," *Fluid Dynamics*, Vol. 14 No. 3, pp. 434–435 (1979).

Crosbie, A. L., and Viskanta, R., "Rectangular Model for Nongray Radiative Transfer," *AIAA Journal*, Vol. 8, pp. 2055–2057 (1970).

Demetriades, S. T., Maxwell, C. D., and Oliver, D. A., "Progress in Analytical Modeling of MHD Generators, II", *Proceedings of 21st Symposium on Engineering Aspects of MHD*, pp. 3.1.1–3.1.20 (1983).

Doss, E. D., and Ahluwalia, R. K., "Three-Dimensional Flow Development in MHD Generators at Part Load," AIAA Paper No. 82–0324 (1982).

Dzung, L. S., "The Magnetohydrodynamic Generator with Hall Effect at the Duct Ends," *Brown Boveri Review*, Vol. 49 No. 6, pp. 211–225 (1962).

Edwards, D. K., and Weiner, M. M., "Comment on Radiative Transfer in Non-isothermal Gases," *Combust. Flame*, Vol. 10 No. 2, pp. 202–203 (1966).

Finkleman, D., and Chien, K. Y., "Semigrey Radiative Transfer," *AIAA Journal*, Vol. 6 No. 4, pp. 755–758 (1968).

Haji-Sheikh, A., and Sparrow, E. M., "Probability Distributions and Error Estimates for Monte-Carlo Solutions of Radiation Problems," *Prog. Heat Mass Transfer*, Vol. 2, pp. 1–22 (1969).

Jeans, J. H., "The Equations of Radiative Transfer of Energy," *Monthly Notices of Royal Astronomical Society*, Vol. 78, pp. 28–36 (1917).

Jeng, San-Mou and Keefer, Dennis, "A Theoretical Evaluation of Laser Sustained Plasma Thruster Performance," *23rd Joint Propulsion Conference*, AIAA Paper No. 87–2166 (1987).

Jeng, S-M., and Keefer, D. R., "Numerical Study of Laser-Sustained Hydrogen Plasmas in a Forced Convective Flow," *22rd Joint Propulsion Conference*, AIAA Paper No. 86–1524 (1986).

Keefer, D. R., Crowder, H. L., and Peters, C. E., "Laser Sustained Argon Plasmas in a Forced Convective Flow," AIAA Paper No. 85–0388 (1985).

Krook, M., "On the Solution of Equations of Transfer, I", *The Astrophysical Journal*, Vol. 122, pp. 488–497 (1955).

Kung, H. C., and Sibulkin, M., "Radiative Transfer in a Nongray Gas Between Parallel Walls," *J. Quant. Spectrosc. Radiat. Transfer*, Vol. 9, pp. 1447–1461 (1969).

Liu, B. L., Schmidt, H. J., Scott, M. H., and Wu, Y. C. L., "Coupled Fluid and Electrodynamic Analysis on MHD Channel with Nonuniform Magnetic Field," *IEEE Trans. on Plasma Science*, Vol. PS-14 No. 4 pp. 444–453 (1986).

Magee, Norman H., Personal Communication. Absorption coefficient data was produced by Mr. Magee at Los Alamos National Laboratory, New Mexico using the program LEDCOP, March 1988.

Mengüc, M. P., and Viskanta, R., "Radiative Transfer in Axisymmetric, Finite Cylindrical Enclosures," *Journal of Heat Transfer*, Vol. 108, pp. 271–276 (1986).

Modest, M. F., "A Simple Differential Approximation for Radiative Transfer in Nongray Gases," *Journal of Heat Transfer*, Vol. 101, pp. 735–736 (1979).

Modest, M. F., and Azad, F. H., "The Differential Approximation for Radiative Transfer in an Emitting, Absorbing and Anisotropically Scattering Medium," *J. Quant. Spectrosc. Radiat. Transfer*, Vol. 23, pp. 117–120 (1980).

Myrabo, Leik N., Kaminski, Deborah A., Rosa, Richard J., Moder, Jeffrey P., "Concept Analysis of a Novel Integrated Directed Energy Power System," Technical Report prepared for Wright Aeronautical Laboratories, AeroPropulsion Laboratory, Subcontract No. 19X–2745909 (1987).

Patch, R. W., "Thermodynamic Properties and Theoretical Rocket Performance of Hydrogen to 100,000 K and 1.01325×10^8 N/m^2," Nasa Rpt. SP–3069 (1971).

Ratzel, A. C., and Howell, J. R., "Two-dimensional Radiation in Absorbing-Emitting Media Using the P-N Approximation," *ASME J. of Heat Transfer*, Vol. 105, pp. 333–340 (1983).

Rosa, Richard J., and Myrabo, Leik N., "Ultra-Performance Closed-Cycle Gas Core Reactors for Orbit Raising," *Progress in Astronautics and Aeronautics*, Vol. 89, Orbit Raising and Maneuvering Propulsion, 1984.

Smith, Theodore F., Fariborz, Alipour-Haghighi, and Phillip, Paul H., "Turbulent MHD Flow of a Radiating Gas," *J. Heat Mass Transfer*, Vol. 24 No. 4, pp. 659–669 (1981).

Sutton, George W., Hurwitz, H., Jr., and Poritsky, H., "Electrical and Pressure Losses in a Magnetohydrodynamic Channel Due to End Current Loops," *Trans. AIEE, Pt. I. Communications and Electronics*, Vol. 80 No. 58 pp. 687–695 (1962).

White, Frank M., *Viscous Flow*, McGraw-Hill, Inc., pp. 641–649 (1974).

Wiebelt, John A., *Engineering Radiation Heat Transfer*, Holt, Rinehart and Winston, Inc. (1966).

Yos, J. M., "Transport Properties of Nitrogen, Hydrogen, Oxygen, and Air to 30,000 K," Research & Development Div., AVCO Corporation, Wilmington, Mass., Technical Memorandum RAD-TM-63-7 (1963).

Yucel, A., and Bayazitoglu, Y., "P-N Approximation for Radiative Heat Transfer in Nongray Medium," *AIAA Journal*, Vol. 21, No. 8, pp.1196–1203 (1983).

Yuen, W. W., and Rasky, D. J., "Application of the P-1 Approximation to Radiative Transfer in a Non-Gray Medium," *Journal of Heat Transfer*, Vol. 103, pp. 182–184 (1981).

1989 National Heat Transfer Conference
HTD-Vol. 106, Heat Transfer Phenomena in Radiation, Combustion and Fires

CORRELATION OF PLUME SPECTRAL RADIANCE MEASUREMENTS WITH PARTICLE SIZE AND COMPOSITION

G. R. Cunnington, Jr. and S. J. Olstad
Lockheed Missiles and Space Company, Inc.
Palo Alto, California

ABSTRACT

Experimental data of the spectral radiance from a particulate laden plume are used to predict the composition and size distribution of the particulate phase of the flow. A scanning broadband radiometer operating over the wavelength range of 2 to 12 micrometers was employed to measure the time-dependent band radiances from a region of a plume located midway between the nozzle exit plane and the downstream shock. An estimate is made of the mass fraction and composition of the solid phase species and their respective mean particle sizes using four spectral bands which do not include any emission contributions from the gaseous products present in the flow. A multiple band radiometric technique was also used to estimate the temperature of the particulate phase of the plume and subsequently, its band emittance characteristics. The solid phase of the flow consisted primarily of carbon with a mean diameter in the range of 1 to 20 micrometers with a trace of silica particulates. The plume was optically thin so multiple scattering phenomena were insignificant.

NOMENCLATURE

A,B,C	detector voltages
b	backscatter fraction
d	particle diameter
E	emission power
F_s	volume fraction of solid
f	forward scatter fraction
l	plume thickness
n	complex refractive index
N	number per unit volume
P	phase function
q	radiant energy flux
Q	efficiency
R	radiometric ratio for temperature
T	temperature
V	voltage, volume
x	size parameter = $\pi d/\lambda$

Greek Letters

ε	emittance
λ	wavelength
σ	coefficient
τ	optical thickness
ω	single scattering albedo

Subscripts

b	blackbody
e	extinction
$\Delta\lambda$	wavelength interval
i	ith filter element
s	scattering
5,6,7,8	number of band filter

Superscripts

-	effective
*	Eq. 8

INTRODUCTION

Thermal radiation from an elevated temperature particulate laden flow may be used as a non-intrusive diagnostic method for estimating mean particle size, solid volume fraction and, in some instances, solid phase composition [1,2]. The radiative constituents of the flow may be categorized into two types of radiation processes. The gaseous species are pure band absorbers (emitters) in the infrared region. The particulates are more nearly continuum emitters which; depending upon size, radiation wavelength and intrinsic optical properties; may be gray or non-gray in nature, both scattering and absorbing and with a strongly varying phase function. In the examination of the particulate phase, spectral regions which do not include any gaseous radiation contribution are selected to simplify the data reduction procedure for media which both scatter and absorb radiation. For the work reported herein we selected four spectral bands in the 5 to 12 micrometer wavelength region which do not contain any gaseous radiation component.

The particulate species anticipated for the combustion induced flow of this study were carbon and silica with a particle diameter range of $1 \leq d \leq 20$ micrometers. Silica in small particles is very weakly absorbing for $0.4 \leq \lambda \leq 6$ micrometers as is seen from Fig. 1 which shows single scatter albedo as a function of particle size and wavelength for both carbon and silica. These data were derived using Mie theory [3] and optical constants from Malitson [4], Champetier [5] and Foster [6]. By selecting spectral bands for $\omega = 1.0$ and $\omega < 1.0$ silica can be detected and its volume fraction quantified. A total of four spectral bands were used to include the influence of the phase functions for both carbon and silica. A two-flux formulation was used to compute the radiative transfer and effective band emittance, assuming an isothermal flow. This assumption is believed to be a reasonable approximation because of the very short distance for cooling from the nozzle exit plane to the measurement region.

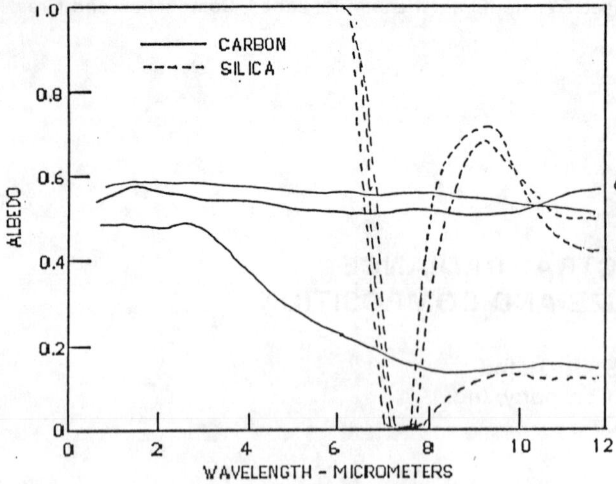

Figure 1. Albedo for Carbon and Silica Spherical Particles

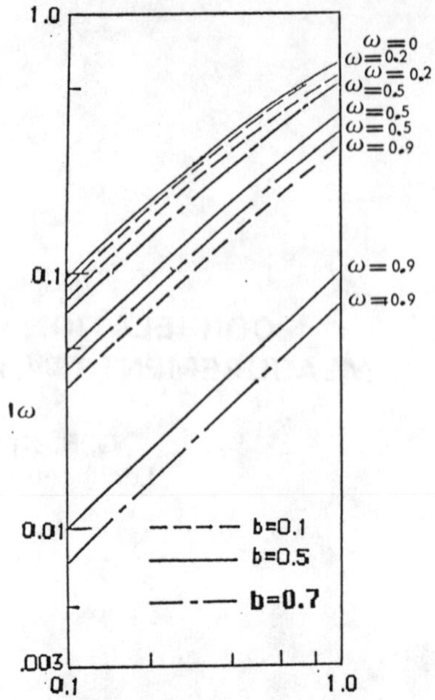

Figure 2. Effective Emittance of a Particle Medium as a Function of Optical Thickness, Albedo and Backscatter Fraction

ANALYTICAL FORMULATION

The monochromatic effective emittance of an isothermal plane layer of an optically thin medium of absorbing and scattering particles is

$$\bar{\varepsilon}_\lambda = \frac{q_\lambda^+}{E_{b\lambda}(T)} \qquad (1)$$

where q_λ^+ is the monochromatic radiant flux in the positive x direction of a two flux model of radiative transfer. This flux is

$$\frac{1}{\sigma_e}\frac{dq_\lambda^+}{dx} = -2(1-\omega f)q_\lambda^+ + 2\omega b q_\lambda^- + 2(1-\omega)n^2 E_{b\lambda}(T) \qquad (2)$$

and

$$\frac{1}{\sigma_e}\frac{dq_\lambda^-}{dx} = -2(1-\omega f)q_\lambda^- + 2\omega b q_\lambda^+ + 2(1-\omega)n^2 E_{b\lambda}(T) \qquad (3)$$

The n^2 term in Eqs. 2 and 3 may be neglected as it approaches unity for low solid fraction (optically thin) media [7]. The terms f and b represent the fractions of energy scattered into the forward and backward directions, respectively. They are derived from the phase function as

$$f = \frac{1}{2}\int_0^1 P(\mu)\,d\mu \quad \text{and} \quad b = \frac{1}{2}\int_{-1}^0 P(\mu)\,d\mu \qquad (4)$$

Equations 1, 2 and 3 may be solved analytically to give an expression for the effective emittance of the particulate medium as

$$\bar{\varepsilon} = 1 - e^{(1-\omega^*)\tau} \qquad (5)$$

where

$$\omega^* \equiv \frac{2\omega b}{1-\omega(1-2b)} \qquad (6)$$

The behavior of effective emittance, calculated using Eq. 5, with optical thickness for several values of ω and b is illustrated in Fig. 2. Decreasing values of both albedo and backscatter fraction increase effective emittance.

The single particle radiative properties of Q_e, Q_s and b are calculated, from Mie theory (Kerker, 1969) assuming spherical geometry, in terms of the size parameter $x = \pi d/\lambda$ and the complex refractive index of the solid material (n of gaseous phase assumed to be unity for this case). The extinction coefficient and albedo are calculated from

$$\sigma_e = Q_e \pi d\, N/4 = 3\, Q_e F_s/2d \qquad (7)$$

and

$$\omega = Q_s/Q_e \qquad (8)$$

The spectral and size influences on ω were seen in Fig. 1 for both carbon and silica particles. Similar dependence of the extinction coefficient and backscatter fraction are illustrated in Figs. 3 and 4. The strong dependences of these radiative properties in the infrared region suggest that multiple spectral band emission data can be used to iteratively solve for size and compositional parameters for mixtures of silica and carbon particles. If the flow consists of a distribution of particle sizes, n(d), the radiative properties must be integrated over this size range using an initially specified size distribution function. In this study a log-normal distribution was assumed. An additional iteration step and more spectral bands are now required to define the size distribution function from the experimental data.

EXPERIMENTAL SYSTEM AND PROCEDURE

Experimental Setup

The test configuration is shown schematically in Fig. 5. The distance between the radiometer steering mirror and the plume target area was 3.6-m which permitted the viewing of a 0.044-m diameter target. The target area was centered 0.050-m from the nozzle exit, as shown in Fig. 6.

A radiometric zero reference target was located on the far side of the plume area in the line-of-sight of the radiometer. This target consisted of a LN_2 cooled copper plate, 0.066-m in diameter and painted black. It provided a radiometric zero reference for the detector. Also, the plate completely subtends the field of view of the radiometer which reduces the amount of background radiation. The target was enclosed in a vacuum chamber with a NaCl window. The window was protected from smoke and particulate impingement with a nitrogen gas purge. A thermocouple was mounted to the LN_2 line exiting the chamber to provide an estimate of the plate temperature.

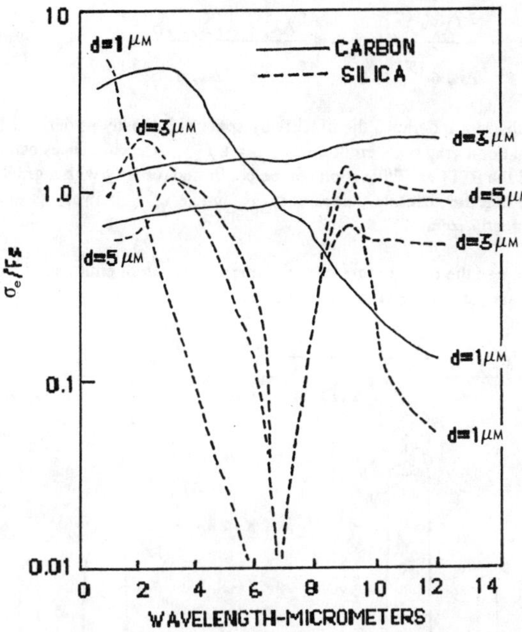

Figure 3. Spectral Extinction Coefficient per Unit Volume for Carbon and Silica

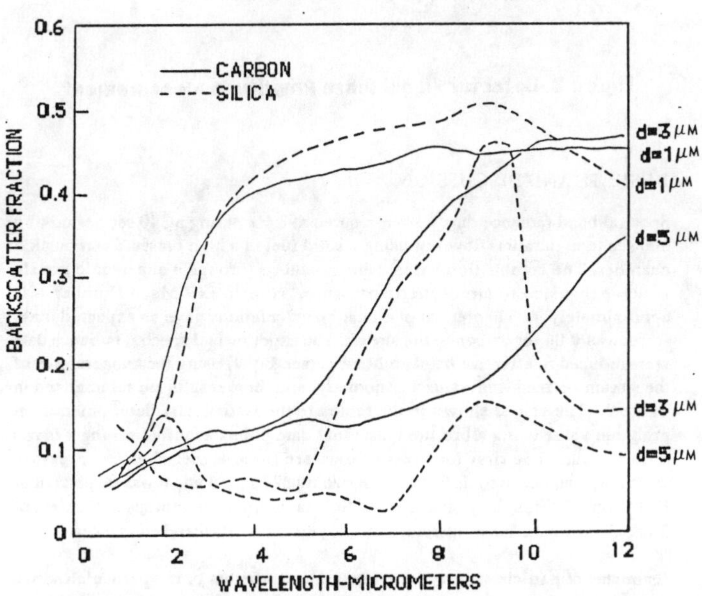

Figure 4. Backscatter Fraction for Carbon and Silica Spherical Particles

<u>Experimental Apparatus</u>

The scanning broadband radiometer is shown schematically in Fig. 7. The flat front-surface steering mirror M1 transfers radiation onto a spherical mirror M2 (0.102-m dia., 0.410-m f), which focuses the target image onto a field stop (FS) via the turning mirror M3. The radiation passing through the FS is refocused with a 90° off-axis ellipsoid mirror (M4) onto a detector element (magnification 6:1).

An aperture stop (AS) is located between the field stop and the ellipsoid mirror. The aperture stop is remotely controlled and can be toggled into and out of the beam path. The purpose of this stop is to attenuate the beam in the event of detector saturation. A reflex viewer is situated in the optical path in front of the field stop and can be rotated into place to assist the user in sighting. The sensor is a HgCdTe detector operating in the photoconductive mode. The sensing element is 0.002-m square and is cooled with liquid nitrogen. The detector signal is passed through a preamp for recording.

An 8-element filter wheel is positioned directly in front of the field stop such that the filters pass through the beam axis as the wheel is rotated. The spectral characteristics of the elements are presented in table 1. The wavelengths λ_1 and λ_2 in the Table are the half-power transmission wavelengths of the filter bandpass, and $\Delta\lambda$ denotes the filter bandwidths at the half-power points. The filter wheel is driven by a DC motor at 8 rps. The detector output is a continuous train of 8 serial pulses per revolution. The pulse amplitudes are proportional to the corresponding band irradiance onto the sensor. The angular position of the filter wheel is sensed by an optical coupler which generates eight sync pulses per revolution, one for each filter element. The sync pulse signals are used to phase the peak-to-reference values of the detector pulses.

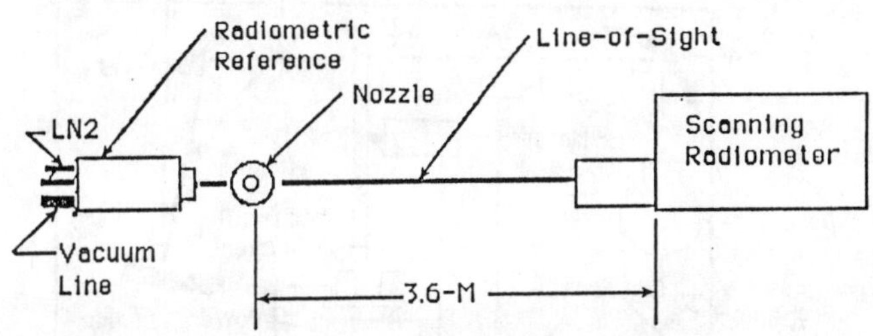

Figure 5. Schematic of Test Setup

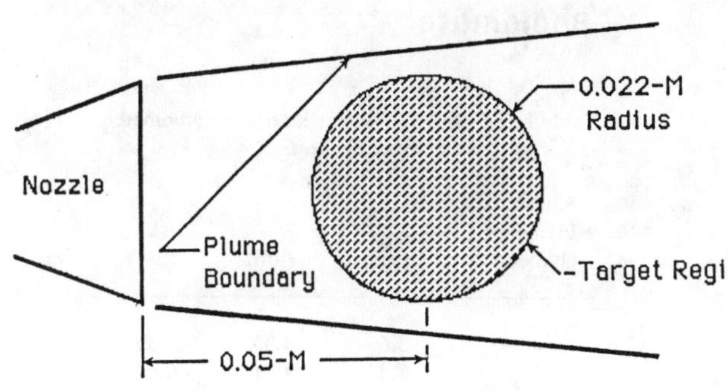

Figure 6. Target Region Geometry

The data were recorded on a Hewlett-Packard model HP-3968A Instrumentation Recorder equipped with FM input boards. To analyze the data, the recorded analog signal was dubbed onto a 1-inch magnetic tape. The data were then digitized at a rate of 8 kHz and transferred onto a computer tape for subsequent analysis.

<u>Data Reduction</u>

Processing of the test data involved three steps: (1) transferring the data onto a VAX computer, (2) pre-processing the digitized data into pulse amplitudes, and (3) converting the voltages to spectral intensity. The data channels recorded during the test (detector, sync. pulse and square wave) were digitized at a rate of 8 kHz, allowing the signals to be reconstructed at a resolution of approximately 1000 points per filter wheel revolution. The digitized data were preprocessed to convert detector voltage versus time signals into overall pulse amplitudes. The analysis of a typical detector signal pulse is represented in Fig. 8. The amplitude V is given by.

$$V = B - \frac{A + C}{2} \tag{9}$$

where A, B and C are the detector signal voltages time averaged across Δt_A, Δt_B, and Δt_C respectively. The signal b is referred to as the pulse signal, and A and C the pre- and post- pulse reference signals. Notice that the reference voltages A and C are not equal; this is because the preamplifier for this test was AC coupled, which causes the baseline voltage to vary with changing peak amplitudes. The changing reference signals make it necessary to average both A and C for the reference measurement.

In this manner, the preprocessing routine steps through the detector pulse data and compute the time averages amplitude and associated uncertainty (standard deviation) of each individual pulse. The pre-processed voltages were converted to spectral intensity $I_{\Delta\lambda}$ (mW/cm^2-µm-sr) using a linear interpolation subroutine. The $I_{\Delta\lambda}$ vs. pulse amplitude was taken from the post test calibration and entered into the subroutine in the form of an eight column array (one column for each of the filtered bandpasses).

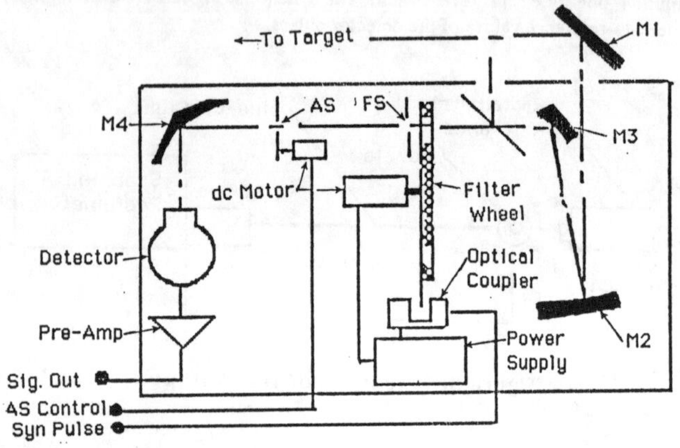

Figure 7. Schematic of the Scanning Broadband Radiometer

Table 1. Spectral Characteristics of the Radiometer Filter Elements

Filter Elements Number	λ_1 (µm)	λ_2 (µm)	$\Delta\lambda$ (µm)
1	3.60	4.95	1.35
2	3.37	4.01	0.64
3	2.06	2.62	0.56
4	4.40	4.90	0.50
5	5.03	5.66	0.63
6	7.36	8.31	0.96
7	9.00	9.55	0.55
8	11.22	11.99	0.77

Temperature and Emittance Determinations

An independent measurement of the plume temperature was not performed for this experiment so an estimate of temperature was inferred from the radiometric data. Proper selection of a radiation thermometry technique requires some knowledge of the spectral emittance of the target. As discussed in the Results section of this paper, the calculated ε_λ values indicate that the target has a relatively constant spectral emittance for the case of $d \geq 3$µm across filter bandpasses 5 and 6. The radiation thermometry method most appropriate for a gray body emitter is the ratio method, whereby the target temperature is determined from the quotient of observed radiances at two different wavelengths. Consider the ratio R(T) of the spectral intensities measured across filter bandpasses 5 and 6 for a target at temperature T:

$$R(T) = \frac{I_{\Delta\lambda,5}(T)}{I_{\Delta\lambda,6}(T)} = \frac{\varepsilon_{\Delta\lambda,5} \cdot I_{\Delta\lambda,5,b}(T)}{\varepsilon_{\Delta\lambda,6} \cdot I_{\Delta\lambda,6,b}(T)} \tag{10}$$

where the subscript b denotes the blackbody spectral intensity as derived from Planck's law. For a gray body emitter ($\varepsilon_{\Delta\lambda,5} = \varepsilon_{\Delta\lambda,6}$), the emittances cancel in the ratio and the R(T) vs. T function can be calibrated directly with a graybody target. Note that the absolute values of $\varepsilon_{\Delta\lambda,5}$ and $\varepsilon_{\Delta\lambda,6}$ can vary as long as they remain nearly equal.

Having estimated the temperature of the plume, the apparent emittance across a given filter bandpass i is determined from the following equation:

$$\varepsilon_{\Delta\lambda,i} = \frac{I_{\Delta\lambda,i}}{I_{\Delta\lambda,i,b}(T)} \tag{11}$$

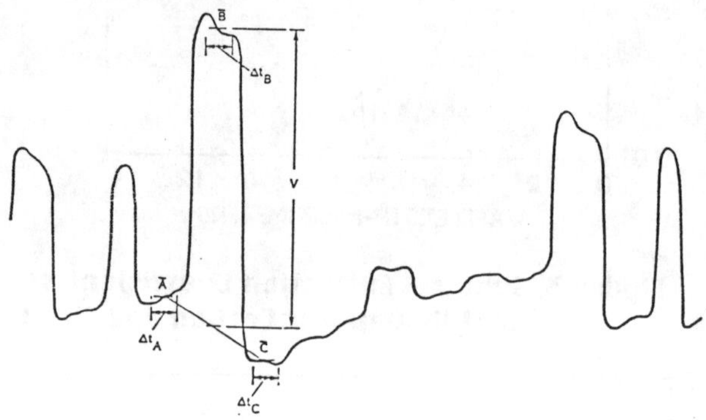

Figure 8. Detector Signal Pulse Amplitude Measurement

RESULTS AND DISCUSSION

Spectral band radiance data were measured at 8 Hz during a 340 sec period of a 600 sec total duration flow test using a solid fuel in a high pressure combustion chamber. The combustion gases were expanded through a supersonic nozzle which exhausted to atmospheric pressure. Nozzle exit Mach Number was approximately 4. Theoretical combustion calculations gave an expected mass fraction of 0.06 for carbon in the stream. The experimental spectral radiance data were reduced to effective band emittances (see Eq. 7) using the temperatures of the stream derived from ratio thermometry, and these results are summarized in Table 2. The values shown in the table are the statistically determined most probable value using all of the individual data points acquired during a given duration run. The first four sets of data are for a higher chamber pressure condition than occurred during the last two runs (P_c for 5 and 6 was 90 percent of P_c for runs 1 through 4). Estimate of the maximum uncertainties in the data of Table 2 are ± 50K for temperature and 16 percent for the band emittances.

Estimates of particle size, volume fraction and whether or not particulate silica was present downstream from the nozzle exit were made using an iterative procedure to deconvolve the emittance results in the four wavelength bands. Initially, calculations were made of the band emittances for both carbon and silica spherical particles as a function of mean particle diameter, using a log-normal size distribution with a one-sigma deviation, and volume fraction. Temperature dependent refractive index data were used for the calculations [5,6]. An initial volume fraction estimate for carbon derived from the theoretical combustion calculation was 3.8×10^{-6}. The upper limit of silica concentration from theoretical considerations was estimated to be a volume fraction of 1×10^{-6}. Figures 9 and 10 illustrate the calculated emittance behavior of carbon and silica particulates, respectively, for the four spectral band passes and three volume fractions. The spectral emittance results were integrated over the Planck function to give a single band emittance ascribed to the band center wavelength which is then used to analyze the experimental results.

The presence of silica in the stream has a minimal effect on emittance for the 5.35 micrometer band. It does not emit at this wavelength ($\omega = 1$), and it has an insignificant effect on extinction at the anticipated low volume fraction.

Experimental data for this wavelength were used to establish a particle size range for the carbon particles present in the stream. If high spectral resolution data were available over this bandpass, Eqs, 5 and 6 would be used to solve for σ_e/F_s, ω and b. However, as only broadband data are available, a further assumption must be made. This was that the volume fraction of carbon was $3.8 \times 10^{-6} \pm 50$ percent. If we simply plot band emittance as a function of particle diameter and solid fraction and make the comparison with the experimental data, a reasonably narrow range of sizes is obtained as is seen in Fig. 11. Using $\varepsilon_{\Delta\lambda} = 0.027$ we arrive at a mean particle diameter of 9.5 to 10.0 micrometers. The 50 percent variation in volume fraction results in a mean diameter range of 7.5 to 12 micrometers.

Table 2. Experimental Results of Particulate Temperature and Spectral Band Emittance

Run No.	Duration of Run (Sec)	Calculated Temp.(a) (K)	$\varepsilon_{\Delta\lambda}$ for Filter Band Center Wavelength (μm) 5.35	7.83	9.27	11.60
1	85	1370	0.027	0.027	0.026	0.028
2	48	1160	0.025	0.025	0.026	0.029
3	45	1140	0.027	0.027	0.027	0.029
4	45	1160	0.027	0.027	0.025	0.029
5	38	1025	0.040	0.040	0.032	0.035
6	46	930	0.050	0.050	0.041	0.042

(a) 5.35 and 7.83 micrometer band center filters used for ratio thermometry.

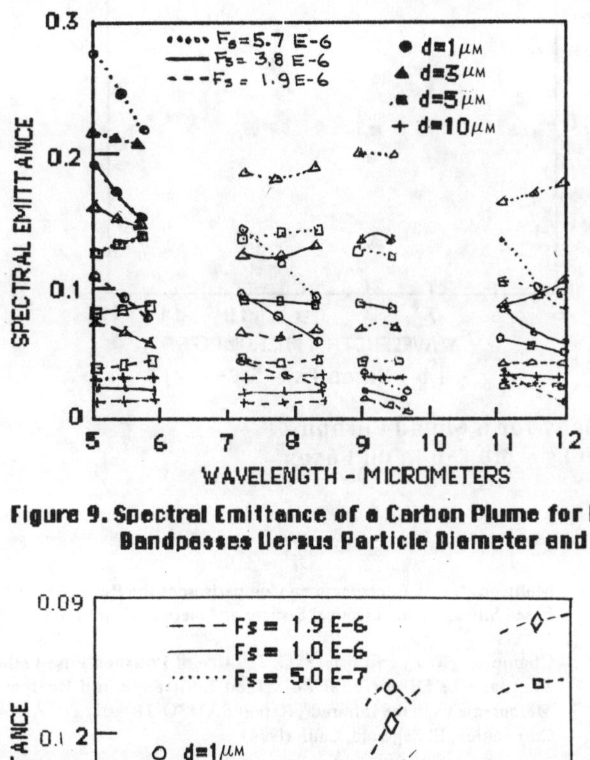

Figure 9. Spectral Emittance of a Carbon Plume for Four Bandpasses Versus Particle Diameter and Fs

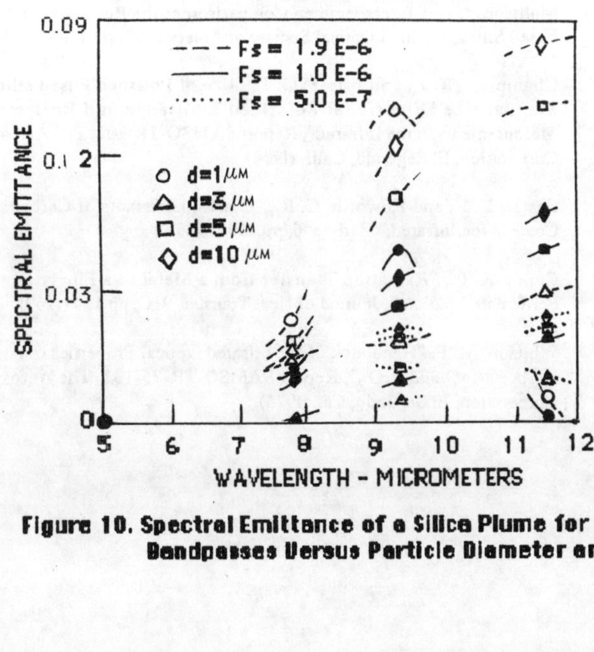

Figure 10. Spectral Emittance of a Silica Plume for Four Bandpasses Versus Particle Diameter and Fs

The next step was to compare the emittance data between the 5.35 and 7.83 micrometer wavelength bands. Again, the presence of silica in small quantities should have little effect on emittance because of the low extinction efficiency for silica in this wavelength region (c.f. Fig. 3). It is seen from Fig. 9 that the emittance for 5 and 10 micrometer particles is nearly equal for these two bands. Also, only for a volume fraction of 3.8×10^{-6} are the emittances nearly equal to the experimental values. Thus, it is estimated that the mean particle diameter for the carbon is 9 to 10 micrometers and the carbon volume fraction is 3.8×10^{-6}.

With this initial determination of the carbon particulate properties, the experimental data are further examined for analysis of silica. Band emittance calculations were made for mixtures of 10 micrometer diameter carbon particles at $F_s = 3.8 \times 10^{-6}$ with varying volume fractions of silica particles over the diameter range of 1 to 10 micrometers. The radiative properties for the mixture were taken as the volume fraction averaged values for σ_e, ω and b: i.e., for example.

$$\omega = \frac{\omega_c F_{s,c} + \omega_s F_{s,s}}{F_{s,c} + F_{s,s}} \tag{12}$$

where subscripts c and s denote carbon and silica, respectively. The band emittances calculated for these conditions are shown in Fig. 12. Comparing these results with the experimental data, it appears that silica is not present in amounts detectable by this experiment. From Fig. 12c for 5 percent volume fraction of silica and Fig. 12d for no silica, it is seen that the experimental data, shown by the horizontal bars on the figures, are consistent with the no silica case and depart from the 5 percent silica case at the 9.27 micrometer band. Thus, the estimate is that the percentage of silica in the plume is less than 5 percent.

The band emittance data for Runs 5 and 6 show a significantly different trend with increasing wavelength than that which is seen for runs 1 through 4. For the lower pressure operation, it appears that the size of the carbon particles shifts to a smaller diameter. Emittance in the two shorter wavelength bands increases strongly and the longer wavelengths band emittance is less than that for the two shorter wavelength bands. It is estimated that the carbon mean particle diameter is 7 to 8 micrometers for these cases.

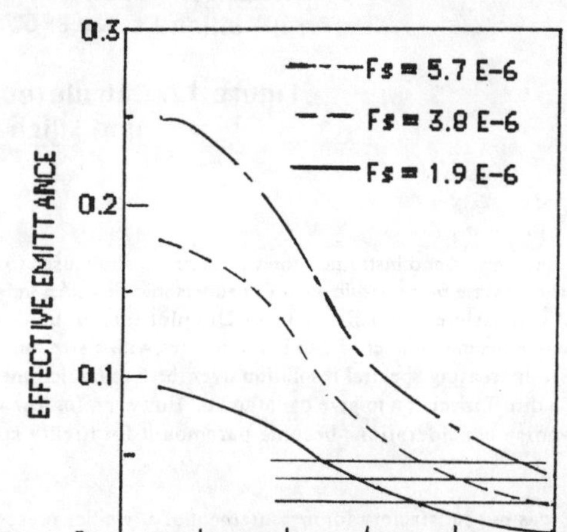

Figure 11. Effective Emittance of Carbon Plume

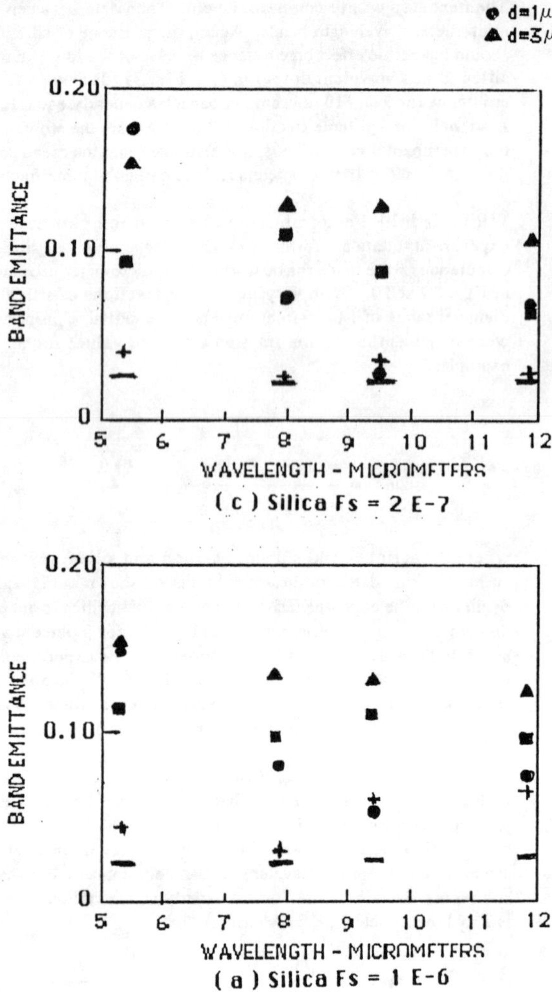

Figure 12. Calculated Band Emittances for a Mixed Carbon and Silica Plume, Carbon Fs = 3.8 E-6 in All Cases

CONCLUSIONS

The experimental method and instrumentation were successfully used to quantify the carbon particle phase presented in a hot gas supersonic flow. An independent measurement of particle size using a laser Doppler technique during the experiment gave a mean diameter of 7 to 8 micrometer with a size range of 5 to 10 micrometer. Increasing spectral resolution over the 5 to 6 micrometer band would permit a direct inversion to give σ_c ω and b. However, for low values of F_s signal-to-noise considerations become paramount for highly spectrally resolved data.

The technique was not satisfactory for measurement of any silica present in the flow. This is probably the result of volume concentrations below the current detectable limit. Increased spectral range, resolution and detector responsivity would enhance this detection capability.

REFERENCES

1. Trinks, H.,. and Hoffman, R. J., "Standard Experimental Exhaust Plume Analysis Procedure", AIAA paper 85-1390, Joint Propulsion Conference, pp 1-11 (1985).

2. Lee, S. C., and Tien, C. L., "Optical Constants of Soot in Hydrocarbon Flames," 18th International Symposium on Combustion, The Combustion Institute, pp 1159-1165 (1981).

3. Kerker, M., "The Scattering of Light", Academic Press, New York (1969).

4. Malitson, I. H., "Interspecimen Comparison of the Refractive Index of Fused Silica," Journal Optical Society of America, 55, pp 1205 (1965).

5. Champtier, R. J., and Friese, G. J., "Use of Polished Fused silica to Standardize Directional Polarized Emittance and Reflectance Measurements in the Infrared", Report SAMSO-TR-202, The Aerospace Corporation, El Segundo, Calif. (1974).

6. Foster, P. J., and Howarth, C. R., "Optical Constants of Carbons and Coals in the Infrared," Carbon, 6, pp 719 (1968).

7. Caren, R. P., "Radiation Transfer from a Metal to a Finely Divided Particulate Medium", Journal of Heat Transfer, 91C, pp 154 (1969).

8. Whitson, M. E. "Handbook of the Infrared Optical Properties of Al_2O_3, Carbon MgO and ZrO_2", Report SAMSO-TR-75-131, The Aerospace Corporation, El Segundo, Ca., (1975).

1989 National Heat Transfer Conference
HTD-Vol. 106, Heat Transfer Phenomena in Radiation, Combustion and Fires

IMPROVING THE PERFORMANCE OF POROUS RADIANT BURNERS THROUGH USE OF SUB-MICRON SIZE FIBERS

T. W. Tong, S. B. Sathel, and R. E. Peck
Department of Mechanical and Aerospace Engineering
Arizona State University
Tempe, Arizona

ABSTRACT

An analysis has been carried out to determine the performance of porous radiant burners (PRB) as a function of fiber size. PRB made with silica or alumina fibers were considered. The radiative properties of the fibers were determined using the electromagnetic wave scattering theory for two different characteristic temperatures, 1000 °C and 1500 °C. The properties were used in a combined-mode heat transfer model to calculate the amount of energy radiated by the PRB. It was found that fibers smaller than order of one micrometer in diameter produced significantly higher radiant output. For a characteristic temperature of 1000 °C, in some cases, the increase in output was as high as 63 and 109 percent for silica and alumina fibers, respectively. For a characteristic temperature of 1500 °C, the corresponding increases were 72 and 150 percent, respectively.

NOMENCLATURE

a	Solid surface area per unit volume of solid
$a_{n,1}$	Coefficients (TM mode)
$a_{n,2}$	Coefficients (TE mode)
$b_{n,1}$	Coefficients (TM mode)
$b_{n,2}$	Coefficients (TE mode)
c	Specific heat
D	Fiber diameter
e_b	Planck's function
G	Non-dimensional emissive power, $\sigma T^4/Qx_3$
h	Heat transfer coefficient
i	Radiant intensity
I	Non-dimensional radiant intensity, i/Qx_3
k	Thermal conductivity
P_1-P_5	Non-dimensional parameters as defined in Table 1
q^r	Radiative heat flux
Q	Heat generation rate per unit volume
Q^r	Non-dimensional radiative heat flux, q^r/Qx_3
Q^r_o	Non-dimensional radiant output, $Q^r(\xi_3)$
Q_e	Extinction efficiency
Q_s	Scattering efficiency
$\bar{Q}_e$	Averaged extinction efficiency
$\bar{Q}_s$	Averaged scattering efficiency
T	Temperature
T_m	Characteristic temperature
u	Velocity
x	Coordinate
x_1-x_3	Lengths defined in Fig. 1
z	Dummy variable in equation (3)
$\bar{z}$	Averaged z

<u>Greek Symbols</u>

α	Size parameter
δ	Unit step function
Θ	Non-dimensional temperature, $(T-T_i)/T_i$
λ	Wavelength of radiation
μ, μ'	Direction cosine
ξ	Non-dimensional coordinate, x/x_3
ρ	Density
σ	Stefan-Boltzmann constant
σ_a	Absorption coefficient
σ_e	Extinction coefficient
$\bar{\sigma}_e$	Averaged extinction coefficient
σ_s	Scattering coefficient
τ	Optical thickness, $(\sigma_a+\sigma_s)x_3$

ϕ Porosity
Φ Incidence angle
ω Single scattering albedo, $\sigma_s/(\sigma_a+\sigma_s)$
$\bar{\omega}$ Averaged single scattering albedo

Superscripts

r Radiation
+ Positive x direction
− Negative x direction

Subscripts

b Blackbody or baseline calculation
e Exit
g Gas
i Inlet
o Output
s Solid

INTRODUCTION

In a recent study on the heat transfer characteristics of porous radiant burners (PRB), Tong and Sathe (1988) found that smaller single scattering albedo (ω) of the porous matrix would result in higher radiant output. Typically, the porous matrix of PRB is made of cylindrical fibers of several microns in diameter. Since the radiation spectrum involved in the application ranges roughly from one to fifteen microns in wavelength, the size parameter ($\alpha \equiv \pi D/\lambda$, where D is the fiber diameter and λ is the wavelength of radiation) of the fibers is of the order of one or slightly larger. It is well known that when $\alpha \ll 1$ the scattering efficiency (Q_s) is small compared to the absorption efficiency (Q_a) and hence, ω is also small because $\omega = Q_s/(Q_s+Q_a)$ (Kerker, 1969). This suggests that it may be possible to increase the radiant output of PRB by using fibers smaller than those currently used. It is the purpose of this work to establish quantitatively if and by how much the radiant output of PRB can be increased by using smaller fibers.

Before the analysis is presented, a brief description of the operation principle of PRB is in order. A premixed gaseous fuel-air mixture enters a non-combustible porous layer such as that illustrated schematically in Fig. 1. By adjusting the mass flow rate appropriately, combustion can be stablized inside the porous layer. The burnt gas heats up the porous matrix via convection. The porous matrix in turn radiates the energy away from the porous layer. Heating is provided to a heat load by means of thermal radiation.

ANALYSIS

The analysis is based on a heat transfer model that treats combustion as a constant heat source (Tong and Sathe, 1988). The theory of electromagnetic wave scattering (Kerker, 1969) is used to determine the radiative properties of the fibers. They are then used as input parameters in the heat transfer model to calculate the radiant output.

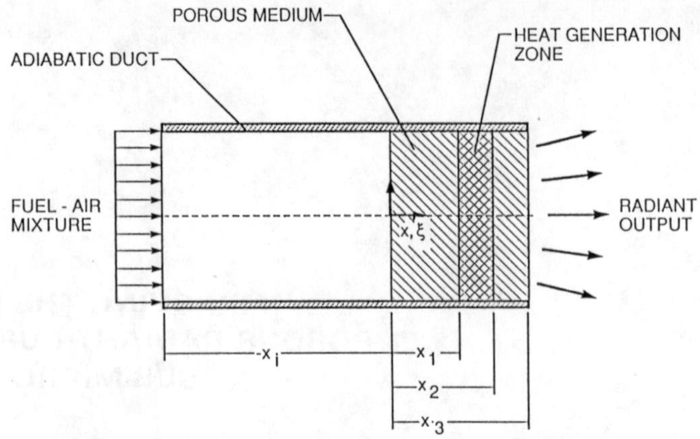

Fig. 1 Schematic diagram of the porous radiant burner.

Radiative Properties of Fibers

This section presents the background on determining the radiative properties of fibers. The fibers are assumed to be circular and infinitely long. For unpolarized incident radiation with wavelength λ, the extinction effeciency Q_e, which is the sum of Q_s and Q_a, and Q_s of a fiber are given by (Kerker, 1969)

$$Q_e = \text{Real}\left[b_{01} + 2\sum_{n=1}^{\infty} b_{n1} + 2\sum_{n=1}^{\infty} a_{n1} + 2\sum_{n=1}^{\infty} b_{n2} + a_{02} + 2\sum_{n=1}^{\infty} a_{n2}\right]/\alpha \qquad (1)$$

$$Q_s = \left[|b_{01}|^2 + 2\sum_{n=1}^{\infty} |b_{n1}|^2 + 2\sum_{n=1}^{\infty} |a_{n1}|^2 + 2\sum_{n=1}^{\infty} |b_{n2}|^2 + |a_{02}|^2 + 2\sum_{n=1}^{\infty} |a_{n2}|^2\right]/\alpha \qquad (2)$$

where a_{n1}, b_{n1}, a_{n2} and b_{n2} are coefficients that depend on the incidence angle Φ, wavelength of radiation λ, fiber diameter D, and the complex refractive index of the fiber. The expressions for a_{n1}, b_{n1}, a_{n2} and b_{n2} can be found in Kerker (1969). The efficiencies averaged over all wavelengths and incidence angles are determined according to

$$\bar{z} = \frac{2}{\pi\sigma T_m^4}\int_0^{\pi/2}\int_0^{\infty} z e_b(T_m)\, d\lambda\, d\Phi \qquad (3)$$

where z represents either Q_e or Q_s, and the

overbar indicates averaged values. Note that Planck's function e_b evaluated at some characteristic temperature T_m has been used as the weighting function in the averaging process. Equations (1) through (3) can be used to compute the averaged efficiencies once D, the complex refractive index of the fiber, and T_m are specified.

The heat transfer model to be employed requires the averaged scattering albedo and the averaged extinction coefficient $\bar{\sigma}_e$ as input properties. The averaged single scattering albedo is

$$\bar{\omega} = \frac{\bar{Q}_s}{\bar{Q}_e} \tag{4}$$

while the averaged extinction coefficient is related to the averaged extinction efficiency by

$$\bar{\sigma}_e = \frac{4\bar{Q}_e(1-\phi)}{\pi D} \tag{5}$$

with ϕ being the porosity.

Heat Transfer

The physical situation considered is depicted in Fig. 1. Assuming constant thermophysical properties, steady and laminar slug flow, and one-dimensional radiative heat transfer, one can write the governing equations in dimensionless form as (Tong and Sathe, 1988)

Energy Equation for the Gas Phase

$$P_1\frac{d\Theta_g}{d\xi} + (1-\phi)P_2(\Theta_g-\Theta_s) = \phi\delta(\xi) + P_3\frac{d^2\Theta_g}{d\xi^2} \tag{6}$$

Energy Equation for the Solid Phase

$$(1-\phi)P_4\frac{d^2\Theta_s}{d\xi^2} + (1-\phi)P_2(\Theta_g-\Theta_s) = \frac{dQ^r}{d\xi} \tag{7}$$

Equation of Radiative Transfer

$$\frac{\mu}{\tau}\frac{\partial I(\xi,\mu)}{\partial \xi} + I(\xi,\mu) = (1-\omega)I_b(\Theta_s) + \frac{\omega}{2}\int_{-1}^{1} I(\xi,\mu')d\mu' \tag{8}$$

Net Radiative Heat Flux

$$Q^r = 2\pi\int_{-1}^{1} I(\xi,\mu')\mu' d\mu' \tag{9}$$

Boundary Conditions for the Gas Phase Energy Equation

$$\Theta_g = 0 \text{ at } \xi = -\xi_i \tag{10a}$$

$$\frac{d\Theta_g}{d\xi} = 0 \text{ at } \xi = \xi_3 \tag{10b}$$

Boundary Conditions for the Solid Phase Energy Equation

$$-\frac{d\Theta_s}{d\xi} = P_5(\Theta_g-\Theta_s) \text{ at } \xi = 0 \tag{11a}$$

$$\frac{d\Theta_s}{d\xi} = P_5(\Theta_g-\Theta_s) \text{ at } \xi = \xi_3 \tag{11b}$$

Boundary Conditions for the Equation of Radiative Transfer

$$I^+(0) = 2\int_0^1 I^-(0,-\mu')\mu' d\mu' \tag{12a}$$

$$I^-(\xi_3) = \frac{G_e(T_e)}{\pi} \tag{12b}$$

where the dimensionless variables are defined as:

$$\xi = x/x_3, \qquad \Theta_g = (T_g-T_i)/T_i,$$

$$\Theta_s = (T_s-T_i)/T_i, \qquad I = i/Qx_3,$$

$$I_b = i_b/Qx_3, \qquad \tau = \sigma_e x_3,$$

$$Q^r = q^r/Qx_3, \qquad G_e = \sigma T_e^4/Qx_3$$

The definitions of the dimensional variables can be found in the Nomenclature. The physical significance and the definitions of the parameters P_1 through P_5 are given in Table 1. Heat generation due to combustion is represented by the term $\phi\delta(\xi)$ in equation (6), where $\delta(\xi)$ is a delta function defined as unity for $\xi_1 \le \xi \le \xi_2$ and zero elsewhere.

Equation (6) will be applied to the entire domain shown in Fig. 1, that is $-\xi_i \le \xi \le \xi_3$. Accordingly, ϕ is defined such that $\phi = 1$ for $\xi < 0$ and $0 \le \phi \le 1$ for $0 \le \xi \le \xi_3$.

Table 1 Significance of P_1 through P_5.

Parameter	Physical Significance
$P_1 = \rho u c_g T_i / Q x_3$	flow enthalpy
$P_2 = h a T_i / Q$	convective energy coupling between gas and solid
$P_3 = k_g T_i / Q x_3^2$	gas conduction
$P_4 = k_s T_i / Q x_3^2$	solid conduction
$P_5 = h x_3 / k_s$	Biot number

Equations (7) and (8) are coupled through equation (9) and they will be used for the porous region. As far as the boundary conditions are concerned, equation (10b) specifies that it is adiabatic for the gas at the exit plane. This results from assuming local one way behaviour at the outflow position which has been shown by Patankar (1980) and Smooke et al. (1983) to be a reasonable approximation for flow situations such as that being considered. Equations (11a) and (11b) equate the convective heat flux between the gas and solid phases to the conductive heat flux. The upstream radiative boundary condition as indicated by equation (12a) is formulated by characterizing the upstream as a gray and perfectly reflecting surface. Equation (12b) assumes that the burner sees a black environment at T_e.

COMPUTATIONS OF THE RADIATIVE PROPERTIES

Both silica and alumina fibers were considered for use as the porous material for the burner. Computations for $\overline{Q}_s$, $\overline{Q}_e$ and $\overline{\omega}_s$ were done by using the trapezoidal rule to approximate the integrals in equation (3). The integration with respect to Φ were done at 1 degree intervals. The integration with respect to λ were performed at 0.05 μm intervals with 1 and 16.5 μm chosen as the lower and upper intergration limits, respectively. This wavelength range contains 98 and 96 percent of the blackbody radiant energy when 1000 °C and 1500 °C, respectively, were employed as T_m. Such a range of characteristic temperatures is representative of the flame temperatures in PRB. For silica fibers the complex refractive index from Malitson (1965) for $\lambda < 7$ μm and from Champtier and Friese (1974) for $\lambda \geq 7$ μm were used. The latter reference also provided the complex refractive index for alumina for all wavelengths.

METHOD OF SOLUTION

Equation (8) was transformed to a set of four first order ordinary differential equations in terms of the moments of the radiant intensity using the P-3 spherical harmonics approximation (Menguc and Viskanta, 1985; Swathi et al., 1987; Tong et al., 1987). Equations (12a) and (12b) were replaced by the Marshak's boundary conditions which required the intensity moments to be conserved. The spherical harmonics method and the Marshak's boundary conditions for solving the equation of radiative transfer have been well documented elsewhere (Davison, 1957; Menguc and Viskanta, 1985; Swathi et al., 1987; Tong et al., 1987). The readers are referred to these references for the mathematical details. As indicated by equation (9), the first moment of the radiant intensity is the net radiative heat flux.

The four equations governing the moments of the radiant intensity and equations (6) and (7) were solved according to the following procedure:

1. Guess $\partial\Theta_g/\partial\xi$ at $\xi = 0$;
2. Solve equation (6) analytically for the region $-\xi_i \leq \xi \leq 0$ using equation (10a) and the boundary condition from step 1;
3. Determine Θ_g at $\xi = 0$ using the solution from step 2;
4. Numerically solve equations (6) and (7) and the four equations resulting from the P-3 spherical harmonics approximation for the region $0 \leq \xi \leq \xi_3$ using equations (10b) (11a) and (11b), the boundary condition from step 3 and the Marshak's boundary conditions;
5. Determine $\partial\Theta_g/\partial\xi$ at $\xi = 0$ using the solution from step 4; and
6. Repeat steps 2 to 5 until convergence is achieved.

The subroutine DVCPR from the IMSL library was used to perform the computation in step 4. Convergence for step 6 was assumed when all the following conditions were met:

1. Fractional change in $\partial\Theta_g/\partial\xi$ at $\xi = 0$ in consecutive iterations was less than 10^{-4};
2. Fractional change in Θ_g at $\xi = 0$ in consecutive iterations was less than 10^{-4}; and
3. An overall energy balance of less than 0.1 percent was achieved.

A relative error tolerance of 10^{-4} was specified for the iterations internal to subroutine DVCPR.

RESULTS AND DISCUSSION

The averaged efficiencies and single scattering albedo for silica and alumina fibers can be found in Tables 2 and 3, respectively. The properties were evaluated at characteristic temperatures of 1000 °C and 1500 °C, as mentioned earlier. The results cover fiber diameters ranging from 0.001 to 5 μm. For both fibers, $\overline{Q}_e$ and $\overline{Q}_s$ first increase and then decrease as the fiber diameter is reduced from 5 μm. This is typical of the oscillatory behavior of the

Table 2 Averaged efficiencies and single scattering albedo of silica fibers: (a) T_m = 1000 °C; (b) T_m = 1500 °C.

(a)

D (μm)	$\overline{Q}_e$	$\overline{Q}_s$	$\overline{\omega}$
5	1.6091	1.5534	0.96540
4	1.6783	1.6261	0.96888
3	1.6990	1.6515	0.97204
2	1.4998	1.4564	0.97107
1	7.5246×10^{-1}	7.2113×10^{-1}	0.95836
0.5	2.3161×10^{-1}	2.1223×10^{-1}	0.91623
0.4	1.3520×10^{-1}	1.1916×10^{-1}	0.88139
0.3	6.0921×10^{-2}	4.8575×10^{-2}	0.79735
0.2	2.1270×10^{-2}	1.2887×10^{-2}	0.60591
0.1	5.6608×10^{-3}	1.4254×10^{-3}	0.25180
0.09	4.8435×10^{-3}	1.0292×10^{-3}	0.21250
0.08	4.1088×10^{-3}	7.1641×10^{-4}	0.17436
0.07	3.4458×10^{-3}	4.7594×10^{-4}	0.13812
0.06	2.8441×10^{-3}	2.9741×10^{-4}	0.10457
0.05	2.2939×10^{-3}	1.7091×10^{-4}	0.07451
0.04	1.7858×10^{-3}	8.6969×10^{-5}	0.04870
0.03	1.3109×10^{-3}	3.6498×10^{-5}	0.02784
0.02	8.6050×10^{-4}	1.0769×10^{-5}	0.01252
0.01	4.2624×10^{-4}	1.3422×10^{-6}	0.00315
0.005	2.1262×10^{-4}	1.6764×10^{-7}	0.00079

(b)

D (μm)	$\overline{Q}_e$	$\overline{Q}_s$	$\overline{\omega}$
5	1.5715	1.5456	0.98352
4	1.6121	1.5879	0.98499
3	1.7392	1.7173	0.98473
2	1.7445	1.7246	0.98845
1	1.1547	1.1402	0.98750
0.5	4.4964×10^{-1}	4.4073×10^{-1}	0.98017
0.4	2.9451×10^{-1}	2.8713×10^{-1}	0.97495
0.3	1.5339×10^{-1}	1.4771×10^{-1}	0.96297
0.2	4.6254×10^{-2}	4.2397×10^{-2}	0.91663
0.1	6.3525×10^{-3}	4.4038×10^{-3}	0.69324
0.09	4.9123×10^{-3}	3.1573×10^{-3}	0.64274
0.08	3.7440×10^{-3}	2.1832×10^{-3}	0.58311
0.07	2.8079×10^{-3}	1.4414×10^{-3}	0.51336
0.06	2.0673×10^{-3}	8.9560×10^{-4}	0.43322
0.05	1.4888×10^{-3}	5.1198×10^{-4}	0.34389
0.04	1.0409×10^{-3}	2.5929×10^{-4}	0.24909
0.03	6.9473×10^{-4}	1.0837×10^{-4}	0.15598
0.02	4.2284×10^{-4}	3.1867×10^{-5}	0.07536
0.01	1.9946×10^{-4}	3.9622×10^{-6}	0.01986
0.005	9.8248×10^{-5}	4.9449×10^{-7}	0.00503

Table 3 Averaged efficiencies and single scattering albedo of alumina fibers: (a) T_m = 1000 °C; (b) T_m = 1000 °C.

(a)

D (μm)	$\overline{Q}_e$	$\overline{Q}_s$	$\overline{\omega}$
5	1.5204	1.5123	0.99467
4	1.5601	1.5529	0.99537
3	1.6663	1.6598	0.99610
2	1.7765	1.7697	0.99616
1	1.3743	1.3690	0.99615
0.5	5.3004×10^{-1}	5.2050×10^{-1}	0.98191
0.4	3.3431×10^{-1}	3.2511×10^{-1}	0.97247
0.3	1.5532×10^{-1}	1.4909×10^{-1}	0.95987
0.2	4.3107×10^{-2}	3.9940×10^{-2}	0.92654
0.1	5.4371×10^{-3}	4.1072×10^{-3}	0.75539
0.09	4.1262×10^{-3}	2.9429×10^{-3}	0.71321
0.08	3.0743×10^{-3}	2.0333×10^{-3}	0.66139
0.07	2.2438×10^{-3}	1.3414×10^{-3}	0.57983
0.06	1.5999×10^{-3}	8.3281×10^{-4}	0.52054
0.05	1.1104×10^{-3}	4.7597×10^{-4}	0.42848
0.04	7.4545×10^{-4}	2.4085×10^{-4}	0.32309
0.03	4.7720×10^{-4}	1.0063×10^{-4}	0.21088
0.02	2.7969×10^{-4}	2.9589×10^{-5}	0.10579
0.01	1.2842×10^{-4}	3.6790×10^{-6}	0.02865
0.005	6.2786×10^{-5}	4.5914×10^{-7}	0.00731

(b)

D (μm)	$\overline{Q}_e$	$\overline{Q}_s$	$\overline{\omega}$
5	1.4586	1.4549	0.99747
4	1.4398	1.4365	0.99772
3	1.5675	1.5646	0.99814
2	1.6946	1.6916	0.99821
1	1.7029	1.7006	0.99864
0.5	8.2748×10^{-1}	8.2334×10^{-1}	0.99499
0.4	5.7336×10^{-1}	5.6939×10^{-1}	0.99307
0.3	3.0259×10^{-1}	2.9990×10^{-1}	0.99111
0.2	9.4844×10^{-2}	9.3476×10^{-2}	0.98557
0.1	9.7215×10^{-3}	9.1463×10^{-3}	0.94084
0.09	7.0249×10^{-3}	6.5131×10^{-3}	0.92715
0.08	4.9227×10^{-3}	4.4725×10^{-3}	0.90854
0.07	3.3233×10^{-3}	2.9330×10^{-3}	0.88256
0.06	2.1424×10^{-3}	1.8107×10^{-3}	0.84514
0.05	1.3035×10^{-3}	1.0290×10^{-3}	0.78941
0.04	7.3666×10^{-4}	5.1839×10^{-4}	0.70373
0.03	3.7653×10^{-4}	2.1568×10^{-4}	0.57281
0.02	1.6992×10^{-4}	6.3201×10^{-5}	0.37194
0.01	5.9670×10^{-5}	7.8375×10^{-6}	0.13135
0.005	2.7939×10^{-5}	9.7769×10^{-7}	0.03499

efficiencies when α is of order one or slightly larger (Kerker, 1969). For the approximations used in carrying out the wavelength integral in equation (3) (i.e. 1 and 16.5 μm for the lower and upper limits, respectively), α ranges from 15.71 to 0.95 for a diameter of 5 μm. Even for a smaller diameter such as 1 μm, α is about order one in most of the spectrum since it ranges from 3.14 to 0.19. Therefore, it is clear that for diameters 1 μm or larger, the efficiencies can be regarded as in the regime where there is oscillatory behavior. When the diameter is less than 1 μm, there is clearly a monotonic decreasing trend for the averaged efficiencies as the diameter is reduced. Since $\overline{\omega}$ is just the ratio of $\overline{Q}_s$ to $\overline{Q}_e$, it follows closely the trend displayed by the averaged efficiencies.

To illustrate the effect fiber size has on the radiant output, calculations have been performed for decreasing fiber size for three different cases: Case (1) ϕ is kept constant;

Case (2) the surface area available for convective heat transfer between the gas and solid phases, i.e. $(1-\phi)a$, is kept constant; and Case (3) both $(1-\phi)a$ and τ are kept fixed. In real situations, all three cases involved increasing the number density of the fibers appropriately. In addition, Case (3) requires increasing the physical thickness of the porous layer.

The input parameters needed to calculate radiant output are shown in Table 4. The parameters in Table 4a are those that are kept constant for all the calculations. Their values are chosen to be representative of PRB and have been used in an earlier study (Tong and Sathe, 1988). Notice that the flame is assumed to be 0.1 cm thick and is situated in the middle of the porous layer $[(\xi_1+\xi_2)/2 = 0.5]$. In Table 4b are the parameters that are varied, and the values indicated are used to establish the baseline results that will serve as the datum for comparison with the results for smaller diameters.

When D is changed, the parameters affected are τ, ω, and P_2 for Case (1), ϕ, τ, ω, and P_2 for Case (2), and ϕ, x_3, ω, and P_2 for Case (3). It follows from equation (5) that the relationship among τ, x_3, and D is

$$\tau = \frac{4\bar{Q}_e(1-\phi)x_3}{\pi D} \qquad (13)$$

Thus, when a new D is considered, the corresponding $\bar{Q}_e$ from Tables 2 or 3 is used in equation (13) to determine τ. Additionally for Case 3, x_3 is adjusted according to the equation to maintain τ constant. Realizing a is equal to 4/D and utilizing the definition of P_2 (see Table 10, one can determine P_2 through

$$P_2 = \frac{P_{2b}D_b}{D} \qquad (14)$$

where subscript b represents values used for the baseline calculation, i.e. values in Table 4b. For $(1-\phi)a$ to remain the same as that for the baseline calculation, ϕ must be related to D according to

$$\frac{1-\phi}{1-\phi_b} = \frac{D}{D_b} \qquad (15)$$

Equation (15) provides the relation for calculating ϕ for Case 2 and Case 3.

Depicted in Figs. 2 and 3 is radiant output for the silica and alumina fibers, respectively. It is presented in a form that the values as read from the graphs correspond to the fraction of the thermal energy generated by combustion. The results for both types of fibers exhibit similar trends.

Table 4 Input parameters for heat transfer calculations.

(a) Parameters that are held constant

	values used for the present study
x_i (m)	1
x_3 (cm)	1
x_2-x_1 (cm)	0.1
$(\xi_1+\xi_2)/2$	0.5
P_1	0.01
P_3	2.5×10^{-4}
P_4	0.02
P_5	5
G_e	2.98×10^{-5}

(b) Parameters that are varied

	values used for the baseline calculation	
	silica fibers	alumina fibers
D (μm)	5	5
ϕ	0.95	0.95
τ	193.6	204.9
ω	0.995	0.965
P_2	795	795

This is not surprising since $\bar{Q}_e$ and $\bar{\omega}$ for both the fibers exhibit similar behavior as D is decreased. For all three cases the radiant output initially increases slowly or even decreases slightly as D is reduced. This is a result of the oscillatory nature of the efficiencies as discussed earlier. The output increases substantially as D is further reduced. However, it reaches a plateau for Case 1 and Case 3 while that for Case 2 drops after reaching a maximum. The plateau occurs when $\bar{\omega}$ is so small that it is no longer an influencing factor and further reduction in fiber size has no impact on the output. This suggests that one only needs to reduce the fiber size to the point where the plateau begins to occur. The drop for Case 2 can be explained by the following consideration. Equation (13) indicates that τ is directly proportional to $\bar{Q}_e(1-\phi)$ and inversely proportional to 1/D. As D is reduced beyond the maximum point, both $\bar{Q}_e$ and $(1-\phi)$ decrease and their product decreases at a faster rate than 1/D increases. Hence, one ends up with smaller τ for decreasing D. As demonstrated by Tong and Sathe (1988), smaller τ results in lower radiant output.

It is quite apparent that there is definite advantage to use fibers that are smaller than order of 1 μm in diameter. Based on the plateau values for Case 1 and Case 3 and the maximum output for Case 2, the

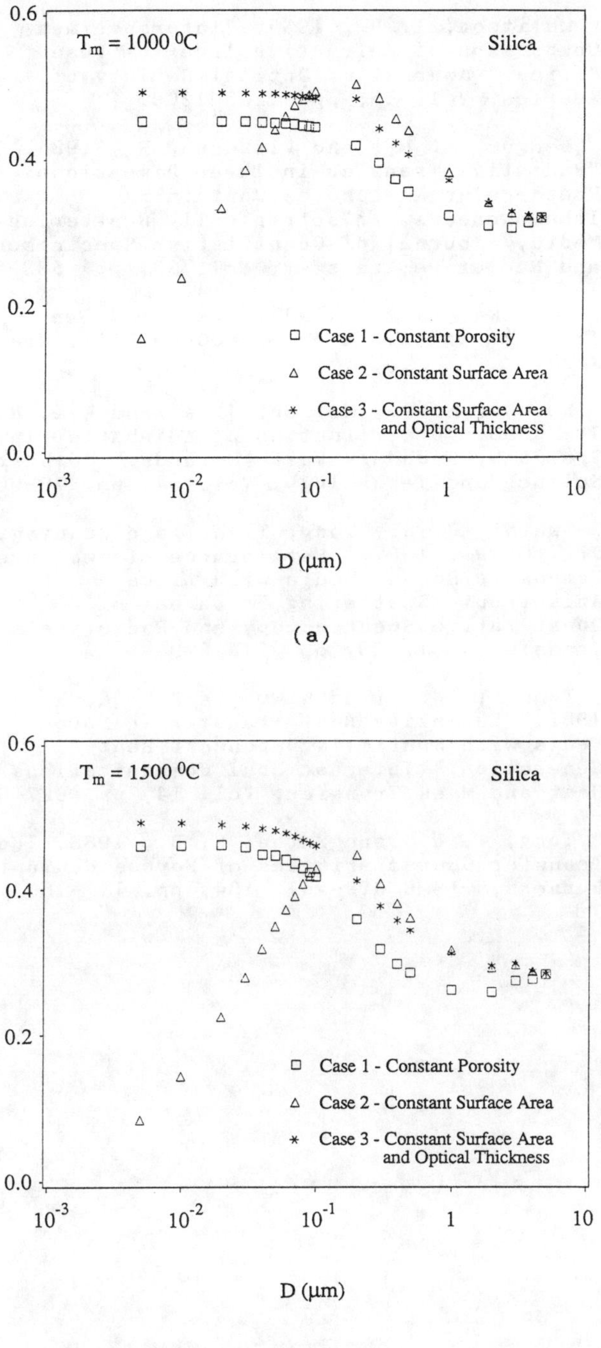

Fig. 2 Radiant output for silica fibers: (a) T_m = 1000 °C; (b) T_m = 1500 °C.

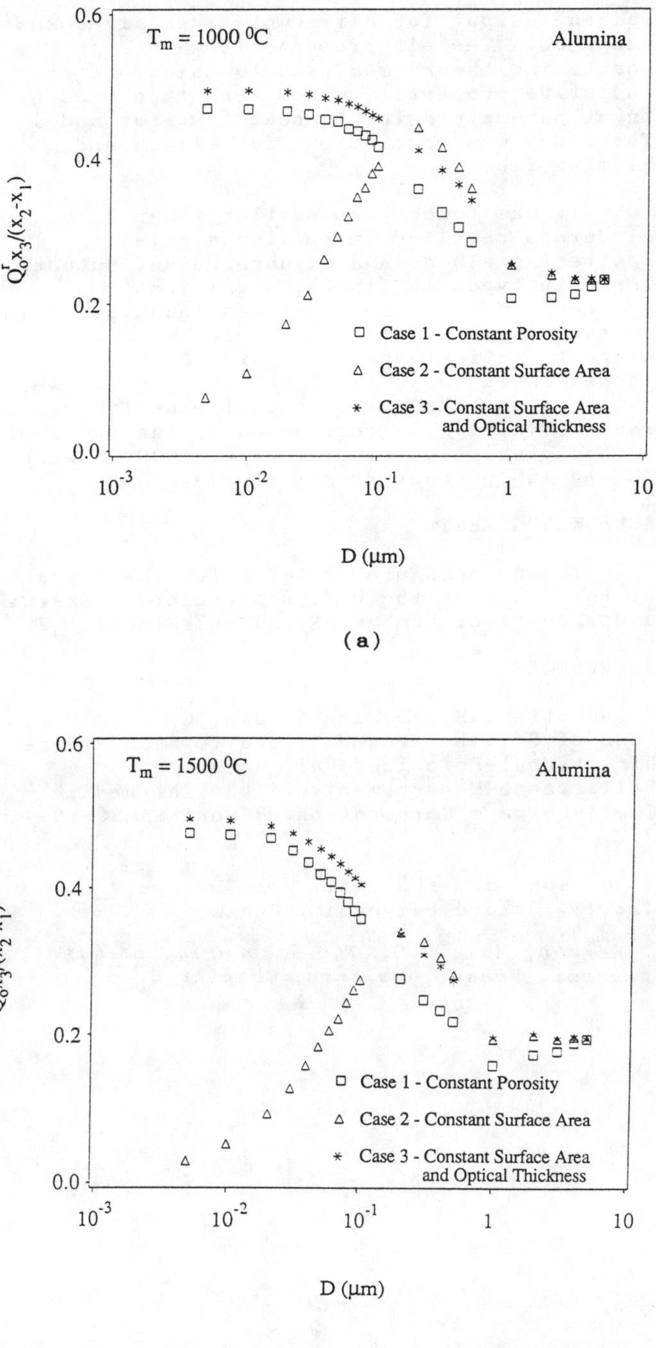

Fig. 3 Radiant output for alumina fibers: (a) T_m = 1000 °C; (b) T_m = 1500 °C.

radiant output when compared to the baseline value increases by 42, 54, and 63 percent, respectively, for silica fibers at T_m = 1000 °C. The corresponding increases are 61, 72 and 57 percent for T_m = 1500 °C. A similar comparison shows 100, 109, and 89 percent increases, respectively, for alumina fibers at T_m = 1000 °C and 140, 150 and 72 percent increases, respectively at T_m = 1500 °C.

CONCLUSIONS

The possibility of using fibers smaller than order of 1 μm in diameter to improve the performance of PRB has been investigated. Based on a one-dimensonal conduction, convection, and radiation heat transfer model with spatially dependent heat generation, radiant output for different fiber sizes was computed. The electromagnetic wave scattering theory was used to calculate the radiative properties which were then used as input parameters in the heat transfer model. The study was carried out for silica and alumina fibers.

It was found that smaller fiber diameters resulted in smaller single scattering albedo and higher radiant output. For both types of fibers, significant increase in radiant output was found possible by reducing the fiber size. For a characteristic temperature of 1000 °C increases as large as 63 and 109 percent were obtained for the silica and alumina fibers, respectively. The corresponding increases at a characteristic temperature of 1500 °C were 72 and 150 percent, respectively

ACKNOWLEDGEMENT

The authors are grateful for the support of this work by the U.S. Department of Energy under contract number DE-FG02-87ER13697.

REFERENCES

Champtier, R. J., and Friese, G. J., 1974, "Use of Polished Fused Silica to Standardize Directional Polarized Emittance and Reflectance Measurements in the Infrared," The Aerospace Corporation, Report SAMSO-TR-202.

Davison, B., 1957, *Neutron Transport Theory*, Oxford University Press.

Kerker, M., 1969, *The Scattering of light*, Academic Press, New York, Chapter 6.

Malitson, I. H., 1965, "Interspecimen Comparison of Refractive Index of Fused Silica," Journal of Optical Society of America, Vol. 55, pp. 1205-1209.

Menguc, M. P., and Viskanta, R., 1985, "Radiative Transfer in Three-Dimensional Rectangular Enclosures Containing Inhomogeneous, Anisotropically Scattering Media," Journal of Quantitative Spectroscopy and Radiative Transfer, Vol. 33, pp. 533-549.

Patankar, S. V., 1980, *Numerical Heat Transfer and Fluid Flow*, McGraw Hill, New York.

Smooke, M. D., Miller, J. A. and Kee, R. J., 1983, "Determination of Adiabatic Flame Speeds by Boundary Value Methods," Combustion Science and Technology, Vol. 34, pp. 79-90.

Swathi, P. S., Tong, T. W., and Cunnington Jr., G. R., 1987, "Reflectance of Two-Layer Composite Porous Media with Linear-Anisotropic Scattering," Journal of Quantitative Spectroscopy and Radiative Transfer, Vol. 38, pp. 273-279.

Tong, T. W., Lin, W. Q., and Peck, R. E., 1987, "Radiative Heat Transfer in Porous Media with Spatially Dependent Heat Generation," International Communications in Heat and Mass Transfer, Vol. 14, pp. 627-638.

Tong, T. W., and Sathe, S. B., 1988, "Heat Transfer Characteristics of Porous Radiant Burners," ASME HTD-Vol. 104, pp. 147-155.

1989 National Heat Transfer Conference
HTD-Vol. 106, Heat Transfer Phenomena in Radiation, Combustion and Fires

MODELING AND ANALYSIS OF HEAT TRANSFER IN A DIRECT-FIRED BATCH REHEATING FURNACE

K. S. Chapman, S. Ramadhyani and R. Viskanta
Heat Transfer Laboratory
School of Mechanical Engineering
Purdue University
West Lafayette, Indiana

ABSTRACT

A mathematical model of a batch reheating furnace has been developed. The furnace was modeled as a well-stirred enclosure with one-dimensional transient heat conduction in the refractory walls and the load. The convective heat transfer rate to the load and refractory was calculated by using existing correlations from the literature, while the radiative heat transfer rate to the load was calculated by considering the radiant energy exchange between the load, combustion gases, and the refractories. The radiant energy exchange was calculated using Hottel's zone method. The directed-flux exchange areas were calculated by using a four-gray gas model to model the combustion gases as a non-gray emitting and absorbing medium.

An extensive parametric investigation has been completed to determine the furnace design and operating characteristics that would lead to optimum furnace efficiency. The parametric investigations included in this paper are used to study the effect of the load and refractory emissivities, load heat capacity, and furnace combustion space on the thermal performance of a directly fired furnace.

NOMENCLATURE

A	area (m^2)
A_{n,T_i}	coefficient used in the weighted sum-of-gray-gases
c	specific heat for a solid (J/kg K)
c_p	specific heat at constant pressure (J/kmole K)
D_{eq}	hydraulic diameter (m)
d_L	load thickness (m)
E_b	blackbody emitted flux at gas temperature (W/m^2)
$\overline{G_iG_j}$	gas to gas radiation total exchange area (m^2)
$\overline{G_jS_i}$	gas to surface radiation total exchange area (m^2)
$\overrightarrow{GS_i}$	gas to surface radiation directed exchange area (m^2)
H	enthalpy flow rate (kW)
ΔH_c	rate of heat release by combustion in gas volume (kW)
h	heat transfer coefficient (W/m^2 K)
$\bar{h}$	effective heat transfer coefficient, accounting for radiation and convective heat transfer (W/m^2 K)
$\Delta h_{f,i}$	heat of formation of substance i (J/kmole K)
k	thermal conductivity (W/m-K)
$\dot{m}$	mass flow rate (kg/s)
$\dot{N}$	molar flow rate (kmole/s)
N	mole number
Q_L	heat transfer rate to the load (kW)
ΔQ_{adv}	net rate of enthalpy transport due to gas flow (kW)
q	heat flux (kW/m^2)
$\overline{S_iS_j}$	surface to surface total radiation exchange area (m^2)
$\overrightarrow{S_iS_j}$	surface to surface directed radiation exchange area (m^2)
T	temperature (K)
V	volume (m^3)
Y	molar oxygen-to-fuel ratio
ε	emissivity
μ	viscosity (Ns/m^2)
ρ	density (kg/m^3)
η	furnace efficiency defined by Eq.(18)
θ	dimensionless temperature
ξ	dimensionless distance
σ	Stefan-Boltzmann constant

τ	dimensionless time

Subscripts

AMB	ambient
a	air
conv	convection
cp	combustion products
f	fuel
g	gases
i	gas volume
init	initial quantity
k	zone index number
L	load
R	refractory
s	surface

INTRODUCTION

Although natural gas has been used as a fuel for many decades, the understanding of combustion kinetics, chemically reacting flow, turbulent transport processes, and heat transfer in industrial and commercial furnaces and equipment is incomplete. In the past, development of direct-fired natural gas furnaces and combustion systems has been mainly empirical. This is because the complex physical and chemical processes involved in flames have eluded detailed analytical description until more recently. During the last two decades, the challenge of ever-increasing unit capacity, improving thermal performance, and the problems of reducing combustion-generated pollution have attracted considerable attention from researchers and designers, resulting in significant progress towards science-based engineering design [Gupta and Lilley, 1985; Gupta et al., 1984; Khalil, 1982].

This paper presents a simplified mathematical model of a batch reheating furnace. The model was developed to identify the furnace design and operating parameters that largely affect furnace performance. The details of the model are provided in the next section, followed by the results from a parametric investigation of some of the more important furnace parameters.

Mathematical models, based on fundamental principles of fluid mechanics, thermodynamics and heat transfer, for predicting the thermal performance of complex combustion systems have proven to be useful design and system optimization tools. Computer based models can provide design and operating information relatively quickly and inexpensively, enabling the designer to explore a wide range of options for improving system performance. They can eliminate the need for costly experimental test stands or greatly assist in the design of new experiments and in the interpretation of experimental data. A whole spectrum ot heating processes, including reheating, heat treatment, melting of metals and non-metals, can benefit from the application of the modeling techniques. The models can be used to predict either the thermal efficiency and the gas consumption, or the design of equipment and construction materials required to achieve an optimum thermal efficiency and heating rate. This information is valuable in estimating the operating and capital costs of the system. Additionally, the mathematical models enable the quality of the heated product to be assessed by predicting both its temperature uniformity and the temperature history of the load as it is heated.

In high temperature directly-fired natural gas combustion systems, radiative transfer from the flame/combustion products and from the walls to the load or sink is usually the dominant mode of heat transfer, representing up to approximately 80% of the heat transferred to the load. Thus, there is a need to develop an understanding and modeling capability to predict radiative transfer in multidimensional combustion systems. Such systems contain selectively absorbing and emitting gases such as CO_2, H_2O, CO, CH_4, etc. and possibly soot and dust particles. Such particles are capable of absorbing, emitting and scattering thermal radiation.

In furnaces and combustion chambers the problem of radiative transfer is complicated by the need to estimate the temperature and concentration of the emitting species (gaseous and particulate, if any). Rigorous treatment of the problem is cumbersome and very difficult. The compromise is the development of simple models followed by a test of their adequacy against either experimental measurements or the predictions from more sophisticated analytical models. Experience has shown that the predictions from models which appear to introduce gross oversimplifications are often surprisingly close to reality. This is because radiation transfer formulations, in general, are tolerant to casual treatment of details, since local imperfections are averaged out in the integration process by calculating the radiation intensity along a line-of-sight in the medium.

TRANSIENT BATCH HEATING FURNACE MODEL DEVELOPMENT

In Hottel's zone method for analyzing radiation heat transfer in furnaces or combustion systems, where significant changes in gas temperature and composition and in sink temperature or emissivity are present, the gas volume and surface areas are subdivided into smaller zones which can be assumed to be isothermal [Hottel and Sarofim, 1967; Hottel and Cohen, 1958]. If there are M gas zones and N surface zones, there will be a total of M+N equations. Energy balances are made on each gas and surface zone separately.

Well-Stirred Furnace Model

The well-stirred furnace model used here to simulate the transient thermal performance is shown in Fig.1. The main assumptions and features of the model are the following:

- The furnace consists of a single gas zone, five surface (wall) zones and one sink.
- The gas composition throughout the gas volume is uniform and equal to that of the combustion products without dissociation.
- The uniform gas temperature is equal to the flue gas temperature calculated from the heat balance on the gas volume.
- The wall and sink surface temperatures are calculated from one-dimensional transient heat conduction equations for each wall.

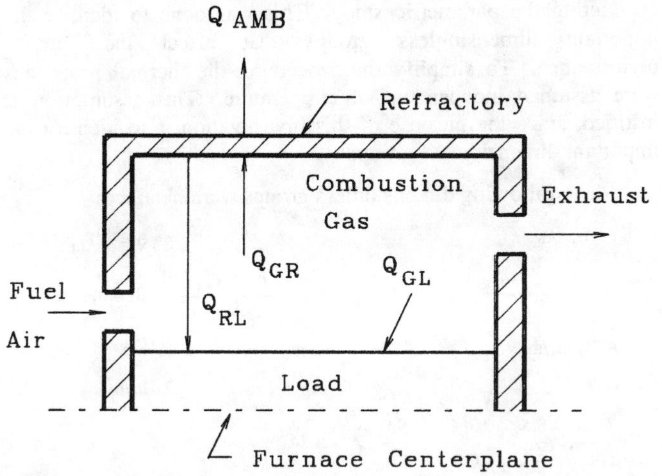

Figure 1 Schematic of the heat exchange occurring within a furnace.

- Radiation heat exchange is calculated using Hottel's zone method.
- The convective heat transfer model allows for the variation of the convective heat transfer coefficient with gas velocity and temperature.
- The bottom of the load is assumed to be adiabatic, i.e., symmetric about the center plane.

Gas energy balance. Neglecting energy stored in the gas, the gas energy balance can be written as

$$H_f + H_a - Q_{loss} = H_{cp} \tag{1}$$

Referring to Fig. 2, which shows the zone indexing scheme, the energy balance on the gas zone can be expressed as

$$H_a + \dot{N}_f \Delta h_{f,f} - H_{cp}(T_g) - \sum_{i=1}^{6} [A_i h_i (T_g - T_{s,i})$$

$$+ \overline{GS}_i \sigma (T_g^4 - T_{s,i}^4)] = 0 \tag{2}$$

The fourth term on the left-hand side of Eq.(2) denotes the convective heat transfer and the fifth term represents the radiative heat exchange between the gas zone and any surface zone i which has a surface temperature $T_{s,i}$. The determination of the gas-to-

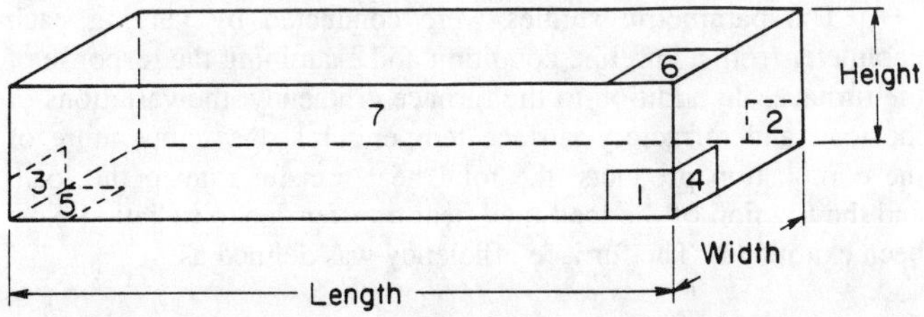

Figure 2 Indexing of the furnace internal surface for the Hottel zone method. The numbers 1-6 represent the six surface zones, and the number 7 represents the single gas zone.

surface total exchange areas, the heat transfer coefficients, and the enthalpy flow rates of the combustion products will be described in a later subsection.

Heat conduction model. Assuming transient, one-dimensional heat conduction in the wall or the load which is being heated, the energy equation may be written as

$$\rho c \frac{\partial T}{\partial t} = \frac{\partial}{\partial n}\left(k \frac{\partial T}{\partial n}\right) + \dot{q} \tag{3}$$

where n is the unit normal into the surface. This equation is applicable to each wall and to the load. The boundary condition on the inside surface of the wall as well as on the load exposed to the combustion products is

$$-k_i \frac{\partial T_i}{\partial n}\Big|_{\text{inner surface}} = q_i = Q_i / A_i \tag{4}$$

where the total (convective plus radiative) heat transfer rate is given by

$$Q_i = A_i h_i (T_g - T_{s,i}) + \overline{GS_i} \sigma (T_g^4 - T_{s,i}^4)$$

$$+ \sum_{j=1}^{6} \overline{S_i S_j} \sigma (T_{s,j}^4 - T_{s,i}^4) \tag{5}$$

Heat transfer from the outside of the furnace wall to the ambient is by convection and radiation and can be expressed as

$$-k_{s,i} \frac{\partial T_i}{\partial x}\Big|_{\text{outer surface}} = \bar{h}_i (T_{s,i} - T_{AMB}) \tag{6}$$

An adiabatic wall (boundary condition) can be obtained from Eq.(6) by setting $\bar{h} = 0$.

Other model parameters. The enthalpy of the combustion products is calculated from

$$H_{cp} = \sum_i \left(\int_{298}^{T_g} c_{p,i} dT + \Delta h_{f,i} \right) \dot{N}_i \tag{7}$$

The temperature dependent molar specific heats were approximated by the functions cited in Van Wylen and Sonntag (1978) and are valid for temperatures between 300 K and 3500 K. The molar flow rates appearing in the foregoing equation depend on the actual air-to-fuel ratio, the stoichiometry of the reaction, the fuel composition, and on the molar flow rate of the fuel, and were calculated by conducting an atom balance on the generalized combustion equation for a hydrocarbon fuel:

$$C_xH_y + YO_2 + 3.76YN_2 \rightarrow N_{CO}CO + N_{CO_2}CO_2$$

$$N_{H_2O}H_2O + N_{N_2}N_2 + N_{O_2}O_2 \tag{8}$$

The convective heat transfer coefficient on a surface exposed to the combustion gases is given by an empirical equation [Lebedev and Sokolov, 1976] based on experimental measurements,

$$Nu = \frac{hD_{eq}}{k_g} = C\left[\frac{\dot{m}_g D_{eq}}{A \mu_g}\right]^{0.75} \tag{9}$$

where C is 0.283, 0.175 or 0.142 for roof, end, or side firing, respectively.

The heat losses from the furnace wall to the ambient is by both convection and radiation and the flux can be approximated as

$$q = \bar{h}(T_s - T_{AMB}) + \varepsilon\sigma(T_s^4 - T_{AMB}^4) \cong \bar{h}(T_s - T_{AMB}) \quad (10)$$

The convective heat transfer coefficient is calculated from empirical correlations for natural convection available in the literature [Incropera and DeWitt, 1985].

Method of Solution of Model Equations

Energy equation for gas temperature. The model equation for the gas temperature, Eq.(2), was solved using the iterative Newton Raphson method:

$$T_{new} = T_{guess} - f(T_{guess})/f'(T_{guess}) \quad (11)$$

The function $f(T_{guess})$ was the left-hand side of Eq.(2), while the first derivative of the function was

$$f'(T_{guess}) = \sum_{i=1}^{6}[A_i h_i T_{guess} + 4\overline{GS_i}\sigma T_{guess}^3] - \partial H_{cp}/\partial T_g \quad (12)$$

Equation (11) was iteratively solved until the difference between T_{new} and T_{guess} was less than some pre-determined error criterion.

The direct exchange areas [Hottel and Sarofim, 1967] were calculated using the Monte Carlo statistical method [Siegel and Howell, 1981]. The method was used because it is simple and yields good accuracy relative to the computational effort. The total exchange areas, $\overline{GS_j}$ and $\overline{S_iS_j}$, were then calculated using the procedure suggested by Hottel and Sarofim (1967). When the weighted sum-of-gray-gases approach was used to evaluate the radiation characteristics of the combustion gases, the directed-flux exchange areas were calculated from the total exchange areas by using a four-gray gas model [Zongyu, et al., 1987; Bartelds, 1979]:

$$\overrightarrow{S_iS_j} = \sum_{n=1}^{4} A_{n,T_i}(\overline{S_iS_j})_n \ ; \ \overrightarrow{GS_j} = \sum_{n=1}^{4} A_{n,T_j}(\overline{GS_j})_n \quad (13)$$

The directed-flux exchange areas were used to replace the total exchange areas in Equations (2), (5), and (12)

Transient heat conduction equation. Equation (3) was discretized by using the fully implicit control volume approach described by Patankar (1980). The boundary heat flux in Equation (4) is the combination of radiation heat exchange with the combustion gases, radiation heat exchange with the other furnace interior surfaces, and convection heat transfer with the product gases. Since the radiation terms make Equation (4) non-linear, the heat flux was linearized by using a first order Taylor series expansion of the heat flux about the surface temperature.

The foregoing numerical method was used to calculate the temperature distribution in each furnace wall by applying the model to each wall separately, assuming the only interaction between the walls to be due to radiation heat exchange.

Integration of the Sub-Models into the Furnace Model. The individual sub-models were integrated into the furnace model. Due to the thermophysical property dependence on temperature assumed for the walls, the determination of the wall temperatures was an iterative process. The iterations were continued until the temperatures in all the furnace walls (and the load) changed by less than some predetermined error tolerance.

DIRECT-FIRED BATCH FURNACE PARAMETRIC INVESTIGATION

The model equations were non-dimensionalized before conducting the parametric study. This was done to identify the important dimensionless groups that affect the furnace performance. To simplify the procedure, the thermal properties were assumed independent of temperature. This assumption is justified, since the purpose of this investigation is to identify the important dimensionless groups.

The following dimensionless groups were obtained:

$\tau = \alpha_L t/d_L$ $\quad \xi = x/d_L$ $\quad \theta = T/T_{f,i}$

$St = \dfrac{hT_{L,init}A_L}{H_f}$ $\quad Bo = \dfrac{\overrightarrow{GS_L}\sigma T_{i,init}^4}{H_f}$ $\quad Bi = hd_L/k_L$

$AF = \dot{m}_a/\dot{m}_f$ $\quad d_R/d_L$ $\quad \bar{h}/h$

k_L/k_R $\quad \alpha_R/\alpha_L$ $\quad h_a/h_f$

$\dfrac{\overrightarrow{GS_i}}{\overrightarrow{GS_L}}, \dfrac{\overrightarrow{S_iS_j}}{\overrightarrow{GS_L}}, \dfrac{\overrightarrow{S_LG}}{\overrightarrow{GS_L}}, \dfrac{\overrightarrow{S_iG}}{\overrightarrow{GS_L}},$ $\quad A_i/A_L$

Although there are several identifiable dimensionless groups, it is apparent that they are not all independent (i.e., the area ratios, A_i/A_L, are directly related to the directed exchange areas, the thermal conductivity ratio is directly related to the thermal diffusivity ratio, etc.). However, it is possible to identify a set of primitive variables that would be appropriate for a parametric investigation. These are shown in Table 1, along with the associated dimensionless groups. The primitive variables were selected so that the associated dimensionless parameters would be independent of each other. Variation of the load thickness, d_L, for instance, affects only Bi. If the load thermal conductivity, k_L, were chosen rather than the load thickness, then both, Bi and the ratio α_R/α_L would be affected. The parameters selected for inclusion in this report were the load and refractory emissivities, the furnace combustion height, and the load heat capacity.

Dimensionless Parameter	Primitive Variable
Bi	d_L
$\bar{h}/h$	$\bar{h}$
d_R/d_L	d_R
α_R/α_L	$(\rho c)_R$, $(\rho c)_L$
AF	$\dot{m}_a$
h_a/h_f	$h_a = f(T_{air,in})$
Directed Exchange	$\varepsilon_R, \varepsilon_L$, Furnace height, A_L

Table 1 Primitive variables appropriate for a parametric investigation. Also shown are the related dimensionless groups.

The parametric studies were conducted by varying each parameter from a baseline condition and examining the response of the furnace. In addition to the furnace efficiency, the variations of the load and refractory surface temperatures, the temperature of the combustion products, the total heat transfer rate to the load, and the fraction of the load total heat transfer due to radiation have been examined. The furnace efficiency was defined as

$$\eta = \frac{\int_{ts}^{te} (\text{Heat transfer rate to the load})dt}{\int_{ts}^{te} (\text{Rate at which energy enters the furnace })dt} \quad (14)$$

where t_s was the time when the combustion gas temperature reached 800°C and the load was inserted into the furnace, and t_e was the time when the load centerplane temperature reached 800°C. The energy entering the furnace was entirely due to the fuel and did not take into account any air preheat, since the preheat was assumed to be from a furnace recuperator. The convective heat transfer rate to the load was calculated by a correlation for a crown firing configuration [Lebedev and Sokolov, 1976]. The baseline operating condition for these studies is shown in Table 2.

	Material	ρc (J/m³K)[1]	k (W/mK)[1]	Thickness (m)	ε[1]
Load	Iron	3770(T-300)	112exp(-0.001T)	0.15	0.3
Refractory	Red Clay Brick	253900	1+0.008(T-475)	0.5	0.8

Gas Properties:

$\Delta h^2_{f,fuel} = -74800\text{J/mole}$ Fuel = CH_4

$\mu(T)^3 = (4.1\times10^{-7})T^{0.666}\text{kg/ms}$ ATF = 20

$\rho(T)^3 = 348T^{-1}\ \text{kg/m}^3$

$k(T)^3 = 0.00028T^{0.81}\,\text{W/mk}$ Non-gray (4 gray gas mode)

$$h^4 = C\left[\frac{\dot{m}D_{eq}}{A\mu}\right]^{0.75}\left(\frac{k}{D_{eq}}\right)$$

Miscellaneous

$T_{AMB} = 300$ K
Furnace Dimensions:
- Length = 5 m
- Height = 1 m
- Width = 1 m

$h_{AMB} = 10\ \text{W/m}^2\text{K}$
Air Preheat = 800 K
Firing Rate:
- Warm-up = 350 kW
- Heating = 500 kW

1 - Incropera and DeWitt, 1985
2 - VanWylen and Sonntag, 1978
3 - White, 1974
4 - Lebedev and Sokolov, 1976

Table 2 Base furnace configuration parameters used for the parametric study

Variation of the Load Heat Capacity

For this study, the load thermophysical properties were held constant, independent of temperature. The load heat capacity was varied from 6.6 MJ/m³K to 0.86 MJ/m³K, while the load thermal conductivity was held constant at 100 W/mK. A heat capacity of 1 MJ/m³K corresponds to a steel mass of 2.3 metric tons. The remaining furnace parameters were as stated in the base furnace configuration (Table 2).

Figure 3 shows the load surface temperature as a function of the load heat capacity and time. As expected, longer periods of time were required to heat the higher heat capacity loads, since large amounts of heat must be stored for a given temperature increase. The gas temperature, shown in Figure 4, exhibits a decreasing trend with an increase in the heat capacity. At low heat capacities, the load surface temperature quickly approaches the gas temperature, resulting in low potential for heat transfer, which, in turn, results in a higher gas temperature. At higher values of load heat capacity, the load surface temperature remains low for a longer period of time, resulting in the opposite heat transfer effect. The final gas temperature is higher than for the low heat capacity case.

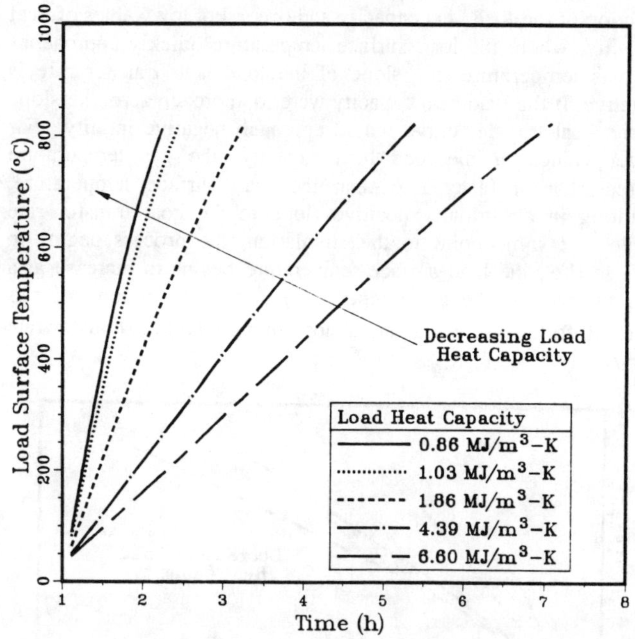

Figure 3 Load surface temperature as a function of time and the load heat capacity.

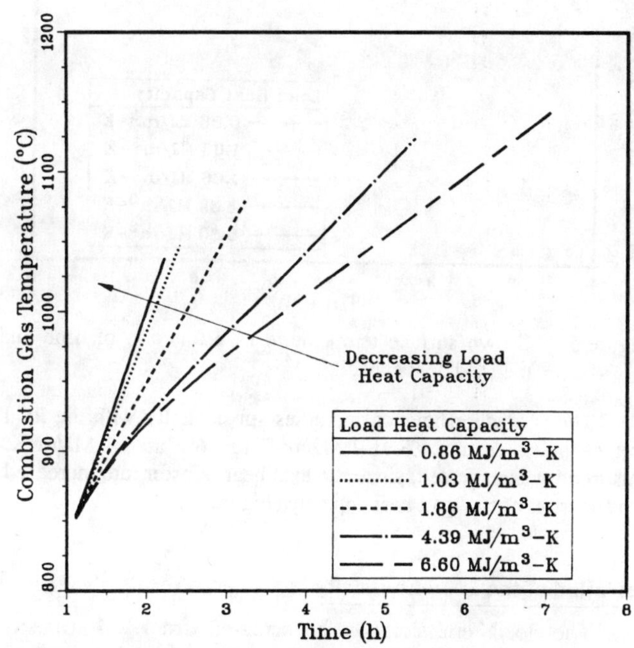

Figure 4 Furnace gas temperature as a function of time and the load heat capacity.

Figure 5 represents the refractory crown surface temperature as a function of the load heat capacity and time. The trends are similar to those exhibited by the gas temperature. As the load heat capacity is increased, the load surface temperature remains lower for longer periods of time, resulting in a larger net radiative heat transfer rate from the refractory to the load, which results in a lower refractory surface temperature. In addition, the gas temperature is lower due to the above described reasons, and, therefore, does not transfer as much heat to the refractory.

Figure 6 shows the total heat transfer rate to the load as a function of the load heat capacity and time. For low values of heat capacity, where the load surface temperature quickly approaches the gas temperature, the slope of the load heat transfer rate is negative. If the load heat capacity were to approach zero, the slope of the heat transfer curve would approach negative infinity. For larger values of the load heat capacity, the gas temperature increases at a faster rate than the load surface temperature, resulting in an initially positive slope to the heat transfer rate curve. At some point in the simulation, the process begins to reverse (i.e., the load surface temperature begins to increase at a rate faster than the gas temperature) and an inflection point is reached, followed by a negative slope in the load heat transfer rate curve.

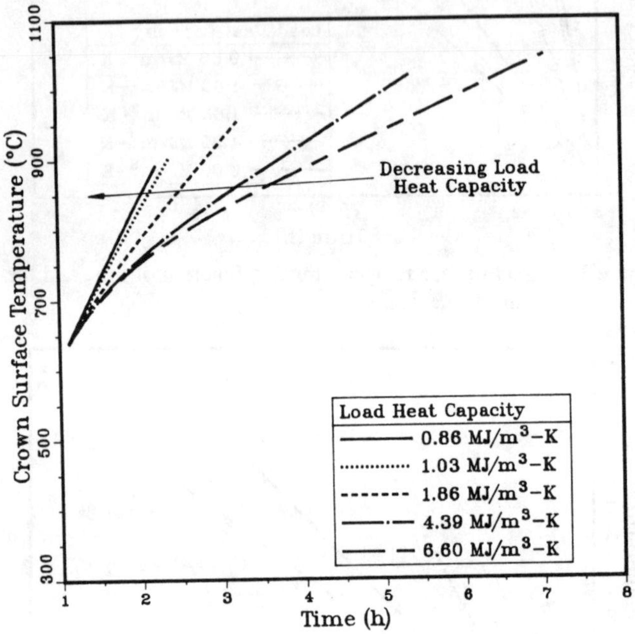

Figure 5 Crown surface temperature as a function of time and the load heat capacity.

The furnace efficiency increases substantially with the load heat capacity, from 24% at 1 MJ/m^3K to 36% at 6.5 MJ/m^3K. This trend was expected, since the load heat transfer rate increased as a function of the load heat capacity (Figure 4).

Variation of the Load Emissivity

The load emissivity was increased from 0.1 (mostly reflective) to 0.995 (nearly black). The remaining furnace parameters were as stated in the base furnace configuration (Table 2).

Figure 7 shows the load surface temperature as a function of the load emissivity. As ε_L is increased from 0.1 to 0.995, more radiant energy is absorbed at the load surface resulting in higher load surface temperatures. Since additional heat is removed from the gas, lower gas temperatures are realized.

The total load heat transfer rate increases as ε_L is increased, due to the increased absorption of incident radiant energy on the load surface (Figure 8). For a large ε_L (close to a black surface), the highest load heat transfer rate is at the initial loading, after which time the load heat transfer rate decreases, due to the rapid rise of the load surface temperature. It is of interest to note that,

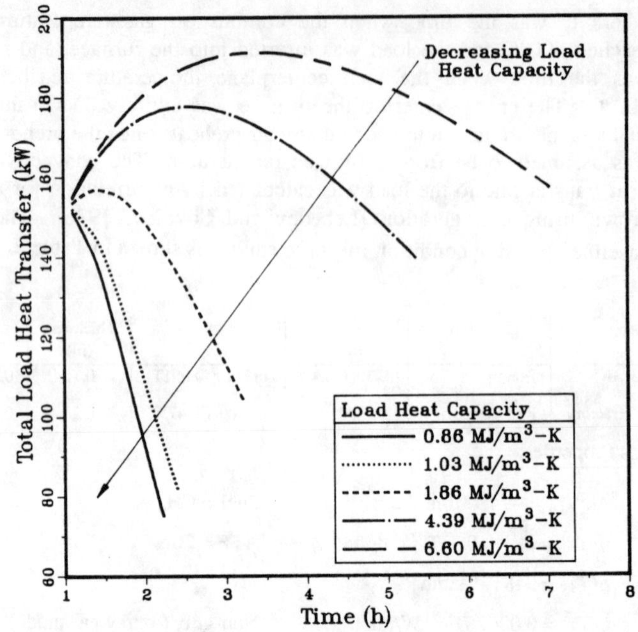

Figure 6 Total load heat transfer coming from the gas and refractory by radiation and convection as a function of time and the load heat capacity.

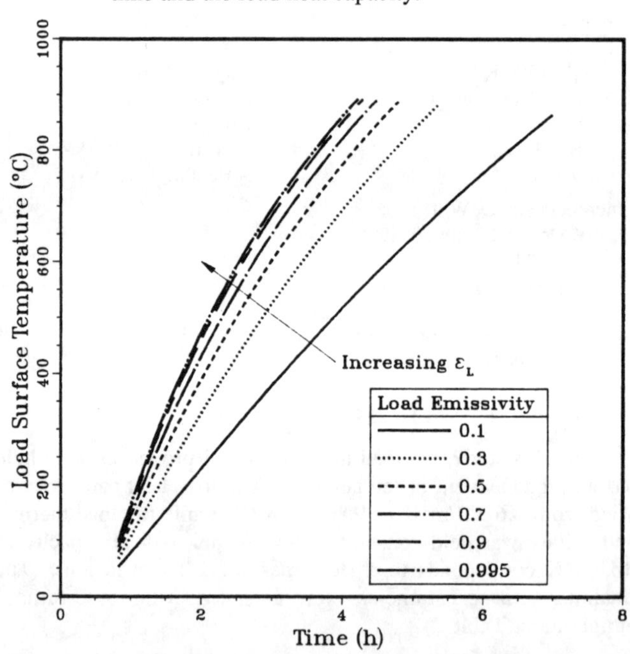

Figure 7 Load surface temperature as a function of time and the load emissivity.

although this is similar in nature to the case of a low load heat capacity, the trend of the heat transfer rate with the load emissivity ε_L is opposite to that of the load heat capacity. When the heat capacity was low, the surface temperature approached the gas and refractory temperatures without an increase in absorbed energy. For the case of a large ε_L, the incident radiant energy is almost entirely absorbed at the surface, and, due to the fairly high load thermal conductivity, the energy absorbed at the load surface is conducted into the load interior.

Since the total heat transfer rate to the load increases with

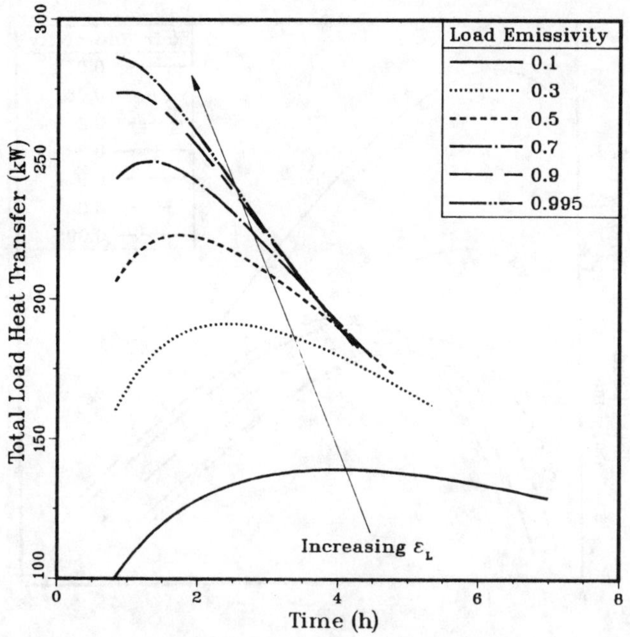

Figure 8 Total load heat transfer coming from the gas and refractory by radiation and convection as a function of time and the load emissivity.

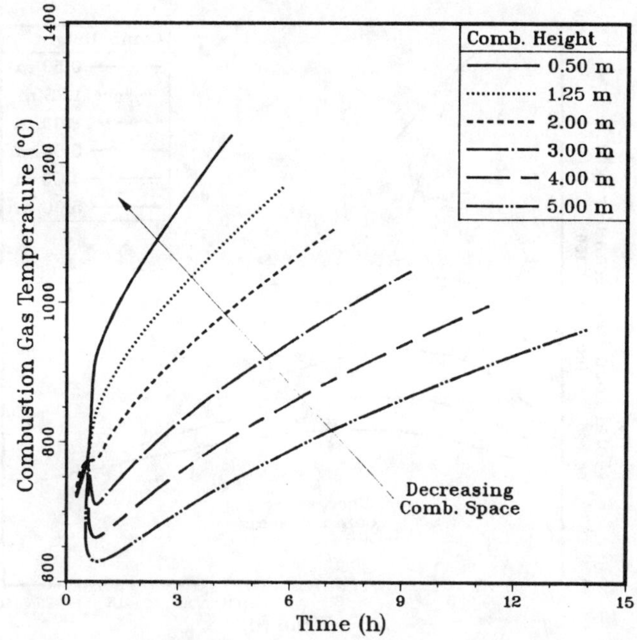

Figure 9 Furnace gas temperature as a function of time and the combustion space above the load.

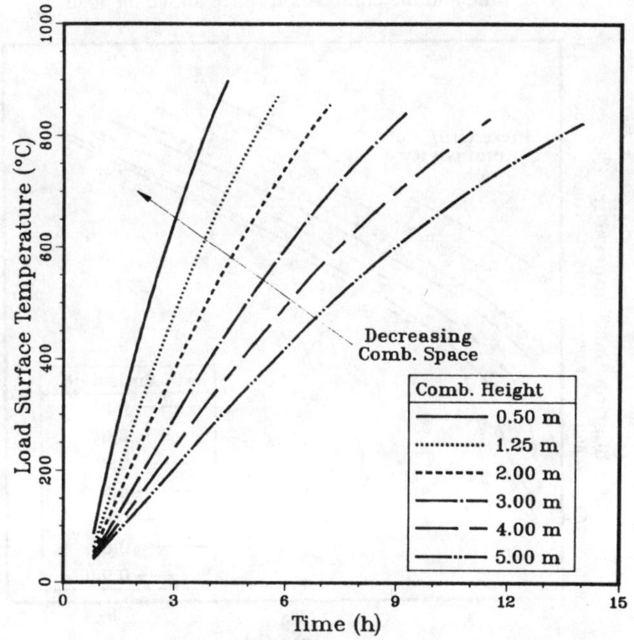

Figure 10 Load surface temperature as a function of time and the combustion space above the load.

ε_L, the furnace efficiency increases as well. The increase in furnace efficiency is substantial increasing from 27% with $\varepsilon_L = 0.1$ to 48% with $\varepsilon_L = 1.0$.

Variation of the Furnace Combustion Space Parameters

The furnace height between the load and crown surfaces was varied from 0.5 m to 5.0 m. The remaining furnace parameters were as stated in the base furnace configuration (Table 2).

Figure 9 shows the gas temperature as a function of time and the furnace height. As the height is increased, the gas temperature is not only reduced to lower values, but also undergoes a severe temperature depression when the load is inserted into the furnace. The overall reduction in gas temperature is due to the increased refractory surface area with which the gas exchanges heat, both by convection and by radiation. The temperature depression, however, is due to the increased radiant heat transfer between the refractory side walls and the load. As the furnace height is increased, the radiant heat exchange from the refractory side walls to the load increases. This, in turn, results in an initial decrease in the refractory temperature. As the refractory temperature decreases, so does the gas temperature.

Figure 10 shows the load surface temperature as a function of the furnace height and time. As can be seen, not only are longer periods of time required to heat the load, but the load surface temperature is lower for higher furnace heights. These trends can be explained by realizing that both the gas and refractory temperatures are lower for higher furnace heights due to greater heat losses to the ambient, resulting in lower heat transfer rates to the load, as confirmed in Figure 11. The resulting effect on the furnace efficiency is strong. The furnace efficiency decreases rapidly from 45% to 12% as the furnace height is increased from 0.5m to 5.0 m.

Variation of Refractory Emissivity

One of the directly fired furnace design features which has recently attracted attention for improving furnace thermal performance is the refractory emissivity. The refractory emissivity was increased from 0.01 (almost perfectly reflecting) to 0.995 (almost perfectly black). The remaining furnace parameters were as stated in the base furnace configuration (Table 2).

The effect of refractory emissivity on the combustion products temperature is shown in Figure 12. The results show that as the emissivity is increased from a value of 0.01 to a value of

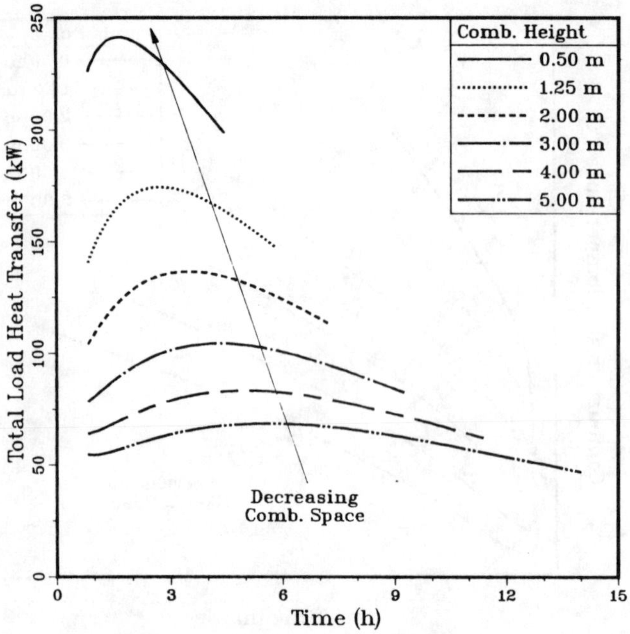

Figure 11 Total load heat transfer coming from the gas and refractory by radiation and convection as a function of time and the combustion space above the load.

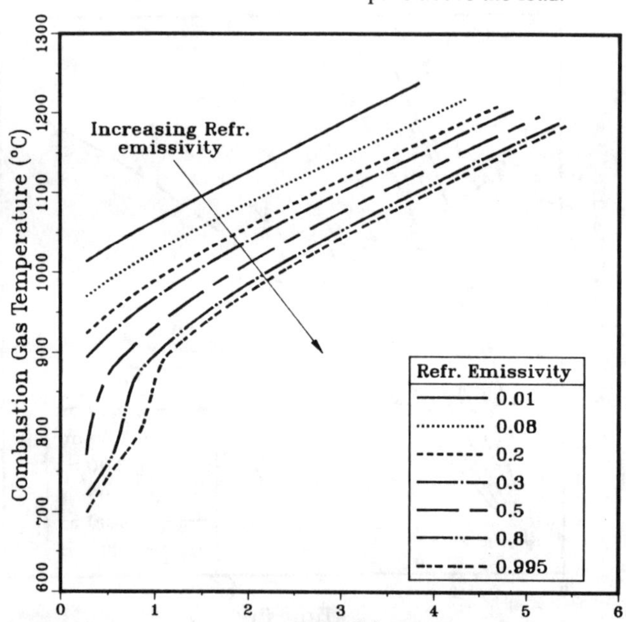

Figure 12 Effect of refractory surface emissivity on the variation of the combustion gas temperature with time.

0.995 the gas temperature decreases. This is understandable, because less radiant energy is reflected from the refractories and re-absorbed by the gas, resulting in a gas temperature decrease. On the other hand, the crown surface temperature reaches its minimum for $\varepsilon_R = 0.01$ because the surface is highly reflecting and absorbs very little of the radiation incident on it. Therefore, the refractory surface temperature must decrease as the refractory emissivity decreases.

The dependence of the total heat transfer rate to the load on the refractory emissivity is illustrated in Figure 13. The results

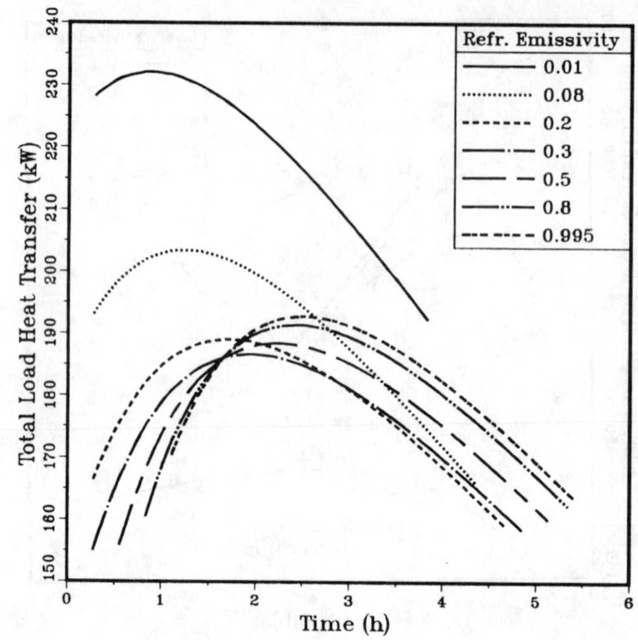

Figure 13 Effect of refractory surface emissivity on the variation of the total heat transfer rate to the load as a function of time.

show that the heat transfer rate to the load, Q_L, is highest for $\varepsilon_R = 0.01$, and that the dependence of the net heat transfer rate on the refractory emissivity is highly nonlinear. Q_L reaches a minimum at $\varepsilon_R = 0.3$, followed by a gradual increase for $\varepsilon_R > 0.3$.

The fractional contribution to the heat transfer at the load surface from the combustion products and the refractory walls is presented in Figures 14 and 15, respectively. The remaining fraction is by convection between the combustion gases and the load surface. The results in Figures 14 and 15 can be explained by considering the variation of the refractory emissivity. As the

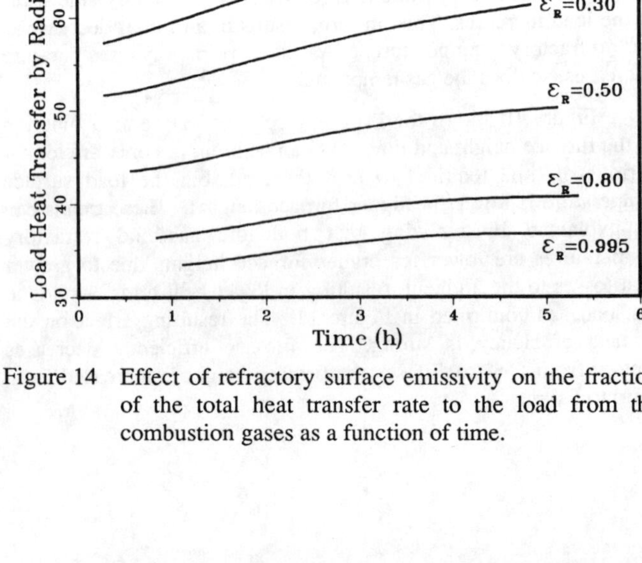

Figure 14 Effect of refractory surface emissivity on the fraction of the total heat transfer rate to the load from the combustion gases as a function of time.

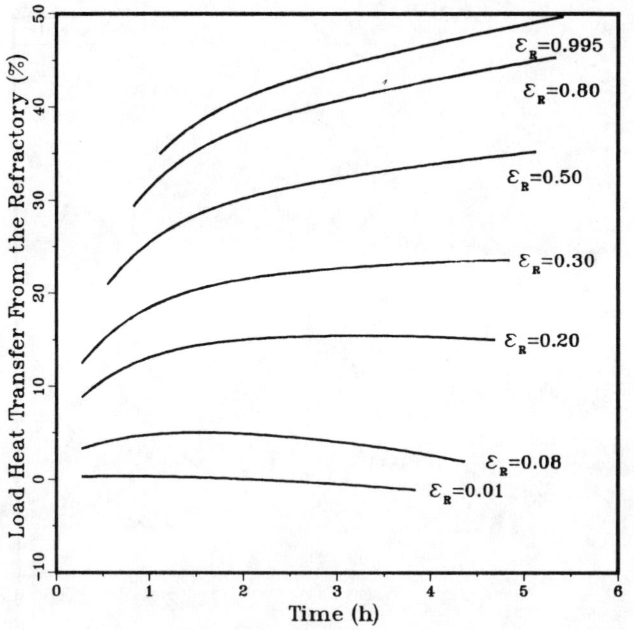

Figure 15 Effect of refractory surface emissivity on the fraction of the total heat transfer rate to the load from the refractory surfaces as a function of time.

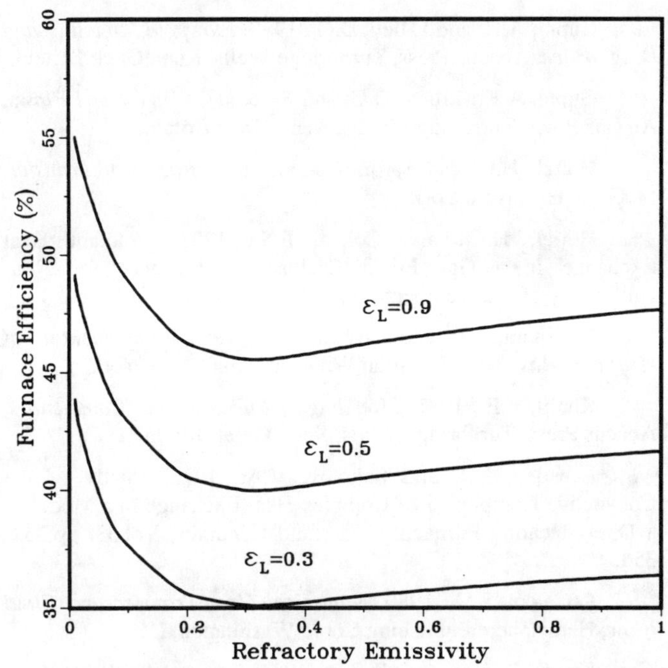

Figure 16 Furnace efficiency as a function of the load and refractory emissivities, showing the minimum location of the furnace efficiency.

refractory emissivity increases to moderately high values, less radiant energy from the gas is reflected by the refractory back to the gas, and is, instead, absorbed by the refractory and either conducted through the refractory to the environment or re-emitted by the refractory. The re-emitted radiation is spectrally continuous, and, since the gas is a selective absorber, a fraction of the emitted radiant energy reaches the load surface, while the remainder is absorbed by the gas. This is in contrast to the case of a perfectly reflecting refractory ($\varepsilon_R = 0$), where the reflected radiant energy retains its spectral band nature and is completely re-absorbed by the gas. For a perfectly absorbing refractory surface the re-emitted radiant energy has a continuous spectral distribution corresponding to the refractory temperature. Only a portion of this re-emitted energy is absorbed by the gas and the rest irradiates the load. Therefore, the refractory contribution to Q_L increases with the refractory emissivity. Also, since the gas absorbs relatively less reflected radiant energy as the refractory emissivity is increased from 0 to 1, the gas temperature, and hence the gas radiation contribution to the Q_L decrease.

Since Q_L is a minimum at a refractory emissivity of 0.3, the furnace efficiency will also be a minimum, as shown in Figure 16. The furnace efficiency, following the same trends as Q_L, increases towards a very high value for a perfectly reflecting refractory, and reaches approximately 36% for an absorbing refractory. Although the minimum furnace efficiency is coincident with the refractory and load emissivities being equal, Figure 16 shows that the load emissivity has little affect on the location of the minimum furnace efficiency. The load emissivity was varied from 0.3 to 0.9, with each case continuing to exhibit a minimum furnace efficiency at a refractory emissivity of approximately 0.3 for the particular furnace design and operating condition.

CONCLUSIONS

The mathematical model of a batch reheating furnace presented in this paper provides an efficient method of identifying the effect of design and operating parameters on the furnace performance. These findings should be qualified by mentioning that the results are based on a particular furnace design and operating condition. In addition, the calculations were performed on the gray basis for the radiation properties of surfaces and with Hottel's sum-of-the-gray gases model for the combustion products.

The major conclusions are summarized as follows:

- The furnace efficiency becomes large as ε_R becomes small. From a practical point of view, however, a nearly reflective refractory surface is not possible with conventional refractories. Even a coating would soon lose its reflective characteristics due to the harsh operating environment. Unless reflective refractory surfaces can be maintained, a nearly black surface is recommended.
- The furnace efficiency is highest when the furnace combustion space is small. Therefore, the furnace should be as short as possible.
- The furnace efficiency is increased substantially by making ε_L approach a black surface. This can easily be accomplished by coating the load surface with a material such as graphite.

ACKNOWLEDGEMENTS

The authors gratefully acknowledge the support of the Gas Research Institute, Chicago, IL (Contract No. 5086-260-1293).

REFERENCES

Bartelds, H., 1979, "Development and Verification of Radiation Models," in Combustion Modelling, AGARD No.CP-275, pp.11-1 to 11-20.

Gupta, A.K. and Lilley, D.G., 1985, *Flowfield Modeling and Diagnostics*, Abacus Press, Turnbridge Wells, Kent, Great Britain.

Gupta, A.K., Lilley, D.G. and Syred, H., 1984, *Swirl Flows*, Abacus Press, Turnbridge Wells, Kent, Great Britain.

Hottel, H.C. and Sarofim. A.F., 1967, *Radiative Transfer*, McGraw-Hill, New York.

Hottel, H.C., and Cohen, E.S., 1958, "Radiant Heat Exchange in a Gas Filled Enclosure: Allowance for Non-Uniformity of Gas Temperature," AIChE Journal, Vol.4, pp.3-14.

Incropera, F.P. and DeWitt, D.P., 1985, *Fundamentals of Heat and Mass Transfer*, John Wiley and Sons, New York.

Khalil, E.E., 1982, *Modeling of Furnaces and Combustors*, Abacus Press, Turnbridge Wells, Kent, Great Britain.

Lebedev, V.I. and Sokolov, V.A., 1976, "Study of the Convective Component of Complex Heat Exchange in a Model of a Direct-Heating Furnace," Glass and Ceramics, Vol.33, pp.352-354.

Patankar, S.V., 1980, *Numerical Heat Transfer and Fluid Flow*, Hemisphere Publishing Corp., Washington, DC.

Siegel, R. and Howell, J.R., 1981, *Thermal Radiation Heat Transfer*, 2nd Edition, McGraw-Hill Book Co., New York.

Van Wylen, G.J. and Sonntag, R.E., 1978, *Fundamentals of Classical Thermodynamics*, Second Edition, John Wiley and Sons, New York.

White, F.H., 1974, *Viscous Fluid Flow*, McGraw Hill, Inc.

Zongyu, L., Barr, P.V., and Brimacombe, J.K., "Computer Simulation of the Slab Reheating Furnace," 1987, in *Proceedings of the International Symposium on the Mathematical Modeling of Metals Processing Operations*, The Metallurgical Society, Warrendale, PA.

1989 National Heat Transfer Conference
HTD-Vol. 106, Heat Transfer Phenomena in Radiation, Combustion and Fires

A 3-D CODE FOR PREDICTING RADIATIVE AND CONVECTIVE HEAT TRANSFER IN BOILERS

G. De Michele, L. Ghiribelli, S. Pasini and A. Tozzi
ENEL — Thermal and Nuclear Research Center
Pisa, Italy

ABSTRACT

A numerical model was developed for predicting the behaviour of the furnace and the convective section of power station boilers. The model is capable of predicting gas aerodynamics, flame heat release and convective and radiative heat transfer for practical furnace geometries, both in the combustion chamber and in the convective section. The model is 3-D both on the gas and steam sides, therefore allowing detailed simulation of the geometry of the convective banks. It can be applied to both gas-, oil- and coal-fired boilers.

This work describes the main features of the algorithm adopted. A simplified procedure based on the original zone method was developed, capable of examining geometries with a large number of zones using less computer time and with smaller storage requirements. An original technique was utilized for the solution of the energy equation in the convective pass, based on direct inversion of the matrix of the coefficients of this equation.

The predictions provided by the model are compared with experimental measurements of gas temperatures, steam temperatures and heat fluxes obtained in a 35 MWe tangentially-fired boiler.

NOMENCLATURE

Symbol	Definition
a	Weighting coefficient
A	Pre-exponential factor
A	Coefficient of algebraic form of finite difference equation
A_j	Area of surface j
cp	Specific heat
d_i	Particle diameter
D	Burner diameter
E_a	Activation energy
E	Black body emissive power
f_v	Soot volume fraction
$g_i g_j$	Gas-gas direct exchange area
$g_i s_j$	Gas-surface direct exchange area
h	Convective heat transfer coefficient
H	Enthalpy
k_c	Kinetic rate coefficient
k_d	Diffusive rate coefficient
k_r	Gas extinction coefficient
K	Exchange coefficient
m	Mass
N'_s	Number of surface zones $\not\subset S_c$
N''_s	Number of surface zones $\subset S_c$
P	Pressure
P	Flow rate
Q	Heat released per unit length
Q_c	Heat release due to combustion
Q_e	Heat exchange by convection
Q_{tot}	Burner heat release
R	Gas constant
s	Curvilinear coordinate starting from the burner
$s_i s_j$	Surface-surface direct exchange area
S	Source term
S_{c1}	Connection surface with the upper cavity
S_{c2}	Connection surface with the convective bank
S_c	$S_{c1} \cup S_{c2}$
t	Time
T	Temperature
U	x-component of velocity
V	y-component of velocity
V	Volume
W	z-component of velocity
W_j	Radiative heat flux leaving surface j
ε	Emissivity
λ	Parameter
ρ_j	Reflectivity of surface j

ρ Density
ϕ Heat flux

Subscripts

CF	Cooling fluid
E	East cell
D	Down cell
F	Fluid
G	Gas
GC	Gas-cooling fluid
GT	Gas-tube
N	North cell
O_2	Oxygen
P	Central cell
r	Index of grey gas
S	South cell
S	Surface
sc	Upper zone
SF	Surface-fluid
Si	Surface i
TF	Tube-fluid
U	Upper cell
V	Volume
W	West cell

Superscripts
* Mean

1. INTRODUCTION

Thanks both to the increase in computational abilities and to the need for more detailed predictions of heat transfer phenomena in industrial furnaces, considerable work has been carried out in this connection in the last five years. Exhaustive reviews of this work can be found in the papers of Richter (1985) and Robinson (1985). Very detailed solutions of heat transfer applied to industrial full scale boilers have appeared recently (Sakai,1984; Cooper,1984; Richter,1985; Gorner,1986; Fiveland,1988). In all these papers attention has been mainly focused on the combustion chamber, neglecting the convective section.

Present deficiencies in understanding the performance of the whole plant were revealed when detailed data were found necessary to evaluate the possibility of using new methods for reducing pollutant emissions in existing power plants (for instance, by injecting calcium sorbents in the upper part of the combustion chamber (De Michele,1988)), or for correctly evaluating the performance of the heat exchange components (specifically superheaters, reheaters and economizers) of large steam generators (Bauver,1987).

The model described in this paper attempts to examine the whole boiler (furnace and convective pass) with the same degree of detail. Radiation modelling in furnace heat transfer is based on a simplified version of the zone method, allowing use of fine grids comparable to the ones used for flow field computation. In the convective region, both convective and radiative heat transfer are considered. The convective banks can be simulated in a very detailed manner, the tube bundles are described following the 3-D flow field discretization, starting from the inlet of the cooling fluid. The energy equation is solved by using an original algorithm for the solution of the resulting eptadiagonal matrix.

2. DESCRIPTION OF THE MODEL

The code developed calculates the performance of a steam generator (gas and steam side), up to the economizer exit.

The code supplies:

- the 3-D distribution of the temperature of the gases;
- the distribution of the radiative and convective heat flux on the walls of the furnace and of the convective banks;
- the distribution of surface temperature on the walls of the furnace and of the convective banks;
- the temperature distribution of the water/steam in the individual sections of the convective banks;
- the partial energy balances (the balance of the furnace calculated at the nose level and at the entrance to the first convective bank, the balance of the single banks and the overall balance).

The structure of the code is modular. The steam generator is divided up into 4 sections, described by separate modules (Fig. 1):

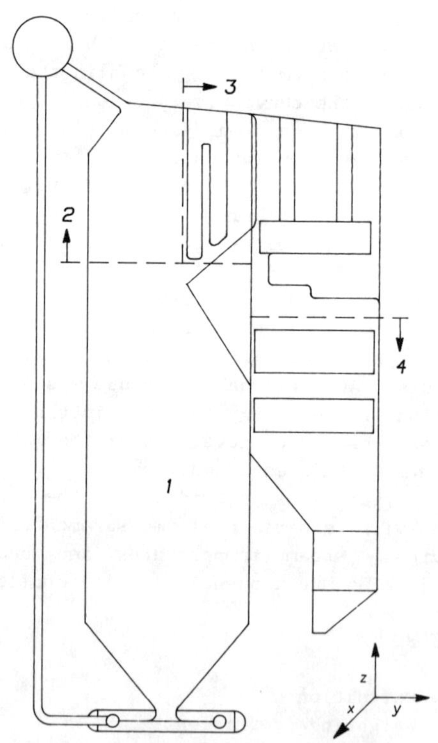

Fig.1 Boiler view
1) Lower furnace
2) Upper furnace
3) 3-D convective section
4) 1-D convective section

1 - The lower furnace, consisting of the combustion chamber up to the nose level.

2 - The upper furnace, consisting of the empty volume in front of the first bank. This section may not be present if there are platens in the boiler occupying the whole of its upper volume.

3 - The 3-D radiative-convective exchange section, including the first part of the convective section (platens, superheaters, reheaters,......) where there are still strong disuniformities in the gas and steam temperatures.

4 - The 1-D convective exchange section.

The coupling between these sections is obtained by means of an iterative process. A description of section 4 will not be found in the present paper, since it is considered insignificant.

The following are the main basic hypotheses common to all the modules:

- the optical properties of the medium can be approximated by means of a grey-gas model;
- the flow-field of the combustion gases (velocity and mass fluxes) is known with the required accuracy;
- the volumetric distribution of the heat released by the combustion is known.

The emissivity of the combustion products is calculated in relation to the composition of the gas, to soot concentration, to mean beam length and to temperature by means of a code (Castellano, 1988), based on a narrow band model in the case of the gas (Ludwig, 1973; Farag, 1976; Hartmann, 1984; Souffiani,1985) and on the approximation of the Mie theory (Hottel,1985) in the case of soot. At present scattering is not modelled. Emissivity is then simulated by means of the weighted gray gas approach, assuming that the weighting coefficients are functions of both temperature and soot concentration (De Michele,1985).

The flow field is determined by means of a finite-difference code that solves the Navier-Stokes equation for a dilatable fluid, using mass-fluxes as primitive variables (De Michele, 1985; Pasini,1988). The density distribution of the gases is determined on the basis of an estimation of the temperature field in the boiler. This estimation is then updated by solving the energy equation. These coupled iterations are repeated until convergence is reached. Normally, just a few iterations are sufficient.

In the case of oil- or gas-fired boilers, flame heat release distribution is obtained by using an empirical flame equation (Cooper, 1981; Godridge, 1983). Heat generation is assumed to decrease exponentially with the distance from the burner quarl, through the equation:

$$(1) \qquad \frac{Q(s)}{Q_{tot}} = \frac{1}{\lambda . D} \exp(-s/\lambda D)$$

where: $\lambda \simeq 1 \div 2.5$.

This model proved to be very efficient in simulating standard high intensity burners (De Michele, 1986). It is no longer suitable for the new generation of low-NOx burners where heat release is greatly dependent on the burner configuration and there is a wide range of variation. A more fundamental approach is then necessary and work is in progress in this field.

In the case of pulverized coal-fired boilers, a simplified combustion model was adopted, based on papers by Smith (1982), Combustion Engineering (1984), Richter (1985), Fiveland (1987), and Wall (1985, 1986). Coal combustion was assumed to take place in three stages: devolatilization, gas-phase combustion of the volatiles and gaseous intermediates, and char combustion. For the coals normally fired the residence time required for devolatilization is very short, less than 100 ms (Smoot, 1985) and particle temperature is very close to the temperature of the surrounding gas. Devolatilization is assumed to be completed in the first downstream burner zone. Flame equation (1) is used for the volatile fraction of the coal particle.

The combustion velocity of a particle with mass m_i and diameter d_i is written as follows:

$$(2) \qquad \frac{dm_i}{dt} = - \frac{k_d . k_c}{k_d + k_c} . P_{O2} . \pi d^2_i$$

with

$$(3) \qquad k_d = 5.1.10^{-7}.T^{0.5}/d_i$$

$$(4) \qquad k_c = A \exp(-E_a / RT)$$

where T is the gas/particle temperature. Constants $A \simeq (200 \div 700 \ kg/m^2.s.atm)$ and $Ea \simeq (20 \div 35 \ kcal/gmole)$ depend on the coal being examined.

The fuel flow is marked by releasing a number of markers (Lagrangian particles) at the individual burners. Each marker is associated to a fraction of fuel combustion enthalpy. The trajectories and thermal histories of the markers are established by using the flow-field already generated. The use of Lagrangian particles seems a reasonable compromise between computational efficiency and accuracy (Fiveland, 1988; Richter, 1985).

By utilizing the empirical flame equation (oil/gas) or the combustion model (coal), it is possible to calculate the volumetric distribution of the combustion heat released by the markers.

3. THE LOWER FURNACE

In this region, calculation of the heat transfer is based on the zone method of analysis (Hottel, 1967). The furnace chamber is divided into N_V gas volumes and N_S surface zones which are assumed to have uniform properties. The volume elements are made up of parallelepipeds and semiparallelepipeds with constant dimensions along the three axes, as shown in Fig. 2. The surface elements are rectangles and triangles.

The energy balance equation for each volume zone is written in terms of direct-exchange areas (Tozzi, 1974). The resulting equations are:

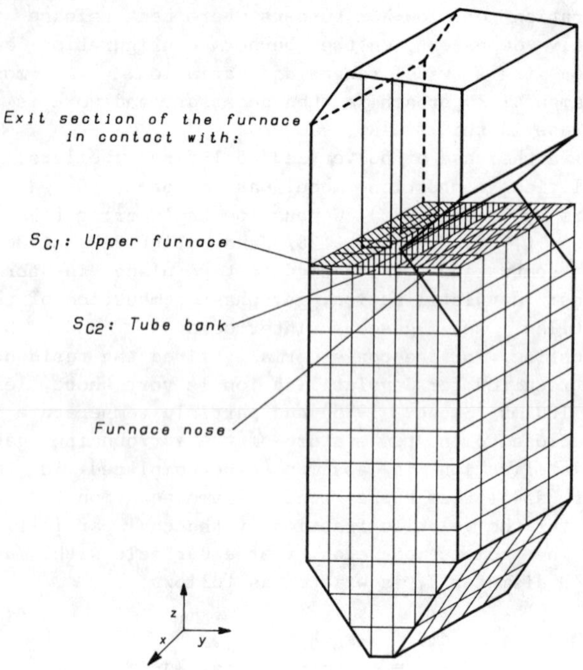

Fig.2 - Discretization utilized for the heat transfer calculation in the furnace.

$$\sum_{r=1}^{N_G}\left[\sum_{j=1}^{N_V} g_i g_j^r \cdot a_r(T_j, f_{vj}) \cdot E_{g,j} + \sum_{j=1}^{N_S} g_i s_j^r \cdot \bar{a}_{r,j} \cdot W_j^r + - 4 K_r \cdot V_i \cdot a_r(T_i, f_{vi}) \cdot E_{g,i}\right] = Q_{c,i} + Q_{e,i} \quad (5)$$

The group $a_{r,j}\, W_j^r$ is calculated directly by explicitly following the reflections up to the second order:

$$\begin{aligned}\bar{a}_{r,j} \cdot W_j^r = {} & a_r(T_s, f_v^*) \cdot \varepsilon_j \cdot E_{s,j} + \frac{\rho_j}{A_j} \sum_{i=1}^{N_V} g_i s_j^r \cdot a_r(T_i, f_{vi}) \cdot E_{g,i} + \\ & + \frac{\rho_j}{A_j} \sum_{i=1}^{N''_S} s_i s_j^r \cdot a_r(T_{sc}, f_v^*) \cdot W_j^r + \sum_{i=1}^{N'_S} s_i s_j^r \cdot \varepsilon_i \cdot \\ & \cdot a_r(T_s, f_v^*) \cdot E_{s,i} + \frac{\rho_j}{A_j} \sum_{i=1}^{N'_S} \frac{\rho_i}{A_i} s_i s_j^r \sum_{k=1}^{N_V} g_k s_i \cdot \\ & \cdot a_r(T_k, f_v^*) \cdot E_{g,k} + \frac{\rho_j}{A_j} \sum_{i=1}^{N'_S} s_i s_j^r \cdot \frac{\rho_i}{A_i} s_k s_i^r \cdot \bar{a}_{r,k}\, W_k^r\end{aligned} \quad (6)$$

In writing the above equations, indirect use has been made of the assumption that the absorptivity of a volume zone equals its emissivity calculated at the temperature of the radiation source. Reflectivity may be a function both of the surface element being considered and of the surface temperature. The downward radiation from the upper region (furnace and convective section) is assessed using estimated values for gas and wall temperatures in these zones.

Once the solution for the whole boiler has been obtained, the calculation can be repeated, with the new temperatures. Usually, since these quantities are small fractions of the energy balance, the iterative calculation can be avoided.

The direct exchange areas $g_i g_j^r$, for the grey gases being considered, are calculated (Sala, 1984) by using the method of the projections of the areas (Becker, 1977) in the case of the contiguous elements, a Montecarlo-type method (Haselgrove, 1961) for the nearby elements, and asymptotic formulas in the case of elements that are far apart.

Due to the kind of geometry which is employed and to the approximation for the representation of the optical properties, it is enough to evaluate and memorize a small number of exchange areas in order to construct a complete system independent of rototranslation operations (De Michele, 1985). In the calculation cycle, the exchange areas are obtained from these basic exchange-areas by means of simple operations on the indexes.

All the weighting factors are evaluated at the temperatures and soot volume fractions of the emitters except the last one for which an average is assumed. When considering emission from the surfaces, a mean soot concentration in the furnace f_v^* is used. In the case of low reflectivity (which is the case normally encountered in power-station boilers where $\rho < 0.4$), this last term makes a negligible contribution to the heat balance equation.

The convective contribution to the energy balance equation is computed from the given velocity field, using a donor-cell scheme. The convective heat-transfer with the walls is calculated by means of a local heat transfer coefficient (Mobsby, 1986). The local surface temperature T_{Si} is related to the heat flux ϕ_i, to the temperature of the coolant and to a preassigned local surface-fluid exchange coefficient h_{SFi}, where:

$$\phi_i = h_{SFi}\,(T_{Si} - T_F) \quad (7)$$

The utilization of this local coefficient h_{SFi} makes it possible to take the differentiated foulings of the tube walls into account.

With the above positions, equations (5) become a system of nonlinear equations in the temperatures of the gas zones. These equations are solved by using a Newton-Raphson method stopped at the second iteration, giving the gas temperatures and the heat fluxes to the furnace wall. The radiative heat flux to the exit section S_c is used as input data for the calculation of the upper furnace and of the convective section.

The use of equation (6), that avoids calculating the total exchange areas together with the great reduction in the number of direct exchange areas to be computed and stored due to symmetry considerations, makes this approach different from the standard zone method and leads to a noticeble increase in computational efficiency. In this way the furnace can be represented by a high number of volume and surface elements with a limited use of memory and computer time. For a parallelepiped furnace subdivided in 10 * 10 * 20 volume elements and using three grey gases, only 10^5 basic exchange areas are computed, instead of more than 10^7 required by the original zone method. Moreover, computation of the total exchange areas is avoided.

4. THE UPPER FURNACE

The upper furnace is simulated in the same way as the lower furnace, with the fluxes from the exit of the lower furnace and from the side surface of the first bank as boundary conditions.

At the end of the program, the flux transmitted towards the lower furnace is again calculated, thereby making it possible to assess the entity of the error of the initial approximate estimation, according to what has already been stated in paragraph 3.

As already mentioned, if there is a considerable difference between the two estimations, it may be necessary to repeat the solution procedure, calculating the lower furnace again with the new value of the imposed flux.

The utilization of an iterative cycle may be necessary even when downward gas recirculations are present across the surface, carrying combustion products from the upper to the lower furnace.

5. THE CONVECTIVE-RADIATIVE SECTION

The stationary energy equation for the flue gases is written disregarding the diffusive transport terms:

$$\frac{\partial(\rho U H)}{\partial X} + \frac{\partial(\rho V H)}{\partial Y} + \frac{\partial(\rho W H)}{\partial Z} = S \tag{8}$$

where S is a source term including the energy exchanges by convection and radiation with the walls and the tube banks.

The radiative gas-gas exchange is disregarded, except for the gas contained in the first bank to which part of the radiation coming from the lower and from the upper furnaces is attributed.

The code calculates the direct exchange areas between the surfaces facing the upper and lower furnaces, the surfaces of the exchanger sections and the walls of the boiler (Fig. 3). The incident radiation fraction corresponding to these exchange areas is attributed to the receiving surfaces, where the residual part is uniformly distributed as a source term of the gas contained in the first exchanger.

Equation (8) is discretized by using a donor-cell type scheme, thus generating the typical eptadiagonal matrix (Patankar, 1980):

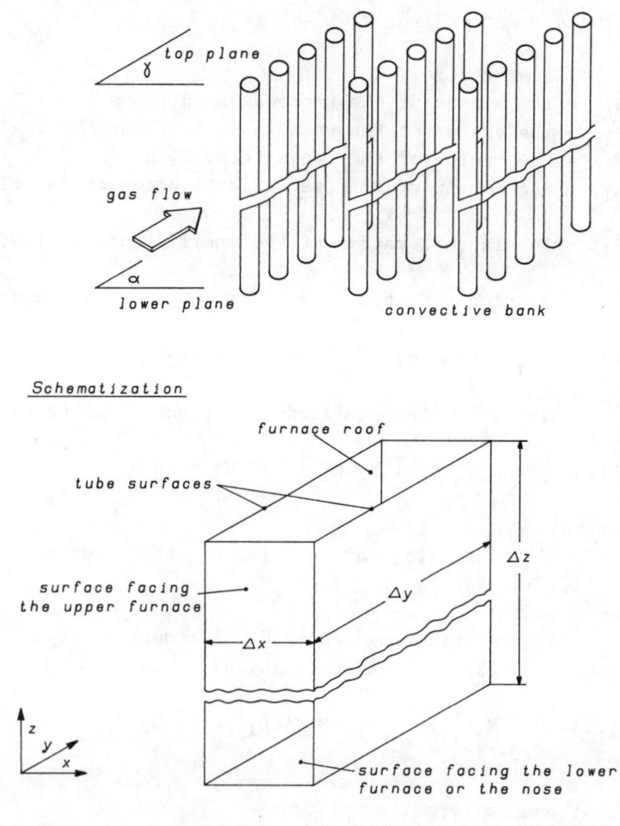

Fig.3 - Convective bank schematization for the calculation of the radiative contribution to heat transfer

$$\begin{aligned} &A_W^{(\alpha)}.H(I-1,J,K) + A_E^{(\alpha)}.H(I+1,J,K) + \\ &A_N^{(\alpha)}\ H(I,J+1,K) + A_S^{(\alpha)}\ H(I,J-1,K) + \\ &A_U^{(\alpha)}\ H(I,J,K+1) + A_D^{(\alpha)}\ H(I,J,K-1) + \\ &A_P^{(\alpha)}\ H(I,J,K) = S(I,J,K) \end{aligned} \tag{9}$$

$$I = 1, NDX\ ;\ J = 1, NDY\ ;\ K = 1, NDZ\ ;\ \alpha = (I,J,K)$$

where $A_{W,E,N,S,U,D}$ represent the input mass fluxes in the generic cell (I,J,K), $A_P = -(A_W+A_E+A_N+A_S+A_U+A_D)$ the output mass flux and NDX, NDY and NDZ are the number of cells according to the three axes.

The first element of equation (9) does not contain unknown terms which are outside the integration volume. Indeed, on the boundaries, either the input mass fluxes are zero and therefore the corresponding A coefficient is also zero, or the enthalpy of the input gas is known. The coefficients of equation (9) are functions of the grid and of the imposed flow-field alone; they are calculated at the beginning of the program and remain constant during the construction of the solution.

The system of equation (9) is solved by means of a method consisting in the use of direct inversion of the matrix that is possible due to the structure of the eptadiagonal matrices.

For plane I = 1, equations (9) can be written in the following form:

$$M_1.H_1 + D_1^+.H_2 = F_1 \qquad (10)$$

Given that:

H_1 = the vector of the unknowns H in plane I = 1
H_2 = the vector of the unknowns H in plane I = 2
F_1 = the vector of the known terms S in the plane I=1
M_1 = the matrix of the coefficients of equation (10) in plane I = 1
D_1^+ =the diagonal matrix of the coefficients $A_E(1,J,K)$

Equation (10) has a formal solution as follows:

$$H_1 = M_1^{-1}.F_1 - M_1^{-1}.D_1^+.H_2 \qquad (11)$$

Similarly, the equations for plane I = 2 become:

$$D_2^-.H_1 + M_2.H_2 + D_2^+.H_3 = F_2 \qquad (12)$$

where:
D_2^- is the diagonal matrix of the coefficients $A_W(2,J,K)$.

By substituting term H_1 of equation (11) in equation (12), we get an equation:

$$D_2^-.M_1^{-1}.F_1 - D_2^-.M_1^{-1}.D_1^+.H_2 + M_2.H_2 + D_2^+.H_3 = F_2 \qquad (13)$$

that can be formally written:

$$\tilde{M}_2.H_2 = \tilde{F}_2 - D_2^+.H_3 \qquad (14)$$

given that:

$$\tilde{M}_2 = M_2 - D_2^-.M_1^{-1}.D_1^+$$
$$\tilde{F}_2 = F_2 - D_2^-.M_1^{-1}.F_1$$

Equation (14) can be formally solved with respect to H_2:

$$H_2 = \tilde{M}_2^{-1}.\tilde{F}_2 - \tilde{M}_2^{-1}.D_2^+.H_3 = \tilde{M}_2^{-1}\ (\tilde{F}_2-D_2^+.H_3) \qquad (15)$$

The process can be repeated as far as plane I = NDX. We therefore have:

$$H_{NDX} = \tilde{M}_{NDX}^{-1}.\tilde{F}_{NDX} \qquad (16)$$

Equation (16) makes it possible to obtain the values of H(NDX,J,K) corresponding to the last plane. The values of the preceding planes are obtained through backward substitution. The values of H being known, the gas temperatures are easily obtained by inverting the relation $H = H(T_G)$.

The procedure indicated is equivalent to solving a series of bidimensional NDX problems, with constant inverse matrices. In order to reduce computer time and to save memory space, the best thing is to assume the direction with the greatest number of cells as the propagation direction of the solution (corresponding to index I in the test).

It has been assumed that three kinds of elements may be present in the region being considered: 1) exchangers, 2) cavities and 3) screen tubes.

The cavities and the screens may be considered to be special cases of the exchangers and are not explicitly taken into consideration in the present paper.

The exchangers are described by specifying the cells crossed and the fraction of exchange-surface connected with the cells in the YZ plane parallel to the banks. This description is carried out by following the tube bundle starting from the point of entrance of the cooling fluid. The intersections between the tube bundles and the grid cells discretize each exchanger in pieces (Fig. 4). Moreover, the number of sections across the gas duct, the diameter of the tubes and the average dimensions along the Y and Z axes are given for each exchanger. The inlet fluid flow and temperature can be imposed or obtained by means of a subroutine which simulates the connections between the exchangers and the logic of the control system.

If the number of sections is greater than NDX, the exchanger is replaced by an equivalent exchanger with NDX sections. The tube bundles are discretized by means of three indices: IS (exchanger), IB (section of the exchanger IS) and IT (piece of section IB). The

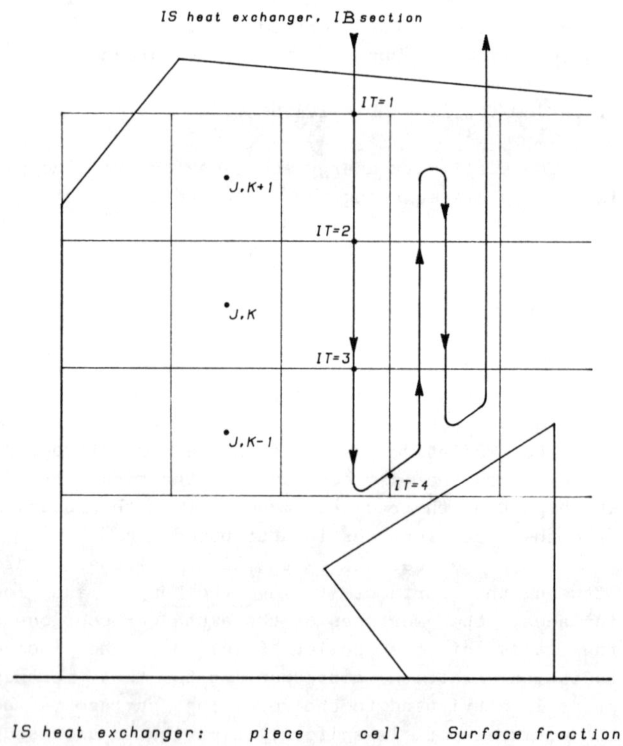

IS heat exchanger:	piece	cell	Surface fraction
IB section	1	J+1, K+1	α_1
	2	J+1, K	α_2
	3	J+1, K-1	α_3

Fig.4 - Discretization of a convective bank and coupling between gas flow and steam flow in the tubes.

indices IS, IB and IT being known, the program makes it possible to identify cell I,J,K which contains segment IT of section IB of exchanger IS.

The direct exchange areas between the volume of gas contained between two tube sections and the corresponding roof of the boiler and the lower and side surfaces are calculated for each exchanger. For this purpose the volume of gas is approximated with a right parallelepiped (Fig. 3). Subsequently, the program calculates the average temperature of the gas contained in the exchanger and the average temperatures of the surfaces of the section and of the walls. These data make it possible to obtain two average radiative exchange coefficients relevant to the side walls and to the roof/floor, by calculating the radiative energy transmitted towards the exchange surfaces and by dividing it by the difference between the average temperature of the gas and that of the surfaces. These coefficients are associated to all the exchange surfaces affected by the exchange in question. The convective exchange coefficients are obtained by the correlations to be found in the literature (Schlunder, 1983), using the temperature and local velocity of the gas. These coefficients, therefore, prove to be a function of the exchanger IS, of the section IB which identifies the coordinate I and of the indices J,K which identify the location in the plane YZ of the individual segment IT of the bank. Adding the two coefficients together, we obtain an overall tube-gas surface exchange coefficient indicated by the symbol $K_{GT}^{(\beta)}$ with:
$\beta = (IS, IB, J(IS,IT), K(IS,IT))$.

Let us consider a segment of bank IB of exchanger IS (Fig. 5). Let us use $T_G(I,J,K)$ to indicate the temperature of the gas in the cell in contact with the segment being considered, $T_F(IT)$ to indicate the temperature of the cooling medium at input, P its flow rate and $T_S(IT + 1/2)$ the average surface temperature of the tube in the segment being considered.

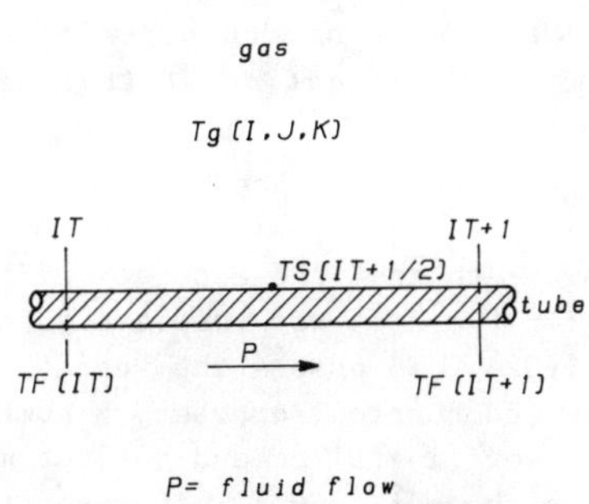

Fig.5 - Schematization of the heat transfer in a piece of a convective bank

The quantity of heat transferred, in stationary conditions, from the contact gas is written:

$$(17) \quad q_{GC} = S(IT) \cdot K_{GT}^{(\beta)} \cdot [T_G(I,J,K) - T_S(IT+1/2)]$$

where S(IT) is the exchange surface.

Equation (17) is applied if the segment being considered exchanges with the cell (I,J,K) alone. Should the segment involve several gas cells, the exchange surface (IT) is distributed among all the cells in contact. Using S(IC) to indicate the partial surfaces and T_G (IC) the temperatures of the cells of the gas involved, we can rewrite (17) as follows:

$$(18) \quad q_{GC} = \sum_{IC} S(IC) \cdot K_{GT}^{(\beta)} \cdot \left[T_G(IC) - T_S(IT+1/2)\right] = \sum_{IC} S(IC) \cdot K_{GT}^{(\beta)} \cdot T_G(IC) - S(IT) \cdot K_{GT}^{(\beta)} \cdot T_S(IT+1/2)$$

where the summation is extended to all the cells in contact and S(IC) = S(IT). For the sake of simplicity, (17) alone will be considered below. The term q_{GC}, with its sign changed, constitutes the source term for the balance equation (9) of the gas. The quantity of heat q_{CF} transferred to the cooler is approximated by means of the equation:

$$(19) \quad q_{CF} = S(IT) \cdot K_{TF}^{(\beta)} \cdot \left[T_S(IT+1/2) - \frac{T_F(IT+1) + T_F(IT)}{2}\right]$$

where $K_{TF}^{(\beta)}$ is the overall heat exchange coefficient between the surface of the tubes and the cooling medium. The values of K_{TF} are assigned at input according to the position, that is to the cells of the gas. We, therefore, have $(\beta) = (I,J,K) = (I(IS,IB), J(IS,IT), K(IS,IT))$. In stationary conditions, we obviously get $q_{GC} = q_{CF}$.

In the program we prefer to substitute the iterative cycle with a temporal cycle, writing a pseudo-transitory equation for the surface temperature:

$$(20) \quad T_S^{n+1}(IT+1/2) = T_S^{n}(IT+1/2) + \frac{\Delta t}{S(IT)} \cdot (q_{GC} - q_{CF})$$

where t is the temporal step and n a number of temporal integration steps. Substituting (17) and (19), equation (20) turns out to be:

$$(21) \quad T_S^{n+1}(IT+1/2) = T_S^{n}(IT+1/2) + \Delta t \left\{ K_{GT}^{(\beta)} \cdot \left[T_G(I,J,K) - T_S^{n+1}(IT+1/2)\right] - K_{TF}^{(\gamma)} \cdot \left[T_S^{n+1}(IT+1/2) - \frac{T_F(IT+1) + T_F(IT)}{2}\right] \right\}$$

The system is closed by the thermal balance equation of the cooler, written as follows:

$$P\cdot\left[\overline{cp}\cdot T_F(IT+1)-\overline{cp}\cdot T_F(IT)\right] = K_{TF}^{(\gamma)}\cdot S(IT)\cdot\left[T_S^{n+1}(IT+1/2)+ - \frac{T_F(IT+1)+T_F(IT)}{2}\right] \quad (22)$$

where
cp is a mean value of the specific heat between 0 and T_F.

Lastly, from (21) and (22), we get:

$$\left\{\left[1+\Delta t\cdot(K_{GT}^{(\beta)}+K_{TF}^{(\gamma)})\right]\cdot\left[1/2+\frac{\overline{cp}\cdot P}{K_{TF}^{(\gamma)}\cdot S(IT)}\right]-\frac{\Delta t}{2}\cdot K_{TF}^{(\gamma)}\right\}\cdot T_F(IT+1) = \left\{-\left[1+\Delta t(K_{GT}^{(\beta)}+K_{TF}^{(\gamma)})\right]\cdot\left[1/2-\frac{\overline{cp}\cdot P}{K_{TF}^{(\gamma)}\cdot S(IT)}\right]+\frac{\Delta t}{2}\cdot K_{TF}^{(\gamma)}\right\}\cdot T_F(IT)+T_S^{n}(IT+1/2)+\Delta t\cdot K_{GT}^{(\beta)}\cdot T_G(I,J,K) \quad (23)$$

Equation (23) can be solved by starting from the feeding point of the bank (IT = 1) where the temperature of the cooling fluid is known.

The walls and the roof of the boiler are dealt with in the same way as the exchangers, with a constant temperature of the cooler being imposed. The convective and radiative exchange coefficients are calculated within each time-step. Subsequently, you update the temperatures of the cooler for all the banks of all the exchangers (eq. 23), the surface temperatures (eq. 21), and the source terms of the balance equation of the gas (eq. 17). The time-step ends with the calculation of the enthalpy and the temperature of the gas and of the energy balance of the gas and of the cooler. The temporal cycle finishes when the overall energy balance is satisfied with the required accuracy.

The use of a direct matrix inversion algorithm allows the exact solution at each time step of the fundamental energy balance equation (8). Its use is favourable if compared with iterative methods (at least if a good degree of accuracy is required) because the inverse matrices $\tilde{M}^{-1}$ are evaluated only once at the beginning of the temporal cycle, whereas at each time step only a limited amount of computation is required (eq. 15 and 16). Moreover, since $\tilde{M}^{-1}$ is function only of the discretization and flow field considered, the same matrices can be used with different steam/gas input conditions and with different fouling coefficients for the convective banks.

The proposed algorithm takes advantage of the peculiar band structure of the matrix and is extremely powerful when the number of cells in one direction greatly exceeds the number in the other two. It can also be easily implemented on a vectorial computer.

6. EXAMPLES OF APPLICATION

The series of codes developed has been tested by comparing their predictions with measurements taken in full-scale power-stations. In the following, a few examples of application will be given.

Gas radiation model

In Fig. 6 the spectral variation of the energy gas is given as a function of the local composition corresponding to three different fuels. The integral data obtained with this code for the range of temperatures, mean beam lengths and composition in question are interpolated using the grey gas formulation.

Fig. 7 shows the weighting coefficients in the case of oil firing with the corresponding values of the total emissivities plotted. The importance of this program is that it makes it possible to obtain total emissivity values for all kind of medium compositions. It also makes it possible to interpret spectral narrow-angle radiometric measurements in boilers in order to obtain mean temperature and soot volumetric fractions.

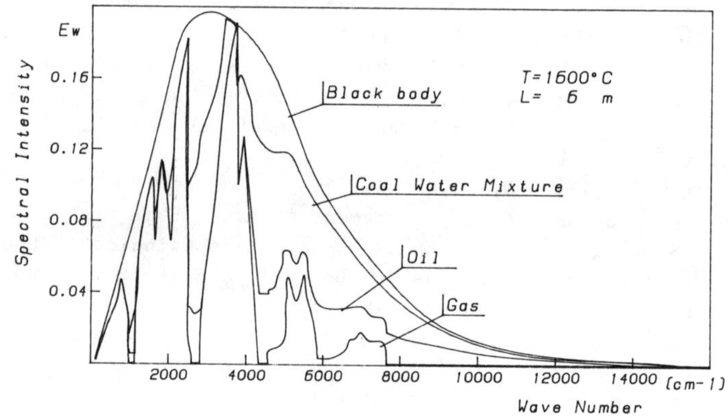

Fig.6 - Spectral variation of gas emission as a function of the fuel fired.

Heat Transfer in the Furnace.

In order to test the accuracy of this model, a small boiler (35 MWe) was completely instrumented (De Michele, 1987). To assess the heat flux distribution inside the combustion chamber, a number of fixed fluxmeters were installed and various measuring ports were made on the boiler wall, in order to take optical and physical measurements. Automatic guides for water-cooled suction pyrometers were installed in front of some of these. A number of tests were conducted with different loads and boiler operating conditions.

In Fig. 8 the predicted heat fluxes are compared with the ones measured in the case of oil-firing and for the test conditions shown in table 1. Two mean values of soot concentration were utilized for the calculation: $f_V = 3\ 10^{-8}$ in the burner zone and $f_V = 0.8 \cdot 10^{-8}$ in the rest of the combustion chamber.

Weighting coefficients for grey-gas approximation.

$$a_n\,(T, f_v) = a_1^n + a_2^n\,T + a_3^n\,T^2 + a_4^n f_v + a_5^n f_v T + a_6^n f_v^2$$

Range of validity : $T = 750\text{-}2250$ [°K]
$L = 0.5\text{-}12$ [m]
$f_v = 0.\text{-}1.10^{-7}$ [m^3/m^3]
$P_{CO_2}/P_{H_2O} \sim 1$

Number of gray gases = 3

N_G	K_n [m^{-1}]	a_1^n	a_2^n	a_3^n
1	1.95296E-02	4.99425E-01	8.02530E-05	2.06552E-08
2	2.69435E-01	1.41653E-01	5.22805E-05	-2.35870E-08
3	2.69685E+00	3.58922E-01	-1.32534E-04	2.93180E-09

N_G	a_4^n	a_5^n	a_6^n
1	-1.00387E-03	-3.71879E-07	1.07487E-06
2	1.25679E-03	3.05830E-07	-1.33073E-06
3	-2.52920E-04	6.60490E-08	2.55860E-07

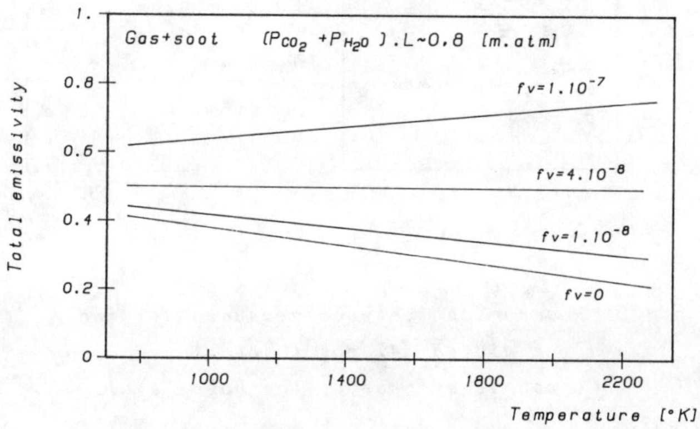

Fig.7 - Gas+soot total emissivity evaluated with the grey-gas model in the case of oil firing and grey gas weighting coefficients.

Table 1 - Furnace operating conditions for the test examined

Oil:		
Higher Heating Value	42354	KJ/Kg
Flow rate	6880	Kg/hr
Combustion Air:		
Flow rate	112600	Kg/hr
Temperature	285	°C
Excess air	18	%
Burners:		
Tilt	+30	°
Number	6.	

This second value was measured with a spectral radiometer; the first one was chosen on the basis of previous experiments and measurements taken in other boilers.

The shape of the predicted heat flux distribution corresponds to the one measured. The location and values of the maximum heat-flux zones are correctly predicted. The temperatures in four traverses in the furnace exit plane are shown in Fig. 9. The maximum discrepancy is of the order of 50 ÷ 70°C, and occurs only in a few points. Most of the values are in close agreement ($\Delta \leq 30$°C) and the shape of the temperature distribution is well reproduced. As for global predictions, the steam production is calculated with an error of less than 5%.

The results reported were obtained with a coarse discretization (6 x 5 x 18) with $\Delta x = 0.8$m; $\Delta y = 0.86$m; $\Delta z = 0.9$m. A greater number of cells is used for the simulation of larger furnaces, in order to keep cell mean dimension between 0.5 and 1 m.

Heat Transfer in the convective section.

This is the newest part of the whole program and tests are still in progress. An example of the calculation is shown in Fig. 10a and b. The boiler considered and the operating conditions are the same as in Figs. 8 and 9.

The schematization adopted and the comparison between the preliminary global predictions and measurements on the steam side are shown in Fig 10a. Fig. 10-b contains the gas temperature distribution in the central YZ plane of the boiler. There are no detailed gas measurements as yet but at least qualitatively the gas temperatures measured with a suction pyrometer in the descending pass of the convective section show the same shape, with higher temperatures near the rear wall of the furnace.

Work is in progress to obtain more detailed gas temperature measurements.

7. CONCLUSIONS

A numerical model has been developed for predicting the behaviour of the whole steam generator (furnace and convective section).

The model is 3-D both on the gas and steam sides.

The structure of the code is modular and the coupling between the different programs is obtained by means of an iterative process.

The steam generator is normally divided into 4 sections:

- lower furnace
- upper furnace
- 3-D convective-radiative exchange section
- 1-D convective exchange section

The first three sections are fully 3-D, the last one is 1-D.

The detailed structure of the convective banks can be simulated.

Original techniques have been used for the numerical solution of the heat transfer equations, both in the radiative and the convective sections.

The various submodels of the code have been thoroughly tested against experimental measurement, giving promising results.

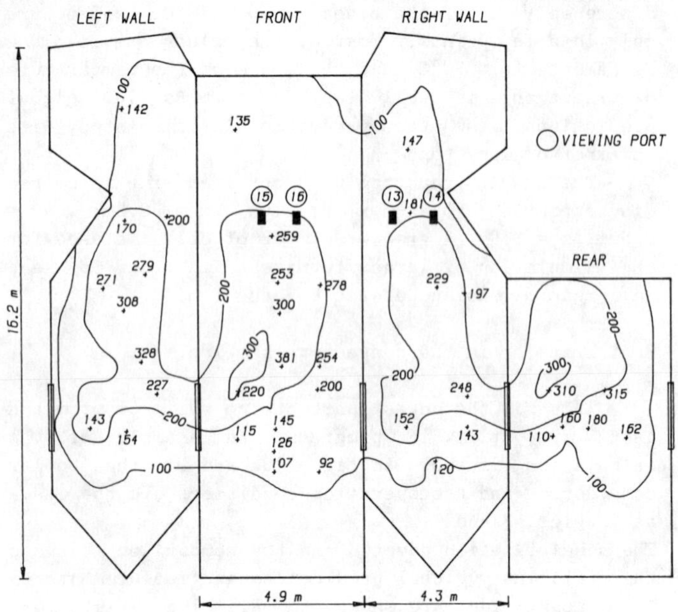

Fig.8 - Comparison between measured (+) and predicted heat fluxes (KW/m²)

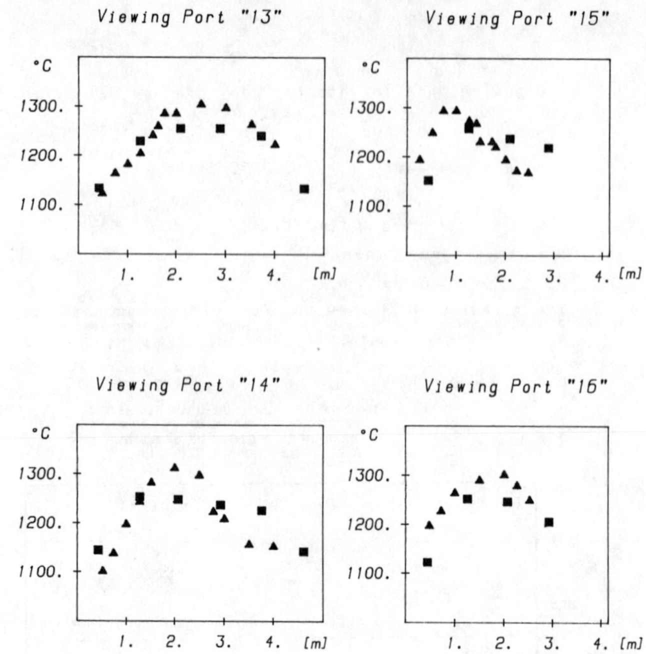

Fig.9 - Comparison between measured (▲) and predicted (■) gas temperatures along four traverses at the nose level.

Santa Gilla Boiler

Load 25 MWe

Heat PickUp

SH_1	6,3 MWt
SH_2	10,5 MWt
Rear Cavity	1,1 MWt

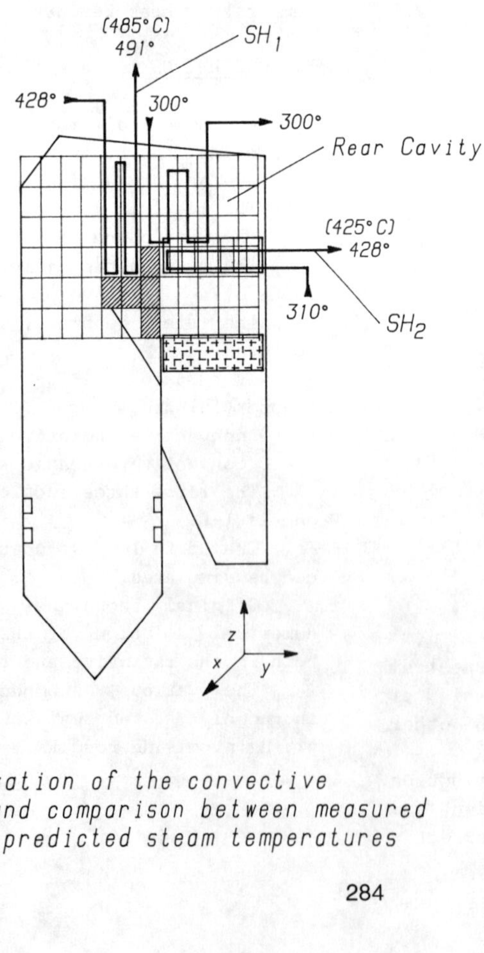

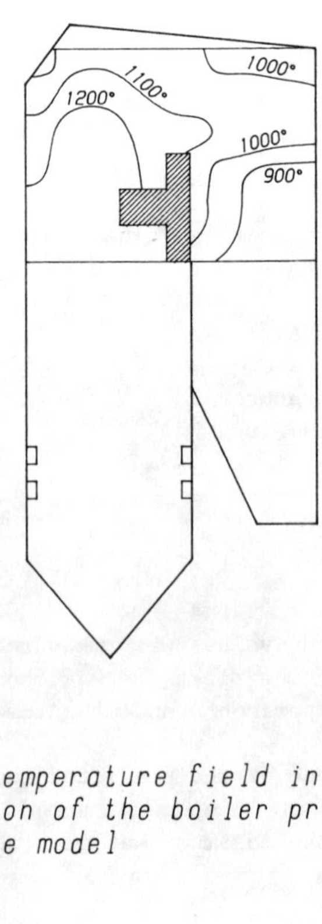

Fig.10 a) Schematization of the convective section and comparison between measured (..) and predicted steam temperatures

b) gas temperature field in the central section of the boiler predicted by the model

REFERENCES

Bauver, W.P., Friedman, C.H., and McGowan J.G., 1987, "Modeling of Flow Maldistribution Effects in Large Steam Generator Heat Exchangers", 24th National Heat Transfer Conference, Pittsburgh, USA.

Becker, H., 1977, "A Mathematical Solution for Gas to Surface Radiative Exchange Area for a Rectangular Parallelepiped Enclosure Containing a Grey-Medium", J. Heat Transfer, 99, 203.

Castellano, L., 1988, "Spectral and Integral Optical Properties Calculation for Homogeneous and Isothermal Combustion Products of Arbitrary Composition", Rept. MATEC MTC/RD1/88.

Combustion Engineering, 1984, "Physics and Chemistry of Carbon Burnup", Eng. Dev. & Code Coordination.

Cooper, S., 1981, "Simple Physical Models of the Spray Flames with Particular Reference to Flame Scaling", I.F.R.F., Doc, nr. F21/Ca/53.

Cooper, S., and Gibb, J., 1984,"High Temperature Heat Transfer: Furnace Performance Assessment", 1s U.K. National Heat Transfer Conf., Leeds.

De Michele, G., Ligasacchi, S., Pasini, S., Tozzi, A.,and Trebbi, G., 1986, "Characterization of a Large HFO-CWS Dual Fuel Burner", 8th Members Conference, IFRF, Noordwijkerhout, Holland.

De Michele, G., Ligasacchi, S., and Pasini, S., 1987, "CWF Firing on an Oil-Designed Utility Boiler", 12th Internat. Conf. on Slurry Technology, New Orleans, USA.

De Michele, G., Pasini, S., and Tozzi, A., 1985, "Aerodynamics and Heat Transfer Modelling in Tangentially-Fired-Boilers", Joint Meeting of the Aerodynamics, Flame Chemistry and Heat Transfer Penels, I.F.R.F., Stockolm, Sweden.

De Michele, G., Pasini, S., and Trebbi, G., 1988, "The RIDOX Project - A Summary Review", Chemistry of Flames/Heat Transfer and Gaseous Fuels Panels, IFRF, Roskilde, Denmark.

Farag, I.H., 1976, "Radiative Heat Transmission from Non-Luminous Gases, Computation Study of the Emissivities of Water Vapor and Carbon Doixide", Sc., D. Thesis, MIT, Cambridge, Mass.

Fiveland, W.A., and Wessel, R.A., 1988, "Numerical Model for Predicting Performance of Three-Dimensional Pulverized-Fuel Fired Furnaces",J. of Engineering for Gas Turbines and Power,110, pp. 117-126.

Fiveland, W.A., Wessel. R.A., and Eskinazi, D. 1987, "Pollutant Model for Predicting Formation and Reduction of Nitric Oxides in Three-Dimensional, Pulverized-Fired Furnaces", 24th National Heat Transfer Conference, Pittsburgh, 1987.

Godridge, A.M., 1983, "Heat Transfer Properties of Large Oil Flames", 7th Members Conference, I.F.R.F., Leenwenhourst, Holland.

Gorner, K., and Zinser, W.,1986,"Prediction of Three-Dimensional Flows in Utility Boiler Furnaces and Comparison with Experiments", ASME 107th Winter Annual Meeting,Anhaeim, USA.

Hartmann, J.M., Levi di Leon, R., and Taine, J., 1984, "Line-by-line and Narrow Band Statistical Model Calculations for H_2O", J.Q.R.S.T., 32(2), pp. 119-122.

Haselgrove, C., 1961, "A Method for Numerical Integration", Math. Comp. 1, 323.

Hottel, H.C., 1985, "Radiation from Carbon Dioxide, Water Vapor and Soot", MIT, Chem. Eng. Dept.

Hottel, H.C., and Sarofim, A.F., 1967, "Radiative Transfer", Mc Graw Hill, New York.

Ludwig, C.B., Malkmus, W., Readon, J.E., and Thomson, J.A.L., 1973, "Infrared Radiation from Combustion Gases", NASA SP-3080, Sci. Tech. Ing. Office, Washington. D.C.

Mobsby, J.A., 1986, "Thermal Modelling of Fossil-Fired Boilers", I. Chem. E. Symposium Series N. 86, pp. 1013-1024.

Pasini, S., Benelli, G., De Michele, G., Tanzini, G., and Tozzi, A., 1988, "Flow Field In Tangentially Fired in Boilers. Comparison Between Physical and Mathematical Models and Furnace Measurements", Poster presented at the 22nd Int. Symposium on Combustion, Seattle, USA.

Patankar, S.V., 1980, "Numerical Heat Transfer and Fluid Flow", Hemisphere Publishing Corporation.

Richter, W., 1985, "Scale-up and Advanced Performance Analysis of Boiler Combustion Chambers", ASME Winter Annual Meeting, Miami Beach, USA.

Robinson, G.F., 1985, "A Three-Dimensional Analytical Model of a Large Tangentially-Fired Furnace", J. of the Inst. of Energy, pp. 116-150.

Sakai, M., Tokuda, K., Ide, Y., Nakashima, F., Asayama, H., and Aiki, H., 1984, "Development of Three-Dimensional Numerical Analysis Method of Boiler Furnace Characteristics - Part 2: Combustion and Heat Transfer Analysis", Technical Review, Mitsubishi Heavy Industries, pp. 24-34.

Sala, R., 1984, "Evaluation of the Direct Exchange Areas Utilized in the Zone Method of Analysis", Report MATC MTC/FL42/84.

Schlunder, E.U., Editor, 1983, "Heat Exchanger Design Handbook", Hemisphere Publishing Company, New York.

Smith, P.J., Smoot, L.D. et al., 1982, "Combustion Processes in a Pulverized-Coal Combustor", EPRI CS-2490-CC M.

Smoot, L.D., and Smith, P.J., 1985, Coal Combustion and Gasification, Plenum Press, New York.

Souffiani, A., Hartmann, J.M., Taine, J., 1985, "Validity of Band-Model Calculations for CO_2 and H_2O Applied to Radiative Properties and Conductive-Radiative Transfer", JQST, 33(3), pp. 243-257.

Tozzi, A., and Trebbi, G., 1974, "Development of a Zone Method of Analysis for Heat Transfer Calculation in Large Boilers", Report ENEL-DSR/G7-4.

Wall, T.F., Phelan, W.J., and Bortz, S., 1985, "Coal Burnout in the IFRF N. 1 Furnace", IFRF doc. n. 20/a/214.

Wall, T.F., Phelan, W.J., and Bortz, S., 1986, "The Prediction and Scaling of Burn-Out in Swirled Pulverized Coal Flames", IFRF Study Rep. F388/a/3.

1989 National Heat Transfer Conference
HTD-Vol. 106, Heat Transfer Phenomena in Radiation, Combustion and Fires

BAND RADIATION FROM A NONHOMOGENEOUS FLUCTUATING MEDIUM

D. A. Nelson
The Aerospace Corporation
El Segundo, California

ABSTRACT

Radiation from a nonhomogeneous fluctuating medium is shown to be well approximated by using results from a homogeneous model. Even greater simplification, based upon Planck function fluctuations, provides approximate results with an error that seldom exceeds 10%. In each case, only local statistical properties are needed as opposed to complex joint, spatial statistics in the exact representations.

NOMENCLATURE

A	band absorption function
c	speed of light
E	emissive power
f	mixture fraction
g	mixture fraction variance
h	Planck constant
k	Boltzmann constant
ℓ	path length
p	probability density function
P	generic symbol denotes all mean properties
q	heat flux
T	temperature
u	$hc\nu/kT$
κ	absorption coefficient
ν	wave number
ρ	density of absorbing species
τ	optical depth
ψ	flux ratio
ω	bandwidth parameter

<u>Subscripts</u>

b	black body function
i,j	zone index
k	band index
ν	spectral quantity

INTRODUCTION

Until recently, little had been done to clarify the role of fluctuations on the radiation output of high temperature gases. In some respects, this is a reflection of the fact that considerable success had been achieved without such explicit consideration (Hottel and Sarofim, 1967). This success, however, still leaves many fundamental questions unanswered. In more recent applications, one cannot be sure that neglect of fluctuations is justified. In particular, buoyant flows tend to show substantially higher scalar fluctuations than those which are momentum dominated (Papanicolaou and List, 1987). Since radiation can be strongly nonlinear in temperature response, a relatively modest increase in fluctuation intensity may produce a disproportionate increase in radiation output. Analytical studies which have focused on radiating media generated by the combustion process are those of Grosshandler and Joulain (1986) and Nelson (1986, 1989). Although different points of view are evident in the radiation models, the treatments of combustion and fluctuation phenomena are quite similar. In Grosshandler and Joulain, the use of a time-dependent sawtooth wave to describe the fluctuations is the equivalent of the uniform probability density function (pdf) used by Nelson. Both studies employ a conserved scalar to determine all other physical properties. Each explores effects of fluctuation intensity, waveform or pdf shape, and mean optical path length.

Grosshandler and Joulain also study spatial effects such as different profiles of the conserved scalar and variations in the statistical properties of the fluctuations. They also consider the effects of soot and select propane and methane as fuels. Nelson takes an instantaneously homogeneous medium to remove spatial effects and concentrate on other fundamental aspects. Combustion properties are varied by choosing CO, H_2, and CH_4 as fuels. The physical path length also fluctuates, since this would occur in any real flame. Grosshandler and Joulain use a total radiation model, whereas Nelson concentrates on band radiation. On the whole, these studies are complementary and fully consistent in the important conclusions. Because of the simpler radiation model, Nelson (1986, 1989) was able to isolate the Planck function as the primary contributor to radiation fluctuations.

Other analytical studies include Cox (1977), Kabashnikov and Kmit (1979), and Kabashnikov and Myasnikova (1985).

Experimental and analytical studies of a number of turbulent, nonpremixed flames have been reported by Faeth and co-workers [Jeng, Lai, and Faeth (1984); Gore, Jeng, and Faeth (1987a); Gore, Jeng, and Faeth (1987b); Gore and Faeth (1988)]. Besides obtaining basic experimental data, they also seek to determine how well fluctuation/radiation interaction can be predicted with standard turbulence and radiation models. They use a narrow-band radiation model and ensemble averaging over a large number of predicted realizations. The very nature of this work makes it very difficult to achieve broad generalizations. Indeed, some of their findings are unexpected, unexplained, and clearly require further study.

The predictive procedure which requires time or ensemble averaging over nonhomogeneous paths could prove to be very difficult to implement in general situations. This would be especially true if the radiation calculations were coupled with the fluid mechanics or if one desired more than simple line-of-sight results.

Nelson (1986, 1989) suggests that a great simplification should be possible. Although much of that work deals with a homogeneous medium, the calculations with a fluctuating path length actually represent a particular, time-averaged nonhomogeneous case. The following shows that the homogeneous results are quite relevant to the nonhomogeneous case and that the simplification previously suggested has full generality.

ANALYSIS

The path through a nonhomogeneous medium can be approximated by a partition into several segments. The properties within each segment are considered uniform. To obtain the radiation emerging from the medium, the contributions from each segment are summed. If a segment is of length ℓ_i, then its spectral contribution can be written

$$q_{\nu i} = E_{\nu b}(T_i) \left[1 - \exp\left(-\rho_i \kappa_{\nu i} \ell_i\right)\right] \exp\left(-\sum_{j=i+1}^{N} \rho_j \kappa_{\nu j} \ell_j\right) \tag{1}$$

To be useful, this must be integrated over a band spectrum. Depending upon the requirements of a particular problem, one could choose either narrow-band, wide-band, or full-spectrum integration. It is convenient to introduce the absorption functions

$$A_\nu(\tau_\nu) = 1 - \exp(-\tau_\nu)$$

$$\omega A(\tau) = \int_{\Delta\nu} A_\nu(\tau_\nu) d\nu$$

Then Eq. (1) can be written

$$q_{\nu i} = E_{\nu b}(T_i) \left[A_\nu \left(\sum_{j=i}^{N} \rho_j \kappa_{\nu j} \ell_j \right) - A_\nu \left(\sum_{j=i+1}^{N} \rho_j \kappa_{\nu j} \ell_j \right) \right]$$

or

$$q_{\nu i} = E_{\nu b}(T_i) \left[A_\nu \left(\sum_{j=i}^{N} \tau_{\nu j} \right) - A_\nu \left(\sum_{j=i+1}^{N} \tau_{\nu j} \right) \right] \qquad (2)$$

To a good approximation, wide-bands can be considered spectrally narrow so that the Planck function can be considered constant when Eq. (2) is integrated. Then, for a band, Eq. (2) becomes upon integration

$$q_{ki} = \omega_{ki} E_{kb}(T_i) \left[A \left(\sum_{j=i}^{N} \tau_{kj} \right) - A \left(\sum_{j=i+1}^{N} \tau_{kj} \right) \right] \qquad (3)$$

This result is not precisely correct because, in general, for a nonhomogeneous medium, it is not possible to separate the bandwidth parameter ω_i from the absorption function $A(\tau)$. However, Eq. (3) is a convenient approximation which satisfies the exact, optically thin limit and is useful at all optical depths.

Note that Eqs. (2) and (3) are analogs in the sense that one can pass from (2) to (3) by the substitutions

$$q_{\nu i} = \frac{dq_i}{d\nu} \rightarrow \frac{q_{ki}}{\omega_{ki}}$$

$$A(\tau_\nu) \rightarrow A(\tau_k)$$

$$E_{\nu b}(T_i) \rightarrow E_{kb}(T_i)$$

However, although Eq. (2) is fully equivalent to Eq. (1), no similar equivalent relationship exists for Eq. (3). In terms of a transmission function, this is expressed by

$$T_\nu(\tau) = 1 - A_\nu(\tau)$$

but

$$T_k(\tau) \neq 1 - A_k(\tau)$$

This is a consequence of nongray absorption.

From Eq. (3), the total emergent radiation from a nonhomogeneous medium is

$$q_k = \sum_i q_{ki} = \sum_{i=1}^{N} \omega_{ki} E_{kb}(T_i) \left[A \left(\sum_{j=i}^{N} \tau_{kj} \right) - A \left(\sum_{j=i+1}^{N} \tau_{kj} \right) \right]$$

In this relation, one must add the condition that

$$\sum_{j=N+1}^{N} \tau_{kj} = 0$$

A rearrangement yields the form

$$q_k = \sum_{i=1}^{N} \omega_{ki} E_{kb}(T_i) A \left(\sum_{j=i}^{N} \tau_{kj} \right) - \sum_{i=2}^{N} \omega_{k(i-1)} E_{kb}(T_{i-1}) A \left(\sum_{j=i}^{N} \tau_{kj} \right) \qquad (4)$$

Equation (4) gives the instantaneous radiation from a line-of-sight path through a nongray, nonhomogeneous medium. It applies to any band model which can be considered narrow in the spectral sense.

The time or ensemble average of Eq. (4) is formally given by (see Appendix)

$$\overline{q_k} = \sum_{i=1}^{N} \overline{\omega_{ki} E_{kb}(T_i) A \left(\sum_{j=i}^{N} \tau_{kj} \right)} - \sum_{i=2}^{N} \overline{\omega_{k(i-1)} E_{kb}(T_{i-1}) A \left(\sum_{j=i}^{N} \tau_{kj} \right)} \qquad (5)$$

It is convenient to compare this with the result based upon mean properties and given by

$$q_k(\bar{P}) = \sum_{i=1}^{N} \omega_k(\bar{T}_i) E_{kb}(\bar{T}_i) A \left(\sum_{j=i}^{N} \tau_{kj}(\bar{\rho}_{kj}, \bar{T}_j) \right) - \sum_{i=2}^{N} \omega_k(\bar{T}_{i-1}) E_{kb}(\bar{T}_{i-1}) A \left(\sum_{j=i}^{N} \tau_{kj}(\bar{\rho}_{kj}, \bar{T}_j) \right) \qquad (6)$$

Equation (6) represents the approach taken in most problem solutions. Consequently, it is of interest to know how and to what extent it differs from Eq. (5).

Nelson (1986, 1989) has studied the homogeneous medium where

$$\bar{q}_k = \overline{\omega_k E_{kb}(T) A(\tau_k)} \qquad (7)$$

Each term in Eq. (5) is of this form. However, the statistics of a typical term in Eq. (5) differ from that of Eq. (7). Clearly, it would be of great value to be able to use results of the simpler problem described by Eq. (7) to evaluate the much more complex situation given by Eq. (5). The feasibility of this requires careful examination of the terms appearing in Eq. (5). To this end, it will be assumed that fluctuations in each path length segment are statistically independent. This can be arranged by choosing segments longer than a correlation length scale. If segments are shorter than the local scale of mean property variations, then Eq. (5) will accurately describe the emergent radiation.

Considering the terms of the second sum, one now has

$$\overline{\omega_{k(i-1)} E_{kb}(T_{i-1}) A\left(\sum_{j=i}^{N} \tau_{kj}\right)} = \overline{\omega_{k(i-1)} E_{kb}(T_{i-1})} \times$$

$$\times \overline{A\left(\sum_{j=i}^{N} \tau_{kj}\right)}$$

because of independence.

In assessing the relationship of the previous work (Nelson, 1986, 1989) to the above averaging, one notes three important factors. First, the value of the total optical depth was relatively unimportant in computing the ratio $\psi = \bar{q}_k / q_k(\bar{P})$; second, the shape of the pdf was also relatively unimportant; and third, $\overline{\omega_k E_{kb}}\ \overline{A(\tau)}$ was a very good approximation of $\overline{\omega_k E_{kb} A(\tau)}$. Taken together, these strongly suggest that the results from the homogeneous study provide a very useful and accurate approximation for each of the terms in the second sum of Eq. (5). The terms in the first sum are somewhat different, for the first and last one has

$$\overline{\omega_{k1} E_{kb}(T_1)\ A\left(\sum_{j=1}^{N} \tau_{kj}\right)}$$

and

$$\overline{\omega_{kN} E_{kb}(T_N)\ A(\tau_{kN})}$$

The last corresponds exactly to the problem of the homogeneous medium. The first is more like the terms in the second sum which have already been discussed. There is only a weak coupling of $\omega_{k1}E_{kb}(T_1)$ to $A(\tau)$ through τ_1. However, the earlier study has already shown that to a good approximation $\omega_k E_{kb}$ can be considered statistically independent of $A(\tau)$, even when their fluctuations are strongly coupled.

Equation (5) can be written in the alternative form

$$\bar{q}_k = \sum_{i=1}^{N} \omega_k(\bar{T}_i) E_{kb}(\bar{T}_i) A \left(\sum_{j=i}^{N} \tau_{kj}(\bar{\rho}_{kj}, \bar{T}_j)\right) \times$$

$$\psi(u_{ki}, \bar{f}_i, \cdots \bar{f}_N, g_i, \cdots g_N)$$

$$- \sum_{i=2}^{N} \omega_k(\bar{T}_{i-1}) E_{kb}(\bar{T}_{i-1}) A \left(\sum_{j=i}^{N} \tau_{kj}(\bar{\rho}_{kj}, \bar{T}_j)\right) \times$$

$$\psi(u_{k(i-1)}, \bar{f}_{i-1} \cdots \bar{f}_N, g_{i-1} \cdots g_N) \qquad (8)$$

where each ψ is the ratio of the averaged term to its value based upon the appropriate average properties.

The foregoing discussion suggests the simple approximation

$$\psi(u_{ki}, \bar{f}_i, \cdots, \bar{f}_N, g_i, \cdots g_N) \approx \psi(u_{ki}, \bar{f}_i, g_i) \qquad (9)$$

where $\psi(u_{ki}, \bar{f}_i, g_i)$ applies to a homogeneous medium and is available for certain cases in Nelson (1986, 1989). Nelson's work also shows that ψ is dominated by Planck function fluctuations and that it can be approximated very well by several different representations, among them

$$\psi \approx \psi_1 = \overline{E_{kb}(T)} / E_{kb}(\bar{T}) \qquad (10a)$$

$$\psi \approx \psi_2 = \overline{\rho_k E_{kb}(T)} / \bar{\rho}_k E_{kb}(\bar{T}) \qquad (10b)$$

$$\psi \approx \psi_3 = \overline{\omega(T) E_{kb}(T)} / \omega(\bar{T}) E_{kb}(\bar{T}) \qquad (10c)$$

Each of these corresponds to a special limit for either homogeneous or nonhomogeneous paths. In particular, Eq. (10a) is appropriate for an optically thick band of fixed, finite bandwidth; Eq. (10b) represents the optically thin limit for all band models; and Eq. (10c) applies to optically thick bands with a temperature-dependent bandwidth. For the homogeneous medium, it was found that $\psi_1 \approx \psi_2 \approx \psi_3$ to within a maximum difference of about 15% for all the conditions examined (Nelson, 1986, 1989). Each of Eqs. (10) is much simpler than the exact expression, because they depend only on local properties. Equation (10b) is more complicated than the other two, because it requires the concentration as well as temperature.

In the following, the ψ factors in Eq. (8) will be determined for representative nonhomogeneous paths and compared to results based on Eqs. (10).

RESULTS AND DISCUSSION

Although the outcomes of numerical experiments can be completely anticipated from the foregoing discussion, they are presented here for concreteness. All computations are for an overall optical depth of unity. However, results are also given for the optically thin and thick limits. In each case, the active region is called zone one and the radiation considered emerges from the boundary of zone N. The number of zones is considered a parameter and the optical depth of each zone is 1./N. Nonhomogeneous variations in mean mixture fraction are considered. The mixture fraction variance is a constant but is taken to be rather large, so that the effects of nonhomogeneity are not suppressed. Profiles of mean mixture fraction are sectionally linear and include monotonic and symmetric forms. These simulate the general features of reacting mixing layers, boundary layers, and jets. Nonsymmetric and nonmonotonic forms are not considered, simply because there are so many possibilities to consider and they are unlikely to reveal anything new. Uniform pdf's of mixture fraction are used, since the form has a weak effect on results (Nelson, 1986, 1989). As before, a logarithmic absorption function is used; but this has no really significant effect other than to point out the minor differences in gray and nongray response to fluctuations.

Table 1 shows the effect of number of zones for a medium homogeneous in the mean at mixture fractions which are fuel lean, stoichiometric, and fuel rich. For a single zone, the medium is instantaneously homogeneous; but with multiple independent zones, this is not the case. In this work, the thermodynamic property variation with mixture fraction has been improved over that used earlier in that temperature is a nonlinear function due to variations in mixture molecular weight. This has had a significant effect for fuel-rich conditions for the H_2/Air flame. Before, the effect of fluctuations was very weak; but now the effect is quite large (for CH_4/Air the change had very little effect on previous results). The general conclusion to be drawn from these results is that multizoning does not have a particularly significant effect on the flux ratio ψ, even for the extreme conditions considered herein. It is safe to say that for a given mean optical depth the precise nature of its fluctuations is relatively unimportant. The last two rows of the tables apply to optically thin and thick limits, respectively, and can be compared to values of ψ, which are for an optical depth of unity. Note that the results given for ratios other than ψ should not depend upon the number of zones. The variations seen here are simply a measure of statistical scatter in the Monte Carlo method and are most often less than 1%.

Further calculations were carried out in which the active zone had the same properties shown in Table 1, but the mean mixture fraction varied from zone to zone in the manner already discussed. Ten zones were used for all cases. For the fuel-lean case of f = 0.009, the mixture fraction at the far boundary was given values of 0.001, 0.030, and 0.060. These represent conditions in which mean zone temperatures and concentrations are strictly decreasing, strictly increasing, and maximizing at about the center of the optical path. In addition, symmetric variations were explored in which centerline values of about 0.035 and 0.065 were used. In the latter case, the mean temperatures and concentrations show a double maximum. Over all these variations, the value of ψ fell between 2.78 and 2.95. These simple variations in the mean mixture fraction qualitatively simulate many of the conditions that are found in mixing layer, boundary layer, and jet flows. The normalized emergent radiation from a zone is remarkably insensitive to details of nonhomogeneous paths and to all fluctuations except those of

Table 1. Effect of Statistically Independent Path Length Segments on Emergent Radiation from a Mean-Homogeneous Fluctuating Medium

	N = 1	2	5	10	20
ψ	2.8762	2.8362	2.8262	2.8226	2.8091
$\overline{E_{kb}}/E_{kb}(\bar{T})$	2.5536	2.5398	2.5407	2.5410	2.5309
$\overline{\rho_k E_{kb}}/\bar{\rho}_k E_{kb}(\bar{T})$ (a)	2.8251	2.8093	2.8103	2.8104	2.7987
$\overline{\omega_k E_{kb}}/\omega_k(\bar{T})E_{kb}(\bar{T})$ (b)	3.0181	3.0006	3.0015	3.0016	2.9886

H_2/Air, $\tau = 1.$, $\bar{f} = 0.009$, $g^{1/2}/\bar{f} = 0.55$, $u_k = 6$

	N = 1	2	5	10	20
ψ	1.9122	1.9372	1.9484	1.9641	1.9689
$\overline{E_{kb}}/E_{kb}(\bar{T})$	1.8274	1.8214	1.8146	1.8230	1.8244
$\overline{\rho_k E_{kb}}/\bar{\rho}_k E_{kb}(\bar{T})$	1.8792	1.8730	1.8659	1.8746	1.8760
$\overline{\omega_k E_{kb}}/\omega_k(\bar{T})E_{kb}(\bar{T})$	2.0036	1.9970	1.9890	1.9984	2.0001

H_2/Air, $\tau = 1.$, $\bar{f} = 0.0283$, $g^{1/2}/\bar{f} = 0.55$, $u_k = 6$

	N = 1	2	5	10	20
ψ	2.1399	2.2403	2.2637	2.2618	2.2763
$\overline{E_{kb}}/E_{kb}(\bar{T})$	1.9447	1.9709	1.9549	1.9433	1.9502
$\overline{\rho_k E_{kb}}/\bar{\rho}_k E_{kb}(\bar{T})$	2.1011	2.1304	2.1127	2.0996	2.1070
$\overline{\omega_k E_{kb}}/\omega_k(\bar{T})E_{kb}(\bar{T})$	2.2502	2.2829	2.2635	2.2483	2.2560

H_2/Air, $\tau = 1.$, $\bar{f} = 0.09$, $g^{1/2}/\bar{f} = 0.55$, $u_k = 6$

(a) optically thin limit

(b) optically thick limit for logarithmic absorption

the Planck function. Indeed, the value of ψ is well approximated by any of the other ratios given in Table 1. These depend only upon the statistics of the active zone, and two of them depend only upon temperature mean and variance. Use of the Planck function ratio has the widest applicability but also the largest error for band radiation with logarithmic absorption. In the table, this error ranges from about 7 to 15%. However, the parameters have been chosen to amplify the error beyond that of greatest relevance. In particular, with $u_k = 6$ the Planck function is more than an order of magnitude below the peak at $u_k = 2.82$. At a temperature of 2000 K, which is close to peak values found in H_2/Air and other flames, the value of u_k at 2.7, 4.3, and 6.3 μm wavelength is 2.63, 1.65, and 1.13, respectively. For these conditions, the error in the use of the Planck function ratio is below 10%.

SUMMARY AND CONCLUSIONS

Band radiation from a nonhomogeneous fluctuating medium can be described by a series of terms, each of which has the same form as that for radiation from a homogeneous medium. The statistics of these terms differ from the homogeneous case through the different realizations which are possible for the optical path. Several variations of the mean mixture fraction along the path were examined with the assumption that the optical path could be divided into statistically independent intervals.

It was found that results were relatively insensitive to the details of the path-length mean property variations and to the number of intervals into which the optical path length was partitioned. When parameter values chosen to emphasize differences were used, the nonhomogeneous results differed from the homogeneous result by an absolute maximum of little more than 3%. From this one can conclude that very accurate calculations of the escape of radiation from a given zone can be made using only local zone statistics; these statistics are applied without appreciable detriment to the optical path length fluctuations in all the other intervals.

With less than about a 10% error, even greater simplification can be made by evaluating only local Planck function fluctuations. This is undoubtedly the most generally useful approach.

APPENDIX

When a nonhomogeneous medium has been partitioned into homogeneous intervals, the instantaneous contribution from a given interval to the escaping band radiation is given by Eq. (3). If $p(\vec{f})$ is the joint pdf for the mixture fraction in all the intervals, then the average value of the first term of Eq. (3) is given by

$$\overline{\omega_{ki}E_{kb}(T_i)\,A\left(\sum_{j=i}^{N}\tau_{kj}\right)} = \int_{\Omega}\omega_{ki}E_{kb}(T_i)\,A\left(\sum_{j=i}^{N}\tau_{kj}\right)\times p(\vec{f}_i)\,d\vec{f}_i \qquad \text{(A-1)}$$

where

$$\vec{f}_i = (f_i, f_{i+1}, \cdots, f_N)$$

is the vector of mixture fraction values from zone i to N and

$$d\vec{f}_i = df_i df_{i+1} \cdots df_N$$

When the intervals are statistically independent

$$p(\vec{f}_i) = p(f_i)p(f_{i+1}) \cdots p(f_N)$$

Equation (A-1) is best evaluated by a Monte Carlo technique guided by physical considerations.

REFERENCES

Cox, G., 1977, "On Radiant Heat Transfer from Turbulent Flames," Comb. Sci. and Tech., Vol. 17, pp. 75-78.

Gore, J. P., and Faeth, G. M., 1988, "Structure and Radiation Properties of Luminous Turbulent Acetylene/Air Diffusion Flames," ASME Journal of Heat Transfer, Vol. 110, pp. 173-181.

Gore, J. P., Jeng, S.-M., and Faeth, G. M., 1987a, "Spectral and Total Radiation Properties of Turbulent Carbon Monoxide/Air Diffusion Flames," AIAA Journal, Vol. 25, pp. 339-345.

Gore, J. P., Jeng, S.-M., and Faeth, G. M., 1987b, "Spectral and Total Radiation Properties of Turbulent Hydrogen/Air Diffusion Flames," ASME Journal of Heat Transfer, Vol. 109, pp. 165-171.

Grosshandler, W. L., and Joulain, P., 1986, "The Effects of Large Scale Fluctuations on Flame Radiation," Dynamics of Reactive Systems. Part II: Modeling and Heterogeneous Combustion, J. R. Bowen, J.-C. Leyer, and R. I. Soloukhin, eds., AIAA, New York, pp. 123-152.

Hottel, H. C., and Sarofim, A. F., 1967, Radiative Transfer, McGraw Hill, New York.

Jeng, S.-M., Lai, M.-C., and Faeth, G. M., 1984, "Nonluminous Radiation in Turbulent Buoyant Axisymmetric Flames," Comb. Sci. and Tech., Vol. 40, pp. 41-53.

Kabashnikov, V. P., and Kmit, G. I., 1979, "Influence of Turbulent Fluctuations on Thermal Radiation," J. Appl. Spectroscopy, Vol. 31, pp. 963-967.

Kabashnikov, V. P., and Myasnikova, G. I., 1985, "Thermal Radiation in Turbulent Flows-Temperature and Concentration Fluctuations," Heat Transfer-Soviet Research, Vol. 17, No. 6.

Nelson, D. A., 1986, "Band Radiation from a Fluctuating Medium," Radiation in Energy Systems, HTD-Vol. 55, ASME, New York.

Nelson, D. A., 1989, "Band Radiation from a Fluctuating Medium," ASME Journal of Heat Transfer, Vol. 111, pp. 131-134.

Papanicolaou, P. N., and List, E. J., 1987, "Statistical and Spectral Properties of Tracer Concentration in Round Buoyant Jets," Int. J. Heat Mass Trans., Vol. 30, pp. 2059-2071.

1989 National Heat Transfer Conference
HTD-Vol. 106, Heat Transfer Phenomena in Radiation, Combustion and Fires

MEASUREMENT OF PULVERIZED COAL FLAME RADIATION INTENSITY BY HIGH-SPEED SPECTROSCOPY

A. Saljnikov and B. Repić
Boris Kidrič Institute of Nuclear Sciences
Institute for Thermal Engineering and Energy Research
Belgrade, Yugoslavia

ABSTRACT

A high-speed spectrophotometer system was developed for the transient measurement of pulverized coal flame radiation. The system is a grating spectrograph, but with a single detector instead of a detector array. Wavelength of the observed band is determined by the position of the four-sided rotationary mirror, and the plane grating is mounted in a completely new arrangement. The system allows the measurement of radiation intensity spectra in a wavelength region from 2.4 to 4.0 μm repeatedly within a period of one second or less. A desk-top combustion system is designed and assembled as to have an ability of varying parameters influencing the pulverized coal combustion.

Experiments performed include the preliminary measurements, the sample transient measurement of the radiation from a constant intensity light source and the transient measurement of the radiation from a pulverized coal flame. Steady-state and transient preliminary measurements have shown that the developed spectrophotometer has a sufficient measuring ability and that it can be used for the measurement. Transient measurement of pulverized coal flame radiation performed on samples of Yugoslav coal powder have produced data of considerable significance which comply well with corresponding results of other authors.

INTRODUCTION

The primary driving force in the pulverized coal combustion is the transfer of the heat from the flame back to the unheated fuel. Because of the nature of suspension, the temperature of products and the speed of combustion reaction, radiation is often the dominant mode of heat transfer. It is also the primary mechanism of energy transfer from the burning pulverized coal cloud to the water-cooled walls of a boiler furnace because of the large enclosure dimensions, relatively low gas velocities, and high temperature. An accurate konwledge of the transient thermal radiative behaviour of the pulverized coal cloud and its constituents it therefore essential before any meaningful degree of flame modeling, heat transfer analysis or radiation prediction can take place.

The thermal emission of a pulverized coal flame is determined by the emission of the triatomic gases CO_2 and H_2O and of the ash and coke particles contained in the stream of these gases. The radiative properties of triatomic gases and particles of the ash dust have been explained in sufficient detail. However, the radiant properties of partially burnt coal particles are not known with any confidence. There are many difficulties involved in acquiring reliable data because the particulate cloud can be made up of varying fractions of unreacted coal, char, ash and soot, each of which has its own optical properties dependent upon wavelength, temperature and particle size.

Numerous researchers have investigated the optical properties of carbonaceous materials. Blokh /1/ has reviewed the results of many experiments on bulk coal, soot and pulverized coal. Using Mie theory, the absorption and scattering coefficients of different coals were calculated as a function of particle size. The optical constants of swelling and nonswelling coals and their respective char products, at room temperature, have been determined by Brewster and Kunitomo /2/.

Viskanta et al. /3/ have recently studied the importance of some of the most influential parameters (mean particle radius, size distribution, temperature, back scatter) for polydispersions of various coals and fly-ash using a numerical model with average spectral properties. As an experimental counterpart, Groshandler and Monteiro /4/ have investigated the absorption and scattering of thermal radiation within a dilute cloud of pulverized coal and char. An experimental investigation of the optical properties of the solid fuels´cokes was performed by Burak /5/. He used the measured values to determine the spectral coefficients of absorption and scattering needed to calculate the thermal emission of coke particles in a pulverized coal flame.

As a natural continuation of the outlined research, one should proceed to the laboratory-scale, later in field, transient measurements of the radiative properties of the burning cloud of pulverized coal. In this report, the development of a new spectrophotometer system for acquiring transient spectral data is outlined, as well as the combustion system designed as to have an ability of varying some of the parameters influencing the pulverized coal combustion. Because of the emphasis on technique, here we report results from just some of the preliminary tests, together with the analysis of the experimental system performance.

EXPERIMENTAL APPARATUS

The apparatus, designed for the transient measurement of the pulverized coal flame radiation, consists of the combustion system and of the high-speed spectrophotometer.

Combustion System (see Fig.1)

Mixing chamber 3 has the function of mixing variable amounts of oxygen and nitrogen as controlled by valves 5. Flowrate of the gases before being mixed is measured by rotameters 4 and 4a. Fine adjustment of the oxygen-nitrogen mixture flowrate, for controlling the elevation of the fluidized bed of coal particles, is accomplished using the valve 5a. Mesh 6 separates the single-phase and the two-phase sections of the flow, also representing the front (lower) boundary of the combustion chamber 8 (glass cylinder). The particulate cloud is ignited by a hot filament electric resistor 7. Thermocouples 9 monitor the mean temperature of different sections of the flow. Dotted circle 10 represents the section of the combustion chamber observed by the spectrophotometer. That section of the cylinder is placed in an enclosure which is flushed with dry nitrogen for minimizing the atmospheric absorption.

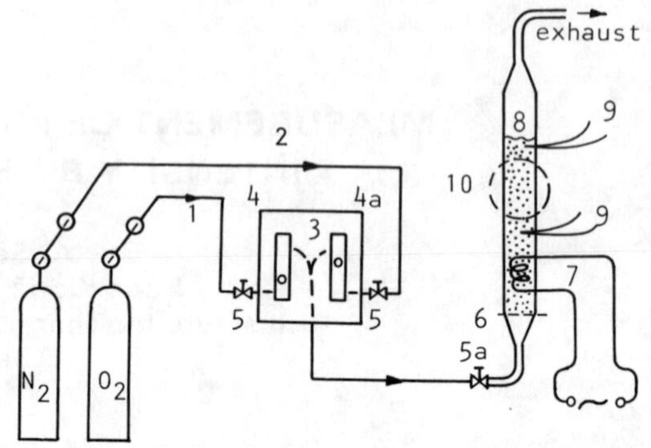

Fig.1 Combustion system - schematic

(1) oxygen supply line
(2) nitrogen supply line
(3) mixing chamber
(4),(4a) rotameter
(5),(5a) valve
(6) mesh
(7) ignitor
(8) combustion chamber
(9) thermocouple
(10) observed section

High-Speed Spectrophotometer

The newly developed spectrophotometer system is a grating spectrograph, but with a single detector instead of a detector array. Wavelength of the observed band is determined by the position of the four-sided rotationary mirror, and the plane grating is mounted in a completely new arrangement. The detector signal is scanned electronically. The spectrophotometer consists of an incident optical system,

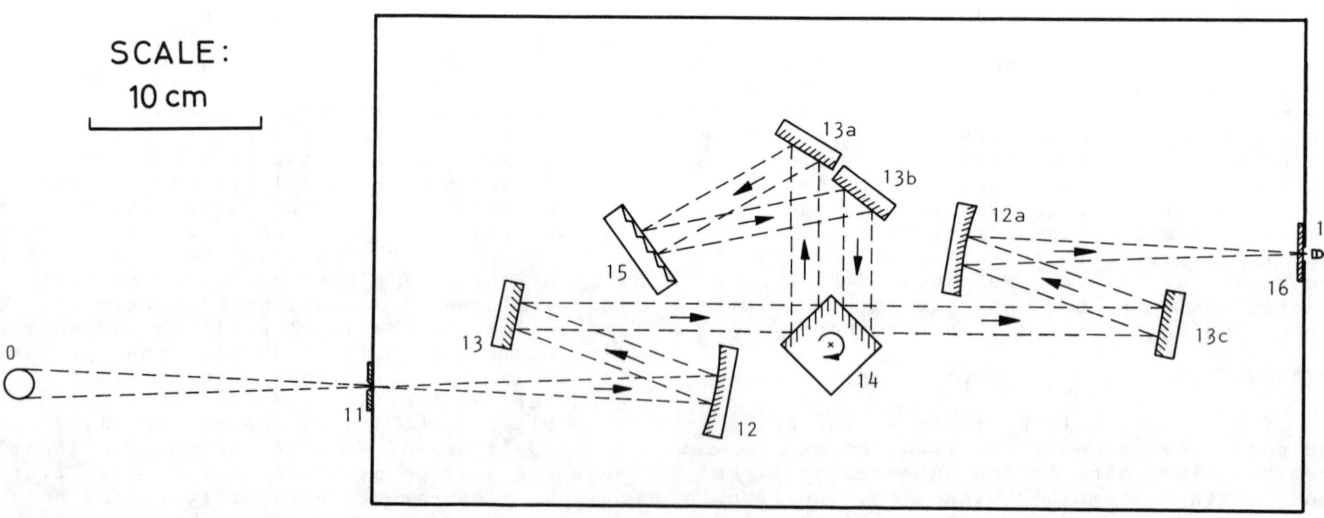

Fig.2 Optical system - schematic

(11) entrance slit
(12),(12a) collimator
(13),(13a),(13b),(13c) mirror
(14) rotationary mirror
(15) diffraction grating
(16) filter
(17) optical sensor

a monochromator system, an electronic system and a digital processing system. The whole measurement course is controlled by a micro-computer.

Figures 2 and 3 show schematic diagrams of the whole optoelectric system. The radiation from the observed section 10 of the flame is projected onto the collimator 12 through the entrance slit 11 of the monochromator system (flushed by nitrogen). To be stressed here is that in this system, as a diffraction element, plane grating is mounted in a completely new arrangement-different from the known ones (Littrow, Ebert-Fastie, or Czerny-Turner /6/).

Collimator 12 transforms the incident beam into a parallel one which is reflected by the stationary mirror 13, rotationary mirror 14 and stationary mirror 13a onto the diffraction grating 15. For the measurable wavelength range of this system (2.0 i.e. 2.4 to 4.0 μm) the ray (beam) coming from mirror 13a is always incident with its complete width onto the diffraction grating. The light is transformed to dispersed spectra by the grating, directed by mirrors 13b, 14 and 13c, and concentrated by the collimator 12a onto the optical sensor 17. By rotating the mirror 14, one can choose the band of the radiation spectrum to be projected onto the detector. All rays (beams) between the two collimators 12 and 12a are parallel ones so all mirrors between them are plane mirrors. Also, the ray (beam) coming from mirror 13a is allowed to travel all the length of the grating when the rotationary mirror 14 is rotated from its starting to its end position. This greatly facilitates the arrangement of mirrors and reduces the cost of the system, which is also achieved by using single detector instead of a detector array. Reduced cost is the main advantage of this system compared to i.e. the one in Ref./7/.

A 150 grooves/mm grating (Bausch & Lomb 35-53-*-760; λ_b=2.0 μm; d=1 mm/150; θ_b=8°38′) is used for the near-infrared measurements. It was adjusted using the harmonics of a He-Ne laser. Only the m=1 spectrum of the grating is considered, for its superior brightness. The focal distance of the collimating mirrors 12 and 12a is 200 mm. The optical sensor is a PbSe photo-conductive cell. The measurable wavelength range of this system is from 2.0 to 4.0 μm. In the present work, the filter 16 has been used with a cut-off wavelength of 2.4 μm. This has, for the time being, narrowed the measurable range to wavelengths from 2.4 to 4.0 μm. Proper baffling for this monochromator remains an unresolved issue.

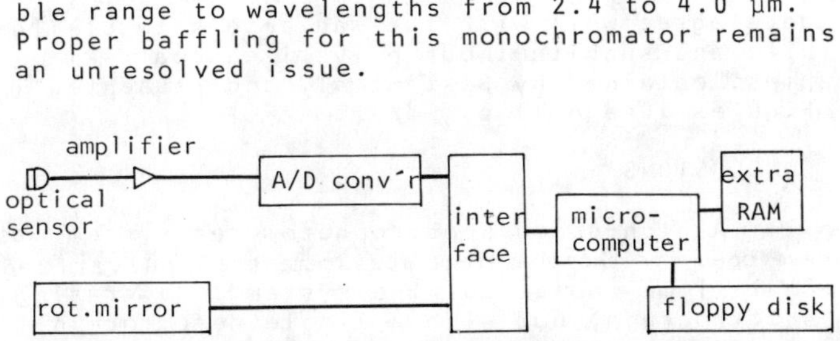

Fig.3 Data processing system - schematic

The amplification and A/D conversion of the detector signal as well as the storage, transfer and processing of the digital data were outlined by Makino et al./7/. The present system measures and stores a 12-bit datum repeatedly in a period of less than 50 μs. That is, if a spectrum of 10 wavelength points could be measured without the mirror rotation by the stepping motor, the period of the spectrum sampling would be about 0.5 ms. However, a stepping motor is used to rotate the mirror, with an interval of about 2.5 ms between two subsequent steps. This means that the time needed to measure and process ten data points is only 50 ms (including the rotation back to the first measuring position). In other words, one can obtain up to 20 spectra, having 10 data points each, within just one second.

RESULTS AND DISCUSSION

Experiments performed include the preliminary measurements, the sample transient measurement of the radiation from a constant intensity light source and the transient measurement of the radiation from a pulverized coal flame.

Preliminary Measurements

In order to check the performance of the developed spectrophotometer, the following test was performed. A constant temperature light source was used, its radiation directed onto the entrance slit of the monochromator. The rotationary mirror position was being changed manually. The detector monitored 9 measuring points approximately equally spaced throughout the 2.4 - 4.0 μm wavelength region. At each measuring position, the polystyrene film calibrator was placed at the optical path in front of the filter. Transmittance values for the film were calculated by simply dividing the voltage readings with and without the film. In this test the output of the electric signal amplifier was sent to the commercial multimeter. In the following measurements the micro-computer controlled system was used.

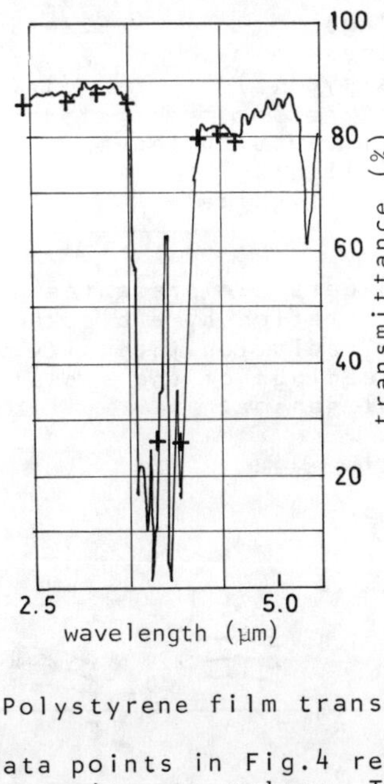

Fig.4 Polystyrene film transmittance

The data points in Fig.4 represent the measured transmittance values. The solid curve is the polystyrene film characteristic provided by its manufacturer. The measured values differ from the standard ones by less than 3% which is considered acceptable. The measurement stability was confirmed by the repeatability of the results of a number of successive measurements. As it is readily observed, data points follow the standard line, including the OH-stretching absorption band gap, closely.

These results show that the spectrophotometer developed has sufficient measuring ability for the future measurement.

<u>Transient Measurement of Pulverized Coal Flame Radiation</u>

There is no published data on thermal radiation characteristics of Yugoslav coal, to date.

In this investigation three types of pulverized Yugoslav coal (one subbituminous, two lignites; see Table 1) were burned in a 50%-50% oxygen-nitrogen flow in the combustion system described previously. The mass loading values of pulverized coal in the levitated state were 0.452, 0.495 and 0.470 gr/cm^3 for Banovići, Bitola and Kolubara coal, respectively. The radiation was monitored by the present measuring system, with data collected at a time increment of one second.

According to the termocouple readings, time values of 1,2 and 3 s correspond to gas temperature values of 740, 760 and 790°C (Banovići subbituminous coal); 720, 730 and 740°C (Bitola lignite); 710, 720 and 730°C (Kolubara lignite), respectively.

The same radiation intensity data are plotted vs. time in Fig.6.

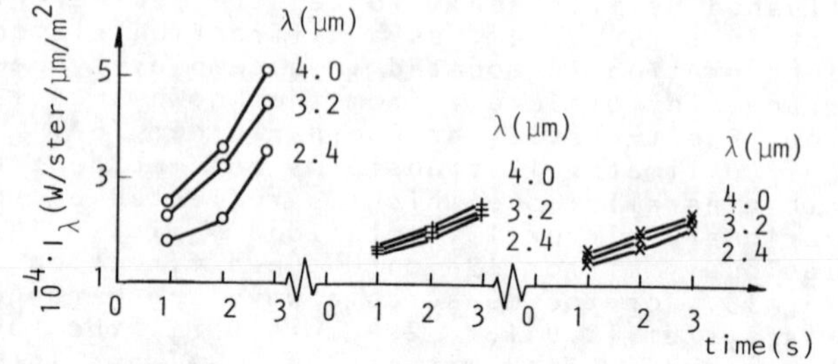

Fig.6 Histories of the pulverized coal combustion

Although the measured signal was so noisy that its fluctuations were almost of the same order of magnitude as the fluctuations of the blackbody source (noise will be relatively reduced by amplifying the signal incident onto the sensor by focusing its vertical component, i.e. image of the slit, too), the data obtained are of considerable significance. Measured data agree well with the radiance data of lignite and subbituminous pulverized coal samples, obtained by Best et al. and presented in Figures 10 and 11 of /8/.

Table 1 - Analyses of coal samples

	moisture - free			moisture - ash - free		
Coal type	o	+	x	o	+	x
Item						
Ash (%)	17.2	27.9	27.5			
Char (%)	61.0	59.6	56.8	52.9	44.1	40.1
C-fixed (%)	43.1	29.5	29.1	52.1	40.9	40.1
Vol.matter (%)	39.0	40.3	43.4	47.1	55.9	59.9
Combustible matter (%)	32.8	72.1	72.5	100.0	100.0	100.0
HEATING VALUE:						
- lower (kJ/kg)	23354	18782	18937	28212	26035	26121
- higher (kJ/kg)	22435	17983	17991	27102	24927	24815
Grinding fineness-remaining on 0.090 mm sieye (%)	73.2	57.6	61.1			

o = Banovići (subbituminous)
\+ = Bitola (lignite)
x = Kolubara (lignite)

Measured data are presented graphically in Fig.5. Calibration by a blackbody was used to derive the radiation intensity values from the voltage readings of the signal generated by the optical sensor.

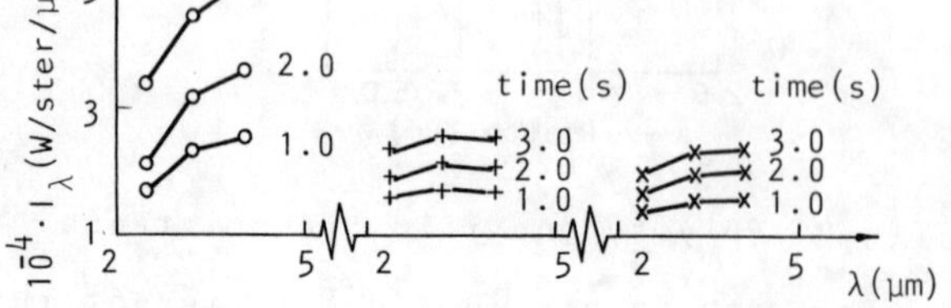

Fig.5 Pulverized coal flame radiation data

CONCLUSIONS

A high-speed spectrophotometer system was developed for the transient measurement of pulverized coal flame radiation. The system is a grating spectrograph, but with a single detector instead of a detector array. Wavelength of the observed band is determined by the position of the four-sided rotationary mirror, and the plane grating is mounted in a completely new arrangement.

Steady-state and transient preliminary measurements have shown that the spectrophotometer developed has sufficient measuring ability and that it can be used for the measurement. Transient measurement of pulverized coal flame radiation performed on samples of Yugoslav coal powder have produced data of considerable significance which comply well with corresponding results of other authors.

ACKNOWLEDGEMENTS

The authors acknowledge prof.Dr Toshiro Makino of Kyoto University for his invaluable guidance and professonal support.

This work was partially supported by the Grant-in Aid for Scientific research of the Ministry of Education, Science and Culture of Japan.

REFERENCES

1. Blokh,A.G., "Problem of Flame as a Disperse System", Heat Transfer in Flames, Scripta Book Co., Washington, D.C., 1974, p.111.

2. Brewster, M.Q., and Kunitomo,T., "The Optical Constants of Coal, Char, and Limestone", Journal of Heat Transfer, Vol.106, Nov.1984, p.678.

3. Viskanta,R., Ungan,A., and Mengüc,M.P., "Predictions of Radiative Properties of Pulverized Coal and Fly-Ash Polydispersions", 20th National Heat Transfer Conference, ASME Paper No.81-HT-24, Milwaukee, Aug.1981.

4. Grosshandler,W.L., and Monteiro,S.L.P., "Attenuation of Thermal Radiation by Pulverized Coal and Char", Journal of Heat Transfer, Vol.104, Nov.1982, p.587.

5. Burak,L.D., "Radiative Characteristics of Coke Particles of a Coal-Dust Flame", Inzhenerno-Fizicheskii Zhurnal, Vol.45, No.2, p.297, August 1983.

6. Kneubühl,F., "Diffraction Grating Spectroscopy", Applied Optics, Vol.8, No.3, March 1969, p.505.

7. Makino,T., Matsuda,S., Hirata,N., and Kunitomo,T., "Measurement of Transient Behaviours in Thermal Radiation Characteristics of Materials by High-Speed Spectroscopy", The 8th International Heat Transfer Conference, San Francisco, August 1986.

8. Best,P.E., Carangelo,R.M., Markham,J.R., and Solomon,P.R., "Extenstion of Emission-Transmission Technique to Particulate Samples Using FT-IR", Combustion and Flame, Vol.66 (1986), p.47.

1989 National Heat Transfer Conference
HTD-Vol. 106, Heat Transfer Phenomena in Radiation, Combustion and Fires

DIFFUSION APPROXIMATIONS FOR INTERACTION OF RADIATION AND CONDUCTION IN FUSED SILICA

Y. H. Pan and W. Aung
University of Maryland
College Park, Maryland

R. Reiss
Howard University
Washington, District of Columbia

ABSTRACT

A numerical procedure for the calculation of temperature response of semitransparent materials has been developed. The method is based on the solution of the one-dimensional energy equation and the equation of transfer for combined conduction and radiation. The use of cold, reradiating and Rosseland approximation is discussed. A novel modified Rosseland approximation is developed to produce accuracy and computation efficiency. The results are compared with experimental data obtained using holographic interferometry, for a fused silica slab subjected to radiant heating on one side. Both boundaries are subject to conduction, natural convection and surface radiation heat losses. The thermal conductivity and specific heat are considered linearly dependent on temperature. It is shown that the modified Rosseland approximation produces acceptable computed results within engineering accuracy. The effects of the thermophysical properties on the computed results are also assessed.

NOMENCLATURE

Bi Biot number: $Bi = \frac{hL}{k^*}$
c specific heat
E_n exponential integral, $n = 1, 2$
G incident radiosity as defined in Eq. (6)
h heat transfer coefficient of free convection
I radiant intensity
k thermal conductivity
L thickness of sample
N_{rc} radiation-conduction parameter: $N_{rc} = \frac{\sigma T_\infty^3 L}{k_\infty^*}$
n index of refraction
q radiant flux
T dimensional temperature
t dimensionless time parameter: $t = \frac{k_\infty^* t^*}{\rho c^* L^2}$
x dimensionless spatial parameter: $x = \frac{x^*}{L}$
β_R Rosseland mean extinction coefficient
ϵ emissivity
θ dimensionless temperature: $\theta = \frac{T}{T_\infty}$
κ absorption coefficient
λ wavelength
μ directional cosine
ρ reflectivity
σ Stefan-Boltzmann constant
τ optical length: $\tau = \int_0^x \kappa \, dx'$
τ_L optical length: $\tau_L = \kappa L$

Subscript

∞ refer to ambient temperature
r refer to reference temperature
0 refer to boundary at x = 0
1 refer to boundary at x = 1
b refer to blackbody assumption

Superscript

$+$ refer to $\mu > 0$
$-$ refer to $\mu < 0$
$*$ refer to dimensional quantity
e refer to external flux
i refer to internal flux

Introduction

The flammability of polymeric materials, of great interest in studies related to fire safety, is governed by complex phenomena that arise in both the gas phase and the condensed phase. In each phase there is a strong coupling between chemical reaction and heat and mass transfer. These effects can have a strong influence on gas phase phenomena such as soot and noxious gas generation, and are subjects of continuing research. In the solid phase, previous investigation (Kashiwagi and Ohlemiller, 1982) have pointed out the importance of heat and mass transfer in the transient heating of polymers prior to ignition; this forms the focus of the present paper.

We consider the radiant heating of a semitransparent polymeric material in the solid phase. The heating may be viewed as caused by a remote source during the pre-ignition stage. When the temperature of the material is sufficiently high, chemical degradation will begin leading to gas evolution. It is important in fire safety research that the gasification rate and the composition of the evolved gases be determined accurately. This requires a quantitative understanding of the coupled heat and mass transfer and chemical reaction process.

Recent studies which have examined the effect of heat transfer include those by Vovelle, Delfan, Reuillon, Bransier and Laraqui (1987). These authors examined the thermal degradation of PMMA subjected to a radiant heat flux. Theoretical calculation were performed using an Arrhenius equation to relate the mass loss rate to the temperature of the materials. The heat transfer model, however, assumed that the incoming radiant energy was absorbed totally on the front surface of the solid. The heat transfer in the solid was therefore caused by conduction alone. This led to the prediction that the maximum temperature occurred at the front surface. The initiation of thermal degradation which was observed (Kashiwagi and Ohlemiller, 1982) to occur in the interior of the solid cannot be predicted by this approach.

We present here an alternative approach towards calculating the heat transfer in the interior of the solid, by considering the material to be semitransparent to thermal radiation. Some of our results have been reported in an earlier paper (Finlayson, Aung, kashiwagi, 1987). The present study represents an effort at obtaining an accurate yet relatively simple theoretical model for combined conduction and radiation in a semitransparent medium. The overall goal is to enable the extension of the model to include effects of chemical reaction.

Combined Conduction and Radiation

Heat radiation problem is usually formulated with an integral equation. Heat conduction, however, is governed by a differential equation. When radiation and conduction occur together, such as in the interaction of heat radiation and conduction in a semi-transparent material, the energy equation becomes a highly non-linear integro-differential equation. There are no general solutions for this problem, and several simplifications such as the opaque, cold, hot, gray, and optically thick models have been investigated.

In the opaque model, the solid is treated as having an infinite absorption coefficient, and the incoming heat flux is totally absorbed on the surface. The heat transfer problem is now one of heat conduction.

The cold model assumes that there is no internal emission of radiation inside the material; i.e., the material is only an absorber of radiation. Such an approach has been employed by Dake, Harleman (1969) and Cobble (1964). This simplification means that the equation of transfer is no longer dependent on the local temperature. The hot (reradiating) model accounts for internal emission inside the material (Stepanov and Bityukov, 1980; Finlayson, Aung, Kashiwagi, 1987). This model should be more accurate, but its shortcoming is that it takes a long time to complete a calculation, even with a supercomputer.

The P-N approximation (Yucel and Bayazitoglu, 1983) is very accurate; however, it is tedious and cumbersome. The two-flux method simplifies the angular dependence of the intensity (Matthews, Viskanta and Incropera, 1985) is sometimes not so accurate. The gray model assumes that the absorption coefficient is constant and independent of wavelength (Stepanov and Kim, 1980; Nelson, 1975).

The optically thick model is valid when the radiation photon mean free path is very small as compared with the characteristic length of the material (Deissler, 1964; Condon 1968). Kowalski (1987) added a heat generation term to modify the Rosseland Approximation. The basic advantage of the optically thick model is that it can change the governing integro-differential equation into a differential equation. However,this model is not accurate near the front boundary, since the optically thick assumption, inherent in the Rosseland approximation, breaks down in the vicinity of x = 0.

All the models mentioned above have their own advantages and disadvantages. The specific objective of this study is to obtain a modified Rosseland approximation which can adequately predict the thermal response of a semi-transparent solid subjected to radiative heating on one surface.

Description of the Problem and Basic Equations

A vertical solid plane exposed to a high temperature radiant flux is analyzed in this study. The solid is in local thermodynamic equilibrium and with negligible scattering. The radiant flux and source temperature are known quantities. The specific heat and thermal conductivity are assumed to be functions of temperature only. The reflectivity is assumed to be wavelength dependent and is determined by the Fresnel equation. The wavelength dependence of the absorption coefficient is approximated by the band model. The general set of boundary conditions allows for the heat balance of conduction, natural convection and surface radiation. The general, transient, non-dimensional energy equation and the equation of transfer with boundary conditions are as follows (Finlayson, Aung, Kashiwagi, 1987; see p.394, Viskanta, 1975):

Energy equation

$$\frac{\partial}{\partial x}\left(k\frac{\partial\theta}{\partial x} - N_{rc}q^r\right) = c\frac{\partial\theta}{\partial t} \tag{1}$$

Boundary conditions

$at \quad x = 0,$

$$-k\frac{\partial\theta}{\partial x}\bigg|_0 + Bi_0(1-\theta_0) + N_{rc}\epsilon_0[f_{opq}(\theta_s)\theta_s^4 - f_{opq}(\theta_0)\theta_0^4] = 0 \tag{2}$$

$at \quad x = 1,$

$$k\frac{\partial\theta}{\partial x}\bigg|_1 + Bi_1(1-\theta_1) + N_{rc}\epsilon_1[f_{opq}(\theta_{s1})\theta_{s1}^4 - f_{opq}(\theta_1)\theta_1^4] = 0 \tag{3}$$

where f_{opq} is a fractional function of the flux of opaque bands to the flux of total bands. The term, θ_{s1}, can be thought to be equal to 1.

Initial condition

$$\theta = 1 \qquad for\ all\ x,\ at\ t = 0. \tag{4}$$

In equation (1),

$$\frac{\partial q^r}{\partial x} = \tau_L(4n^2\theta^4 - G) \tag{5}$$

where the spectral incident radiation for azimuthally symmetric radiation, G, is defined as (see p.268, Ozisik, 1973):

$$G = 2\pi[\int_0^1 I^+(x,\mu)d\mu + \int_0^1 I^-(x,\mu)d\mu] \tag{6}$$

Equation of Transfer

$$\mu\frac{dI}{dx} + \tau_L I = \tau_L I_b \tag{7}$$

The boundaries for I are affected by both external and internal flux (Finlayson, Aung, Kashiwagi, 1987; p.279, Ozisik, 1973):

$$I_0^+(\mu) = \left[I_e + \rho_i J_0(\mu) + \rho_i^2 J_1(\mu) exp(\frac{-\tau_L}{\mu})\right]/\beta(\mu) \tag{8}$$

$$I_1^-(\mu) = \left[\rho_e I_e exp(\frac{-\tau_L}{\mu}) + \rho_i J_1(\mu) + \rho_i^2 J_0(\mu) exp(\frac{-\tau_L}{\mu})\right]/\beta(\mu) \tag{9}$$

where the parameters $J_0(\tau)$, $J_1(\tau)$ and $\beta(\tau)$ are defined as:

$$J_0(\mu) = \int_0^1 \frac{\tau_L}{\mu} I_b(x') exp(\frac{-\tau_L x'}{\mu}) dx'$$

$$J_1(\mu) = \int_0^1 \frac{\tau_L}{\mu} I_b(x') exp(\frac{-\tau_L(1-x')}{\mu}) dx'$$

$$\beta(\mu) = 1 - \rho_e^2 exp(\frac{-2\tau_L}{\mu})$$

In equation (8), the first term, I_e represents the intensity contribution from the external flux, and the remaining terms are due to energies reflected at the boundary at x = 0 and x = 1, respectively, as a result of energies arriving from within the solid. Terms in equation (9) have similar meaning, except that there is no term representing the direct contribution from external flux.

The objective of the "hot" or reradiating model is to solve the above coupled system of equations (1)-(9) without any further simplification. This model has been described in detail by Finlayson, Aung and Kashiwagi (1987). Since it is necessary to calculate the internal emission energy for every grid point inside the slab in every time step in a finite difference scheme, experience has shown that the time needed for the calculation has been excessive. Therefore, for further research including effects of polymer degradation, it is unrealistic to use the reradiating model. The use of the Rosseland diffusion approximation can lead to a great simplification in the treatment of radiative transfer, since this approach changes the "integro-differential" form of equation of transfer into a "diffusion" or differential type of equation.

Diffusion Approximation

A medium is said to be optically thick if the radiation photon mean free path is very small compared with the characteristic dimension of the medium (Siegel and Howell, 1981). For the optically thick case, the radiation energy emitted by locations removed from the immediate vicinity of the location P is greatly attenuated before reaching the location P. In this situation, the energy transfer depends only on the conditions in the immediate surrounding of node P and can be described in terms of the local gradients. This approximation was first derived by Rosseland (1936). This approach is applied in the present study as shown in the following.

Derivation of Conventional Rosseland Approximation

The derivation of the conventional Rosseland approximation is outlined here for completeness and for the purpose of definition of terms. Defining (Goldstein and Howell, 1968):

$$\epsilon_s = \frac{1}{\tau_L}$$

equation (7) can be written as:

$$\epsilon_s \mu \frac{dI}{dx} + I = I_b \tag{10}$$

For $\tau >> 1$, I and I_b can be expanded:

$$I = I^{(0)} + \epsilon_s I^{(1)} + \epsilon_s^2 I^{(2)} + ... \tag{11}$$

$$I_b = I_b^{(0)} + \epsilon_s I_b^{(1)} + \epsilon_s^2 I_b^{(2)} + ... \tag{12}$$

Substituting equations (11) and (12) into (10), and collecting terms of the same ϵ_s order, we have:

$$order\quad 0: \quad I^{(0)} = I_b^{(0)}$$

$$order\quad \epsilon_s: \quad I^{(1)} = I_b^{(1)} - \mu\frac{dI^{(0)}}{dx}$$

Neglecting the high order terms,

$$I = I^{(0)} + \epsilon_s(I_b^{(1)} - \mu\frac{dI_b^{(0)}}{dx})$$

$$= I^{(0)} + \epsilon_s I_b^{(1)} - \mu\epsilon_s\frac{dI_b^{(0)}}{dx}$$

$$I = I_b - \frac{\mu}{\tau_L}\frac{dI_b}{dx} \tag{13}$$

But

$$q^r = 2\pi\int_{\mu=-1}^{1} I(x,\mu)\,\mu d\mu \tag{14}$$

Substitute equation (13) into equation (14), and since $I_b = I_b(\theta)$ only, we obtain

$$q^r = -\frac{4\pi}{3\tau_L}\frac{dI_b[\theta(x)]}{dx} \tag{15}$$

Equation (15) represents the conventional Rosseland approximation which gives the radiative heat flux in term of the gradient of the radiative intensity. Near the front boundary of solids, this approximation is not valid. Consequently, the solution is incorrect near the wall. Goldstein and Howell (1968) developed a method which used the asymptotic expansion method to linearize the temperature near the wall to match the temperature calculated from the diffusion equation which is valid only far from the front surface. In deriving their method, the above authors assumed the surface to be black and the boundary temperature to be constant; further considered the gray model and a fixed radiation conduction parameter.

Derivation of Modified Approaches

To improve the conventional Rosseland approximation, we can add an "external flux" into the radiative heat flux equation, i.e., equation (15), to modify the Rosseland Method. The external flux can be taken as the incoming radiation, which is the quantity relevant to the cold model. Thus the modified approach can be viewed as the "cold + diffusion" approach. The radiation term in the energy equation is (see p.273 in Ozisik, 1973):

$$\frac{dq^r(\tau)}{d\tau} = 4\pi I_b(\tau) - 2\pi\Big[I_0^+(\mu)E_2(\tau) + I_1^-(\mu)E_2(\tau_1 - \tau) + \int_0^{\tau_1} I_b(\tau')E_1(|\tau - \tau'|)d\tau'\Big] \quad (16)$$

The term, $4\pi I_b(\tau)$, is the emission flux from the location under consideration. The first two terms in the square brackets in equation (16) represent the absorbed fluxes from the front boundary and rear boundary, respectively. The last term in the bracket gives the fluxes absorbed from all other points within the slab. To simplify equation (16), we expand $I_b(\tau')$ at $\tau' = \tau$,

$$I_b(\tau') = I_b(\tau) + (\tau' - \tau)\frac{dI_b(\tau')}{d\tau'}\Big|_\tau + \frac{(\tau' - \tau)^2}{2!}\frac{d^2 I_b(\tau')}{d\tau'^2}\Big|_\tau + ...$$

It can be easily shown, based on the theory of exponential integral (cf. appendix, Ozisik, 1973), that

$$\int_0^\infty E_1(\tau')d\tau' = 1$$

$$\int_0^\infty \tau' E_1(\tau')d\tau' = \frac{1}{2}$$

$$\int_0^\infty \tau'^2 E_1(\tau')d\tau' = \frac{2}{3}$$

Letting τ, τ_1, $\tau - \tau' >> 1$ and using the above relations, we obtain the following equations:

$$\int_0^\tau I_b(\tau')E_1(\tau - \tau')d\tau' = I_b(\tau) - \frac{1}{2}\frac{dI_b(\tau)}{d\tau} + \frac{1}{3}\frac{d^2 I_b(\tau)}{d\tau^2} \quad (17)$$

$$\int_\tau^{\tau_1} I_b(\tau')E_1(\tau' - \tau)d\tau' = I_b(\tau) + \frac{1}{2}\frac{dI_b(\tau)}{d\tau} + \frac{1}{3}\frac{d^2 I_b(\tau)}{d\tau^2} \quad (18)$$

Substituting equations (17) and (18) into (16), we obtain

$$\frac{dq^r(\tau)}{d\tau} = -2\pi[I_0^+(\mu)E_2(\tau) + I_1^-(\mu)E_2(\tau_1 - \tau)] - \frac{4\pi}{3}\frac{d^2 I_b(\tau)}{d\tau^2} \quad (19)$$

Substituting equations (8) and (9) into (19) and neglecting the terms $\rho_i J_0$, $\rho_i J_1$, $\rho_i^2 J_0$ and $\rho_i^2 J_1$, we can formulate the following derivative radiation term in the energy equation for the modified Rosseland approximation which is essentially a "Cold + Rosseland" model.

$$\frac{dq^r(\tau)}{d\tau} = \frac{-2\pi}{\beta(\mu)}\Big[I_e E_2(\tau) + \rho_e I_e exp(\frac{-\tau_1}{\mu})E_2(\tau_1 - \tau)\Big] - \frac{4\pi}{3}\frac{d^2 I_b(\tau)}{d\tau^2} \quad (20)$$

The first group of terms in equation (20), those inside the square brackets, consist of the contributions due to external flux, I_e, and can be thought as from cold model. The complete derivation of cold model is in the (p.382, Viskanta, 1975). The last term is from the conventional Rosseland approximation. This equation can be substituted into equation (1) thereby transforming it from an "integro-differential" equation into a "diffusion" equation.

Numerical Solution Method

Control Volume Scheme

A control volume scheme with uniform mesh is adopted in this study. The Crank-Nicolson scheme is used to maintain second order accuracy. The discretization of the energy equation leads to the following finite difference equation:

$$c\Delta x(\theta_P' - \theta_P^o) = \frac{1}{2}\Delta t\Big[\frac{k_e(\theta_E' - \theta_P')}{\Delta x} - \frac{k_w(\theta_P' - \theta_W')}{\Delta x}\Big] + \frac{1}{2}\Delta t\Big[\frac{k_e(\theta_E^o - \theta_P^o)}{\Delta x} - \frac{k_w(\theta_P^o - \theta_W^o)}{\Delta x}\Big] + \int_t^{t+\Delta t}\int_w^e \frac{\partial q^r}{\partial x}dxdt$$

The nomenclature is explained in figure 1. The interface conductivities of k_e and k_w are are well discussed in (p.46, Patankar, 1980).

$$k_e = \frac{2k_P k_E}{k_E + k_P}$$

Near Boundary Control Volume

To get more accurate results, the control volume of adjacent to a boundary uses a regular control volume instead of half control volume (see p.69 Patankar, 1980). This arrangement is shown in figure 2. The typical boundary face 1 is actually located at the boundary point. With such an arrangement, there is no need for extra discretization of equation for the near boundary control volume.

Linearization

The surface radiation term is linearized by using Taylor series expansion and the equation is as follows:

$$\theta_0^4 = 4\theta_0'^3\theta_0 - 3\theta_0'^4$$

where θ_0' is the surface temperature at the previous iteration.

Rosseland Mean Extinction Coefficient

To transform I_b into θ in the last term of equation (20), we introduced Rosseland mean extinction coefficient β_R (see p.25, Ozisik, 1973):

$$\frac{1}{\beta_R} = \int_0^\infty \frac{1}{\beta_\lambda}\frac{\partial I_{b\lambda}(T)}{\partial I_b(T)}d\lambda$$

where β_λ is equal to κ_λ in our non-scattering assumption. The Rosseland mean extinction coefficient can be evaluated by utilizing the table of fractional function of the second kind.

$$\frac{1}{\beta_R} = \sum_{m=1}^{M}\frac{1}{\beta_m}[f^*_{0-\lambda_m}(T) - f^*_{0-\lambda_{m-1}}(T)]$$

where M is the number of bands, and f^* is from the table of fractional function of the second kind. The quantity β_m is the mean value of β_λ in the wavelength interval from λ_{m-1} to λ_m.

After introducing β_R and substituting equation (20) into equation (1), we obtain the following relation:

$$c\frac{\partial\theta}{\partial t} = (k + \frac{4N_{rc}\theta^3}{3\beta_R})\frac{\partial^2\theta}{\partial x^2} - \frac{2\pi}{\beta(\mu)}\Big[I_e E_2(\tau) + \rho_e I_e exp(\frac{-\tau_1}{\mu})E_2(\tau_1 - \tau)\Big] \quad (21)$$

The external flux (the last term of the above equation) can be considered as a source term in the energy equation, and needs to be calculated only once because it is a function of location only.

Note that equation (21) is a non-steady, non-linear partial differential equation. We need to compute, at each time step in the solution procedure, the values of β_R, k and c at every node point and $I_b(0)$, $I_b(1)$ at the boundaries at x = 0 and x = 1, respectively.

Physical Properties

In order to analyze the combined radiation and conduction problem, the index of refraction, absorption coefficient, thermal conductivity, and specific heat have to be known functions of temperature and wavelength. The density of the solid is assumed to be constant throughout this study. To focus on developing an efficient and accurate heat transfer model, we consider the semi-transparent material to be fused silica at this time; this is because of its stable and relatively well quantified physical properties, and negligible chemical degradation in the temperature range of interest. Further research will involve more complicated material such as PMMA.

Refractive-Index and Reflectivity

The refractive index of fused silica are functions of temperature and wavelength (Malitson, 1965); however the magnitude value of $\frac{dn}{dt}$ is approximately 10^{-6}, hence the dependence of temperature can be neglected. The value of the wavelength dependent refractive index is obtained from a four-term Sellmeier equation (Brixner, 1967).

The reflectivity is related to the refractive index at both parallel and perpendicular directions, and is determined by the Fresnel equation (Siegel and Howell, 1981 and Viskanta, 1975). It can be simplified into Snell's law for normal incident radiation. The Snell's law is:

$$\rho_\lambda = \left(\frac{1-n_\lambda}{1+n_\lambda}\right)^2$$

Absorption Coefficient

The absorption coefficient of fused silica is considered a function of wavelength only. The value of κ_λ is assumed to be 0 cm^{-1} when its wavelength is smaller than 2μ, and infinite when larger than 10μ. The absorption coefficient is related to transmittance by the following equation (Viskanta, 1975).

$$t_\lambda = (1-\rho_\lambda)^2 \frac{exp(-\kappa_\lambda L)}{1-\rho_\lambda^2 exp(-2\kappa_\lambda L)}$$

The value of the transmittance is obtained from IR spectrum experiment, and length of the sample is 0.5 inch. The absorption coefficient is given explicitly by the above equation as:

$$\kappa_\lambda = \frac{-1}{L}\, ln\left(\frac{(1-\rho_\lambda)^2 + \sqrt{(1-\rho_\lambda)^4 + 4t_\lambda^2\rho_\lambda^2}}{2\rho_\lambda^2 t_\lambda}\right)$$

Thermal Conductivity

Data on thermal conductivity for fused silica are obtained from product information supplied by Dynasil Corporation. This property is almost a linear function of temperature from 0 to 450 $°C$, and is given by the following equation:

$$k = 0.703 + 0.297\frac{\theta}{\theta_r}$$

Specific Heat

The specific heat of fused silica is approximated by the following dimensional equation with an uncertainty only of 0.4% in the range 298 to 2000 $°K$ (Men and Sergeev, 1973):

$$c^* = 931.6 + 0.256T - 2.40 \times 10^{-7}\ T^{-2}$$

Results

Figure 3 gives a comparison of the modified ("Cold + Rosseland") model, reradiating model and experiment results for fused silica. The experimental results are obtained from holographic interferometry details of which will be reported in a separate paper. The temperature profiles are displayed vs. positions at different times. The sample thickness is 3.1 cm. The irradiated boundary is subjected to a heat flux of 4.89 $\frac{W}{cm^2}$. The temperature of the flux source is 1636 $°K$. Five groups of temperature profiles (30, 70, 170, 290, 340 seconds after the beginning of heating) are recorded.

The results show that the modified Rosseland model is acceptable but not as accurate as the reradiating model, especially near the irradiated boundary at later times. Note that the reason for the temperature peak to occur inside the slab is due to natural convection losses from the boundary. It is obvious from the figure that, at later times, the calculated temperature profiles are steeper than the experimental results. The discrepancies may be in part due to inaccuracy in specifying the thermal conductivity in the theoretical models, since a higher thermal conductivity leads to a flatter temperature profile. To insure the accuracy of the calculation of the modified Rosseland approximation, the total energy balance was examined and verified as correct during the calculation.

The overprediction by the modified Rosseland model for the temperature profile near the front surface is apparently caused by too much energy being retained at the front boundary. It is intuitively feasible to further modify this model by taking $f_{opq}(\theta_0) = 1$ in equation (1) to increase the dissipation energy from the front boundary. This means that the front boundary acts as a blackbody emitter in the radiation exchange between the source and itself. This truncated model represents a convenient means of reducing the thermal energy entering the solid.

As shown in figure 4, the truncated model is very accurate. The temperature profiles of the truncated model is almost identical to those given by the reradiating model. The total amount of computer (CDC CYBER 205) time for most computations is about 100 seconds for this model, as compared with 2000 seconds for the reradiating model. It should be noted that there is a crossover between the truncated model and the reradiating model near the irradiated boundary. This is because of the excessive amount of emission energy from the boundary at x = 0. The temperature profiles of the truncated model is also slightly flatter than temperatures calculated using the reradiating model; this is because in the modified or truncated model, the thermal conductivity of the material has been augmented by a contribution involving the Rosseland mean extinction coefficient, β_R [see terms in the first set of parentheses on the right hand side of equation (21)].

Figure 5 shows the maximum temperature vs. time at different fluxes. Six incident fluxes are included in the figure. In each case, steady state is reached in about 5000 seconds after the beginning of heating. Beyond this point, the incoming flux is equal to the energy loss from the boundary, approximately.

Figure 6 shows heat conduction from the front boundary for a radiative flux of 2 $\frac{W}{cm^2}$. The change in the sign of conducted heat is caused by a change in the direction of the temperature gradient. Initially, the maximum temperature is located at the front boundary itself, and heat conduction occurs from the boundary into the solid and is in the positive x-direction. Later, the increasing natural convection loss from the boundary due to higher wall temperature results in a slower rise in boundary temperature as compared with that inside the solid, and the location of the maximum temperature shift to the inside of the solid. Heat conduction at the boundary therefore reverses direction. This figure clearly indicates that but for a very short duration at the beginning of the heating period, conduction transfer heat occurs in the negative direction. This result is totally different from that given by the opaque model. The front boundary of opaque model absorbs all the flux, and always has a maximum temperature.

As shown in the figure 3, there is a temperature peak inside the slab, and the reason is the natural convection loss from the front boundary. The heat transfer coefficient has been approximated by the equation (Ozisik, 1985):

$$h = 1.42(\frac{T_{surface} - t_{\infty}}{height\ of\ slab}) \quad (22)$$

To examine in detail this heat convection loss, we have artificially increased and decreased the heat transfer coefficient value computed using the above equation. The results are shown in figure 7. The radiative flux is 4.89 $\frac{W}{cm^2}$, and the irradiation period is 340 seconds. The solid curve in the figure is the original curve based on equation (22). The dashed curve above is the result when the heat transfer coefficient is artificially decreased by a factor of 10. The curve underneath the solid curve is for the case when the heat transfer coefficient is increased to ten times of the original value. As expected, a small heat transfer coefficient leads to small heat losses to the ambient, and gives higher temperatures in the solid. What is more important is that the temperature peak shifts to the front boundary. As for the higher heat transfer coefficient, the resultant temperature profile is lower and the temperature peak tends to shift to the inside of the slab. The difference between the peak temperature and the surface temperature also tends to increase.

Figure 8 gives a comparison of the truncated model and reradiating model at different values of thermal conductivity. The heat flux is 4.89 $\frac{W}{cm^2}$ and elapsed time is 300 seconds after beginning of heating. There are five groups of curve (according to the different thermal conductivities) in the figure. The solid lines represent the reradiating model, and the dashed lines the truncated model. The thermal conductivity k_r is the value for fused silica at temperature 25 $°C$. We artificially increase or decrease the thermal conductivity to change the radiation-conduction parameter to examine its effect on the truncated model. For the higher thermal conductivity cases (10 and 100 times of the original value), the truncated model gives higher but acceptable temperature profiles. Heat conduction dominates the energy transfer processes for these two cases, and the temperature profiles are very flat. For $k = 0.1kr$, the accuracy of the truncated model appears still acceptable; however, for $k = 0.01kr$, the truncated model is not acceptable when compared against the reradiating model.

Conclusion

In the present study a radiative-conductive heat transfer calculation scheme has been developed by adding the cold model to the Rosseland approximation. Although the results of the modified Rosseland approximation is acceptable, it is still not accurate enough when compared with the reradiating model. The main cause of this inaccuracy still lies in the inaccuracy of the optically thick assumption near the front boundary. The diffusion approximation has apparently failed to provide a proper thermal energy balance at the boundary. The truncated model artificially increase the emitted energy from the irradiated boundary, and leads to a significant improvement. The predicted temperature profile is almost identical to the reradiating model. The computation time for the truncated model is only about 5% of the reradiating model. Definitely, considering the accuracy and computation time, the truncated model is an acceptable engineering model for the combined radiation and conduction heat transfer problem.

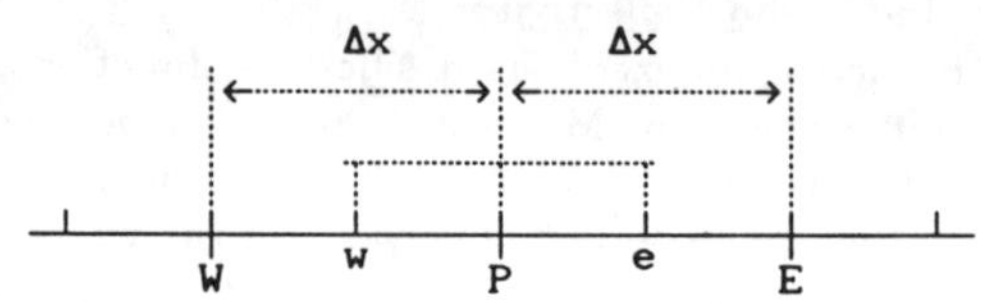

Fig. 1 Nomenclature for finite difference scheme

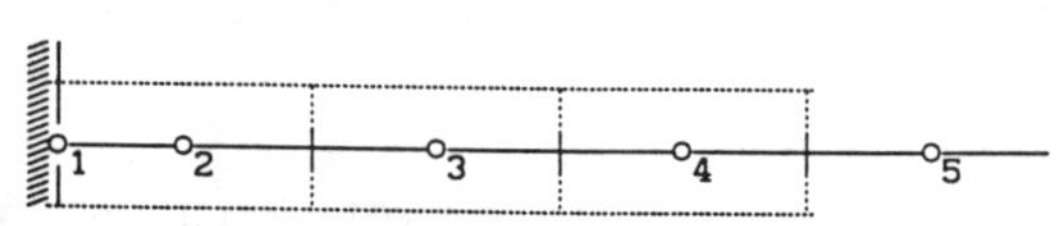

Fig. 2 Near boundary control volume

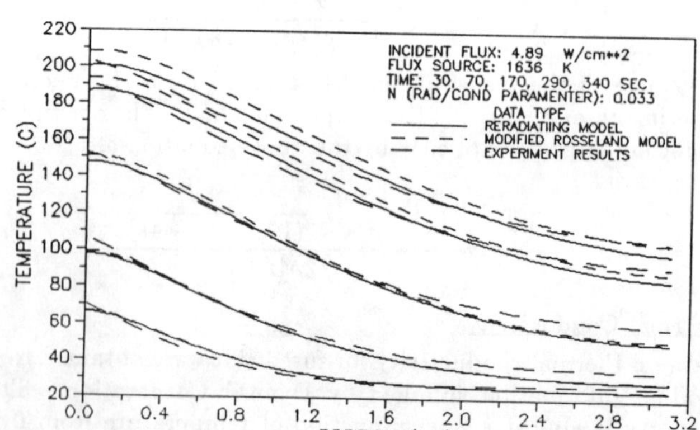

Fig. 3 Comparison of profiles for experiment results, reradiating and modified model

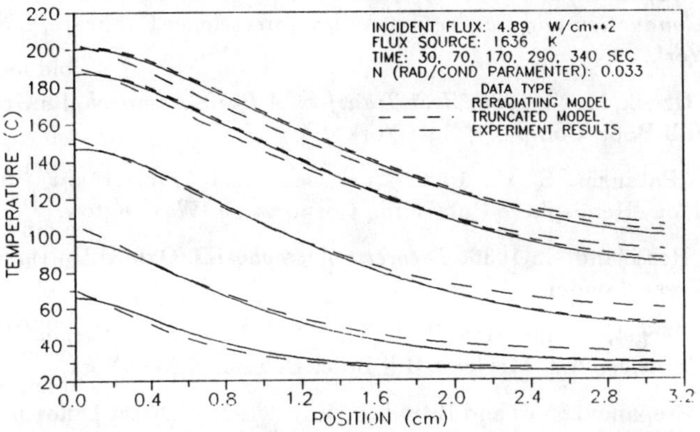

Fig. 4 Comparison of profiles for experiment results, reradiating and truncated model

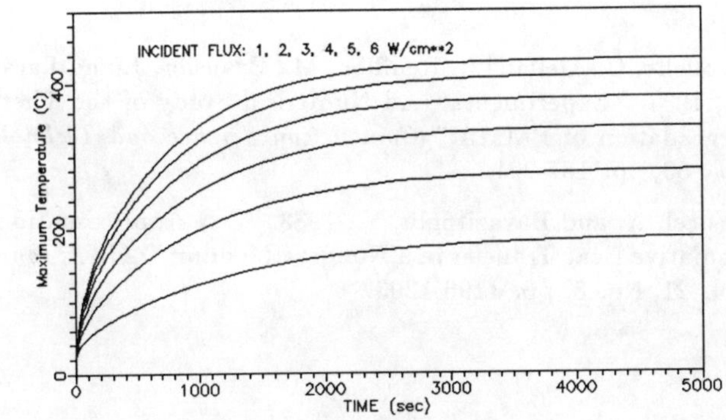

Fig. 5 Maximum temperature vs. time for different fluxes

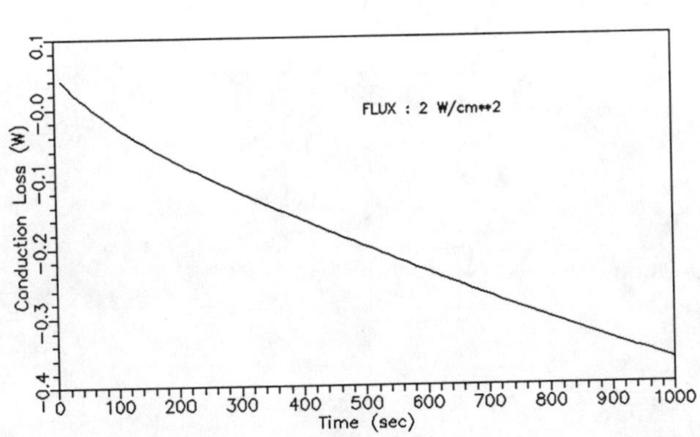

Fig. 6 Conduction heat loss from the front boundary

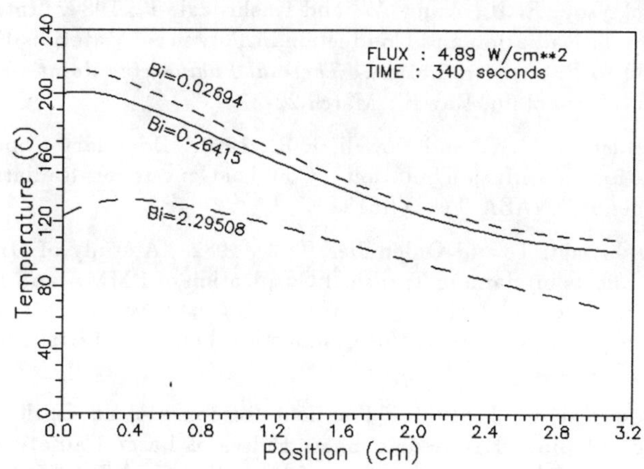

Fig. 7 Comparison of effect of Biot number at front surface

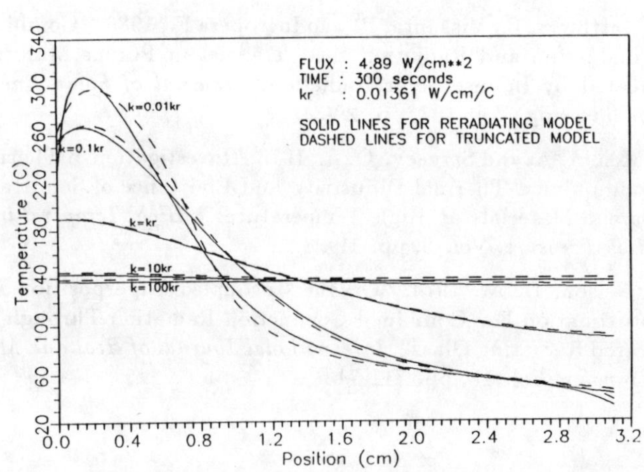

Fig. 8 Comparison of effect of thermal conductivity

REFERENCES

Brixner, B., 1967, "Refractive-Index Interpolation for Fused Silica," *Journal of the Optical Society of America*, Vol 57, pp. 674-676.

Cobble, M. H., 1964, "Irradiation into Transparet Solids and the Thermal Trap Effect," *Journal of the Franklin Institute*, Vol. 278, No. 6, pp. 383-393.

Condon, E. U., 1968, "Radiative Transport in Hot Glass," *Journal of Quant. Spectrosc. Radiation Transfer*, Vol. 8, pp. 369-385.

Dake, J. A.and Harleman, D. R., 1969, "Thermal Stratification in Lakes: Analytical Laboratory Studies," *Water Resource Research*, Vol. 5, pp. 484-495.

Deissler, R. G., 1964, "Diffusion Approximation for Thermal Radiation in Gases with Jump Boundary Conduction," *Journal of Heat Transfer*, Vol. 86, pp. 240-246.

Finlayson, E. U., Aung, W. and Kashiwagi, T., 1987, "Interaction of Radiation and Conduction in Polymeric Materials Exposed to External radiation," *Thermal Engineering Joint Conference*, Honolulu, Hawaii: March 22-27.

Goldstein, M. E. and Howell, J. R., 1969, "Boundary Conditions for the Diffusion Solution of Coupled Conduction-Radiation Problems," NASA TN D-4618.

Kashiwagi, T. and Ohlemiller, T. J., 1982, "A study of Oxygen Effects on flaming Transient Gasification of PMMA and PE During Thermal Irradiation," *Nineteenth Symposium (International) on Combustion*, the Combustion Institute, Pittsburgh, pp. 815-823.

Kowalski, G. J., 1987, "Transient Response of an Optically Thick Medium Exposed to Short Pulses of Laser Radiation," Fundamentals and Applications of Radiation Heat Transfer: ed. A. M. Smith and P. S. Smith, pp.67-74.

Malitson, I. H., 1965, "Interspecimen Comparison of the Refractive Index of Fused Silica," *Journal of the Optical Society of America*, Vol 55, pp. 1205-1210.

Matthews, L., Viskanta, R. and Incropera F., 1985, "Combined Conduction and Radiation Heat Transfer in Porous Materials Heated by Intense Solar Radiation," *Journal of Solar Energy Engineering*, Vol. 107, pp. 29-34.

Men, A. A. and Sergeev. O. A., 1973, "Investigation of Thermal conductivity, Thermal Diffusivity, and Emittance of Semitransparent Materials at High Temperatures," *High Temperature - High Pressure*, Vol. 5, pp. 19-28.

Nelson, D. A., 1974, "On the Uncoupled Superposition Approximation For Combined Conduction-Radiation Through Infrared Radiating Glass", *International Journal of Heat and Mass Transfer*, Vol. 18, pp. 711-713.

Ozisik, M. N., 1973, *Radiative Transfer and Interactions with Conduction and Convection*, Wiley-Interscience Publication, New York.

Ozisik, M. N., 1985, *Heat Transfer: A Basic Approach*, McGraw-Hill Book Company, New York.

Patankar, S. V., 1980, *Numerical Heat Transfer and Fluid Flow*, Hemisphere Publishing Corporation, Washington.

Rosseland, S., 1936, *Theoretic Astrophysics*, Oxford University Press, London.

Siegel, R. and Howell, J. R., 1981, *Thermal Radiation Heat Transfer*, 2nd, McGraw-Hill Book Company, New York.

Stepanov, S. V. and Bityukov, V. K., 1979, "Direct Differential Methods in the Theory of Radiative and Radiative-Conductive Heat Transfer," *High Temperature*, Vol. 17, pp. 348-359.

Stepanov, S. V. and Kim, D. M., 1980, "Radiation-Conduction Heat Transfer in a Plane Layer of Slightly Absorbing Medium," *Journal of Heat Transfer*, Vol. 102, pp. 388-390.

Viskanta, R. and Anderson, E. E., 1975, "Heat Transfer in Semitransparent Solids," *Advance in Heat Transfer*, Vol 11, pp. 316-441.

Vollelle, C., Delfau, J., Reuillon, M., Bransier, J. and Laraqui, N., 1987, "Experimental and Numerical Study of the Thermal Degradation of PMMA," *Combustion Science and Technology*, Vol. 53, pp. 187-201.

Yucel, A. and Bayazitoglu, Y., 1983, "P-N Approximation for Radiative Heat Transfer in a Nongray Medium," *AIAA Journal*, Vol. 21, No. 8, pp. 1196-1203.

1989 National Heat Transfer Conference
HTD-Vol. 106, Heat Transfer Phenomena in Radiation, Combustion and Fires

SCALES OF POOL FIRES

V. S. Arpaci and A. Selamet
Department of Mechanical Engineering
and Applied Mechanics
The University of Michigan
Ann Arbor, Michigan

ABSTRACT

A fundamental dimensionless number for pool fires,

$$\Pi_\beta = \left(\frac{\sigma_\beta}{1 + \sigma_\beta}\right) Ra_\beta$$

is proposed. Here σ_β and Ra_β denote a flame Schmidt number and a flame Rayleigh number. The sublayer thickness of a turbulent pool fire is shown in terms of Π_β to be

$$\frac{\delta_\beta}{\ell} \sim \Pi_\beta^{-1/3}$$

ℓ being an integral scale. The fuel consumption in a turbulent fire expressed in terms of δ_β leads to

$$\frac{m'}{\rho\, D\, Ra^{1/3}} \sim \frac{B}{(1 + B)^{2/3}}$$

ρ being density, D mass diffusivity, Ra the usual Rayleigh number, and B is the transfer number.

1. INTRODUCTION

So far the buoyancy-driven diffusion flame above a horizontal fuel in general or the pool fire in particular has remained mathematically untractable. A hybrid (analytical + dimensional) attempt, first neglects the effect of buoyancy, considers a one-dimensional (heat plus species) formulation in the direction of gravity, and obtains the fuel consumption in terms of an assumed stagnant film thickness. Next, it relates the fuel consumption to a power of the usual Grashof number (see, for example, Glassman 1987). A more realistic approach follows from some qualitative arguments and suggests, in terms of some dimensionless numbers, the prediction of large scale fires from the results of laboratory experiments (see,[1] for example, Blinov and Khudyakov 1959, 1961; Emmons 1961, 1973; Hottel 1961; Williams 1969; de Ris and Orloff 1972; Orloff and de Ris 1972; de Ris, Kanury and Yuen 1973; and Kanury 1975). The present study subscribes the philosophy of this approach. Moreover, it follows rigorous dimensional arguments with no reference to any empiricism which often is the rule rather than exception in studies on fires. Accordingly, following a recent preliminary study by Arpaci and Selamet (1985a), it proposes a basic dimensionless number and develops a model in terms of this number for turbulent pool fires. The model throws further light on experimental literature and the proposed dimensionless number is expected to be helpful in the presentation of future experimental results.

The study consists of seven sections: following this Introduction, Section 2 illustrates the theory of laminar

[1] For other pertinent works, refer, for example, to Eckert and Jackson 1951; Emmons 1956; Lees 1958; Akita and Yumoto 1965; Marxman 1965; Corlett 1968, 1970; Hall 1973; de Ris 1979; Kinoshita 1981; Babrauskas 1983; McCaffrey 1983; Mizner and Eyre 1983; Gengembre *et al.* 1984; Delichatsios 1987; Lockwood and Corlett 1987; Schneider and Kent 1987.

pool burning in terms of the proposed dimensionless number; Section 3 demonstrates the general nature of this number by developing a theory for turbulent pool fires in terms of the same number; Section 4 briefly reviews the recent literature on gas radiation; Section 5 modifies the physics of pool fires in terms of this literature; Section 6 provides some comparison with experimental data; and Section 7 concludes the study with some final remarks.

2. LAMINAR POOL FIRE. A DIMENSIONLESS NUMBER

Although the objective of the study is turbulent pool fires over a stagnant liquid fuel, first a basic dimensionless number is conveniently developed from laminar considerations.

On dimensional grounds, the balance of momentum (including the longitudinal and transversal inertial forces, viscous and buoyant forces)

$$U\frac{U}{\ell} + \nu\frac{U}{\delta^2} \sim g\left(\frac{\Delta\rho}{\rho}\right) - v_w\frac{U}{\delta}, \tag{1}$$

U being a characteristic longitudinal velocity, ℓ and δ longitudinal and transversal lengths characterizing the directions of flow and diffusion, respectively (Fig. 1). Also, the balance of the first Schvab-Zeldovich (heat+oxidizer) property including the transversal gain is

$$U\frac{B}{\ell} - v_w\frac{B}{\delta_\beta} \sim D\frac{B}{\delta_\beta^2} \tag{2}$$

where v_w is the wall value of the transversal velocity; δ_β the boundary layer thickness of β; $\beta = -b/B$; $Le=\alpha/D=1$, α and D being thermal and mass diffusivities, respectively; b and the transfer number B (Busemann 1933; Spalding 1963, 1979) are defined as,

$$b = \left[\left(Y_O - Y_{O\infty}\right)Q/\nu_O M_O + h\right] / h_{fg}, \tag{3}$$

and

$$B = b_\infty - b_w, \tag{4}$$

or, explicitly,

$$B = \left(Y_{O\infty}Q/\nu_O M_O - h_w\right) / h_{fg}. \tag{5}$$

Here, Y_O is the mass fraction of the oxidizer; $Y_{O\infty}$ its free-stream value; Q the heat released by ν_F moles of fuel; M_O the molecular weight of oxidizer; ν_O and ν_F the stoichiometric coefficients of oxidizer and fuel, respectively; h the specific enthalpy; and h_{fg} the heat of evaporation (see Spalding 1953, 1954a,b, 1960, 1963, 1979 for an integral approach). In terms of the wall mass balance,

$$\rho v_w \sim \rho D\frac{B}{\delta_\beta}, \tag{6}$$

equations (1) and (2) may be rearranged as

$$U\frac{U}{\ell} + \nu\frac{U}{\delta^2} \sim g\left(\frac{\Delta\rho}{\rho}\right) - D\,B\frac{U}{\delta\,\delta_\beta}, \tag{7}$$

$$U\frac{B}{\ell} \sim D(1+B)\frac{B}{\delta_\beta^2}, \tag{8}$$

and employing the Squire assumption for buoyancy-driven flows,

$$\delta \sim \delta_\beta, \tag{9}$$

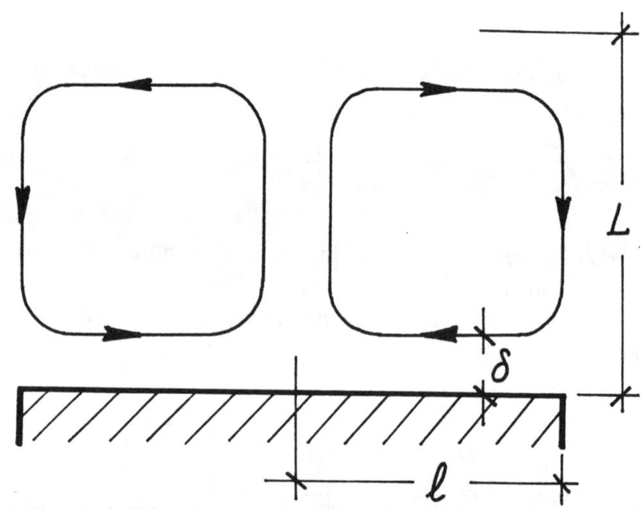

Fig. 1 Characteristic length scales

equation (7) may further be rearranged as

$$U\frac{U}{\ell} + \nu\left(1+\frac{B}{\sigma}\right)\frac{U}{\delta_\beta^2} \sim g\left(\frac{\Delta\rho}{\rho}\right), \tag{10}$$

where

$$\sigma = \nu/D$$

is the Schmidt number. For notational convenience, let

$$\nu_\beta = \nu\,(1 + B/\sigma) \quad \text{and} \quad D_\beta = D\,(1 + B)\,. \qquad (11)$$

Then, equations (8) and (10) are reduced to

$$U\,\frac{B}{\ell} \sim D_\beta\,\frac{B}{\delta_\beta^2}\,, \qquad (12)$$

$$U\,\frac{U}{\ell} + \nu_\beta\,\frac{U}{\delta_\beta^2} \sim g\left(\frac{\Delta\rho}{\rho}\right) \qquad (13)$$

which are identical in form to the balance of thermal energy and that of momentum for buoyancy-driven boundary layers.

A dimensionless number essential for laminar pool fires may now be obtained by coupling equations (12) and (13). Since velocity is a dependent variable for buoyancy-driven flows, its elimination between these equations yields

$$\frac{\ell}{\delta_\beta^4}\left(1 + \frac{D_\beta}{\nu_\beta}\right) \sim \frac{g}{\nu_\beta\,D_\beta}\left(\frac{\Delta\rho}{\rho}\right) \qquad (14)$$

which, in terms of a flame Schmidt number,

$$\sigma_\beta = \nu_\beta\,/\,D_\beta \qquad (15)$$

and a flame Rayleigh number,

$$Ra_\beta = \frac{g}{\nu_\beta\,D_\beta}\left(\frac{\Delta\rho}{\rho}\right)\ell^3\,, \qquad (16)$$

may be rearranged as

$$\frac{\ell}{\delta_\beta} \sim \Pi_\beta^{1/4}\,, \qquad (17)$$

where

$$\Pi_\beta \sim \left(\frac{\sigma_\beta}{1 + \sigma_\beta}\right) Ra_\beta \qquad (18)$$

is a fundamental dimensionless number for pool fires. Also, in view of

$$\frac{\Delta\rho}{\rho} \sim \frac{\rho_\infty - \rho_f}{\rho_f} = \frac{T_f - T_\infty}{T_\infty} \sim \frac{h_{fg}}{c_p T_\infty}\,, \qquad (19)$$

(see, Kim, de Ris, and Kroesser 1971 for a more involved model on $\Delta\rho/\rho$) the Rayleigh number may be more appropriately written as

$$Ra_\beta = \frac{g\,h_{fg}\,\ell^3}{\nu_\beta\,D_\beta\,c_p T_\infty}\,. \qquad (20)$$

Now, in terms of a (fuel) mass transfer coefficient h_β,

$$Sh_\beta = \frac{h_\beta\,\ell}{D} \sim \frac{\ell}{\delta_\beta}\,, \qquad (21)$$

Sh_β being a flame Sherwood number. Then, the fuel consumption from a fuel of width ℓ,

$$\frac{m'}{\rho\,D} = \frac{m''_w\,\ell}{\rho\,D} = Sh_\beta\,B = B\,\frac{\ell}{\delta_\beta}\,, \qquad (22)$$

(m''_w being the fuel consumption per unit area) may be written in terms of equation (17) as

$$\frac{m'}{\rho\,D} \sim B\,\Pi_\beta^{1/4}\,, \qquad (23)$$

or, explicitly,

$$\frac{m'}{\rho\,D} \sim B\left(\frac{\sigma_\beta}{1 + \sigma_\beta}\right)^{1/4} Ra_\beta^{1/4}\,, \qquad (24)$$

or, in terms of the usual Rayleigh number for mass transfer,

$$Ra = \frac{g}{\nu\,\alpha}\left(\frac{\Delta\rho}{\rho}\right)\ell^3\,, \qquad (25)$$

as

$$\frac{m'}{\rho\,D\,Ra^{1/4}} \sim B\left(\frac{\sigma_\beta}{1 + \sigma_\beta}\right)^{1/4}\left(\frac{\nu\,D}{\nu_\beta D_\beta}\right)^{1/4}. \qquad (26)$$

Noting equations (11) and (15),

$$\frac{\nu\,D}{\nu_\beta\,D_\beta} = \frac{1}{(1 + B/\sigma)\,(1 + B)}\,, \qquad (27)$$

$$\sigma_\beta = \sigma\left(\frac{1 + B/\sigma}{1 + B}\right), \qquad (28)$$

and assuming

$$\sigma \sim 1\,,$$

equation (26) may be reduced to

$$\frac{m'}{\rho\, D\, Ra^{1/4}} \sim \frac{B}{(1 + B)^{1/2}}, \tag{29}$$

or, to an equality,

$$\frac{m'}{\rho\, D\, Ra^{1/4}} = \frac{C_1\, B}{(1 + C_o\, B)^{1/2}}, \tag{30}$$

where C_o and C_1 are numerical constants to be determined from an analytical or computational study, or from experimental data. The foregoing dimensional arguments provide a complete understanding of the physical foundations of laminar pool fires. Extension of these arguments to turbulent pool fires, in view of recent developments on turbulent natural convection (Arpaci and coworkers 1985b, 1986a,b,c, 1987a, 1988a, 1989) presents no difficulties.

3. TURBULENT POOL FIRE

Let the turbulent fire over a horizontal solid fuel be controlled by a sublayer next to the fuel. The thickness of this layer may be obtained by invoking isotropy for turbulence. Accordingly, letting

$$\ell \to \delta_\beta \tag{31}$$

in equation (14) gives the thickness of sublayer,

$$\frac{1}{\delta_\beta^3}\left(1 + \frac{1}{\sigma_\beta}\right) \sim \frac{g}{\nu_\beta\, D_\beta}\left(\frac{\Delta\rho}{\rho}\right),$$

or,

$$\delta_\beta \sim \left(\frac{1 + \sigma_\beta}{\sigma_\beta}\right)^{1/3} \left[\frac{\nu_\beta\, D_\beta}{g\,(\Delta\rho/\rho)}\right]^{1/3}. \tag{32}$$

An alternative form of this scale may be expressed in terms of the buoyant production of turbulent energy $\mathcal{P}_\beta$ rather than the buoyant force. Note

$$\mathcal{P}_\beta \sim v\, g\left(\frac{\Delta\rho}{\rho}\right) \tag{33}$$

and obtain the isotropic velocity v from equation (12) by satisfying equation (31),

$$v \sim \frac{D_\beta}{\delta_\beta}. \tag{34}$$

Elimination of v between equations (33) and (34) gives

$$\mathcal{P}_\beta \sim \left(\frac{D_\beta}{\delta_\beta}\right) g\left(\frac{\Delta\rho}{\rho}\right), \tag{35}$$

and the elimination of $(\Delta\rho/\rho)$ between equations (32) and (35) yields

$$\delta_\beta \sim \left(\frac{1 + \sigma_\beta}{\sigma_\beta}\right)^{1/4} \left(\frac{\nu_\beta\, D_\beta^2}{\mathcal{P}_\beta}\right)^{1/4} \tag{36}$$

the microscale identical in form to that recently discovered by Arpaci (1986a) for buoyancy-driven flows. For $\sigma_\beta \to \infty$, equation (36) is reduced to the microscale

$$\delta_\beta \sim \left(\frac{\nu_\beta\, D_\beta^2}{\mathcal{P}_\beta}\right)^{1/4}, \qquad \sigma_\beta \to \infty \tag{37}$$

discovered by Batchelor (1959) for a passive contaminant in isotropic turbulent flow.

Returning back to the original form of δ_β given by equation (32), and relative to ℓ,

$$\frac{\ell}{\delta_\beta} \sim \Pi_\beta^{1/3} \tag{38}$$

which shows the universal nature of Π_β (valid under any flow condition). Then, following the steps leading to equation (30), the fuel consumption in a turbulent fire is found to be

$$\frac{m'}{\rho\, D\, Ra^{1/3}} = \frac{C_1\, B}{(1 + C_o\, B)^{2/3}}, \tag{39}$$

where C_o and C_1 are the numerical constants to be determined from experimental data (note that C_o and C_1 of equation 39 are not related to those of equation 30). Foregoing dimensional arguments leading to equations (30) and (39) provide the fuel consumption in a laminar and turbulent fire over a horizontal solid fuel provided the effect of radiation remains insignificant. Under the influence of radiation, these equations need to be modified. Before proceeding to these modifications, a brief review of the recent literature on gas radiation is pertinent.

4. RADIATIVE HEAT FLUX

As is well known, the primary effect of thermal radiation on high temperature problems is on the heat flux. Following some dimensional arguments, Arpaci and Tabaczynski (1982), (1984) have shown the radiative heat flux for an arbitrary optical thickness to be

$$q^R \sim -\left[\frac{\eta\,\tau}{1 + 3\,\tau^2\,(1 - \alpha\omega)/(1 - \omega)}\right] 4\,E_b\,, \quad (40)$$

where $\eta=(\kappa_P/\kappa_R)^{1/2}$ is the geometrically weighted nongrayness, κ_P and κ_R being Planck and Rosseland mean absorption coefficients, respectively; $\tau=\kappa_M\ell$ the optical thickness, $\kappa_M=(\kappa_P\kappa_R)^{1/2}$ being the geometrically weighted absorption coefficient, ℓ a characteristic length; $\omega=\kappa_s/\kappa_e$ the albedo of single scattering, κ_s and κ_e being scattering and extinction coefficients, respectively; $\omega=0$ corresponds to no scatter, $\alpha=0$ to isotropic scattering, and $0<\alpha<1$ to a degree weighted anisotropic scattering. However, the respective behavior of radiation as a sink and as a diffusion for small and large optical thicknesses is obscured in equation (40) and needs to be taken into account. Furthermore, the boundaries have not been included in the foregoing arguments and should also be incorporated into equation (40).

Following Arpaci and coworkers (1968, 1970a,b, 1987b, 1988b) the sink behavior of thin gas near a boundary is

$$\frac{dq_x^R}{dx} = 4\,\kappa_P\left[\left(E_b - E_{b\infty}\right) - \frac{\epsilon_w}{2}\left(E_{bw} - E_{b\infty}\right) E_2\right], \quad (41)$$

E_2 being the exponential integral of order two. On the boundary,

$$\left(\frac{dq_x^R}{dx}\right)_w = 4\,\kappa_P\left(1 - \frac{\epsilon_w}{2}\right)\left(E_{bw} - E_{b\infty}\right) \quad (42)$$

or, in terms of $\xi=x/\ell$,

$$u_R''' = \left(\frac{dq_\xi^R}{d\xi}\right)_w \quad (43)$$

and

$$u_R''' = 4\,\eta\,\tau\left(1 - \epsilon_w/2\right)\left(E_{bw} - E_{b\infty}\right),\ \tau < 1\,. \quad (44)$$

For not so large values of optical thickness (say $\tau \le 1$), the combination of equations (40) and (43) yields

$$u_R''' \sim \frac{4\,\eta\,\tau\left(1 - \epsilon_w/2\right)\left(E_{bw} - E_{b\infty}\right)}{1 + 3\,\tau^2\,(1 - \alpha\omega)/(1 - \omega)}\,,\quad \tau \le 1 \quad (45)$$

which shows the behavior of radiation as a thermal sink.

Also shown by Arpaci and coworkers (1968, 1969, 1970b, 1987b, 1988b), the thick gas flux near a boundary is

$$q_x^R = -\frac{4}{3\kappa_R}\left(1 - \rho_w\,E_3 - \frac{3}{2}\,E_4\right)\frac{dE_b}{dx}\,, \quad (46)$$

ρ being the wall reflectivity, E_3 and E_4 exponential integrals of order three and four. On the boundary,

$$q_w^R = \frac{\epsilon_w}{2}\left(-\frac{4}{3\kappa_R}\frac{dE_b}{dx}\right), \quad (47)$$

and far from the boundary

$$q_x^R = -\frac{4}{3\kappa_R}\frac{dE_b}{dx} \quad (48)$$

which is the Rosseland flux. Rearrange equations (47) and (48) in terms of temperature as

$$q_w^R = \frac{\epsilon_w}{2}\left(\frac{16\,\sigma\,T_w^3}{3\kappa_R}\right)\frac{dT}{dx} \quad (49)$$

and

$$q_{x=\ell}^R = -\left(\frac{16\,\sigma\,T_\infty^3}{3\kappa_R}\right)\frac{dT}{dx}\,. \quad (50)$$

The diffusive behavior of equations (47) through (50) is well-known. However, in terms of Fig. 2, the contribution

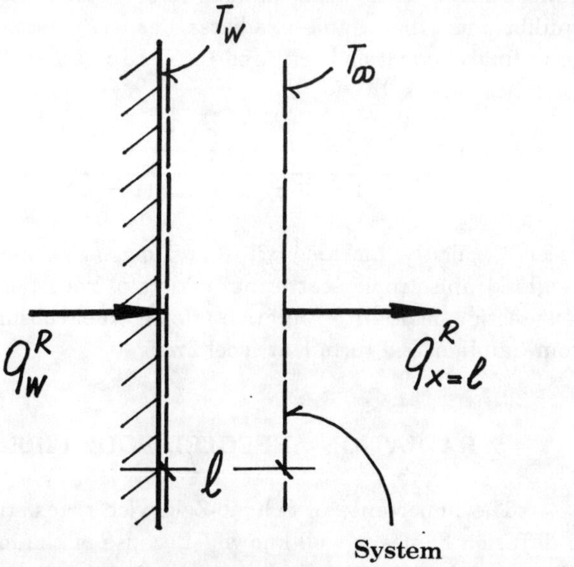

Figure 2. System for radiative balance of thick gas

of thick gas may also be reduced to a sink. The First Law applied to the system shown in the figure gives

$$q_w^R - q_{x=\ell}^R - u_R''' = 0 \,. \tag{51}$$

Let the temperature gradient across a characteristic length ℓ (say a boundary layer thickness) be constant. Then, in terms of equations (49), (50), and (51)

$$u_R''' = \frac{16\,\sigma\, T_\infty^3}{3\kappa_R}\left(1 - \frac{\epsilon_w}{2}\frac{T_w^3}{T_\infty^3}\right)\frac{dT}{dx}\,, \tag{52}$$

which may be rearranged, for a reasonably large optical thickness (say $\tau > 1$), in view of $\xi = x/\ell$ and equations (40) and (51), as

$$u_R''' \sim \frac{16\eta\tau\sigma T_\infty^3\,[1 - (\epsilon_w/2)(T_w/T_\infty)^3]\,(T_w - T_\infty)}{1 + 3\,\tau^2\,(1 - \alpha\omega)/(1 - \omega)}\,,\ \tau > 1\,. \tag{53}$$

The foregoing general considerations may now be applied to radiation affected pool fires.

On dimensional grounds, there are two domains in pool fires over a fuel; the first domain is between the (laminar) flame sheet or the location of the maximum temperature of (turbulent) flame and the fuel; the second domain is between the flame sheet (or maximum temperature) and the ambient. In general,

$$T_f \gg T_w \sim T_\infty\,, \tag{54}$$

and, the radiation from fuel and ambient can **quantitatively** be neglected relative to that from the flame. Also, under the assumption of local radiative equilibrium, the flame radiates as a black body. Accordingly, equations (45) and (53) are reduced, for any optical thickness, to

$$u_R''' \sim \frac{\eta\,\tau\,E_{bf}}{1 + 3\,\tau^2\,(1 - \alpha\omega)/(1 - \omega)} \tag{55}$$

which explicitly includes all (emission, absorption and weighted anisotropic scattering) effects of radiation. The next section deals with radiation affected fuel consumption from (laminar and turbulent) pool fires.

5. RADIATION-AFFECTED POOL FIRE

The importance of Schvab-Zeldovich transformation to diffusion flames is well-known. Because of the intrinsic difference between radiation and conduction (or any diffusion), this transformation no longer applies to radiation-affected flames. Consequently, one has to follow a computational approach by numerically and simultaneously solving the mass, momentum, thermal energy and species equations, or, introducing some intuitive arguments, modify the transfer number by recognizing the sink and/or diffusion effects) of radiation. Both of these effects lower the flame temperature, and have been already incorporated into the heat of combustion and the latent heat of evaporation by implicitly lowering these properties by fractional γ and ψ, respectively (see, for example, Kanury 1975). In view of the developments in Section 4 on radiation, γ and ψ may now be explicitly related to all (emission, absorption and weighted scattering) effects of radiation.

Let

$$m''\,Q^R = u_R''' \tag{56}$$

which appears to be reasonably supported by experiments (see Fig.4 in Kanury 1975). Then

$$\gamma = \frac{Q^R}{Q} = \frac{u_R'''}{m''\,Q}\,. \tag{57}$$

A thermal energy balance on the fuel surface

$$m''\,h_{fg} = -\,k\left(\frac{\partial T}{\partial y}\right)_w + Q_w^R \tag{58}$$

may be rearranged as

$$m''\,h_{fg} = -\,k\,(1 + \psi)\left(\frac{\partial T}{\partial y}\right)_w\,, \tag{59}$$

or,

$$m''\left(\frac{h_{fg}}{1 + \psi}\right) = -\,k\left(\frac{\partial T}{\partial y}\right)_w\,, \tag{60}$$

where

$$\psi = \frac{Q_w^R}{Q_w^K} = \frac{Q_w^R}{-\,k\,(\partial T/\partial y)_w}\,. \tag{61}$$

Equations (60) and (61) may respectively be interpreted as an increased conductivity or a reduced heat of evaporation. The intrinsic nature of radiation as an increased effective conductivity for $\tau > 1$ is well-known. Then, in terms of equations (57) and (61), the transfer number becomes

$$B = \left[\frac{Y_{O\infty}\,Q\,(1 - \gamma)}{\nu_O M_O} - c_p(T_b - T_\infty)\right]\frac{(1 + \psi)}{h_{fg}}\,.$$

6. EXPERIMENTAL SUPPORT

On fundamental grounds, the proposed theory leads to the well-known exponents 1/4 and 1/3 for buoyancy-driven laminar and turbulent flows, respectively. Experimental data and previously proposed empirical models on fires support the same exponents. The present theory proposes a transfer number dependence which is different than the existing empirical models which are usually based on the theory of stagnant flame. The experimental data on small size fires (see Corlett 1968, 1970, de Ris and Orloff 1972) appears to correlate well with the proposed theory against the transfer number as shown in Fig. 3. The explicit form of the correlation is

$$\frac{m'}{\rho\, D\, Ra^{1/3}} = \frac{0.12\, B}{(1 + 0.16\, B)^{2/3}}$$

for turbulent fires.

7. CONCLUDING REMARKS

In this study a fundamental dimensionless number

$$\Pi_\beta \sim \left(\frac{\sigma_\beta}{1 + \sigma_\beta}\right) Ra_\beta$$

is introduced for (laminar and turbulent) diffusion flames in general and pool fires in particular. Neglecting the effect of σ_β, the fuel consumption for laminar and turbulent pool fires are given in terms of Ra_β. By definition,

$$\sigma_\beta = \left(\frac{\text{Viscous force}}{\text{Inertial force}}\right)\left(\frac{\text{Flow of B}}{\text{Diffusion of B}}\right).$$

Since the inertial/viscous force ratio is usually small for buoyancy-driven flows, neglecting inertial effects, let

$$\sigma_\beta \to \infty$$

and

$$\left(\frac{\sigma_\beta}{1 + \sigma_\beta}\right) \to 1\,.$$

However, for large turbulent fires, inertial effects may become important and need to be taken into account through σ_β.

The most important outcome of this study is the discovery of a diffusion (or dissipation) scale δ_β for pool fires. It is shown relative to an integral scale ℓ that

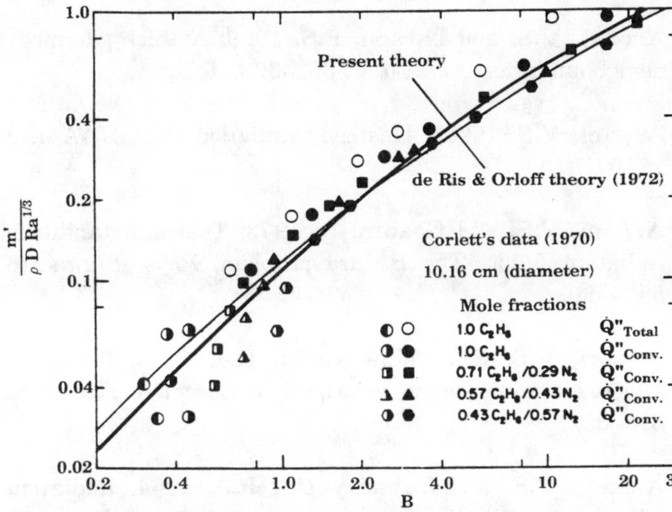

Figure 3. Experimental data versus theory (Original figure from de Ris and Orloff, 1972)

$$\frac{\delta_\beta}{\ell} \sim \Pi_\beta^{-n},$$

the exponent of Π_β assuming fractional values $n = 1/4$ and 1/3 for laminar and turbulent fires, respectively.

Recent developments on the transport and properties of radiation so far do not appear to be explicitly incorporated into radiation affected pool fires. The intuitive approach involving implicit fractional reduction in the heat of combustion and the latent heat of evaporation for the radiation effect is extended by interpreting the fractional reductions in terms of the explicit emission, absorption and weighted scattering effects of radiation.

The experimental data on radiating pool fires lacks sufficient information on the emission, absorption and weighted effects of radiation, and may not yet be correlated with the proposed theory in the present study. With the assumption of small radiation effect, an excellent correlation between some of the experimental data and the proposed theory has been established.

REFERENCES

Akita, K. and **Yumoto, T.** 1965, Heat transfer in small pools and rates of burning of liquid methanol, *Tenth Symposium (International) on Combustion*, pp. 943–948. The Combustion Institute, Pittsburgh.

Arpaci, V.S. 1968, Effect of thermal radiation on the laminar free convection from a heated vertical plate, *Int. J. Heat Mass Transfer* **11**, 871–881.

Arpaci, V.S. and **Larsen, P.S.** 1969, A thick gas model near boundaries, *AIAA J.* **7**, 602–606.

Arpaci, V.S. 1970b, Unsteady radiation slip, *AIAA J.* **8**, 1910–1913.

Arpaci, V.S. and **Gözüm, D.** 1973, Thermal stability of radiation fluids: The Bénard problem, *Phys. Fluids* **16**, 581–588.

Arpaci, V.S. and **Tabaczynski, R.J.** 1982, Radiation-affected laminar flame propagation, *Combust. Flame* **46**, 315–322.

Arpaci, V.S. and **Tabaczynski, R.J.** 1984, Radiation-affected laminar flame quenching, *Combust. Flame* **57**, 169–178.

Arpaci, V.S. and **Selamet, A.** 1985a, Pool fire, in *Heat Transfer in Fire and Combustion Systems, The 23rd ASME National Heat Transfer Conference*, HTD Vol. **45**, 153–157, Denver, Colorado.

Arpaci, V.S. and **Kabiri-Bamoradian, K.** 1985b, A theory for turbulent natural convection, *The 23rd ASME National Heat Transfer Conference*, Paper No. 85-HT-24, Denver, Colorado.

Arpaci, V.S. 1986a, Microscales of turbulence and heat transfer correlations, *Int. J. Heat Mass Transfer* **29**, 1071–1078.

Arpaci, V.S. 1986b, Two thermal microscales for natural convection and heat transfer correlations, *ASME HTD-60*, 117–121.

Arpaci, V.S. 1986c, Two thermal microscales and a heat transfer model, *39th APS Meeting*, Colombus, Ohio.

Arpaci, V.S. and **Dec, J.E.** 1987a, A Theory for buoyancy-driven turbulent flows, *The 24th ASME National Heat Transfer Conference*, Paper No. 87-HT-5, Pittsburgh, Pennsylvania.

Arpaci, V.S. 1987b, Radiative entropy production – Lost heat into entropy, *Int. J. Heat Mass Transfer* **30**, 2115–2123.

Arpaci, V.S. 1989, Microscales of turbulence and heat transfer correlations, to appear in *Ann. Rev. Num. Fluid Mechanics and Heat Transfer*.

Babrauskas, V. 1983, Estimating large pool fire burning rates, *Fire Technology* **19**, 251–261.

Batchelor, G. K. 1959, Small-scale variation of convected quantities like temperature in a turbulent fluid, *J. Fluid Mech.* **5**, 113–133.

Blinov, V.I. and **Khudyakov, G.N.** 1959, Certain laws governing diffusive burning of liquids, (Reviewed by H.C. Hottel), *Fire Res. Abstr. Rev.* **1**, pp. 41–44. National Academy of Sciences – National Research Council, Washington, D.C.

Blinov, V.I. and **Khudyakov, G.N.** 1961, Diffusion burning of liquids, *T-1490 a-c, AD-296-762*. Academv of Sciences, Moscow.

Busemann, A. 1933, *Der Wärme-und Stoffaustausch*. Springer, Berlin.

Corlett, R.C. 1968, Gas fires with pool-like boundary conditions, *Combust. Flame* **12**, 19–32.

Corlett, R.C. 1970, Gas fires with pool-like boundary conditions: Further results and interpretation, *Combust. Flame* **14**, 351–360.

Delichatsios, M.A. 1987, Air entrainment into buoyant jet flames and pool fires, *Combust. Flame* **70**, 33–46.

de Ris, J. 1979, Fire radiation – A review, *Seventeenth Symposium (International) on Combustion*, pp. 1013–1016. The Combustion Institute, Pittsburgh.

de Ris, J., **Kanury, A.M.**, and **Yuen, M.C.** 1973, Pressure modeling on fires, *Fourteenth Symposium (International) on Combustion*, pp. 1033–1044. The Combustion Institute, Pittsburgh.

de Ris, J. and **Orloff, L.** 1972, A dimensionless correlation of pool burning data, *Combust. Flame* **18**, 381–388.

Eckert, E.R.G. and **Jackson, J.W.** 1951, Analysis of turbulent free-convection boundary layer on flat plate, *NACA TR 1015*.

Emmons, H.W. 1956, The film combustion of liquid fuel, *Z. angew. Math. Mech.* **36**, 60–71.

Emmons, H.W. 1961, *The Use of Models in Fire Research*, Publication 786, pp. 50–67, 121–123. National Academy of Sciences – National Research Council, Washington, D.C.

Emmons, H.W. 1973, Heat transfer in fire, *J. Heat Transfer* **95**, 145–151.

Gengembre, E., **Cambray, P.**, and **Bellet, J.C.** 1984, Turbulent diffusion flames with large buoyancy effects, *Combust. Sci. Tech.* **41**, 55–67.

Glassman, I. 1987, *Combustion* (2nd Edn), pp. 281–282. Academic Press, New York.

Hall, A.R. 1973, Pool burning, *Oxidation and Combustion Reviews* **6**, 169–225.

Hottel, H.C. 1961, The Use of Models in Fire Research, Publication 786, National Academy of Sciences – National Research Council, Washington, D.C., p.32.

Kanury, A.M. 1975, Modeling of pool fires with a variety of polymers, *Fifteenth Symposium (International) on Combustion*, pp. 193–202. The Combustion Institute.

Kim, J.S., **de Ris, J.**, and **Kroesser, F.W.** 1971, Laminar free convective burning of fuel surfaces, *Thirteenth Symposium (International) on Combustion*, pp. 949–961. The Combustion Institute, Pittsburgh.

Kinoshita, C.M. 1981, Stagnant-film combustion of an absorbing-emitting gas, *Combust. Flame* **43**, 291–302.

Larsen, P.S. and **Arpaci, V.S.** 1988a, Two models for natural convection, *ASME Proc. 25th National Heat Transfer Conference, HTD – Vol.96*, Vol.**2**, 207–211, Houston, Texas.

Lees, L. 1958, Convective heat transfer with mass addition and chemical reactions, in *Combustion and Propulsion – Third AGARD Symposium*, p. 451. Pergamon Press, London.

Lockwood, R.W. and **Corlett, R.C.** 1987, Radiative and convective feedback heat flux in small turbulent pool fires with variable pressure and ambient oxygen, in *Heat and Mass transfer in Fires*, ASME/AIChE – 24th National Heat Transfer Conference, HTD-Vol. **73** (Edited by A.K. Kulkarni, Y. Jaluria), pp. 421–426. Pittsburgh, PA.

Lord, H. and **Arpaci, V.S.** 1970a, Effect of nongray thermal radiation on laminar forced convection over a heated horizontal plate, *Int. J. Heat Mass Transfer* **13**, 1737–1751.

Marxman, G.A. 1965, Combustion in the turbulent boundary layer on a vaporizing surface, *Tenth Symposium (International) on Combustion*, pp. 1337–1349. The Combustion Institute, Pittsburgh.

McCaffrey, B.J. 1983, Momentum implications for buoyant diffusion flames, *Combust. Flame* **52**, 149–167.

Mizner, G.A. and **Eyre, J.A.** 1983, Radiation from liquefied gas fires on water, *Combust. Sci. Tech.* **35**, 33–57.

Orloff, L. and **de Ris, J.** 1972, Cellular and Turbulent Ceiling Fires, *Combust. Flame* **18**, 389–401.

Schneider, M.E. and **Kent, L.A.** 1987, Measurements of gas velocities and temperatures in a large open pool fire, in *Heat and Mass transfer in Fires*, ASME/AIChE – 24th National Heat Transfer Conference, HTD-Vol. **73** (Edited by A.K. Kulkarni, Y. Jaluria), pp. 37–47. Pittsburgh, PA.

Selamet, A. and **Arpaci, V.S.** 1988b, Entropy production in radiation-affected boundary layers, *The AIAA Thermophysics, Plasmadynamics and Lasers Conference*, Paper No. AIAA-88-2640, San Antonio, Texas.

Spalding, D.B. 1953, The combustion of liquid fuels, *Fourth Symposium (International) on Combustion (Combustion and Detonation Waves)* , pp. 847–864. The Williams and Wilkins Co., Baltimore.

Spalding, D.B. 1954a, Mass transfer in laminar flow, *Proc. Roy. Soc.* **A221**, 78–99.

Spalding, D.B. 1954b, Mass transfer from a laminar stream to a flat plate, *Proc. Roy. Soc.* **A221**, 100–104.

Spalding, D.B. 1960, A standard formulation of the steady convective mass transfer problem, *Int. J. Heat Mass Transfer* **1**, 192–207.

Spalding, D. B. 1963, *Convective Mass Transfer*. Edward Arnold, London.

Spalding, D. B. 1979, *Combustion and Mass Transfer*. Pergamon Press, Oxford.

Williams, F.A. 1969, Scaling mass fires, *Fire Research Abstracts and Reviews* **11**, 1–23.

1989 National Heat Transfer Conference
HTD-Vol. 106, Heat Transfer Phenomena in Radiation, Combustion and Fires

COMBINED NATURAL CONVECTION AND FORCED FLOW THROUGH SMALL OPENINGS IN A HORIZONTAL PARTITION, WITH SPECIAL REFERENCE TO FLOWS IN MULTICOMPARTMENT ENCLOSURES

M. Epstein and M. A. Kenton
Fauske and Associates, Inc.
Burr Ridge, Illinois

ABSTRACT

Estimates of the magnitude of buoyancy-driven exchange flows through openings in partitions that separate compartments are needed to assess the movement of toxic gases and smoke through buildings. An experiment using water and brine as a substitute for a light gas moving in a dense gas was designed to measure combined forced and buoyancy-driven exchange flow through a single opening in a horizontal partition. No theoretical treatment exists for this configuration. The same apparatus was used to determine the magnitude of the forced flow required to purge the opening of the oppositely directed buoyant component (i.e., the "flooding" limit). Finally, combined forced and buoyancy-driven flows in a multicompartment enclosure were measured. It has been demonstrated that the combined forced and buoyancy generated flows in the multicompartment structure can be predicted by a direct application of the results of the study of exchange flow through a single opening.

NOMENCLATURE

C_D Opening orifice coefficient.

D Diameter of circular opening (aperture or tube).

g Acceleration due to gravity.

L Axial length of opening (partition thickness or tube length).

q Flooding or purging flow rate.

Q Buoyancy-driven exchange flow rate.

Q_{BF} Buoyancy-driven component of combined convection.

Q_{cc} Purely buoyancy-driven exchange flow rate across a partition.

Q_u One-way buoyancy-driven or supply flow rate.

S Length of a side for a square opening.

u Velocity of fluid within opening.

W Width of rectangular opening.

z Vertical distance measured from center of opening in wall.

z_n Vertical location of plane of zero velocity in opening in wall.

$\Delta\rho$ Density difference, $\rho_H - \rho_L$.

ρ_L Density of light fluid.

ρ_H Density of heavy fluid.

$\bar{\rho}$ Mean density, $(\rho_H + \rho_L)/2$.

Subscripts

1,2,3 Compartment numbers in three-compartment apparatus.

INTRODUCTION

The present paper is a continuation of the work recently published by Epstein (1988). The previous paper dealt with buoyancy-driven exchange flow through an opening in a horizontal partition that separates two otherwise closed compartments. The fluid in the upper compartment is more dense than that in the lower compartment. The density difference induces a buoyancy-driven, countercurrent downflow of the heavier fluid and an upflow of the lighter fluid through the opening. A universal correlation between Froude number (dimensionless exchange flow rate) and opening aspect ratio was obtained. Epstein (1988) also investigated the buoyancy-driven flow with two openings in the horizontal partition. Two openings were observed to give rise to three different flow configurations involving both one-way and countercurrent flows within the openings. Indications were that a purging (or flooding) mechanism accounted for the observed transition between one-way and countercurrent flow in an opening, much as has been observed in vertical two-phase flows.

In the present paper we report an experimental study which provides an empirical formula for the one-way flooding flow rate that is necessary to prevent countercurrent exchange flow within a single opening in a horizontal partition. Also experimental results for forced fluid flow and corresponding reverse natural convection flow through a single opening in a horizontal partition are presented. An empirical correlation is developed that yields the magnitude of the buoyant component of this combined flow in terms of the pure countercurrent flow rate (Epstein, 1988) and the flooding flow rate for the opening. We shall see later in the paper, via an experimental study of natural and forced convection in a model three compartment structure, that this correlation for combined flow in a single opening may represent the key to prediction of the convection patterns within complex building configurations.

As in the previous study (Epstein, 1988) and in the study of Steckler, et.al (1986) on the movement of smoke fronts during fires in multiroom structures, brine and water were used to simulate buoyancy-driven gas movement between compartments. This so-called salt water modeling technique is known to be valid so long as molecular viscosity and thermal diffusivity are not important parameters. Steckler, et.al (1986) examined the nondimensional (scaled) governing equations of mass, momentum, and energy transfer for free turbulent flow and concluded that the molecular transport terms are negligible in comparison with the other terms when the Reynolds number is larger than 10^4 and, therefore, that the equations are essentially identical for hot and cold gas and water-brine flows. It is noteworthy in this regard that Prahl and Emmons (1975) experimentally confirmed the applicability to gas exchange flows across vertical partitions of data obtained with liquid-liquid systems for Reynolds numbers as low as 10^3. One can arrive at the same conclusion for buoyancy induced flows across horizontal partitions by comparing the brine-water data of Epstein (1988) with the hot gas flow data of Brown (1962).

There are two reasons for examining the exchange flow through a small opening in a horizontal partition in some detail. First, the rate of movement of radiological or toxic gases, or smoke in tall industrial or commercial buildings under emergency conditions is mainly a function of leakage between floors as opposed to walls. Secondly, from a fundamental point of view, there is a big difference between horizontal and vertical openings (passages) with respect to the relative motion of the two fluids. The pattern of flow in the vertical passage (horizontal partition) is chaotic and may not amenable to analysis, whereas the two fluids in a horizontal passage (vertical partition) are observed to be stably stratified. The application of simple momentum flux equations including an empirically determined flow coefficient to the superposed layers appears justified for the horizontal passages (Steckler, et.al, 1984; Prahl and Emmons, 1975; and Brown and Solvason, 1962). A theoretical and experimental study of stratified, combined forced flow and natural convection across a rectangular opening in a vertical partition has been reported by Shaw (1971). To the present authors' knowledge, the phenomenon of combined forced and natural exchange flow through a small opening in a horizontal partition has not been dealt with in the literature.

COMBINED NATURAL CONVECTION AND FORCED FLUID FLOW THROUGH AN OPENING IN A VERTICAL PARTITION

As a means of introducing our empirical approach to simultaneous forced flow and reverse natural convection flow through an opening in a horizontal partition, let us first consider the simpler case of such a combined flow in an opening in a vertical partition. As mentioned in the Introduction, previous experimental studies for this case have shown that the lateral compartments exchange mass via a horizontal stratified flow which can be treated semiempirically by standard hydraulic practice. Shaw's theory (1971) of combined flow through an opening in a vertical partition is expressed in terms of an excess supply pressure acting on the natural convection flow, which is not a convenient form for the purposes of the present study. As we shall see later on, the following treatment of combined horizontal flow through an opening in a vertical partition will serve to guide our experimental approach to combined vertical flow through an opening in a horizontal partition.

Consider the two-compartment configuration connected by a common opening in Fig. 1. For the sake of simplicity, we first focus our attention on a rectangular opening of width W and height H. Compartment 1 contains the heavier fluid of density ρ_H while compartment 2 is filled with the lighter fluid of density ρ_L. The contents of each compartment are assumed to be well mixed. Note that the compartments are not sealed. Fluid is supplied to compartment 1 and extracted from compartment 2 at a volumetric flow rate Q_u.[1] We are interested in determining the buoyancy-driven leakage back-flow Q_{BF} from compartment 2 to compartment 1 against the forced fluid flow Q within the opening.

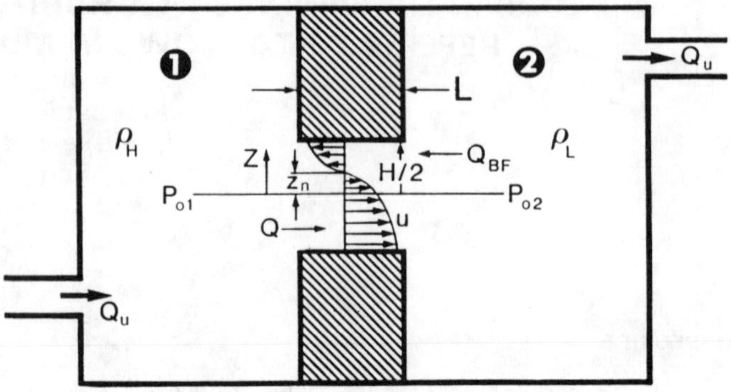

Fig. 1 Schematic representation of combined buoyantly-driven and forced flow through a rectangular opening in a vertical partition.

The pressure P_1 in compartment 1 at a level z above the midplane is

$$P_1 = P_{01} - \rho_H g z \tag{1}$$

and the pressure P_2 in compartment 2 at this elevation is

$$P_2 = P_{02} - \rho_L g z \tag{2}$$

Note that the pressures P_{01} and P_{02} at the midplane are not equal owing to the excess pressure in compartment 1 due to the supply flow. The pressure difference between compartments at level z is

$$P_2 - P_1 = P_{02} - P_{01} + (\rho_H - \rho_L) g z \tag{3}$$

The flow velocity u within the opening at level z is obtained by using the Bernoulli equation. Thus

$$u = \sqrt{\frac{2(P_2 - P_1)}{\bar{\rho}}} = \sqrt{\frac{2\Delta\rho g z}{\bar{\rho}} - \frac{2(P_{01} - P_{02})}{\bar{\rho}}} \tag{4}$$

Strictly speaking, the use of an average density in Eq. (4) is only valid when $\Delta\rho/\bar{\rho} << 1$. However, this form of the Bernoulli equation permits a useful degree of simplicity in the analysis that follows and is sufficiently accurate for many situations of practical interest. The buoyant, back-flow through the opening, Q_{BF}, is obtained by integrating u over the vertical distance z from the "neutral plane". The neutral plane is defined by the condition that the velocity u is zero, namely [see Fig. 1 and Eq. (4)]

$$z_n = \frac{P_{01} - P_{02}}{g\Delta\rho} \tag{5}$$

Integrating from the neutral plane to the top of the opening at z = H/2 we obtain the total back flow:

$$Q_{BF} = C_D W \int_{z_n}^{H/2} u dz = \frac{2 C_D W}{3} \left(\frac{2\Delta\rho g}{\bar{\rho}}\right)^{1/2} \left(\frac{H}{2} - z_n\right)^{3/2} \tag{6}$$

1. The subscript u pertains to unidirectional or one-way flow.

where the pressure difference $P_{01} - P_{02}$ has been eliminated in favor of the neutral height via Eq. (5). The quantity C_D is the discharge (or orifice) coefficient of the opening, to be determined by experiment.

In the limit $Q_u = 0$, there is no net volume flow across the opening and fluid exchange between compartments takes place via purely buoyancy-driven countercurrent flow. Under this condition, the zero velocity plane coincides with the midplane of the opening. Thus the purely buoyancy-driven volumetric flow rate through half of the opening, Q_{cc}, is obtained by setting $z_n = 0$ in Eq. (6). The result is

$$\frac{Q_{cc}}{\sqrt{W^2 H^3 g\Delta\rho/\bar{\rho}}} = \frac{C_D}{3} \tag{7}$$

Equation (6) can be transformed into a more compact form by normalizing it with respect to the purely countercurrent exchange rate to obtain

$$\frac{Q_{BF}}{Q_{cc}} = \left(1 - \frac{2\,z_n}{H}\right)^{3/2} \tag{8}$$

Now the flow rate Q from compartment 1 to compartment 2 (see Fig. 1) is, after some simplification,

$$Q = C_D\, W \int_{-\frac{H}{2}}^{z_n} u\,dz = \frac{2\,C_D\,W}{3}\left(\frac{2\Delta\rho g}{\bar{\rho}}\right)^{1/2}\left(\frac{H}{2} + z_n\right)^{3/2} \tag{9}$$

or, in dimensionless form,

$$\frac{Q}{Q_{cc}} = \left(1 + \frac{2\,z_n}{H}\right)^{3/2} \tag{10}$$

A volumetric flow balance on, say, compartment 1 in Fig. 1 requires that

$$Q_u = Q - Q_{BF} \tag{11}$$

Combining Eqs. (8), (10), and (11) gives the following expression for the location of the neutral plane within a rectangular opening in inverted form:

$$\frac{Q_u}{Q_{cc}} = \left(1 + \frac{2\,z_n}{H}\right)^{3/2} - \left(1 - \frac{2\,z_n}{H}\right)^{3/2} \tag{12}$$

The purging or flooding flow rate $Q_u = q$ necessary to render $Q_{BF} = 0$ is achieved when the neutral height is positioned at the top of the opening, i.e. $z_n = H/2$. Thus, from Eqs. (7) and (12),

$$q = 2^{3/2}\, Q_{cc} = \frac{2\,C_D}{3}\sqrt{\frac{2\,W^2\,H^3\,g\Delta\rho}{\bar{\rho}}} \tag{13}$$

is the flooding condition for a rectangular opening in a wall.

When the quantity z_n/H lies somewhere between 0 and 1/2, which is the range of physical interest, the right-hand-side of Eq. (12) can be approximated by the simple linear function

$$\frac{Q_u}{Q_{cc}} \simeq 2^{5/2}\,\frac{z_n}{H} \tag{14}$$

to within an accuracy of 5 percent. Using Eqs. (13) and (14), Eq. (8) can be rewritten in terms of Q_{cc} and q as follows:

$$Q_{BF} = Q_{cc}\,(1 - Q_u/q)^{3/2} \tag{15}$$

For combined flow through a <u>circular</u> opening in a wall; the theoretical results corresponding to Eqs. (7), (8), and (10) are

$$\frac{Q_{cc}}{\sqrt{D^5 g\Delta\rho/\bar{\rho}}} = \frac{C_D}{2}\int_0^1 \sqrt{y(1 - y^2)}\,dy = 0.24\,C_D \tag{16}$$

$$\frac{Q_{BF}}{Q_{cc}} = 2.083\,(1 - z_n/R)^{3/2}\int_0^1 y^{1/2}\sqrt{1 - [(1 - z_n/r)y + z_n/R]^2}\,dy \tag{17}$$

and

$$\frac{Q}{Q_{cc}} = 2.083\,(1 + z_n/R)^{3/2}\int_0^1 y^{1/2}\sqrt{1 - [z_n/R - (1 + z_n/R)y]^2}\,dy \tag{18}$$

By setting $z_n/R = 1$ in Eq. (17), integrating the result, and using Eq. (16), we get the flooding flow rate for circular openings in walls:

$$q = 3.143\,Q_{cc} = 0.75\,C_D\sqrt{D^5\,g\Delta\rho/\bar{\rho}} \tag{19}$$

The integrals of Eqs. (17) and (18) were evaluated numerically for the entire range of z_n/R values of interest. The resulting numerical data for Q_{BF}, Q, and Q_u, the latter quantity determined from Eq. (11), fit well the expression

$$Q_{BF} = Q_{cc}\,(1 - Q_u/q)^{1.6} \tag{20}$$

Equations (15) and (20) express the buoyantly-driven volumetric flow rates, Q_{BF}, through openings in vertical walls in the presence of oppositely directed forced flows Q_u. We note that the buoyant flow $Q_{BF} \rightarrow 0$ as the flooding limit $Q_u = q$ is approached, and that the condition of purely buoyantly-driven countercurrent flow is reached when $Q_u = 0$.

It is apparent from our previous work (Epstein, 1988) that it is not possible to derive from hydraulic theory alone similar functional relationships between Q_{BF} and Q_u which are applicable to openings in horizontal partitions, in view of the mingling of the heavy and light streams in the vertical orientation. For purposes of developing an empirical correlation for combined flow across a horizontal partition, it will be assumed here that the nature of the variation of Q_{BF} with Q_u is similar to that suggested by Eqs. (15) and (20). Thus it is believed to be reasonable to proceed by first correlating measurements of the purely buoyantly-driven countercurrent flow rate Q_{cc} and the flooding flow rate q and then determine if the experimental results for Q_{BF} conform closely to the equation $Q_{BF} = Q_{cc}\,(1 - Q_u/q)$, where m is some constant exponent. In this regard Epstein (1988) has already provided an empirical correlation for the buoyantly-driven countercurrent exchange rate Q_{cc} in the form of the Froude number $Q_{cc}/(D^5 g\Delta\rho/\bar{\rho})$ versus the length-to-diameter ratio of the opening. His result is

$$\frac{Q_{cc}}{(D^5 g\,\Delta\rho/\bar{\rho})^{1/2}} = \frac{0.055\,[1 + 400\,(L/D)^3]^{1/6}}{\left\{1 + 0.00527[1 + 400\,(L/D)^3]^{1/2}[(L/D)^6 + 117\,(L/D)^2]^{3/4}\right\}^{1/3}} \tag{21}$$

This equation gives the volumetric flux of either fluid within a single opening connecting two closed compartments (i.e., pure countercurrent flow) for L/D's in the range 0.01-20 and $\Delta\rho/\bar{\rho}$'s in the range 0.025-0.17. While the above correlation is based mainly on data with round openings, the experimental evidence suggests that Eq. (21) can be applied with reasonable accuracy to square openings of characteristic dimension S (length of a side) by making the identification D = 1.1 S. Based on the theory presented in the foregoing for flow through openings in vertical partitions, we tentatively assume that the appropriate effective diameter for a rectangular opening of width W and length H is $D = 1.1\ (H^3 W^2)^{1/5}$ [see Eq. (7)].

In the following section, experimental results are presented for the volumetric flooding flow rate q, and q is correlated in terms of the dimensionless Froude number.

AN EMPIRICAL CORRELATION FOR THE FLOODING (PURGING) CONDITION

Separate experiments were performed on forced flow through a single opening to obtain a definitive relation between the flooding flow rate q, above which buoyancy-driven exchange flow is prevented, and the opening length-to-diameter ratio, L/D. The same apparatus was employed in this study as in the pure exchange flow experiments described by Epstein (1988), except that fluid was supplied or extracted from the tank via a tube that penetrated one of the walls of the tank at the bottom of the lower compartment (see Figs. 2 and 3). The tank is constructed of plexiglass and is 0.546 m square and 0.762 m deep. A horizontal plexiglass partition divides the tank into a lower compartment and an upper compartment, each of height 0.33 m.

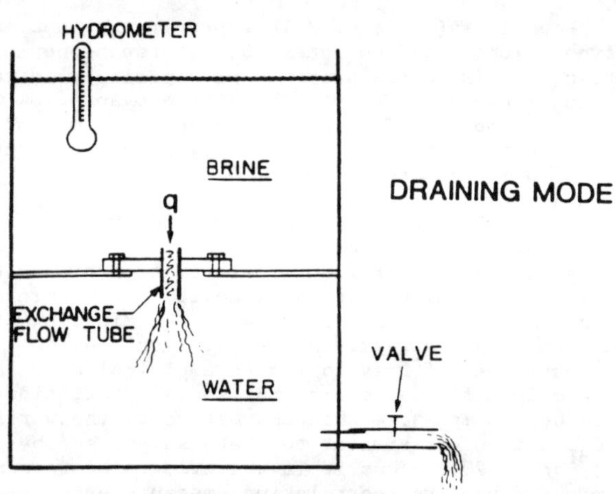

Fig. 2 Schematic diagram of the apparatus for the determination of the flooding flow rate via a quasi-steady draining technique.

In the "extraction tests", the fresh water in the lower compartment was allowed to drain from the tank and enter the outlet tube under gravity. A valve was installed in the outlet tube to control the draining rate. In this manner a quasi-steady downward flow of brine was established in the opening connecting the two compartments (Fig. 2). At low draining rates the fresh water was observed to flow upwards through the opening against the imposed downward brine flow. In this case the draining rate was not strong enough to overcome the buoyant force due to the density difference $\Delta\rho$ across the opening. At high draining rates the fresh water was prevented from entering the opening. This situation was evident by the absence of any signs of fresh water above the opening, either as a plume or in the form of individual packets (drops) of fluid. The flooding or purging condition was defined by an intermediate draining rate that represented a boundary between these countercurrent and cocurrent regimes of flow.

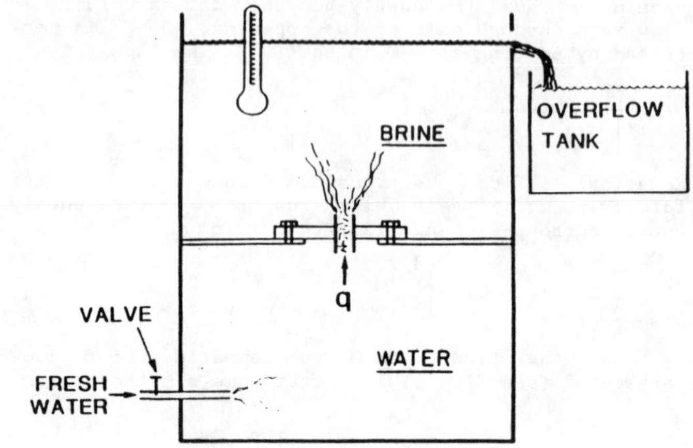

Fig. 3 Schematic diagram of the apparatus for the determination of the flooding flow rate in steady-state by water injection.

For openings of low L/D, the flooding condition was detected by adjusting the flow so that the rate of upflow of fresh water was reduced to a very slow rate of release of fluid packets into the upper compartment. With the tube-like openings (high L/D) it was found possible to adjust the draining rate so that the brine-water mixing zone filled the tube without any net upflow of water. The draining rate that produced this condition of incipient buoyant flow was defined and recorded as the flooding rate.

The aim of the second group of experiments on flooding was to determine the flooding condition in the presence of a steady-state forced-convection flow. This was accomplished by pumping fresh water through the control valve and into the lower compartment (see Fig. 3). An equivalent volumetric flow of brine was discharged from the tank at the free surface. In these experiments the observed flooding conditions were the same as those described above for the draining experiment, except that in this case it was the upwards forced flow of water that prevented the countercurrent buoyant flow of brine through the opening.

The experimental results from the draining and steady-state experiments are brought together in Fig. 4 where, again, the results are plotted in the form of a Froude number versus L/D. The measured flooding flow rates and experiment parameters are presented elsewhere in tabular form (Epstein, 1989). The range of length scales and density differences investigated here may not appear to be wide enough to establish the generality of Fig. 4. However, previous comparisons with data on buoyantly driven flows in similar but much larger geometries indicate that the Froude number versus L/D relations for exchange flows show little dependence on density difference and can be developed with small scale simulations (Epstein, 1988 and Prahl and Emmons, 1975). Results from experiments performed for the same conditions indicated that q was reproducible to within ± 8.0 percent. The two data points obtained with square openings were plotted using the notion of the effective diameter discussed previously and blend nicely with the round opening results. Aside from the moderate scatter, the dimensionless flow data can be regarded as lying on a single curve. Interestingly enough, the shape of the curve is similar to the Q_{cc} versus L/D trend reported by Epstein (1988) for exchange flow in a closed system. Presumably, the regimes of flow encountered in his study of pure countercurrent flow are also operative under flooding conditions when there is no net flow of one of the fluids within the opening.

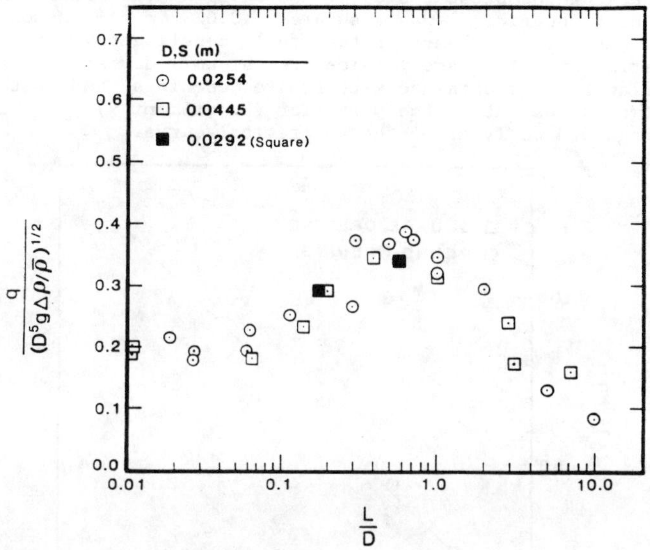

Fig. 4 Experimental results for the critical purging (or flooding) flow rate into a single opening in a horizontal partition.

For purposes of computer studies of natural convection in buildings, one is tempted to inquire if a formula can be used to estimate q for cases of interest. The function

$$\frac{q}{(D^5 g\ \Delta\rho/\bar{\rho})^{1/2}} = \frac{0.19\ [1 + 4 \times 10^3\ (L/D)^3]^{1/9}}{\left[1 + 5.091 \times 10^{-2}\ (L/D)^{16/7}\ [1 + 4 \times 10^3\ (L/D)^3]^{4/9}\right]^{1/4}} \quad (22)$$

was developed and found to pass through the data in Fig. 4 with a relative error of less than about 20%.

A CORRELATION FOR COMBINED FLOW THROUGH A SINGLE OPENING

The main intention of this portion of the study is to determine the effects of forced-fluid motion on natural convection across a single opening in a horizontal partition. In this section we will develop a correlation for combined forced flow and natural convection across a single opening. In the next section we will show how this correlation may be used to treat multiple openings and systems with more than two compartments.

The apparatus used for the combined convection experiments is the same as that already described and shown in Fig. 2 for measuring flooding rates by the draining technique. Combined convection conditions were obtained simply by adjusting the draining rate with the control valve to preselected values below the corresponding flooding rate for the experimental conditions chosen. The buoyancy-driven upflow of water Q_{BF} against the forced downflow of brine was obtained by measuring the rate of dilution of the brine in the upper compartment with a hydrometer (see Epstein, 1988). At high draining rates, but below the flooding rate, the countercurrent flow of water into the upper brine compartment was weak and it was difficult to accurately detect density changes during the course of a run without draining a significant fraction of the brine inventory. Errors resulting from this difficulty account for much of the scatter in the data. The scatter at forced flow rates Q_u greater than about 60% of the flooding rate q was so large that no meaningful data could be obtained. From a practical point of view this situation is not serious because the buoyant component of the flow is negligible compared to Q_u for $Q_u \gtrsim 0.6$ q (see below). The buoyancy-driven components Q_{BF} determined from the experiments are available in tabular form (Epstein, 1989). The estimated experimental uncertainty in the reported values is between 10 and 30 percent, which again is mainly due to uncertainties in reading the hydrometer.

The theory given previously for combined flow within an opening in a vertical partition suggests the correlation scheme shown in Fig. 5 where Q_{BF}/Q_{cc} is plotted against $1 - Q_u/q$. We note from the figure that a power law interpolation equation does a reasonable job of correlating a wide variety of data obtained as part of this study for combined natural and forced convection flow. It can be seen from Fig. 5 that the externally supplied flow rate Q_u need only exceed a value equivalent to about 60 percent of the flooding rate q to essentially "wipe away" the natural convection component Q_{BF}. An expression for the solid, correlating line in the figure, which gives Q_{BF} to an accuracy of 25 percent, is

$$Q_{BF} = Q_{cc}\ (1 - Q_u/q)^{2.3} \quad (23)$$

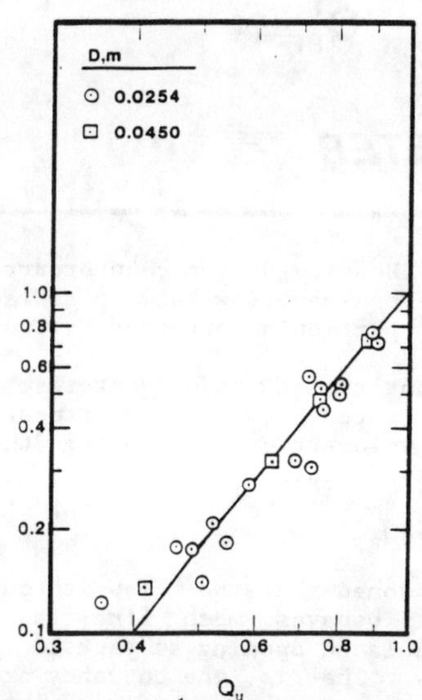

Fig. 5 The natural convection component Q_{BF} of combined natural convection and forced flow across circular openings in a horizontal partition; correlation of data in terms of Q_{cc}, Q_u, and q.

The only difference between this expression and Eqs. (15) and (20) for the combined effect of natural and forced flow through openings in walls is in the value of the exponent. The increase in the exponent compared to the stratified case is not surprising considering that two fluids flowing in opposite directions through an opening in a vertical partition remain essentially unmixed while countercurrent flow in openings in horizontal partitions promotes vigorous mixing.

COMBINED FLOW AND MULTIPLE OPENINGS

It is interesting to inquire as to whether Eq. (23) for a single opening can be used to predict the measured exchange flow rates for the two-opening system reported in the previous paper (Epstein, 1988). The flow pattern of interest is shown schematically in Fig. 6. While in these experiments an externally supplied flow was not used, a natural unidirectional flow Q_u is present in the system owing to the difference in density between the light fluid that enters opening 1 and the heavier fluid above opening 2. We are particularly interested in the situation where the unidirectional flow is not strong enough to prevent the buoyant downflow of brine at opening 1, as illustrated in Fig. 6. The exchange flow rates for the flow configuration illustrated in this figure are more difficult to deal with than the exchange rates associated with a one-way loop flow, where Q_u is too strong for combined free and forced convection to develop in openings 1 and 2.

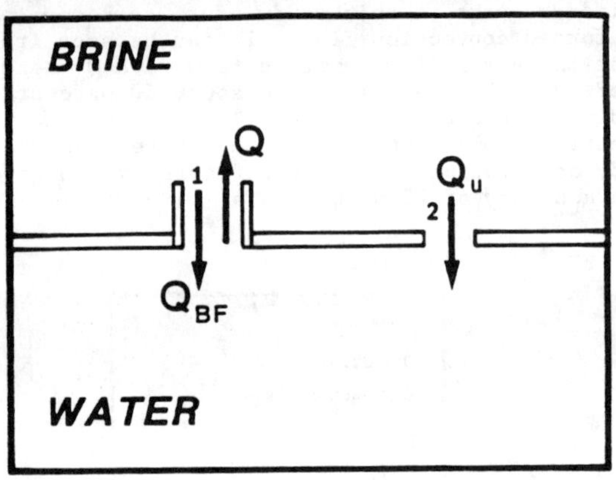

Fig. 6 Density-driven countercurrent and one-way flow between sealed compartments connected by two openings.

The quantity of major interest is the net exchange flow rate between compartments, namely Q. Clearly, the volumetric flows through the openings are related by the formula

$$Q = Q_u + Q_{BF} \tag{24}$$

It can be reasoned that the flow through opening 1 (short tube) behaves much like the combined flow through an isolated opening subject to an externally supplied flow. That is, the buoyancy-driven component of the flow can be replaced by Eq. (23). Consequently, Eq. (24) becomes

$$Q = Q_u + Q_{cc}\ (1 - Q_u/q)^{2.3} \tag{25}$$

where we now identify Q_u with the one-way natural loop flow in the system. A second and major assumption in the development is that the unidirectional loop flow Q_u that passes through both openings and appears in Eq. (25) can be evaluated by application of the Bernoulli equation, as if the unidirectional flow was the only flow in the system. The Bernoulli model for the determination of Q_u for the two-opening configuration shown in Fig. 6, for substitution into Eq. (25), can be found in Epstein (1988).

The measured exchange rates Q for the experiments involving two openings and the flow pattern shown in Fig. 6 are plotted in Fig. 7 in the coordinate system suggested by Eq. (25). Five of the data points are taken from Table 3 of Epstein's (1988) paper, Experiments 1-4 and 9. Four additional experiments were performed with the two-opening configuration as part of the present study, including three runs in which opening 1 was a square of 0.0292 m side dimensions. The exchange rates fall consistently on a straight line and, aside from a small deviation for flow results obtained with square openings and with two holes cut in the partition (Experiment 9), can be regarded as lying on the theoretical curve.

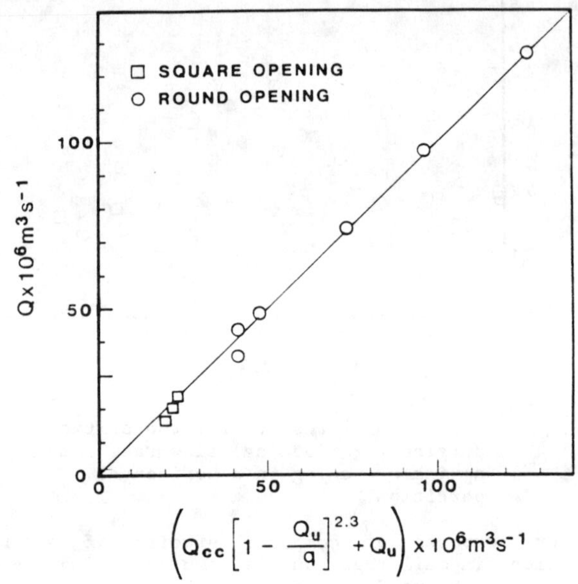

Fig. 7 Measured exchange rates across a horizontal partition with two openings, unidirectional flow in one opening and countercurrent in the other, plotted against Eq. (25).

In view of our success in applying Eq. (23) to two openings, we suggest that combined flow problems involving more than two openings and two compartments can be treated with this equation. The recommended procedure for solving such problems is as follows: The volumetric flow is calculated by first assuming one-way flow throughout the system. In this manner a Q_u is calculated for each opening using the Bernoulli equation and the appropriate orifice discharge coefficients. Then each opening is checked for countercurrent natural convection flow by calculating q and comparing it with Q_u. If openings are found such that $Q_u < q$ then the exchange flow rate in these openings are recomputed using Eq. (25). The Q_u's to be input in Eq. (25) are those already obtained from the initial, purely unidirectional analysis.

The experimental apparatus was modified to test this computational scheme, as shown in Fig. 8. As seen, a vertical wall with an opening at its center divides the upper compartment of the previous apparatus (see Fig. 3) into two rooms (compartments 2 and 3), resulting in a three-room configuration. The volume of the upper room is 0.0984 m³ while the volume of each of the lower rooms is 0.0481 m³. The cutting of two openings in the horizontal partition ensured fluid mechanical communication between all three compartments and increased the number of openings to three. At the beginning of each experiment, the brine solution of known density was poured into the tank until all three compartments were filled. Flow of fresh water into the lower room (compartment 1) was then initiated by opening the valve on the supply tube. From then onward, the densities of the brine solutions in the upper compartments were carefully monitored. The fresh water flow was maintained constant throughout the data run at levels which ranged from 400 to 3000 ml min^{-1}. The brine solution flowed out of the tank via a weir arrangement at the free surface in compartment 3. In this manner the height of the liquid in the tank was maintained constant. The experimental volumetric flow rate was determined by measuring the volume of brine collected in the overflow tank as a function of time. Run times ranged from 1 to 5 hours, increasing with lower supply flow rates. Hydrometer data were collected at various intervals by temporarily turning off the supply flow and sealing the openings. During these "off periods" two density measurements were made of the mixtures in the upper compartments. The brine-water mixtures were mechanically stirred between hydrometer readings to obtain a "feeling" for the magnitude of brine-water stratification or, equivalently, the rapidity of mixing within the compartments. Density differences due to stratification (slow mixing) were found to be small on the scale of the density change measured over the course of an experiment. Some of the experiments were repeated with hydrometer readings taken at different intervals to ensure that the results are not sensitive to the density measurement technique nor to the effects of brine-water stratification. The apparatus shown in Fig. 8 can be regarded as a laboratory tool for simulating the fluid mechanical response of an actual building atmosphere, as represented by the brine, to an essentially point source release of hot gases, which can be identified with the injection of the less dense fresh water.

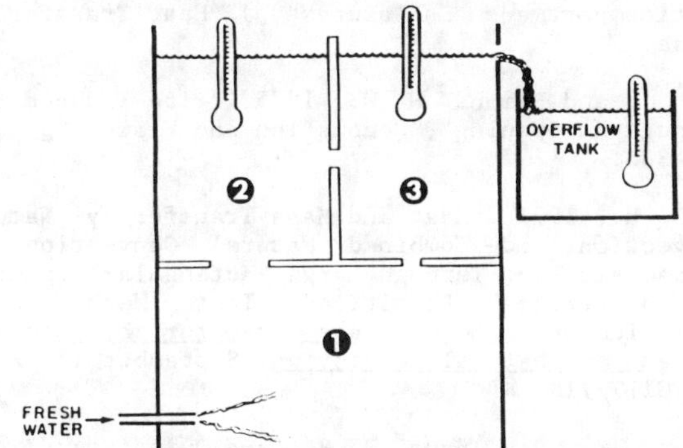

Fig. 8 Three-compartment test configuration. Numbers identify compartments.

The measured dilution histories within the upper compartments are shown in Fig. 9 for a case in which all the compartments were connected by circular openings of 0.025 m in diameter and the flow rate was 2000 ml min^{-1}. During the first 8 minutes of the test, the flows from compartment 1 through the openings in the horizontal partition were unidirectional and upward. The flow through the vertical partition was unidirectional from compartment 2 to compartment 3. For the rest of the 1.28-hour experiment the flow through the opening connecting compartments 1 and 2 was unidirectional and downward. The flow through the vertical partition also changed direction and was unidirectional from compartments 3 to 2. The flow directions are indicated in Fig. 9. It is evident that the flow path of least resistance is from compartment 1 to compartment 3. Thus compartment 2 experiences the lowest rate of dilution by the incoming fresh water, resulting in a counterclockwise loop flow. The onset of this buoyancy-driven convective loop in the system at 8 minutes was predicted by the model. Also, good agreement is seen between predicted and experimental density versus time curves for compartments 2 and 3. Actually the calculational method was utilized in planning the experiment program as it was useful to know beforehand the flow patterns that would be observed for each set of experimental conditions. Thus, the theory was used to prepredict the experimental results.

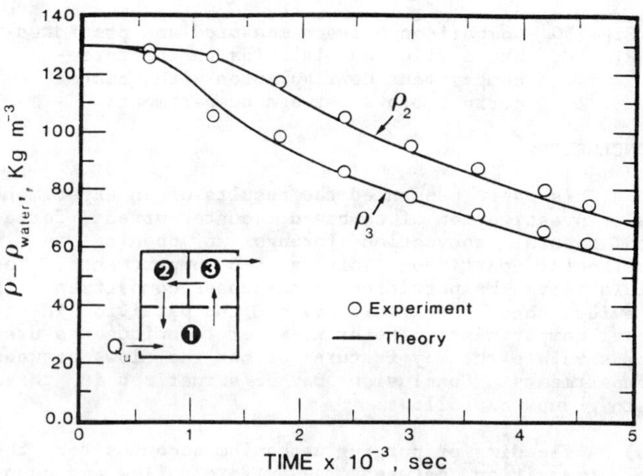

Fig. 9 Comparison between measured and predicted brine dilution histories in the three-compartment configuration under conditions where unidirectional flow prevails.

It should be mentioned here that a discharge coefficient C_D = 0.71 was used in applying the Bernoulli flow model to predict the volumetric, unidirectional flows through all the openings. This value was deduced from separate experiments on pure buoyancy-driven exchange flow through the opening in the vertical partition; it agrees reasonably well with the value of approximately 0.68 measured by Prahl and Emmons (1975) and by Steckler, et.al (1984) for flows through doors and windows.

The theory led us to believe that, if the 0.025-m diameter openings used in the experiment described above were changed so that the openings cut in the horizontal partition were substantially larger, say, 0.06 m in diameter, while the opening in the vertical partition was reduced in size to, say, 0.0095 m, countercurrent flow would occur in the openings in the horizontal partition. Indeed, this was observed in the experiments with openings of these sizes, beyond approximately 10 min. Fig. 10 shows the results for a fresh water injection rate of 400 ml min^{-1}. The agreement between theory and experiment is good and typical of the agreement obtained with all the tests peformed in the three-compartment configuration. The predicted density versus time results obtained with a purely unidirectional Bernoulli model are represented by the dashed lines in Fig. 10. These curves indicate the magnitude of the error incurred by ignoring the effects of countercurrent flow within the openings in the horizontal partition.

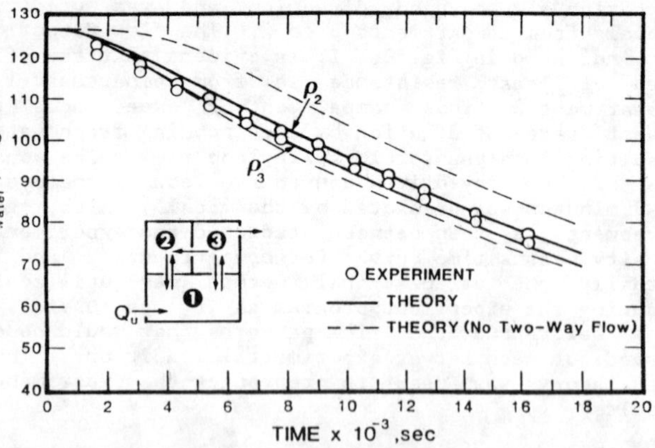

Fig. 10 Comparison between measured and predicted brine dilution histories in the three-compartment configuration with counter-current flows between compartments.

CONCLUSIONS

This paper presented the results of an experimental investigation of combined, countercurrent, forced and natural convection through an opening in a horizontal partition joining two compartments. The fluid above the partition in the upper compartment is heavier than the fluid below the partition in the lower compartment. The brine-water technique was used to simulate the key features of hot gas flows between compartments. Conclusions may be summarized in three main groups as follows:

(a) A flooding or purging mechanism accounts for the transition between one-way forced flow and countercurrent combined flow in an opening. An empirical formula was obtained for calculating the flooding flow rate q that is necessary to prevent countercurrent flow within an opening in a horizontal partition.

(b) The combined flow regime of buoyancy-driven flow in opposition to the externally supplied flow within an opening lies between pure countercurrent buoyant flow Q_{cc} in a sealed system and the flooding condition. Correlation of the buoyancy component of the combined flow, Q_{BF}, was achieved with a function $Q_{BF}(Q_{cc},q)$ suggested by the application of hydraulic theory to openings in walls that interpolates between the limits of pure countercurrent flow and the flooding condition.

(c) A methodology was developed for predicting the exchange flow rates within multicompartment buildings. Briefly, the Bernoulli equation is applied to flooded openings ($Q_u > q$) and the function $Q_{BF}(Q_{cc}, q)$ is used in the remaining openings for which $Q_u < q$. The natural and forced convection flow patterns and fluid densities within three compartments of the model-scale apparatus were measured experimentally and compared with calculations. The results indicate that the interpolation function $Q_{BF}(Q_{cc},q)$ in combination with the Bernoulli model can be used to predict convection in multilevel/multicompartment buildings.

Whether or not the influence of stratification within compartments, an effect which was either eliminated by mixing or was found to be negligible in the absence of mixing in the present experiments, is sufficient to render invalid the extrapolation of present results to full-scale buildings may be a matter for further study. Our feeling is that the question as to what determines the magnitude of the stratification will be very difficult to deal with. In actual buildings the stratification is affected not only by the heat loads resulting from walls and internal structures, but also by the flow between compartments. It is believed, however, that the methodology developed in this paper for well mixed compartments is applicable to actual buildings in that it will provide reasonable estimates of volume-averaged temperature and gas species concentrations.

ACKNOWLEDGEMENTS

The authors are grateful for Electric Power Research Institute (EPRI) funding and the support provided by Dr. Edward L. Fuller, EPRI Project Manager. The assistance of Mr. Alan S. La Croix and Mr. Verden R. Fletcher in conducting the experiments is also gratefully acknowledged.

REFERENCES

Brown, W. G., 1962, "Natural Convection Through Rectangular Openings in Partitions-2. Horizontal Partitions", Int. J. Heat Mass Transfer, 5, pp. 869-878.

Brown, W. G. and Solvason, K. R., 1962, "Natural Convection Through Rectangular Openings in Partitions - Vertical Partitions", Int. J. Heat Mass Transfer, 5, pp. 859-867.

Epstein, M., 1988, "Buoyancy-Driven Exchange Flow Through Small Openings in Horizontal Partitions", J. Heat Transfer, 110, pp. 885-893.

Epstein, M., 1989, "Combined Natural Convection and Forced Flow Through Small Openings in a Horizontal Partition, With Special Reference to Flows in Multicompartment Enclosures", J. Heat Transfer, in press.

Prahl, J. and Emmons, H. W., 1975, "Fire Induced Flow Through an Opening", Combustion and Flame, 25, pp. 369-385.

Shaw, B. H., 1971, "Heat and Mass Transfer by Natural Convection and Combined Natural Convection and Forced Air Flow Through Large Rectangular Openings in a Vertical Partition", Inst. Mech. Engrs. Symposium on Heat and Mass Transfer by Combined Forced and Natural Convection, September 15, Paper No. C117/71.

Steckler, K. D., Baum, H. R., and Quintiere, J. G., 1984, "Fire Induced Flows Through Room Openings - Flow Coefficients", 20th Symposium (Int.) on Combustion, The Combustion Institute, pp. 1591-1600.

Steckler, K. D., Baum, H. R., and Quintiere, J. G., 1985, "Salt Water Modeling of Fire Induced Flows in a Multiroom Enclosure", 21st Symposium (Int.) on Combustion, The Combustion Institute, pp. 143-149.

1989 National Heat Transfer Conference
HTD-Vol. 106, Heat Transfer Phenomena in Radiation, Combustion and Fires

AN EXPERIMENTAL INVESTIGATION TO IMPROVE AND CONTROL THE REPEATABILITY OF THE HEAT SOURCE USED IN THE AIRCRAFT AND ORDNANCE SAFETY PROGRAM

M. V. Gunaji
Lockheed-ESC
Las Cruces, New Mexico

D. L. Fenton
Kansas State University
Manhattan, Kansas

L. J. Linley
NASA Johnson Space Center
Las Cruces, New Mexico

ABSTRACT

The Naval Weapons Center (NWC) at China Lake subjects various ordnance devices, such as bombs and rockets, to conditions representative of a fire caused by an accidental fuel spill on an aircraft carrier deck. This paper describes the investigation of various techniques to improve the control and the repeatability of the heat source (pool fire) used in the Aircraft and Ordnance Safety Program at China Lake. Only those techniques that would not alter the structure of the pool fire were considered. Using wind screens with a porosity of 37 percent outside the pool radius, appeared to be the most promising technique because the screens would reduce the wind velocity and still allow the fire to entrain air. Several wind profile tests were performed to evaluate the effectiveness of the screens. To eliminate the uncertainty in scaling, full-scale tests were performed using screen heights of 3.6 m and 6.0 m. The ratio of the wind velocities inside to outside the screens (K_s) decreased with increased wind velocities for the shorter screens, but was nearly constant (K_s = 0.18 for a wind velocity up to 8.2 m/s) for the 6.0-m-high screen. Pool fire tests were performed both inside the wake of the screens and outside to compare the repeatability of the pool fires. For these tests, a simulated ordnance instrumented with thermocouples was exposed to the fire under different wind conditions. When the ordnance was tested inside the screens with wind velocities outside the screens up to 4.5 m/s, the average exposure temperature was of the order required by the Military Standard. This wind velocity is a 5-fold increase above the current 0.9 m/s wind velocity limit.

NOMENCLATURE

K_s - Fraction of Wind Velocity
V_i - Velocity inside the screens (m/s)
V_o - Velocity outside the screens (m/s)

INTRODUCTION

The Naval Weapons Center (NWC) at China Lake conducts an Aircraft and Ordnance Safety Program for in-service air-launched weapons to evaluate the thermal sensitivity of ordnance devices. The program involves subjecting various ordnance devices, such as bombs and rockets, to conditions representative of a fire caused by an accidental fuel spill on an aircraft carrier deck.

The ordnance device is exposed to the heat flux, which is largely determined by the presence and velocity of wind. The wind affects the flame shape, the temperature in the flame plume, and the ordnance device by moving the flame away from the device. As tests conducted under different wind conditions are not comparable, NWC limits testing to those days when the velocity of the wind is 0.9 m/s or less as required by MIL-STD-1648A (Department of Defense, 1977).

In an effort to produce repeatable and comparable test results, NASA White Sands Test Facility investigated various techniques to control the affects of the wind in a way that does not alter the thermal structure of the fire.

The most important fire characteristics are the burn rate of the fuel and the thermal environment in the flame plume. These characteristics are sensitive to atmospheric conditions, such as wind, ambient temperature, and humidity. Wind seems to be the most important because it causes the flame to tilt, thus affecting the amount of energy feedback to the fuel and the burn rate (Fu, 1972). Additionally, flame bending causes a change in radiation and temperature levels in and around the fire.

Welker and Sliepcevich (1965) studied the effect of wind on a liquid pool fire flame. Their experimental study resulted in a correlation between the wind velocity and the flame-tilt angle; however, their study was limited to pool diameters of 0.6 m and wind velocities less than 1.8 m/s. The extent of flame bending was found to be dependent on the Reynolds number, the Froude number, and the ratio of the fuel density to that of ambient air.

To abate the wind effects on the fire, wind barriers have been investigated. Some barriers did protect the fire from wind, but affected the thermal environment in the fire. Yoshimura et al. (1983) describes a facility to obtain a repeatable pool fire environment in support of the Lightweight Air-Transportable Resistant Container Program at Sandia National Laboratories, NM. In this technique, the fire was enclosed to control the wind; however in doing so, the air entrained by the fire was found to be affected. This technique is not appropriate for the present investigation because the effects of the enclosure on the pool fire radiation mechanisms are unknown. Also, the cost of repairing the damage to the enclosure, resulting from an explosion of a test article, is prohibitively expensive.

The aerodynamic jet technique was studied by Tozer (1983). In this technique, the wind was diverted away from the pool fire by an air jet that was directed upward at some angle from the vertical. Tozer found that although this technique offered some wind protection, fuel vapors were entrained into the jet and the pool fire structure was altered.

The use of wind screens (for example, snow fences) has been recognized and used for many years. Snow fences have been used to control snow drifts by reducing the wind velocity, thus causing the snow to deposit in front of and behind the fence. Tall fences have been reported to be more efficient than short fences in controlling the wind velocity, and the wind velocity behind a fence has been reported to decrease as the height of the fence is increased (Gray, 1977).

Wind screens have been used successfully to subdue gale-force winds on mountain bridges and to prevent vehicle accidents caused by high-velocity crosswinds (Nordlin and Bourget, 1969).

Horn et al. (1971) performed wind-tunnel simulation studies of barrier-flow phenomena, and concluded that the barrier height is of primary importance in reducing the wind velocity in the wake of the screens. Based on their study, they recommended a porosity of 37 percent as an optimum for the wind barrier that was erected along a 1.8 km erosion test section on the Rocket Sled Test Track at Holloman Air Force Base, Holloman, NM.

Based on the above discussion and the investigation of other methods to control the wind velocity (Gunaji, 1987), it was determined that the most promising method is the use of wind screens. The advantage to this approach would be that the screens would provide a wind barrier and allow the fire to entrain air. Several criteria used in the design and location of the snow fences (Gray, 1981) were extended for use as a wind barrier, for example, the porosity of the fence material and the optimum distance from the area to be protected. The advantage of having the screens in two concentric circles would be to reduce the wind velocity in stages, and thus reduce wind loading on the screens. Also, the flow rate per unit area of the screens would be the same and a barrier would be provided for any wind direction. The screens would be located at sufficient distances from the edge of the fire to prevent melting of the polyethylene, and the screen height would be sufficient so that the reattachment length would be beyond the area to be protected.

EXPERIMENTAL

Wind profile tests were conducted to determine the decrease in wind velocity across the screen in absence of a pool fire. These tests were conducted using 3.6-m-high and 6.0-m-high screens to investigate the effect of the screen in decreasing the wind velocity.

Pool fire tests were conducted both in the wake created by the screens and outside the screens. The pool fire tests were conducted using the simulated ordnance device and either Jet-A or JP-5 jet fuel.

TEST SYSTEM DESCRIPTION

Figure 1 shows the schematic of the test system. The test system consisted of inner and outer screens set up in circular configurations on a concrete deck. The radii of the inner and outer screens from the center of the concrete pad were 14 m and 22.1 m, respectively. The wind-screen material selected for the test was 1.5-mm-thick high-density polyethylene and was commercially available in 1.2-m x 30-m rolls. The screen had a mesh size of 5 mm x 23 mm and a porosity of 37 percent. The screens were supported by I-beams and attached to the support structure with weather-resistant wire ties.

A 6.1-m-diameter fuel-containment pond was located concentric to the inner and outer screens. A sand berm was used, and the containment area in the sand berm was lined with a polyethylene sheet. A simulated ordnance was supported on three stands 1.2 m above the pool, which is the standard per MIL-STD-1648A. The simulated ordnance device is shown in Figure 2. The thickness of the fuel layer above the water layer in the containment pond ranged from 30 mm to 50 mm for various tests, resulting in a 7- to 8-minute burn time. A gasoline-filled plastic bag was ignited with a remotely operated nichrome-wire silicone igniter. The burning gasoline ignited the fuel and the flame spread over the entire pond area within 60 to 120 seconds.

INSTRUMENTATION

The sensors used to measure the wind velocity were Gill UVW anemometers with a worst-case error reading of ± 2 percent. For the wind profile tests, four anemometers were used to measure the wind velocity in the wake created by the inner and outer screens. Figure 1 shows the locations of the anemometers. One anemometer (D) was set up outside the perimeter of the test system approximately 3 m above the ground to record the ambient wind velocity. For each test, four anemometers (A, B, C, and E) were placed on concentric circles of 6.1 m or 10.0 m diameters to record wind velocity at heights where the effects of wind on the fire were to be minimum; that is, 1.2 m and 2.4 m.

For the pool fire tests, only the anemometer outside the perimeter of the test system was used. The simulated ordnance device used in the pool fire tests was a 2.3-m-long stainless steel pipe with an outside diameter of 0.08 m, an inside diameter of 0.076 m, and 0.3-cm-thick walls. The device was filled with powdered copper to simulate the worst case of heat transfer and was instrumented with eight sheathed, 9.1-m-long stainless steel Chromel-Alumel (Type K) ungrounded thermocouples, 1.6 mm in diameter, positioned symmetrically along its axis.

Two thermocouples, one 5.1 cm below the simulated ordnance and the other 1.3 cm from the end, were used to measure the flame temperature. The thermocouple positioned below the simulated ordnance device was shielded by fiberfrax insulation to minimize the radiative contribution from the ordnance. Thermocouples were also used to measure the liquid fuel and the ambient air temperatures. The fuel regression rate was measured by using a variable reluctance differential pressure transducer. Video data were taken for all tests.

RESULTS AND DISCUSSIONS

Wind Profile Tests

To evaluate the effectiveness of the screens, as a means for protecting the fire up to 2.4 m in height, wind velocities were measured in protected areas created by the screens and in unprotected areas outside of the screens. The variation of the wind velocity is shown in Figures 3 and 4. The reduction in wind velocity across the screens was evaluated as the ratio of the measured velocity inside the screen to the measured velocity outside the screen (K_S). Table 1 gives the results of the wind profile tests.

Figure 5 shows the variation of wind velocity with time obtained from the anemometer (D) outside the screen and one of the inner anemometers (A) inside the screens. The correlation between the changes in the wind velocity measured outside and inside the screens indicates that basic structure of the wind remains intact with the use of the screens.

The value of K_S for tests 1 through 6 determined with anemometers at 1.2 m high decreased as the velocity outside the screen increased. The decrease in the K_S value indicates that the effectiveness of the 3.6-m-high screen increases as the upstream velocity increases. Because the ordnance item is located at a height of approximately 1.2 m, a lower value of K_S is preferred at that height than the 0.34 average obtained for tests 1 through 6. To achieve a lower value of K_S at heights of 1.2 m and 2.4 m, the height of the screens was increased to 6.0 m. Table 1 shows that the K_S values obtained with the 6.0-m-high screens had less variability and that the average value was 0.18. The results show that the screens are an effective way of controlling the wind velocity. The 6.0-m-high screen is more effective at decreasing the wind velocity than the 3.0-m-high screen.

Pool Fire Tests

Pool fire tests were conducted to evaluate the repeatability of the pool fire in the wake created by the screens and to compare the results with pool fire tests conducted without the screens. The results of the pool fire tests are shown in Table 2. Tests 2 and 4 were conducted during relatively calm wind conditions (less than 2 m/s), while tests 1, 3, and 5 were conducted during windy conditions (greater than 2 m/s). Tests 6 and 7 were conducted without the screens. Test 6 was conducted during windy conditions, and test 7 was conducted during relatively calm conditions.

The fuel regression rate, the rate at which the liquid fuel level drops, was determined from the time the fire was fully developed (spread over the entire pool area) to the time where the fuel was consumed. The pressure of the water level plus the fuel layer gave a differential pressure value corresponding to the weight of these fluids. As the fuel burned, the pressure level decreased in proportion to the fuel regression rate. At the end of the test, all the fuel was burned leaving only water. The pressure of the water alone was checked to determine the amount of water evaporation by the burning fuel. Within the limits of the instrumentation, the data indicate no difference in the measured fuel regression rates (change in the burn rate) during windy (tests 1, 3, and 5) and relatively calm (tests 2, 4, and 7) conditions.

Figure 6 shows the variation of the average temperature inside the simulated ordnance for each of the tests. The ceramic fiber insulation at the three support stands was found to mask the fire impingement onto the ordnance. Masking reduced temperatures at these locations and lowered the average temperatures overall. The temperatures at the masked locations were lower than those at the other locations. Also, the temperature at the masked locations continued to increase, even after the fire was out, because of the conductive heat transfer within the simulated ordnance.

By qualitatively comparing the temperatures at different locations in the simulated ordnance device for tests 1 and 3, the temperatures at most locations in test 3 are higher than those in test 1, even though the ambient wind velocity was 4.3 m/s for the former and 3 m/s for the latter. The higher temperature is attributed to the sintering of powdered copper in the simulated ordnance device during test 1, causing higher peak temperatures in later tests. This effect was not observed in subsequent tests because sintering of the copper powder was essentially complete after the first test.

The average peak temperature and the rate of temperature increase inside the simulated ordnance device is given in Table 2. Qualitatively, a constant heat-flux incident on the simulated ordnance device would result in a constant rate of temperature increase. Table 2 shows that the rate of temperature increase in the tests conducted inside the screens is nearly constant (1.9 ± 0.14 K/s) for wind velocities up to 4.5 m/s outside the screens, indicating a constant heat flux to the simulated ordnance device. The only exception is test 1 which shows a rate of increase in temperature of 0.6 K/s probably because of the sintering of powdered copper.

Military Standard MIL-STD-1648A requires an average flame temperature of at least 1145 K as measured by all valid thermocouples at the test item. Figures 7a through 7d show the flame-temperature profiles for the locations 5.1 cm beneath the center of the ordnance and 1.3 cm from the end of the ordnance. Tests 4 and 5 were performed inside the screens, and tests 6 and 7 were performed outside the screens. The flame temperatures at the center and end of the simulated ordnance were considerably lower for tests 6 and 7. The average flame temperature at the center of the simulated ordnance for tests 4 and 5 was 1295 ± 93 K and 1071 ± 211 K; for tests 6 and 7 the temperature was 700 ± 309 K and 790 ± 376 K. The average fire temperature at the end of the simulated ordnance for tests 4 and 5 was 1127 ± 131 K and 943 ± 171 K; for tests 6 and 7 the temperature was 523 ± 146 K and 432 ± 101 K.

The above average flame temperatures at the center and the end of the simulated ordnance for tests conducted inside the screen area under different wind velocities, are representative of those required by the MIL-STD-1648A. Similar tests conducted with no screens do not meet the requirements of MIL-STD-1648A. Thus, the screens proved to be a good method for improving the control and repeatability of the pool fire tests under variable wind conditions.

SUMMARY

The results of the wind profile tests indicate that the 6.0-m-high screens were more effective than the 3.6-m-high screens. A reduction of 82 percent in wind velocities up to 8.2 m/s was achieved with the 6.0-m-high screens.

The results of the pool fire tests indicate that there was no variation in regression rates for tests conducted inside the screens, which implies that the burning rate of the fire was not affected by the screens. The average ordnance temperature, coupled with the rate of temperature increase, qualitatively indicates that the thermal environment, engulfing the simulated ordnance device, was not affected in the presence of the screens and the wind. The flame temperatures that were measured meet the MIL-STD-1648A requirement at wind velocities of 4.5 m/s as compared with those required at 0.9 m/s.

ACKNOWLEDGMENTS

The authors would like to acknowledge Carl Wright and Bruce Havenor, who performed most of the tests from which the data presented in this paper were obtained. In addition, the authors wish to thank Jim Dugan for his efforts and assistance in choosing the instrumentation, and Dr. Harold Beeson for reviewing the paper.

REFERENCES

Department of Defense. "Criteria and Test Procedures for Ordnance Exposed to an Aircraft Fuel Fire." In *Military Standard* MIL-STD-1648A: 1 January 1977.

Fu T. T. *Aviation Fuel Fire Behavior Study.* Technical Report AGFSRS 72-2, U. S. Naval Civil Engineering Laboratory, Calif. 1972.

Gray, D. M. and D. H. Male, eds. *Handbook of Snow: Principles, Processes and Management and Use.* N.Y.: Pergamon Press, 1981, pages 636 - 646.

Gunaji M. V. *An Experimental Investigation to Improve and Control the Repeatability of the Heat Source used in the Aircraft and Ordnance Safety Program.* Technical Report TR-432-001, White Sands Test Facility, New Mexico, 1987.

Horn J. D., G. S. Campbell, A. L. Wallis, and R. C. Mcintyre. *Wind Tunnel Simulation and Prototype Studies of Barrier Flow Phenomena.* Technical Report ECOM-5416, U. S. Army Electronics Command, New Jersey, 1971.

Nordlin, E. F., and Bourget, Louis. *Wind Screen Subdues Gale Force Winds on a Mountain Bridge.* Report No. MXR 636357, Division of Highways CA Research, Calif. 1969.

Tozer, B. L. *An Investigation to Improve JP-5 Fuel Pool Fire Test Repeatability.* NWC Technical Memorandum 5205, Naval Weapons Center, China Lake, Calif. 1983

Welker J. R., C. M. Sliepcevich. *The Effect of Wind on Flames.* NBS Contract CST 1142, National Bureau of Standards, NTIS AD-474-283, Washington D. C., 1965.

Yoshimura, H. R., W. Gill, G. C. Allen, W. L. Uncapher, J. Moya, D. D. Cline, R. G. Eakes, J. O. Davis. "A Description of the Sandia Controlled Fire Facility." In *7th International Symposium on Packaging and Air Transportation of Radioactive Materials Summaries*, CONF-830528-SUMM, page 449. New Orleans: Sandia National Laboratories, May 15, 1983.

Table 1. Variation of the Fraction of Wind Velocity (K_s) and the Velocity Outside and Inside the Screens

Test No.	Anemometer Height m	Velocity[a] Inside Screen V_i (m/s)	Velocity Outside Screen V_o (m/s)	$K_s = V_i/V_o$
		Screen Height 3.6 m		
1	1.2	0.5 ± 0.3[b]	0.9 ± 0.4	0.51
2	1.2	0.5 ± 0.3	1.2 ± 0.8	0.41
3	1.2	0.5 ± 0.3	1.4 ± 0.7	0.34
4	1.2	0.4 ± 0.3	1.3 ± 0.7	0.30
5	1.2	0.6 ± 0.4	2.3 ± 1.0	0.26
6	1.2	0.7 ± 0.5	3.0 ± 1.2	0.24
7	2.4	0.4 ± 0.3	1.0 ± 0.4	0.38
8	2.4	0.6 ± 0.4	1.3 ± 0.6	0.43
9	2.4	1.1 ± 0.6	3.2 ± 1.1	0.34
10	2.4	0.9 ± 0.6	2.2 ± 1.1	0.41
		Screen Height 6.0 m		
1	2.4	1.3 ± 0.8	7.9 ± 2.0	0.16
2	2.4	1.3 ± 0.8	8.2 ± 2.1	0.19
3	2.4	1.2 ± 0.7	6.8 ± 1.7	0.17
4	2.4	1.1 ± 0.6	7.2 ± 1.6	0.16
5	2.4	0.8 ± 0.5	5.2 ± 1.2	0.16
6	1.2	0.2 ± 0.2	1.3 ± 0.6	0.17
7	1.2	0.4 ± 0.2	1.9 ± 0.7	0.19
8	1.2	0.5 ± 0.4	2.5 ± 1.1	0.20
9	1.2	0.7 ± 0.4	3.3 ± 1.0	0.21
10	1.2	0.7 ± 0.5	4.1 ± 1.2	0.18
11	1.2	0.8 ± 0.5	5.0 ± 1.2	0.17
12	1.2	0.8 ± 0.5	4.5 ± 1.1	0.17
13	1.2	0.9 ± 0.5	4.5 ± 1.0	0.20
14	1.2	0.7 ± 0.5	4.2 ± 1.1	0.17
15	1.2	0.5 ± 0.4	2.3 ± 1.0	0.19

a Average of 4 anemometers
b Standard deviation

Table 2. Pool Fire Test Results

Test No.	Pool Diameter (m)	Fuel Depth (mm)	Duration of Fire (min)	Average Fuel Regression Rates (mm/min)	Wind Velocities (m/s)	Time for Fire to Spread (+ 30 sec at end) (sec)	Ordnance: Average Peak Temperature (K)	Ordnance: Rate of Temperature Increase (K/s)
1	6.1	30	9.6 ± 0.5	5.0 ± 1.5	3.0 ± 0.7[a]	48	316	0.6
2	6.1	34	7.8 ± 0.5	4.4 ± 1.7	1.3 ± 0.5	104	1089	2.0
3	6.1	37	9.2 ± 0.5	5.5 ± 1.6	4.3 ± 1.0	57	853	1.9
4	6.1	47	7.8 ± 0.5	6.0 ± 2.2	1.1 ± 0.4	120	866	1.7
5	6.1	47	7.1 ± 0.5	6.6 ± 2.8	4.5 ± 1.14	58	713	2.0
6[b]	6.1	–	–	–	4.3 ± 1.4	50	450	1.1
7[b]	6.1	43	8.0 ± 0.5	5.3 ± 2.2	1.7 ± 0.7	80	561	1.8

a Standard deviation
b Tests conducted outside the screens

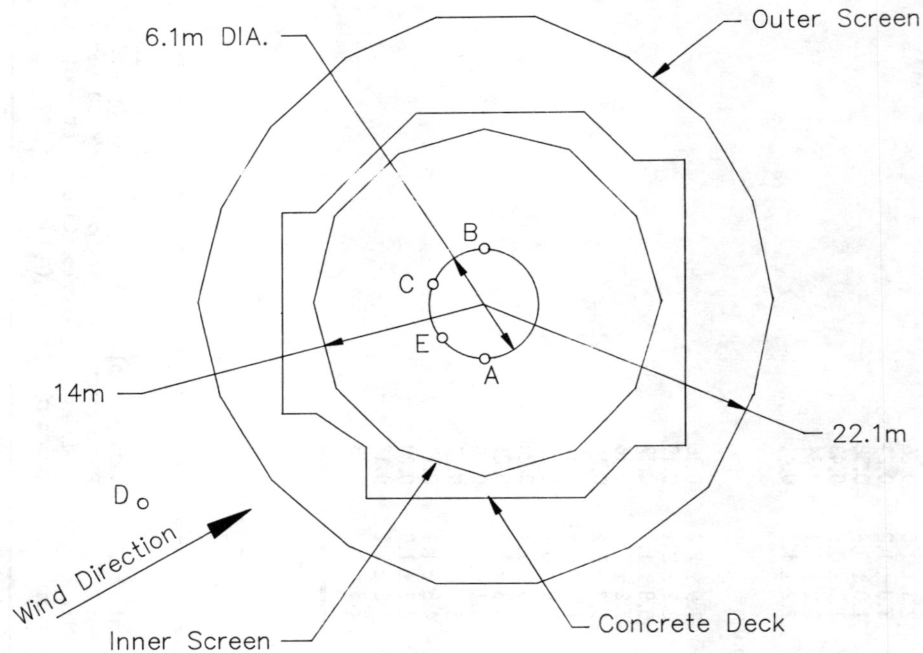

Fig. 1 Test system

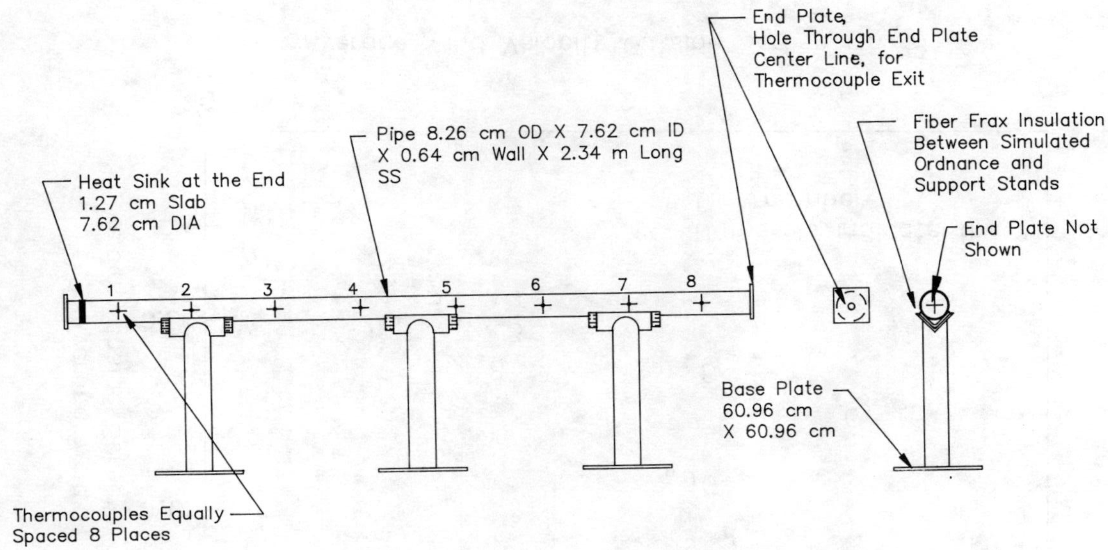

Fig. 2 Simulated ordnance device

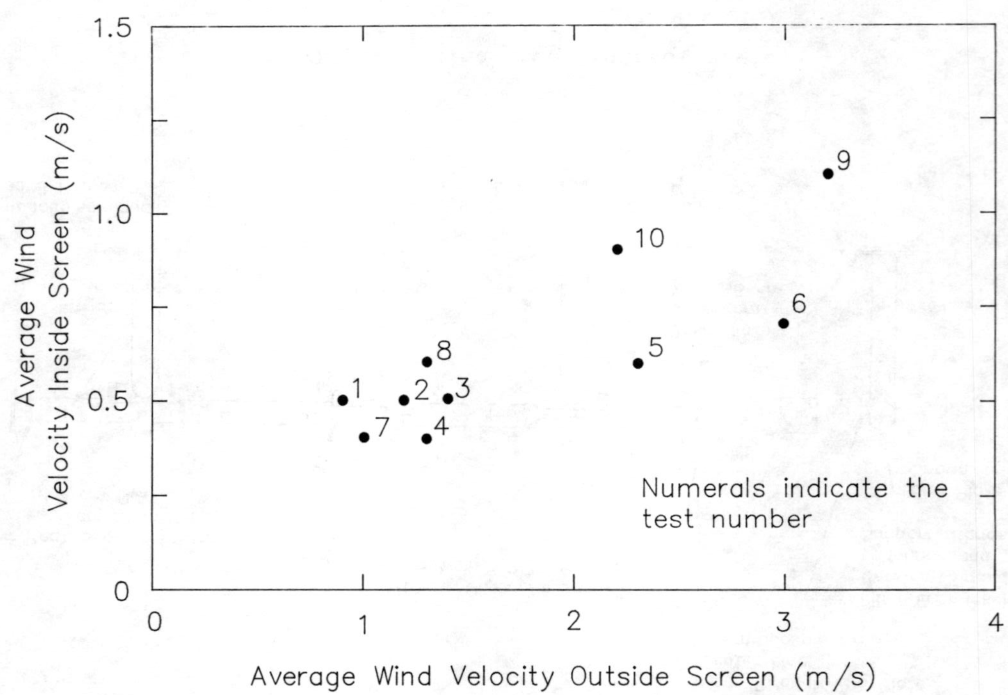

Fig. 3 Variation of the wind velocity for the 3.6-m-high screens

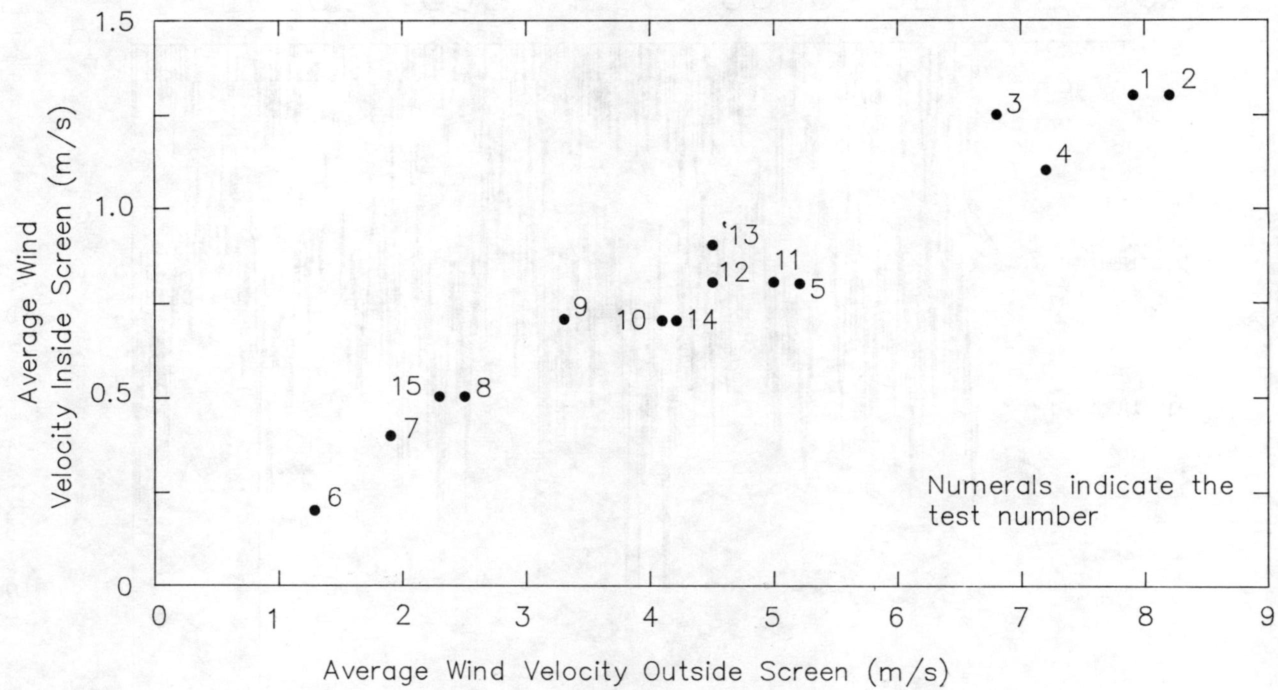

Fig. 4 Variation of the wind velocity for the 6.0-m-high screens

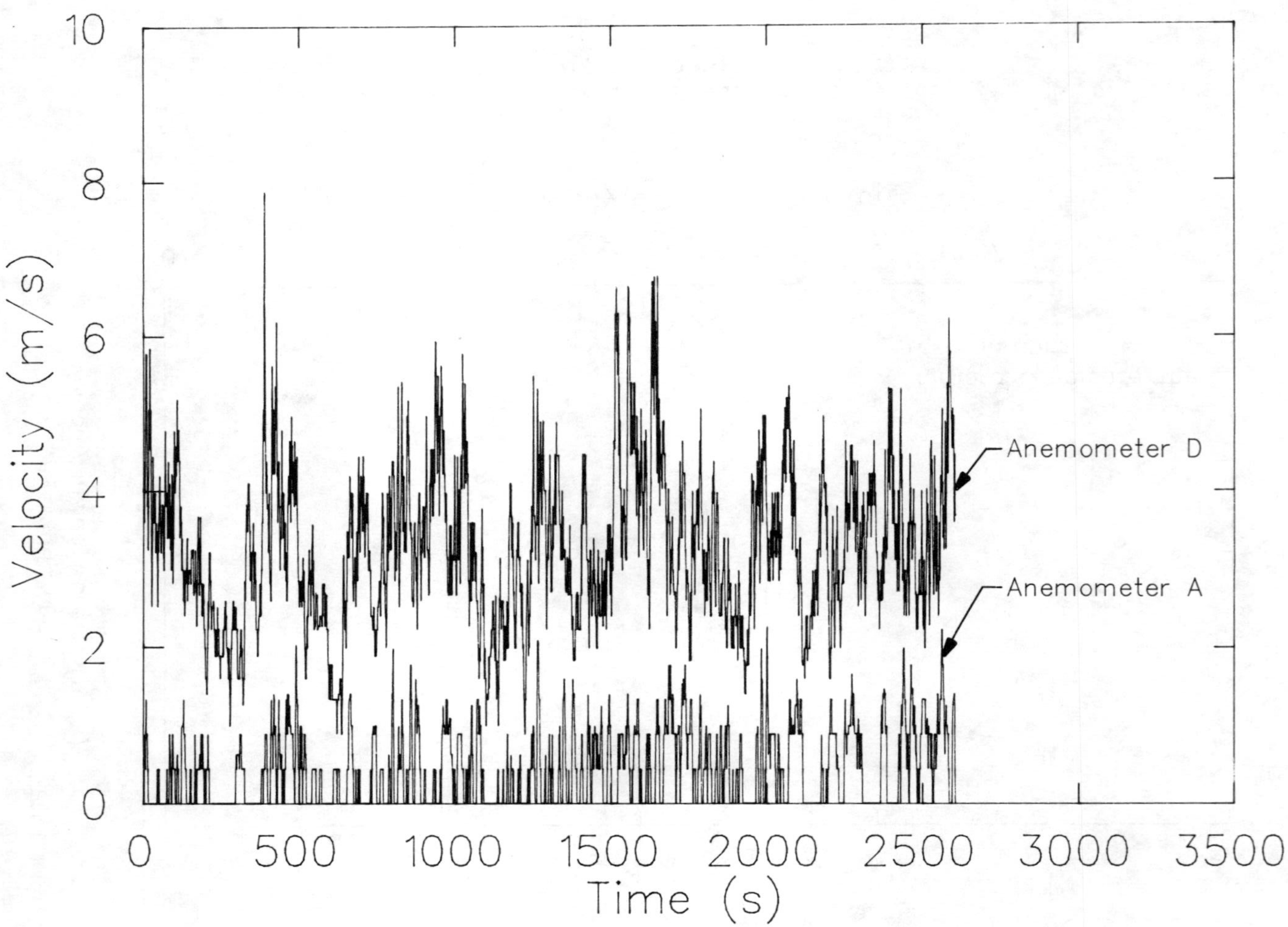

Fig. 5 Variation of wind velocity with time

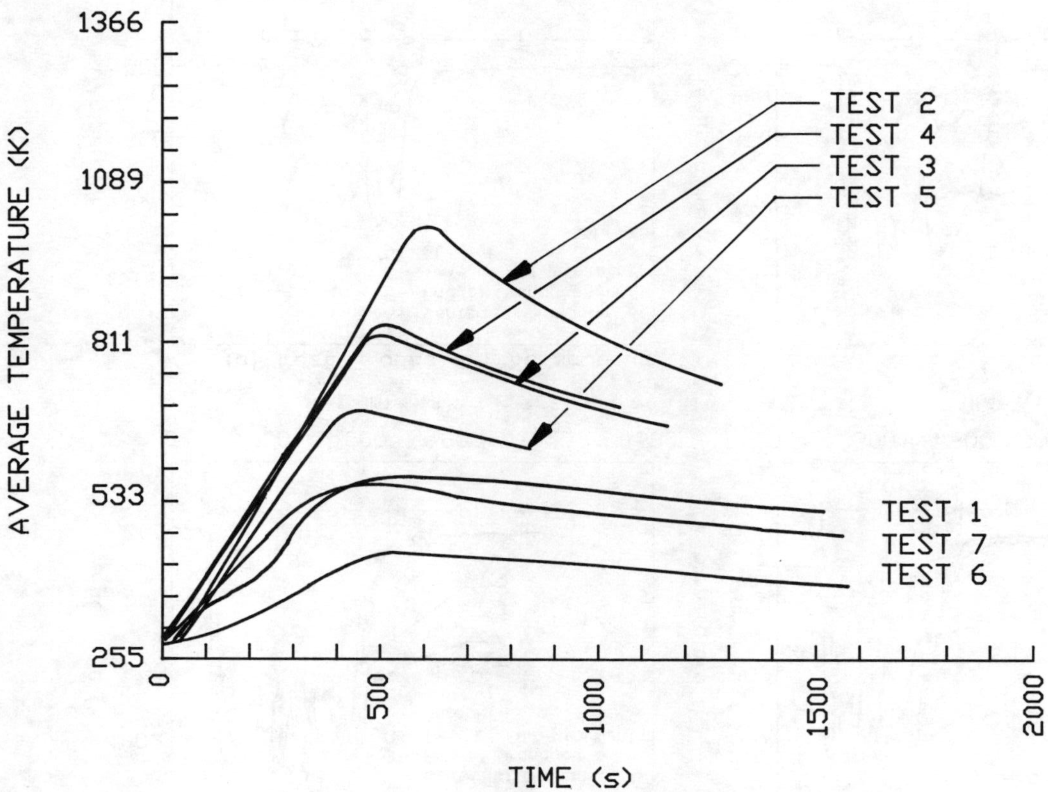

Fig. 6 Variation of the average temperature inside the simulated ordnance

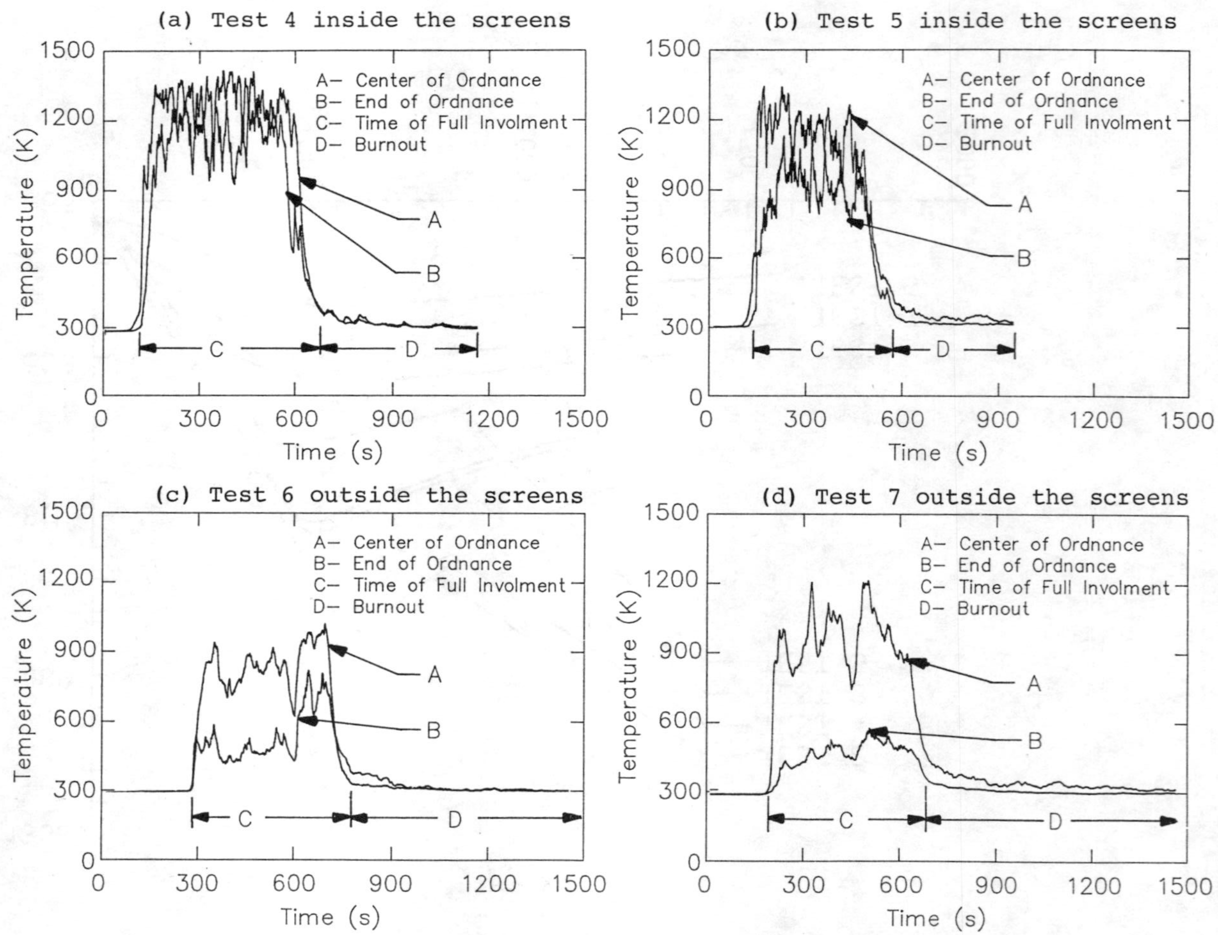

Fig. 7 Variation of temperature over time

1989 National Heat Transfer Conference
HTD-Vol. 106, Heat Transfer Phenomena in Radiation, Combustion and Fires

FIRE ENGULFMENT OF HORIZONTAL CYLINDRICAL TANKS

G. V. Hadjisophocleous, A. C. M. Sousa and J. E. S. Venart
Fire Science Centre and Department of Mechanical Engineering
University of New Brunswick
Fredericton, New Brunswick, Canada

ABSTRACT

The present work deals with the mathematical modeling of the response to heating resulting from fire engulfment of horizontal cylindrical tanks partially filled with a liquefied gas. The numerical model developed considers both radiation and convection heat transfer from the flames to the tank wall to determine the fire heat flux. The conduction equation is then solved, yielding the wall temperature distribution enabling the calculation of heat transfer to the liquid and vapor zones. Finally, the equations governing free convection are solved numerically, employing the control volume approach and a boundary-fitted coordinate system. Radiation heat transfer from the vapor-wetted wall to the vapor and liquid interface is determined using the net radiation method for an enclosure filled with an isothermal gas. Comparison of predicted results with available experimental data presented demonstrates the capability of the developed model to predict the response of LPG tanks to fire conditions. Simulations performed to study the effect of the filling level on the tank response are presented and analyzed.

NOMENCLATURE

A	geometric configuration factor;
c	specific heat capacity;
E	emissive power;
g	acceleration of free fall;
h	heat transfer coefficient;
J	Jacobian of coordinate transformation; radiosity;
k	thermal conductivity;
K	flame absorption coefficient;
L_r	reference length
P	non-dimensional pressure
q	heat flux
r	radial coordinate
R	radius
S	source term
t	time and non-dimensional time
T	temperature and non-dimensional temperature
u,v	velocity and non-dimensional velocity components in Cartesian coordinates
U,V	contravariant velocity components
x,y	non-dimensional Cartesian coordinates

Greek letters

α	thermal diffusivity
α,β,γ	metric coefficients
β	coefficient of thermal expansion
Γ	diffusion coefficient
ϵ	emissivity of surface
θ	plane angle
υ	kinematic viscosity
ξ,η	boundary fitted coordinates
ρ	density; reflectivity
σ	Stefan-Boltzman constant
τ	transmissivity
Φ	general variable

Subscripts

b	blackbody
conv	convection
e	fire boundary
g	gas
i	tank exterior surface zone
t	turbulent
v	vapour
w	wall

Dimensionless Numbers

Nu	Nusselt number
Pr	Prandtl number
Ra	Rayleigh number

INTRODUCTION

Pressure liquefied gases such as LPG (liquefied petroleum gas) and LNG (liquefied natural gas) are usually stored and transported in cylindrical or spherical tanks of various sizes. The filling level of these tanks with use is also variable, ranging from empty to the filling limits set by regulatory agencies. During fire engulfment, the response of the tank and its contents to heating depends on factors such as tank size, filling level, fire temperature and exposure, and location and size of any pressure relief valves. In order to understand the complex thermohydraulics of tank contents, extensive research efforts have been devoted to the experimental and mathematical modeling to the subject. Experiments are being carried out at various research centres such as the Health and Safety Executive, United Kingdom (Moodie et al., 1988), the Federal Institute of Material Testing, West Germany, (Droste and Schoen, 1988; Schoen and Droste, 1988), Davis Engineering, Canada (Birk, 1988) and the University of New Brunswick, Canada (Venart et al., 1988). The facilities at the University of New Brunswick consist of a small-scale tank with sophisticated data acquisition and data processing systems.

Most researchers involved in the mathematical modeling of tanks engulfed by fire use a zone model approach. With these methods, tank contents are subdivided into various zones and conservation laws and empirical relations are then used to model the various heat transfer and thermodynamic processes taking place in each zone. Zone models, to name a few, have been developed by Birk (1988), Sousa et al. (1985), Hunt and Ramskill (1987), Beynon et al. (1988) and Aydemir et al. (1988).

The present work represents a first step in developing a field model to study the response of LPG tanks to fire conditions. In order to make this project feasible, the following simplifying assumptions were necessary:

(i) the interface liquid/vapor doest not allow mass transfer and is static;
(ii) no boiling occurs at the tank walls;
(iii) the vapor is considered isothermal in determining the radiation heat transfer from the vapor-wetted wall to the liquid interface; and
(iv) flow is two-dimensional and incompressible;
(v) the Boussinesq approximation is valid;
(vi) the fluid is Newtonian; and
(vii) viscous dissipation is considered to be negligible.

Due to these assumptions the developed model can only be used to predict the tank response during the initial stages of heating (up to the first valve opening since during this period boiling is somewhat suppressed by pressurization).

The objectives of the present work are:
(a) develop a numerical model to study free convection in partially-filled horizontal cylinders;
(b) realistically model the engulfing fire;
(c) compare the predictions with experimental data; and
(d) study the effect of filling level on the tank behavior.

THE MATHEMATICAL MODEL

The system considered consists of the engulfing fire, the tank shell and the tank contents: liquid and vapor. Heat is transferred from the flames to the exterior tank surface by both radiation and convection; it is then conducted through the tank wall in the radial and circumferential directions and finally enters the tank contents by convection and radiation. The mathematical model developed consists of four sub-models; the fire model, the wall conduction model, the vapor space radiation model and the free convection model. These models are discussed in the subsequent sections.

Fire model

Radiation model. Radiation heat transfer is modelled following the approach used by Birk and Oosthuizen (1982), in which the fire is assumed to occupy a rectangle surrounding the cylindrical tank. The tank exterior surface is divided into surface zones and the configuration factors, A_{ij}, between each surface zone (A_i) and the imaginary walls enclosing the fire (A_e) are computed by numerically integrating the relation,

$$A_{ie} = \int_{A_i} \int_{A_e} \frac{\cos\theta_i \cos\theta_j \, e^{-\int K dR}}{\pi R^2} dA_j \, dA_i \qquad (1)$$

where R is the distance between the centers of the surface zones, K is the mean absorption coefficient of the flames, θ_i is the angle between the normal vector at a surface element and the radius vector connecting the two elements, and subscript "j" refers to areas of the fire enclosure wall.

Using the configuration factors and realizing that the zones on the cylinder can only "see" part of the enclosure walls and part of the surrounding gas, the energy balance equations are written as:

$$Q_i = A_i e_i E_i - \alpha_i (A_{ei} E_e + A_{gi} E_g) \qquad (2)$$

where Q_i is the heat transfer rate, e_i, the surface emmissivity, E_i, the emmissive power, and a_i, the surface absorptivity. The subscripts "g" and "e" stand for gas and fire enclosure wall, respectively.

Equation 2 is solved directly to give the radiation heat flux at the various surface zones of the tank surface.

Convection model. The convection heat flux, q_{conv}, from the hot combustion gases to the exterior tank wall is given by:

$$q_{conv} = h\,(T_w - T_g) \tag{3}$$

The convective heat transfer coefficient, h, is calculated based on heat transfer measurements for forced convection to cylinders in cross flow of air, as presented in Kreith and Bohn (1986). For Reynolds numbers (Re) less than 70,800, a mean heat transfer coefficient is obtained using the relation for the Nusselt number (Nu):

$$Nu = c\,Re^m\,Pr^n \tag{4}$$

where c=0.26, m=0.6, n=0.36 and Pr is the Prandtl number.

For Reynolds numbers greater than 70,800, the heat transfer coefficient varies along the tank surface as obtained from Kreith and Bohn (1986). The Reynolds number is calculated based on the tank exterior diameter and a gas velocity determined from measured fuel evaporation rates for large diameter pool fires (Blinov and Khudiakov, 1959).

Wall conduction model

The wall conduction model, using the control volume approach, numerically solves the transient conduction equation in cylindrical coordinates (r,θ)

$$\rho c\frac{\partial T}{\partial t} = \frac{1}{r}\frac{\partial}{\partial r}\left(rk\frac{\partial T}{\partial r}\right) + \frac{1}{r}\frac{\partial}{\partial \theta}\left(\frac{k}{r}\frac{\partial T}{\partial \theta}\right) \tag{5}$$

where ρ is the density, k is the thermal conductivity and c is the specific heat of the shell.

The boundary conditions employed for the solution of equation 5 are the fire heat flux at the exterior surface, calculated by the fire model, and the heat flux to the tank contents at the interior surface. The latter boundary condition is computed based on the solution of the liquid and vapor temperatures of the previous time step. For the vapor-wetted wall, the heat leaving the shell has radiative and convective components, while heat transfer to the liquid is only by convection.

Vapor space radiation model

This model calculates the radiative component of heat transfer from the vapor-wetted wall to the vapor and to the liquid/vapor interface, using the net radiation method for an enclosure filled with an isothermal gas (Siegel and Howell, 1981). For this purpose the vapor-wetted wall and the interface are divided into surface zones, each having a uniform temperature. The gas space is considered as a single-volume zone with uniform radiative properties. Configuration factors are calculated for each surface zone and the energy balance equations are solved yielding the radiation heat flux at each zone. An energy balance on the entire enclosure and vapor space is also calculated which gives the heat absorbed by the vapor.

Free convection model

The equations governing two-dimensional free convection are the continuity equation; the two momentum equations and the energy equation. They are simplified using the assumptions previously stated and normalized using, as reference parameters, liquid properties, tank diameter and ambient temperature and then transformed from the Cartesian coordinate system (x,y) to the boundary-fitted coordinate system (ξ,η). The Cartesian velocities (u,v) are maintained, however, as the dependent variables (Maliska and Raithby, 1984). The resulting equations, which exhibit a form similar to their Cartesian counterparts; are:

$$\frac{\partial U}{\partial \xi} + \frac{\partial V}{\partial \eta} = 0 \tag{6}$$

$$J\frac{\partial \Phi}{\partial t} + \frac{\partial (U\Phi)}{\partial \xi} + \frac{\partial (V\Phi)}{\partial \eta} =$$

$$\Gamma\left[\frac{\partial}{\partial \xi}\left(\frac{\dot\alpha}{J}\frac{\partial \Phi}{\partial \xi} - \frac{\dot\beta}{J}\frac{\partial \Phi}{\partial \eta}\right) + \frac{\partial}{\partial \eta}\left(\frac{\dot\gamma}{J}\frac{\partial \Phi}{\partial \eta} - \frac{\dot\beta}{J}\frac{\partial \Phi}{\partial \xi}\right)\right] + S(\xi,\eta) \tag{7}$$

where:

$$\dot\alpha = x_\eta^2 + y_\eta^2 \qquad \dot\beta = x_\xi x_\eta + y_\xi y_\eta$$

$$\dot\gamma = x_\xi^2 + y_\xi^2 \qquad J = x_\xi y_\eta - x_\eta y_\xi$$

$$U = u y_\eta - v x_\eta \qquad V = v x_\xi - u y_\xi$$

Φ, Γ and $S(\xi,\eta)$ have values as shown in Table 1, and ν and α are the kinematic viscosity and thermal diffusivity of the liquid, respectively.

Table 1. Symbols used in governing equations.

Equation	Φ	Γ	$S(\xi,\eta)$
x-momentum	u	$Pr\nu_t/\nu$	$-y_\eta\frac{\partial P}{\partial \xi} + y_\xi\frac{\partial P}{\partial \eta}$
h-momentum	v	$Pr\nu_t/\nu$	$-x_\xi\frac{\partial P}{\partial \eta} + x_\eta\frac{\partial P}{\partial \xi} + JRa\,(T - T_m)$
energy	T	α_t/α	0

The turbulent viscosity, γ_t, and thermal diffusivity, α_t, are obtained using the prescribed eddy viscosity model developed by Thompson et al. (1987).

The governing equations are discretized using the control volume approach and solved following the SIMPLEC method (Van Doormall and Raithby, 1984). Details of the solution procedure can be found in Hadjisophocleous (1989).

Major steps of the solution method

The four sub-models discussed in this section are combined to form the complete numerical model. The solution of each sub-model depends on the solution of the other sub-models. This dependency is introduced either in the form of boundary conditions or as a heat source or sink in the governing equations. The major steps of the solution procedure are:

(i) read and initialize all variables;
(ii) calculate the heat flux distribution at the tank exterior surface due to the fire using the fire model;
(iii) solve the conduction equation for the tank wall to obtain the temperature distribution of the shell;
(iv) calculate the heat flux from the wall to the tank contents by convection and radiation;
(v) employ the SIMPLEC method to solve free convection in the vapor and liquid regions;
(vi) calculate the tank pressure (i.e., the saturation pressure corresponding to the mean interface temperature);
(vii) proceed to the next time step, return to 2, and continue until the tank pressure has reached the set pressure for the valve to open.

RESULTS AND DISCUSSION

The developed numerical model is used to study the response of LPG tanks to a variety of conditions. The first simulation performed is an attempt to examine the predictability of the model by comparing its results with experimental data. The data used for the comparisons have been obtained by HSE of the United Kingdom (Moodie et al., 1988). A series of simulations is also carried out to investigate the effect of filling level on tank response to a fire environment. In the following section, the results obtained from these simulations are presented and analyzed.

Comparisons of predicted solution with experimental data

A numerical simulation has been carried out to compare the numerical predictions to experimental data obtained by the HSE (Moodie et al., 1988). The tank considered has a diameter of 1.694 m and is 80% filled with liquid propane. The fire temperature is extracted from the experimental data; however, an accurate representation cannot be easily made as the temperature varies with location and time. Variations of up to 200 K can be seen in the data. From the video record of the test, it is apparent that a strong wind was present during the test period, which sometimes caused the fire to expose the top of the tank. For this simulation, the fire shape is assumed to occupy the space between the wall enclosing the tank; thus the fire completely engulfs the tank during the numerical simulation. The flames absorption coefficient is chosen as 1.5 m^{-1} (Steward, 1988).

The tank pressure predicted by the model is compared with the pressure recorded during the experiment (Figure 1) and they are in excellent agreement. The numerical model predicts that the pressure relief valve opens in 305 seconds; during the test, the valve opened at 312 seconds.

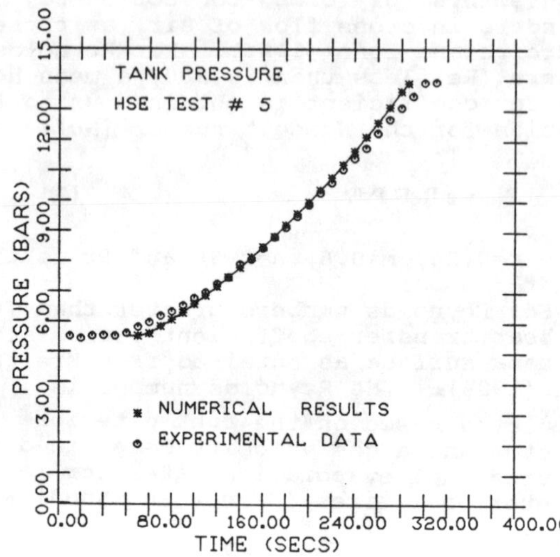

Figure 1. Comparison of predicted with experimental pressure.

Figures 2a and 2b indicate comparisons of liquid temperatures at the liquid top and bottom, respectively. In general, the predicted temperatures compare well with the experimental data, especially during the first 180 seconds of the test. Towards the end of the simulation run, the model seems to overpredict the liquid temperatures at the top of the tank. The temperatures at the bottom of the tank, however, are in good agreement (Figure 2b).

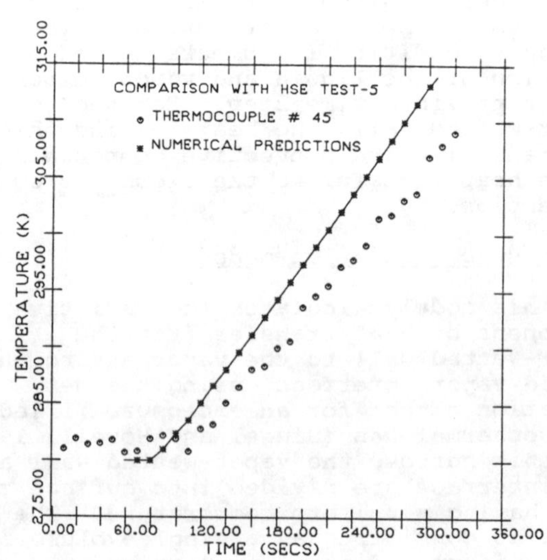

a)

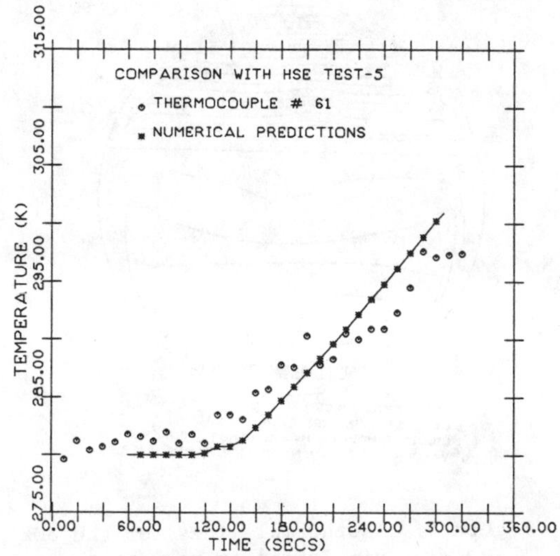

b)

Figure 2. Comparison of predicted with experimental temperatures at a) the top and b) the bottom of the liquid.

The predicted temperatures of the vapor and vapor-wetted wall are much higher than the experimental ones. A possible reason for this may be that, during the test, the vapor-wetted wall was not engulfed completely by the flames due to the wind. Hence, the temperature of this wall segment remained low due to radiation heat transfer to the ambient air. This may also explain why the model overpredicts the liquid temperatures, as the predicted radiation from the wall to the interface is much smaller than the actual radiation. Another factor influencing the liquid and vapor temperature variations between the experimental data and the predicted data is the amount of impurities in the experimental fluid, which is not accounted for in the numerical model.

Effect of filling level on tank response

A series of numerical tests were carried out to study the effect of filling level on tank behavior. The diameter of the tank for all these simulations was kept constant at 1.694 m. Fire conditions were also the same for all tests. Fire tests were conducted with filling levels of 20%, 40%, 60%, 80% and 98%. The solutions obtained from these simulations are presented and anlyzed in the following sections.

Flow and temperature field development. The flow development in both the liquid and vapor regions, for all simulations, follows a similar pattern. The velocity vectors of the 60%-filled case are shown in Figure 3.

After application of the fire heat flux, the temperature of the fluid adjacent to the tank wall increases. This causes the fluid to move upward due to buoyancy, creating a boundary layer flow. In the liquid region, the fluid continues to move upwards until it

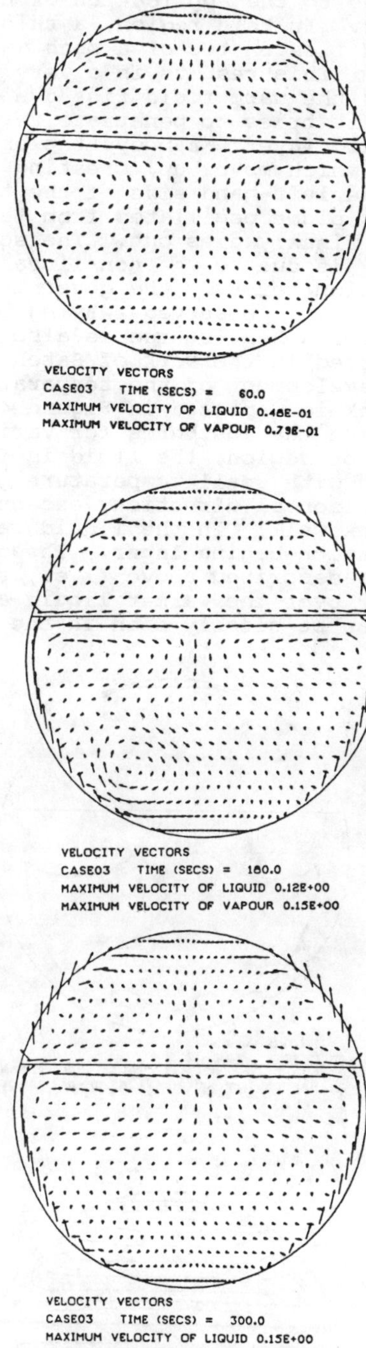

Figure 3. Velocity vectors for the 60% simulation at various times.

reaches the liquid-vapor interface where it then flows towards the tank interior. At the tank centre, it meets the stream coming from the other side and both streams move downward and diffuse in the bulk liquid. A similar behavior is seen in the vapor region. These motions generate two recirculating eddies in both regions. As time proceeds, some hydrodynamic instabilities occur at the tank bottom, due to the application of heat flux from below. In this region, a thin layer of cold fluid lies on top of a much warmer layer of liquid. This results in a very unstable system; as the warm fluid rise due to buoyancy, it tries to penetrate through the colder fluid layer. A buoyant plume is generated which moves up, creating a smaller eddy in the left-hand side of the tank. With time, this plume oscillates from left to right and finally dies out. The solution unsymmetry is due to the non-linearities in the advection terms of the governing equations. This observed swaying behavior of a plume over a heat source is also similar to that reported in the work of Satoh (1988).

The development of the temperature field in the tank is depicted in Figure 4, illustrating the iostherms for various times. In the vapor region, the fluid is fully stratified with small temperature gradients in the horizontal direction, except adjacent to the tank wall. In the liquid region, stratification beging later. The unstable condtiions described above (i.e., the thin cold layer over the warmer liquid at the tank bottom) can be clearly seen in the figure.

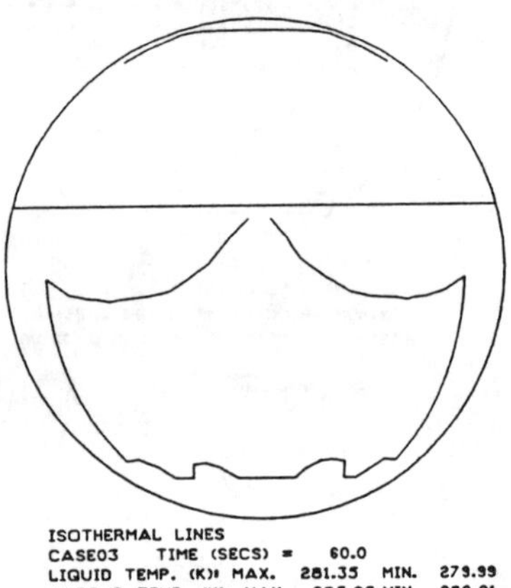

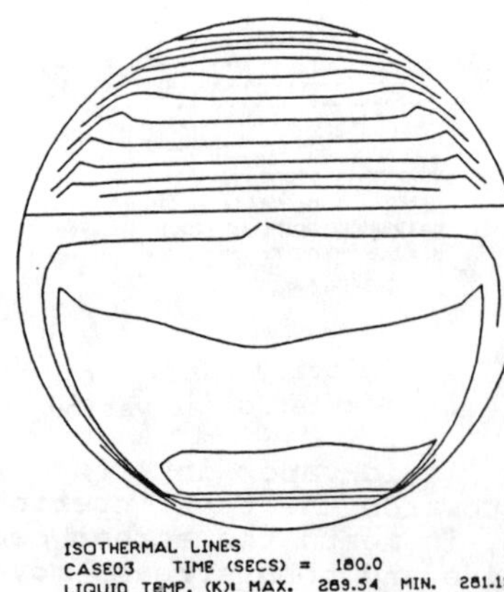

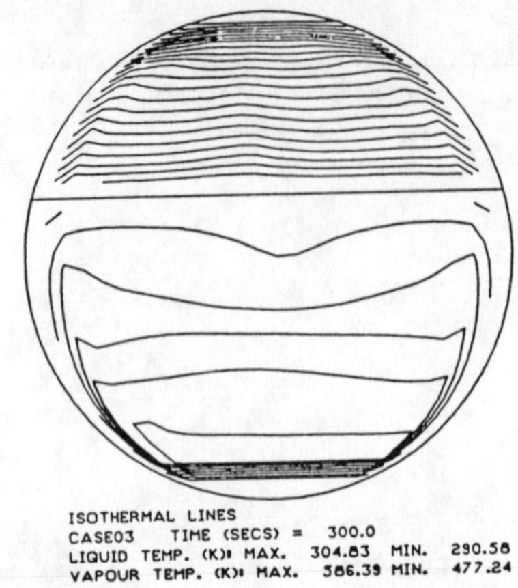

Figure 4. Isothermal lines for the 60% simulation at various times.

For better visualization of temperature field development in the liquid region, temperature profiles along the vertical diameter are plotted as function of time with the vertical axis starting from the tank bottom and extending up to the interface (Figure 5).

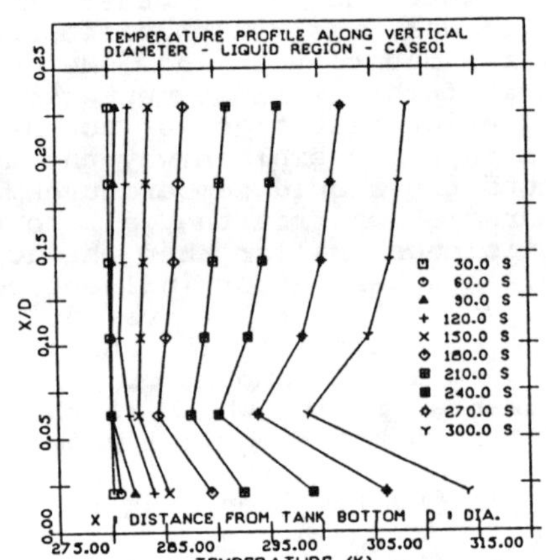

a)

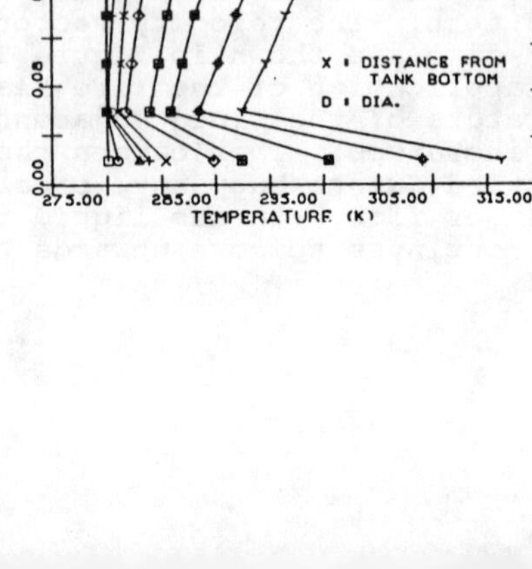

b)

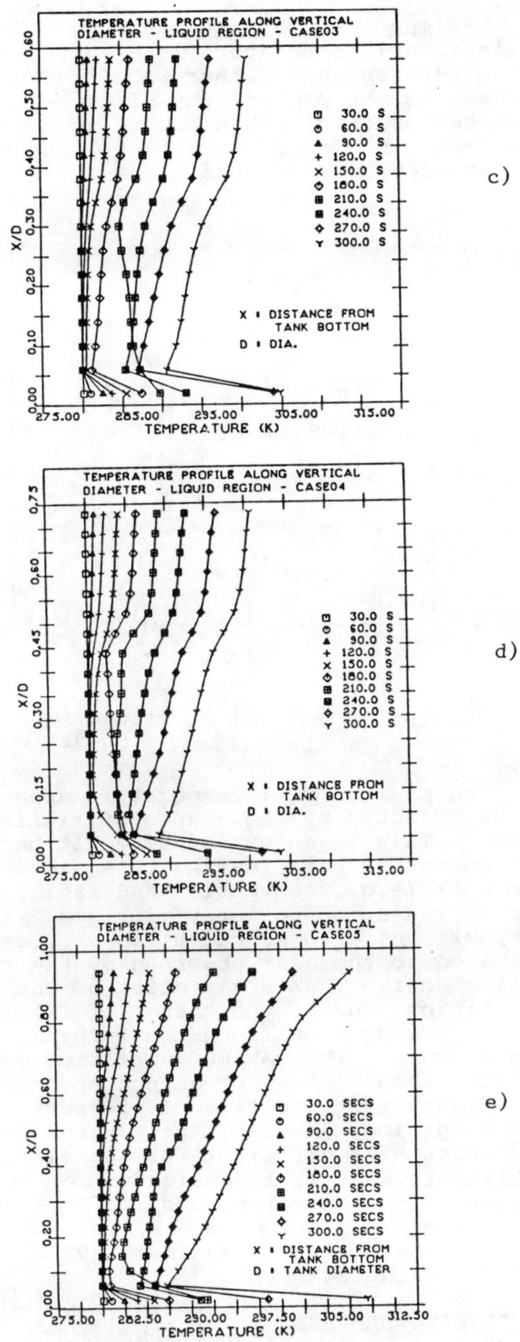

Figure 5. Temperature profiles along the vertical diameter in liquid for a) 20%, b) 40%, c) 60%, d) 80% and e) 98% simulation.

A distinct thermal stratification layer exists in all but the 98%-fill case. The fluid within the stratified layer has an almost uniform temperature, with negligible gradients in the vertical or horizontal direction. Within the bulk liquid layer, the temperature decreases continuously from top to bottom and reaches its minimum value just above the first control volume at the tank bottom. A gradient reversal exists for all cases at the tank bottom, due to the heat input from below. The temperature profiles of the 98%-fill cases has a different shape to that of the other cases. The liquid appears to be fully stratified but no distinction can be made between the stratified and bulk layer. An increase of the temperature gradient, adjacent to the liquid-vapor interface, is observed, mainly due two reasons; higher liquid velocities and shorter length of the interface. As a result, hot liquid from the tank wall reaches the tank centre before its temperature decreases and transfers its heat downward with its motion.

The mean temperatures of the vapor region, averaged over all control volumes for all cases, are plotted in Figure 6.

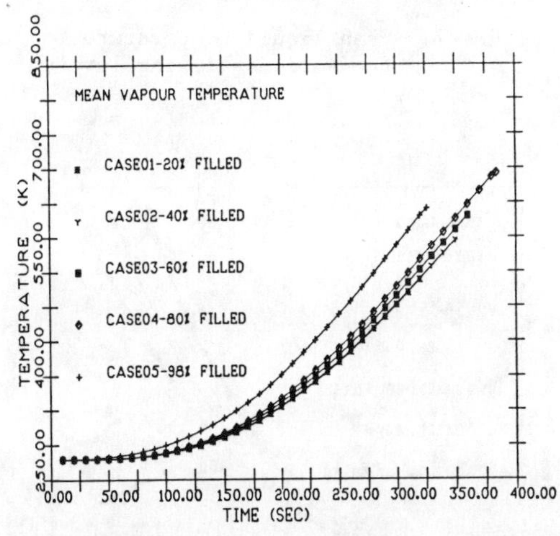

Figure 6. Mean vapour temperature for the various filling levels.

The results, as expected, indicate that the mean vapor temperature of the 98%-fill case is higher than that of the other cases. As the filling level decreases, the mean vapor temperature decreases. No apparent differences, however, exist between the mean vapor temperature of the 40%- and 20%-fill cases. Similarly, the mean liquid temperatures for all cases are shown in Figure 7. In the liquid region, the mean temperature is maximum when the filling level is 20% and decreases as the filling level increases. The variations for filling levels above 60% are, however, very small.

Tank pressure. The predicted pressures for the various simulations performed are shown in Figure 8 as a function of time.

The pressure history exhibits a similar behavior for all cases. During the initial time steps, the pressure remains relatively constant and as time progresses, it begins to increase in an exponential fashion and quickly reaches the pressure setpoint for the relief valve. This behavior is mainly due to the exponential nature of the surrounding tank and also to the fact that the mean interface temperature is almost constant during the first time steps because the warm

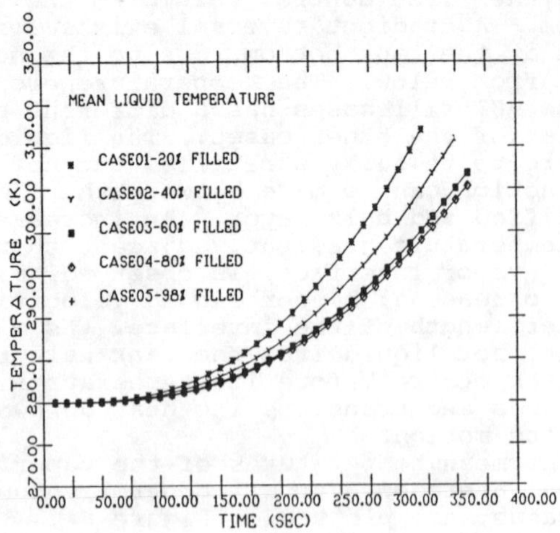

Figure 7. Mean liquid temperature for the various filling levels.

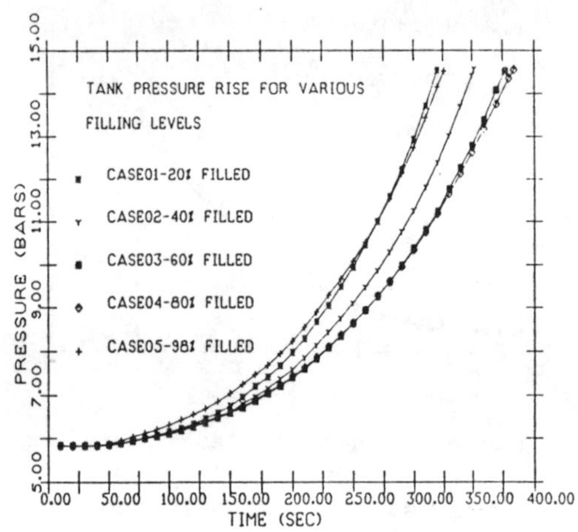

Figure 8. Tank pressure for the various filling levels.

liquid at the walls has not yet moved. As time proceeds, hot fluid moves to the interior along the interface causing its temperature to increase and hence the increase in tank pressure. The development of the stratified layer in the liquid enhances the rise of the interface temperature and thus tank pressure. Tank pressure, in the 20%-fill case, rises to the specified limit within 320 seconds. The time period required for the valve to open for the other cases increases as filling level increases, except for the 98%-fill case, where the valve opens at 326 seconds. This behavior is to be anticipated since the liquid temperature increases faster as the mass of the liquid decreases. The time needed for the pressure to increase to the setpoint for the 98%-fill case is less, however, due to the very short length of the interface, which enhances the temperature of the interface.

<u>Maximum tank wall temperature</u>. The maximum wall temperature is a very important parameter, which together with tank pressure, influences the probability of tank explosion. This temperature, which occurs at the top of the tank, is plotted in Figure 9 for all filling levels simulated.

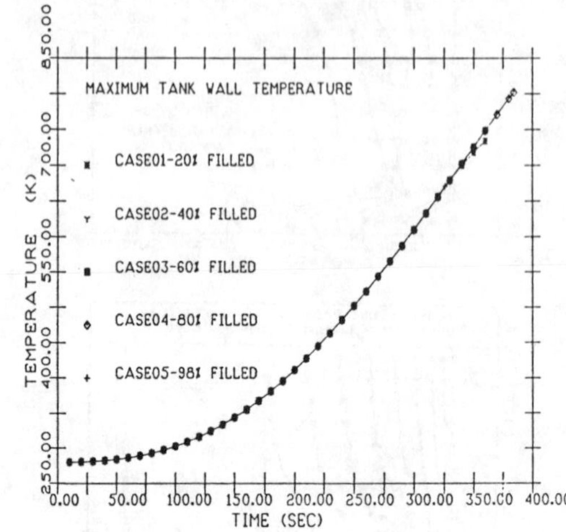

Figure 9. Maximum tank wall temperature for the various filling levels.

The maximum wall temperature appears not to be affected by the change of filling level. This is an expected result because the amount of heat leaving this section of the tank (e.g., convection and radiation to vapor, radiation to interface, and circumferential conduction) is very small compared to the heat absorbed by the tank wall itself. Thus a variation of these quantities, due to the change of filling level, yields a small change in tank wall temperature. The maximum variation predicted is less than 1 K, which is not apparent in the figure. At the time of pressure relief valve opening, however, the maximum wall temperature is different for the different filling level. It is analogous to the time required for the valve to open; therefore, it has its minimum value (656 K) for the 20%-fill case and its maximum (793 K) for the 80%-fill simulation.

CONCLUDING REMARKS

A numerical model has been developed to study the response of a LPG tank to fire conditions. A numerical simulation performed in order to compare predictions with experimental data indicated that, despite the difficulties in accurately modeling the engulfing fire, the developed model can reasonably predict the time to first valve opening. The model seems, however, to overpredict the vapor and liquid temperatures, which maybe expected due to the variation in the fire exposure during the test.

The developed numerical model has been used to simulate tank response to fire engulfment for a variety of filling levels. The predicted results lead to the following conclusions:

(i) the time for the first pressure relief valve opening increases as the filling level increases, except for the 98%-fill case;

(ii) the vapor region, for all cases, becomes fully stratified very early in the transient and remains in a pseudo-steady state condition thereafter.

(iii) two distinct thermal regions develop in the liquid region; a stratified and bulk zone. The two zones cannot, however, be distinguished in the 98%-fill simulation.

(iv) the thickness of the liquid stratified layer is approximately equal to 33% of the distance from the tank bottom to the interface at the time the pressure relief valve opens.

(v) the temperature difference between the mean temperature of the liquid stratified layer and the bulk layer increases with the filling level; and

(vi) the maximum vapor-wetted wall temperature is not significantly affected by the variation of the filling level.

ACKNOWLEDGEMENTS

This work was supported by Transportation Development Centre, Transport Canada and the Natural Sciences and Engineering Research Council of Canada.

REFERENCES

Aydemir, N.U., Magapu, V.K., Sousa, A.C.M., and Venart, J.E.S., 1988, "Thermal Response Analysis of LPG Tanks Exposed to Fire", J. Hazardous Materials, Vol. 20, pp. 239-262.

Beynon, G.V., Cowley, L.T., Small, L.M., and Williams, I., 1988, "Fire Engulfment of LPG Tanks: HEATUP, a Predictive Model", J. Hazardous Materials, Vol. 20, pp. 227-238.

Birk, A.M., 1985, "Fire Test to Study the Significance of a Burning Relief Valve Flare from a Tank-car Engulfed in Fire", Transport Canada Report No. TP6694E, Ottawa, Ontario.

Birk, A.M., 1988, "Modelling the Effects of a Torch Type Fire Impingement on a Rail or Highway Tank Car", J. Hazardous Materials, Vol. 20, pp. 197-225.

Birk, A.M., and Oosthuizen, P.H., 1982, "Model for the Prediction of Radiant Heat Transfer to a Horizontal Cylinder Engulfed in Flames", ASME 82-WA/HT-52.

Blinov, V.I., and Khudiakov, G.N., 1959, (Reviewed by Hottel, H.C.), "Certain Laws Governing Diffusive Burning of Liquids", Fire Research, Abstracts and Review, Vol. 1, pp. 41.

Droste, B., and Schoen, W., 1988, "Full Scale Fire Tests with Unprotected and Thermal Insulated LPG Storage Tanks", J. Hazardous Materials, Vol. 20, pp. 41-53.

Hadjisophocleous, G.V., 1989, "Numerical Modelling of LPG Horizontal Cylindrical Tanks Subjected to Heating Environments", Ph.D. Thesis, University of New Brunswick, Fredericton, N.B.

Hunt, D.L.M., and Ramskill, P.K., 1987, "A Description of ENGULF - a Computer Code to Model the Thermal Response of a Tank Engulfed in Fire", Technical Report, No. SRD R 354, United Kingdom Atomic Energy Authority.

Kreith, F., and Bohn, M.S., 1986, Principles of Heat Transfer, 4th Edition, Harper and Row Publishers, New York.

Maliska, C.P., and Raithby, G.C., 1984, "A Method for Computing Three Dimensional Flows Using Non-Orthogonal Boundary Fitted Coordinates", Int. J. Numerical Methods in Fluids, Vol. 4, pp. 519-537.

Moodie, K., Cowley, R.B., Denny, R.B., Small, L.M., and Williams, I., 1988, "Fire Engulfment Tests on a 5 Tonne Tank", J. Hazardous Materials, Vol. 20, pp. 55-71.

Satoh, K., 1988, "A Numerical Study of Dynamic Plume Behavior (Natural Swaying Motion of the Plume above a Line Heater," JSME International Journal, Series II, Vol. 31, pp. 402-409.

Schoen, W., and Droste, B., 1988, "Investigation of Water Spraying Systems for LPG Storage Tanks by Full Scale Fire Tests", J. Hazardous Materials, Vol. 20, pp. 73-82.

Siegel, P, and Howell, J.R., "Thermal Radiation Heat Transfer", 2nd Edition, Hemisphere Publication Corporation, Washington, D.C.

Sousa, A.C.M., Wentzell, P.R., and Venart, J.E.S., 1985, "Modelling of LPG Tanks Exposed to Fire", Heat and Technology, Vol. 20, pp. 1-23.

Steward, F.R., 1988, Personal communication.

Thompson, C.P., Wilkes, N.S., and Jones, I.P., 1987, "Numerical Studies on Buoyancy-Driven Turbulent Flow in a Rectangular Cavity", Int. J. Numerical Methods in Engineering, Vol. 24, pp. 89-99.

Van Doormall, J.P., and Raithby, G.D., 1984, "Enhancements of the SIMPLE Method for Predicting Incompressible Fluid Flows," Num. Heat Transfer, Vol. 7, pp. 147-163.

Venart, J.E.S., Sousa, A.C.M., Steward, F.R., and Prasad, R.C., 1984, "Experiments on the Physical Modelling of LPG Tank Cars Under Accident Conditions", Proc. Int. Conf. Storage and Transport of LPG and LNB, Vol. 1, Royal Flemish Soceity of Engineers, Brugge, pp. 385-390.

1989 National Heat Transfer Conference
HTD-Vol. 106, Heat Transfer Phenomena in Radiation, Combustion and Fires

HOLOGRAPHIC INTERFEROMETRY TEMPERATURE MEASUREMENT OF LIQUID PHASE UNDER SPREADING FLAME CONDITION

A. Ito and D. Masuda
Department of Energy Engineering
Oita University
Oita, Japan

K. Saito
Department of Mechanical Engineering
University of Kentucky
Lexington, Kentucky

ABSTRACT

This study was undertaken as an effort to improve our understanding of flame spread mechanism over subflash point liquids. Studies of flame spread over liquids are of current interest due to our fundamental curiosity about its mechanism and practical concern about accidental occurence of fires on liquid fuels. A holographic interferometry technique was applied to obtain detail temperature profiles in liquid phase for pulsating and uniform spread over subflash point liquids. Subsurface liquid convection motion which is comprised of two different circulation zones, was found in the liquid phase for both pulsating and uniform spread regions. In addition, flame spread behavior was recorded by a video camera and spread rate and pulsation frequency were then deduced. Based on those measurements, the flame spread rate was calculated, and which compared favorably with the experimental results.

INTRODUCTION

A series of experimental studies were performed to understand the flame spread mechanism on alcohols (Roberts,1959; Burgone et al., 1968a,1968b,1968c; Glassman et al.,1969; Akita,1974; Newman,1979; Hirano et al.,1980) and hydrocarbon fuels (Glassman et al.,1969; Mackinven et al.,1970). Based on these results, Glassman and Dryer (1980) made a review which includes explanations on mechanism of pulsating spread and uniform spread over subflash point liquids. In this paper, we call the region where pulsating spread occurs "pulsating region" and the region where uniform spread occurs "uniform region". According to them, flame spread in pulsating region could be a result of coupled gas phase/liquid phase convection effects, since it occurs only when subsurface liquid convection has been generated. While flame spread in uniform region occurs when a flame conducts sufficient heat to the liquid underneath and just ahead of it to volatilize some fuel. This additional fuel entering the gaseous fuel-air mixture just ahead of the flame brings that mixture within the lean flammability limit and the flame is able to propagate.

To understand the heat transfer mechanism through gas phase, Hirano et al. (1980) measured the induced air velocity above a methanol surface using high speed Schlieren and his unique hot gas tracer device with changing initial liquid temperature between -5 ^{o}C and 30 ^{o}C. They found that when the initial liquid temperature (T_∞) was less than 5 ^{o}C, there was no induced air flow above the liquid surface in the area preceeding the flame's leading edge, while for 5 $^{o}C < T_\infty < T_c$ there was induced air flow which was small at the lower temperature and became larger with increasing T_∞ (where T_c is the closed cup flash point and is 11 ^{o}C for methanol). They have suggested that for $T_\infty < 5$ ^{o}C, flame spread is governed by gas phase conduction and subsurface liquid convection, while for 5 $^{o}C < T_\infty < T_c$ it is governed by gas phase convection and liquid phase conduction.

Yet, two questions still remained to be answered: (1) a boundary temperature between the uniform spread and the pulsating spread over methanol is 1 ^{o}C from Akita's data (1974), while gas phase velocity measurement by Hirano et al.(1980) resulted in a critical temperature of 5 ^{o}C, below which there is no horizontal gas flow. There is a need for consistent explanation for the existence of the boundary temperature between the uniform spread and the pulsating spread. (2) In uniform region, if heat was given to the liquid ahead of the flame's leading edge either through gas phase or liquid phase, a temperature gradient must exist in the liquid phase. This temperature gradient may generate subsurface liquid convection, however no such flow has been observed in uniform region.

To answer these two questions, a detail and instantaneous temperature mapping in the liquid phase is essential. To satisfactorily obtain this mapping, a high space/time resolution and simultaneous collection of spatially distributed signals are required, since the phenomenon is highly time dependent. The conventional thermocouple point by point measurement technique, however, is incapable of satisfying these requirements, because of the effects of the wire disturbances on the flow and the limited spatial resolution of the thermocouple bead (Newman,1979; Ito and Kashiwagi,1988).

A holographic interferometer connected to a photographic recording system can detect a sudden and sensitive temperature change occurring in a distributed area instantaneously. By applying this technique, detail temperature profiles in the subsurface liquid convection current were investigated. Basic principles of this technique were explained by Ito and Kashiwagi (1987), and applicability of the technique was successfully demonstrated for 1-decene by Kashiwagi and Kashiwagi (1986) and for polymethylmethacrylate (PMMA) by Ito and Kashiwagi (1988). We tested three different alcohols (methanol, ethanol and n-propanol) in a rectangular tray with a temperature control unit and detailed discussions are made for the n-propanol results. In addition, flame behavior was recorded by a video camera and an average flame spread rate was deduced from the video film, in the same manner as was done by Saito et al.(1985). The results obtained were plotted as a function of T_∞ and then compared to other published data.

The subsurface liquid convection current was found to be governed mainly by surface tension and viscous forces, but little by gravity (Sirignano, 1972 and Torrence, 1971). It should be mentioned that, apart from a thin layer of surface tension driven flow (STF), a round circular convective zone (CCZ) in which gravity may play a role, was always found underneath the flame's leading edge. In the following sections, we will present the experimental method and interferograms which reveal a detailed temperature profiles in the liquid phase.

EXPERIMENTAL APPARATUS AND PROCEDURE

Schematics of experimental apparatus, which include a tray of 40 cm length, 1 cm height and four different widths (0.5, 1.0, 2.0, and 5.0 cm) with a rectangular cooling unit, and a video camera are shown in Fig. 1a. In the figure, x coordinate was fixed to the flame spread direction, y to the tray width direction and z to the vertical direction. Two long sides of the tray were made of Pyrex of 0.1cm thickness. Fuel was uniformly ignited at one end of the tray by a pilot flame. A video camera system was used to record the flame spread phenomena. Each test could then be reviewed on a video cassette recorder as was previously done by Saito et al.(1985). The video camera had a built-in digital clock (0.1 s accuracy) and could be filmed along with each test to allow a reliable method for measurement of pulsation frequency and spread rate, as well as detail observation of flame spread behavior.

A schematic diagram of the optical system used for holographic interferometry is shown in Fig. 1b. Lenses and mirrors are arranged on an optical bench and a 5 mW He-Ne laser was used as the light source. A diffraction image was enhanced by enhancing the reference light so that a real time method (Ito and Kashiwagi,1987) could apply. Interferogram for a reference (25 °C) condition was photographed with 1 s exposure time, while interferogram for spreading flame was photographed with 1/500 s exposure time, as it was fast enough to "freeze" the temperature profiles. Each interferogram was taken at a 0.1 s interval by a motor driven camera and the corresponding visible flame location was determined from the video film. To

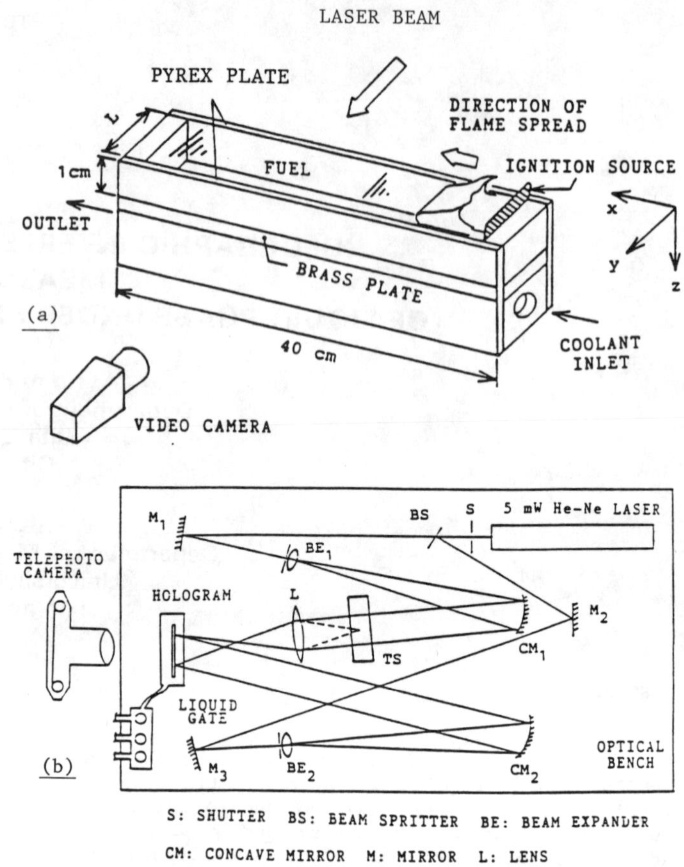

Fig.1 Shematic diagram of the apparatus: (a) tray and video camera, and (b) optical system used for holographic interferometry.

acccurately resolve each fringe in the interferogram, a uniform/short path length through the sample (liquid and Pyrex), and two dimensionality in temperature profiles in gas and liquid are required. To obtain a uniform/short path length, the narrower tray width is more advantageous, so that a tray with 0.5 cm and 2.0 cm width was used, while to ensure two dimensionality in temperature profiles in gas and liquids, tray with widths (0.5 cm, 2.0 cm and 5.0 cm) were used. Each result was then compared to the corresponding flame shape recorded by a video film. The flame shape was indeed distorted near the side wall due to heat loss for the smallest tray width. However, a reasonable degree of two dimensionality was sustained for the tray with a width larger than 2.0 cm. Effects of tray width on spread rate were studied by Mackinven et al.(1970) for n-decane. They found that, for a tray width larger than 60 cm, the spread rate became independent of width. However, we decided not to use a wide tray, since the spread mechanism wouldn't be significantly affected. To satisfy the "deep pool" condition (Williams,1985), a 1 cm fuel thickness was decided to be used based on the experimental results by Isoda (1971) and our exploratory tests on the spread rate measurements over various fuel layer thicknesses. Therefore all the detail measurements were done with a tray of 2 cm width and with 1 cm fuel thickness.

Assuming the laser beam emitted along y-direction (normal to x-z plane in Fig. 1a) moves straight and temperature variation along y-direction is negligible, the relationship between fringe number, N and temperature, T can be described as follows:

$$\left| L \int_{T_\infty}^{T} \frac{dn_1}{dT}\, dT + 2xl \int_{T_\infty}^{T} \frac{dn_s}{dT}\, dT \right| = N\lambda \qquad (1)$$

N=1,2,3

where L: the inner width of tray, dn/dT: thermooptic coefficientmaterial (n_1: refractive index of liquid and n_s: refractive index of Pyrex), l: thickness of Pyrex wall, λ: the wave length of light, and T_∞: the initial liquid temperature. Detail discussion of Eq.(1) is given by Ito et al.(1989). The thermooptic coefficient of Pyrex is one order smaller than that of liquid. The thickness of the Pyrex wall is 0.1 cm which is small compared to the width of the tray (2 cm). Therefore, the second term of the left-hand side in Eq.(1) can be neglected.

Thermooptical coefficients for alcohols and hydrocarbon liquids under 25 °C were investigated experimentally by Hauf and Grigull (1970) and theoreticlly by Murphy and Alpert (1976). However, their temperature dependence is unknown. Therefore, we measured the thermooptical coefficients of four different alcohols (methanol, ethanol, n-propanol and n-butanol) as a function of temperature between 5 °C and 80 °C using an Abbe refractometer. Special care was taken to prevent evaporation of the liquid during the measurement. The measured values were approximated by a quadratic expression. The thermooptical coefficients were then obtained by taking derivatives and favorably compared with those at 25 °C by Hauf and Grigull (1970) in Table 1. The reference temperature for the interferogram was obtained by a 50 μm wire diameter copper-constantan thermocouple. The estimated error for the temperature reading from the interferogram of n-propanol was within 3 % (for detail discussion of the error analysis, see Ito et al., 1989).

Table 1 Thermooptical coefficient for five different alcohols

Liquid fuels	-dn/dT ℃⁻¹ (λ=632.8nm)	
	Hauf&Grigull(1970) At 25°C	This work [Temperature range °C]
Methyl alcohol	4.0E-4	(3.7+0.0105T)E-4 [5~60]
Ethyl alcohol	4.0E-4	(3.8+0.0106T)E-4 [5~70]
n-Propyl alcohol	------------	(3.8+0.0137T)E-4 [5~80]
Iso propyl alcohol	4.15E-4	------------------------
n-Butyl alcohol	------------	(4.0+0.0122T)E-4 [5~80]

RESULTS AND DISCUSSION

Effects of Ambient Air Temperature and Tray Width

To check whether or not our experimental apparatus could provide data agreeable with those of others, flame spread behavior over ethanol was recorded on a video film. Average spread rate $\bar{V}$ (tray length divided by time for flame to propagate from one end to the other) was then measured and plotted as a function of T_∞ (Fig. 2). Our results favorably compare to those of Matsumoto (1980) for ethanol.

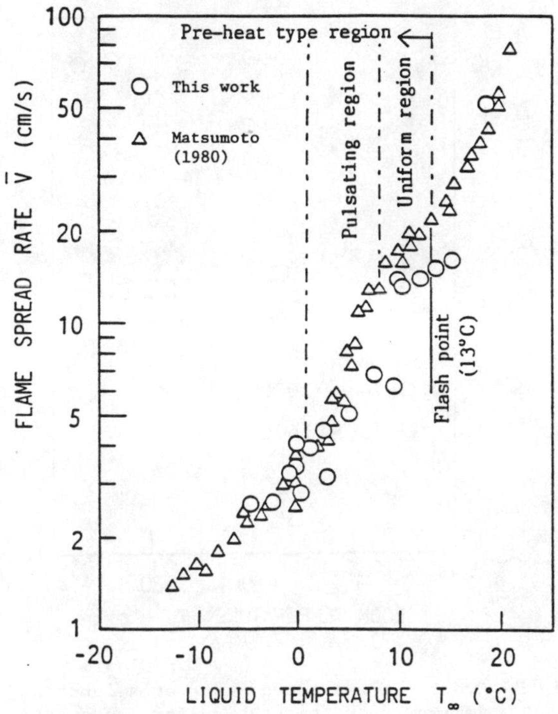

Fig.2 Average flame spread rate as a function of the initial liquid temperature for ethanol.

Ideally, it is better to perform spread tests under the same liquid and ambient gas temperatures. With n-propanol whose closed cup flash point, T_c is 25 °C (Glassman and Dryer,1980), it is relatively easy to satisfy this condition. A series of detailed measurements were, therefore, performed for n-propanol. The dependence of flame spread rate on ambient air temperature, T_o was investigated for two different T_∞'s (T_∞ = 16 °C and 20 °C) as shown in Fig. 3, in which temperature range the average spread rate was found to be nearly independent on the ambient air temperature. Average spread rate was also plotted as a function of the initial liquid temperature (Fig. 4) and tray widths for three different T_∞'s (Fig. 5). The trend in our results is in agreemet with previous studies by Akita (1974) and Mackinven et al.(1970).

Interferograms of STF and CCZ for n-Propanol

Two different series of interferograms (each series contains 4 photographs, (a) through (d)) for pulsating region under T_∞ = 16 °C are shown in Figs. 6 and 7. Illustrations of the iso-thermal lines for the interferogram of case (b) for Fig. 6 and case (c) for Fig. 7 are also included in each figure. Spread rate of gas phase flame front (V) and that of subsurface

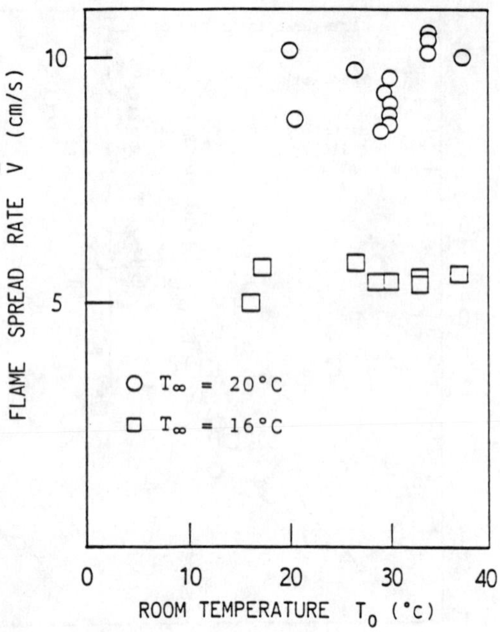

Fig.3 Average flame spread rate as a function of ambient room temperature for n-propanol.

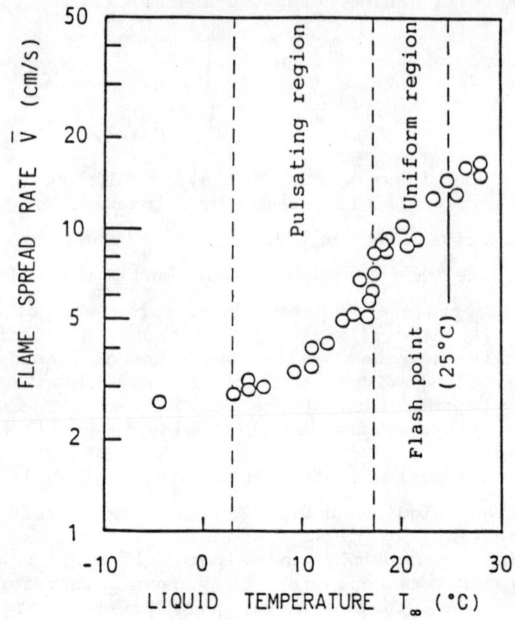

Fig.4 Average flame spread rate as a function of initial liquid temperature for n-propanol.

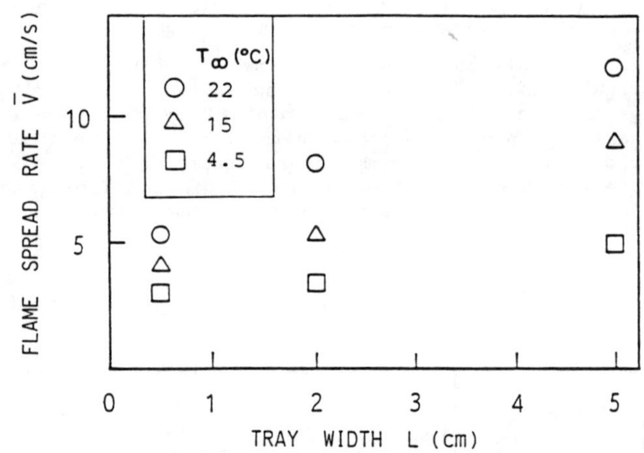

Fig.5 Average flame spread rate as a function of tray width for n-propanol.

liquid convection current (S) for the Fig. 6 case are shown in Fig. 8a and those for Fig. 7 case are shown in Fig. 8b. It can be seen from Fig. 8 that V and S in Fig. 8a are in a steady state condition with nearly the same spread rate, and the liquid surface temperature close to the flame edge is below T_c. Contrary in Fig. 8b, V is in a rapid acceleration mode, S remains nearly constant, and the liquid surface temperature exceeds T_c by 7 °C, suggesting that the flame spread occurs in the premixed gas phase. The last observation is consistent with previous results by Roberts (1959) and Mackinven et al.(1970). Our interferogram results also agree with the interpretation given by Glassman and Dryer (1980) that the pulsating spread is the result of an appearance of gas phase spread and liquid phase spread alternately.

Based on the series of interfelograms obtained, the process of pulsating flame spread is schematically illustrated in Fig. 9 which includes steps (a) through (e). The steps (a) and (b) correspond to the condition indicated in Figs. 6 and 8a, while steps (c) and (d) correspond to the condition in Figs. 7 and 8b. For step (c) surface temperature of subsurface liquid convection exceeds T_c by 7 °C (the gaseous fuel-air mixture over the fuel surface is within flammability limit), therefore flame can spread in the formed premixed gas layer. Once the flame's leading edge reaches the front of STF, it slows down and stays there momentarily as shown in step (d). A new circulation zone CCZ is then formed in the front location of the old circulation zone (step(e)). The small CCZ grows until a new STF is generated. The time interval of repeating the processes (a) through (e) was found to be exactly the same as that of flame pulsation measured from the video film.

For uniform region, two interferograms and illustrations of the corresponding isothermal lines are shown in Fig. 10a for T_∞ = 19 °C and Fig. 10b for 22.8 °C. It was found that the subsurface liquid convection was distinguished into the two different patterns by a critical initial liquid temperature ($T_{\infty,c}$ = 17 °C) which distinguishes uniform region and pulsating region. In uniform region (17 °C < T_∞ < T_c), CCZ and STF were always found to be close together (the distance (δ) between head of STF and that of CCZ is less than 1 cm). While at T_∞ = $T_{\infty,c}$, the onset of pulsating spread, a sudden increase of δ to 1.8 cm occurred. For 4 °C < T_∞ < 17 °C, δ remained

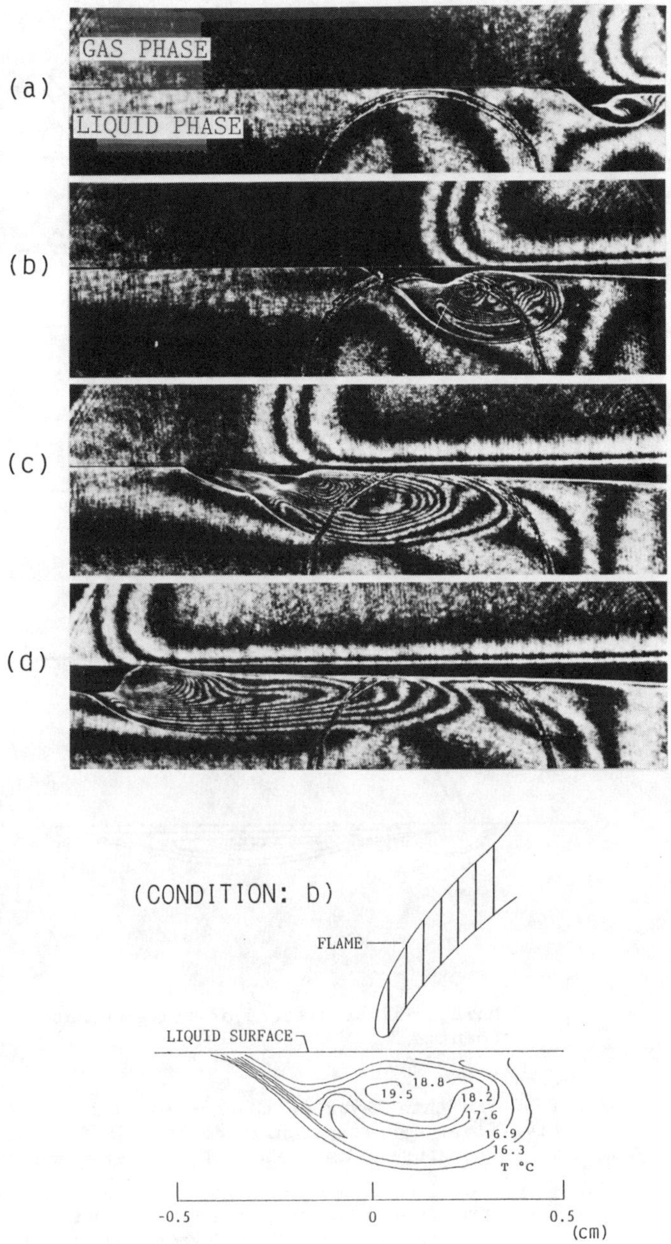

Fig.6 Four different interferograms (a) through (d) and the iso-thermal lines for condition (b) for n-propanol under T_∞ = 16 °C.

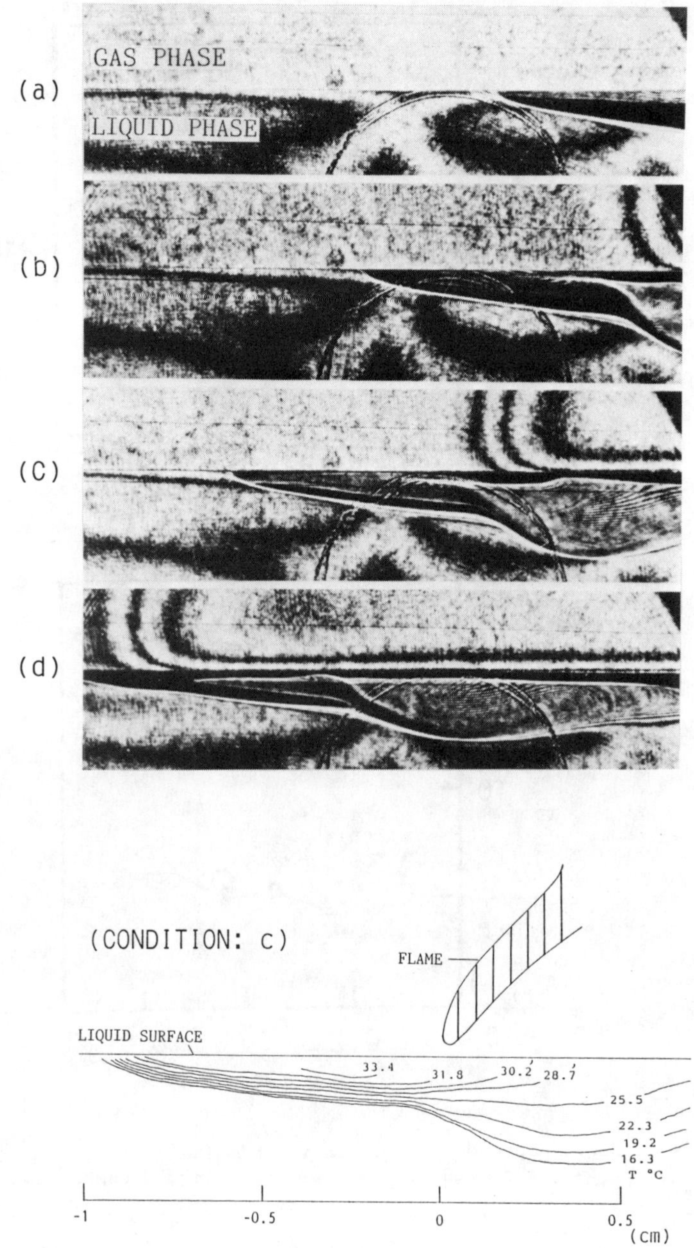

Fig.7 Four different interferograms (a) through (d) and the iso-thermal lines for condition (c) for n-propanol under T_∞ = 16 °C.

approximately to be 2 cm which is more than twice as large as δ for uniform region. The relative distance between the head of the subsurface liquid convection current and the flame's leading edge was measured from the interferograms and plotted as a function of T_∞ (Fig.11). However, neither STF nor CCZ was found in the liquid phase for $T_\infty > T_c$, which is consistent with the results of Akita (1974). Since there exists a lean flammability mixture over the fuel surface, spread is able to occur in the gas phase without having any heat input from the flame to the fuel.

It should be emphasized that, for liquids whose T_∞ is lower than T_c, the subsurface liquid convection current was always found in the liquid phase. This result is discrepant from that of Akita's (1974) which did not show any convective currents in methanol in uniform region. Two reasons were thought: (1) even if a small convective motion was generated in the liquid subsurface, the size of the motion is so small that a distorted liquid layer near the side window or hot gas layer near the flame edge could mask the flow from the streak shadowgraph technique. Holographic

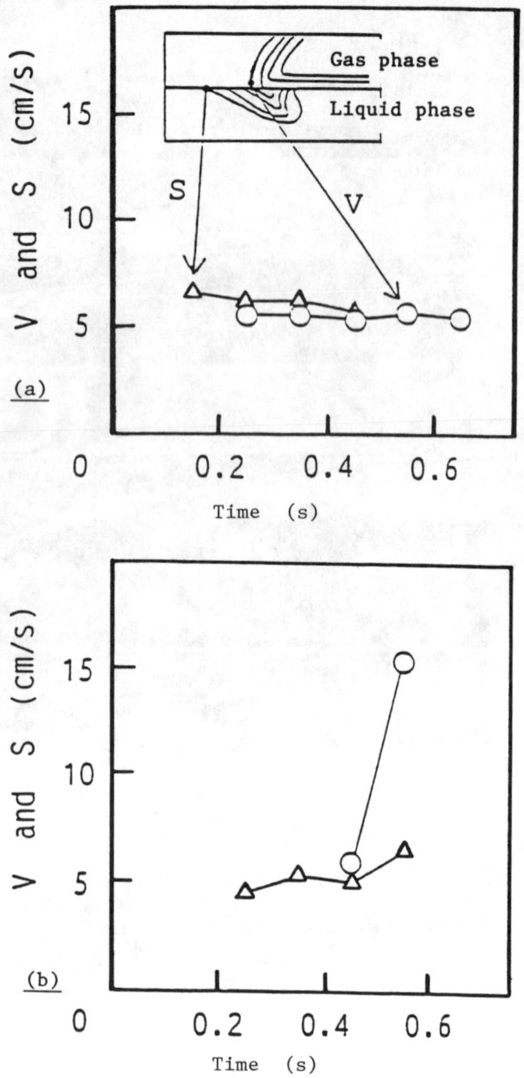

Fig.8 Spread rate of the visible flame front, V and the head of the subsurface liquid convection flow, S for n-propanol under $T_\infty = 16\ ^\circ C$. ((a) is for Fig. 6 case and (b) for Fig. 7 case).

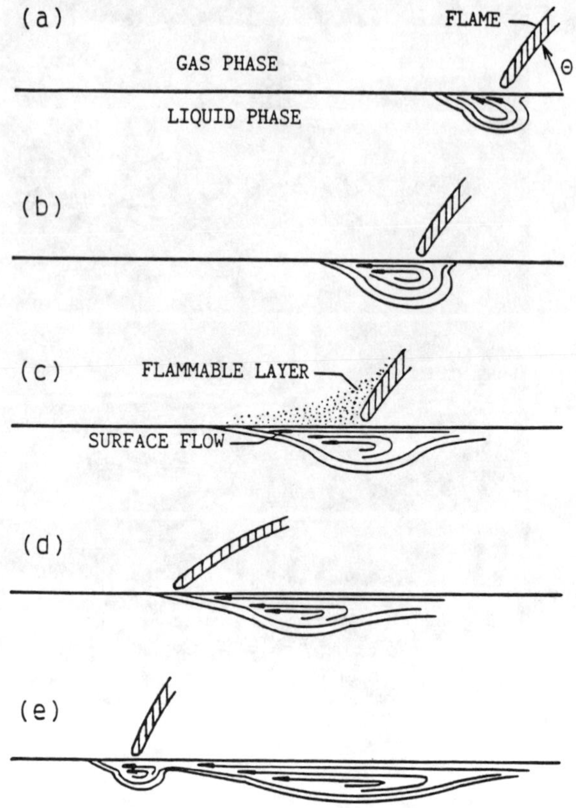

Fig.9 Schematic illustrations of flame pulsation mechanism.

interferometry, however, was sensitive enough to detect the subsurface liquid convection current in uniform region and pulsating region. (2) different liquid fuels may have different flame spreading mechanisms over the subflash point liquids. To check this posibility, interferograms for methanol and ethanol were obtained under the subflash point liquid conditions using the same method as for n-propanol. As a result, we always found subsurface liquid convection current for all three alcohols.

Our interferogram results and gas phase velocity measurement results by Hirano et al.(1980) suggest that for uniform region, both gas phase and liquid phase convections are important for preheating the liquid ahead of the flame. A series of interferograms taken for uniform region revealed that a thin layer of subsurface flow whose temperature was above T_c was always ahead of the flame's leading edge (two representative interferograms are shown in Fig. 10). Apart from the thin layer of the hot subsurface liquid convective flow, a cold circular convective flow whose maximum temperature was below T_c always appeared underneath the flame's leading edge (see Fig. 10). The two zone structure appears to be governed by two different mechanisms, since these two circulations are connected together and have the same rotational direction (counter-clockwise). A possible explanation for the two zone structure is that the hot subsurface liquid convective flow is a surface tension-driven flow, while the cold circular convective flow is a buoyancy-driven flow. Spread rate calculation for each current will appear in a later section.

Effects of Flame Inclination

Flame inclination angle θ, defined in Fig. 9, is approximately the same (60°) for steps (a) through (c); but smaller (30°) for step (d) and larger (75°) for step (e). The smaller θ (30°) may be attributed to flame propagation in premixed gas mixtures, since the color of the flame is so pale. Flame propagates faster near the fuel surface where the premixed gas satisfies the lean flammability limit. With increasing distance from the fuel surface, the concentration of the fuel vapour becomes thinner, so that the spread rate

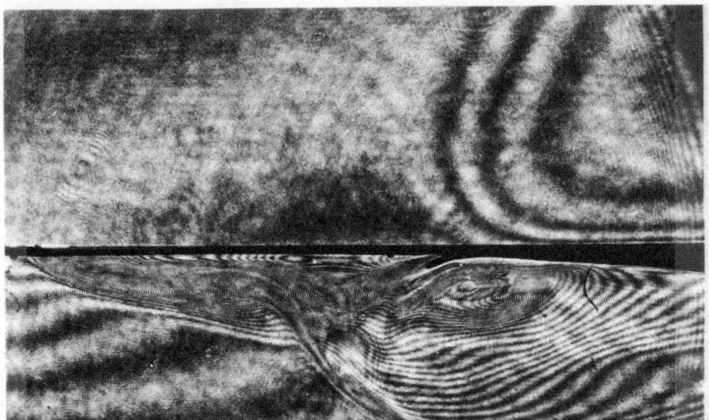

Fig.10 (a) Interferogram obtained for n-propanol under T_∞ = 19 °C and its iso-thermal lines

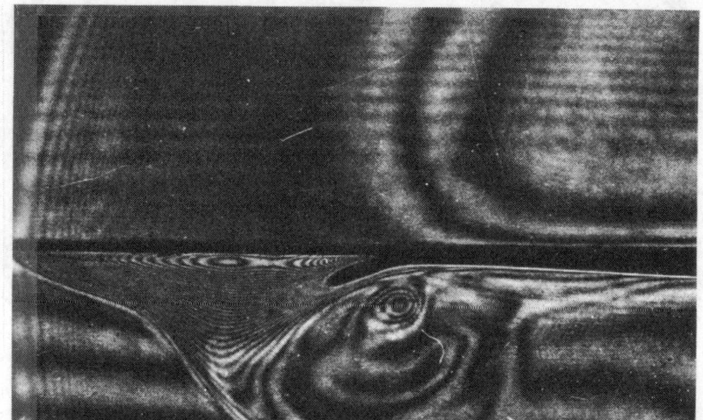

Fig. 10 (b) Interferogram obtained for n-propanol under T_∞= 22.8 °C and its iso-thermal lines

decreases and the inclination angle becomes small. The appearance of both STF and CCZ may be inter-related to a change in the flame inclination angle. The flame with small θ provides more uniform heat flux on the fuel surface either by conduction or convection. However, flame with large θ provide a concentrated heat flux just beneath the flame's leading edge and a very small amount of heat flux in the other area. This concentrated heat flux distribution contributes to form a large temperature gradient in the liquid which generates a new circulation zone under the flame edge as shown in Fig. 9e.

Spread Mechanism in Uniform Region

Based on our interferogram results, we suggest a modified explanation for the uniform spread over subflash point liquid. While for the mechanism of pulsating spread, the interpretation made by Glassman and Dryer (1980) was found to be in agreement with our results.

In the uniform region, surface temperature just ahead of the flame's leading edge is always found to be higher than T_c which suggests that the flammable gaseous fuel-air mixture is already formed above the fuel surface. The hot circulation zone was always found in the liquid just ahead of the flame's leading edge as seen in Fig. 10. This hot circulation zone could be made either by the induced air convection or by the subsurface liquid convection current. The distance between head of the hot circulation zone and the flame's leading edge (δ) was large at the low temperature and decreased with increasing T_∞ (Fig. 11) and the maximum temperature of the hot circulation zone slightly decreased with increasing T_∞ (Fig. 10). However the induced air velocity measured by Hirano et al. (1980) increased with increasing T_∞. Therefore, the gas phase convection is unlikely to be responsible for the hot circulation zone. From the interferogram for n-propanol, we estimated $\partial T/\partial x$ and the depth of the heated layer (h) for the two circulation zones at T_∞ =

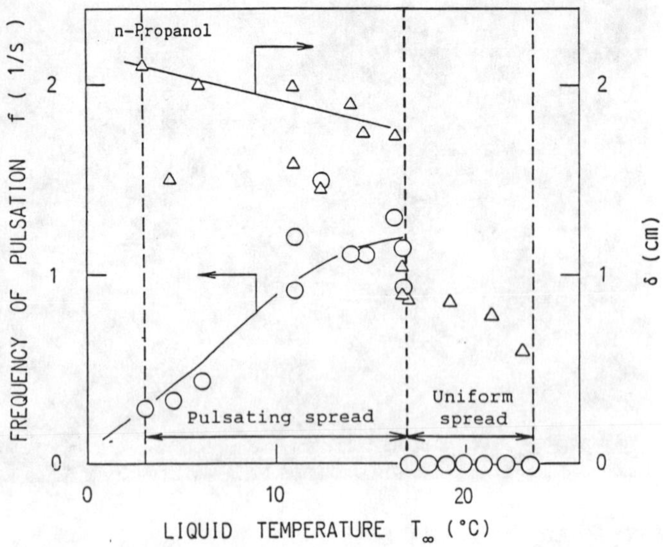

Fig.11 Frequency of flame pulsation and δ as a function of the initial liquid temperature. Where δ is the distance between the head of subsurface liquid convection and the flame's leading edge.

Table 2 Experimentally determined $(\partial T/\partial x)$ and h

T_∞ [°C]	$(\partial T/\partial x)_s$ [°C/cm]	h_s [cm]	$(\partial T/\partial x)_c$ [°C/cm]	h_c [cm]
19.0	17.4	0.15	3.5	0.4
22.3	21.7	0.15	4.7	0.35

19 °C and 22.8 °C. The spread rates were then calculated by the following equations (Williams,1985) and using the experimental values listed in Table 2.

For STF,

$$V_s = h_s(\partial T/\partial x)_s(\partial \sigma/\partial T)/\mu \qquad (2)$$

and for CCZ,

$$V_c = (gh_c^3)(\partial T/\partial x)_c(\partial \rho/\partial T)/\mu \qquad (3)$$

where subscripts s and c denote STF and CCZ, respectively. Other symbols used are σ: surface tension force, μ: viscosity of n-propanol at the corresponding liquid temperature, g: the acceleration due to gravity and ρ: the density of n-propanol.

The calculation resulted that V_s = 9.53 [cm/s] and V_c = 8.11 [cm/s] for T_∞ = 19 °C case, and V_s = 11.9 [cm/s] and V_c = 7.3 [cm/s] for T_∞ = 22.8 °C case. Since STF was generated from the head of CCZ and they are jointly moving in the same direction, the spread rate of the head of STF will be equal to $V_s + V_c$. For the T_∞ = 19 °C case, the measured average spread rate,

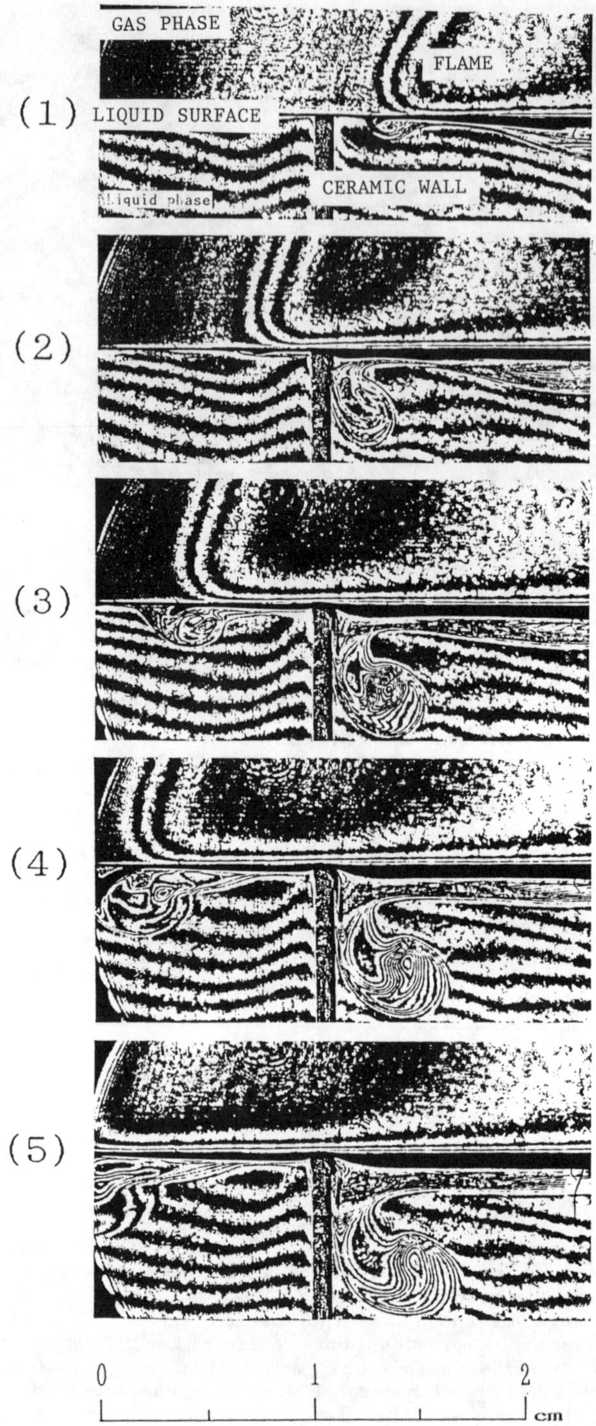

Fig. 12 Interferograms obtained for n-propanol in 0.1 sec/frame (1 to 5 in order) with 1/250 exposure time under T_∞ = 19 °C with a ceramic wall partition in the tray.

$\bar{V}$ = 12 [cm/s] and the local spread rate (V) measured at the corresponding location where the interferogram was taken was V = 15 [cm/s], while the calculated spread rate was 17.64 [cm/s]. For the T_∞ = 22.8 °C case, $\bar{V}$ = 16 [cm/s] and V = 19 [cm/s], while the calculated spread rate was 19.2 [cm/s]. It appears that the slight difference between the calculated spread rate and measurement is due to the observed irregularity of isothermal lines for CCZ which results in the uncertainty in h^3 for CCZ. In general, however, our calculation is favorably compared with the experimental results.

With decreasing T_∞ the maximum surface temperature of STF slightly increased, so that the temperature gradient along the fuel surface becomes larger. At $T_\infty = T_{\infty,c}$ (17 °C for n-propanol), heat from the flame cannot sustain the fuel surface temperature above T_c, because of a large heat loss to the cold liquid. There are no flammable gaseous mixtures available for the flame to propagate under this condition. Figure 4 shows a discontinuous decrease occurred for the spread rate at $T_\infty = T_{\infty,c}$, which is attributed to a change of spread mechanism from the subsurface liquid convection control to a combination of the subsurface liquid convection and premixed gas control.

Further experiment was performed to reconfirm that the spread rate in the uniform region is controlled by the subsurface liquid convection current. If the flame spread rate is controlled by the subsurface liquid convection current, the spread could be stopped by stopping the current. A thin ceramic wall which is chemically inert and has a small thermal diffusivity, was inserted into the liquid normal to the spread direction and was held vertical to the liquid surface (see Fig. 12). This ceramic partition separates the liquid into two section along the flame spread direction, so that the subsurface liquid convection (if any) can be stopped by the partition. Spread test was performed for T_∞ = 19 °C and its interferogams were taken at 0.1 second interval and a set of these which includes five interferograms is shown in Fig. 12. As a result, it was proved that when the current hits the ceramic wall, the current was stopped by the wall, and the gas phase spread was also stopped. When sufficient heat was given to the liquid from the flame, a new subsurface liquid current was regenerated in the liquid phase beyond the wall, which was followed by flame spread in the gas phase.

SUMMARY AND CONCLUSIONS

Holographic interferometry technique was applied to obtain detail temperature profiles in alcohols with subflash point and the following conclusions were obtained.

(1) Flame spread rate in the uniform region is controlled by the subsurface liquid convection current which is comprised of the surface tension-driven current and the buoyancy-driven current. The calculated flame spread rate was found to be in good agreement with the experimental results.

(2) The pulsating flame spread phenomenon was understood as a result of coupled gas phase and liquid phase convection effects which is in agreement with interpretation previously made by Glassman and Dryer (1980).

ACKNOWLEDGEMENTS

Technical discussions with F.A. Williams concerning spread rate calculation is greatly acknowledged. One author (KS) wishes to acknowledge partial support from National Bureau of Standards under Grant (#60NANB7D0739) and from Center for Robotics and Manufacturing Systems, University of Kentucky under Grant (#204002).

REFERENCES

Akita, K. 1972, "Some Problems of Flame Spreading Along a Liquid Surface", Fourteenth (International) Symposium on Combustion, The Combustion Institute, p.1075.

Burgoyne, J.H., Roberts, A.F. and Quinton, P.G., 1968a, "The Spread of Flame Across a Liquid Surface, Part I: The Induction Period", Proceedings of Royal Society, A308, 39.

Burgoyne, J.H. and Roberts, A.F., 1968b, "The Spread of Flame Across a Liquid Surface, Part II: Steady State conditions, Proceedings of Royal Society, A308, 55.

Burgoyne, J.H. and Roberts, A.F., 1968c, "The Spread of Flame Across a Liquid Surface, Part III: A Theoretical Model', Proceedings of Royal Society, A308, 69.

Glassman, I., Hansel, J. and Eklund, T., 1969, "Hydrodynamic Effects in the Flame Spreading, Ignitability and Steady Burning of Liquid Fuels", Combustion and Flame, 13, 199.

Glassman, I. and Dryer, F.L., 1980, "Flame Spreading Across Liquid Fuels", Fire Safety Journal, 3, 123.

Hauf, W. and Grigull, U., 1970, "Optical Methods in Heat Transfer ", in Advances in Heat Transfer Series 6, Academic Press, p.268.

Hirano, T., Suzuki, T., Mashiko, I., and Tanabe, N., 1980, "Gas Movements in Front of Flames Propagation Across Methanol", Combustion Science and Technology, 22, 83.

Ito, A. and Kashiwagi, T., 1987, "Measurement Technique for Determining the Temperature distribution in a Transparent Solid Using Holographic Interferometry", Applied Optics, 26, 954.

Ito, A. and Kashiwagi, T., 1988, "Characterization of Flame Spread Over PMMA Using Holographic Interferometry Sample Orientation Effects", Combustion and Flame, 71, 189.

Ito, A., Choudhury, S.K., and Fukano, T., 1989, "A Technique for Measuring an Unsteady Temperature Distribution in a Liquid Interferometry", Transaction of JSME, submitted.

Isoda, H., 1971, "Liquid Motion Generated by a Heated Wire", Kikaino-Kenkyu, 23, 45 (in Japanese).

Kashiwagi, T. and Kashiwagi, T., 1984, "A Study of The Radiative Ignition Mechanism of A Liquid Fuel Using High Speed Holographic Interferometry", Nineteenth (International) Symposium on Combustion, The Combustion Institute, p.1511.

Mackinven, R., Hansel, J. and Glassman, I., 1970, "Influence of Laboratory Parameters on Flame Spread Across Liquid Fuels", Combustion Science and Technology, 1, 293.

Matsumoto, Y and Saito, T., 1980, "Propagation of Pool Burning", Transactions of JSME, 46(B), 998 (in Japanese).

Murphy, C.G. and Alpert, S.S., 1974, Am. J. Phys., 39, 834.

Newman, J.S., 1979, "Laser Doppler Velocimeter Measurements of the Gas and Liquid Flow Fields Induced by Flame Propagation over a Liquid Fuel Surface", MSE Thesis, Mech. and Aerospace Engineering Dept.,Princeton

University, Princeton, N.J. 08544.

Roberts, A.F., 1959, "Spread of Flames on a Liquid Surface", Ph.D Thesis, Imperial College, London, England.

Sirignano, W.A., 1972, "A Critical Discussion of Theories of Flame Spread Across Solid and Liquid Fuels", Combustion Science and Technology, 6, 95.

Saito, K., Quintiere, J.G. and Williams, F.A., 1986, "Upward Turbulent Flame Spread, Proceedings of the First International Symposium on Fire Safety Science, Hemisphere Pub., p.75.

Torrance, K., 1971, "Surface Flows Preceding Flame Spread Over a Liquid Fuel", Combustion Science and Technology, 3, 3.

Williams, F.A., 1985, Combustion Theory, 2nd ed., Chap.12, Benjamin/Cummings.

1989 National Heat Transfer Conference
HTD-Vol. 106, Heat Transfer Phenomena in Radiation, Combustion and Fires

HEAT LOSS TO THE INTERIOR OF A FREE BURNING VERTICAL PMMA SLAB AND ITS INFLUENCE ON ESTIMATION OF HEAT OF PYROLYSIS

A. K. Kulkarni and C. I. Kim
Department of Mechanical Engineering
The Pennsylvania State University
University Park, Pennsylvania

ABSTRACT

Conduction heat transfer into a vertical free burning PMMA (polymethylmethacrylate) slab as a function of time and distance from the leading edge for clear and black PMMA, and its contribution to the known values of heat of pyrolysis were estimated. The heat conduction into the pyrolyzing surface was deduced from the measured temperatures in the interior of burning slabs. The surface conduction heat flux decreased from 14 kW/m^2 for clear PMMA and 13 kW/m^2 for black PMMA at t=300 s to almost 8 kW/m^2 at around t=1000 s and then reached a plateau, indicating a substantial degree of unsteadiness in the early part of combustion of slabs. It is concluded that, when estimating burning rates in numerical calculations one has to be careful in selecting a value for the heat of pyrolysis because of the generally unsteady nature of heat conduction into the solid interior.

NOMENCLATURE

A, B = constants in Eq. (1) and (2)
c_p = heat capacity, kJ/kg·K
h_p = heat of pyrolysis, kJ/kg
k = thermal conductivity, kJ/m·s·K
$\dot{m}''$ = burning rate per unit area, kg/m^2s
$\dot{q}_c''$ = heat conduction to the interior at the pyrolyzing surface, kW/m^2
$\dot{q}''_{CONV}$ = convective heat feedback from gas phase to the pyrolyzing surface, kW/m^2
$\dot{q}''_{RAD\ FL}$ = radiative heat feedback from flames to the pyrolyzing surface, kW/m^2
$\dot{q}''_{RERAD}$ = radiation loss by surface to atmosphere, kW/m^2
T = temperature, K
T_w = temperature of pyrolyzing surface, K
T_∞ = ambient temperature, K
x = distance from the leading edge of the sample, m
y = distance from the surface to the interior of the sample, m
y_w = location of the pyrolyzing surface, m

INTRODUCTION

Several experimental studies have been conducted on burning of vertical slabs or cylinders of polymethylmethacrylate (PMMA) in fire and combustion research. PMMA is preferred in many wall fire studies primarily because it is an ashless burning solid, with not very sooty flames, can be available in a form that does not drip very much, and many of its properties have been thoroughly investigated. One of these properties, which is extremely important in fire spread and growth studies, is h_p, often referred to as the heat of pyrolysis. It involves heat of depolymerization and melting, latent heat of pyrolysis and gasification, and the sensible energy to heat the solid from room temperature to vaporization temperature. Values for h_p between 1512 to 1884 kJ/kg have been reported in literature (Magee and Reitz, 1974 and Tewarson and Pion, 1976).

Most of the experimental and analytical studies on burning of PMMA slabs or cylinders have used a constant value for h_p for further calculation of burning rates, energy analysis, spread rate, etc. A constant value of h_p is appropriate if the temperature of pyrolyzing surface, T_w, remains constant during the combustion and if the

PMMA slab behaves like a thermally thick solid. For PMMA burning in a vertical wall configuration with natural convection in air, T_w is found to remain almost constant. For a fuel to burn like a thermally thick solid, the burning must be steady state, i.e., the surface regression rate must be equal to the thermal wave penetration rate. (Without any surface vaporization, a semi-infinite solid with a constant surface temperature is an inherently unsteady problem of heat conduction.) Therefore, the question is, whether a PMMA slab burns like a thermally thick solid under the commonly studied experimental or analytical conditions. Most studies have used slabs or vertical solid cylinders of dimension 1.3 cm to 2.5 cm perpendicular to the burning surface; a few studies have used thicker slabs. A sample that burns like a thermally thick solid will have a heat loss at the pyrolyzing surface in the interior of the solid that is time independent and this energy flux component of the overall energy balance is usually included in the so called heat of pyrolysis/gasification, h_p. If a sample does not burn like a thermally thick solid, then one must be careful in using a value of h_p in heat and mass transfer calculations.

Recently Jackson (1988) attempted to estimate time-dependent heat of gasification for PMMA by exposing 9 cm diameter, 2.5 cm thick disks in an inert atmosphere. His heat of gasification value at near-steady state condition for high heat flux (40 kW/m^2) was close to 1600 kJ/kg. He indeed discovered that the value of heat of gasification is highly transient initially, for example, with an external heat flux of 10 kW/m^2, the heat of gasification decreased from as much as 55,000 kJ/kg at the beginning of gasification (which was at t = 13 min from the instant the radiation was turned on) to 5,500 kJ/kg in the first 1000 s, and it took almost 1000 s longer to reach the near-steady state value of 4,000 kJ/kg. At higher external radiation fluxes, the steady state was achieved earlier and the heat of gasification was considerably lower; the lowest value being 1600 kJ/kg. However, all these tests were carried out under non-burning, non-flaming conditions with a controlled flow of nitrogen gas over the sample.

The objective of the present experimental investigation is to measure conduction heat transfer into a vertical free and fully burning PMMA slab (having a nonpropagating fire), as a function of time and distance from the leading edge for clear and black PMMA and compare the change in its magnitude to the known values of h_p.

Heat conduction in the solid interior of a burning PMMA slab or cylinder has been measured in some experimental studies, notably by Sibulkin and others (1974, 1975), Fernandez-Pello and Williams (1974), and more recently by Ito and Kashiwagi (1988). However, in all of these studies, conduction heat loss to the interior was reported at the pyrolysis (or vaporization) front of a spreading flame, which is a fundamentally different configuration compared to the present experimental condition. In the present investigation, the emphasis is on finding out whether a fully burning slab or cylinder has a steady heat conduction loss to the interior, and if not , how much error will be introduced by assuming it to be a steady burning slab through the use of a fixed value of h_p. In a flame-spread situation, the interior heat conduction loss at the pyrolysis front may or may not be a constant depending on the history of heating of the unburned fuel before the pyrolysis front reaches there, and therefore its unsteadiness and magnitude will strongly depend on the type of flame spread. This point is discussed further in view of the present experimental results. The importance of the present investigation lies in the fact that several analytical and experimental studies on burning PMMA have used a constant value of h_p in their analysis and data reduction (for example, Fernandez-Pello 1977 and 1978, and Quintiere and Harkleroad 1986), and the question is whether it was a reasonable assumption.

EXPERIMENTS

A schematic of the apparatus and measurement scheme is shown in Figure 1. Samples of clear and black PMMA were in the form of 1.9 cm thick, 7.5 cm wide, and 15 cm high slabs, all cut from a single large sheet for each type of PMMA. Samples were ignited with a flat

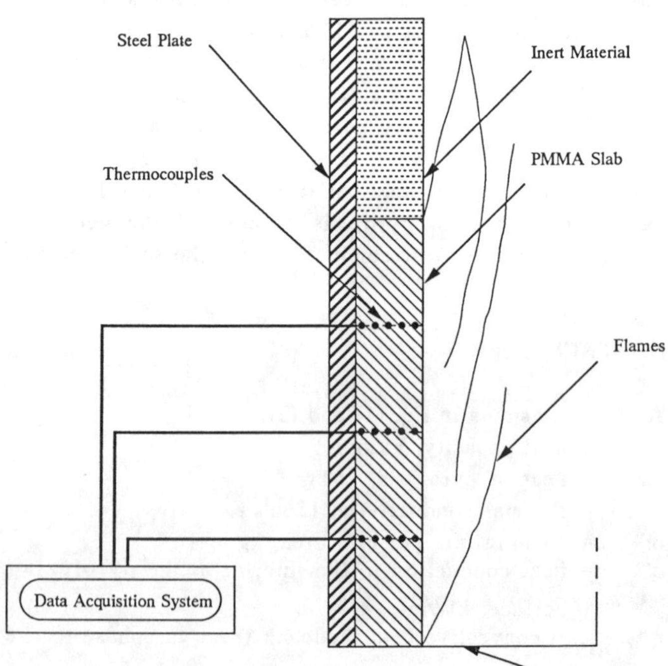

Figure 1. Schematic of Experimental Setup

flame burner using an identical, predetermined ignition procedure for all the test runs. The ignition procedure involved heating the sample sequentially at three different vertical locations for ten seconds each, repeated four times. The front surface of the samples was fully involved before t = 300 s. Samples were mounted on a vertical steel plate with side walls to prevent flames from wrapping around. Each sample was instrumented with fifteen, type K, 0.3 mm thermocouples. Thermocouples were embedded in the slab from behind in blind holes and glued using a PMMA solvent to ensure close thermal contact of thermocouples with the solid interior. The thermocouple junctions were at 2, 5, 8, 11, and 14 mm from the front (burning) surface at 3, 7.5, and 12 cm from the leading edge. Temperature was recorded at every four-second intervals from all thermocouples for approximately 1000 s of total burning time using an automatic data acquisition system.

Once the temperature (T) vs. depth (y) distribution was measured, the heat conduction to the interior at the pyrolyzing surface, $\dot{q}_c''$, was deduced as,

$$\dot{q}_c'' = -k \left.\frac{\partial T}{\partial y}\right|_{y=y_w} \qquad (1)$$

where k is the thermal conductivity of PMMA, for which a value of $2.688\text{x}10^{-4}$ kJ/m·s·K was used (Fernandez-Pello, 1979). Since the pyrolyzing surface regressed as the sample continued to burn, the location of the pyrolyzing surface, $y = y_w$, had to be defined carefully. As mentioned earlier, it has been observed that the temperature of pyrolyzing surface of a free burning PMMA slab, T_w, remains approximately constant. We conducted separate experiments to measure T_w of clear and black PMMA using fine (50 μm) thermocouples and they were found to be in a sufficiently narrow temperature range to be assumed constant as follows,

$T_w = 639.2 \pm 5$ K for clear PMMA
$T_w = 638.5 \pm 5$ K for black PMMA

The small difference between black and clear PMMA surface temperature is expected because the surface temperature is mainly dictated by the pyrolysis chemistry of the solid (with other conditions of heat flux being the same between the two types of PMMA). There is very little difference between the chemical composition of clear and black PMMA (which contains a very small amount of carbon soot for color). Therefore, the overall temperature and burning characteristics of the two types of PMMA should not be drastically different. The measured temperature at various locations was curvefitted to an exponential function as,

$$T(y) = A\, e^{-By} \qquad (2)$$

where A and B are constants. This curve was then extrapolated to $T = T_w$ to find the position of the pyrolyzing surface, $y = y_w$. The conduction heat loss then becomes,

$$\dot{q}_c'' = kAB\, e^{-By_w} \qquad (3)$$

In computing the heat flux at the pyrolyzing surface from the indepth solid temperature measurements, one has to be careful to avoid several possible errors. Beck (1962) discussed a similar situation and pointed out the "temperature disturbance" or loading error caused by the difference in the thermal conductivities of the thermocouple material and material it is embedded in, and because of the thermocouple leads not being in an isothermal plane. In the present experiment, there is a possibility for this type of error, however, it could not be estimated by the parametric charts given by Beck (1962) because the ranges of the charts are not suitable for the present conditions. The error is probably not very significant because the interior temperature gradients are not very large and the thermocouples are staggered. Singh and Dybbs (1976) also discussed error in temperature measurements due to condition along the leads which depended on the knowledge of the contact resistance at the thermocouple junction. In the present experiments, although it is difficult to estimate the contact heat resistance quantitatively, the resistance is expected to be very small. Each thermocouple was installed in place using a solvent in several steps, so that when it was completely dried, the thermocouple was solidly embedded inside and it could be separated only by cutting the wires. Therefore, the local temperature measurement is expected to reflect the actual interior temperature accurately.

The largest error in the estimation of temperature gradient at the surface is perhaps due to the extrapolation of the temperature data to the burning surface (defined as the location where $T = T_w$ given above). If a higher order scheme is used (such as a fourth order polynomial or a fourth order polynomial in the exponent of e), all the five data points are connected, however, the behavior of the function beyond the data range can be very erratic and a root may be nonexistant at $T = T_w$. An exponential functional form shown in Eq.2 was used because, in the absence of any significant indepth radiation, the temperature variation can be relatively closely simulated by the exponential function. The indepth radiation is mostly stopped at the bubbling, darkened surface before it reaches the first thermocouple.

It should be noted here that, the emphasis of the present investigation is on pointing out the transient nature of the apparent value of the heat of pyrolysis and, therefore, the <u>relative</u> magnitudes of the transient values of h_p, and conclusions based on these results, will not be

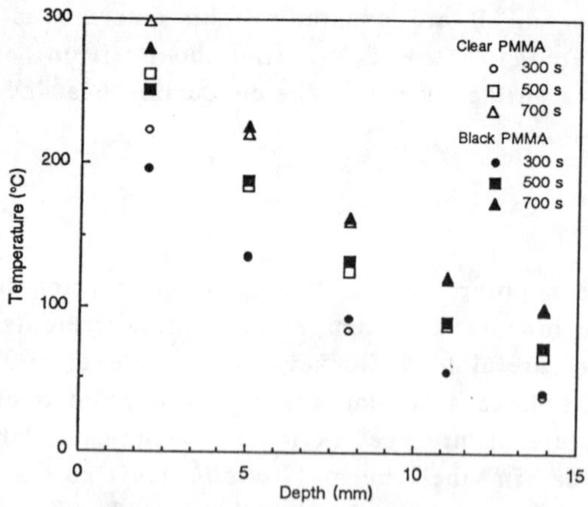

Figure 2. Measured Temperature Distribution in the Solid Interior of 1.9 cm Burning PMMA Slabs

greatly affected by the possible errors discussed earlier.

RESULTS

Figure 2 shows a typical temperature distribution inside the solid as a function of the distance from the original front surface location at three different instants. As the time increases from 300 s. to 700 s., temperature continues to rise. After considering surface regression which is approximately $8.0x10^{-3}$ mm/s after the sample is fully involved (Kulkarni and Sibulkin, 1982), the rise at a given location will appear to be slower with respect to the pyrolyzing surface. When the data for clear and black PMMA are compared, it is apparent that the temperature of clear PMMA is greater at the first location, but the difference is very small deeper in the solid. This can be explained based on the indepth absorption of flame radiation incident at the surface. Recent measurements by Saito (1988) on 1.2 cm PMMA slabs show that the optical depth (based on the attenuation of incident radiation by a factor of $1\text{-}e^{-1}$) is approximately 4 mm for clear PMMA and approximately 0.5 mm for black PMMA for a radiation source temperature of 1400 K. This means that in the case of clear PMMA, a substantial amount of incident radiative energy can be absorbed in the first several millimeters of the surface. If there is a thermocouple junction embedded in the surface, the indepth radiation will be intercepted by the thermocouple junction and a corresponding temperature rise will be apparent. However, it must be noted that the above-mentioned optical depths were for a non-pyrolyzing PMMA surface; but in the present case the surface is pyrolyzing. A closer examination of the pyrolyzing surface of burning samples using floodlights showed that the surface is rough due to boiling action and it is blackened by the deposition of soot. In this situation, the indepth absorption of radiation at the surface for a clear PMMA sample will not be as much as the non-pyrolyzing PMMA. Although an exact quantitative determination of indepth absorption is not possible here under flaming condition, this aspect is important and should be explored in the future. However, it may further be noted that none of the previous experimental studies on the measurement of interior heat conduction using thermocouples took this effect into account quantitatively. In the case of the black PMMA practically all the heat transfer at the first thermocouple at y = 2 mm is via conduction.

Figure 3 shows temperature-time history at the thermocouple location nearest to the burning surface (2 mm in interior) for a clear PMMA sample. The temperature data taken at 4 s intervals, were averaged over 40 s for clarity in plotting in this and the subsequent figures. The ignition procedure involved heating of the sample using a line burner at various locations in a particular sequence from t = 0 to 120 s and the samples appeared to be fully involved at around 300 s. This figure was plotted to check if the entire front surface of the sample was heated and ignited simultaneously, i.e., within a reasonably short interval of time. It is clear from Figure 3 that the temperature history at the three vertical locations (of 3 cm, 7.5 cm, and 12 cm from the bottom edge) is almost identical during a large part of the burn period (up to 800 s). This confirms that the interior temperature profiles had a negligible dependence on the height, and that the ignition procedure was satisfactory.

At t > 800, the thermocouple at 3 cm starts measuring a higher temperature. This happens because the local mass burning rate is greater nearer to the leading edge, which roughly varies as $x^{-1/4}$ (Kim et al., 1971) for small-scale laminar fires such as those in the present experiments. Consequently, the thermocouple at 3 cm is very close to the pyrolyzing surface at around 800 s where it perhaps "sees" the flame radiation through the front surface, and finally it starts protruding out (based on a visual observation) at around t = 1000 s.

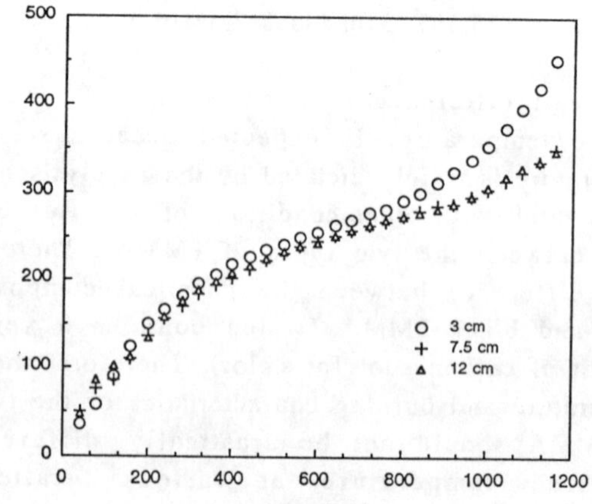

Figure 3. Temperature-Time History for Clear PMMA at a Depth of 2mm

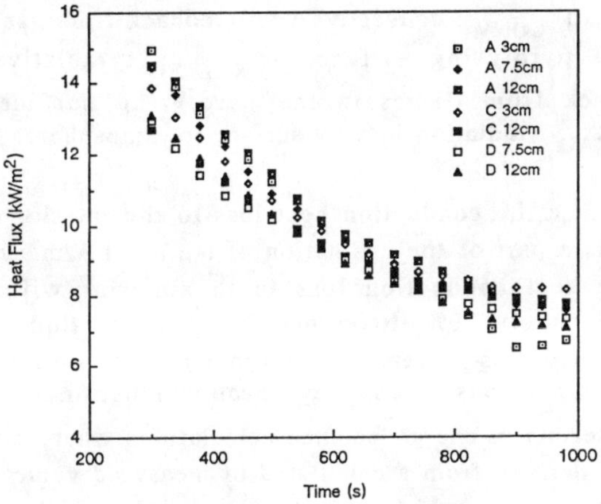

Figure 4. Conduction Heat Flux Derived from Temperature Measurements in Clear PMMA at Three Different Distances from the Leading Edge for Test Runs A, C, and D.

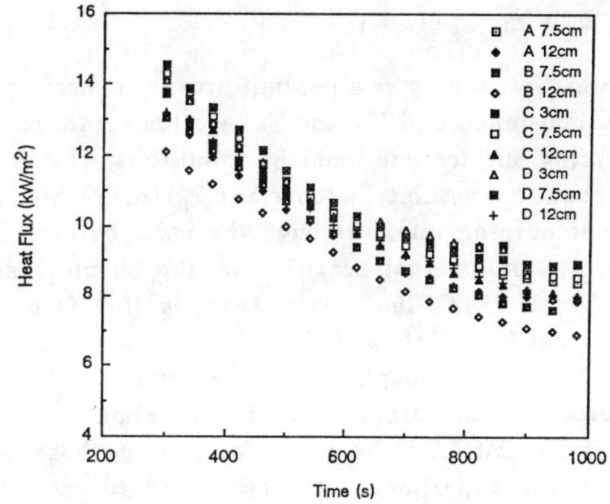

Figure 5. Conduction Heat Flux Derived from Temperature Measurements in Black PMMA at Three Different Distances from the Leading Edge for Test Runs A, B, C, and D.

Figures 4 and 5 show conductive heat flux to the interior derived from the temperature measurements for three test runs with clear PMMA and four test runs with black PMMA respectively, each at three different vertical locations (except for a few of those where the thermocouples malfunctioned). Based on the accuracy of temperature measurement (±2 °C), repeatability of time determination (±10 s), location of the thermocouples in interior, and most important, the scatter of the derived results of heat flux, the accuracy of the heat flux results is estimated at around 10% at early times and up to 20% at later times.

The difference between heat fluxes at different heights is within the error of estimation of the heat flux and, therefore, the effect of vertical position is seen to be not significant in the present experiments. The surface conduction heat flux decreases with time and starts to reach a plateau at around 1000 s. There is a decrease from 14 kW/m^2 for clear PMMA and 13 kW/m^2 for black PMMA at t = 300 s to almost 8 kW/m^2 at around t=1000 s at which time the heat flux starts to approach a plateau. This indicates a substantial degree of unsteadiness in the early part of combustion of slabs. Similar unsteadiness was observed, but not quantitatively measured, by Sibulkin and Lee (1974) in the combustion of PMMA cylinders.

DISCUSSION

As mentioned in the Introduction, previous studies on the measurement of conductive heat loss into the interior of burning PMMA samples were aimed at estimating the heat flux at the pyrolysis front of a spreading fire in various orientations. In case of a downward spreading fire, the unburned fuel is heated from room temperature to pyrolysis temperature in a relatively short distance and therefore the conductive heat flux into the surface is high. (The short-distance heating is caused by the closeness of the blue flame tip to the fuel surface.) On the other hand, for upward spreading fire, the flames hang directly over the unburned fuel well before it reaches the pyrolysis temperature, therefore, the unburned fuel receives more heat before it reaches the pyrolysis temperature, and consequently, the conductive heat flux at the pyrolysis edge is relatively smaller. This phenomenon was clearly demonstrated recently by Ito and Kashiwagi (1988) who used holographic interferometry to probe temperature in the immediate interior of the surface (with a spatial resolution of 50 μm). They showed that the interior conductive heat flux at the pyrolysis edge is as high as 70 kW/m^2 for downward spreading fires and 28 kW/m^2 for upward spreading fires. The interior conductive heat flux for downward flame spread at the pyrolyzing edge, derived from thermocouple data of Sibulkin and Lee (1974) for PMMA cylinders and from those of Fernandez-Pello and Williams (1974) for flat sheets of PMMA, indicated values of 55 kW/m^2 and 73 kW/m^2 respectively.

In the present configuration, the fuel is systematically and slowly heated over a relatively large period (120 s). Consequently, the present results show a lower heat flux (around 14 kW/m^2) at ignition. The point to be made here is that the conduction heat flux to the interior is not constant in time and even at the ignition instant it can not be the same for various configurations. It basically depends on the amount of preheating the sample undergoes prior to ignition. Since the present experiments were conducted with identical ignition procedure for all test runs, the results may be further analyzed with self-consistency. As shown by the present data, a single value of conductive heat loss to the interior has no meaning because it is dependent on time as well as history prior to ignition.

If a slab burns like a thermally thick solid, the conduction heat loss to the interior will be given by,

$$\dot{q}''_c = \dot{m}'' \, c_p (T_w - T_\infty) \quad (4)$$

where $\dot{m}''$ is burning rate per unit area, c_p is heat capacity of the substance and T_w and T_∞ are temperatures at the pyrolyzing surface and ambient conditions, respectively. This equation assumes a constant c_p for PMMA, and a constant burning rate. Although the mass burning rate is a function of external heat flux on the burning surface (Saito et al, 1988), the present investigation is made for no external heat flux, and therefore, constant $\dot{m}''$ should be a reasonable assumption. Kulkarni and Sibulkin (1982) measured a local burning rate of approximately 9.5×10^{-3} kg/m^{2}s for PMMA slabs of similar size as those used in the present experiment. Assuming $\dot{m}''$ to be 9.5×10^{-3} kg/m^{2}s, c_p = 2.1 kJ/kg·K, T_w = 366 °C and T_∞ = 25 °C, $\dot{q}''_c$ is found to be 6.8 kW/m^2 which is not very different from the value reached at the plateau in Figures 4 and 5. However, it must be noted that the slab thickness used in the present experiments was 1.9 cm with a 3 mm steel backing plate and therefore there was always some heat loss at the back surface of the samples.

There is a small difference between the computed conduction heat flux for clear and black PMMA as seen from the comparison of Figures 4 and 5. At early times, the clear PMMA has a higher heat flux than the black PMMA, but later the trend is reversed. In any case, the difference is smaller than experimental errors and the scatter of the data, so no attempt is made to explain this phenomenon in terms of the small difference in temperatures and/or the difference in indepth radiation absorption characteristics of the two materials.

Tewarson and Pion (1976) reported that for PMMA the heat loss at the surface was approximately 21 kW/m^2. Their heat loss term was defined as the difference between the flame heat transfer to the surface and the heat flux required to pyrolyze the mass at the surface, in the absence of any external radiation. Thus, it included the heat flux to the interior by conduction and the surface reradiation. Assuming T_w = 366°C and emissivity equal to 1.0, the radiation loss is approximately 9 kW/m^2 and therefore the present range of measurements of heat conduction to the interior are generally consistent with the measurements of Tewarson and Pion (1976).

A final point, and perhaps the most important outcome of the present investigation, may be noted as follows. Assuming a typical burning rate of 9.5×10^{-3} kg/m^{2}s and a value reported in the literature for h_p of 1600 kJ/kg, the net heat flux responsible for the burning rate is found to be, $\dot{m}'' \times h_p$ = 15.2 kW/m^2. The definition of h_p often assumes, and includes, a constant conductive heat loss into the interior, as given by the energy balance at the surface. In the absence of any external heat flux, the energy flux required for pyrolysis is given by

$$\dot{m}'' h_p = \dot{q}''_{CONV} + \dot{q}''_{RAD\,FL} - \dot{q}''_{RERAD} \quad (5)$$

where, $\dot{q}''_{CONV}$: convective heat feedback from gas phase to the pyrolyzing surface, $\dot{q}''_{RAD\,FL}$: radiative heat feedback from flames to the pyrolyzing surface, and $\dot{q}''_{RERAD}$: radiation loss by surface to atmosphere.

Since the conduction heat loss to the interior of the solid is a part of the estimation of h_p, a 5 kW/m^2 change in the heat conduction loss to the interior will cause approximately 30% difference in the calculation of $\dot{m}''$. Therefore, it is extremely important to take into account the instantaneous value of $\dot{q}''_c$ (heat conduction loss into the interior) in energy balance calculations where burning rate is derived from a calculated or measured value of net heat flux to the surface and a constant value for h_p.

SUMMARY AND CONCLUSION

Conduction heat loss into the solid interior was determined for vertical, fully burning slabs of clear and black PMMA by measuring instantaneous temperature profiles inside the solid. Substantial unsteadiness was observed in the early part of combustion of slabs. Therefore, it is concluded that the thermally thick solid assumption is inappropriate in the early stage (first several hundred seconds) of the fire for combustion of PMMA slabs or cylinders of the typically used dimensions of 1 or 2 cm. When the conduction heat loss into the interior reaches a plateau, the thermally thick slab assumption is reasonably in agreement with experimental results. Therefore, it is suggested that when estimating burning rates, instantaneous value of the conduction heat loss into the interior should be considered to avoid substantial errors.

ACKNOWLEDGEMENT

Partial support for this work was provided by Center for Fire Research, National Institute of Standards and Technology, Grant No. 60NANB8D0849.

REFERENCES

Beck, J. V., May 1962, "Thermocouple Temperature Disturbances in Low Conductivity Materials," ASME Journal of Heat Transfer, pp.124-132.

Fernandez-Pello, A. C., 1979, "Flame Spread in a Forward Forced Flow," Combustion and Flame 36, pp.63-78.

Fernandez-Pello, A. C., 1977, "Upward Laminar Flame Spread under the Influence of Externally Applied Thermal Radiation," Comb. Sci. and Tech. Vol.17, pp.87-98.

Fernandez-Pello, A. C., 1978, "A Theoretical Model for the Upward Laminar Spread of Flames over Vertical Fuel Surfaces," Combustion and Flame 31, pp.135-148.

Fernandez-Pello, A. C. and Williams, F. A., 1974, "Laminar Flame Spread over PMMA Surfaces," Fifteenth Symp. (Int'l) on Comb., The Combustion Institute, Pittsburgh, PA, pp.217-231.

Ito, A. and Kashiwagi, T., 1988, "Characterization of Flame Spread over PMMA Using Holographic Interferometry Sample Orientation Effects," Combustion and Flame 71, pp.189-204.

Jackson, J. L., October 1988, "Direct Measurement of Heat of Gasification for Polymethylmethacrylate," NISTIR 88-3809, U. S. Department of Commerce, National Institute of Standards and Technology, Center for Fire Research, Gaithersburg, MD.

Kim, J. S., deRis, J., and Kroesser, F. W., 1971, "Laminar Free-Convective Burning of Fuel Surfaces," Thirteenth Symp. (Int'l) on Comb., The Combustion Institute, Pittsburgh, PA, pp.949-961.

Kulkarni, A. K. and Sibulkin, M., 1982, "Burning on a Vertical Fuel Surface with Finite Chemical Reaction Rate," Combustion and Flame 44, pp.187-199.

Magee, R. S. and Reitz, R. D., 1974, "Extinguishment of Radiation Augmented Plastic Fires by Water Sprays," Fifteenth Symp. (Int'l) on Comb., The Combustion Institute, Pittsburgh, PA, pp.337-347.

Quintiere, J. and Harkleroad, M., 1986, "Wall Flames and Implications for Upward Flame Spread," Comb. Sci. and Tech. Vol.48, pp.191-222.

Saito, K., 1988, Personal Communication.

Saito, K., Delichatsios, M. A., Venkatesh, S., and Alpert, R. L., 1988, "Measurements and Evaluation of Parameters Affecting the Preheating and Pyrolysis of Noncharring Materials," Fall Technical Meeting, Eastern Section of the Combustion Institute, Clearwater Beach, Florida, 5-7 December 1988, p.69-1.

Sibulkin, M. and Hansen, A. G., 1975, "Experimental Study of Flame Spreading over a Horizontal Fuel Surface," Comb. Sci. and Tech. Vol.10, pp.85-92.

Sibulkin, M. and Lee, C. K., 1974, "Flame Propagation Measurements and Energy Feedback Analysis for Burning Cylinders," Comb. Sci. and Tech. Vol.9, pp.137-147.

Singh, B. S. and Dybbs, A., August 1976, "Error in Temperature Measurements Due to Conduction Along the Sensor Leads," ASME Journal of Heat Transfer, pp.491-495.

Tewarson, A. and Pion, R. F., 1976, "Flammability of Plastics - I. Burning Intensity," Combustion and Flame 26, pp.85-103.

1989 National Heat Transfer Conference
HTD-Vol. 106, Heat Transfer Phenomena in Radiation, Combustion and Fires

HEAT AND MASS TRANSFER IN LIVE FUEL DURING THE PROCESS OF DRYING, PYROLYSIS, AND IGNITION

J. Mardini, A. S. Lavine and V. K. Dhir
University of California, Los Angeles
Los Angeles, California

E. B. Anderson
USDA Forest Fire Laboratory
Riverside, California

ABSTRACT

A framework has been developed to study the problem of drying, pyrolysis, and ignition of live fuel. The motivation for this work is to ultimately be able to model the propagation of fires in Southern California wildlands. The model considers the coupling between the fuel and its surrounding environment. The time to reach either flaming or glowing ignition is predicted. It is found that the presence of a "pilot species" with a low spontaneous ignition temperature often causes flaming ignition to occur before glowing ignition otherwise would.

NOMENCLATURE

A_s = fuel surface area
c_p = specific heat at constant pressure
E_p = activation energy in Arrhenius equation
F_{12} = shape factor from fuel to flame
g = mass transfer coefficient
h = heat transfer coefficient
h_{fg} = enthalpy of vaporization of water
K_p = pre-exponential constant in Arrhenius equation
m = mass
$\dot{m}$ = mass flow rate
p = pressure
Q_p = energy for pyrolysis, per unit mass of pyrolyzate
R = gas constant for mixture in environment
R_u = universal gas constant
t = time
T = temperature

Greek Symbols

ε = effective flame-fuel emissivity
ω = mass fraction
σ = Stefan-Boltzmann constant

Subscripts

a = air
c = char
e = environment
f = solid fuel
fl = flame
i = into environment control volume
o = out of environment control volume
s = surface of fuel
v = volatiles
w = water vapor

INTRODUCTION

In this paper, a model has been developed to study the burning of live fuel from the early stages of drying, through pyrolysis, until ignition occurs. Ignition refers to either glowing ignition of the solid fuel, or flaming ignition of the combustible gases. Specifically, the model predicts the length of time required for ignition to occur when a piece of live wood or foliage (the "fuel element") is exposed to a distant flame. The flame does not come into direct contact with the fuel element. The motivation is to understand the role that combustible gases play in the ignition process. As the fuel element is heated, volatiles are released from it starting at low temperatures. As the temperature increases, pyrolysis occurs and additional combustible gases are released. In live fuels, there is no clear distinction between release of volatiles and pyrolysis, and in this paper we use the word "volatiles" to refer to any gases released from the structure of the fuel, and "pyrolysis" to refer to the process by which this occurs.

The hypothesis of this paper is that the volatiles may, under certain circumstances, cause flaming ignition to occur before the solid fuel would otherwise undergo glowing ignition. When this happens, the volatiles produce an additional heat source which may cause the fire to propagate. This paper presents a framework for analyzing this problem up to the time of ignition. Unfortunately, there is insufficient information available concerning the various combustible volatiles which are released from plants as a function of temperature. Therefore, it is not yet possible to solve this problem in a quantitatively correct way. Instead, this paper explores the sensitivity of the results to the various relevant parameters. Work is presently underway to obtain the needed experimental data and related information to improve and validate the model.

The motivation for this work is to ultimately be able to model the propagation of fires in Southern California wildlands. Chamise is typical of Southern California fuels. It can be described as a large bush with many thin branches covered with clusters of small needles.

Considerable research has been conducted on the heat and mass transfer from live fuel during fires. Susott (1980, 1982) investigated experimentally the release of volatiles from a variety of forest fuels. He used ether and benzene-ethanol solutions to extract the volatiles. The ether extractives are of importance because of their high heats of combustion, volatility, and low limits of flammability. He found that the ether extractives typically constitute between 5 and 9% of the dry sample. Typical Southern California fuels (chamise and manzanita) were shown to have a high rate of combustible volatile production below 300 °C. Preliminary results of Anderson (1988) show light hydrocarbons such as methane, ethane, ethene, and acetylene being released from chamise at temperatures as low as 50 °C.

Tinney (1965) conducted an analytical and experimental investigation of the heating and burning of wooden dowels in air. In his model, Tinney solved the heat conduction equation simultaneously with a first order Arrhenius equation for the pyrolysis process. Tinney changed the constants in his Arrhenius equation in such a way as to agree with the experimental results for temperature and weight loss. He found that the pyrolysis constants changed at a certain time corresponding to a given percentage weight loss. His pre-exponential constant ranged from 6×10^7 to 2×10^9 s^{-1} and his activation energy from 30 to 43 kcal/mole. Roberts (1970) reviewed the literature, and suggested typical values of 30 kcal/mole for the activation energy and 7×10^7 s^{-1} for the pre-exponential constant. However, there were variations in both constants, due to sample size and structural and autocatalytic effects. Roberts review also indicated that the first several percent of weight loss may not be governed by the same rate equation as the remainder of the volatiles. Havens et al. (1972) developed a model which predicts the transient temperature distribution and volatile product evolution rates in heating and decomposing wood specimens.

Gandhi and Kanury (1986) developed a model of ignition of solid fuels. The model addressed the heat and mass transfer in the fuel, coupled to exothermic reactions occurring in a natural convection boundary layer. The assumption of a natural convection boundary layer is not expected to be accurate in actual forest conditions, where a wind is usually present.

The new contribution made by this work is to consider the coupling between the fuel and its surrounding environment, in order to predict when flaming ignition will occur, for conditions representative of Southern California wildlands.

PROBLEM STATEMENT AND ASSUMPTIONS

A representative fuel element is considered. This fuel element can be thought of as a thin twig with clusters of needles attached to it, as is found on chamise. The fuel element is idealized as a long cylinder of diameter d, at a fixed distance L from a flame (see Fig. 1a). The flame does not contact the fuel element. There are assumed to be many geometrically identical fuel elements, and each one is associated with the finite volume of air surrounding it (see Fig. 1b). The model addresses the heat and mass transfer in the fuel element and its surrounding environment. As the fuel is heated, the physical phenomena which occur are evaporation of water, pyrolysis, and finally ignition -- either glowing ignition of the solid fuel or flaming ignition of the combustible volatiles. The basic assumptions in the model are as follows:

a) There is radiation heat transfer between the adjacent control volumes, but all are assumed to have the same temperature, so this does not cause any net exchange of heat.

b) A lumped capacity approach is used for the energy balances on the fuel element and surrounding environment. This is not highly accurate, even in the fine fuels considered here: the Biot number based on an effective heat transfer coefficient for radiation is of order one, and the mass transfer Biot number is of order 0.1. However, since the model focuses on the interaction between the fuel and environment, a simple approach was chosen for each of these two control volumes individually.

c) Consistent with assumption b), the internal diffusion resistance for mass transfer of water is neglected, and water is assumed to be available at the surface of the fuel element, at its saturation pressure corresponding to the fuel temperature.

d) The mass fraction of liquid water in the fuel is taken to be 10%, and the mass of char remaining after pyrolysis is complete is 30% (Susott, 1982).

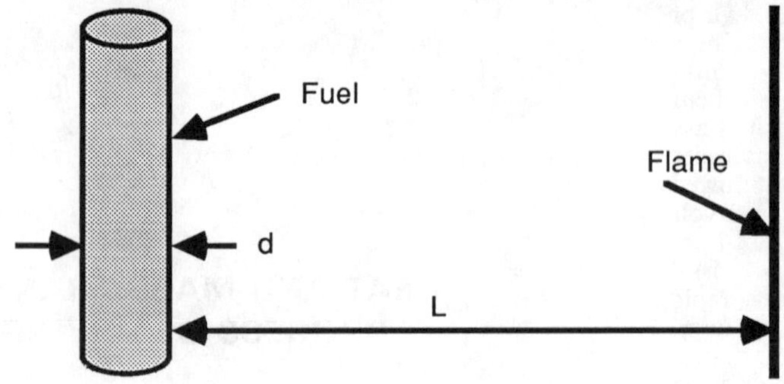

Figure 1a: Side View of Flame and a Single Fuel Element

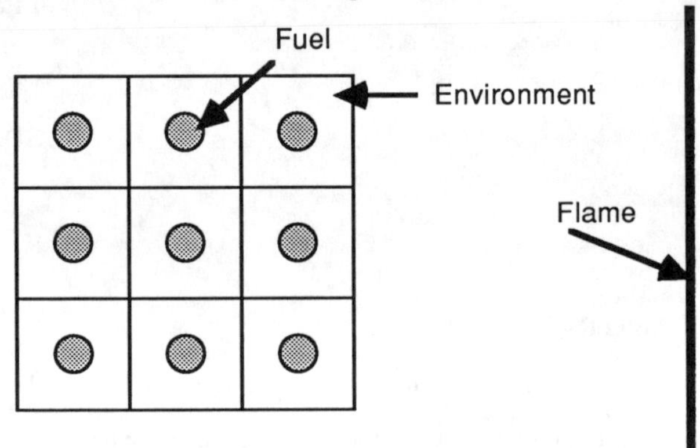

Figure 1b: Top View of Fuel Element, Environment, and Flame Geometry

e) It is not within the domain of this work to solve for the air flow caused by the combined effect of wind and natural convection. In developing the governing equations, it has been assumed that a known mass flow rate of air enters the environment control volume, which affects the mass and energy balances. However, in all cases considered here, the inlet mass flow rate of air is taken to be zero -- this is exactly equivalent to the assumption that all control volumes have identical temperatures and species concentrations, so that air flow between control volumes does not cause any net exchange of heat or mass.

f) The effect of the air flow on the exchange of heat and water vapor between the fuel and environment is characterized by heat and mass transfer coefficients. Different values of these coefficients are considered.

g) If the fuel temperature reaches 100 °C while there is still liquid water present in the fuel, then the fuel temperature will be held fixed at 100 °C until all the water evaporates.

h) Some of the water vapor in the environment will of course condense if its partial pressure would otherwise exceed its saturation pressure. The condensate is assumed to leave the environment control volume (fall downward).

i) The gases in the environment control volume (air, water vapor, and volatiles) are well-mixed.

j) As water vapor and volatiles enter the environment, some of the resulting mixture leaves the environment control volume (rises out of it) to maintain atmospheric pressure. Thus, the rate at which mass exits the environment control volume is greater than the rate at which it enters.

k) The gases in the environment control volume are assumed not to radiate or attenuate radiation from the distant flame. This assumption is particularly good because the environment and fuel element temperatures generally remain very close up to the time of ignition.

l) Based on comments by Susott (1980), some small fraction of the combustibles (less than 8% of the initial fuel mass) is assumed to be a single species with a low spontaneous ignition temperature (below the temperature at which glowing ignition begins) and a low lean limit of flammability in air. Two examples which will be considered are acetylene and diethyl ether (see Table 1 for properties of these species). This species will be referred to as the pilot species.

m) The pyrolysis process is described using a first order Arrhenius model (Roberts, 1970). Again based on Susott (1980), it is assumed that the pilot species is the first species to be released by pyrolysis. The constants in the Arrhenius equation are allowed to be different for the pilot species and the remainder of the volatiles. The constants for the pyrolysis of the pilot species will be varied in the model.

n) The volatiles are assumed to convect into the environment as rapidly as they are released from the structure of the fuel (Kung, 1972).

o) The pyrolysis process is assumed to be endothermic (Kanury, 1972).

p) As volatiles enter the environment, and mass leaves the environment to maintain atmospheric pressure, the fuel-air ratio of the pilot species is monitored. If it falls between the lean and rich limits for combustion, and the environment temperature exceeds the spontaneous ignition temperature, flaming ignition occurs.

q) Glowing ignition is assumed to begin at a fixed temperature of 450 °C (Hshieh and Richards, 1989).

r) The environment gases are ideal gases, with specific heats which are independent of temperature.

A set of differential equations describing the heat and mass transfer for the fuel element and the surrounding environment will now be presented. These equations describe the time variation of the fuel and environment temperatures, mass of solid fuel remaining, and mass of water vapor, volatiles, and air in the environment. These equations are solved at discrete time intervals using a fourth order Runge-Kutta method. In addition to the physical phenomena described by the differential equations, condensation is accounted for between successive time intervals. This will be discussed after the differential equations are presented.

Equations Describing Heat and Mass Transfer

The rate of evaporation of water vapor from the fuel to the environment is expressed as (see Nomenclature):

$$\dot{m}_w = gA_s(\omega_{ws} - \omega_{we}) \tag{1}$$

where ω_{ws} is found from saturation conditions at T_f. When all the water has left the fuel, $\dot{m}_w$ is set to zero.

The rate of change of water vapor in the environment is due to evaporation from the fuel and the flow in and out of the control volume:

$$\frac{dm_{we}}{dt} = \dot{m}_w + \dot{m}_i\omega_{wi} - \dot{m}_o\omega_{we} \tag{2}$$

The rate of release of volatiles from the fuel is expressed as:

$$\dot{m}_v = K_p(m_f - m_c)\exp(-E_p/R_uT_f) \tag{3}$$

The pyrolysis constants are first set to the values for the pilot species. After the assumed fraction of the pilot species has left the fuel, the constants are changed to the values for the remaining volatiles. Pyrolysis is complete when the mass of the fuel, m_f, reaches the mass of the char, m_c. Thereafter, $\dot{m}_v$ is set to zero.

The rate of change of volatiles in the environment is given by:

$$\frac{dm_{ve}}{dt} = \dot{m}_v + \dot{m}_i\omega_{vi} - \dot{m}_o\omega_{ve} \tag{4}$$

The rate of change of air in the environment is given by:

$$\frac{dm_{ae}}{dt} = \dot{m}_i\omega_{ai} - \dot{m}_o\omega_{ae} \tag{5}$$

Since the instantaneous fuel mass is required in Eq. (3), it is determined by keeping track of the rates at which water vapor and volatiles leave the fuel. That is:

$$\frac{dm_f}{dt} = -\dot{m}_w - \dot{m}_v \tag{6}$$

In the above equations:

$$\omega_{we} = \frac{m_{we}}{m_e} = m_{we}\frac{RT_e}{pV_e}, \quad \omega_{ve} = \frac{m_{ve}}{m_e} = m_{ve}\frac{RT_e}{pV_e}, \tag{7}$$

$$\omega_{ae} = \frac{m_{ae}}{m_e} = m_{ae}\frac{RT_e}{pV_e}$$

where p=1 atm and R is the gas constant for the mixture in the environment. The outgoing mass flow rate is determined from a global mass balance on the environment control volume:

$$\dot{m}_o = \dot{m}_i + \dot{m}_w + \dot{m}_v + \frac{pV_e}{RT_e^2}\frac{dT_e}{dt} \tag{8}$$

The last term is $-dm_e/dt$, with the assumption that the pressure remains atmospheric at all times.

The energy balance on the fuel includes: a) radiation from the distant flame, b) convection to the surrounding environment, c) latent heat of vaporization of water, and d) energy required for pyrolysis. Thus the energy equation for the fuel is given by:

$$m_fc_{pf}\frac{dT_f}{dt} = \sigma\epsilon F_{12}A_s(T_{fl}^4 - T_f^4) + hA_s(T_e - T_f) - \dot{m}_wh_{fg} - \dot{m}_vQ_p \tag{9}$$

The energy balance on the surrounding environment includes: a) convection from the fuel, b) sensible heat of the incoming water vapor, c) sensible heat of the incoming volatiles, and d) sensible heat of the incoming air. (In addition, condensation affects the environment temperature, but this is accounted for separately from the differential equation below.) Thus the sensible energy equation for the surrounding environment is given by:

$$\overline{m_ec_{pe}}\frac{dT_e}{dt} = hA_s(T_f - T_e) + \dot{m}_wc_{pw}(T_f - T_e) + \dot{m}_vc_{pv}(T_f - T_e) + \dot{m}_ic_{pi}(T_i - T_e) \tag{10}$$

Equations (1)-(10) can be solved for the dependent variables m_{we}, m_{ve}, m_{ae}, T_f, and T_e. The initial conditions are $T_f = T_e =$ 300 K, $m_{ve} = 0$, m_f based on a density of 700 kg/m^3, and m_{we} and m_{ae} based on 70% relative humidity. The quantity $\overline{m_ec_{pe}}$ is calculated by summing up the instantaneous contributions from water vapor, volatiles, and air. The values chosen for the constants in these equations are discussed in the Appendix and tabulated in Table 2. Thus Eqs. (1)-(10) can be solved at each time step. Then, as mentioned previously, condensation is accounted for separately. This will now be described.

Condensation

At the end of a time step, the mass of water vapor in the environment may exceed the mass corresponding to saturated conditions. This will only occur for temperatures less than 100 °C. If this occurs, the extra mass is assumed to condense. There is a corresponding increase in the environment temperature due to the release of latent heat, i.e. $\Delta T_e = m_{cond}h_{fg}/\overline{m_ec_{pe}}$. The water is then assumed to leave the environment control volume (it falls to the ground).

Summary of the Model

The system of equations (1)-(10), together with the additional consideration for condensation, constitute the model for drying, pyrolysis, and ignition of live fuels. These equations predict the temperatures and masses of the various species in the fuel and surrounding environment as a function of time. At the end of each time step, the environment temperature and pilot species-air ratio are tested to determine if ignition would occur. Flaming ignition occurs if the pilot species-air mass ratio is between the lean and rich flammability limits, and the environment temperature exceeds the spontaneous ignition temperature (see Table 1). Glowing ignition occurs if the fuel temperature exceeds 450 °C. Thus the time required for flaming or glowing ignition to occur can be predicted as a function of the various parameters. An assessment can be made of the importance of combustible volatiles in bringing about ignition.

RESULTS AND DISCUSSION

Before considering the detailed results of the model, it is helpful to discuss some general considerations. First of all, it must be recognized that if it were not for the fact that the pilot species has a spontaneous ignition temperature below the glowing ignition temperature, ignition would always occur first in the solid fuel. (This might then cause piloted ignition of the combustible volatiles.) The hypothesis of this paper is that the pilot species may, under certain circumstances, cause flaming ignition to occur before the fuel reaches the glowing ignition temperature. The model in fact shows that, for the conditions considered here, flaming ignition does often occur before glowing ignition. Note that the conditions under which flaming ignition does not occur first are either a) the mixture is still too lean when the glowing ignition temperature is reached, or b) the mixture is too rich before the spontaneous ignition temperature is reached. The following comments can be made concerning the differences between acetylene and diethyl ether (see Table 1). Since acetylene has a lower lean flammability limit and a higher spontaneous ignition temperature, it has a tendency to reach the lean limit first, and then undergo flaming ignition when the ignition temperature is reached. Diethyl ether has the opposite tendency of reaching the ignition temperature first and undergoing flaming ignition as soon as the lean flammability limit is reached. For this reason, diethyl ether tends to ignite sooner than acetylene. Also, since acetylene has a lower rich flammability limit, it has a greater tendency to become rich, and therefore undergo glowing ignition rather than flaming.

It is found that the heat transfer coefficient has little effect on the fuel temperature, because radiation is usually the dominant term in the fuel energy equation. The heat transfer coefficient does affect how closely the environment temperature tracks the fuel temperature, but the effect on time to ignition is fairly minor, and will not be discussed.

The model was run for a variety of conditions. Table 2 shows the parameter values used for the "base case," as well as the ranges of values used for the parameters that were varied. The chosen values are justified in the Appendix. In the results presented below, all variables are held fixed at their "base" values, except as noted. All cases considered here have zero flow of air into the environment control volume, which is exactly equivalent to assuming that adjacent control volumes are identical, so that there is no net exchange of heat or mass between them.

Figures 2 and 3 show the fuel and environment temperatures as a function of time, for diethyl ether and acetylene, for three different fuel element diameters. The points where the curves end correspond to ignition. The type of ignition is indicated with a symbol at the end of the curve -- a star for flaming ignition and a circle for glowing ignition (see Legend below Fig. 6). Note that the fuel element spacing stays fixed, so the largest diameter fuel element corresponds to a smaller environment control volume. In all cases, the temperature of the environment tracks that of the fuel within several degrees. The portion of the curve below 373 K (100 °C) corresponds to evaporation of the water; the temperature cannot exceed 373 K until all the water evaporates. The evaporation process takes much longer for the largest diameter fuel because it contains more water, and because the environment is smaller and therefore becomes saturated more rapidly. For both diethyl ether and acetylene, the 0.32 cm diameter fuel element undergoes flaming ignition (as does the 0.95 cm case for diethyl ether). Glowing ignition occurs for the two larger diameters for acetylene, and the largest diameter for diethyl ether. There are two reasons for this. First of all, the fuel heats up more slowly for the larger diameters, so there is more time for volatiles to be released. Second, there is more water in the fuel, which displaces the air in the environment. Both of these phenomena contribute to the mixture exceeding the rich limit, so that glowing ignition occurs.

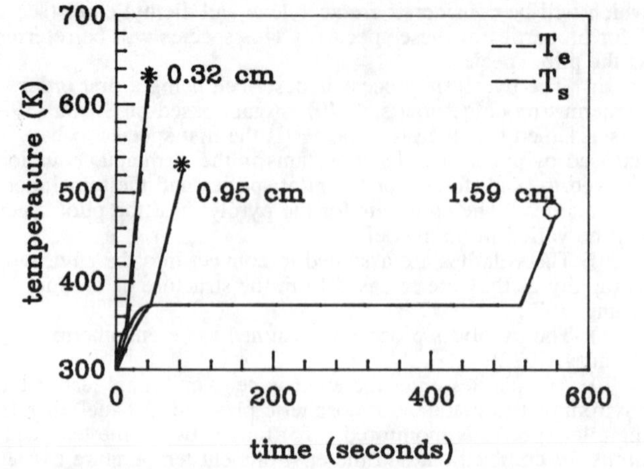

Figure 2: Fuel and Environment Temperatures vs. Time for Diethyl Ether

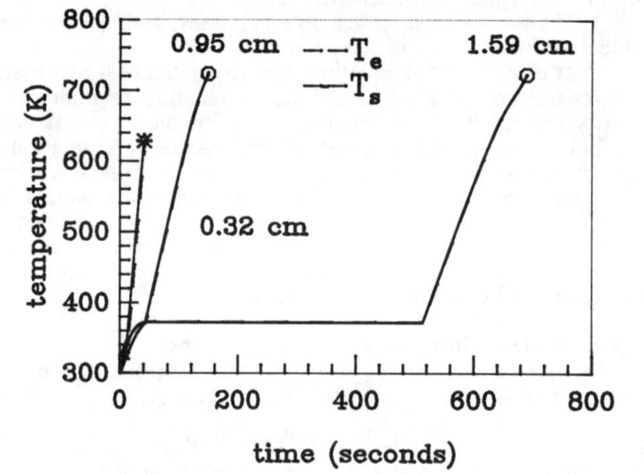

Figure 3: Fuel and Environment Temperatures vs. Time for Acetylene

Figures 4 and 5 show the percent weight loss of the solid fuel as a function of time, corresponding to Figures 2 and 3. The first 10% of the weight loss is due to evaporation of water. As explained previously, it takes longer to dry out the larger fuel elements because there is more water in the fuel and a smaller environment. After all the water has evaporated, the loss of volatiles occurs very slowly at first, due to the low temperatures (nearly horizontal portions of curves). For the cases which persisted to high temperatures (particularly acetylene for the two larger diameters), the rate of weight loss does increase rapidly near the end.

Figure 6 demonstrates the effect of the pre-exponential constant for the pilot species, K_p, on the time to ignition, for diameters of 0.32 cm and 0.95 cm. Note that in Figs. 6 through 9, the symbols indicate the type of ignition, and the line type distinguishes between diethyl ether and acetylene (see Legend below Fig. 6). For diethyl ether (dashed line), flaming ignition occurs in all cases. As usual, the ignition temperature is reached first, and then the lean flammability limit. Therefore, as the rate constant increases, the time to ignition decreases slightly. For acetylene (solid line), for the smaller diameter case, the results are essentially identical to those for diethyl ether. However for the 0.95 cm diameter, the results differ. For the lower values of K_p, flaming ignition occurs. The lean limit is reached first and then the ignition temperature, so that ignition occurs at essentially the same time, independent of K_p. The time to ignition is slightly larger than for diethyl ether because the acetylene has to reach 578 K to ignite. For larger values of K_p, the rich flammability limit of acetylene is exceeded, due to the combination of a larger fuel mass and a smaller air mass. Therefore glowing ignition occurs at a significantly later

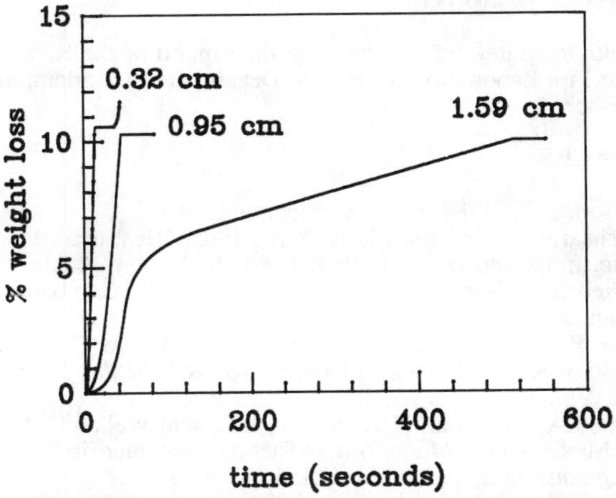

Figure 4: Percentage Weight Loss vs. Time for Diethyl Ether

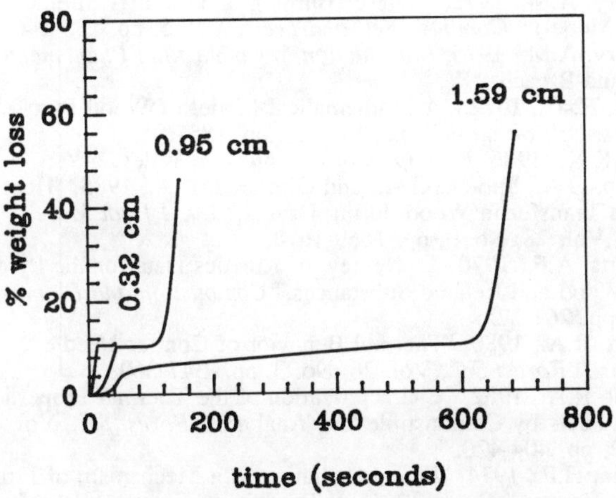

Figure 5: Percentage Weight Loss vs. Time for Acetylene

time than for flaming ignition, again independent of K_p.

Figure 7 shows the effect of the amount of the pilot species present in the fuel (as a fraction of the initial fuel mass), for fractions less than 0.01 (1%). For the range 1% to 8% (beyond the range of the figure), there is no significant change. For very small fractions (less than roughly 0.002 for both diethyl ether and acetylene), the lean limit is never reached, and glowing ignition occurs. For somewhat larger fractions, there is enough of the pilot species to reach the lean limit, and flaming ignition occurs, with a corresponding decrease in the time to ignition. For still higher fractions (greater than about 0.0044), acetylene exceeds the rich limit, and returns to glowing ignition.

The effect of the product of shape factor times effective emissivity is shown in Fig. 8. Clearly, as this product decreases, the fuel heats up more slowly, and therefore takes longer to ignite. For diethyl ether, flaming ignition always occurs. For acetylene, however, glowing ignition occurs for values of εF_{12} below about 0.2. The glowing ignition occurs because the pilot species has more time to be released, and therefore reaches the rich limit before the ignition temperature is reached.

The effect of fuel element spacing is presented in Fig. 9. Consider first the behavior for the larger fuel element spacings shown in the figure (spacings greater than about 0.06 m). For both fuels, flaming ignition occurs. The time to ignition increases with increasing spacing, because it takes longer to reach the lean limit. As the spacing is increased still further (beyond the range shown in the figure), the mixture does not reach the lean limit even when all of the pilot species has been released from the fuel, so that glowing ignition occurs. For small spacing (less than about 0.06 m), the behavior is more complicated, and in general the time to

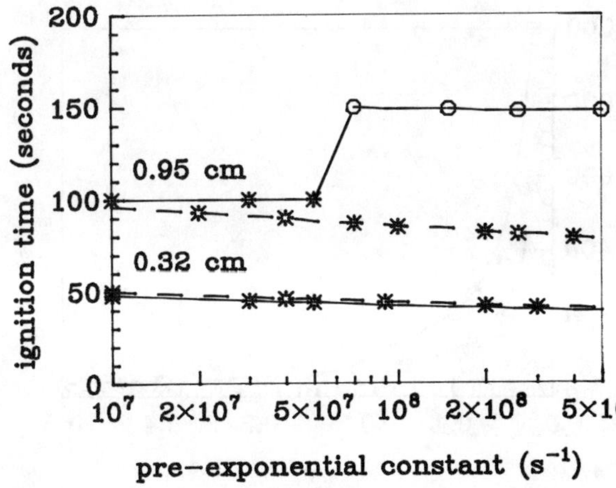

Figure 6: Effect of Pre-Exponential Constant on Ignition Time

Legend for Figures 6–9
-- Diethyl Ether
— Acetylene
◇ Glowing Lean
○ Glowing Rich
⋇ Flaming

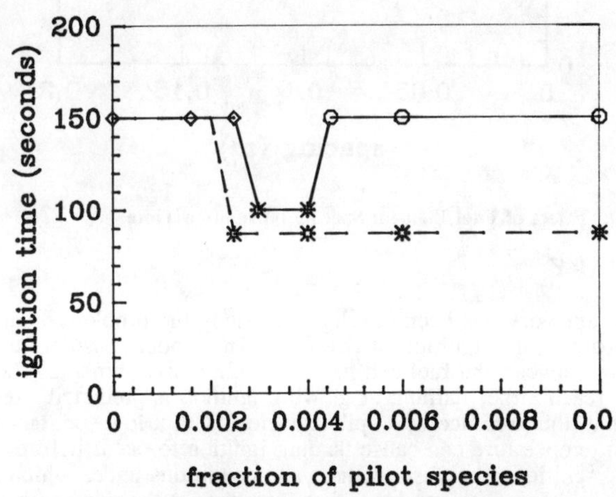

Figure 7: Effect of Fraction of Pilot Species on Ignition Time

ignition is higher than for the larger spacings. For these very small spacings, the behavior is dominated by the length of time required for the water to evaporate. The solid fuel has to stay at 100 °C for an extended period before it dries out, thus lengthening the time to ignition. The time needed to dry out (and hence the time to ignition) varies nonmonotonically with spacing for the following reasons. As the spacing increases, the environment stays cooler, heat is lost more rapidly from the fuel, and it tends to heat up more slowly. At the same time, as the spacing increases, the environment stays drier, which increases the evaporation rate. These two competing effects are the cause of the observed nonmonotonic behavior. It is also noted that acetylene undergoes glowing ignition due to the mixture becoming too rich at the small spacings.

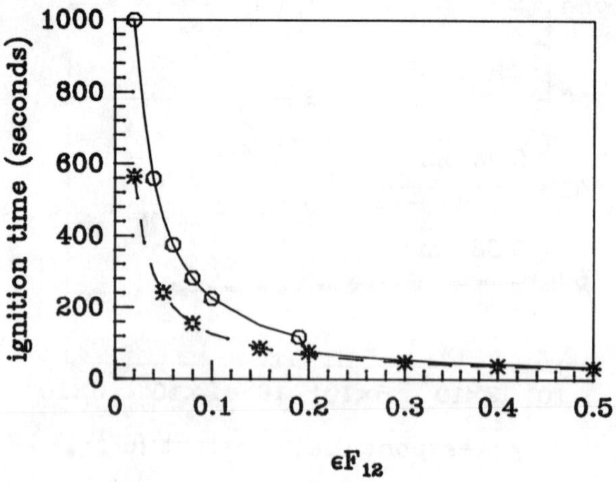

Figure 8: Effect of Flame Radiation on Ignition Time

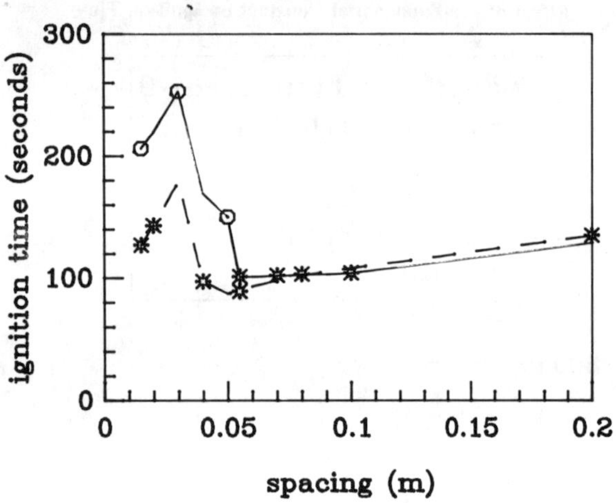

Figure 9: Effect of Fuel Element Spacing on Ignition Time

SUMMARY

A framework has been developed to study the problem of drying, pyrolysis, and ignition of live fuel. The model considers the coupling between the fuel and its surrounding environment. The time to reach either flaming or glowing ignition is predicted. It is found that the presence of a "pilot species" with a low spontaneous ignition temperature can cause flaming ignition to occur before glowing ignition otherwise would. The conditions under which flaming ignition does not occur first are either a) the mixture is still too lean when the glowing ignition temperature is reached, or b) the mixture is too rich before the spontaneous ignition temperature is reached. The effects of various parameters were explored, such as the fuel element diameter and spacing, the pre-exponential constant, the fraction of the pilot species present in the fuel, the shape factor-emissivity product, the lean and rich flammability limits, and the spontaneous ignition temperature. Some complex interactions between these parameters were observed, and results were obtained which would not have been immediately obvious without the aid of the model. Although there is insufficient experimental data with which to define some of the model parameters, the model is useful in suggesting qualitative phenomena which may occur in ignition of live fuels typical of Southern California.

ACKNOWLEDGEMENT

The authors gratefully acknowledge the support of the Riverside Forest Fire Laboratory of the U.S. Department of Agriculture Forest Service.

REFERENCES

Anderson, E.B., 1988, personal communication.

Blackshear, P.L., Jr., and Murty, K.A., 1965, "Heat and Mass Transfer to, from, and within Cellulosic Solids Burning in Air," *Tenth Symposium (International) on Combustion,* The Combustion Institute, pp. 911-923.

Gandhi, P.D. and Kanury, A.M., 1986, "Criterion for Spontaneous Ignition of Radiantly Heated Organic Solids," *Combust. Sci. and Tech.*, Vol. 50, pp. 233-254.

Havens, J.A., Hashemi, H.T., Brown, L.E., and Welker, J.R., 1972, "A Mathematical Model of the Thermal Decomposition of Wood," *Combust. Sci. and Tech.*, Vol. 5, pp. 91-98.

Hshieh, F.Y. and Richards, G.N., 1989, "Factors Influencing Chemisorption and Ignition of Wood Chars," *Combustion and Flame*, Vol. 76, pp. 37-47.

Kanury, A.M., 1972, "Rate of Burning of Wood (A Simple Thermal Model)," *Combust. Sci. and Tech.*, Vol. 5, pp. 135-146.

Kanury, A.M., 1975, *Introduction to Combustion Phenomena,* Gordon and Breach, NY.

Kung, H.-C., 1972, "A Mathematical Model of Wood Pyrolysis," *Combustion and Flame*, Vol. 18, pp. 185-195.

Kuo, K.K., 1986, *Principles of Combustion,* Wiley, NY.

Plumb, O.A., Spolek, G.A., and Olmstead, B.A., 1985, "Heat and Mass Transfer in Wood during Drying," *Int. J Heat Mass Transfer*, Vol. 28, No. 9, pp. 1669-1678.

Roberts, A.F., 1970, "A Review of Kinetics Data for the Pyrolysis of Wood and Related Substances," *Combustion and Flame*, Vol. 14, pp. 261-272.

Susott, R.A., 1980, "Thermal Behavior of Conifer Needle Extractives," *Forest Sci.*, Vol. 26, No. 3, pp. 347-360.

Susott, R.A., 1982, "Characterization of the Thermal Properties of Forest Fuels by Combustible Gas Analysis," *Forest Sci.*, Vol. 28, No. 2, pp. 404-420.

Telisin, H.P., 1974, "Flame Radiation as a Mechanism of Fire Spread in Forests," Chap. 29, *Heat Transfer in Flames,* eds. Afgan and Beers, Scripta, Washington, D.C.

Tinney, E.R., 1965, "The Combustion of Wooden Dowels in Heated Air," *Tenth Symposium (International) on Combustion,* The Combustion Institute, pp. 925-930.

APPENDIX

This Appendix describes how the various constants used in the equations were determined.

Fuel Masses and Geometry. The fuel element is thought of as a long thin twig with clusters of needles attached to it. The diameter of this twig was varied in the calculations, from 0.32 to 1.59 cm (1/8" to 5/8"). The length is not a significant parameter, since all masses, volumes, and surface areas scale with the length. For the purposes of calculating the initial mass and specific heat of the fuel, the initial composition of the fuel by volume was assumed to be 85% solid wood, 10% liquid water, and 5% trapped gases (e.g. air and water vapor) (Plumb et al., 1985). The specific heat of the fuel was assumed to remain constant, although the mass obviously decreases. On a weight basis, the fuel element was taken to be 10% water and 60% volatiles. The remaining 30% is left as char after pyrolysis is complete.

Environment Masses and Geometry. The volume of the environment was varied in the calculations based on an assumed twig spacing of 3 to 10 cm. The air and water vapor masses corresponding to these volumes were calculated assuming a relative humidity of 70% at an ambient temperature of 300 K.

Convection Coefficients. The heat and mass transfer coefficients can vary greatly depending on whether there is natural convection due to the heating of the twig or forced convection due to an ambient wind. In the case of natural convection, the coefficients depend strongly on the orientation of the twig relative to gravity and the twig vertical dimension. In the case of forced convection, the coefficients depend strongly on the wind speed and orientation relative to the wind. Thus it is possible to have a substantial range of heat and mass transfer coefficients. For instance, forced convection due to a 2.2 m/s (5 mph) wind in cross flow over a 1.59 cm diameter cylinder yields h=30 W/m^2-K, g=0.03 kg/m^2-s whereas a 8.9 m/s (20 mph) wind over a 0.32 cm diameter cylinder yields h=140 W/m^2-K, g=0.11 kg/m^2-s. Natural convection typically yields smaller values. The range of heat transfer coefficients from 2 to 100 W/m^2-K was used in the calculations, but did not yield significant differences in the time to reach ignition. The mass transfer coefficient was held fixed at 0.09 kg/m^2-s (Blackshear and Murty, 1965), because it only affects how quickly the water evaporates, which was not of major interest in this study.

Flame Radiation. The flame temperature was held fixed at 1200 K (Telisin, 1974). The other variable which affects flame radiation is the product of the shape factor and the effective flame-fuel emissivity. Assuming the flame is a plane, and therefore does not see itself, the effective flame-fuel emissivity is:

$$\varepsilon = \frac{\varepsilon_1 \varepsilon_2}{\varepsilon_1 + \varepsilon_2(1 - \varepsilon_1)F_{12}}$$

where 1 stands for fuel and 2 stands for flame. Note that the emissivity of wood and vegetation is typically between 0.8 and 1.0. Taking the fuel emissivity to be close to 1.0, the effective emissivity is close to the flame emissivity. The flame emissivity can be approximated by ε_2 = 1-exp(-0.16D), where D is the flame depth in meters (Telisin, 1974). Clearly, this can vary between 0 and 1 depending on D. The shape factor F_{12} can vary greatly depending on the distance between the flame and the fuel, and their relative sizes. The shape factor from the fuel to the flame has a maximum value of 0.5, corresponding to a flame which appears infinite to the fuel element and no blockage by other fuel elements. There is no minimum value for the shape factor. In the limit of a distant flame (again with no blockage), the shape factor is L/2πR, where L is the flame width and R is the distance between the flame and the fuel. Thus, the product of shape factor and effective emissivity can vary from a maximum of 0.5 down to arbitrarily small values. The range from 0.02 to 0.5 was considered in the calculations.

Pyrolysis Constants. As explained in the literature review, there is evidence that points to pyrolysis constants of $K_p = 7 \times 10^7$ s^{-1} and E_p = 30 kcal/mole for temperatures over about 230 °C (Roberts, 1970). However, there is also evidence (Tinney, 1965) of a very broad range of pyrolysis constants, particularly the pre-exponential constant K_p, especially in larger fuels (greater than 0.95 cm diameter). Furthermore, there are indications that the first several percent of the pyrolyzate, which comes off at temperatures lower than 230 °C, has a higher pre-exponential constant. In the calculations, the activation energy was held fixed at 30 kcal/mole for both the pilot species and the remainder of the volatiles. The pre-exponential constant was varied from 10^7 to 5×10^8 s^{-1} for the pilot species, and held fixed at 7×10^7 s^{-1} for the remainder of the volatiles. The pyrolysis process was assumed to be endothermic with $Q_p = 3.3 \times 10^5$ J/kg (Kanury, 1972 and Gandhi and Kanury, 1986).

Table 1: Properties of "Pilot" Species

	Acetylene	Diethyl Ether
Chemical Formula	C_2H_2	$(C_2H_5)_2O$
Lean flammability limit	0.023 g/g air	0.049 g/g air
Rich flammability limit	0.75 g/g air	2.4 g/g air
Spontaneous ignition temperature	578 K	458 K
Specific heat	2.04 kJ/kg-K	1.45 kJ/kg-K
Molecular weight	26.0	74.1

From Kanury (1975), except for the rich limit of acetylene, from Kuo (1986).

Table 2: Parameters Used in Model

	Base Case	Range
Fuel Element		
Diameter	0.95 cm	0.32 to 1.59 cm
Density for determining initial mass	700 kg/m^3	
Specific heat	3.0 kJ/kg	
Initial water mass	10% of fuel mass	
Final char mass	30% of fuel mass	
Initial pilot species mass	8% of fuel mass	0 to 8%
Initial fuel temperature	300 K	
Environment		
Fuel element spacing	5 cm	1 to 20 cm
Relative humidity	70%	
Initial environment temperature	300 K	
Heat transfer coefficient	100 W/m^2-K	2 to 100 W/m^2-K
Mass transfer coefficient	0.09 kg/m^2-s	
Inlet mass flow rate	0	0
Flame Radiation		
Flame temperature	1200 K	
Product of shape factor and effective emissivity	0.15	0.02 to 0.5
Pyrolysis		
Pre-exponential constant for pilot species	7×10^7 s^{-1}	10^7 to 5×10^8 s^{-1}
Pre-exponential constant for remainder	7×10^7 s^{-1}	
Activation energy for pilot species	30 kcal/mole	
Activation energy for remainder	30 kcal/mole	
Specific energy required for pyrolysis	330 kJ/kg	

1989 National Heat Transfer Conference
HTD-Vol. 106, Heat Transfer Phenomena in Radiation, Combustion and Fires

RADIATIVE FLUX MEASUREMENTS IN A SOOTY POOL FIRE USING A MULTIHEAD TRANSPIRATION RADIOMETER

L. Matthews
Engineering Research Center
A. Harris and G. Garcia
Department of Mechanical Engineering
New Mexico State Unversity
Las Cruces, New Mexico

ABSTRACT

Sooty pool fires are dominated by thermal emission processes which represent approximately 70% to 80% of the total heat transfer to an object engulfed by the flames. The multihead transpiration radiometer is a device which was designed to measure the radiative flux coming from five different directions in the fire. With 180 degree field of view and the rugged design of the radiometer, it is ideally designed for use in such a severe environment. The design, calibration and operation of the radiometer is discussed.

INTRODUCTION

Large sooty pool fires have been studied for a number of years by a number of investigators [1-5]. Of late, interest has focused on fires caused by transportation accidents which could occur during the movement of volatile or potentially hazardous material. These fires, as shown in Figure 1, generate large amounts of soot which makes measuring heat flux very difficult.

Convection and radiation heat transfer play important roles in the net heat transfer to any engulfed object. To understand the fire environment as well as the heat transfer mechanisms, radiative flux field must be measured. The transpiration radiometer is a device which can measure radiative flux in a sooty pool fire environment. The multihead transpiration radiometer is rugged, stable, repeatable and relatively temperature insensitive. The important aspects of the radiometer (its design, operation and calibration) in a pool fire environment are described in the following sections.

BACKGROUND

Radiation is the dominant mode of heat transfer to an object in a sooty pool fire. Birk and Oosthuizen [1] and Russel and Canfield [2] show that convection contributes approximately 20 percent of the total heat flux to an object in the fire. de Ris [3] and Pagni and Bard [4] also state that radiation is the dominant mode of heat transfer within large scale sooty pool fires. The aforementioned authors reached the following conclusions:

1. As an initial approximation, the combustion gases can be assumed to be nonscattering and act as a gray medium.

2. The surroundings act as a low-temperature blackbody absorber. The total heat flux within the fire is primarily thermal radiation.

3. Scaling of the thermal response of fires will be difficult if not impossible.

4. Luminous radiation is characteristic of sooty pool fires. Usually this type of fire is highly nonhomogeneous with steep temperature and species gradients.

Felske and Tien [6] showed similar results from their analysis, namely that soot emission dominates gas species emission and that scattering is nonexistent with the gray gas assumption being a good one.

The realization that sooty pool fires are thermal radiation dominated has pushed the

Fig. 1 30FT by 60FT Pool Fire Test

development of transducers that measure radiative fluxes while being unaffected by convection in the fire. The primary transducer discussed here is the multihead transpiration radiometer.

EQUIPMENT DESCRIPTION

The concept of a transpiration radiometer was first described by Moffat [7]. The basic idea being that a sensing surface could be made which was porous to allow a gas to flow through and one junction of a differential thermocouple to be attached. Moffat used the transpiration gas for several purposes.

1. The transpiration gas blows the convective boundary layer off of the sensor thereby eliminating the convective component to the flux measurement.

2. Reduce conduction in the sensing surface to the axial direction only. Radial conduction was eliminated.

3. Keep contaminants off of the sensing surface.

4. The transpiration gas acts as the reference temperature for the second thermocouple junction.

Basically the sensor has one junction of a differential thermocouple attached to it and another junction in the transpiration gas flow stream, (before the gas reaches the sensor) which acts as a steady, constant reference temperature. As the sensor heats up a greater and greater temperature differences exist between the two thermocouple junctions which can be calibrated against heat flux.

Considerable work, redesign and characterization was needed to make the concept viable for a sooty pool fire environment. Inherent problems in designing such an instrument for a pool fire environment include:

1. Coating of the sensing area by soot.

2. Removing the convection component of the total heat flux.

3. Maintaining a 180 degree field of view.

4. Obtaining an accurate calibration that includes the fire blackbody spectrum.

5. Choosing a sensor surface that is repeatable, stable and sensitive to small changes in the fire environment.

6. Keep the effect of the sensor on the fire environment to a minimum.

7. Surviving high temperatures present in the fire (1300 K to 1500 K).

The transpiration radiometer was an ideal candidate to measure radiative flux over an 180 degree field of view in a fire. Although several commercial radiometers have been tried in a sooty pool fire, none have met with the success of the transpiration radiometer. Reduced field of view, soot build up, temperature requirements and stability problems caused all commercially available designs tested to date to fail in the sooty pool fire environment.

A single head sensor was designed first [5] and used in several fires at Sandia National Laboratories. Refer to Figures 2 and 3. Although this design worked well it was clear that a more efficient (space wise) design was needed to be able to measure heat fluxes coming from multiple directions. Hence the need for the multiple sensor transpiration radiometer.

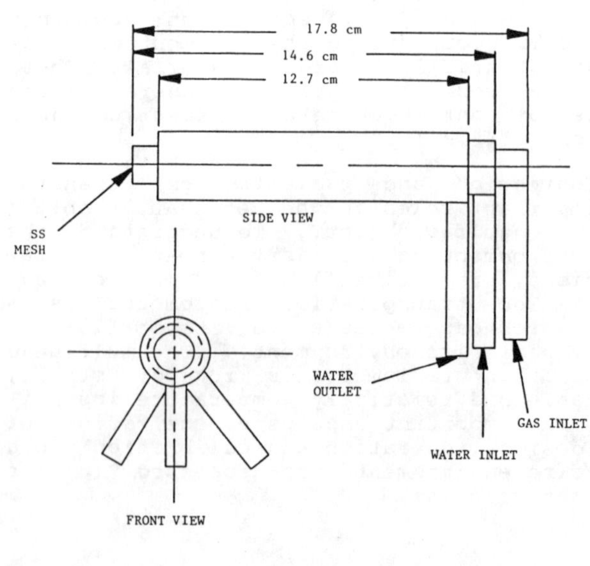

Fig. 2 Single Head Radiometer

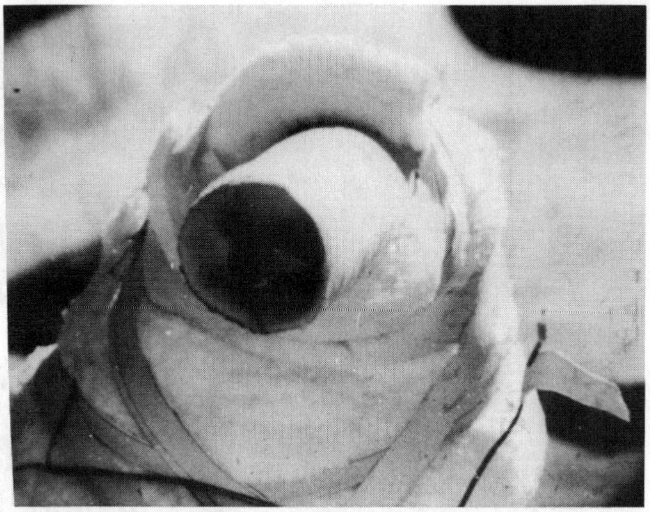

Fig. 3 Insulated Radiometer

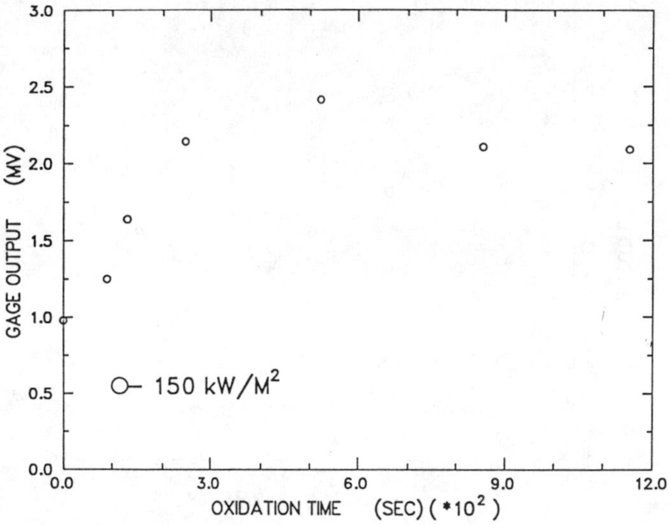

Fig. 4 Absorptivity Changes with Oxidation Time

A 304 stainless steel mesh with 85 micrometer diameter wire and a thickness of 0.25 mm was used. When holding the mesh up to the sun no light appears to pass through, however the transpiration gas (nitrogen in this case) can readily pass through. To increase the sensitivity of the mesh it was oxidized in air. The absorptivity of the mesh was increased and the surface was made more stable by the process. Stability was improved because the stainless steel mesh was heated during the oxidation process to a temperature well beyond any that should be seen during normal operation. At operating mesh temperatures of approximately 200 C the mesh should not continue to oxidize. Figure 4 shows how the mesh sensitivity changes with oxidation.

To achieve the desired convection effects, Moffat described a blowing parameter which must be kept above a certain value,

$$BP = \left[\frac{m'' \; Cp \; R^2}{4 \; k \; t}\right]^{0.5}$$

where

Cp= specific heat of the gas

k= mesh thermal conductivity

m"= mass flux flow rate

R= sensor disk radius

t= sensor disk thickness.

The blowing parameter should be kept above 3, according to theory, to assure convection free operation of the sensor. As a test of this concept the new multisensor radiometer was calibrated with BP equal to 3, 2 and 1 in an environment at 800 K with gas velocity of approximately 3 m/sec. A 5% to 10% variation in output was noticed between BP equal to 3 and 2. A 5% to 15% variation was noticed between BP equal to 3 and 1. This verifies the need to keep the blowing parameter in the range of 3 or higher. Of course, the higher BP is the lower will be the sensitivity of the gage.

Figure 5 shows how the transpiration radiometer sensitivity, measured here as the gage output voltage, changes with the blowing parameter. Obviously the best situation for maximum sensitivity is to use as low a BP as feasible while still maintaining a convection free sensor. From measurements taken during the latest tests the BP=3 suggestion by Moffat is as low as the radiometer should be taken. Useful descriptions of radiometers such as circular foil heat flux gages normally include the flux level of the gage at an output of 10 mV. Using this type of description the transpiration radiometers built to date are measuring 500 kW/m^2 or higher for an output of 10 mV. Most sooty pool fires emit radiative fluxes up to only 200-300 kW/m^2 with lower values around 10-50 kW/m^2. For this reason the BP needs to be kept as low as possible and the sensor surface needs to be made as absorptive as possible.

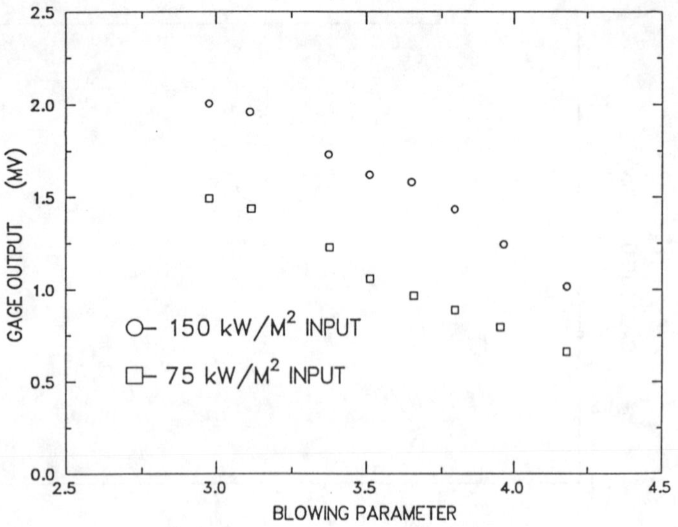

Fig. 5 Gage Sensitivity versus Gas Flow Rate

Fig. 6 Multihead Radiometer Sections

The multisensor transpiration radiometer design required that five sensors be located on a compact water cooled body with a self contained gas flow metering capability. As shown in Figures 6 and 7 the radiometer was designed and machined into four segments. The lower segment is a nitrogen supply header and a cooling water collection header. The next segment distributes the cooling water to the rest of the radiometer. Segment number three contains the flow metering nozzles and four of the flux measuring sensors. The top segment turns the water flow for return back through the radiometer and also houses the last flux measuring sensor.

Flow metering is performed by placing flow nozzles in the radiometer in each nitrogen supply line to the sensors. Each nozzle passes a calibrated flow of nitrogen to each sensor for a given back pressure. Considerable testing has shown that the nozzles are very repeatable and essentially unaffected by temperature changes. These nozzles allow the flow to be easily metered at the radiometer without the need for a flowmeter external to the radiometer. As can be seen, the multisensor transpiration radiometer is very compact and self contained.

Fig. 7 Blackbody Calibrator

CALIBRATION

Atkinson, Strange and Moffat [9] have described several aspects of calibrating transpiration radiometers. The radiometer mesh was modeled as an isothermal imperfect heat exchanger which compensated for most of the errors associated with the concept of the mesh being a perfect heat exchanger. This model helped bring the theoretical results in agreement with the calibration data. One of the major contributions of their paper was the

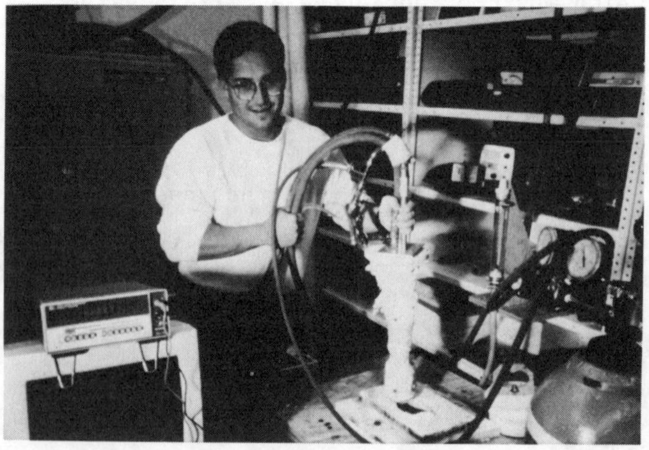

Fig. 8 Radiometer Ready for Calibration

description of the need to calibrate the radiometer in an environment similar to the one in which it was being used. For the current project accurate calibration of the multihead transpiration radiometer required a blackbody cavity that operated at temperatures close to those seen in a sooty pool fire. As described earlier, the fire is luminous without significant spectral line contributions from the combustion process. To keep the spectrum used for calibration similar to that seen in the fire, the cavity shown in Figures 7 and 8 was constructed.

Radiant lamps or possibly a solar furnace could be used for calibration purposes however the spectrum mismatch with the fire could lead to errors in the radiometer calibration curves. In addition, since the radiometer monitors flux from 5 different directions, see Figure 6, a cavity is required which can provide a uniform flux over essentially a 360 degree angle. The cavity shown in Figures 7 and 8 has an inner steel coil through which hot air is passed and then exhausted from the coil at the bottom of the cavity. The air then mixes with the radiometer transpiration gas that has passed through the stainless steel meshes and leaves through the cavity top. This helps keep the cavity walls at a uniform temperature and provide convective heating for the radiometer body. In this way a realistic environment for calibration can be provided for the radiometer.

During calibration the cavity wall temperatures were monitored with thermocouples. The maximum variation around the cavity was 3 degrees out of an average of 750 degrees Kelvin. The multihead radiometer was tested with low coolant water flow and high coolant water flow which showed that the calibration changed in direct proportion to the radiometer body temperature. This was easily explained because the nitrogen transpiration gas flow increased in temperature thereby heating the free stream differential thermocouple junction which caused a lower temperature differential to exist between the free stream junction and the stainless steel mesh. Further tests showed, however, that if the multihead radiometer was well insulated as shown in Figure 8 and at least 5 gpm of coolant water was passed through it, then the radiometer body did not heat up significantly. In all calibration cases with full water flow the body temperature never exceeded 35 degrees centigrade. Of course calibrating the radiometer in an environment similar to the fire temperature environment further reduces the effect of body temperature changes on the calibration.

Figure 9 shows a calibration curve for one of the multihead transpiration radiometers. Each radiometer demonstrated a high degree of stability and repeatability. After performing an uncertainty analysis on the calibration procedure, the radiometer calibration curves are given with a (+-)5% uncertainty band.

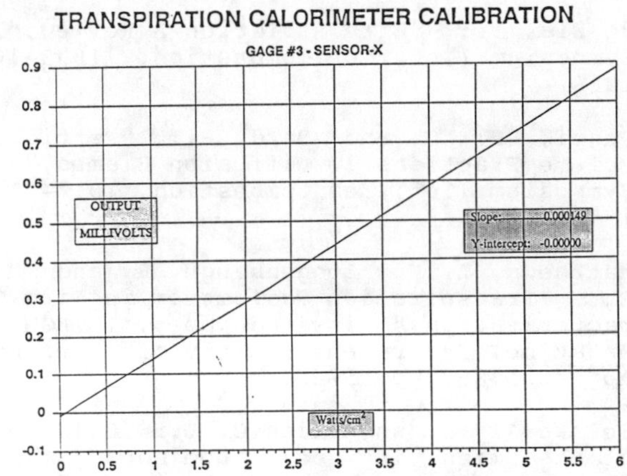

Fig. 9 Calibration Curve

CONCLUSION

A multihead transpiration radiometer has been described which can be used to measure radiative flux in severe environments such as a sooty pool fire. The radiometer is rugged, dependable and unaffected by the soot generated in a fire (meaning that the soot will not coat and therefore adversely affect the sensor) as well as the convective currents which can occur.

The radiometer can easily be calibrated for different environments such as engine combustors or rocket exhaust plumes. The radiometer flux measuring level can be easily raised to accommodate high flux level environments by simply changing the transpiration gas flowrate and recalibrating. The major drawbacks to the radiometer are the low sensitivity to read low flux levels (say below 50 kW/m^2) and response time, which is typically in the 2-5 second range for a zero to full scale reading. Future improvements to the design will reduce the impact of these drawbacks.

Acknowledgment: The authors would like to thank Dr. M. E. Schneider, Mr. L. A. Kent and Mr. R. Longenbaugh of Sandia National Laboratories for their contributions to the design of the multihead transpiration radiometer. In addition, part of this effort was funded by the National Science Foundation through grant # RII-8704253.

REFERENCES

1. Birk, A. M. and Oosthuizen, P. H.: "Model for the Prediction of Radiant Heat Transfer to a Horizontal Cylinder Engulfed in Flames," ASME Paper No. 82-WA/HT-52, 1982.

2. Russel, L. H. and Canfield, J. A.: "Experimental Measurement of Heat Transfer to a Cylinder Immersed in a Large Aviation-Fuel Fire," ASME Paper No. 73-HT-2, 1973.

3. de Ris, J.: "Fire Radiation-A Review,"17th Symposium (Int.) on Combustion, 1003-1016, 1978.

4. Pagni, P. J. and Bard, S.: "Particulate Volume Fractions in Diffusion Flames," 17th Symposium (Int.) on Combustion, 1017- 1028, 1978.

5. Matthews, L. K., Longenbaugh, R. and Mata, R.: "Transpiration Radiometer in Fire Characterization," invited paper, INTECH Magazine, Instrument Society of America, August 1986, pp.55-60.

6. Felske J. D. and Tien C. L.: "Calculation of the Emissivity of Luminous Flames," Combustion Science and Technology, 7, 25-31, 1973.

7. Moffat, R. J. et al.: "Development of a Transpiration Radiometer," Instrument Society of America, Conference Proceedings, Paper No. 613, 1971.

8. Longenbaugh, R. and Matthews, L. K.: "Experimental and Theoretical Analysis of the Radiation Transfer Inside a Sooty Pool Fire," Sandia National Laboratories Report, SAND 86-0083.

9. Atkinson, W. H., Strange, R. R. and Moffat, R. J.: "Development of Porous Plug Radiometers for Use in Advanced Gas Turbine Programs," AIAA/ASME/SAE/ASEE 24th Joint Propulsion Conference, paper # AIAA-88-3040, July 11-134, 1988, Boston, Mass.

1989 National Heat Transfer Conference
HTD-Vol. 106, Heat Transfer Phenomena in Radiation, Combustion and Fires

THE RADIATIVE-CONVECTIVE PARTITIONING OF HEAT TRANSFER TO STRUCTURES IN LARGE POOL FIRES

J. T. Nakos and N. R. Keltner
Thermal Test and Analysis Division 7537
Sandia National Laboratories
Albuquerque, New Mexico

I. ABSTRACT

This paper explores the relative contributions of radiative and convective heat transfer to structures (like shipping containers) in large pool fires. Results include measurements of the temperature of two structures (called "calorimeters") in the fire, the flames surrounding the calorimeters, the total heat flux to the calorimeters and the radiative component of the heat transfer at a few locations using transpiration radiometers. These measurements will be compared with calculations. Agreement between the measured radiative heat flux and the calculated radiative heat flux is good. The convective contribution was calculated from the total and radiative parts and was found to be about 10-20% of the total.

II. NOMENCLATURE

h = convective heat transfer coefficient, W/m^2-C
q = heat flux, kW/m^2
T = temperature, C
ε = emissivity
σ = Stefan-Boltzman const., 5.6693×10^{-12} W/cm^2-K^4

Subscripts

c = convective heat flux
er = emitted radiative heat flux
f = flame temperature value
ir = incident radiative heat flux
neti = net flux to surface 'i'
refl = reflected heat flux
si = surface 'i'
t = total = radiative + convective parts

III. INTRODUCTION

This paper explores the radiative/convective partitioning of heat transfer to two calorimeters in a JP-4 (an aviation fuel) pool fire. This is an important consideration in the design of any structural assembly that can be subjected to an accidental fire, because the total heat transfer to that structure and its final temperature will determine whether it can perform its intended function.

In this paper, the general physics of the convective heat transfer mechanism to structures in a turbulent pool fire will be discussed. More complicated phenomena, such as the effects of a participating medium and accelerating flows, are not considered. The convective part of the total heat transfer from a recently completed experiment will be presented and analyzed. Experimental data from a flat plate and a cylinder in cross flow will be analyzed. These are two common configurations of radioactive material shipping containers, so they were used as the modelling configurations for these heat transfer studies. The experimental data from the test includes information from transpiration radiometers and thermocouples mounted on a 3.05 m high vertically oriented flat plate (the "plate calorimeter") and a 1.42 m diameter cylinder (the "cylindrical calorimeter").

The experiments presented in this paper were performed in a fire test facility located at Sandia National Laboratories in Albuquerque, N.M. Tests have been run over a period of several years in a 9 m x 18 m rectangular pool with JP-4 jet fuel floating on water. A variety of shipping containers have been tested simultaneously with other experiments designed to help characterize the fire environment.

IV. LITERATURE REVIEW

The magnitude of the radiative and convective components is important in understanding the total heat transfer to a structure in a fire. The thermal mass, size and shape of the structure will affect that heat transfer. Neill et al. (1970) have studied fires of methanol, acetone, hexane, JP-4, cyclohexane and benzol in circular pans 30.48 cm (12"), 45.72 cm (18") and 60.96 cm (24") in diameter and a cluster burner consisting of 9, 15.24 cm (6") diameter circular pans arranged with one pan in the center and eight pans located 45 degrees apart on 30.48 cm centers. They estimated the convective component to a 11.37 cm diameter, 22.86 cm long vertical cylinder to be from 10-61% of the total heat transfer. From Russell and Canfield (1973), the measured rates of the convective component to a 21.67 cm diameter, 30.48 cm long horizontal cylinder in cross flow ranged from 17-63% of the total in a 2.44 m by 4.88 m (8 ft x 16 ft) pool using JP-5 aviation fuel. Thomas et al. (1965) have estimated the convective part of the total flux to the fuel surface to be from 1/2 of the total at the edge to 1/5 in the center. These measurements were made in wood and ethyl alcohol fires 91 cm in diameter at a location 1-2 cm above the fuel surface. Hamann et al. (1980) modelled the response of a railcar/cask system assuming constant values of the heat transfer coefficient h of 56.8 W/m^2-C and 28.4 W/m^2-C. They obtained the best agreement with experimental data using the lower value for h. Gregory et al. (1987) analyzed the convective heat transfer from a 1.4m diameter flat plate oriented vertically. They found the convective part to be 7% of the total and estimated h to be about 9 W/m^2-C.

There have been many convective heat transfer correlations used in the analysis of fires. Some are standard forced convection correlations taken from heat transfer papers (Russell and Canfield (1973)) and several studies estimate values of the convective heat transfer coefficient from data taken in a plasma-jet (Ahmed (1968)) and a pool fire (Russell and Canfield (1973)).

The data from Russell and Canfield (1973), which was from turbulent JP-5 pool fire experiments, indicated an average heat transfer coefficient of about 56.1 W/m^2-C (9.88 Btu/hr-ft^2-F). A correlation was presented and was used to estimate heat transfer coefficients of various gases, presumably those that might be present in a JP-5 fire. The calculated values ranged from 15.3 W/m^2-C to 92.0 W/m^2-C. This range spans the measured value, but spans it by a rather large amount. For example, if dry air were assumed as the gas present, the calculated value of the heat transfer coefficient would be 15.9 W/m^2-C (2.80 Btu/hr-ft^2-F); this is 72% below the measured value. Therefore, depending on which gas was assumed present in the fire, the calculated heat transfer coefficient could be significantly different than the actual value.

The highly turbulent free stream conditions may augment the heat transfer coefficient when compared with forced convective heat transfer correlations from flat plates cited in the literature. For example, Drysdale (1985) presented correlations for forced convection that show the transition from laminar to turbulent flow occurs at a Reynolds number of $3x10^5$. However, the free stream flow in our experiments is turbulent at Reynolds numbers near $4x10^4$. This Reynolds number is very close to the value cited by Gregory et al. (1987), i.e. 39,300.

The point to be made is that standard correlations may not be adequate for convective heat transfer calculations in fires. This was also pointed out in Shepherd (1987). One reason that the correlations may not be adequate is that the exact composition of the gas is not known, and in fact changes with time. This would cause property value estimates to be in error. Second, many of the existing correlations are for high Reynolds number turbulent flow at relatively small differences in temperature, whereas in a pool fire the flow is turbulent at low Reynolds numbers and has very high temperature differences. (Ahmed (1968) has shown that standard correlations do need modifications if high temperature differences are present.) Because the convective heat transfer contribution can be more than half of the total, errors in convective coefficients can conceivably have a significant effect on the final results.

V. DESCRIPTION OF CONVECTION IN FIRES

Figure 1 shows several side views of a vertical flat plate. Such a situation would occur if a cylindrical shipping container were oriented upright or if a rectangular container were oriented with one side vertical.

View A shows a situation in which the plate is colder than the surrounding gas and the free stream gas is moving vertically upward, as in a fire. This situation occurs during the early stages of a fire when the object is colder than the surrounding gas. Kent and Schneider (1987) and Schneider and Kent (1989) have shown that the vertical velocities range from an average of about 4 m/s at 2.8 m above the initial pool surface to about 12 m/s at 6.8 m above the surface. Their results also indicate that the flow in these large JP-4 fires is highly turbulent. As a point of comparison, the maximum velocity estimated from natural convection theory from a vertical flat plate 2 m from the leading edge is 2-3 m/sec. This is for a gas at 925C (1700F) and the plate at 38C (100F). In this case, the natural tendency of the fluid near the plate to fall downward is opposed by the "free stream" velocity. Depending on the elevation, the velocities may be roughly the same magnitude. This situation may create several scenarios, two of which are these: 1) the average velocity near the plate may be reduced below the free stream value, which could cause reduced heat transfer or 2) the turbulence may be increased near the plate which could augment the heat transfer mechanism.

View B shows the situation during the middle stages of the fire, after the plate has been heated close to the surrounding gas temperature. Even though there is a free stream velocity flowing upward, there will be little net convective heat transfer because the surface and free stream are at roughly equal temperatures.

Lastly, View C shows the situation that would occur at the end of the fire or during the fire when the winds "tilt" the fire away from the object. In this case the object (650C or 1200F) is hotter than the surrounding gas and there is no free stream velocity. Therefore, the plate will cool only by natural convection.

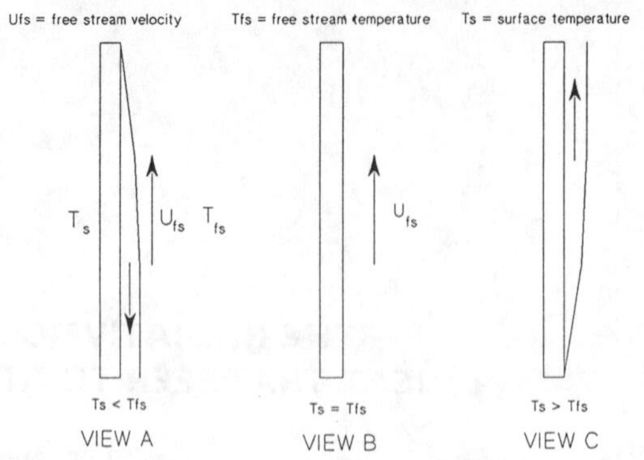

FIGURE 1, CONVECTION PROCESSES IN FIRES

In summary, there are three distinctive convective regimes, depending on the temperature of the structure. These regimes correspond roughly to the early times of the fire, the middle part when the object has reached temperatures close to the flame temperatures and lastly to the end of the fire when the flames have died due to the lack of fuel (or to high winds). During the early regime the free stream is turbulent and moving upward while there is a natural convective force driving gas near the plate downward. The convective heat transfer may be highest during this period. During the middle part of the fire there will be little net convective heat transfer due to the diminished temperature difference. Lastly, after the flames have died, the convective mechanism is purely natural, because the free stream flow has abated.

VI. EXPERIMENT DESCRIPTION

The majority of the experimental data to be presented was gathered from what is called the "plate calorimeter". The plate calorimeter is essentially a heavily instrumented pair of vertically oriented flat steel plates. One plate is 0.48 cm (3/16") thick and is made of 304 stainless steel and the other is 1.91 cm (3/4") thick mild steel. Both plates are about 30.5 cm wide and 305 cm long. They are mounted side by side. Thermocouples were mounted on the front and back faces of the plates and just in front of the plates to measure the flame temperature. In addition, there were two transpiration radiometers mounted on the front face of the plates to measure the incident radiative component of the heat flux. Figures 2 and 3 show an overall view and a sectional view of the plate calorimeter.

Limited data will be presented from a "1.4 m diameter cylindrical calorimeter"; it is described in Gregory, et al. (1987). Figure 4 shows a side view of the 1.4 m diameter cylindrical calorimeter. The calorimeter is 1.4 m diameter and 6.4 m long. The walls are 3.18 cm (1-1/4") thick and are made of A517 steel

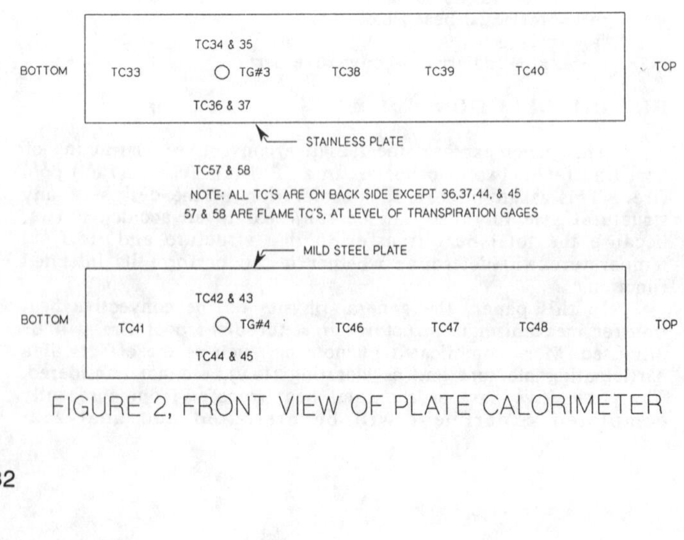

FIGURE 2, FRONT VIEW OF PLATE CALORIMETER

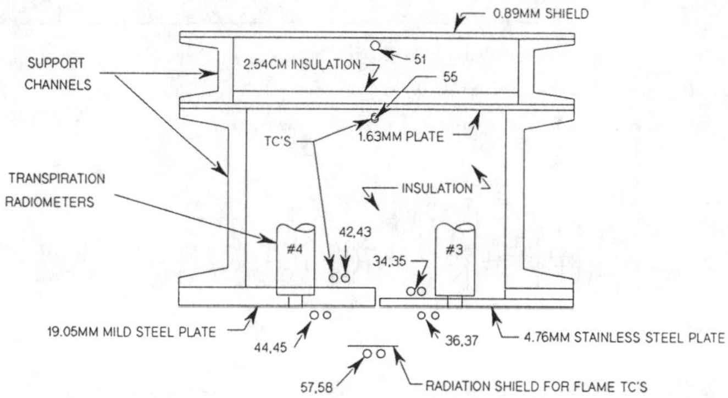

FIGURE 3, SECTION THROUGH PLATE CALORIMETER

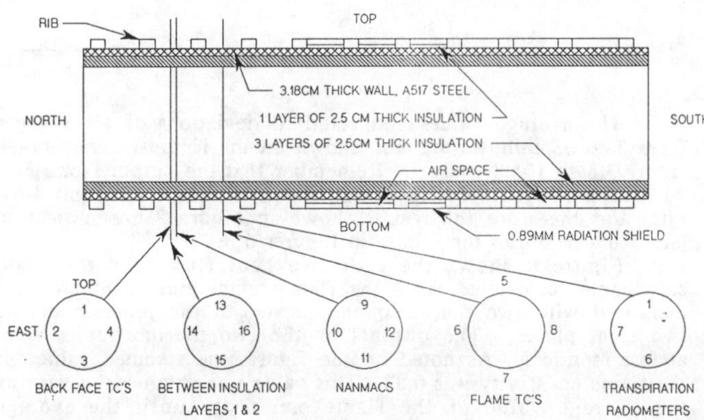

FIGURE 4, SIDE VIEW OF 1.4M DIA. CALORIMETER

with strengthening ribs every 61 cm. Mounted in the 1.4 m calorimeter were transpiration radiometers and thermocouples. Measurements included flame temperatures, temperatures on the backface of the 3.18 cm thick walls and pencil probe "eroding" thermocouples used to measure outer surface temperatures on the calorimeter.

The layout of the pool is shown in Figure 5. The long axis of the pool is oriented in the east-west direction. The test unit, a shipping container, was located on the east side and the 1.4 m diameter calorimeter was located on the west side. The long axis of the test unit was oriented east-west while the long axis of the 1.4 m calorimeter was oriented north-south. The plate calorimeter was placed between the test unit and the 1.4 m cylindrical calorimeter. It was placed north of the centerline of the pool. The side of the calorimeter that contained the transpiration gages faced to the south, into the fire. As indicated in Figure 5, there were other experiments in the pool, but those results will not be presented here. The entire 9m x 18m area of the pool was filled with JP-4 fuel to a depth of about 20 cm (8"). This amounted to about 33,000 liters (8,800 gal) of fuel.

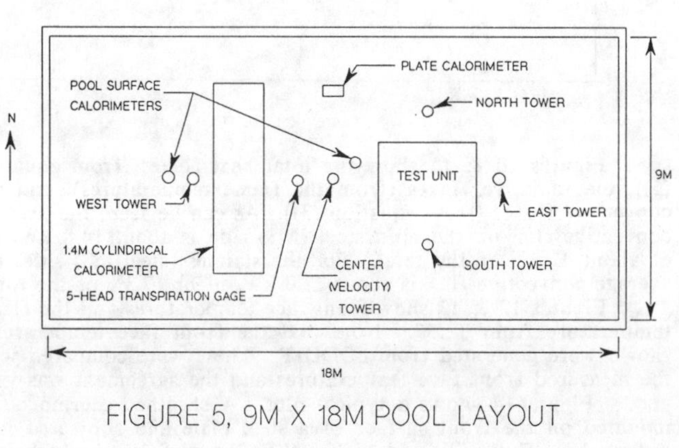

FIGURE 5, 9M X 18M POOL LAYOUT

VII. ANALYSIS

Calculation of the convective heat transfer component has its basis in the following energy balance:

$$q_{net} = q_{ir} - q_{er} + q_c - q_{refl}, \tag{1}$$

where q_{net} = net absorbed heat flux, q_{ir} = incident radiant flux, q_{er} = emitted radiative flux, q_c = convective heat flux, and q_{refl} = reflected radiative flux. An energy balance for surface 'i' can be written as follows:

$$q_{neti} = \sigma\varepsilon_f T_f^4 - \sigma\varepsilon_{si} T_{si}^4 + h_{si}(T_f - T_{si}) - (1-\varepsilon_{si})\sigma\varepsilon_f T_f^4, \tag{2}$$

The above relation assumes that the shape factors are equal to 1.0 and that the surface absorptivities equal the surface emissivities. Subscripts 'net' refer to net absorbed values, 'f' to flame temperature values, and 'si' to 'surface i'.

Solving (2) for the convective component gives:

$$h_{si}(T_f - T_{si}) = q_{neti} + \sigma\varepsilon_{si} T_{si}^4 - \sigma\varepsilon_{si}\varepsilon_f T_f^4. \tag{3}$$

Equation (3) can be used to estimate the convective heat transfer to surface 'i' from the measured flame temperatures, surface temperatures and the absorbed heat flux. The net absorbed heat flux was estimated from the backface temperature measurements by using an inverse heat conduction algorithm called SODDIT (Blackwell et al. (1987)). SODDIT also estimates the front face temperature and so these values can be compared with measured values from the "eroding" thermocouples and 1.59 mm (1/16") diameter sheathed thermocouples.

Writing equation (2) for the stainless and mild steel plates, subtracting one from the other and rearranging gives:

$$h = [q_{net1} - q_{net2} - \sigma\varepsilon(T_{s2}^4 - T_{s1}^4)]/[T_{s2} - T_{s1}]. \tag{4}$$

This equation assumes that the heat transfer coefficients for both plates are the same, which is not true if the plate temperatures are significantly different. It was also assumed that the flow fields on the two plates were the same, for example, similar to that shown in either View A or B in Figure 1. Equation (4) can be used to estimate the convective heat transfer coefficient, as long as the temperature difference $[T_{s2} - T_{s1}]$ is not zero and as long as the coefficients h_{s1} and h_{s2} are equal. In equation (4), ε_{s1} and ε_{s2} were made the same by painting both steel surfaces with a high absorptivity paint, Pyromark Black. Similarly, equation (3) for each surface can be used to estimate the convective heat transfer coefficient h. There is at least one advantage in using the equation (4) over equation (3), to calculate h. Flame temperatures are used in equation (3), as are assumed values for flame emissivity. Flame temperatures are measured, but flame emissivity values are not. In addition, flame temperatures fluctuate considerably and are measured only at a point several inches out from the plate surface. As a result, that flame temperature measurement may not be an accurate indication of the effective temperature of the radiation source to the plate surface. The use of equation (4) eliminates the need for flame temperatures or emissivities, which could have large errors, and so is presumably more accurate. However, the use of (4) to determine h could also have problems, including possibly a different frequency content of the temperature data and total absorbed heat flux data (q_{net1} and q_{net2}) for the two thicknesses of the plates. In addition, if the temperature difference $[T_{s2} - T_{s1}]$ becomes small, the coefficients h_{s1} and h_{s2} are not equal or the flow fields are not the same, the calculations for h would not be valid.

The total heat flux incident on the front surface can be estimated from the sum of the radiative and convective parts ($q_t = q_{ir} + q_c$), which, from equation (2), gives the following relation:

$$q_t = q_{neti} + q_{er} + q_{refl} = q_{neti} + \sigma\varepsilon_{si} T_{si}^4 + (1-\varepsilon_{si})\sigma\varepsilon_f T_f^4 \tag{5}$$

The net absorbed flux, q_{neti}, is calculated from SODDIT. The convective, radiative and total heat fluxes are calculated separately in a small program written for these calculations.

Note that, as defined, the total flux can be less than the radiative flux. The fluxes are defined as positive if they go into the surface, negative if they are outgoing.

From the ratio of the total and convective parts of the heat transfer, the convective contribution can be obtained as a function of time during the fire. Lastly, the calculated radiative contribution can be compared with the measured radiative contribution obtained from the transpiration gage data.

VIII. RESULTS

Data from three of the transpiration radiometers are shown in Figures 6-8. In Figures 6 & 7, the radiative heat fluxes incident on the mild steel (MS) and stainless steel (SS) sides of the plate calorimeter are shown. One would expect that the output of gages #4 & #3 (Figures 6 & 7) would be close, since they are only about 0.6 m apart. This is the case until about the 27 minute time, when the output of gage #3, Figure 7, starts to drop. Perhaps the gage surface was affected by soot buildup, because there was a considerable negative offset after the fire was out. Some of the gages did have soot buildup after the test, even though the buildup should have been minimal due to the transpiration gas flow. The exact cause(s) of this offset are not known at this time. However, gage #4, Figure 6, shows very little offset at either the beginning or the end of the test.

Figure 8 shows the heat flux (looking upward) on the top of the 1.4 m cylindrical calorimeter. The two other radiometers on the cylindrical calorimeter produced data with large offsets at the end of the test, and so those data are not shown. The output from the gage on the top of the 1.4 m calorimeter was modified because the actual transpiration gas flow rate during the test was not the same as that used in the calibration of the gage. Because the gage output is highly sensitive to the flow rate of the gas (Matthews et al. (1986)), the output at the lower flow rate was higher than the calibration values. Gage #1, Figure 8, showed some offset after the test, but not as much as gage #3.

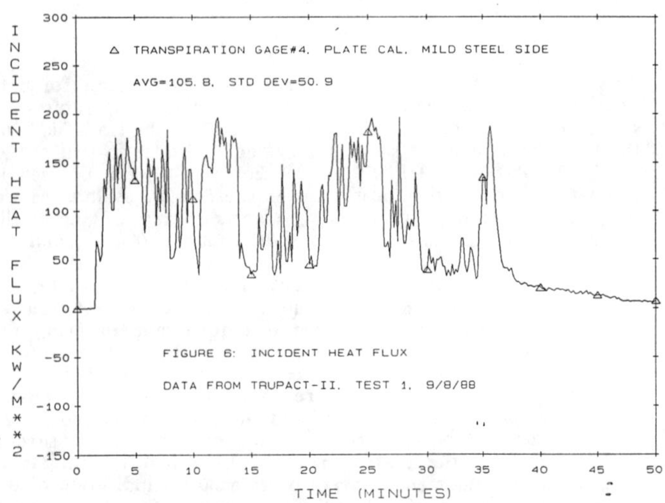

FIGURE 6: INCIDENT HEAT FLUX

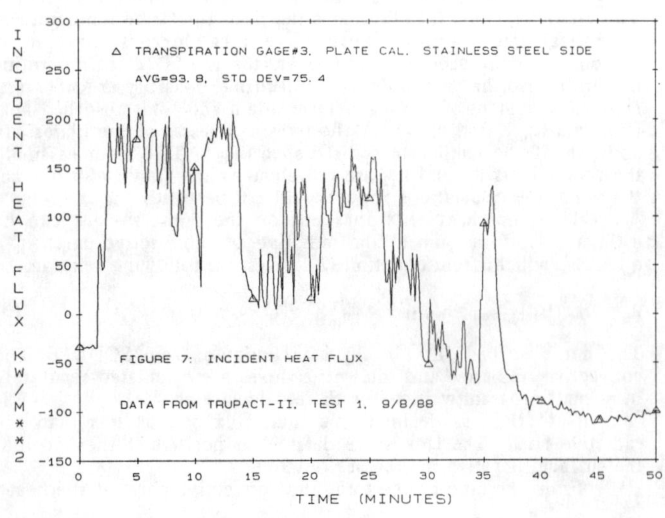

FIGURE 7: INCIDENT HEAT FLUX

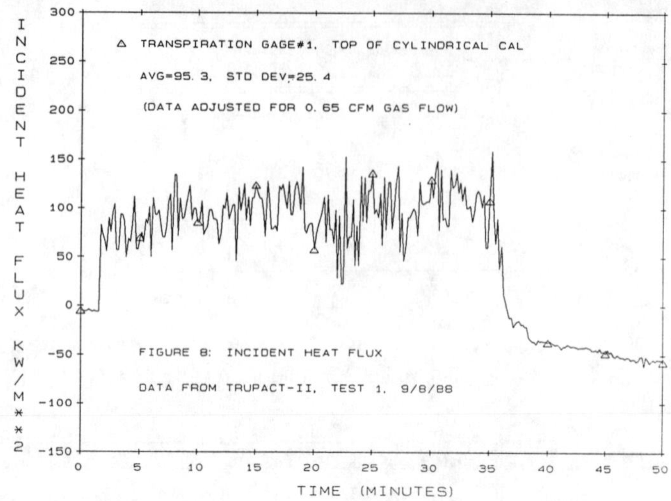

FIGURE 8: INCIDENT HEAT FLUX

The average values and standard deviations of all 3 gages from 3 to 33 minutes are also shown on the figures. They range from 93.8 to 105.8 kW/m². Remember that the output from gage #1 was modified to account for the low transpiration gas flow rate, and therefore the results shown in Figure 8 are subject to larger uncertainties than those in Figures 6 & 7.

Figure 9 shows the radiative heat flux near the plate calorimeter calculated from the flame temperatures, which were measured with two thermocouples located at the junction of the two steel plates. The outputs of the two thermocouples were almost identical. As noted on the figure, the assumed value for the flame emissivity was 0.82. This value was obtained by varying the average value of the flame emissivity until the average radiative heat flux calculated by the equation $\sigma \epsilon_f T_f^4$ was the same as that obtained from the transpiration radiometer shown in Figure 6. As can be seen, the average values shown in Figures 6 and 9 are almost identical. In addition, if one overlays these two plots, the calculated and measured radiative heat fluxes show good agreement.

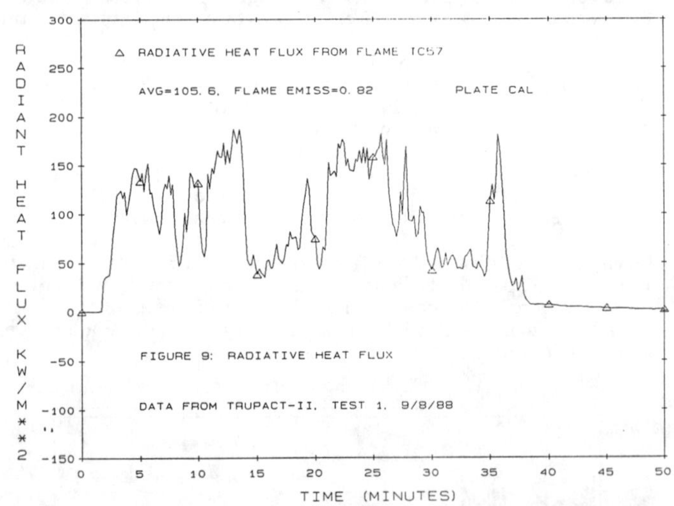

FIGURE 9: RADIATIVE HEAT FLUX

Figures 10 & 11 show the total heat fluxes (from equation (5)), the radiative fluxes (from the flame temperatures), and the convective fluxes (from equation (3)). As can be seen, the average convective flux on the mild steel (MS) side is about 14.8 kW/m², or about 12.3% of the total. For the stainless steel (SS) side, the average convective flux is about 22.0 kW/m² or 17.3% of the total.

Figures 12 & 13 show front face temperatures and the flame temperature from TC57. Note that the front face temperatures shown were generated from SODDIT. These were compared with the measured front face temperatures and the agreement was very good. Figure 14 shows a typical plot. A sheathed thermocouple mounted on the front surface of a steel plate and subjected to a radiant heat flux will read higher than the surface to which it is

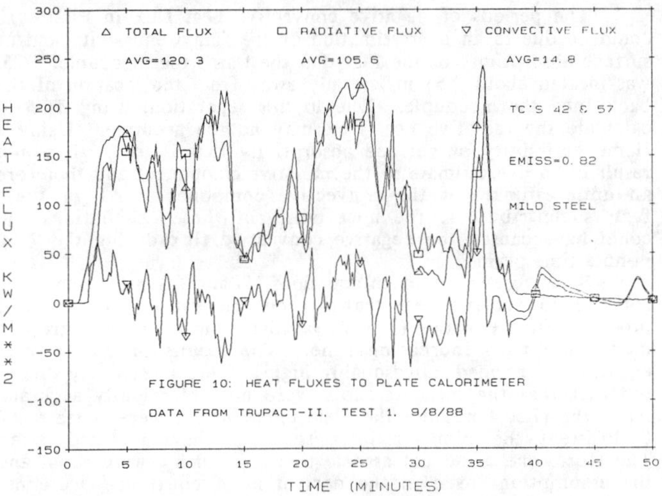

FIGURE 10: HEAT FLUXES TO PLATE CALORIMETER
DATA FROM TRUPACT-II, TEST 1, 9/8/88

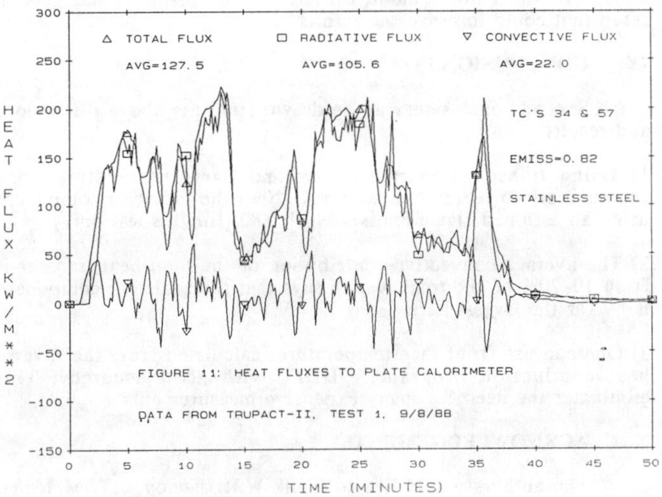

FIGURE 11: HEAT FLUXES TO PLATE CALORIMETER
DATA FROM TRUPACT-II, TEST 1, 9/8/88

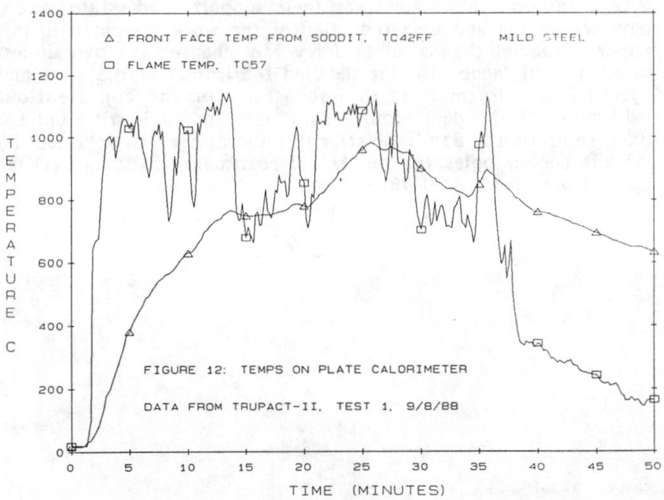

FIGURE 12: TEMPS ON PLATE CALORIMETER
DATA FROM TRUPACT-II, TEST 1, 9/8/88

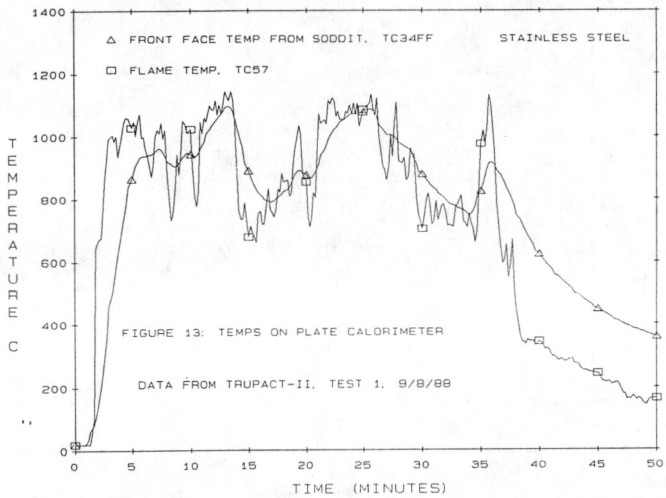

FIGURE 13: TEMPS ON PLATE CALORIMETER
DATA FROM TRUPACT-II, TEST 1, 9/8/88

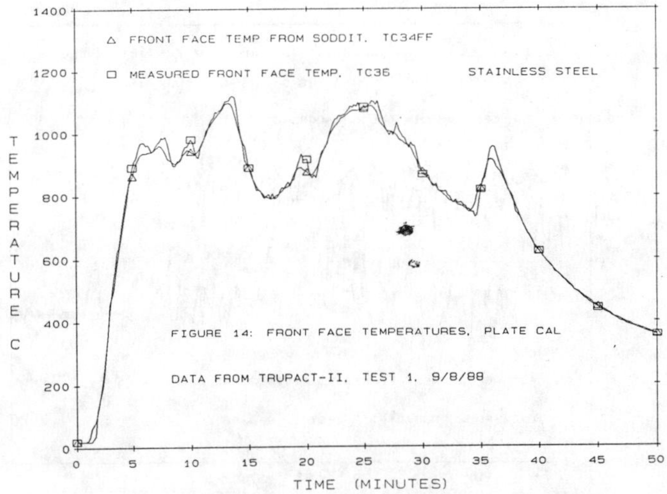

FIGURE 14: FRONT FACE TEMPERATURES, PLATE CAL
DATA FROM TRUPACT-II, TEST 1, 9/8/88

attached. This trend is evident from the data in Figure 14. As a result, it was decided that the calculated front face temperatures are preferred, and these were used in Figures 12 and 13.

It is instructive to compare the heat flux plots in Figures 10 and 11 with the temperature plots in Figures 12 and 13. Note in Figure 12 that from the very beginning of the test to a time of about 13-14 minutes, the flame temperature was higher than the surface temperature of the mild steel plate. As a result, the convective flux, defined as $h(T_f - T_s)$, would be positive. This scenario was discussed in Section IV and shown in Figure 1, View A. During 14-20 minutes, the flame and plate temperatures are close, and so the convective flux should be small (View B). From about 21-26 minutes the flames are again hotter than the plate and so the convective flux should be positive. From about 26-34 minutes the two temperatures are close and so the convective flux should be small. Lastly, the spike at about 35 minutes again brings the flame temperature above the plate temperature and so the convective flux should be positive. A close scrutiny of the plot in Figure 10 shows the general trends just outlined.

Similar trends are difficult to establish on the stainless steel side (Figures 11 and 13) because the plate temperature and flame temperatures were much closer throughout the test, except at the very beginning. The stainless steel plate is much thinner than the mild steel plate and so responds faster.

Figures 15 and 16 show heat flux data from the 1.4m diameter cylindrical calorimeter. Figure 15 shows the calculated radiative heat flux from TC5 assuming a flame emissivity of 0.82. The average of this plot is about 116.4 kW/m^2, which is 20% above the average value of 95.3 kW/m^2 from the radiometer data shown in Figure 8. This discrepancy is not surprising because the radiometer data were corrected for the low transpiration gas flow rate in a very approximate manner and the correction may not be accurate. However, the general trends in Figures 8 and 15 are similar.

The calculated total, radiative and convective fluxes on the top of the cylindrical calorimeter are shown in Figure 16. As can be seen, the average value of the convective part is 15.1 kW/m^2 or about 11.5% of the total. This contribution is on the low side of those reported in Russell and Canfield (1973) for a small cylinder (17-63%).

Figure 17 shows the flame temperature and front face temperature on the top of the 1.4m diameter calorimeter. As can be seen, the flame temperature is greater than the surface

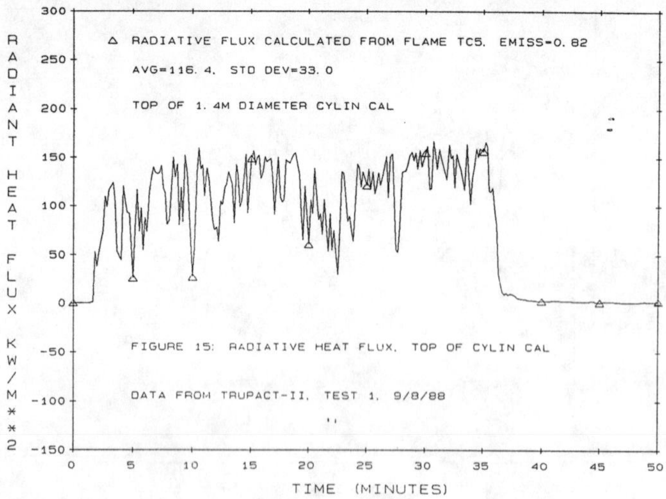

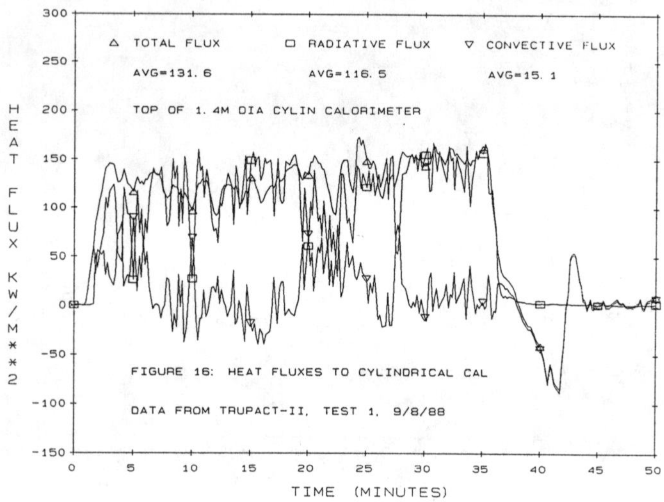

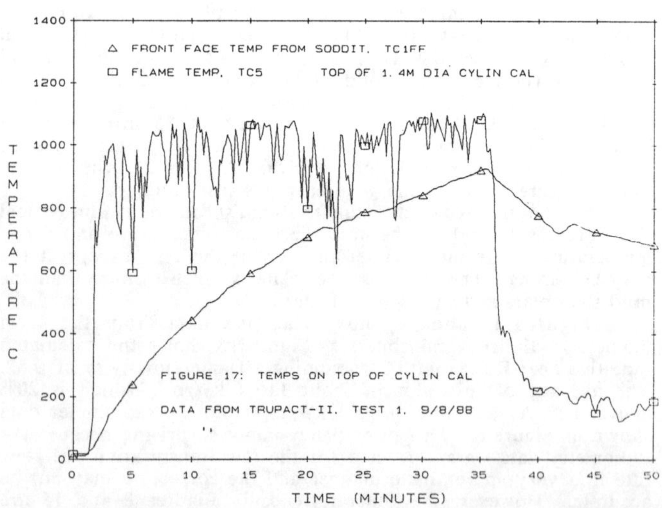

temperature through most of the test. As a result, the convective flux should be positive for most of the test, and this is generally the case (see Figure 16). However, during the 7-20 minute time, the convective component seems more negative than positive. The convective flux also seems to decrease towards the end of the test, as would be expected since the temperature difference decreases steadily. Lastly, note that the convective flux drops to almost -100 kW/m² after the fire is over. This would be expected because the hot cylinder would generate a natural convective heat loss to the cooler environment, as shown in View C, Figure 1.

The periods of negative convective heat flux in Figure 16 could be due to an overestimation of the flame emissivity and/or surface absorptivity or the fact that the flame thermocouple, TC5, was located about 0.55 m laterally away from the location of the back face thermocouple. Due to this separation, using TC5 to calculate the radiative heat flux may not be accurate. Using a flame emissivity or surface absorptivity that is too high would result in an overestimate of the radiative component and therefore an underestimate of the convective component. All of these factors contribute to the uncertainty of these calculations and could have caused the negative convective flux during the 7-20 minute time period.

Some results, based on equations (3) and (4), were generated for the convective heat transfer coefficient h. The results produced from equation (3) were oscillatory in nature and grew in magnitude with increasing time. The results produced from equation (4) seemed unreasonably high and are most likely due to the fact that the measurements were not sufficiently accurate. After the first 4 minutes the plate temperatures were at least 400 C different, the thinner stainless steel plate having heated faster. Therefore, one could not assume that h_{s1} and h_{s2} were equal and the assumptions used in the derivation of equation (4) are not valid. In the future, modifications will be made to the general setup that could improve the results.

IX. CONCLUSIONS

Several conclusions can be drawn from the above discussions and results:

1) Using transpiration radiometers and flame temperatures, one can compare to reasonable accuracy (20%) the radiative component using an assumed flame emissivity of 0.82 (for this test only).

2) The average convective contribution of the total heat transfer is from 10-20% of the total, which says that the radiative component is by far the largest part.

3) One can use front face temperatures calculated from the inverse heat conduction program SODDIT with good accuracy. This eliminates the need for many expensive measurements.

X. ACKNOWLEDGEMENTS

The authors would like to thank W.R. Jacoby, J.T. Meloche, B.G. Strait and L.A. Kent for their support in the fabrication, data acquisition and data reduction of the work presented in this paper. Special thanks go to Jerry Meloche for the overall test setup, to Bill Jacoby for the detailed fabrication of the plate and cylindrical calorimeters, to Bob Strait for the conscientious calibration of the data acquisition system and to Larry Kent for data reduction. Sandia National Laboratories is operated by AT&T Technologies for the U.S. Department of Energy (DOE) under Contract DEAC0476DP00789.

XI. REFERENCES

Ahmed, A.M., "Forced Convective Heat Transfer to Cooled Cylinders at Low Reynolds Numbers and with Large Temperature Differences", Canadian Armament Research and Development Establishment Technical Report 588/68, Sept., 1968.

Blackwell, B.F., Douglass, R.W., Wolf, H., "A Users Manual for the Sandia One-Dimensional Direct and Inverse Thermal (SODDIT) Code", SAND85-2478, May, 1987.

Drysdale, D., An Introduction to Fire Dynamics, J. Wiley & Sons, 1985, pg 56.

Gregory, J.J., Keltner, N.R., and Mata, R., Jr., "Thermal Measurements in Large Pool Fires", Heat and Mass Transfer in Fires, ASME-HTD-VOL. 73, pp 27-36, 1987. (To be published in the Journal of Heat Transfer.)

Hamann, J.E., Klein, D.E., Pope, R.B., and Yoshimura, H.R., "Modelling of Pool Fire Environments Using Experimental Results of a Two-Hour Test of a Railcar/Cask System", 6th International Symposium on the Packaging and Transportation of Radioactive Materials (PATRAM), 1980, pp 1081-1089.

Kent, L.A., and Schneider, M.E., "The Design and Application of Bi-Directional Velocity Probes for Measurements in Large Pool Fires", ISA Transactions, Vol. 26, No. 4, 1987.

Matthews, L., Longenbaugh, R., and Mata, R., "Transpiration Radiometer in Fire Characterization", InTech, ISA, August, 1986.

Neill, D.T., Welker, J.R., Sliepcevich, C.M., "Direct Contact Heat Transfer from Buoyant Diffusion Flames", J. Fire & Flammability, Vol.1, October, 1970, pp 289-301.

Russell, L.H., Canfield, J.A., "Experimental Measurement of Heat Transfer to a Cylinder Immersed in a Large Aviation-Fuel Fire", J. Heat Transfer, August, 1973, pp 397-404.

Schneider, M.E., and Kent, L.A., "Measurements of Gas Velocities and Temperatures in a Large Open Pool Fire", Fire Technology, Vol. 25, No. 1, Feb., 1989, pp 51-80.

Shepherd, J.E., "Analysis of Diffusion Flame Tests", Sandia National Laboratories Report SAND86-0419, 1987.

Thomas, P.H., Baldwin, R., and Heselden, A.J.M., "Buoyant Diffusion Flames: Some Measurements of Air Entrainment, Heat Transfer, and Flame Merging", Tenth Symposium (International) on Combustion, The Combustion Institute, 1965, pp 983-996.

1989 National Heat Transfer Conference
HTD-Vol. 106, Heat Transfer Phenomena in Radiation, Combustion and Fires

ENCLOSED LIQUID/GAS SYSTEM HEATED NONUNIFORMLY FROM ABOVE: VARIABLE PROPERTIES AND VAPORIZATION EFFECTS

D. N. Schiller
University of California at Irvine
Irvine, California

B. Abramzon
Ben-Gurion University of the Negev
Negev, Israel

W. A. Sirignano
University of California at Irvine
Irvine, California

ABSTRACT

A computational study has been made of transient heat transfer and fluid flow in a cylindrical enclosure containing a two-layer gas-and-liquid system heated nonuniformly from above. The effects of variable density and thermophysical properties, vaporization, and two-dimensional wall conductivity have been considered. Natural and thermocapillary convection, radiative heat transfer, thermal inertia and conduction of the walls bounding the enclosure, and the magnitude of the gravity field play important roles in the development of the temperature and velocity fields. The constant property assumption with the Boussinesq approximation significantly under-predicts gas temperatures immediately below the heater, over-predicts velocities in the majority of the gas phase, and under-predicts velocities in the liquid phase and immediately above the liquid surface. Variable density is important in the gas phase, whereas in the liquid phase variable viscosity is most important.

NOMENCLATURE

c_p	c'_p/c_{p*}, specific heat
$\mathcal{D}_{ij}$	$\mathcal{D}'_{ij}/\mathcal{D}_*$, binary diffusion coefficient
g	$g't_*/U_*$, gravitational acceleration (≥ 0)
h	$h'/c_{p*}T_*$, enthalpy
h'_o	formation enthalpy
h'_1	$h' - h'_o$, excess enthalpy (dimensional)
h_1	$h'_1/c_{p*}T_*$, excess enthalpy (nondimensional)
h'_{fg}	latent heat of vaporization
$\bar{h}_{\text{conv}}$	heat transfer coefficient [W/m^2·K]
H'	height of gas or liquid phase
k	k'/k_*, thermal conductivity
L_*	characteristic length
M	molecular weight
P_D	$P'_D/(\Delta P)_*$, dynamic pressure
P_G	P'_G/P_*, global pressure
P_{sat}	saturated vapor pressure
q_R	q'_R/q_*, radiative heat flux
q_*	$k_{*,g}T_*/L_*$, reference heat flux
r	r'/L_*, radial coordinate
R'	radius of enclosure
$\mathcal{R}'$	universal gas constant
t	t'/t_*, time
Δt	$\Delta t'/t_*$, time step
T	T'/T_*, temperature
U_*	L_*/t_*, reference velocity
u	u'/U_*, r-component of velocity
v	v'/U_*, y-component of velocity
$\vec{V}$	velocity vector
y	axial coordinate
Y	mass fraction of species
Bo	static Bond number
Ec	Eckert number
Eu	Euler number
Pe	Péclet number
Pr	Prandtl number
Re	Reynolds number
Sc	Schmidt number

Greek symbols

α	thermal diffusivity
β	volume expansion coefficient
χ	mole fraction
δ	δ'/L_*, wall thickness
μ	μ'/μ_*, dynamic viscosity
ν	kinematic viscosity
ψ	$\psi'/(\rho UL)_*$, stream function
ρ	ρ'/ρ_*, density
σ	surface-tension coefficient

Subscripts

A	air
F	fuel vapor

g	gas phase
h	heater
i	species index (e.g., air, fuel vapor)
l	liquid phase
lg	gas/liquid interface
o	initial condition
w	wall
$*$	reference quantity

Superscripts

$'$	dimensional quantity

Unless otherwise specified, a variable without a superscript or subscript represents a nondimensional quantity. All reference quantities are evaluated at 300 K, which is taken as the initial temperature of the fluid, T'_o.

INTRODUCTION

Fire safety problems of ignition and flame spread above a pool of flammable liquid arise in many accident situations where a flammable liquid spills in the vicinity of a hot ignition source. Another prototypical hazard situation might be the rupture of a fuel tank when hot engine parts or exhaust gases may appear in close proximity of liquid fuel. Heat and mass transport which control ignition delay and flame spread rates are complicated by multiple energy and mass transport processes, phase change, and chemical reaction.

The literature of liquid pool burning theory has for the most part neglected the gas and liquid phase coupling, and concerned itself solely with liquid motion (Torrance, 1971; Torrance and Mahajan, 1975). However, the most interesting scientific situation (and one of very practical importance) involves not-so-volatile fuels whereby the coupling between the liquid and gas phases is of critical importance. That is, heat from the gas phase is required to vaporize some portion of the liquid fuel in order to provide a combustible mixture in the gas phase. Convective heat and mass transport between the liquid and gas phases is critical and often dominant during the preignition period. Other modes of heat transfer, such as thermal radiation from the ignition source, may also play a very important role in the liquid heating. Temperature gradients in the gas and liquid phases, as well as concentration gradients in the gas mixture, create natural convection currents. Temperature gradients on the liquid surface initiate "surface-tension-driven convection" which can be a critical factor in the ignition delay and flame spread phenomena. On account of nonuniform heating from above due to the ignition source or the flame, liquid motion will be driven by both surface-tension gradients and by buoyancy. These driving forces generally support in concert surface fluid motion away from the heat source. Therefore this motion tends to delay ignition as heat from the source is convected rather than concentrated. On the other hand, the motion supports flame spread as the convection assists the preheating process ahead of the flame. The gravity effect on ignition and flame spread is, in general, opposite to that of surface tension. As gravity decreases, the fluid mixing due to natural convection in both phases is retarded, and the liquid at the surface is heated and vaporized more rapidly. Therefore the reduction in gravity should lead to shorter ignition delay and slower flame spread. However, the final result depends on the relative strengths of buoyancy versus surface tension.

Recent theoretical studies (Aggarwal et al., 1986; Abramzon et al., 1987) indicate that surface tension dominates buoyancy only below a Grashof number of 10^4. These studies were based on a computational model which included the effects of black-body radiation, surface tension, gravitational level, and one-dimensional wall conductivity in a cylindrical enclosure containing a two-layer gas-and-liquid system heated nonuniformly from above. The present paper adds the effects of variable density and thermophysical properties, vaporization, and two-dimensional wall conductivity to the computational study. Variable density is important in the gas phase because during preignition heating, high temperature gradients are expected near the ignition source. In such a situation, the use of the constant density approach along with the Boussinesq approximation may produce very serious quantitative errors in the analysis of the phenomena. The incorporation of variable thermophysical properties in the liquid phase is also very important since the viscosity and volume expansion coefficient of many liquids vary considerably with temperature. Vaporization is included in the computational model in order to study the transport of fuel vapor in the gas phase.

The geometry of the problem is shown in Figure 1. Here, the cylindrical enclosure of radius R' and height $H' = H'_g + H'_l$ contains a two-layer gas/liquid system. The initial temperature of the system is uniform at $T' = T'_o$. At time zero ($t' = 0$), the temperature of a small circular spot (heater) of radius R'_h at the center of the top cover increases suddenly to $T' = T'_h \gg T'_o$ and then remains constant. The total pressure in the gas phase is initially 1 atm. The initial condition for the mass fraction of fuel vapor in the gas phase is $Y_F(r, y, 0) = 0$.

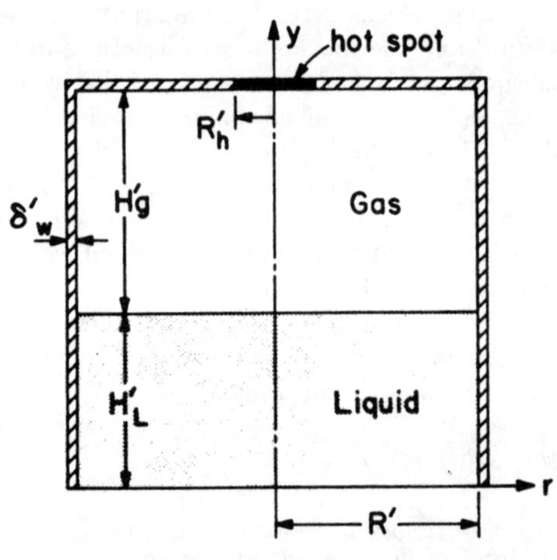

Fig. 1 Geometry of the problem

STATEMENT OF THE PROBLEM

The mathematical formulation of the axisymmetric problem is based on the following assumptions:

1. The gas/liquid interface remains flat and horizontal, since the static Bond number, $Bo = (\rho'_l - \rho'_g)g'R'^2/\sigma'$, in this study is $Bo \simeq 770$.

2. All of the solid surfaces as well as the gas/liquid interface are assumed opaque and black, and gas radiation is neglected.

3. The "hot spot" has no thermal contact with the remaining part of the top cover (i.e., the sides of the heater are insulated). The bottom of the liquid phase is maintained at the constant temperature $T' = T'_o$.

4. Chemical reactions are neglected.

5. The energy equation is simplified by the following approximations: (*i*) low Mach number flow, (*ii*) work by body forces is neglected, (*iii*) no volumetric heat sources, and (*iv*) heat transferred by interdiffusion due to the difference in the specific heats of components is neglected.

These assumptions differ from those made in the previous study by Abramzon et al. (1987) in that:

1. In the previous study, constant properties with the Boussinesq approximation were used. In the present model all properties may vary.

2. In the previous study, the aluminum walls bounding the enclosure were taken to be so thin that the temperature drop across the thickness was negligible for relatively short-time simulations (one-dimensional wall conductivity was assumed). The top and side walls were insulated. In order to model relatively long experiments which use acrylic containers, the explicit one-dimensional wall conductivity model (Abramzon et al., 1987) has been replaced in the present study by an explicit two-dimensional calculation.

3. Liquid vaporization and multicomponent diffusion in the gas phase are now included in the numerical model.

The nondimensional conservation equations of mass, species, momentum, and energy follow:

Continuity equation

$$\frac{\partial \rho}{\partial t} + \frac{1}{r}\frac{\partial}{\partial r}(r\rho u) + \frac{\partial}{\partial y}(\rho v) = 0 \tag{1}$$

Species equation

$$\frac{\partial}{\partial t}(\rho Y_i) + \frac{1}{r}\frac{\partial}{\partial r}\left[r\left(\rho u Y_i - \frac{\rho \mathcal{D}_{ij}}{Sc_* Re_*}\frac{\partial Y_i}{\partial r}\right)\right] + \frac{\partial}{\partial y}\left(\rho v Y_i - \frac{\rho \mathcal{D}_{ij}}{Sc_* Re_*}\frac{\partial Y_i}{\partial y}\right) = 0 \tag{2}$$

Momentum equation in radial direction

$$\begin{aligned}\frac{\partial(\rho u)}{\partial t} &+ \frac{1}{r}\frac{\partial}{\partial r}\left[r\left(\rho u^2 - \frac{\mu}{Re_*}\frac{\partial u}{\partial r}\right)\right] \\ &+ \frac{\partial}{\partial y}\left(\rho u v - \frac{\mu}{Re_*}\frac{\partial u}{\partial y}\right) = -Eu_*\left(\frac{\partial P_D}{\partial r}\right) \\ &+ \frac{1}{Re_*}\left\{\frac{1}{r}\frac{\partial}{\partial r}\left[r\left(\tau_{rr} - \mu\frac{\partial u}{\partial r}\right)\right]\right. \\ &+ \left.\frac{\partial}{\partial y}\left(\tau_{yr} - \mu\frac{\partial u}{\partial y}\right) - \frac{\tau_{\theta\theta}}{r}\right\}\end{aligned} \tag{3}$$

where

$$\tau_{rr} = \mu\left(2\frac{\partial u}{\partial r} - \frac{2}{3}\vec{\nabla}\cdot\vec{V}\right) \tag{4}$$

$$\tau_{yr} = \mu\left(\frac{\partial v}{\partial r} + \frac{\partial u}{\partial y}\right) \tag{5}$$

$$\tau_{\theta\theta} = \mu\left(2\frac{u}{r} - \frac{2}{3}\vec{\nabla}\cdot\vec{V}\right) \tag{6}$$

Momentum equation in axial direction

$$\begin{aligned}\frac{\partial(\rho v)}{\partial t} &+ \frac{1}{r}\frac{\partial}{\partial r}\left[r\left(\rho u v - \frac{\mu}{Re_*}\frac{\partial v}{\partial r}\right)\right] \\ &+ \frac{\partial}{\partial y}\left(\rho v^2 - \frac{\mu}{Re_*}\frac{\partial v}{\partial y}\right) = -Eu_*\left(\frac{\partial P_D}{\partial y}\right) \\ &+ \frac{1}{Re_*}\left\{\frac{1}{r}\frac{\partial}{\partial r}\left[r\left(\tau_{ry} - \mu\frac{\partial v}{\partial r}\right)\right]\right. \\ &+ \left.\frac{\partial}{\partial y}\left(\tau_{yy} - \mu\frac{\partial v}{\partial y}\right)\right\} + g(1-\rho)\end{aligned} \tag{7}$$

where $\tau_{ry} = \tau_{yr}$ and

$$\tau_{yy} = \mu\left(2\frac{\partial v}{\partial y} - \frac{2}{3}\vec{\nabla}\cdot\vec{V}\right) \tag{8}$$

The Boussinesq approximation may be obtained by replacing the last term in equation (7) by $g(T - 1.0)$ for the gas phase and $g\beta_* T_*(T - 1.0)$ for the liquid phase (where T is evaluated at the center of the v-momentum finite difference control volume). For a constant property simulation, the physical properties of the gas are calculated at the temperature $T' = 0.5(T'_o + T'_h)$, while the liquid properties are taken at $T' = T'_o$.

Energy equation

$$\frac{\partial(\rho h_1)}{\partial t} + \frac{1}{r}\frac{\partial}{\partial r}\left[r\left(\rho u h_1 - \frac{k/\bar{c}_p}{Pr_* Re_*}\frac{\partial h_1}{\partial r}\right)\right] + \frac{\partial}{\partial y}\left(\rho v h_1 - \frac{k/\bar{c}_p}{Pr_* Re_*}\frac{\partial h_1}{\partial y}\right) = Eu_* Ec_* \frac{P_*}{(\Delta P)_*}\frac{\partial P_G}{\partial t} \tag{9}$$

The above equations are derived in detail by Schiller (1988). Physical properties ($\rho, \mu, k, \bar{c}_p$) for vapor-air mixtures are calculated using the standard additive rules for an ideal gas.

The dimensionless Reynolds, Schmidt, Prandtl, Euler, and Eckert numbers are defined for each phase (with the exception of the Schmidt number) using reference quantities as follows:

$$Re_* = U_* L_*/\nu_* \qquad Sc_* = \nu_*/\mathcal{D}_* \qquad Pr_* = \nu_*/\alpha_*$$

$$Eu_* = \frac{(\Delta P)_*}{\rho_* U_*^2} \qquad Ec_* = \frac{U_*^2}{c_{p_*} T_*}$$

Note that the above numbers are constant, since they are defined with reference quantities evaluated at 300 K, although they differ between the gas and liquid phase. Since the properties are highly variable (especially density), there is some arbitrariness in the selection of the reference parameters. Based on the results of the previous study (Abramzon et al., 1987), these values have been chosen as follows: $L_* = R' = 5$ cm; $t_* = 5$ s; thus, $U_* = 1$ cm/s.

The hydrodynamic boundary conditions are as follows:

1) Zero velocity on all solid walls,

$$u = v = 0 \tag{10}$$

2) Symmetry of the velocity field at the axis $r = 0$,

$$u = 0, \qquad \frac{\partial v}{\partial r} = 0 \tag{11}$$

3) Balance of the stresses and mass flux and continuity of the radial velocity on the gas/liquid interface,

$$\left(\frac{\partial u}{\partial y}\right)_l = \frac{\mu_{*,g}}{\mu_l \mu_{*,l}} \left[\mu_g \left(\frac{\partial v}{\partial r} + \frac{\partial u}{\partial y}\right)_g + \left(\frac{T_*}{\mu_{*,g} U_*}\right)\left(\frac{\partial \sigma'}{\partial T'}\right)_* \left(\frac{dT}{dr}\right)_{lg}\right] \tag{12}$$

$$u_g = u_l \tag{13}$$

$$v_g = -\frac{1}{Sc_* Re_*}\left[\left(\frac{\mathcal{D}_{ij}}{1 - Y_{F,lg}}\right)\left(\frac{\partial Y_F}{\partial y}\right)_{lg}\right] \tag{14}$$

$$v_l = \frac{\rho_{*,g}\rho_g}{\rho_{*,l}\rho_l} v_g \tag{15}$$

Here v_g is the mass-average vaporization velocity based on the Stefan flow assumption (Edwards et al., 1979). The mass fraction of fuel vapor at the liquid surface, $Y_{F,lg}$, is determined by

$$Y_{F,lg} = \frac{\chi_{F,lg} M_F}{\chi_{F,lg} M_F + (1 - \chi_{F,lg}) M_A} \tag{16}$$

where

$$\chi_{F,lg} \simeq \frac{P'_{\text{sat}}(T'_{lg})}{P'_G} \tag{17}$$

When vaporization is not included in the analysis, the boundary condition simply becomes $v_g = v_l = 0$. Note also that a zero flux boundary condition for Y_F is used at the walls and along the centerline in the gas phase.

The thermal boundary conditions are as follows:

1) Constant temperature of the hot spot ($r < R_h$, $y = H'_g/L_*$),

$$T = T_h \tag{18}$$

2) Constant temperature (300 K) at the bottom of the liquid phase ($y_l = 0$),

$$T = 1 \tag{19}$$

3) Heat balance and continuity of the temperature on the gas/liquid interface,

$$\left(\frac{\partial T}{\partial y}\right)_l = \frac{k_{*,g}}{k_l k_{*,l}}\left[\left(k\frac{\partial T}{\partial y}\right)_g + q_R - \frac{\rho'_g v'_g h'_{fg}}{q_*}\right] \tag{20}$$

$$T_g = T_l \tag{21}$$

4) Transient heat conduction equation governing the two-dimensional temperature distribution in the side and top walls using constant properties for the walls,

$$\frac{\partial T_w}{\partial t} = \frac{\alpha_w}{Pe_{*,w}}\left(\frac{\partial^2 T_w}{\partial r^2} + \frac{1}{r}\frac{\partial T_w}{\partial r} + \frac{\partial^2 T_w}{\partial y^2}\right) \tag{22}$$

Here $Pe_{*,w} = Pr_{*,w} Re_{*,w}$.

Equation (22) is solved explicitly using the total heat fluxes from the gas and liquid phases from the beginning of the time step for the inner wall boundary condition and Newton's law of cooling for the outside wall boundary condition. The external heat transfer coefficient, $\overline{h}'_{\text{conv}}$, is taken as (Lienhard, 1981)

$$\overline{h}'_{\text{conv}} = \frac{k'}{L'}\left[0.678 Ra^{1/4}\left(\frac{Pr}{0.952 + Pr}\right)^{1/4}\right] \tag{23}$$

where

$$Ra = \frac{g' L'^3 (|T'_{w,b} - T'_e|/T'_e)}{\nu' \alpha'} \tag{24}$$

is the Rayleigh number of the air and T'_e is the outside environment temperature (assumed 300 K). In equations (23) and (24), properties are of air evaluated at $T' = 0.5(T'_e + T'_{w,b})$. Also, $T'_{w,b}$ is the boundary temperature of the wall and L' is the length of the wall. Although equation (23) is a correlation for vertical walls, it is also used in the present study for the external heat transfer for the top wall of the cylinder. Little error is introduced here as $\overline{h}'_{\text{conv}}$ is approximately 5 W/(m^2·K), which leads to external heat fluxes of 10–50 W/m^2 in this study; this is nearly a full order of magnitude less than the total heat fluxes on the inner walls (100–300 W/m^2).

METHOD OF SOLUTION

Pressure Splitting

The total pressure (P') is divided into three parts in the computational model: global pressure (P'_G), dynamic pressure (P'_D), and the hydrostatic pressure ($\rho_* g'_y y' = -\rho_* g' y'$). Hence

$$P' = P'_G(t) + P'_D(r, y) - \rho_* g' y' \tag{25}$$

Following Paolucci (1982), the global pressure (used only in the gas phase) is calculated as

$$P'_G = \frac{1}{V'}\iiint\left(\rho' T' \mathcal{R}' \sum_i Y_i/M'_i\right) dV' \tag{26}$$

where V' is the volume of the gas phase. Since $P'_G \sim 10^5$ N/m^2 while the typical maximum values of the other two terms in the total pressure equation are of order 1 or 10^{-2} N/m^2,

$$\begin{aligned} P'_D(r,y)\text{max} &\sim \rho_* |\vec{V}|^2_{\text{max}} \sim 10^{-2}\ \text{N/m}^2 \\ (\rho_* g' y')\text{max} &\sim 1\ \text{N/m}^2 \end{aligned}$$

we can make the approximation $P' \simeq P'_G$. Thus the global pressure is used to calculate density in the gas phase using the ideal gas law at each cell location. The global pressure increases during the heating and vaporization processes.

Numerical Method

In order to include variable properties in the problem, the numerical method was changed from a modified SMAC method (Abramzon et al., 1987) to the SIMPLE algorithm (Patankar, 1980) with the SIMPLEC modification (Van Doormaal and Raithby, 1984). This method uses primitive variables (u, v, p, h), a staggered mesh, and the hybrid-differencing scheme. The continuity equation is satisfied by solving the pressure correction equation in the SIMPLE algorithm.

The gas and liquid phase solutions are time-split in that the solution at time t represents the gas phase at time t and the liquid phase at time $t + \frac{1}{2}\Delta t$. The sequential calculation procedure for each time step is as follows:

- Determine wall temperatures using explicit scheme:
  1. Calculate total heat fluxes (radiative + conductive) to the walls in the gas phase based on temperatures at time t.
  2. Calculate conductive heat fluxes to the walls in the liquid phase using temperatures at time $t + \frac{1}{2}\Delta t$.
  3. Calculate wall temperatures based on above heat flux data. These wall temperatures are then used as boundary conditions for the gas and liquid phase solutions below.
- Determine gas phase solution at time $t + \Delta t$:
  1. Calculate gas phase properties at beginning of time step. Only values of c_p are updated during the internal iteration cycle. This helps stabilize the solution. Values of c_p are updated in order to update values of enthalpy.
  2. Perform internal iteration loop for gas phase (SIMPLE algorithm). Boundary values of wall temperatures (from explicit solution) and liquid temperatures and velocities (from time $t + \frac{1}{2}\Delta t$) are used [see equations (13), (14), and (21)].
- Determine liquid phase solution at time $t + \frac{3}{2}\Delta t$:
  1. Calculate liquid phase properties at beginning of time step. Only values of c_p are updated during the internal iteration cycle.
  2. Perform internal iteration loop for liquid phase. Boundary values of T_w, v_g, and *gradients* of T_g and u_g (from time $t + \Delta t$) are used [see equations (12), (15), and (20)].

The species, momentum, and energy (excess enthalpy) equations are said to be converged when their normalized residuals are reduced to less than 10^{-5}. Typically, the residuals of these equations are reduced by 5 orders of magnitude (from initial values of 10^{-3}–10^{0} to 10^{-8}–10^{-5}) after only 2 or 3 internal iterations of the algorithm. At the beginning of the time step, the initial pressure correction equation residual

$$\vec{\nabla}\cdot(\rho\vec{V}) - \frac{\partial\rho}{\partial t}$$

is usually of the order 10^{-2} or 10^{-3}, and the pressure correction residual is then reduced by one order of magnitude after 3–5 internal iterations. Frequently the pressure correction residual decreases by less than 0.1% from one iteration to the next after a few internal iterations. This is because the fluid properties are fixed at the beginning of each time step (they are not subsequently altered during the iteration cycle for the time step). The pressure correction equation is said to be converged when this residual "freeze" occurs, or if the value of the pressure correction equation is reduced by one order of magnitude to below approximately 10^{-3}. Experience with the computer code has indicated that this convergence criterion is acceptable.

Unless otherwise specified, the calculations were performed with a 32 x 32 uniform mesh in each phase. Since 2 boundary nodes are used in each direction, there are 30 x 30 internal cells in this mesh. Four internal cells were used for the thickness of each wall, with the cells aligned longitudinally with the cells from the adjacent fluid domain. The time step was typically 0.01 s and required approximately 50–80 s of CPU time per time step on an Apollo Domain Series 3000 computer for the two-phase calculations. Results using the 32 x 32 mesh differed by approximately 10% with those using a 22 x 22 mesh, whereas the 32 x 32 mesh results differed by less than 1% compared to those obtained from a 42 x 42 mesh.

RESULTS AND DISCUSSION

The flow field in the enclosure will be presented in terms of the streamlines and the fluid velocities. In polar coordinates, the nondimensional stream function $\psi(r, y)$ is defined by

$$\psi = -\int_0^r \rho r v dr, \qquad \psi = r\int_0^y \rho u dy \tag{27}$$

Verification of Computer Code

The computer code was tested in three ways: (*i*) a single-phase, constant property, planar geometry comparison with the benchmark solution of De Vahl Davis and Jones (1981), (*ii*) comparisons with experimental results for the single-phase (air only) problem, and (*iii*) constant-property comparisons with numerical results obtained with the modified SMAC method (Abramzon et al., 1987).

The boundary conditions for the De Vahl Davis comparison are insulated top and bottom with a given temperature difference between the two isothermal side walls. For a Rayleigh number of 10^6 with a 1°C temperature difference between the side walls and a 42 x 42 mesh, we obtain values of $|\psi|_{max}$ which are 3.1% higher than that obtained by De Vahl Davis. The locations of the local maxima agree within 1% to those cited by De Vahl Davis. Results using a 32 x 32 mesh are nearly identical to the above.

Qualitatively, single-phase computational results have agreed very well with experiments (Ross et al., 1989). Two-phase simulations without the effects of surface tension or vaporization also agree qualitatively with experimental results (Ross et al., 1989). No quantitative experimental data have been collected to date.

Constant property comparisons were made with the previous results (Abramzon et al., 1987) for a variety of mesh sizes. The test cases used were for a water/air two-layer system and an air-only system with a 2 cm radius, 1000 K hot spot in a 10 cm radius cylindrical container with thin aluminum walls. The aspect ratio of the gas and liquid phases was 1. Using the same grid, time step, properties, reference velocity, and one-dimensional wall calculation scheme for the comparisons, the results are qualitatively similar for each of the two tests. However, quantitatively there were significant differences in the results between the two numerical methods. In particular, values of gas and liquid phase velocities predicted by the SIMPLE algorithm were 50–200% greater than those found by the modified SMAC method. We believe that the present code which uses the SIMPLE algorithm is more accurate than the modified SMAC code mainly because the latter uses the upwind approximation for the convective terms even at small cell Péclet numbers. This produces considerably large numerical diffusion which tends to reduce the field gradients. In contrast, the present SIMPLE method uses the hybrid scheme, in which the convective terms at cell $|Pe| \leq 2$ are approximated by the central-difference scheme. Results from the previous study (Abramzon et al., 1987) showed that the cell Péclet numbers were less than 2 over most of the container volume. Furthermore, the modified SMAC code yielded 15–20% errors in the maximum vortex intensity compared to the benchmark solution of De Vahl Davis and Jones (1981).

Preliminary Remarks

As was done by Abramzon et al. (1987), small time analytical solutions were used for the temperature of the inner walls of the gas phase and liquid surface, and for the liquid surface radial velocity. This was necessary because the values for these boundary conditions, when derived from equations (12) and (20) (with an analogous boundary condition equation for the inner walls), are not an explicit function of time. Therefore when radiation is included in the analysis, initially unrealistic values will be calculated unless the mesh size is very fine near these boundaries. For example, for the initial temperature gradient immediately below the gas/liquid interface, equation (20) gives (neglecting the inital conductive heat flux and vaporization in the gas phase)

$$\left(\frac{\partial T}{\partial y}\right)_l \approx \left(\frac{k_{*,g}}{k_l k_{*,l}}\right)\frac{q'_R}{q_*} \tag{28}$$

Using the above equation for a typical radiative heat flux of 860 W/m^2 (obtained from the results to follow using a 2 cm diameter, 800 K heater) close to the centerline of the liquid surface, we would predict a liquid surface temperature of 305.2 K for a 32 x 32 grid (note the boundary nodes lie directly on the liquid surface). This would be the surface temperature after the first time step, *regardless* of the magnitude of the time step! Using the analytical solution (Carslaw and Jaeger, 1959)

$$T'_{lg}(t) \simeq T'_o + 2\frac{q'_R}{k_{*,l}}\sqrt{\frac{\alpha_{*,l}t'}{\pi}} \tag{29}$$

we would obtain this temperature only after 6.34 seconds. By comparing equations (28) and (29), the time at which the computational liquid surface temperature equals the analytical prediction (denoted by t'_α) is given by

$$t'_\alpha = \frac{\pi(\Delta y')^2}{4\alpha_{*,l}} \tag{30}$$

where $\Delta y'$ is the distance from the liquid surface to the first internal temperature node below the liquid surface. A similar variable may be defined for the analytical cut-off time for the inner wall temperatures (this is considerably smaller since the wall grid is typically finer).

Similarly, the analytical solution for the liquid surface velocity is given by (Carslaw and Jaeger, 1959)

$$u'_{lg} \simeq \frac{\sigma'_T}{2\rho_{*,l}k_{*,l}\sqrt{Pr_{*,l}}}\left(\frac{dq'_R}{dr'}\right)t' \tag{31}$$

where $\sigma'_T = (\partial\sigma'/\partial T')_* = -9.2\times10^{-5}$ N/m·K for liquid n-decane, and the cut-off time for the analytical solution, t'_ν, is

$$t'_\nu = \frac{2(\Delta y')^2\sqrt{Pr_{*,l}}}{\nu_{*,l}} \tag{32}$$

For liquid n-decane with a uniform 32 x 32 grid, $t'_\alpha = 6.34$ s and $t'_\nu = 4.56$ s. The analytical solutions given by equations (29) and (31) are used only for $t' < t'_\alpha$ and $t' < t'_\nu$, respectively.

Variable Property Effects

The effects of variable properties will be demonstrated for a 10 cm height and diameter plexiglass container ($\delta'_w = 4$ mm) with $R'_h = 1$ cm, $T'_h = 800$ K, and $H'_g = H'_l = 5$ cm. This configuration was selected in order to compare results with planned experimentation (not included in this paper). Vaporization is not included in this comparison.

Three simulations were used for this comparison: (*i*) constant properties with the Boussinesq approximation, (*ii*) variable density (no Boussinesq approximation) with the other properties (μ, c_p, and k) held constant, and (*iii*) all properties variable. Figure 2 shows a comparison of the ψ-contours in the gas phase for these cases at $t' = 500$ s. Since the constant properties are evaluated at $T' = 0.5(T'_o + T'_h)$ in the gas phase, the value of ρ is 0.545 (300/550) throughout the gas phase for the constant property simulation. As shown by the isotherm plot in Figure 3 (only the variable property isotherms are shown as those for the other two simulations are very similar except for close to the heater), approximately 85% of the gas phase is between 300 K and 315 K (the volume-averaged temperature is 315 K). Hence $\rho \approx 1.0$ throughout most of the gas phase. Therefore if the predicted velocities were equal for all cases, then one would expect the values of ψ predicted by a simulation with variable density to be approximately 80% greater than those predicted by the constant property run in the region where $T' \leq 315$ K. Thus Figure 2 shows that the velocities predicted by the Boussinesq approximation are greater than those predicted by the variable density runs for the upper vortex, while they are slightly lower for the lower vortex near the liquid surface. Figure 4, which shows three u-velocity profiles at various heights in the gas phase, clearly indicates this result.

By comparing radial velocities with the variable property simulation, Figure 4 also shows that the effects of variable density are most important near the heater, whereas the effects of the other properties are most important for the liquid surface. Variable properties are also important for the prediction of the temperature profiles just below the heater in the gas phase. As shown by Figure 5, both the constant property and variable-density-only simulations under-predict the temperatures in this region.

Plots of radial velocity in the liquid phase show that the variable-density-only and constant property simulations predict nearly identical velocities throughout most of the liquid phase. However, the subsurface velocities are approximately 12% lower for these simulations compared to the variable property run for $t' = 500$ s. There is a slightly larger difference in the prediction of the maximum liquid surface velocity (1.98 cm/s for the variable property simulation versus 1.63 cm/s for the other two runs). Figure 6 shows ψ-contours and isotherm contours in the liquid phase for the variable property simulation. The corresponding plots for the variable-density-only and constant property simulations are nearly identical to each other and are also qualitatively very similar to those in Figure 6. The only significant difference is the maximum value of $|\psi|$ in the upper vortex (12% lower for these two runs compared to the variable property case). The above results are consistent with what one would expect by considering the variation of properties with temperature (see the appendix). Variable density is most important in the gas phase

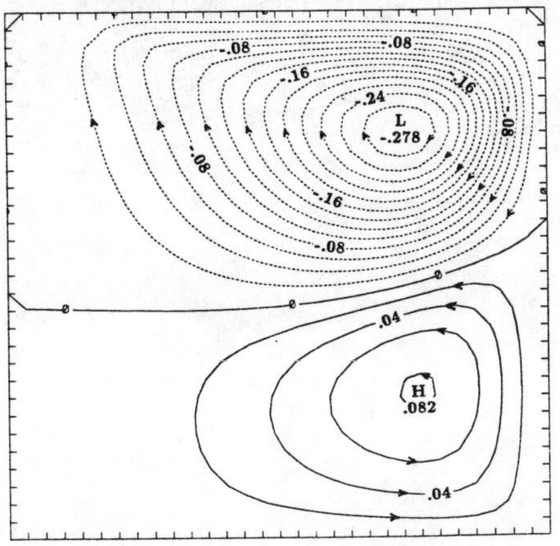

(a) All properties constant

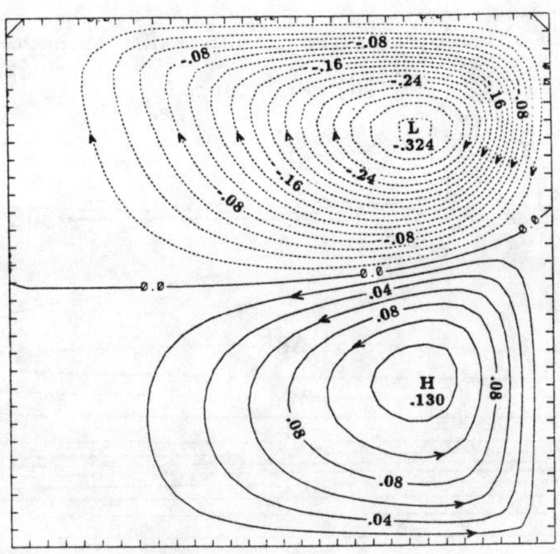

(b) Variable density only

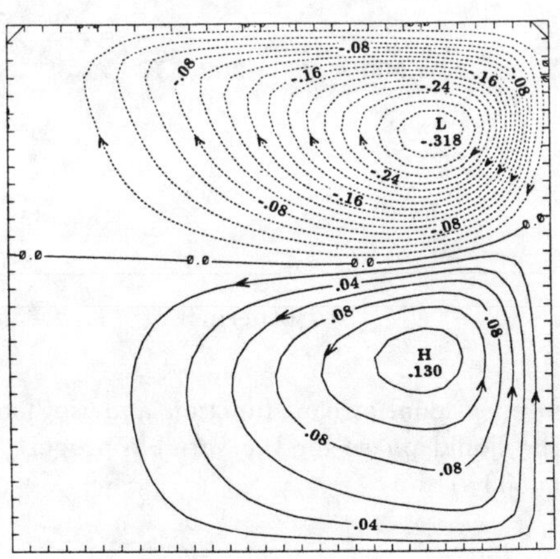

(c) All properties variable

Fig. 2 Nondimensional stream function contours in the gas phase at $t' = 500$ s. The left side of the plots corresponds to the axis of symmetry of the cylinder

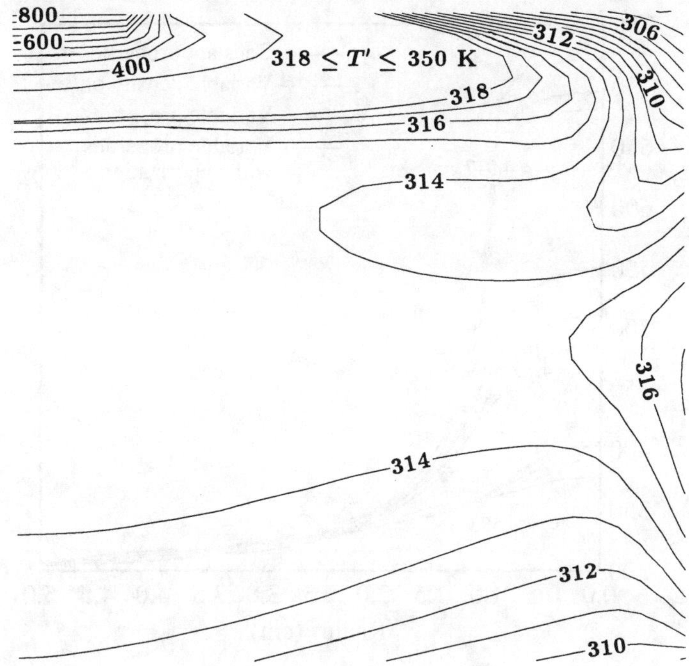

Fig. 3 Isotherm contours in the gas phase for the variable property simulation ($t' = 500$ s)

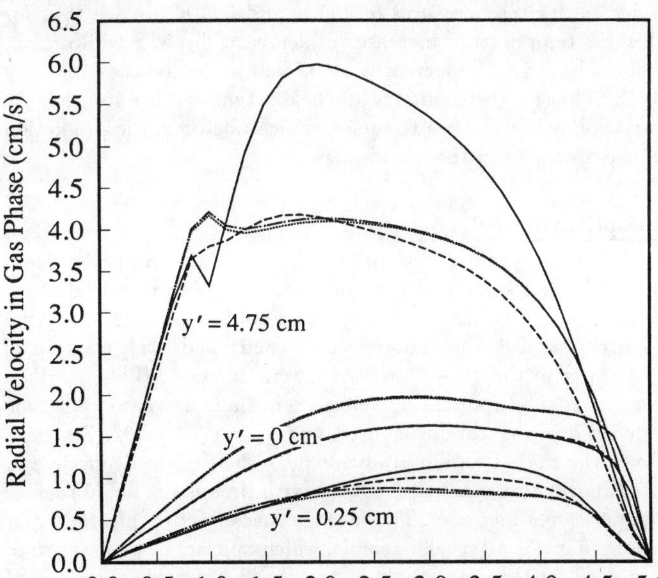

Fig. 4 Radial velocity profiles at various heights in the gas phase ($t' = 500$ s); —— constant properties, ---- variable density only, —— variable properties, —— variable properties with vaporization

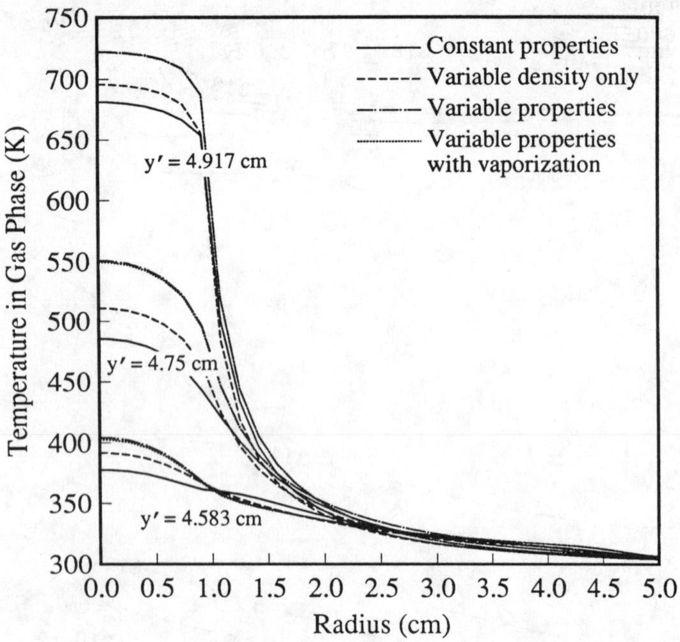

Fig. 5 Temperature profiles just below the heater in the gas phase ($t' = 500$ s)

near the heater because here the temperatures and temperature gradients are quite large. In reduced gravity, conduction becomes more dominant and these gradients are significantly reduced (Aggarwal et al., 1986; Abramzon et al., 1987). Hence the variation of density becomes very important over a larger domain in the gas phase in reduced gravity. By comparing properties of liquid n-decane for temperatures of 300 K and 303 K, we see that for this 1% temperature increase, μ' decreases by 5%, while c_p' increases by 0.5%, k' decreases by 0.7%, and ρ' decreases by only 0.3%. Thus for the geometry and heater temperature studied, the variation of viscosity is the most critical factor in the prediction of the flow field in the liquid phase.

Vaporization Effects

Including vaporization in the above variable property simulation did not significantly change the temperature or radial velocities (see Figure 4) in the gas or liquid phase. As shown by Figures 7 and 8, the vaporization velocity and mass fraction of fuel vapor in the gas phase were maximum along the centerline. Note that the vaporization velocities initially decrease with time since the initial condition used was $Y_F(r, y, 0) = 0$. For later times the mass fraction gradients near the surface decrease and the vaporization velocities increase with time as the liquid surface temperatures increase. The average mass fraction of fuel vapor is less than 2% after 500 seconds, which appears to be lower than expected for the area-weighted average vaporization velocities of $\sim 0.7\ \mu$m/s. Note that although the maximum vaporization velocities are 7–10 μm/s, the average velocity is much less due to condensation at the outer radii. This condensation is caused by the surface-tension-driven flow and the no-slip condition which sweeps fuel vapor immediately above the interface from the centerline toward the outer radii. At larger radii, the liquid surface temperatures and thus mass fractions of fuel vapor are lower than those at the centerline. The values of Y_F predicted by the simulation are all less than the lean flammability limit of n-decane in air, which is cited as 0.033–0.038 by Coward and Jones (1952).

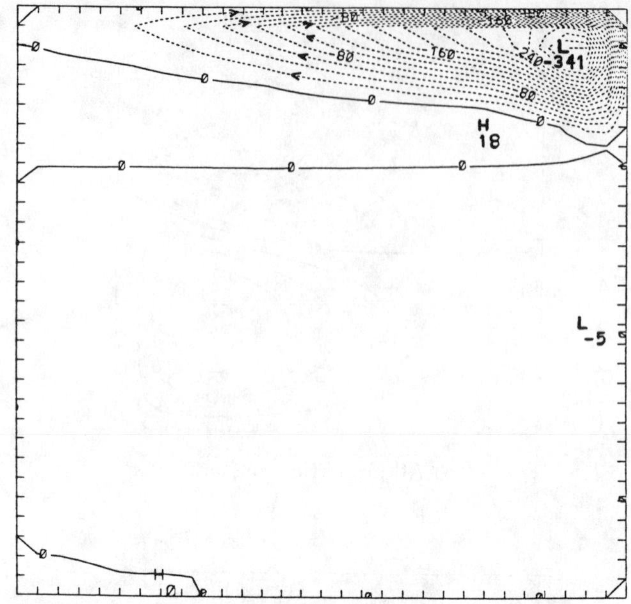

(a) ψ-contours

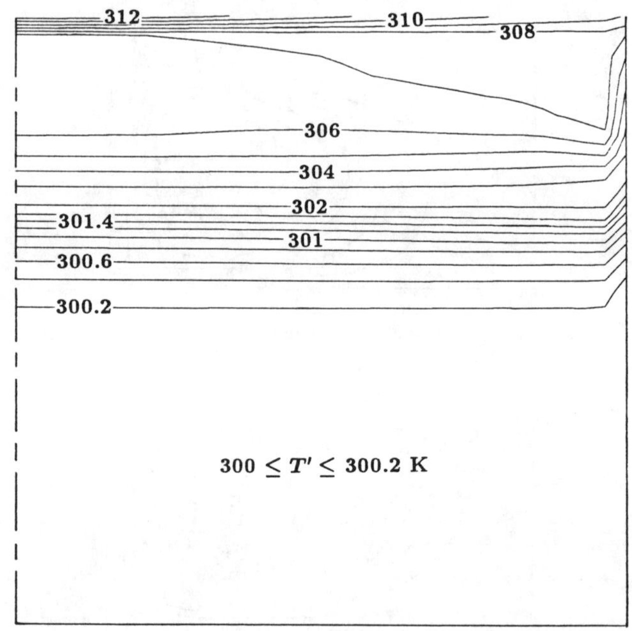

(b) Isotherms

Fig. 6 Nondimensional stream function and isotherm contours in the liquid phase for the variable property simulation ($t' = 500$ s)

Figure 8 also shows that a boundary layer develops in time above the liquid surface with respect to the concentration of fuel vapor (see the curves for $r' = 0$ and $r' = 2.5$ cm). This figure, as well as plots of ψ-contours in the gas and liquid phases, also indicates that the flow becomes quasi-steady after approximately

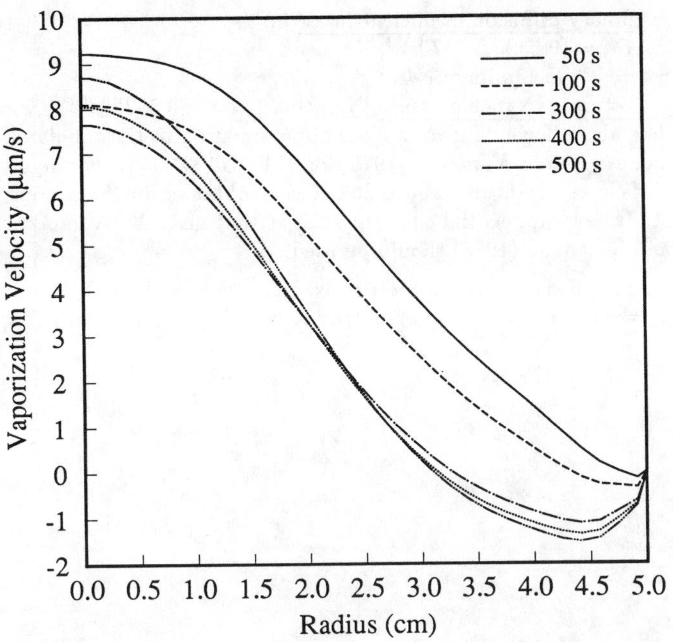

Fig. 7 Vaporization velocity (v_g' at $y = 0$) profiles at different times

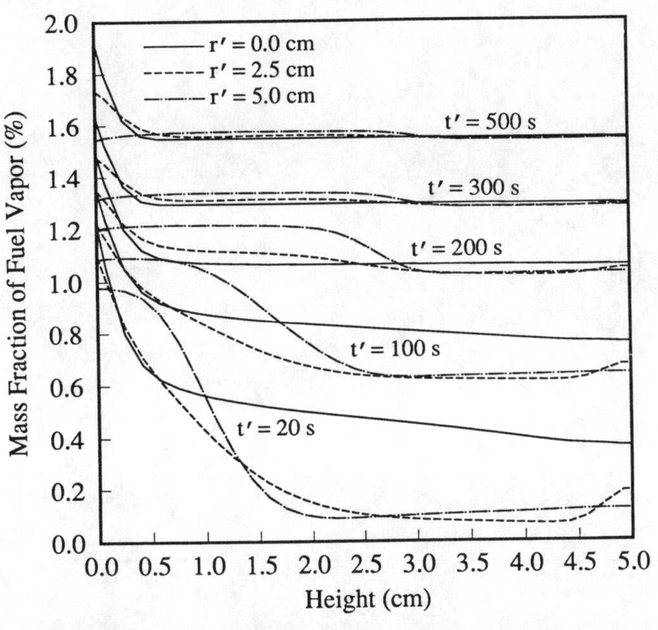

Fig. 8 Mass fraction of fuel vapor for different times at various radial locations in the gas phase

200–300 seconds. (The ψ-contour plots for the simulations without vaporization indicated that the flow field for the constant property simulation became quasi-steady about 100 seconds later than the other simulations.)

In the future, the use of a finer mesh near the gas/liquid interface and double precision for the calculations should be investigated. Double precision may be required because velocities (especially in the axial direction) vary by 4–5 orders of magnitude in both the gas and liquid phases. The effects of radiative absorption in the liquid phase should also be studied.

CONCLUSIONS

A computational study was made of the transient heat transfer and fluid flow in a cylindrical enclosure containing a two-layer gas-and-liquid system heated nonuniformly from above. Results indicate that the constant-property assumption with the Boussinesq approximation significantly (*i*) over-predicts velocities and under-predicts temperatures near the heater in the gas phase, and (*ii*) under-predicts velocities in the liquid phase and immediately above the liquid surface. The effects of variable density are significant in the gas phase, whereas in the liquid phase variable viscosity is most important.

When vaporization is included in the study, the mass fraction of fuel vapor in the gas phase reaches only $\sim$ 2% after 500 seconds. The vaporization velocities are maximum at the centerline and range from 7–10 μm/s with an area-weighted average of $\sim$ 0.7 μm/s. The fuel vapor profile immediately above the liquid surface is a boundary layer type. The flow patterns for the simulations studied become quasi-steady after 200–300 seconds.

ACKNOWLEDGEMENTS

Research at UCI was conducted in support of NASA Grant No. NAG 3-627 under the technical monitoring of Dr. H. Ross. The authors are grateful to Dr. Ross for many stimulating technical discussions.

REFERENCES

Abramzon, B., Edwards, D. K., and Sirignano, W. A., 1987, "Transient, Stratified, Enclosed Gas and Liquid Behavior with Concentrated Heating from Above," *J. Thermophysics*, Vol. 1, No. 4, pp. 355–364.

Abramzon, B., and Sirignano, W. A., 1988, "Droplet Vaporization Model for Spray Combustion Calculations," AIAA preprint No. 88-0636, presented at the 26th AIAA Aerospace Sciences Meeting, Reno, Nevada.

Aggarwal, S. K., Iyengar, J., and Sirignano, W. A., 1986, "Enclosed Gas and Liquid with Nonuniform Heating from Above," *Int. J. Heat Mass Transfer*, Vol. 29, No. 10, pp. 1593–1604.

Carslaw, H. S. and Jaeger, J. C., 1959, *Conduction of Heat in Solids*, 2nd ed., Oxford University Press, New York.

Coward, H. F., and Jones, G. W., 1952, "Limits of Flammability of Gases and Vapors," Bulletin 503, *Bureau of Mines.*

De Vahl Davis, G. and Jones, I. P., 1981, "Natural Convection in a Square Cavity: A Comparison Exercise," *Numerical Methods in Thermal Problems*, edited by R. W. Lewis et al., Pineridge Press, Swansea, Wales, pp. 553–572.

Edwards, D. K., Denny, V. E., and Mills, A. F., 1979, *Transfer Processes: An Introduction to Diffusion, Convection, and Radiation*, Hemisphere Publishing Corp., New York.

Lienhard, J. H., 1981, *A Heat Transfer Textbook*, Prentice-Hall, Inc., New Jersey.

Paolucci, S., 1982, "On the Filtering of Sound From the Navier-Stokes Equations," *Sandia National Laboratories Rep.* SAND82-8257.

Patankar, S. V., 1980, *Numerical Heat Transfer and Fluid Flow*, McGraw-Hill, New York.

Ross, H. D., Schiller, D. N., Disimile, P., and Sirignano, W. A., 1989, "Behavior in Normal and Reduced Gravity of An Enclosed Liquid/Gas System with Nonuniform Heating From Above," AIAA preprint No. 89-0070, presented at the 27th AIAA Aerospace Sciences Meeting, Reno, Nevada.

Schiller, D. N., 1988, "Science Experiments Plan for Pool Fires: Preignition Processes," M.S. Thesis, University of California, Irvine.

Torrance, K. E., 1971, "Subsurface Flows Preceding Flame Spread Over a Liquid Fuel," *Comb. Sci. Tech.*, Vol. 3, pp. 113–143.

Torrance, K. E. and Mahajan, R. L., 1975, "Surface Tension Flows Induced by a Moving Thermal Source," *Comb. Sci. Tech.*, Vol. 10, pp. 125–136.

Touloukian, Y., 1970, *Thermophysical Properties of Matter*, Vol. 11, Plenum Press, New York.

Van Doormaal, J. P. and Raithby, G. D., 1984, "Enhancements of the SIMPLE Method for Predicting Incompressible Fluid Flows," *Numerical Heat Transfer*, Vol. 7, pp. 147–163.

Vargaftik, N. B., 1975, *Tables on the Thermophysical Properties of Liquids and Gases*, 2nd. ed., Hemisphere Publishing Corp., Washington, D.C.

APPENDIX

Physical Properties

Physical properties of air, n-decane liquid and vapor, and air-vapor mixtures (e.g., binary diffusion coefficient) were taken from Vargaftik (1975), Touloukian (1970), Lienhard (1981), and Abramzon and Sirignano (1988). Curve-fits were then made for the anticipated required temperature ranges. Note that $T = T'/(300\text{ K})$.

For air, the following correlations were used:

heat capacity [J/kg·K]:
If $T' < 315$ K: $c'_p = 1003.0$
If $315 \leq T' \leq 530$ K: $c'_p = 1003.0 + 42.87(T - 1.050)$
If $T' > 530$ K: $c'_p = 1033.7 + 72.03(T - 1.767)$
thermal conductivity [W/m·K]: $k' = 0.02614T^{0.8003}$
dynamic viscosity [kg/m·s]: $\mu' = 1.853 \times 10^{-5}T^{0.6862}$

For liquid n-decane, the following correlations were used:

density [kg/m³]: $\rho' = 724.42 - 234.785(T - 1.0)$
heat capacity [J/kg·K]: $c'_p = 2214.86 + 1056.42(T - 1.0)$
thermal conductivity [W/m·K]: $k' = 0.1381 - 0.0982(T - 1.0)$
dynamic viscosity [kg/m·s]: $\mu' = 7.794 \times 10^{-4}T^{-5.4009}$
latent heat of vaporization [J/kg]:
If $T' \leq 333$ K: $h'_{fg} = 355,311 - 140,100(T - 1.0)$
If $T' > 333$ K: $h'_{fg} = 339,900 - 150,000(T - 1.11)$

For n-decane vapor, the following correlations were used:

heat capacity [J/kg·K]: $c'_p = 239.83 + 1532.7T - 58.01T^2 - 28.65T^3$
thermal conductivity [W/m·K]: $k' = -6.004 \times 10^{-3} + 1.138 \times 10^{-2}T + 6.708 \times 10^{-3}T^2$
dynamic viscosity [kg/m·s]: $\mu' = 5.64 \times 10^{-6} + 5.31 \times 10^{-6}(T - 1.0)$

The above formula was derived for n-octane. No sufficient tabulated data was found in the literature for n-decane gas.

binary diffusion coefficient in air [m²/s]:
$\mathcal{D}'_{ij} = (0.5533/P'_G)T^{1.583}$
where P'_G has units of N/m².

saturated vapor pressure [N/m²]: $P'_{\text{sat}} = 180.36T^{19.2039}$

The above formula gives a close approximation to the tabulated values given by Vargaftik (1975) for $273 \leq T' \leq 363$ K. For liquid surface temperatures above 363 K (not obtained in the present study) an exponential curve-fit such as that given by Abramzon and Sirignano (1988) should be used.

For plexiglass, the properties used were (Edwards et al., 1979): $\rho'_w = 1185$ kg/m³, $c'_w = 1470$ J/kg·K, and $k'_w = 0.21$ W/m·K.

1989 National Heat Transfer Conference
HTD-Vol. 106, Heat Transfer Phenomena in Radiation, Combustion and Fires

SIMPLIFIED CALCULATIONS OF TOTAL FLUX AND ENTRAINMENT USING EXPERIMENTAL RESULTS FROM A MEDIUM-SCALE POOL FIRE

E. J. Weckman
Department of Mechanical Engineering
The University of Western Ontario
London, Ontario, Canada

ABSTRACT

Simultaneous, temporally resolved velocity and temperature data, measured in the continuous flame region of a medium-scale pool fire, were incorporated in a simplified two-dimensional control volume analysis to estimate mass, momentum and energy fluxes in this region of the fire. The total fluxes, as well as the air entrainment distribution were examined for heights up to one pool diameter above the surface of the fuel. The analysis was extended to investigate methods for incorporation of the temperature dependence of the thermophysical properties in regions of the fire where the temperature field fluctuated with time and position. The effects of including mean "turbulent" flux terms in the mass, momentum and energy balances also were estimated. The analysis has provided estimates of entrainment near the base of the fire and indicated the extent of "turbulent" transport through the continuous flame zone.[1]

NOMENCLATURE

c_p specific heat at constant pressure
D diameter
g gravitational acceleration
$\dot{m}$ mass flux
Q heat release rate of the fire
r radial coordinate
R radius of cylindrical control volume
T temperature
U axial (streamwise) velocity
u' fluctuating axial velocity
V radial velocity
v' fluctuating radial velocity
z axial coordinate
Z distance from liquid fuel surface

Greek

α entrainment coefficient
γ''' rate of heat release per unit volume
ρ mass density
χ_{rad} fractional heat loss by radiation

Subscripts

cl centreline
e edge of pool
ent entrained
∞ ambient

Superscripts

— time-averaged portion

[1]Keywords: Pool Fire, Flux, Entrainment

INTRODUCTION

Mathematical models of fire behaviour are needed to predict characterisitics of fires in various environments and under varying ambient conditions. Although much effort has gone into the development of detailed models for the behaviour of pool fires, the success of their application has been limited by poor understanding of the complex physical processes (i.e., buoyancy, turbulent transport and entrainment) which occur at the base of the fire. This has been due, in part, to a lack of detailed experimental data, taken in the continuous flame zone of these fires (i.e., up to one pool diameter above the surface of the fuel), that might be used to develop and test submodels for the parameters important to fire development.

Dimensional analyses of the fire have indicated that a large number of non-dimensional parameters are required to fully characterize pool fire behaviour (Emmons, 1961; Emori and Saito, 1983). They have also pointed to the difficulty of defining representative values of temperature, velocity and the thermophysical fluid properties (ρ, c_p and μ) in regions of the fire where temperature fields fluctuate in space and time. Zone models include more complex mathematical formulations to describe the behaviour of a fire in various environments (Emmons, 1979) and have been used to demonstrate the influence of some external parameters on overall fire behaviour (Ndubizu *et al.*, 1983). However, they also are limited by inconsistent descriptions of, for example, the entrainment mechanisms and radiation in the fire. (Ndubizu *et al.*, 1983; Beyler, 1986).

Figure 1: The experimental system showing the overflow pan burner, the velocity and temperature measurement probes and the screen enclosure.

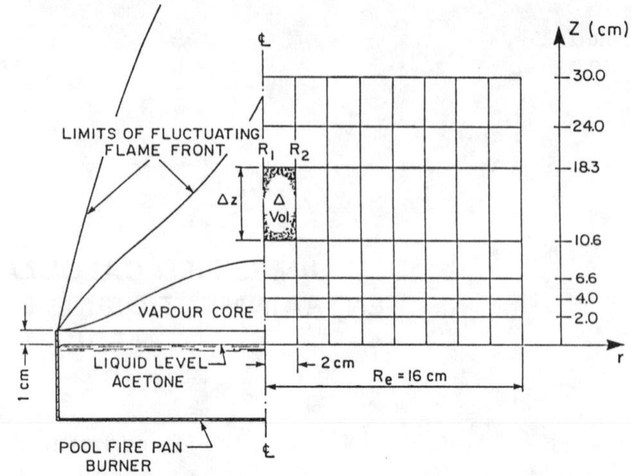

Figure 2: Schematic of the pool fire flow field showing the time-averaged position of the flame front and the discretization mesh used in the analysis. Measurements were taken at the control volume nodes.

In field models, the coupled continuity, momentum, energy and species conservation equations are solved with appropriate boundary conditions. Field models based on point source plume theory have been used to predict temperatures and velocities in the convection column high above a fire (e.g., Nielsen and Tao, 1965). In the lower regions of the fire, it is necessary to simplify the full equation set due to incomplete specification of fire boundary conditions and lack of models for buoyancy, entrainment and turbulent transport in the continuous flame zone (Tamanini, 1977).

Models of fire behaviour can be developed further by including more complete descriptions of buoyancy and turbulence in the flow field and of air entrainment near the base of the fire. One approach is to make detailed experimental measurements in the continuous flame zone. Through these, realistic boundary conditions can be specified for the fire together with values for the, as yet, unquantified terms in the equations (i.e., $\overline{u'v'}$, $\overline{u'T'}$) (Tamanini, 1977; Gengembre *et al.*, 1984).

The present paper reports an investigation in which detailed measurements of velocity and temperature have been combined in a control volume analysis to estimate mass, momentum and energy flux, and entrainment throughout the continuous flame zone of a medium-scale pool fire (Weckman, 1987). The analysis was extended to examine methods by which the temperature dependence of the thermophysical properties could be included for regions where temperature fluctuated with position and time. The effects of measured correlations of fluctuating axial velocity and temperature were estimated in calculations of total mass, momentum and energy fluxes. The simplified analysis provided estimates of air entrainment into the base of the fire and of the extent of "turbulent" transport through the continuous flame zone.

THE EXPERIMENTS

In the experiments, a 30.5 cm acetone pool fire was established on a modified overflow-type pan burner (see Figure 1). The burner was operated with constant fuel feed (2.19 cm^3/s) and a rim height of 1.0 cm. The total heat release rate of the fire was 53.3 kW.

The pool burner was mounted on a two-way traversing mechanism designed to cause minimal obstruction to air flow into the fire. It was centered under a natural draft fume hood and was positioned more than one pool diameter (30.5 cm) above the laboratory floor to minimize the effects of surrounding surfaces on the behaviour of the fire (McCaffrey, 1983; Crauford, 1984). To shield the fire from drafts in the main laboratory, the entire system was enclosed by panels of coarse wire grating covered in aluminum window screen. Monitoring and diagnostic equipment were operated from outside the enclosure.

Velocity measurements were performed using a one component laser Doppler anemometer (LDA) operated in the forward-scatter mode. In the flame, naturally occurring soot particles were used as seed, while ambient (entrained) air was artificially seeded using TiO_2 particles. Temperature was measured using 50 μm Pt-Pt10%Rh thermocouples. Data acquisition was performed with an IBM/AT compatible microcomputer directly interfaced to both the LDA counter processor and thermocouple amplifier circuits. This allowed independent or simultaneous sampling of velocity and temperature data. After a full time-series of data was collected (on average 40,000 independent measurements), it was analysed to provide mean and rms values of velocity and temperature, as well as joint or independent probability functions and correlation coefficients of the variables. More complete descriptions of the experimental system and measurements may be found in Weckman *et al.* (1986) and Weckman (1987).

Radial profiles of axial and radial velocity and temperature were taken at 2.0 cm intervals from the pool centreline to beyond the outside edge of the pool (16.0 cm from the centreline). They were repeated at heights ranging from 2.0 to 30.0 cm above the fuel surface (see Figure 2). These provided a detailed picture of the continuous flame zone of the fire, from the vapour core (2.0–6.6 cm above the fuel) through the highly fluctuating region (6.6–18.3 cm above the fuel) to the top of the continuous flame zone (18.3–30.0 cm above the fuel). Representative radial profiles of measured velocities and temperatures for positions 10.0 and 30.0 cm above the pool surface are given in Table 1. In Figure 3, it can be seen that scaled centreline temperature data were comparable to data and correlations reported in the literature.

CONTROL VOLUME ANALYSIS

The experimental data were incorporated into a simplified model of the fire flow field in order to examine mass, momentum and energy flux through the continuous flame zone.

Discretization Of The Basic Equations'

The general equations for conservation of mass, momentum and energy in cylindrical coordinates were formulated for a two-dimensional steady flow with only small variations in thermophysical properties between horizontal planes in the flow field (Currie, 1974). It was assumed that the radial pressure gradient

Table 1: Representative profiles of measured data for positions Z=10.0 cm and Z=30.0 cm

R	$\overline{U}$	$\sqrt{u'^2}$	$\overline{T}$	$\sqrt{T'^2}$	$\frac{\overline{u'T'}}{\sqrt{u'^2}\sqrt{T'^2}}$
cm	cm/s	cm/s	K	K	
			z = 10.0 cm		
16	3.7	6.5	295	0	.0004
14	6.7	9.3	295	11	.1302
12	14.2	18.1	336	99	.6535
10	45.7	42.1	534	239	.6694
8	70.3	54.1	725	267	.6891
6	111.1	58.6	1027	243	.5998
4	125.8	48.6	1109	178	.4108
2	115.2	43.5	1003	175	.4283
0	109.4	40.2	919	137	.4138
			z = 30.0 cm		
16	9.0	12.7	325	64	.0430
14	23.0	28.6	374	124	.4969
12	42.2	46.8	485	193	.5641
10	75.7	69.9	583	221	.6444
8	127.5	82.7	730	238	.6226
6	212.4	97.8	928	250	.5854
4	272.6	104.1	1072	223	.4255
2	333.8	93.2	1225	154	.2731
0	331.9	90.4	1240	137	.2800

was small (McCaffrey, 1983), that viscous terms could be neglected, that heat transfer by conduction was minimal and that heat loss by radiation could be estimated as a fraction, χ_{rad}, of the chemical heat release within a given volume of the flow. The simplified forms of the equations became for mass

$$\frac{1}{r}\frac{\partial}{\partial r}(\rho r\tilde{V}) + \frac{\partial(\rho\tilde{U})}{\partial z} = 0, \tag{1}$$

for momentum in the z-direction

$$\tilde{V}\frac{\partial(\rho\tilde{U})}{\partial r} + \tilde{U}\frac{\partial(\rho\tilde{U})}{\partial z} = (\rho_\infty - \rho)g, \tag{2}$$

for momentum in the r-direction

$$\tilde{V}\frac{\partial(\rho\tilde{V})}{\partial r} + \tilde{U}\frac{\partial(\rho\tilde{V})}{\partial z} = 0, \tag{3}$$

and for thermal energy

$$\tilde{V}\frac{\partial(\rho c_p T)}{\partial r} + \tilde{U}\frac{\partial(\rho c_p T)}{\partial z} = (1 - \chi_{rad})\gamma''', \tag{4}$$

where $\tilde{U}, \tilde{V}$ and T are general streamwise and radial velocities and temperature, respectively, and γ''' is the rate of heat release per unit volume of the fire.

The region above the pool was discretized using rectangular, toroidal control volumes as shown in Figure 2. The nodes of the control volumes were placed at the measurement positions and equations (1) to (4) were integrated over each volume. Representative steps in the analysis are outlined below for the continuity equation and the corresponding forms of the equations are written for momentum and energy conservation.

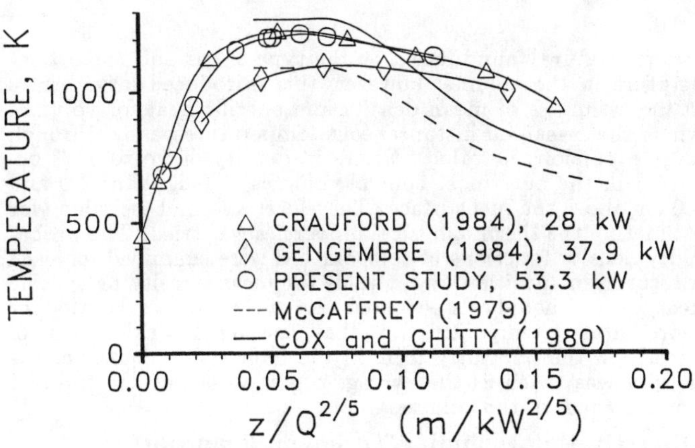

Figure 3: Comparison of temperature data from the present study with data (Crauford, 1984; Gengembre, 1984) and correlations (McCaffrey, 1979; Cox and Chitty, 1980) reported in the literature.

Integrating equation (1) from 0 to 2π and from R_{N-1} to R_N and noting that for a cylindrical volume $R \neq f(z)$ gives

$$\rho\tilde{V}(R_N)2\pi(R_N - R_{N-1}) + \frac{\partial}{\partial z}\left[\int_{R_{N-1}}^{R_N}(\rho\tilde{U})2\pi r\,dr\right] = 0. \tag{5}$$

Assuming top-hat profiles of $\tilde{U}$ for each control volume and integrating between two horizontal planes in the flow field, the continuity equation for each volume became,

$$\left[\rho_N U_N \pi(R_N^2 - R_{N-1}^2)\right]_{top} = \left[\rho_N U_N \pi(R_N^2 - R_{N-1}^2)\right]_{bottom} + \frac{\rho_{N_{top}} + \rho_{N_{bottom}}}{2} V_N 2\pi R_N (Z_{top} - Z_{bottom}) - \frac{\rho_{N-1_{top}} + \rho_{N-1_{bottom}}}{2} V_{N-1} 2\pi R_{N-1}(Z_{top} - Z_{bottom}). \tag{6}$$

Following similar reasoning, the momentum equation became

$$\left[\rho_N U_N^2 \pi(R_N^2 - R_{N-1}^2)\right]_{top} = \left[\rho_N U_N^2 \pi(R_N^2 - R_{N-1}^2)\right]_{bottom} + \left[\rho_\infty - \frac{(\rho_{top} + \rho_{bottom})}{2}\right] g\pi(R_N^2 - R_{N-1}^2)(Z_{top} - Z_{bottom}), \tag{7}$$

and the balance of thermal energy

$$\left[\rho_N c_{p_N} U_N T_N \pi(R_N^2 - R_{N-1}^2)\right]_{top} = \left[\rho_N c_{p_N} U_N T_N \pi(R_N^2 - R_{N-1}^2)\right]_{bottom} + \frac{\rho_{N_{top}} + \rho_{N_{bottom}}}{2} c_p V_N T_N 2\pi R_N (Z_{top} - Z_{bottom}) - \frac{\rho_{N-1_{top}} + \rho_{N-1_{bottom}}}{2} c_p V_{N-1} T_{N-1} 2\pi R_{N-1}(Z_{top} - Z_{bottom}) + (1 - \chi_{rad})\gamma''' \pi(R_N^2 - R_{N-1}^2)(Z_{top} - Z_{bottom}). \tag{8}$$

Using the measurements of mean temperature and velocity, the fluxes of mass, momentum and energy were calculated for each of the eight volumes in a given horizontal plane, then the total flux for that plane calculated as the sum of the individual values. In the discretized equations, $\tilde{U_N}$ and T_N were taken as the average of the mean values at the R_N and R_{N-1} nodes of a volume and the radial velocity was taken as the mean inflow velocity at a given radius, averaged between the top and bottom horizontal edges of the volume. Initially, density and specific heat at constant pressure were evaluated as functions of temperature, found based on the type of gas and mean temperature in the volume; however, this introduced large errors in the estimates of mean density and specific heat for volumes where the measured instantaneous temperature passed through large excursions in value. Errors in density of up to 35% occurred in the outermost control volumes at heights of 2.0 and 4.0 cm above the fuel surface. Therefore, a second technique for estimating the thermophysical properties was tried. The probability density functions of temperature were examined for each measurement position and a value for mean density or specific heat was estimated as a weighted sum of the values evaluated for representative temperatures at that position. For calculation of density in the buoyancy term in the momentum equation, the density was taken as the average of the values at the top and bottom edges of the volume.

Discretization Including Turbulent Transport

To estimate the effects of the mean "turbulent" flux terms in the equations of mass, momentum and energy conservation, equations (1), (2) and (4) were written in boundary layer form for a steady, turbulent, axisymmetric flow, including the same body forces and energy source terms as in equations (2) and (4) and neglecting triple correlations of the fluctuating quantites (McCaffrey, 1983). They became for mass

$$\frac{\partial}{\partial z}\left(\overline{\rho}\overline{U}+\overline{\rho' u'}\right)=0, \tag{9}$$

for momentum

$$\frac{\partial}{\partial z}\left(\overline{\rho}\overline{U}^2+2\overline{U}\,\overline{\rho' u'}+\overline{\rho}\overline{u'^2}\right)=(\rho_\infty-\overline{\rho})g, \tag{10}$$

and for thermal energy

$$\frac{\partial}{\partial z}\left(\overline{\rho c_p}\overline{U}\overline{T}+\overline{T}\,\overline{\rho' u'}+\overline{\rho}\overline{u'T'}+\overline{U}\,\overline{\rho' T'}\right)=(1-\chi_{rad})\gamma''', \tag{11}$$

where the overbar denotes a time-averaged quantity and the $'$ indicates a fluctuating component. Using the perfect gas law for small changes of pressure, redefining density based upon the mean temperature, $\overline{\rho}=\frac{\rho_\infty T_\infty}{\overline{T}}$ and integrating as above, the equations for each control volume became for mass

$$\left[\overline{\rho}_N\overline{U}_N\left\{1+\frac{\overline{T'^2}}{\overline{T}^2}-\frac{\overline{u'T'}}{\overline{U}\,\overline{T}}\right\}_N\pi(R_N^2-R_{N-1}^2)\right]_{top}=$$
$$\left[\overline{\rho}_N\overline{U}_N\left\{1+\frac{\overline{T'^2}}{\overline{T}^2}-\frac{\overline{u'T'}}{\overline{U}\,\overline{T}}\right\}_N\pi(R_N^2-R_{N-1}^2)\right]_{bottom}+$$
$$\frac{\rho_{N_{top}}+\rho_{N_{bottom}}}{2}\overline{V}_N 2\pi R_N(Z_{top}-Z_{bottom})-$$
$$\frac{\rho_{N-1_{top}}+\rho_{N-1_{bottom}}}{2}\overline{V}_{N-1}2\pi R_{N-1}(Z_{top}-Z_{bottom}), \tag{12}$$

for momentum

$$\left[\overline{\rho}_N\overline{U}_N^2\left\{1+\frac{\overline{T'^2}}{\overline{T}^2}+\frac{\overline{u'^2}}{\overline{U}^2}-2\frac{\overline{u'T'}}{\overline{U}\,\overline{T}}\right\}_N\pi(R_N^2-R_{N-1}^2)\right]_{top}=$$
$$\left[\overline{\rho}_N\overline{U}_N^2\left\{1+\frac{\overline{T'^2}}{\overline{T}^2}+\frac{\overline{u'^2}}{\overline{U}^2}-2\frac{\overline{u'T'}}{\overline{U}\,\overline{T}}\right\}_N\pi(R_N^2-R_{N-1}^2)\right]_{bottom}+$$
$$\left[\rho_\infty-\overline{\rho}\left\{1+\frac{\overline{T'^2}}{\overline{T}^2}\right\}\right]_N g\pi(R_N^2-R_{N-1}^2)(Z_{top}-Z_{bottom}), \tag{13}$$

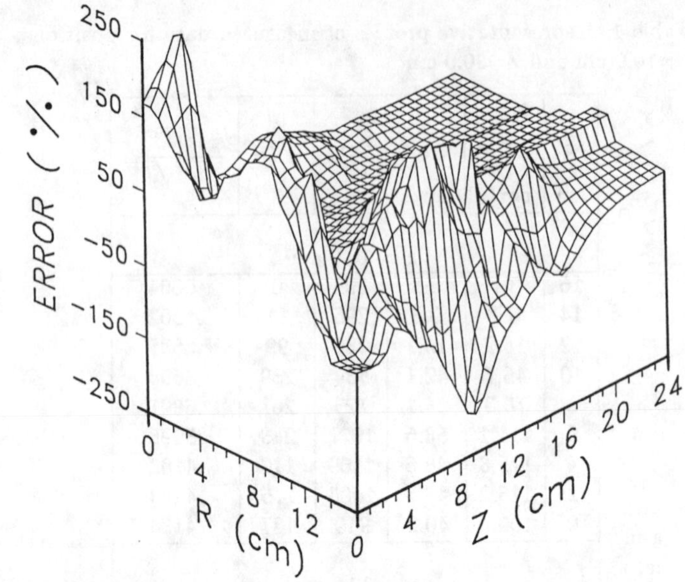

Figure 4: Surface map of % error in total mass flux entering a volume with respect to measured mass flux leaving the volume.

and for thermal energy

$$\left[\overline{\rho}_N c_{p_N}\overline{U}_N\left\{(\overline{T}-T_\infty)-T_\infty\left(\frac{\overline{T'^2}}{\overline{T}^2}\right)+T_\infty\left(\frac{\overline{u'T'}}{\overline{U}\,\overline{T}}\right)\right\}_N\pi(R_N^2-R_{N-1}^2)\right]_{top}=$$
$$\left[\overline{\rho}_N c_{p_N}\overline{U}_N\left\{(\overline{T}-T_\infty)-T_\infty\left(\frac{\overline{T'^2}}{\overline{T}^2}\right)+T_\infty\left(\frac{\overline{u'T'}}{\overline{U}\,\overline{T}}\right)\right\}_N\pi(R_N^2-R_{N-1}^2)\right]_{bottom}+$$
$$\frac{\rho_{N_{top}}+\rho_{N_{bottom}}}{2}c_p\overline{V}_N T_N 2\pi R_N(Z_{top}-Z_{bottom})-$$
$$\frac{\rho_{N-1_{top}}+\rho_{N-1_{bottom}}}{2}c_p\overline{V}_{N-1}T_{N-1}2\pi R_{N-1}(Z_{top}-Z_{bottom})+$$
$$(1-\chi_{rad})\gamma'''\pi(R_N^2-R_{N-1}^2)(Z_{top}-Z_{bottom}). \tag{14}$$

The mean entrainment terms were included but fluctuations in the velocity of the entrained air were neglected. Measured values of $\overline{u'T'}, \overline{T'^2}$ and $\overline{u'^2}$ such as those shown in Table 1 were substituted into these equations in the same manner as for the analysis described above. Mass, momentum and energy fluxes were recalculated for each control volume and the summations performed for various heights above the pool surface. These were compared to values calculated without "turbulent" terms to provide a quantitative measure of the relative importance of "turbulent" transport for the continuous flame zone.

RESULTS AND DISCUSSION

Mass Flux

In Figure 4 are surface maps of % error in total mass flux entering each control volume with respect to measured mass flux leaving the volume. The figure indicates that the largest errors occur close to the fuel surface, shown as the large positive peaks in the figure. These can be accounted for through errors in measured velocities in this region, since the magnitude of velocity is small, less than 2 cm/s, and problems were encountered with LDA measurements due to the large refractive index gradients at

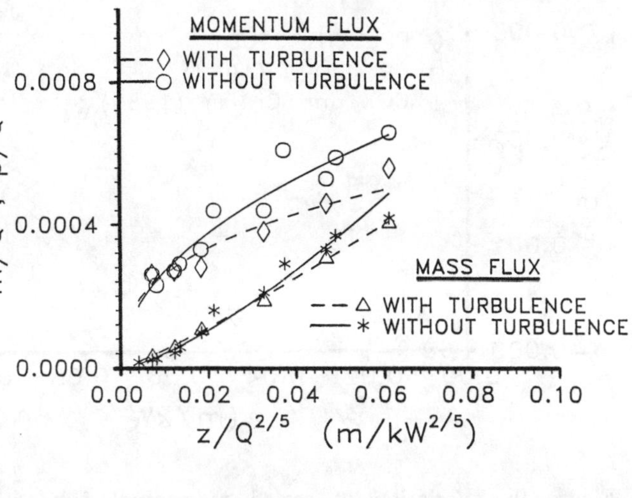

Figure 5: Scaled mass, m/Q (kg/s-kW) and momentum, $p/Q^{6/5}$ (kg-m/s^2-kW$^{6/5}$) flux versus scaled distance above the fuel surface, $z/Q^{2/5}$ (m/kw$^{2/5}$), with and without the inclusion of measured mean "turbulent" flux terms.

the edges of the fire (Weckman and Sobiesiak, 1987). The peaks seen in the figure for the region 6.0 to 14.0 cm above the pool surface and 1.0 to 4.0 cm in from the edge of the pool cannot be attributed to errors in measured parameters, since both velocity and temperature measurements were repeatable to within 10% in these regions of the fire. Rather, this is the region of the fire in which the flame front fluctuates in both space and time, so the error may be attributable to poor estimates of mean density, which varied by more than 20% depending on the temperatures taken as representative for a given control volume. Even with large errors in the balance of mass for individual volumes, total mass flux into and out of each vertical measurement plane balanced to within 15% throughout the continuous flame zone. This value of total flux was used to determine the effect of "turbulent" terms on mass flux in the fire (see Figure 5). Throughout the continuous flame zone, the total mass flux with "turbulent" terms included was approximately 15% less than that calculated without taking them into account. This agrees well with previous estimates that at 30 cm above a pool burner, the decrease in mass flux, with the inclusion of the "turbulence" terms, was 4% on the fire centreline, increasing to 14% near the fire edges (Crauford, 1984). Although these terms may be important in some regions of the fire, they do not appear to have a large effect on predictions of total mass flux exiting the continuous flame zone of a medium-scale pool fire.

In Figure 6, the values of total mass flux are compared to data and correlations found in the literature. Agreement between data is excellent. The fires were of similar size; however, they used different fuels and consequently had substantially different heat release rates. Agreement between any two correlations is poor and none represents the data adequately. This may be due to differences amongst researchers of experimental technique, fire size and fuel type used in determination of total mass flux or to inadequacies in the factors used to scale mass flux and distance above the fuel surface. The wide variation in results points again to the need for further work in development of a general expression for prediction of mass flux through the continuous flame zones of pool fires (Beyler, 1986).

Momentum Flux

In Figures 7 and 8, surface maps of constant % error are plotted for two different methods of calculating momentum flux across individual control volumes. Close comparison indicates the sensitivity of momentum flux balances to the estimation of mean density, $(\rho_{top} + \rho_{bottom})/2$, in the buoyancy term. In Figure 7, the mean density was estimated based on the average

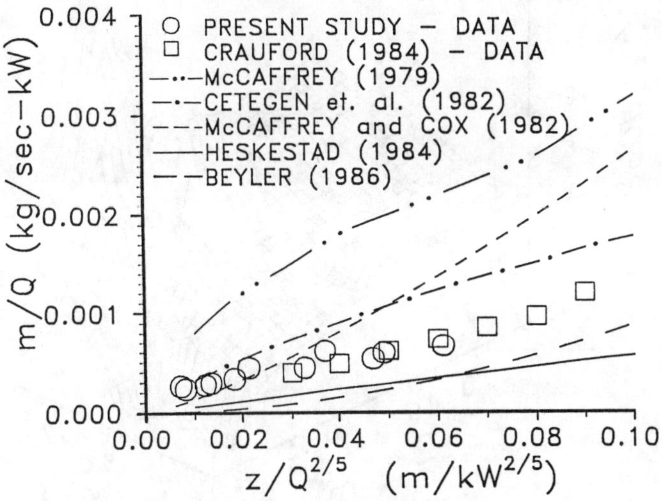

Figure 6: Variation of scaled mass flux, m/Q, with scaled distance through the continuous flame zone, $z/Q^{2/5}$, as compared to data of Crauford (1984) and correlations of McCaffrey (1979), Cetegen *et al.* (1982), McCaffrey and Cox (1982), Heskestad (1984) and Beyler (1986).

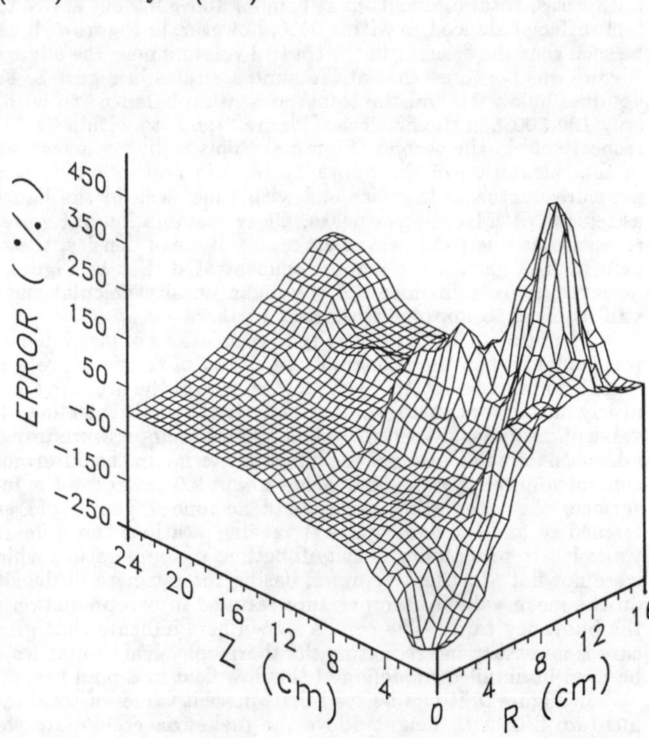

Figure 7: Surface maps of constant % error between measured and calculated momentum flux into and out of each control volume with ρ estimated based on the mean values of temperature.

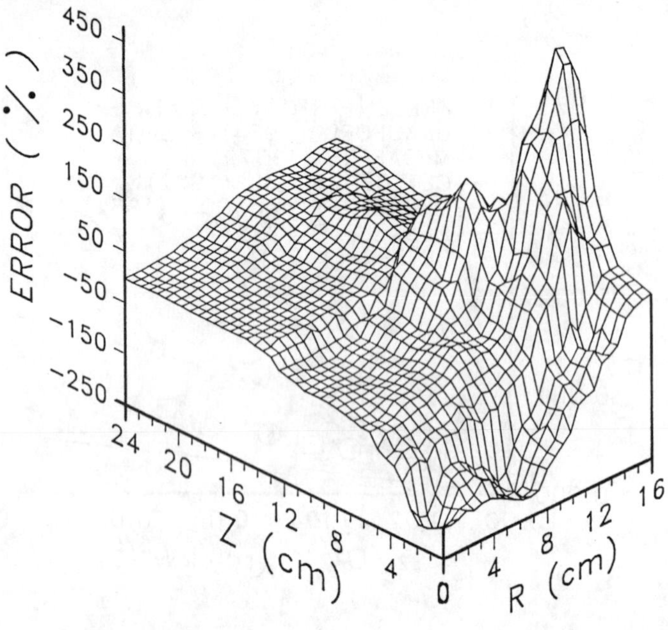

Figure 8: Surface maps of constant % error between measured and calculated mass flux into and out of each control volume with ρ estimated based weighted values of mean temperature.

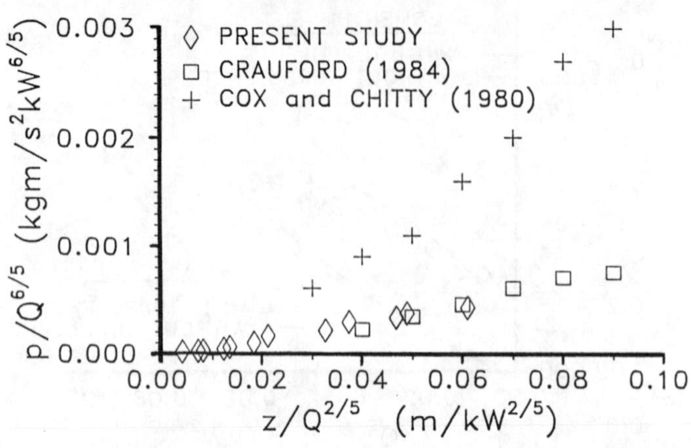

Figure 9: Variation of scaled momentum flux, $p/Q^{6/5}$ (kg-m/s^2-kW$^{6/5}$), with scaled distance through the continuous flame zone, $z/Q^{2/5}$, as compared to results of Crauford (1984) and Cox and Chitty (1980).

of the mean temperatures at the nodes of each control volume, while in Figure 8, it was estimated as a weighted sum of the densities evaluated for representative temperatures at a position. In both cases, total momentum at heights above 9.0 cm above the fuel surface balanced to within 25%; however, in Figure 7, it can be seen that the % error in the control volumes near the edges of the fire was 1-2 times that at the same positions in Figure 8. For volumes below 9.0 cm, the total momentum balanced to within only 100-300% in the first case (Figure 7), but to within 75-25% respectively in the second (Figure 8). This is due to inaccuracy in the calculation of the buoyancy term in regions where temperature fluctuates in space and with time, seen in the figures as regions with large error peaks. Close examination of Figure 8 reveals that due to improved estimates of mean density, the region with large % error is more concentrated than in Figure 7; however, more refinement of the mean density calculations is still required to improve agreement further.

The reason for the sensitivity to estimations of mean density was evident when the probability functions of temperature were examined in more detail. In the central core of the fire, they were nearly flat, so for control volumes close to the pool centreline, the value of mean density estimated using mean temperature proved adequate for calculation of the buoyancy term. In the outermost control volumes, particularly below about 9.0 cm above the fuel surface, the alternating presence of acetone vapour, cold entrained air and hot flame gases at varying positions and different times led to probability density functions of temperature which were not flat. In these volumes, basing the estimate of density on the mean value of temperature resulted in overprediction of the buoyancy term. The results shown here indicate that great care is necessary in estimating the thermophysical properties to be used in detailed modeling of the flow field in a pool fire.

In Figure 9, it can be seen that, present values of total momentum flux with height above the fuel surface compare well with results of Crauford (1984), but there are significant deviations from those of Cox and Chitty (1980). These are due to their use of a constant density and an assumed Gaussian profile of velocity, neither of which accurately represents the fire flow field in the lowest regions.

Although the remaining errors in balance of momentum flux through the flow field were thought to arise mainly from further errors in estimation of mean density, it was decided to examine the contribution of the mean "turbulent" terms in the momentum equation (10) (see Figure 6). Inclusion of these terms decreased the total flux by more than 30% in the lowest regions of the fire, but this difference decreased to less than 2% by 30.0 cm above the pool surface. These values are in reasonable agreement with Crauford (1984) who suggested that, at 30.0 cm above the fuel, momentum flux decreased by about 7% with inclusion of the "turbulent" flux terms in the governing equations. As was found in the case of mass flux, inclusion of the "turbulent" terms in the momentum calculations does not greatly alter estimates of total momentum flux through the continuous flame zone of the fire.

Table 2: Fraction of chemical heat release and radiation with height above the fire (estimated from Tamanini, 1983).

Position Z (cm)	Fractional Heat Release	Radiative Fraction
2	0.15	0.006
4.0	0.2	0.022
6.6	0.35	0.051
10.0	0.45	0.081
18.3	0.6	0.135
24	0.65	0.168
30	0.7	0.197

Energy Flux

The terms in the energy balance were estimated using mean densities and mean specific heats based on mean temperatures weighted as in the corrected momentum calculations. The chemical heat release and radiation in a volume of the flow field were based upon published results of energy distribution in a 30.8 cm diameter propane pool fire (Tamanini, 1983). Values used are in Table 2. Total flux across all horizontal planes balanced to within 30% except from 6.0 to 9.0 cm above the fuel surface.

In Figure 10, the total energy fluxes, calculated with and without inclusion of the "turbulent" terms are compared with results from the literature. The inclusion of the "turbulent" terms appears to decrease total energy flux by more than 50%; how-

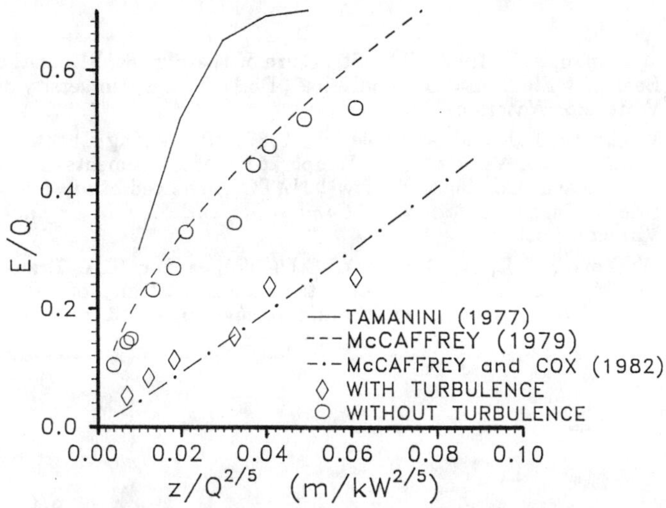

Figure 10: Variation of scaled energy flux, E/Q (kW/kW), with scaled distance through the continuous flame zone, $z/Q^{2/5}$, as compared to results of Tamanini (1977), McCaffrey (1979) and McCaffrey and Cox (1982).

ever, this is likely an indication of inaccuracies in the estimation of thermophysical properties, of the total heat release and of the radiation loss at various heights in the fire than a true physical effect occurring in the continuous flame zone. It is interesting to note that data excluding turbulence agrees with that of McCaffrey (1979) while that including turbulence agrees with that of McCaffrey and Cox (1982). The reasons for this are not clear, but in general, until better measurements of radiative fraction and total heat released by combustion are obtained through the continuous flame zone, a simplified balance of total energy likely will not provide further understanding of the flow field.

Air Entrainment

In many models of fire behaviour, empirical relationships are required to quantify air entrainment into the fire. For characterization of thermal plumes, the entrainment relation has been expressed as a coefficient, α, relating the local gradient of mass flux at a given height, $(d\dot{m}/dz)_{local}$, to the centreline mass flux at the same height (Crauford, 1984),

$$\alpha = \frac{(d\dot{m}/dz)_{local}}{2\pi R \rho \overline{U}_{cl}}. \tag{15}$$

Also, α may be calculated as the ratio of mean radial velocity at the edge of the fire to mean axial velocity on the centreline of the fire. In Figure 11, values of α from measured velocity ratios are compared to those calculated using equation (15). Near the fuel surface, $0.015 < z/Q^{2/5} < 0.03$, there is a high entrainment coefficient, decreasing to $z/Q^{2/5} \simeq 0.03$. Similar trends were observed by Toner *et al.* (1987); however, since they measured only total air entrained into the fire up to a given height, it is difficult to deduce local values of α from their results. At values of $z/Q^{2/5} > 0.035$, measured and calculated values of α are comparable with those measured by Crauford (1984). Except in the lowest region of the fire, local values of α are lower than the average values, $\alpha = 0.07$ - 0.16, reported by other investigators (Cox and Chitty, 1980; Tamanini, 1977); however, the average value, to the top of the continuous flaming region is 0.09, comparable to their values. This is reasonable since these values are found by averaging entrainment over the entire flaming portion of the fire. Unfortunately, the use of a single value for α does not truly account for any significant variations in local entrainment, and therefore mass flux, with height above the fuel surface.

A promising method for calculation of total mass of air entrained into a fire up to a specified height has been proposed by

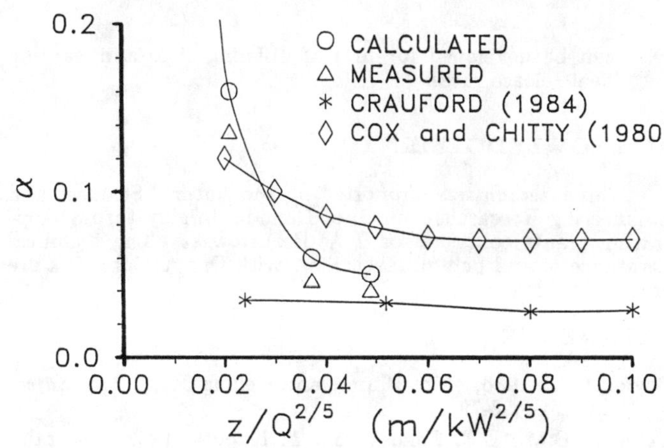

Figure 11: Calculated and measured entrainment coefficients, α, compared with values reported by Crauford (1984) and Cox and Chitty (1980). Vertical distance is scaled as, $z/Q^{2/5}$.

Delichatsios (1986). Using dimensional scaling with fire Froude number and fuel stoichiometric ratio, dependencies of entrained mass flux on fire size and geometry are eliminated in order to develop universal entrainment laws for each region in the fire. Scaled data from the present study agree well with that reported by Delichatsios (1986) suggesting that scaling of entrainment in this manner could provide correlations which are more universal than the calculations of α discussed above. To date, such laws cannot be tested since there presently are insufficient detailed data relating to entrainment into fire plumes.

CONCLUSIONS

This simplified analysis of total mass, momentum and energy flux and air entrainment at various heights in the continuous flame zone of a 30 cm diameter acetone pool fire has indicated several very important considerations for the development of more accurate models of flow field behaviour in the flaming region of a pool fire.

1. Temporal and spatial variations of the thermophysical properties must be accounted for in a detailed model of the fire flowfield, since they contribute large errors to calculations of mass, momentum and energy flux through the continuous flame zone.

2. Inclusion of fluctuating velocity-temperature correlations ("turbulent" flux terms) in the mass and momentum equations decreases the value of these fluxes by less than 15% throughout the continuous flame zone, so need not be considered in a detailed model of that region.

3. Inclusion of these same terms in the energy equation decreases total energy flux by more than 50%; however, it is thought that larger errors in the prediction of energy flux arise through incorrect approximations of the radiative fraction and total heat released by combustion in this region of the fire. To correct this, better measurements of these two variables are required in fires of varying diameter and heat release rate.

4. In general, poor correlation between the data and existing correlations for mass, momentum and energy flux, as well as air entrainment, indicate the necessity of collecting and analyzing more time-resolved measurements of velocity and temperature throughout the fire before universal laws for scaling of transport and entrainment phenomena

can be developed for fires of differing sizes and varying heat release rates.

ACKNOWLEDGEMENTS

This research was supported by the Natural Sciences and Engineering Research Council of Canada, largely through operating grants to Professor T.A. Brzustowski. The technical assistance of and helpful discussions with Dr. A. Sobiesiak are gratefully acknowledged.

REFERENCES

Beyler, C.L., 1986, "Fire Plumes and Ceiling Jets", *Fire Safety J.*, Vol. 11, pp.53-75.

Cetegen, B.M., E.E. Zukoski and T. Kubota, 1982, "Entrainment and Flame Geometry of Fire Plumes", *National Bureau of Standards Report NBS-GCR-82-402*, Washington, D.C..

Cox, G. and R. Chitty, 1980, "A Study of the Deterministic Properties of Unbounded Fire Plumes", *Comb. and Flame*, Vol. 39, pp. 191-209.

Crauford, N.L., 1984, "The Structure of an Unconfined Buoyant Diffusion Flame", Ph.D. Thesis, University of Southhampton, England.

Currie, I.G., 1974, *Fundamental Mechanics of Fluids*, McGraw-Hill, New York, p.422.

Delichatsios, M.A., 1986, "Air Entrainment into Buoyant Jet Flames and Pool Fires", *Comb. and Flame* Vol. 70, pp. 33-47.

Emmons, H.W., 1961, "Some Observations on Pool Burning", *International Symposium on the Use of Models in Fire Research*, Publication 786, NAS/NRC Washington, D.C..

Emmons, H., 1979, "The Prediction of Fires in Buildings", *Seventeenth Symposium on Combustion*, The Combustion Institute, Penn., pp. 1101-1111.

Emori, R.I. and K. Saito, 1983, "A Study of Scaling Laws in Pool and Crib Fires", *Comb. Sci. Tech.*, Vol. 31, pp. 217-231.

Gengembre, E., Cambray, P., Karmed, D. and Bellet, J.C., 1984, "Turbulent Diffusion Flames with Large Buoyancy Effects", *Comb. Sci. Tech.*, Vol. 41, pp. 55-67.

Heskestad, G., 1984, "Engineering Relations for Fire Plumes" *Fire Safety J.*, Vol. 7, pp. 25-32.

McCaffrey, B.J., 1979 "Purely Buoyant Diffusion Flames: Some Experimental Results", *National Bureau of Standards Report NBSIR-79-1910*, Washington, D.C..

McCaffrey, B.J., 1983 "Momentum Implications for Buoyant Diffusion Flames", *Comb. and Flame*, Vol. 52, pp. 149-167.

McCaffrey, B.J. and G. Cox, 1982 "Entrainment and Heat Flux of Buoyant Diffusion Flames", *National Bureau of Standards Report NBSIR-82-2473*, Washington, D.C..

Ndubizu, C.C., D.E. Ramaker, P.A. Tatem and F.W. Williams, 1983, "A Model of Freely Burning Pool Fires", *Comb. Sci. Tech.*, Vol. 31, pp. 223-247.

Nielsen, H.J. and L.N. Tao, 1965, "The Fire Plume Above A Large Free-Burning Fire", *Tenth (10th) Symposium (International) on Combustion*, The Comb. Inst., Penn., pp. 965-972.

Tamanini, F., 1977, "Reaction Rates, Air Entrainment and Radiation in Turbulent Fire Plumes", *Comb. and Flame*, Vol. 30, pp. 85-101.

Tamanini, F., 1983, "Direct Measurements of the Longitudinal Variation of Burning Rate and Product Yield in Turbulent Diffusion Flames", *Comb. and Flame*, Vol. 51, pp. 231-243.

Toner, S.J., E.E. Zukoski and T. Kubota, 1987, "Entrainment, Chemistry and Structure of Fire Plumes", *National Bureau of Standards Report NBS-GCR-87-528*, Washington, D.C..

Weckman, E.J., 1987, "The Structure of the Flowfield Near the Base of A Medium-Scale Pool Fire", Ph.D. Thesis, University of Waterloo, Waterloo, Canada.

Weckman, E.J. and A. Sobiesiak, 1987, "Processing Effects in Simultaneous Velocity and Temperature Measurements in an Intermittent Combusting Flow", NATO Advanced Study Institute on *Instrumentation for Combustion and Flow in Engines*, Vimiero, Portugal.

Weckman, E.J., A. Sobiesiak, C.D. Tropea and T.A. Brzustowski, 1986, "LDA Measurements in the Base Region of a Pool Fire", *Laser Anemometry in Fluid Mechanics*, Vol. 3, Portugal.

1989 National Heat Transfer Conference
HTD-Vol. 106, Heat Transfer Phenomena in Radiation, Combustion and Fires

EFFICIENT RETRIEVAL OF TEMPERATURE AND CONCENTRATION PROFILES FROM ABSORPTION MEASUREMENTS WITH ERROR

M. Ravichandran and F. C. Gouldin
Cornell University
Sibley School of Mechanical
and Aerospace Engineering
Ithaca, New York

ABSTRACT

The widely used methods for obtaining the inverse of the Radon transform, the underlying reconstruction problem in the determination of temperature and species concentration profiles from absorption measurements, have been developed for medical imaging applications and are not well-suited for limited data problems. The application of a new method, that is especially suited for limited data problems involving smooth distributions, to this retrieval problem has been previously reported. Performance of the new method with error containing absorption data is studied with regard to the stability and accuracy of the retrieved temperature and concentration profiles. Test results demonstrating the superiority of this new method over the convolution backprojection method, a widely used benchmark for the discrete data inverse Radon transform problem, are presented. Strategies to speed up the computationally intensive sections of the algorithm used to implement the new method have been identified and performance statistics demonstrating the substantial improvements that have been effected are reported. The problem of using optically thick absorption paths, which may result if the absorption frequency is to be employed as an optimization parameter for a wide range of measurement section sizes, and a possible method to resolve it have also been studied.

INTRODUCTION

Absorption spectroscopy techniques being non-intrusive are eminently suited for combustion and other hostile flow measurements requiring the determination of temperature and species concentration simultaneously over an extended region of the flow cross-section. Except in the trivial case of spatially homogeneous flows, such a determination requires the solution of two or more reconstruction problems to obtain the molecular absorption coefficient distributions at different optical frequencies from the corresponding line-of-sight absorption measurements.

The appropriate unifying framework for a large class of such reconstruction problems in the general 2-D case is the Radon (1917) transform. Even though its exact mathematical inverse is known, the practical *discrete* data problems are ill-posed and optimal solution strategies have been known to be problem or application specific (Herman and Rowland, 1973). Most past studies of reconstruction problems involving the simultaneous retrieval of temperature and concentration profiles, *e.g.*, Emmerman *et al.* (1980) and Ray and Semerjian (1983), have, however, involved straightforward applications of methods whose development has been largely motivated by medical imaging applications. In these applications, the distributions of interest are characterized by high contrast, and a large number of measurements are typically employed.

A new method that is especially suited for limited data problems involving smooth distributions has been reported in Ravichandran and Gouldin (1988a) and is termed the ***finite domain direct inversion*** (*FDDI*) method. Its application to the simultaneous retrieval of temperature and concentration profiles from absorption measurements including the use of other optimization parameters such as absorption frequencies has been reported in Ravichandran and Gouldin (1988b). Test results obtained by using synthetically generated absorption measurements show that the *FDDI* method is superior to the convolution backprojection (*CBP*) method, a widely used method for solving the discrete data, inverse Radon transform problem which is also typically used as a benchmark for evaluation (Ravichandran and Gouldin, 1988 a&b).

Here we address certain practical aspects of applying the *FDDI* method to the problem of retrieving temperature and concentration profiles from absorption measurements. First, the effect of pseudo-measurement error on the stability and the accuracy of the retrieved temperature and concentration profiles is studied. Test results obtained by applying the *FDDI* method to retrieve temperature and NO concentration profiles from synthetic absorption data containing random measurement error are presented; comparison with the *CBP* method once again demonstrates the superiority of the *FDDI* method. The *FDDI* method is intrinsically better equipped to cope with error containing data, because it permits the rational choice of a special parameter that controls the trade off between stability and accuracy of the reconstruction which is inevitable in the case of error containing data. Second, strategies to speed up the computationally intensive sections of the algorithm implementing the *FDDI* method have been identified with a view to making it computationally competitive with the *CBP* method. Substantial improvements have been effected by vectorizing appropriate sections of the computer code, and relevant statistics are presented. Lastly, possible approaches to the problem of using optically thick absorption paths, which may result if the absorption frequency is to be employed as an optimization parameter for a wide range of conditions and measurement section sizes, have been explored. A method is suggested which approximates the effect of spatial variations in absorption line width and line shape.

PROBLEM FORMULATION

The equation of radiative transfer with particulate scattering can be written as

$$\frac{dI_\nu}{dt} = -(k_\nu + k_\nu^s)I_\nu, \tag{1}$$

where I_ν is the specific intensity at frequency ν, k_ν, is the volumetric absorption coefficient, k_ν^s is the volumetric scattering coefficient and t is the path length. Emission and multiple scattering contributions are ignored.

For small optical depths and after compensating for particle scattering effects, the absorption measurement along a line-of-sight can be related to the molecular absorption coefficient using Eq. (1) as (Ravichandran and Gouldin, 1986):

$$M(\nu;b) = \int_{\Delta\nu} I_{\nu'}(a)\left(1 - \int_a^b k_{\nu'} dt\right)\tau_{\nu'}\, d\nu', \tag{2}$$

where $M(\nu;b)$ is the measured intensity at $t = b$, $I_{\nu'}(a)$ is the intensity at ν' of the light source at $t = a$, $\tau_{\nu'}$ represents the frequency response of the measurement system, $\Delta\nu$ is the measurement slit width centered at ν, and $\int_a^b k_\nu dt$ is referred to as the optical depth.

For measurement paths defined in Fig. 1, Eq. (2) can be recast in terms of the underlying reconstruction problem as (Ravichandran and Gouldin, 1988b)

$$\begin{aligned} M_p(\nu_{mn};s,\theta) &= \int_{-\infty}^{+\infty} n_m^*(s\cos\theta - t\sin\theta, s\sin\theta + t\cos\theta)\, dt \\ &= \int_{-\infty}^{+\infty}\int_{-\infty}^{+\infty} n_m^*(x,y)\, \delta(s - x\cos\theta - y\sin\theta)\, dx\, dy = \mathcal{R}n_m^*, \end{aligned} \tag{3}$$

where $M_p(\nu_{mn};s,\theta)$ is the line integral (along a line-of-sight defined by (s,θ)) which represents the pseudo-measured quantity that is derivable from the measurement in Eq. (2), ν_{mn} is the characteristic absorption frequency at which there is a transition from a certain state m to another state n, $n_m^*(x,y)$ is the distribution of the equilibrium population of state m, and $\mathcal{R}$ denotes the Radon transform. Further, $n_m^*(x,y)$ for a heteronuclear diatomic molecule can be related to its temperature $T(x,y)$ and concentration $n(x,y)$ as[1] (Incropera, 1974)

$$n_m^*(x,y) = n(x,y)\, F_v[T(x,y)]\, F_J[T(x,y)]; \tag{4}$$

$$F_v(T) = \exp(-vT_v/T)\,[1 - \exp(-T_v/T)],$$

where T_v is the characteristic temperature for vibration and v is the vibrational quantum number of state m; and

$$F_J(T) = \frac{(2J+1)\exp[-J(J+1)T_{rot}/T]}{T/T_{rot}},$$

where T_{rot} is the characteristic temperature for rotation and J is the rotational quantum number of state m.

Thus if the absorption coefficient distributions $n_{m1}^*(x,y)$ and $n_{m2}^*(x,y)$ can be reconstructed from line integrals $M_p(\nu_1;s,\theta)$ and $M_p(\nu_2;s,\theta)$, measurements at frequencies corresponding to different rotational lines,[2] the required distributions $n(x,y)$ and $T(x,y)$ can be determined using Eq. (4). The measurements are assumed to have sufficient spectral resolution for observing individual rotational lines.

SOLUTION METHOD

With a finite number of measurements the reconstruction problem of interest outlined above is mathematically ill-posed. And in the case of limited data problems (typical of combustion/high temperature applications), it is necessary and desirable to recognize and exploit all *a priori* information such as smoothness of the molecular absorption coefficient profiles as well as finiteness of domain and constraints on the distribution.

The practical problem of reconstructing the absorption coefficient distributions from a finite number of line-of-sight measurements distributed over a number of angular views can be stated as (Ravichandran and Gouldin, 1988b):

> Given $p^\ell(s_i,\theta_j)$ for $j = 1,\ldots,N$ $i = 1,\ldots,M_j$ and $p^\ell(s,\theta) = \mathcal{R}[f^\ell(x,y)]$,
> determine $f^\ell(x,y)$ subject to $(n,T) \in Q$.

$p^\ell(s,\theta) \equiv M_p(\nu_\ell;s,\theta)$, called a projection, and $f^\ell(x,y) \equiv n_{m\ell}^*(x,y)$ for $\ell = 1,2$; N is the total number of views and M_j is the number of projections in view j. Q represents the *domain* as well as the *range* constraints on the temperature and concentration distributions.

The *FDDI* method has been shown to be well suited for solving the above discrete data reconstruction problems (Ravichandran and Gouldin, 1988a&b), and the method is summarized below.

[1] Electronic excitation is not considered here because it does not materially affect the characteristics of the reconstruction problem of interest.

[2] States $m1$ and $m2$ differ only in their J numbers, while having the same electronic and vibrational states.

FDDI Method

Formulation. The discrete data reconstruction problem defined above is first transformed into a linear system of equations by converting it into a nearby continuous problem using consistent basis functions for the distribution[3] f as well as the measurements[3] p.

The distribution f is represented by axisymmetric 2–D basis functions $b(x,y)$ $(b(x,y) \equiv b(\sqrt{x^2+y^2}))$ located on the nodal points (x_k,y_k) of a triangular grid in the x–y plane (see Fig. 2):

$$f(x,y) \approx \sum_{k=1}^{K} f_k\, b(x - x_k, y - y_k). \tag{5}$$

K denotes the number of grid points inside the measurement domain Ω, which in this case is assumed to be the (scaled) unit circle without loss of generality. Since $p = \mathcal{R}f$ (from Eq. (3)), p is implicitly represented by the 1–D basis function defined as $\psi(s) = \mathcal{R}b(x,y)$. The reconstruction problem referred to above can then be converted into a linear system, *viz.*,

$$C\vec{f} = \vec{p}, \tag{6}$$

where $C \in \mathbb{R}^{M_N \times K}$; $\vec{f} \in \mathbb{R}^K$; and $\vec{p} \in \mathbb{R}^{M_N}$. M_N represents the total number of projections, *i.e.*, $M_N = \sum_{j=1}^{N} M_j$. This system is actually an assemblage of N sub-systems, one for each angular view (j), *viz.*,

$$C_j\vec{f} = \vec{p}_j, \quad j = 1,\ldots,N, \tag{7}$$

where

$$C_j \equiv \{c_{ik}^{[j]}\} \in \mathbb{R}^{M_j \times K}, \text{ and } c_{ik}^{[j]} = \psi(s_i - x_k\cos\theta_j - y_k\sin\theta_j);$$

$$\vec{f} = \{f_1,\ldots,f_K\}^T;$$

$$\vec{p}_j \in \mathbb{R}^{M_j}, \text{ with } \vec{p}_j = \{p(s_1,\theta_j),\ldots,p(s_{M_j},\theta_j)\}^T.$$

$b(x,y)$ and $\psi(s)$ are generated from the "Kaiser" window, a window commonly used in signal processing (Rabiner and Gold, 1975), using the known relationship between Radon and Fourier transforms (see **Appendix**).

Unconstrained Solution. To reduce the impact of measurement error the system of equations in Eq. (6) is purposefully made overdetermined with $M_N > K$. The resulting least squares problem, *i.e.*, $\min\|C\vec{f} - \vec{p}\|$, is solved through the *singular value decomposition (SVD)* of the C–matrix. A "regularization" approach is used to account for the ill-conditioning of the C–matrix due to the ill-posedness of the finite data reconstruction problem, and the solution is given by (see Ravichandran and Gouldin (1988a) for details)

$$\vec{f}_{LS}^{(\lambda)} = \sum_{i=1}^{K} \frac{u_i^T\vec{p}\,\sigma_i}{\sigma_i^2 + \lambda^2}\, v_i. \tag{8}$$

The u_i's ($u_i \in \mathbb{R}^{M_N}$) and the v_i's ($v_i \in \mathbb{R}^K$) are referred to as the left and right singular vectors, respectively; and the σ_i's ($\sigma_i \in \mathbb{R}$) are the singular values such that $\sigma_1 \geq \ldots \geq \sigma_K \geq 0$; and λ ($\lambda \in \mathbb{R}$) is the regularization parameter. It should be noted that Eq. (8) represents the solution to the following "damped" least squares problem

$$\min\|C\vec{f} - \vec{p}\|^2 + \lambda^2\|\vec{f}\|^2. \tag{9}$$

The damping is provided by design to minimize the impact of noise in the form of instability in the solution. Noise can be present either explicitly as random measurement error or implicitly as representation error (manifested in the C–matrix). The extent of damping required or the optimal value of λ is determined using a plot of the residual norm ($\|C\vec{f}_{LS}^{(\lambda)} - \vec{p}\|$) and the solution norm ($\|\vec{f}_{LS}^{(\lambda)}\|$).

Constrained Solution. The *domain* and *range* constraints on the distribution of f are used to arrive at the constrained least squares formulation.

[3] To simplify the notation, the superscript denoting the measurements at two frequencies is omitted while describing the *FDDI* method.

The fact that the measurement domain is physically bounded is incorporated by appropriately modifying the projection matrix C so that Eq. (6) is replaced by

$$D\vec{f} = \vec{p}, \tag{10}$$

where $D \in \mathbb{R}^{M_N \times K}$ is explicitly computable for a given grid, measurement domain, and experiment design in terms of the orientation of the lines-of-sight (see Ravichandran and Gouldin (1988a) for details).

The constraints on the *range* of f can in general be represented as

$$f_{\min} \leq f \leq f_{\max}.$$

Taking $f_{\min}$ to be defined at $(x_{l'}, y_{l'})$, $l' = 1, \ldots, L'$; and $f_{\max}$ to be defined at $(x_{l''}, y_{l''})$, $l'' = 1, \ldots, L''$; the constraints can be expressed in matrix form as

$$G\vec{f} \equiv \begin{bmatrix} G' \\ -G'' \end{bmatrix} \vec{f} \geq h \equiv \begin{bmatrix} \hat{f}_{\min} \\ -\hat{f}_{\max} \end{bmatrix}, \tag{11}$$

where $G' \in \mathbb{R}^{L' \times K}$, $G'' \in \mathbb{R}^{L'' \times K}$, $G \in \mathbb{R}^{L \times K}$, $\hat{f}_{\min} \in \mathbb{R}^{L'}$, $\hat{f}_{\max} \in \mathbb{R}^{L''}$, and $h \in \mathbb{R}^{L}$. It is implied that $L = L' + L''$. Elements of the matrices G' and G'' represent the appropriate 2–D basis function contributions corresponding to the locations where f is constrained.

The constrained problem then becomes

$$\textit{Minimize } \|D\vec{f} - \vec{p}\| \textit{ subject to } G\vec{f} \geq h.$$

This has been referred to as the PROBLEM LSI (denoting linear least squares with linear inequality constraints) and is solved by a series of problem transformations (Lawson and Hanson, 1974). The solution after incorporating the regularization approach of the unconstrained problem is given by

$$\vec{f}_{LSI}^{(\lambda)} = \vec{f}_{LS}^{(\lambda)} + V\left\{\Sigma_S^{(\lambda)}\right\}^{-1} \hat{z}. \tag{12}$$

$\vec{f}_{LS}^{(\lambda)}$ is the unconstrained solution corresponding to Eq. (10); and V, $\Sigma_S^{(\lambda)}$, $\hat{z}$ are defined in the **Appendix**.

Application of the *FDDI* Method to the Retrieval Problem

Use of *a priori* Information. The formulation of the constrained problem (above) presupposes the availability of range constraints on f^1 and f^2. However, the constraints that are available, as stated earlier while defining the discrete data reconstruction problem, relate to the range of temperature and concentration distributions. While the non-negativity constraints on f^1 and f^2 are always applicable, the specific constraint distributions for f^1 and f^2 have to be derived from the constraints on the range of $n(x,y)$ and $T(x,y)$. This has been dealt with in detail in Ravichandran and Gouldin (1988b), and is excerpted in the **Appendix**.

Minimizing Representation Error. It has been noted earlier that there is an implicit noise in the reconstruction process due to the representation error. This error is unavoidable because we are in fact solving the *nearest* continuous problem that exists in the subspace of the chosen basis functions. Two separate strategies have been identified in Ravichandran and Gouldin (1988b) to improve representability, *viz.*, Use of shifted distributions and Judicious choice of the absorption frequencies.

Using "shifted distributions" involves subtracting an appropriate function g from the distribution f with the objective of minimizing the component of the resulting function f^ℓ $(f^\ell \equiv f^\ell - g^\ell)$ that resides in the null space of the 2–D basis functions used to represent the distribution. The measurements (in p^ℓ) are also to be suitably modified.

Choice of "the absorption frequency or the J–value" has been found to have a strong influence on the reconstruction error. Also, since f^ℓ reconstructions corresponding to 2 J–values are required, the existence of a preferred range for the optimal pair has been identified in Ravichandran and Gouldin (1988b).

ABSORPTION MEASUREMENTS—SYNTHETIC DATA GENERATION

In order to test the performance of the *FDDI* method with error containing data, synthetic absorption data are generated using phantom temperature and concentration distributions.[4]

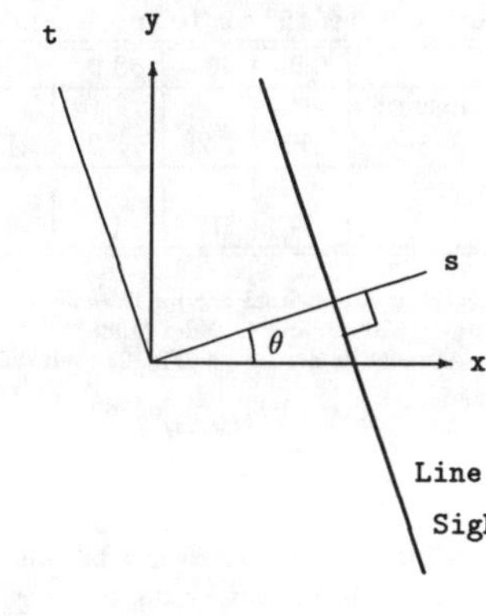

Fig. 1 Coordinate system—Absorption measurement line-of-sight defined by constant θ and s is shown.

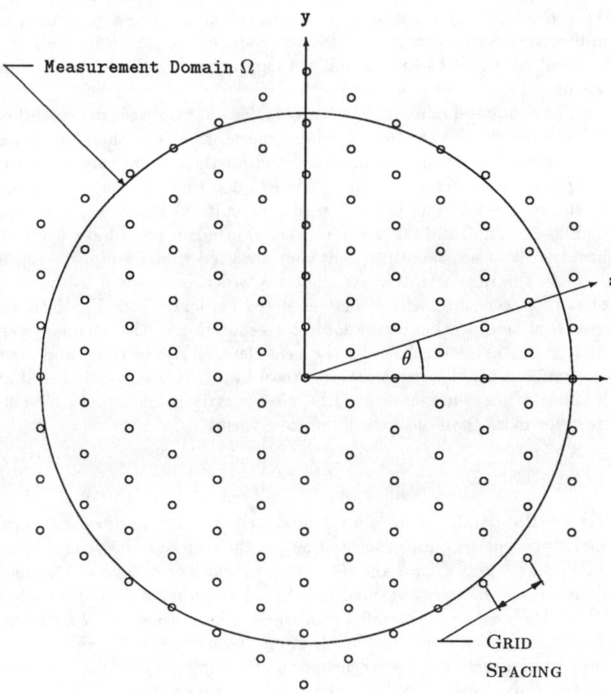

Fig. 2 Triangular grid representing the location of the 2–D basis functions.

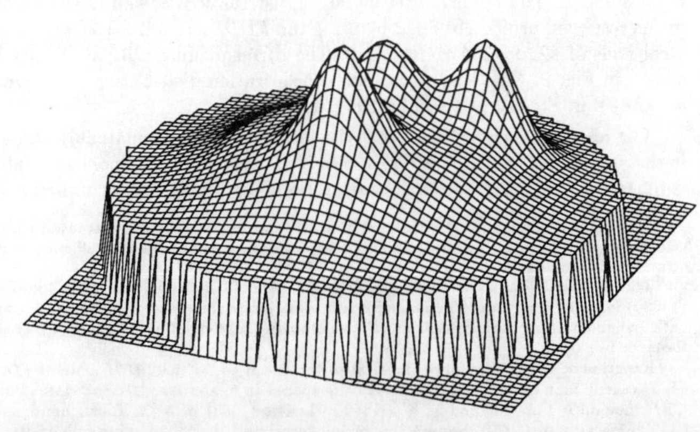

Fig. 3 Temperature distribution phantom—Original. Maximum temperature = 1144°K.

[4] These distributions have been used as "test phantoms" in Ravichandran and Gouldin (1988b). They are constructed by superposing a sum of 3 Gaussians on experimentally observed axisymmetric temperature and NO concentration profiles reported in Anand and Gouldin (1985) and by normalizing the composite distributions.

Table I: Experiment design[a] for 102 measurements in 6 views.

View Angle θ_j (°)	8.95	40.9	68.95	100.9	128.95	160.9
Spacing between lines-of-sight[b] in view j—Δs_j	.12	.126	.12	.126	.12	.126
No. of measurements[b] in view j—M_j	17	17	17	17	17	17

[a]This experiment design is an optimal one for the chosen number of measurements and has been used in Ravichandran and Gouldin (1988b).

[b]The lines-of-sight intersect the s-axis (see Fig. 2) such that

$$s_i = \left[i - \frac{(M_j+1)}{2}\right] \cdot \Delta s_j \quad i = 1, \ldots, M_j.$$

First, the f^ℓ distributions are constructed using the phantom profiles in Fig. 3 and 4 using Eq. (4). Second, simulated (error free) measurements $p(s_i, \theta_j)$ are computed from the f^ℓ distributions using the experiment design given in Table I. The grid for locating the 2–D basis functions consists of 97 nodal points in the domain, i.e., K=97 (this is identical to the one shown in Fig. 2). Lastly, random errors (a certain fraction (say ϵ) of the maximum) were introduced in the synthetically generated absorption data. This was done by constructing an error vector using a uniform random number generator suitably scaled to produce a sequence in the range [-ϵ, +ϵ] and adding it to the normalized measurement vector (whose elements are in [0,1]).

The simulated random error vector referred to above is dependent on the "seed" used in the random number generator. It was therefore necessary to determine if several simulated experiments were necessary to include a representative error domain. This was done by studying the sensitivity of the reconstructions to the "seed." Results of this study are reported in Fig. 5. The residual norm *vs.* solution norm plots have been chosen because these are the same ones that are used to determine the optimum value of the regularization parameter λ which is sensitive to the presence of random measurement errors (as noted earlier). From Fig. 5, it can be seen that the location of the foot (corresponding to the optimum value of λ) is much more sensitive to the error level (ϵ) than the "seed" used in the random number generator.[5] Accordingly, it was concluded that the number of measurements used in a set was large enough to obviate the need for more than one simulated experiment.

TEST RESULTS

Temperature and NO concentration distributions were retrieved from synthetic absorption data generated as per the procedure described above at 2 J–values,[6] *viz.*, $J_1 = 8$ and $J_2 = 12$ for various error levels. The distributions f^1 and f^2 were obtained first by solving for the shifted distributions $\boldsymbol{f}^1$ and $\boldsymbol{f}^2$ with the modified absorption data using the *FDDI* method described earlier (see Ravichandran and Gouldin (1988b) for details). A simple unidirectional constraint of the form $\boldsymbol{f}^\ell \geq$ Const. is also imposed on the shifted distributions. The regularization parameter that was used with different error levels for the 2 J–values is given in Table II. The results quantified in terms of the error distance measures (defined in the **Appendix**) including a comparison with the *CBP* method[7] are given in Figs. 6 and 7. The retrieved temperature distribution as well as the error in the retrieved profile obtained by using the *FDDI* method for a simulated error level of 4% are shown in Fig. 8. The corresponding *CBP* profiles are shown in Fig. 9. The associated NO concentration distribution and error are shown in Fig. 10 (*FDDI*) and Fig. 11 (*CBP*).

The results clearly indicate that the *FDDI* method is remarkably stable in the presence of error. The *FDDI* retrieved profiles are also more accurate compared to the *CBP* ones in terms of both the profile averaged distance

[5]It should be noted that the parameters used in generating the reconstructions applicable to the plots in Figs. 5(a) and 5(b) were identical except for the differences in error simulation aspects under study.

[6]This pair of J–values is the optimal choice given in Ravichandran and Gouldin (1988b) for error free data. The characteristic variation in the efficacy of reconstruction with error containing data for different J–values was found to be similar to that reported for error free data.

[7]Experiment design is almost identical to the one used for the *FDDI* method (Table I) except that the 6 views are are equally spaced in θ, and $\Delta s_j \equiv \Delta s = .118$. The *CBP* method is implemented as described in Lewitt (1983) with the Ramachandran-Lakshminarayanan (1970) discrete convolving kernel and the use of cubic spline interpolation for the convolved projections. Shifting functions identical to the ones in the *FDDI* method are also employed.

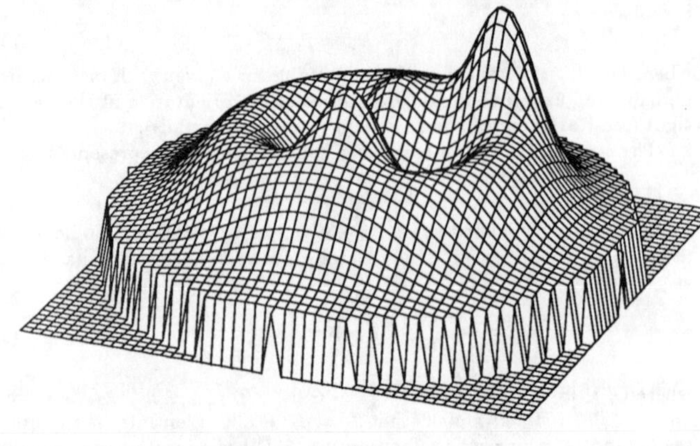

Fig. 4 NO concentration distribution phantom—Original. Maximum NO concentration = 1.07 ppm.

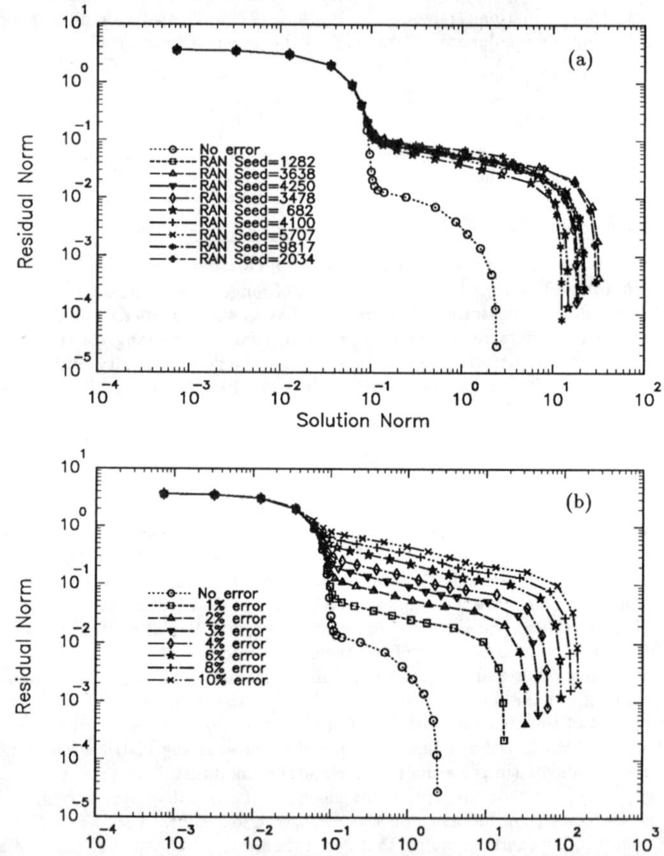

Fig. 5 Residual norm ($\|D\vec{f}_{LS}^{(\lambda)} - \vec{p}\|$) *vs.* solution norm ($\|\vec{f}_{LS}^{(\lambda)}\|$) for a range of values of the regularization parameter λ (from $10\sigma_1$ to $\sigma_K/10$: σ_1 and σ_K are the extreme singular values of the D-matrix.). *FDDI* reconstructions of f using synthetic data generated from phantom temperature and NO concentration profiles in Figures 3 and 4 with $J = 8$. (a) Plots for different data sets having an error level (ϵ) = 2% of maximum. (b) Plots with different error levels from $\epsilon = 0\%$ to $\epsilon = 10\%$ with the same seed used in each case for generating the data with random error.

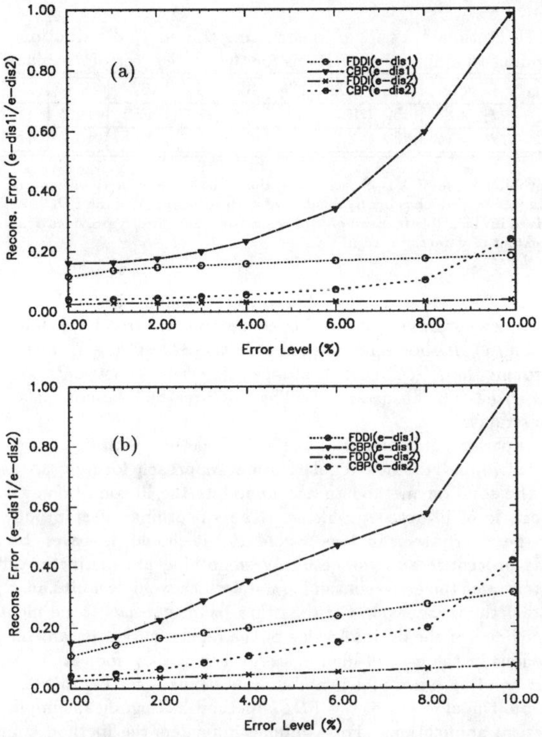

Fig. 6 Error (e–dis1/e–dis2) in the temperature and concentration profiles retrieved by the *FDDI* and the *CBP* methods for different error levels in the absorption data; $J_1 = 8$, $J_2 = 12$. (a) Temperature. (b) NO concentration.

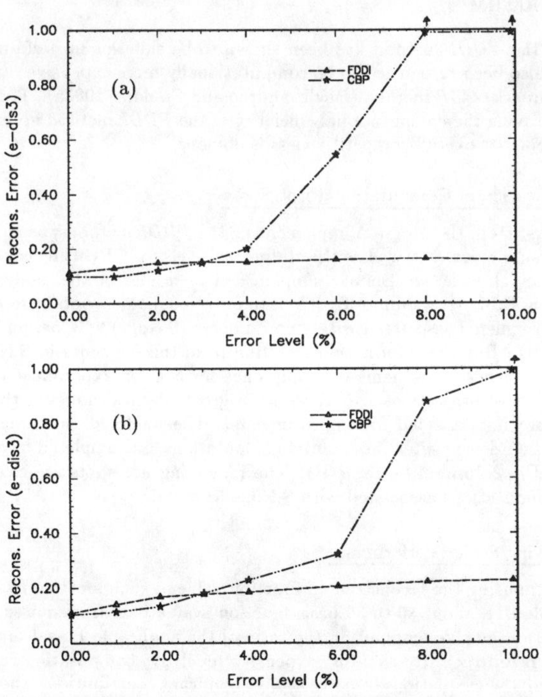

Fig. 7 Error (e–dis3) in the temperature and concentration profiles retrieved by the *FDDI* and the *CBP* methods for different error levels in the absorption data; $J_1 = 8$, $J_2 = 12$. (a) Temperature. (b) NO concentration. Maximum deviation represented by e–dis3 has been normalized in each case with a reference value which is equal to the maximum in the original distribution. *CBP* maximum deviations at high error levels are very large due to the occurrence of some non-physical values in the reconstructed f^1 and f^2 distributions.

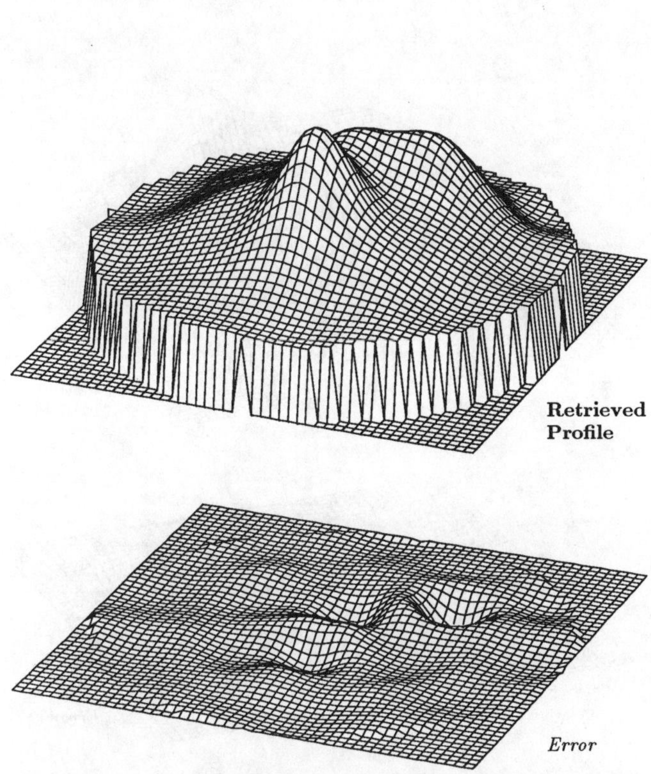

Fig. 8 Retrieval of temperature distribution by the *FDDI* method using 102 measurements (with random error) distributed over 6 viewing angles at 2 frequencies. Error level in the simulated measurements = 4% of maximum. *Error*, defined as $T_{retr.} - T$, has been plotted to the same scale as the retrieved profile.

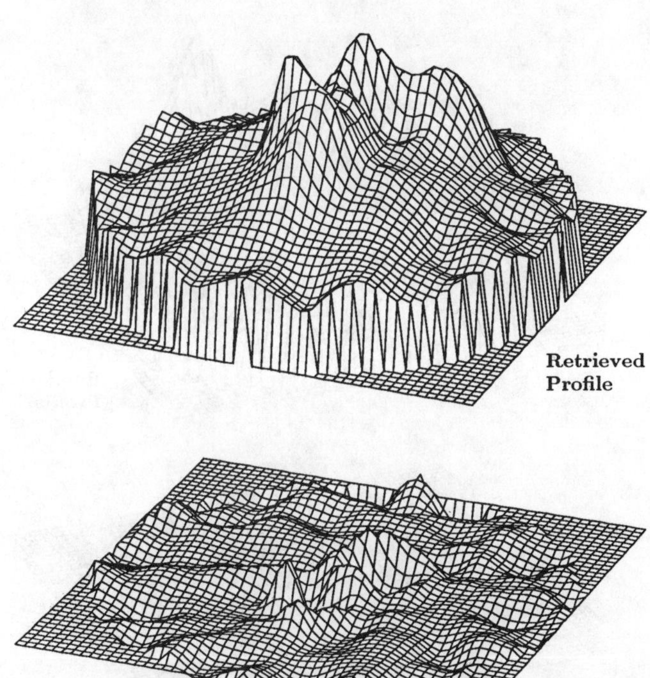

Fig. 9 Retrieval of temperature distribution by the *CBP* method using 102 measurements (with random error) distributed over 6 viewing angles at 2 frequencies. Error level in the simulated measurements = 4% of maximum. *Error*, defined as $T_{retr.} - T$, has been plotted to the same scale as the retrieved profile.

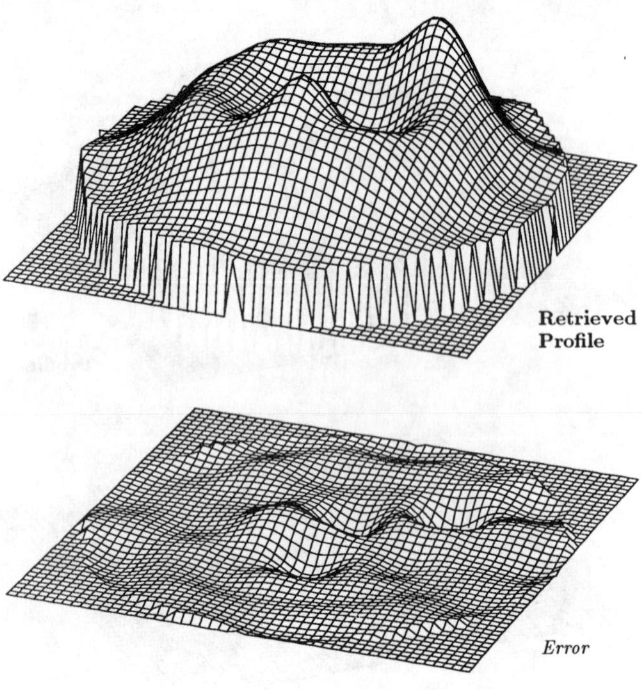

Fig. 10 Retrieval of NO–concentration distribution by the *FDDI* method using 102 measurements (with random error) distributed over 6 viewing angles at 2 frequencies. Error level in the simulated measurements = 4% of maximum. *Error*, defined as $n_{\text{retr.}} - n$, has been plotted to the same scale as the retrieved profile.

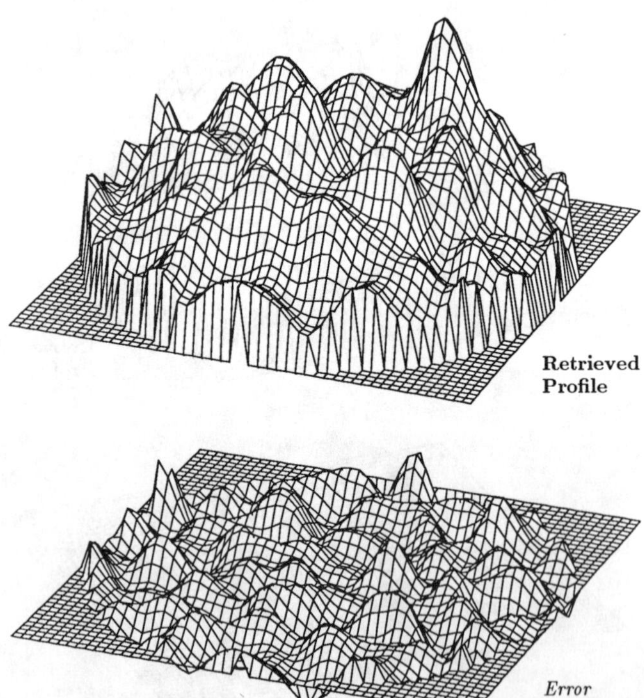

Fig. 11 Retrieval of NO–concentration distribution by the *CBP* method using 102 measurements (with random error) distributed over 6 viewing angles at 2 frequencies. Error level in the simulated measurements = 4% of maximum. *Error*, defined as $n_{\text{retr.}} - n$, has been plotted to the same scale as the retrieved profile.

Table II: Optimal[a] λ's for reconstructing shifted $\boldsymbol{f}^{\ell}$ distributions corresponding to different error levels for the chosen pair of J–values.

Error Level		0%	1%	2%	3%	4%	6%	8%	10%
λ^2	$J_1 = 8$	.50	3.3	5.5	7.6	9.3	12.1	14.2	17.2
	$J_2 = 12$	.20	2.2	4.7	6.4	7.7	10.8	13.4	15.4

[a]Optimum value of λ is chosen using the method given in Ravichandran and Gouldin (1988a), which is briefly mentioned in the description of the *FDDI* method. λ^2 is given in the Table (instead of λ), because its value directly provides a measure of the extent of damping used in Eq. (9).

measures, e–dis1 and e–dis2 (see Fig. 6). Only the maximum deviation (e–dis3) is somewhat larger for the *FDDI* method for error levels lower than 3% (see Fig. 7). It should be noted that in these "test" cases, the potential for improving the *FDDI* reconstructions using range constraints on $T(x, y)$ and $n(x, y)$ has not been tapped. The *CBP* method cannot make use of such constraints.

The superior performance of the *FDDI* method with error containing data is attributed to the "regularization" approach forming an integral part of the solution method to accommodate the ill-conditioning that is characteristic of ill-posed problems. The smoothing effect produced by this can be clearly seen in Figs. 8 and 10. It should, however, be noted that this procedure and the resulting smoothing are distinctly different from filtering of the error containing data which would require an *a priori* estimate of the error level. On the other hand, it needs to be mentioned that the choice of the optimal value of the regularization parameter is not very obvious in the case of high error levels when the foot in the residual norm *vs.* solution norm plot is not sharply defined (see Fig. 5(b)).

The built-in stability of the *FDDI* method has significant implications for practical applications. For example, it renders the method eminently suited for control applications where a meaningful trade off exists between speed of measurement and accuracy.

IMPROVING THE COMPUTATIONAL EFFICIENCY OF THE *FDDI* ALGORITHM

While the *FDDI* method has been shown to be superior in performance, it has also been recognized to be computationally more expensive than the (benchmark) *CBP* method (Ravichandran and Gouldin, 1988a). The value of improving the computational efficiency of the *FDDI* method in terms of enhancing its overall competitiveness is obvious.

Identification of Computational Hot Spots

As a first step, the program implementing the *FDDI* method was analyzed at the subprogram as well as the statement levels for CPU time consumed using an "Interactive Debug Sampler." The results of the analysis are given in Table III (column 3). The computational "hot spots" are clearly in the segments associated with step numbers (S.No.'s) 2, 3, 5a, and 6a of Table III. But the information resulting from the segments in S.No.'s 1, 2, and 3 needs to be generated only once for a given experiment design, *i.e.*, a set of projections and views for a given physical domain; they are therefore not targeted for improvements. The modified run times that would be relevant after these initial computations are completed have been clocked in column 4 of Table III. The remaining hot spots are clearly in the computations associated with S.No.'s 5a and 6a.

Improving Scalar Performance

For computing the reconstructed function value at each location (S.No. 6a of Table III), about 30 to 36 basis function evaluations are required. Each evaluation involves computing the ratio of the applicable Bessel functions (see **Appendix**). By avoiding expensive repetitive computations and restricting the evaluations to numerically significant contributions, the scalar code associated with S.No. 6a of Table III has been substantially improved (see column 5).

Vectorization

The maximum potential for vectorization lies in the other hot spot, *viz.*, S.No. 5a of Table III. Vectorization was initially done using the vectorizing compiler which identifies vectorizable segments. Special attention was then paid to sections of the code which are known hot spots, but had been

Table III: CPU Time (in milliseconds) Comparisons[a] for a Test Problem—102 Projections/97 Nodes, Non-negativity Constraint at the Nodes.

1	2	3	4	5	6	7
S.No.	Description	*FDDI*[b]	*FDDI*[c]	*FDDI*[d]	*FDDI*[e]	*CBP*[f]
1	Grid generation	250[b]	10[c]	10	10	–
2	Computing the projection (D) matrix	7487[b]	187[c]	187	187	–
3	Computing the SVD of the D–matrix	2260[b]	40[c]	40	40	–
4	Computing the constraint matrix	72	72	72	72	–
5a	Solving the constr. least squares problem for the 2–D basis function coeffs.	2119	2119	2119	372[e]	–
5b	Computing the convolved projections	–	–	–	–	19[f]
6a	Computing the fn. values[g] from the basis fn. coeffs.	1875	1875	810[d]	810	–
6b	Computing the fn. values[g] by backprojection	–	–	–	–	513[f]
	Total	14063[b]	4303[c]	3238[d]	1491[e]	532[f]

[a]Performance statistics have been obtained on an IBM® 3090 machine at the Cornell National Supercomputing Facility.
[b]Full run with scalar code; information for S.No.'s 1, 2 and 3 is computed *ab initio*.
[c]Solution only run with scalar code; data already generated w.r.t S.No. s 1, 2, and 3 is assumed to be read from secondary storage.
[d]Solution only run with improved scalar code.
[e]Vectorized code as elaborated in text.
[f] *CBP* method is implemented as indicated earlier (see footnote 7).
[g]The reconstructed function values are computed on a grid with 50 x 50 subdivisions.

confined to the scalar mode by the conservative vectorizing compiler. Selective compiler directives have been incorporated in such sections to override the objections *of* or the inefficient vectorizing choices made *by* the vectorizing compiler so as to achieve maximum vectorization while preserving the integrity of the code, *i.e.*, obtaining the same reconstruction results as with the scalar code. The improvement in performance is reflected in the run time statistics after vectorization that are presented in column 6 of Table III.

Comparison with the *CBP* Method

The computational performance of the *CBP* method is shown alongside that of the *FDDI* algorithm in column 7 of Table III. Considering that the *CBP* method does not accommodate constraints on the f^t distributions to be reconstructed, most of the contribution due to S.No. 5a must be removed from the run time of the *FDDI* method for a valid comparison. After doing so, the present implementation of the *FDDI* method would be half as fast as the *CBP* method. While it is to be noted that the algorithm implementing the *CBP* method may not be the most optimal one, the *FDDI* method can be considered to be competitive with the *CBP* method in an order of magnitude sense. More improvements in the section of the code represented by S.No. 6a in Table III are considered to be feasible, and would enable improving the competitiveness of *FDDI* method further.

THE "OPTICALLY THICK" PROBLEM

An integral equation (Eq. (2)) which formed the basis of the underlying characteristic reconstruction problem (in Eq. (3)) was derived earlier under the assumption of small optical depths. In practical situations, large measurement sections would entail large physical path lengths, thereby limiting the choice of the absorption frequency in order to satisfy this assumption. On the other hand, the ability to work with large optical depths would not only enhance the absorption signal with obviously favorable implications for the S/N ratio, but also permit improved reconstructions resulting from a wider choice of absorption frequencies.

Without the assumption of small optical depths, the absorption measurement along a line-of-sight can be related to the molecular absorption coefficient by the original form of Eq. (2) as

$$M(\nu_0; b) = I_{\nu_0}(a) \int_{\Delta\nu} \exp\left[-\int k_\nu dt\right] \tau_\nu \, d\nu, \tag{13}$$

where ν_0 is the absorption frequency. The source intensity $I_{\nu_0}(a)$ is assumed to be a weak function of ν.

The equilibrium population n_m^*, the reconstruction of whose distribution is our underlying motivation, can be related to the absorption coefficient k_ν through the associated line shape function g_ν centered at ν_0 (Ravichandran and Gouldin, 1986) and Eq. (13) can then be written as

$$M(\nu_0; b) = I_{\nu_0}(a) \int_{\Delta\nu} \exp\left[-\Gamma \int n_m^* g_\nu dt\right] \tau_\nu \, d\nu, \tag{14}$$

where Γ is a catch-all for terms independent of t and ν.

The problem of extracting the required integral $\int n_m^* dt$ from Eq. (14) is not trivial because of the spatial dependence of g_ν (primarily through the temperature dependence). Consider the case of Doppler broadening; the normalized Doppler-broadened shape g_ν is defined as (Yariv, 1971)

$$g_\nu = \frac{2(\ln 2)^{0.5}}{\pi^{0.5}\Delta\nu_D} \cdot \exp\left[-\frac{4\ln 2(\nu-\nu_0)^2}{\Delta\nu_D{}^2}\right]; \tag{15}$$

$$\Delta\nu_D = 2\nu_0\sqrt{\frac{2\kappa T}{Mc^2}}, \tag{16}$$

where M is the molecular weight, c is the velocity of light, κ is Boltzmann's constant, and T is the temperature.

As a first step towards solving for $\int n_m^* dt$, the Gaussian line shape for g_ν is approximated by a triangle whose base is proportional to $\Delta\nu_D$, and whose height is proportional to $1/\Delta\nu_D$. Eq. (14) can then be recast as

$$M(\nu_0; b) = I_{\nu_0}(a) \int_{\Delta\nu} \exp\left[-\Gamma_1 \int \frac{n_m^*}{\sqrt{T}}(\Gamma_2 - \frac{|\nu-\nu_0|}{\sqrt{T}})dt\right] \tau_\nu \, d\nu, \tag{17}$$

where Γ_1 and Γ_2 are catch-alls for terms independent of t and ν. The next step is to find an effective temperature $T_{\text{eff.}}$ such that Eq. (17) can be written as

$$M(\nu_0; b) = I_{\nu_0}(a) \int_{\Delta\nu} \exp\left[-\int \frac{n_m^*}{\sqrt{T}} dt\right]^{\Gamma_1(\Gamma_2 - \frac{|\nu-\nu_0|}{\sqrt{T_{\text{eff.}}}})} \tau_\nu \, d\nu. \tag{18}$$

Eq. (18) can then be solved to obtain $\int(n_m^*/\sqrt{T})dt$, which can then be used to solve the reconstruction problem. Since the temperature as well as the n_m^* distributions are not known to start with, $T_{\text{eff.}}$ would have to be solved for iteratively using an initial guess. In practice, one hopes that just one or two iterations may be sufficient; and in control applications, the "previous" value is likely to be good enough.

SUMMARY AND CONCLUDING REMARKS

The suitability of the *FDDI* method for the problem of retrieving temperature and species concentration measurements from a limited number of absorption measurements has previously been established (Ravichandran and Gouldin, 1988b). Here we have examined certain important practical issues relating to the application of the *FDDI* method to this problem.

The robustness of the the *FDDI* method in coping with absorption data containing random errors has been demonstrated through the stability of the profiles retrieved from simulated measurements based on test phantoms. The fact that the *FDDI* retrieved profiles are also much more accurate than the (benchmark) *CBP* ones proves that the trade off between stability and accuracy, which is integral to the *FDDI* solution method, is not far from being optimal. It is important to note that the value of the "regularization" parameter, which controls this trade off, is internally determined from the absorption data without requiring *a priori* estimates of the measurement error.

Even though the *FDDI* method has a computationally intensive structure, it has been shown that its computational competitiveness in practice can be enhanced substantially by intelligent use of primary and secondary computer storage as well as by vectorizing the relevant sections of the algorithm implementing the method.

Recognizing the value of using the absorption frequency as an optimization parameter in obtaining improved reconstructions on the one hand and the need for accommodating a wide range of section sizes in practical measurement situations on the other, we have proposed a method for rendering the resultant "optically thick" problem mathematically tractable.

ACKNOWLEDGEMENTS

This work has been supported by the National Science Foundation (Grant CBT-8714040); and this research was conducted using the Cornell National Supercomputing Facility, a resource of the Center for Theory and Simulation in Science and Engineering at Cornell University, which is funded in part by the NSF, New York State, and the IBM Corporation and members of the Corporate Research Institute.

REFERENCES

Anand, M. S., and Gouldin, F. C., 1985, "Combustion efficiency of a premixed continuous flow combustor," *J Engineering for Gas Turbines and Power Trans ASME*, Vol. 107, pp. 695–705.

Emmerman, P. J., Goulard, R., Santoro, R. J., and Semerjian, H. G., 1980, "Multiangular absorption diagnostics of a turbulent Argon-Methane jet," *J. Energy* Vol. 4, p. 70.

Herman, G. T., 1980, *Image Reconstructions from Projections: The Fundamentals of Computerized Tomography*, Academic Press, New York.

Herman, G. T. and Rowland, S. W., 1973, "Three methods of reconstructing objects from x rays: A comparative study," *Computer Graphics and Image Processing* Vol. 2, pp. 151–178.

Incropera, F. P., 1974, *Introduction to Molecular Structure and Thermodynamics*, John Wiley and Sons, New York.

Lawson, C. L. and Hanson, R. J., 1974, *Solving Least Squares Problems*, Prentice Hall, Englewood Cliffs, New Jersey.

Rabiner, L. R. and Gold, B., 1975, *Theory and Application of Digital Signal Processing*, Prentice Hall, Englewood Cliffs, New Jersey.

Radon, J., 1917, "Über die Bestimmung von Funktionen durch ihre Integralwerte längs gewisser Mannigfaltigkeiten," *Berichte Sächsische Akademie der Wissenschaften, Leipzig, Math. Phys. Kl.* Vol. 69, p. 262.

Ramachandran, G. N. and Lakshminarayanan, A. V., 1970, "Three dimensional reconstruction from radiographs and electron micrographs: Application of convolutions instead of Fourier transforms," *Proc. Nat. Acad. Sci. USA* Vol. 68, pp. 2236–2240.

Ravichandran, M. and Gouldin, F. C.,1986, "Determination of temperature and concentration profiles using (a limited number of) absorption measurements," *Combustion Science and Technology* Vol. 45, pp. 47–64.

Ravichandran, M. and Gouldin, F. C., 1988a, "Reconstruction of smooth distributions from a limited number of projections," *Applied Optics* Vol. 27, pp. 4084–4097.

Ravichandran, M. and Gouldin, F. C., 1988b, "Retrieval of asymmetric temperature and concentration profiles from a limited number of absorption measurements," *Combustion Science and Technology* Vol. 60, p. 231.

Ray, S. R. and Semerjian, H. G., 1983, "Laser tomography for simultaneous concentration and temperature measurement in reacting flows," *Progress in Astronautics and Aeronautics, AIAA* Vol. 92, p. 301.

Vincenti, W. G. and Kruger, C. H., 1965, *Introduction to Physical Gas Dynamics*, John Wiley and Sons, New York.

Yariv, A., 1971, *Introduction to Optical Electronics*. Holt, Rinehart and Winston, New York.

APPENDIX

1-D and 2-D Basis Functions—*FDDI* Method

Based on the known relationship between Radon and Fourier transforms, *viz.*, $\mathcal{F}_1(\mathcal{R}f) = \mathcal{F}_2(f)$, $b(x,y)$ and $\psi(s)$ are generated from the "Kaiser" window as follows:

$$b(x,y) \equiv b(r) = \mathcal{F}_2^{-1}[W(R)] = \begin{cases} \dfrac{2\pi C_0^2}{I_0(2\pi H_0 C_0)} \dfrac{J_1\{2\pi C_0(r^2 - H_0^2)^{0.5}\}}{2\pi C_0(r^2 - H_0^2)^{0.5}} & \text{if } |r| \geq H_0 \\ \dfrac{2\pi C_0^2}{I_0(2\pi H_0 C_0)} \dfrac{I_1\{2\pi C_0(H_0^2 - r^2)^{0.5}\}}{2\pi C_0(H_0^2 - r^2)^{0.5}} & \text{if } |r| \leq H_0, \end{cases} \tag{19}$$

$$\psi(s) = \mathcal{F}_1^{-1}[W(R)] = \begin{cases} \dfrac{2C_0}{I_0(2\pi H_0 C_0)} \dfrac{\sin\{2\pi C_0(s^2 - H_0^2)^{0.5}\}}{2\pi C_0(s^2 - H_0^2)^{0.5}} & \text{if } |s| \geq H_0 \\ \dfrac{2C_0}{I_0(2\pi H_0 C_0)} \dfrac{\sinh\{2\pi C_0(H_0^2 - s^2)^{0.5}\}}{2\pi C_0(H_0^2 - s^2)^{0.5}} & \text{if } |s| \leq H_0. \end{cases} \tag{20}$$

$\mathcal{F}_1$ and $\mathcal{F}_2$ are one- and two-dimensional Fourier transforms; $W(R)$ is the "Kaiser" window; R is the Fourier space variable; J_1 is the Bessel function of the first kind–order 1; I_ℓ is the modified Bessel function of the first kind–order ℓ; C_0 is the cut off frequency of the window function; and H_0 is the influence length parameter. By definition, $b(r)$ and $\psi(s)$ are band-limited; they can be rendered numerically space-limited by an appropriate choice of the non-dimensional product[8] $C_0 H_0$. The cut off frequency C_0 is related to the grid spacing.[9]

Definitions of Terms Used in *FDDI* Method—Constrained Solution

Singular value decomposition of the D-matrix gives

$$D = U\Sigma V^T \equiv [\underbrace{U_1}_{K} : \underbrace{U_2}_{M_N - K}] \begin{bmatrix} \Sigma_S \\ 0 \end{bmatrix} V^T, \tag{21}$$

where $\Sigma_S \in \mathbb{R}^{K \times K}$ is a diagonal matrix of the singular values;

$$\Sigma_S^{(\lambda)} = \Sigma_S + \lambda^2 \Sigma_S^{-1};$$

and $\hat{z}$ is the solution to the PROBLEM LDP (Least Distance Programming), *viz.*,

$$\textit{Minimize } \|z\| \textit{ subject to } \tilde{G}z \geq \tilde{h},$$

where $\tilde{G} = GV\left\{\Sigma_S^{(\lambda)}\right\}^{-1}$; and $\tilde{h} = h - GV\left\{\Sigma_S^{(\lambda)}\right\}^{-1} U_1^T \vec{p}$.

Derived Range Constraints on f^1 and f^2

The constraints on the range of $n(x,y)$ and $T(x,y)$ can in general be expressed as

$$\begin{aligned} n_{\min} &\leq n \leq n_{\max}, \\ T_{\min} &\leq T \leq T_{\max}. \end{aligned} \tag{22}$$

The constraint distributions for f^1 and f^2 are derived as

$$\left.\begin{aligned} f_{\min}^{\ell} &= n_{\min} \cdot \min_{T_{\min} \leq T' \leq T_{\max}} \{F_v(T') F_{J\ell}(T')\} \\ f_{\max}^{\ell} &= n_{\max} \cdot \max_{T_{\min} \leq T' \leq T_{\max}} \{F_v(T') F_{J\ell}(T')\} \end{aligned}\right\} \ell = 1, 2. \tag{23}$$

Constraints that are applicable to the ratio f^1/f^2 are derived as

$$\begin{aligned} \frac{f^1}{f^2} &\leq \exp\left[\frac{T_{rot}}{T_{\min}}\{J_2(J_2+1) - J_1(J_1+1)\}\right] \cdot \frac{(2J_1+1)}{(2J_2+1)}, \\ \frac{f^1}{f^2} &\geq \exp\left[\frac{T_{rot}}{T_{\max}}\{J_2(J_2+1) - J_1(J_1+1)\}\right] \cdot \frac{(2J_1+1)}{(2J_2+1)}. \end{aligned} \tag{24}$$

Without any loss of generality, it is assumed that $J_2 > J_1$ in Eq. (18).

Error Distance Measures

Three error distance measures often used in the reconstruction literature (Herman, 1980), are defined as follows:

$$\begin{aligned} \text{e-dis1} &= \|f_{\text{recons.}} - f\| / \|f - \{\overline{f}\}_\Omega\|, \\ \text{e-dis2} &= \{|f_{\text{recons.}} - f|\}_\Omega / \{|\overline{f}|\}_\Omega, \\ \text{e-dis3} &= \max_\Omega\{|f_{\text{recons.}} - f|\}, \end{aligned} \tag{25}$$

where

$$\{\overline{f}\}_\Omega = \int_\Omega f(x,y)\, dx\, dy \Big/ \int_\Omega dx\, dy.$$

e-dis1 is a *normalized root mean square distance measure.* A large difference in a few places causes the value of e-dis1 to be large. e-dis2 is a *normalized mean absolute distance measure.* It emphasizes the importance of a lot of small errors rather than of a few large errors. e-dis3 represents the maximum deviation and is a *worst case distance measure.* The integrals in Eq. (19) are typically approximated by summations involving a finite number of function evaluations.

[8] A default value of ≈ 1.6 is suggested by Ravichandran and Gouldin (1988a).

[9] A limiting value of 1.5 to 1.6 times the 1-D sampling frequency of 1/(2×Grid Spacing) is considered to be appropriate by Ravichandran and Gouldin (1988a).

1989 National Heat Transfer Conference
HTD-Vol. 106, Heat Transfer Phenomena in Radiation, Combustion and Fires

EFFECT OF DROPLET INTERACTIONS ON IGNITION DELAY TIMES

W. R. Laster and K. Annamalai
Department of Mechanical Engineering
Texas A&M University
College Station, Texas

ABSTRACT

The ignition and evaporation of a spherical cloud of droplets in a hot quiescent atmosphere is examined numerically using transient group combustion analysis. Ignition delay times are calculated as a function of cloud radius, ambient temperature, drop size and droplet number density. The ignition temperature for a cloud of drops was found to be less than that obtained for a single drop. The results indicated an interaction between chemical and physical effects resulting in the possibility of an optimal interdrop spacing for ignition. The model results indicate that for interdrop spacings of less than 5 the ignition and evaporation of a cloud of drops is confined to a thin layer at the surface of the cloud. For drops spaced farther apart thermal penetration from the hot ambient is possible resulting in vaporization within the cloud.

NOMENCLATURE

A	Preexponetial Factor
B	Transfer Number for Evaporation
C_p	Specific Heat
d	Droplet Diameter
D	Diffusion Coefficient
Ea	Activation Energy
F_B	Blowing Correction
G	Group Combustion Number
h	Convective Heat Transfer Coefficient
h_m	Convective Mass Transfer Coefficient
h_T	Total Entholpy
hc	Heating Value
h_L	Latent Heat of Vaporization
h_m	Convective Mass Transfer Coefficient
l	Inter Drop Spacing
$\dot{m}$	Mass Flow Rate
$\dot{m}'''_{F,p}$	Vapor Production Rate Per Unit Volume
$\dot{m}'''_{F,c}$	Vapor Consumption Rate Per Unit Volume
$\dot{m}_{s,c}$	Combustion Rate of Single Drop in the Cloud
n	Number of Drops Per Unit Volume
Nu	Nusselt Number
n_F, n_{o2}	Reaction Order with Respect to Fuel and Oxidizer
n_T	Total Reaction Order
$\dot{q}$	Convection Heat Transfer to Drop
r	Radius
Rc	Cloud Radius
Sh	Sherwood Number
T	Temperature
t	Time
Y	Mass Fraction
$Y_{F,w}$	Fuel Mass Fraction at the Droplet Surface (Saturation)
λ	Thermal Conductivity
ν_o	Stoichiometric Mass of Oxygen
ρ	Density
τ	Ignition Delay

Subscript

c	Cloud
c,o	Cloud at Time t=0
F	Fuel

1	Liquid
O_2	Oxidizer
∞	Freestream

INTRODUCTION

The time required to ignite fuel air mixtures is of practical importance in the design of combustion systems. For liquid fuels the ignition delay time is composed of a chemical portion due to chemical kinetics and a physical portion due to evaporation and mixing. The understanding of the ignition process in liquid fuel spray requires a model which includes both processes. Many models exist for individual droplet ignition but these results are applicable only to very dilute clouds, a situation which rarely occurs in practical combustion systems. In practical combustion systems ignition often occurs near the injector where the spray is very dense. It has been observed that when the droplets are sufficiently close together, a diffusion flame will form around a cloud of droplets. This behavior is called group combustion. The purpose of this paper is to examine the ignition behavior of dense sprays including the effects of group combustion.

LITERATURE REVIEW

The earliest work on group combustion was conducted by Suzuki and Chiu (1971). They obtained group combustion results for a spherical cloud of liquid propellants. Later Labowsky and Rosner (1978) obtained a criteria for incipient group combustion by applying the analogy of porous char combustion to the group combustion of droplets. Chiu and Liu (1977) used a quasi-steady (QS) model for spherical, cylindrical and planar clouds of droplets. Using a perturbation analysis they obtained a power series solution for the total group combustion rate and flame location.

The effects of group combustion are related to the group combustion number, G, which can be defined as, (Annamalai, 1985)

$$G=3(\text{Fuel Volume Fraction})(\text{Cloud Radius}/\text{Droplet Radius})^2$$

If the G number exceeds a critical value G_c, then the group flame stands outside the cloud. If $G < G_c$, then the burning of individual droplets inside the cloud is possible. The existence of a group combustion mode has been confirmed by the liquid fuel spray experiments performed by McCreath and Chigier (1973). The common observation among these experiments was the detection of a sheath flame around the outer periphery of the liquid spray, indicating group combustion rather than individual drop combustion.

Group combustion analysis has been applied to the problem of evaporation of a cloud of drops. Correa and Sichel (1982; 1983) used an asymptotic technique to obtain the propagation rate of a vaporization wave for a very dense cloud (G=12,600). Bellan et al. (1983) used the "bubble" model to investigate the evaporation of a cloud of drops enclosed in an adiabatic vessel.

The QS formulation by Sichel and Palaniswamy (1984) for the ignition and extinction included a sheath combustion criteria in terms of vaporization wave thickness ratio. They provided explicit results for ignition, extinction and cloud burning time. Considering a cloud of dodecane drops, Annamalai et al. (1984) accounted for chemical reaction both in the cloud and outside the cloud and obtained numerical results for the steady state ignition temperature of the cloud. The QS ignition analysis of a cloud of drops shows that the ignition of a cloud of drops can occur at significantly lower temperatures than that of an isolated droplet.

The term group combustion as defined so far implies the formation of a group flame. More generally, group combustion is a result of droplet interactions with or without the formation of a group flame. These studies are useful in determining the behavior of drops in the core of a liquid spray. Annamalai (1984) considered the group combustion of a cloud of drops as well as the group evaporation of a cloud of drops. As long as the mass fraction of O_2 is negligible at the surface of each droplet in the cloud, the mass burning rates of a cloud with finite chemistry and infinite chemistry are identical. He compared the mass burning rate for individual particle combustion with that for the cloud and determined a correction factor. The burning rate of a typical cloud of drops can be as low as 10% of the burning rate obtained with an individual droplet combustion model at G=100. Annamalai (1985) compared the group burning rate to the sum of individual drop burning rates at various G numbers and found that for $G < 0.3$, the results for the group burning rate are 90% of the rate predicted with an individual particle combustion model. Recently group combustion models have been extended to a cloud of char and coal particles (Annamalai, 1987; 1988; Ryan, 1988).

OBJECTIVE

This study will concentrate on the ignition of a cloud of fuel drops in a hot quiescent atmosphere. This problem has been studied previously using quasi-steady state analysis (QS). The results have indicated cloud ignition temperatures significantly lower than would be expected for a single drop. Correa and Sichel (1983) studied a 10cm radius cloud of saturated octane drops containing 10^9 drop/m^3 with a drop diameter of 200 microns. They found ignition occurred at temperatures as low as 500 K. Annamalai (1984) studied a 1 cm diameter cloud of dodecane drops containing 10^8 drops/m^3 with a drop diameter of 100 microns (G=7). It was found that for this case ignition was possible for temperatures as low as 700 K. The G number for both of these studies was greater than G_c indicating a significant group combustion effect.

The limitation of the steady state or quasi-steady state analysis is that although they can provide minimum ignition temperatures, this temperature requires an infinite amount of time for ignition to occur. In a practical combustion system ignition is a transient process and must be accomplished in a few milliseconds. In order to obtain ignition delay times it is necessary to perform a transient analysis. This paper presents the results of a transient group combustion model applied to a cloud of liquid droplets.

MODEL FORMULATION

The system modeled for this study is shown in Figure 1. A spherical cloud of radius, R_c, containing n droplets of a uniform diameter, d, is considered for this analysis. The interior of the cloud is essentially a two phase zone consisting of both gas and liquid phases while outside the cloud is a single phase zone. The cloud is assumed initially at a

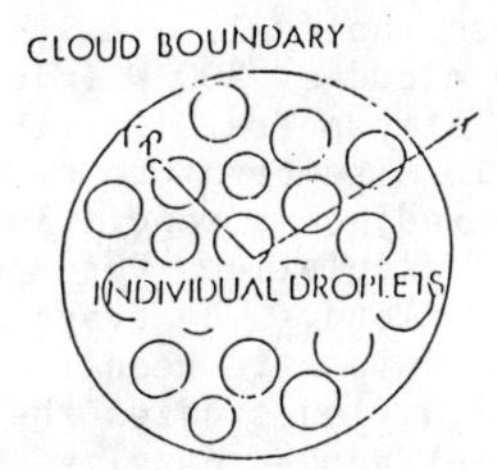

Figure 1: Schematic of a cloud of droplets

uniform temperature, $T_{c,o}$, and is suddenly immersed in a hot quiescent atmosphere of T_∞. For this case a thermal wave propagates from ambient to the cloud interior. As the thermal wave penetrates the cloud the gas temperatures adjacent to the drops increase, resulting in heating and evaporation of the liquid. The gas vapors produced by the drops in the cloud are carried to the single phase zone outside the cloud by both convection and diffusion. If the local mixture conditions outside the cloud are appropriate for ignition, the local heat generation rate will exceed the heat loss rate producing a temperature rise. Ignition is defined as the location where the gas temperature rise produces a 10 K "bulge" above the gas temperature in the adjacent zone. To model the ignition process the transient equations for conservation of mass, species and energy were employed.

The assumptions used in the present analysis are:

(i) The drops are treated as point mass and heat sources. This assumption is justified when the volume occupied by the drops is less than 5% of the cloud volume which corresponds to an interdrop spacing (l/d) of approximately 2.5.

(ii) The mass and heat transfer relationships for the drop and adjacent gas are governed by steady state relationships for a single drop. In order for this assumption to be valid, it is essential that the time necessary to establish a QS profile in the boundary layer adjacent to the drop is shorter than the lifetime of a drop. This condition is satisfied as long as the thermal capacity of the liquid is high compared to that of the gas.

(iii) Spherical symmetry is assumed. The profiles for gas temperature and mass fraction are functions of cloud radius only.

(iv) The gas phase oxidation is modeled by a single step global Arrhenius reaction.

Conservation equations for mass, species and energy are solved for the gas and liquid phases. These equations are:

Gas Phase

Mass

$$4\pi r^2 \frac{\partial \rho}{\partial t} + \frac{\partial \dot{m}}{\partial r} = \dot{m}'''_{F,p} 4\pi r^2 \qquad r<Rc \qquad (1)$$

$$= 0 \qquad r>Rc$$

where $\dot{m}'''_{F,p} = \dot{m}_{s,c} n$

Species

Fuel

$$4\pi r^2 \frac{\partial(\rho Y_F)}{\partial t} + \frac{\partial}{\partial r}(\dot{m} Y_F) - \frac{\partial}{\partial r}\left(4\pi\rho D r^2 \frac{\partial Y_F}{\partial r}\right) =$$

$$(\dot{m}'''_{F,p} - \dot{m}'''_{F,c}) 4\pi r^2 \qquad r<Rc \qquad (2)$$

$$= -\dot{m}'''_{F,c}\, 4\pi r^2 \qquad r>Rc$$

The fuel consumption rate is obtained by global reaction theory as

$$\dot{m}'''_{F,c} = A\rho^{n_T} Y_{02}^{n_{02}} Y_F^{n_F} \exp\left(-\frac{Ea}{RT}\right) \qquad (3)$$

Oxidizer

$$4\pi r^2 \frac{\partial(\rho Y_{02})}{\partial t} + \dot{m} \frac{\partial(\dot{m} Y_{02})}{\partial r} - \frac{\partial}{\partial r}\left(4\pi\rho D r^2 \frac{\partial Y_{02}}{\partial r}\right) =$$

$$-\dot{m}'''_{F,c}\, \nu_o\, 4\pi r^2 \qquad (4)$$

Energy

$$\frac{\partial}{\partial t}(\rho h_T) + \frac{\partial}{\partial r}(\dot{m} h_T) - \frac{\partial}{\partial r}\left(4\pi\lambda r^2 \frac{\partial T}{\partial r}\right)$$

$$= \dot{m}'''_{F,c} h_c + \dot{q}'''_m - \dot{q}'''_{conv} - \dot{m}'''_{F,p} h_L \qquad r<Rc$$

$$= \dot{m}'''_{F,c} h_c \qquad r>Rc \qquad (5)$$

where $\dot{q}'''_{conv} = n\, \dot{q}_{conv}$

$$\dot{q}'''_m = n\dot{m}_{s,c} c_p (T_1 - T_{ref})$$

Liquid Phase

Mass

$$\rho_1 \pi \frac{d^2}{2} \frac{d(d)}{dt} = -\dot{m}_{s,c} \qquad (6)$$

where

$$\dot{m}_{s,c} = h_m \pi d^2 (Y_{F,w} - Y_F) F_B$$

where

$$h_m = Sh \frac{\rho D}{d}$$

and

$$F_B = \ln(1+B)/B$$

$$B = \frac{(Y_{F,w}-Y_F)}{(1-Y_{F,w})}$$

Energy

$$\rho_l c_{pl} \pi \frac{d^3}{6} \frac{dT_l}{dt} = (\dot{q}_{conv} - \dot{m}_{sc} h_L) \qquad (7)$$

where

$$\dot{q}_{conv} = h \pi d^2 \, (T_g - T_l) \, F_B$$

$$h = \frac{Nu \, d}{\lambda}$$

Boundary and Initial Conditions

$$\text{At} \quad t = 0, \; T = T_\infty \qquad R_c < r < \infty$$

$$T = T_{c,o} \qquad 0 < r \le R_c$$

$$\dot{m} = 0$$

$$\text{At} \quad t > 0, \; T = T_\infty \qquad r = \infty$$

$$\dot{m} = \frac{\partial T}{\partial r} = \frac{\partial Y_i}{\partial r} = 0 \qquad r = 0$$

The conservation equations are solved using an explicit upwind finite difference scheme. The problem is nodalized using 60 nodes in the radial direction, with half of the nodes initially outside the cloud. A study of the effect of grid size revealed that increasing the number of grid points above 60 did not change the calculated delay time. When all the drops within a single layer are evaporated the cloud radius is allowed to shrink. Details on the numerical method are given in Ryan (1988), and Annamalai et al. (1988b).

RESULTS

Two different fuels were studied, octane and dodecane. The properties for these fuels were obtained from Correa and Sichel and Annamalai and are given in Table 1. The one step kinetics for dodecane were obtained from Williams (1981) and the kinetics for octane came from Barnett and Hibbard (1959). These values were chosen to match the assumptions of Correa and Sichel and Annamalai. These are more representative of upper and lower limits than differences between the fuels. All cases studied were an initially cold cloud at 300 K injected into a hot quiescent ambient atmosphere. For this model ignition was defined as a 10 K temperature rise above the gas temperature in the adjacent zone. In addition to the ignition studies, calculations were performed for a purely evaporating cloud. The evaporating cloud was compared with the asymptotic results of Correa and Sichel. For the ignition studies the effect of temperature, droplet number density, cloud radius and drop diameter were investigated.

TABLE 1
Properties of Octane and Dodecane

	Octane C_8H_{18}	Dodecane $C_{12}H_{26}$
Heat of Combustion	44,770 kJ/kg	44,360 kJ/kg
Critical Pressure	24.8 atm	17.9 atm
Critical Temperature	568.8 K	659 K
Boiling Point	399 K	490 K
Heat of Vaporization	300 kJ/kg	256 kJ/kg
Density	707 kg/m^3	749 kg/m^3
Activation Energy	10^5 kJ/K	1.5×10^5 kJ/K
Arrhenius Constant	9.9×10^6	3.8×10^8
n_{o2}	1	1
n_F	1	1
n_T	2	1

EVAPORATION STUDIES

The model first was applied to the case of a purely evaporating cloud. The 10 cm radius cloud of octane drops studied by Correa and Sichel was used for the evaporation studies. The cloud was assumed to be initially saturated with octane vapor at 358 K (Y_F=0.625). The free stream temperature was set to 500 K and the mass fraction of octane vapor in the free stream was 0.23. The difference between this study and that of Correa and Sichel was that they assumed a steady state temperature profile in the gas outside the cloud as an initial condition while this analysis assumed a constant temperature gas outside of the cloud as an initial condition. This should result in a higher initial evaporation rate for the present analysis. This is indeed the case as this analysis predicts that the cloud radius will have decreased to 81% of its initial value in 46 seconds compared with 73 seconds for the Correa and Sichel case. The radial temperature profiles for an evaporating cloud predicted by both analyses at the time when the cloud radius has decreased to 81% its initial value are compared in Figure 2. The difference between the two profiles is that Correa and Sichel assumed a steady state temperature distribution for the gas outside the cloud at all times. The results of the transient analysis indicate that the time required to obtain a steady state temperature profile in the gas is greater than the evaporation time of the cloud. However, the results of this analysis confirmed Correa and Sichel's assumption that for very dense clouds evaporation is confined to the outmost layer of drops.

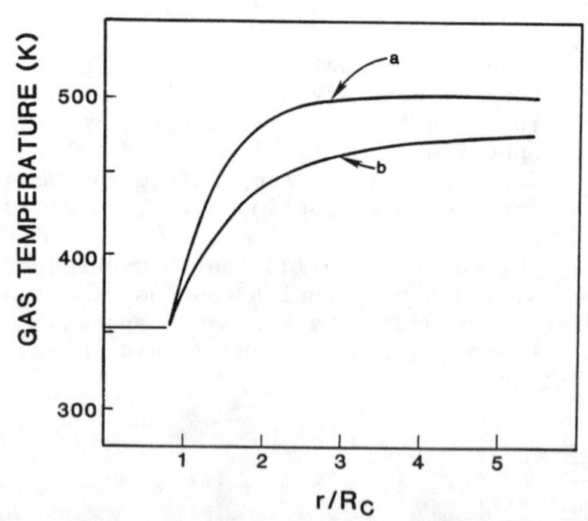

Figure 2: Gas temperature profile for an evaporating octane cloud (R_c = 10 cm, d = 200 micron, n = 10^9 drops/m^3, time corresponding to an instant at which cloud radius = 81% of Initial Value)
a) Present work
b) Correa and Sichel

IGNITION STUDIES

Effect of Temperature

Figure 3 presents ignition delay times for a cloud of octane drops and a cloud of dodecane drops as a function of ambient temperature. Both clouds contained $1x10^{10}$ drops/m^3. The initial cloud radius was held fixed at 1 cm and the droplets were 100 microns. Assuming chemical control, the delay time vs reciprocal temperature plot should be linear (dashed lines). This is true for both octane and dodecane at low temperatures (long delay times). At high temperatures, however, the physical effects become important as evidenced by Figure 3. This shift from chemical to physical control is due to the fact that the chemical delay is a strong function of temperature. At low temperatures the chemical delay is significantly lower than the physical delay and therefore it dominates. At high temperatures the chemical delay is extremely short causing physical effects to become more important. The fact that dodecane has a longer ignition time at high temperatures indicates a greater physical delay. A calculation of the apparent activation energy from Figure 3 results in a value of 66,560 kJ/mole for octane and 114,060 kJ/mole for dodecane. These values are lower than the chemical activation energies used in the program (100,000 and 150,000 KJ/mole; respectively) indicating that although the physical effects at low temperature are small compared to chemical effects they are not negligible. From the figure it can be seen that for dodecane the ignition delay curve deviates from chemical control at lower temperatures than octane. This indicates that physical delays are more important for dodecane. This

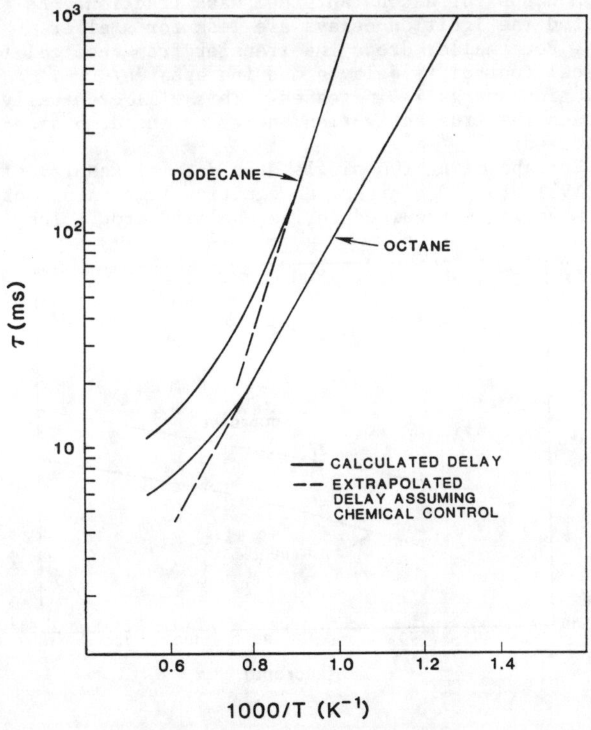

Figure 3: Effect of ambient temperature on the ignition delay times for a cloud of drops (R_c = 1 cm, d = 100 micron, n = 10^{10} drops/m^3)

is reasonable since dodecane has a higher boiling point than octane resulting in a lower evaporation rate. In addition a higher activation energy is used for dodecane in this analysis, which requires a higher gas temperature to initiate the reaction. This increases the physical delay because the fuel vapors must diffuse to a higher temperature region in the gas before the reaction can begin.

Effect of Droplet Density

A study of the effect of droplet number density (or interdrop spacing) was performed and the results are presented in Figure 4. For this case the initial cloud diameter was held fixed at 1 cm, the droplet diameter at 100 microns and the ambient temperature at 1200 K. From this figure it is noted that dodecane and octane produce quantitatively different behavior. As the number of drops are increased, the ignition delay for dodecane decreases until a minimum ignition time is reached at a number density of $1x10^{10}$ drops/m^3, or an inter drop spacing of approximately 5. For closer drop spacings an increase in the ignition delay is observed with an increase in droplet number density. The number density of drops at which of this minimum ignition time was relatively insensitive to ambient temperature. This minimum ignition delay was not observed for octane. For octane drops the ignition delay continues to decrease as the number of drops is increased. At droplet spacings closer than 5, however, the ignition delay remained essentially constant. The closest inter drop spacing studied was approximately 2.5, where the drops are almost touching.

An understanding of this different behavior for dodecane and octane can be obtained from an examination of the temperature profiles at ignition. Figure 5 shows the temperature profiles at ignition for the dodecane cloud with droplet number densities of $1x10^9$ and $5x10^{10}$ drops/m^3. Figure 6 shows the corresponding temperature profiles for the octane cloud. In both cases for very dense clouds there is virtually no penetration of the hot ambient inside of the cloud. All of the evaporation occurs at the outer layer of droplets. As the droplet number density is decreased, the hot ambient is able to penetrate

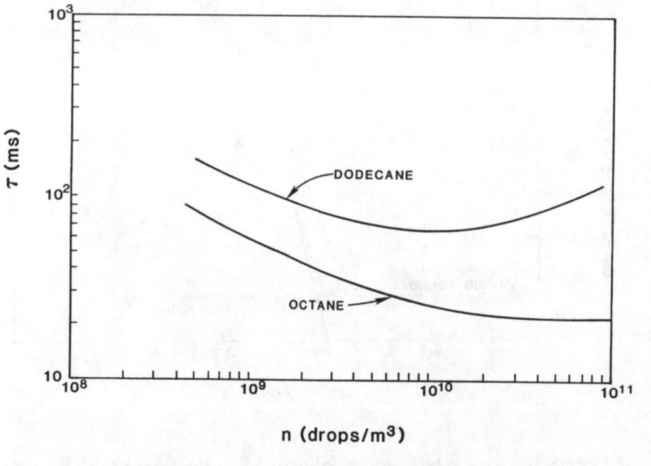

Figure 4: Effect of droplet number density on ignition delay (T_∞ = 1200 K, d = 100 micron, R_c = 1 cm)

farther inside of the cloud and evaporation occurs not only at the outmost layer but also farther in the cloud. The physical ignition delay for the cloud of drops is due to the evaporation time and the time required to form a flammable mixture. As the number of drops is increased, the thermal penetration is reduced and the increased delay is due to the fact that the fuel vapors must travel farther from the cloud before they encounter sufficiently high temperatures for the reaction to occur. The minimum delay does not occur for octane because of the lower activation energy used in this study. The octane vapors react at a lower temperature or nearer to the cloud surface. The increased slope of the gas temperature outside the cloud in the octane case is due to an increased reaction rate at low temperature. The existence of the minimum is strongly tied to the kinetics. If the high temperature kinetics used for dodecane are used in the octane analysis, a minimum ignition occurs.

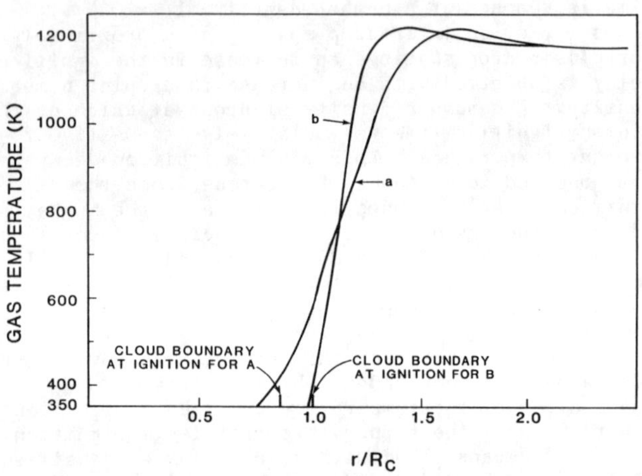

Figure 5: Gas temperature profile at ignition for a cloud of dodecane drops (T_∞ = 1200 K, d = 100 micron, R_c = 1 cm)
a) n = 10^9 drops/m^3
b) n = $5x10^{10}$ drops/m^3

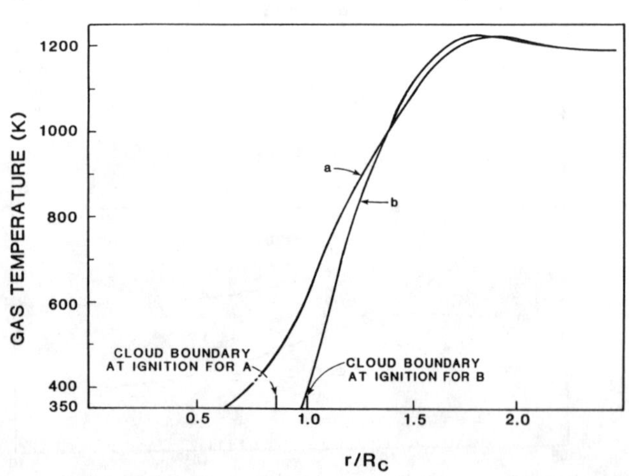

Figure 6: Gas temperature profile at ignition for a cloud of octane drops (T_∞ = 1200 K, d = 100 micron, Rc = 1 cm)
a) n = 10^9 drops/m^3
b) n = $5x10^{10}$ drops/m^3

Effect of Cloud Size

The cloud diameter for this study was varied between 10 cm and 0.2 cm. It was found that the ignition delay times decreased as the cloud radius decreased for both octane and dodecane. Although smaller clouds ignite quicker, a lower ignition temperature was possible for larger clouds. This is due to the fact that at low temperatures smaller clouds will vaporize without burning. As the fuel vapors diffuse outward from the cloud to high temperature regions where ignition is possible this concentration becomes diluted. This dilution is greater for smaller clouds. The amount of fuel vapor necessary for ignition is never present at the proper temperature. The effect of increasing ignition temperature with decreasing drop size has been observed in individual droplet experiments (LeMott, et al; Saitoh, et al.). These results indicate that a dense cloud of drops exhibits qualitatively the same behavior as a single large drop having the same radius as the cloud.

Effect of Drop Diameter

Figure 7 shows a comparison of the ignition delay times for both dodecane and octane as a function of the drop diameter. The initial fuel volume fraction in the cloud was held constant for these calculations at $5.2x10^{-4}$. The temperature was 1200 K and a 1 cm cloud was studied. The result for dodecane and octane are quantitatively similar. As the drop size is decreased, the ignition delay is decreased for both cases. Smaller drops evaporate faster reducing the evaporation portion of the ignition delay time. Figure 8 shows a comparison of the ignition delay time as a function of temperature between 40 micron and 100 micron drops for a constant fuel mass fraction. As expected the ignition delays are less for smaller drops. For smaller drops the transfer from chemical to physical control is delayed and the apparent activation energy is increased. This will eventually approach the true activation energy as the drop size is reduced.

Extrapolating the single drop data of LeMott, et al. (1971) to 100 microns, an ignition temperature of 1425 K would be required for an isolated drop. The

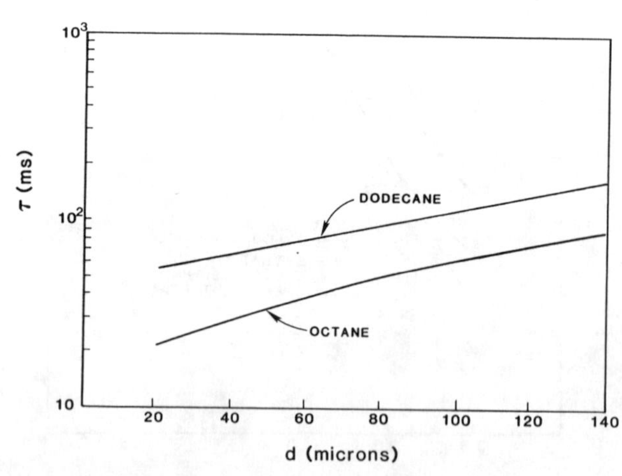

Figure 7: Effect of drop size on ignition delay (T_∞ = 1200 K, R_c = 1 cm)

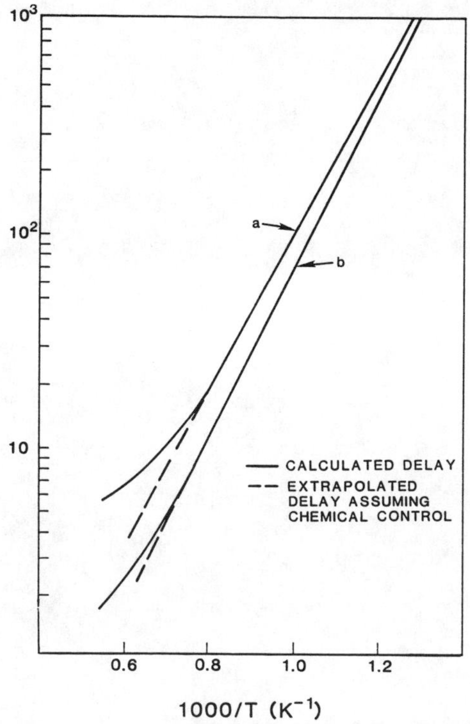

Figure 8: Ambient temperature dependence of ignition delay for octane drops (R_c = 1 cm, n = 10^{10} drops/m^3)
a) d = 100 microns
b) d = 40 microns

present analysis shows that a cloud of 100 micron drops can easily be ignited at temperatures as low as 1000 K. The higher temperature for a single drop is due to the rapid dilution of the fuel vapors as they diffuse to the high temperature ambient. In addition ratio of heat loss to heat generation is greater for smaller diameters. At low temperatures the single drop will vaporize without burning.

CONCLUSIONS

The ignition and evaporation of a cloud of droplets has been examined numerically using a transient group combustion model. The results indicate an interaction between chemical and physical processes resulting in an optimal ignition case. For the dodecane cloud studied there is an optimal interdrop spacing for ignition. This optimal condition does not exist for octane. The existence of the optimal drop spacing is strongly tied to the kinetic parameters assumed in the analysis and may not represent a difference between the fuels. The slower kinetics assumed for dodecane is responsible for this minimum. The model results indicate that for interdrop spacings of less than 5, the ignition and evaporation of a cloud of drops is confined only to a thin layer at the surface of the cloud. For drops spaced farther apart thermal penetration from the hot ambient is possible and vaporization can occur over many layers within the cloud interior. A cloud of drops can ignite at much lower temperatures than is possible with a single drop.

REFERENCES

Annamalai, K., Madan, A. and Mortada, Y.I., "Ignition of a Cloud of Droplets", ASME Paper No. 84-WA/HT-18, 1984.

Annamalai, K., "The Evaporation and Combustion of a Cloud of Droplets", in Heat Transfer in Fire and Combustion Systems, C. K. Law, ed, ASME, New York, pp. 27-35, 1985.

Annamalai, K., and Ramalingam, S., "Distinctive Burning Characteristics of Carbon Particles", Combustion and Flame, 70, pp. 307-332, 1987.

Annamalai, K., Ramalingam, S., Dahdah, T., and Chi, D., "Group Combustion of a Cylindrical Cloud of Char Particles", Journal of Heat Transfer, 110, pp. 190-200, 1988.

Annamalai, K., Ryan, W. and Ramalingam, S., "Group Combustion of Coal Particles" Annual Report DE-FG22-85 PC 80528, Pittsburgh Energy Technology Center - Department of Energy Pittsburgh, PA. 1988b.

Barnett, H.C. and Hibbard, R.R., "Basic Considerations in the Combustion of Hydrocarbon Fuels With Air" NACA Report 1300, 1959.

Bellan, J. and Cuffel, R., "A Theory of Nondilute Spray Evaporation Based Upon Multiple Drop Interactions", Combustion and Flame, 51, pp. 55-67, 1983.

Chiu, H.H., and Liu, F.M., "Group Combustion of Liquid Droplets", Combustion Science and Technology, 17, pp. 127-142, 1977.

Correa, S.M. and Sichel, M., "The Boundary Layer Structure of a Vaporizing Fuel Cloud", Combustion Science and Technology, 28, pp. 121-130, 1982.

Correa, S.M. and Sichel, M., "The Group Combustion of a Spherical Cloud of Monodisperse Fuel Droplets", 19th Symposium (International) on Combustion (Combustion Institute, Pittsburgh: 1983).

Labowsky, M. and Rosner, D.E., "Group Combustion of Droplets in Fuel Clouds. I Quasi-Steady Predictions", in Evaporation-Combustion of Fuels, Advances in Chemistry Series, J.T. Chang, ed., American Chemical Society, Washington, DC, No. 166, pp. 63-79, 1978.

LeMott, S.R., Peskin, R.L. and Levine, D.G., "Effect of Fuel Molecular Weight on Particle Ignition", Combustion and Flame, 16, pp. 17-27, 1971.

McCreath, C.G., and Chigier, N.A., "Liquid Spray Burning in the Wake of a Stabilizer Dirk", Proceedings 14th Symposium (International) on Combustion, The Combustion Institute, Pittsburgh, PA, pp. 1355-1363, 1973.

Ryan, W.R., "Group Pyrolysis, Ignition, and Combustion of a Spherical Cloud of Coal Particles", M.S. Thesis, Texas A&M University, 1988.

Saitoh, T., Ishiguro, S. and Niioka, T., "An Experimental Study of Droplet Ignition Characteristics Near the Ignitable Limit", Combustion and Flame, 48, pp. 27-32, 1982.

Sichel, M. and Palaniswamy, S., "Sheath Combustion of Sprays", Proceedings 20th Symposium (International) on Combustion, The Combustion Institute, Pittsburgh, PA, pp. 1789-1798, 1984.

Suzuki, T. and Chiu, H.H., "Multi-Droplet Combustion of Liquid Propellants", Proceedings of the 9th International Symposium on Space Technology and Science, AGNE Publishing, Tokyo, Japan, pp. 145-154, 1971.

Williams, F.A., "A Review of Flame Extinction" Fire Safety Journal, 3, pp 113-175, 1981.

1989 National Heat Transfer Conference
HTD-Vol. 106, Heat Transfer Phenomena in Radiation, Combustion and Fires

EXPERIMENTAL STUDY OF THE THERMAL STRUCTURE OF A PLANAR SPRAY FLAME

M. Queiroz
Brigham Young University
Provo, Utah

S. C. Yao
Carnegie-Mellon University
Pittsburgh, Pennsylvania

ABSTRACT

The effect of gas-phase turbulence intensity and fuel-vapor concentration on the thermal structure of an array of burning droplet streams has been investigated. Particular attention has been focused on the relationship between the dynamic motion of the flame front and the flame thermal structure. Bare microthermocouples, digitally compensated for thermal inertia effects, were used to perform dynamic measurements of gas-phase temperatures in a dilute flame of 400 μm nominal diameter hexane droplets, injected through nine droplet streams oriented in a plane and horizontally separated by a distance of 4 mm. Higher fuel-vapor concentration in the gas phase generated higher flame temperatures at the average location of the premixed-gas flame for flames having a strong premixed-gas ignition mechanism. Furthermore, the essential features in both the average and fluctuating temperature profiles as well as the pdf surfaces were unchanged as the fuel-vapor concentration varied. The increase in the turbulence intensity promoted higher temperature fluctuations by widening the intermittent combustion zone of the flame. It also changed the dynamics of the drifting motion of the flame as well as the instantaneous temperature profile across the flame, causing the disappearance of the bimodal shapes in the pdf surfaces.

INTRODUCTION

Knowledge of the thermal structure of flames plays an important role in the fundamental understanding of flame propagation, and in the development of suitable combustion models because of the strong interrelation between the burning process and the temperature field. In attempting to obtain useful information about the thermal structure of flames, yet maintaining low experimental costs, many researchers have used microthermocouples to measure temperatures in turbulent premixed [1-5], diffusion [6-8], and spray flames [10-12]. Generally, the local gas-phase temperatures are measured using bare microthermocouples compensated for thermal inertia effects. For turbulent gaseous flames the information obtained usually includes time-averaged temperatures (which may be corrected for radiation effects), temperature fluctuations (rms of the temperature signal), probability density functions (pdf) of fluctuating temperature, and power spectral density (psd) of the temperature fluctuations. The thermal inertia compensation is usually electrically performed, using a simple RC circuit on the thermocouple output [1-8]. This paper focuses on the use of digitally-compensated thermocouple measurements to investigate the thermal structure of a dilute spray flame in terms of the parametric effect of fuel-vapor concentration and gas-phase turbulence intensity.

Using the techniques described above, different thermal structures have been identified in turbulent premixed Bunsen-burner flames by Yoshida and co-workers. In one study of a 10 mm burner [1,2], a structure was found that was different from that of a wrinkled laminar-flame model (WLF) with significant contributions on probability density functions from intermediate temperatures in the intermittent combustion zone. In another set of experiments with a 40 mm burner [3,4], the pdf showed well-defined bimodal shapes, and the maximum temperature fluctuations were close to those estimated from WLF models. The effect of fuel equivalence ratio and turbulence intensity on the thermal structure of the flame was also investigated [3]. With increasing equivalence ratio, the spatial temperature peak appeared at higher temperature values; however, the essential features of the thermal structure were not changed. It was also observed that the pdf profiles were not influenced by the turbulence intensity of the unburnt mixture. Mean and fluctuating temperatures were found to increase with equivalence ratio; nevertheless, the influence of turbulence intensity could not be observed.

Lenz and Gunther [6] have made microthermocouple temperature measurements in a turbulent free-jet diffusion flame and obtained similar information on the thermal structure of the flame. The results of these measurements give detailed information about the alternating effect of turbulent mixing and reaction on the thermal structure of diffusion flames. Ballantyne and Moss [7] have discussed the use of fine thermocouples to analyze the thermal structure of diffusion flames. Measurements made in an open turbulent methane flame were compared with simple model predictions with the intention of identifying the interaction between the measurement technique and the structure of the flame. Digital compensation of the thermocouple signal was also discussed, although apparently not implemented. While uncertainties do surround aspects of the detailed measurements, it was observed that the thermocouple technique appears to offer valuable insights into the temperature structure of turbulent flames, bearing in mind the difficulty, in general, of obtaining temperature information in flames. Ballantyne and Moss's discussion of thermocouple performance in terms of detailed local flame structure demonstrates the capabilities of this instrumentation technique. Roberts and Moss [8] have proposed a model exploiting the mechanistic interpretation of the flame front dynamics in terms of wrinkled or flapping flame behavior. The model is capable of almost complete reconstruction of the measured temperature pdf, using values of mean and variance obtained from compensated microthermocouple temperature signals.

Although microthermocouples have been extensively used in turbulent gaseous flames and their performance compared favorably with state-of-the-art laser techniques [9], their use to study the thermal structure of spray flames has usually been limited to measurement of average gas-phase temperature information in locations where the liquid droplets are almost completely vaporized [10,11]. In some cases where additional information from the temperature measurements was sought [12], the influence of droplets impinging on the thermocouple bead was not addressed and the parametric effects on the thermal structure of the flame were not investigated.

Queiroz, et al. [13] performed a preliminary study on the microthermocouple performance in the presence of liquid fuel droplets in dilute spray flames. The technique was found to be appropriate for a dilute hexane spray flame with a low frequency of droplet collision if attention is focused on locations in the flame where the average temperature is above 350°C. Digital compensation of the thermocouple output has also been successfully demonstrated [21], as well as the possibility of correcting the average temperature information for radiation effects, provided reliable information on radiation properties is available [14,15].

The objective of the exploratory study presented here is to reveal the thermal structure of a dilute spray flame, using microthermocouples digitally compensated for thermal inertia effects to perform gas-phase temperature measurements. Particular attention is focused on the effect of the large-scale, low-frequency, drifting motion of the flame front [16] on the thermal structure of the flame. The parametric effect of fuel-vapor concentration and turbulence intensity on these results was also investigated.

EXPERIMENTAL METHODS

The spray flame used in this thermal-structure study has been previously described by Queiroz and Yao [16]. Nine vertical streams of hexane fuel droplets with a constant lateral spacing of 4 mm were injected at a total rate of 37.2 cm^3/min and mixed with air to form the combustible mixture. The airflow mixed the droplet streams to produce a homogeneous, well-dispersed spray at the point of ignition. The rectangular cross section of the dispersed spray measured approximately 5 by 45 mm at the ignition point. This geometric configuration was selected because of the facility with which the flame front motion and thermal structure could be documented, using sequential photography and microthermocouples, respectively.

The oblique flame was ignited by placing a small hydrogen pilot near the edge of the outermost stream (Fig. 1). The pilot had the shape of a candle flame and was 3 mm wide and 6 mm long. The region of thermal influence of the small pilot flame was determined by taking temperature measurements around it with the airflow, but without the droplets. The temperature measurements revealed that the limit of the lateral region of thermal influence of the pilot flame was very close to the edge of its visible light. At 5 mm away from the hydrogen source in the horizontal direction, the average temperature was 47°C. The influence of the pilot flame downstream of its location was broader because of the convective action of the gas flow. At 10 mm downstream from the hydrogen source, the average temperature value was 840°C. The temperature dropped to 470°C at a location 20 mm downstream of the hydrogen source.

Initial liquid fuel temperatures between 8°C and 43°C were obtained, using a heat exchanger in the fuel supply line, to indirectly investigate the effect of fuel-vapor concentration in the gas phase. Two different levels of air turbulence were used to investigate the effect of this parameter on the thermal structure of the flame. A perforated plate and a fine wire screen produced mean velocities of 1.30 and 1.00 m/s and turbulence intensity levels of 17 and 2.5%, respectively.

Temperature measurements were made with uncoated, 50μm wire diameter, Pt-Pt/10% Rh thermocouples with bead diameters of approximately 100μm. The average thermocouple time constant used in the compensation of the thermal inertia effects was found to be 40 ms, based upon measurements using a modified step-response method [14]. This average value compares favorably with similar measurements performed by Yule et al. [15] on thermocouples of approximately the same size and having comparable aerodynamic conditions. The maximum and minimum temperature values on the pdf surfaces were checked to determine if the selected time constant value was overcompensating the temperature data. This procedure was followed based on the work of Yoshida and Gunther [3]. Data points at locations where the average air velocity was substantially higher than values at the ignition point were not used in order to minimize flame stretching effects.

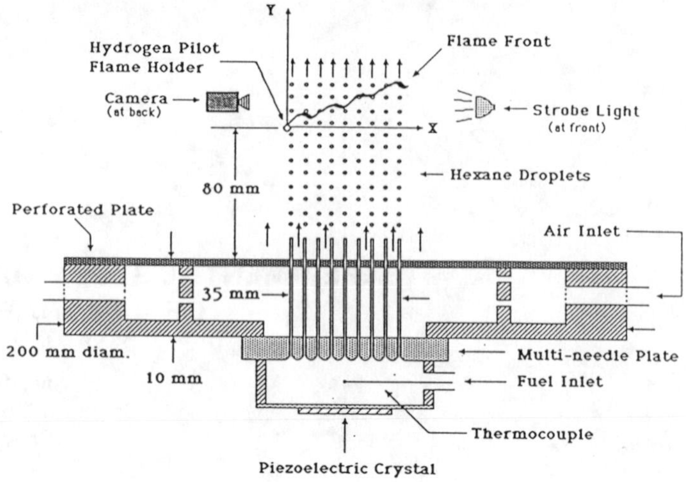

Fig.1 Schematic of the droplet generator, not drawn to scale.

A schematic of the thermocouple probe and the system used to process the thermocouple output is shown in Fig. 2. An electronic ice-point junction was used in the thermocouple output to obtain the thermocouple voltage with respect to the voltage at the freezing point of water as given in thermocouple tables. The voltage signal of the thermocouple was amplified before sampling, and the dynamics of the amplifier used was found not to distort the sampled signal in the frequency window of interest. A Nicolet digital oscilloscope and an IBM-AT computer were used to measure the amplified voltage signal from the thermocouple and to perform data reduction calculations.

Temperatures at different vertical locations in the flame were measured by moving the thermocouple along the spray plane in small vertical increments at the horizontal location of x = 14 mm. 4096 data points were sampled at each location. Source of errors associated with the temperature measurements are the typical ones mentioned in the open literature [1-11,15]. Conduction losses through the leads, radiation losses, catalytic effects on the surface of the wires, soot deposition, and variations on time constant in the flame are the most important ones. In these experiments, conduction losses were minimized by the probe design suggested by Yule et al. [15]. Radiation corrections (5 to 10% maximum correction at the highest temperatures) were not performed, although the possibility of correcting the average temperature information for radiation effects, provided reliable information on radiation properties is available, has been demon-

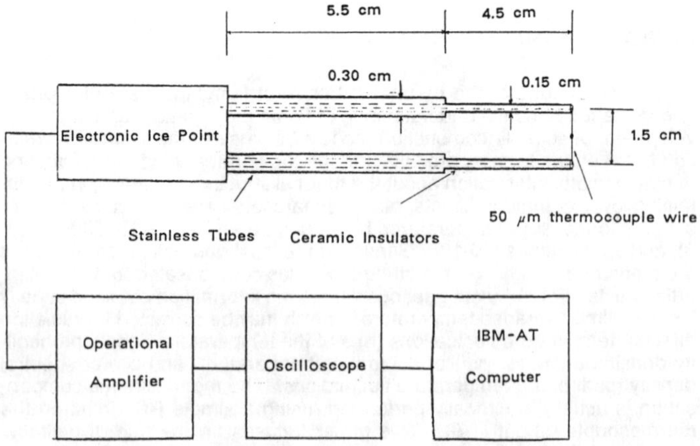

Fig.2 Schematic of the thermocouple probe and data acquisition system.

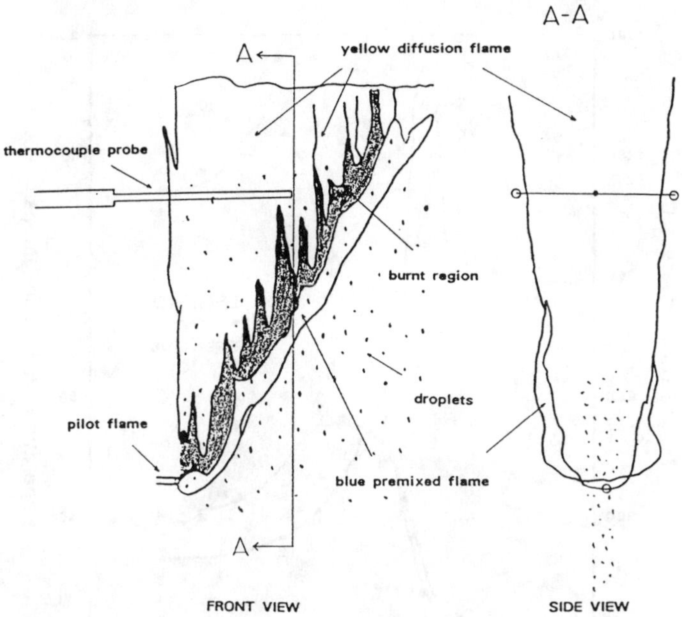

Fig.3 Schematic of the thermocouple probe in the flame, not drawn to scale.

strated [14,15]. The amount of catalytic effects on the temperature measurements is questionable [22] and thermocouple probes are usually left uncoated to maximize the frequency response. Soot deposition was not observed to be a problem in the regions of the flame where temperature measurements were made in this study. Variations of time constant through the flame is the most important source of error for the measurement of temperature fluctuations. Time constants can be measured in the flame as a function of location or it can be assumed constant (based on local time constant measurements in the flame and inspection of the temperature pdf's for overcompensation or undercompensation). In the present study, the former method was employed, although the digital compensation method used here is a definite plus compared to electrical methods because the same raw data can be compensated more than once without redoing the experiments. Root-mean-square noise in the temperature signal was measured to be about 10 ^{o}C and the relative error in average temperature was estimated to be $\pm$ 10%. Figure 3 shows a schematic of the thermocouple probe in the flame (not drawn to scale).

RESULTS AND DISCUSSION

Physical Phenomena

Two types of ignition mechanism for spray flames, depending mainly on the amount of fuel vapor present in the gas phase, have been described in the literature [17-20]. A premixed-gas ignition mechanism exists when the amount of droplet evaporation after the fuel injection is enough to maintain a premixed-gas flame at the ignition source. This premixed-gas flame, characterized by a blue, smooth, continuous flame front, then serves as the ignition source for the vaporizing droplets. Under this ignition mechanism, the influence of the droplet spatial distribution on spray ignition and flame propagation phenomena is somewhat reduced, because the premixed-gas flame exists to ignite any droplet or group of droplets flowing through it. The second mechanism is called a relay ignition mechanism. This mechanism is present when sufficient vaporization does not occur ahead of the ignition source to produce the continuous blue gaseous flame to ignite the droplets. Instead, the thermal front from an ignited droplet near the ignition source moves towards the neighboring droplet or droplets, igniting them. This mechanism is identified with a yellow, very rough, and brushy flame front. In this second ignition mechanism, the influence of the spatially random droplet distribution on ignition and flame propagation is strong, resulting in a more violent drifting motion of the flame front when compared to the first ignition mechanism.

Also affecting the thermal structure of the flame is the behavior of the flame front characterized by an intermittent combustion zone where, in this geometric configuration, the flame moves up and down with the ignition point acting as a pivot point. In this intermittent zone, the flame front presents different dynamic characteristics as spray conditions are varied. The flame front motion exists mainly because of aerodynamic interaction of the flame front with the flow field, and the characteristics of this motion depend on the variation of important spray parameters such as fuel-vapor concentration and turbulence intensity in the gas phase [16].

Effect of Fuel-Vapor Concentration

This parameter was indirectly varied by changing the initial temperature of the liquid fuel. Different levels of fuel-vapor concentration were associated with different liquid initial fuel temperatures of 43°C, 25°C, and 8°C. Furthermore, this study was performed under two levels of gas-phase turbulence intensity (17% and 2.5%) with mean velocities of 1.30 and 1.00 m/s, respectively. The fuel-vapor concentration in the gas phase was estimated using single droplet evaporation rate and droplet number density, assuming that the direct interaction of droplets in the dilute spray was negligible. In general, the gas phase was lean in fuel vapor.

Low Turbulence - Figure 4 shows the profiles of mean temperature (T_m), rms of temperature fluctuations (T_{rms}), and the fluctuation intensity (I_{int}) given as a percentage in terms of $I_{int} = T_{rms}/T_m$. This figure represents the case of low turbulence intensity (2.5%) and 25°C initial fuel temperature. A strong premixed-gas ignition mechanism prevailed for this condition.

T_m gradually increases until the thermocouple crosses the average location of the premixed-gas flame (y = 17 mm). The average temperature at this location is about 1080°C. Locally the average temperature gradient decreases beyond the average position of the premixed-gas flame, primarily because of the evaporative cooling of droplets downstream due to their ignition delay and possibly because of the lateral air entrainment. T_m increases again when the average location of the diffusion flame is approached (y = 23 mm), where it reaches the maximum value of about 1350°C, and finally starts to decrease for positions inside of the diffusion flame envelope. An overall average temperature gradient of 79°C/mm is obtained from the average temperature profile. This gradient is defined as the change in average temperature between the highest average temperature in the flame and the low temperature of 100°C, divided by the distance in which the change occurred.

T_{rms} increases sharply through the premixed-gas flame intermittent combustion zone, because in this region the temperature-time signals frequently indicate both high (burnt gases) and low (unburnt gases) temperatures. T_{rms} peaks at y = 13 mm and then decreases to a constant value of about 200°C. Inside of the intermittent combustion zone, T_m rises from 20% to 80% of its final value, and I_{int} decreases from about 160% down to 25%. For the case under consideration, the intermittent combustion zone is about 10 mm long. One of the distinctive characteristics of this spray flame is noticed in the development of the temperature fluctuation profiles. For axially symmetric gaseous flames [1-4], the temperature fluctuations, in contrast to the present flame, decay almost to negligible amounts inside the flame. This is mainly due to the air entrainment and to the exposure of the thermocouple to the lateral reaction zones which are also moving in a direction perpendicular to the spray plane.

Figure 5 shows the thermal profile for the case of low fuel-vapor concentration (initial fuel temperature of 8°C), and low turbulence intensity (2.5%). The general trend of this thermal profile is similar to the previous one. This flame also has a strong premixed-gas ignition mechanism and a separate diffusion flame region. At the average location of the premixed-gas flame, a lower average temperature, as compared to the case in Fig. 4, is recorded (900°C compared to 1080°C). This is because the fuel-vapor equivalence ratio is reduced due the lower initial fuel temperature (lower droplet evaporation rates), causing a lower premixed-gas flame temperature. At the average location of the diffusion flame, T_m was only 80°C lower (1270°C compared to 1350°C) than in the previous case. At this location T_m would be expected to be nearly the same as the first case, since diffusion flames burn at stochiometric conditions; however, this small difference may have been caused by experimental error.

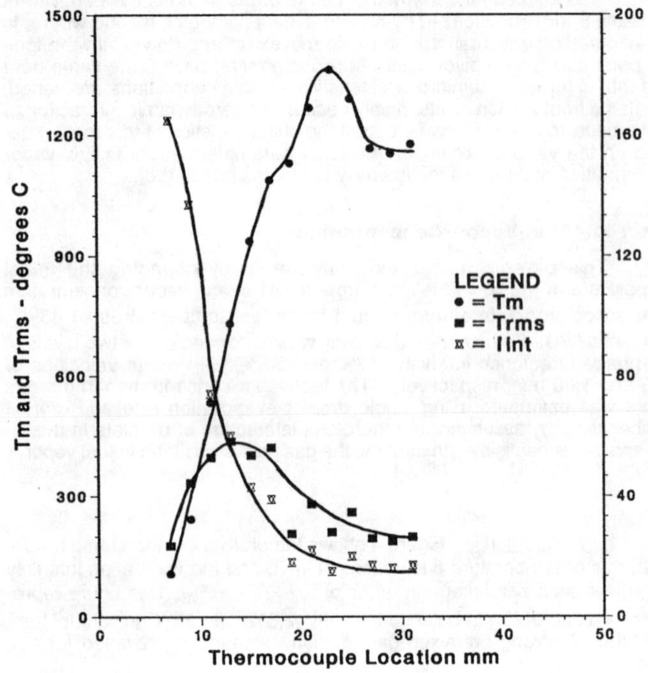

Fig.4 Temperature profile at x = 14 mm for the hexane spray flame with low turbulence, 25 °C initial temperature of the liquid fuel, and 400 μm average droplet size.

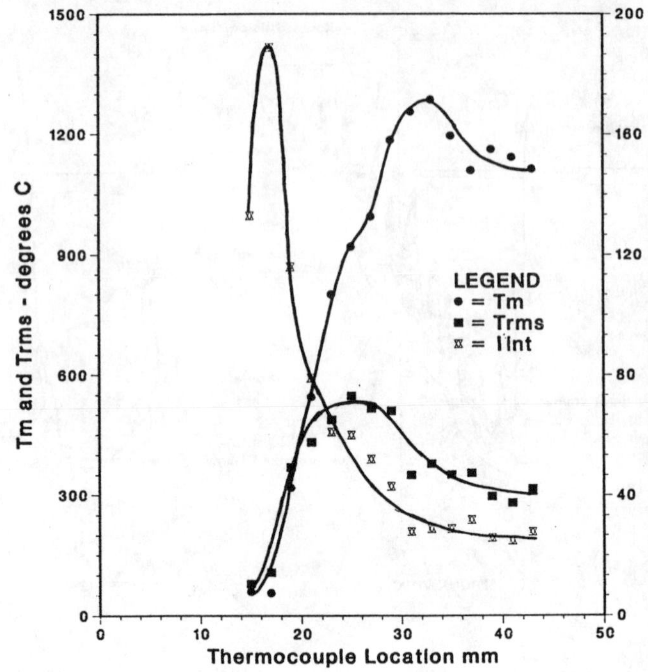

Fig.5 Temperature profile at x = 14 mm for the hexane spray flame with low turbulence, 8 °C initial temperature of the liquid fuel, and 400 μm average droplet size.

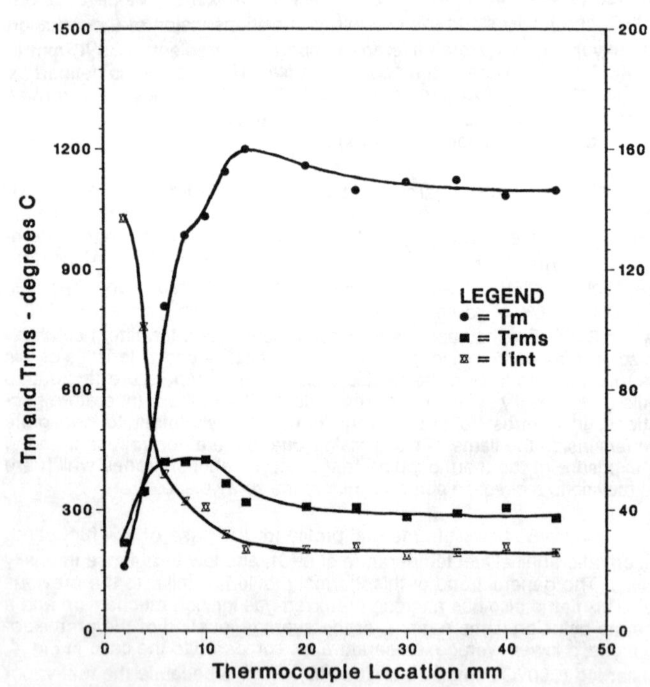

Fig.6 Temperature profile at x = 14 mm for the hexane spray flame with high turbulence, 43 °C initial temperature of the liquid fuel, and 400 μm average droplet size.

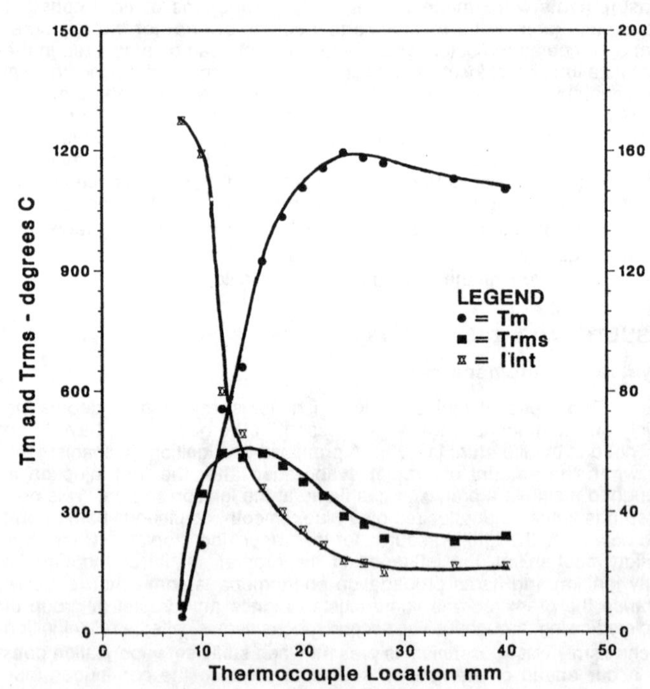

Fig. 7 Temperature profile at x = 14 mm for the hexane spray flame with high turbulence, 25 °C initial temperature of the liquid fuel, and 400 μm average droplet size.

The cause for the increase in T_{rms} when the fuel temperature was lowered from 25°C to 8°C is not well understood. Although the flame does present a slightly higher flame front fluctuation intensity when the fuel temperature is lowered [16], the same trend in the T_{rms} profiles is not observed in the high turbulence tests, or in similar parametric studies in turbulent premixed flames [3]. One possibility is that the temperature profiles were overcompensated in the 8°C case, but not enough evidence exists for such a claim. The overall average temperature gradient for this case was approximately 74°C/mm.

High Turbulence - The effect of fuel-vapor concentration on the thermal structure of the flame was also investigated at high turbulence intensity. Figures 6 and 7 show the temperature profiles for liquid initial fuel temperatures of 43°C, and 25°C, respectively.

For the 43°C case (Fig. 6), despite the increased diffusivity due to the higher turbulence level, a strong premixed-gas ignition mechanism was present because of the high droplet vaporization rate associated with the high initial temperature of the fuel. The premixed-gas flame was followed by the diffusion flame, as can be inferred from the average temperature profile. A sharp overall average temperature gradient can be seen (approximately 87°C/mm), due to the reduced flame front fluctuations [16]. T_{rms} is high (425°C) in the intermittent combustion zone, and it then drops to a constant value of about 300°C.

Because the strength of the premixed-gas ignition mechanism was substantially reduced when the fuel temperature was decreased to 25°C, the continuous bluish gas flame disappeared and the existence of the average location of the premixed-gas flame for this second high turbulence case was not detected by the thermocouple (Fig. 7). As observed in studying photographs of the flame for this spray condition, the premixed-gas flame strength is reduced to the point that the blue flame is only visible in a small region around the individual streams. The diffusion flame starts right after the premixed-gas flame, and essentially no gap between the two flames exists. T_m attains a maximum value of about 1200°C, when the average location of the diffusion flame is reached. As the flame fluctuation intensity increased, compared with the previous high turbulence case, the thickness of the intermittent combustion zone increased, causing a lower overall temperature gradient (about 72°C/mm). The T_{rms} profile follows the general trend with maximum values comparable (within experimental error) with the ones in Fig. 6.

Under the spray conditions studied here, the effect of fuel-vapor concentration in the thermal structure of the flame is strongly related to the spray ignition mechanisms through their effect on the flame front dynamic behavior. As compared to the relay ignition mechanism, the premixed-gas ignition mechanism causes higher average temperatures, lower temperature fluctuations, and steeper overall average temperature gradients across the flame. Furthermore, for flames with a strong premixed-gas ignition mechanism, the effect of higher fuel-vapor concentrations on the thermal profile is similar to the observations of Yoshida and Gunther in turbulent premixed flames [3]. Although with increasing fuel-vapor concentration the average temperature across the premixed flame peaks at higher values, the essential features in both the average and fluctuating temperature profiles are unchanged.

Probability Density Functions - In addition to the global information on the thermal structure of the flame obtained by analyzing variations of time-averaged temperatures and rms of temperature fluctuations, further study of the phenomenon is possible by extracting more information from the time-resolved temperature signals in the form of probability density functions (pdf's) of the temperature fluctuations. This further analysis may provide information on the instantaneous temperature profile across the flame as well as on the dynamic behavior of the flame front. Figure 8 is the pdf surface for the case of 25°C initial fuel temperature and 2.5% turbulence intensity. The thermal profile for this case is shown in Fig. 4.

At the locations upstream of the flame front, the pdf shows a sharp peak at the low temperature side. At this location, since the average gas-phase temperature is below the threshold value (350°C) for the operation of a microthermocouple in a dilute spray flame of liquid hexane fuel [13], one needs to carefully interpret the results obtained, due to the possible effect of droplet impaction and wetting on the thermocouple (although the qualitative information obtained is as expected). As the thermocouple moves downstream, the probability of ambient temperature decreases and a long tail to high temperatures appears, indicating the existence of high temperature spikes when the flame arrives. Further downstream (y = 10 mm), the pdf has two peaks at the low temperature (about room temperature) and at the high temperature (about 700°C), resulting in a bimodal distribution. One may say that the bimodal pdf shape corresponds to the large scale flapping motion of the flame front passing by the measuring point detecting regions of burnt and unburnt gas, respectively. The fact that the bimodal pdf curve is not dominated by near-delta-function peaks suggests a flame front with a smooth instantaneous temperature profile, as observed by Roberts and Moss [8]. The detection of the burnt gas region by the thermocouple is shown by the tail of the bimodal curve extending to high temperatures. This observation is consistent with the wrinkled laminar-flame model, although in this case the bimodal pdf is not dominated by near-delta functions at essentially fixed temperatures, as is the case with the thermal structure of some turbulent gaseous flames [3]. The low temperature spike then starts to disappear, and the probability of high temperature increases. Further downstream, the probability of low temperatures becomes minimal, and the high temperature peaks at about 1450°C at the location where the thermocouple crosses the average location of the diffusion flame front (y = 23 mm). The shifting of the pdf surface towards the low temperatures after this vertical location is associated with the drop in average temperature inside the diffusion flame envelope. The mixing of burnt gases with side-entrained ambient air can be observed by the broad shape of the pdf surface downstream of the average location of the diffusion flame in the spray plane.

Figure 9 shows the pdf surface for the case of 8°C initial fuel temperature and 2.5% turbulence intensity, corresponding to the thermal profile of Fig. 5. The general shape of this surface is similar to the one in Fig. 8. The steeper instantaneous temperature profile from the unburnt to burnt gases in the flame front can be inferred from the well-defined bimodal shapes of the pdf surface in the intermittent combustion zone. The probability of intermediate temperature is reduced, supporting this statement. Also, for this spray condition the probability of having low temperatures persists well into the flame.

The effect of fuel-vapor concentration on the pdf surfaces at high turbulence intensity is shown in Figs. 10 and 11, corresponding to the temperature profiles of Figs. 6 and 7, respectively. The sudden drop in the projection of the pdf profiles shown in Figs. 10 is not related to any physical phenomena in the flame, but rather to a deficiency of the plotting subroutines used to generate the pdf surfaces.

The general trend of the development of these surfaces with thermocouple position is similar to the ones for the surfaces shown in Figs. 8 and 9. Two major differences exist, due to the presence of gas-phase turbulence, which may be related to a different instantaneous temperature profile across the flame and to different flame front dynamics (different ignition mechanisms). These differences are indicated by the overall smoothness of the pdf surfaces (smoother instantaneous temperature profile across the flame), and by the different sizes of intermittent combustion zones (different flame front dynamics) where the pdf surfaces change from the low to the high temperature side.

Effect of Gas-Phase Turbulence Intensity

The effect of turbulence intensity on the thermal structure of the flame can be observed by comparing Figs. 4 and 7 and their corresponding pdf surfaces in Figs. 8 and 11, since these tests have approximately the same conditions except for the gas-phase turbulence intensity level (2.5 and 17%, respectively).

A study of Figs. 4 and 7, complimented with flame photographs for both flame condition, reveals that the increased level of the turbulence intensity reduced the premixed-gas ignition mechanism substantially. This increased the flame intermittent combustion zone [16] and reduced the gap between the premixed flame and diffusion flames to the point that, generally, the average location of the premixed flame was not detected under the high turbulence intensity condition. The larger intermittent combustion zone and the increase in the flame fluctuation intensity sustained high values of T_{rms} longer, and caused higher levels of T_{rms}, respectively. Increase in transport properties due to turbulent diffusion accounts for the lower value of T_m at the diffusion flame average location for the high turbulence case. The overall temperature gradient was approximately the same (79°C/mm compared to 73 °C/mm), despite the larger intermittent combustion zone for the high turbulence case.

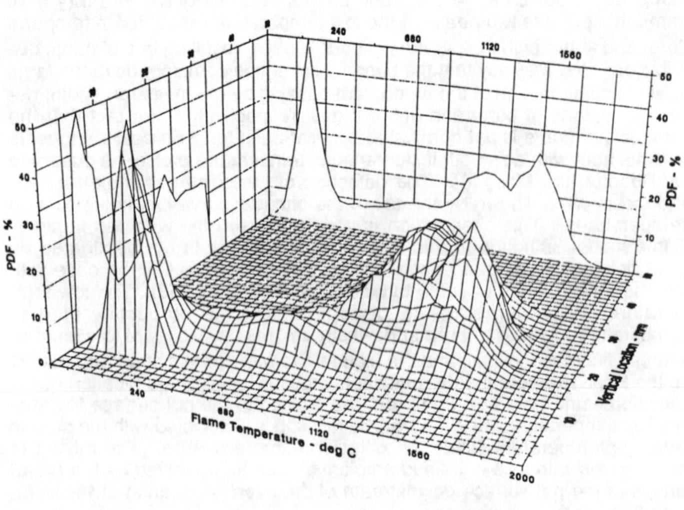

Fig. 8 Pdf surface for the temperature profile shown in Fig. 4.

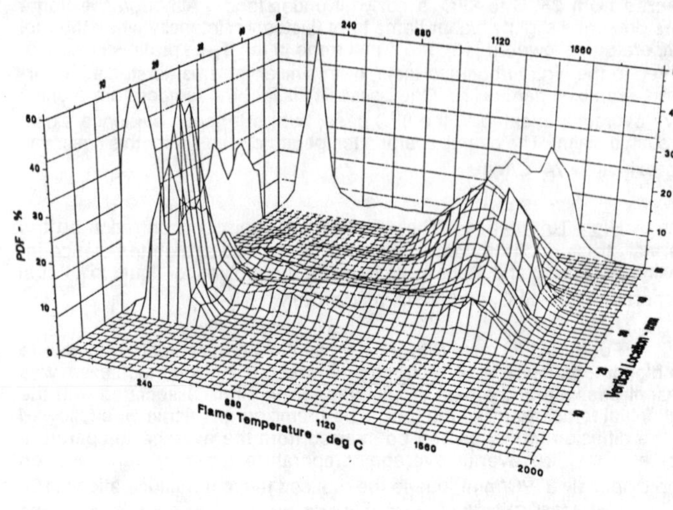

Fig. 9 Pdf surface for the temperature profile shown in Fig. 5.

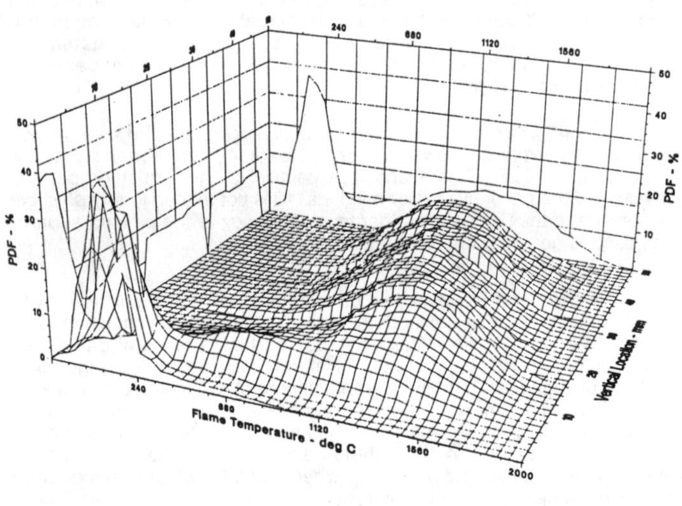

Fig. 10 Pdf surface for the temperature profile shown in Fig. 6.

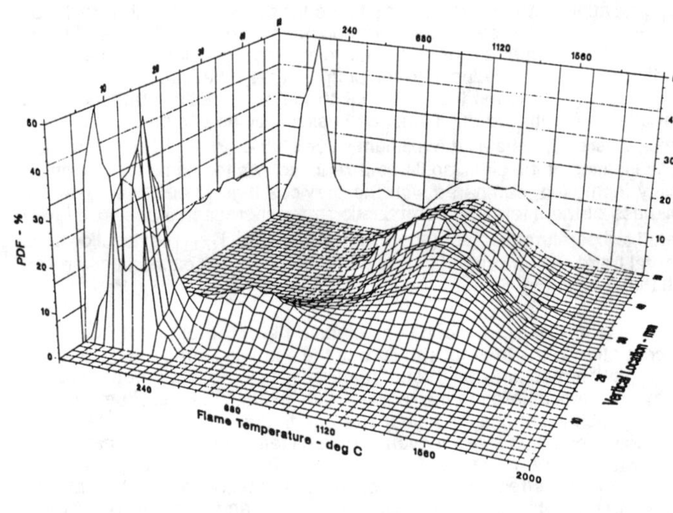

Fig. 11 Pdf surface for the temperature profile shown in Fig. 7.

For the pdf surfaces, the major differences between Figs. 8 and 11 are the bimodal shapes for the low turbulence case, and the smoother surface for the high turbulence case. It is believed that the influence of turbulence in the disappearance of the bimodal pdf's is two-fold. It first increases the diffusion of heat, causing the instantaneous temperature profile through the flame to be less steep. Second, its aerodynamic interaction with the flame causes the flame front to fluctuate more with perhaps different dynamic characteristics.

CONCLUSIONS

The effect of the flame front dynamic behavior on the thermal structure of a dilute spray flame was explored in the present study. Thermocouple measurements for flames with different fuel-vapor concentrations and turbulence intensities were studied to identify the effect of these spray parameters on the flame thermal structure. The following was concluded from this study:

1. Good agreement was found between the previous observations on the combustion dynamics of the spray flame front and its thermal structure. Overall average temperature gradient, values of T_{rms}, and shape of the pdf surfaces were found to be related to the intermittent combustion zone where the flame front presented a different drifting motion as spray conditions varied [16].

2. The spray ignition mechanisms, which were found to affect the drifting motion of the flame [16], also influenced its thermal profile. The particular ignition mechanism that exists for a given spray flame condition is a function of the fuel-vapor concentration at the ignition point, which is affected by the initial temperature of the liquid fuel, as well as the turbulence intensity level in the gas phase. Compared to the premixed-gas ignition mechanism, the relay ignition mechanism caused lower average temperatures, higher temperature fluctuations, and also less steep overall average temperature gradients across the flame.

3. Under the experimental conditions of the present study, higher fuel-vapor concentrations generated higher premixed-gas flame temperatures as observed by higher values of T_m at the end of the intermittent combustion zone for flames having a strong premixed-gas ignition mechanism. Furthermore, taking into account experimental errors, the essential features in both the average and fluctuating temperature profiles as well as the pdf surfaces were unchanged when the fuel-vapor concentration varied.

4. The increase in the turbulence intensity level promoted higher T_{rms} values by widening the intermittent combustion zone. It also changed the dynamics of the drifting motion of the flame as well as changing the instantaneous temperature profile across the flame, causing the disappearance of the bimodal shapes in the pdf surfaces.

ACKNOWLEDGEMENT

The authors appreciated the support of the Army Research Office under grant No. DAAG 29-84-k-007 during this research effort. The preparation of this manuscript was done while the first author was supported by a grant from the Advanced Combustion Engineering Research Center (ACERC) at Brigham Young University. Funds for this center are received from the National Science Foundation, The State of Utah, 23 industrial participants, and the U.S. Department of Energy.

REFERENCES

1. Yoshida, A., and Tsuji, H., "Characteristic Scale of Wrinkles in Turbulent Premixed Flames," *Nineteenth Symposium (International) on Combustion*, The Combustion Institute, Pittsburgh, PA, pp. 403-411, (1982).

2. Yoshida, A., and Tsuji, H., "Measurements of a Fluctuating Temperature and Velocity in a Turbulent Premixed Flame," *Seventeenth Symposium (International) on Combustion*, The Combustion Institute, Pittsburgh, PA, pp. 945-956, (1979).

3. Yoshida, A., and Gunther, R., "Experimental Investigation of Thermal Structure of Turbulent Premixed Flames," *Combustion and Flame*,

4. Yoshida, A., "An Experimental Study of Wrinkle Laminar Flame," *Eighteenth Symposium (International) on Combustion*, The Combustion Institute, Pittsburgh, PA, pp. 931-939, (1981).

5. Toong, T.Y., and Chang, C., "Thermal Structure of Turbulent Premixed Rod-Stabilized V-Flames," *Proceedings of Eastern Section Meeting*, The Combustion Institute, Pittsburgh, PA, paper No. 19, (1986).

6. Lenz, W., and Gunther, R., "Measurements of Fluctuating Temperature in a Free-Jet Diffusion Flame," *Combustion and Flame*, 37:63-70, (1980).

7. Ballantyne, A., and Moss, J.B., "Fine Wire Thermocouple Measurements of Fluctuating Temperature," *Combustion and Flame*, 17:63-72, (1977).

8. Roberts, P.T., and Moss, J.B., "A Wrinkled Flame Interpretation of the Open Turbulent Diffusion Flame," *Eighteenth Symposium (International) on Combustion*, The Combustion Institute, Pittsburgh,

9. Chandran, S.B.S., Komerath, N.M., Grissom, W.M., Jagoda, J.I. and Strahle, W.C., "Time Resolved Thermometry by Simultaneous Thermocouple and Rayleigh Scattering Measurements in a Turbulent Flame," *Combustion Science and Technology*, 44:47-60, (1985).

10. Onuma, Y., Ogasawara, M., and Inove, T., "Further Experiments on the Structure of a Spray Combustion Flame," *Fifteenth Symposium (International) on Combustion*, The Combustion Institute, Pittsburgh,

11. Onuma, Y., and Ogasawara, M., "Studies on the Structure of a Spray Combustion Flame," *Fifteenth Symposium (International) on Combustion*, The Combustion Institute, Pittsburgh, PA, pp 453-465, (1975).

12. Ballal, D.R., and Lefebvre, A.H., "Flame Propagation in Heterogeneous Mixtures of Fuel Droplets, Fuel Vapor and Air," *Eighteenth Symposium (International) on Combustion*, The Combustion Institute, Pittsburgh, PA, pp. 321-328, (1980).

13. Queiroz, M., Dietvorst, J., Yao, S.C., and Chigier, N.A., "Digitally Compensated and Processed Temperature Measurements in a Thin Spray Flame Using a Micro-Thermocouple," *Proceedings of Eastern Section Meeting*. The Combustion Institute, Pittsburgh, PA, paper No. 72, (1986).

14. Queiroz, M., "Combustion Dynamics and Thermal Structure of a Planar Spray Flame," Ph.D. Thesis, Carnegie Mellon University, June, (1987).

15. Yule, A.J., Taylor, D.S., and Chigier, N.A., "On-Line Digital Compensation and Processing of Thermocouple Signals for Temperature Measurement in Turbulent Flames," *AIAA Sixteenth Aerospace Sciences Meeting*, Huntsville, AL, (1978).

16. Queiroz, M., and Yao, S.C., "A Parametric Exploration on the Dynamic Behavior of Flame Propagation in Planar Sprays." Accepted for publication in *Combustion and Flame*, (1989).

17. Burgoyne, J.L., Cohen, L. "The Effect of Drop Size on Flame Propagation in Liquid Aerosols," *Proceedings of Royal Society of London*, a225(1162):375-392, (1954).

18. Hayashi, S., Kumagai, S., and Sakay, T., "Propagation Velocity and Structure of Flame in Droplet-Vapor-Air Mixtures," *Combustion Science & Technology*, 15:169-177, (1976).

19. Hayashi, S., Ohtani, T., Iinuma, K., and Kumagai, S., "Limiting Factor of Flame Propagation in Low-Volatility Fuel Clouds," *Eighteenth Symposium (International) on Combustion*, The Combustion Institute, Pittsburgh, PA, pp. 361-367, (1981).

20. Ashgriz, N., and Yao, S.C., "Mechanism of Flame Propagation in Sprays," *Proceedings of Eastern Section Meeting*, The Combustion

21. Son, S.F., Queiroz, M., and Wood, C.G., "Digital Compensation of Thermocouples for Thermal Inertial Effects," *Proceedings of Western Section Meeting*, The Combustion Institute, Pittsburgh, PA, paper No.

22. Baker, H.D., Ryder, E.A., and Baker, N.H., "Temperature Measurements in Engineering," Volumes 1 and 2, Omega Engineering, Inc., Stamford, CT.

1989 National Heat Transfer Conference
HTD-Vol. 106, Heat Transfer Phenomena in Radiation, Combustion and Fires

EVAPORATION OF A MOVING LIQUID DROPLET: SOLUTIONS FOR INTERMEDIATE REYNOLDS NUMBERS

L. J. Huang and P. S. Ayyaswamy
Department of Mechanical Engineering and Applied Mechanics
University of Pennsylvania
Philadelphia, Pennsylvania

ABSTRACT. In this paper, we have examined the quasi-steady evaporation of an n-heptane droplet moving in the intermediate Reynolds number regime (Re = O(100)). The droplet environment is taken to consist of its own vapor and air. The ambient pressure is taken to be one atmosphere. The droplet trajectory and hence the instantaneous terminal velocity of the droplet have been determined by a balance of the drag and body forces. Variable thermophysical properties, liquid phase motion and heating, and the effect of variations in droplet size have been included in this investigation. A numerical procedure that is based on the hybrid difference scheme has been employed. We provide histories of droplet radius, mass, velocity, Reynolds number and liquid heating fraction. The total drag, friction drag, pressure drag, and evaporation drag histories are also individually predicted.

INTRODUCTION. Estimates of the rates of the evaporation of droplets in high temperature convective environments are important to a large number of industrial processes and applications. Typically, fuel sprays consist of a spectrum of droplets of various sizes and velocities. The smaller droplets (diameter < 30 μm) essentially follow the ambient flow field, and will be in the low translational Reynolds number regime (Re < 1) during most of their lifetime [Gogos and Ayyaswamy (1988), Gogos et al. (1986)]. The bigger droplets (diameter - 100 μm), because of their large inertia and momentum, may be in the intermediate Reynolds number flow regime (Re=O(100)) for significant portion of their lifetime. They penetrate deep into the spray core. The bigger droplets determine the overall behavior of the spray combustion process since they constitute most of the spray mass. The smaller droplets have a lesser impact since they last over a comparatively shorter period of time. The predominant mode of droplet vaporization of relevance to industrial applications is, therefore, convective evaporation at intermediate Reynolds numbers. At such a Reynolds number, we may envision thin boundary layers for the transport of momentum, heat, and mass, both on the outside and on the inside of the droplet surface. Prakash and Sirignano (1980) have examined the convective evaporation of n-hexane, n-decane and n-hexadecane droplets at 10 atmospheres ambient pressure by a boundary-layer analysis. This work invokes the constant-property assumption and all processes except liquid heating are regarded as quasi-steady. The velocity and pressure distribution outside the boundary layer are assumed to be those for potential flow over a sphere; these assumptions preclude the determination of the correct total drag coefficient governing droplet motion [Ayyaswamy (1989)]. Also, at Re=O(100), there will be a recirculating wake in the rear stagnation-point region. In this region the elliptic effects are important, and the boundary layer assumptions are not valid. Dwyer and Sanders (1984) have numerically examined the convective evaporation of dodecane droplets at ambient pressures of 1, 5, and 25 atmospheres. Their model includes the effects of transients in both phases as well as the transient deceleration of the droplet under the influence of its own drag. The study also assumes constant properties. The solution has been obtained by employing a numerical technique that uses an adaptive grid to track the shrinking droplet. The drag coefficients and Nusselt numbers obtained in this numerical study are found to be lower than those predicted by existing quasi-steady correlations [Renksizbulut and Yuen (1983)]. We note that constant property assumptions may not be valid in the study of droplet evaporation. Haywood et. al. (1988) have used a finite volume numerical technique to model the evaporation in air (800 K, 1 atm) of a n-heptane droplet with an initial Reynolds number of 100. Variable thermophysical properties, liquid phase motion and heating, and transient variation in the droplet size and velocity are included in this analysis. This study makes a very important contribution to the literature on droplet evaporation. However, in this paper, the conservation of momentum equation does not seem to take into account the presence of body forces. The effect of this omission on the results obtained is difficult to predict; in particular, the predicted temporal variations may not be correct. The results provided in [Haywood et. al. (1988)] indicate that with appropriate correction for the effects of variable properties and liquid phase heating, quasi-steady correlations may be used to predict the transient histories of drag coefficients, Nusselt, and Sherwood numbers when the ambient pressure is not very high.

In this paper, we have examined the quasi-steady evaporation of an n-heptane droplet moving in the intermediate Reynolds number regime (Re = O(100)). The environment is taken to consist of its own vapor and air. The ambient pressure is taken to be one atmosphere. The droplet trajectory has been determined by a balance of the drag and body forces. With the knowledge of the instantaneous terminal velocity of the droplet, we are able to provide results for the transient behavior. Variable thermophysical properties, liquid phase motion and heating, and the effect of variations in droplet size have been included in this study. A numerical procedure that is based on the hybrid difference scheme has been employed. In the hybrid scheme, at the start of every iteration, based on current guess solutions, the coefficients for the convective and diffusive terms are estimated at each grid point using the Central Difference Scheme. The Central Difference Scheme is retained where the coefficient for the diffusion contribution is larger, and the Upwind Difference Scheme is employed when the coefficient for convection becomes dominant.

We provide histories of droplet radius, mass, velocity, Reynolds number and liquid heating fraction. The total drag, friction drag, pressure drag, and evaporation drag histories are also individually predicted.

FORMULATION OF THE VAPORIZATION PROBLEM. Consider a single component liquid drop of initial radius R_o translating at an initial velocity U_o, in an infinite expanse of a hot, insoluble gaseous species (air) of mass fraction Y_1. The drop is initially cold at a temperature T_o, and vaporizes to a gaseous species whose mass fraction is Y_F. The ambient temperature T_∞ and the vapor mass fraction $Y_{F,\infty}$ are taken to be prescribed. In the formulation of the problem it is assumed that: (i) the droplet maintains a spherical shape, (ii) the flow field is laminar and axisymmetric, (iii) thermal radiation, viscous dissipation, Soret and Dufour effects and pressure diffusion are negligible, (iv) thermodynamic equilibrium prevails at the gas/liquid interface, and (v) air is insoluble in the liquid droplet.

The governing equations and boundary conditions are:

Trajectory equation:

$$\frac{dV}{dt} = \frac{R_o^2 g}{U_o \alpha_\ell} - \frac{3}{16}\frac{C_D V Pe_{\ell,t}}{\bar{\rho}_\ell R^2} \qquad (1)$$

where

$$Pe_{\ell,t} = \frac{U_\infty 2R}{\alpha_\ell}, \qquad V = \frac{U_\infty}{U_o}, \qquad R^* = \frac{R}{R_o}$$

and the density ρ_ℓ has been scaled by $\rho_{g,o}$. The drag coefficients are:

$$C_D = C_F + C_P + C_E \qquad (2)$$

where,

$$C_F = \frac{8}{Re_{g,\infty}}\int_0^\pi (\tau_{r\theta}\sin\theta - \tau_{rr}\cos\theta)_s \sin\theta \, d\theta \qquad (3)$$

$$C_P = \int_0^\pi p_s \sin 2\theta \, d\theta \qquad (4)$$

$$C_E = 2\int_0^\pi \rho_s \left(v_r^2 \sin 2\theta - 2 v_r v_\theta \sin^2\theta\right)_s d\theta \qquad (5)$$

The nondimensional equations for the gaseous phase (Figure 1) are given by (subscript g refers to mixture, 1 to air, and F to vapor):

$$\nabla \cdot (\rho_g \mathbf{u}_g) = 0 \qquad (6)$$

$$Re_{g,\infty}\rho_g \mathbf{u}_g \cdot \nabla \mathbf{u}_g = -\frac{1}{2} Re_{g,\infty} \nabla p_g + 2\nabla \boldsymbol{\sigma} \qquad (7)$$

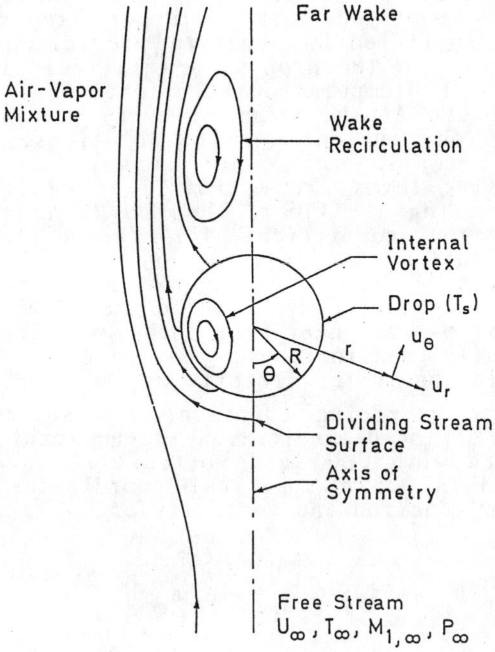

Fig. 1. Schematic of the problem.

$$Pe_{gt,\infty}\rho_g \, \mathbf{u}_g \bullet \nabla T_g = 2 \, \nabla \bullet (k_g \nabla T_g) \quad (8)$$

$$Pe_{gm,\infty}\rho_g \, \mathbf{u}_g \bullet \nabla Y_g = 2 \, \nabla \bullet (\rho_g D_g \nabla Y_g) \quad (9)$$

where

$$\boldsymbol{\sigma} = \mu_g[\nabla \mathbf{u}_g + (\nabla \mathbf{u}_g)^T - \frac{2}{3} (\nabla \bullet \mathbf{u}_g) \, \mathbf{I}] \quad (10)$$

The velocity $\mathbf{u}_g$ has been scaled by U_∞, the pressure p_g, the temperature T_g and mass fraction m_1 have been non-dimensionalized as follows:

$$p_g^* = \frac{(p_g(\theta) - p_\infty)}{\frac{1}{2}\rho_g U_\infty^2}, \qquad Re_{g,\infty} = \frac{U_\infty 2R}{\nu_g}$$

$$T_g^* = \frac{T_g}{T_\infty}, \qquad Y_1^* = \frac{m_1}{m_{1,\infty}}$$

The Peclet and Prandtl numbers are:

$$Pe_{g,t} = \frac{U_\infty 2R}{\alpha_g}, \quad Pe_{g,m} = \frac{U_\infty 2R}{D_g}, \quad Pr_g = \frac{\nu_g}{\alpha_g}$$

where D_g is the mass diffusivity.

The non-dimensional equations for the liquid phase are:

$$\nabla \bullet \mathbf{u}_\ell = 0 \quad (11)$$

$$Re_\ell \, \mathbf{u}_\ell \bullet \nabla \mathbf{u}_\ell = -\frac{1}{2} Re_\ell \, \nabla p_\ell + 2 \, \nabla^2 \mathbf{u}_\ell \quad (12)$$

$$\frac{\partial T_\ell}{\partial t} + \frac{1}{2} Pe_{\ell,t} \, \mathbf{u}_\ell \bullet \nabla T_\ell = \nabla^2 T_\ell \quad (13)$$

where the velocity $\mathbf{u}_\ell$ has been scaled by U_∞, temperature T_ℓ, time t, pressure p_ℓ, Reynolds number Re_ℓ are

$$t^* = \int_0^t \frac{\alpha_\ell \, dt}{R^2}, \qquad T_\ell^* = \frac{T_\ell}{T_\infty}$$

$$p_\ell^* = \frac{(p_\ell(\theta) - p_\infty)}{\frac{1}{2}\rho_\ell U_\infty^2}, \qquad Re_\ell = \frac{U_\infty 2R}{\nu_\ell}$$

Initial and boundary conditions governing the evaporation problem are:

<u>The initial conditions:</u>

$$T_\ell = T_o \quad \text{at} \quad t = 0 \quad (14)$$

<u>At the interface (r=1):</u>

Continuity of tangential velocity

$$u_{g,\theta} = u_{\ell,\theta} = u_\theta \quad (15)$$

Continuity of shear stress

$$r \frac{\partial}{\partial r} \frac{u_{g,\theta}}{r} + \frac{1}{r} \frac{\partial u_{g,r}}{\partial \theta} = \frac{\mu_\ell}{\mu_g} \left[r \frac{\partial}{\partial r} \frac{u_{\ell,\theta}}{r} + \frac{1}{r} \frac{\partial u_{\ell,r}}{\partial \theta} \right] \quad (16)$$

Continuity of mass flux

$$\rho_g(u_{g,r} - \dot{R}) = \rho_\ell(u_{\ell,r} - \dot{R}) \quad (17)$$

where $\dot{R}$ is the regression rate of the drop, μ_ℓ and μ_g have been scaled by $\mu_{g,\infty}$, and ρ_g by $\rho_{g,\infty}$.

Impermeability of the air

$$\frac{2 \, D_{g,s}}{Pe_{gm,\infty}(1-E)} \frac{\partial Y_1}{\partial r} = u_{g,r} \Big|_{r=1} \quad (18)$$

where the parameter E referred to as the evaporation parameter is given by $E = 1 - m_{1,s}/m_{1,\infty}$, and it is a function of the thermodynamic conditions p_∞, T_∞ and T_s. E varies from 0 to 1.

Temperature Continuity

$$T_g = T_\ell \quad (19)$$

Heat Continuity

$$Ja_g \frac{\partial T_g}{\partial r} - \frac{Pe_{g,t}}{2}\left[u_{g,r} + \frac{\rho_g \bar{u}_{g,r}}{\rho_\ell - \rho_g}\right] = q_s \quad (20)$$

where the Jakob number Ja_g is given by $Ja_g = c_{pg} T_\infty/\lambda$, and the non-dimensional heat flux is given by $q_s^* = q_s R/\rho_g \alpha_g \lambda$. The latent heat of evaporation is λ and the specific heat is c_p. The quantity $\bar{u}_{g,r}$, is the average value of $u_{g,r}$. The normalized mass fraction at the interface:

$$Y_1 = Y_s \quad (21)$$

Far-stream inlet conditions ($r \to \infty$, $0 \le \theta \le \pi/2$):

Uniform velocity:

$$\mathbf{u}_g = \mathbf{j} \quad (22)$$

where $\mathbf{j}$ is the unit vector. The bulk mixture conditions are

$$T_g = p_g = 0 \ , \quad Y_1 = 1\text{-}E \quad (23)$$

Far-stream outlet conditions ($r \to \infty$, $\pi/2 \le \theta \le \pi$):

$$\frac{\partial \mathbf{u}_g}{\partial y} = \frac{\partial T_g}{\partial y} = \frac{\partial Y_1}{\partial y} = 0 \quad (24)$$

where y is the y-coordinate.

Axi-symmetric conditions at $\theta=0$ and $\theta=\pi$:

$$u_{g,\theta} = \frac{\partial u_{g,r}}{\partial \theta} = \frac{\partial p_g}{\partial \theta} = \frac{\partial T_g}{\partial \theta} = \frac{\partial Y_1}{\partial \theta} = 0 \quad (25)$$

$$u_{\ell,\theta} = \frac{\partial u_{\ell,r}}{\partial \theta} = \frac{\partial p_\ell}{\partial \theta} = 0 \quad (26)$$

The definitions of Nusselt and Sherwood Numbers are

$$Nu_\infty = \frac{1}{2}\int_0^\pi \left(\frac{2k_s}{1-T}\frac{\partial T}{\partial r}\right)_{s,\theta} \sin\theta \, d\theta \quad (27)$$

$$Sh_\infty = \frac{1}{2}\int_0^\pi \left(\frac{2\rho_s}{Y_{F,\infty}-Y_F}\frac{\partial Y_F}{\partial r}\right)_{s,\theta} \sin\theta \, d\theta \quad (28)$$

SOLUTION PROCEDURE. A numerical procedure that is based on the hybrid difference scheme will be employed to solve the coupled set of governing equations subject to appropriate boundary and interface conditions. The scheme is a combination of a second-order accurate central difference scheme (CDS) and a first-order accurate upwind difference scheme (UDS) [Sundararajan and Ayyaswamy (1984)]. The CDS is employed near the drop-surface in conjunction with a small step-size. Far away from the drop, UDS is employed with a larger step-size. Thus, sufficiently accurate predictions of transport to the drop at a relatively lower level of computational effort are made possible. At the start of every iteration, based on current guess solutions, the coefficients for the convective and diffusive terms are estimated at each grid point using the CDS. The CDS is retained where the coefficient for the diffusion contribution is larger, and the Upwind Difference Scheme is employed when the coefficient for convection becomes dominant. The step-size near the drop surface is carefully monitored so as to get a CDS formulation in the vicinity of the drop.

The governing equations are solved in terms of the dimensionless stream function ψ (scaled with $U_\infty R^2$) and vorticity ζ (scaled with U_∞/R). In spherical coordinates the stream function and vorticity are

$$\rho_g \mathbf{u}_g = \nabla \mathrm{x} \left[-\frac{\psi_g}{r \sin\theta} \mathbf{e}_\phi\right]$$

$$\zeta_g \mathbf{e}_\phi = \nabla \mathrm{x} \, \mathbf{u}_g \quad (29)$$

and

$$\mathbf{u}_\ell = \nabla \mathrm{x} \left[-\frac{\psi_\ell}{r \sin\theta} \mathbf{e}_\phi\right]$$

$$\zeta_\ell \mathbf{e}_\phi = \nabla \mathrm{x} \, \mathbf{u}_\ell \quad (30)$$

RESULTS AND DISCUSSION. The computations have been made for an n-heptane droplet evaporating in air at 800 K and 1 atmosphere pressure. The droplet initial radius R_o = 0.2 mm, T_o = 300 K, and $Re_{g,o}$ = 100 based on freestream thermophysical properties. These are the same conditions examined by Haywood et al. (1989).

The Reynolds number, radius, and liquid heating fraction histories are shown in Fig. 2. The droplet spends a significant portion of its lifetime at Reynolds numbers where convection effects are dominant, and liquid heating persists for considerable portion of the droplet lifetime. Since the Reynolds Number continuously decreases with time, a boundary layer analysis cannot be used to describe the evaporation process for the entire lifetime of the drop. Liquid heating is seen to be very important and cannot be ignored in analyzing droplet evaporation or combustion. A comparison with the results of Haywood et al. (1989) shows that their results overpredict the decrease in transient Reynolds Number. This is probably due to the neglect of the body force term in the trajectory equation in their analysis.

Fig. 3 shows the variation of the various drag coefficients. The total drag coefficient C_D decreases very slightly during the early transient period, and then

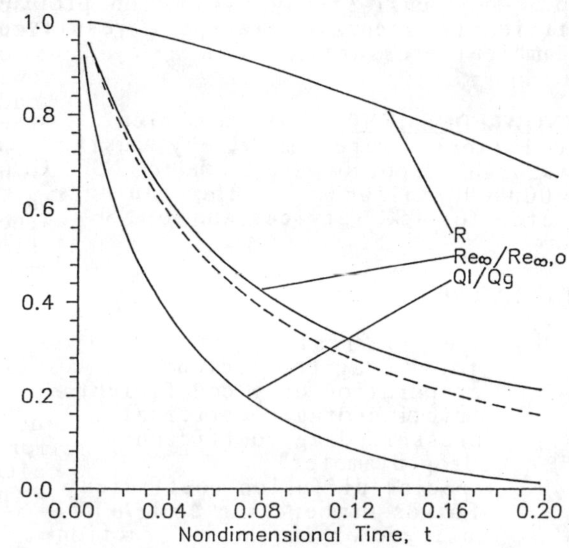

Fig. 2. Droplet radius, Reynolds number, and liquid heating fraction time histories; dot line - $Re_\infty/Re_{\infty,o}$ from [Haywood etal. (1989)].

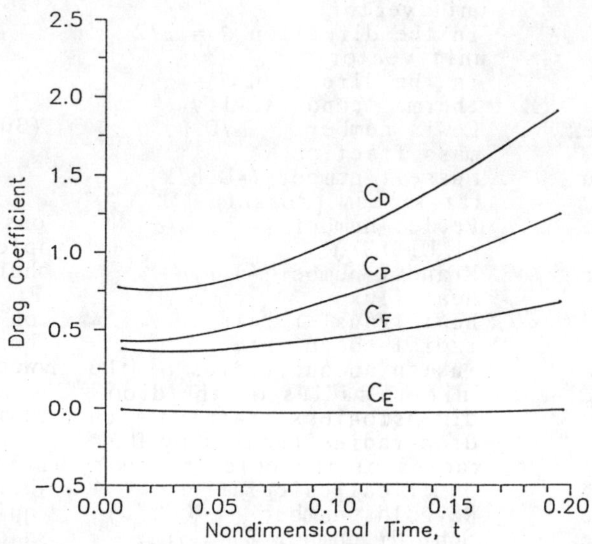

Fig. 3. Drag coefficients time histories.

continuously increases with time. The total drag force on an evaporating drop in intermediate Reynolds Number flow is essentially made up of the pressure and friction drag force components. The evaporation drag does not play a significant role. This is unlike in the case of a slowly moving evaporating drop [Sadhal and Ayyaswamy (1983); Gogos and Ayyaswamy (1988)]. Both C_P and C_F increase with time. The vorticity is convected away from the drop and there is very little pressure recovery in the rear of the drop. The blowing due to evaporation reduces friction drag but C_F is noted to increase because of the decrease in droplet Reynolds Number with time. The early rapid reduction in C_D lasting for a very short time is explained by Haywood et al. (1989) as due to the modification of the flow field and transport properties in the vicinity of droplet surface with the onset of evaporation.

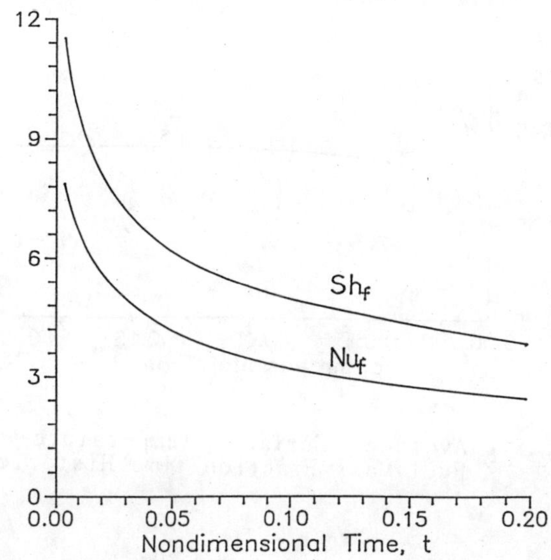

Fig. 4. Nusselt and Sherwood number time histories.

The transient variations in heat and mass transfer are shown in Figure 4. Both the Sherwood and Nusselt Numbers rapidly decrease during the early transient period. This feature is due to (i) blowing associated with evaporation (thick boundary layers), and (ii) variable property effects (colder boundary layers) [Renksizbulut and Yuen (1983)]. With increasing time, the droplet deceleration decreases the effects of convection, and a gradual decrease in both Sh and Nu are noted.

The variations in the average surface temperature and fuel mass fraction are shown in Fig. 5. Both the surface temperature and mass fraction rapidly increase during the early transient, resulting in significant heat and mass transfer. This period corresponds to the liquid heat-up of the droplet. We conclude that the often invoked assumption of constant surface temperature (neglect of liquid heating) in the evaluation of droplet evaporation rate is likely to lead to significant errors.

The transient variation of the local Nusselt Number with angular position is shown in Figure 6. With increasing time, the stagnation point heat transfer (θ = 0) decreases directly as a result of the weakening of the convective effects. It is also noted that the neglect of wake effects (contribution from θ > 120°) may lead to errors in estimating evaporation rates.

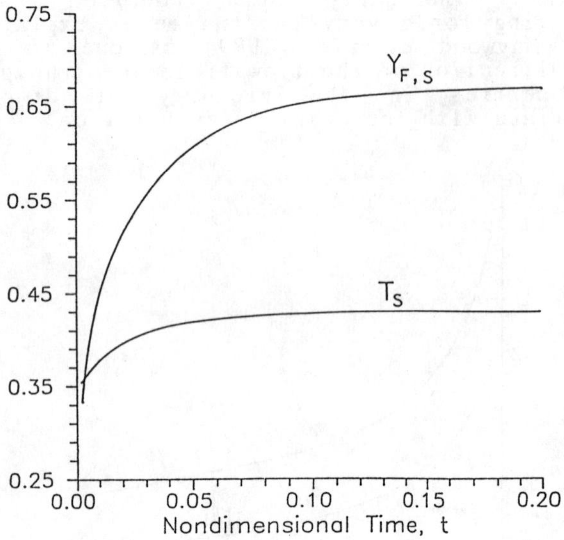

Fig. 5. Average Surface Temperature and Fuel Mass Fraction time Histories.

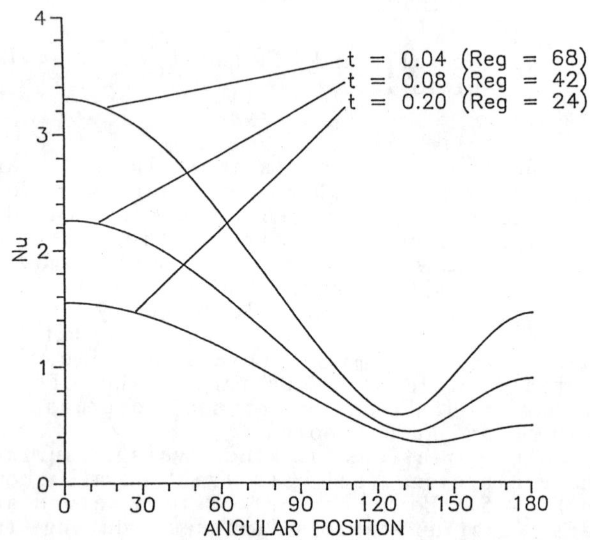

Fig. 6. Local Nusselt number time history.

CONCLUSIONS. In estimating droplet evaporation rate, liquid heating plays an important role. A boundary layer analysis cannot be used to describe the evaporation during the entire droplet lifetime due to the deceleration experienced by the drop. The assumptions of constant surface temperature and constant properties are likely to lead to errors in the estimations of heat and mass transfer to an evaporating drop. The quasi-steady assumption provides sufficiently accurate transport results at low ambient pressures.

ACKNOWLEDGEMENT. The numerical calculations were made by using the Pittsburgh Supercomputer under NSF Grant ECS-0000000/8515068. The authors are grateful for PSC services and to NSF.

NOMENCLATURE.

c_p	specific heat
C_D	total drag coefficient
C_E	evaporation drag coefficient
C_F	friction drag coefficient
C_P	pressure drag coefficient
D	drop diameter
D_g	species diffusion coefficient
Eo	Eötvös number (= g $\Delta\rho$ D^2/σ)
$\mathbf{e}_r$	unit vector in the r-direction
$\mathbf{e}_\theta$	unit vector in the θ-direction
$\mathbf{e}_\phi$	unit vector in the ϕ-direction
E	Evaporation parameter (= $1-m_{1,s}/m_{1,\infty}$)
g	gravity
h	heat transfer coefficient
h_m	mass transfer coefficient
Ja_g	Jakob number (= c_{pg} T_∞/λ)
I	unit tensor
i	unit vector in the direction θ = $\pi/2$
j	unit vector in the direction θ = π
k	thermal conductivity
Le	Lewis number (= α/D_g)
m	mass fraction
Nu	Nusselt number (=D h/k)
p_∞	far-stream pressure
Pe	Peclet number (= U_∞ D/α or U_∞ D/D_g)
Pr	Prandtl number (= ν/α)
q	heat flux
Q	heat transfer rate
r	radial coordiante
R	instantaneous radius of the drop
R_o	initial radius of the drop
R	dimensionless rate of change of drop-radius (scaled by U_∞)
R_∞	radius of the outer boundary in the numerical calculation.
Re	Reynolds number (= U_∞ D/ν)
Sc	Schmidt number (= ν/D_g)
Sh	Sherwood number (= D h_m/D_g)
t	time
T	temperature
T_b	instantaneous bulk temperature of the drop
T_s	surface temperature of the drop
u	velocity
U_0	initial velocity of drop
U_∞	far-stream translational velocity
u_r, u_θ	velocity components
x	x-coordinate in the direction θ = $\pi/2$
y	y-coordinate in the direction θ = π
Y	normalized mass fraction
z	transformed radial coordinate (= ℓn r)

Greek Symbols

α	thermal diffusivity
$\Delta\rho$	density difference between drop and gaseous phase
ζ	vorticity
τ	stress
θ	polar angle
λ	latent heat of evaporation
μ	dynamic viscosity
ν	kinematic viscosity
ρ	density
σ	surface tension
ϕ	azimuthal angle
ψ	stream function

Subscripts

av	average
f	film conditions
F	fuel component
g	gas-phase
ℓ	liquid-phase
m	mass transfer
p	pressure
s	drop-surface
t	thermal
0	at initial time; stagnant drop
1	air
∞	far-stream

Superscripts

-	average
*	dimensionless quantity

REFERENCES.

1. P. S. Ayyaswamy, "Combustion Dynamics of Moving Droplets," Chapter 20 in Encyclopedia of Environmental Control Technology, vol. 1 : Thermal Treatment of Hazardous Wastes, Ed: P.N. Cheremisinoff, pp. 479-532, Gulf Publishing Co., Houston, TX, (1989).
2. H. A. Dwyer and B. R. Sanders, "Droplet Dynamics and Vaporization with Pressure as a Parameter", ASME Paper 84-WA/HT-20, New Orleans, LA (1984).
3. G. Gogos, S. S. Sadhal, P. S. Ayyaswamy, and T. Sundararajan, "Thin Flame Theory of the Combustion of a Moving Liquid Drop: Effects due to Variable Density", J. Fluid Mech., 171, 121-144 (1986).
4. G. Gogos, and P. S. Ayyaswamy, "A Model for the Evaporation of a Slowly Moving Liquid Droplet", Combustion and Flame, 74, 111-129 (1988).
5. R. J. Haywood, R. Nafziger, and M. Renksizbulut, "A Detailed Examination of Gas and Liquid Phase Transient Processes in Convective Droplet Evaporation", In Press, J. Heat Transfer, (1989).
6. S. Prakash and W. A. Sirignano, "Theory of Convective Droplet Vaporization with Unsteady Heat Transfer in the Circulating Liquid Phase", Int. J. Heat Mass Transfer, 23, 253-268 (1980).
7. M. Renksizbulut and M. C. Yuen, "Experimental Study of Droplet Evaporation in a High Temperature Air Stream," J. Heat Transfer, 105, No. 2, 384-388 (1983).
8. M. Renksizbulut and M. C. Yuen, "Numerical Study of Droplet Evaporation in a High Temperature Air Stream," J. Heat Transfer, 105, No. 2, 389-397 (1983).
9. S. S. Sadhal and P. S. Ayyaswamy, "Flow Past a Liquid Drop with a Large Non-Uniform Radial Velocity," J. Fluid Mech., 133, 65-81 (1983).
10. T. Sundararajan and P. S. Ayyaswamy, "Hydrodynamics and Heat Transfer Associated with Condensation on a Moving Drop: Solutions for Intermediate Reynolds Numbers", J. Fluid Mech., 149, 33-58 (1984).

1989 National Heat Transfer Conference
HTD-Vol. 106, Heat Transfer Phenomena in Radiation, Combustion and Fires

A MODEL OF THE ENHANCEMENT OF COAL COMBUSTION USING HIGH INTENSITY ACOUSTIC FIELDS

S. Yavuzkurt, M. Y. Ha, and G. H. Koopmann
Department of Mechanical Engineering

A. Scaroni
Fuel Science Program
The Pennsylvania State University
University Park, Pennsylvania

ABSTRACT

A model for the enhancement of coal combustion in the presence of high intensity acoustics has been developed. A high intensity acoustic field induces an oscillating velocity over pulverized coal particles otherwise entrained in the main gas stream, resulting in increased heat and mass transfer. The augmented heat and mass transfer coefficients, expressed as space- and time-averaged Nusselt and Sherwood numbers for the oscillating flow, were implemented in an existing computer code (PCGC-2) capable of predicting various aspects of pulverized coal combustion and gasification.

Increases in the Nusselt and Sherwood numbers of about 45, 60 and 82.5% at sound pressure levels of 160, 165 and 170 dB for 100 μm coal particles were obtained due to increases in the acoustic slip velocity associated with the increased sound pressure levels. The main effect of the acoustic field was observed during the char combustion phase in a diffusionally controlled situation. A decrease in the char burnout length (time) of 15.7% at 160 dB and 30.2% at 170 dB was obtained compared to the case with no sound for the 100 μm coal particles.

NOMENCLATURE

A	=	Preexponential factor
c	=	speed of sound
C_{og}	=	oxidant concentration
D	=	diameter
D_v	=	Diffusivity of the vapor in the surrounding fluid
E	=	activation energy
h	=	heat transfer coefficient
h_m	=	mass transfer coefficient
k	=	thermal conductivity
L_p	=	sound pressure level
Nu	=	Nusselt number (hD/k)
P_{rms}	=	root mean square sound pressure
R	=	gas constant
R_c	=	kinetic reaction
R_{prim}	=	radius of primary air and coal injection tube
Re_o	=	steady Reynolds number (U_oD/μ)
Re_1	=	acoustic Reynolds number (U_1D/μ)
Sc	=	Schmidt number ($\mu/\rho D_v$)
Sh	=	Sherwood number (h_mD/D_v)
SPL	=	sound pressure level
t	=	time
T	=	temperature
U_o	=	steady slip velocity
U_1	=	peak velocity of acoustic field
U_a	=	acoustic velocity
V_1	=	particle peak velocity
Y_{jm}	=	stoichiometric coefficient
γ	=	specific heat
ε	=	velocity ratio
η	=	entrainment factor
μ	=	viscosity
ρ	=	density
ρ_c	=	reaction rate of char
τ	=	relaxation time
ω	=	angular frequency

Subscripts

c	=	char
m	=	mass transfer
g	=	gas
p	=	particle
prim	=	primary
oc	=	oxidant concentration
rms	=	root mean square
t	=	space- and time-averaged value
0	=	steady
1	=	acoustic

INTRODUCTION

In a pulverized coal combustor, coal particle and char are entrained in the main gas flow for a significant length of time. The slip velocity between the entrained particles and the gas is very low, leading to low heat and mass transfer to and from the particles. This results in long combustion times for the particles, particularly if the combustion is diffusionally controlled.

When a high intensity acoustic field is applied to the gas

flow, it superposes high velocity oscillations on the main flow. If the frequency of this acoustic field is sufficiently high (about 2000 Hz for 100 μm particle) the particles are not entrained in the acoustically induced flow oscillations. This creates a periodically varying slip velocity between the particles and the hot gas, leading to higher convective heat and mass transfer to and from the particles, thereby enhancing the combustion.

Zinn [1] and Rudinger [2] indicated the advantages of a slip velocity between the flow and a liquid or solid fuel in the presence of acoustic oscillations on the rate of diffusion of oxidant molecules and on increasing the convective heat transfer. Zinn et al. [3] and Faeser [4], respectively, reported promising results on the combustion of coal by using pulsating combustion with a Rijke type combustor and crushed coal combustors with high intensity sound produced by acoustic drivers. Zikikou et al. [5] and Poinsot et al. [6] studied the combustion characteristics of a gas flame such as vortex shedding, growth, interaction and burning by the mode selective excitation of high and low frequencies in the presence of acoustic fields generated by loud speakers. Yavuzkurt et al. [7] observed an increase in the combustion rate of coal particles in a flat flame in the presence of high intensity acoustic fields, as inferred from the increased light intensity of the flame in an increase in the flame width. Rawson [8] showed a substantial increase in water evaporation rates in experiments with a vertical sonic water spray evaporator in the presence of a high intensity acoustic field. At 152 dB an average of 33% increase in the evaporation rate was observed. The effects of an oscillating flow field with and without a steady component on the heat and mass transfer from single spherical particles were investigated by Mori et al. [9] and Larsen and Jensen [10]. However, no detailed models currently exist for the combustion of pulverized coal particles in the presence of acoustic fields. This study is being carried out to address this issue.

The objective of the present study is to model the enhancement in the rate of pulverized coal combustion in the presence of high intensity acoustic fields. Initially, the dimensionless space- and time-averaged heat and mass transfer coefficients (Nusselt and Sherwood numbers) were calculated as a function of the ratio of the peak acoustic velocity to the steady slip velocity, the Prandtl number and the Schmidt number. Ha et al. [11] obtained these results from a computer code using the two-dimensional, unsteady, conservation of mass, momentum and energy equations in spherical coordinates for a spherical particle exposed to a steady slip velocity and an oscillating acoustic velocity.

The resulting Nusselt and Sherwood number correlations are implemented in an existing computer program (PCGC-2) which is a steady, axisymmetric code capable of modeling pulverized coal combustion and gasification, and the combustion of coal water slurry fuels, in cylindrical combustors. This code has been developed at Brigham Young University during the past 10 years [12-17]. The addition of acoustic effects to this code enables it to calculate the effects of the acoustic field on combustion behavior.

This paper presents the char burnout history for coal particles with diameters between 80-120 μm exposed to sound pressure levels of 160-170 dB and compares the results to similar cases with no acoustic field. The temperature of the particles and the gas, and the Nusselt and Sherwood numbers along particle trajectories under sound and no sound conditions are presented.

THEORY

In a travelling wave acoustic field with a sound pressure of P_{rms} and acoustic impedance $\rho_g c$, where ρ_g is the density and c is the speed of sound, the local acoustic velocity U_1 is given by:

$$U_1 = \frac{\sqrt{2}P_{rms}}{\rho_g\,(\gamma R T_g)^{0.5}} \tag{1}$$

where

$$P_{rms} = 10^{\frac{L_p - 94}{20}} \tag{2}$$

The unit of P_{rms} is N/m^2 and Lp is in dB.

This acoustic field creates an oscillating slip velocity. The boundary layer formed over the particles is reversed with every half-period of the wave if the coal particles in the combustor are not entrained in the acoustic field.

The entrainment factor η is obtained by equating the particle drag inertial forces for a smooth sphere [8].

$$\eta\ (=\frac{V_1}{U_1}) = \frac{1}{(1 + \omega^2\tau^2)^{1/2}} \tag{3}$$

where U_1 and V_1 represent, respectively, the amplitudes of the velocity of the surrounding medium and the particle velocity as defined by:

$$U_a = U_1 \sin\omega t \tag{4}$$

and

$$V_a = V_1 \sin t(\omega t - \phi) \tag{5}$$

Here, ω is the angular frequency of the acoustic wave, t is the time and φ is the phase difference between the particle and gas motion and τ represents the particle relaxation time defined by:

$$\tau = \frac{\rho_p D^2}{18\mu} \tag{6}$$

Thus, full entrainment corresponds to $\eta = 1$ while $\eta = 0$ means no entrainment. In equation (6), ρ_p is the particle density, D is the mean diameter of the particles and μ is the dynamic viscosity of the medium.

Figure 1 shows the particle entrainment for coal particles in air at 1500K at several sound frequencies as an example. The entrainment η becomes 0.0053 for 100 μm coal particles at 2000 Hz frequency as shown in Figure 1. Thus, if the frequency of the acoustic field is sufficiently high, the entrainment of the individual coal particles is minimized. Thus, hot gas will blow back and forth over the coal particles, thereby increasing the convective heat and mass transfer substantially.

The acoustic field given with equations (1) and (2) is superposed on a steady flow of velocity. Then, dimensionless space- and time-averaged, quasi steady state heat and mass transfer coefficients, normalized by their steady flow values, as a function of the ratio $\varepsilon\ (=\frac{U_1}{U_o}$ or $\frac{Re_1}{Re_0})$ for a spherical particle can be expressed as follows:

$$\frac{Nu_t - 2}{Nu_o - 2}\ (\text{or } \frac{Sh_t - 2}{Sh_o - 2}) = F(\varepsilon) \tag{7}$$

Summarizing the experimental data compared with quasi-steady analysis by Mori et al. [9] and Larsen and Jensen [10] and the numerical analysis using the unsteady, two-dimensional computer

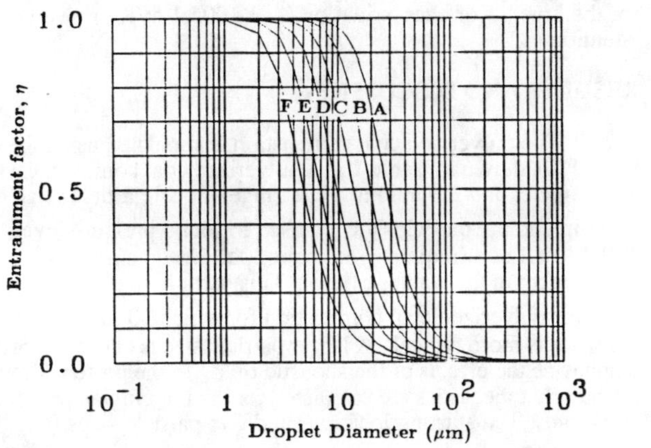

Figure 1. Coal particles entrainment factor vs. particle diameter in air at 1500K for sound frequencies: (A) 250 Hz, (B) 500 Hz, (C) 1000 Hz, (D) 2000 Hz, (E) 4000 Hz, and (F) 8000 Hz.

code by Ha et al. [11] for the heat and mass transfer over spherical particles in the presence of acoustic fields, F in equation (7) can be expressed as follows:

$$F = 1.0 + 9.608\chi 10^{-3}\varepsilon - 0.109608\varepsilon^2 \text{ for } 0 \le \varepsilon \le 1 \quad (8)$$

$$F = 0.7628\varepsilon^{0.5} + 0.1372 \text{ for } \varepsilon > 1 \quad (9)$$

and as $U_o \to 0$ and $\varepsilon \to \infty$, equation (9) gives asymtotically the following:

$$F = 0.7628\varepsilon^{0.5} \quad (10)$$

which is the same expression given by Larsen and Jensen [10]. The correlations (solid line) given in equations (8) and (9) are shown in Figure 2 and compared with the quasi-steady analysis of Mori et al. [9] at $0 \le \varepsilon \le 1$ and the numerical data of Ha et al. [11]. As shown in Figure 2 these equations (8) and (9) represent well the previous results. These correlations are implemented in the PCGC-2 code in order to simulate the pulverized coal combustion in the presence of an acoustic field.

Limitations and ranges of validity of these equations are discussed in detail in [11]. This reference shows that under the conditions of this study they are valid. PCGC-2 uses the following correlations for the steady Nusselt or Sherwood numbers (Nu_o or Sh_o) in equation (7):

$$Nu_o - 2 \text{ (or } Sh_o - 2) = 0.6\, Re^{1/2}\, Pr^{1/3} \text{ (or } Sc^{1/3}) \quad (11)$$

The PCGC-2 code, which was developed at Brigham Young University [12-17] over several years, uses an Eulerian flow field for the gas phase and a Lagrangian approach for the particle phase. Smoot [13] has provided a detailed review of PCGC-2, including the various aspects of the hydrodynamics of the gas flow, turbulence modeling, gaseous combustion, particle dispersion, coal devolatilization, heterogeneous char reaction, radiation, and pollutant formation. Also, he has explained the capability of the code to simulate pulverized coal combustion, and provided comments on the limitations of the code.

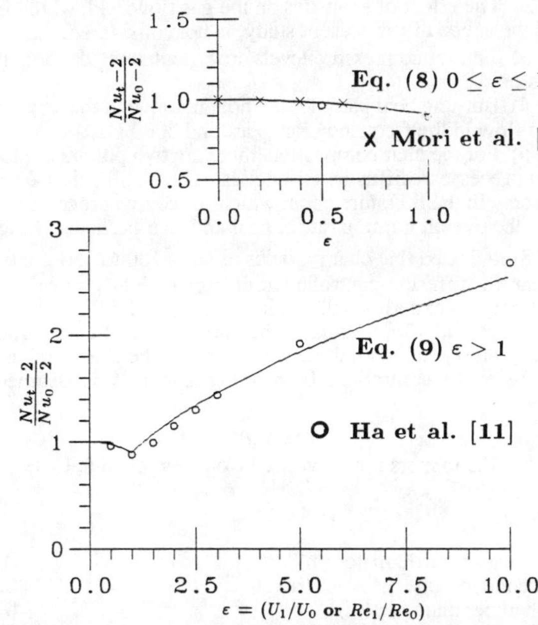

Figure 2. Space- and time-average Nusselt number normalized by its steady value as a function of Reynolds number ratio.

The gas phase equation includes a two-equation (k-ε) turbulence model for closure with the assumption at most gas-phase reactions are mixing-limited, so that gaseous combustion is modeled with a statistical probability density function based on the mixture fraction.

For the particle phase equations, PCGC-2 assumes that the reacting particles are composed of specified amounts of moisture, raw coal (char and volatiles) and ash for pulverized coal combustion. A specified amount of water is included in the case of combustion of coal water slurry fuels. During the pyrolysis, the raw coal (i.e., the dry, ash-free portion of the virgin coal) undergoes devolatilization to form volatiles and char by one or more reactions of the form [14]:

$$\text{(Raw Coal)j} \to Y_{jm} \text{ (Volatiles)} + (1-Y_{jm}) \text{ (Char)} \quad (12)$$

where Y_{jm} represents the stoichiometric coefficients of the volatiles which react further in the gas phase.

The char reacts heterogeneously with gaseous oxidants that diffuse to the particle surface from the bulk gas as follows [13]:

$$\rho_c = C_{og}/[1/h_m) + (1/R_c)] \quad (13)$$

where

$$R_c = AT^n \exp(-E/RT_p) \quad (14)$$

C_{og} in equation (12) is the oxidant concentration in the bulk gas and h_m is the oxidant mass transfer coefficient to the particle surface. The following assumptions are used in the present calculation:

1) The calculations are made using the PCGC-2 code developed by BYU [17] during the last 10 years and are evaluated by comparing them with experimental data. Thus, the purpose of the present research is not to improve upon the limitations of PCGC-2 but to implement the effects of the acoustic field on heat and mass transfer coefficients contained in the code in order to investigate the effects of high intensity acoustic fields on the combustion of coal-water slurry fuels.

2) The effect of acoustics on the gas flowfield, which is beyond the scope of the present study, is not considered.

3) The sound pressure levels are not attenuated along the combustor.

4) Burning coal particles are not entrained in the acoustically induced flow in the frequency range around 2000 Hz.

5) For the char combustion, there are two potential rate-limiting processes: diffusion or kinetics. There still exists a controversy in the literature about which of the two processes controls the overall burning rate of coal and char particles. Mitchell et al. [18] indicated that char particles of sizes 100 to 150 μm burn at rates near the diffusion-controlled limit from both theoretical calculations and experimental results. Walsh et al. [19] and Srinvasachar et al. [20] calculated the burnout rate of char particles of coal-water slurry fuel with the assumption that the char combustion was diffusionally controlled. In the present study, it is assumed that the char combustion is diffusionally controlled, which is achieved by assigning large values to the preexponential factor in the expression for the kinetics of char oxidation (A = 1005.1 m/K s) given by equation (12).

TABLE 1
Parameters for Pulverized Coal Combustion Case

Geometry	
Primary tube diameter (m)	0.022
Secondary tube diameter (m)	0.084
Chamber diameter (m)	0.203
Chamber length (m)	3.000
Feed Rates	
Primary gas (kg/s)	6.228E-03
Secondary gas (kg/s)	0.019
Coal in primary (kg/s)	2.853E-03
Inlet Gas Properties	
Primary swirl number	0.000
Primary turbulent intensity	15.0
Primary temperature (K)	300.0
Primary mole fractions:	
AR	0.046
H_20	0.035
N_2	0.725
0_2	0.194
Secondary swirl number	2.000
Secondary turbulent intensity	18.0
Secondary temperature (K)	589.0
Secondary mole fractions:	
AR	0.009
N_2	0.781
0_2	0.210
Reactor Parameters	
Reactor pressure (N/sq.m)	8.600E+04
Side wall temperature (K)	1000.0
Particle Parameters	
Particle solid density (kg/cu.m)	1340.0
Heat of formation, daf (J/kg)	5.611E+05
Mass mean particle diameter (m)*	1.000E-04
Initial analysis:	
raw coal	0.931
ash	0.069
Elemental analysis (daf):	
C	0.724
H	0.047
O	0.218
N	0.012

*Continuous distribution simulated with 5 discrete sizes

RESULTS AND DISCUSSION

The pulverized coal combustion is simulated using the data of BYU [10] shown in Table 1 in a pulverized coal combustor with 20 cm diameter and 3.0 m long. The range of coal particles size is 80-120 μm and the particles are exposed to sound pressure levels of 160-170 dB at a frequency of about 2000 Hz (to ensure no entrainment in the acoustic field). This particle size is somewhat large for pulverized coal combustion for which 50 μm average diameter is more typical. A larger particle size is chosen in order to emphasize the effects of the acoustic field. In the present simulation, 50 particle trajectories are calculated for the Lagrangian particle equations. Ten starting locations and five particle sizes (80, 90, 100, 110, and 120 μm) at each starting location are used. The initial locations of the ten particles are equally distributed between r/R_{prim} = 0 and r/R_{prim} = 1 where r is the distance from the centerline and R_{prim} is the radius of primary air and coal injection tube.

Figure 3 shows the particle trajectories as a function of distance along the combustor for 100 μm particles that are introduced at the centerline (r/R_{prim} = 0), halfway between the centerline and the primary tube wall (r/R_{prim} =0.5), and near the primary tube wall (r/R_{prim} = 1) with no sound and with sound pressure levels of 160 and 170 dB. At the centerline particles under sound pressure levels of 160 and 170 dB follow almost the same trajectories as in the absence of an acoustic field, whereas the particles released near the primary tube wall move further towards the wall of the combustor with increasing sound pressure levels. The particles which are introduced at the centerline deviate from this line downstream due to sudden expansion of the combustor geometry at injection which creates deviations in streamlines of the flow.

Figures 4(a) and (b) show the temperature of the gas and the particles as a function of distance along the combustor for 100 μm coal particles which are released at the centerline (r/R_{prim} = 0) and near the primary tube wall (r/R_{prim} = 1) with no sound and sound pressure levels of 170 dB. Figure 4(c) is enlarged in order to observe the temperature rise of the gas and coal particles during the pyrolysis stage. Due to the increased heat and mass transfer the coal particles introduced at r/Rprim = 0 with no sound and sound pressure level of 170 dB reach a particle temperature of 1000 K at about 0.14 m from the inlet of the combustor whereas the coal particles with no

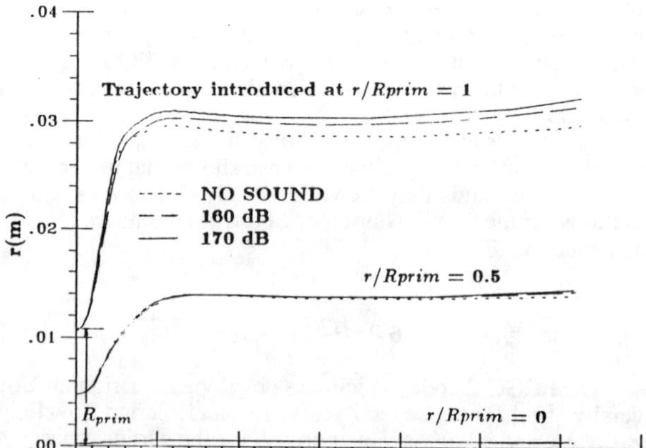

Figure 3. Particle trajectories as a function of distance along the combustor for 100 μm coal particles introduced at r/R_{prim} = 0.0, 0.5 and 1.0

sound reach 1000 K at about 0.17m as shown in Figure 4. The peak at the gas temperature around x = 0.1 m occurs due to the combustion of volatiles, most of which are released and burn in the gas phase before x=0.1 m. Between x=0.1 and 0.2 m the char combustion is not taking place at a high rate, i.e. Tp < Tg. Due to energy loss with radiation to the wall (1000 K) the gas cools.

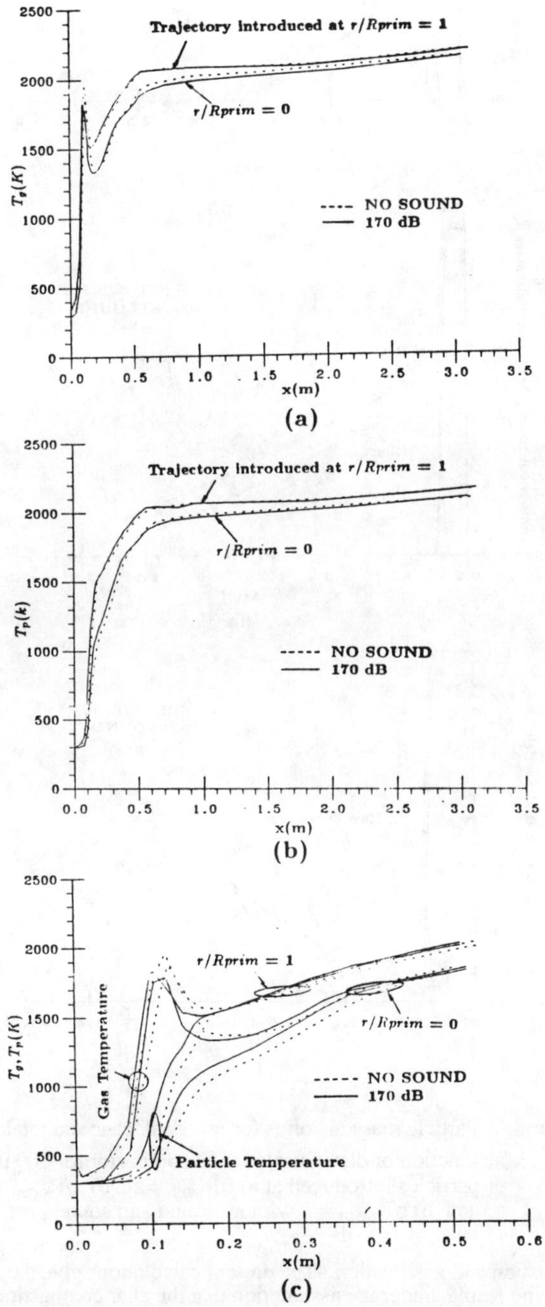

Figure 4. Temperature of the gas and the particle as a function of distance along the combustor for 100 μm coal particles introduced at r/R_{prim} = 0.0 and 1.0 with no sound and SPL = 170 dB

Figure 5 shows the steady slip velocity (U_o) with no sound and the acoustically induced slip velocity (U_1) at sound pressure levels of 160, 165 and 170 dB between the coal particles and the surrounding gas as a function of distance along the combustor for 100 μm particles released at the centerline. This acoustically induced slip velocity U_1 (peak value) is calculated using equation (1) along the particle trajectories. As shown in Figure 5 the steady slip velocity U_o is almost zero at x = 0.5 m when no acoustic field exists, indicating that the coal particles are entrained in the surrounding gas field. When the surrounding gas temperature increases from the inlet (x = 0 m) to x = 0.2 m as shown in Figure 4, the acoustically induced slip velocities reach levels of about 20, 40, and 70 m/s at sound pressure levels of 160, 165 and 170 dB. When the gas temperature decreases in the region just downstream of the ignition location followed by an increase of the gas temperature as shown in Figure 4, it is observed that the acoustic slip velocities follow this trend due to an approximate dependence of the acoustic velocities on $T_g^{0.5}$.

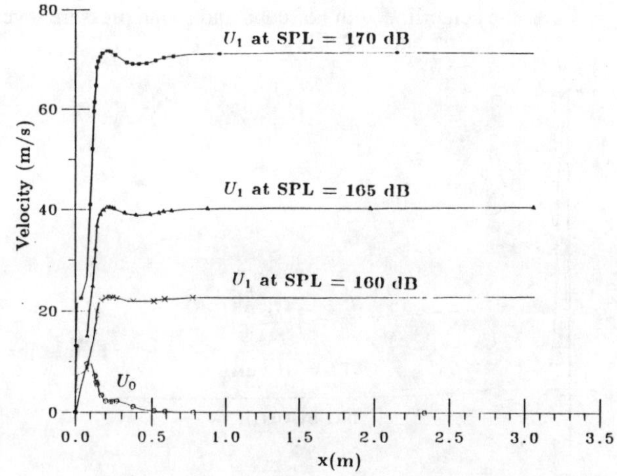

Figure 5. Steady slip velocity U_o with no sound and the acoustically induced slip velocity U_1 at sound pressure levels of 160, 165 and 170 dB

Higher oscillating slip velocities induced by the high intensity acoustic field result in the increased heat and mass transfer as shown in Figures 6 and 7 and as explained by Ha et al. [11]. Figures 6 and 7, respectively, show the Nusselt and Sherwood numbers for 100 μm particles released at the centerline with no sound and with sound pressure levels of 160, 165 and 170 dB. As shown in these figures the Nusselt and Sherwood numbers with no sound are almost equal to 2, which corresponds to the pure conduction value, at x = 0.5 m since the steady slip velocity U_o is almost zero at this location (Figure 5). An increase in the Nusselt and Sherwood numbers of about 45% at 160 dB, 60% at 165 dB and 82.5% at 170 dB is observed in these figures. Augmentation of the Nusselt and Sherwood numbers is the cause of the increase in the particle temperature shown in Figure 4 which may enhance the onset of pyrolysis.

Figure 8 shows particle mass histories for the raw coal (dry ash free basis), char (dry ash free basis) and the total (raw coal on a dry basis) as a function of distance along the combustor for 100 μm particles released at the centerline (r/R_{prim} = 0), halfway between the centerline and the primary tube wall (r/R_{prim} = 0.5), and the primary tube wall (r/R_{prim} = 1) with no sound and with a sound pressure level of 165 dB. Devolatilization starts at about 15 cm and is completed very rapidly. The gas phase products formed during this devolatilization burn in the bulk gas phase, resulting in the sharp increase in the gas temperature as shown in Figure 4(a). While char is being formed by devolatilization, it is also being depeleted by the oxidations reactions. The peak value in the amount of char occurs at

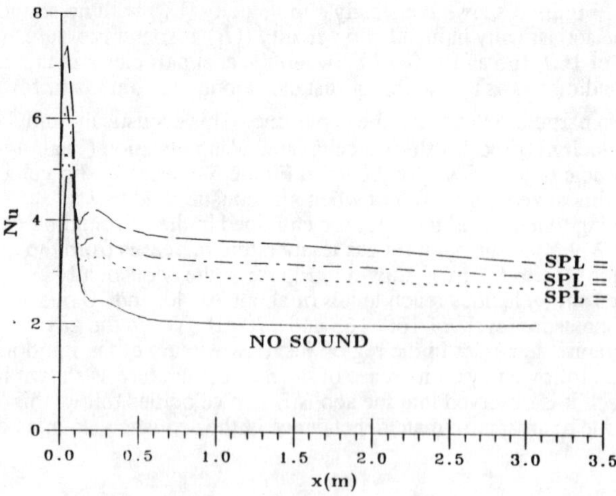

Figure 6. Nusselt number for 100 μm coal particles introduced at the centerline with no sound and sound pressure levels at 160, 165 and 170 dB

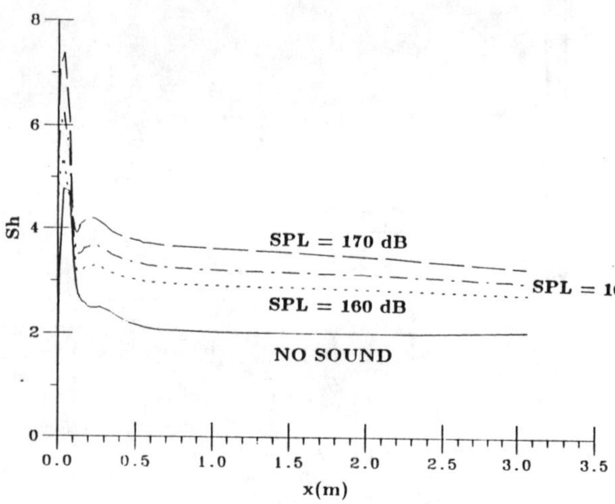

Figure 7. Sherwood number for 100 μm coal particles introduced at the centerline with no sound and sound pressure levels of 160, 165, and 170 dB

the time of complete devolatilization.

As shown in Figure 8, the significant effects of the high intensity acoustic field can be observed, especially during the char burning phase. A decrease in the char burnout length of 19.2% for particles released at $r/R_{prim} = 0$, 21% at $r/R_{prim} = 0.5$ and 25.4% at $r/R_{prim} = 1$ is obtained compared to the case with no sound. This is especially due to augmented mass transfer of oxidant.

Figure 9 shows char burnout histories as a function of distance along the combustor for 100 μm particles introduced at $r/R_{prim} = 0.5$, with no sound and sound pressure levels of 160, 165, and 170 dB in order to examine the effect of the sound pressure level. A decrease in the char burnout length of 15.7% at 160 dB, 21% at 165 dB, and 30.2% at 170 dB sound pressure level is obtained compared to the case with no sound. This is a significant improvement as far as combustor size is concerned.

SUMMARY AND CONCLUSIONS

The space- and time-averaged heat and mass transfer coefficients (Nusselt and Sherwood numbers) including the effects of the acoustic field superposed on a steady slip velocity are implemented in an existing computer code PCGC-2, which is a steady, axisymmetric code capable of modeling pulverized coal

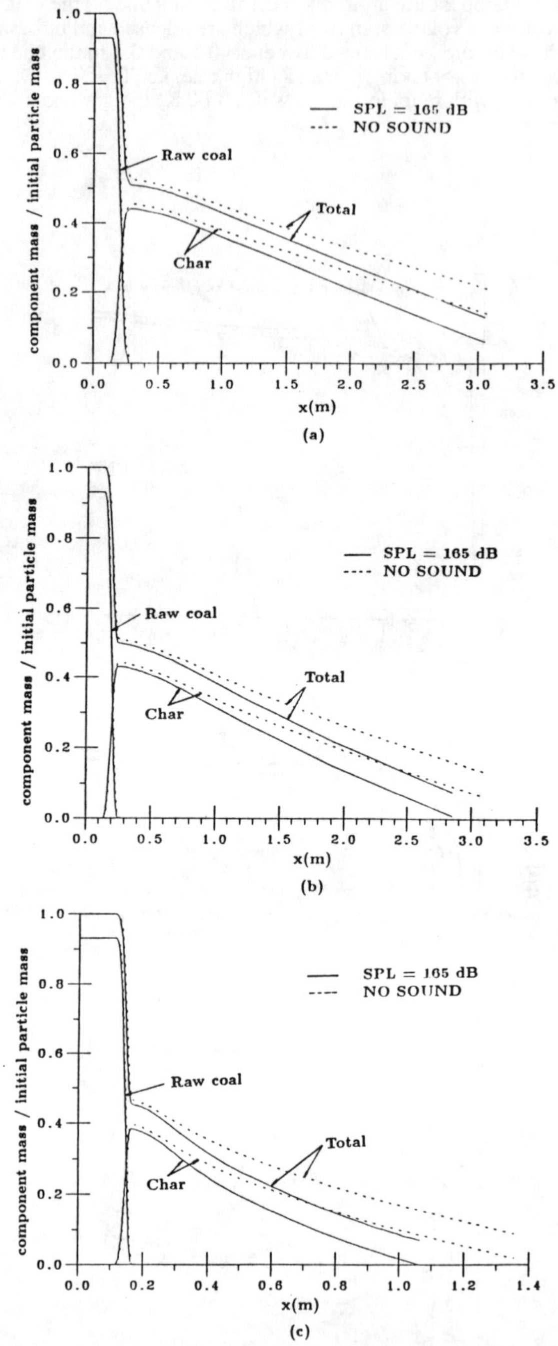

Figure 8. Particle mass histories for raw coal, char and total as a function of distance along the combustor for 100 μm particles introduced at a) $r/R_{prim} = 0$, b) $r/R_{prim} = 0.5$ and c) $r/R_{prim} = 1$ with no sound and sound pressure level of 165 dB

combustion and gasification. The present calculations give the following results under the assumption that the char combustion is diffusionally controlled.

Coal particles of 100 μm are almost entrained by the surrounding gas field at x = 0.5 m in the example combustor, with almost a zero steady slip velocity. Peak slip velocities of about 20, 40 and 70 m/s are induced on the entrained coal particles at sound pressure levels of 160, 165 and 170 dB, respectively. This happens

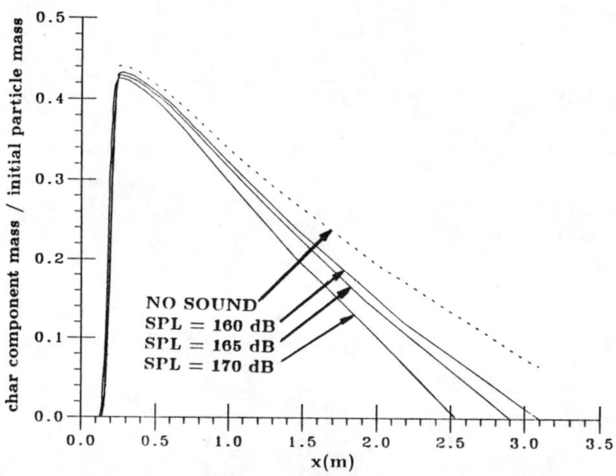

Figure 9. Char burnout histories as a function of distance along the combustor for 100 μm particles introduced at r/R_{prim} = 0.5 with no sound and sound pressure level of 160, 165 and 170 dB

at frequencies of around 2000 Hz since 100 μm coal particles are not entrained in the acoustic field at high frequencies. This induced slip velocity results in an increase in the Nusselt and Sherwood numbers of about 45, 60 and 82.5% at sound pressure levels of 160, 165, 170 dB, respectively.

A decrease in the char burnout length of 19.2, 21 and 25.4% for 100 μm coal particles released at the centerline, halfway between the centerline and the primary tube wall and near the primary tube wall, respectively, is obtained at a sound pressure level of 165 dB compared to the case with no sound due particularly to increased mass transfer in the char burning phase as a consequence of the acoustically induced slip velocity. Increasing the sound pressure level results in a decrease in the char burnout length of 15.7% at 160 dB, 21% at 165 dB and 30.2% at 170 dB compared to the case with no sound for the 100 μm coal particles introduced halfway between the centerline and the primary tube wall.

The high intensity acoustic field induces an oscillating acoustic slip velocity over the coal particles which augments the heat and mass transfer rates. Increased heat and mass transfer enhances the combustion rate of the pulverized coal particles, especially in the char burning phase if the process is diffusionally controlled. Because char burning is the slowest step in the combustion of the pulverized coal, application of high intensity acoustic fields is expected to result in combustors which are up to 20% shorter for sound pressure levels of 165 dB than conventional combustors with the same capacity.

ACKNOWLEDGMENTS

This work was supported by the U.S. Department of Energy, Morgantown Energy Technology Center, under contract DE-RA21-86MC23257.

REFERENCES

1. Zinn, B.T., "State of the Art and Research Needs of Pulsating Combustion", ASME Paper 84-WA/NCA-19, 1984.

2. Rudinger, G., "Effect of Velocity Slip on the Burning Rate of Fuel Particles", J. of Fluids Engineering, pp. 321-326, 1975.

3. Zinn, B.T., Carvalho, Jr., J.A., Miller, N., and Daniel, B.R., "Pulsating Combustion of Coal in a Rijke Type Combustor", 19th Symposium (International) on Combustion, The Combustion Institute, pp. 1197-1203, 1982.

4. Faeser, R.J., "Acoustic Enhancement of Pulverized Coal Combustion", ASME Paper 84-WA/NCA-18, 1984.

5. Zikikou, S., Candal, S., Poinsot, T., Trouve, A. and Esposito, E., "High Frequency Combustion Oscillations Produced by Mode Selective Acoustic Excitation", 21st Symposium (International) on Combustion, The Combustion Institute, pp. 1427-1434, 1986.

6. Poinsot, T.J., Trouve, A.C., Veynante, D.P. , Candel, S.M., and Esposito, E.J., "Vortex-driven Acoustically Coupled Combustion Instabilities", J. Fluid Mechanics, Vol. 177, pp. 265-292, 1987.

7. Yavuzkurt, S., Ha, M.Y., Reethof, G., Koopmann, G. and Scaroni, A.W., "Effect of Acoustic Field on the Combustion of Coal Particles in a Flat Flame Burner", The 12th Annual ASME Energy-Sources Technology Conference and Exhibition, Fossil Fuel Combustion Symposium, Houston, TX, 1989.

8. Rawson, S.A., "An Experimental Investigation of the Influence of High Intensity Acoustics on Heat and Mass Transfer Rates from Spheres as Related to Coal-Water Slurry Fuel Combustion Enhancement", M.S. Thesis, The Pennsylvania State University, 1988.

9. Mori, Y., Imabayashi, M., Hijikata, K. and Yoshida, Y., "Unsteady Heat and Mass Transfer from Spheres", Int. J. Heat Mass Transfer, Vol. 12, pp. 571-585, 1969.

10. Larsen, P.S. and Jensen, J.W., "Evaporation Rates of Drops in Forced Convection with Superposed Transverse Sound Field", Intl. J. Heat and Mass Transfer, Vol. 21, pp. 511-517, 1978.

11. Ha, M.Y., Yavuzkurt, S., Koopmann, G. and Scaroni, A.W., "Modeling the Effects of High Intensity Acoustic Fields on the Enhancement of Heat and Mass Transfer from Spheres", paper accepted for presentation at the session on Interfacial Heat and Mass Transfer, 26th ASME/AICHE National Heat Transfer Conference, Philadelphia, PA, 1989.

12. Smith, P.J., "Theoretical Modeling of Coal and Gas Fired Turbulent Combustion and Gasification Processes", Ph.D. Thesis, Brigham Young University, 1979.

13. Smoot, L.D., "Pulverized Coal Diffusion Flames: A Perspective Through Modeling", Eighteenth Symposium (International) on Combustion, The Combustion Institute, pp. 1185-1202, 1981.

14. Smith, P.J., Fletcher, T.H., and Smoot, L.D., "Model for Pulverized Coal-Fired Reactors", Eighteenth Symposium (International) on Combustion, The Combustion Institute, pp. 1285-1293, 1981.

15. Smith, P.J., Hill, S.C., and Smoot, L.D., "Theory for NO Formation in Turbulent Coal Flames", Nineteenth Symposium (International) on Combustion, The Combustion Institute, pp. 1263-1270, 1982.

16. Hill, S.C., Smooth, L.D. and Smith, P.J., "Prediction of Nitrogen Oxide Formation in Turbulent Coal Flames", Twentieth Symposium (International) on Combustion, The Combustion Institute, pp. 1391-1400, 1984.

17. Smith, P.J., Fletcher, T.H., and Smoot, L.D., "Prediction and Measurement of Entrained Flow Coal Gasification Process", Volume II, User's Manual for a Computer Program for Two-Dimensional Coal Gasification or Combustion (PCGC-2)" Final Report to DOE/METC under Contract No. DE-AC21-81MC16518, 1985.

18. Mitchell, R.E., Tichenor, D.A. and Hencken, K.R., "Overall Burning Rates of Coal Chars from Simultaneous Measurements of Particle Sizes, Temperature and Velocity in Burning Polydisperse Pulverized Coal Streams", presented at the 23rd National Heat Transfer Conference, K11 Committee on Heat Transfer in Fire and Combustion Systems, Denver, CO., pp. 37-45, 1985.

19. Walsh, P.M., Zhang, M., Farmayan, W.F. and Beer, J.M., "Ignition and Combustion of Coal-Water Slurry in a Confined Turbulent Diffusion Flame", Twentieth Symposium (International) on Combustion, The Combustion Institute, pp. 1401-1407, 1984.

20. Srinivasachar, S., Toqan, M.A., Teare, J.D. and Beer, J.M., "A Mechanistic Model of Coal-Water Fuel Combustion", Twenty First Symposium (International) on Combustion, The Combustion Institute, pp. 379-388, 1986.

1989 National Heat Transfer Conference
HTD-Vol. 106, Heat Transfer Phenomena in Radiation, Combustion and Fires

ON THE CONVECTIVE THERMAL IGNITION PROCESS AND IGNITION CRITERIA OF THE REACTIVE SOLID PARTICLE

T.-T. Yang, H.-Y. Li, and G.-G. Wang
Department of Power Mechanical Engineering
National Tsing Hua University
Hsinchu, Taiwan

ABSTRACT

This work uses the boundary layer theory to study the thermal ignition process of the reactive solid particle. The research also includes an investigation of the criteria of the convective ignition of the solid fuel. The theoretical model explores the 2-D unsteady boundary layer equations in the gaseous phase. The governing equations of mass, momentum, energy, and species in gas phase are first transformed to ordinary differential equations through series expansion with respect to the azimuthal angle. The equations are then quasilinearized to be the initial value problem and solved through Runge-Kutta method. The minimum heat flux from gas phase to the fuel is evaluated as the most suitable criterion for the convective thermal ignition. This analysis indicates the existence of the critical gas velocity, at which the ignition condition is optimal. Moreover, The results shows that the ignition position may not fall at the stagnation point zone. The influences of the fuel pyrolysis rate and the gas phase chemical reaction rate on the ignition process are also discussed in detail.

NOMENCLATURE

A	preexponential factor
C_p	specific heat of gas mixture, cal/g-K
D	mass diffusivity, cm^2/s
E	activation energy, cal/mol
K	thermal conductivity, cal/cm-s-K
L	heat of gasification, cal/g
$\dot{m}_p''$	pyrolysis rate, g/cm^2-s
$\dot{m}_R'''$	volumetric chemical reaction rate, g/cm^3-s
n	stoichiometric ratio
Pr	Prandtl number
Q	heat of combustion, cal/g
q	heat flux, cal/cm^2-s
r	radial coordinate
R_1	fuel particle radius, cm
R_2	distance from surface to symmetric axis
Re	Reynolds number, $\rho u R_1/\mu$
t	time, s
T	temperature, K
u	tangential velocity, m/s
v	radial velocity, m/s
x	peripheral coordinate
y	normal coordinate
Y	mass fraction
α	thermal diffusivity, cm^2/s
η	nondimensional position, define as Eq.(18)
μ	dynamic viscosity, g/cm-s
ν	kinematic viscosity, cm^2/s
ρ	density, g/cm^3
σ	azimuthal angle, define as Eq.(16)
ψ	dimensionless stream function

Subscripts

e	edge of boundary layer
f	fuel
i	solid phase initial value
o	oxygen
p	pyrolysis
R	reaction
s	solid
w	fuel surface
∞	free stream

INTRODUCTION

The phenomena of flame spreading, flame stabilization, and other combustion problems are closely related to the ignition process of solid fuel. In addition, a comprehensive understanding of the mechanism of ignition will benefit the design of combustors and the prevention of hazardous fires. The competing influences among diffusion, reaction, and the flow around the fuel have to be thoroughly analyzed to yield a true description of the ignition mechanism. Moreover, the ignition condition is not a conserved property; it depends not only on the fuel itself, but also on the properties of the surroundings. Due to all these complexities, the development of generalized ignition criterion is extremely difficult.

According to Birk (1980) there exists three modes of ignition for solid fuels: gas phase ignition, solid phase ignition, and heterogeneous ignition. Among these three catagories, the gas phase ignition is frequently encountered for most condensed fuel burnt in the moderate range of Reynolds number. The ignition criteria previously proposed for the gas phase ignition were summarized by Annamalai and Durbetaki(1977) and Ganhdi and Kanury(1986) as specifications of the gas temperature (Fernandez-Pello and Law, 1982a), the rate of gas temperature rise (Price and Bradely, 1966), the rate of energy generated (Kumar, 1983), and the rate of chemical reaction (Kashiwagi and Summerfield, 1973).

Ganhdi and Kanury(1986) developed a one-dimensional mathematical model and suggested that the criterion of the gradient reversal of the gas temperature profile at the solid-gas interface is most suitable for the radiantly induced ignition of organic solids. The problem of combustion of solids in a hot gas stream is the most common case and the problem is somewhat similar to the case they examined.

On the aspect of two-dimensional theoretical model, Kashiwagi et al.(1971) used the boundary layer theory to analyze the ignition of a PBAA plate heated by hot air. Similarity theory was assumed to be valid. Kashiwagi and Summerfield(1973) dealt with the same problem but replaced the similarity approach by the non-similar approach. The model predicted the ignition delay time and the ignition position for the fuel plate. Sibulkin et al. (1981, 1983) also used the same approach to analyze the burning of PMMA plate and discussed the differences between the similar and non-similar approaches. Fernandez-Pello and Law (1982a, 1982b) used the boundary layer approximation and large asymptotics to study the ignition and extinction in the stagnation point of a fuel particle. Birk(1980) investigated the convective ignition of a cylindrical nitrocellulose-based propellant. He concluded that the ignition occurred in the gas phase and his model could predict the relation among the ignition delay, ignition position, velocity, temperature, and gas concentration, although the model is still limited to the stagnation point zone. Generally there exist very few ignition models for the solid fuels as compared with models for premixed combustible mixtures.

The purpose of this paper is to improve the previous models and to reexamine the suitability of the ignition criteria. The model is developed for an unsteady, reactive boundary layer and an appropriate numerical scheme is given which can be applied to the whole boundary layer before the separation point. Since the problem analyzed is at points away from the stagnation zone, the present analysis is non-similar in nature. The correlation among the ignition condition, the heat flux, and the distributions of the gas temperature, concentrations, velocity are thoroughly investigated. The ignition criteria as well as the ignition mechanism are studied.

THEORETICAL ANALYSIS

The overall sketch of the theoretical model is shown in Fig. 1. Initially, the hot air with a temperature of T_∞, oxygen concentration of $Y_{O\infty}$, and velocity of u_∞ flows over a spherical solid fuel particle and forms the momentum and the thermal boundary layers. The solid fuel is heated and the pyrolyzed vapor diffuses to the boundary layer in which the chemical reaction occurs. During the reaction, part of the released energy is transferred back to the fuel surface and thus both the pyrolysis rate and the chemical reaction rate are enhanced. When the runaway chemical reaction occurs somewhere in the boundary layer, the fuel is ignited and the flame starts to spread over the fuel surface.

Due to the chemical reaction and the large temperature variation during the ignition process, the mathematical formulation is complicated and highly nonlinear. The following assumptions, are made in order to simplify the problem.

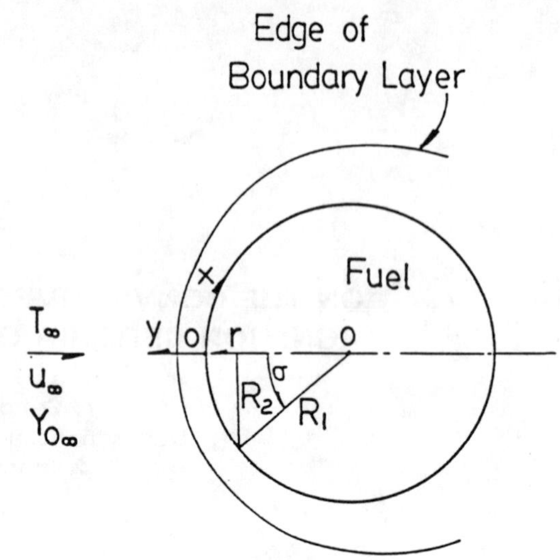

Fig. 1 The configuration and the coordinates of the model.

(1) The boundary layer equations for the gaseous phase are regarded as quasi-steady, even though the solid phase is treated as unsteady.

(2) The boundary layer is laminar.

(3) The pyrolysis rate of the fuel $\dot{m}_p''$ obeys the Arrhenius law, and is expressed as,

$$\dot{m}_p'' = \rho_s A_p \exp(-E_p/R\,T_w) \qquad (1)$$

(4) The chemical reaction is described by a one-step and second-order Arrhenius kinetics. The volumetric chemical reaction rate $\dot{m}_R'''$ is given as,

$$\dot{m}_R''' = A_R \rho^2 Y_o Y_F \exp(-E_R/R\,T) \qquad (2)$$

(5) The gas mixture obeys the ideal gas law.

(6) The volumetric pyrolysis inside the solid fuel is negligible and the specific heat, the conductivity, and the density are constant.

The origin of the coordinate system of gaseous phase equations is at the front stagnation point of the particle, as depicted in Fig. 1. The x-axis is along the fuel surface and the y-axis is perpendicular to the x-axis. The origin of the formulation of the solid phase is at the center of the fuel.

The equation of conservation of mass is given as

$$\frac{\partial(R_2^j \rho u)}{\partial x} + \frac{\partial(R_2^j \rho v)}{\partial y} = 0 \qquad (3)$$

where R_2^j is the radius of the contour of fuel surface. For two-dimensional flows j equals to zero whereas in axisymmetric flows j equals to one.

The equations of conservation of oxygen and fuel are composed of convective, diffusive, and chemical reaction terms, as interpreted in the following,

$$\rho u\frac{\partial Y_o}{\partial x} + \rho v\frac{\partial Y_o}{\partial y} = \frac{\partial}{\partial y}\left(\rho D\frac{\partial Y_o}{\partial y}\right) - n\rho^2 Y_o Y_F A_R \exp(-E_R/RT) \qquad (4)$$

$$\rho u\frac{\partial Y_F}{\partial x} + \rho v\frac{\partial Y_F}{\partial y} = \frac{\partial}{\partial y}(\rho D\frac{\partial Y_F}{\partial y})$$

$$-\rho^2 Y_o Y_F A_R \exp(-E_R/RT) \qquad (5)$$

The equation of conservation of momentum is

$$\rho u\frac{\partial u}{\partial x} + \rho v\frac{\partial u}{\partial y} = \rho_e u_e\frac{du_e}{dx} + \frac{\partial}{\partial y}(\mu\frac{\partial u}{\partial y}) \qquad (6)$$

where u_e is the velocity of air external to the momentum boundary layer and can be calculated from the potential flow theory as,

$$u_e = \frac{3}{2} u_\infty \sin\frac{x}{R_1} \qquad (7)$$

The energy conservation equation is given as,

$$\rho u C_p\frac{\partial T}{\partial x} + \rho v C_p\frac{\partial T}{\partial y} = \frac{\partial}{\partial y}(K\frac{\partial T}{\partial y})$$

$$+ Q\rho^2 Y_o Y_F A_R \exp(-E_R/RT) \qquad (8)$$

The equation of state is given as,

$$\rho T = \rho_\infty T_\infty \qquad (9)$$

where it is assumed that the pressure change is small across the y- direction.

The conservation equation for energy in the solid phase is given as,

$$\frac{1}{r}\frac{\partial^2}{\partial r^2}(rT_s) = \frac{1}{\alpha_s}\frac{\partial T_s}{\partial t} \qquad (10)$$

where it is assumed that there is no volumetric pyrolysis in the solid.

Since the distributions of velocity, temperature, and concentrations are symmetric at the stagnation point, the boundary conditions at x = 0 are

$$u = 0 \qquad (11a)$$

$$\frac{\partial T}{\partial x} = 0 \qquad (11b)$$

$$\frac{\partial Y_o}{\partial x} = 0 \qquad (11c)$$

$$\frac{\partial Y_F}{\partial x} = 0 \qquad (11d)$$

On the fuel surface the non-slip condition is fulfilled and the fuel is blown out perpendicular to the surface, hence the boundary conditions on the fuel surface are

$$u = 0 \qquad (12a)$$

$$v = v_w = \frac{\rho_s V_s}{\rho_w} = \frac{\rho_s}{\rho_w} A_p e^{-E_p/RT_w} \qquad (12b)$$

$$T = T_w \qquad (12c)$$

$$q = K_w\frac{\partial T}{\partial y}\Big|_w = K_s\frac{\partial T_s}{\partial r}\Big|_{R_1} + \rho_s v_s L \qquad (12d)$$

$$-\rho_w D_w\frac{\partial Y_o}{\partial y}\Big|_w + \rho_w v_w Y_{ow} = 0 \qquad (12e)$$

$$-\rho_w D_w\frac{\partial Y_F}{\partial y}\Big|_w + \rho_w v_w Y_{Fw} = \rho_s v_s \qquad (12f)$$

The boundary conditions on the edge of the boundary layer are

$$u = \frac{3}{2} u_\infty \sin\frac{x}{R_1} \qquad (13a)$$

$$T = T_\infty \qquad (13b)$$

$$Y_o = Y_{o\infty} \qquad (13c)$$

$$Y_F = 0 \qquad (13d)$$

The initial condition and the boundary conditions in the solid phase are

$$t = 0 : \quad T_s = T_i \qquad (14)$$

$$t \geq 0 \quad r = 0 : \quad \frac{\partial T_s}{\partial r} = 0 \qquad (15a)$$

$$r = R_1 : \quad T_s = T_w \qquad (15b)$$

The following dimensionless variables are introduced.

$$\sigma = \frac{x}{R_1} \qquad (16)$$

$$R_2 = R_1 \sin\sigma \qquad (17)$$

$$\eta = (\tfrac{3}{2}Re)^{\frac{1}{2}} \frac{1}{R_1} \int_0^y \frac{\rho}{\rho_\infty} dy \qquad (18)$$

where Howarth transformation is used to reduce the compressible equation into incompressible form.

In addition, a stream function, ψ, with j = 1 is defined as,

$$R_2\rho u = \mu_\infty \frac{\partial(\psi R_2)}{\partial y} \qquad (19)$$

$$R_2\rho v = -\mu_\infty \frac{\partial(\psi R_2)}{\partial x} \qquad (20)$$

With the defined variables and the assumptions, equations (4) to (6) and equation (8) can be transformed into

$$\frac{\partial\psi}{\partial\eta}\frac{\partial Y_o}{\partial\sigma} - (\frac{\partial\psi}{\partial\sigma} + \psi\cot\sigma)\frac{\partial Y_o}{\partial\eta} = (\tfrac{3}{2}Re)^{\frac{1}{2}} Pr^{-1}\frac{\partial^2 Y_o}{\partial\eta^2}$$

$$- (\tfrac{3}{2}Re)^{-\frac{1}{2}} \frac{\rho_\infty^2 R_1^2 n}{\mu_\infty} \frac{\rho}{\rho_\infty} Y_o Y_F A_R \exp(-E_R/RT) \qquad (21)$$

$$\frac{\partial\psi}{\partial\eta}\frac{\partial Y_F}{\partial\sigma} - (\frac{\partial\psi}{\partial\sigma} + \psi\cot\sigma)\frac{\partial Y_F}{\partial\eta} = (\tfrac{3}{2}Re)^{\frac{1}{2}} Pr^{-1}\frac{\partial^2 Y_F}{\partial\eta^2}$$

$$- (\tfrac{3}{2}Re)^{-\frac{1}{2}} \frac{\rho_\infty^2 R_1^2}{\mu_\infty} \frac{\rho}{\rho_\infty} Y_o Y_F A_R \exp(-E_R/RT) \qquad (22)$$

$$\frac{\partial\psi}{\partial\eta}\frac{\partial^2\psi}{\partial\sigma\partial\eta} - (\frac{\partial\psi}{\partial\sigma} + \psi\cot\sigma)\frac{\partial^2\psi}{\partial\eta^2} = (\tfrac{3}{2}Re)\frac{\rho_\infty}{\rho}\sin\sigma\cos\sigma$$

$$+ (\tfrac{3}{2}Re)^{\frac{1}{2}}\frac{\partial^3\psi}{\partial\eta^3} \qquad (23)$$

$$\frac{\partial\psi}{\partial\eta}\frac{\partial T}{\partial\sigma} - (\frac{\partial\psi}{\partial\sigma} + \psi\cot\sigma)\frac{\partial T}{\partial\eta} = (\tfrac{3}{2}Re)^{-\frac{1}{2}} Pr^{-1}\frac{\partial^2 T}{\partial\eta^2}$$

$$+ (\tfrac{3}{2}Re)^{-\frac{1}{2}} \frac{\rho_\infty^2 R_1^2 Q}{\mu_\infty C_p} \frac{\rho}{\rho_\infty} Y_o Y_F A_R \exp(-E_R/RT) \qquad (24)$$

Since the profiles of variables are axisymmetric, the stream function, temperature, and concentration profiles are further expanded with respect to the azimuthal angle, σ, as

$$\psi = (\tfrac{3}{2}Re)^{\frac{1}{2}} \; (\sigma f_o(\eta) + \sigma^3 f_1(\eta) + O(\sigma^5)) \qquad (25)$$

$$\frac{T-T_\infty}{T_\infty} = G_o(\eta) + \sigma^2 G_1(\eta) + O(\sigma^4) \qquad (26)$$

$$\frac{Y_o - Y_{o\infty}}{Y_{o\infty}} = H_o(\eta) + \sigma^2 H_1(\eta) + O(\sigma^4) \qquad (27)$$

$$\frac{Y_F}{Y_{o\infty}} = I_o(\eta) + \sigma^2 I_1(\eta) + O(\sigma^4) \qquad (28)$$

Substituting the above expressions into equations (21) to (24) and expanding the triangular functions into azimuthal angle, the following ordinary differential equations, through the comparison of the coefficients of azimuthal angle, are obtained.

$$f_o''' + 2f_of_o'' - (f_o')^2 = -1-G_o \qquad (29)$$

$$f_1''' + 2f_of_1'' - 4f_o'f_1' + 4f_o''f_1 - \tfrac{1}{3}f_of_o'' = \tfrac{2}{3}(1+G_o) - G_1 \qquad (30)$$

$$Pr^{-1}G_o'' + 2f_oG_o' + \{(\tfrac{3}{2}Re)^{-1} \frac{\rho_\infty^2 R_1^2 Y_{o\infty} Q}{\mu_\infty C_p T_\infty} A_R \exp[\frac{-E_R}{R\,T_\infty(1+G_o)}]\} \frac{1+H_o}{1+G_o} I_o = 0 \qquad (31)$$

$$Pr^{-1}G_1'' - 2f_o'G_1 + 2f_oG_1' + 4f_1G_o' - \tfrac{1}{3}f_oG_o' + \{(\tfrac{3}{2}Re)^{-1} \frac{\rho_\infty^2 R_1^2 Y_{o\infty}^2 Q}{\mu_\infty C_p T_\infty} A_R \exp[\frac{-E_R}{R\,T_\infty(1+G_o)}]\}$$
$$(\frac{(1+H_o)I_1 + I_oH_1}{1+G_o} + \frac{(1+H_o)I_oG_1E_R/RT_\infty}{(1+G_o)^3} - \frac{(1+H_o)I_oG_1}{(1+G_o)^2}) = 0 \qquad (32)$$

$$Pr^{-1}H_o'' + 2f_oH_o' - \{(\tfrac{3}{2}Re)^{-1} \frac{\rho_\infty^2 R_1^2 n Y_{o\infty}}{\mu_\infty} A_R \exp[\frac{-E_R}{R\,T_\infty(1+G_o)}]\} \frac{1+H_o}{1+G_o} I_o = 0 \qquad (33)$$

$$Pr^{-1}H_1'' - 2f_o'H_1 + 2f_oH_1' + 4f_1H_o' - \tfrac{1}{3}f_oH_o' - \{(\tfrac{3}{2}Re)^{-1} \frac{\rho_\infty^2 R_1^2 n Y_{o\infty}}{\mu_\infty} A_R \exp[\frac{-E_R}{RT_\infty(1+G_o)}]\}$$
$$(\frac{(1+H_o)I_1 + I_oH_1}{1+G_o} + \frac{(1+H_o)I_oG_1E_R/RT_\infty}{(1+G_o)^3} - \frac{(1+H_o)I_oG_1}{(1+G_o)^2}) = 0 \qquad (34)$$

$$Pr^{-1}I_o'' + 2f_oI_o' - \{(\tfrac{3}{2}Re)^{-1} \frac{\rho_\infty^2 R_1^2 Y_{o\infty}}{\mu_\infty} A_R \exp[\frac{-E_R}{RT_\infty(1+G_o)}]\} \frac{1+H_o}{1+G_o} I_o = 0 \qquad (35)$$

$$Pr^{-1}I_1'' - 2f_o'I_1 + 2f_oI_1' + 4f_1I_o' - \tfrac{1}{3}f_oI_o' - \{(\tfrac{3}{2}Re)^{-1} \frac{\rho_\infty^2 R_1^2 Y_{o\infty}}{\mu_\infty} A_R \exp[\frac{-E_R}{RT_\infty(1+G_o)}]\}$$
$$(\frac{(1+H_o)I_1 + I_oH_1}{1+G_o} + \frac{(1+H_o)I_oG_1E_R/RT_\infty}{(1+G_o)^3} - \frac{(1+H_o)I_oG_1}{(1+G_o)^2}) = 0 \qquad (36)$$

Similarly, the boundary conditions are transformed and given as follows,

$$f_o'(0) = 0 \qquad (37a)$$

$$f_1'(0) = 0 \qquad (37b)$$

$$f_o(0) = \frac{-\rho_s R_1}{2\mu_\infty}(\tfrac{3}{2}Re)^{-\frac{1}{2}} \; A_P \; e^{-E_P/RT_W} \qquad (37c)$$

$$f_1(0) = \frac{-\rho_s R_1}{24\mu_\infty}(\tfrac{3}{2}Re)^{-\frac{1}{2}} \; A_P \; e^{-E_P/RT_W} \qquad (37d)$$

$$G_o(0) = \frac{T_W - T_\infty}{T_\infty} \qquad (37e)$$

$$G_1(0) = 0 \qquad (37f)$$

$$H_o'(0) + 2Prf_o(0)\;(1+H_o(0)) = 0 \qquad (37g)$$

$$H_1'(0) + 2PrH_1(0)f_o(0) = 0 \qquad (37h)$$

$$I_o'(0) = \frac{2Prf_o(0)(1-Y_{o\infty}I_o(0))}{Y_{o\infty}} \qquad (37i)$$

$$I_1'(0) + 2Prf_o(0)I_1(0) = 0 \qquad (37j)$$

$$f_o'(\infty) = 1 \qquad (37k)$$

$$f_1'(\infty) = -\tfrac{1}{6} \qquad (37l)$$

$$G_o(\infty) = 0 \qquad (37m)$$

$$G_1(\infty) = 0 \qquad (37n)$$

$$H_o(\infty) = 0 \qquad (37o)$$

$$H_1(\infty) = 0 \qquad (37p)$$

$$I_o(\infty) = 0 \qquad (37q)$$

$$I_1(\infty) = 0 \qquad (37r)$$

NUMERICAL METHOD

The governing equations of gaseous and solid phases are coupled and nonlinear. A value of surface temperature is first assumed to decouple the gaseous equations from the equations of the solid. The higher order equations of gas are transformed into a set of second order differential equations and then

Table 1 The fuel properties and the related parameters adopted in the theoretical model (Kashiwagi and Summerfield, 1973).

K_s=0.0003 cal/cm-s-K	α_s= 3×10^{-4} cm^2/s
ρ_s=1.54 g/cm^3	R_1=0.2 cm
T_∞=1600K	u_∞=200 cm/s
Pr=0.7	ρ_∞=0.000221 g/cm^3
K_∞=0.000239 cal/cm-s-K	μ_∞=0.000563 g/cm-s
$Y_{o\infty}$=0.23	L=120 cal/g
Cp=0.298 cal/g-K	n=6
$A_RQ=12\times10^{12}$ $cal\text{-}cm^3/g^2\text{-}s$	E_R=16000 cal/mol
$A_p=10^3$ cm/s	E_p=12000 cal/mol

quasilinearized. Through the process of quasilinearization (Lee, 1968), these sets of boundary type equations are transformed into the initial value problems and can be easily solved through Runge-Kutta method. Finally, the gas phase solutions and the solid phase conditions are matched on the fuel surface. The fuel properties and the related parameters adopted are shown in Table 1 (Kashiwagi and Summerfield, 1973). The accuracy of the numerical solutions was verified by comparing the results with and without the fourth order terms of azimuthal angle of the temperature profile. The error was found to be less than 1 K, therefore the fourth order terms are neglected through the following analysis.

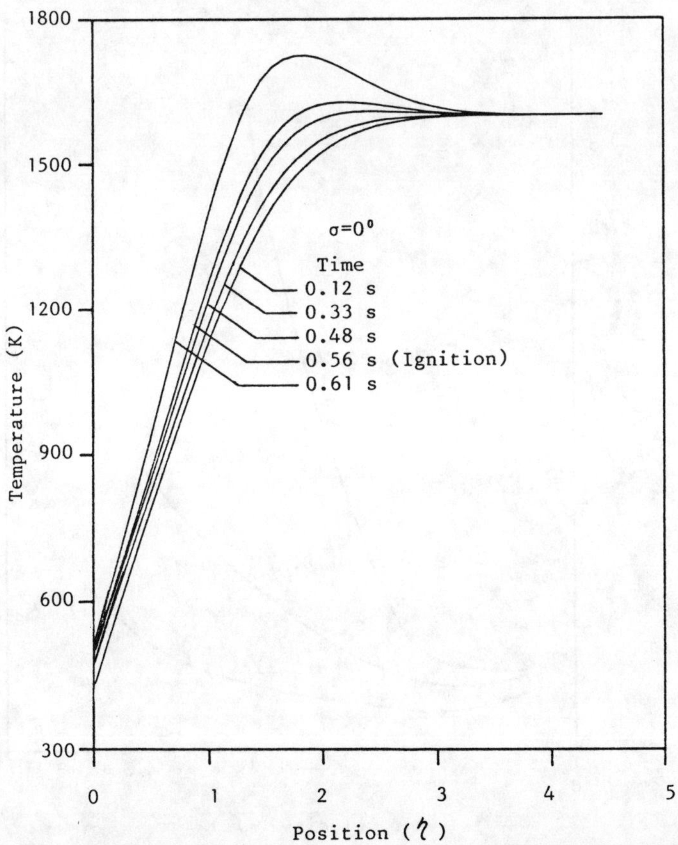

Fig. 2 The temperature profiles at various times around the igntion time. (T_∞ = 1600 K, u_∞ = 2 m/s, R_1 = 0.2 cm, $Y_{O\infty}$ = 0.23)

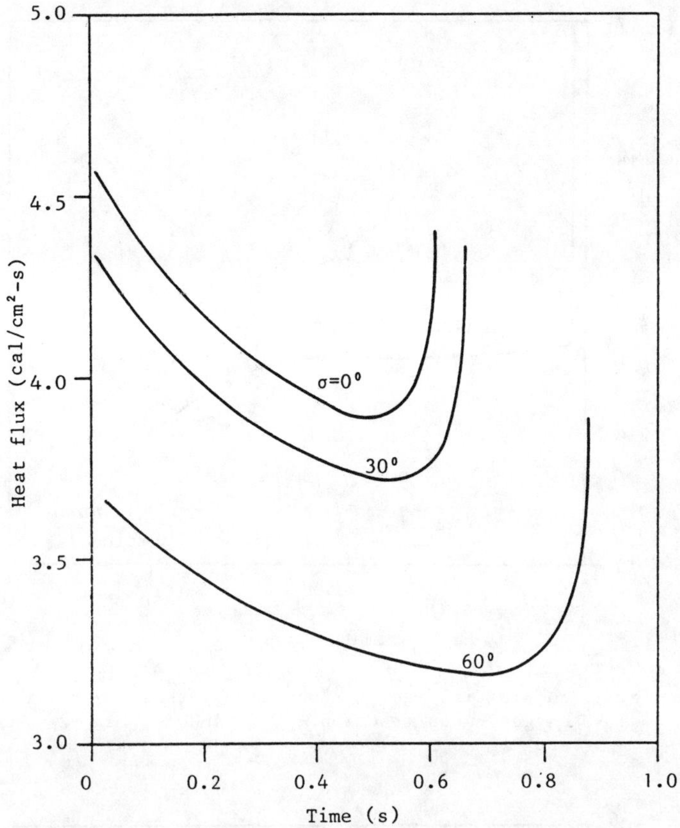

Fig. 3 The histories of the heat flux to the solid fuel at the azimuthal angles of 0°, 30°, and 60°. (T_∞=1600 K, u_∞ =2 m/s, R_1=0.2 m/s, $Y_{O\infty}$ =0.23)

RESULTS AND DISCUSSION

Ignition criterion

It is reported that the major variables related to the ignition criteria are temperature and concentration of fuel vapor in the gas phase (Mutoh et al. 1979). Since the fuel vapor concentration in this study is within the rich flammability limit, the major variable discussed is the gas temperature. Figure 2 shows the temperature profiles prior to and right at the ignition time. Before ignition, the gas temperature near the stagnation point rises monotonically from the cool wall temperature to the free stream temperature of 1600 K. At 0.48 s, the first small peak appears near the edge of the boundary layer and propagates inward quickly. The corresponding heat fluxes from the gas phase to the fuel surface at the azimuthal angles of 0°, 30°, 60°, are shown in Fig. 3. The heat flux from the gas phase to the solid wall, as given in equation (12d), is at its minimum value at 0.565 s and then starts to increase. At that moment the reaction rate is high enough to elevate the gas temperature. As the gas temperature is significantly higher than the free stream temperature the heat flux and the gas temperature begin to runaway. Therefore, it is concluded that the inverse point of the slope of the heat flux to the fuel, dq/dt = 0, is the suitable criterion for the convective ignition of the

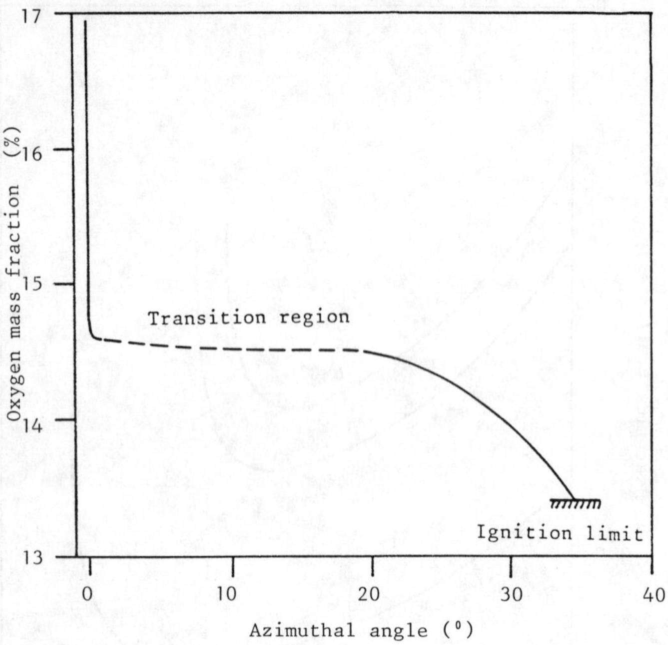

Fig. 4 The initial ignition position versus free stream oxygen mass fraction. (T_∞ =1600 K, u_∞ =2 m/s, and R_1 = 0.2 cm).

reactive solid. This criterion should fulfill the physical situation of ignition more precisely. In addition, it covers the catagory of the criteria of null heat transfer, maximum temperature, null temperature gradient, and the critical chemical reaction rate.

<u>Ignition mechanism</u>

In the previous theoretical models the ignition is often assumed to begin at the stagnation point. However, this study shows that the ignition point may be blown down to the downstream location since the Damkohler number is small. Figure 4 shows that the ignition point is near the stagnation point for the oxygen mass fraction is higher than 14.6%. It moves from 0 azimuthal angle to 35 azimuthal angle while the oxygen mass fraction drops from 14.6% to 13.5%. This phenomena has been reported in experimental studies(Birk, 1980; Yang and Wang, 1989).

Figure 5 shows that the ignition delay is a strong function of the oxygen mass fraction at 2 m/s flow velocity of the N_2/O_2 mixture and is only slightly affected as the flow velocity is 0.3 m/s. When the flow velocity is 0.9 m/s the ignition delay is minimum as the oxygen mass fraction is greater than 15%.

It is then determined that the gas phase chemical reaction and the solid fuel pyrolysis are the two determining steps for convective ignition. When the flow rate is low, the rate of heat-up is not substantial and the boundary layer thickness is relatively thick. This condition prolongs the transport process of the pyrolyzed product to the free stream. Therefore, the rate-determining step is the solid phase pyrolysis.

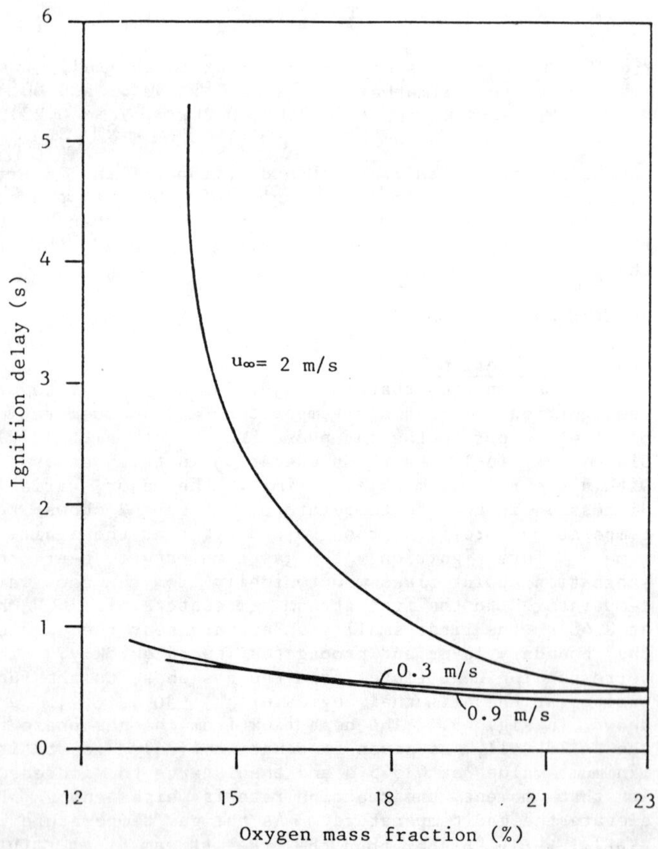

Fig. 5 The effect of air velocity on the ignition delay under verious oxygen mass fractions. (T_∞ = 1600 K, R_1 = 0.2 cm)

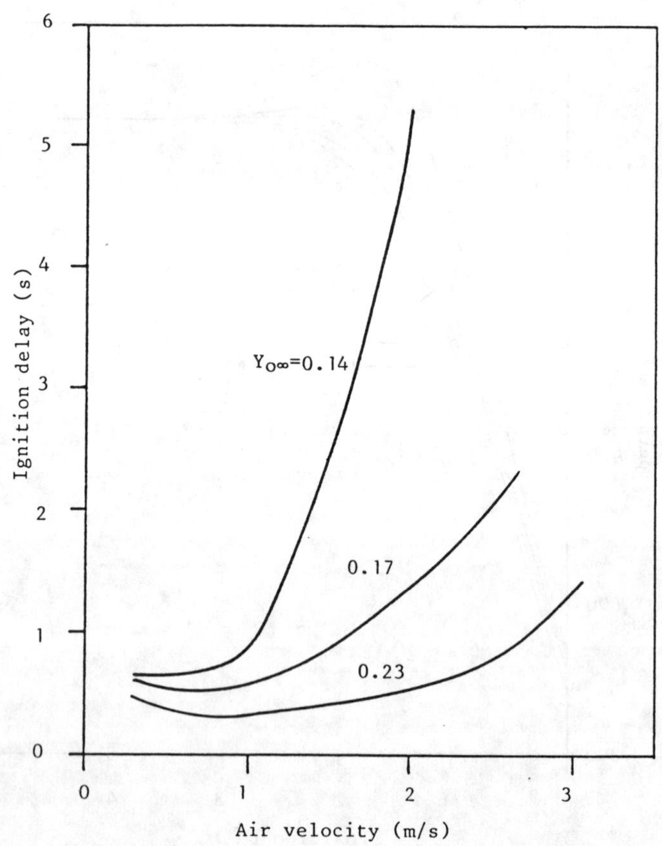

Fig. 6 The effect of oxygen concentration on the ignition delay at various air velocity. (T_∞ = 1600 K, R_1 = 0.2 cm)

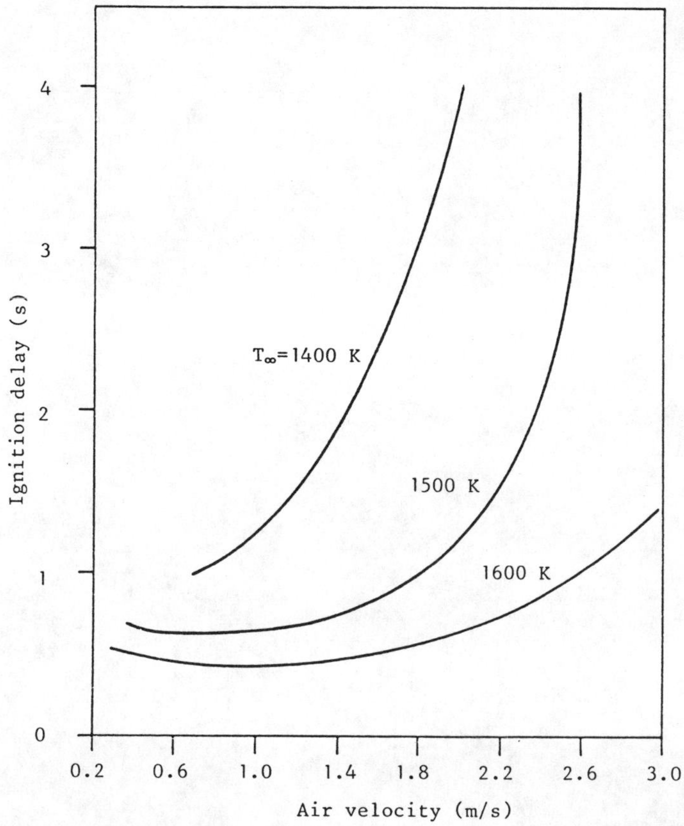

Fig. 7 The effect of temperature on the ignition delay at various air velocity.
($Y_{O\infty}$ = 0.23, R_1 = 0.2 cm)

Therefore, the ignition delay time decreases gradually as the air flow rate increases, as shown in Fig. 6. Under high air flow rate the pyrolyzed fuel vapor is more easily swept downstream and the rate-determining step becomes the gas phase chemical reaction. Hence, the ignition delay time increases monotonically with respect to the air velocity. There are the critical velocities, 0.9 m/s and 0.5 m/s, which correspond to the optimal ignition conditions for oxygen mass fractions of 23% and 17%, respectively. For the case of 14% oxygen mass fraction the whole range indicated in Fig. 6 is controlled by the slow gas phase chemical reaction rate and the corresponding critical velocity does not exist. The effect of air temperature on the ignition delay shows a similar trend as depicted in Fig. 7.

CONCLUSIONS

A two-dimensional, unsteady model is developed to study the convective thermal igntion process of the reactive solid particle in the region between the stagnation point and the separation point. The condition of the minimum heat flux from the gas phase to the fuel is suggested as the ignition criterion. Once this criterion is fulfilled the local gas temperature and the reaction rate are high enough to sustain the combustion flame.

The analysis indicates that the ignition process is mainly controlled by the rate of gas phase reaction and the rate of solid pyrolysis and the latter is closely related to the physical heat-up of the fuel particle. There exists a critical air velocity, at which the ignition time is a minimum. Below this critical air velocity pyrolysis of the fuel is the rate-determining factor of the ignition process. When the ignition process is controlled by the gas phase chemical reaction the ignition may occur downstream of the stagnation point and the ignition delay time increases as the air velocity increases.

ACKNOWLEDGEMENTS

The authors wish to acknowledge the support of this study by the National Science Council of the Republic of China, under Contract No. NSC78-0401-E007-02.

REFERENCES

Annamalai, K. and Durbetaki, P., 1977, "A Theory on Transition of Ignition Phase of Coal Particles," Combustion and Flame, Vol. 29, 193-208.

Birk, A, 1980, Convective Ignition of Propellant Cylinders in A Developing Cross Flow Field, Ph.D. Thesis, Princeton University.

Fernandez-Pello, A.C. and Law, C.K., 1982a, "On the Mixed-Convective Flame Structure in the Stagnation Point of A Fuel Particle," 19th Sym. (Int.) on Combustion, 1037-1044.

Fernandez-Pello, A.C. and Law, C.K., 1982b, "A Theory for The Free- Convective Burning of A Condensed Fuel Particle,"Combustion and Flame, Vol. 44, 97-112.

Ganhdi, P.D. and Kanury, A.M., 1986, "Criterion for Spontaneous Ignition of Radiantly Heated Organic Solids," Combust. Sci. and Tech., Vol. 50, 233-254.

Kashiwagi, T., Macdonald, B.M., Isoda, H., and Summerfield, M., 1971, "Ignition of a Solid Polymeric Fuel in A Hot Oxidizing Gas Stream," 13th Sym. (Int.) on Combustion, 1073-1086.

Kashiwagi, T. and Summerfield, M., 1973, 14th Sym. (Int.) on Combustion, 1235-1247.

Kumar, R.K., 1983, "Gas Phase Ignition of A Composite Solid Propellant Subjected to Radiant Heating," Combust. Sci. and Tech., Vol. 30, 273-288.

Lee, E.S., 1968, Quasilinearization and Invariant Imbeding, Academic Press, New York, 73-81.

Mutoh, N., Hirano, T., and Akita, K., 1979, "Experimental study on Radiative Ignition of Polymethylmethacrylate," 17th Sym. (Int.) on Combustion, 1183-1190.

Price, E.W., Bradely, Jr., H.H., and Dehority, G.L., 1966, "Theory of Ignition of solid Propellants,"AIAA J., Vol. 4, 1153-1161.

Sibulkin, M., Kulkarni, A.K., and Annamalai, K., 1981, "Effects of radiation on the burning of vertical fuel surfaces," 18th Sym. (Int.) on Combustion, 611-617.

Sibulkin, M., Kulkarni, A.K., and Malary, S.F., 1983, "Nonsimilar calculation for free convection diffusion flames," Combustion and Flame, Vol. 50, 59-64.

Yang, J.T. and Wang, G.G., 1989, "The Experimental Study of The Convective Ignition of PMMA Particle," to be presented on 26th ASME/AIChE National Heat Transfer Conference, Pittsburgh, Pennsylvania, U.S.A.

1989 National Heat Transfer Conference
HTD-Vol. 106, Heat Transfer Phenomena in Radiation, Combustion and Fires

TURBULENT REACTING GAS JETS IN LIQUIDS AT ELEVATED TEMPERATURES AND PRESSURES

S. H. Chan and T. R. Shen
Department of Mechanical Engineering
University of Wisconsin — Milwaukee
Milwaukee, Wisconsin

ABSTRACT

The analytical work proposed recently has been applied to construct new equilibrium state relationships of the liquid metal combustion simulation reactants: hydrogen chloride gas as oxidant and ammonia aqueous solution as fuel. The state relationships have been employed to predict the structure of sonic, turbulent, reacting HCl gas jets submerged in ammonia aqueous solutions. Fuel temperature and ambient pressure effects have been examined. It is found that either raising fuel temperature or lowering ambient pressure increases evaporation, deters entrainment rate, delays the completion of reaction and extends greatly the total jet penetration length. Good agreement of penetration lengths with reported experimental data over a wide range of fuel concentration, temperature and ambient pressure confirms the validity of the theoretical model proposed in the recent paper.

NOMENCLATURE

a	gravitational acceleration (m/s^2)
D	injector diameter (m)
d	jet diameter at the end of the expansion region (m)
f	mixture fraction
g	square of the mixture fraction fluctuations
k	turbulent kinetic energy (J/kg)
P	probability density function
r	radial distance from axis of symmetry (m)
S_ϕ	source term in governing equations
T	temperature (K)
u	axial velocity (m/s)
v	radial velocity (m/s)
x	axial distance (m)
Y_i	mass fraction of species i
ϵ	dissipation rate of turbulent kinetic energy (J/(kg·s))
α	void fraction
μ	molecular viscosity (kg/(m·s))
μ_t	turbulent viscosity (kg/(m·s))
$\mu_{eff,\phi}$	effective viscosity (kg/(m·s))
ρ	density (kg/m^3)
σ	turbulent Schmidt or Prandtl number
γ_i	activity coefficient of species i

<u>Subscripts</u>

aq	aqueous solution
c	crystal
g	vapor phase
l	liquid phase
o	injector exit condition
∞	ambient condition

<u>Superscripts</u>

(‾)	time averaged quantity
(~)	mass averaged quantity

INTRODUCTION

Liquid metal combustion has been a subject of recent interest, particularly for under water propulsion systems (Hughes et al., 1983). In these power systems, thermal energy is produced in a combustor filled with a molten metal fuel bath, into which a choked submerged jet of reacting gas oxidant is injected. The reaction forms a complex, multiphase, turbulent diffusion flame. The combustor is unique in that power density is high and combustion products are condensable so that all products are containable in reactant vessel itself. Thus no discharge of products is necessary during operation. Therefore the propulsion system is unaffected by the surrounding back pressure, well suited for under water applications.

In our prior theoretical study (Chan et al., 1988), two parts were proposed for the analysis of liquid metal combustion. First a formulation and code development for chemical equilibrium computations of general multiphase reacting systems was proposed. The reactants in the system can be ideal, nonideal, electrolytic or nonelectrolytic mixtures. The second part constitutes physical models for the prediction of turbulent structure of liquid metal diffusion flame. The former is for the construction of equilibrium state relationships which are pre-requisite information necessary for flow computation of the latter. Applications were made to predict the reaction of a gaseous hydrogen chloride (oxidant) jet submerged in an ammonia aqueous solution (fuel), where limited experimental data was available for comparison (Cho et al., 1987). Since then, much more extensive experimental data have been reported over a wide range of system pressure and fuel temperature conditions (Cho et al., 1988). The objective of the present study is to continue our prior study to examine the pressure and temperature effects on turbulent structure of such diffusion reacting jets. This, together with the prior study which focused on fuel concentration effects, constitutes a better verification of the theoretical work proposed previously (Chan et al., 1988).

It is noted that the reaction of HCl gas/NH_3 aqueous solution simulates well many characteristics of liquid metal combustion - multiphase, exothermic reaction, fuel vaporization, condensation of products and condensation of excess reactants (Cho et al., 1987). In fact analysis of HCl gas/NH_3 solution reaction is by no means simpler than liquid metal reaction since the former also involves an additional process, namely, dissolution of HCl gas in ammonia solution, which is not found in liquid metal combustion. Therefore successful prediction of HCl/NH_3 reaction will lend further validity of the analytical work proposed for liquid metal combustion where little data is yet available.

THEORY

Details of theoretical model and formulation have been described previously (Chan et al., 1988). They are briefly summarized below.

The method of calculation of chemical equilibrium was based on the principle of minimization of Gibbs free energy subject to the constraints: the conservation of elements and balance of charges. The latter is needed since ions are present in electrolyte solutions. With the introduction of proper partial molal (or molar) enthalpy and chemical potential expressions for imperfect gases, solutes and solvent of nonideal and electrolytic solutions, as well as proper activity coefficient expressions and thermodynamic data, the system of nonlinear equations is solved by a modified Newton-Raphson iteration scheme.

Like the last paper, the equilibrium code developed is now applied to the reacting pair of gaseous HCl as oxidant and aqueous NH_3 solutions as fuel. The possible species considered in liquid phase after reaction are: H^+, NH_4^+, Cl^-, $NH_{3(aq)}$ and $H_2O_{(l)}$. The gas phase species allowed are $H_2O_{(g)}$, $NH_{3(g)}$ and $HCl_{(g)}$, while $NH_4Cl_{(c)}$ crystal is the one considered in solid phase. Chemical equilibrium composition and thermodynamic properties after mixing for ten representative experimental conditions will be constructed later. Those conditions before mixing (Cho et al., 1988) are summarized in Table 1, which lists the concentration of ammonia fuel in aqueous solution ($Y_{NH3,\infty}$), the solution (ambient bath) temperature (T_∞) and the ambient bath pressure (P_∞). In all test runs, the choked HCl gas jets are discharged vertically through the bottom of ammonia aqueous solution through a nozzle diameter of 1 mm (i.e., D = 1 mm). The HCl gas was supplied by a constant source with a stagnation pressure of 0.377 MPa (40 psia) and a stagnation temperature of 293.15 K.

The primary variable in the first group (case No. A through E) is the ammonia concentration in aqueous solution (i.e., fuel concentration variation). It's effects were studied in the last paper. Fuel bath temperature and system pressure are the variables in the second (cases E to H) and the third (cases I and J) group, respectively. Their effects are the subject of present interest.

The computation of reaction zone structure of turbulent HCl gas jets submerged in aqueous ammonia solution uses the model and the method described in the previous paper (Chan et al., 1988), it follows closely the method developed by Lockwood and Naguib (1975) and Chen and Faeth (1983), etc. The method uses the locally homogeneous flow approximation for multiphase flow, the conserved scalar method for mass transfer and the K-ϵ-g model for turbulence. The transport equations solved are summarized in Table 2; they include the Favre-averaged, parabolic forms of continuity, momentum, mixture fraction (f), turbulence kinetic energy (k), dissipation rate of turbulent kinetic energy (ϵ) and the square of the mixture fraction fluctuations (g). Other mean scalar properties not solved by the transport equations are determined by the equilibrium state relationship weighted by a clipped Gaussian profile of probability density function. The later is governed by two parameters, the variance and the most probable value, both of which are in turn determined by f and g solved at each local point in the flow. More details can be found in (Chan et al., 1988).

THEORETICAL RESULTS AND COMPARISON WITH EXPERIMENTAL DATA

The equilibrium states achieved after reaction have been calculated as a function of mixture fraction f for experimental conditions listed in Table 1. The mixture fraction is so defined that f = 1 designates the pure HCl gas at the end condition of expansion region (see later) while f = 0 designates pure aqueous ammonia solution at bath conditions. Two representative cases, E and H, are presented in Figs. 1 and 2, respectively, to show effects of initial fuel temperature on equilibrium state relationships. In each set of figures, the equilibrium mixture void fraction, mass fractions of liquid, vapor and solid phases, mixture temperature and density are presented in part A, such as Fig. 1A. The detailed mass fractions of all individual species, including electrolytes, are given in part B, like Fig. 1B. Similarly, for pressure effects, the equilibrium state relationships for cases I and J are shown in Figs. 3 and 4, respectively.

Once the equilibrium state relationships have been constructed, the structure of the reacting HCl gas jets submerged in NH_3 aqueous solutions can be readily computed using the flow model summarized above. To account for underexpansion effects, the sonic jets discharged from nozzle is assumed to underexpand isentropically. By mass and energy balances, the jet's temperature, density and velocity at the end of the underexpansion region have been calculated and listed in Table 1. Those end conditions are found to satisfy the conservation of momentum within 2.6 to 10.6%. They serve as the initial conditions of computation. The underexpansion length is neglected as compared to the plume length. Thus numerical computations begins at injector's exit and marches downstream. The rest of computational details should be referred to the previous study (Chan et al., 1988).

Figures 5 and 6, respectively, present the predicted mean transport values along the center line axis of vertical HCl gas jets injected into the ammonia solution pool for experimental conditions of cases E and H. They include the mean centerline velocity (u/u_o), mixture fraction (f), temperature (T), void fraction (α) and mass fractions of gas (Y_g), liquid (Y_l) and solid (Y_c) phases. For cases I and J, they are shown in Figs. 7 and 8 respectively.

To assess bath temperature effects, the species mass fraction profiles of cases E and H shown in Figs. 1B and 2B are examined and compared first. The fuel temperature has the most significant effect on fuel vapor, namely ammonia and water vapors. As temperature increases from 14°C to 59.5°C, the span of excess $NH_{3(g)}$ is greatly extended to lower value of mixture fraction, namely from f = 0.2 ~ 0.07 to f = 0.2 ~ 0.01. Similarly, the water vapor range is extended from f = 1 ~ 0.07 to f = 1 ~ 0.01. The combined effect of $NH_{3(g)}$, $H_2O_{(g)}$ and the oxidant $HCl_{(g)}$ can be examined by comparing Figs. 1A and 2A. It is seen that the gas phase mass fraction and the mixture void fraction have accordingly been extended from f = 1 ~ 0.07 to f = 1 ~ 0.01 when the bath temperatures increases from 14°C to 59.5°C.

The extension of scalar property range in these state relationship figures (Figs. 1 and 2) should stretch those scalar property profiles in the flow structure diagram (Figs. 5 and 6) further downstream. This is obvious if one notices that the scalar properties near the nozzle exit correspond roughly to those given in the state relationships at f nearly equal to 1. As one travels downstream, the scalar properties in flow structure diagram correlate with those in state relationship diagram with decreasing f values. Therefore, when bath temperature increases, equilibrium fuel vapor and void fraction appear at lower values of f, which in turn causes significant change in jet structure and substantial elongation of total penetration length as shown in Figs. 5 and 6. The calculated penetration lengths, which are arbitrarily taken as the length where void fraction reduces to 0.01, are also listed in Table 1 for ease of comparison.

The above elongation of the total penetration length is predominantly an effect due to lower subcooling of the liquid bath; the flow with lower subcooling needs to take a longer distance to reach the "cut-off" condition. Another interesting phenomenon we noticed from the detailed printouts of the computations is that vaporization in the jet significantly affects the entrainment rate, namely, higher fuel temperatures trigger more evaporation but less entrainment rate of the ambient fuel. Comparing the two cases in Figs. 5 and 6, the one with higher temperature clearly shows higher values of Y_g and f. Higher Y_g values mean larger mass fraction attributed to gaseous phase due to more evaporation, and higher f values imply less entrainment in view of the definition of the mixture fraction. Physically what happens is as follows. Since NH_3 vapor is lighter than ambient aqueous solution and HCl gas in the main jet stream, as more NH_3 evaporates from the jet interphase boundary, the jet mixture density reduces near the boundary. Consequently turbulent viscosity and entrainment rate of ambient fuel should reduce accordingly (Spalding, 1977). The resulting f and u/u_o profiles along the jet axis should then decrease at a rate slower for case H (Fig. 6) with higher ambient temperature. They in turn prolong the void fraction profile and delay appearance of liquid (Y_l) and crystal (Y_c) mass fraction profiles to axial locations of larger x/D values. Therefore by referring to Figs. 5 and 6, when T_∞ increases from 14 to 59.5°C, the plume penetration length increases by an order of magnitude because of the entrainment and subcooling effects.

As to pressure effects, let us examine case I (P_∞ = 50.5 kpa) and case J (P_∞ = 84.4 kpa). Their state relationships are compared in Figs. 3 and 4 and their flow structures in Figs. 7 and 8. The comparison shows that, like temperature effects, lower ambient pressure produces larger spans of NH_3 and H_2O vapor regions in state relationship diagrams and prolongs f and α profiles in flow structure diagrams. In fact, either lowering pressure or raising temperature yields the same net effect, namely, reducing subcooling of fuel. Thus the above discussion about temperature effects is equally applicable here which needs not be reiterated.

Finally the calculated reacting jet penetration lengths are compared with the experimental data (Cho, 1988) and summarized in Table 1. The total penetration lengths were found experimentally to be oscillating in nature, and both their ranges of oscillation and their mean values are listed in the table. The comparison shows that the calculated results somewhat overpredict the mean penetration length in cases E to H but slightly underpredict cases I and J. Nevertheless, most of them, including cases A through D, are within the oscillating range of the plume length. The comparison of ten cases indicated in the table is considered to be rather extensive since variation of experimental conditions in fuel concentration, fuel temperature and ambient pressure have been covered in cases A ~ E, E ~ H and I ~ J, respectively. In addition, reaction of turbulent submerged HCl gas jets in ammonia solutions is in reality rather complex. It involves a dissolution process (HCl gas dissolving in aqueous solution) not found in metal combustion, and many processes found in liquid metal combustion, including heat of reaction, evaporation, condensation and solubility effects. Thermal radiation and immiscibility effects which may be important in liquid metal combustion are however not important in NCl-NH_3 system. Separate effort is being under way to model thermal radiation/turbulence interaction for future liquid metal combustion computation. The immicibility effect can be satisfactorily accounted by the present model as to be reported later. The satisfactory agreement in all cases of computations of $HCl_{(g)}$/$NH_{3(aq)}$ reaction seems to attest the validity of the theoretical work proposed for the liquid metal combustion, at least for the realistic modelling of the global characteristics of the flow structure.

CONCLUSIONS

New equilibrium state relationships for the reactants of gaseous hydrogen chloride and aqueous ammonia solution have been constructed. They have been used to study effects of fuel temperature and ambient pressure on structure of turbulent, diffusion, reacting HCl jets in ammonia aqueous solutions. It is found that raising the fuel temperature or lowering the ambient pressure increases vaporization of fuel liquid to the jet, reduces the entrainment flux of the liquid, delays the completion of reaction and greatly extends the total jet penetration length. Satisfactory agreement with extensive experimental data lends further support to the validity of our prior theoretical work proposed for the equilibrium computation of general multiphase reacting systems and for modeling the structure of turbulent, diffusion jets in reacting liquids.

TABLE 1. Experimental Conditions and Comparison of Theoretical and Experimental Results

Case No.	$NH_{3(aq)}$ $Y_{NH_3,\infty}$ %	T_∞ °C	HCl Initial Conditions P_∞ kpa	T_o °C	ρ_o kg/m^3	$\tilde{u}_o$ m/s	dx10^3 m	Plume Length (x/D) Results Experimental Range	Experimental Mean	Theoretical Mean
A	0	20	101.3	- 72.15	2.208	382.2	1.087	3 ~ 4.5	3.9	3.8
B	10.6	20	101.3	- 72.15	2.208	382.2	1.087	8 ~ 13	11.3	9.9
C	19	20	101.3	- 72.15	2.208	382.2	1.087	12 ~ 22	17.3	20
D	25.6	20	101.3	- 72.15	2.208	382.2	1.087	30 ~ 48	43	45
E	12	14	101.3	- 72.15	2.208	382.2	1.087	5 ~ 11	7.5	9.5
F	12	45	101.3	- 72.15	2.208	382.2	1.087	12 ~ 24	20.4	26
G	11.9	50	101.3	- 72.15	2.208	382.2	1.087	24 ~ 37	31.4	37.7
H	11.9	59.5	101.3	- 72.15	2.208	382.2	1.087	57 ~ 73	63.9	85.0
I	12	25	50.5	-108.11	1.342	452	1.283	22 ~ 32	27.1	23.0
J	12	25	84.4	- 82.05	1.937	403.3	1.13	9 ~ 20	15.9	13.3

TABLE 2. Governing Equations

$$\frac{\partial \bar{\rho}\tilde{u}}{\partial x} + \frac{1}{r}\frac{\partial}{\partial r}(r\bar{\rho}\tilde{v}) = 0$$

$$\bar{\rho}\tilde{u}\frac{\partial \phi}{\partial x} + \bar{\rho}\tilde{v}\frac{\partial \phi}{\partial r} = \frac{1}{r}\frac{\partial}{\partial r}\left(r\mu_{eff,\phi}\frac{\partial \phi}{\partial r}\right) + S_\phi$$

ϕ	$\mu_{eff,\phi}$	S_ϕ
$\tilde{u}$	$\mu + \mu_t$	$a(\rho_\infty - \bar{\rho})$
$\tilde{f}$	$(\mu/Sc) + (\mu_t/\sigma_t)$	0
k	$\mu + (\mu_t/\sigma_k)$	$\mu_t(\frac{\partial \tilde{u}}{\partial r})^2 - \bar{\rho}\varepsilon$
ε	$\mu + (\mu_t/\sigma_\varepsilon)$	$[C_{\varepsilon 1}\mu_t(\frac{\partial \tilde{u}}{\partial r})^2 - C_{\varepsilon 2}\bar{\rho}\varepsilon]\frac{\varepsilon}{k}$
g	$(\mu/Sc) + (\mu_t/\sigma_g)$	$C_{g1}\mu_t(\frac{\partial \tilde{f}}{\partial r})^2 - C_{g2}\bar{\rho}\frac{\varepsilon g}{k}$

WHERE: $\mu_t = C_\mu \bar{\rho} k^2/\varepsilon$, $C_\mu = 0.09$, $C_{\varepsilon 1} = 1.44$,

$C_{\varepsilon 2} = 1.84$, $C_{g1} = 2.80$, $C_{g2} = 1.84$,

$\sigma_k = 1.0$, $\sigma_\varepsilon = 1.3$, $\sigma_f = 0.7$,

$\sigma_g = 0.7$, $Sc = 0.7$,

$$\bar{\rho} = [\int_0^1 P(f)/\rho(f)\, df]^{-1}$$

ACKNOWLEDGMENTS

This research was sponsored by the Office of Naval Research, Contract No. N00014-85-K-0752 and N00014-89-J-1267, under the technical management of Drs. G. Roy, R. S. Miller and L. Parnell.

REFERENCES

Chan, S. H., Janke, P. J. and Shen, T.R., 1988, "Equilibrium Computations of Multiphase Nonideal Electrolytic Systems and Structure of Turbulent Reacting Dissolving Jets," Presented at the Twenty-second Sym. (Int.) on Combustion, Aug 14-18, 1988, Seattle, Washington; also in Proceedings (in press).

Chen, L. D. and Faeth, G. M., 1983, "Structure of Turbulent Reacting Gas Jets Submerged in Liquid Metals," Combustion Science and Technology, Vol. 11, pp. 111-131.

Cho, D. H., Armstrong, D. R., and Bova, L., 1987, "Dynamic Behavior of Reacting Gas Jets Submerged in Liquids," Presented at the National Heat Transfer Conference, Pittsburgh, Pennsylvania, August 9-12, 1987, Paper No. 87-HT-67.

Cho, D. H., Armstrong, D.R., and Bova, L., 1988, "Heat Release Effects of Reacting Gas Jets in Liquids," ASME Proceedings of the 1988 National Heat Transfer Conference, (H. R. Jacobs, ed.), Vol. 1, pp. 27-34.

Hughes, T. G., Smith, R. B., and Kiley, D. H., 1983, "Stored Chemical Energy Propulsion System for Underwater Applications," J. of Energy, Vol. 7, pp. 128-133.

Lockwood, F. C., and Naguib, A. S., 1975, "The Prediction of the Fluctuations in the Properties of Free, Round-Jet, Turbulent, Diffusion Flames," Combustion and Flame, Vol. 24, pp. 109-124.

Spalding, D. B.: GENMIX: A General Computer Program for Two-Dimensional Parabolic Phenomena, Pergamon Press, Oxford, 1977.

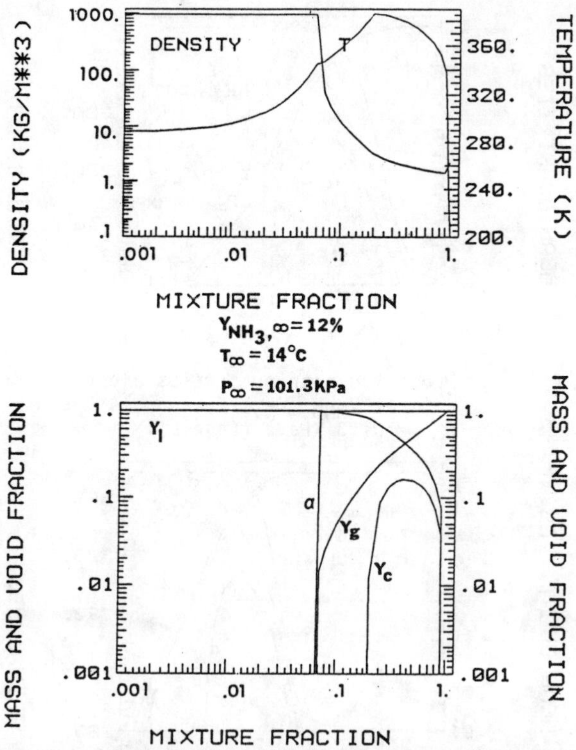

Fig. 1A. Equation of state (density, temperature, void fraction and mass fractions) for HCl-NH_3-H_2O system at $Y_{NH3,\infty}$ = 12%, T_∞ = 14°C and P_∞ = 101.3 KPa. (Case E).

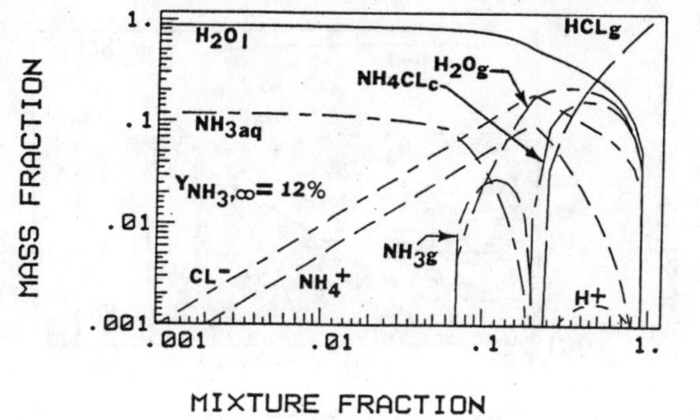

Fig. 1B. Equilibrium composition of HCl-NH_3-H_2O system at $Y_{NH3,\infty}$ = 12%, T_∞ = 14°C and P_∞ = 101.3 KPa. (Case E).

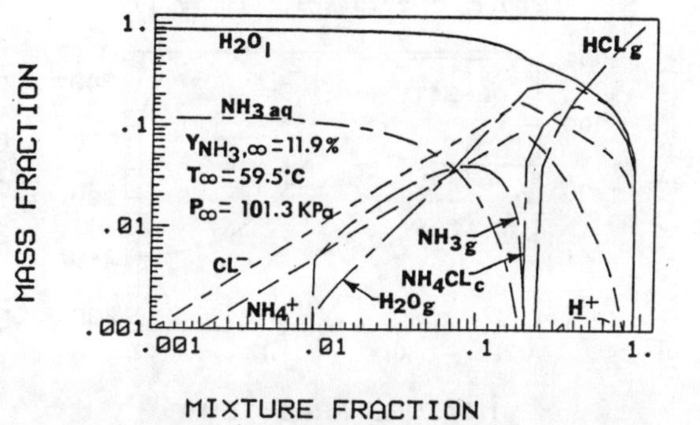

Fig. 2B. Equilibrium composition of HCl-NH_3-H_2O system at $Y_{NH3,\infty}$ = 11.9%, T_∞ = 59.5°C and P_∞ = 101.3 KPa. (Case H).

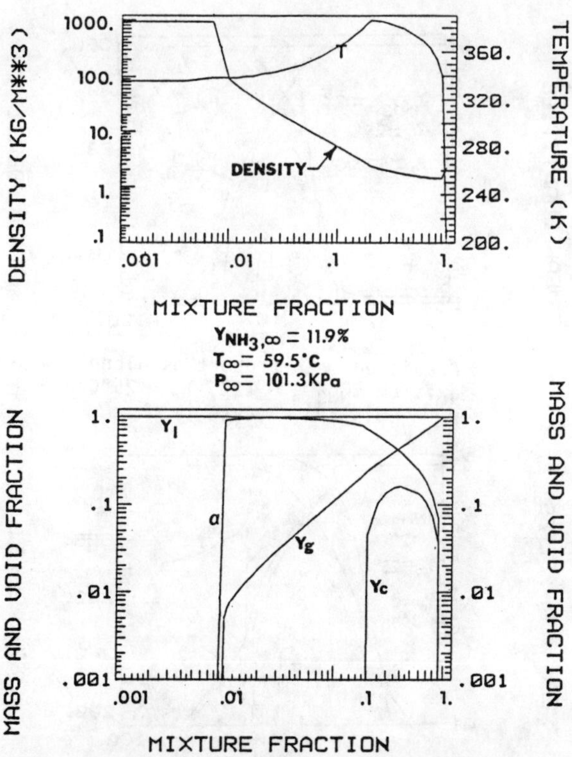

Fig. 2A. Equation of state (density, temperature, void fraction and mass fractions) for HCl-NH_3-H_2O system at $Y_{NH3,\infty}$ = 11.9%, T_∞ = 59.5°C and P_∞ = 101.3 KPa. (Case H).

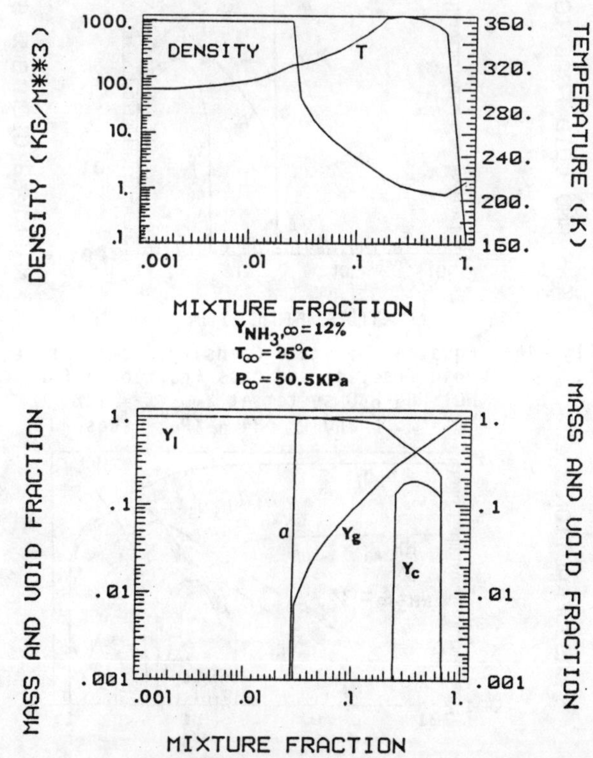

Fig. 3A. Equation of state (density, temperature, void fraction and mass fractions) for HCl-NH_3-H_2O system at $Y_{NH3,\infty}$ = 12%. T_∞ = 25°C and P_∞ = 50.5 KPa. (Case I).

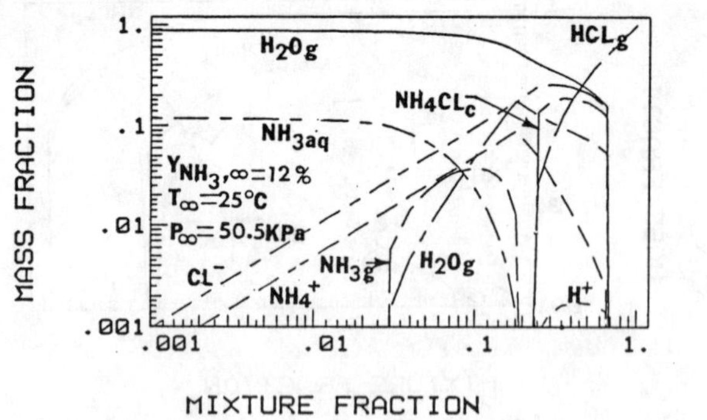

Fig. 3B. Equilibrium composition of $HCl-NH_3-H_2O$ system at $Y_{NH3,\infty}$ = 12%, T_∞ = 25°C and P_∞ = 50.5 KPa. (Case I).

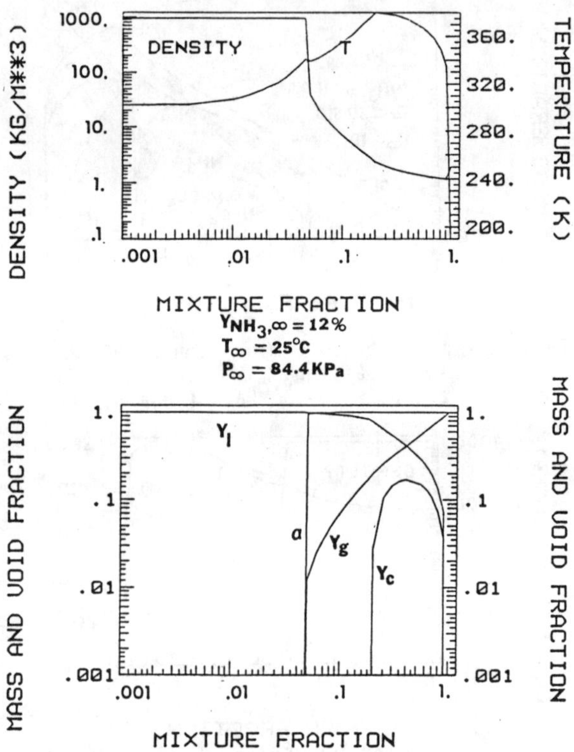

Fig. 4A. Equation of state (density, temperature, void fraction and mass fractions) for $HCl-NH_3-H_2O$ system at $Y_{NH3,\infty}$ = 12%. T_∞ = 25°C and P_∞ = 84.4 KPa. (Case J).

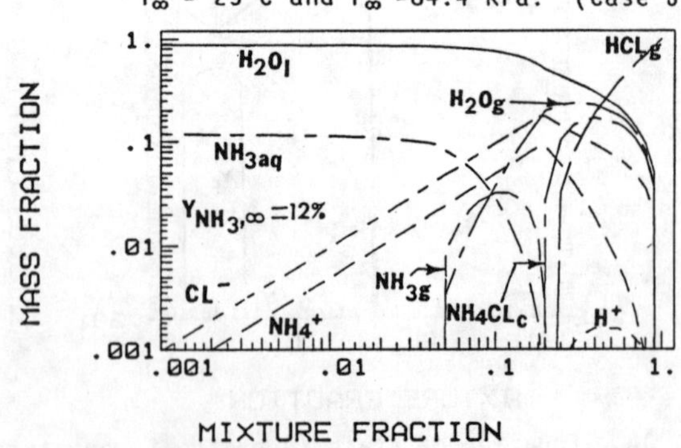

Fig. 4B. Equilibrium composition of $HCl-NH_3-H_2O$ system at $Y_{NH3,\infty}$ = 12%, T_∞ = 25°C and P_∞ = 84.4 KPa. (Case J).

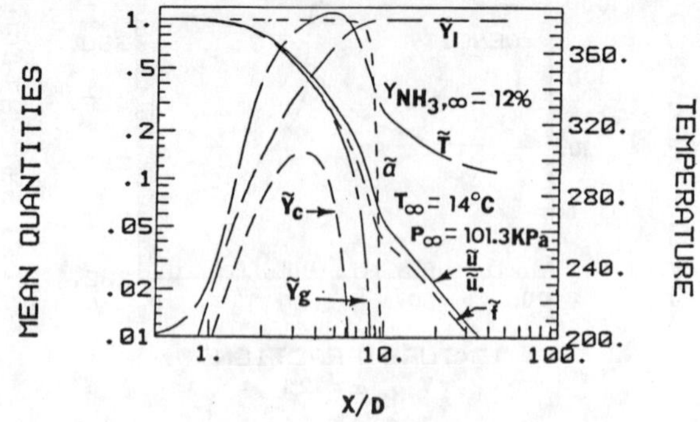

Fig. 5. Predicted mean properties along the jet axis at $Y_{NH3,\infty}$ = 12%, T_∞ = 14°C and P_∞ = 101.3 KPa. (Case E).

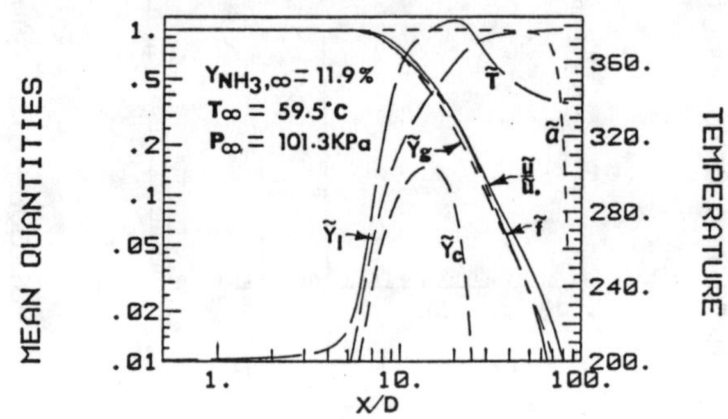

Fig. 6. Predicted mean properties along the jet axis at $Y_{NH3,\infty}$ = 11.9%, T_∞ = 59.5°C and P_∞ = 101.3 KPa. (Case H).

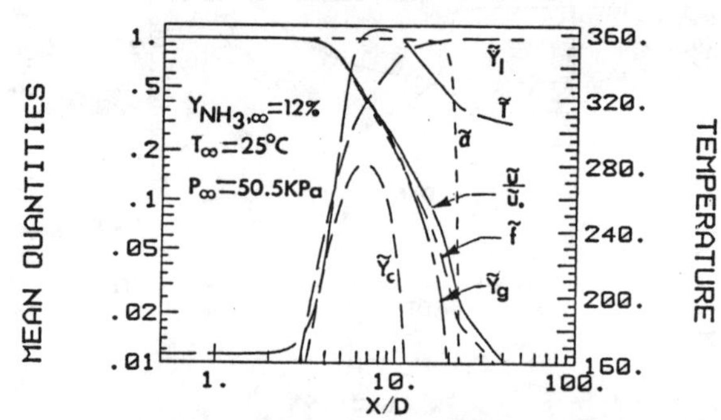

Fig. 7. Predicted mean properties along the jet axis at $Y_{NH3,\infty}$ = 12%, T_∞ = 25°C and P_∞ = 50.5 KPa. (Case I).

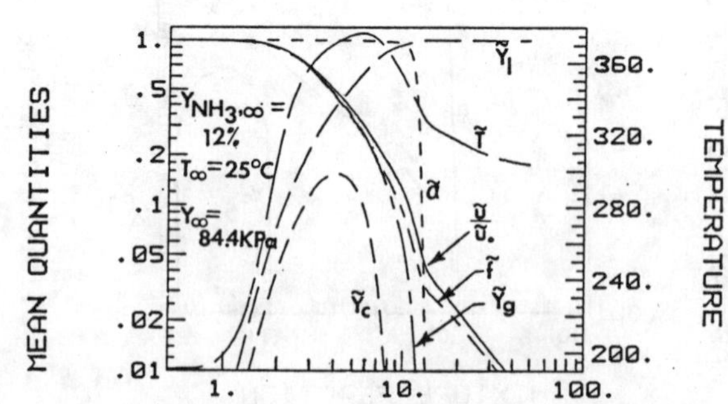

Fig. 8. Predicted mean properties along the jet axis at $Y_{NH3,\infty}$ = 12%, T_∞ = 25°C and P_∞ = 84.4 KPa. (Case J).

1989 National Heat Transfer Conference
HTD-Vol. 106, Heat Transfer Phenomena in Radiation, Combustion and Fires

A NUMERICAL ANALYSIS OF COMBUSTION AND HEAT TRANSFER IN POROUS RADIANT BURNERS

S. B. Sathe, R. E. Peck and T. W. Tong
Arizona State University
Department of Mechanical and Aerospace Engineering
Tempe, Arizona

ABSTRACT

A numerical study of combustion and multimode heat transfer in porous radiant burners is performed. Burner characteristics such as flame speeds, radiant outputs and efficiencies are investigated using a one-dimensional conduction, convection, radiation, and premixed flame model. The porous medium is assumed to emit, absorb, and scatter radiant energy. Non-local thermal equilibrium between the solid and gas is accounted for by introducing separate energy equations for the gas and the solid phase. Combustion is described by a one-step global mechanism. The effect of the optical depth, scattering albedo, solid thermal conductivity, upstream environment reflectivity, and interphase heat transfer coupling on the burner performance are studied. It was revealed that for maximizing the radiant output the optical depth should be about ten and the flame should be stabilized near the center of the porous medium. Also, low solid thermal conductivity, low scattering albedo, and high inlet environment reflectivity produced a high radiant efficiency.

NOMENCLATURE

a surface area per volume of solid [1/m]
c specific heat capacity at constant pressure [J/kg-K]
h heat transfer coefficient [W/m²-K]
h_k species specific enthalpy [J/kg]
i radiant intensity [W/m²-sr]
k thermal conductivity [W/m-K]
L porous layer length [m]
p pressure [N/m²]
P_1 dimensionless heat transfer coefficient, $haL/\sigma T_i^3$
P_2 dimensionless thermal conductivity, $k_s/\sigma T_i^3 L$
q^r radiant heat flux [W/m²]
q_c reaction enthalpy flux [W/m²]
Q_c dimensionless enthalpy flux, $q_c/\sigma T_i^4$
Q^r dimensionless radiant heat flux, $q^r/\sigma T_i^4$
r reflectivity
R universal gas constant [J/mole-K]
S dimensionless flame speed, $u(0)/u_{ad}$
T temperature [K]
u axial velocity [m/s]
V diffusion velocity [m/s]
W molecular weight [kg/mole]
$\bar{W}$ average molecular weight [kg/mole]
x axial coordinate [m]
Y mass fraction

Greek Symbols

ϵ emissivity
η_o output radiative efficiency, $q^r(L)/q_c$
η_t total radiative efficiency, $\{q^r(L)-q^r(0)\}/q_c$
θ dimensionless temperature, $(T-T_i)/T_i$
μ direction cosine in Eqn. 5
ξ dimensionless axial coordinate, x/L
ρ gas density [kg/m³]
σ Stefan Boltzmann constant [W/m²-K⁴]
σ_a, σ_s, σ_e absorption, scattering and extinction coefficients [1/m]
τ optical depth, $\sigma_e L$
ϕ porosity
$\dot{\omega}$ mass generation rate [moles/m³-s]
ω scattering albedo, σ_s/σ_e

Superscripts

$+$ positive direction
$-$ negative direction
r radiation

Subscripts

a apparent
ad adiabatic

b blackbody
c combustion
d diffuse, downstream
e exit plane
g gas
i inlet plane
k index for species
s solid
u upstream

INTRODUCTION

A promising development in gas-burner technology is the concept of porous radiant burners (PRB). These devices operate by stabilizing a premixed flame inside or near a non-combustible porous medium. The enthalpy of combustion released in the gas phase heats the porous matrix which then emits thermal radiation to a heat load. In small-scale applications PRB have already shown performance gains over conventional open-flame burners in the form of higher efficiencies, lower NO_x emissions, and more uniform heating (GRI, 1984). In order to exploit the potential benefits of this technology for the myriad of industrial heating applications more detailed information about the factors controlling burner performance is needed. The goal of the present work is to provide a better understanding of the link between combustion and heat transfer in PRB and the influence of these processes on the radiant output and burner efficiency.

Recently, Tong and Sathe (1988) performed a detailed study of the heat transfer characteristics of PRB. Combustion was represented as a heat generation zone inside the porous medium. It was shown that the properties of the solid matrix such as the thermal conductivity, the optical depth, and the scattering albedo significantly affected the radiative output from the burner. More realistic combustion models were employed by Yoshizawa *et al.* (1988) and Chen *et al.* (1987) to reveal the structure of radiation-controlled flames in inert porous media. Both studies, however, neglected the important effect of radiation scattering in their analysis and fixed the flame zone at or near the middle of the porous medium. Their results demonstrated the significance of radiative heat transfer in promoting energy feedback ahead of the reaction zone yielding higher than adiabatic flame speeds. Sathe *et al.* (1989) studied numerically the effects of multimode heat transfer and combustion on the flame stability in a finite porous layer. It was shown that stable combustion could be maintained in two different spatial domains. The radiative characteristics of the porous matrix such as the optical depth and scattering albedo were also shown to have a considerable effect on flame stability. In some related work DesJardin and Kendall (1987) investigated the effect of stoichiometry on the liftoff behavior and radiant output of a radiant surface burner using a simple surface model to account for the radiant heat flux emanating from a porous burner.

The aim of the present study is to investigate the effect of combustion and multimode heat transfer on the radiative output and efficiency of PRB. The study is based on a one-dimensional, premixed methane-air flame model wherein the flame thickness is much thinner than the length of the porous medium. Combustion is modeled using a single-step, irreversible reaction. The solid and gas phases are assumed not to be in local thermal equilibrium. And, the solid is assumed to emit, absorb, and scatter radiant energy. The computational domain is extended beyond the porous region on either side to accurately model flames close to the edges of the porous region.

MATHEMATICAL ANALYSIS

The schematic diagram of the problem under consideration is shown in Fig. 1. A premixed methane-air gas mixture enters an adiabatic duct containing an inert porous layer of length L. The flame may be located either inside or outside the porous layer and the flame thickness is much less than L. The flow is assumed to be one-dimensional, steady, and laminar. The viscous effects are neglected so that the momentum equation can be integrated and replaced by the condition that the thermodynamic pressure is a constant, which is the case for most PRBs. The solid matrix is assumed to be gray and to emit, absorb, and scatter radiant energy. Gaseous radiation is neglected compared to solid radiation. Non-local thermal equilibrium between the gas and the solid phase is accounted for by considering separate energy equations for the two phases. All the properties of the solid are assumed to be homogeneous and constant.

The methane-air combustion is modeled using the one-step global reaction (Reitz and Bracco, 1982) for an equivalence ratio of one so that the only species involved are CH_4, O_2, N_2, CO_2 and H_2O. The reaction rate constant is given by $k_c = 1.5\text{x}10^8 \exp(-125600/RT_g)[CH_4]$ moles/m³-s. The thermodynamic and transport properties of the

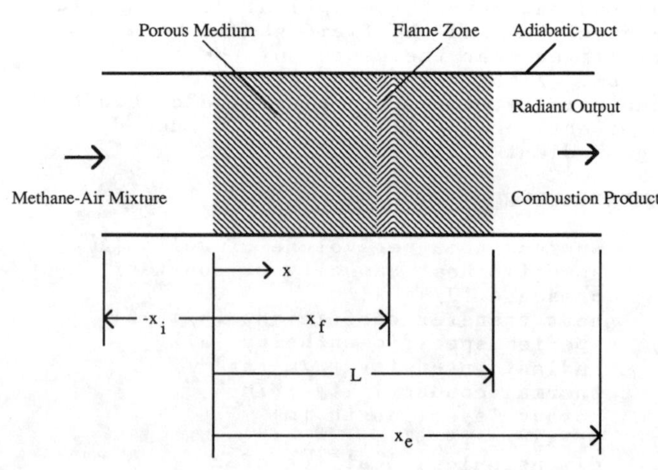

Fig. 1 Schematic diagram of the porous radiant burner.

gases are evaluated using the CHEMKIN (Kee *et al.*, 1980) and TRANFIT (Kee *et al.*, 1986) codes, respectively, from Sandia National Laboratory.

The governing equations are written by incorporating convection, conduction, radiation, and combustion mechanisms. A more detailed description of the heat transfer model may be found in Tong and Sathe (1988). The governing equations are then:

Continuity Equation

$$\frac{d}{dx}(\rho u \phi) = 0, \tag{1}$$

where ϕ is the porosity so that $\phi < 1$ for $0 < x < L$ and $\phi = 1$ elsewhere. All the symbols are identified in the nomenclature.

Species Conservation Equation

$$\rho u \phi \frac{dY_k}{dx} + \frac{d_k}{dx}(\rho \phi Y_k V_k) = \phi \dot{\omega}_k W_k \quad (k = 1, K), \tag{2}$$

where the diffusion velocity V_k is the sum of the diffusion velocities due to mole fraction gradients and the thermal gradients (Kee *et al.*, 1986).

Gas Energy Equation

$$\rho u \phi c_g \frac{dT_g}{dx} + (1-\phi) ha(T_g - T_s) + \phi \sum_{k=1}^{K} \rho Y_k V_k c_k \frac{dT_g}{dx}$$
$$= -\phi \sum_{k=1}^{K} \dot{\omega}_k h_k W_k + \frac{d}{dx}(\phi k_g \frac{dT_g}{dx}) \tag{3}$$

Solid Energy Conservation

$$\frac{d}{dx}[(1-\phi) k_s \frac{dT_s}{dx}] + (1-\phi) ha(T_g - T_s) = \frac{dq^r}{dx} \tag{4}$$

Radiative Transfer Equation (RTE)

$$\mu \frac{\partial i(x,\mu)}{\partial x} + (\sigma_a + \sigma_s) i(x,\mu) = \sigma_a i_b(T_s)$$
$$+ \frac{\sigma_s}{2} \int_{-1}^{1} i(x,\mu') d\mu' \tag{5}$$

where μ' is the dummy variable of integration.

Equation of State

$$\rho = \frac{\overline{W}p}{RT_g} \tag{6}$$

Boundary Conditions

The gas temperature and the species concentrations are specified at the inlet section ($x = -x_i$), whereas vanishing gradients are imposed at the exhaust section ($x = x_e$)

$$Y_k = Y_{k,i}, \; T_g = T_i \; \text{at} \; x = -x_i$$
$$\frac{dY_k}{dx} = \frac{dT_g}{dx} = 0 \; \text{at} \; x = x_e. \tag{7}$$

The boundary conditions for the solid temperature are written assuming that the solid loses heat convectively to the gas (Tong and Sathe, 1988) so that

$$-k_s \frac{dT_s}{dx} = h(T_g - T_s) \; \text{at} \; x = 0, \text{ and}$$
$$k_s \frac{dT_s}{dx} = h(T_g - T_s) \; \text{at} \; x = L. \tag{8}$$

The upstream boundary condition for the radiant intensity is written assuming that the upstream environment can be characterized by a gray diffusely emitting and reflecting surface. On the downstream side, the burner is assumed to be seeing a black environment at T_b = 298 K. Thus, we have

$$i^+(0) = \epsilon i_b(T_i) + 2r_d \int_0^1 i^-(0,-\mu')\mu' d\mu'$$
$$\text{at } x = 0, \text{ and} \tag{9}$$
$$i^-(L) = i_b(T_b) \; \text{at} \; x = L.$$

The radiative heat flux appearing in the solid energy equation is given by

$$q^r = 2\pi \int_{-1}^{1} i(x,\mu')\mu' d\mu'. \tag{10}$$

Upon non-dimensionalizing the above equations, the dimensionless quantities relevant to this study are: $\xi = x/L$, $\theta = (T-T_i)/T_i$, $P_1 = haL/\sigma T_i^3$, $P_2 = k_s/\sigma T_i^3 L$, $\omega = \sigma_s/\sigma_e$, and $\tau = \sigma_e L$ where $\sigma_e = \sigma_a + \sigma_s$. The symbols have been described in the nomenclature.

SOLUTION PROCEDURE

The radiative transfer equation (RTE) was transformed into a set of ordinary differential equations using the spherical harmonics approximation, and the Marshak's formulation was used to model the boundary conditions (Davison, 1957; Ozisik, 1973). The reader is referred to these references for the details on the spherical harmonic method. An efficient numerical scheme was devised to solve the governing equations by modifying PREMIX (Kee *et al.*, 1986). The PREMIX code has been applied successfully to solve the governing equations pertaining to flame propagation in premixed gaseous systems. The code employs finite-difference techniques with adaptive gridding and solves the set of nonlinear governing equations by the damped Newton-Raphson method. The principal modification to the code was the addition of the solid energy equation and the differential equations resulting from the spherical harmonic approximation for the RTE. As noted earlier, the computational domain was extended on both the upstream and downstream sides of the porous layer in order to accurately model flames situated close to the edge of the porous medium. The flame location was fixed numerically by setting the gas temperature at a particular point in the domain, and the governing equations were then solved so that the flame speed was an eigenvalue. A word should be added about the spherical harmonics approximation. Considerable convergence difficulties were observed when the order of the spherical harmonics method was increased and only results for the P-3 approximation could be obtained. Nevertheless, a separate study using the earlier code (Tong and Sathe, 1988) revealed that for optical thicknesses greater than or equal to one, the difference in the radiative outputs obtained by the P-3 and the P-11 approximation did not change by more than one percent.

The governing equations were solved using the IBM 3090 computer. The time required to solve for one set of parameters ranged widely from approximately 100-500 cpu seconds, depending on the initial guess provided. Since the Newton-Raphson method was used, very good initial guesses were required to obtain convergence, otherwise divergence resulted. A relative convergence criterion of 10^{-4} was specified for the Newton's iteration. Computations were continued on successively finer grids until the flame speed changed by no more than 0.1 percent. An overall energy balance of approximately 0.5 percent was achieved for most of the cases, with an energy balance of less than 1 percent in the worst case. Approximately 100 grid points were required. Reducing the grid refinement criterion had a negligible effect on the results but required more computer time. Also, care was taken to keep the computational domain large enough so that flame speed was not affected by the length of the computational domain (Smooke *et al.*, 1983). The accuracy of the numerical code was checked by comparing the results with laminar flame speed and adiabatic flame temperature data of Reitz and Bracco (1982) by letting the porosity be close to one.

RESULTS AND DISCUSSION

The above model is used to calculate results for a wide range of properties of the porous layer that may be encountered in PRB. In particular, the effect of the extinction coefficient (σ_e), scattering coefficient (σ_s), solid thermal conductivity (k_s), heat transfer coupling (ha), and the upstream environment reflectivity (r_d) on the flame speeds, radiant outputs, and radiant efficiencies are investigated. The porous layer is assumed to be 5 centimeters long and have a porosity (ϕ) of 0.95. A stoichiometric fuel-air mixture is allowed to enter the duct with a uniform velocity and inlet temperature, T_i = 298 K. Variations were made in the following baseline property values to produce the results presented below: $\sigma_e = 20\ m^{-1}$, $\sigma_s = 10\ m^{-1}$, k_s = 10 W/m-K, ha = 2.0×10^9 W/m³-K, and $r_d = 0$. The thermal conductivity of the solid, k_s was intended to be that for non-metallic solids, such as alumina and silica encountered in PRBs. The value of ha was estimated using the experimental data for h for cylinders in cross flow and a was computed assuming infinitely long fibers of 10 μm diameter (Tong and Sathe, 1988). The corresponding relevant non-dimensional parameters are: $\tau = 1$, $\omega = 0.5$, $P_1 = 8.35 \times 10^7$, $P_2 = 6.67$.

Figure 2 presents the computed burning velocities for different flame locations, ξ_f, obtained by varying the extinction coefficient ($\sigma_e = \sigma_a + \sigma_s$) from 20 m^{-1} to 200 m^{-1} so that the total optical thickness of the layer ($\tau = \sigma_e L$) ranges from 1.0 to 10.0 while the scattering albedo ($\omega = \sigma_s/\sigma_e$) is kept constant at 0.5. The flame location is defined as the position where the reaction rate is a maximum. The upstream (inlet) and downstream (outlet) edges of the porous medium are denoted by vertical dashed lines at $\xi = 0$ and $\xi = 1$, respectively. The flame speeds are normalized by the adiabatic flame speed which prevails a few millimeters beyond either side of the porous layer. It was revealed in an earlier study (Sathe *et al.*, 1989) that the flame speed variation near the edge of the porous medium is controlled by conduction; whereas, radiation heat transfer becomes a contributing factor for flame propagation in the interior. When the flame approaches the porous medium from the downstream side, the flame speed increases abruptly because of the higher thermal

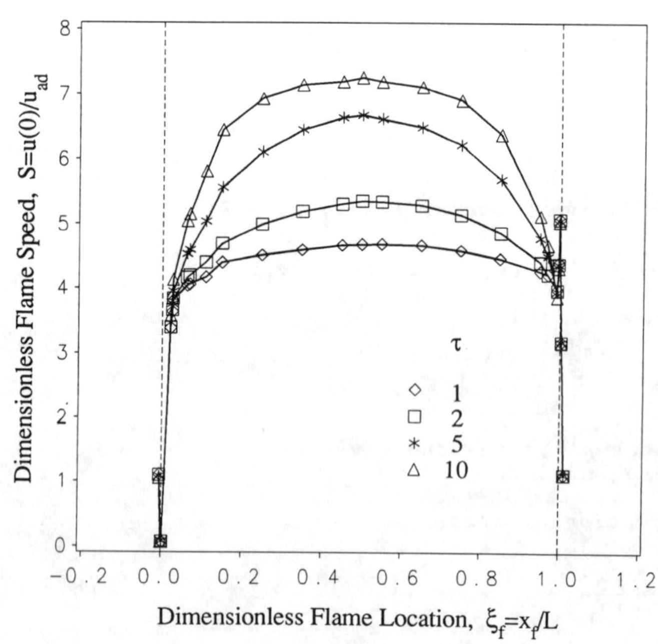

Fig. 2 Effect of the burner optical thickness on the flame speed for different flame locations.

conductivity encountered in the preheat zone until a local maximum is observed at the edge of the porous medium. As the flame enters the porous region the burning velocity is reduced because of increasing conduction and radiation losses downstream. The radiation loss accounts for the local minimum in flame speed that is lower for higher optical depths, τ. As the flame is shifted toward the center of the porous region, the lengthening amount of material radiating in the post-flame zone directs the radiation upstream causing the flame temperature and speed to increase. Beyond the symmetry axis this radiant energy feedback escapes across the inlet plane producing lower flame speeds. A broad maximum in the flame speed near the center of the porous layer was also observed by Yoshizawa *et al.* (1988). As the flame approaches the upstream edge, the flame speed drops drastically due to increased radiative loss and the preferential distribution of the conductive flux in the downstream direction due to greater resistance to upstream conduction. After this minimum the flame speed returns to the adiabatic value as there is no radiation loss from the flame and the conductivity value is that of the fuel-air mixture. Because the flame location is important from the standpoint of radiant energy exchange and determining the maximum allowable heat release, it should be emphasized that only those portions of the velocity profile with positive slope would be expected to maintain stable combustion.

The effect of the optical thickness on combustion-heat transfer behavior deep in the porous region is illustrated in Fig. 3. It is evident that the reactant enthalpy flux Q_c (proportional to the flame speed) increases as τ is increased from 1 to about 10 due to enhanced radiation feedback across the flame in the upstream direction. This explanation is substantiated by the radiative energy fluxes at the upstream and downstream edge of the flame zone. The quantities $Q^r{}_u$ and $Q^r{}_d$ in the figure are the radiant fluxes on the upstream and downstream edge of the flame zone. The edge of the flame zone is taken to be the location where the reaction rate is 1/1000 of its peak value. However, as τ is increased further beyond 10, the diminishing radiative energy feedback across the flame zone lowers the reactant enthalpy flux. This trend is consistent with results obtained by Chen *et al.* (1987). The output radiative efficiency, defined as the ratio of the radiant output (radiant heat flux at $\xi = x/L = 1$) and the reactant enthalpy flux, reaches a maximum at about $\tau \approx 5$ and then decreases very gradually.

Shown in Fig. 4 are the upstream and downstream radiative fluxes (at $\xi = x/L = 0$ and $\xi = x/L = 1$, respectively) resulting from different flame locations and optical thicknesses. The upstream and downstream radiative fluxes depend on the heat generation rate, which is proportional to the flame speed, and the optical depths between the flame and the edges of the porous medium. For a fixed τ the upstream radiative flux increases as the flame moves upstream from the outlet plane due to an increase in flame speed and because there is less shielding between the flame and the upstream edge. The heat flux continues to increase due to the

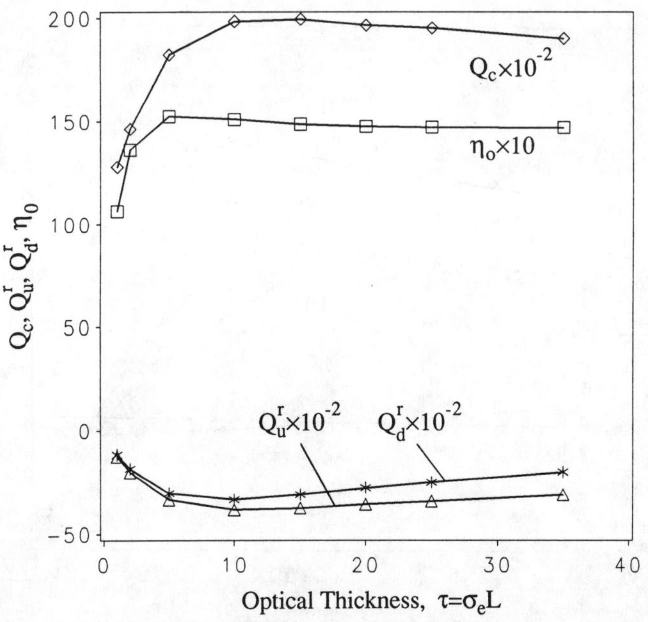

Fig. 3 Effect of the optical depth on the radiation feedback, flame speed, and radiative output efficiency for $x_f/L = 0.5$.

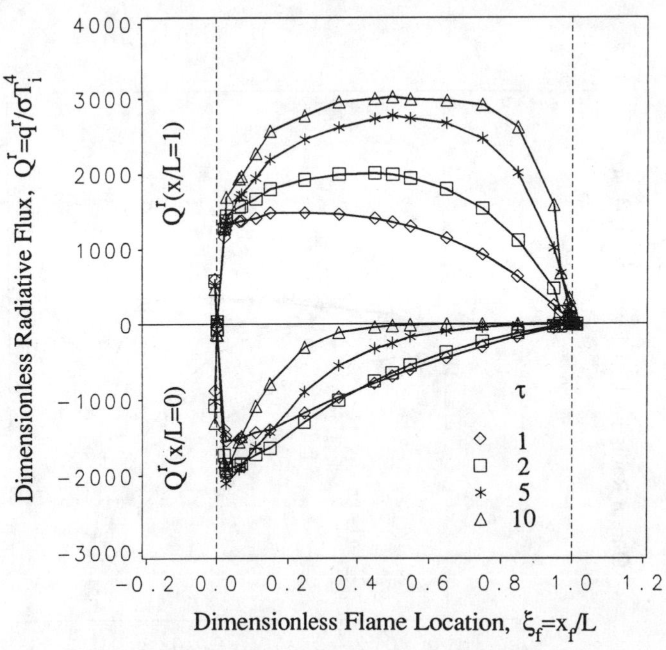

Fig. 4 Radiant output (downstream radiative flux) and upstream radiative flux for varying optical thicknesses and flame locations.

reduced shielding until the sharp drop in flame speed near the inlet occurs. When the flame is near the downstream edge, the shielding effect dominates, so that the upstream radiative flux is higher for lower τ. However, when the flame is outside the porous layer and near the upstream edge, there is no shielding effect involved and the upstream radiant flux is largest for the largest τ due to greater emission. Similarly, the downstream radiative flux increases when the flame is moved upstream due to an increase in the flame speed and due to an increase in the extent of porous material at elevated temperatures. As the flame moves further upstream, the radiative output decreases eventually because the reduction in flame speed and shielding on the downstream side outweighs the increase in the amount of high-temperature porous material. The shielding effect is more important for larger optical thicknesses; hence, the peak in the radiant output occurs at more downstream positions. For the flame located upstream of the porous region, due to the shielding effect, the radiant output is larger for a smaller τ. These results are consistent with those of Tong and Sathe (1988) where the peak in the radiant output shifted downstream with increasing τ even when the flow speed was held constant.

The radiative output efficiencies for different flame locations and optical depths are shown in Fig. 5. The efficiencies are nearly zero when the flames are at the outlet plane and increase quickly at higher τ to values ranging from 12-15 percent as the flame is moved upstream. It is interesting to note that the highest efficiencies are achieved when the flame is located outside the porous medium near the upstream edge and that the efficiency at this position increases with decreasing optical depth. However, from the recent study of Sathe *et al.* (1989) the flame outside the porous layer near the upstream edge is unstable since the slope of the curve of flame speed versus distance is negative. Negligible increase in efficiencies are observed in the upstream half (which is the stable flame region) of the porous material when τ is increased above 5. These findings suggest that to maximize both the radiant output and efficiency the flame should be located at the center of the porous material, but the maximum radiant output occurs when $\tau \approx 10$ and the maximum radiant output efficiency occurs when $\tau \approx 5$.

Figure 6 shows the effect of the upstream environment (at $\xi = -x_i/L = -\xi_i$) reflectivity and the scattering albedo on the burner characteristics for a flame located at $x_f/L = 0.5$. A reduction in albedo increases the flame speed due to stronger emission from the solid. It must be noted that when there is no emission or absorption by the solid, i.e. when $\omega = 1$, the flame speed exceeds the adiabatic value because the thermal conductivity of the solid matrix is much greater than the gas. Both the radiant output and the efficiency also increase with a decrease in the albedo; hence, a strongly emitting-absorbing medium is preferred to enhance the burner performance. The results demonstrate that scattering by the solid is an important mechanism in the radiative transport and cannot be neglected. Results for flame locations other than at the center of the solid indicated similar trends when

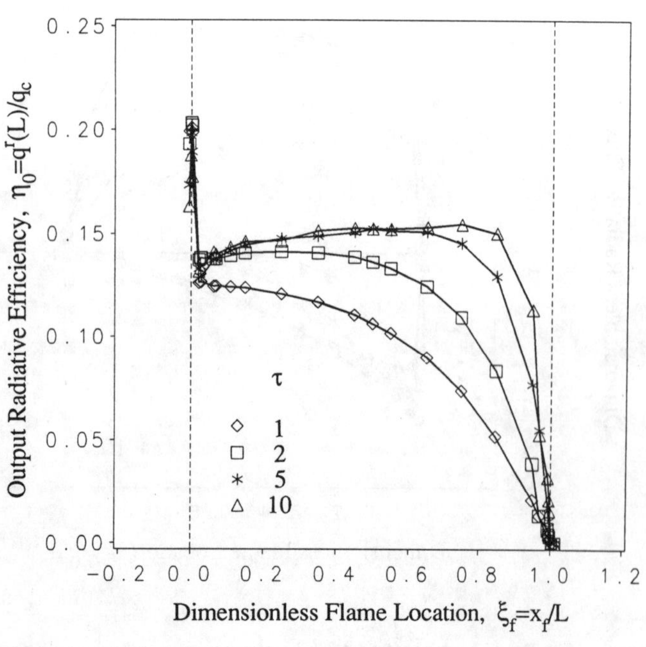

Fig. 5 Variation of the radiative output efficiencies for different optical thicknesses and flame locations.

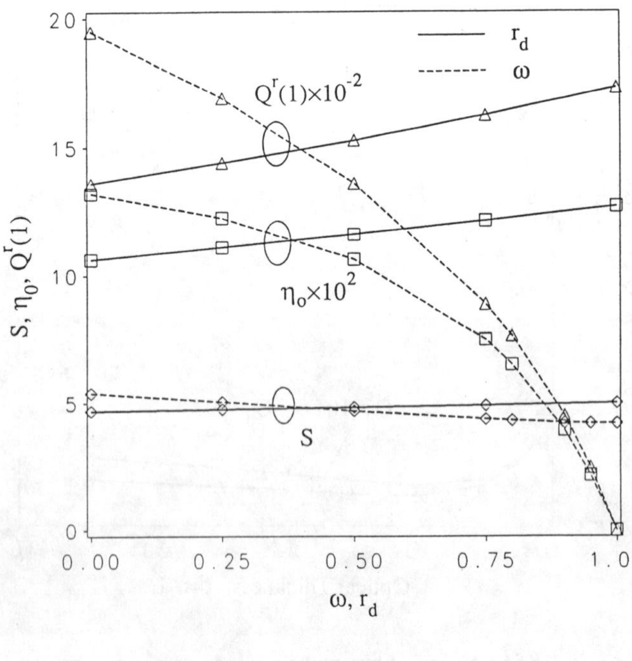

Fig. 6 Burner characteristics for different scattering albedos and upstream reflectivities when $x_f/L = 0.5$.

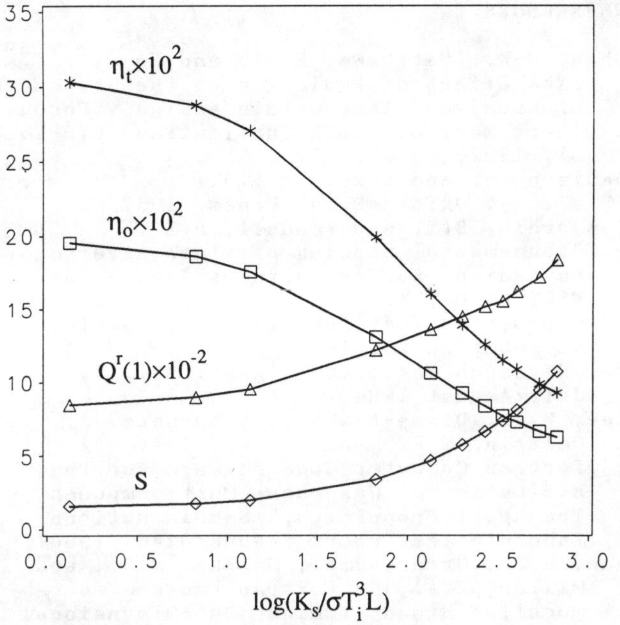

Fig. 7 Effect of the solid thermal conductivity on the burner characteristics when $x_f/L = 0.5$.

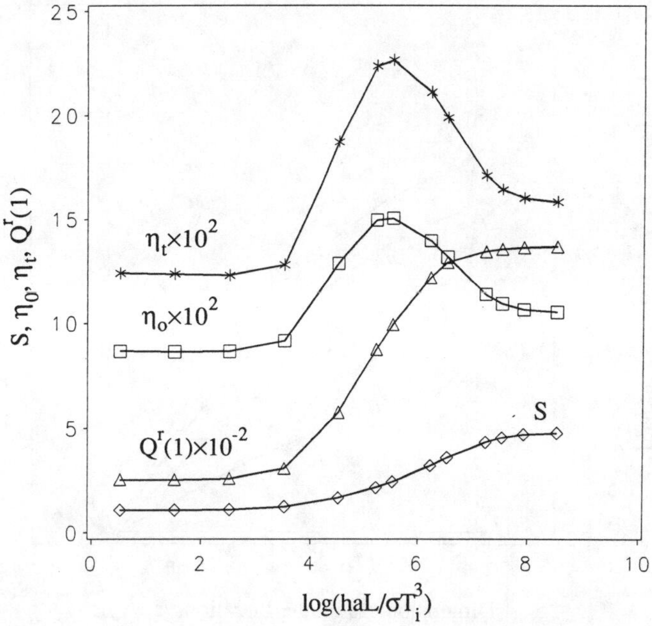

Fig. 8 Effect of the heat transfer coupling between the solid and the gas on the burner characteristics when $x_f/L = 0.5$.

the albedo was varied and are not shown here. An increase in the upstream reflectivity reduces the radiation loss from the flame in the upstream direction thereby raising the flame temperature and accelerating the flame speed. The radiant output increases when the reflectivity is increased because the flame speeds are higher and less radiation escapes in the upstream direction. This latter effect also improves the output efficiency and suggests that a large reflectivity is desired for the best burner performance. These results are consistent with the findings of Tong and Sathe (1988).

Shown in Fig. 7 are the burner characteristics for different thermal conductivities of the solid. An increase in the thermal conductivity increases the flame speed due to an enhancement in the preheating of the incoming reactants. However, the increase in the solid conductivity does not increase the radiant output to the same extent as the flame speed so that the radiative efficiency actually declines. This result is due to the fact that for the same flowrate and heat release rate, an increase in the solid conductivity produces lower post-flame temperatures and higher pre-flame temperatures, thereby reducing the radiative output (Tong and Sathe, 1988). Thus, the radiant output can be increased by increasing the conductivity and concurrently sacrificing on the radiant efficiency.

The effect of the heat transfer coupling between the solid and the gas is reported in Fig. 8. A stronger heat transfer coupling results in a better pre-heating effect when the solid conductivity is higher than that of the gas. Hence, the flame speed increases as the heat transfer coupling is increased.

Once the coupling is large enough so that the gas and solid are in local thermal equilibrium, i.e. $P_1 \geq 8 \times 10^7$, the rise in flame speed becomes negligible. On the other hand, when the heat transfer coupling is weak, very little energy is transferred from the gas to the solid so that the flame speed approaches the adiabatic flame speed. This asymptotic approach to the adiabatic flame speed has served as an additional check of the code. The radiative output increases as the heat transfer coupling is enhanced because of higher solid temperatures, and no appreciable increase occurs once the solid and gas have reached thermal equilibrium. The radiant efficiency is observed to pass through a maximum as the heat transfer coupling is enhanced.

Shown in Fig. 9 is the apparent emissivity of the burner, defined as $\epsilon_a = q^r/\sigma T^4$ at $\xi = 1$. An attempt is being made in the industrial sector to characterize the emissivity of the porous burner as a gauge of its performance (DesJardin and Kendall, 1987). It is seen that as the porous medium becomes strongly emitting and absorbing (albedo becomes smaller), the apparent emissivity increases. This result is due to the fact that as albedo decreases, q^r increases whereas the temperature rise is not appreciable. The emissivity is also a strong function of the flame location. Thus, a unique value of emissivity cannot be assigned to a particular burner unless the exact flame location for a given firing rate is known.

CONCLUSIONS

The effect of combustion and multimode

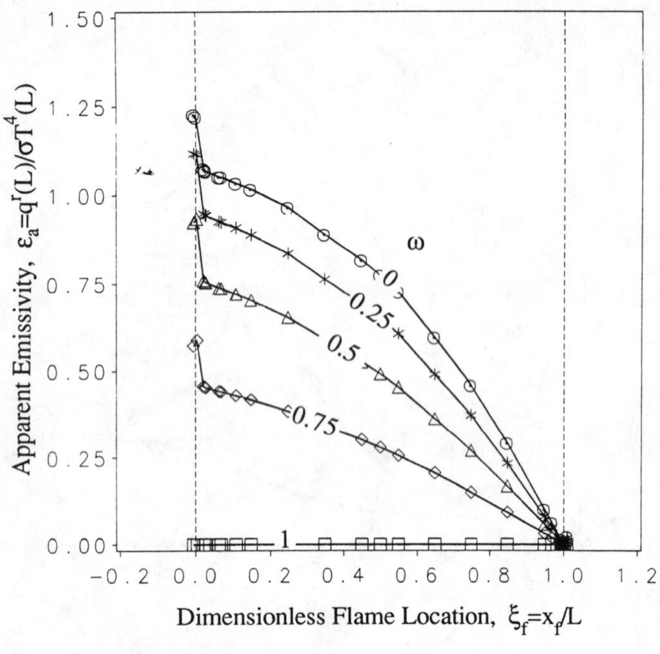

Fig. 9 Apparent emissivity of the burner for different scattering albedos.

heat transfer on the performance characteristics of PRB has been studied numerically. The model incorporated emission, absorption, and scattering of radiant energy by the porous matrix and accounted for non-local thermal equilibrium between the gas and solid phases. The computational domain was extended on either side of the porous medium to accurately model the flames located near the edge of the porous medium. It was revealed that for maximizing the radiant output, the optical depth should be approximately ten with the flame located near the center of the porous medium, and the heat transfer coupling should be high enough to ensure local thermal equilibrium between the gas and the solid. The solid-phase conduction and scattering had a significant effect on the burner performance. Low solid conductivities and scattering albedos are desirable to increase the radiant efficiency.

ACKNOWLEDGMENT

The authors gratefully acknowledge the support of this research provided by the Office of Basic Energy Sciences, U.S. Department of Energy under contract number DE-FG02-87ER13697.

REFERENCES

Chen, Y-K., Matthews, R.D., and Howell, J.R., "The Effect of Radiation on the Structure of Premixed Flame within a Highly Porous Inert Medium," ASME Publication, HTD-Vol. 81, 1987, pp. 35-42.

Davison, B. and Sykes, J., *Neutron Transport Theory*, Oxford Univ. Press, 1957.

DesJardin, S.T. and Kendall, R.M., "Fundamental Studies of Flame Attachment on Radiant Surface Burners," paper no. WSCI 87-81, Western States Section/The Combustion Institute meeting, Honolulu, November 22-25, 1987.

Gas Research Institute Digest, Vol. 7, July/August 1984.

Kee, R.J., Dixon-Lewis, G., Warnatz, J., Coltrin, M.E., and Miller, J.A., "A Fortran Computer Code Package for the Evaluation of Gas-phase Multicomponent Transport Properties," Sandia National Laboratories Report SAND86-8246, 1986.

Kee, R.J., Grcar, J.F., Smooke, M.D., and Miller, J.A., "A Fortran Program for Modeling Steady Laminar One-Dimensional Premixed Flames," Sandia National Laboratories Report SAND85-8240, 1985.

Kee, R.J., Miller, J.A., and Jefferson, T.H., "CHEMKIN: A General-Purpose Problem-Independent, Transportable, Fortran Chemical Kinetics Code Package," Sandia National Laboratories Report SAND80-8003, 1980.

Ozisik, M.N., *Radiative Heat Transfer and Interactions with Conduction and Convection*, Wiley-Interscience, 1973.

Reitz, R.D. and Bracco, F.V., "Toward the Formulation of a Global Local Equilibrium Kinetics Model for Laminar Hydrocarbon Flames," in *Numerical Methods in Laminar Flame Propagation*, N. Peters/J. Warnatz (eds.), Vieweg, 1982, pp. 130-151.

Sathe, S.B., Peck, R.E., and Tong, T.W., "Flame Stabilization and Multimode Heat Transfer in Inert Porous Media: A Numerical Study," paper no. WSCI89-37, Western States Section/The Combustion Institute meeting, Pullman, WA, March 20-21, 1989.

Smooke, M.D., Miller, J.A. and Kee, R.J., "Determination of Adiabatic Flame Speeds by Boundary Value Methods," *Comb. Sci. Techn.*, Vol. 34, 1983, pp. 79-90.

Tong, T.W. and Sathe, S.B., "Heat Transfer Characteristics of Porous Radiant Burners," ASME Publication, HTD-Vol. 104, 1988, pp. 147-155.

Yoshizawa, Y., Sasaki, K. and Echigo, R., "Analytical Study of the Structure of Radiation Controlled Flame," *Int. J. Ht. Mass Transf.*, Vol. 31, 1988, pp. 311-319.

1989 National Heat Transfer Conference
HTD-Vol. 106, Heat Transfer Phenomena in Radiation, Combustion and Fires

MATHEMATICAL MODELING OF INTERACTING VAPORIZING FUEL DROPLETS

C. Kleinstreuer, H. Chiang and T.-Y. Wang
Department of Mechanical and Aerospace Engineering
North Carolina State University
Raleigh, North Carolina

ABSTRACT

A basic understanding of the effects of interacting vaporizing droplets on the transfer processes in dense-dispersed flows are of importance for the improved design of fuel droplet combustion and other spray systems. After a brief review of basic modeling approaches suitable for dense (combustion) sprays, model developments and/or preliminary results are presented for a finite element microscale analysis of several spheres/droplets in a linear array and for a boundary-layer analysis of dynamically interacting vaporizing droplets on a one-dimensional trajectory. Both computer simulation models have been validated with experimental data sets. Results discussed include: individual drag coefficient and local Nusselt number correlations for three interacting spheres/droplets (finite element solution) as well as transient changes of the individual droplet radii, Reynolds numbers, vaporization rates and inter-droplet distances for three interacting vaporizing droplets (approximate solution).

NOMENCLATURE

A	surface area of drop
C_D	drag coefficient
d	diameter
D	diffusion coefficient
h	enthalpy of the gaseous mixture
k	thermal conductivity
L	latent heat of vaporization
$\overline{Nu}$	spatially averaged Nusselt number
p	pressure
Pr	Prandtl number
q	heat flux
R	radius of the drop
Re_{do}	initial Reynolds number; $Re_{do} = wd_o/\nu_\infty$
r	distance from axis of symmetry to the drop surface
T	temperature
t	time
u	velocity component in x-direction
v	velocity component in y-direction
w	drop velocity
x	streamwise coordinate measured along surface from forward stagnation point
Y	vapor mass fraction
y	coordinate normal to surface
z	axial coordinate

Greek Symbols

α	thermal diffusivity
ρ	density
λ	gas constant
μ	absolute viscosity
ν	kinematic viscosity
θ	dimensionless temperature (gas-phase), angle
θ_s	angle of flow separation measured from forward stagnation point
τ	dimensionless time; $\tau_g = 4t\nu_g/d^2$ whereas $\tau_\ell = \alpha_\ell t/R_o^2$

Subscripts

b	boiling condition
e	edge condition
g	gas-phase condition
i	interface condition
j	index for particle; j^{th} sphere/droplet
o	initial condition
p	particle, sphere/droplet
∞	ambient condition
ℓ	liquid-phase condition
s	single sphere

I. INTRODUCTION

A basic understanding of the physics of interacting spherical particles is of importance for the improved design of a number of dispersed flow systems. Examples include: fuel droplet vaporization, particle suspension flows, flue gas desulfurization, spray cooling, spray coating, and biosprays. Of special interest are the coupled droplet dynamics, heat transfer, and mass transfer in combustion sprays.

In general, dense sprays are characterized by dynamic droplet and droplet streamer interactions, Reynolds numbers of O(100), initial liquid phase heating, possible internal droplet circulation, rapid droplet vaporization, turbulence effects, and strong coupling between the liquid-phase and the gas-phase quantities. This is followed by a dilute spray region where droplet interactions are negligible, but the other coupled effects somewhat prevail. Additional complexities may arise from swirl flow, droplet collisions, wall effects, soot formation, trace gas absorption with chemical reactions, solid-slurry fuel droplets and multicomponent fuels as well as secondary atomization, oscillating droplets, non-spherical vaporization, and random motion. Thus, it is necessary to accommodate large changes in time and length scales during droplet life times, and to consider simultaneous momentum, heat and mass transfer in the continuous gas coupled with the dynamics of discrete interacting solid or liquid particles. Most of these aspects eluded in the past theoretical and experimental analysis on a microscale (cf. Brauner & Schmidt-Laine, 1988; Oran & Boris, 1987; Faeth, 1987; Kuo, 1986; Bracco, 1985; Law, 1984; Sirignano, 1983; Faeth, 1983; and Chigier, 1983). A detailed and accurate simulation of some of the coupled transfer processes for three interacting vaporizing fuel droplets is presented here, and the results may contribute towards improved global spray modeling and ultimately better designs of combustion systems.

Traditionally, dispersed flows, in particular dilute (droplet) sprays, are modeled employing a locally homogeneous flow (LHF) method or a separated flow analysis (cf. Kuo, 1986 or Faeth, 1987). While the LHF method concentrates on the effect of mixing like in a single phase, deterministic or stochastic separated flow analyses highlight finite interphase transport rates (DSF method) or dispersed phase/turbulence interactions (SSF method). In any case, (the) coupled gas-phase transport equations for the mean velocity field, mixture fraction, and turbulence properties ($k - \varepsilon - g$) have to be solved and empirical correlations, state relationships, probability density functions and additional source terms have to be known, in order to describe dispersed jets, particle-laden flows or combustion sprays. Faeth (1987) suggested that the computationally efficient LHF melthod may be better suited for modeling global dense spray characteristics. O'Rourke & Bracco (1980) were some of the first who treated interacting droplets using a statistical model to study pulsating combustion sprays.

An alternative approach, employing the cylindrical cell model proposed by Tal & Sirignano (1982), has been carried out by Tong & Chen (1988). They found average Nusselt number correlations for three spheres with and without blowing (or surface velocity) in a cylindrical hot gas stream for $10 \le Re_d \le 150$ using a finite difference method with a hybrid or upwind scheme for the convective terms.

In contrast to depicting whole sprays or at least (linear) arrays of droplets, numerous theoretical/experimental studies have focused on a single vaporizing droplet with the ultimate goal of representing an entire dispersed-flow system by one equivalent droplet in a hot gas stream (cf. Abramizon & Sirignano, 1988; Prakash & Krishna, 1985; Fernandez-Pello, 1983; Aggarwal et al., 1984; and Prakash & Sirignano, 1980). Needless to say, such approximate single droplet solutions, although interesting, are too idealized to contribute to the understanding of complicated interaction effects and hence may not be useful for predictive design applications of practical sprays. Recently, the dynamics of two and three interacting vaporizing droplets have been published by Raju & Sirignano (1987) and by Kleinstreuer et al. (1988), respectively. Raju & Sirignano (1987) employed a finite difference method developed by Dwyer & Sanders (1984) and a grid-transformation scheme developed by Thompson (1976) to solve the full transport equations for laminar axisymmetric flow ($Re_d \le 100$) past two vaporizing droplets in tandem. They showed that interaction effects are very significant in the evaluation of such systems. Ramachandran et al. (1989a,b) developed a finite element microscale analysis for thermal axisymmetric flow past closely-spaced spheres/droplets. Their results, briefly discussed in Section 4.1, form the bases for effective drag/interaction coefficients and a thermal near-wake model used in the present boundary-layer analysis.

Experimental results up to 1976, typically for single drops or spheres, are summarized in Clift et al. (1978). They provide graphs and correlations for the separation angle, the dimensionless wake length, the dimensionless wake volume and the Strouhal number versus Reynolds number for flow around a solitary sphere. One of the earliest experimental two-sphere investigations was done by Rowe & Henwood (1961) providing individual drag coefficients. More recently, Tsuji et al. (1982) used condensed milk to visualize the longitudinal and transverse flow patterns occuring from the interaction of two closely-spaced spheres in a moderate to high Reynolds number stream. They published individual drag coefficients, flow separation angles, and oscillating wake schematics for different sphere spacings and relatively high free-stream Reynolds numbers.

Complementary to these somewhat detailed two-sphere flow measurements, numerous global experiments with single droplet streamers can be found in the combustion literature (e.g., Sangiovanni & Kesten, 1977; Sangiovanni & Labowsky, 1982; and Rah et al., 1986). Here, the overall droplet evaporation rate and burning time for different atomizer conditions (initial velocity, frequency, size distribution, etc.) are of interest rather than the microstructure of interacting vaporizing droplets. Nevertheless, these measurements provide valuable data points for comparison with end results of multiple droplet-streamer studies.

2. ANALYSIS

2.1 Governing Equations and Boundary Conditions

The general conservation equations for total mass, momentum, enthalpy and mass of species j in terms of fluxes (written for an inertial coordinate frame) are (cf. Kleinstreuer 1990):

$$\frac{\partial \rho}{\partial t} + \mathrm{div}(\rho\vec{v}) = 0 \tag{1}$$

$$\partial(\rho\vec{v})/\partial t + \mathrm{div}(\rho\vec{v}\vec{v}) = -\mathrm{grad}\,(p) + \mathrm{div}(\vec{\vec{\tau}}) + \sum_{j=1}^{n} \rho_j \, \phi_j \tag{2}$$

$$\partial(\rho h)/\partial t + \mathrm{div}(\rho\vec{v}h) = -\,\mathrm{div}(\vec{j}_h) + S_h \tag{3}$$

and

$$\partial \rho_j/\partial t + \mathrm{div}(\rho_j \vec{v}) = -\,\mathrm{div}(\vec{j}_{m,j}) + S_{m,j}\ ;$$

$$j = 1,\ 2,\ \ldots(n-1) \qquad (4)$$

where $h = cT + \Sigma(\rho_j h_j)$ and $\vec{\Phi}_j$ is the external force acting on species j. The stress tensor $\bar{\bar{\tau}}$ and the flux vectors $\vec{j}_h$ and $\vec{j}_{m,j}$ have to be specified for a given case study. A finite element solution of the specific form of equations (1) to (4) describing laminar axisymmetric flow past three closely-spaced spheres is discussed in Sections 2.2 and 4.1. For the approximate analysis of multiple droplets injected into a hot gaseous ambient, Newton's second-law of motion will be used to track dynamically interacting droplets of mass m_p, viz:

$$d(m_p\vec{v})_k/dt = \sum_n \vec{F}_n^k\ ; \qquad k = 1,\ 2,\ \ldots \qquad (5)$$

where the external forces acting on particle/drop k include drag, virtual mass, interaction, etc. For example, the time-dependent drag/interaction force is computed from the solution of equation (2) which, in turn, depends on (1), (3), (4), (5) and the interfacial conditions discussed below.

At gas-liquid interfaces, convective diffusive mass fluxes and heat fluxes have to be conserved. In addition, new "free surface" boundary conditions have to be invoked in order to calculate the changing shape of (the) axisymmetric droplets. Thus,

$$\text{(mass conservation)}\ (\rho v)_i\ (Y_i - 1) = \left(\rho D\ \frac{\partial Y}{\partial y}\right)_i \qquad (6)$$

and

$$\text{(energy conservation)}\ (\rho v)_i\ AL + \left(k_\ell\ A\ \frac{\partial T_\ell}{\partial r}\right)_i$$

$$= \left(k_g\ A\ \frac{\partial T}{\partial y}\right)_i \qquad (7)$$

The surface temperature T_i and the vapor mass fraction Y_i are related via the Clausius-Clapeyron equation.

$$Y_i = \exp\left[\frac{L}{\lambda}\left(\frac{1}{T_b} - \frac{1}{T_i}\right)\right] \qquad (8)$$

One of the most important (unknown) variables is the "blowing" velocity $v_i = v_i[R(t),\ \theta,\ t]$ which couples all transport equations and boundary conditions, including the receding droplet surface.

2.2 Microscale Analysis

In order to obtain detailed and accurate results for thermal flow past interacting particles in a linear array, equations (1) through (4) for laminar axisymmetric flow are solved for the gas-phase. The related dimensionalized equations, boundary and compatibility conditions, and finite element solution method are discussed in Ramachandran et al. (1989a,b). Comparisons with accepted data sets and submodels developed from the microscale analysis are given in Sections 2.3 and 4.1.

2.3 Approximate Analysis

In order to extend the complex microscale study (Sect. 2.2) to dynamically interacting vaporizing droplets and to have a simplified model for use in dense spray investigations, a boundary-layer analysis using reduced forms of equations (1) through (4), coupled with equation (5), is proposed. The key assumptions for the approximate analysis are laminar axisymmetric boundary-layer flow, spherical droplets, constant liquid-phase properties, single component fluid, and non-collisional droplet interactions. As indicated in Section 1, these are very restrictive assumptions when compared with actual dense sprays. However, some effects, such as internal droplet circulation and free convection, may be negligible in practicle spray applications and other assumptions, such as laminar boundary layers and droplet sphericity are quite realistic. Furthermore, the present model can be readily extended to include multiple droplet-streamer interactions, trace gas uptake and chemical reactions.

Considering n spherical droplets on a one-dimensional trajectory, the dynamics of the jth droplet is described by Newton's second law of motion (cf. Eq. (5) and Fig. 1):

$$\frac{dw_j}{dt} = -\frac{3}{4}\,\frac{C_{D,j}}{d}\,\frac{\rho_a}{\rho_p}\,|\,w_j\,|\ w_j;\ j = 1,\ 2,\ \ldots,\ n \qquad (9)$$

subject to

$$z_j = z_{j,o}\ \text{ and }\ w_j = w_{j,o}\ \text{ at } t = 0. \qquad (10)$$

With n = 3, the dimensionless droplet distance is at any time step

$$d_{j,j+1} = (z_j - z_{j+1})\ /\ d\ ; \qquad j = 1,\ 2 \qquad (11)$$

where the velocity of the jth droplet is

$$w_j = \frac{dz_j}{dt} \qquad (12)$$

The $C_{D,j}$'s represent the effective drag forces on the jth particle due to its motion and the presence of the other (two) particles (cf. Sect. 2.2 and Ramachandran et al., 1989a). Specifically, for $10 \le Re_d \le 200$ and $2 \le d_{ij} \le 6$,

$$\frac{C_{D,1}}{(C_{Ds})_1} = 1 - 0.096\ (Re_d)_1^{\,0.2475}\ d_{12}^{\,-0.965}$$

$$\cdot\ e^{0.4764/d_{23}} \qquad (13)$$

$$\frac{C_{D,2}}{(C_{Ds})_2} = 1 - (Re_d)_2^{\,0.1593}\ \Big(0.2932\ d_{12}^{\,-0.4876}$$

$$+\ 0.1341\ d_{23}^{\,-0.4242}\Big) \qquad (14)$$

and

$$\frac{C_{D,3}}{(C_{Ds})_3} = 1 - 0.325\ [\ln\ (Re_d)_3 + 1]^{0.603}$$

$$d_{23}^{-0.385}\ e^{-0.282/d_{12}} \tag{15}$$

where the standard drag correlation for the solitary droplet is (White, 1974):

$$(C_{Ds})_j = \frac{24}{(Re_d)_j} + \frac{6}{1 + \sqrt{(Re_d)_j}} + 0.4$$

The effective approach stream temperature is taken to be

$$T_{\infty,k+1} = 0.2\ T_{i,k} + 0.8\ T_{\infty,k}\ ; \quad k = 1,\ 2 \tag{16}$$

based on the work by Ramachandran et al. (1989b).

The gas-phase fluid mechanics, liquid-phase heating and droplet vaporization is described by (cf. Wang & Kleinstreuer, 1989)

$$\text{(Continuity)} \quad \frac{\partial(r\rho u)}{\partial x} + \frac{\partial(r\rho v)}{\partial y} = 0 \tag{17}$$

$$\text{(Momentum)} \quad \rho\left(u\frac{\partial u}{\partial x} + v\frac{\partial u}{\partial y}\right) = \rho_e u_e \frac{du_e}{dx} + \frac{\partial}{\partial y}\left(\mu\frac{\partial u}{\partial y}\right) \tag{18}$$

$$\text{(Energy)} \quad \rho\left(u\frac{\partial h}{\partial x} + v\frac{\partial h}{\partial y}\right) = \frac{\partial}{\partial y}\left(k\frac{\partial T}{\partial y}\right) \tag{19}$$

$$\text{(Species)} \quad \rho\left(u\frac{\partial Y}{\partial x} + v\frac{\partial Y}{\partial y}\right) = \frac{\partial}{\partial y}\left(\rho D\frac{\partial Y}{\partial y}\right) \tag{20}$$

$$\text{(Conduction)} \quad \frac{\partial T_\ell}{\partial t} = \frac{\alpha_\ell}{r^2}\frac{\partial}{\partial r}\left(r^2\frac{\partial T_\ell}{dr}\right) \tag{21}$$

The compatibility conditions at the gas-liquid interface are given as equations (6) and (7).

The associated gas-phase boundary conditions are

$$\underline{\text{at } y = 0:} \quad u = 0,\ v = v_i,\ T = T_i,\ Y = Y_i \tag{22}$$

and

$$\underline{\text{at } y \to \infty:} \quad u = u_e(x),\ T = T_\infty,\ Y = Y_\infty = 0 \tag{23}$$

The liquid phase conditions are

$$T_\ell(t = 0) = T_o \quad \text{for all } r \tag{24}$$

and

$$\left.\frac{\partial T_\ell}{\partial r}\right|_{r=0} = 0\ , \quad \left.\frac{\partial T_\ell}{\partial r}\right|_{r=R(t)} \sim q_i \tag{25}$$

The mass fraction at the interface is a function of the droplet surface temperature following equation (8). The time rate of change of the droplet radius can be expressed as

$$\frac{d}{dt}R(t) = \frac{1}{2\rho_\ell}\int_0^{\theta_s} (\rho v)_i\ \sin\theta\ d\theta \tag{26}$$

The transformation and solution of equations (17) - (21) subject to (22) - (25) with equations (6) - (8) are given in Wang & Kleinstreuer (1989). Their experimentally verified computer code appears as a subroutine in the present model for the calculation of the interfacial conditions and hence the droplet size at each time level.

3. BOUNDARY-LAYER SOLUTION

Keller's box method has been extended to solve the system of gas-phase boundary-layer equations for momentum, heat and mass transfer simultaneously (cf. Wang & Kleinstreuer, 1989). A different implicit finite difference scheme has been used to solve the liquid-phase transient conduction equation. After initialization of the program with an educated guess for the droplet surface temperature, the preliminary interfacial mass fraction can be obtained from equation (8) and the transformed versions of equations (17) - (20) subject to conditions (22) and (23) which are solved at each time step. Now the interfacial conditions (6) and (7) have to be fulfilled which yields a boundary condition, i.e., (25), for the solution of equation (21). Then the new droplet surface temperature is being used in the solution of the gas-phase equations until, iteratively, convergence has been achieved, i.e., $|(T_i^{new} - T_i^{old}) / T_i^{new}| \leq 10^{-4}$. The new droplet radius is obtained from equation (26). Thus, the gas-side equations are solved for quasi-steady-state conditions (cf. $\Delta t = O(1\ \mu sec)$), although the individual droplet radii are changing with time.

Simultaneously, at any given time level, equations (9) - (12) are solved where the $C_{D,j}$-expressions (13) - (16) yield the effective approach stream Reynolds number for each droplet. The computational flow chart for a multiple-droplet iterative procedure at each time step is given in Fig. 2.

The accuracy of the computer simulation model has been checked in two ways: (a) comparison of the predictive results with published experimental and theoretical data sets for specific case studies, and (b) testing of the mesh-independence of the results (cf Wang & Kleinstreuer, 1989).

4. RESULTS AND DISCUSSION

4.1 Boundary-Layer Model Verification

The predictive capabilities of the core of the present computer simulation model has been successfully verified in the past with special case studies of mixed convection heat transfer of a sphere with blowing (cf. Kleinstreuer & Wang, 1988) and single droplet vaporization in a convective gas stream (Wang & Kleinstreuer, 1989). Because of the lack of experimental or analytical results for three interacting particles, comparisons are made for two spheres/drops in tandem. Thus, the correlations (13) to (15) are reduced for a two-sphere system by eliminating the effect of d_{23}, so that

$$\frac{C_{D,1}}{(C_{Ds})_1} = 1 - 0.096\, Re_{d,1}^{\,0.2475}\, d_{12}^{\,-0.965} \tag{27}$$

and

$$\frac{C_{D,2}}{(C_{Ds})_2} = 1 - 0.2932\, Re_{d,2}^{\,0.1593}\, d_{12}^{\,-0.4876} \tag{28}$$

A comparison of the computer predictions with both experimental data (Rowe & Henwood, 1961) and numerical results (Zapryanov & Toshev, 1986) are shown for different particle spacings in Table 1. While Table 1 summarizes individual drag coefficient ratios for closely-spaced spheres, Figs. 3 and 4 show for two interacting vaporizing droplets a comparison of the present results with the finite difference solution by Raju & Sirignano (1987). The input data used are listed in Table 2. During the initial stage ($\tau_g \leq 10$), the amount of fuel evaporating is low and the droplet size reduction is minor but the droplet Reynolds number decreases rapidly due to drag/interaction effects (Fig. 3a,b). The second droplet approaches the first one very quickly as shown in Fig. 4. As indicated in Table 1, the effective drag force is much lower for the second particle being in the wake of the leading particle and hence the faster moving second droplet catches up with the first droplet. The comparison of different system characteristics (Figs. 3a,b and 4) with the numerical results by Raju & Sirignano (1987) is very good for the two shrinking droplets and for the transient Reynolds number of the leading droplet. However, Re_d for the second droplet and hence the droplet spacing differ. Since experimentally and theoretically verified drag/interaction coefficients have been employed in the present analysis, the discrepancies may be due to blowing effects, near-wake effects, and/or the use of different property values, not fully reported by Raju & Sirignano (1987) (cf. Table 2).

TABLE 1

Comparison of drag coefficient ratio ($C_{D,j}/C_{Ds}$) for flow past two tandem spheres; $Re_d = 40$

	$d_{12} = 1.5$			$d_{12} = 2$		$d_{12} = 4$	
	Tal et al. (1984)	Rowe & Henwood (1961)	Present Method	Zapryanov & Toshev (1986)	Present Method	Zapryanov & Toshev (1986)	Present Method
1st sphere	0.880	0.800	0.840	0.880	0.877	0.920	0.937
2nd sphere	0.530	0.540	0.567	0.620	0.624	0.740	0.732

Keeping in mind that less than half of the surface area is being considered in the (attached) boundary-layer analysis, the accuracy of the approximate solution is somewhat surprising. However, Ramachandran et al. (1989b) evaluated the regional contributions to the total amount of heat transferred over the entire (spherical) surface. As can be seen from Table 3, based on the piecewise averaging of the local Nusselt numbers for each interacting sphere ($k = 1, 2, 3$) at various distances d_{ij} and Reynolds numbers $Re_d = 100$ and 200, the boundary-layer (BL) part of each sphere contributes at least 80% to $\overline{Nu}_k$ (cf. $k = 2$, $d_{12} = 2$, $d_{23} = 4$ and $Re_d = 200$). The near-wake (NW) contributions stay below 10%. For small intersphere distances, a recirculation zone (RZ) is formed ($\tau_{wall} \leq 0$) which reduces the attached boundary-layer region and may contribute measurably to Nu_k.

TABLE 2.

Input Data Sets for Figures 3 to 8

	Raju & Sirignano (1987) Figs. 3a,b & 4	Present Method (n-decane)
Operational Data		
d_o (μm)	50	50
$w_{j,o}$ (m/sec)	25	25
T_∞ (K)	1000	1000
T_o (K)	350	300
p (atm)	10	10
Fuel Properties		
T_b (K)	573	565
ρ_ℓ (kg/m^3)	NR	730
α_ℓ (m^2/sec)	NR	7.2×10^{-8}
D (m^2/sec)	NR	1.155×10^{-8}
c_p (cal/gm-mole K)	NR	0.539
L (cal/gm mole)	NR	66
MW (gm/gm mole)	192	142.3
k_ℓ (cal/m-sec-K)	NR	2.84×10^{-6}

NR ≅ not reported

TABLE 3

Relative Contributions to $\overline{Nu}_k$ from various Flow Regions

d_{ij}	Sphere #, k	$Re_d = 100$			$Re_d = 200$		
		RZ	BL	NW	RZ	BL	NW
$d_{12} = 2$, $d_{23} = 4$	1	---	94.5	5.5	---	92.0	8.0
	2	6.9	89.4	3.7	13.5	80.9	5.6
	3	---	93.0	7.0	---	91.1	8.9
$d_{12} = 4$, $d_{23} = 2$	1	---	93.1	6.9	---	90.7	9.3
	2	---	93.4	6.6	---	90.6	9.4
	3	7.1	87.1	5.8	9.5	84.0	6.5
$d_{12} = 4$, $d_{23} = 4$	1	---	93.1	6.9	---	91.8	8.2
	2	---	93.7	6.3	---	93.3	6.7
	3	---	93.4	6.6	---	93.6	6.4

4.2 *Three-Droplet Dynamics*

The transient changes in droplet radius, mass and Reynolds number for three dynamically interacting droplets are depicted in Figs. 5a - 5c. While the transient characteristics of droplets ② and ③ stay very similar, the first droplet shrinks faster and decelerates more rapidly. As a result, the second and third droplet in tandem catch up with the leading droplet (cf. Fig. 6). Ejecting the third droplet closer to the second droplet (i.e., $d_{12} = 6$ but $d_{23} = 2$) causes measurable upstream effects as can be concluded from the solid curves in Fig. 7 relative to Fig. 6; all three droplets may coalesce if their lifetimes permit. When the second droplet is ejected closely to the leading droplet ($d_{12} = 2$) and the third droplet at $d_{23} = 6$, the last droplet becomes solitary, leaving the formation, while the first two droplets become a rapidly approaching pair (Fig. 8). Needless to say, systematic variations of the initial and operational conditions allows the user of the new computer simulation model to map out optimal scenarios in terms of maximum vaporization without droplet collisions.

Presently, both modeling approaches, i.e., the microscale finite element analysis and the approximate boundary-layer analysis are being extended. The main goal is to study vortex shedding, gas-phase turbulence, and non-sphericity effects on interacting vaporizing droplets and multiple closely-spaced droplet streamers. The anticipated, experimentally verified, modeling results will improve the physical understanding and design applications of dense (fuel) droplet spray systems.

ACKNOWLEDGEMENT

This work has been supported in part by the Department of Energy, Office of Basic Energy Science; Grant No. DE-FG05 87 ER 13728.

The authors gratefully acknokwledge the excellent wordprocessing and assistance of Mrs. M. A. Sullivan.

REFERENCES

Abramzon, B. and Sirignano, W. A., 1988, "Droplet Vaporization Model for Spray Combustion Calculations," Combustion Processes at the AIAA 26th Aerospace Science Meeting, Reno, Nevada, January.

Aggarwal, S. K.; Tong, A. Y., and Sirignano, W. A., 1984, "A Comparison of Vaporization Models in Spray Calculations," *AIAA J.*, **22**, pp. 1448-1457.

Bellan, J. and Harstad, K., 1987, "Analysis of the Convective Evaporation of Nondilute Clusters of Sprays," *Int. J. HMT*, **30**, pp. 125-136.

Bracco, F. V., 1985, "Modeling of Engine Sprays," SAE Tech. Paper Series No. 850394.

Brauner, C. M. and Schmidt-Laine, C. (eds) (1988) Mathematical Modeling in Combustion, Martinus Nighoff Publ., Dordrecht, The Netherlands (NATO SAO).

Chigier, N., 1983, "Drop Size and Velocity Instrumentation," *Prog. Energy and Comb. Sci.*, **9**, pp. 155-177.

Clift, R., Grace, J. R. and Weber, M. E., 1978 *Bubbles, Drops and Particles,* Academic Press, New York.

Dwyer, H. A. and Sanders, B. R., 1984, "Detailed Computation of Droplet Dynamics," presented at the 20th Int. Symp. on Comb., August 12-17.

Faeth, G. M., 1983, "Evaporation and Combustion of Sprays," *Prog. Energy Comb. Sci.*, **9**, pp. 1-76.

Faeth, G. M. (1987) "Mixing, Transport and Combustion in Sprays," *Prog. Energy Comb. Sci.*, **13**, pp. 293-345.

Fernandez-Pello, A. C., 1983, "Theory of the Mixed Convective Combustion of a Spherical Fuel Particle," *Comb. and Flame*, **53**, pp. 23-32.

Finlayson, B. A., and Olson, J. W., 1987, "Heat Transfer to Spheres at Low to Intermediate Reynolds Numbers," *Chem. Eng. Comm.*, **58**, pp. 431-447.

Kleinstreuer, C., Wang, T-Y, and Chiang, H. 1989, "Analysis of Closely-Spaced Evaporating Fuel Droplets," Proceedings of 5th Int. Miami Symp. on Multi-Phase Transport & Particulate Phenomena, Dec. 12-14, 1988, Miami Beach; T. N. Veziroglu (ed) CERI, Univ. of Miami, Coral Gables, FL.

Kleinstreuer, C., 1990, *Engineering Fluid Dynamics - An Interdisciplinary Systems Approach*, Springer Verlag, New York (to be issued).

Kleinstreuer, C. and Wang, T. Y., 1988, "Heat Transfer between Rotating Spheres and Flowing Power-law Fluids with Suction and Injection," *Int. J. Heat and Fluid Flow*, **9**, pp. 328-333.

Kuo, K. K., 1986, *Principles of Combustion,* Wiley-Interscience, N.Y., NY.

Law, C. K., 1984, "Heat and Mass Transfer in Combustion: Fundamental Concepts and Analytical Techniques," *Prog. Energy Comb. Sci.*, **10**, pp. 295-318.

Oran, E. S. and Boris, J. P. (eds) 1987, "Numerical Simulation of Reactive Flow," Elsevier, N.Y.

O'Rourke, P. J. and Bracco, F. V., 1980, "Modeling of Drop Interactions in Thick Sprays and a Comparison with Experiments," Stratified Charge Auto. Engineers Conf. The Inst. of Mech. Eng., London.

Prakash, S. and Krishan, G., 1984, "Convective Droplet Vaporization with Transient Non-convective Liquid-phase Heating, *20th Symp. on Combustion*, pp. 1735-1742.

Prakash, S. and Sirignano, W. A., 1980, "Theory of Convective Droplet Vaporization with Unsteady Heat Transfer in a Circulating Liquid Phase, *Int. J. Heat Mass Transfer*, **23**, pp. 253-268.

Rah, S. C.,; Sarofim, A. F. and Beer, J. M., 1986, "Ignition and Combustion of Liquid Fuel Droplets: Part I: Impact on Pollutant Formation," *Comb. Sci. Tech.*, **48**, pp. 273-284.

Raju, M. S. and Sirignano, W. A., 1987, "Unsteady Navier-Stokes Solution for Two Interacting Vaporizing Particles, AIAA 25th Aerospace Sciences Meeting, Reno, Nevada.

Ramachandran, R. S., Wang, T. Y., and Kleinstreuer, C., 1989a, "Fluid Flow Characteristics of Three Closely-spaced Spheres, *ASME J. Fluid Engineering*, (under review).

Ramachandran, R. S., Kleinstreuer, C. and Wang, T. Y., 1989b, Forced Convection Heat Transfer of Interacting Spheres, *Numerical Heat Transfer*, Part A, **15**.

Ramachandran, S. and Kleinstreuer, C., 1986, "Laminar Mass Transfer in the Wake Region of Closely-spaced Drops, *Int'l. Comm. Heat & Mass Transfer*, **13**, 4; pp. 389-402.

Rowe, D. N. and Henwood, G. A., 1961, "Drag Forces in a Hydraulic Model of a Fluidized Bed - Part I, *Trans. Inst. Chem. Engr.*, **39**, pp. 43-54.

Sangiovanni, J. J. and Kesten, A. S., 1977, "Effect of Droplet Interaction on Ignition in Monosized Droplet Streams," 16th Symp. on Combustion, The Combustion Institute, Pittsburgh, pp. 577-592.

Sangiovanni, J. J. and Labowsky, M., 1982, "Burning Times of Linear Fuel Droplet Arrays: A Comparison of Experiment and Theory," *Comb. and Flame*, **47**, pp. 15-30.

Sirignano, W. A., 1983, "Fuel Droplet Vaporization and Combustion Theory," *Prog. Energy Comb. Sci.*, **9**.

Sirignano, W. A., 1986, "The Formulation of Spray Combustion Models: Resolution Compared to Droplet Spacing," *ASME J. Heat Transfer*, **108**, pp. 633-639.

Tal, R. and Sirignano, W. A., 1982, "Cylindrical Cell Model for the Hydrodynamics of Particle Assemblages at Intermediate Reynolds Numbers," *AIChE J.*, **28**, pp. 233-237.

Tal, R., Lee, D. N., and Sirignano, W. A., 1984, "Heat and Momentum Transfer around a pair of Spheres in Viscous Flow, *Int. J. Heat Mass Transfer*, **27**, pp. 1953-1962.

Thompson, J. F., Thames, F. C. and Mastin, C. W., 1977, "TOMCAT - A Code for Numerical Generation of Boundary-fitted Curvilinear Coordinate Systems on Fields Containing Any Number of Arbitrary Two-dimensional Bodies," *J. Comp. Phys.*, **24**, pp. 274-302.

Tong, A. Y. and Chen, S. J., 1988, "Heat Transfer Correlations for Vaporizing Liquid Droplet Arrays in a High-Temperature Gas at Intermediate Reynolds Number," *Int. J. Heat & Fluid Flow*, **9**, (2), pp. 118-130.

Tsuji, Y., Morikawa, Y. and Terashima, K., 1982, "Fluid-dynamic Interaction Between Two Spheres," *Int. J. Multiphase Flow*, **8**, pp. 71-82.

Wang, T. Y. and Kleinstreuer, C., 1989, "Laminar Thermal Boundary-layer Flow Past a Vaporizing Spherical Droplet," *Num. Heat Transfer*, (under review).

White, F. M., 1974, *Viscous Fluid Flow*, McGraw-Hill, New York.

Zaprynov, Z. D., and Toshev, E. T., 1986, "Hydrodynamics and Heat Transfer Around Two Separated Spherical Particles," Proc. of the 8th Int. Conf. on Heat Transfer, San Francisoc, CA, Vol. 5, pp. 2549-2553.

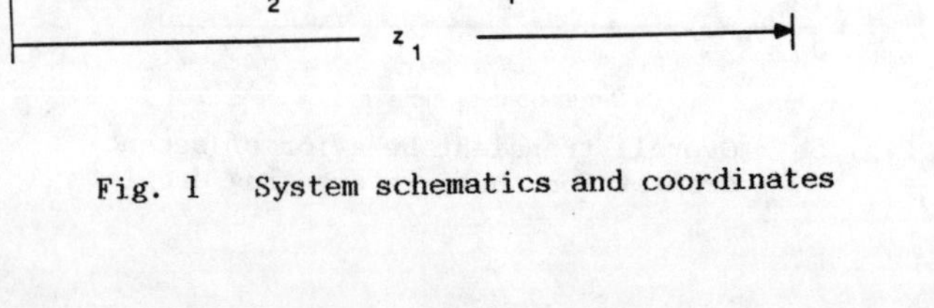

Fig. 1 System schematics and coordinates

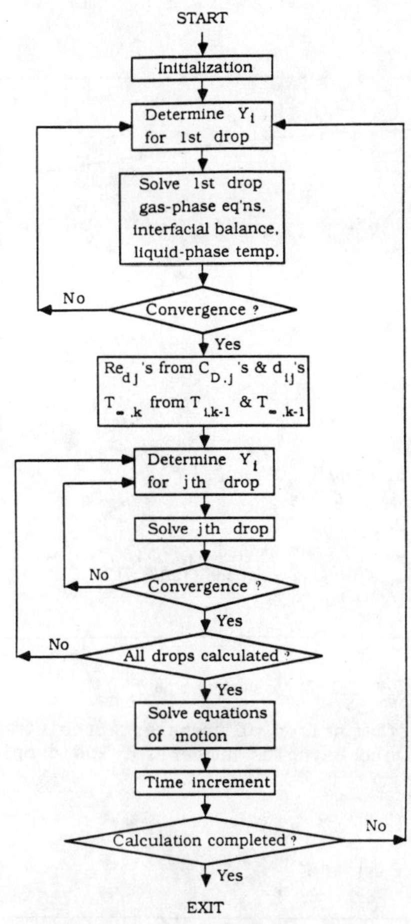

Fig. 2 Computational flow chart

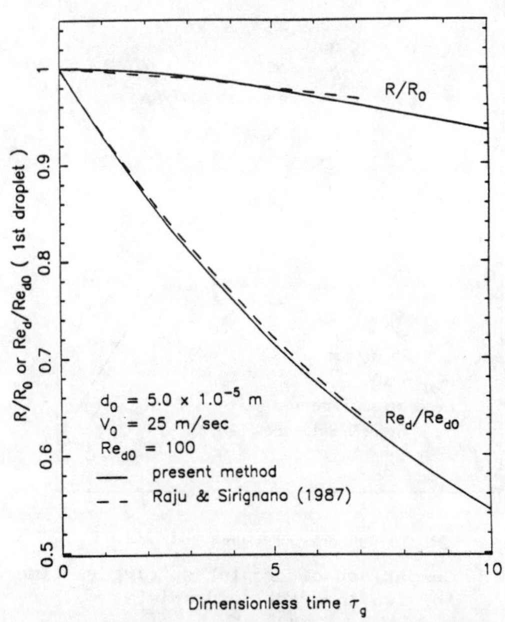

Fig. 3a Comparison of transient droplet radius and Reynolds number for 1st droplet

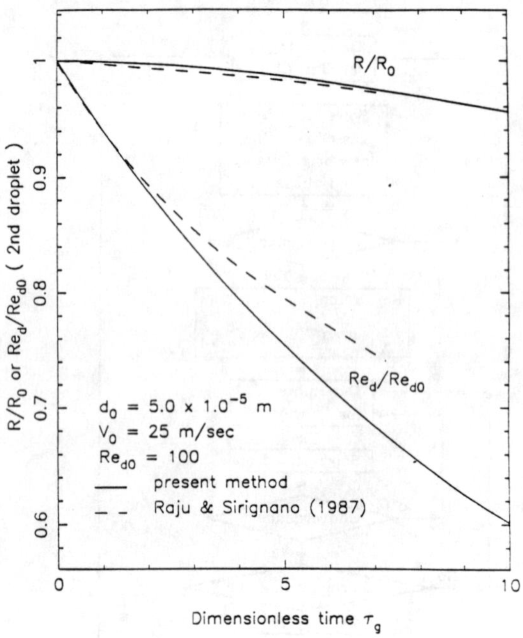

Fig. 3b Comparison of transient droplet radius and Reynolds number for 2nd droplet

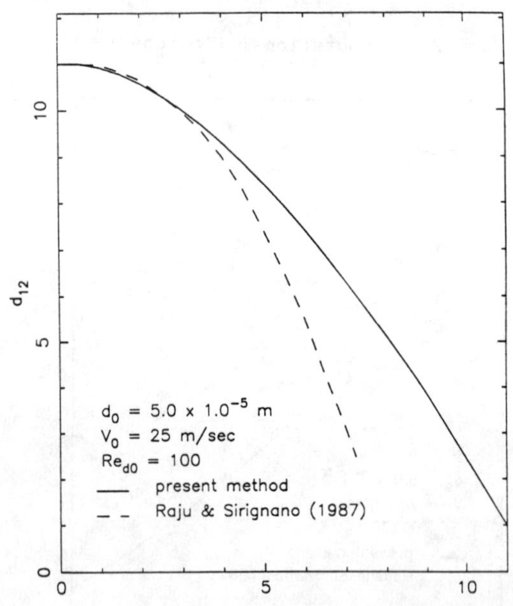

Fig. 4 Comparison of droplet spacing vs time for two interacting droplets

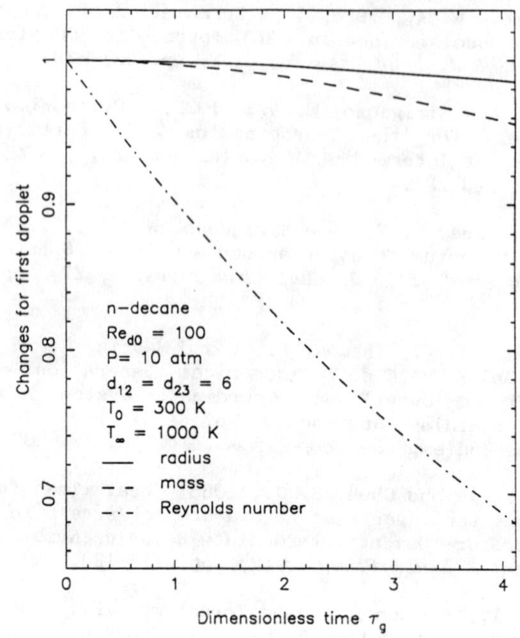

Fig. 5a Overall transient behavior of leading droplet for three interacting droplets

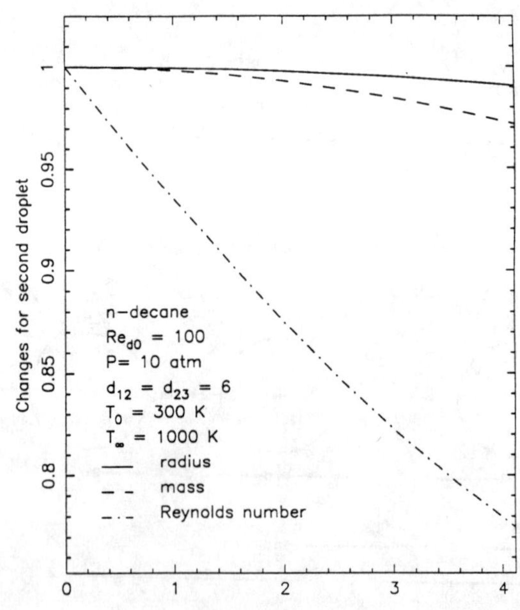

Fig. 5b Overall transient behavior of second droplet for three interacting droplets

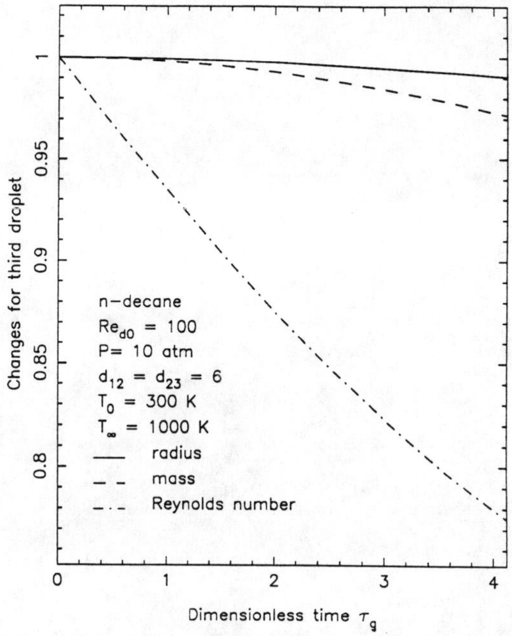

Fig. 5c Overall transient behavior of third droplet for three interacting droplets

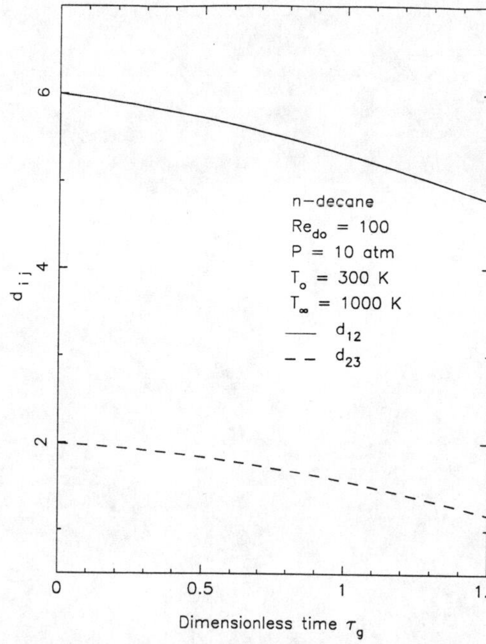

Fig. 7 Droplet spacings vs time for three interacting droplets

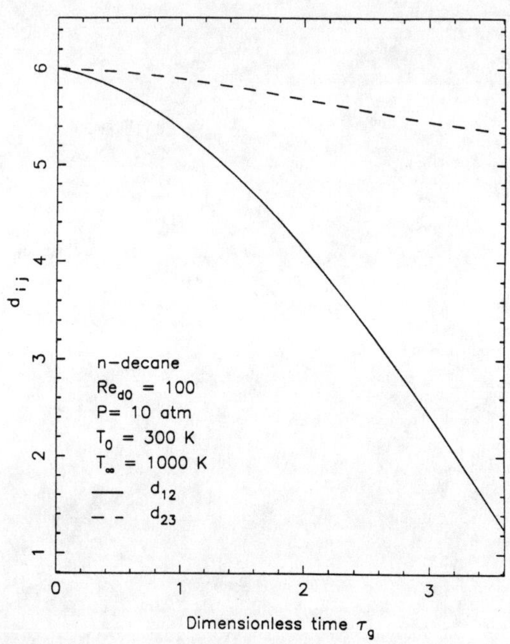

Fig. 6 Droplet spacings vs time for three interacting droplets

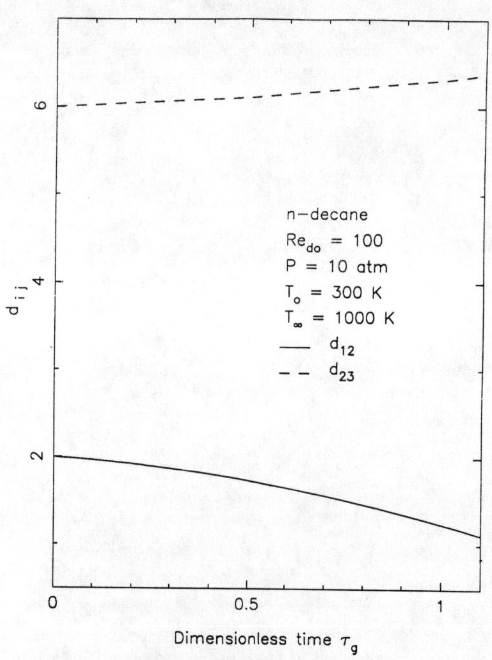

Fig. 8 Droplet spacings vs time for three interacting droplets

1989 National Heat Transfer Conference
HTD-Vol. 106, Heat Transfer Phenomena in Radiation, Combustion and Fires

NUMERICAL MODELING OF DIFFUSION AND PREMIXED TURBULENT REACTING FLOWS

A. A. Boretti
Dipartimento di Energetica
Firenze, Italy

ABSTRACT

This paper presents a model for turbulent, compressible, reacting, gas flows in which the reactants have been partially mixed prior to reaction. A numerical method for the simulation of steady state, two-dimensional flows is then described. The method is applied to the computation of both diffusion and premixed turbulent reacting flows. The computed temperature distributions compare favorably with experimental data.

NOMENCLATURE

AE = Arrhenius activation enthalpy
AV = artificial viscosity
C = constant
$d_{i,j}$ = Kronecker delta
dt = time step
D = wall term
D = diffusion vector
e = kinetic energy dissipation rate
E = wall term
f = function
f = unknown vector
F = compressibility term
G = compressibility term
h = enthalpy
$\mathbf{H}_{opt}$ = optimization matrix
H_{fu} = chemical heat release
K = turbulence kinetic energy
k_f = reaction rate constant
p = pressure
P = turbulence kinetic energy production
PF = Arrhenius prefactor
q = heat flux vector
R_t = turbulence Reynold number
s = stoichiometric ratio
Sc = Schmidt number
t = time
T = residual
u = velocity
x = axial coordinate
Y = mass fraction
α = specific heat ratio
ρ = density
μ = viscosity coefficient
τ = viscous stress tensor
θ = conserved scalar
$\emptyset$ = rate of chemical reaction

Subscripts
o = total
i = inviscid
fu = fuel
ox = oxidizer
t = turbulent
v = viscous

Superscripts
m = time level

INTRODUCTION

The design of future combustion concepts requires accurate numerical methods for the simulation of liquid or gaseous fuels in a highly turbulent air stream (Mularz et al., 1985). These chemical reacting turbulent flows involve fluid mechanics, chemistry and turbulence-combustion interaction. The present approach is devoted to the solution of the full turbulent reacting flow problem within an engineering accuracy. Only averaged quantities are considered, and the reactive gas is described as a mixture of three species: fuel, oxidizer and products, with combustion controlled by a single step, irreversible chemical reaction.

The paper presents a numerical method for the solution of the conservation equations governing steady, reacting, turbulent viscous flows in two-dimensional geometries. The reliability of a numerical method is a function of both the mathematical modeling of the physical process and the solution algorithm. These two features are then taken into account in order to obtain an efficient numerical method.

The governing equations are written here in Favre-averaged form. No simplification is used in deriving the energy conservation equation as usually performed (Bray, 1980, Bilger, 1980). The closure is achieved by using a first order model. The K-e turbulence model adopted here includes low Reynolds number terms, so that the equations are valid all over the laminar, transition and turbulent region. Furthermore, the model includes a pressure-density gradient term to better simulate variable density effects. The modified eddy-break up mixing controlled reaction rate expression adopted here introduces a concentration gradient dependence other than the classical concentration dependence. The influence of the kinetics is taken into account. The combustion model has been developed for imperfectly premixed flows, but it is useful in cases where the combustion is neither fully premixed nor entirely diffusion controlled, as it is likely to occur under many circumstances even if the fuel and oxidizer enter the combustion chamber in separate streams. These two models allow to simulate fluid mechanics turbulence, chemistry and turbulence-combustion interaction.

The solution is obtained by using a pseudo-unsteady method with improved perturbations propagation properties. Pseudo-unsteady methods with artificial equations are quite common in turbomachinery applications (Viviand, 1984), but up to now they have received little use in computing reacting flows, even if they are faster and simpler than classical pressure iteration methods (Hirt et al., 1976). The equations are discretized in space by using a finite-volume formulation (Couston, 1975). An explicit, multi-stage, dissipative, Runge-Kutta algorithm is then used to advance the flow equations in the pseudo time. Multi-stage schemes for the numerical solution of ordinary differential equations are usually designed to give a high degree of accuracy, but in a pseudo-unsteady solution these schemes are selected only for their properties of stability and damping (Schmidt and Jameson, 1982). The four-stage scheme adopted here allows a Courant number of about 2.6 with only negligible changes with reference to the Lax-Wendroff algorithm.

The method is applied to the computation of both diffusion and premixed turbulent reacting flows.

FLOW MODEL

The basic variables are the gas density, velocity and enthalpy. Their conservation equations are written in Favre-averaged form as follows

$$\bar{\rho}_{,t} + (\bar{\rho}.\tilde{u}_k)_{,k} = 0$$

$$(\bar{\rho}.\tilde{u}_i)_{,t} + (\bar{\rho}.\tilde{u}_k.\tilde{u}_i)_{,k} = - \bar{p}_{,i} + [\bar{\tau}_{ik} - \overline{\rho.u''_k.u''_i}]_{,k}$$

$$(\bar{\rho}.\bar{\bar{h}})_{,t} + (\bar{\rho}.\tilde{u}_j.\bar{\bar{h}})_{,j} = \bar{p}_{,t} + \overline{u''_j.p_{,j}}$$
$$+[- \bar{q}_j - \overline{\rho.u''_j.h''}]_{,j} + [\bar{\tau}_{ij}.\tilde{u}_i - \overline{\rho.u''_i.u''_j}.\tilde{u}_i]_{,j}$$
$$+\overline{u''_{i,j}.\tau_{ij}} + \overline{\rho.u''_i.u''_k}.\tilde{u}_{i,k} + \tilde{\varnothing}_{fu}.H_{fu}$$

where (Hirsch et al., 1985)

$$\bar{\bar{h}}_0 = \tilde{h} + \tilde{u}_i.\tilde{u}_i/2 = \tilde{h}_0 - \overline{\rho.u_i''.u_i''}/(2.\bar{\rho})$$

The state equation is written as follows

$$p = (\alpha - 1)/\alpha.\rho.h$$

while the constitutive relations are

$$\tau_{ij} = \mu.(u_{i,j} + u_{j,i} - 2/3.u_{k,k}.d_{ij})$$

and

$$q_j = - Sc_h^{-1}.\mu.h_{,j}$$

The previous equations have to be closed by introducing a turbulence and a chemical reaction rate model.

TURBULENCE MODEL

The closure is achieved by using a first order model. The K-e turbulence model adopted here includes low Reynolds number terms (Boretti and Martelli, 1988c) and a pressure-density gradient term (Jones and McGuirk, 1987). The Reynolds stresses and the turbulent heat flux vector are expressed as follows

$$\overline{\rho.u''_i.u''_j} = -\mu_t.(\tilde{u}_{i,j}+\tilde{u}_{j,i}-2/3.d_{ij}.\tilde{u}_{1,1}) +2/3.d_{ij}.K$$

$$\overline{\rho.u''_k.h''} = - Sc_{ht}^{-1}.\mu_t . \tilde{h}_{,i}$$

where

$$\mu_t = \bar{\rho}.c_\mu.f_\mu.K^2/e$$

The Favre-averaged conservation equations for K and e are written as follows

$$(\bar{\rho}.K)_{,t}+(\bar{\rho}.\tilde{u}_i.K)_{,i} = [Sc_K^{-1}.(\mu_t+\mu).K_{,i}]_{,i}$$

$$+ P + F - \bar{\rho}.e + D$$

$$(\bar{\rho}.e)_{,t} + (\bar{\rho}.\tilde{u}_i.e)_{,i} = [Sc_e^{-1}.(\mu_t+\mu).e_{,i}]_{,i}$$

$$+ e/K.[C_{e1}.f_1.(P+G)-C_{e2}.f_2.\bar{\rho}.e]+E$$

where

$$P = - \overline{\rho.u''_i.u''_j} \cdot \tilde{u}_{i,j}$$

the wall terms D and E are (Boretti and Martelli, 1988 c)

$$D = - 2.\mu.[(K)^{1/2}{}_{,k}]^2$$

$$E = - 2.\mu.[(e)^{1/2}{}_{,k}]^2$$

and the compressibility terms F and G are (Jones and Mc Guirk, 1987)

$$F = - \mu_t/\bar{\rho}^2.\bar{\rho}_{,i}.\bar{p}_{,i}$$

$$G = - \mu_t/\bar{\rho}^2.\bar{\rho}_{,i}.\bar{p}_{,i}$$

The values of the previous constants are C_μ=0.09, C_{e1}=1.43, C_{e2}=1.92, Sc_K=1.0, Sc_e=1.3, while the functions are expressed as

$$f_\mu = \exp[-2.5/(1+R_t/50)]$$

$$f_1 = 1.0$$

$$f_2 = 1.0 - 0.33.\exp[-R_t2]$$

where

$$R_t = \rho.K^2/(\mu.e)$$

Finally,

$$\overline{u''_j.p_{,j}} = - \mu_t/\bar{\rho}^2.\bar{\rho}_{,i}.\bar{p}_{,i}$$

$$\overline{u''_{i,j}.\tau_{ij}} = - e$$

The intoduction of this K-e turbulence model improves considerably the mathematical model previously used (Boretti and Martelli, 1988 b) where a simpler and less reliable generalized mixing length model was adopted.

COMBUSTION MODEL

The reaction rate for fuel is a function of some of the previous variables plus two reactive variables, the fuel mass fraction and the conserved scalar

$$\Theta = Y_{fu} - Y_{ox}/s$$

Their conservation equations are written in Favre-averaged form as follows

$$(\bar{\rho}.\tilde{Y}_{fu})_{,t} + (\bar{\rho}.\tilde{u}_k.\tilde{Y}_{fu})_{,k} =$$

$$(Sc_Y^{-1}.\mu.(\tilde{Y}_{fu})_{,k} - \overline{\rho.u''_k.Y_{fu}''})_{,k} - \tilde{\varnothing}_{fu}$$

$$(\bar{\rho}.\tilde{\Theta})_{,t}+(\bar{\rho}.\tilde{u}_k.\tilde{\Theta})_{,k} = (Sc_\Theta^{-1}.\mu.\tilde{\Theta}_{,k}-\overline{\rho.u''_k.\Theta''})_{,k}$$

where

$$\overline{\rho.u''_k.\Theta''} = - Sc_{\Theta t}^{-1}.\mu_t \cdot \tilde{\Theta}_{,i}$$

$$\overline{\rho.u''_k.Y_{fu}''} = - Sc_{Yt}^{-1}.\mu_t \cdot (\tilde{Y}_{fu})_{,i}$$

The problem is completely closed when a formula for the averaged rate of chemical reaction for fuel is supplied. This quantity is related to both mean values and fluctuations of reactant concentrations, density and reaction rate constant (Libby and Williams, 1980)

$$\tilde{\varnothing}_{fu} = \bar{\rho}.K_f.(\bar{\rho}.\tilde{Y}_{fu}.\tilde{Y}_{ox} + \overline{\rho''.Y''_{fu}.Y''_{ox}})$$

$$+ \overline{\rho''.(\rho.k_f)''.Y''_{ox}}.\tilde{Y}_{fu}$$

$$+ \overline{\rho''.(\rho.k_f)''.Y''_{fu}}.\tilde{Y}_{ox}$$

$$+ \overline{\rho''.(\rho.k_f)''.Y''_{fu}.Y''_{ox}}$$

A modelled expression can be derived from classical approaches as follows. Consider first a reaction controlled by the turbulent mixing. In a diffusion flame, as a result of the conserved scalar approach (Bilger, 1980), the reaction rate for fuel is proportional to the viscous dissipation of the fluctuations of the conserved scalar. Then, we may suppose that

$$\tilde{\varnothing}_{fu} = C_{DF}.\{[K^3/e^2.\tilde{\Theta}_{,k}.\tilde{\Theta}_{,k}]\}^{1/2}.\bar{\rho}.e/K$$

or

$$\tilde{\varnothing}_{fu} = C_{DF}.\bar{\rho}.\tilde{\Theta}.e/K$$

where C_{DF} is a proportionality coefficient.

In the particular case $\Theta \geq 0$, these expressions can be rewritten in terms of Y_{fu} as

$$\tilde{\varnothing}_{fu} = C_{DF}.\{[K^3/e^2.(\tilde{Y}_{fu})_{,k}.(\tilde{Y}_{fu})_{,k}]\}^{1/2}.[\bar{\rho}.e/K]$$

or

$$\tilde{\varnothing}_{fu} = C_{DF}.\bar{\rho}.\tilde{Y}_{fu}.e/K$$

In a premixed flame, as a result of the progress variable approach (Bray, 1980), the reaction rate for fuel is proportional to the viscous dissipation of the fluctuations of the progress variable. Then, we may suppose

$$\tilde{\varnothing}_{fu} = C_{PF}.\{[K^3/e^2.(\tilde{Y}_{fu})_{,k}.(\tilde{Y}_{fu})_{,k}]\}^{1/2}.[\bar{\rho}.e/K]$$

or

$\tilde{\O}_{fu} = C_{PF}.\bar{\rho}.\tilde{Y}_{fu}.e/K$

where C_{PF} is a proportionality coefficient, while no expression can be given in terms of θ.

If we consider both expressions, we may take here

$$\tilde{\O}_{fu} = \min [C_A.\tilde{Y}_{fu},$$

$$C_A.\tilde{Y}_{ox}/s,\ C_B.K/e.\{K.[(\tilde{Y}_{fu})_{,k}]^2\}^{1/2}].\bar{\rho}.e/K$$

where C_A and C_B are empirical constants. Their typical values are found to be $C_A \simeq 4.0$ and $C_B \simeq 0.4$. This expression can be regarded as a simple improvement of the model proposed by Magnussen and Hjertager (1976). Further improvement can be added according to the available experimental data.

When the kinetic is considered, we may suppose

$$\tilde{\O}_{fu} = -PF.\bar{\rho}^2.\tilde{Y}_{fu}.\tilde{Y}_{ox}.\exp[-AE/\tilde{h}]$$

where AE is the Arrhenius activation enthalpy and PF is the Arrhenius prefactor. For kinetically influenced processes, the actual reaction rate can then be taken as the smaller between the previous expressions.

The introduction of this reaction rate expression further improves the matematical model previously used (Boretti and Martelli, 1988 b) where an expression valid only for perfectly premixed flows was adopted, the influence of kinetic was disregarded and the turbulence-combustion interaction was considered in a very simplified way.

BOUNDARY CONDITIONS

The physical flow domain is generally limited from inlet, exit, solid, and symmetry boundaries. A number of boundary conditions equal to the number of conservation equations minus one is imposed at the inlet (ρ, u_i, K, e, Y_{fu}, θ are known), while only one condition is imposed at the exit (p is known). Along the solid boundaries the no-slip condition requires $u_i = K = e = 0$, while the heat flux vector or the temperature and the diffusion vector for species have to be specified. (Usually no heat flux and diffusion conditions are adopted.) On a plane of symmetry, the normal derivatives of all the flow parameters vanish. The only exceptions to this specification occur for those flow parameters which must be set equal to zero, like the normal velocity component. More details about the particular boundary conditions are given in the following applications.

NUMERICAL METHOD

The numerical solution of the previous equations is described in details elsewhere (Boretti and Martelli, 1988 a,b,c, Boretti, 1989) and is only outlined here.

When the previous physical time dependent equations are written in the following vector form

$$\mathbf{f}_{,t} = \mathbf{T}^+(\mathbf{f})$$

the solution of the steady problem is obtained by solving the artificial time dependent equations

$$\mathbf{f}_{,t} = \mathbf{H}_{opt}.\mathbf{T}^+(\mathbf{f}) = \mathbf{T}(\mathbf{f})$$

where $\mathbf{H}_{opt}$ is a matrix defined in order to optimize the rate of convergence towards the steady state (Boretti and Martelli, 1988 a).

The equations are discretized in space by using a finite-volume formulation (Boretti and Martelli, 1988 c).

The equations are then integrated in time by using an explicit, multi stage, dissipative, Runge-Kutta algorithm (Boretti, 1989). The solution at the iteration m+1, i.e., at the local time t=(m+1).dt, is expressed as a function of quantities evaluated at the times (m).dt and (m*).dt with the following scheme

$$\mathbf{f}^{(0)} = \mathbf{f}^m$$

$$\mathbf{f}^{(1)} = \mathbf{f}^{(0)} - 1/4.dt.[\mathbf{T}_i(\mathbf{f}^{(0)}) + \mathbf{T}_v(\mathbf{f}^{m*}) + \mathbf{D}(\mathbf{f}^{(0)}) - \mathbf{AV}.\mathbf{D}(\mathbf{f}^{m*})]$$

$$\mathbf{f}^{(2)} = \mathbf{f}^{(0)} - 1/3.dt.[\mathbf{T}_i(\mathbf{f}^{(1)}) + \mathbf{T}_v(\mathbf{f}^{m*}) + \mathbf{D}(\mathbf{f}^{(0)}) - \mathbf{AV}.\mathbf{D}(\mathbf{f}^{m*})]$$

$$\mathbf{f}^{(3)} = \mathbf{f}^{(0)} - 1/2.dt.[\mathbf{T}_i(\mathbf{f}^{(2)}) + \mathbf{T}_v(\mathbf{f}^{m*}) + \mathbf{D}(\mathbf{f}^{(0)}) - \mathbf{AV}.\mathbf{D}(\mathbf{f}^{m*})]$$

$$\mathbf{f}^{(4)} = \mathbf{f}^{(0)} - dt.[\mathbf{T}_i(\mathbf{f}^{(3)}) + \mathbf{T}_v(\mathbf{f}^{m*}) + \mathbf{D}(\mathbf{f}^{(0)}) - \mathbf{AV}.\mathbf{D}(\mathbf{f}^{m*})]$$

$$\mathbf{f}^{(m+1)} = \mathbf{f}^{(4)}$$

where **D** is a dissipative term approximating second order derivatives, and **AV** is a vector of viscosity coefficients, depending on the density gradients. The terms referred to the iteration m* are updated only at particular iterations and taken constants between two updatings. Viscous and inviscid components of the residual have been separated for convergence purposes. The scheme allows a Courant number of about 2.6.

It is noticeable that the flow parameters not specified by a boundary condition are evaluated along the solid boundaries by using the same numerical process as inside the passage, by solving their conservation equations over half control volumes, while a simpler extrapolation from the interior points is preferred along the inlet and outlet boundaries and symmetry planes (Boretti and Martelli, 1988 c).

RESULTS

Calculations were performed for a partially premixed turbulent reacting flow. Figure 1 shows schematically the flow domain. In this configuration, the mixing of two parallel streams in a constant area duct is considered. One stream is made up of hot gases, the other of a fresh mixture of air and methane. Due to the hot jet, a flame is ignited and stabilized in the duct. The inlet duct, with a 100x100 mm^2 cross- section, is split into two parts: the upper one, with a cross-section 80 x 100 mm^2, is assigned to the fresh air and methane mixture, with a velocity of 65 m/s, a temperature of 580 K, and a mixture ratio of 0.8; the lower one, with a cross-section of 20 x 100 mm^2, is assigned to the pilot flame made up of hot gases, with a velocity of 130 m/s and a temperature of 2000 K. The walls are insulated. Experimental results are presented by Dupoirieux (1986).

Calculations were performed on a 35x45 computational grid, nonuniform but orthogonal. Conventional steady solutions were obtained in about 2.10^3 iterations with a total CPU time of about 1.5 h on a VAX 8800. Figures 2,3 show a comparison of predicted and measured transverse temperature profiles at axial locations x=42 mm and x=122 mm. The agreement in both sections lies within engineering accuracy. The introduction of the influence of kinetics in the evaluation of the reaction rate and a better calibration of the model constants lead to a better accuracy than the one previously obtained (Boretti and Martelli, 1988 c) even if at x=122 mm the temperatures in the mixing layer are still underestimated.

The method has then been applied to a turbulent diffusion reacting flow. Figure 4 shows schematically the flow domain. Experimental and theoretical results are presented by Lin (1987). In this configuration, there is a central jet of fuel, to a large extent methane, with a velocity of 21.3 m/s and a temperature of 300 K, and a concentric jet of oxidizer, with a velocity of 34.3 m/s and a temperature of 589 K. The outer radius of the fuel jet is 8 mm, the inner and outer radius of the air jet are respectively 11.1 and 28.6 mm. The radius of the combustor is 101.6 mm. The wall temperature is 1140 K.

Calculations were performed on a 45x45 computational grid, nonuniform but orthogonal. Conventional steady solutions were obtained in about 3.10^3 iterations with a total CPU time of about 3 h on a VAX 8800. Figures 5,6 show a comparison of predicted and measured transverse temperature profiles at axial locations x=95 mm and x=398 mm.

Even if the accuracy appears to be not optimal, the proposed method allows a better accuracy than the method of Lin (1987), despite the single-step chemical reaction adopted here. This is substantially due to the introduction of the fuel concentration gradient dependence in the mixing controlled reaction rate expression.

CONCLUSIONS

This paper has presented a numerical method for the solution of the conservation equations governing steady, reacting, turbulent viscous flows in two-dimensional geometries. The method is an improvement of a previous method in both the reliability of the mathematical modeling of the physical process and the efficiency of the solution algorithm.

The governing equations have been written in the full Favre-averaged form, without simplification in deriving the energy conservation equation. The proposed K-e turbulence model includes low Reynolds number and compressibility effects. The proposed reaction rate expression includes both the influence of turbulence and kinetic . The resulting flow model allows to simulate fluid mechanics turbulence, chemistry and turbulence-combustion interaction within an engeneering accuracy.

The solution has been obtained by using a pseudo-unsteady method with improved perturbations propagation properties. An explicit, multi-stage, dissipative, Runge-Kutta algorithm has been used to advance the flow equations in the pseudo time for its properties of stability and damping. This results in a good computational efficiency of the numerical solution algorithm.

The method has been applied to the computation of both diffusion and premixed turbulent reacting flows with a good agreement, even if some uncertainties still remain expecially in the evaluation of the reaction rate and in the influence of the numerical viscosity.

ACKNOWLEDGEMENTS

The author would like to express his indebtedness to NASA Le.R.C., Cleveland, Ohio, and in particular to Dr. L.Povinelli, for the support to the present research activity.

REFERENCES

Bilger, R.W., 1980, "Turbulent Flows with Nonremixed Reactants," in Libby, P.A., and Williams, F.A.,Turbulent Reacting Flows, Springer-Verlag, New York.

Boretti, A.A. and Martelli, F., 1988a, "Application of a Fast Pseudo Unsteady Method to Steady Compressible Turbulent Reacting Flows," Proceedings of the First National Fluid Dynamic Congress, Cincinnati.

Boretti, A.A. and Martelli, F., 1988b, "Numerical Modeling of Turbulent Combustion in Premixed Flows," Proceedings of the National Heat Transfer Conference, Houston.

Boretti, A.A. and Martelli, F., 1988c, "Accuracy and Efficiency of Time Marching Approach for Combustor Modelling," paper presented at the "AGARD Fluid Dynamics Panel Symposium on Validation of Computational Fluid Dynamics", Lisbon.

Boretti, A.A., 1989, "An Explicit Runge-Kutta Method for Turbulent Reacting Flows Calculations", paper to be presented at the "INRIA Third International Conference on Numerical Combustion", Antibes.

Bray, K.N.C., 1980, "Turbulent Flows with Premixed Reactants," in Libby, P.A., and Williams, F.A., Turbulent Reacting Flows, Springer-Verlag.

Couston, M., 1975, "Methode de calcul de l'ecoulement inter aubes pseudo-tridimensionnel en regime transsonique," PhD Thesis, VKI-University of Brussels.

Dupoirieux, F., 1986, "Numerical calculations of turbulent reactive flows and comparison with experimental results", ONERA TP 1986-80.

Hirt, C.W., Cloutman, L.D., and Romero, N.C., 1976, "SOLA-ICE: A Numerical Solution Algorithm for Transient Compressible Fluid Flows," Los Alamos Laboratory Report LA-6236.

Hirsch, C., and Deconinck, H., 1985, "Description of Various Flow Models from Navier-Stokes to Potential Flow Models," in Ucer, Stow, and Hirsch, Thermodynamics and Fluid Mechanics of Turbomachinery, Martinus Nijhoff Publishers, Boston.

Libby, P.A., and Williams, F.A., 1980, "Fundamental Aspects," in Libby, P.A., and Williams, F.A., Turbulent Reacting Flows, Springen-Verlag.

Jones, W.P., and McGuirk, J.J., 1979, "Mathematical Modelling of Gas Turbine Combustion Chambers", Combustor Modelling, AGARD CP 275.

Lin, C.-S., 1987, "Numerical Calculations of Turbulent Reacting Flow in a Gas-Turbine Combustor," NASA TM 89842.

Magnussen, B.F., and Hjertager, B.H., 1976, "On Mathematical Modeling of Turbulent Combustion with Special Emphasis on Soot Formation and Combustion," Proceedings 16th Symposium (International) on Combustion, Pittsburg.

Mularz, E.J., and Claus, R.W., "Combustion Research for gas Turbine Engines", Proceedings 7th International Symposium on Air Breathing Engines, Bejing.

Peyret, R., and Taylor, T.D., 1983, Computational Methods for Fluid Flow, Springer-Verlag, New York.

Satofuka, N., 1985, "Numerical Solution of Fluid Dynamic Equations Using Rational Runge-Kutta Methods", Proceedings of the 7th INRIA International Conference on Computing Methods in Applied Sciences and Engineering, Versailles.

Schmidt, W. and Jameson, A., 1982, "Recent Developments in Finite-Volume Time-Dependent Techniques for Two and Three Dimensional Transonic Flows," Computational Fluid Dynamics, VKI LS 1982-04.

Viviand, H., 1984, "Pseudo-Unsteady Systems for Inviscid Flow Calculation," ONERA TP No. 1984-69.

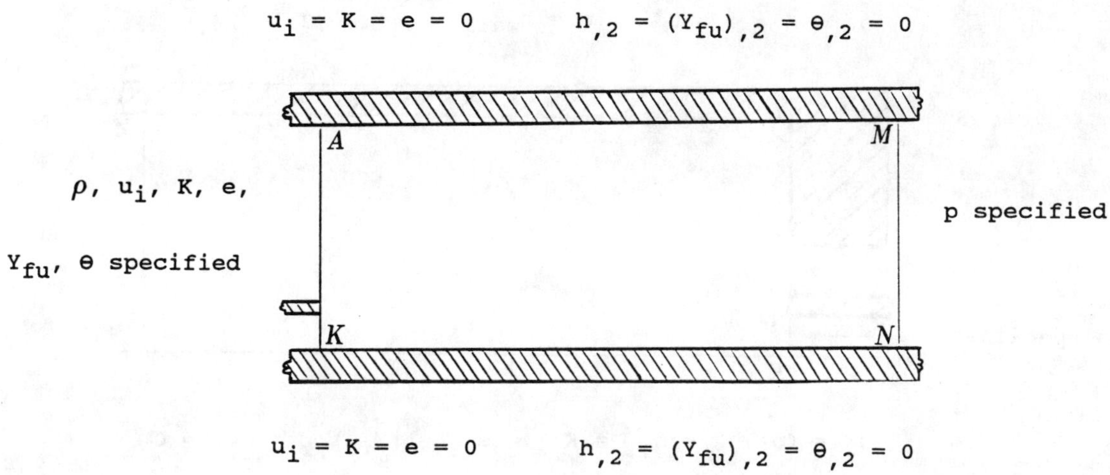

Figure 1 - Premixed turbulent reacting flow: physical domain.

theoretical, present method
theoretical, Boretti and Martelli (1988 b)
experimental, Dupoirieux (1986)

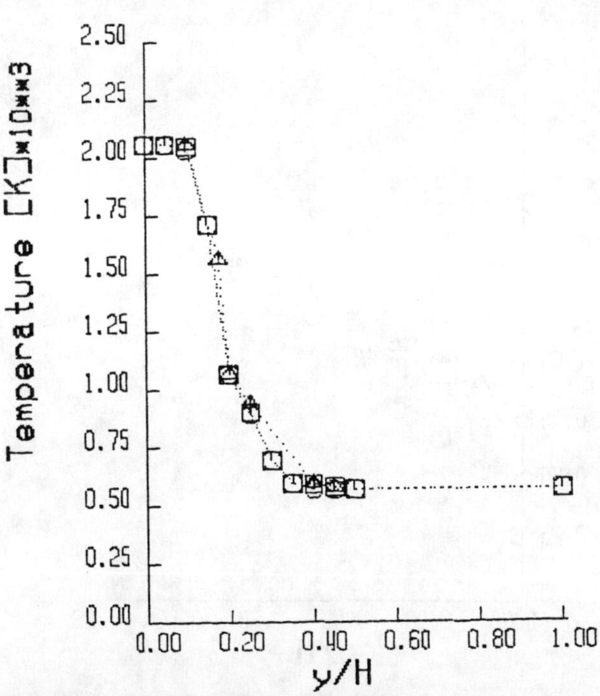

Figure 2 - Premixed turbulent reacting flow: comparison of predicted and measured temperature distributions.

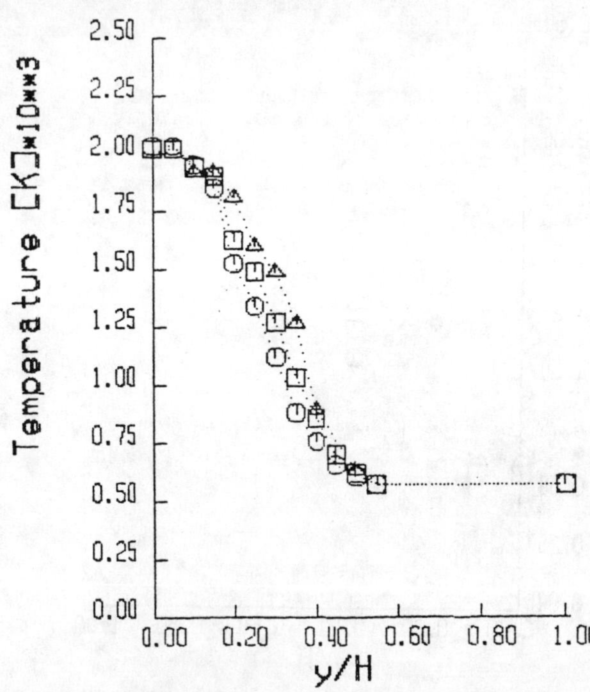

Figure 3 - Premixed turbulent reacting flow: comparison of predicted and measured temperature distributions.

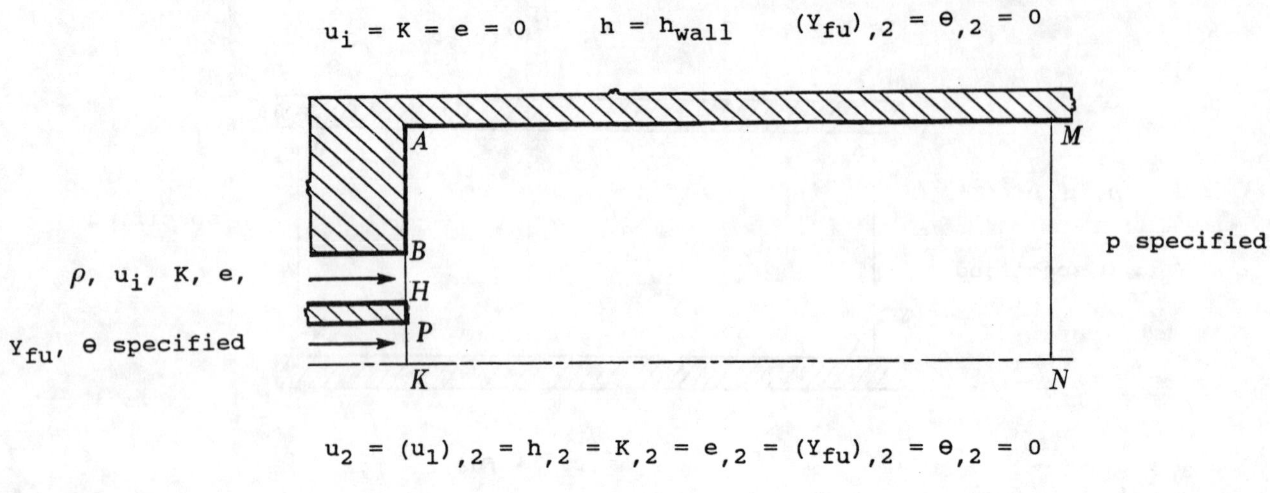

Figure 4 - Diffusion turbulent reacting flow: physical domain.

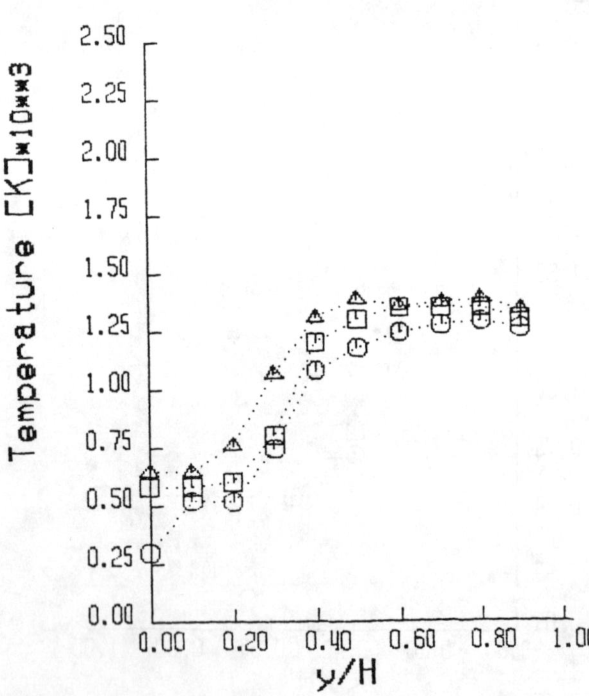

Figure 5 - Diffusion turbulent reacting flow: comparison of predicted and measured temperature distributions.

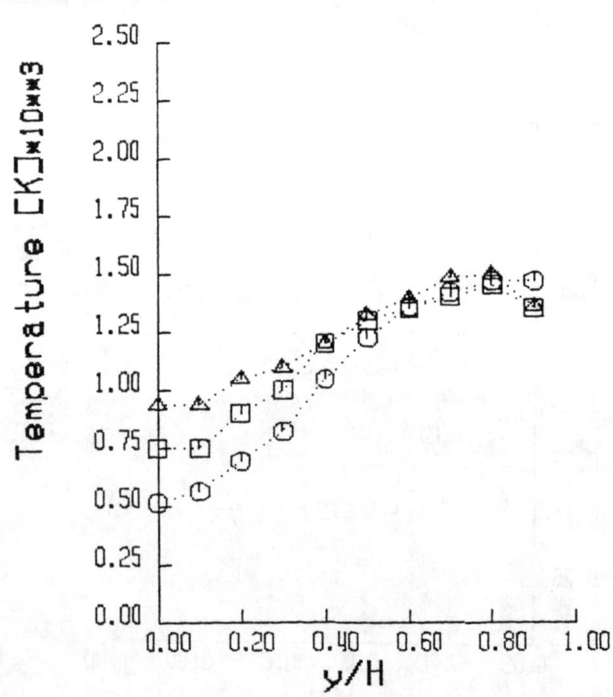

Figure 6 - Diffusion turbulent reacting flow: comparison of predicted and measured temperature distributions.

1989 National Heat Transfer Conference
HTD-Vol. 106, Heat Transfer Phenomena in Radiation, Combustion and Fires

COMBUSTION OF METHANE IN A RIJKE PULSATING COMBUSTOR

G. A. Brinckman and D. L. Miller
Department of Mechanical Engineering and Mechanics
Drexel University
Philadelphia, Pennsylvania

ABSTRACT

This paper presents the results of initial investigations of burning methane in a Rijke pulsating combustor. The combustor consists of a vertical tube with decoupling chambers attached at both ends. Methane was delivered through a homemade nozzle which was oriented axially into the combustor and located one quarter of the distance from the bottom opening of the tube (L/4, where L is the tube length). The interaction between the combustion process and the oscillating velocity and pressure fields produces pulsating combustion, which is accompanied by a loud, deep audible tone. This study demonstrated that pulsating combustion can be achieved under both self-aspirating and forced mode conditions for a variety of operating ranges. Initial studies revealed the dependence of the sound pressure level on independent variables such as fuel flow rate, combustor tube length, location of the fuel nozzle about the L/4 position and equivalence ratio. The experiments revealed that pulsating combustion could be achieved within a range of fuel flow rates from as low as 0.5 L/min to as high as 7 L/min. It was shown that increased sound pressure levels could be attained with the exit of the fuel injector nozzle located above the theoretically optimum location of L/4. These studies indicated that the combustor can be operated under a variety of conditions.

NOMENCLATURE

a Constant Value

c Sonic Speed

L Combustor Tube Length

n Harmonic Number

p Pressure

P Pressure Waveform Amplitude

q Instantaneous Heat Release Rate

Q Fuel Flow Rate

R Ideal Gas Constant

t Time

T Temperature

u Fluid Velocity

U Velocity Waveform Amplitude

x Heat Source Position Relative to Bottom of Tube

X Nozzle Exit Position Relative to L/4

Greek

ø Equivalence Ratio

ϕ Phase

ω Angular Frequency

γ Specific Heat Ratio

Subscripts

o reference level

n nth mode

Superscripts

\- mean value

' acoustic perturbation

INTRODUCTION

A research program has been developed to investigate the combustion characteristics of a Rijke-type pulsating combustor. This paper describes the results of a study in which methane was burned in the combustor while varying several independent parameters that affect the combustor's performance. The research was prompted by a desire to understand the mechanisms which control pulsating combustion. There is a lack of information about what mechanisms control pulsating combustion and how they affect its operation. Although this has not affected the development of some applications, such as rocketry, space heating, water heating and food drying, further development is hindered by the lack of fundamental knowledge and fear that venturing into uncharted operational regimes will end in failure.

Interest in pulsating combustion exists because it offers the following advantages over conventional technologies: higher combustion intensities, higher combustion efficiencies, lower NO_X and other pollutant formation, higher thermal efficiencies, higher mass and heat transfer rates, better air/fuel mixing, which reduces the amount of excess air required, reduced slagging and cleaner heat transfer areas (Carvalho et al., 1984). These advantages are a direct result of the highly turbulent flow field created within the combustor by oscillating pressure and velocity fields. The oscillatory pressure and velocity fields established inside of the combustor are believed to improve the rates of mass, momentum and heat transfer processes (Severyanin, 1982). The oscillating flow fields are produced when the heat released by the combustion process excites one or more of the combustor's natural acoustic modes.

Pulsating combustion conditions exist when the combustion driven pressure and velocity oscillations are spontaneously excited by the heat released from the combustion process. The interaction between the combustion process and the flow oscillations results in periodic combustion. The energy supplied to maintain the oscillations must compensate for the energy lost due to viscous dissipation, heat transfer and radiation of acoustic energy out of the system (Zinn, 1985). In order to drive and maintain the pulsations, the heat addition process must satisfy Rayleigh's criterion. Rayleigh's criterion suggests that the periodic release must be in phase with the pressure oscillations (Rayleigh, 1945). This is expressed mathematically as:

$$\int_{cycle} p'q'dt > 0, \tag{1}$$

where the integral is evaluated over one cycle of oscillation.

Equation (1) can be derived from conservation considerations and can be mathematically applied to a continuously distributed heat source, which can be regarded as the superposition of many concentrated heat sources (Carvalho et al., 1989). The heat release rate term depends on the total instantaneous velocity of the convected flow. Carvalho et al. (1989) and Zinn et al. (1979) assumed an expression for q' of their analysis for a combustion process such that:

$$q' = a\,|\bar{u} + u'|, \tag{2}$$

where a is a constant. Neglecting damping, the acoustic velocity and acoustic pressure associated with the nth harmonic of an open-ended tube are given as:

$$u' = -U_n \cos\left(\frac{n\pi x}{L}\right) \sin(\omega t), \tag{3}$$

$$p' = P_n \cos\left(\frac{n\pi x}{L}\right) \sin(\omega t + \phi), \tag{4}$$

where ω is the angular frequency which is expressed as:

$$\omega = \frac{n\pi c}{L} . \tag{5}$$

The value of c can be estimated by applying the equation to determine the acoustic speed in a perfect gas, which is:

$$c = \sqrt{\gamma RT}, \tag{6}$$

where R is the ideal gas constant of air.

Three major classes of pulsating combustors have been developed to date. These are the Schmidt tube, the Helmholtz resonator and the Rijke tube (Zinn, 1985). Each can be categorized according to the acoustics involved in their operation. The Schmidt tube exhibits the acoustics of the quarter-wave tube, which consists of a pipe closed at one end and open at the other end. The acoustics of the Helmholtz resonator resemble that of the Schmidt tube. The Rijke tube combustor exhibits the acoustics of a half-wave tube (open-ended tube).

The acoustic Rijke tube was discovered in 1859. Rijke observed that placing a heated metal gauze at the L/4 position in a vertical tube open at both ends spontaneously excited the natural acoustic modes of the tube (Rayleigh, 1945). The resulting oscillations continued as long as the metal gauze remained hot and positioned at the L/4 location. In Rijke pulsating combustion, the hot metal gauze is replaced by a combustion process.

Each combustor possesses regions where pressure and/or velocity oscillations are significant. The combustion process in the Schmidt tube and the Helmholtz resonator occurs in the vicinity of the inlet section where only the magnitude of the pressure oscillations is significant. However, the combustion process in the Rijke tube occurs at the L/4 position, where the amplitude of both the pressure and the velocity oscillations is significant. This is important because observations indicate that the velocity oscillations play a more significant role in the enhancement of the mixing processes (Zinn, 1986). Another advantage offered by the Rijke tube combustor is that it is self-aspirating, eliminating the need for mechanical valves, which are required by the other two combustor types. Therefore, the configuration based on the Rijke tube was chosen, designed and constructed to study pulsating combustion.

The initial objective of the research program is to characterize the combustor based on the sound pressure level produced within the unit during pulsating combustion. It was previously mentioned that pulsating combustion could be sustained if the energy supplied to the oscillations by the driving mechanisms was sufficient to overcome the dissipation of energy caused by viscous interactions and acoustic radiation from the system. Therefore, one could argue that the amplitude of the acoustic pressure wave is proportional to the energy available to overcome dissipation. Determining the effect of independent parameters such as fuel delivery rate, stoichiometry, combustor length and location of the fuel nozzle exit location on the sound pressure level may suggest either that the damping phenomena or the effectiveness of the driving mechanisms or both are influencing the flow oscillations. Therefore, measuring the sound pressure level provides information about the overall mechanisms which drive the oscillations.

EXPERIMENTAL SECTION

The facility consists of a main combustor unit, a fuel and air supply system, temperature and pressure instrumentation and the capability to support exhaust analysis equipment. The combustor, when fully assembled, consists of five flanged sections, each of which is 51.8 cm in length with a 10.2 cm internal diameter. The sectioned design allows for the study of the effect of the combustor tube length on pulsating combustion performance. The overall length of the combustor with the five sections assembled is 2.63 m.

The tube is made from a Nickel alloy which is capable of withstanding temperatures up to 2300 K and is more resistant to corrosion than stainless steel. This is important because future studies may include burning halogenated and sulfonated hydrocarbons. Each section contains an array of instrument ports where probes may be inserted into the tube to measure temperatures, acoustic pressures and to extract exhaust gas samples for analysis. A quartz viewing window is located in the second section of the combustor near the L/4 location to allow for visual observation of the flame under operating conditions. A schematic diagram of the combustor is shown in Figure 1.

The combustion air is supplied by an in-house compressor. The flow rate is monitored by a rotameter and the delivery pressure is controlled by a pressure regulating valve. The air is supplied through a manifold to the lower decoupling chamber located at the bottom end of the combustor tube. The products leave the combustor via an upper decoupling chamber which leads to an exhaust stack.

The air inlet manifold consists of a one inch diameter inlet and four one inch diameter outlets leading to the inlet decoupling chamber. This is necessary to assure that the flow of air entering the decoupling chamber does not issue directly into the entrance of the combustor and interfere with the velocity oscillations, which would alter the acoustically open-ended boundary condition. This configuration effectively reduces the velocity of the air delivered to the combustor and uniformly distributes the air entering the decoupling chamber.

The decoupling chambers provide a pseudo open-ended boundary condition which is required by Rijke pulsating combustors,

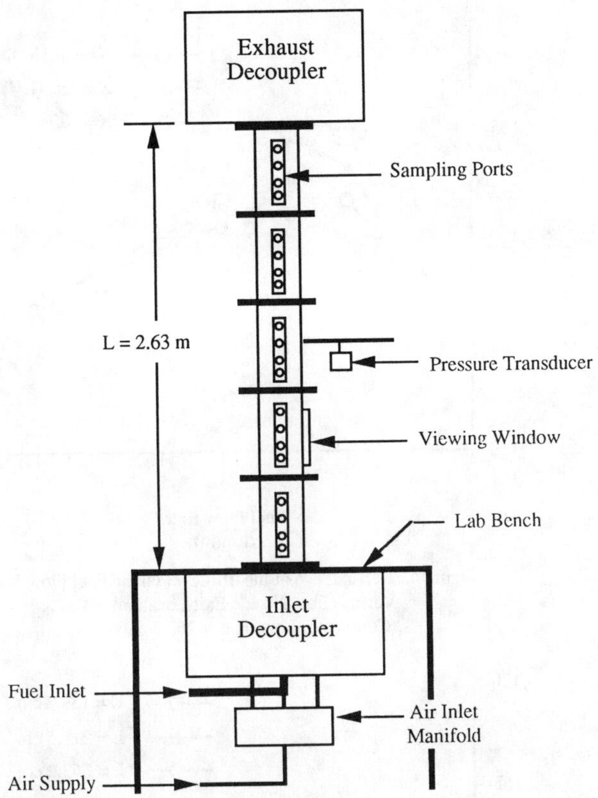

Figure 1. Schematic Diagram of the Rijke Tube Pulsating Combustor and Support Systems.

i.e., p' = 0 at the entrance and the exit of the combustor. If the open-ended boundary condition is not satisfied, the acoustic configuration of the system changes and the theoretically optimum location for the combustion zone for driving the oscillations will not be at the L/4 location.

While the ultimate plan is to operate with liquid fuels, for characterization studies, a gaseous fuel system was assembled utilizing the fuel line and the flow control devices purchased for liquid fuel operation. The fuel was supplied by a nozzle fabricated to provide a flame similar to that produced by a Fischer burner. The nozzle produces a premixed flame by entraining combustion air approximately six inches below its exit plane.

The amplitude of the pressure oscillations produced within the combustor were measured with an Omega PX-242 piezoelectric pressure transducer located at the midpoint of the combustor tube where the amplitude of the fundamental mode of the pressure oscillation is a maximum. The signal is analyzed by a Norland Prowler digital oscilloscope which provides amplitude and frequency information of the pressure waveform. The measured pressure amplitude is converted to sound pressure level, or decibel level, measured in decibels, by the following equation:

$$\text{S.P.L.(decibels)} = 10 \log_{10} (P^2/P_o^2), \qquad (7)$$

where P is the measured pressure amplitude and P_o equals 2 x 10^{-5} Pa, which is the zero decibel reference level.

Experiments were conducted to determine the effect of varying independent variables such as combustor length, combustion zone location and fuel flow rate on the sound pressure level while operating under self-aspirating conditions. The tests were performed with tube lengths of 159.7 cm (three flanged sections), 211.5 cm (four flanged sections) and 263.2 cm (five flanged sections). The fuel nozzle position was varied +/-10.16 cm about the L/4 location while varying the fuel flow rate to determine the effect of the fuel flow rate on the sound pressure level.

The fuel flow rate was varied within the range in which pulsating combustion could be sustained. The effect of the fuel nozzle exit location on the sound pressure level was determined by varying its position +/-20 cm relative to the L/4 location. Finally, similar tests were conducted while operating under forced mode conditions to determine the effect of the equivalence ratio on the decibel level.

RESULTS AND DISCUSSION

Initial tests of the combustor were conducted while operating under self-aspirating conditions. The upper and lower decoupling chambers were removed and the combustor was allowed to entrain air from the surrounding environment. Data were taken and some general observations were made while the combustor was operating under pulsating conditions. The tests demonstrated that the nozzle design produced conditions conducive to pulsating combustion. Pulsating combustion was achieved almost instantaneously after the fuel was ignited. Once pulsating conditions were achieved, the flame appeared to be extremely agitated, deep blue in color, and only a few inches in length, as opposed to a long, yellow-orange, laminar flame present during non-pulsating conditions.

There existed a minimum fuel flow rate below which the flame would blow out several seconds after ignition. During these few seconds, pulsating conditions existed, but the highly turbulent environment produced by the oscillating flow field extinguished the flame. There existed also a maximum fuel flow rate for each set of conditions above which the pulsations were damped. As the flow rate of the fuel was increased, the flame grew in length until the pulsations were damped, whereupon the flame became laminar in nature. It is known that the presence of a long flame causes a shift in the phasing between the pressure and velocity standing waves, thus changing the phasing requirements for driving the oscillations (Putnam, 1964).

Pulsating combustion was accompanied by a deep, audible tone. The waveform recorded by the oscilloscope was sinusoidal, as opposed to a straight line produced by non-pulsating conditions. "Proper" pulsating combustion was characterized by a high and relatively constant amplitude of the sine wave generated on the oscilloscope.

The fundamental mode of the pressure oscillations was excited for each length of the combustor tube used for this study. The frequency of the waveform measured by the oscilloscope corresponded to the calculated value of the fundamental frequency, which was based on combustor tube length and the average temperature of the flow field. The sonic speed was calculated using equation (6) and thus the frequency was determined knowing that the wavelength of the fundamental mode of the oscillation is equal to twice the length of the combustor tube.

It was shown that the intensity of the pulsations could be varied by varying the amount of energy supplied to the system, which is dictated by the amount of fuel delivered to the system. Figure 2 shows the effect of the fuel flow rate on the rms sound pressure level while varying the location of the fuel nozzle about the L/4 position, where L = 159.7 cm. For this set of conditions, the minimum and maximum flow rate of fuel which sustained pulsating combustion were 0.5 L/min and 6 L/min, respectively. Figure 3 and figure 4 show the same relationship between the decibel level and the fuel flow rate but with a combustor tube length of L = 211.5 cm and L = 263.2 cm, respectively. Each curve displays the same trend. The sound pressure level initially increases with increasing fuel flow rate until it reaches a peak, where it gradually decreases to below the initial level. For each case, the peak sound pressure level increased as the fuel nozzle was traversed through the L/4 location, indicating that the theoretical L/4 position is not absolute and that the optimum location for the combustion zone varies with operating conditions.

The study also showed a dependence of the sound pressure level on the length of the combustor tube. Figure 5 shows the effect of varying the combustor tube length on the sound pressure level

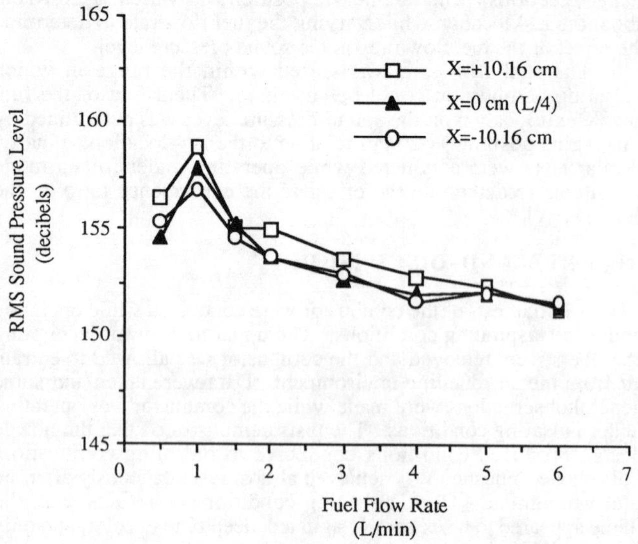

Figure 2. Dependence of the dB Level on the Fuel Flow Rate at Various Fuel Nozzle Exit Locations (Combustor Length, L = 159.7 cm)

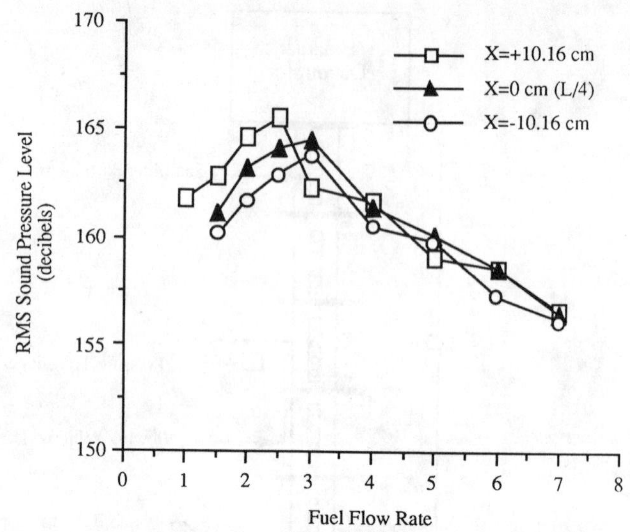

Figure 4. Dependence of the dB Level on the Fuel Flow Rate at Various Fuel Nozzle Exit Locations (Combustor Length, L = 263.2 cm)

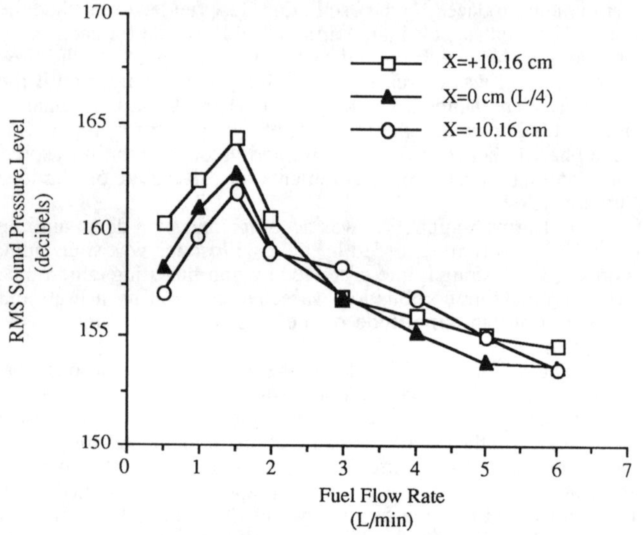

Figure 3. Dependence of the dB Level on the Fuel Flow Rate at Various Fuel Nozzle Exit Locations (Combustor Length, L = 211.5 cm)

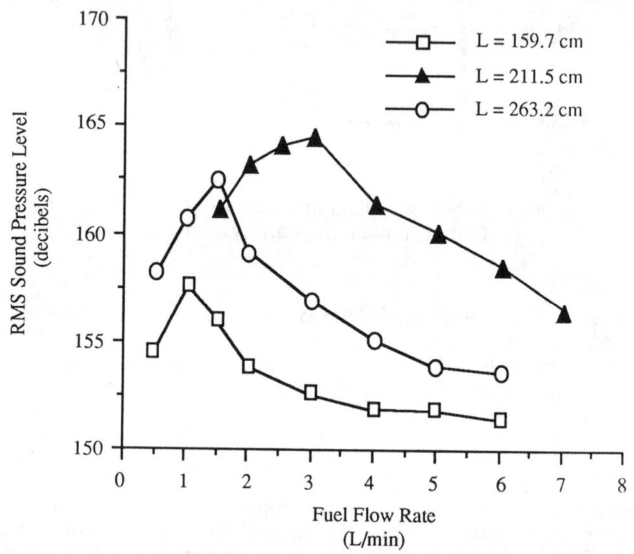

Figure 5. Dependence of the dB Level on the Fuel Flow Rate for Various Combustor Tube Lengths. (Fuel Nozzle Exit located at L/4 for each case)

with the exit of the fuel nozzle located at L/4 for each case. As the length of the combustor is increased, the maximum sound pressure level attained increases and the corresponding fuel flow rate also increases. Graphically, this is represented by a shift in the curves up and to the right as the combustor length is increased. The maximum fuel flow rate which produced pulsating combustion conditions increased also with increasing combustor length.

Figure 6 indicates more clearly the dependence of the sound pressure level on the nozzle exit position at various fuel flow rates for a combustor length of 211.5 cm. The location of the fuel nozzle exit was varied up to +/-20 cm about the L/4 position in order to acquire an extended view of the dB level profile. The figure shows that there is a slight dependence on varying the fuel nozzle exit location about the L/4 position. It shows that for a given fuel flow rate there appears to be an nozzle location above L/4 for which the sound pressure level reaches a maximum. The results for combustor tube lengths of 159.7 cm and 263.2 cm exhibited the same trends as those indicated in Figure 6.

Several tests were conducted with the combustor operating under forced mode conditions. The upper and lower decoupling chambers were attached and a predetermined amount of air was supplied to the combustor. Figure 7 shows the dependence of the sound pressure level on the nozzle exit position and the equivalence ratio at a fuel flow rate of 3.0 L/min. The figure shows that pulsating combustion is possible for both fuel rich ($\phi > 1$) and fuel lean ($\phi < 1$) conditions. The plot for each value of ϕ exhibits the same trend. The decibel level increases as the nozzle exit position moves towards the L/4 position, reaches a peak and decreases. As the equivalence ratio increases (by decreasing the supply of air and holding the fuel flow rate constant) for a given nozzle exit position, the decibel level increased, indicating a possible dependence on the air supply rate. It is known that the sound pressure level exhibits a

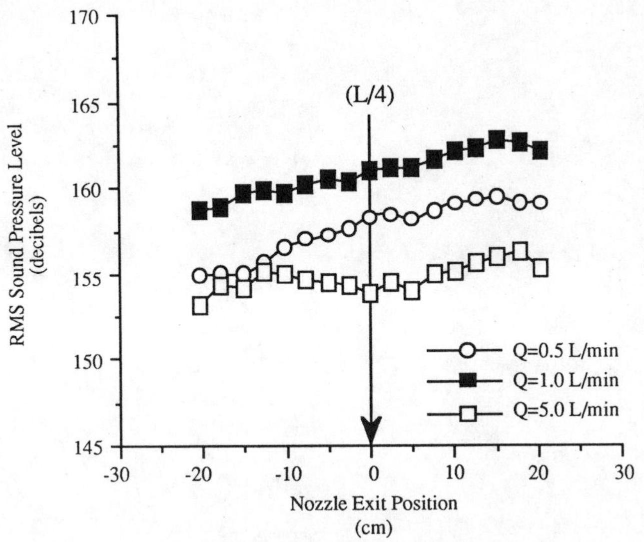

Figure 6. Dependence of the dB Level on the Fuel Nozzle Exit Position for Various Fuel Flow Rates (Combustor Length, L = 211.5 cm)

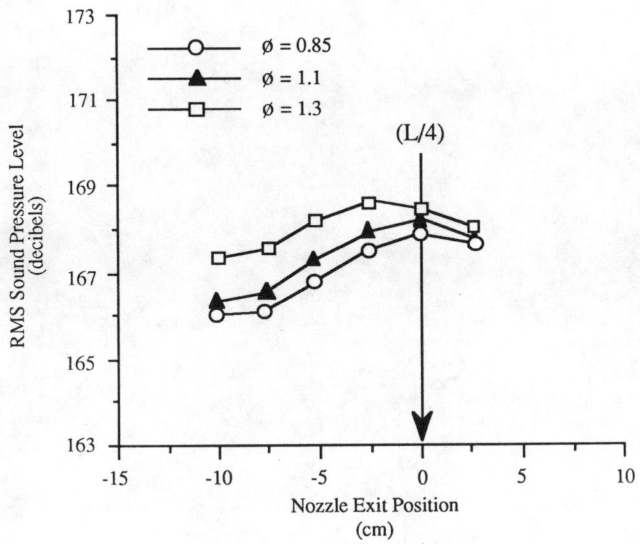

Figure 7. Dependence of the dB Level on the Fuel Nozzle Exit Position for Various Stoichiometries (Combustor Length, L = 263.2 cm)

dependence on the magnitude of the upward steady flow produced within the tube (Romie, 1956). Therefore, the flow of the combustion air will have an effect on the flow field created by the oscillating velocity field. However, further studies will have to be conducted to determine the extent of this dependence.

SUMMARY AND CONCLUSIONS

The Rijke pulsating combustor developed for this research program has been built and is operational. Initial tests were conducted with the combustor operating under both self-aspirating and forced mode conditions while burning methane. The data presented show that pulsating combustion conditions can be achieved for a variety of fuel flow rates, equivalence ratios, nozzle exit positions and combustor lengths. Initial tests showed the effect of varying these independent variables on the sound pressure level encountered inside the combustor. The tests under self-aspirating conditions showed that there is a range of fuel flow rates for which pulsating combustion can be sustained.

It was shown that pulsating combustion can be achieved with the heat source located about the geometric quarter length of the tube from its entrance. The unit displayed a great deal of flexibility with regard to this operational parameter. When operated under forced mode conditions, pulsating combustion could be sustained for a range of equivalence ratios. However, each produced nearly the same dB level, indicating that operating under fuel lean conditions may be a viable alternative to operating under fuel rich conditions without sacrificing performance.

REFERENCES

Carvalho, J. A., Jr., Wang, M. R., Miller, N., Daniel, B. R. and Zinn, B. T., 1984, "Controlling Mechanisms and Performance of Coal Burning Rijke Type Pulsating Combustors." *Twentieth Symposium (International) on Combustion*, The Combustion Institute, pp. 2011 - 2017.

Carvalho, J. A., Jr., Ferreira, M. A., Bressan, C., Ferreira, J. I. G., 1989, "Definition of Heater Location to Drive Maximum Amplitude Acoustic Oscillations in a Rijke Tube," *Combustion and Flame*, Vol. 76, pp. 17 - 27.

Putnam, A. A., 1964. "General Considerations of Autonomous Combustion Oscillations," Nonsteady Flame Propogation, Markstein, G. H., ed, Pergamon Press, New York, pp. 183 - 198.

Lord Rayleigh, 1945, The Theory of Sound, Vol. II, Dover Publication, New York.

Romie, F. E., 1956. "Heat Transfer to Fluids Flowing with Velocity Pulsations in a Pipe." Ph. D. Thesis, University of California.

Severyanin, V. S., March, 1982, "Application of Pulsating Combustion in Industrial Installations." *Proceedings of the Symposium on Pulse Combustion Applications,* Vol. 1, Paper No. 7, Atlanta, GA.

Zinn, B. T., August 1985 "Pulsating Combustion." *Mechanical Engineering*, pp. 36 - 41.

Zinn, B. T., Feb., 1986 "Rijke Pulse Combustors for Liquid Fuels," *The ECUT Pulse Combustion Research Program - A Milestone Report, Sandia National Laboratories*, pp. 57 - 64.

Zinn, B. T., Daniel, B. R., Sheshadri, T. S., 1979, *Proceedings of the Symposium on Pulse Combustion Technology for Heating Applications*, Argonne National Laboratory, pp. 239 - 248.

1989 National Heat Transfer Conference
HTD-Vol. 106, Heat Transfer Phenomena in Radiation, Combustion and Fires

DIAGNOSIS OF THERMAL AND FLUID DYNAMIC PERFORMANCE OF A BOILER

F. W. Bauer, R. V. Nayak, C. R. Worsley, and T. P. Tonden
Stone & Webster Engineering Corporation
W. Kohanik
Public Service Electric and Gas Company
Boston, Massachusetts

ABSTRACT

A mathematical model of a boiler was developed for use as a diagnostic tool and for optimized corrective design in problem resolution. The model was developed as a series of individual detailed models based upon first principles, which are used for specific study purposes. The data from these models can be reduced to equations, matrices, or families of curves and input into an overall performance program, which can be used for establishing the interrelationship of the various components for transient studies. These programs are described in a series of technical papers ([1-7]), which have been included in the appendix to this paper. This paper describes the application of these methods to a boiler of very unique design to determine a reason for excessive wall tube burnouts and to analyze manufacturers' proposals to correct the problem, in which solutions differed widely. The burnouts were found to be the result of departure from nucleate boiling at low loads, which was exacerbated by the change in boiler operation from designed base loading to cycling service. The boiler model was used to evaluate alternative solutions, to select the lowest cost acceptable proposal, and to recommend ways to reduce cost and improve performance. The model was also used to select alternative fuels.

This paper is intended to demonstrate methods of applying the model and approach to problem solution and offers the attached documents for description of the theory.

INTRODUCTION

This paper describes the analytical approach used and the results achieved in the assessment of the thermal and fluid dynamic performance of a 320-MW, wet bottom, coal-fired utility boiler to determine the causes of excessive furnace wall tube failures. The arithmetic and first-principle applications described in the preceding referenced documents are not repeated herein, except to demonstrate their use in a complex problem. The analysis evaluated and determined the technical viability of boiler redesigns provided by two competitive boiler manufacturers.

The failures were demonstrated to be a direct result of changing system load demands. The boiler, originally designed for base-load operation, is not presently suited for cyclic operation. Reducing load increases the occurrence of tube failures as a result of the departure from nucleate boiling (DNB) phenomenon. In the first performance analysis using unit operating records, boiler performance test data, and computer modeling to determine design deficiencies, it was determined that tube failures were caused by flow distribution in both the waterwalls and the furnace wall radiant superheater tubes and by flow instability due to low mass velocities and high heat burn in the superheater tubes. At lower loads, these deficiencies exacerbated the problem. Low-mass velocities in the radiant superheater tubes caused high tube metal temperatures, resulting in high-temperature oxidation and scale buildup inside the tubes, which led to tube failure. In the water-cooled lower furnace, sidewall tubes have a very short heated length compared to their total length, and therefore operate close to DNB conditions. Substituting ribbed tubes for the original smooth tubes in the lower furnace provided some safety margin; however, because the minimum flow requirement is not met at low loads in the sidewall tubes, ribbed tubes are ineffective in preventing DNB.

Therefore, the existing design proved unsuitable for reliable cyclic operation at low loads and required modification. The proposed solution by the original equipment manufacturer (OEM), upgrading tube materials and converting the primary superheater to all upflow walls, provided an effective solution to many problems and was predicted to provide reliable operation and reduced maintenance costs. The calculated model assessment showed that the unit should perform reliably in cyclic mode in the 78 to 320-MW range if the tubes were upgraded to 304H. Although these modifications would permit cyclic operation over the load range, the boiler would not be suitable for on-off cycling because of the steam-cooled radiant superheater furnace walls. Alternative proposals by other manufacturers required removal of the primary superheater from the furnace walls and redesign to conventional convective sections. While these extensive revisions were better suited to cyclic operation or two-shift cyclic mode, the cost was too high for this single advantage.

Boiler mathematical modeling methods developed for this boiler provided a description of the problems causing the failures and a low-cost study of alternative solutions with a high degree of confidence in the results. This approach is better suited and less expensive than the traditional approach of diagnosing the causes for excessive furnace and convection surface tube failures by extensive field testing and trial-and-error field modifications.

The specific causes of wall tube characteristic failures appeared to be determined best by development of a boiler-unique thermodynamic model of furnace performance of sufficient specific location accuracy to determine waterwall and primary superheater heat transfer flux density and resulting two-phase natural circulation flow and to use this data to anticipate DNB. A utility seeking to analyze the boiler performance of two almost identical units contracted to analyze only one and to use the history of the second as background information.

Placed into service between 1960 and 1962, each unit consisted of twin pressurized, wet bottom, natural circulation divided furnaces which were front-fired with pulverized coal. Each single reheat boiler was designed to produce 2,060,000 lb/hr of steam at 2,450 psig and 1,100°F/1,050°F to produce 320 MW of electricity. Each boiler is operated at 2,200,000 lb/hr main steam flow at 1,050°F/1,050°F. The boiler is an unusual design in which the steam generating waterwalls form the lower portion of the furnace and the primary superheater surface forms furnace walls above the waterwalls. In the primary superheater, tubing is integral with the waterwalls in an upflow/downflow circuitry on all four furnace walls, from 10 ft above the centerline of the top burner row to the top of the furnace. A simplified circuitry diagram of the superheater furnace is shown on Figure 1. The reheater furnace is similar. Attemperation is provided by a condenser-type heat exchanger header at the drum outlet and biased firing of the two furnaces.

These boilers originally experienced numerous waterwall tube failures due to insufficient water mass flow. Tubes in the burner zone were replaced with spiral tubing, which reduced the problem to an acceptable level. Recently, the unit has been operated in cyclic mode, with the load varying between 140 and 320 MW. The unit has reported increasing occurrences of furnace wall radiant superheater tubing failures, which have increased unit maintenance costs. Moreover, changing load demands require modifications to the boiler so that it will operate reliably in cycling mode. Reducing the boiler load to the desired minimum load of 70 MW results in a sharp increase in radiant superheater tube failures; therefore, the minimum load is restricted to 140 MW.

Bids were solicited from the OEM and another boiler vendor to recommend changes so the boiler could be operated reliably in a cycling mode in the load range of 70 to 320 MW. The manufacturers offered radically different design modifications at radically different costs. Evaluation of the two proposals with regard to their appropriateness and reasonableness required a more complete analysis to determine the best solution for the owner. A zone analysis-type mathematical model[1] of the furnace was used, along with boiler history, field testing, and specialist review, to analyze tube failure causes and to evaluate the manu-

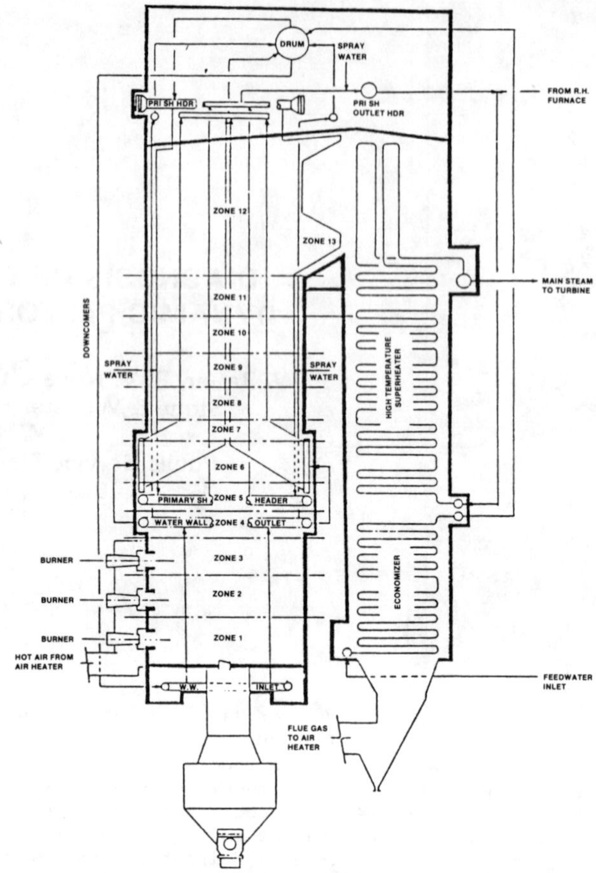

FIGURE 1

SUPERHEATER FURNACE ARRANGEMENT

facturers' proposed corrective modifications. The OEM's offer was determined to be acceptable with some recommended modifications, and was substantially lower in cost; therefore, it is the only offering described herein.

The boiler mathematical modeling methods described in this paper have been developed, and routinely used, to diagnose and correct boiler performance combustion-related problems at relatively low cost. The analyses are based upon fundamental heat transfer methods, zones, and finite-element applications and do not rely on previously manufacturer-developed test data, except as a check to ensure accurate results. The illustrative example shows how these methods are applied to diagnose and correct unique boiler problems. The methods can also be utilized to study, e.g., the causes of a boiler aperture deposit problem, and the effect of changing or blending fuels, or the capability of burning lower cost coal fuels.

BOILER PERFORMANCE ANALYSIS PROCEDURE

A normal approach to the diagnosis of excessive furnace and convective surface tube failures includes field testing and analytical methods. Solutions developed and implemented on a best-effort basis often represent less than optimum corrective measures, thus requiring subsequent corrections. It may be necessary to develop an alternative such as the selected option, in which a mathematical model of the boiler is adjusted to agree with actual boiler performance.

This approach reduces the amount of field data needed, identifies and corrects inaccurate field test data, and improves the confidence level of performance problem diagnosis. The model is modified to reflect recommended changes, and these recommendations are tested in the computer rather than on the boiler.

In the example, three operational characteristics were suspected of contributing to radiant superheater wall tube failures: improper heat transfer that caused high heat flux rates, spot absorptions, and uneven flow distribution that caused high tube wall temperatures. The reported waterwall tube failures were the result of low water circulation rates, which caused DNB. Changing the tubes in the lower furnace area previously to rifled tubing reduced the problem considerably. The mathematical model enabled a review of both failure areas. Although interrelated, the heat transfer and flow distribution analyses were performed separately to permit greater flexibility in investigating alternatives.

After the mathematical model was developed for the existing boiler configuration and adjusted to suit actual boiler conditions and fuels at various plant loads, proposed plant corrections were incorporated and calculations performed to analyze performance of the modified boiler.

Data Acquisition

The accuracy of and confidence in diagnostic methods depend upon valid input data. Design drawings, manufacturer-predicted performance (see Table 1), and plant operating test data (see Table 2) can be obtained from plant records to provide the data needed to develop a thermal performance model for the boiler.

Heat Transfer Analysis

The computerized analytical model of boiler performance developed for a furnace and combustion system consists of the following three calculation programs:

- Basic Combustion Model. Boiler operating data and fuel ultimate analysis are used to calculate fuel

TABLE 1

MANUFACTURER-PREDICTED BOILER PERFORMANCE

Furnace		Reheat		Superheat
Load		1/2		1/2
Steam	M lb/hr	985	2,060	1,075
Pressure-Drum	psig	2,602		2,607
Superheater Out	psig		2,456	
Temperature				
Reheater/Superheater Out	°F	1,050		1,100
Feedwater	°F		525	
Feedwater Leaving Economizer	°F	590		613
Air Entering Unit	°F		150	
Air Leaving Air Heater	°F	639		608
Gas Leaving Furnace	°F	2,005		1,940
Gas Leaving SH or RH	°F	932		1,075
Gas Leaving Economizer	°F	685		665
Gas Leaving AH, Corr	°F	280		275
CO_2 at Boiler Outlet	%		16.1	
Excess Air - Boiler Exit	%		15	
Wet Gas - Air Heater Exit	M lb/hr		2,810	
Air Entering Unit	M lb/hr		2,620	
Draft				
Boiler Loss	in. H_2O	0.25		
Superheater or Reheater	in. H_2O	2.26		2.06
Economizer	in. H_2O	1.38		2.10
Air Heater	in. H_2O	3.90		3.90
Flues	in. H_2O	1.18		1.18
Total	in. H_2O	8.97		9.49
Air Pressure Loss				
Air Heater	in. H_2O	3.55		3.55
Ducts	in. H_2O	0.10		0.10
Air Pressure in Windbox	in. H_2O	4.60		4.60
Air Pressure - Total	in. H_2O	8.25		8.25
Water Pressure - Through Economizer	psi	50		50
Fuel Burned	M lb/hr		205	
Liberation, Efficient Volume	Btu/hr/cf	14,150		14,150
Liberation, Total Volume	Btu/hr/cf	74,400		74,400

Boiler		
Waterwalls	sq ft	31,250
Radiant SH, Projected	sq Ft	34,122
Convective Superheater	sq ft	16,024
Economizer	sq ft	79,740
Air Heater	sq ft	96,000
Reheater	sq ft	714,000
Total furnace volume	cu ft	88,550
Projected furnace envelope	cu ft	188,332
	sq ft	38,556

Fuel		Bit Coal
Proximate Analysis		
Moisture	%	5.0
Volatile Matter	%	36.0
Fixed Carbon	%	49.0
Ash	%	10.0
Ash Fusion Temperature	°F	2100
Ultimate Analysis		
Ash	%	10.0
S	%	3.0
H_2	%	4.5
C	%	72.1
H_2O	%	5.0
N_2	%	1.2
O_2	%	4.2
HHV	Btu/lb	13,000

Heat Losses		
Dry Gas	%	2.88
H_2 and Fuel Moisture	%	3.74
Moisture in Air	%	0.07
Unburned Combustion	%	0.35
Radiation	%	0.15
Unaccounted for	%	1.50
Total Losses	%	8.69
Efficiency	%	91.31

Number of Units 2 Design Pressure 2,735
Description: Foster Wheeler pulverized coal fired, divided twin furnaces, natural circulation, wet bottom.

TABLE 2

SUMMARY OF SUPERHEATER FURNACE TEST DATA

Load (MWG)	320	224	187	78
Feedwater				
Flow (lb/hr)	1,154,000	844,000	649,000	263,000
Pressure (psig)	2,910	2,785	2,785	1,985
Inlet Temperature (°F)	530	485	470	400
Outlet Temperature °(F)	596	569	520	520
Primary Superheater Outlet Temperature (°F)	777	780	773	762
PSH Intermediate Header Temperature (°F)	-	694	-	-
Main Steam				
Flow (lb/hr)	2,204,000	4,582,000	1,353,400	606,000
Pressure (psig)	2,433	2,414	2,385	1,635
Temperature (°F)	1,024	982	998	956
Gas Temperature (°F)				
Economizer Outlet	705	618	652	563
Air Preheater Outlet	274	257	222	205
Air Temperature (°F)				
Air Preheater Inlet	130	142	102	98
Air Preheater Outlet	632	566	576	516
Total Air (%)	118	123	128	235
Attemperator Flow (lb/hr)	0.0	62,000	117,800	52,000

Fuel (Virginia High Volatile Coal)

Ultimate Analysis - (% wt)	
Carbon	75.48
Hydrogen	4.72
Oxygen	4.42
Nitrogen	1.06
Sulfur	0.97
Ash	6.90
Moisture	5.95
HHV (Btu/lb)	13,576

and air flow requirements, as well as flue gas flow and constituents for various boiler loads and operating conditions. This program also determines adiabatic flame temperature and the specific heat of resulting flue gases. The results are necessary inputs to the furnace zone analysis program.

- Furnace Performance Calculations. A furnace zone analysis program is applied to determine heat transfer in a furnace. The furnace is divided into 13 combustion zones, with each zone small enough that the gas within is well mixed and at a uniform temperature. The results of the basic combustion model, furnace geometry, flame and wall emissivities, and slag and scale thicknesses comprise program input data. The temperature leaving each zone is determined by balancing the next input and the heat released in the combustion process with the heat transferred to the walls and the areas above and below the zone. This program determines the gas temperature profile, heat transfer flux or surface heat absorption, temperatures of the tube walls for each tube type, and average flue gas retention time in each zone.

  The gas temperature profile obtained from the boiler zone analysis is subsequently utilized in the burn time program, and the fuel particle burning profile is used to make corrections, if necessary, to the initial heat release profile in the furnace.

- Fuel Burning Time. The burn time program calculates the time necessary to completely burn a fuel particle in a furnace. Actual fuel particle size distribution is used, if available; otherwise, a Rosin-Rammler fuel size distribution is assumed. Some of the fuel particles are too large for complete combustion in the furnace; this provides the basis for calculating the amount of unburned carbon leaving the furnace. Boiler efficiency and loss of ignition then can be determined.

Although these programs were simplified to the accuracy necessary for the application described, they are phenomenological models of the processes which occur in a boiler. Results obtained from the model are compared to boiler tests conducted to ensure their validity.

Flow Distribution Analysis

The furnace waterwall circuits are analyzed to determine if any circuit circulation problems are evident and any waterwall tubes are in danger of overheating due to DNB. The calculations utilize close approximations of the circuit. The point at which the boiler waterwall circuit will operate for a given heat input and specific volume difference between the cold and hot legs of the circuit is calculated for various steam quantities. These points are compared to published data concerning minimum mass flow required to determine the potential of DNB.

Steam flow in the radiant superheater tubes is analyzed by calculating pressure drops in various parts of the steam circuit. Based on these calculated pressure drops, the potential for steam flow imbalances in the various parallel circuits is determined:

- Waterwall Mass Flow Rates. Mass flow rate, defined as pounds per square foot of flow area per hour, is an important criterion in determining the adequacy of tube cooling. For each inlet temperature, pressure, and heat flux to tubes, there is a maximum permissible steam by weight (SBW) in the waterwall tubes which is dependent upon mass flow. This is the point at which tubes may overheat or burn out due to DNB. Designs which would result in operation at steam analog in excess of this limit could be subject to tube failures due to metal temperatures above material limitations. The permissible quality for a given heat flux, mass flux, and pressure can be increased using rifle-bore tubing to ensure a continuous water film adjacent to the tube metal surface. However, the quality of the mixture is not the only criterion of proper flow conditions. Minimum allowable internal velocities are also important, especially for rifle-bore tubes, to ensure efficiency. The relationship between the anticipated density and pressure drop of the flowing fluid must also be considered.

- Radiant Superheater Mass Flow Rates. Mass flow rates directly affect the temperature drop across the steam film layer at the inside diameter of the tube and determine mean tube metal temperatures. Tube metal temperatures and mass fluxes are utilized to select tube diameters, thicknesses, and materials of construction.

  Boiler manufacturers currently use mass flow rates of approximately 700,000 lb/hr-sq ft as standard guidelines for radiant superheaters at full-load operation.

Material Evaluation

A tube material evaluation is made of both the furnace waterwall tubes and the radiant superheater tubes to ascertain their suitability at the temperatures involved, from a high-temperature oxidation standpoint, and to determine the required minimum wall thickness.

PERFORMANCE ANALYSIS OF EXISTING BOILER DESIGN

Existing boiler circuitry and operating data can be analyzed to develop an understanding of past and present operating problems so that proposed unit modifications can be evaluated properly. Since some boiler features are uncommon, similarly designed units can be investigated. In the example, these also exhibited similar high frequencies of tube failures and difficulties in cycling.

Boiler Performance

The mathematical model of furnace performance reflected current boiler performance for the load range of 78 to 320 MW for the superheater furnace (see Table 2). In the example, the boiler was found to operate close to the original predicted performance. One difference, however, was that the main steam system, designed to operate at 1,100°F, was currently operating at 1,050°F. Reduced load boiler performance was close to that originally predicted. The heat absorptions in the water and steam circuits calculated by the model and obtained from field data were consistent over the load range.

The average furnace exit gas temperature (FEGT) was calculated for both the entrance to and exit from the furnace aperture. The furnace heat absorption distribution between waterwall and steam tubes is given in Table 3. The heat flux profile calculated along the height of the furnace is shown on Figures 2, 3, and 4 for the waterwall, upflow, and downflow radiant superheater tubes, respectively. This heat absorption profile is typical of a wet-bottom furnace.

TABLE 3

SUPERHEATER FURNACE PERFORMANCE

Load (MW)	320	300	224	187	78
Temperature (°F)					
Furnace Exit Gas	2,420	2,350	2,170	2,055	1,545
Aperture Outlet	2,160	2,115	1,935	1,835	1,450
Furnace Heat Absorption (MMBtu/hr)					
Waterwall	480	435	385	335	115
Radiant Superheater	230	210	170	135	40
Total	710	645	555	470	155

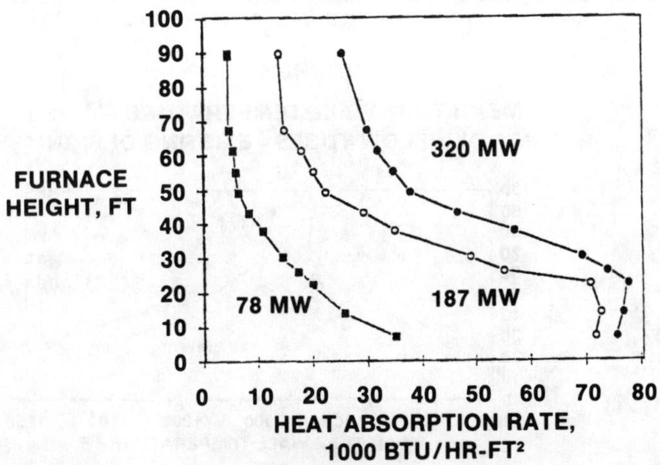

FIGURE 2

HEAT ABSORPTION RATE - WATERWALL TUBE - PRESENT DESIGN

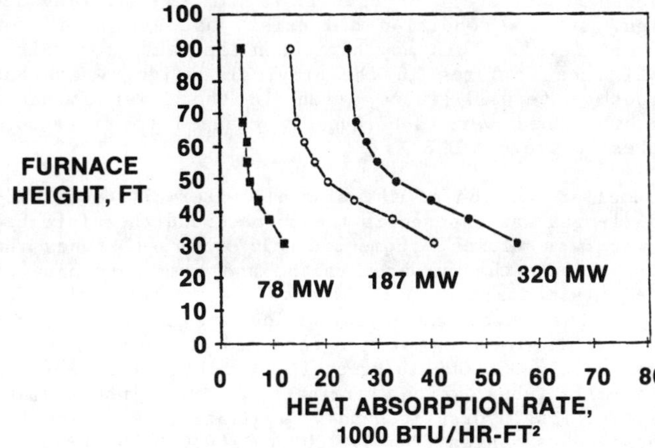

FIGURE 3

HEAT ABSORPTION RATE - STEAM DOWNFLOW TUBES - EXISTING DESIGN

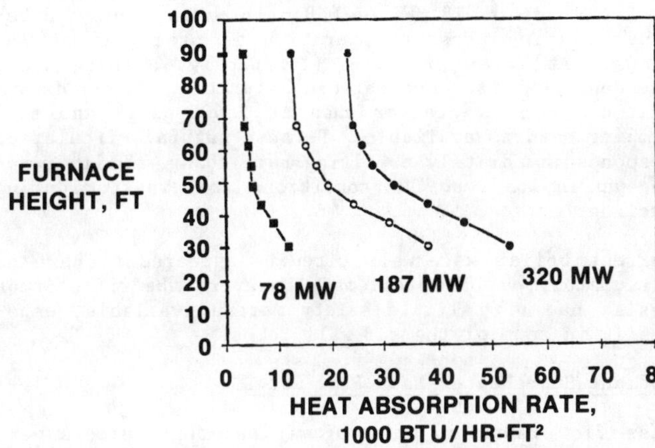

FIGURE 4

HEAT ABSORPTION RATE - STEAM UPFLOW TUBES - EXISTING DESIGN

Waterwall Mass Flow Rates

The waterwall circuits of the superheater furnace were analyzed to determine whether circulation problems were evident. Mass flux in the waterwall tubes was calculated by balancing the friction and other flow-related losses with the pumping head available due to density difference. At the full load of 320 MW, calculated mass flux was approximately 500,000 lb/hr-sq ft, and calculated recirculation ratio was nearly 4.5. Based on the heat flux to the waterwall tubes along the furnace height and the local steam quality and mass flux, it was concluded that mass flow in the front wall tubes, which run straight up to the upper header, was sufficient to prevent DNB. Similarly, the rear wall tubes and the front and rear wall tubes which form the division walls had sufficient safety margin to prevent DNB.

However, the mass flux of 500,000 lb/hr-sq ft in water-cooled smooth tubes in a natural circulation unit was marginal. At this mass flux, smooth tubes could be in DNB regime with conditions of 15 percent SBW and

75,000 Btu/hr-sq ft of heat flux. In the existing design, such a condition did exist just above the top burner level. This may have been the cause for waterwall tube failures in the original design, which had smooth waterwall tubes, even in the lower furnace. Smooth tubes were subsequently changed to rifle-bore tubes to prevent DNB.

A smaller portion of the sidewall waterwall tubes' total length was exposed in the furnace,and therefore had lower mass fluxes. Some sidewall waterwall tubes run straight up the furnace length, and some terminate at the intermediate header. The calculated mass flux for the tubes which run straight up the furnace length without terminating at the intermediate header is approximately 350,000 lb/hr-sq ft at full load. For the sidewall tubes which terminate at the intermediate header, less hydraulic head is available. Mass flow in these tubes ranges from 170,000 to 200,000 lb/hr-sq ft. Although these tubes are ribbed, the advantage of ribbed tubes is not available unless the mass flow is above a minimum level of approximately 200,000 lb/hr-sq ft. Therefore, the sidewall tubes which terminate at the intermediate header have a marginal safety factor to prevent DNB.

At low loads of 78 MW, no DNB problem is predicted because sliding pressure operation is practiced at low loads. At lower pressures (1,650 psig in this case), the density difference between saturated water and saturated steam is greater than at 2,600 psig, and more pumping head is available. Because natural circulation responds immediately to firing rate changes while ramping up in load, no DNB conditions are expected during cycling.

Current boiler waterwall circuits appeared to have no circulatory or DNB problems. However, the rifled-tube design had very little safety margin available, especially in some of the sidewall tubes.

Radiant Superheater Mass Flow Rates

Mass flow rates of existing radiant superheaters were substantially lower than accepted guidelines. Full load rates were as low as 430,000 lb/hr-sq ft in the downflow tube and as high as 650,000 lb/hr-sq ft in the upflow tubes. These rates were proportionately lower at lower loads.

Radiant superheater tube failures occurred primarily in the area immediately above the radiant intermediate header and midway between the header and the roof. Since steam temperatures at the intermediate header were relatively low and furnace temperatures above the burner zone were high, the heat flux to the radiant superheater tubes was high in this zone. Since the mass flux was low, mean tube wall temperatures in this zone were expected to be high.

Steam Flow Instability and Maldistribution

The present design of headers, feeder lines, and wall tubes in the superheater circuit resulted in a pressure-drop distribution in the heated and unheated circuit of equal order of magnitude. For uniform flow in the parallel heated circuits, the pressure drop in unheated circuits should be an order of magnitude lower than in the heated circuits. Any unequal distribution due to unsymmetrical arrangement of unheated circuit elements does not affect flow in the parallel heated circuit elements. In the existing design, the transfer risers in the radiant superheater circuit have a higher pressure drop than in the furnace tubes. These risers, of unequal length and connected to an end-connected outlet header, cause unequal flows in the furnace tubes.

The existing radiant superheater had upflow/downflow steam circuitry. Downflow of steam in heated tubes can easily lead to stagnation and even flow reversal under certain conditions. When the hydraulic head, due to elevation of the radiant superheater tube, exceeds the friction loss due to low flow in some tubes, it first causes flow stagnation and, upon further heating, results in flow reversal. From a tube failure point of view, flow stagnation is the worst condition. The existing design of radiant superheater tubes yields low mass flow velocities, and at loads below 135 MW, the pressure drop in the downflow tubes becomes less than the hydraulic head in the tube. Therefore, at loads below 135 MW, flow stagnation and reversal can occur in these tubes. In the example, the utility cycled the boiler down to 78 MW (gross), and this practice resulted in a high rate of radiant superheater tube failures. Currently, the lowest load on the boiler is limited to 145 MW. It is likely that repeated occurrences of stagnation in tubes led to rapid buildup of inside scale in these tubes due to high-temperature oxidation, which further exacerbated this condition.

Material Evaluation

The existing design was evaluated to determine whether materials used in the radiant superheater and waterwall tubes could withstand predicted tube metal wall temperatures. The unheated circuit elements were not evaluated. The boiler zone analysis program calculated the mean wall temperature along the length of the tubes. The mean tube wall temperature profiles for the downflow and upflow tubes for the entire load range are presented on Figures 5 and 6, respectively. The allow-

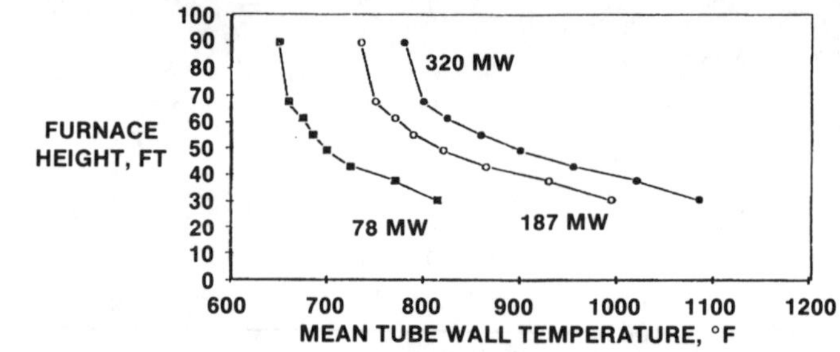

FIGURE 5

MEAN TUBE WALL TEMPERATURE - STEAM DOWNFLOW TUBES - EXISTING DESIGN

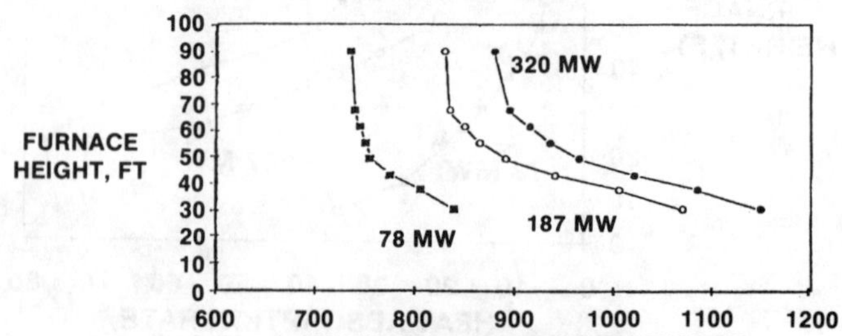

FIGURE 6

MEAN TUBE WALL TEMPERATURE - STEAM UPFLOW TUBES

able tube metal temperature of 1,050°F for the existing tube material (SA213-T11) of the downflow tubes and 1,125°F for the upflow tube material (SA213-T22) were predicted to be exceeded above the intermediate headers at loads greater than 187 MW. The model predicted the documented tube failures near this scissor area. The mean tube wall temperatures at loads less than 187 MW were below the maximum allowable temperature.

The high tube wall temperatures were predicted for tubes covered with 1/2 in. of slag and 20 mils of scale. The boiler zone model also was used to predict the tube wall temperatures of clean tubes. As expected, high spot heat absorption rates could cause tube wall temperatures to approach or exceed the maximum allowable temperature in areas up to the midpoint of the furnace. The conditions leading to high tube wall temperatures would be aggravated during startup and rampup operating modes. Clean tubes during startup and overfiring required by both operating modes produce the most sensitive combination of conditions, high heat and low mass fluxes.

Table 4 presents the allowable outer tube wall temperatures for materials used in the existing design and compares them with predicted temperatures. The predicted temperatures were calculated for high spot heat flux and upset steam flow due to flow imbalance. The radiant superheater tube materials near the scissors area, SA213-T11 and SA213-T22, are not adequate under certain upset conditions. However, the outer tube metal temperatures, which were calculated using average values for steam flow and heat flux with minimal amount of inside scale (5 mils), did not exceed the allowable temperature limit. This suggests that cycling of the unit, which generally results in upset conditions, caused scale buildup in these tubes. Once the scale thickness reached 30 to 35 mils, as reported from plant samples, the outer tube wall temperatures exceeded allowable limits, even under normal operating conditions. Scale buildup inside the tubes also caused higher mean tube wall temperatures. The minimum wall thickness required by the ASME Code for these increased temperatures exceeded that provided by the original design (see Table 4), thus contributing to the higher incidence of tube failures.

TABLE 4

TUBE MATERIAL EVALUATION - EXISTING DESIGN

Component	Tube Description (Outside diameter x mean wall) (in.)	Tube Material	Maximum Outer Temperature Expected (°F)	Tube Temperature Recommended (°F)	Minimum Wall Required (in.)
Waterwall					
Furnace basket	3 x 0.3	SA-210-A1	870	950	0.370
Above furnace basket	3 x 0.3	SA-210-A1	780	950	0.278
Radiant Superheater					
Downflow - All Walls:					
Below El 180 ft-11 1/4 in.	1 1/2 x 0.281	SA213-T11	1,230	1,050	0.689
Above El 180 ft-11 1/4 in.	1 1/2 x 0.281	SA213-T11	1,230	1,050	0.689
Upflow					
Front Wall:					
Below El 216 ft-10 1/2 in.	1 1/2 x 0.281	SA213-T11	1,325	1,050	-
Above El 216 ft.-10 1/2 in.	1 1/2 x 0.313	SA213-T22	900	1,125	0.146
Rear Wall:					
Below El 189 ft.-8 in.	1 1/2 x 0.281	SA213-T11	1,325	1,050	-
Above El 189 ft.-8 in.	1 1/2 x 0.313	SA213-T22	950	1,125	0.164
Sidewall:					
Below El 196 ft-6 in.	1 1/2 x 0.281	SA213-T11	1,325	1,050	-
Above El 196 ft-6 in.	1 1/2 x 0.313	SA213-T22	900	1,125	0.146

NOTES:

1. At maximum mean wall temperature as defined by ASME Boiler Code.
2. Five mils inside scale; full-load flux; no slag deposit.
3. No inside scale; full-load flux; no slag deposit.
4. From Babcock & Wilcox Steam Table 3, p. 29-11.

A number of tube failures occurred in the roof and aperture areas. An evaluation of these tubes showed that the high tube metal temperature could have resulted from either flow stagnation or initial inside scale formation from feedwater leakage in the condenser attemperator, both of which are likely causes. It is believed that the initial cause of internal scale formation is not high bulk steam temperature or high heat flux rate. Leaking feedwater is converted into steam, and impurities in the feedwater remain as scale. Internal scale acts as an insulating layer to the transfer of heat from the furnace gas to the steam and raises the tube metal temperature. As the tube metal temperature increases, so does the rate of internal scale formation. As scale thickness increases, so does the tube metal temperature; the cycle progressively gets higher with time, eventually causing tube failure.

PERFORMANCE ANALYSIS OF OEM-PROPOSED REDESIGN

The OEM analyzed present performance and operating problems of the boilers in detail and recommended design changes to correct the problems. Similar analyses were performed by the OEM for other projects featuring partially steam-cooled furnaces with downflow/upflow superheater circuitry. As a result of the investigations, the OEM proposed redesign of the radiant superheaters. The OEM did not recommend more extensive changes in overall boiler design, including conversion to a more typical all watercooled furnace with platen superheaters, because a complete design overhaul would be economically prohibitive.

The proposed redesign consisted of changing the primary superheater tube configuration in the walls - maintaining the same total heat absorption surface area, but increasing the number of tubes, reducing the diameter, and increasing tube wall thickness, resulting in a high steam flow mass velocity to increase mass flux rates. The downflow/upflow circuitry was converted to an all upflow arrangement. The all upflow design utilized a two-pass arrangement which uses unheated downcomers to transfer steam. The first pass consisted of the front and rear walls, and the second pass, the sidewalls. The second major redesign feature was upgraded tube materials selection and panelization of the radiant wall tubing. Finally, the addition of 8 ft of refractory was recommended in the lower furnace. The mathematical model prepared for the existing boiler was modified to reflect these changes for analysis of the modified boiler.

The study reviewed recommended modifications for the following areas:

- Boiler performance
- Mass flow rates - waterwall
- Mass flow rates - radiant superheater
- Steam flow instability and maldistribution
- Material evaluation

Boiler Performance

The predicted boiler performance for the proposed OEM redesign is shown in Table 5 for high (320 MW) and low (78 MW) loads.

Boiler performance remained close to current operating performance; however, total absorption increased slightly due to increased absorption by the redesigned radiant superheater surfaces. This resulted in lower gas temperatures leaving the furnace; therefore, final reheat and superheat temperature might not be attained. The recommended addition of an 8-ft-high refractory in the furnace is necessary if no other surface changes are made to increase FEGT or to increase absorption of the final reheater or superheater surfaces. The refractory also will reduce corrosion of waterwalls in the lower furnace.

TABLE 5

PREDICTED FURNACE PERFORMANCE - OEM REDESIGN (WITH REFRACTORY)

Load	320	78
Percent MCR	100	24
Furnace Heat Absorption (MMBtu/hr)		
Waterwall	465	105
Radiant superheater	250	45
Total	715	150
Furnace Exit Gas Temperature (°F)		
Before aperture	2,410	1,580
After aperture	2,150	1,480

The average FEGT is given for both the entrance to and exit from the furnace aperture. The furnace heat absorption distribution between waterwall and steam tubes was unchanged (see Table 5). The heat flux profiles along the height of the furnace are shown on Figures 7, 8, and 9 for the waterwall and radiant superheater tubes, respectively. The heat flux values are typical of a wet bottom furnace.

Waterwall Mass Flow Rates

The redesign did not change the waterwall circuits. However, the OEM recommended the addition of 8 ft of refractory in the lower furnace, just above the floor, to cause higher heat fluxes in the upper furnace. The present front and rear wall design of ribbed tubes in the lower furnace had sufficient safety margin that the

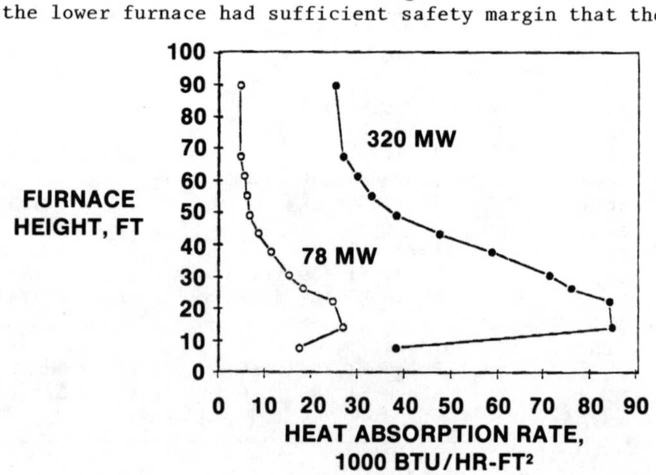

FIGURE 7

HEAT ABSORPTION RATE - WATERWALL TUBES - OEM REDESIGN

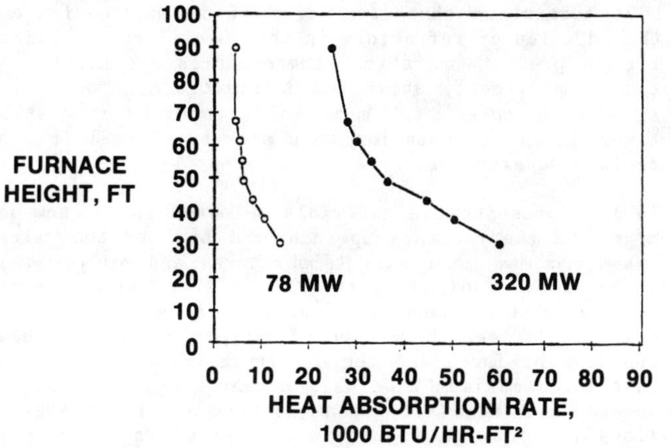

FIGURE 8

HEAT ABSORPTION RATE - FIRST PASS STEAM TUBES - OEM REDESIGN

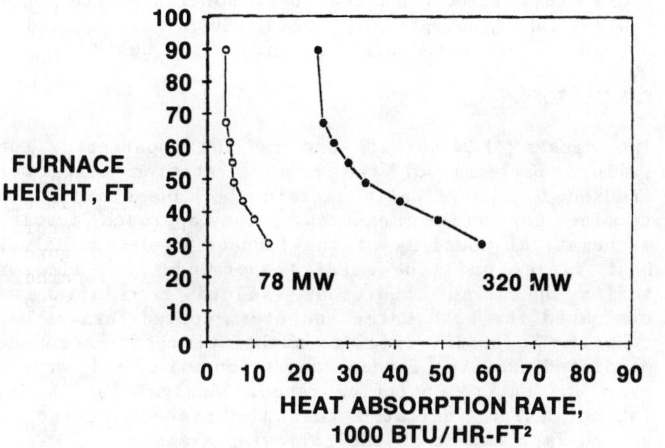

FIGURE 9

HEAT ABSORPTION RATE - SECOND PASS STEAM TUBES - OEM REDESIGN

higher heat fluxes will not cause problems. The division wall smooth tubes will not be affected detrimentally by increased heat flux in the upper furnace. However, the sidewall water-cooled tubes have the potential of being unsafe by this recommended change. The sidewall tubes will not absorb sufficient heat to generate the necessary amount of steam. Hydraulic head, then, is insufficient to maintain mass flux above 150,000 lb/hr-sq ft. The effectiveness of ribbed tubes is lost at such low mass flux (velocity) and could lead to DNB conditions in the tubes. Lower mass velocity also causes flow distribution problems. However, flow maldistribution does not lead to catastrophic failures, since flow by natural circulation eventually corrects itself.

To improve waterwall circulation, forced circulation should be evaluated and compared to other solutions, such as making changes to riser tubes from the sidewall intermediate header to reduce pressure drop in the circuit.

The OEM also suggested constant pressure boiler operation at lower loads. A calculation was performed to determine the sidewall tube circulation at 78 MW with drum pressure at 1,660 psig as currently practiced and 2,400 psig as proposed by the OEM. The mass flow through the tubes at 1,660 psig is approximately twice that at 2,400 psig. Since the mass velocity is low for sidewall tubes, it was recommended that the boiler be operated in a sliding pressure mode to maintain the safety margin in the waterwall circulation.

There are several other advantages to operating the boiler in the current dual pressure-sliding pressure mode, i.e., extended control range (high steam temperatures at low loads) of both the main and reheat steam, improved steam flow distribution in the radiant superheater at lower loads due to higher specific volume of the steam at lower pressure, better turbine-temperature match (minimal variation in first-stage shell temperature), reduced thermal stresses in the turbine, and improved overall power plant heat rate. From a boiler standpoint, constant pressure cyclic operation is advantageous because it reduces thermal stresses on the thick-walled boiler components, resulting in lower cyclic life expenditure. A constant-pressure operation mode keeps the drum pressure essentially constant over the entire load range, and a higher boiler ramp can be obtained because stored energy in the drum is not lost. With sliding pressure operation, as drum pressure decreases at lower loads, some stored energy in the drum is utilized; while ramping up, this additional amount of energy must be supplied before higher load can be achieved. This requires extra time and lowers the ramp-up rate. Generally, ramp-up rate is limited by other boiler components, e.g., superheater outlet header or superheater tubes; therefore, sliding pressure operation alone may not restrict ramp rate.

Radiant Superheater Mass Flow Rates

The boilers were operating with inadequate mass flow rates for proper tube cooling. Low mass flux rates, combined with high heat absorption rates, were causing tube failures due to higher than allowable tube metal temperatures.

The redesign raised mass fluxes at full load to over 1,000,000 lb/hr-sq ft. At lower loads the mass flow rate is proportionately lower. These values were slightly higher than, but generally consistent with, current boiler manufacturer guidelines. High mass flow rates are beneficial to a cycling unit because they maintain uniform flow in parallel circuits, even at low loads. At full load this results in slightly lower main steam pressure (approximately 20 psi), which affects turbine performance.

Steam Flow Instability and Maldistribution

Steam flow uniformity and stability in the furnace wall tubes are functions of pressure drop in the tubes in comparison to pressure drop in header and transfer pipes. The revised design reduced the tube size (OD) and increased the tube wall thickness of the radiant superheater tubes, significantly reducing the flow area and increased mass velocity and pressure drop in the tubes. The sizes of headers, downcomers, and transfer pipe in the new design were sufficiently increased to cause a very small pressure drop compared to the furnace wall radiant superheater tubes. At full load, this ratio of pressure drops exceeded 50 to 1 (furnace tube pressure drop to header pressure drop). Even at 78 MW, the radiant wall tube pressure drop in each pass was sufficient (approximately 3 psi) to cause uniform flow in the steam-cooled tubes. The transfer pipes, which feed the inlet headers, were suitably redesigned and rearranged to ensure uniform steam distribution in the inlet headers.

Flow stagnation and reversal cannot occur in the upflow steam circuits since this is contrary to the natural law that heated fluids (steam) become less dense and flow upward. Although flow maldistribution and stagnation can conceivably occur if headers are not properly sized and feeder/transfer pipes are incorrectly arranged, this did not appear to be the case in the proposed redesign.

Material Evaluation

Stainless steel tubing (SA213-TP304H) was selected for the first 10 ft of radiant surface for the first and second-pass tubing. First-pass tubing was switched to T22 from this point to the roof, where it was switched back to 304H. Second-pass tubes are 304H for the entire length. The maximum allowable temperatures for T22 and 304H steels are 1,125°F and 1,400°F, respectively. Tube wall temperatures are not expected to exceed these levels. The boiler zone model was used to predict tube wall temperatures of clean tubes. High spot absorption rates due to clean tubes do not cause tube wall temperatures to exceed the maximum allowable temperature.

The boiler zone analysis program calculated the mean tube wall temperature along the length of the tubes for the proposed redesign. The mean tube wall temperature profile is presented on Figures 10 and 11 for the first and second-pass radiant superheater tubes, respectively. Mean tube wall temperatures are predicted to increase over current operation as a result of improved

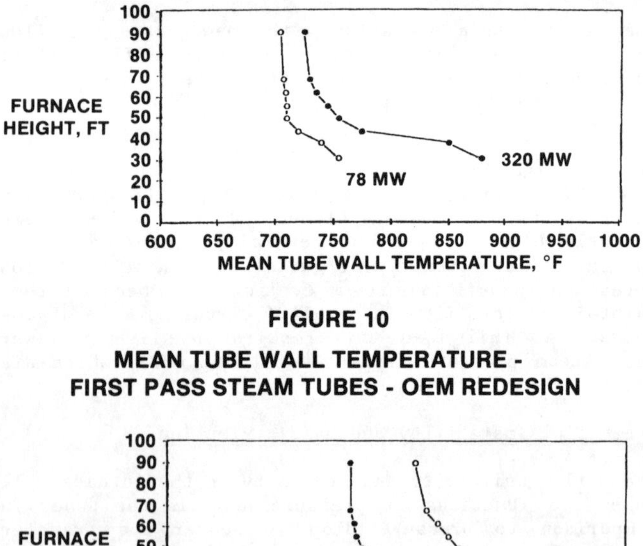

FIGURE 10

MEAN TUBE WALL TEMPERATURE - FIRST PASS STEAM TUBES - OEM REDESIGN

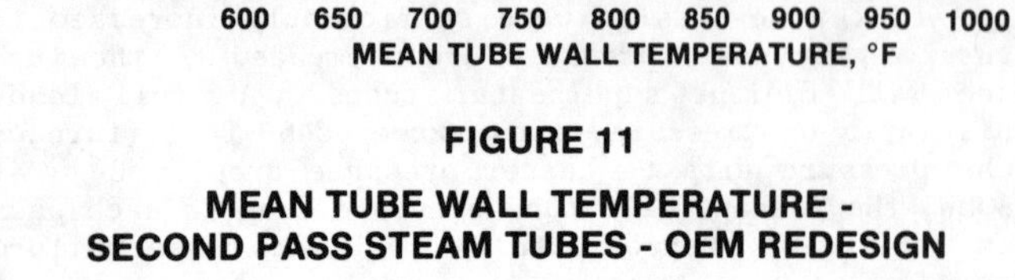

FIGURE 11

MEAN TUBE WALL TEMPERATURE - SECOND PASS STEAM TUBES - OEM REDESIGN

heat absorption characteristics of the new design and the addition of refractory in the lower furnace. Similar to present operation, temperatures are highest in the area directly above the intermediate header. The effects of these high mean wall temperatures must be taken into consideration when selecting metal for the radiant superheater.

Table 6 presents the materials selected in the new design for the radiant superheater tubes and the calculated maximum mean wall temperatures and minimum wall thicknesses required by the ASME Boiler Code. The selection of tube materials and minimum tube wall thicknesses will meet the projected extreme conditions these tubes might face, and the design is conservative. The calculated maximum tube wall temperatures are based on upset steam flow condition (50 percent of the average flow) and/or high spot heat absorption rate (130 percent of average heat flux).

The radiant superheater tube wall temperatures were lower in the new design because no inside scale was considered. In time, there will be some scale buildup, but it is not expected to approach the 30- to 35-mil thickness in the present tubes. A 5-mil-thick scale will raise the maximum outer wall temperature to approximately 1,200°F in the lower portion of the superheater tubes, suggesting that TP304H is required to provide an adequate safety margin in the design.

CONCLUSION

The causes of waterwall and radiant superheater tube failure problems of the example station units were studied to specifically explain the causes and to determine corrective measures. The approach involved mathematical modeling of the furnace to determine wall heat fluxes and tube metal temperatures for numerous boiler operating conditions. Fluid circulation was evaluated for both water and steam-cooled furnace wall tubes. The boiler-specific mathematical furnace model predicted existing furnace operation with good accuracy over the entire operating range. Analysis of the existing design indicated that the present design required improvement in the following areas:

- Mass flow in the waterwalls.
- Mass in the radiant superheater.
- Radiant superheater tube material.
- Radiant superheater steam flow circuitry.

The need for design improvements primarily arises from a change in the mode of boiler operation. These units performed well in their original design for base-load operation; however, when operated in a load-following (cyclic) mode the incidence of tube failures increased dramatically. Aware of redesign requirements, the OEM recommended changes to the existing design to make the unit suitable for cyclic operation. The changes involved converting the upflow/downflow steam circuitry of the upper furnace walls to all upflow circuitry with two-pass flow. The steam-cooled tube material was upgraded to 304H stainless steel, and steam mass flow was increased by reducing tube diameter and increasing tube wall thickness. The attemperation system was changed to spray-type.

The design changes also were evaluated and found to be satisfactory in reducing steam tube failures. Redesign improves steam flow maldistribution by reducing pressure losses in the feeder and riser tubes and feeding the headers at quarter points instead of at the ends. Steam flow stagnation and flow reversal are corrected

TABLE 6

TUBE MATERIAL EVALUATION - OEM REDESIGN

Component	Tube Description (Outside diameter x mean wall) (in.)	Tube Material	Maximum Outer Temperature Expected (°F)	Tube Temperature Recommended (°F)	Minimum Wall Required (in.)
Waterwall					
Furnace basket	3 x 0.3	SA-210-A1	900	950	0.433
Above furnace basket	3 x 0.3	SA-210-A1	850	950	0.323
Radiant Superheater					
First-Pass:					
Below El 165 ft	1 3/8 x 0.34	TP213-304H	1105	1400	0.125
Above	1 3/8 x 0.34	TP213-T22	825	1125	0.122
Roof	1 3/8 x 0.34	TP213-304H	825	1400	0.12
Second Pass:					
Below El 165 ft	1 3/8 x 0.13	TP213-304H	1165	1400	0.132
Above	1 3/8 x 0.13	TP213-304H	970	1400	0.122
Aperture	1 3/8 x 0.13	TP213-304H	970	1400	0.122

NOTES:

1. At maximum mean wall temperature as defined by ASME Boiler Code.
2. Five mils inside scale; full load flux; no slag deposit.
3. No inside scale; full-load flux; no slag deposit.
4. From Babcock & Wilcox Steam Table 3, p. 29-11.

by upflow steam circuitry. Furnace modeling indicated that the new tube materials were suitable and of sufficient margin to prevent tube failures from high-temperature oxidation. The OEM was not requested to address the waterwall circuitry, which also requires design modifications. The analysis indicated that some sidewall tubes have a very high pressure drop compared to the total available pumping head and are liable to enter into a DNB regime. This problem can be corrected by modifying the riser tube configuration or providing forced circulation. Calculations also indicated the desirability of operating the unit in sliding-pressure mode at lower loads to improve waterwall tube mass flow at lower loads and to prevent DNB problems.

The following design modifications are desirable to prevent furnace tube failures during the cyclic mode of operation:

- Change to all upflow steam circuitry for the upper furnace walls.
- Increase steam mass flow by reducing tube diameter and/or increasing tube wall thickness.
- Upgrade tube material to T22 and TP304 to reduce high-temperature oxidation.
- Modify steam header and feeder tube size and arrangement for even steam distribution in various heater parallel circuits.
- Change the lower sidewall water-cooled tube arrangement, including riser tube arrangements, to improve mass flow in these tubes, or convert to forced circulation flow.
- Operate the boiler in a dual-pressure mode or in a sliding-pressure mode at lower loads.

REFERENCES

1. Bauer, F.W. and Traub, J.W. Boiler and Steam Generator Zone Analysis Applications. 8th International Symposium on Coal Slurry Fuels Preparation and Utilization, Orlando, FL, May 1986.

2. Bauer, F.W. and Nayak, R.V. Engineering Application of Fuel Particle or Droplet Burntime Calculations. Coal Technology '86, 9th International Coal Utilization and Transportation Exhibition and Conference, Pittsburgh, PA, November 1986.

3. Nayak, R.V., Bauer, F.W., and Traub, J.W. Erosion Considerations in Boiler Conversion Studies - A Mathematic Model to Evaluate Boiler Derating. 12th International Conference on Slurry Technology, New Orleans, LA, March-April 1987.

4. Nayak, R.V., Bauer, F.W., and Tonden, T.P. Mineral Matter in Coal - Origin, Identification, High Temperature Transformation and Boiler Erosion. Coal Quality Journal, April 1987.

5. Bauer, F.W., Tonden, T.P., Nayak, R.V., and Hebert, C.T. Combustion Calculation Methods and Their Application in Boiler Analysis. ASME Winter Annual Meeting, Boston, MA, December 1987.

6. Bauer, F. W., Nayak, R. V., and Hernandez, J. Boiler Plant Performance Model for Utility Plant Engineering use. To be presented at ASME Winter Meeting, San Francisco, CA, December 1989.

7. Bauer, F. W., Nayak, R. V., King, R, and Hernandez, J., Convective Heat Exchanger Performance by Finite Element Analysis. To be presented at 1989 ASME Joint Power Generation Conference, Dallas, TX, October 1989.

1989 National Heat Transfer Conference
HTD-Vol. 106, Heat Transfer Phenomena in Radiation, Combustion and Fires

A MODEL FOR SINGLE PARTICLE COMBUSTION OF A BITUMINOUS COAL CHAR

R. Sahu
Development Engineer
Kinetics Technology International Corporation
Monrovia, California

ABSTRACT

A model for combustion of single cenospheric bituminous coal char particles at high temperatures is proposed. The primary purpose of the present study was to determine the effect of the internal particle structure on the combustion of such particles. Only one heterogenous reaction involving carbon oxidation to carbon monoxide is considered. The parameters used for characterizing the internal geometry and the reactivity of the particle were obtained from experiments. The effects of particle size, density, emissivity, reaction rate, void fraction, and initial void size on particle temperature and burntime are examined. Although the description of the particle internal structure is kept simple, it is concluded that the model adequately describes the essential processes occuring during combustion even though some simplifying assumptions were made for the sake of mathematical tractability.

INTRODUCTION

The size and efficiency of boilers used in pulverized coal combustion depends largely on the time taken for char burnout. Upon injection into the hot-zone, where temperatures are in the range 1000 K-2000 K, pulverized coal particles undergo a series of transformations which depend on many parameters such as the rank of the coal and the temperature and gas composition in the hot-zone. First, the volatiles are emitted and they burn rapidly, in the order of tens of milliseconds. The emission of volatiles in the case of high volatile coals is often accompanied by swelling of the particle, resulting in a highly porous, low density, carbonaceous char particle, the cenosphere. Cenospheric char combustion is characterized by a complex series of physical and chemical steps involving diffusion of oxidant to the char surface, reaction at its external surface, further diffusion of oxidant into the interior of the porous char, internal reaction, and finally, outward diffusion of the products of reaction (Gavalas, 1981). Depending on many factors such as the size of the char, the composition and temperature of the ambient, and the char reactivity, one or more of the above processes can become rate limiting. The internal pore structure of the char plays a significant role during combustion because it affects and, in turn, is affected by the reaction and diffusion inside the particle.

This paper will describe a mathematical model and the results from the numerical implementation of that model, that takes into account the above factors affecting char combustion at high temperatures. The parameters characterizing the internal porosity of the char and its reactivity were obtained from experiments (Sahu, 1988). The effects of char size, density, radiation properties, reactivity, porosity, and pore size on char burnout time are investigated.

Although there are many models of coal and char combustion in the literature, the effect of the internal structure of cenospheric chars on their burnout time has not been adequately investigated. The present model assumes that the char particle is spherical and has internal spherical symmetry. From measurements made in the laboratory, it is seen that pore sizes in such chars can vary by as much as four orders of magnitude. However, as a first step in modelling the internal structure, a relatively simple pore model consisting of initially monodisperse spherical voids is used. This model has been extended to include the more realistic case of micro, transition, or macro pores (Sahu, 1988). The results from the simple model proposed here show adequately the importance of the internal structure.

The mass and energy balance equations on the particle are solved to determine the particle mass and temperature at each time step. The temperature variation inside the particle is neglected (Field, 1967). The diffusion equation is then solved inside the particle to determine the local oxygen concentration as a function of radius. Once that is known, the local reaction rate and the total carbon flux and mass loss from the particle are then determined. It is assumed that the heterogeneous reaction is $2C + O_2 \rightarrow 2CO$ since it is the dominant one at high temperatures (Mitchell, 1986). In the external gas phase, species balance equations for oxygen, carbon monoxide, and nitrogen are used to determine the oxygen concentration at the particle surface. The gas phase is assumed to be quasi-steady with respect to the solid phase combustion. Gas phase reactions are not considered (*i.e.*, carbon monoxide oxidation is assumed to take place far from the particle). As expected, the internal and external processes are strongly coupled *i.e.*, while the internal oxygen concentration profile in the particle obviously depends on the external oxygen concentration, the latter is dependent on the reaction rate inside the particle, and therefore on the internal oxygen profile. Thus, iterative procedures were employed to determine all quantities at a particular time step before marching to the next one.

The calculation of the particle radius at any time step includes not only the effect of surface reaction but also that of particle crumbling. When the void fraction exceeds a critical value, it is seen that the char particle can no longer maintain its structural integrity and it crumbles into many small fragments. Usually, since the reaction rate is highest near the particle surface (as the oxygen concentration is highest there), the critical void fraction is first reached near the particle edge. When that happens, the particle crumbles to a radius where the void fraction is below the critical level. Obviously this affects the particle mass balance. It also affects the energy balance, in two ways. Firstly, the energy balance is a strong function of the particle radius. Secondly, the energy balance is affected by whether or not the shed fragments burn close to the particle. In the former case, some heat may be transferred back to the particle, thereby increasing its temperature..

A key feature of the model is that all thermodynamic and transport properties are assumed to be functions of temperature. This is a significant improvement over models that assume constant or average property values over wide ranges of temperature variation.

The presence of ash has been neglected in this model because: (*i*) the bituminous char does not have as much ash as many other coals and chars and thus it is not as significant; and (*ii*) the main purpose of the modelling effort was to test the effect of the internal structure as the particle burns. Without knowing the details of ash behaviour inside the particle, its inclusion would obscure the interpretation of results. At the present moment our knowledge of ash behavior does not justify its inclusion in this model.

Since the reaction can take place inside the particle, the kinetic parameters used in the model are the intrinsic Arrhenius parameters derived from single particle experiments (Sahu *et al*, 1988). These are based on the total surface area per unit mass of the particle (*e.g.* N_2-BET area). The reaction kinetics will be assumed to be first order with respect to local oxygen concentration.

EQUATIONS

Internal Morphology

As in any continuum model, the initial porosity will be assumed to be constant throughout the particle. However, as combustion proceeds, variations in the radial distribution of porosity will develop. In this model, the initial porosity is assumed to be due to monodisperse spherical voids present in the particle. The radius of the void and the void volume fraction (or porosity) are the two parameters that characterize the initial porosity. Let a_0 be the initial radius of the spherical voids and let there be λ voids per unit particle volume. Then, accounting for pore overlap (Gavalas, 1985), the initial void fraction in a solid of infinite extent is

$$\epsilon_{init} = 1 - e^{-\lambda \frac{4}{3}\pi a_0^3} \tag{1}$$

The void fraction of a finite particle will be slightly larger due to edge effects. This correction is neglected. Equation (1) determines λ once a_0 and ϵ_{init} are specified. The void fraction at any later time depends on the local carbon recession, $q(r,t)$, defined later by equation (7), and is given by

$$\epsilon(r,t) = 1 - e^{-\lambda \frac{4}{3}\pi [a_0 + q(r,t)]^3} \tag{2}$$

Once $\epsilon(r,t)$ is known, the local surface area, $S(r,t)$, is given by

$$S(r,t) = [1 - \epsilon(r)]\, 4\pi\lambda (a_0 + q(r,t))^2 \tag{3}$$

Diffusion

The diffusion equation inside the particle is

$$\frac{1}{r^2}\frac{\partial}{\partial r}\left(r^2 \mathcal{D}_e \frac{\partial c}{\partial r}\right) = \frac{1}{\rho_C} A_i e^{-E_i/RT_P} S(r) c \tag{4}$$

where c is the local oxygen concentration, $\mathcal{D}_e$ is the effective diffusivity, ρ_C is the density of carbon, R is the universal gas constant, T_P is the particle temperature, A_i and E_i are the intrinsic Arrhenius parameters, and $S(r)$ is the internal surface area per unit volume given in (3). The boundary conditions for this equation are

$$r = 0 : \frac{\partial c}{\partial r} = 0 \tag{5}$$

and

$$r = r_p : c = c_p \tag{6}$$

The first condition is from spherical symmetry. r_p and c_p are the particle radius and the oxygen concentration at the particle surface, respectively. c_p is determined from the gas phase equations given later.

The local carbon recession, $q(r,t)$, is determined by

$$q(r,t) = \int_0^t \frac{A_i e^{-E_i/RT_p(t')} c(r,t')}{\rho_c} dt' \tag{7}$$

This important equation determines the depth of carbon burned at any given radial location as a function of time. It is assumed that the void radius grows by the amount q with time.

To take into account the effect of crumbling mentioned earlier, the radius of the particle is calculated as follows:

$$\begin{aligned} &\text{for} \quad \epsilon_{ext} < \epsilon_{crit} : r_p = r_0 - q|_{ext} \\ &\text{for} \quad \epsilon_{ext} \geq \epsilon_{crit} : r_p = r_{p_{old}} - \left.\frac{d\epsilon/dt}{d\epsilon/dr}\right|_{ext} \Delta t \end{aligned} \tag{8}$$

where the subscript ext refers to the external surface of the particle and ϵ_{crit} denotes the critical value of the void fraction at which crumbling starts. ϵ_{crit} was assumed to be 0.8 in the model. r_0 is the initial particle radius. If the model is run at fine enough time intervals the magnitude of the shedding is small and the radius appears to decrease smoothly.

The effective diffusivity in equation (4) is calculated from the void fraction using

$$\mathcal{D}_e(r) = \frac{1}{\tau} D\epsilon(r) \tag{9}$$

where, τ is a tortuosity factor, commonly taken as 2, and D is the bulk diffusion coefficient at the particle temperature. Strictly, D should be the binary diffusion coefficients of the various gases present in the particle. But in this paper the D value used is that of a mixture of nitrogen and oxygen.

Mass and Energy Balance

The particle mass balance can be written in terms of the intrinsic reaction rate as:

$$\frac{dm_C}{dt} = -A_i e^{-E_i/RT_p} c S(r) \tag{10}$$

where m_C is the mass of carbon.

The particle energy balance can be written as:

$$\frac{d}{dt}\{m_C H_C\} = \text{Energy Generation Rate} - \text{Energy Loss Rate} \tag{11}$$

where H_C is the enthalpy of carbon. The energy generation rate is given by

$$-\frac{dm_C}{dt}\Delta H_{comb}(T_p) \tag{12}$$

where ΔH_{comb} is the enthalpy of combustion of the heterogenous reaction at the particle temperature. The rate of mass loss can also be rewritten using the carbon mass flux per unit external area at the particle surface, N_p, as

$$-\frac{dm_C}{dt} = 4\pi r_p^2 N_p \tag{13}$$

The energy generation term can then be written in terms of the enthalpies and the mass fluxes of the gas phase components (O_2=1: CO=2: N_2=3) as

$$-4\pi r_p^2 \sum_{i=1}^{3} N_i H_i \Big|_{r=r_p} \tag{14}$$

In equation (14), N_i and H_i are, respectively, the mass flux away from the particle and specific enthalpy of the gas phase component i. From stoichiometry we have

$$N_{1p} = -\frac{4}{3}N_p, \quad N_{2p} = \frac{7}{3}N_p, \quad N_{3p} = 0 \tag{15}$$

There are two heat loss terms accounting for conduction and radiation. The heat loss by conduction is given by

$$e_{cond} = -4\pi r_p^2 k \left.\frac{dT}{dr}\right|_{r=r_p} \tag{16}$$

where k is the thermal conductivity of the gas phase. The heat loss due to radiation from the particle to its surroundings is given by

$$e_{rad} = 4\pi r_p^2 \sigma \epsilon_o \{T_p^4 - T_\infty^4\} \tag{17}$$

where σ is the Stefan-Boltzmann constant of radiation, ϵ_o is the emissivity of the particle, and T_∞ is the temperature of the wall. The unsteady term on the left hand side of equation (11) can be expanded using the chain rule. Also the enthalpy H_C of carbon can be written as

$$H_C = H_{C0} + \int_{T_0}^{T_p} c_{pC}\, dT \tag{18}$$

where H_{C0} is the enthalpy of carbon at some reference temperature T_0. Using the chain rule, equation (13), and equation (18), the left hand side of equation (11) becomes

$$(m_C c_{pC})\frac{dT_p}{dt} - 4\pi r_p^2 N_p H_C \tag{19}$$

Therefore combining equations (14),(16),(17) and (19), the particle energy balance becomes

$$(m_C c_{pC})\frac{dT_p}{dt} = 4\pi r_p^2 \left[N_p H_C - \sum_{i=1}^{3} N_i H_i \Big|_{r=r_p} + k\left.\frac{dT}{dr}\right|_{r=r_p} - \sigma\{\epsilon_o T_p^4 - \epsilon_\infty T_\infty^4\} \right] \tag{20}$$

Defining

$$e_p \equiv \left[-k\frac{dT}{dr} + \sum_{i=3}^{3} N_i H_i \right]_{r=r_p} \tag{21}$$

and

$$\gamma_p \equiv \frac{e_p}{N_p} \tag{22}$$

the energy balance becomes

$$(m_C c_{pC})\frac{dT_p}{dt} = 4\pi r_p^2 \left[N_p H_C - N_p \gamma_p - \sigma\{\epsilon_o T_p^4 - \epsilon_\infty T_\infty^4\} \right] \tag{23}$$

Transport in the gas phase is described by the Stefan-Maxwell equations:

$$\frac{dY_i}{dr} = \sum_j \frac{(Y_i N_j - Y_j N_i)}{\rho_g \mathcal{D}_{ij}} \tag{24}$$

where Y_i is the mass fraction of species i, N_i is the mass flux of i, ρ_g is the gas density and $\mathcal{D}_{ij}$ is the binary diffusion coefficient of component i into component j. Here, in our 3-component system: species 1 = O_2, 2 = CO, and 3 = N_2. Since these molecules are similar in size, the binary diffusion coefficients $\mathcal{D}_{ij}$ are assumed to be equal. This simplifies (24) to

$$N_i = -\rho_g \mathcal{D}\frac{dY_i}{dr} + NY_i \tag{25}$$

where N is the total mass flux $\sum_i N_i$ which satisfies

$$r^2 N = r_p^2 N_p \tag{26}$$

where the subscript p, as before, denotes values at the surface of the particle. Loewenberg *et al.* (1987) have shown that the equations (25) for the various species can be combined to give

$$\frac{d(Y_1/4 + Y_2/7)}{dY_3} = \frac{Y_1/4 + Y_2/7}{Y_3} \tag{27}$$

From this, along with boundary conditions at r→∞ and the fact that $\sum_i Y_i = 1$, Y_2 and Y_3 are obtained in terms of Y_1 as

$$Y_3 = \frac{3/28 Y_1 + 1/7}{1/Y_{3\infty}(Y_{1\infty}/4 + Y_{2\infty}/7) + 1/7} \tag{28}$$

$$Y_2 = 1 - Y_1 - Y_3 \tag{29}$$

where the subscript ∞ denotes values far from the particle. The energy balance for the gas phase is:

$$r^2\{-k\frac{dT}{dr} + \sum_i N_i H_i\} = r_p^2 e_p \tag{30}$$

From the stoichiometric relations (27) and overall mass balance (26),

$$r^2 k \frac{dT}{dr} = r_p^2 \left(\frac{N_p}{3} (h(T) - e_p) \right) \tag{31}$$

with $h(T) = \{7H_2 - 4H_1\}/3$. The boundary conditions are:

$$r = r_p : T = T_p; r \to \infty : T \to T_\infty \tag{32}$$

Integrating (31) gives

$$r_p N_p = \int_{T_\infty}^{T_p} \frac{kdT}{\gamma_p - h(T)} \tag{33}$$

This is an algebraic expression for mass flux at the surface. Combining (33) with the mass balance for oxygen (31) yields

$$r^2 \rho_g \mathcal{D} \frac{dY_1}{dr} = r_p^2 N_p \{Y_1 + \frac{4}{3}\} \tag{34}$$

Dividing (34) by (31) gives

$$\frac{\rho_g \mathcal{D}}{k} \frac{dY_1}{dT} = \frac{Y_1 + 4/3}{h(T) - \gamma_p} \tag{35}$$

Assuming that the temperature and mass fraction of oxygen far from the particle are T_∞ and $Y_{1\infty}$ equation (35) can be integrated to yield

$$ln \frac{Y_{1\infty} + 4/3}{Y_{1p} + 4/3} = \int_{T_\infty}^{T_p} \frac{kdT}{\rho_g \mathcal{D} \{\gamma_p - h(T)\}} \tag{36}$$

This is a quadrature for the energy flux at the particle surface. Equations (33) and (36) were solved numerically to obtain T_P and Y_{1p} as a function of time. The oxygen concentration, c_p, can then be easily obtained from its mass fraction Y_{1p}.

Solution Procedure

The particle was radially discretized into fifty shells of equal radial thickness. Conditions were assumed to be uniform inside each shell. As the radius decreased with time, the number of remaining shells correspondingly decreased. The position of the external surface was calculated as a function of time by interpolating between shells. Hence, the resolution in calculating the particle radius was not limited by the number of shells chosen.

The overall carbon flux for the particle was the sum of all the fluxes at the various radii. This flux, in conjunction with the gas phase, determined the new external oxygen concentration. This was used to recalculate new internal profiles of the various quantities. This iterative process was continued until convergence was achieved at each time step. Then the overall energy balance was used to calculate the new particle temperature and a new particle radius was also calculated. This was repeated over as many time steps as needed to achieve burnout.

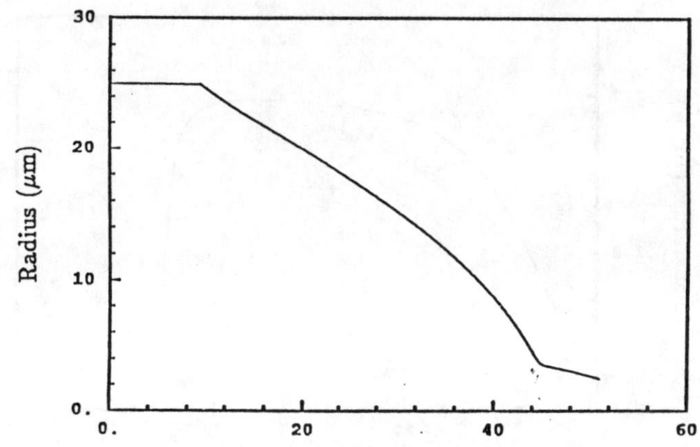

Fig. 1a Particle radius versus time.

RESULTS

Figure 1a shows the size of the particle as a function of time. A typical particle, having an initial radius of 25μm and an initial void fraction of 0.1 consisting of 0.1μm diameter voids, was used. The density of the solid carbon was assumed to be 2.0g/cc. The initial particle and wall temperatures were both 1500K and the critical void fraction was 0.8. The wall and the particle emissivities were 0.9 and 0.8 respectively. The figure shows three distinct regions. At first, the external radius of the particle decreases very slowly because the surface void fraction is below the critical void fraction and the radius can only

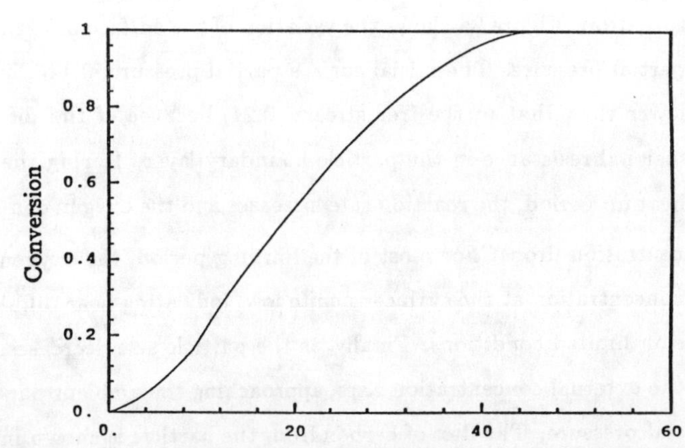

Fig. 1b Carbon conversion versus time.

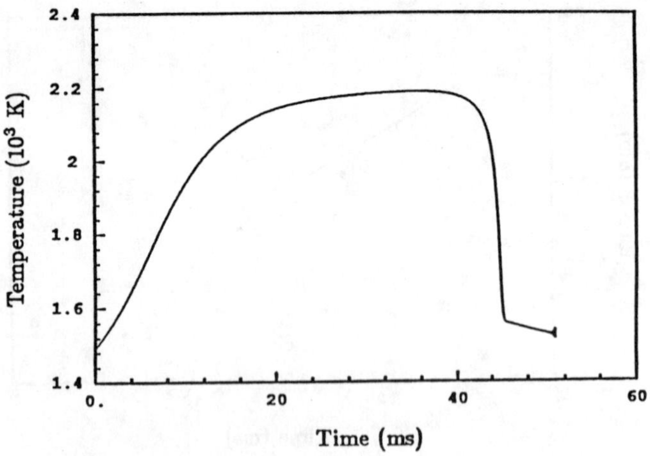

Fig. 1c Particle temperature versus time.

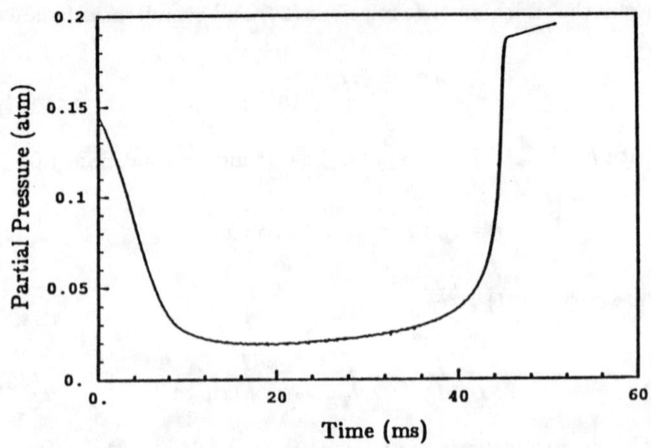

Fig. 2a Surface oxygen partial pressure versus time.

change by reaction. When the surface void fraction does reach the critical value, crumbling starts and the radius reduces as a faster rate. As the particle burns, however, oxygen penetrates further and more of the outer shell reaches the critical value and is shed. This causes the radius to decrease faster. Finally, when the particle becomes quite small, there is complete oxygen penetration and it burns in the kinetic limited regime. The shedding thickness becomes smaller and the radius decrease is arrested. Eventually, the entire particle reaches the critical void fraction and disintegrates. This happens when the radius is around 3μm. Therefore the final radius is not zero. Figure 1b shows the variation of conversion versus time. The particle temperature is shown in Figure 1c. The heat up period, the almost isothermal phase, the rapid decrease and finally the kinetic regime are all shown. In the kinetic regime the particle is in thermal equilibrium with its environment. In fact the kinetic regime is seen because of the preceeding temperature drop. Figure 2a shows the variation of the surface oxygen partial pressure. The initial surface partial pressure, 0.145, is lower than that in the free stream, 0.21, because of the diffusional resistance in the particle boundary layer. During the heat up period, the reaction rate increases and the oxygen concentration drops. For most of the burning period, the oxygen concentration at the surface is quite low, indicating near diffusion limited conditions. Finally, as the particle size decreases, the external concentration rises, approaching the ambient partial pressure. The flux of carbon from the particle is shown in Figure 2b. The fluctuations are due to the numerical method employed and the choice of each time step interval. The temporal resolution in various runs was 0.1 milliseconds. Figure 2c shows the surface and total void fractions of the particle as a function of time. Since the critical void fraction is chosen as 0.8, the surface void fraction increases to 0.8 and then remains constant throughout most of the combustion. The drop at the start of the kinetic regime is due to computational inaccuracies. The total void fraction starts from the value of the initial void fraction (0.1). At first it rises because the radius is not changing and there is some reaction. As soon as shedding begins, it starts decreasing because the particle loses its most porous external shell. However, the thickness of the outer layer that crumbles decreases with time, and the particle void fraction decreases to a relative minimum. Eventually, the particle size becomes small enough that oxygen penetrates throughout the particle. The accelerating reaction then increases the particle void fraction monotonically till the critical void fraction is reached when the entire particle crumbles.

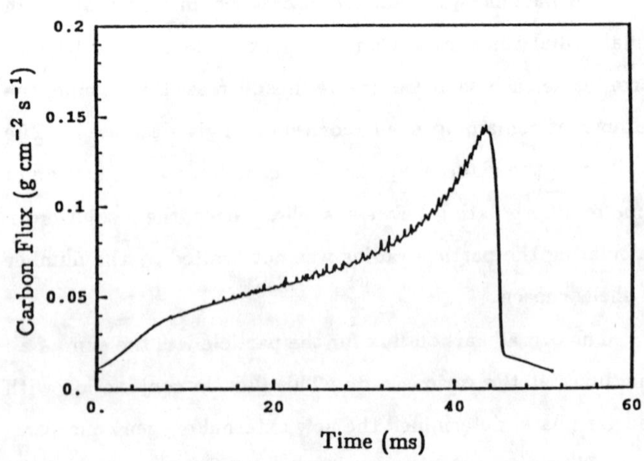

Fig. 2b Carbon flux versus time.

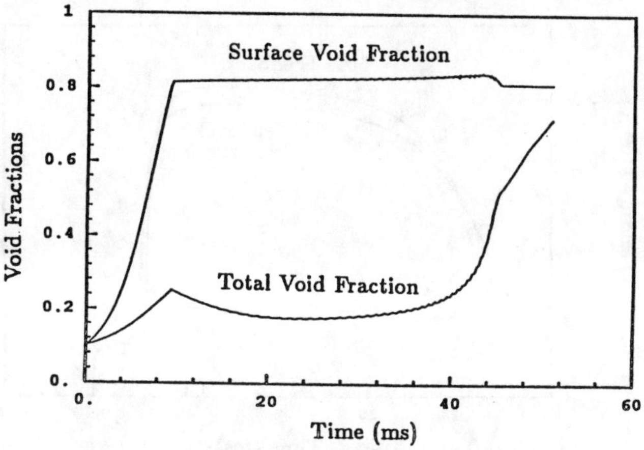

Fig. 2c Total and surface void fractions versus time.

The effect of particle radius on the combustion temperature plot is shown in Figure 3a. The burn times again scale approximately as the square of the size because of the aforementioned shrinking core type of behavior. The peak temperature is almost the same for the two particle sizes examined. The initial void fraction and void radius, were, in both cases, 0.1 and 0.05μm, respectively. The wall and initial particle temperatures were both 1500K. The intrinsic pre-exponential factor was 10^5 kg $m^{-2}s^{-1}$ and the intrinsic activation energy was 179740 J/mole (43000 cal/mole). Figure 3b shows that the carbon density affects only the burn time and not the peak temperature. The lower density particle also heats more rapidly due to its lower thermal inertia. The influence of carbon emissivity is shown in Figure 3c. Again, the change in the temperature is small and the higher the emissivity, the lower the final particle temperature. In both cases the wall emissivity was assumed to be 0.9. In Figure 3d, the influence of

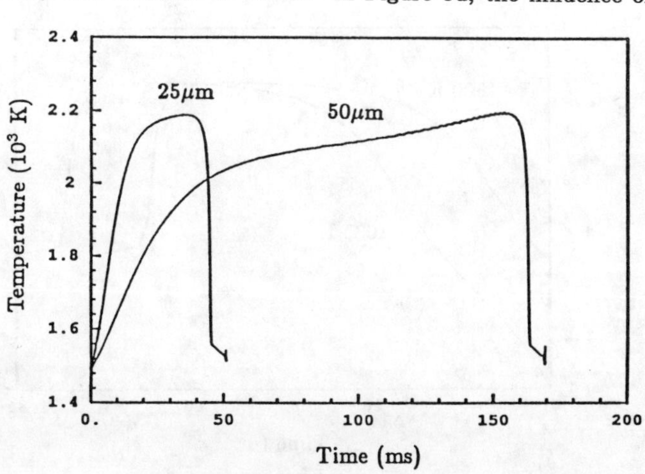

Fig. 3a Effect of initial particle size (radii).

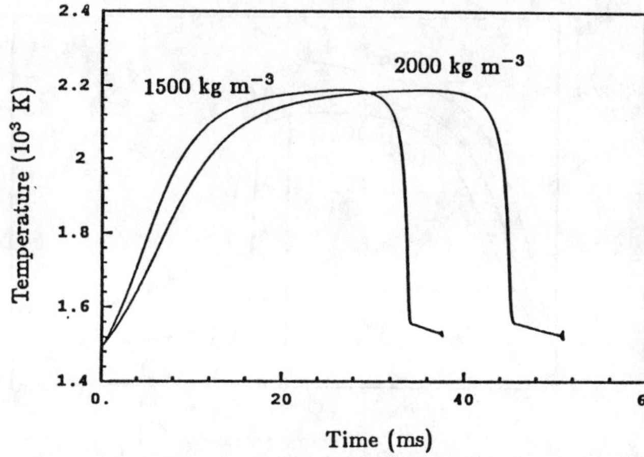

Fig. 3b Effect of carbon density.

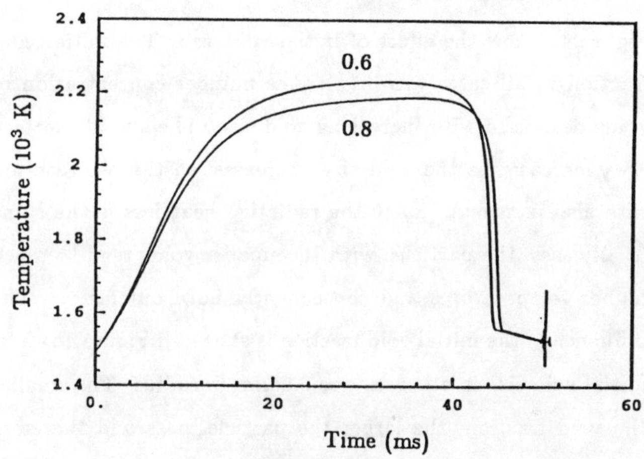

Fig. 3c Effect of carbon emissivity.

the pre-exponential factor is demonstrated. The nature of the variation is as expected. The initial rate of rise is not commensurate with the rise in reaction rate, but the duration of time the particle spends in the final kinetic regime is dramatically increased with decreasing reaction rate. The curves in

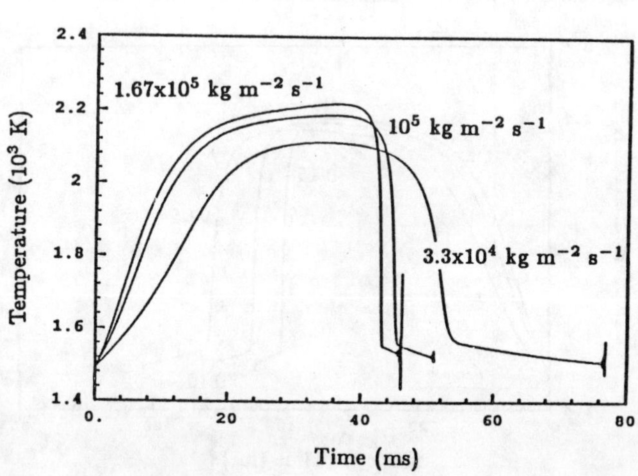

Fig. 3d Effect of intrinsic Arrhenius pre-exponential factor.

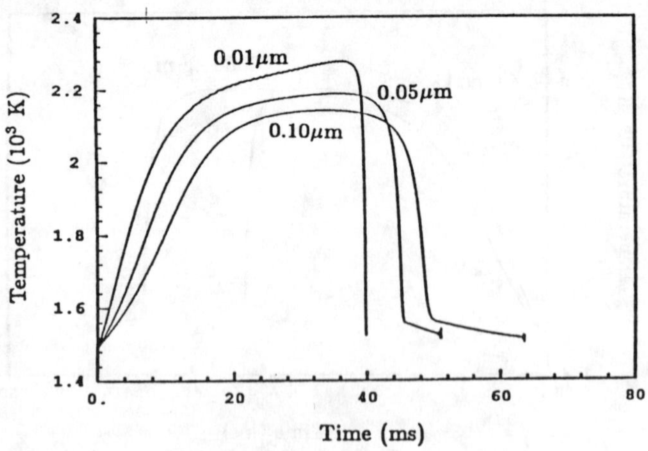

Fig. 4a Effect of initial void radius.

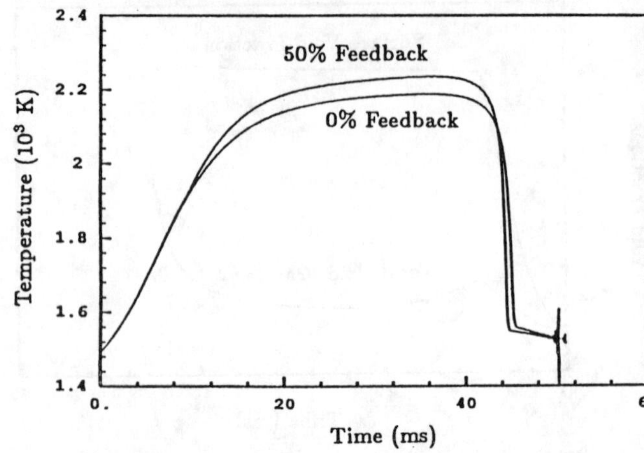

Fig. 4c Effect of energy feedback from fragmentation.

Figure 4a show the effect of initial void size. The initial void fraction in all cases was 0.1, so the number concentration of voids decreased with increasing void size. The specific surface area increases as the void size decreases, so the net reaction rate also increases. Since the radiative heat loss is the same in all cases, the particles with the smaller voids tend to reach higher temperatures and consequently burn out faster. The influence of the initial void fraction is shown in Figure 4b. The initial void size in all cases was 0.05μm in radius. The smaller the void fraction, the larger the particle mass and therefore the longer the burn time. The particle temperature does not vary significantly with void fraction. In all the cases discussed above, the energy released by the core particle and the burning fragments was completely fed back to the particle. However, not all the energy of combustion is necessarily available for the remaining particle. Burning fragments expelled from the particle may burn far from the particle. To simulate such a situation, various fractions of energy released by the exterior crumbling shell were fed back to the particle and the effect on its combustion behavior noted. Figure 4c shows two cases involving no (0%) and 50% feedback. The particle burns at almost 100K higher in the second case. The final comparision (Figure 4c) involves various wall temperatures. The heat up period is longer when the wall is cool, but the latter part of the combustion is similar in both cases.

Figure 5 shows the oxygen concentration profiles inside the particle as a function of time. It is for a case with a pre-exponential factor of 10^5 kg $m^{-2}s^{-1}$, initial void radius of 0.05 μm, and initial void fraction of 0.1. The first profile is labelled 0 ms. It shows that after the first time step, the radius of the particle is still 25 μm and the external oxygen partial pressure is 0.12. There is no penetration below 10 μm radius, implying

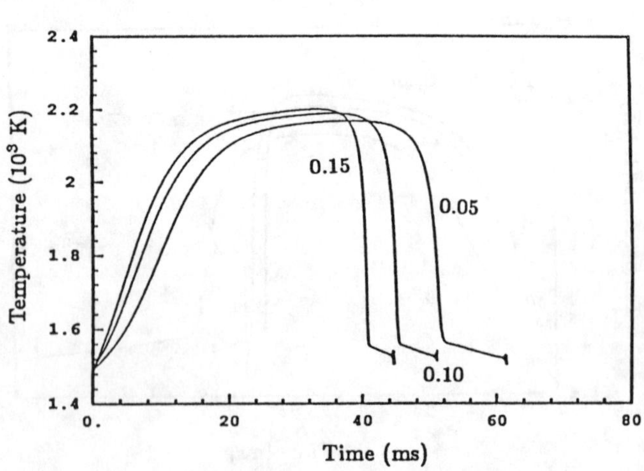

Fig. 4b Effect of initial void fraction.

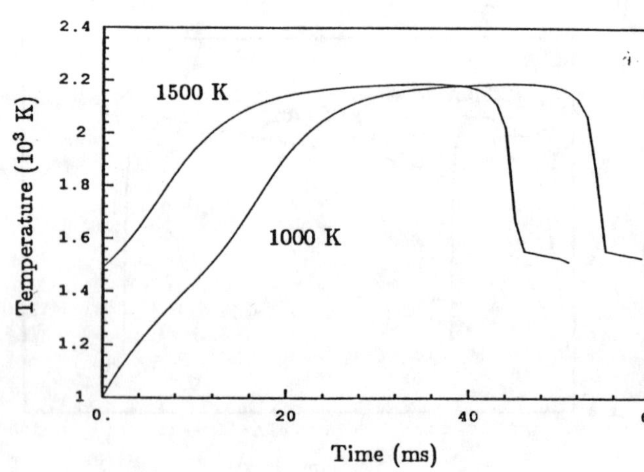

Fig. 4d Effect of wall temperature.

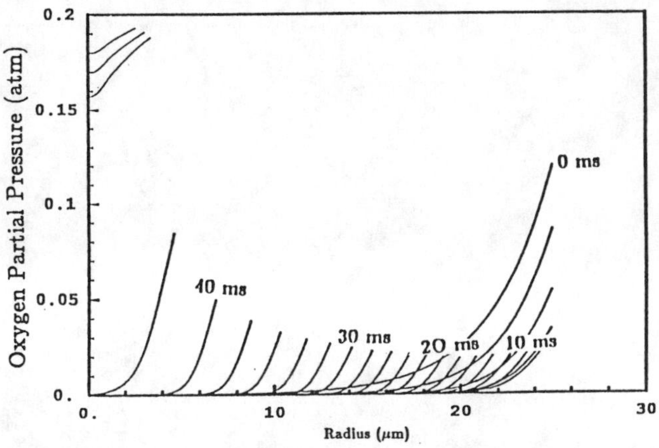

Fig. 5 Internal oxygen partial pressure profiles versus time.

that the oxygen is confined to an outer layer 15 μm thick. The profiles are drawn every 2 ms and labelled every 10 ms for clarity. Twenty milliseconds after the start of combustion, the profile shows that the particle radius is around 18 μm and the penetration depth is only 3 μm. The surface partial pressure is 0.025. As combustion continues, the external concentration rises, and the radius decreases. The penetration depth remains around 3 μm. At about 42 ms after the start of combustion, the concentration at the center of the particle becomes non-zero for the first time and the particle is in the kinetic regime. Although there is some drop in concentration inside the particle, the penetration of oxygen is not limited to a layer on the outside.

CONCLUSIONS

A model has been proposed that treats the effect of particle internal geometry on the combustion process. The effects of important parameters such as char size, char reactivity, pore size, porosity, and char emissivity are investigated. The model shows that significant oxygen penetration takes place into the interior of the char particle during its burning history. However, there are significant diffusion resistances, both in the outer boundary layer and inside the particle. Clearly, for the major part, the particle burns neither in a diffusion-limited nor in a kinetic-limited mode. Therefore theoretical analyses with such assumptions are very approximate.

REFERENCES

1. Field, M.A., Gill, D.W., Morgan, B.B., and Hawksley, P.G.W., Combustion of Pulverized Coal, BCURA, 1967, 314.
2. Gavalas, G.R., "Analysis of Char Combustion Including the Effect of Pore Enlargement", Combust. Sci. Tech., Vol. 24, 1981, 197.
3. Gavalas, G.R., Loewenberg, M., Bellan, J., and Clayton, R.M., "Structure and Combustion of Cenosphere Particles", AIChE Winter Annual Meeting, Chicago, 1987.
4. Loewenberg, M., Bellan, J. and Gavalas, G.R., Chem. Engg. Comm., Vol. 58, 1987, 89.
5. Mitchell, R.E., "Experimentally Determined Overall Burning Rates of Coal Chars", Combined Canadian and Western States Section Technical Meeting of the Combustion Institute, Banff, Canada, 1986.
6. Sahu, R., Combustion of Bituminous Coal Chars, PhD thesis, Calif. Inst. of Tech., 1988.
7. Sahu, R., Northrop, P.S., Flagan, R.C. and Gavalas, G.R., Combust. Sci. Tech., Vol. 60, 1988, 215.

1989 National Heat Transfer Conference
HTD-Vol. 106, Heat Transfer Phenomena in Radiation, Combustion and Fires

COMPENSATION OF THERMOCOUPLES FOR THERMAL INERTIA EFFECTS USING A DIGITAL DECONVOLUTION

S. F. Son and M. Queiroz
Mechanical Engineering Department
Brigham Young University
Provo, Utah

C. G. Wood
Mechanical Engineering Technology Department
Weber State College
Ogden, Utah

ABSTRACT

A digital technique which uses a fast Fourier transform (FFT) algorithm has been implemented to compensate thermocouple measurements for thermal inertia effects. The digital compensation technique is more accurate, more flexible, and more easily implemented than the traditional method of electrical compensation. The proposed method is described in detail and an analysis is made of its advantages and accuracy. Errors that may occur using this method are quantified by comparing results with the analytical compensation of a generated temperature signal. Realistic noise is also added to the generated signal to evaluate the method in a more practical environment. Digital filtering is implemented to minimize the effects of noise on the compensated signal. Finally, temperature measurements made in a turbulent spray flame are compensated and the effect of varying the average time constant on the compensated temperature measurements is presented. The probability density function (pdf), the power spectral density (psd), the autocorrelation, and the average and rms temperature are also presented to fully illustrate the compensation technique.

1. INTRODUCTION

Current mathematical models of practical combustion systems commonly require the prior knowledge of some fluctuating characteristic scalar such as temperature or concentration. Assuming fast reaction, negligible thermal radiation, low Mach number, and equality of transport coefficients, the specific enthalpy and concentrations may be linearly related. The significance of this is that if the probability density function (pdf) for a single passive scalar is known the highly non-linear time-averaged reaction rate, which is otherwise difficult to calculate, can be straightforwardly determined [1]. Further, spatially- and temporally-resolved scalar measurements are needed to gain a better understanding of the phenomenon involved, to validate current models and to indicate possible improvements in future numerical models of combustion systems.

Micro-thermocouples are an attractive alternative to laser systems to measure average and fluctuating temperatures in turbulent combustion because of their low cost. The thermocouple has been and continues to be used extensively as a combustion diagnostic in turbulent gaseous flames [2-13] and in turbulent spray flames [14-19]. Thermocouples are frequently used to calibrate and document the accuracy of laser techniques [13]. Chandran et al. [2] compared thermocouple signals, compensated electronically for thermal inertia effects using an average time constant, with simultaneous molecular Rayleigh scattering measurements and found that the two fluctuating signals compared favorably. Average temperatures measured with a thermocouple have also been shown to compare favorably with laser Raman spectroscopy [7]. However, the major difficulty with using thermocouples to measure high frequency temperature fluctuations continues to be the compensation of the signal for thermal inertia effects.

Currently, most investigators use a version of the electronic RC compensating circuit first developed by Shepard and Warshawsky in 1952 [20]. Although this technique has been proven useful in the study of fluctuating temperatures, it has severe limitations. The most significant is that, as the time constant increases, the maximum frequency that may be compensated accurately decreases substantially because of the increased electrical noise from the compensator. Furthermore, higher frequencies are attenuated more as the time constant increases [1]. In fact, the noise level in electronic compensation has been found to be proportional to the time constant of the compensating network and to depend on the setting of the electronic low-pass filter that is commonly used [3]. Also, in electrical compensation a pre-amplifier and linearizer before the compensator are also necessary, further increasing the noise level. Pre-amplifiers, such as the transformer or the AC types, provide higher signal-to-noise benefits, but attenuate the low frequency components of the temperature signal [12,20]. Certainly this is not acceptable in fluctuating temperature measurement where a significant part of the signal is in the lower frequency range. Furthermore, physical limitations of the electronic compensator prevent its use beyond a certain maximum frequency.

Although digital compensation does not completely overcome all the limitations of electrical compensation, it reduces the noise level because fewer electronic components are used, allows easy estimation of the signal-to-noise level from the digitally calculated power spectral density (psd) of the sampled temperature signal, is more flexible, and is generally more accurate. Digital signal processing also simplifies the procedure of applying additional corrections to the temperature data such as radiation heat transfer, and conduction heat transfer to the leads. Another advantage is that various time constants may be used to compensate the same data file without resampling. Further, the low cost and speed of modern computer equipment and the use of efficient algorithms, such as the fast Fourier transform (FFT) which is used in this study, make digital compensation attractive in terms of time efficiency and cost.

Yule et al. [5] were among the first to use digital techniques in thermocouple compensation to increase accuracy and flexibility in fluctuating temperature measurement; however, very little detail of the implementation of the technique used is readily available. A major step in thermal inertia compensation is calculating the derivative at each point in the measured temperature. If a compensation procedure involves the calculation of derivatives by direct numerical methods, some potential numerical problems often arise. Moreover, an assumption must be made about the derivative of the first and last points, the method is computationally inefficient, and digital smoothing is often necessary because the derivative is sensitive to any error present, especially quantization error [21]. Another approach is to perform the compensation utilizing the FFT algorithm to perform a digital deconvolution that avoids these problems. This digital technique is commonly used in other applications such as systems analysis [22]. No assumption must be made about the derivative of the first and last points, it is computationally efficient, and digital filtering may be implemented simultaneously with the compensation technique. In order to apply this method, as with other methods of thermal inertia compensation, the nature of the frequency response function of the thermocouple must be known. If it can be assumed that the thermocouple is a first-order system, the time constant must be determined.

Usually an average time constant is used in digital compensation, as is done with electrical compensation; however, in a few studies an instantaneous time constant has been used which involves the simultaneous measurement of velocity using laser velocimetry [8,9]. Possible error in this approach is due to the accretion of seed particles on the probe, and the uncertainties involved in the Nusselt correlation that may be used to calculate the heat transfer coefficient to the thermocouple. Data acquisition and reduction would also be likely to be more difficult and time consuming. Although efforts to instantaneously compensate thermocouples are commendable, the possible small increase in accuracy may not justify the added cost and complexity for many investigations. The use of an average time constant greatly simplifies the measurement and data reduction because the additional simultaneous velocity measurement is not necessary and the required number of calculations are significantly reduced. Further, Cambray compared the instantaneous thermal inertia compensation using the two-wire method [4] with that of using an average time constant and found that, even in cases of large time constant fluctuations, compensation with an average time constant does not lead to significant differences. Chandran's favorable comparison of a thermocouple compensated using an average time constant and Rayleigh scattering temperature measurements is also encouraging [2]. The average time constant may be measured *in situ* by electrical heating methods [1,5,6] or by the two-wire method described by Cambray [4].

The objective of this paper is to describe and demonstrate a digital deconvolution technique that has been implemented to accurately and quickly compensate thermocouples for thermal inertia effects. An effort was also made to identify and quantify errors involved by demonstrating the technique applied to a generated temperature signal and by compensating temperature measurements in a dilute spray flame composed of an array of six vertical droplet streams. Probability density functions (pdf's), power spectral densities (psd's), autocorrelations, and average and rms temperature profiles are presented to demonstrate that the compensation technique can be used to obtain physically meaningful data.

2. COMPENSATION TECHNIQUE

This section describes the proposed digital compensation technique which is essentially a digital deconvolution using the FFT algorithm. The measured temperature signal is first transformed by the FFT to the frequency domain where each term in the Fourier series is individually compensated by dividing it by the thermocouple transfer function evaluated at discrete frequencies. The inverse fast Fourier transform (IFFT) is then performed to return the compensated data to the time domain.

The commonly assumed first-order transfer function is used where, in the case of thermocouples, radiation to the surroundings, conduction to the leads, and catalytic effects (if using uncoated probes) are neglected. If radiation or conduction to the leads is significant, it may be possible to apply approximate corrections after the compensation for thermal inertia is performed. For the above assumptions it is easily shown by conservation of energy applied at the thermocouple bead that [22],

$$mC_p \frac{dT_b(t)}{dt} = hA(T_g(t) - T_b(t)) \tag{1}$$

where $T_g(t)$ and $T_b(t)$ are the gas-phase and bead temperatures, m and C_p are the mass and the specific heat of the thermocouple, h is the convective heat transfer coefficient, and A is the thermocouple surface area. Rearranging, it follows that

$$T_g(t) = (1 + \tau \frac{d}{dt}) T_b(t) \tag{2}$$

where

$$\tau = \frac{mC_p}{hA} \tag{3}$$

is the so-called time constant.

Note that the measured voltage signal must be transformed into a temperature signal before compensation may be performed because the derived equations and τ are in terms of temperature and because the relationship between voltage (emf) and temperature for a thermocouple is not linear. In digital compensation this is easily accomplished by evaluating a polynomial fit to the emf curve of the thermocouple. In electronic compensation the same operation may be performed using an electronic linearizer before compensation is performed. However, digital linearization is usually more accurate than electronic linearization. In this study, the maximum error in the linearization process was less than 1°C compared to 10°C reported by Ballantyne and Moss [11].

Equation (2) can also be transformed to the frequency domain to form the transfer function, **X**(jω), which can be written in either complex or phasor notation:

$$\mathbf{X}(j\omega) = \frac{T_b(j\omega)}{T_g(j\omega)} = \frac{1}{1+j\omega\tau} = \frac{1}{\sqrt{1+\omega^2\tau^2}} \angle \mathrm{Tan}^{-1}\omega\tau \tag{4}$$

where the usual substitution of jω for the differential operator has been made. ω is the angular frequency (rad/s), and $j = \sqrt{-1}$. Figure 1 illustrates the plot of Equation (4) (the thermocouple response) and the inverse of the equation (the compensator response) for a cutoff frequency of 10 Hz.

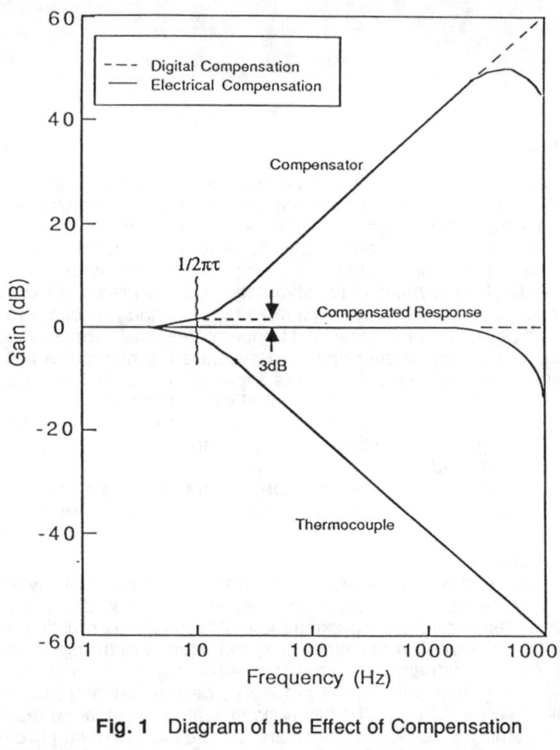

Fig. 1 Diagram of the Effect of Compensation

In the frequency domain, the first order differential equation may now be solved by algebraic manipulation. The solution of Equation (4) may be written as,

$$T_g(j\omega) = \frac{T_b(j\omega)}{\mathbf{X}(j\omega)}. \tag{5}$$

Equation (5) is the central idea of the proposed compensation method. The sampled bead temperature data is first transformed to the frequency domain by the FFT algorithm. While in the frequency domain, Equation (5) can be applied to produce $T_g(j\omega)$. The IFFT is then used to efficiently return the compensated data, $T_g(j\omega)$, to the time domain. This operation of reversing the effect of a transfer function is known as deconvolution.

The proposed method described above is straightforward; however, because a finite discrete Fourier transform is performed instead of a continuous and infinite transform, some important precautions must be taken to insure that the discrete transform closely approximates the continuous transform. The FFT differs from the integral Fourier transform in that it treats a finite data set as if it were a periodic function extending from $t = -\infty$ to $t = +\infty$. If $T_b(t)$ does not begin and end with the same value, a discontinuity is introduced when the function is repeated, causing higher frequencies to be introduced which cannot be adequately described by the limited number of terms in the series. In fact, because of this discontinuity the derivative which is part of the compensation will fail to converge, at some points at least. This effect is known as ringing in the time domain and in the frequency domain leakage occurs which is evidenced by frequency components bleeding into adjacent frequencies. Gibb's phenomenon, which is overshooting at a discontinuity, is also present.

De Levie et al. [23] demonstrated a method to virtually eliminate errors due to ringing and Gibb's phenomenon. He showed that since both Fourier transformation and differentiation are associative, any analytical function P(t) may be subtracted from the original data $T_b(t)$, such that the remainder R(t) is zero valued and has zero-valued derivatives at its beginning and ending, thus eliminating both Gibb's phenomenon and the convergence problem. The analytical function P(t), can then be differentiated explicitly and the remainder, R(t) by Fourier transformation. The two derivatives can then be added together to yield the derivative of the original function.

The simplest analytical function, P(t), called by Hayes et al. [24] the "translation-rotation" (T-R) method, is a line drawn from the first and last points. De Levie et al. [23] showed that a polynomial curve fit through the first and last few points would produce superior results. This is because the polynomial fit will force not only the end points to zero but also their derivatives, which is especially important if the deconvolution involves a derivative. Since the derivative and the other operations in the thermal inertia compensation operator described in Equation (2) are linear, both P(t) and R(t) can be compensated separately and added together to obtain the desired result which will converge and be free of Gibb's phenomenon. P(t) is compensated analytically, R(t) is compensated by using the FFT algorithm and the sum is the compensation of the original data set.

Another important point is the performance of any data reduction algorithm in the presence of noise. In measurement systems it is often found that white or near-white noise exists; that is, noise exists at approximately the same magnitude at all frequencies of interest. In many systems some type of electronic filtering is used to enhance the signal-to-noise ratio of the output or, if digital data acquisition is used, to prevent aliasing. Digital filtering cannot be used to prevent aliasing, but has been shown to have significant advantages over electronic methods when used to enhance the signal-to-noise ratio [24]. These methods are easily applied to thermocouple measurements when the signal is digitally recorded, and can be simultaneously implemented with the compensation if the FFT algorithm is used. The FFT algorithm that was used in this study is based on one originally written by N. Brewer of Lincoln Laboratories [25]. Further detail on the implementation of this compensating method can be found elsewhere [26].

3. APPLICATION TO A GENERATED TEMPERATURE SIGNAL

The algorithm is applied to a generated temperature signal so that an analytical compensation can be performed to quantify the error present in the compensation method. Further, the method previously described to eliminate Gibb's phenomenon and to ensure convergence is implemented. Noise is then added and the resulting errors evaluated. A digital low-pass filter is implemented to reduce noise error, taking into consideration the frequency content of the sampled signal. Aliasing is avoided by sampling the data at more than twice the highest frequency components of the measured signal.

3.1 Gibb's Phenomenon and Convergence

The bead temperature function, $T_b(t)$, used to evaluate the proposed technique was chosen to be:

$$T_b(t) = a\sin(\omega_1 t) + b\sin(\omega_2 t) + c. \qquad (6)$$

Applying Equation (4) to the above equation and letting $\tau = 0.2$ s, $a = 500$ °C, $b = 150$ °C, $c = 1500$ °C, $\omega_1 = 2$ rad/s, and $\omega_2 = 20$ rad/s, it follows that

$$T_g(t) = 538.52\ \sin(2t+0.38) + 618.47\ \sin(20t+1.32) + 1500. \qquad (7)$$

If an integer number of periods of sampled sine waves is chosen, ringing errors do not occur and error is due mainly to the round-off error of the computer used, assuming that no measurement noise is present. Figure 2 shows the sampled signal (Equation (6)) and the digitally compensated signal (using the proposed digital deconvolution). The digital compensation was found to be nearly identical to the analytical compensation predicted by Equation (7).

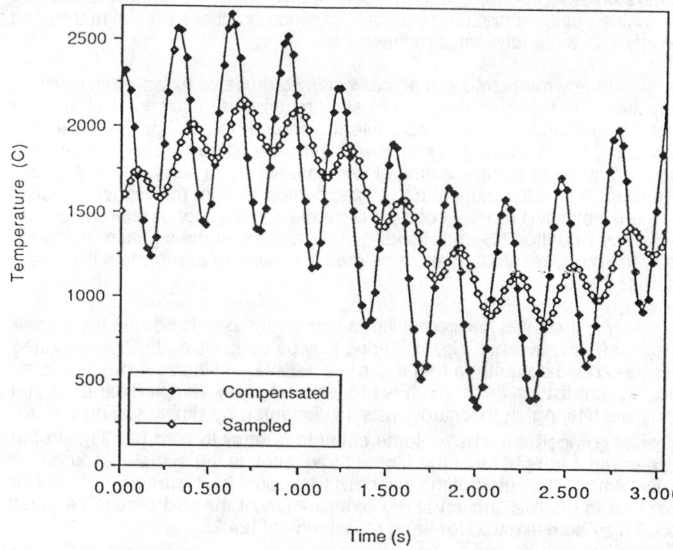

Fig. 2 Compensated Analytical Temperature Signal With No Ringing Error

To study Gibb's phenomenon and convergence, a more realistic signal was generated, using Equation (6) to produce a discontinuity of approximately 500 °C between the first and last points. Figure 3 shows the ringing effect when the proposed method is directly applied to the new $T_b(t)$. The ringing effect seems to affect the extremes of the data file the most; however, it should be noted that all of the data are affected by this error. Figure 3 also illustrates the improved compensation obtained by applying the T-R method. It was observed that there is still some slight deviation that seems to affect the extremes of the data set. This is probably due to the fact that with the T-R method the slope at the beginning and end of R(t) is not forced to zero. A third degree least-squares curve fit of the first and last few points was then used as P(t). As illustrated in Figure 3, the use of the polynomial instead of the T-R method improves the compensation to within acceptable limits of error.

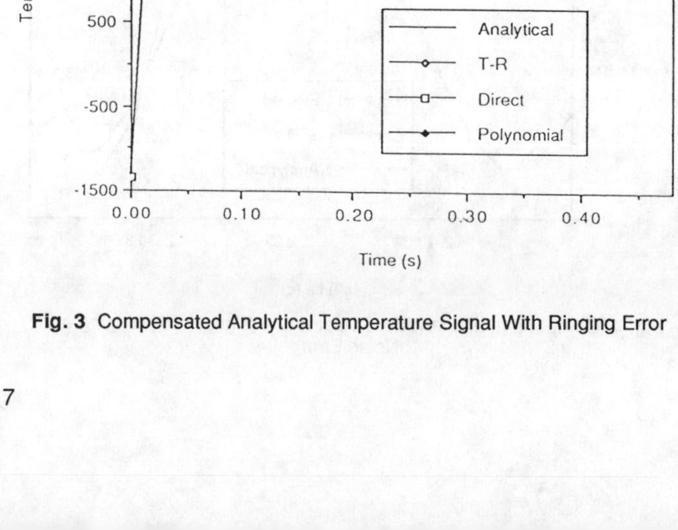

Fig. 3 Compensated Analytical Temperature Signal With Ringing Error

3.2 Noise Error Reduction

Any practical data reduction method must perform well in the presence of noise. In order to simulate a white noise level of +/- 1, +/- 5, and +/- 10 °C, the generated signal (Equation (6)) is summed with generated random numbers and the resulting data set is compensated using the proposed algorithm. Figure 4 illustrates the effect of +/- 10 °C noise in the generated signal on the compensation. It also shows the improvement that is obtained using a digital low-pass filter which produces results that match closely with the analytical compensation.

In any measured signal, particularly attenuated signals that must be amplified, the signal-to-noise level at all measured frequencies should be carefully monitored and suitable filtering should be used or significant errors could result even in systems with very low noise levels. The error is mainly due to the compensation of low signal-to-noise signals at higher frequencies. In both electrical and digital compensation, the higher frequencies are amplified more to compensate the greater signal attenuation, as shown by Equation (4) and Figure 1. This can cause the signal-to-noise ratio of the compensated signal to be greatly reduced if appropriate filtering is not used.

An electrical low-pass filter is commonly used to control the signal-to-noise ratio; however, digital filtering may be used instead. Digital filtering has the added advantage that it is more flexible and it also reduces the effect of quantization error which is present in digitally sampled data. Digital low-pass filtering, in this study, was implemented by simply setting the frequency components above some cutoff frequency to zero [24,25]. In this generated example case the highest frequency in the signal was known *a priori* so that the cutoff of the low-pass filter could be accurately set. This is not true in general; however, by examination of the psd a suitable cutoff point may be estimated for each measurement [24,25].

Table 1 summarizes the results obtained from compensating the analytical function (Equation (6)) using the proposed method. The tabulated average error shows that applying the polynomial method to a signal with no noise virtually eliminates the Gibb's phenomenon and ringing. The slight residual error may be due in part to what is known as wraparound error which occurs because of the assumed periodicity of the FFT and to errors in the curvefit. This wraparound effect may be minimized by padding the sampled data with zeros or by simply discarding a few of the end points equal to the short time that the first-order response function is not nearly zero in the time domain [25]. It is recommended that the later method be used since the end points are also where the error due to Gibb's phenomenon and ringing (if present) are the most prevalent. The effect of noise is seen to be greatly reduced by appropriate digital filtering. Further, it is observed from the results shown in Table 1 that, although the average error of the deviation from the analytical compensation varies substantially, the compensated average and the rms temperatures do not seem to be nearly as sensitive to noise error, Gibb's phenomenon, or ringing.

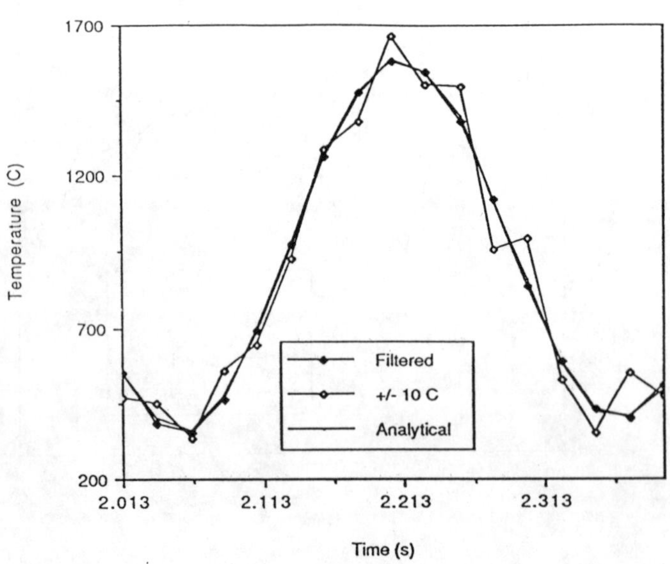

Fig. 4 Compensated Analytical Temperature Signal With Ringing and Noise Error

Table 1 Summary of Results

(All Values °C)	No Added Noise			With Added Noise (Polynomial Applied)			
	Direct FFT Comp.	R-T Applied	Poly. Applied	+/- 1 °C	+/- 5 °C Noise	+/- 10 °C Noise	Filtered +/- 10 °C
Error in Average	1.19	3.1 e-2	1.6 e-2	7.0 e-3	2.2 e-2	4.7 e-2	4.8 e-2
Error in rms	8.19	5.0 e-2	4.6 e-3	5.7 e-2	1.64	6.70	0.27
Average Error	55.44	1.59	0.15	7.24	36.08	72.14	7.45

To fully demonstrate the usefulness of this compensation technique, the method was applied to temperature measurements made in a spray flame. The following sections describe the experimental apparatus and the results obtained.

4. EXPERIMENTAL APPARATUS

4.1 Data Collection and Reduction System

Figure 5 is a schematic of the data collection and reduction system used. The micro-thermocouples used were 50 µm diameter Pt-Pt 10% Rh (Type S). An electronic ice point is used to reference the voltage signal to the freezing point of water so that the measured voltage can be accurately converted to temperature. The system features a sample-and-hold circuit (Metrabyte SSH-4) to make simultaneous two-point measurements and a heat-up circuit to measure the average time constant *in-situ*. The system is also designed to collect data from hot-wire anemometer probes which are used to measure boundary conditions of the turbulent airflow.

To reduce electrical noise, all cables used are shielded and are less than ten feet long. The high precision instrumentation amplifier (AD624B, 500 gain) is battery-powered and is enclosed in a grounded metal box. The amplifier was chosen because of its accuracy (+/- 0.35%), dynamic response (up to 50 kHz at a gain of 500), low-noise (0.2 µV/p-p), and low drift characteristics (.25 µV/C input, 10 µV/C output). A 12-bit A/D converter (Metrabyte DAS-20) and an Epson Equity III Plus computer were used to collect and reduce the data.

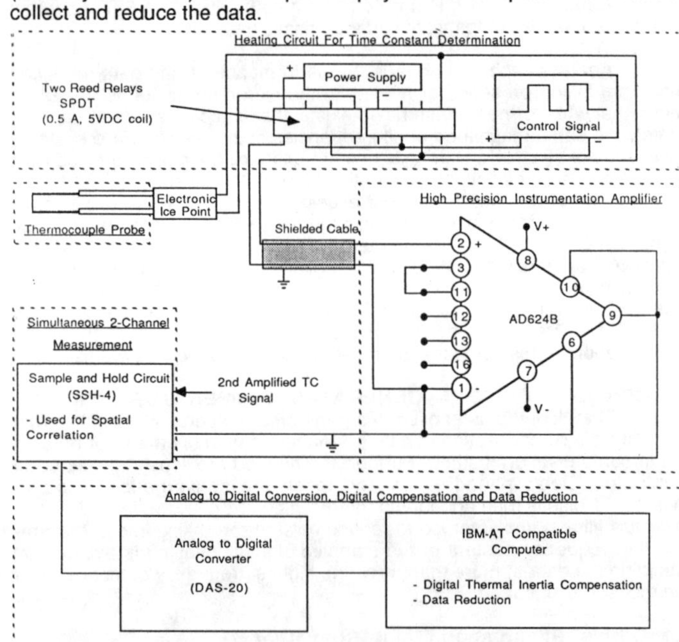

Fig. 5 Data Aquisition and Reduction System

4.2 Spray Flame Configuration

A schematic of the spray generator and air system is illustrated in Figure 6. The spray flame consists of an array of six vertical streams of hexane at a total flow rate of 28.1 cc/min. This arrangement was chosen because of its simple geometry, which makes it easily accessible to probes and photographic documentation, and also because it is easier to numerically model compared to more complex flames [27]. Furthermore, average

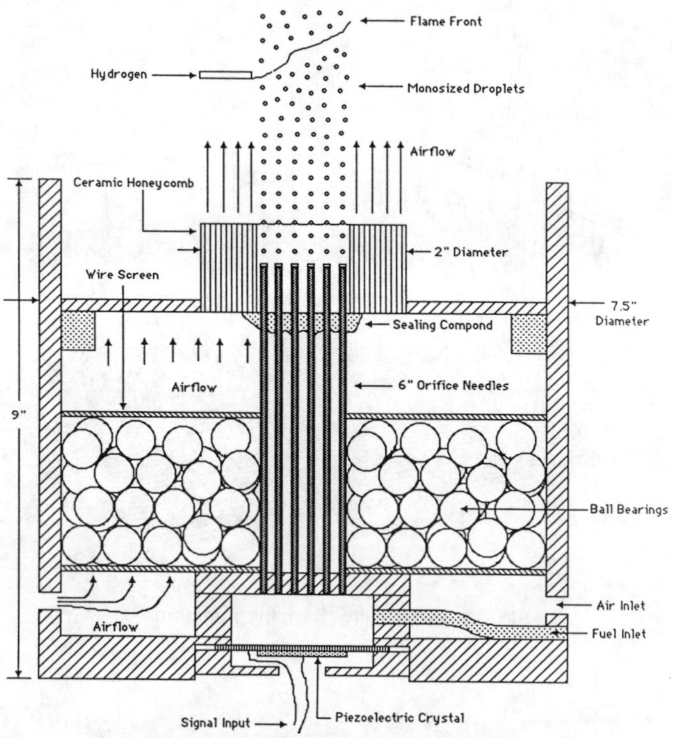

Fig. 6 Spray Generator and Airflow System

flame speeds may be estimated from the angle of the flame front [18]. This system also features an airflow system that allows the nearly monosized droplets to be formed with minimal interference from the air stream. To improve uniform droplet formation, precision-made, 150 μm diameter optical apertures are used as orifices [18]. The operational details of the spray generator are described in more detail elsewhere [28]. The system also includes a large exhaust hood to minimize flame stretching effects.

The temperature of the air entering the test section was approximately room temperature (23 °C). The initial droplet size, temperature and velocity were 280 μm, 23 °C and 4.40 m/s respectively. The spray was ignited by a small hydrogen pilot flame, as shown in Figure 6, near one edge of the array of droplets. The pilot flame is visibly small (3 mm diameter, 6 mm base to tip) and its thermal effect on the airflow without droplets has been fully documented [18]. The average air velocity, which was measured with a hot wire anemometer at the ignition point with no droplets present, was 1.2 m/s and the turbulence intensity level was 3.4%.

Two flame regimes were observed: a blue, smooth, continuous premixed-gas type flame and a yellow, very rough, and brushy flame front which is characteristic of a diffusion flame. The premixed-gas type flame was established by the initial evaporation and air entrainment before the ignition source was reached. Following this premixed region, a dark region exists due to further evaporation of the droplets, ignition delay, and air entrainment [19].

5. COMPENSATED TEMPERATURE MEASUREMENTS

Temperature measurements made in the spray flame described above were compensated to illustrate the thermal inertia compensation method proposed in this paper. A preliminary study showed that above approximately 350°C, gas phase temperature measurements are not significantly affected by hexane droplets impinging on the thermocouple bead [17]. In this study, the thermocouple was positioned between the innermost streams and the droplet dispersion was such that droplet impingement was unlikely.

A major advantage of digital compensation is that the same data may be reduced using various average time constants. In a high temperature, rapidly fluctuating environment an accurate average time constant is difficult to obtain so several time constants may be used to verify that the conclusions are not dependent on the uncertainty of this measurement. For this study the effect of the time constant on the pdf, psd, and autocorrelation is shown. Unless otherwise stated an average time constant of 25 ms was used in all of the measurements. Time constants have been measured in this flame using the heating circuit shown (Figure 5). These measured time constants ranged from 20 ms to 50 ms. Also, Equation 3 may be evaluated directly to obtain an average time constant. Using the measured cold flow velocity and at a temperature of 1600 °C, a time constant of 32 ms was obtained. Further, the pdf's obtained produce physically reasonable results using a time constant of 25 ms; that is, the compensated temperature is less than the adiabatic flame temperature and above room temperature. Ballantyne and Moss [11] also used 50 μm thermocouples and used similar values for the time constant.

The noise level was measured by imbedding a thermocouple in a large cup of wax. Electrical and quantization errors were measured to be less than 35 μV at room temperature (less than +/-2 °C) and from calculated power spectral density plots, it was seen to be essentially white noise. From examination of the psd of the measurements, a cutoff frequency corresponding to a minimum signal-to-noise ratio of 3:1 was determined for each measurement point in the flame. Measurements were also made at various locations and sampling frequencies to ensure that aliasing had not occurred. Figure 7 shows a section of the measured and compensated signal in the spray flame. The average time constants used for the compensated signal in Figure 7 were 20, 25, and 30 ms. The signal is clearly seen to be amplified and shifted to the left which is the expected effect of compensation, since with a first-order measuring system the measured signal is attenuated and experiences a phase lag.

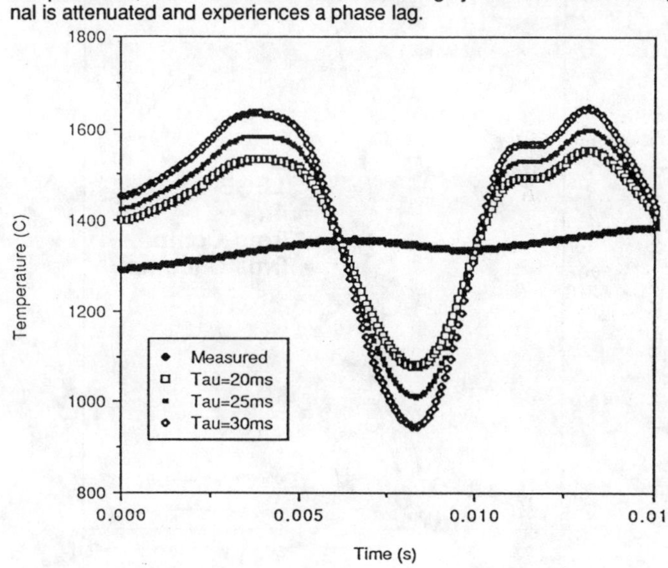

Fig. 7 Compensated temperature Measurements Using Tau = 20, 25, and 30 ms

The autocorrelation and psd were also estimated using the FFT algorithm. The autocorrelation was calculated as described by Van Atta and Chen [32]. They also showed that this method resulted in correlations nearly identical with those calculated directly [32]. Briefly, the method consists of transforming the data to the frequency domain using the FFT algorithm, multiplying this result by its conjugate, and then returning this product to the time domain using the IFFT. To estimate the psd, the data is multiplied by the Hanning window function to minimize leakage, and then transformed to the frequency domain using the FFT algorithm where it is multiplied by its conjugate. For the temperature measurements in this study, the products of the transforms were averaged over 256 records. The data was reduced in blocks of 2048, which for 256 records corresponds to a total time interval of 40.33 seconds. The maximum frequency resolution of the psd for the temperature measurement was 6.35 Hz.

Figure 8 shows the profiles of mean temperature (T_m) and rms of temperature fluctuation (T_{rms}) of a single vertical traverse between the inner two streams. Both uncompensated and compensated T_{rms} profiles, using an average time constant of 25 ms, are shown. The mean temperatures were corrected for radiation losses using a procedure described by Gupta [29]. Montgomery et al. [30] applied a numerical model to evaluate the accuracy of this approach compared with using an instantaneous compensation. It was found that this approximation was, for the worst case considered, within 50 °C, and was usually within 10 °C. Furthermore, although instantaneous correction of temperature measurements seems possible the computational time seems prohibitive for practical use. Also, radiation properties over the entire temperature range of interest are difficult to find in the literature.

T_m increases rapidly until a mean temperature of 1045°C, where the average temperature gradient decreases slightly. This decrease in gradient is similar to the temperature profile obtained in a lifted flame burning gaseous fuel[6]. T_m then continues to increase until a maximum of 1843°C is reached in the diffusion flame region at z=29 mm measured from the hydrogen flame holder. T_{rms} increases through the intermittent combustion zone, because large temperature fluctuations occur between the burnt and unburnt gases. T_{rms} reaches a maximum near z = 22 mm. The T_{rms} for the uncompensated temperatures is qualitatively the same but significantly lower due to thermal inertia of the thermocouple. The average temperature measurements were seen to be identical, except for the radiation correction, as is to be expected when an average time constant is used.

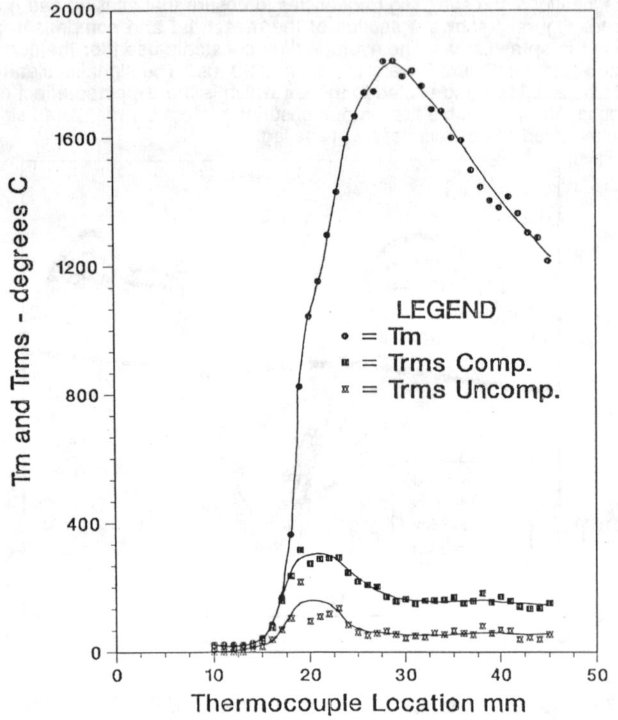

Fig. 8 Vertical Tm and Trms Profiles

Figure 9 is the pdf surface calculated from the same compensated temperature measurements shown in Figure 8. At the lower locations the pdf shows a sharp peak near room temperature. In the low temperature region the effect of droplet impingement may be significant, however, the results seem qualitatively correct. As the thermocouple is positioned higher in the flame, a high temperature tail is developed indicating the occasional presence of the flame front. Further downstream, near the intermittent region of the blue premixed zone, a somewhat bimodal distribution is observed which corresponds to large-scale movement of the flame front passing the measurement point. In the diffusion region, the pdf again assumes a more Gausian type profile at higher temperatures indicating a higher probability of hot burnt gases. The accuracy of the average time constant used for compensation can be roughly evaluated by examination of the pdf of the compensated signal [31]. All the temperatures must certainly lie between the adiabatic flame and ambient temperatures. If these limits are exceeded, then the temperature signal has probably been overcompensated. This approach is limited, however, because these high and low boundaries do not exist at every point in the flame. Currently, the best approach seems to be to measure the time constant *in situ* [1,5,6].

Figures 10, 11, 12 are the results of a parametric study of the effect of varying the time constant on the autocorrelation, psd, and pdf, respectively. These were obtained using the raw data at the z = 29 mm location. Figure 10 shows the autocorrelation, measured at z = 29 mm, for time con-

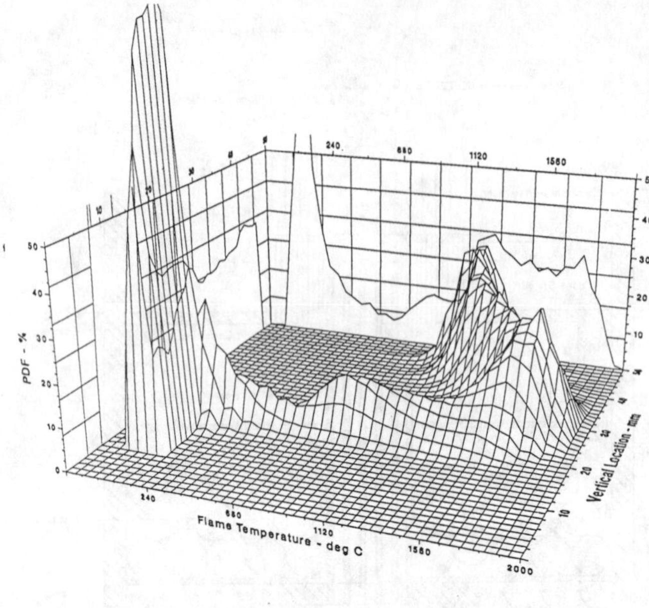

Fig. 9 Probability Density Functions

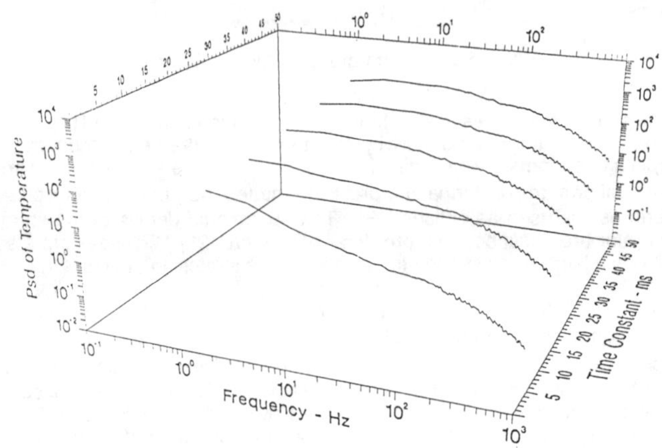

Fig. 10 Autocorrelation at Various Time Constants

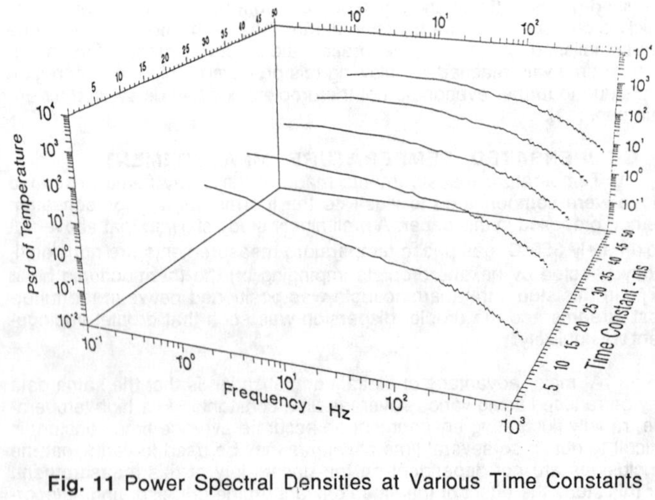

Fig. 11 Power Spectral Densities at Various Time Constants

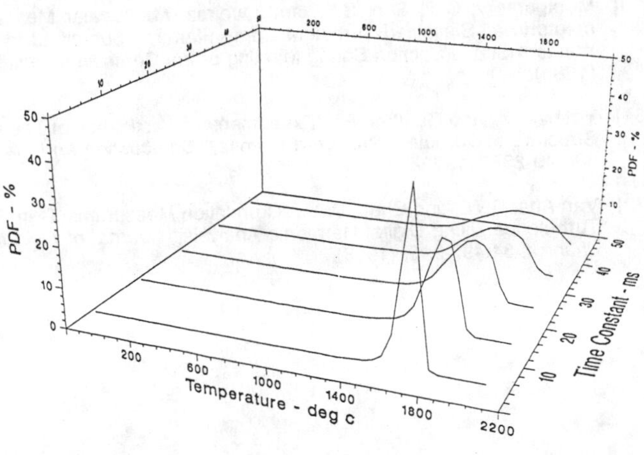

Fig. 12 Probability Density Function at Various Time Constants

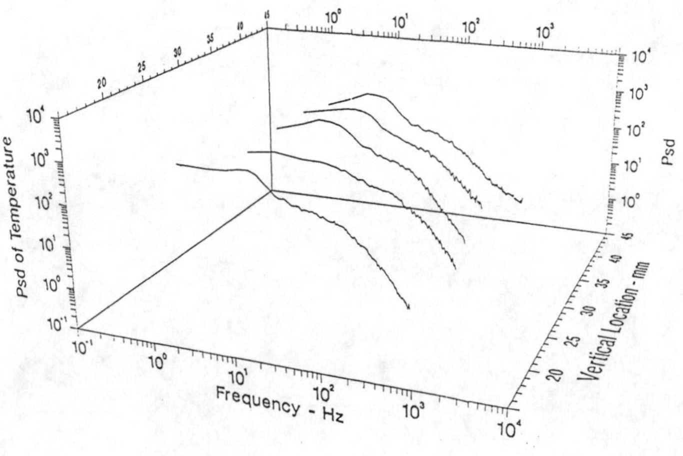

Fig. 13 Power Spectral Densities at Five Vertical Locations

stants of τ = 5 , 15, 25, 35, and 45 ms. The autocorrelation begins at the value of one and quickly approaches zero, as is expected. Very near the zero time separation, the slope is nearly zero and then decreases rapidly. Further, it is seen that as the time constant increases the correlation decreases more rapidly to zero. Likewise, for the same point, Figure 11 shows the psd for the same time constant variation. As the time constant is raised the magnitude is amplified more, especially at higher frequencies. Figure 12 shows the effect of varying the time constant on the pdf at the same point. As the time constant is increased the pdf widens and consequently becomes shorter.

Figure 13 shows the psd's obtained at selected vertical locations in the flame. The maximum frequency that was resolved was about 1200 Hz. Hinze [33] suggested that turbulent temperature fluctuations could occur at frequencies as high as 5000 Hz; however, since the measured cold-flow turbulence intensity level was only about 3.4%, higher frequency fluctuations may not have been present at measurable magnitudes. As expected of turbulent temperature measurements, all of the psd's show a similar slope at the higher frequencies. On several of the psd's, a small peak is seen near 30 Hz which may be due to some characteristic fluctuation frequency of the flame front past the thermocouple. From these results it is seen that the temperature measurements provide significant physical insight and are, in the very least, qualitatively correct.

6. CONCLUSIONS

A digital technique to compensate thermocouple measurements for thermal inertia effects has been presented and its usefulness was demonstrated by compensating a generated temperature signal and actual temperature measurements in a spray flame. The effect of noise in the signal was examined by adding random noise to the generated signal at three noise levels. A low-pass digital filter was used to maintain an adequate signal-to-noise ratio at all compensated frequencies. Temperature measurements made in the spray flame were compensated using three different time constants. The average temperature, rms temperature, autocorrelation, psd, and the pdf of temperature fluctuation in the turbulent spray flame are presented. The following are the major conclusions of this study:

1. The compensation technique was found to be accurate, computationally efficient and easily implemented. Gibb's phenomenon and convergence were identified as potential problems and methods of nearly eliminating these effects were demonstrated.

2. Electrical and quantization noise was found to be a significant source of error that must be carefully managed. Digital low-pass filtering was implemented simultaneously with the proposed algorithm to maintain an adequate signal-to-noise ratio in the presence of electrical and quantization noise.

3. Because of the difficulty in obtaining an accurate time constant, a significant advantage of digital compensation is that the same measured data may be compensated using various time constants to verify that the conclusions made are independent of this uncertainty.

4. Measurements made in a spray flame illustrate the usefulness of this technique to measure different parameters related to turbulent temperature profiles in a harsh combustion environment (e.g., average temperature (T_m), temperature fluctuation (T_{rms}), autocorrelation, psd, and pdf of temperature fluctuation), thus revealing the thermal structure of the flame.

Acknowledgements

This work was sponsored by the Advanced Combustion Engineering Research Center (ACERC). Funds for this center are received from the National Science Foundation, The State of Utah, 26 industrial participants, and the U.S. Department of Energy.

References

[1] Lockwood, F.C. and Moneib, H., "Fluctuating Temperature Measurements in a Heated Round Free Jet," *Combustion Science and Technology*, 22:63-81, (1980).

[2] Chandran, S.B.S., Komerath, N.M., Grissom, W.M., Jagoda, J.I. and Strahle, W.C., "Time Resolved Thermometry by Simultaneous Thermocouple and Raleigh Scattering Measurements in a Turbulent Flame," *Combustion Science and Technology*, 44:47-60, (1985).

[3] Lenz, W. and Gunther, R., "Measurements of Fluctuating Temperature in a Free-Jet Diffusion Flame," *Combustion and Flame*, 37:63-70, (1980).

[4] Cambray, P., Vachon, M. Maciaszed, T., and Bellet, J.C., "Instantaneous Compensation of Fine Wire Thermocouple Inertia in a Buoyant Turbulent Diffusion Flame," The 23rd ASME National Heat Transfer Conference, Denver, Colorado, (1985).

[5] Yule, A.J., Taylor, D.S. and Chigier, N.A., "On-Line Digital Compensation and Processing of Thermocouple Signals for Temperature Measurement in Turbulent Flames," AIAA Sixteenth Aerospace Sciences Meeting, (1978).

[6] Lockwood, F.C. and Moneib, H.A., Fluctuating Temperature Measurements in Turbulent Jet Diffusion Flame," *Combustion and Flame* , 47:291-314, (1982).

[7] Schoenung, S.M., and Mitchell, R.E., "Comparison of Raman and Thermocouple Temperature Measurements in Flames," *Combustion and Flame* ,35:207-211, (1979).

[8] Heitor, M.V., Taylor, A.M.K.P. and Whitelaw, J.H., "Simultaneous Velocity and Temperature Measurements in a Premixed Flame," *Experimental Fluids*, 3:323-339, (1985).

[9] Yoshida, A., "Structure of Opposed Jet Premixed Flame," Presented at the Twenty-Second (International) Symposium on Combustion, The Combustion Institute, Seattle, WA, paper 152, (1988).

[10] Katsuki, M., Mizutani,Y. and Matsumoto, Y., "An Improved Thermocouple Technique for Measurement of Fluctuating Temperatures in Flames," *Combustion and Flame*, 67:27-36, (1987).

[11] Ballantyne, A. and Moss, J.B., "Fine Wire Thermocouple Measurements of Fluctuating Temperature," *Combustion Science and Technology*, 17:63-72, (1977).

[12] Lockwood, F.C. and Odidi, A.O.O., "Measurement of Mean and Fluctuating Temperature and of Ion Concentration in Round Free-Jet Turbulent Diffusion and Premixed Flames," Fifteenth Symposium (International) on Combustion, The Combustion Institute, Pittsburgh, PA, 561-571, (1976).

[13] Laurendeau, N.M., "Temperature Measurements by Light-Scattering Methods," 24th ASME National Heat Transfer Conference, Pittsburgh, PA, August, (1987).

[14] Cameron, C.D., Brouwer, J., and Samuelsen, G.S., "A Model Gas Turbine Combustor With Wall Jets and Optical Access for Turbulent Mixing, Fuel Effects, and Spray Studies," Presented at the Twenty-second (International) Symposium on Combustion, The Combustion Institute, Seattle, WA, paper 50, (1988).

[15] Hirano, T., Hisamoto, D., and Kawamata, M., "Behavior of Fuel Droplets Across Propagating Spray Flames," 24th ASME National Heat Transfer Conference, Pittsburgh, PA, (1987).

[16] Khalil, E.E. and Whitelaw, J.H., "Aerodynamic and Thermodynamic Characteristics of Kerosene-Spray Flames," Fifteenth Symposium (International) on Combustion, The Combustion Institute, Pittsburgh, PA, (1976).

[17] Queiroz, M., Dietvorst, J., Yao, S.C. and Chigier, N.A., "Digitally Compensated and Processed Temperature Measurements in a Dilute Spray Flame Using a Micro-Thermocouple," Proceedings of the Combustion Institute Eastern Section Meeting, The Combustion Institute, Puerto Rico, (1986).

[18] Queiroz, M., "Combustion Dynamics and Thermal Structure of a Planar Spray Flame," Ph.D. Thesis, Carnegie Mellon University, (1987).

[19] Queiroz, M. and Yao, S.C., "A Parametric Exploration of the Dynamic Behavior of Flame Propagation in Planar Sprays," *Combustion and Flame*, in press, (1989).

[20] Shepard, C.E. and Warshawsky, I., "Electrical Techniques for Compensation of Thermal Time Lag of Thermocouples and Resistance Thermometer Elements," NACA Technical Note 2703, (1952).

[21] Dantzig, J.A., "Improved Transient Response of Thermocouple Sensors," *Review of Scientific Instruments*, 56:723-25, (1985).

[22] Doebelin, E.O., *Measurement Systems Application and Design*, McGraw-Hill, New York (1983).

[23] De Levie, R., Sarangapani, S., Czekaj, P., and Benke, G., "Numerical Differentiation by Fourier Transformation as Applied to Electrochemical Interfacial Tension Data," *Analytical Chemistry*, 50:110-115, No. 1, (1978).

[24] Hayes, J.W., Glover, D.E., and Smith, E.S., "Some Observations on Digital Smoothing of Electroanalytical Data Based on the Fourier Transformation," *Analytical Chemistry*, 45:277-284, No. 2, (1973).

[25] Press, W.H., Flannery, B.P., Teukolsky, S.A. and Vetterling, W.T., *Numerical Recipes, The Art of Scientific Computing*, Cambridge University Press, (1986).

[26] Son, S.F., "Thermal Structure of an Array of Burning Droplet Streams," M.S. Thesis, Mechanical Engineering Dept., Brigham Young University, Provo, UT (In progress).

[27] Rangel, R.H., and Sirignano, W.A., "Unsteady Flame Propagation in a Spray With Transient Droplet Vaporization," Presented at the Twenty-second (International) Symposium on Combustion, The Combustion Institute, Seattle, WA, paper 194, (1988).

[28] Ashgriz, N., and Yao, S.C., "Development of Multi-Orifice Impulsed Spray Generators for Heterogeneous Combustion Experiments," ASME/JSME Thermal Engineering Joint Conference Proceedings, 2:433-439, Book no. 100158-B, (1981).

[29] Gupta, A.K. and Lilley, D.G., *Flowfield Modeling and Diagnostics*, Abacus Press, (1985).

[30] Montgomery, C.J., Son, S.F., and Queiroz, M., "Scalar Measurements in a Simplified Turbulent Spray Flame," Submitted to the Western States Section Spring Meeting of the Combustion Institute (1989).

[31] Yoshida, A. and Gunther, R., "Experimental Investigation of Thermal Structure of Turbulent Premixed Flames," *Combustion and Flame*, 38:249-258, (1980).

[32] Van Atta, C.W., and Chen, W.Y., "Correlation Measurements in Grid Turbulence Using Digital Harmonic Analysis," *Journal of Fluid Mechanics*, 34:497-515, (1968).

1989 National Heat Transfer Conference
HTD-Vol. 106, Heat Transfer Phenomena in Radiation, Combustion and Fires

LIQUID JET IMPINGEMENT HEAT TRANSFER ON A UNIFORM FLUX SURFACE

X. Liu and J. H. Lienhard V
Department of Mechanical Engineering
Massachusetts Institute of Technology
Cambridge, Massachusetts

ABSTRACT

Convective heat transfer to an impinging liquid jet is investigated. A circular jet of subcooled liquid impinges on a heating surface maintained at a uniform heat flux. An integral method is used to obtain analytical predictions of the temperature distribution in the liquid film and the local Nusselt number. As a result of the radial development of the viscous and thermal boundary layers and of the film thickness, several characteristic heat transfer regions appear. The variation of Nusselt number is different in each region.

Experiments were performed to test the predictions of the theory. Subcooled water jets leaving a circular orifice cooled an electrically heated sheet under atmospheric conditions. The radial temperature distribution was measured and the Nusselt number was computed as a function of the jet Reynolds number. Agreement between the predictions and the experiments is generally good. Several factors are identified which may cause differing behavior outside a laboratory environment.

NOMENCLATURE

Roman Letters

A function defined in equation (10).
C_2, C_3, C_4 constants of integration.
c_p liquid specific heat capacity.
d jet diameter.
g_m local mass transfer coefficient from liquid surface to environment.
$h(r)$ local thickness of liquid sheet.
$\hat{h}$ local heat transfer coefficient from liquid surface to environment.
h_{fg} latent heat of vaporization.
h_{r_1} liquid sheet thickness at $r = r_1$.
k thermal conductivity of the liquid.
m_w, m_∞ water vapor mass fractions at liquid surface, in environment.
Nu_r Nusselt number, $qr/k(T_w - T_f)$.
Nu_x Nusselt number, $qx/k(T_w - T_f)$.
p, s parameters defined in equation (19).
q wall heat flux.
r radius measured from point of jet impact.
r_o radius at which δ reaches the surface of the liquid sheet.
r_1 radius at which δ_t reaches the surface of the liquid sheet.
r_2 radius at which T_w reaches the liquid saturation temperature.
Re_d Reynolds number of the jet, $u_f d/\nu$.
Sc Schmidt number for water vapor in air.
$T(r, y)$ liquid temperature distribution.
T_f jet temperature at impingement.
$T_{f4}(r)$ jet surface temperature distribution in region 4.
T_{sat} liquid saturation temperature.
$T_w(r)$ wall temperature distribution.
$u(r, y)$ radial velocity distribution in liquid film.
u_f velocity of impinging jet.
u_{max} local maximum film velocity (liquid surface velocity), equal to u_f in region 2.
y distance normal to the wall.

Greek Letters

δ viscous boundary layer thickness.
δ_t thermal boundary layer thickness.
η normalized velocity just outside the boundary layer, u/u_f.
θ jet impingement angle.
μ dynamic viscosity.
ν kinematic viscosity.
ρ density.

INTRODUCTION

Jet impingement cooling has broad application, such as controlled cooling of metals or non-metals in manufacturing processes, as a result of its easy implementation and high efficiency. Many investigations have dealt with convective heat transfer by submerged jet impingement (gas jets in gases or liquid jets in liquids), but studies of free liquid jets (travelling through gases) are much less common, particularly those presenting theoretical analyses of the heat transfer by liquid jets. The reason may be the complexity of the local phenomena.

When a liquid jet strikes a flat surface, it spreads radially in a thin film. This film is responsible for convective heat removal form the surface. Following Watson (1964), we subdivide the jet impingement flow into four regions:

1. *The stagnation zone.*
2. *The boundary layer region.* In this region, the viscous boundary layer thickness is less than the liquid sheet thickness, so that the liquid on the surface is unaffected by wall friction.
3. *The fully viscous sheet.* In this region, the boundary layer thickness is the same as the sheet thickness. The viscous influence extends through the entire liquid film, from the wall to the free surface.
4. *The hydraulic jump.* An abrupt increase of liquid sheet thickness occurs, and in this region the liquid velocity is much lower than in the upstream region.

Watson analyzed the hydraulic jump by an inviscid theory and regions 2 and 3 by a similarity solution of the boundary layer equations. Sharan (1984) applied the momentum integral method to analyze regions 2 and 3 and obtained results in good agreement with those of Watson.

McMurray *et al.* (1966) studied convective heat transfer to impinging *plane* jets from uniform heat flux walls. They observed the heat transfer to be subdivided into an impingement zone and a zone of uniform parallel flow. To fit their data, they based heat transfer correlations on the stagnation flow in the impingement zone and on the flat plate boundary layer in the uniform parallel flow zone. They obtained results in the form

$$\mathrm{Nu}_x = f(\theta, \eta, \mathrm{Re}_x, \mathrm{Pr}) \quad (1)$$

where θ is the angle of impingement and η is jet velocity normalized with that just outside the boundary layer, u/u_f.

Chaudhury (1964) solved the axisymmetric energy equation for a constant wall temperature condition in the fully viscous region, where a similarity solution of the momentum equation is available. The solution was expressed as a series expansion.

Metzger *et al.* (1974) experimentally studied the effects of Prandtl number on heat transfer by liquid jets. They used a uniform surface temperature boundary condition at the test surface. They presented only surface average values of the Stanton number, determined from measurement of the total heat flux, the test surface temperature, and the jet adiabatic wall temperature. No local measurements or analytical results were given. Their correlations are based on data for oil and water; for water, their correlation represents 95% of the data for disk radii of up to 6.6 jet diameters to within ±25%. The possible effect of evaporation resulting from apparently high liquid surface temperatures was not documented.

Carper *et al.* (1978,1986) studied the heat transfer from a rotating disk struck by a liquid jet. The papers show that disk rotation enhances the heat transfer

In the present paper, an integral method is used to predict the Nusselt number for constant heat flux and to examine the distance required to reach liquid saturation temperatures (relevant to the boiling jet problem). In a corroborating experiment, the wall temperature distribution was measured for positions up to 43 jet diameters from the point of impact and the local values of the Nusselt number were calculated. The analytical results are compared to the experimental data. Although the predictions of the theory are not simple, we hope that simplified correlations for convective heat transfer by jet impingement can be developed on the basis of these more complicated results.

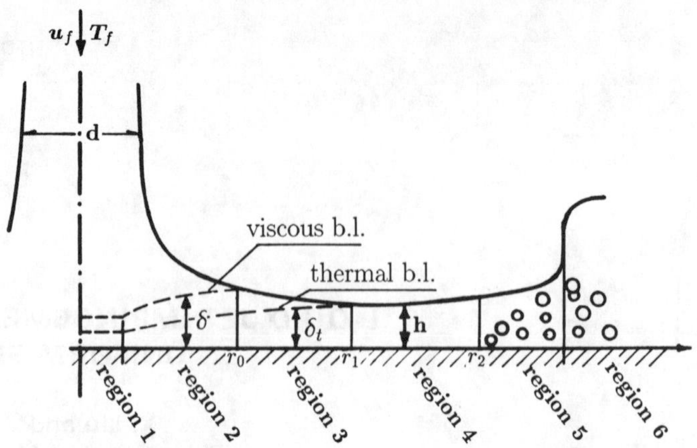

Figure 1: Region map for the downward-flowing jet.

Heat Transfer Regions

The heat transfer regions are more complicated than those of the flow field because more parameters are involved. We shall assume that the Prandtl number is greater than unity, Pr > 1, as is the case for most liquids. The heat transfer regions can then be subdivided as follows (*see* Figure 1).

1. *The stagnation zone.*
2. $\delta < h$ *region:* Within this region, the velocity outside the viscous boundary layer is undisturbed and approximately equal to the jet velocity, u_f.
3. $\delta = h$ *and* $\delta_t < h$ *region:* Within this region, the temperature outside the thermal boundary layer is not affected by the heat transfer, but the thermal boundary layer is affected by the viscous retardation of the momentum boundary layer.
4. $\delta = h, \delta_t = h$, *and* $T_w < T_{sat}$ *region:* In this region, the thermal boundary layer has reached the surface of the liquid sheet, and the temperature of the liquid surface increases with radius. Our analysis below shows that this region cannot exist for Pr > 4.859.
5. *The boiling region:* This region may include regions of nucleate boiling, burnout, and dry surface if the heat flux is high enough. Since boiling might occur at any wall temperature beyond T_{sat}, depending on heater finish and other factors, we will refer to the entire region for which $T_w > T_{sat}$ as the boiling region. We avoid all detailed consideration of nucleation incipience in the present work.
6. *The hydraulic jump:* In this region the heat transfer deteriorates dramatically. If the jet is directed upward (Figure 2), the hydraulic jump is very different than for the downward-directed jet. In that case, the jump is associated with a Rayleigh-Taylor instability. This instability will be discussed in a separate paper.

In the present paper, we do not treat region 5 or region 6. It must be noted that the all of the above regions may not exist at the same time, and that the last two may occur in a different sequence. For example, boiling incipience may take place in any region, and it can change the region map. The hydraulic jump may even occur in region 2 for low heat flux.

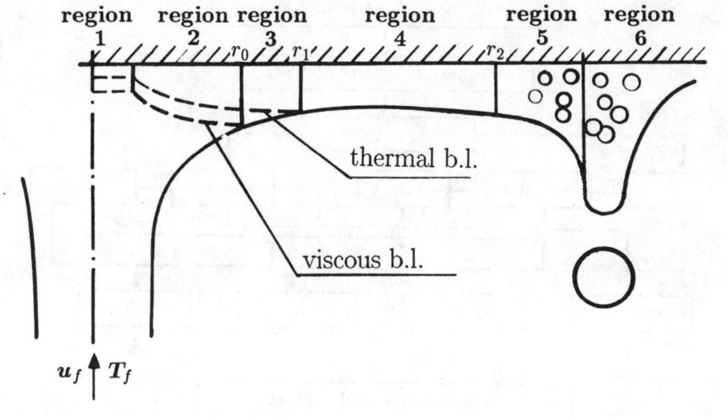

Figure 2: Region map for the upward-flowing jet.

We note in passing that for Pr < 1, region 2 and region 3 will instead be:

2. $\delta < h$ *region:* Since Pr < 1, the viscous boundary layer is thinner than the thermal boundary layer.
3. $\delta_t = h$ *and* $\delta < h$ *region:* In this region, the thermal boundary layer has reached the surface while the viscous boundary layer has not.

For the constant wall temperature problem, the region map would be simpler since we know in advance whether or not boiling occurs. In the present paper, we consider the regions 2, 3, and 4 without boiling for a uniform wall heat flux with Pr > 1.

ANALYSIS

Region 2

Sharan (1984) shows that treating the flow as purely radial is a very good approximation. We assume that in the regions beyond region 1, a purely radial flow occurs. In these regions, the integral energy equation is given by

$$\frac{d}{dr}\int_0^{\delta_t} ru(T-T_f)\,dy = \frac{q}{\rho c_p}r \tag{2}$$

where T is the liquid temperature profile, T_f is the subcooled incoming jet temperature, q is the uniform wall heat flux, u is the radial liquid velocity profile, and other terms are defined in the nomenclature. We approximate the profiles in equation (2) as:

$$T - T_w = (T_f - T_w)\left[\frac{3}{2}\frac{y}{\delta_t} - \frac{1}{2}\left(\frac{y}{\delta_t}\right)^3\right] \tag{3}$$

$$u = u_{max}\left[\frac{3}{2}\frac{y}{\delta} - \frac{1}{2}\left(\frac{y}{\delta}\right)^3\right] \tag{4}$$

for T_w the wall temperature and $u_{max} = u_f$ the constant liquid surface speed in region 2. For Pr > 1, we neglect $3\delta^3/280\delta_t{}^3$ relative to $3\delta/20\delta_t$ and integrate to obtain

$$\frac{27}{80}\mathrm{Re}_d\mathrm{Pr}\frac{1}{\mathrm{Nu}_r^3}\left(\frac{r}{d}\right)^2\frac{r^2}{\delta d} = \frac{1}{2}\left(\frac{r}{d}\right)^2 + C_2 \tag{5}$$

where $\mathrm{Nu}_r = qr/k(T_w - T_f)$ and $\mathrm{Re}_d = u_f d/\nu$.

For $d/2 \ll r$, Sharan shows that the initial boundary layer thickness can be neglected. We have the same situation for the thermal boundary layer and may use Sharan's expression for δ

$$\delta = 2.679\left(\frac{rd}{\mathrm{Re}_d}\right)^{1/2} \tag{6}$$

Thus, we find

$$\mathrm{Nu}_r = 0.632\,\mathrm{Re}_d^{1/2}\mathrm{Pr}^{1/3}\left(\frac{r}{d}\right)^{1/2} \tag{7}$$

At $r = r_o$, the viscous boundary layer reaches the surface of the liquid sheet, which is the border of region 2. Sharan gives

$$r_o = 0.1773\,\mathrm{Re}_d^{1/3}d \tag{8}$$

Region 3

The significant difference between this region and region 2 is that in this region the velocity at the liquid surface, u_{max}, is not constant. Watson (1964) and Sharan (1984) show

$$u_{max} = \frac{A}{hr} = \frac{1}{5}\frac{u_f d^2}{hr} \tag{9}$$

where

$$A = \frac{u_f d^2}{5} \tag{10}$$

and

$$h = 0.1713\left(\frac{d^2}{r}\right) + \frac{5.147}{\mathrm{Re}_d}\left(\frac{r^2}{d}\right) \tag{11}$$

In this region, the energy equation is again equation (2), but u_{max} is now a function of r and h and $\delta = h$. Integration with equations (3), (4), and (9), neglecting higher order terms as before, gives

$$\frac{3}{100}\mathrm{Re}_d\mathrm{Pr}\frac{1}{\mathrm{Nu}_r}\left(\frac{\delta_t}{h}\right)^2\frac{r}{d} = \frac{1}{2}\left(\frac{r}{d}\right)^2 + C_3 \tag{12}$$

With

$$\mathrm{Nu}_r = 3r/2\delta_t \tag{13}$$

from equation (3), substitution of equation (11) into equation (12) and subsequent rearrangement then yields

$$\mathrm{Nu}_r = \frac{0.407\,\mathrm{Re}_d^{1/3}\mathrm{Pr}^{1/3}(r/d)^{1/3}}{\left[0.1713\left(\frac{d}{r}\right)^2 + \frac{5.147}{\mathrm{Re}_d}\frac{r}{d}\right]^{2/3}\left[\frac{1}{2}\left(\frac{r}{d}\right)^2 + C_3\right]^{1/3}} \tag{14}$$

From equation (8), at $r = r_o$

$$\mathrm{Nu}_r = 0.632\,\mathrm{Re}_d^{1/2}\,\mathrm{Pr}^{1/3}\left(\frac{r_o}{d}\right)^{1/2} \tag{15}$$

Solving equations (14 - 15) for C_3 produces

$$C_3 = \frac{0.267(d/r_o)^{1/2}}{\left[0.1713\left(\frac{d}{r_o}\right)^2 + \frac{5.147}{\mathrm{Re}_d}\frac{r_o}{d}\right]^2\mathrm{Re}_d^{1/2}} - \frac{1}{2}\left(\frac{d}{r_o}\right)^2 \tag{16}$$

At $r = r_1$, the border of region 3, we have $\delta_t = h$. To locate r_1, we use equations (13) and (12) to find

$$\frac{2}{100}\mathrm{Re}_d\mathrm{Pr}\left(\frac{h}{d}\right) = \frac{1}{2}\left(\frac{r}{d}\right)^2 + C_3 \tag{17}$$

Then, using equation (11) and rearranging, we get

$$\left(\frac{r_1}{d}\right)^3 + p\left(\frac{r_1}{d}\right) + s = 0 \tag{18}$$

where we define

$$p = \frac{-2C_3}{(0.2058\,\mathrm{Pr} - 1)} \qquad s = \frac{0.00686\,\mathrm{Re}_d\mathrm{Pr}}{(0.2058\,\mathrm{Pr} - 1)} \tag{19}$$

Since C_3 is usually less than zero, equation (18) has real, positive roots when

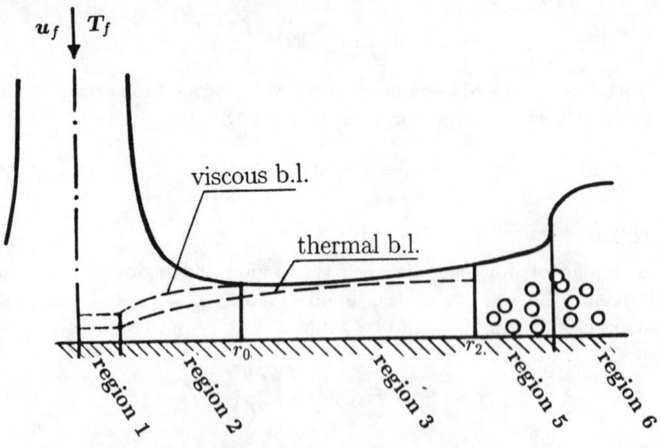

Figure 3: Region map for Pr > 4.859.

$$s < 0 \tag{20}$$

This condition is satisfied only for

$$\mathrm{Pr} < 4.859 \tag{21}$$

This is a very interesting result. It means that for Pr > 4.859 the thermal boundary layer can never reach the surface of the liquid sheet before subcooled boiling occurs: the increase of h owing to viscous retardation is more rapid than the growth of the thermal boundary layer.[1] In this case, region 4 cannot exist. The situation is shown schematically in Figure 3.

For Pr < 4.859, equation (18) yields

$$\frac{r_1}{d} = \left\{-\frac{s}{2} + \left[\left(\frac{s}{2}\right)^2 + \left(\frac{p}{3}\right)^3\right]^{1/2}\right\}^{1/3} + \left\{-\frac{s}{2} + \left[\left(\frac{s}{2}\right)^2 - \left(\frac{p}{3}\right)^3\right]^{1/2}\right\}^{1/3} \tag{22}$$

With this and equation (11), we can calculate the liquid sheet thickness, h_{r1}, at $r = r_1$. The wall temperature can be determined from the expression

$$T_w = T_f + \frac{q\,r}{k\,\mathrm{Nu}_r} \tag{23}$$

Since boiling might occur for any wall temperature which exceeds the saturation temperature, depending on heater surface finish and other factors, we exclude any case for which $T_w > T_{sat}$ from subsequent considerations.

Region 4

For Pr < 4.859, after the thermal boundary layer reaches the surface, the surface temperature increases with radius. If we neglect the heat transfer from the liquid surface, the energy equation may be written

$$\frac{d}{dr}\int_0^h ruT\,dy = \frac{q}{\rho c_p}r \tag{24}$$

We assume the temperature profile to be

$$T - T_w = (T_{f4} - T_w)\left[\frac{3}{2}\frac{y}{h} - \frac{1}{2}\left(\frac{y}{h}\right)^3\right] \tag{25}$$

where the surface temperature in region 4, T_{f4}, is a function of r. Substituting equations (4), (9), and this temperature distribution into equation (24), we obtain

$$A\left(\frac{5}{8}T_{f4} + \frac{13\,qh}{140\,k}\right) = \frac{q}{\rho c_p}\frac{r^2}{2} + C_4 \tag{26}$$

The constant may be found by evaluating this expression at $r = r_1$, where $T_{f4} = T_f$ and $h = h_{r1}$:

$$C_4 = A\left(\frac{5}{8}T_f + \frac{13\,q\,h_{r1}}{140\,k}\right) - \frac{q}{\rho c_p}\frac{r_1^2}{2} \tag{27}$$

Since $T_{f4} = T_w - 2qh/3k$,

$$\mathrm{Nu}_r = \frac{0.25\,(r/d)}{\frac{1}{\mathrm{Pr}\mathrm{Re}_d}\left(1 - \frac{r_1^2}{r^2}\right)\left(\frac{r}{d}\right)^2 + 0.130\frac{h}{d} + 0.0371\frac{h_{r1}}{d}} \tag{28}$$

and

$$T_w = T_f + \frac{q}{k}\left[0.518\,h + \frac{4\alpha(r^2 - r_1^2)}{u_f\,d^2} + 0.149\,h_{r1}\right] \tag{29}$$

From equation (29) we can find r_2, the point at which T_w exceeds the saturation temperature, and beyond which subcooled boiling might occur. We exclude consideration of the region $r > r_2$ from this paper.

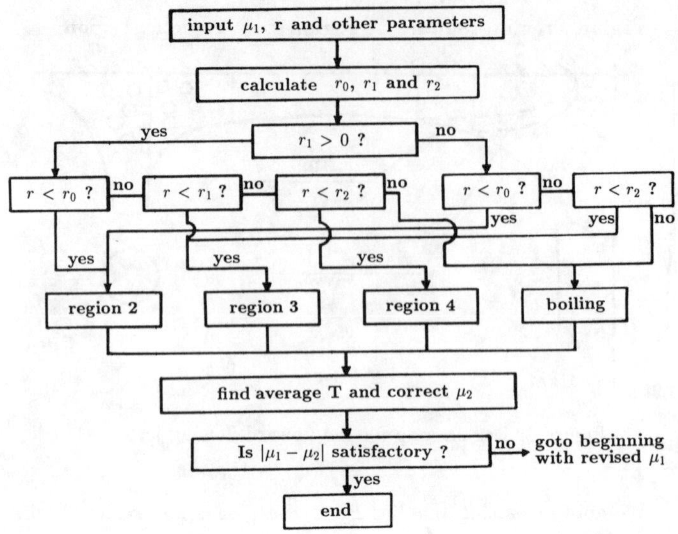

Figure 4: Calculation procedure for integral predictions.

Calculations

To use these results in calculating Nu_r and $T_w(r)$, a numerical iteration is most expedient. In particular, the viscosity varies strongly with temperature for most liquids, and properties must be evaluated at the local average temperature across the film. The some of the different expressions for each region are also quite complex. The flowchart in Figure 4 shows the general procedure for using the analytical results of this section to determine the distribution of Nusselt number and wall temperature. This procedure was used in obtaining the calculated results presented in Section 4 below.

EXPERIMENTS

To verify the predictions of the preceding sections, experiments were performed with water jets under atmospheric conditions. The radial wall temperature distributions beneath a jet striking a uniformly electrically heated surface were measured, and the data were used to calculate the local values of the Nusselt number. In this section, the

[1]The precise value of Pr will be affected by the various minor approximations inherent in the integral solutions of the transport equations. Historically, integral solutions have shown good agreement with exact solutions in spite of the approximations involved.

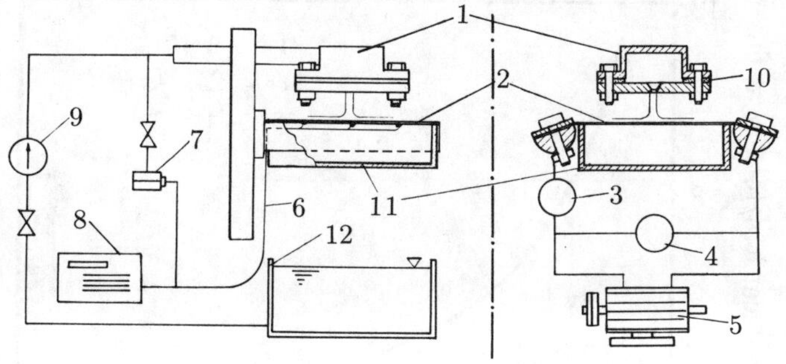

Figure 5: Experimental Apparatus: 1. orifice and plenum; 2. test surface; 3. ammeter; 4. voltmeter; 5. generator; 6. thermocouples; 7. pressure transducer; 8. digital voltmeter; 9. pump; 10. gasket; 11. plastic box; 12. water tank.

experimental procedure is described. The data and the theory are compared in the next section.

Experimental Apparatus

The experimental apparatus is shown schematically in Figure 5. An orifice plate faces vertically downward toward an electrically heated plate. The plate temperature was measured for water jets of differing diameter and speed hitting the plate when heated at different fluxes.

The electric heater is made of stainless steel sheet 0.10 mm thick. The variation in thickness of the steel plate is about ±3%. The working dimensions of the sheet are 7.8 cm by 15.2 cm. This sheet is clamped at either end onto the flat surfaces of two 5.1 cm diameter hemispherical copper rods which span the entire 15.2 cm width of the heater sheet. The rods are connected to a low voltage, high current generator, thus causing Joule heating of the steel sheet uniform to ±6%. Conductive heat loss to the rods was estimated to be negligible as a result of the thinness of the sheet and the low temperature differences involved.

The sheet temperature distribution was measured by twelve evenly-spaced thermocouples along a radial line parallel to the copper rods and passing through the point of jet impact. The thermocouples were each attached to the underside of the heater sheet with glue and were electrically separated from it by a thin wafer of mica. The thickness of the mica sheets is less than 0.015 mm and the thermal conductivity is not less than 5 W/(mK); the thermal resistance of the mica is negligible compared to that of the natural convection and the thermalcouple temperatures were within ±0.01°C of the plate surface temperatures. The back of the heater and the thermocouples were covered with an air filled plastic box which kept them dry and isolated them from room air currents. Free convection heat losses through the back of the heater were estimated to be negligible ($\approx 10 \text{W/m}^2$) because the heater temperatures were quite low.

Subcooled water, near room temperature, was supplied to the orifice through a high pressure pump. The rise in water supply temperature was slow compared to the time required for measurement of the radial temperature profile, and the inlet water temperature was thus effectively constant for each profile. The plenum behind the orifice is supplied by a 2.54 cm ID pipe. The plenum was 5.08 cm in diameter and 3.18 cm deep. The orifice plate covered this plenum; the orifice was a small hole with countersunk entry drilled in the center of the plate. The gauge pressure upstream of the plenum was recorded with a pressure transducer, and the jet velocities were computed from it using the standard jet relations. The jet struck the sheet halfway between the two rods and 2.8 cm from one edge. The precise effects of this azimuthal asymmetry on the flow field may warrant further study.

Pump pressure fluctuations were damped using an accumulator upstream of the orifice plenum. However, turbulence present in the inlet pipe and plenum still led to some minor fluctuations in the jet surface. These fluctuations gave rise to intermittent splattering downstream in the liquid sheet on the heater plate. The total volume of liquid leaving the surface as droplets was small compared to that remaining in the film. The average effects of this splattering are discussed further in the following section. Errico and Middleman (1986) have discussed the mechanisms of splattering in detail.

Table 1: Measured Nusselt number for water under atmospheric conditions.

$q = 12.4 \times 10^4$, $d = 2.58$ mm				
	Re_d			
r/d	3.54×10^4	4.50×10^4	5.27×10^4	6.24×10^4
3.38	436	440	479	544
7.75	599	651	713	787
11.6	590	748	831	897
15.5	544	665	767	880
19.4	499	584	692	763
23.3	462	528	599	660
27.1	385	436	569	504
31.0	313	413	507	549
34.9	289	389	449	492
38.8	250	261	398	399
42.6	211	256	361	282

$q = 7.73 \times 10^3$, $d = 3.05$ mm		
	Re_d	
r/d	4.03×10^4	5.89×10^4
3.28	424	484
6.56	549	669
9.84	667	781
13.1	434	853
16.4	417	718
19.7	394	640
23.0	403	731
26.2	401	578
29.5	336	536
32.8	307	432
36.1	175	308

$q = 17.2 \times 10^4$, $d = 2.58$ mm	
	Re_d
r/d	41.5×10^4
3.88	489
7.75	686
11.6	752
15.5	686
19.4	604
23.3	528
27.1	491
31.0	437
34.9	394
38.8	314
42.6	233

Experimental data were obtained for orifice diameters from 2.58 to 3.05 mm, heat fluxes from 7.73×10^3 to 17.2×10^4 W/m^2, and jet velocities in the range 12 to 24 m. The measured values of Nu_r and T_w are presented in Table 1 as a function of r/d for various q and Re_d. Experimental uncertainties in q and Re_d were estimated to be about 2%. The experimental uncertainty in Nu_r drops from about 13% at the smallest r/d to 1% at the largest r/d; it is associated primarily with the measurement of temperature difference, which is quite small near the point of jet impact.

RESULTS AND COMPARISONS

In this section, we consider the predictions of Section 2 in detail, and we discuss the results of the experiments performed to test the predictions. Experimental agreement with the predictions is generally good, although several effects present in real jets were discovered which can complicate the application of the theory to some non-laboratory systems.

The analytical predictions are compared to data for $q = 12.4 \times 10^4$ W/m^2 and $d = 2.58$ mm (our largest single data set) in Figure 6a. The radial variation of the Nusselt number, Nu_r, is shown for several values of the Reynolds number, Re_d, for r/d up to 43. The observed behavior may be related to the heat transfer regions defined in the analysis.

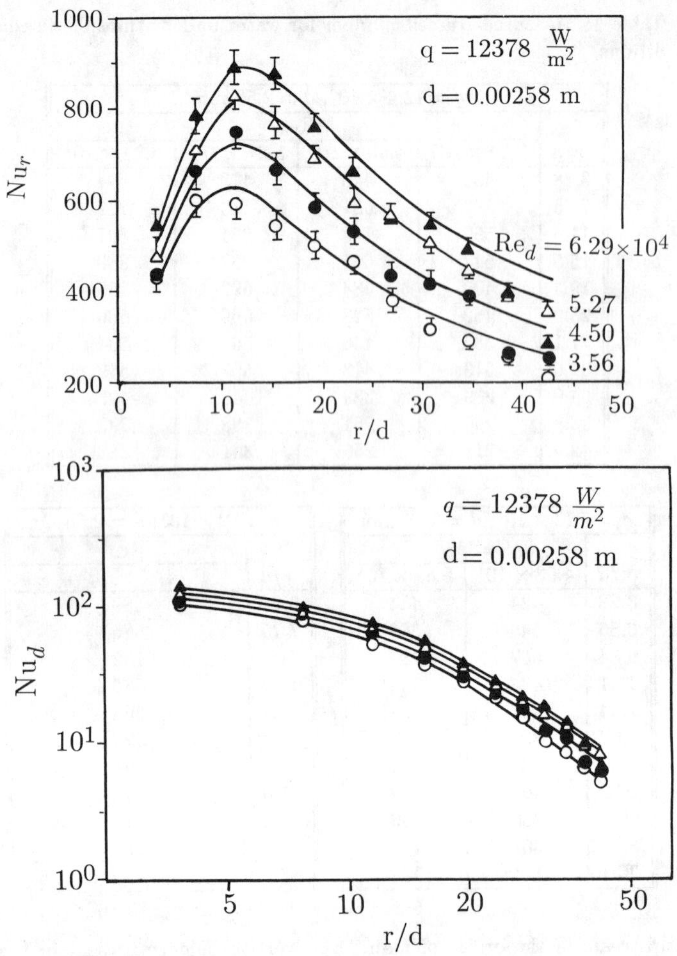

Figure 6: (a) Distribution of Nu_r: theory, —— ; data, ▲ $Re_d = 6.29 \times 10^4$, △ $Re_d = 5.27 \times 10^4$, • $Re_d = 4.50 \times 10^4$, ∘ $Re_d = 3.56 \times 10^4$. (b) Distribution of Nu_d: theory, —— ; data, (symbols as before).

The Nusselt number increases over small values of r/d in regions 1 and 2 and the beginning of region 3; after reaching a maximum, it decreases for larger r/d in region 3. The boundaries of these regions change with Reynolds number, but we may cite typical values. Region 2 ends at $r = r_o$: for $Re_d = 3.54 \times 10^4$, $r_o/d = 5.81$; for $Re_d = 6.29 \times 10^4$, $r_o/d = 7.05$. For cases in which $Pr < 4.859$, the boundary of region 3, r_1, would be on the order of $78d$, beyond the range of this experiment, but the present flows all had $Pr > 4.859$. Since the inlet water temperature was 15°C, saturation temperatures would not be reached until $r_2 \approx 270d$, well beyond the experimental domain. The hydraulic jump, which was also beyond experimental range, would probably occur first in this situation.

If the Nusselt number is instead based on jet diameter, rather than the radius along the sheet, a monotonic decay is seen (Figure 6b). This shows directly the increase in surface temperature. The difference in magnitude of the Nusselt number among different values of the Reynolds number decreases at larger r/d, although this is not obvious on the logarithmic coordinates of Figure 6b.

For the data of Figure 6, the agreement between experiment and theory was about 3% for $r/d < 35$ and about 15% for $r/d > 35$. The agreement deteriorates at the largest radii, for reasons we now examine.

In our analysis, the evaporation from the liquid surface was neglected. This assumption should be carefully checked in any practical situation, since evaporative heat loss can become a significant fraction of the wall heat flux. Evaporation will tend to cool the liquid at

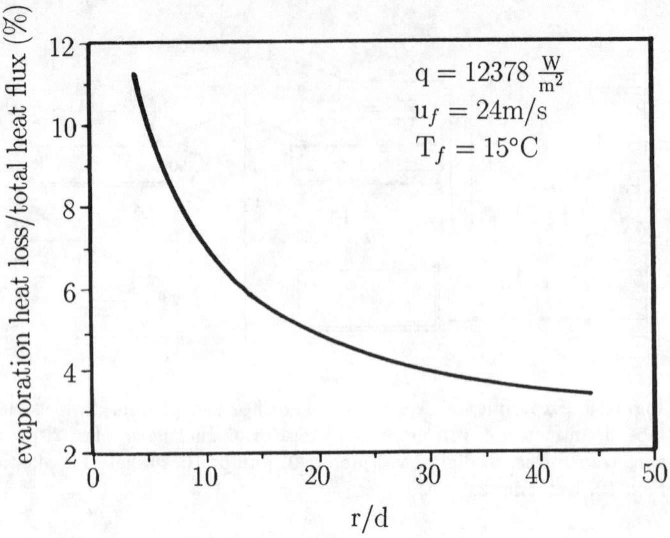

Figure 7: Estimated evaporative heat loss for the data of Figure 6 at $Re_d = 6.24 \times 10^4$.

the surface and reduce the rate of wall temperature increase (lowering Nu_r), even if only a very small amount of the liquid film is vaporized. The rate of evaporation increases rapidly with the surface temperature of the liquid; in dry air it will rise by a factor of 29 as the surface temperature increases from 5°C to 65°C. Some estimates are in order for the present data.

The evaporative heat flux from the liquid surface can be calculated as

$$q_e = h_{fg}\, g_m (m_w - m_\infty) \tag{30}$$

where h_{fg} is the latent heat of vaporization, m_w is the mass fraction of water vapor at the liquid surface, m_∞ is the mass fraction of water in the environment, and the mass transfer coefficient, g_m, is calculated from the Lewis analogy

$$g_m = \frac{\hat{h}}{c_p}\left(\frac{Pr}{Sc}\right)^{2/3} \tag{31}$$

with $\hat{h}$ the estimated local heat transfer coefficient from the moving liquid to the air. For 15°C water, a wall heat flux of 12.4×10^4 W/m², and $u_f = 24$ m/s, the ratio of evaporative heat flux, q_e, to wall heat flux, q, is shown in Figure 7.

The heat flux by evaporation decreases rapidly with r/d. For $r/d > 20$, it is less than 5% of the wall heat flux, and it is never more than about 12% . The rapid decrease in q_e is due in part to the rapid slowing of the liquid surface velocity in region 3. While the evaporation is greatest in region 2, it has no effect on wall temperature until the surface and wall thermal boundary layers meet further downstream. The overall effect of evaporative cooling probably remains negligible in the present case because the overall reduction in liquid bulk temperature is only a few percent of the increase produced by wall heating. However, the same estimation procedure shows, for example, that for a surface temperature of 45°C, evaporation would become comparable to a wall flux of 12.4×10^4 W/m² and cannot be neglected.

A direct comparison between the analytical prediction and the data is given in Figure 8, where different symbols are used for small ($r/d < 35$) and large radius ($r/d > 35$) data for $d = 2.58$ mm and for the data at $d = 3.08$mm. The agreement is excellent for the $d = 2.58$ mm at $r/d < 35$, but shows larger errors further out. The $d = 3.08$mm shows larger error at all radii. Two major factors may be responsible for the deviation of the data from the prediction. One is surface evaporation, as discussed, and the other is splattering of liquid from the moving film. While evaporation tends to raise the Nusselt number, splatter decreases it by reducing the volume of liquid within

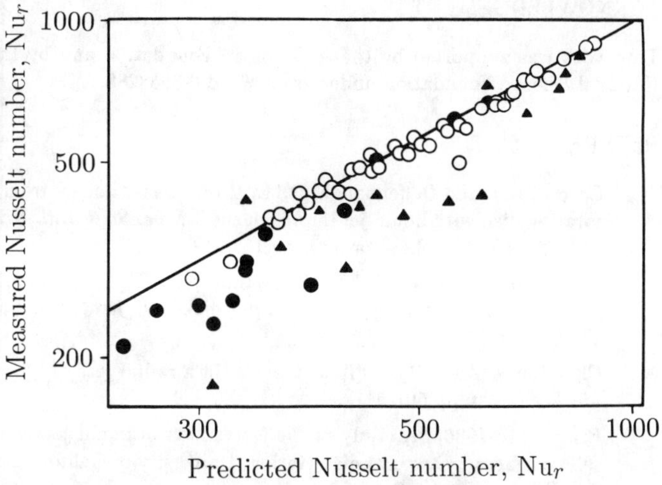

Figure 8: Comparison of predicted and measured Nusselt number: o, $d = 2.58$ mm, $r/d < 35$; •, $d = 2.58$ mm, $r/d > 38$; ▲, $d = 3.08$ mm.

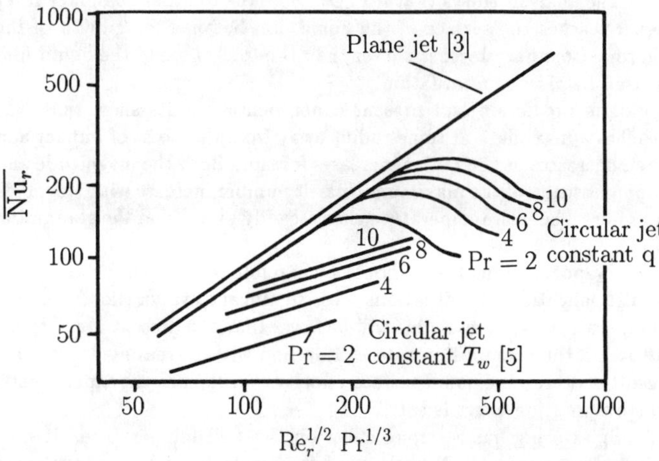

Figure 10: Comparison of predicted average Nusselt number for uniform flux to correlation for uniform wall temperature in circular jet impingement and to correlation for uniform flux plane jet impingement.

the film and producing an associated increase in the rate of rise of liquid bulk temperature. The effect of splatter accumulates in the film and is more serious downstream than upstream. In the present case, splattering is the dominant error.

The splattering of impinging liquid jets is driven by minor irregularities or unsteadiness in the incoming jet (in this case, by turbulence in the jet plenum) as discussed by Errico and Middleman (1986). The deviation of the measured and predicted Nusselt number in the region far from the center suggests that splatter may cause a decrease in the [time average] Nusselt number of more than 20% . No theoretical analysis predicts the effects of splatter; further experiments are needed in which splatter is both well controlled or eliminated.

The larger orifice ($d = 3.08$ mm) produced greater splatter and was run at lower heat flux. Both factors may contribute to the greater errors seen for that data.

Figure 9 shows a comparison of our prediction to the *plane* jet data of McMurray *et al.* (1966). The most significant difference is that, for the axisymmetric jet, Nu_r reaches a maximum and then decreases with radius, but that for the plane jet it does not. This is because

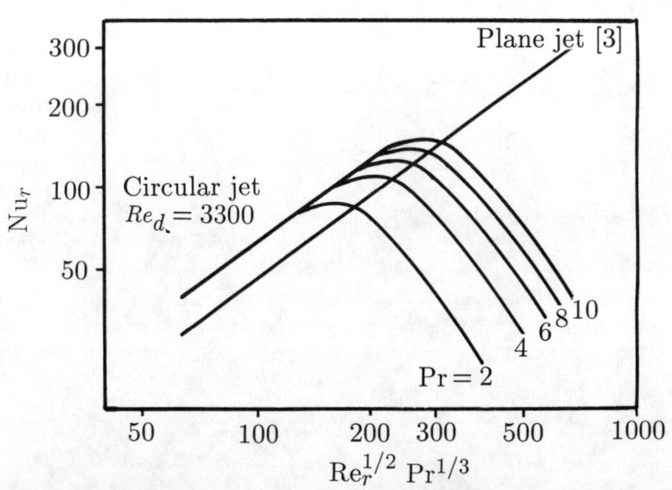

Figure 9: Comparison of predicted local Nusselt number for circular jets to local Nusselt number correlation for plane jets.

the liquid layer thickness produced by the plane jet of McMurray *et al.* was so large that the viscous boundary layer never reached the liquid surface. Thus, the differences are not surprising: McMurray's experiment remains in region 2 throughout. Before the boundary layer reaches the surface, the plane and axisymmetric jets show the same variation with Reynolds number, although for the axisymmetric jet the coefficient is 0.632 while for the plane jet it is 0.47, 25% lower.

Figure 10 compares the *average* Nusselt number for our uniform wall heat flux experiments to that for the uniform wall temperature axisymmetric jet experiments of Metzger *et al.* (1974). The average Nusselt number for a uniform wall heat flux is generally larger than that for a uniform wall temperature. Average Nusselt numbers for the uniform wall flux plane jet data of McMurray *et al.* are also shown. Although the local Nusselt number for the plane jet is smaller than for the axisymmetric jet in region 2 and part of region 3, the average Nusselt number for the plane jet is generally larger than that for the axisymmetric jet. In region 2, the average Nusselt number for the plane jet is twice as large as the local Nusselt number, but that for the axisymmetric jet is 4/3 the local one. This switch in size relates to the different areas involved in planar and axisymmetric averaging.

Observations in a separate apparatus show that, after the jet strikes the plate, the laminar flow formed in center region becomes wavy at some distance from center, finally becoming turbulent at a larger radius. The transition points for surface waves and for turbulence, and the heat transfer coefficient in turbulent region need further investigation. The divergence of our data and theory at large r/d may be related to such effects, although a turbulent region could not be observed directly in these experiments.

CONCLUSIONS

The heat transfer phenomena beneath an impinging circular jet can be divided into different regions by taking account of the development of the viscous and thermal boundary layers and the possible occurrence of nucleate boiling or a hydraulic jump. These regions may appear in different combinations, depending the jet Reynolds number, temperature levels, wall heat flux, and liquid physical properties. The radial variation of Nusselt number, Nu_r, is interpreted in terms of the development of the thermal boundary layer, and an integral analysis is presented for each of the convective regions.

The analysis shows that if $Pr > 4.859$, the thermal boundary layer never reaches the surface of the liquid film because the growth of the thermal boundary layer is slower than the thickening of the liquid film caused by viscous retardation.

The prediction and present experimental results show that Nu_r reaches a maximum at some radius away from the point of impact and then decreases as the radius increases further. Both the magnitude and radial position of the maximum Nusselt number increase with Reynolds number. The wall temperature rises steadily away from the stagnation point.

Evaporative heat loss should be carefully examined for any practical application. In situations with significant evaporation, the wall temperatures at large radius will be lower than the present predictions, although the evaporation rate will decline with increasing radius as a result of decreasing liquid surface velocity. For the present experiments, the evaporation effect is small.

Splattering has a strong effect on the wall temperature. It may cause decreases in the Nusselt number of up to 20% . The downstream flow is more seriously affected than the upstream flow. In most real systems, which are likely to be noisy and thus have splatter, this effect will reduce the efficiency of the jet in cooling the surface and increase the wall temperatures above the present predictions.

Nevertheless, the agreement between the theory and the present data is generally good, and confirms the trends predicted. Further experiments, in which the level of splatter is directly controlled, and a theoretical analysis of the role of splatter, are both needed. The possible effects of turbulence and of flow asymmetries also need further study.

The average Nusselt number for a constant heat flux is generally larger than for a constant wall temperature. Although the local Nusselt number for the plane jet is smaller than that for an axisymmetric jet in region 2 and part of region 3, the average Nusselt number for a plane jet is generally larger than that for an axisymmetric jet.

ACKNOWLEDGEMENTS

This work was supported by the A. P. Sloan Foundation and by the National Science Foundation under grant #CBT-8858288.

REFERENCES

Carper, H.J. and Deffenbaugh, D.M., 1978, "Heat transfer from a rotating disk with liquid jet impingement," *Proc. Sixth Intl. Heat Transfer Conf.*, Vol.4, Toronto, pp.113-118.

Carper, H.J., Saaveda, J.J., and Suwanprateep, T., 1986, "Liquid jet impingement cooling of a rotating disk," *J. Heat Transfer*, Vol.108, pp.541.

Chaudhury, Z.H., 1964, "Heat transfer in a radial jet," *J. Fluid Mech.*, Vol.20, pp.501-511.

Errico, M., 1986, "A study of the interaction of liquid jets with solid surfaces," Doctoral dissertation, University of California at San Diego, Chapter 3.

McMurray, D.C., Myers, P.S., and Uyehara, O.A., 1966, "Influence of impinging jet variables on local heat transfer coefficients along a flat surface with constant heat flux," *Proc. Third Intl. Heat Transfer Conf.*, Vol.2, Chicago, pp.292-299.

Metzger, D.E., Cammings, K.N., and Ruby, W.A., 1974, "Effects of Prandtl number on heat transfer characteristics of impinging liquid jets," *Proc. Fifth Intl. Heat Transfer Conf.*, Vol.2, Tokyo, pp.20-24.

Sharan, A., 1984,"Jet-disc boiling: burnout predictions and application to solar receivers," Master's thesis, University of Houston.

Watson, E.J., 1964, "The radial spread of a liquid over a horizontal plane," *J. Fluid Mech.*, Vol.20, pp.481-499.

1989 National Heat Transfer Conference
HTD-Vol. 106, Heat Transfer Phenomena in Radiation, Combustion and Fires

A TECHNIQUE FOR MEASURING LOCAL HEAT TRANSFER DUE TO AN IMPINGING LIQUID SPRAY

C. J. Davenport
Alcan International Limited
Banbury, United Kingdom

ABSTRACT

This paper describes a technique for measuring local heat transfer coefficients produced by a liquid spray impinging on a surface. The method uses microencapsulated chiral nematic liquid crystals to measure the transient response of a metal sheet when suddenly exposed to the spray.

The method is discussed and the tolerances in coefficient distribution evaluated. Data are presented on the heat transfer coefficients produced by a solid water cone spray impinging on a cylindrical surface. It was found that the heat transfer coefficient in the region of direct impingement was essentially uniform but decreased rapidly outside this region.

NOMENCLATURE

A	(m^2)	Surface area of plate
Bi	-	(hd/k), Biot Number
Bi_{tr}		Ratio heat transfer by convection to lateral conduction
Cp	(J/kgK)	Specific heat capacity of plate material
d	(m)	Thickness of plate
F_o	-	$(kt/\rho Cpd^2)$ Fourier Number
h	(W/m^2K)	Heat Transfer coefficient between spray and surface
k	(W/mK)	Plate material thermal conductivity
q	(W)	Heat Flow between spray and surface
t	$(secs)$	Time
tm	$(secs)$	Time from shutter-raise signal to contour appearance
t_{lag}	$(secs)$	Time from shutter-raise signal to spray arrival
T	$(°C)$	Instantaneous surface temperature
T_{cont}	$(°C)$	Temperature at which the liquid crystal changes appearance
T_{start}	$(°C)$	Starting temperature of the plate
T_w	$(°C)$	Water temperature
x	(m)	Transverse distance on plate
δ	(m)	Tolerance in contour position
ρ	(kg/m^3)	Plate Material density

INTRODUCTION

Liquid sprays are used for a wide variety of industrial cooling operations and are capable of producing very high heat transfer coefficients. In some applications, sprays are used not merely to remove heat but to control temperature distribution. For example, the work rolls of metal rolling mills are usually spray-cooled with individual sprays which are often separately controlled to allow the cooling distribution to be adjusted in order to give accurate control of the roll thermal expansion. This is necessary for the accurate control of the dimensions and flatness of the rolled strip. When sprays are used in this way, it is desirable to know the distribution of heat transfer coefficient produced by a given array and how this varies as members of the array are switched. This information is useful in the design of sprays systems.

This paper describes a method which may be used to measure such a distribution of coefficient for heat transfer, without boiling, between a surface and a single spray or a combination of sprays.

DESCRIPTION OF THE TECHNIQUE

The method employs the transient response of a metal plate to the sudden impingement of the spray system under investigation. Temperature sensitive liquid crystals were used to measure the resulting time-varying temperature distribution on the surface. Local heat transfer coefficients were then derived from these observations.

Liquid crystals have been used by Jones and others (Refs. 1 and 2) in transient heat transfer measurements. In their investigations, relatively low heat transfer coefficients due to gases flowing over surfaces were measured. Plastic components with a coating of liquid crystals were used to detect the time-response of the surface to the sudden onset of flow The plastic pieces were treated as semi-infinite bodies and the transient heat transfer by conduction in the plastic taken as one-dimensional. This enabled the use of well-established analytical expressions for deriving the heat

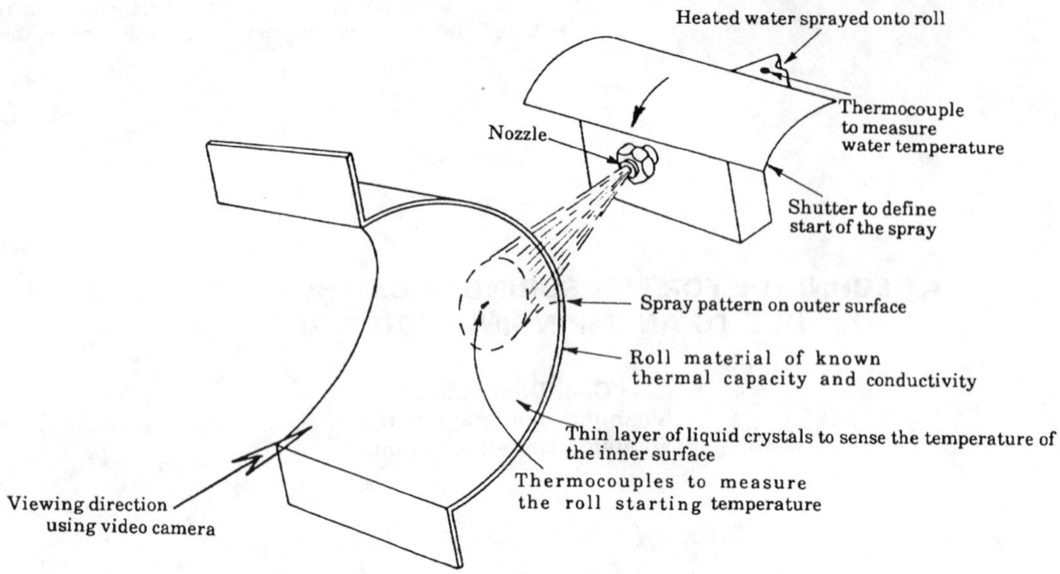

Fig. 1 - Schematic of apparatus

transfer coefficients from the temperature/time observations at the surface. The technique of Jones *et al* cannot be employed directly to the case of liquid sprays since the spray would quickly remove the liquid crystal layer. The method reported here therefore used a thin metal plate with the liquid crystal layer on the opposite side to the liquid spray as shown in the schematic (Fig.1).

The cylindrical surface representing the work roll of a rolling mill, was formed from 2mm thick aluminum sheet and the nozzle was connected to a pump circulating heated water. The flow from the nozzle could be interrupted by a shutter, controlled by a pneumatic cylinder. An indicator light showed when the shutter-raise signal had been given. Thermocouples were attached to the inside surface of the cylinder to measure its starting temperature and the water temperature was measured using a thermocouple fixed into the pipe immediately behind the nozzle. Both these measurements together with the shutter status indicator light were displayed within view of the video camera referred to in Fig.l.

The inside surface of the cylinder was coated with a thin layer of black paint followed by a layer of chiral nematic liquid cyrstals. The liquid crystals were in a micro-encapsulated form with a capsule diameter of approximately 10 microns. An essentially mono-capsule layer was deposited on the surface. The liquid crystals have the property of changing their appearance over a narrow temperature range. The range was centred on 42.3°C for the particular material used in this case. When the temperature was higher or lower than the critical range, very little light was reflected. Within the range, the crystals reflect light. Effectively, they indicated isothermal lines on the cylinder surface.

To measure the coefficient distribution, the starting temperature of the roll was set by flooding cold water down the outside surface. This water was then turned off and the shutter raised to direct spray onto the roll while recording the progress of the temperature contours on the sensitive side of the roll using a video camera. These time-dependant contours were then used to derive the distribution of heat transfer coefficient.

The pattern developed over a time period of between 100 millisecs and 2 secs. This was faster than could be observed with the naked eye but was well within the response capability of the liquid crystals themselves. This has been measured by Ireland (Ref.3) to be in the order of 3 millisecs. The liquid crystals were illuminated by a tungsten filament lamp held in the same position for all tests. They were calibrated against a thermocouple attached to the surface, using very slow and uniform heating / cooling of the surface. This thermocouple was itself calibrated against a standard thermocouple in a constant temperature calibration apparatus. The same thermocouple was used to measure the initial surface temperature.

In order to establish the starting time for each pattern the metal plate was replaced by a transparent plastic one of the same diameter to enable visualisation of the flow pattern. The development of the pattern following the raising of the shutter was video recorded in the same way as the liquid crystal patterns. The time lag between the shutter-raise signal and the arrival of the spray on the surface could therefore be determined within about 20 millisecs. These time-lag values for each region of the pattern were then subtracted from the times measured during the heat transfer tests.

DERIVATION OF HEAT TRANSFER COEFFICIENTS.

The heat transfer coefficient h between the surface and the fluid was defined by:

$$q = h\,A\,(T_w - T) \qquad (1)$$

Under the following assumptions :

a. Negligible temperature drop through the thickness of the cylinder wall.
b. Negligible lateral heat conduction within the wall.

The heat flow can also be expressed in terms of the rate of increase in the temperature of the surface :

$$q = A\,d\,\rho\,C_p \frac{\partial T}{\partial t} \qquad (2)$$

Equating (1) and (2) and integrating between 0 and (t_m - t_{lag}) and T_{start} and T_{cont} leads to the following expression for h:

$$h = \frac{\rho C_p d}{(t_m - t_{lag})}\,\ell n \left[\frac{T_w - T_{start}}{T_w - T_{cont}}\right] \qquad (3)$$

This is an approximate value for h suitable for generating a starting value for a more rigorous calculation.

The calculation method used removed assumption a) by calculating the temperature distribution through the wall thickness using a one-dimensional explicit transient procedure for a postulated h value, at first as calculated by equation (3), and comparing the computed inner surface temperature with T_{cont}. The h value was then adjusted iteratively until the computed surface temperaure was equal to T_{cont}.

The temperature drop across the wall was typically 1 to 2°C with a water temperature of 50°C.

ACCURACY CONSIDERATIONS

The thickness of cylinder wall and its thermal diffusivity were chosen to give acceptable sensitivity to coefficient differences without incurring excessive lateral heat conduction. This was done using the analytical solution for one dimensional transient conduction in a slab of material with surface convection from one surface (Ref.4 article 3.1 1) This series solution is plotted in Ref.4 as a family of curves of temperautre against Fourier Number for different Biot numbers. For acceptable sensitivity of measured time to heat transfer coefficient, two conditions must be satisfied

$$Fo = \frac{kt}{\rho\, C_p\, d^2} > 1$$

and

$$Bi = \frac{hd}{k} < 1$$

For h values in the order of 50,000 W/m2K and measured times,t, of about 0.2 secs, this implied :

$$\left(\frac{k}{\rho C_p}\right)\frac{1}{d^2} > 5 \ \ /sec$$

$$\frac{k}{d} > 50{,}000\ W/m^2K$$

These conditions are satisfied by common alloys of copper and aluminum in workable gauges. They are not satisfied by plastic materials whose thermal conductivity is too low.

LATERAL CONDUCTION

The position of a temperature contour depends not only on heating through the thickness but also on heat conducted laterally within the wall. This will be most noticeable when the heat transfer coefficient is low but the gradient across the surface is large. It depends on the ratio of heat transfer by convection at the surface to the transfer by lateral conduction. This ratio is similar to the Biot number except that the two rates are in perpendicular directions. It will be termed Bi_{tr}.

An estimate of Bi_{tr}, may be obtained by considering the surface to be divided into squares of side δ (see Fig.2). δ is the effective tolerance in the contour position.

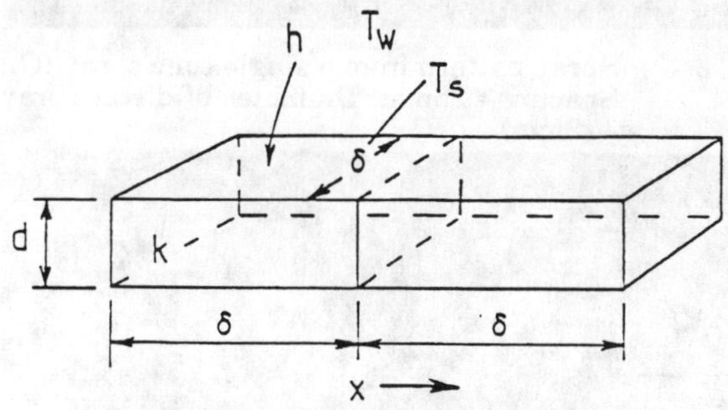

Fig. 2 - Error due to lateral conduction

By neglecting the temperature variation through the plate thickness, equation (3) may be used to estimate temperature gradient as a function of heat transfer coefficient gradient at the instant when the critical liquid crystal temperature T_{cont} is reached. This allows calculation of the lateral heat conduction, using the Fourier conduction equation. Comparison with the surface convection heat transfer rate gives :

$$Bi_{tr} = \frac{h^2 \delta}{k\, d \dfrac{\partial h}{\partial x}} \quad \frac{1}{\ell n\left(\dfrac{T_w - T_{start}}{T_w - T_{cont}}\right)} \qquad (4)$$

If a value of Bi_{tr}, of 10 is used to ensure that convection predominates over lateral conducton, then an expression for δ can be written as:

$$\delta = \frac{10\,k\,d}{h^2}\left(\frac{\partial h}{\partial x}\right)\ell n\left(\frac{T_w - T_{start}}{T_w - T_{cont}}\right) \qquad (5)$$

If T_w, T_{cont}, and T_{start} take approximate values of 50,42 and 20°C respectively, the value of the logarithmic term in (5) is about 1.3. For a 2mm thick aluminum alloy plate (d =0.002m, k=170 W/mK), equation (5) can be simplified to :

$$\delta = 4.5\,\frac{\left(\frac{\partial h}{\partial x}\right)}{h^2} \qquad (6)$$

δ represents the scale of lateral conduction occurring at the instant the contour appears at the relevant point on the surface, and was used to estimate the error in contour position due to lateral conduction.

Equation (3) was used to estimate the errors arising from the measurements of T_w and T_{start} (± 0.5°C) , T_{cont} (±0.5°C) and also of t_m and t_{lag} (± .02secs). Errors in aluminium properties and plate thickness were neglected compared with those in temperatures and times. The largest errors arose from the time measurements and since these appear as time differences in equation (3), the errors are highest for the highest heat transfer coefficients. They are summarised below :

Coefficient KW/m^2K	Percent Deviation
50	33
25	20
18	15
14	12

The errors in measuring the higher coefficients could be reduced by the use of high speed film or videography for which the effective frame times would be much shorter.

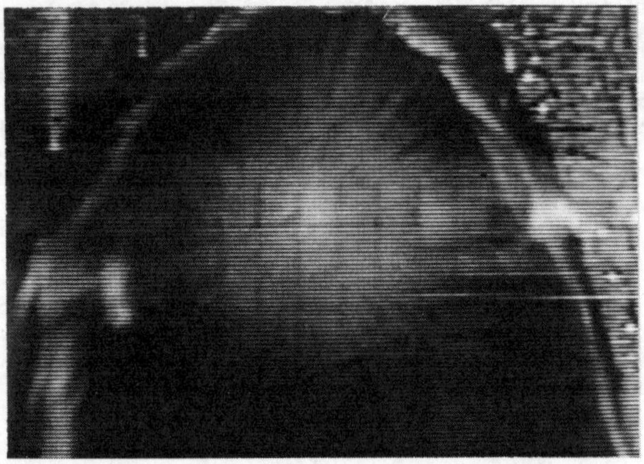

Fig. 3 - Spray pattern from a single cone spray (Grid Spacing - 25mm. Diameter of direct spray - 112mm)

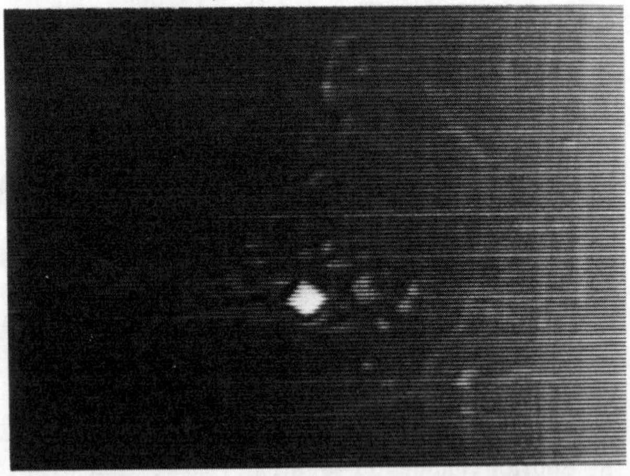

Fig. 4 - Liquid crystal pattern 1.46 secs after shutter raised.

RESULTS FOR A SINGLE CONE SPRAY

Fig.3 shows the spray pattern produced by a single cone spray impinging onto a cylindrical surface and Fig.4 shows a temperature contour for the same spray as measured by the liquid crystal layer.

Fig.5 is a plot of heat transfer coefficient distribution across the centre of the spray pattern for three different spray intensities. The tolerances on the maximum values are also indicated. Equation (6) was used to estimate the error in the contour position due to lateral conduction and gave a tolerance of ± 4mm where the coefficient was 14,000 W/m^2K increasing to ±8mm where the coefficient was 5000 W/m^2K. The width of the visible contour was about 5mm so that the error due to lateral conduction was similar to that involved in tracing the contour.

Jeschar et al (Ref.5) have measured heat transfer coefficients in the centre of spray patterns from cone sprays, mostly for surface temperatures in excess of 100°C. The values they obtained at the lowest

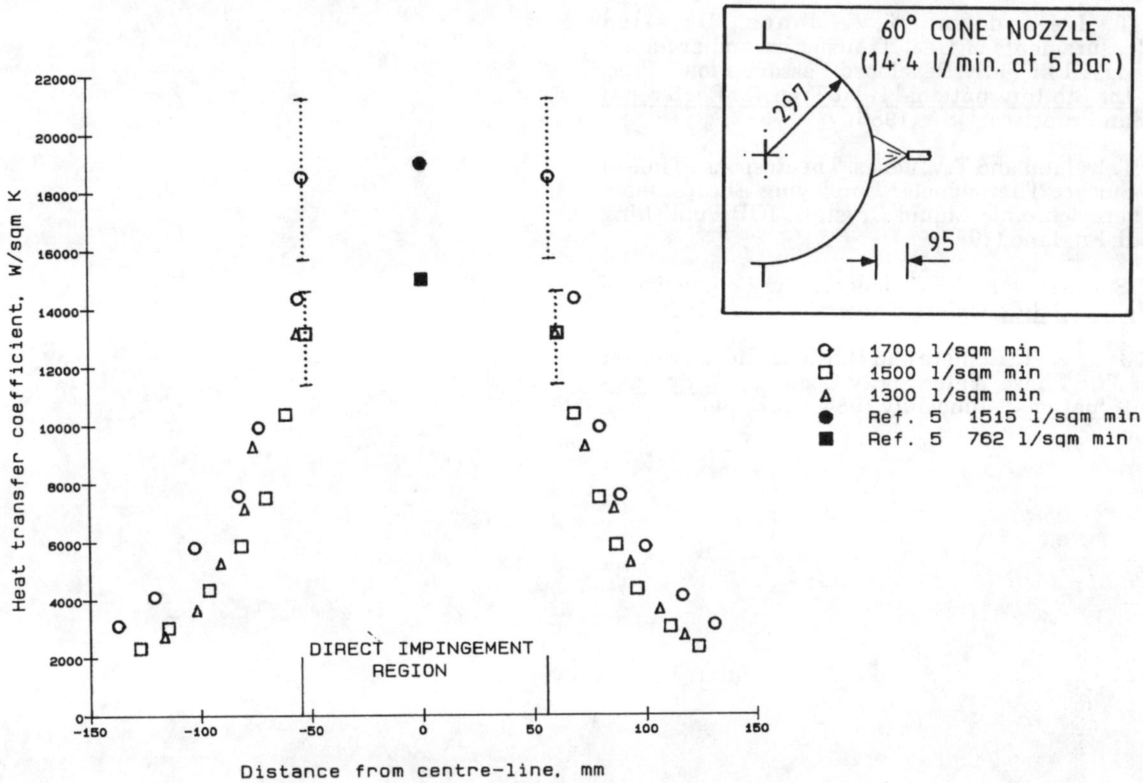

Fig. 5 - Heat transfer coefficient distribution for a single cone spray

temperatures for two intensities are shown in Fig.5 for comparison.

There are no points plotted near the centre of the pattern. This is because the first change of appearance of the liquid crystal layer was over a complete circular area, indicating that the heat transfer coefficient was constant over this area, within the tolerance of the measurements. This area coincided with the direct impingement area calculated from the spray angle and the distance from the surface. The coefficients declined very rapidly outside this area even though visualisation of the spray pattern (Fig.3) indicated very active fluid flow in this region. They were only 50% of their maximum values at a distance from the centre-line of about 1.45 times the direct area radius. This indicated the direct impingement of droplets was a much more effective heat transfer mechanism than tangential flow at the same mass velocity. It implies that in order to maintain a uniform heat transfer coefficient over a surface it is not sufficient to cover the surface with turbulent coolant, but that direct impingement should be maintained over as much of the surface as possible.

When a liquid spray is directed against a rotating cylinder, it is expected that the cylinder surface velocity will influence the heat transfer coefficient. This effect is not included in the technique reported here. However the effect of rotation in the region of direct impingement is expected to be small when the spray pressure is such that liquid droplet velocities are large compared with surface velocity. This will be true in many engineering applications, for example when cooling rolls in a hot rolling mill.

The technique may be applied to the investigation of heat transfer coefficient distribution due to an array of sprays and to the effects of switching combinations of sprays. Such coefficient distributions are of great relevance to the control of the shape of hot rolls during metal rolling.

CONCLUSIONS

A technique has been described which enables measurement of heat transfer coefficient distributions due to liquid sprays to accuracies varying from ± 12% for coefficients of 1400 W/m^{2}K to ±33% for coeffficients of 50000 W/m^{2}K.

When applied to the case of a single cone spray on a cylindrical surface, the coefficient was found to be essentially constant in the region of direct impingement but dropped off rapidly outside this region, being 50% of its maximum value at about 1.45 times the radius of direct impingement.

REFERENCES

1. T.V. Jones and C.M.B Russell , Efficiency of Rectangular Fins, ASME National Heat Transfer Conference, Orlando, Florida, Paper 80 HT-121,(Jul.27-30 1980).

2. P.T. Ireland and T.V. Jones, Detailed Measurements of Heat Transfer on and around a Pedastel in Fully Developed Passage Flow., Proc. of the 8th International Heat Transfer Conference , San Fransisco, USA (1986).

3. P.T. Ireland and T.V. Jones, The Response Time of a Surface Thermometer Employing Encapsulated Thermochromic Liquid Crystals, IOP Publishing Ltd, England,(1987).

4. H.S. Carslaw and J.C. Jaeger, The Conduction of Heat in Solids, Oxford University Press, 1959.

5. R.Jescher, U.Reiners and R.Scholz, Heat Transfer in Two Phase Water Spray Cooling., Wärme Gas International, June/July 1984, pp229-308.

1989 National Heat Transfer Conference
HTD-Vol. 106, Heat Transfer Phenomena in Radiation, Combustion and Fires

EVAPORATION OF DROPLETS SPRAYED INTO AN IMMISCIBLE LIQUID

A. Ullmann and R. Letan
Mechanical Engineering Department
Ben-Gurion University of the Negev
Beer-Sheva, Israel

ABSTRACT

The present work relates to dilute spray systems where interaction between droplets or bubbles does not take place. The analysis of a single droplet-bubble projects in such cases on the whole spray system.

Photographs of colored droplets evaporating into colorless bubbles have illustrated splashing of the liquid inside the bubble, and the non-regular instantaneously varying shapes of the evaporating bubble.

Close to the spray nozzle the droplet accelerates from rest towards its terminal velocity. Further up, as the fraction of vapor increases the bubble accelerates due to its decreasing density. Finally, the shape distortion and the large drag balances the reduced density, and a constant velocity is reached. Integration of the velocity yields the bubble path and the height needed for complete evaporation.

A theoretical time-dependent velocity including terms of "added mass" and "history" is presented and numerically solved for a bubble evaporating in an immiscible liquid. Integration of the velocity with respect to time yields the path of the bubble. The drag coefficient employed in the model relates to the bubble size and shape expressed in its aspect ratio.

Experiments of other researchers were utilized for comparison and validation of the theoretical model. Those experiments conducted with butane and pentane bubbles evaporating in water were presented against the theoretical velocity and path curves. The comparison with the experimental path showed good agreement.

NOMENCLATURE

A - surface area of droplet-bubble
a - axis normal to flow
b - axis parallel to flow
C_D - drag coefficient
C_{Ds} - drag coefficient of a sphere
C_{Dsd} - drag coefficient of a spheroid
C_{Dd} - drag coefficient of a droplet
d - diameter
d_e - equivalent diameter (sphere of the same mass)
d_o - initial diameter
E - aspect ratio, b/a
F - force
F_b - buoyancy force
F_D - drag force
H - height of droplet above nozzle
M - mass of droplet
R - radius of droplet-bubble
R_o - initial radius
$\bar{R}$ - nondimensional radius, R/R_o
s - time variable
T - temperature
t - time
U - velocity
U_o - initial velocity
Γ - density ratio, ρ_1/ρ or $\bar{\rho}/\rho$
Δ_A - "added mass" coefficient
Δ_a - drag correlating factor
Δ_H - "history" coefficient
μ - viscosity
ρ - density of surrounding fluid
ρ_1 - initial density of the droplet
$\bar{\rho}$ - average density of the evaporating droplet
σ - surface tension

INTRODUCTION

Researchers who studied condensing bubbles and evaporating droplets in an immiscible liquid used semi-empirical and theoretical models to predict the droplet-bubble velocity and path. The correct prediction of bubble velocity is most significant in calculations of convective heat and mass transfer coefficients over the phase-changing bubble. The velocity is also of importance in estimates of the condenser or evaporator height.

Sideman and Taitel (1964) empirically obtained the velocity of pentane bubbles evaporating in water. The instantaneous height and surface area of the rising droplet were measured and correlated with time. The velocity was then obtained by differentiation of the height correlation.

Simpson et al (1974) worked with butane droplets evaporating in water. The authors concluded that velocity of the butane droplet-bubble was constant (25 cm/s) up to a diameter ratio of 2, with respect to the initial size. Above that value velocity varied as the square root of the diameter ratio.

Moalem et al. (1980) theoretically studied a droplet of freon, R-113, in a cycle process of evaporation and condensation in water. The freon droplets being heavier than water, rises up and falls down. It accelerates and decelerates. The authors calculated its instantaneous velocity by using an averaged density of the phase-changing droplet, and an "added mass" term in the equation of motion. The term of "history" has not been used by those researchers in their analysis.

Lerner et al. (1987) studied accelerating-decelerating bubbles condensing in an immiscible liquid both experimentally and theoretically. The size of the phase-changing bubble and its velocity were solved simultaneously.

Ullmann and Letan (1988) have shown that the motion of an accelerating droplet is closely related to the wake separetion and its shedding. These phenomena are responsible for oscillations of velocity which have not been yet fully understood. The time-dependent velocity represents an average over these oscillations over short intervals of time.

Letan (1988) has shown the experimental methods which can be applied to study the motion of droplets, bubbles and sprays changing phase.

MOTION OF AN EVAPORATING DROPLET

Let us analyze the motion of an evaporating droplet in an immiscible liquid. In the above studies liquid droplets were sprayed into an immiscible liquid at a velocity close to zero. The droplets evaporated in the surrounding liquid at a temperature above the saturation temperature of the droplet. The droplet moved up, and increased in size. Its average density decreased and its velocity increased. Droplets lighter than the medium, like butane or pentane in water, move unidirectionally up. Droplets heavier than the medium, like R-113 in water, fall down at first, and as the vapor makes them lighter, change direction and rise up. The analysis presently is concerned only with the rising-up bubbles. The range of sizes involves droplets of an initial diameter of 2-4 mm. With the change of phase the diameter increases 5-7 times the initial size, e.g. up to 25-30 mm. Bubbles of that size are very little affected by iterfacial forces. Therefore the bubbles shape into spherical caps or ellipsoidal caps ($E_o \geq 40$), as shown in Fig. 1.

The photographs of Fig. 1 were obtained in our studies with pentane droplets evaporating in water (Ullmann, 1987). The whole process of evaporation with the size and shape variations is illustrated in 24 photographs. The liquid pentane droplets were red-colored before injection. Upon evaporation the vapor became colorless, and the non-evaporated liquid in the bubble showed red, illustrating the spread-out of the liquid.

Sideman and Taitel (1964) assumed the liquid to accumulate at the bottom of the evaporating droplet. Simpson et al. (1974) observed the liquid splashing from side to side over the whole surface area as the bubble oscillatedon its way up. Our studies (Ullmann, 1987) have confirmed the phenomenon of splashing. Our findings were qualitative and obtained by visualization of the colored liquid in the droplet. The photographs of Fig. 1 : 8, 9, 14, 23, 24, show the colored liquid in side-positions at some stages of the oscillation. At other stages of the oscillating bubble the splashing liquid appears at the top of the bubble, and the vapor at the bottom: photographs 17, 20, 22. A close look at the photographs reveals the colored liquid film spread all over the surface of the bubble.

Fig. 1 shows also the shape variation of the injected spherical droplet to a spheroidal droplet-bubble, and finally to a vapor cap.

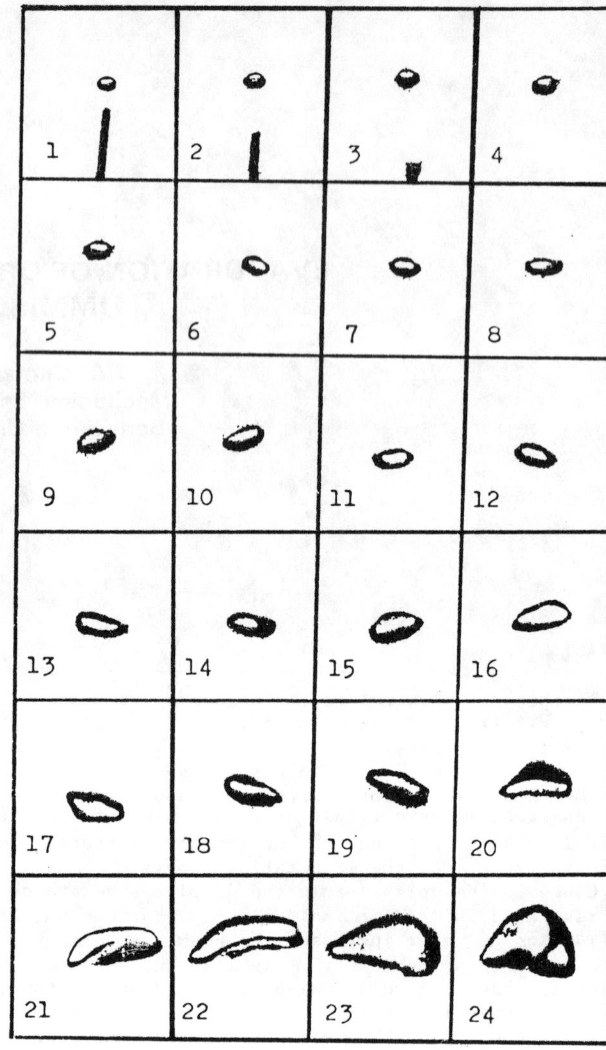

Fig. 1 Evaporation of a pentane bubble in water.

THE THEORETICAL TIME-DEPENDENT VELOCITY

The velocity of an evaporating droplet can be calculated if the initial radius of the injected droplet, R_o, and the instantaneous radius R, of the bubble are known. That knowledge can be achieved either experimentally or theoretically. The theoretical approach demands simultaneous solution of both the radius and the velocity (Lerner et al., 1987). That can be done, provided the mechanism of heat transfer to the evaporating droplet is established. Our present formulations would relate to velocity at defined radii of the evaporating droplet.

The equation of motion of an accelerating droplet is expressed as,

$$\sum F = M\,(dU/dt) \tag{1}$$

Forces of drag and buoyancy are exerted on an accelerating bubble or droplet (lighter than the surrounding liquid),

$$F_b = (1/6).\pi\,.\,\bar{d}_e.\,(\rho - \rho_p)\,.\,g \tag{2}$$

The drag forces, including the terms of "added mass"

and "history", as postulated by Odar and Hamilton (1964) were formulated in the following way:

$$F_D = (1/8).C_{Dd}.\pi . d_e^2.\rho.U^2 + (1/12).\Delta_A.\pi.d_e^3.\rho.(dU/dt) + (3/2).\Delta_H.d_e^2.(\pi.\rho.\mu)^{1/2} . \int_0^t U_{(s)}.(t-s)^{-1/2}.ds \qquad (3)$$

Combining these equations yields,

$$dU/dt = (9/(2\Gamma + \Delta_A)).((2/9).g.(1 - \Gamma) - ((1/(6.d_e).C_{Dd}.U^2) - (2.\Delta_H/d_e).(\mu/\pi.\rho)^{1/2}. \int_0^t U_{(s)}.(t - s)^{-1/2}.ds \qquad (4)$$

where Δ_A is the "added mass" coefficient, Δ_H is the "history", and $\Gamma = \bar{\rho}/\rho$ is the density ratio.

When the droplet-bubble reaches its terminal state, the acceleration term $dU/dt \rightarrow 0$. That takes place as $(t - s) \rightarrow \infty$ in the above equation. Then the terminal velocity of the vapor bubble is,

$$U_T = ((4/3).d_e.(\rho - \bar{\rho}_p).g/(C_{Dd}.\rho))^{1/2} \qquad (5)$$

The drag coefficient, C_{Dd}, at the bubble equivalent diameter and in steady motion is obtained as follows,

$$C_{Dsd} = C_{Ds} .\Delta a \qquad (6)$$

where, C_{Dsd}, is the drag coefficient of the spheroidal bubble. C_{Ds}, is the drag coefficient of a sphere of the same cross section normal to the direction of flow as that of the spheroid ($d_s = 2a$). Δa is correction coefficient. It varies with the Reynolds number, and depends on the aspect ratio, $E = b/a$, of the spheroid (axis b parallel to flow, axis a normal to flow).

The drag coefficient of the bubble in terms of the equivalent diameter, C_{Dd}, can be substituted for C_{Dsd}, by using the aspect ratio, E, of the bubble (Clift et al., 1978),

$$C_{Dd} = C_{Ds}.\Delta a . (1/E^{2/3}) \qquad (7)$$

Reynolds number for C_{Ds} remains based on $d = 2a$, but also may be recalculated as,

$$Re(2a) = U. (d_e . E^{-1/3}).\rho/\mu \qquad (8)$$

where,

$$d_e = 2R = 2.(a^2.b)^{1/3} \qquad (9)$$

The terms of "added mass" and "history" in the equation of motion are not easily determened. The method proposed by Odar and Hamilton (1964) has been also adopted in the present work, using the nondimensional acceleration group M_A,

$$M_A = (dU/dt).d_e.(1/U^2) \qquad (10)$$

in the "added mass" coefficient,

$$\Delta_A = 2.1 - 0.132 M_A^2.(1/(1 + 0.12 M_A^2) \qquad (11)$$

and in the "history" coefficient,

$$\Delta_H = 0.48 + 0.52.M_A^3 (1/(1 + M_A)^3) \qquad (12)$$

These coefficients were introduced into the equation of motion making it an integral, nonlinear equation. It had to be solved for the instantaneous velocity, U(t), of the accelerating bubble. A numerical solution was sought.

The initial conditions for a bubble accelerating from rest are,

$$\text{at } t = 0 : U = 0, \; M_A \rightarrow \infty \; \Delta_A = \Delta_H = 1$$

$$dU/dt = 2g.(1 - \Gamma)/(1 + 2\Gamma) \qquad (13.1)$$

$$\text{where,} \quad \Gamma = \rho_1/\bar{\rho} \qquad (13.2)$$

Practically the droplet is not injected at rest. Therefore the initial conditions of the numerical solution were adjusted to the experimental conditions,

$$\text{at } t = 0: \; U = U_o, \; \Delta_A = \Delta_H = 1$$

$$M_A = (dU/dt).(d_e/U_o^2) \qquad (14)$$

$$dU/dt = 2g.(1 - \Gamma)/(1 + 2\Gamma) - 3C_{Dd}.U_o^2/4R_o(1 + 2\Gamma)$$

The initial density of the droplet, ρ_1, changes in the process to the average density, $\bar{\rho}$,

$$\bar{\rho} = \rho_1.\bar{R}^3 \qquad (15)$$

The large bubbles at the later stages of the evaporation process, as shown in Fig. 1, photographs 16-24, are less affected by interfacial forces. At $E_o \geq 40$, the aspect ratio $E = b/a$ is calculated for $E_o = 40$, and used for the drag coefficient.

COMPARISON WITH EXPERIMENTAL RESULTS

The formulated model is compared with experimental results by Nazir (1972), and by Sideman and Taitel (1964). In those studies pentane and butane droplets evaporating in water were experimented with. The physical properties of pentane and butane were computarized by us to be incorporated in the numerical solutionof the instantaneous velocity and height.

In the experiments conducted by Simpson and Nazir, Sideman and Taitel the path of the evaporating bubble was cine-photographed. These measurements were utilized by us for comparison with our numerical solutions in Figs. 2-7.

The definitions of the equivalent diameter adopted in the above mentioned works differed a little. Sideman and Taitel (1964) considered the evaporating droplet as a spheroid, and calculated the equivalent diameter accordingly,

$$d_e^3 = d_1^2 . d_2 \quad \text{where} \quad d_1 > d_2 \quad (d_1 = a, \; d_2 = b) \qquad (16)$$

$$\text{and} \quad E = d_2/d_1 \qquad (17)$$

$$\text{yield,} \quad d_e = d_1 . E^{1/3} \qquad (18)$$

In Nazir's (1972) study the equivalent diameter was defined by,

$$\bar{d}_e = (d_1 + d_2)/2 \qquad (19)$$

Substituting again for $d_2 = E.d_1$ yields,

$$\bar{d}_e = d_1.(1 + E)/2 \qquad (20)$$

The ratio of these two differently defined diameters if differentiated yields a maximum deviation of about 6% at $E = 0.5$. Therefore, practically there is no need to recalculate the Nazir's results.

Sideman and Taitel (1964) correlated their experimental results for the surface area of the droplet, A(t), and for the height, H(t), as

$$A = K.t^{m} + A_{o} \tag{21}$$

$$H = B.t^{p} + H_{o} \tag{22}$$

The constants were obtained by experiment. We have used the surface area, A, to calculate the diameter of the droplet.

The experimental results of Nazir (1972) for the evaporating butane droplets were presented in his thesis as,

$$U = 0.177.(d/d_o)^{1/2} \quad \text{for } d/d_o > 2 \text{ in brine} \tag{23.1}$$

$$U = 0.152.\ (d/d_o) \quad \text{for } d/d_o > 2.7 \text{ in water} \tag{23.2}$$

Up to the above specified ratio of diameters the velocity of the evaporating butane droplets was found to be constant, 25 cm/s.

The experimental height measurements of Nazir (1972) are presented in Figs. 2-4, and compared with our theoretical curves.

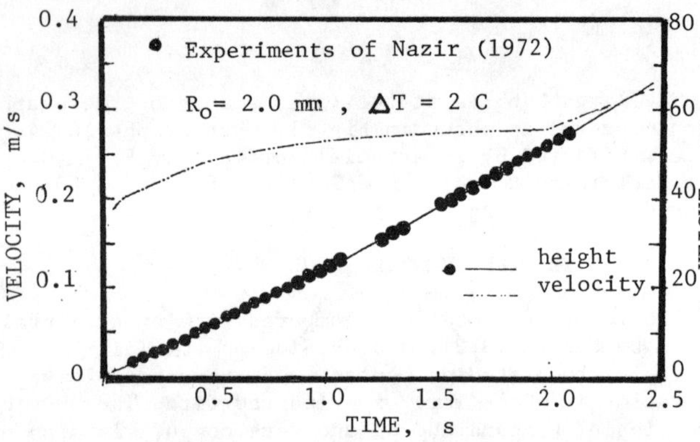

Fig. 2 Evaporation of butane droplets in water. Comparison of theoretical curves with experiments of Nazir (1972).

The experimental points of instantaneous velocity measured by Nazir (1972) are presented only in Fig. 3, and compared with the theoretical velocity curve.

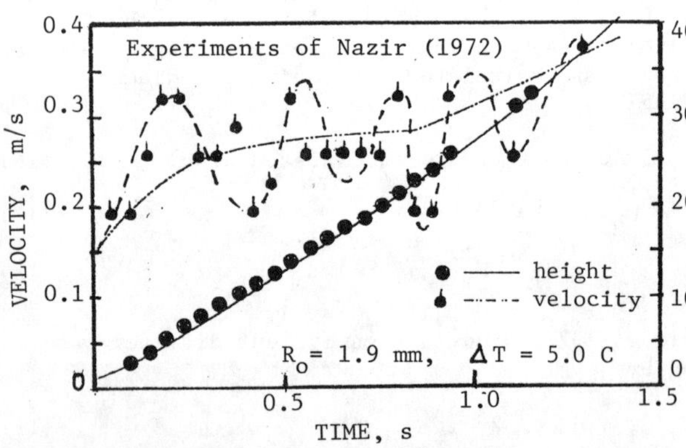

Fig. 3 Evaporation of butane droplets in water. Comparison of theoretical curves with experiments of Nazir (1972).

The measured points appear to periodically oscillate around an average velocity which agrees with the theoretical value. The same pattern was observed in our previous works on motion of droplets in a liquid (Ullmann and Letan, 1988). This pattern of oscillations could be analyzed and understood only in conjunction with the wake movement and configuration.

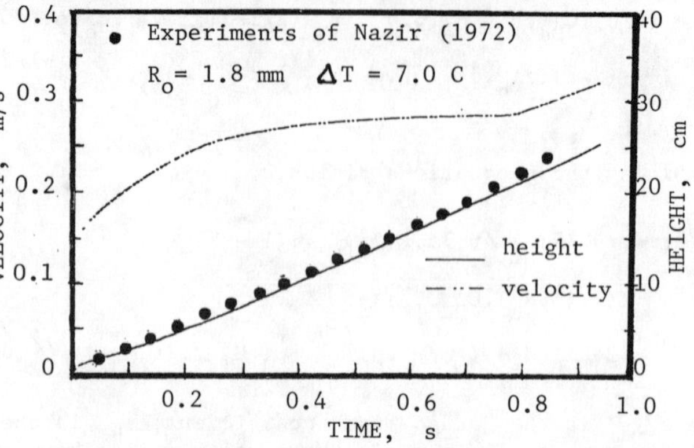

Fig. 4 Evaporation of butane droplets in water. Comparison of theoretical curves with experiments of Nazir (1972).

The experimental results of Sideman and Taitel (1964) are compared with our theoretical solutions in Figs. 5-7. Experiments at different temperature difference were selected for the illustrations. Each of the figures illustrates the theoretical time-dependent velocity and the path of the evaporating droplet against the measurements of instantaneous height

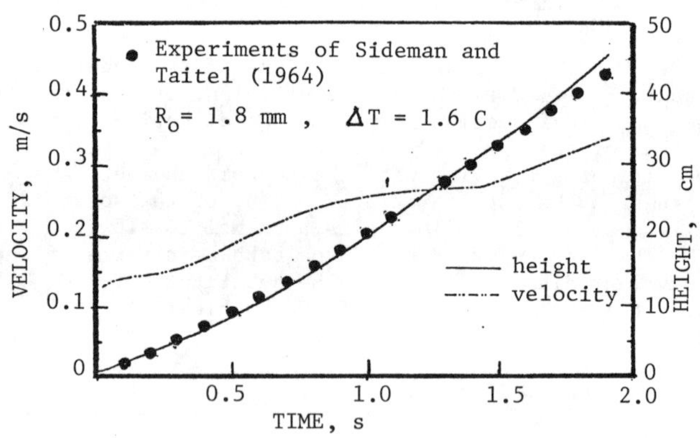

Fig. 5 Evaporation of pentane droplets in water. Comparison of theoretical curves with experiments of Sideman and Taitel (1964).

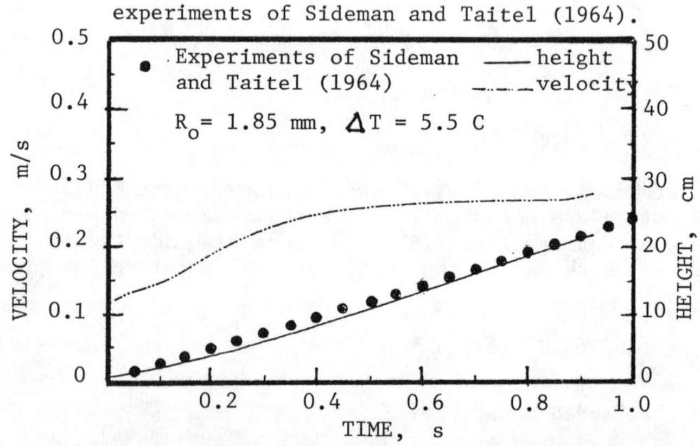

Fig. 6 Evaporation of pentane droplets in water. Comparison of theoretical curves with experiments of Sideman and Taitel (1964).

The velocity curves illustrated in Figs. 2-7 show how and where the evaporating droplet accelerates. The initial acceleration is due to the rising up from

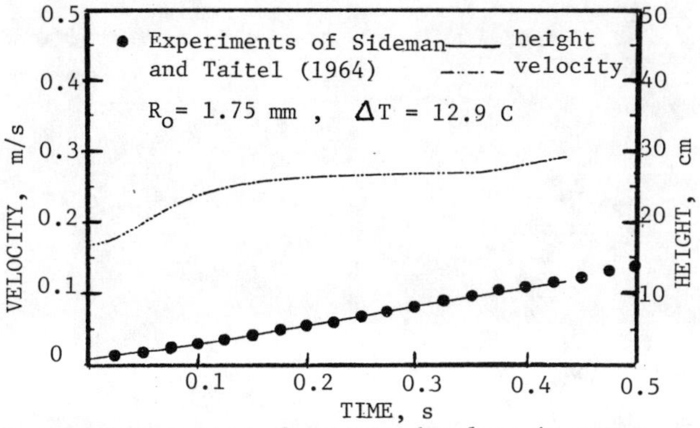

Fig. 7 Evaporation of pentane droplets in water. Comparison of theoretical curves with experiments of Sideman and Taitel (1964).

rest, and due to the reduced density by vapor. At a later stage the acceleration almost ceases, and the velocity levels off to about 0.26-0.28 m/s in the butane-water system, and 0.24-0.26 m/s in the pentane-water system. The almost constant velocity appears simultaneously with the shape distortion of the larger bubbles. The large cross section of the distorted bubble, in the direction normal to flow makes the drag forces to increase, and balance the higher buoyancy forces caused by the reduced density of the larger fraction of vapor in the bubble.

The graphical presentations, Figs. 2-7, show theoretically and experimentally the height of the center of the bubble above the nozzle at t = 0. The time t = 0 is recorded at the instant of full appearance of the bubble at its nozzle.

The calculated curves gradually change up to E_o= 40. The sharp change at this point is arbitrarily imposed because at $E_o > 40$ the interfacial forces are negligible and E = const. The velocity of the bubble varies with its density only.

E_o is calculated as,

$$E_o = (\rho - \bar{\rho}).g.d_e^2/ \sigma = (\rho - \rho_1 / \bar{R}^3).g.(d_o.\bar{R})^2 / \sigma \tag{24}$$

Using the physical properties of the system in the above equation yields for butane: $\bar{R}$ = 3.8 at E_o= 40. The velocity at these conditions is about 0.27 m/s, and agrees with the experimental results.

If the velocity were measured at a high frequency we might have obtained the full oscillations around the time-dependent average velocity calculated by use of the theoretical model. The experimental instantenous height points show a good agreement with the theoretical curves, because the path is the integrated average velocity, and does not reveal the deviations caused by the oscillations. Thus, the theoretically calculated path of the evaporating bubble is validated by experiment.

SUMMARY

The theoretical model derived for prediction of the time-dependent velocity and the bubble path in the process of evaporation accounts for the acceleration of the surrounding fluid by use of "added mass" and "history" coefficients in the equation of motion.

The model does not predict the instantaneous oscillations of velocity. An average time-dependent velocity can be calculated by the theoretical model.

The theoretical curves of instantaneous height were compared with the experimental results of previous researchers in systems of butane-water and pentane-water. The illustrated good agreement validates the proposed model.

REFERENCES

Clift, R., Grace, J.R., and Weber, M.E., 1978, Bubbles, Drops and Particles, Academic Press, N.Y.

Lerner, Y., Kalman, H., and Letan, R., 1987, "Condensation of an Accelerating-Decelerating Bubble: Experimental and Phenomenological Analysis", ASME Journal of Heat Transfer, Vol. 109, pp. 509-517.

Letan, R., 1988,"Visualization of Droplets, Bubbles and Sprays", Proceedings, 25th National Heat Transfer Conference,Houston, Vol. 2, pp. 565-572.

Moalem-Maron, D., Sokolov, M., and Sideman, S., 1980, "A Closed Periodic Condensation-Evaporation Cycle of an Immiscible, Gravity Driven Bubble", Int. Journal of Heat & Mass Transfer, Vol. 23, pp. 1417-1423.

Nazir, M., 1972, Ph.D. Thesis, University of Strathclyde, Glasgow.

Odar, F., and Hamilton, W.S., 1964, Journal Fluid Mechanics, Vol. 18, pp. 302-314.

Sideman, S., and Taitel, Y., 1964, "Direct-Contact Heat Transfer with Change of Phase: Evaporation of Drops in an Immiscible Liquid Medium", Int. Journal of Heat and Mass Transfer, Vol. 7, pp. 1273-1289.

Simpson, S., Beggs, H.C., and Nazir, M., 1974, Desalination, Vol. 15, pp. 11-23.

Ullmann, A., and Letan, R., 1988, "Motion of an Accelerating Droplet", Proceedings, 25th National Heat Transfer Conference, Houston,Vol. 2, pp. 409-414.

Ullmann, A., 1987, "Motion of Drops and Bubbles in an Immiscible Medium", M.Sc. Thesis, Ben-Gurion University, Beer-Sheva.

1989 National Heat Transfer Conference
HTD-Vol. 106, Heat Transfer Phenomena in Radiation, Combustion and Fires

INFLUENCE OF INITIAL NONUNIFORM SPRAY PATTERNS ON QUASI-MONODISPERSE DROPLET DIFFUSION FLAMES

J. B. Greenberg
Faculty of Aerospace Engineering
Technion-Israel Institute of Technology
Haifa, Israel

ABSTRACT

The influence of initial nonuniform spray patterns on combustion is examined using a Burke-Schumann type of diffusion flame in which the fuel is supplied as a cloud of droplets having a velocity different to that of the surrounding environment. The spray of droplets is described using a sectional approach and the concept of a quasi-monodisperse spray. Dimensionless parameters associated with vaporization and droplet drag are identified. Nonuniformity in initial conditions is modeled through functions describing initial distributions of sectional mass fractions and/or velocity.

The effects of these nonuniformities on the downstream evolutionary structure of the spray with consequent influence on flame height and shape are illustrated in a discussion of results computed from the analytical solution presented herein.

NOMENCLATURE

c	L/R
$\bar{C}$	sectional vaporization constant
D_g	diffusion coefficient for gases
H	flame height normalized with respect to gaseous Burke-Schumann flame height
L	radius of inner duct
m	$m_{fuel} - m_{O_2}/\nu$
m_i	mass fraction of species i
m_d	mass fraction of droplets
$m_{totfuel}$	total initial fuel mass fraction
Pe	Peclet number (Eq. (6))
Q_p	integrals defined in Eq. (13)
R	radius of outer duct
V	dimensionless initial mass fraction of oxygen
v	velocity component of gaseous environment
$v_d, \tilde{v}_d$	droplets velocity (normalized with respect to v)
x,y	radial and axial coordinates, respectively

GREEK SYMBOLS

γ, γ_d	normalized mass fractions for gases and droplets
$\delta, \tilde{\delta}$	initial fraction of liquid fuel
Δ	nondimensional vaporization parameter defined in Eq. (7)
Δ_1	nondimensional drag parameter defined in Eq. (8)
ξ, η	nondimensional coordinates
τ	time factor associated with drag coefficient
ν	stoichiometric coefficient
$\underline{\psi}$	initial nonuniform distribution vector (Eq. (14))
$\underline{\phi}=(\phi_{\gamma_d}, \phi_{v_d})$	parameters of nonuniformity

INTRODUCTION

It is a well established fact that combustion efficiency can be rather significantly impaired owing to distortion of fuel spray patterns brought on by even slight nozzle lip imperfections. The resulting nonuniform spray patterns can result in poor mixing between fuel and air, hence lowering combustion efficiency and increasing emitted pollutants. Some recent experimental work has been performed, using a variety of configurations of a model aerating nozzle, to try to analyze the main factors that produce nonuniform patterns in fuel sprays [1,2] albeit in noncombusting systems. It is probably a truism to add that such distortion of the fuel patternation is probably not due to a steady state process but rather the result of some time dependent mechanism involved in the inherent flow instability that is at work during atomization of the liquid fuel.

Nevertheless, much experimental research is still carried out in both noncombusting and combusting spray systems based on the assumption of an axisymmetric spray and its steady state [3-7]. Indeed much relevant and useful information on spray combustion can be gleaned from such studies, with data thus obtained providing both the essential input necessary for a deeper understanding of the physics involved as well as motivation for the development of more sophisticated measuring techniques.

On the other hand the mathematical analysis of spray combustion, involving a complex description of aerothermochemistry of two phase flows with heat, mass and momentum transfer between the phases, is still in its infancy. Large computer codes are often called upon to provide some insight into the combustion of realistic spray systems [see, e.g. Ref. 8] but do not always furnish a fundamental understanding of the relative importance of each of the participating processes.

Much of the original analytical work on spray combustion was confined to examining the burning of individual droplets, see relevant references in Williams [9]. However, it is known that a cloud of droplets does not necessarily burn as the sum of its individual members. The concept of 'group combustion' is often employed and various models of the burning of clouds of droplets, both in quiescent and nonstationary environments, have been proposed [9-27]. But even some of the most advanced of these models are restricted by the assumption of a monodisperse homogeneously distributed spherical cloud of droplets [17-21]. Recently the author [28] has reexamined the classical Burke-Schumann diffusion flame [29] with the fuel being supplied as a cloud of droplets. Tambour's sectional approach [22-26] was adopted to model the spray. The solution domain was taken to be situated in the so-called far-field zone [30,31] in which momentum transfer between the phases no longer exists. The results are indicative of the way in which, at the expense of some plausible simplifying assumptions, tractable analytical solutions can be found that provide, at least, a qualitative description of spray combustion as well as delineating the realms of relevance of the various physico-chemical processes at work. In the current paper such a philosophy has been adopted to try to make a preliminary assessment, within the framework of a simplified model, of the influence of initial nonuniform spray patterns on diffusion flames. The nonuniformities are specified using functions describing the initial distributions of sectional mass fractions and/or velocity. Analytic solutions will be presented for cylindrical flames, and effects of various initial distribution functions on flame characteristics will be discussed.

ASSUMPTIONS OF THE MODEL

The diffusion flame under consideration is formed as a result of diffusive intermixing of two gaseous streams exiting from an inner and outer duct, see Fig. 1. The outer duct contains air whilst a mixture of inert gas, gaseous fuel and fuel droplets flows continuously from the inner one. The velocity of the gases present will be assumed approximately constant. However, the droplets will be presumed to be injected into the inner duct at a large distance upstream such that upon reaching its exit they have already undergone initial breakup and partial vaporization although their velocity will be different from that of the host fluid in which they are found. Nevertheless, the spray velocity will be unidirectional and parallel to the axial coordinate. This locates the solution domain towards the end of the near-field and in the vicinity of the far-field [30,31].

It will be further assumed that the volume of the liquid cloud is sufficiently small and/or the spray sufficiently fine such that the effects of drag due to the environment are felt by the droplets, but not vice versa. The diffusion coefficients of the gases are taken as equal, and the Lewis number is set equal to unity.

As per Burke and Schumann [29] the flame front is defined as the locus of points at which the gaseous fuel and oxidant are in stoichiometric ratio and where instantaneous chemical reaction occurs.

GOVERNING EQUATIONS

The sectional approach [22] is herein adopted to model the fuel spray. As details of the development have appeared elsewhere [22,32] they will not be repeated here. It suffices to say that the concept of a quasi-monodisperse spray will be utilized, although a multisectional polydisperse spray model can also be accommodated at the expense of algebraic complexity.

In nondimensional form the spray equations are:

$$v_d \frac{\partial v_d}{\partial \eta} = (Pe \cdot \Delta_1)(1 - v_d) \tag{1}$$

$$v_d \frac{\partial \gamma_d}{\partial \eta} = -(Pe \cdot \Delta)\gamma_d - \gamma_d \frac{\partial v_d}{\partial \eta} \tag{2}$$

whereas the equation for mass transfer in the gaseous phase is given by:

$$\frac{\partial\gamma}{\partial\eta} = \frac{1}{\xi}\frac{\partial}{\partial\xi}\left(\xi\frac{\partial\gamma}{\partial\xi}\right) + (Pe\cdot\Delta)\gamma_d \quad (3)$$

where, in the above, the non-dimensionalization has been performed with the aid of:

$$(\xi,\eta,c) = \left(\frac{x}{R}, \frac{yD_g}{vR^2}, \frac{L}{R}\right) \quad (4)$$

$$(\gamma,\gamma_d,V) = (m, m_d, m_{O_2}(x,0)/\nu)/m_{totfuel} \quad (5)$$

where $m_{totfuel}$ is the total mass fraction of fuel (i.e. droplets + vapor) at the exit of the inner duct, and the dimensionless parameters have appeared:

$$Pe = vR/D_g \quad (6)$$

$$\Delta = R\bar{C}/v \quad (7)$$

$$\Delta_1 = R/\tau v \quad (8)$$

Here τ is a relaxation time associated with the assumed linear drag law. The expression $(Pe\cdot\Delta)$ represents a sort of Damkohler number for vaporization (ratio of characteristic times for evaporation and diffusion) whilst $(Pe\cdot\Delta_1)$, is a drag coefficient.

The function $\gamma(\xi,\eta)$ is defined as:

$$\gamma(\xi,\eta) = \gamma_{g,fuel} - \gamma_{O_2} \quad (9)$$

Following Burke and Schumann the flame will be located at all those points for which $\gamma=0$.

In the current analysis all dimensionless parameters will be assumed constant. Whilst recognizing the limitations imposed by such a step it can be justified (see Ref. [28]), even if only on pragmatic and tentative grounds. Also, following Williams [9] it is possible to solve for the temperature field using a quantity analogous to γ. The details of the development are parallel and, hence, omitted here.

BOUNDARY CONDITIONS

The boundary conditions for the problem at hand are:

(i) For $\eta = 0$:

$$v_d = \begin{cases} v_d(\xi,0) & 0 \le \xi \le c \\ 0 & c \le \xi \le 1 \end{cases}$$

$$\gamma_d = \begin{cases} \delta(\xi) & 0 \le \xi \le c \\ c & c \le \xi \le 1 \end{cases}$$

$$\gamma = \begin{cases} 1-\delta v_d & 0 \le \xi \le c \\ -V & c \le \xi \le 1 \end{cases}$$

(ii) For $\xi = 0,1$, $\eta > 0$: $\partial\gamma/\partial\xi = 0$

(iii) As $\eta \to \infty$: $\partial\gamma/\partial\eta = 0$.

Boundary conditions (i) describe the situation at the exit of the two concentric tubes. The spray's velocity and mass fraction distributions are given as arbitrary functions of ξ. The function γ is accordingly defined. However, the concentration of oxygen remains uniform in the outer tube. Boundary conditions (ii) refer to the axisymmetry assumed here. Essentially the spray's radial nonuniformity is supposed to be symmetric. (The truly three dimensional effects introduced via circumferential and nonsymmetric nonuniformities will be reported on at a later stage.) The outer wall is impermeable to mass transfer. Finally, boundary condition (iii) describes the equilibrium attained far downstream.

SOLUTION OF THE GOVERNING EQUATIONS

The equations for the liquid phase are solved first. The solution for the velocity of the spray is found from the implicit equation:

$$(Pe\cdot\Delta_1)\eta = v_d(\xi,0) - v_d(\xi,\eta) + \ln\left[\frac{1-v_d(\xi,0)}{1-v_d(\xi,\eta)}\right]$$

$$0 \le \xi \le c \quad (10)$$

Substituting for $\partial v_d/\partial\eta$ in Eq. (2) enables the following solution for γ_d to be extracted:

$$\gamma_d(\xi,\eta) = \delta(\xi)\frac{v_d(\xi,0)}{v_d(\xi,\eta)}$$

$$\exp\left[-(Pe\cdot\Delta)(\eta+(v_d(\xi,\eta)-v_d(\xi,0))/(Pe\cdot\Delta_1))\right] \quad (11)$$

Armed with this solution Eq. (3) can be tackled. The use of a Hankel transform then yields:

$$\gamma(\xi,\eta) = (1+V)c^2 - V - 2\int_0^c \xi\delta(\xi)v_d(\xi,0)d\xi + 2Q_o(\eta)$$

$$+ \sum_{p>0} 2\frac{J_o(p\xi)}{J_o^2(p)}\left\{e^{-p^2\eta}\left[\frac{c(1+V)}{p}J_1(pc) - \int_0^c \xi J_o(p\xi)\delta(\xi)v_d(\xi,0)d\xi\right] + e^{-p^2\eta}\int_0^\eta e^{p^2\eta}Q_p(\eta)d\eta\right\} \quad (12)$$

where:

$$Q_p(\eta)=(Pe\cdot\Delta)\int_0^c \gamma_d(\xi,\eta)\xi J_o(p\xi)d\xi, \qquad p \geq 0 \tag{13}$$

where p assumes the values of all the positive roots of $J_1(p)=0$ and J_o and J_1 are Bessel functions of the first kind. For arbitrary initial distributions of the spray's velocity and mass fractions the integrals Q_p must evaluated numerically. The flame front is deduced by setting the lefthandside of Eq. (12) to zero.

COMPUTED RESULTS AND DISCUSSION

In this section typical computed results will be presented. The basic data utilized referred to the combustion of the fuel n-decane, $C_{10}H_{22}$, that reacts according to the global chemical reaction:

$$C_{10}H_{22} + \frac{31}{2}O_2 \rightarrow 10CO_2 + 11H_2O \tag{14}$$

For all results to be described the value of V was taken to be 0.306. The total amount of fuel entering through the inner duct was 20% with the fraction of droplets δ between zero and one. The ratio of the radii of the two ducts, c, was given the value 1/6. Values of the nondimensional parameters $Pe\Delta$ and $Pe\Delta_1$ were either computed from experimental data (e.g. Chigier [33]) or preassigned. Other relevant data will be enumerated where appropriate.

The following form of assumed spray velocity and mass fraction distribution exiting from the inner tube was used:

$$\underline{\psi}(\xi,0) = \tilde{\underline{\psi}}(1-(\xi/c)^{\phi_\psi}) \qquad 0 \leq \xi \leq c \tag{15}$$

where:

$$\left.\begin{aligned} \underline{\psi}(\xi,0) &= (\delta(\xi),\ v_d(\xi,0)) \\ \tilde{\underline{\psi}} &= \left[\tilde{\delta},\tilde{v}_d\right] \text{ (constants)} \end{aligned}\right\} \tag{16}$$

$$\text{and } \underline{\phi}_\psi = 0 \leq \phi_{\gamma_d},\ \phi_{v_d} \leq \infty$$

If the parameters of nonuniformity ϕ are infinitely large the initial distributions are uniform. The smaller they are the narrower the profiles becomes about $\xi=0$. For all calculations the flame height was normalized relative to the height of the classical gaseous Burke-Schumann flame. In Fig. 2 the effect of the Damkohler number on flame height is illustrated for various initial uniform percentages of liquid fuel. Notice that as the vaporization coefficient becomes infinitely large (i.e instantaneous vaporization of the liquid fuel) the standard Burke-Schumann flame is recaptured. At the other extreme, for low vaporization rates the flame height is primarily determined by the amount of fuel vapor initially available for combustion, since the droplets evaporate too slowly to contribute to the combustible mixture and actually go to waste beyond the flame region. However, for values of the vaporization Damkohler number greater than about 30 the rate at which vapors are produced by the spray is such as to enhance the flame height, in one instance, by as much as almost 40% relative to the Burke-Schumann flame height.

Effects of spray nonuniformity are shown in Fig. 3. Curve (a) relates to a uniform mass fraction distribution but nonuniform velocity distribution, (b) the converse, and (c) both distributions nonuniform. A wide variety of values of the parameters of nonuniformity, ϕ, have been considered. For the data used here all the flames are somewhat taller than that having initial uniform distributions only. The case of combined nonuniformities leads to an increase of about 12% in the flame height, when $\phi=2$. If only the velocity distribution is nonuniform the most striking effect also occurs for $\phi=2$ with almost a 15% increase involved.

An illustration of the downstream evolution of the spray's velocity profile is given in Fig. 4. The profile is initially parabolic with droplets located radially in the range $0\leq\xi<0.12$ having a velocity greater than that of the host environment. Note that the initially slower droplets (i.e. those between $0.12\leq\xi\leq c$) accelerate rather rapidly to "catch up" with the surroundings. The deceleration of the initially faster droplets is much more gradual.

Evolution of the mass fraction distribution can be seen in Fig. 5. Once again the initial profile is initially parabolic. The concentration of droplets originally situated in $0\leq\xi<0.06$ actually increases at the two downstream stations shown. (Subsequently, further downstream they decrease.) This phenomenon was pointed out by De La Mora and Rosner [34] and is the result of the second term on the righthandside of Eq. (2). For values of $\partial v_d/\partial\eta<0$ and under certain operating conditions, this term will effectively serve to outbalance the depletion due to evaporation described by the other term on the RHS of Eq. (2). (This possibility of an increase in γ_d is also predictable from the solution, Eq. (11).) However, droplets initially found in $0.06\leq\xi\leq c$ behave in a more (intuitively) predictable fashion with the mass fractions simply decreasing monotonically in their downstream journey.

Finally, flame shapes are presented in Fig. 6 for different values of the nonuniformity parameter. The flame height decreases the more uniform the initial distributions become. In addition the flame thickness is also affected; the flames become fatter the more pronounced the nonuniformity.

The following possible qualitative interpretation of the results illustrated in Figs. 2-6 is suggested. The flame height (and shape) will depend on both the initial amount of fuel vapor available for combustion and the rate at which vapors can become available locally downstream. Reference to Fig. 2 (in which the droplet mass fraction and velocity distributions are uniform) and its description amply illustrate this point. Now in Fig. 3 as ϕ tends to zero the Burke-Schumann flame is recovered. However as ϕ becomes large uniform initial conditions are attained with the flame being slightly lower than that of Burke-Schumann. For values of ϕ=0(1) the differences in flame height come to the fore. For case (a) the mass fraction of droplets is uniform. Hence, less vapor is initially available than in cases (b) and (c). The flames in the latter cases are thus anticipated to be lower in height due to the earlier availability of fuel vapors. Now compare cases (b) and (c). Although identical mass fractions of fuel vapor are initially present, in case (b) the entire cloud of droplets issues from the inner tube at a speed twice that of the surroundings. Hence, more droplets are likely to traverse the flame region before managing to deposit their vapors than in case (c) where a portion of the cloud is travelling at a velocity less than that of the surroundings. An initial droplet velocity $v_d<1$ implies a longer residence time for vaporization in the pre-flame region than if $v_d>1$. A similar rationale is applicable to Fig. 6. The more nonuniform the initial droplet velocity distribution the greater the portion of the cloud containing droplets moving more slowly than the surroundings. These slower portions deposit their vapors before reaching the flame region. Thus the more nonuniform the initial distribution the higher the flame. The same sort of argument can also be applied to the behavior of the flame width.

CONCLUDING REMARKS

Effects of nonuniform spray velocity and mass fraction distributions on diffusion flame characteristics have been examined within the framework of a simplified model. The nonuniformities lead to differing qualitative (and quantitative) sorts of behavior in different parts of the cloud of droplets with subsequent influence on both flame height and shape. A more detailed study of this subject, including asymmetric and time-dependent nonuniformities, is currently under way and will be reported on in the future.

ACKNOWLEDGEMENT

The support of the Technion V.P.R. Fund - Seniel Ostrow Research Fund is gratefully acknowledged.

REFERENCES

1. Rosfjord, T.J. and Russell, S., "Influences of Fuel Spray Circumferential Uniformity", AIAA Paper AIAA-87-2135, AIAA/SAE/ASME/ASEE 23rd Joint Propulsion Conference, San Diego, California, 1987.

2. Rosfjord, T.J. and Eckerle, W.A., "Alterations of Fuel Spray Distribution by Nozzle Airflow", AIAA Paper, AIAA-88-3140, AIAA/SAE/ASME/ASEE 24th Joint Propulsion Conference, 1988.

3. Presser, C., Gupta, A.K. and Semerjian, H.G., "Aerodynamic Effects on Fuel Spray Characteristics: Air-Assist Atomizer", Winter Annual Meeting of ASME, Chicago, 1988.

4. Presser, C., Gupta, A.K., Semerjian, H.G., "Droplet/Air Interaction in a Swirl Stabilized Spray Flame", 2nd ASME/JSME Thermal Engineering Conference, Vol. 1, 1987, pp. 73-83.

5. Mao, C.-P., Wang, C. and Chigier, N.A., "An Experimental Study of Air-Assist Atomizer Spray Flames", Twenty-First Symposium (International) on Combustion, The Combustion Institute, 1987, pp. 655-673..

6. Bachalo, W.D., Houser, M.J. and Smith, J.N., "Evolutionary Behavior of Sprays Produced by Pressure Atomizers", AIAA Paper, AIAA-86-0296, 24th Aerospace Sciences Meeting, 1986.

7. Edwards, C.F., "Investigation of Spray Flame Structure in an Near-Axisymmetric Optical Access Research Furnace", Spring Meeting of the Western States Section, The Combustion Institute, 1988.

8. Cloutman, L.D., Dukowicz, J.J., Ramshow, J.D. and Amsden, A.A., "CONCHAS-SPRAY-A Computer Code for Reactive Flows with Fuel Sprays", LA-9294-MS, May 1982, Los Alamos National Laboratory, Los Alamos, New Mexico.

9. Williams, F.A., Combustion Theory, 2nd Edition, The Benjamin Cummings Publishing Co., Menlo Park, California, 1985.

10. Chiu, H.H. and Liu, T.M., "Group Combustion of Liquid Droplets", Combustion Science and Technology, Vol 17, 1977, pp. 127-142 .

11. Chiu, H.H., Kim, H.Y. and Croke, E.J., "Internal Group Combustion of Liquid Droplets", Nineteenth Symposium (International) on Combustion, The Combustion Institute, 1982, pp. 971-980.

12. Labowsky, M. and Rosner, D.E., "Group Combustion of Droplets in Fuel Clouds: Quasi-Steady Predictions", Ch. 4 in Evaporation-Combustion of Fuels, American Chemical Society, 1978, pp. 63-79.

13. Labowsky, M., "Calculation of the Burning Rates of Interacting Fuel Droplets", Combustion Science and Technology, Vol 22, 1980, pp. 217-226.

14. Correa, S.M. and Sichel, M., "The Group Combustion of a Spherical Cloud of Monodisperse Fuel Droplets", Nineteenth Symposium (International) on Combustion, The Combustion Institute, 1982, pp. 981-991.

15. Choudhury, P.R. and Gerstein, M., "Analysis of a Fuel Spray Subjected to Coupled Evaporation and Decomposition", Nineteenth Symposium (International) on Combustion, The Combustion Institute, Pittsburgh, 1982, pp. 993-997.

16. Bellan, J. and Cuffel, R., "A Theory of Nondilute Spray Evaporation Based Upon Multiple Drop Interactions", Combustion and Flame, Vol. 51, 1983, pp. 55-67.

17. Bellan, J. and Harstad, K., "Evaluation of the Importance of the Relative Velocity During Evaporation of Drops in Sprays", Int. J. Heat Mass Transfer, Vol.29, 1986, pp. 647-651.

18. Bellan, J. and Harstad, K., "Analysis of the Convective Evaporation of Nondilute Clusters of Drops", Int. J. Heat Mass Transfer, Vol. 30, 1987, pp. 125-136.

19. Bellan, J. and Harstad, K., "The Details of the Convective Evaporation of Dense and Dilute Clusters of Drops", Int.J. Heat Mass Transfer, Vol. 30, 1987, pp. 1083-1093.

20. Bellan, J. and Harstad, K., "Ignition of NonDilute Clusters of Drops in Convective Flows", Combustion Science and Technology, Vol 53, 1987, pp. 75-87.

21. Bellan, J. and Harstad, K., "Evaporation, Ignition and Combustion of Nondilute Clusters of Drops", Collected Papers in Heat Transfer 1988 presented at Winter Annual Meeting of the ASME, HTD-Vol. 104, 1988, pp. 121-129.

22. Tambour, Y., "A Sectional Model for Evaporation and Combustion of Sprays of Liquid Fuels", Israel J. of Technology, Vol. 18, 1980, pp. 47-56.

23. Tambour, Y., "Transient Mass and Heat Transfer From a Cloud of Vaporizing Droplets of Various Size Distributions: A Sectional Approach", Chem. Eng. Commun., Vol. 44, 1986, pp. 183-196.

24. Tambour, Y., "Vaporization of Polydisperse Fuel Sprays in a Laminar Boundary Layer: A Sectional Approach", Combustion and Flame, Vol. Vol. 58, 1984, pp. 103-114.

25. Tambour, Y., "A Lagrangian Sectional Approach for Simulating Droplet Size Distribution of Vaporizing Fuel Sprays in a Turbulent Jet", Combustion and Flame, Vol. 60, 1985, pp. 15-28.

26. Tambour, Y., "Coalescence of Vaporizing Kerosene Fuel Sprays in a Turbulent Jets", Atomisation and Spray Technology, Vol. 1, 1985, pp. 125-146. See also: "Simulation of Coalescence and Vaporization of Kerosene Fuel Sprays in a Turbulence Jet: A Sectional Approach", AIAA/ASME/ASEE 21st Joint Propulsion Conference, Monterey, 1985. AIAA Paper No. 85-1315.

27. Tambour, Y., Greenberg, J.B. and Albagli, D., "Vaporization and Coalescence of Sprays in a Three-Dimensional Axisymmetric Stagnation Flow", ICLASS-85, 3rd. International Conference on Liquid Atomisation and Spray Systems, Institute of Energy, London, VIA/2/1-VIA/2/15.

28. Greenberg, J.B., "The Burke-Schumann Flame Revisited - with Fuel Spray Injection", Combustion and Flame, in press.

29. Burke, S.P. and Schumann, T.E.W., "Diffusion Flames", Ind. and Eng. Chemistry, Vol. 20, No. 10, 1928, pp. 998-1004.

30. Chigier, N.A., Energy, Combustion and Environment, McGraw Hill (1981).

31. Greenberg, J.B. and Tambour, Y., "Far-Field Coalescence Effects in Polydisperse Spray Jet Diffusion Flames", Twenty-First Symposium (International) on Combustion, The Combustion Institute, 1986, pp. 655-663.

32. Greenberg, J.B., Albagli, D. and Tambour, Y., "An Opposed Jet Quasi-Monodisperse Spray Diffusion Flame", Combustion Science and Technology, Vol 50, 1986, pp. 255-270.

33. Chigier, N., "Atomization and Spray Research for Gas Turbine Engines", Collection of Papers of the 30th Israel Annual Conference on Aviation and Astronautics, 1989, pp. 120-124.

34. De La Mora, J.F. and Rosner, D.E., "Inertial Deposition of Particles Revisited and Extended: Eulerian Approach to a Traditionally Lagrangian Problem", Physico-Chemical Hydrodynamics, Vol. 2, 1981, pp. 1-21.

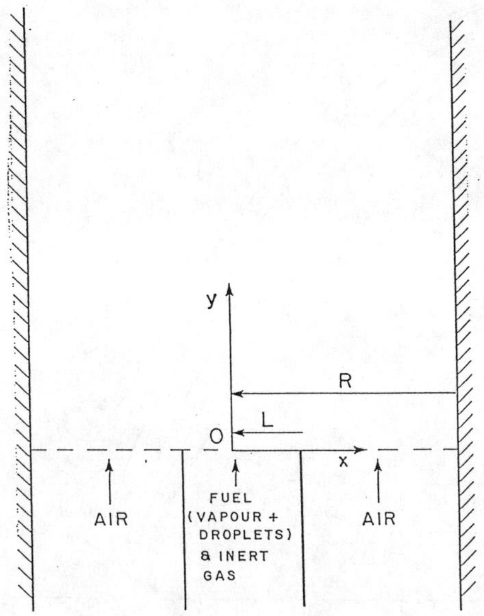

Fig. 1. Configuration for diffusion flame formation.

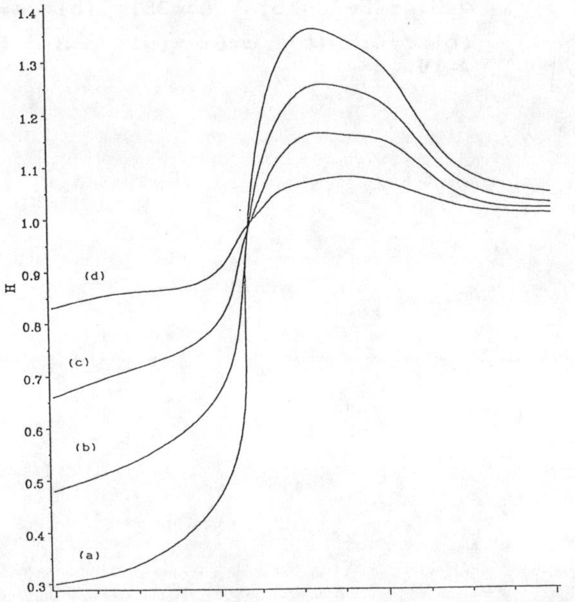

Fig. 2. Effect of Damkohler number on normalized flame height for various initial percentages of liquid fuel; $v_d=1$, $Pe\Delta_1=1$, $\phi_{\gamma_d}=\phi_{v_d}=\infty$, flame height normalized relative to classical Burke-Schumann flame height = 2.58×10^{-2}; (a) $\tilde{\delta}=0.8$; (b) $\tilde{\delta}=0.6$; (c) $\tilde{\delta}=0.4$; (d) $\tilde{\delta}=0.2$.

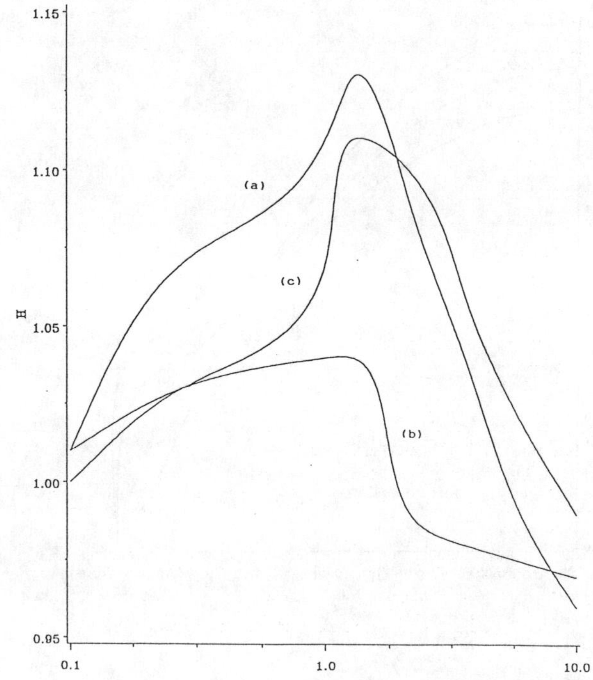

Fig. 3. Effect of initial nonuniform spray velocity and mass fraction distributions on flame height; $v_d=2$; $\delta=0.4$; $Pe\Delta_1=15$; $Pe\Delta=35$; flame height normalized as in Fig. 2. (a) $\phi_{\gamma_d}=\infty$; $\phi=\phi_{v_d}$ (b) $\phi=\phi_{\gamma_d}$; $\phi_{v_d}=\infty$ (c) $\phi=\phi_{\gamma_d}=\phi_{v_d}$.

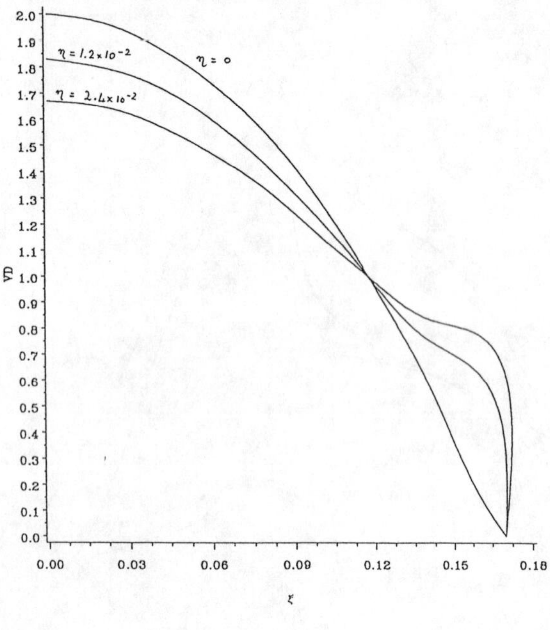

Fig. 4. Spray velocity profile evolution for initial nonuniform velocity distribution; $v_d=2$; $\delta=0.4$; $Pe\Delta_1=15$; $Pe\Delta=35$; $\phi_{\gamma_d}=\phi_{v_d}=2$.

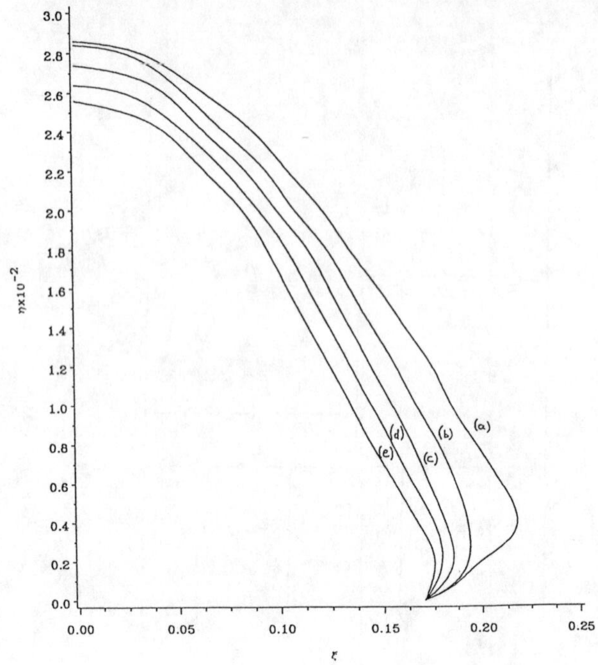

Fig. 6. Effect of initial spray nonuniform velocity and mass fraction distributions on flame shape; $v_d=2$; $\tilde{\delta}=0.4$; $Pe\Delta_1=15$; $Pe\Delta=35$; (a) $\phi=2$; (b) $\phi=4$; (c) $\phi=6$; (d) $\phi=8$; (e) $\phi=10$.

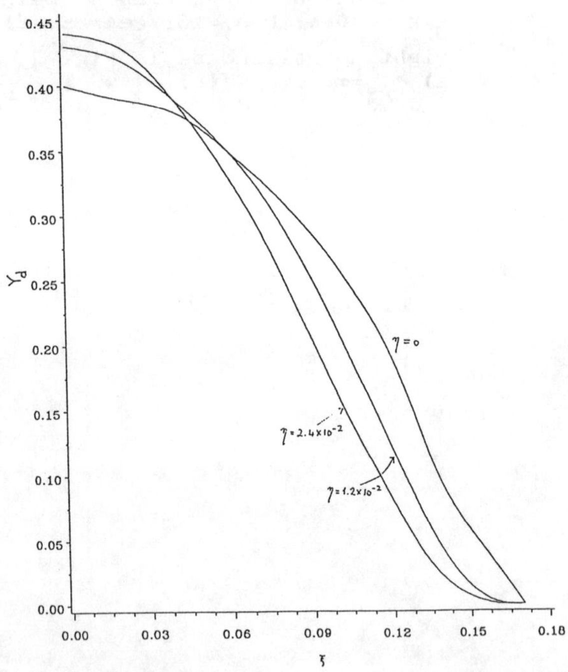

Fig. 5. Spray mass fraction profile evolution for initial nonuniform mass fraction distribution; $v_{d_o}=2$; $\tilde{\delta}=0.4$; $Pe\Delta_1=15$; $Pe\Delta=35$; $\phi_{\gamma_d}=\phi_{v_d}=2$.

1989 National Heat Transfer Conference
HTD-Vol. 106, Heat Transfer Phenomena in Radiation, Combustion and Fires

INFLUENCE OF LIQUID VISCOSITY ON PRESSURE — SWIRL ATOMIZER PERFORMANCE

S. K. Chen and A. H. Lefebvre
Thermal Science and Propulsion Center
Purdue University
West Lafayette, Indiana

J. Rollbuhler
NASA-Lewis Research Center
Cleveland, Ohio

ABSTRACT

Measurements of mean drop size and dropsize distribution are carried out over wide ranges of liquid injection pressure, using seven different simplex swirl nozzles to achieve a wide variation in the length/diameter ratio of the final discharge orifice. Mixtures of water and glycerine-water are used to provide a twelve to one variation in liquid viscosity. For liquids of low viscosity it is found that mean drop size diminishes with increase in discharge orifice diameter. Reduction in the length/diameter ratio of the discharge orifice produces a finer spray, especially at low levels of liquid injection pressure. The influence of liquid viscosity on mean drop size diminishes with increase in injection pressure. Increase in liquid viscosity generally leads to a broader distribution of drop sizes in the spray, while increase in discharge orifice length/diameter ratio produces a more uniform dropsize distribution.

NOMENCLATURE

A,B	constants in Eq. (2)
A_p	total inlet ports area, m^2
D	drop diameter, m
D_s	swirl chamber diameter, m
d_o	discharge orifice diameter, m
FN	Flow Number $= [\dot{m}_L/(\Delta P_L \rho_L)^{0.5}], m^2$
l_o	length of final discharge orifice
$\dot{m}$	flowrate, kg/s
P	pressure, Pa
ΔP_L	injection pressure differential across nozzle, Pa
Q	fraction of total liquid volume contained in drops of diameter less than D
q	Rosin - Rammler size distribution parameter
SMD	Sauter mean diameter, m
t	film thickness in final discharge orifice, m
ρ	density, kg/m^3
μ	dynamic viscosity, kg/ms
σ	surface tension, kg/s^2
θ	nominal spray cone half angle, degrees
θ_m	mean spray cone half angle, degrees

Subscripts

A	air
L	liquid

INTRODUCTION

Pressure-swirl atomizers are used extensively for injecting liquid fuel into a wide range of aircraft, marine, and industrial gas turbine combustors. The various designs employed include simplex, duplex, dual-orifice, and spill-return nozzles. The simplest form of pressure-swirl atomizer is the so-called "simplex" nozzle illustrated in Fig. 1. Fuel is fed into a swirl chamber through tangential ports that give the fuel a high angular velocity, thereby creating an air-cored vortex within the swirl chamber. The outlet from the chamber is the final discharge orifice, and the rotating liquid flows through this orifice under both axial and radial forces to emerge from the atomizer in the form of a hollow conical sheet whose cone angle is determined by the relative magnitude of the radial and axial components of velocity at exit. As the conical sheet expands its thickness diminishes and, due to the combined effects of liquid sheet instability and interaction with the surrounding air or gas, it disintegrates into ligaments which collapse into multitudinous drops in the form of a hollow cone spray.

Previous studies have shown that the liquid properties of most importance to pressure-swirl atomization are surface tension and viscosity. However, for the hydrocarbon fuels employed in gas turbines it is found that wide ranges in viscosity are accompanied by only slight changes in surface tension. For example, the surface tensions of diesel oil and aviation gasoline differ by only about ten percent, but the viscosity of diesel oil is sometimes an order of magnitude higher than that of aviation gasoline. For this reason, when considering the effect of variations in fuel type on atomizer performance, emphasis is usually placed on the role of viscosity.

Published data on the effect of viscosity on mean drop size are usually expressed as SMD $\alpha\ \mu_L^b$. Values for b of 0.16, 0.20, 0.118, 0.16, and 0.215 have been reported by Jasuja (1979), Radcliffe (1960), Dodge and Biaglow (1985), Jones (1982), and Knight (1955) respectively. However, according to Kennedy (1985) viscosity has no effect on mean drop size. Kennedy attributes this, and other differences between his results and those of other workers, to the larger Weber numbers resulting from his use of nozzles of exceptionally high flow number. However, Jones (1982) using large industrial nozzles of much higher flow number than those employed by Kennedy, found the effects of surface tension and viscosity on mean dropsize to be fully consistent with all previous observations made on small-scale nozzles.

The experimental work described in this paper is focused on small simplex nozzles of the type that are widely used in liquid fuel-fired furnaces and small gas turbines, and also serve as "primary" or "pilot" atomizers for the dual-orifice and hybrid airblast atomizers employed in large gas turbine engines. For such nozzles it seems reasonable to assume that the effects of liquid viscosity on atomization are very dependent on the size and geometry of the final discharge orifice, since this is where the liquid attains its highest velocity and, therefore, where the effects of viscosity and viscous losses are most pronounced. In order to assess the extent to which the effects of viscosity on atomization are modified by variations in the geometrical features of the discharge orifice, tests are carried out over wide ranges of liquid viscosity, using a range of nozzles which exhibit appreciable differences in regard to final discharge orifice diameter and length/diameter ratio.

EXPERIMENTAL

Seven simplex nozzles conforming to the basic design shown in Fig. 1 were specially manufactured for this program by the Delavan Company, all having the same flow number of $8.10^{-8} m^2$. Five of these nozzles have the same discharge orifice diameter, d_o, of 0.457 mm, but the length/diameter ratio of the discharge orifice is different for each nozzle, the actual values being 0.5, 1.0, 2.0, 3.0, and 4.0. The two additional nozzles both have a discharge orifice length, l_o, of 0.457 mm, but d_o is 0.356 mm for one nozzle and 0.559 mm for the other. These seven nozzles allow the effects of liquid viscosity and nozzle operating conditions to be examined for several values of d_o, l_o, and l_o/d_o, in a manner that permits the separate effects of these geometrical parameters on atomization performance to be readily determined.

The apparatus used to measure spray characteristics is shown schematically in Fig. 2. Liquid is supplied to the nozzle from a nitrogen-pressurized tank. The amount of liquid ejected from the nozzle is controlled by a hand-operated valve located in the bypass line. A number of valves, pressure gauges, and flow meters are located in the nozzle feed line to provide the desired nozzle operating conditions in terms of liquid injection pressure and mass flow rate.

Mean drop sizes are measured using a Malvern model 2600 particle size analyzer. This instrument is based on the Fraunhofer diffraction theory of a collimated laser beam scattered by moving drops. The principle of operation of the Malvern instrument has been reviewed by Swithenbank et al. (1977) and Dodge and Cerwin (1984). The instrument was fitted with a 300 mm focal lens to provide 16 size bins in the range from 1.5 to 564 μm. All measurements were carried out at a distance of 150 mm downstream of the nozzle with the laser beam passing through the centerline of the spray. As the dropsize distributions obtained conformed reasonably well to the Rosin-Rammler (1933) distribution parameter, Rosin-Rammler data reduction was used throughout the investigation.

The three liquids employed were water and two different mixtures of glycerine and water, to give viscosities of 0.001, 0.006, and 0.012 kg/ms. The addition of glycerine to water caused its surface tension to decline slightly (6 percent maximum) while increasing the viscosity by a factor of twelve. Thus, the results shown in Figs 3 to 17 may be considered as demonstrations of the influence of viscosity on atomization performance for a sensibly constant value of surface tension.

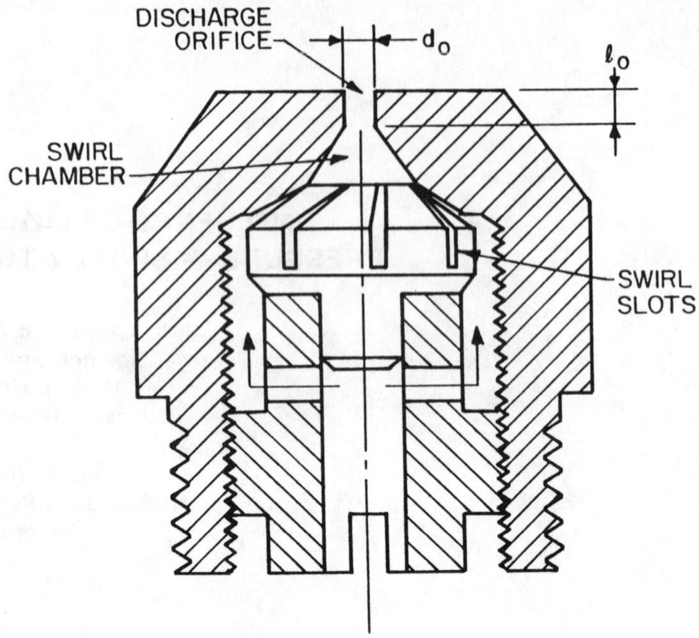

Fig. 1. Simplex swirl atomizer.

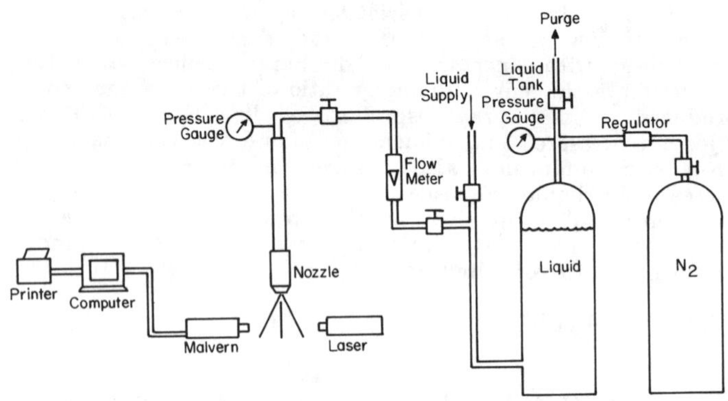

Fig. 2. Apparatus for drop size measurements.

RESULTS

This investigation yielded a large amount of experimental data on the effects of liquid injection pressure, liquid viscosity, and discharge orifice length/diameter ratio on Sauter mean diameter (SMD) and dropsize distribution. Limitations on space do not allow more than a small fraction of these data to be included in this paper. However, the results presented in Figs. 3 to 17 are considered typical of those obtained over broad ranges of nozzle operating conditions.

Mean Drop Size

Figure 3 shows the influence of liquid injection pressure, ΔP_L, on mean drop size for three nozzles having a constant discharge orifice length of 0.457 mm and discharge orifice diameters of 0.356, 0.457 and 0.559 mm. The liquid employed is water. Figure 3 shows that for all three nozzles the SMD declines with increase in ΔP_L, in accordance with all previous observations on simplex nozzles. The relationship between

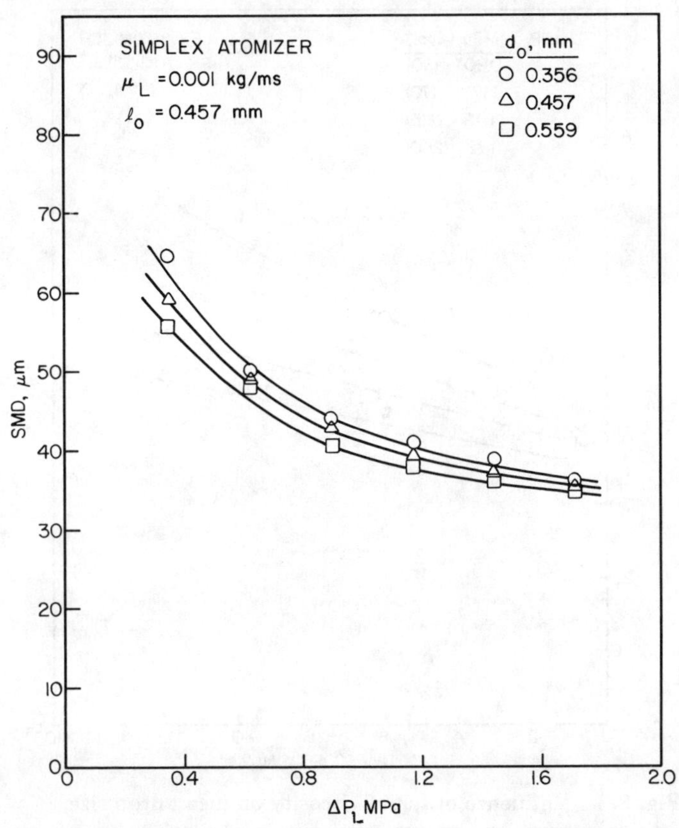

Fig. 3. Influence of injection pressure on mean drop size for a liquid of low viscosity.

mean drop size and liquid injection pressure can be expressed as SMD $\alpha\ \Delta P_L^{-0.33}$.

Figure 3 also shows that the mean drop size is reduced slightly by an increase in discharge orifice diameter. This result is probably due to the influence of d_o on spray cone angle. Taylor (1948) was the first to show that for nonviscous liquids the spray cone angle is determined by the geometrical parameter $A_p/D_s d_o$. According to Rizk and Lefebvre (1985), for viscous liquids an appropriate relationship for spray cone angle is the following:

$$2\theta_m = 6(A_p/D_s d_o)^{-0.15}\ (\Delta P_L d_o^2 \rho_L/\mu_L^2)^{0.11} \tag{1}$$

This dimensionally-correct equation provides a useful indication of the effects of injection pressure, liquid properties, and nozzle dimensions on spray angle. It shows that an increase in d_o causes the spray cone angle to widen. This in turn causes the mean drop size to decline. Thus the reduction in mean drop size with increase in discharge orifice diameter, as shown in Fig. 3, may be attributed directly to the increase in spray cone angle which accompanies an increase in discharge orifice diameter.

Figure 4 contains similar data to those presented in Fig. 3 but for a liquid of higher viscosity ($\mu_L = 0.012$ kg/ms). Two differences between Figs. 3 and 4 may be observed; one is a stronger dependence of mean drop size on liquid injection pressure (SMD $\alpha\ \Delta P_L^{-0.50}$), and the other an almost negligible dependence of mean drop size on discharge orifice diameter. This latter effect is also apparent in Fig. 5 in which mean drop size is plotted against liquid viscosity for three different values

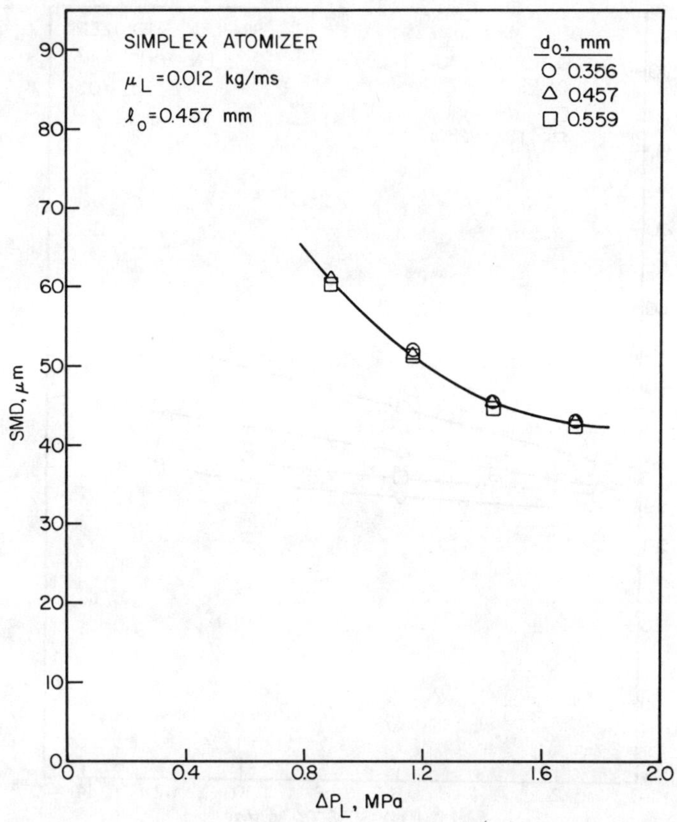

Fig. 4. Influence of injection pressure on mean drop size for a liquid of high viscosity.

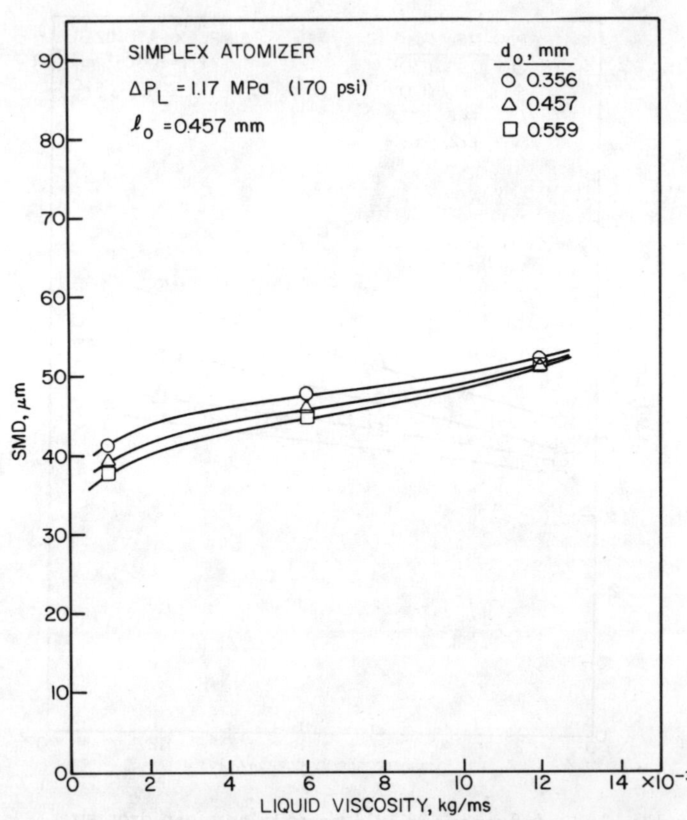

Fig. 5. Influence of liquid viscosity on mean drop size.

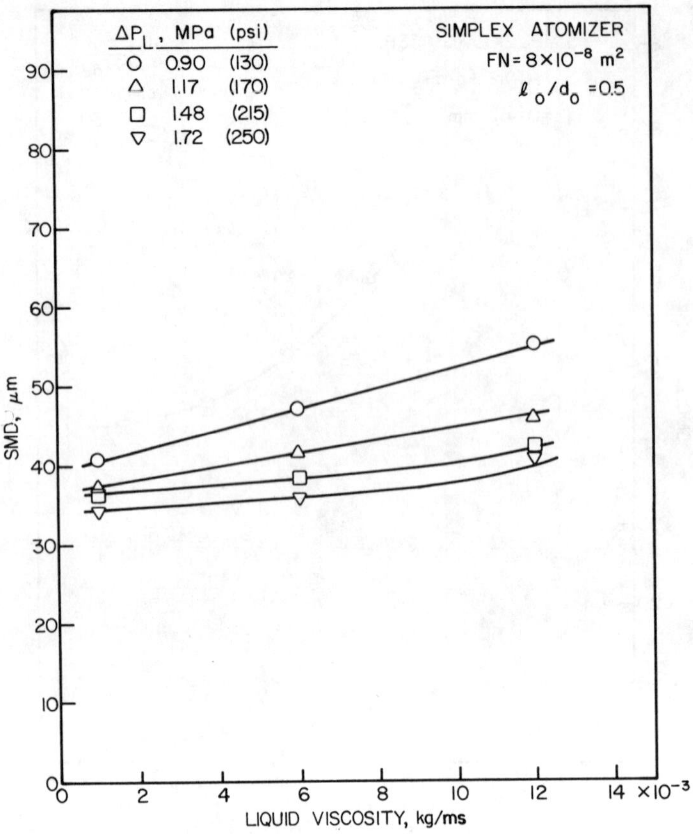

Fig. 6. Influence of liquid viscosity on mean drop size.

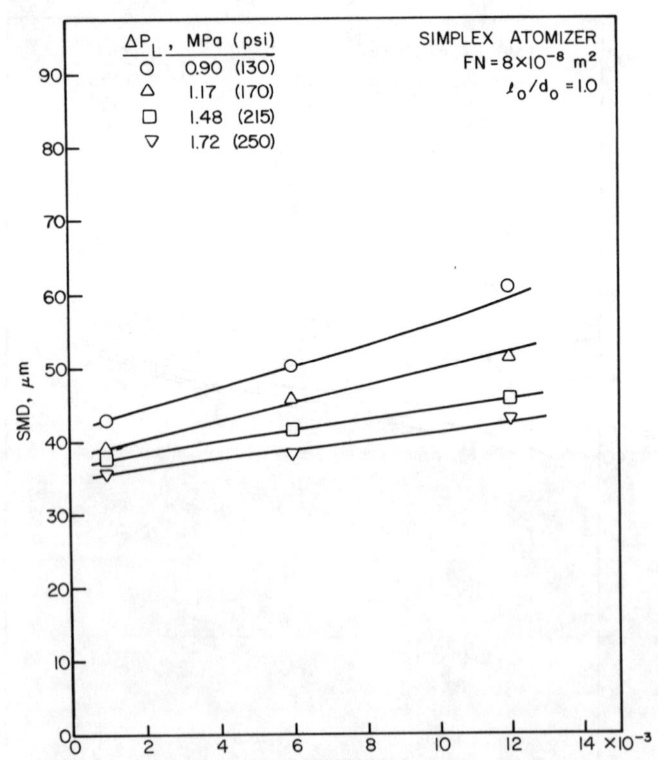

Fig. 7. Influence of liquid viscosity on mean drop size.

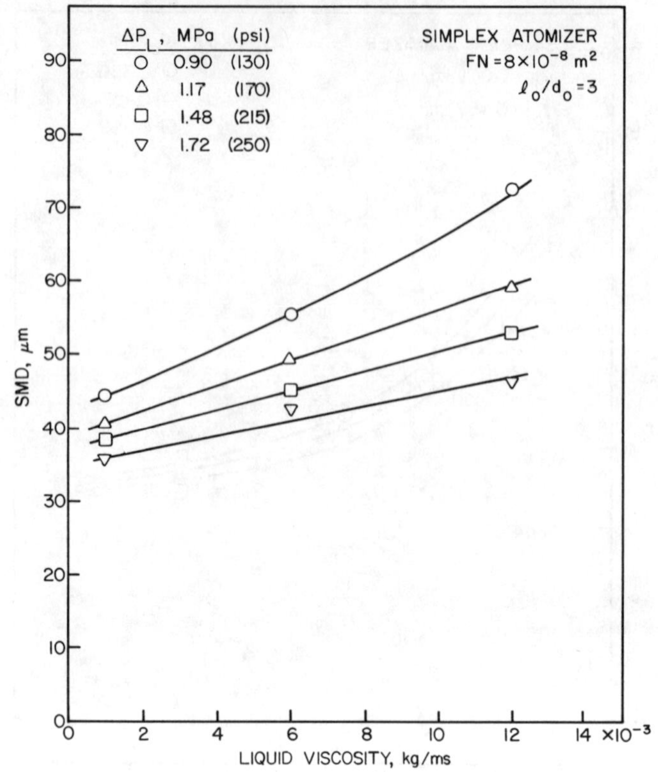

Fig. 8. Influence of liquid viscosity on mean drop size.

of discharge orifice diameter. This figure shows that the influence of d_o on SMD diminishes with increase in μ_L, becoming negligibly small at the highest value of liquid viscosity. Whether or not this insensitivity of SMD to variations in d_o is due to the fact that, for liquids of high viscosity, spray cone angle is relatively unaffected by changes in d_o, will be investigated in a subsequent phase of this experimental program.

An important parameter influencing atomization performance is the length/diameter ratio of the final discharge orifice. As viscous losses tend to be high in this region of high flow velocity, normal practice is to make the l_o/d_o ratio as small as possible without creating a sharp edge that would be difficult to manufacture consistently and would also be susceptible to damage during handling. The influence of discharge orifice length/diameter ratio on mean drop size is illustrated in Figs. 6 through 9, in which SMD is plotted against liquid viscosity for values of l_o/d_o of 0.5, 1.0, 3.0, and 4.0 respectively. The four lines drawn in these figures correspond to liquid injection pressures of 0.90, 1.17, 1.48 and 1.72 MPa (130, 170, 215, and 250 psi). Comparison of Figs. 6 through 9 reveals some interesting features. First and foremost it is clear that reduction in l_o/d_o leads to improved atomization. The figures show that the SMD diminished continuously over the range of l_o/d_o from 4.0 to 0.5. An optimal value of l_o/d_o was not encountered, so presumably further improvement in atomization quality could be gained by further reduction in l_o/d_o below 0.5. This finding is consistent with the results of previous studies by Jones (1982), Elkotb (1987), and Chen et al. (1989) on the effects of simplex atomizer geometry on spray droplet size. Elkotb found that atomization quality improved continuously as the l_o/d_o ratio was reduced in steps from 2.82 through 2, 1.6, and 1.2, to 0.43. Jones observed that reduction in l_o/d_o, over the range from 0.9 to 0.1, caused a slight reduction in mean drop size. Similar effects were noted by Chen et al. who reported a progressive reduction in mean drop size as l_o/d_o was systematically reduced from 4.0 to 0.5.

Inspection of Figs. 6 through 9 reveals that an increase in l_o/d_o causes the four lines drawn to represent the four different levels of injection pressure to become spaced further apart, thereby indicating that the SMD becomes more dependent on ΔP_L with increase in l_o/d_o. Presumably this is due to the fact that as l_o/d_o goes up a higher proportion of the total liquid injection pressure differential is employed in overcoming the frictional losses in the final discharge orifice, so consequently less energy is available for atomization.

Another interesting feature of Figs. 6 through 9 is that the dependence of mean drop size on liquid viscosity diminishes with increase in injection pressure. This effect is noticeable in all four figures, but is especially apparent in Fig. 9. It is a result that is predicted by the following equation for the mean drop sizes produced by pressure-swirl atomizers (Lefebvre, 1987).

$$\mathrm{SMD} = A\left[\frac{\sigma \mu_L^2}{\rho_A \Delta P_L^2}\right]^{0.25} [t\cos\theta]^{0.25} + B\left[\frac{\sigma \rho_L}{\rho_A \Delta P_L}\right]^{0.25} [t\cos\theta]^{0.75} \quad (2)$$

where t, the liquid film thickness in the discharge orifice, is obtained by (Simmons and Harding, 1980) as

$$t = \frac{0.48FN}{d_o \cos\theta} \quad (3)$$

In Eq. (2) mean drop size is given as the sum of two separate terms, only one of which contains liquid viscosity. A ncteworthy feature is that this term exhibits a stronger dependence on injection pressure than the term without viscosity. This means that although an increase in injection pressure reduces both terms on the right hand side of Eq. (2), it reduces the first term, which contains viscosity, by a larger

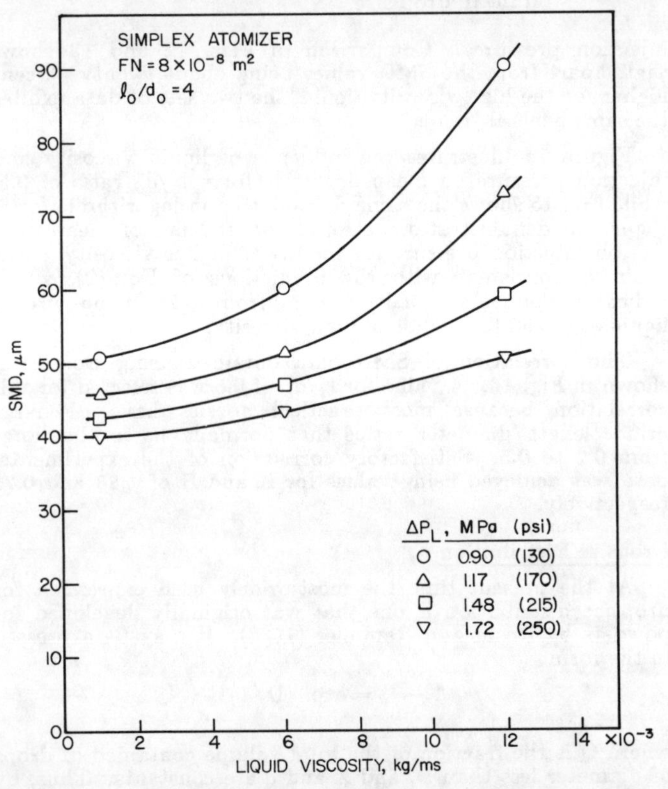

Fig. 9. Influence of liquid viscosity on mean drop size.

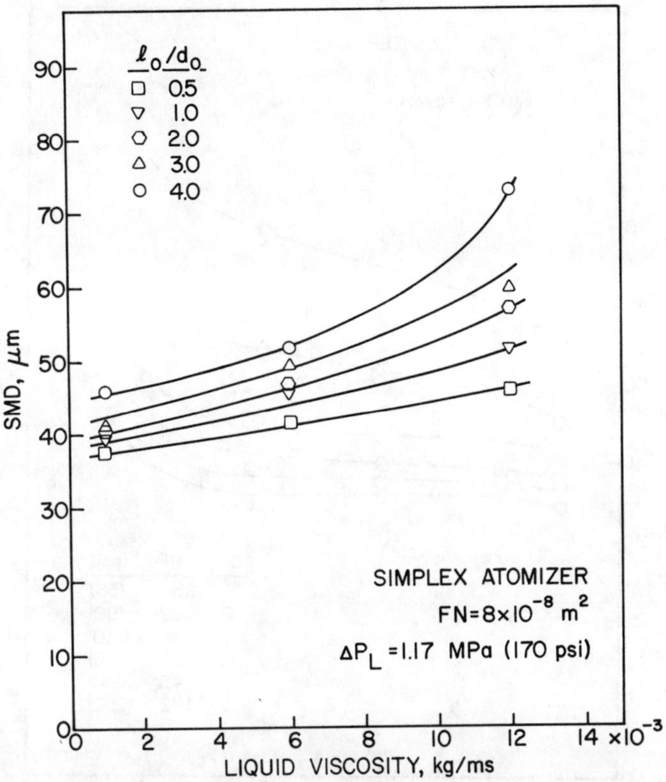

Fig. 10 Influence of liquid viscosity on mean drop size.

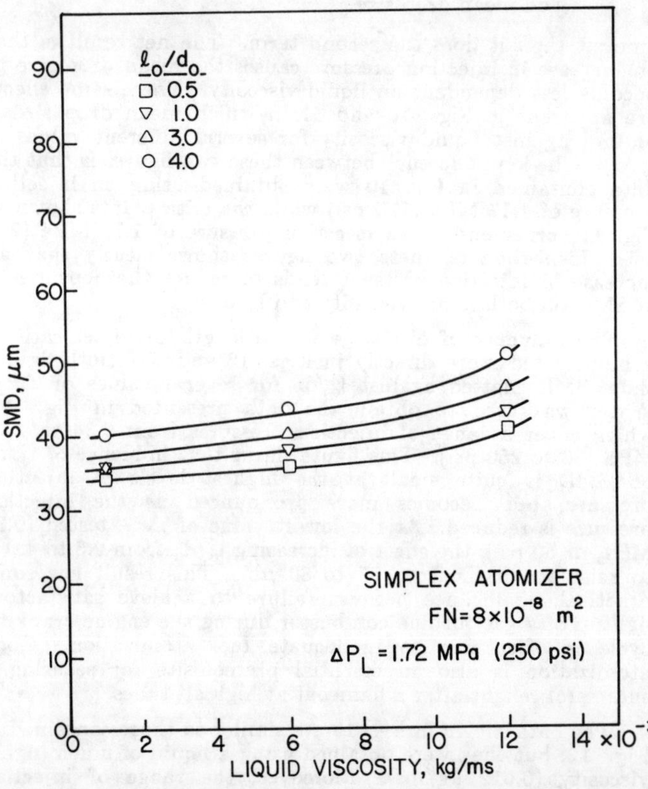

Fig. 11. Influence of liquid viscosity on mean drop size.

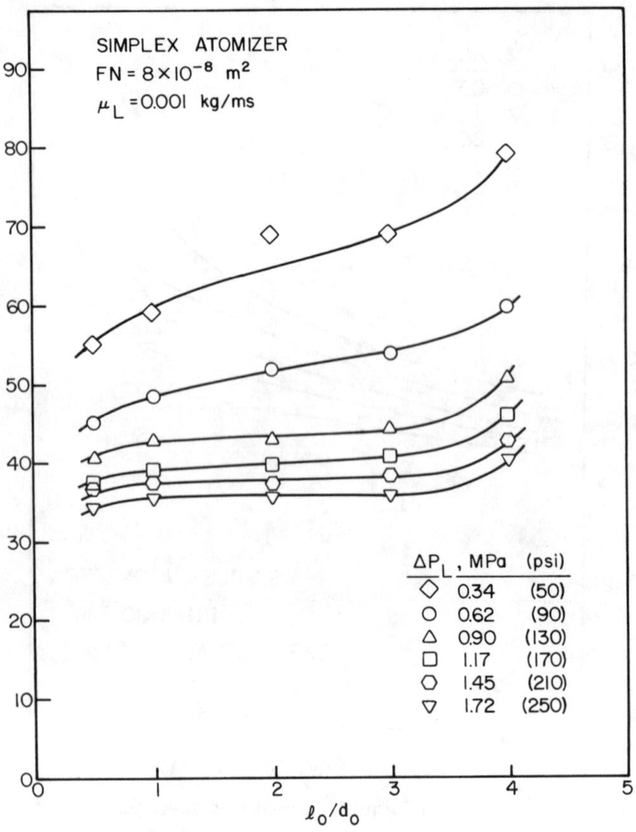

Fig. 12. Influence of discharge orifice length/diameter ratio on mean drop size.

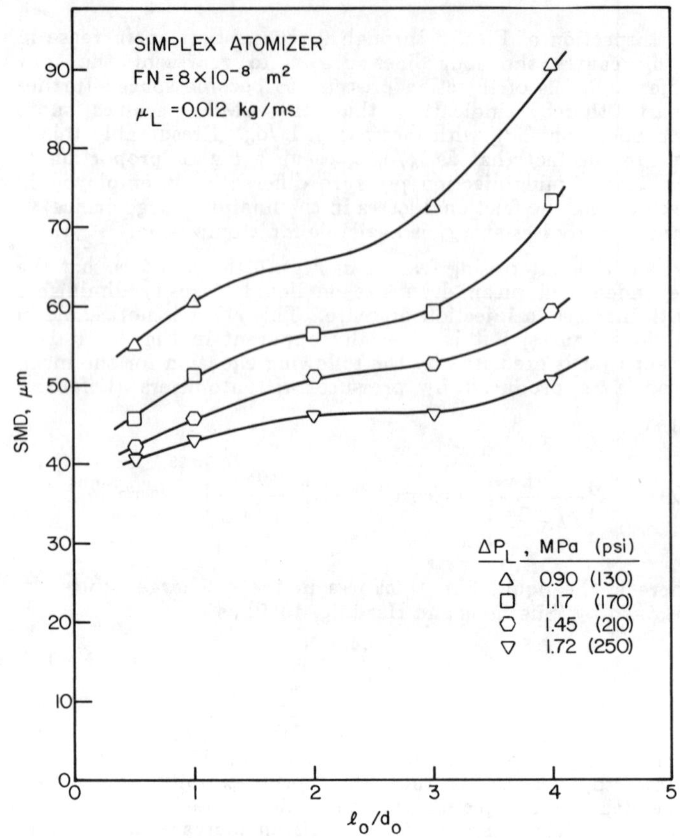

Fig. 13. Influence of discharge orifice length/diameter ratio on mean drop size.

amount than it does the second term. The net result is that an increase in injection pressure causes the mean drop size to become less dependent on liquid viscosity. These same effects are apparent in Figs. 10 and 11 in which mean drop size is plotted against liquid viscosity for several different values of l_o/d_o. The key difference between these two figures is that the data contained in Fig. 10 were obtained using an injection pressure of 1.17 MPa (170 psi) while the data points shown in Fig. 11 correspond to an injection pressure of 1.72 MPa (250 psi). Inspection of these two figures shows clearly that an increase in injection pressure tends to reduce the dependence of SMD on both liquid viscosity and l_o/d_o.

The influence of discharge orifice length/diameter ratio is demonstrated more directly in Figs. 12 and 13. Both figures show SMD plotted against l_o/d_o for several values of ΔP_L. Water was used to obtain the data presented in Fig. 12, which cover a range of injection pressures from 0.34 to 1.72 MPa (50 to 250 psi). This figure shows that influence of l_o/d_o on SMD is quite small at the highest levels of injection pressure, but becomes more pronounced as the injection pressure is reduced. At the lowest value of ΔP_L tested (0.34 MPa, or 50 psi), the effect of increasing l_o/d_o from 0.5 to 4.0 is to raise the SMD from 55 to 80 μm. This result has some practical significance because failure to achieve satisfactory lightup of a gas turbine combustor during the engine cranking cycle is often due to inadequate fuel atomization. Good atomization is also an essential prerequisite for securing a successful relight after a flameout at high altitudes.

The data shown in Fig. 13 are similar to those contained in Fig. 12, but they were obtained using a liquid of much higher viscosity (0.012 kg/ms). Moreover, the range of injection pressures covered at this high viscosity level was somewhat lower due to practical difficulties in achieving consistent and reliable measurements of mean drop size at the lowest injection pressures. Comparison of Figs. 12 and 13 shows that, apart from the SMD values being about twenty percent higher for the high viscosity liquid, the two sets of data exhibit the same general trends.

Figure 14 illustrates the influence of liquid viscosity and injection pressure on mean drop size for a l_o/d_o ratio of 0.5, while Fig. 15 shows the same data plotted in logarithmic form. Figure 15 demonstrates a stronger dependence of mean drop size on injection pressure for liquids of higher viscosity. This again is consistent with the predictions of Eq. (2), which indicates exponents for ΔP_L varying from -0.25 for nonviscous liquids to -0.50 for liquids of high viscosity.

The correlation of SMD data obtained using Eq. (2) is shown in Fig. 16. A value for l_o/d_o of 0.5 was selected for this correlation because most practical nozzles have discharge orifice length/diameter ratios that normally lie in the range from 0.2 to 0.5. Satisfactory correlation of the experimental data was achieved using values for A and B of 1.83 and 0.71 respectively.

Dropsize Distribution

At the present time the most widely used expression for dropsize distribution is one that was originally developed for powders by Rosin and Rammler (1933). It may be expressed in the form

$$1 - Q = \exp-(D/X)^q \tag{4}$$

where Q is the fraction of the total volume contained in drops of diameter less than D, and X and q are constants. Thus, by applying the Rosin-Rammler relationship to sprays, it is possible to describe the dropsize distribution in terms of the two parameters X and q. The exponent q provides a measure

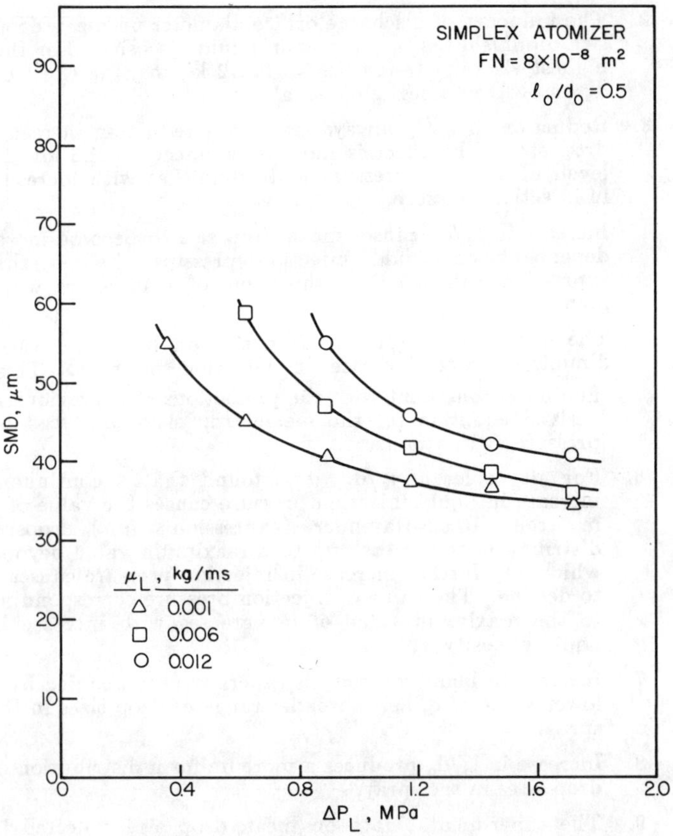

Fig. 14. Influence of injection pressure on mean drop size for liquids of different viscosities.

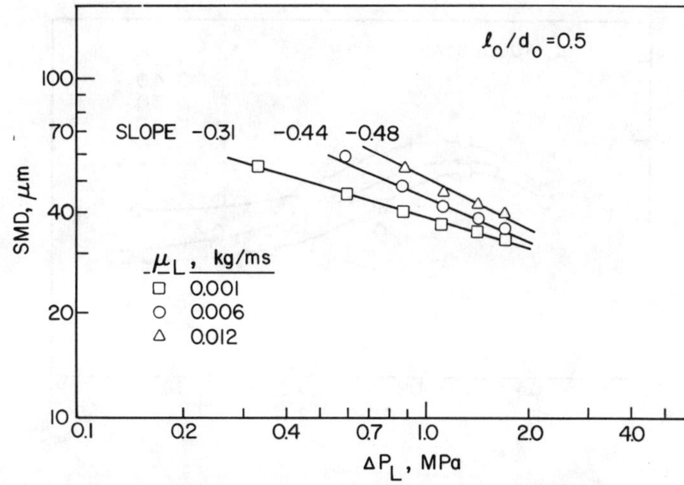

Fig. 15. Influence of injection pressure on mean drop size for liquids of different viscosities.

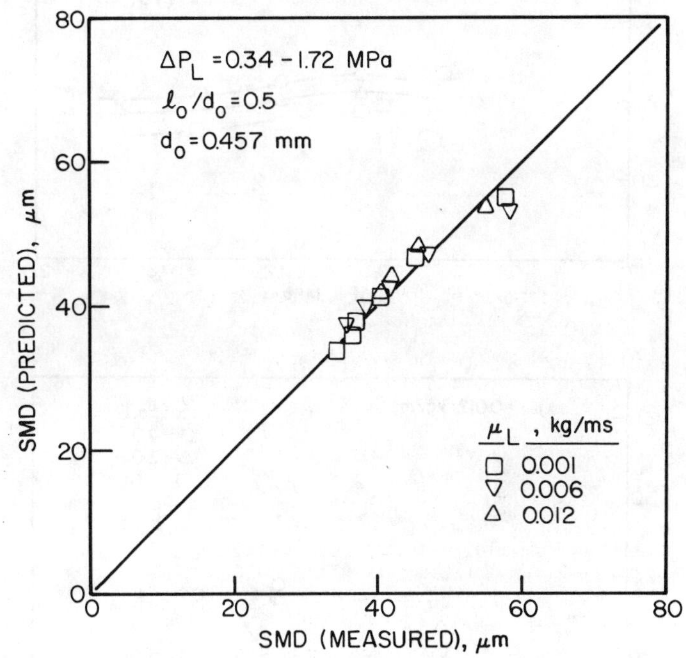

Fig. 16. Correlation of drop size data.

of the spread of drop sizes. The higher the value of q, the more uniform is the spray.

The variation of q with injection pressure and discharge orifice length/diameter ratio is illustrated in Fig. 17. Inspection of this figure reveals some interesting features. It shows that q increases slightly with increase in ΔP_L, reaches a maximum, and then declines slightly with further increase in ΔP_L. This is believed to be due to the combined effect of two conflicting factors. Previous work has shown that an increase in injection pressure, by producing more droplets in the small size range, leads to a broader distribution of drop sizes in the spray, i.e. to a lower value of q (Zhao et al., 1986). However, in a separate series of tests it has been found that an increase in injection pressure also causes the spray cone angle to widen. This tends to reduce the range of drop sizes in the spray, i.e. increase the value of q, for two reasons, (1) the drops are now being produced from a thinner sheet initially, and (2) the spray is dispersed throughout a larger volume of gas which reduces the risk of droplet coalescence. From inspection of Fig. 17 it would appear that the increase in spray cone angle tends to dominate at low values of injection pressure, causing q to increase with increase in ΔP_L. With continuous increase in injection pressure, q attains a maximum value and then slowly declines. The decline in q is due to the fact that, at high levels of ΔP_L, the spray cone angle is much less susceptible to variations in ΔP_L. This allows the influence of injection pressure per se on drop-size distribution to become paramount, with the result that q is diminished by an increase in ΔP_L.

Comparison of the results presented in Fig. 17 for three different values of viscosity shows that, in general, increase in viscosity leads to a lower value of q, i.e. to a broader range of drop sizes in the spray. Also of interest in this figure is that an increase in discharge orifice length/diameter ratio causes the dropsize distribution to become more uniform. For the nozzles employed in acquiring the data points in Fig. 17, the variation in l_o/d_o was accomplished by maintaining d_o constant and varying l_o. It would appear therefore, that increasing the length of the final discharge orifice provides more time and opportunity for the flow striations caused by the use of a finite number of swirl ports to dampen out. In consequence, the annular sheet of liquid created at the nozzle exit has a more uniform thickness, so that when it disintegrates it forms drops of a more uniform size.

CONCLUSIONS

From extensive measurements of mean drop size and dropsize distribution in the sprays produced by several simplex nozzles, all conforming to the same basic design but having

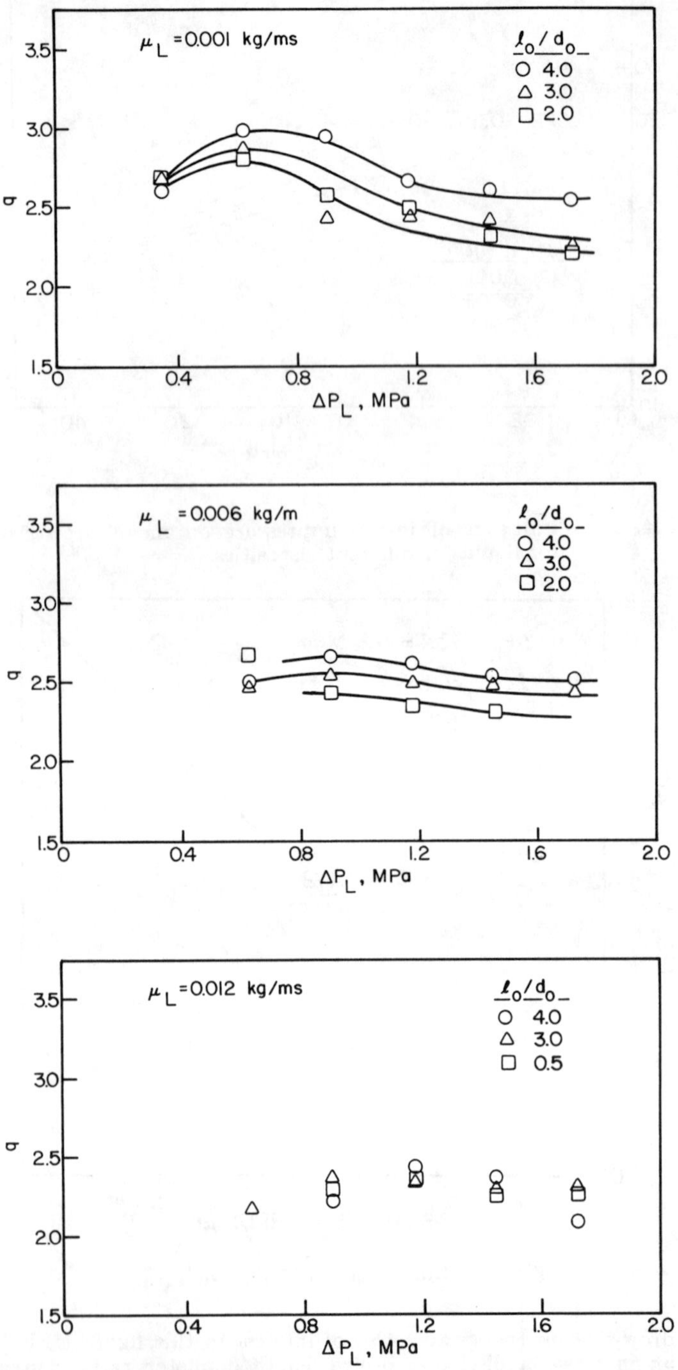

Fig. 17. Influence of injection pressure, viscosity, and l_o/d_o on Rosin - Rammler dropsize distribution parameter.

different values for the length, diameter, and length/diameter ratio of the final discharge orifice, the following conclusions are drawn:

1. For liquids of low viscosity, the mean drop size in the spray declines slightly with increase in discharge orifice diameter, d_o. This is attributed to the beneficial effect on atomization quality of the increase in spray cone angle which accompanies an increase in d_o.

2. The influence of discharge orifice diameter on mean drop size diminishes with increase in liquid viscosity. For the highest viscosity tested ($\mu_L = 0.012$ kg/ms) the effect of d_o on SMD was negligibly small.

3. Reduction in l_o/d_o always leads to a reduction in mean drop size. The effect is most pronounced at the lowest levels of injection pressure and diminishes with increase in injection pressure.

4. Increase in l_o/d_o causes mean drop size to become more dependent on liquid injection pressure, i.e. in the expression SMD $\alpha \Delta P_L^{-x}$, the value of x increases with increase in l_o/d_o.

5. The dependence of mean drop size on liquid viscosity diminishes with increase in injection pressure. This finding is consistent with the predictions of a previously-derived equation for the mean drop sizes produced by pressure-swirl atomizers.

6. For all values of l_o/d_o, it is found that a continuous increase in liquid injection pressure causes the value of q in the Rosin-Rammler expression for dropsize distribution to increase up to a maximum value beyond which any further increase in injection pressure causes q to decline. The value of injection pressure corresponding to the maximum value of q increases with increase in liquid viscosity.

7. Increase in liquid viscosity is generally accompanied by a lower value of q, i.e. a broader range of drop sizes in the spray.

8. Increase in l_o/d_o produces a more uniform distribution of drop sizes in the spray.

9. The experimental data on mean drop size collected in this study for an l_o/d_o of 0.5 are correlated with good accuracy by the equation:

$$SMD = 1.83\left[\frac{\sigma \mu_L^2}{\rho_A \Delta P_L^2}\right]^{0.25}[t\cos\theta]^{0.25} + 0.71\left[\frac{\sigma \rho_L}{\rho_A \Delta P_L}\right]^{0.25}[t\cos\theta]^{0.75}$$

REFERENCES

Chen, S.K., Lefebvre, A.H. and Rollbuhler, J., 1989, "Influence of Geometrical Features on the Performance of Pressure-Swirl Atomizers," *Proceedings of ASME Fossil Fuels Symposium*, Ed. S.N. Singh, Houston, pp. 79-84.

Dodge, L.A., and Biaglow, J.A., 1985, "Effect of Elevated Temperature and Pressure on Sprays from Simplex Swirl Atomizers," *ASME J. Eng. Power*, Vol. 108, No. 1, pp. 209-215.

Dodge, L.A., and Cerwin, S.A., 1984, "Extending the Applicability of Diffraction-Based Drop Sizing Instruments," in *Liquid Particle Size Measurement Techniques*, ASTM STP 848, ed J.M. Tishkoff, R.D. Ingebo and J.B. Kennedy.

Elkotb, M.M., Rafat, N.M. and Hanna, M.A., 1987, "The Influence of Swirl Atomizer Geometry on the Atomization Performance," *Proceedings of the 1st International Conference on Liquid Atomization and Spray Systems*, Tokyo, pp. 109-115.

Jasuja, A.K., 1979, "Atomization of Crude and Residual Fuel Oils," *ASME J. Eng. Power*, Vol. 101, No. 2, pp. 250-258.

Jones, A.R., 1982, "Design Optimization of a Large Pressure-Jet Atomizer for Power Plant," *Proceedings of the 2nd International Conference on Liquid Atomization and Spray Systems*, Madison, Wisconsin, pp. 181-185.

Kennedy, J.B., 1985, "High Weber Number SMD Correlations for Pressure Atomizers," *ASME J. Eng. Power*, Vol. 108, No. 1, pp. 191-195.

Knight, B.E., 1955, "Communication on the Performance of a Type of Swirl Atomizer," by A. Radcliffe, *Proc. Inst. Mech. Eng.*, Vol. 169, p. 104.

Lefebvre, A.H., 1987, "The Prediction of Sauter Mean Diameter for Simplex Pressure-Swirl Atomizers," *Atomization and Spray Technology*, Vol. 3, No. 1, pp. 37-51.

Radcliffe, A., 1960, "Fuel Injection," *High Speed Aerodynamics and Jet Propulsion*, Vol. XI, Section D., Princeton University Press.

Rizk, N.K. and Lefebvre, A.H., 1985, "Prediction of Velocity Coefficient and Spray Cone Angle for Simplex Swirl Atomizers," *Proceedings of the 3rd International Conference on Liquid Atomization and Spray Systems*, London, pp. III C/2/1-16.

Rosin, P. and Rammler, E., 1933, "The Laws Governing the Fineness of Powdered Coal," *J. Inst. Fuel*, Vol. 7, No. 31, pp. 29-36.

Simmons, H.C., and Harding, C.F., 1980, "Some Effects of Using Water as a Test Fluid in Fuel Nozzle Spray Analysis," ASME Paper 80-GT-90, presented at ASME Gas Turbine Conference, New Orleans.

Swithenbank, J., Beer, J.M., Taylor, D.S., Abbot, D. and McCreath, G.C., 1977, "Experimental Diagnostics in Gas Phase Combustion Engines," *Prog. Aeronaut. Astronaut.*, B.T. Zin (Ed.), AIAA, Vol. 53, p. 421.

Taylor, G.I., 1948, "The Mechanics of Swirl Atomizers," *Seventh International Congress of Applied Mechanics*, Vol. 2, Part 1, pp. 280-285.

Zhao, Y.H., Hou, M.H. and Chin, J.S., 1986, "Dropsize Distributions from Swirl and Airblast Atomizers," *Atomization and Spray Technology*, Vol. 2, No. 2, pp. 3-15.

1989 National Heat Transfer Conference
HTD-Vol. 106, Heat Transfer Phenomena in Radiation, Combustion and Fires

MEASUREMENTS OF SPRAY FLAME PROPERTIES IN A JET-ENGINE COMBUSTOR

A. M. Attya, M. A. Habib and M. R. Taha
Cairo University
Faculty of Engineering, Mechanical Power Department
Cairo, Egypt

ABSTRACT

Detailed measurements of the mean temperature and species concentration of carbon dioxide, carbon monoxide and oxygen have been obtained in an actual jet-engine combustor called straight flow combustion chamber manufactured by Rolls Royce company. This type of experiment provides useful information which are essential to facilitate understanding of the spray combustion in this real combustor where the three dimensional nature of the primary zone causes intense combustion followed by dilution where the flow pattern can be a strong function of geometry. The combustor was equipped with a twin-fluid swirl atomizer fuelled with kerosene fuel. The scheme of measurements has been arranged to study the effect of air-to-fuel ratio and droplet size on the flame behaviour. Four flame configurations are identified and correspond to a small droplet size of 40 μm Sauter mean diameter with high (260 kg/h) and low (165 kg/h) combustion air. The large droplet size of 50 μm was used with the same two flow rates of combustion air. In all experiments the mass flow rate of fuel was kept constant at 2.06×10^{-3} kg/h and the flames were stabilized by a swirler having a swirl number of 1.31.

The results have revealed a very intense combustion region in the recirculation zone and around the centerline. This main flame region has been characterized by a high temperature of around 1600 °C and by 14% CO_2 volume percentage corresponding to small droplet size and high combustion air. As the droplet size increases the main flame region spreads and lowered in temperature down to 1400 °C. The increase in the combustion air is found to reduce the size of the reaction zone and to enhance the rate of oxidation of CO to CO_2 in the case of the low atomizing air.

1. INTRODUCTION

The confined spray flame combustion is relevant to many applications ranging in size from simple domestic flames to jet engines and industrial furnaces. Therefore the investigation of spray flames is important to understand their behaviour and improve their combustion characteristics to save fuel. However, spray combustion involves too many complex phenomena involving simultaneous heat, mass and momentum transfer and chemical reaction beside the complicated geometry required for certain applications. Therefore, the problem has been attacked from different points of view both experimentally and analytically. These researches include Williams [1], Law [2], Chigier [3], Sangiovanni and Dodge [4], Nicholls et al. [5], Attya and Whitelaw [6,7], Elbanhawy and Whitelaw [8], Styles and Chigier [9], Noyce et al. [10], Yule et al. [11] and Mellor et al. [12, 13, 14].

In general the previous work provides useful information which are limited by the geometry of the combustor, type of fuel atomization and the number of the investigated parameters. The available information are considerable but in arrangements which are less suited to calculation methods and are sometimes far from gas-turbine practice. The confined spray flame arrangements of [6] and [7] provide useful information on droplet velocity, temperature and species concentrations for different sets of preheated and unpreheated kerosene spray flames, but the combustor geometries were far from actual gas turbine combustors. The experiments of [3], [9] and [11] have reported measurements in unconfined combusting and non-combusting sprays emerging from burners with pressure, pressure-swirl and twin fluid atomizers. Mellor et al. [12, 13 and 14] have obtained measurements concerning the emissions from and within the Allison J-33 and film-cooled gas turbine combustors. Noyce et al. [10] have presented detailed measurements of gas analysis with a simplified gas turbine combustor to investigate the effects of a number of variables on combustor emissions at idling conditions.

Experiments of the type described above provide information which are essential to the improved understanding of the flows in the combustors but are not completely representative of the flow properties of real combustors where the three-dimensional nature of the primary zone causes intensive combustion which is followed by dilution where the flow pattern of air streams can be a strong function of the geometry of the combustor. To overcome this defficiency the present work has been done in an actual combustor called straight-flow jet engine combustion chamber manufactured by Rolls Royce and fuelled with kerosene fuel through a twin fluid swirl atomizer. The measurements include mean temperature and species concentration of carbon dioxide, carbon monoxide and oxygen aiming at the understanding of the flame structure inside such a complex geometry. The investigation of the effect of air-to-fuel ratio and droplet size on the flame behaviour is also considered. The measurements also provide a complete set of data to help in improving the modeling of gas turbine combustion.

2.FLOW CONFIGURATION AND MEASUREMENT TECHNIQUES

The experiments were carried out in an actual jet engine combustor called straight flow gas turbine combustion chamber manufactured by Rolls Royce company. To facilitate the understanding of the spray combustion inside such a complicated geometry, the dilution air holes along the surface of the combustion chamber were closed. This is justified by the fact that the objectives of the present study are restricted to the main combustion zone downstream of the entrance section of the combustion chamber. Figure 1 gives a detail drawing of the combustor and shows the tapping points prepared for the measurements. The combustion air was supplied to the combustion chamber from a 15 kW centrifugal fan. At the entry of the combustor, the air was divided into two parts; primary stream and secondary stream. The primary air was passed to the combustion domain from a swirler having a swirl number of 1.31. The secondary air was admitted from four rows of holes drilled in the front conical section surrounding the swirler. The total amount of the combustion air was measured by a calibrated orifice plate located downstream of the fan. The ratio between the flow rates of the primary and secondary air streams is dependent on the geometry, size and number of the inlet holes of the secondary air stream and the diameter and angle of the primary air swirler. In order to measure this ratio it was essential to modify the geometry upstream the entrance section, and as a result the ratio between the flow rates of the primary and secondary air streams would be influenced. However the primary air percentage was obtained via a semi-empirical calculation procedure described by Attya, Habib and Taha[15] and found to be 46.6%.

The combustion chamber was surrounded with a cooling air jacket to remove the heat from the combustor wall. The cooling air was supplied from another fan and admitted to the cooling jacket through an air distributer to achieve a uniform distribution of air surrounding the combustor walls.

The combustion chamber was equipped with a twin fluid swirl atomizer illustrated in Fig.2. The

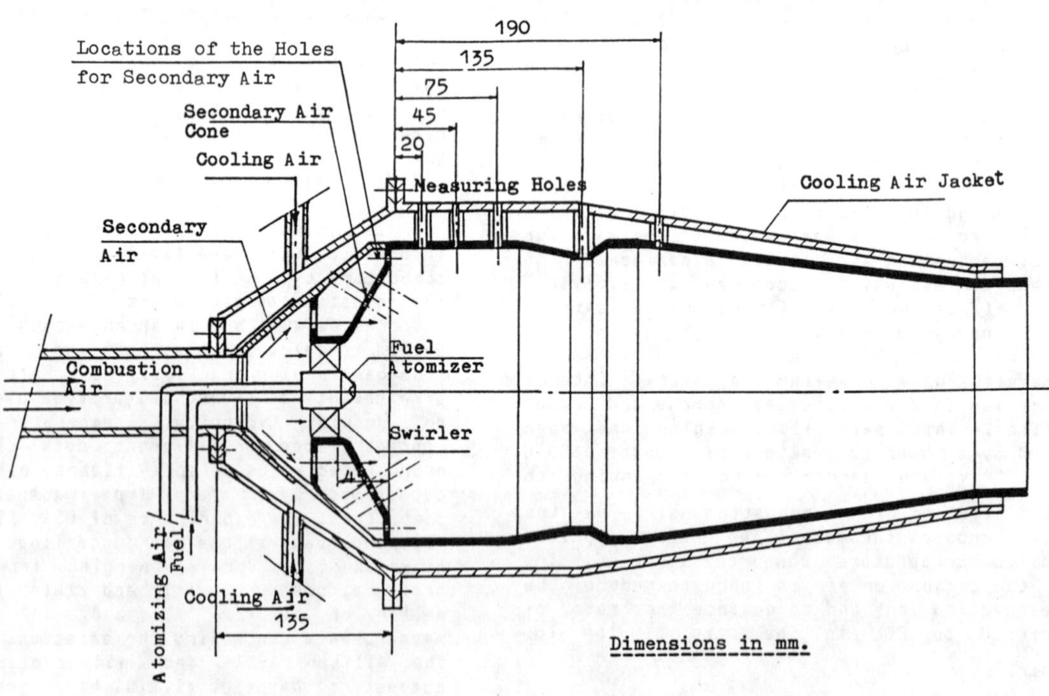

Fig. 1 Geometry of the Gas Turbine Combustor.

atomizer gives a spray with a cone angle of nearly 110 degrees. The fuel used was kerosene which was pumped from a tank by a 1.5 kW, two stage radial pump. The fuel flow rate was measured and cotrolled by a calibrated rotameter. The fuel flow rate was kept constant throughout the experiments at 2.06 kg/h. The changes in the fuel flow rate were kept below 2% via a valve located in the by-pass line of the fuel. The atomizing air was supplied from a centrifugal compressor, controlled by a regulator and measured by a rotameter.The atomizing air had a swirl number of 0.5 in the same direction of the swirl of the combustion air. The estimation of the spray Sauter mean diameter was based on the mass ratio of the atomizing air to fuel as given by Fraster et al.[16]. The complete layout of the test rig is shown in Fig.3.

The operating conditions were arranged to study the effect of spray mean droplet diameter and combustion air flow rate on the spray flame structure. Table 1 summarizes these test conditions. The measurements were taken at five axial stations located at X = 6.5, 9.0, 12.0, 18.0 and 23.5 cm from the swirler exit. The measurements include the mean temperature and species concentrations of CO_2, CO, and O_2. The temperatures were measured by a Platinum-Platinum 13% Rhodium fine wire thermocouple having 80 μm diameter. The thermocouple wires were mounted on support wires of 500 μm diameter and having the same material of the thermocouple wires. These wires were carried in a twin bore alumina insulator of 2.5 mm external diameter. The thermocouple length between the hot junction and the supporting wires ensured small temperature gradient and thus the conduction loss is minimized. The radiation errors associated with the measured temperature were corrected experimentally according to the method described by Attya [17].

Concentration measurements of CO_2, CO and O_2 were made on the dry basis analysis. Gas samples were extracted from the combustion domain by a water cooled stainless steel probe of 6 mm external diameter having a tip of less than 4 mm long similar to that described by many investigators [12, 17 and 18]. The sample was sucked through the probe by a diaphram pump where it was quenched ,as it enters the probe, by the cooling water to stop any further reactions. The quenching rate (dT/dt) was estimated according to [17], [18] and [19] to be $2.5x10^{-3}$ K/s which is high enough to prevent any further chemical reactions. The sample was then passed through a water separator, filter and magnesium perchlorate to remove the condensed vapour

TABLE (I): Operating Conditions of the Experiments

	Run 1	Run 2	Run 3	Run 4
Combustion Air Flow Rate (kg/h)	165.0	260.0	165.0	260.0
Atomizing Air Flow Rate (kg/h)	3.05	3.05	2.75	2.75
Fuel Mass Flow Rate (kg/h)	2.06	2.06	2.06	2.06
Droplet Sauter Mean Diameter (μm)	40	40	50	50

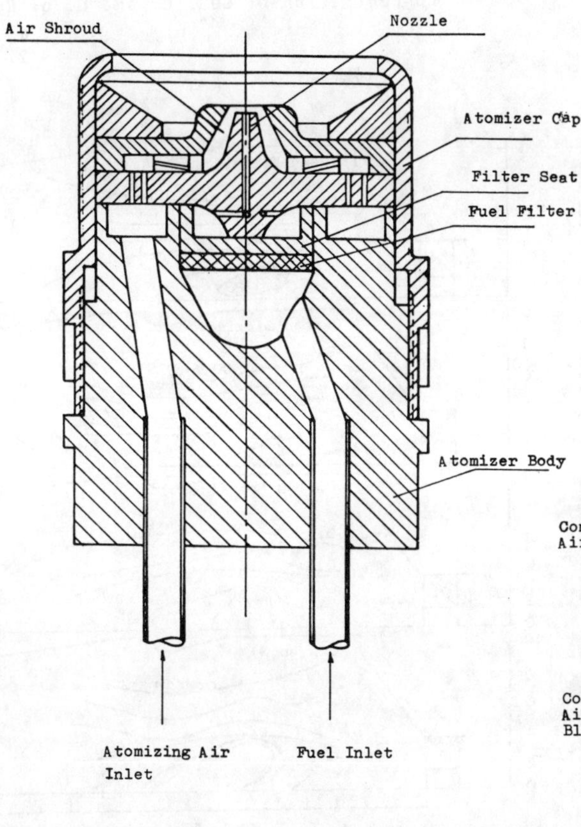

Fig.2 Detailed Drawing of the Fuel Atomizer.

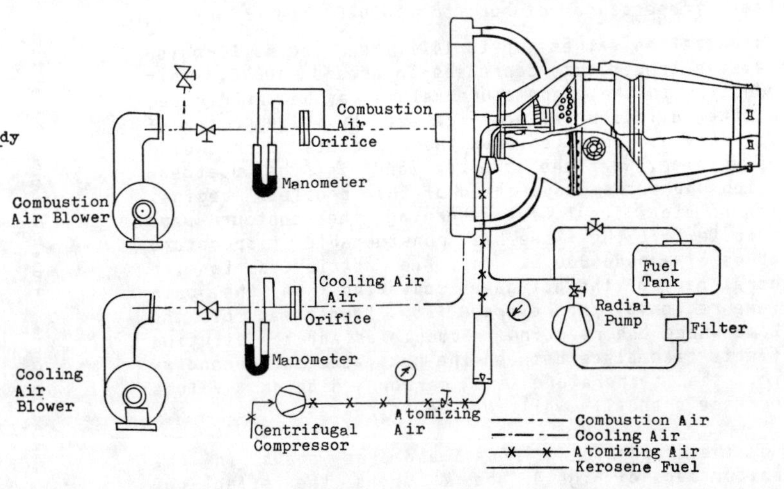

Fig.3 Layout of the Test Rig.

and soots from the sample. The sample flow rate was metered by a rotameter and kept in the range of nearly 60 cm3/ min. which is suitable for the measuring instruments. The carbon dioxide and carbon monoxide were measured using infra-red gas analysers (model Uras 2T manufactured by Hartmann & Braun Co.) while oxygen concentration was measured by a paramagnetic analyser [Servomax O_2 Analyser OA 570]. The analysis of the instrumentation techniques indicated that the uncertainties associated with the concentration measurements are less than ± 6%.

3. RESULTS AND DISCUSSION

The present experimental data include contour maps of the measured mean temperature, carbon dioxide, carbon monoxide and oxygen as presented in Figs 4 to 7 and correspond to the experimental conditions described in Table 1. The present results also include the radial profiles of mean temperature, carbon dioxide, carbon monoxide and oxygen for runs 1, 2, 3 and 4 for comparison purposes as shown in Figs 8 to 11.

The general features of the four flames are similar as shown from the contour maps. The results indicate that the main flame region is located in the core of the combustor around the centerline. This region is characterized by the presence of the recirculation zone which is expected to be in this region around the centerline for this high swirl number of 1.31. In this region it is expected that the incoming air and fuel droplets are heated up by the hot gases in the reverse flow region. Thus the fuel droplets are evaporated and burnt in a short distance from the swirler exit. However, the contours of the temperature and CO_2 concentrations are characterized by peak values off (but not far from) the centerline. This indicates an annular combustion zone and resembles the flame structure obtained by Owen et al.[20]. The occurrence of the annular combustion zone is attributed to the high swirl of the fuel jet caused by the swirling atomizing air.

The main flame region is characterized by high flame temperature of up to 1580°C and high CO_2 concentration values (up to 14%). Near the wall region the mean temperature decreases to around 400°C. This reduction in the temperature values may be attributed to the dilution effects caused by high rate of secondary air injected from the air ports surrounding the swirler. Near the atomizer and in the upstream region where the evaporation of the droplets begins, with no significant fuel burning the contour maps indicate that the flame has considerable temperature values of around 800°C. This is attributed to the entrainment of the hot gases convected from the main flame region by the reversed flow. Downstream of the flame where the reaction is completed and the dilution effects take place between the hot gases and secondary air, the temperature and carbon dioxide values decrease gradually while oxygen increases.

The comparison of runs 1 and 3 as shown in the contour maps of Figs 4 and 6 shows the effect of decreasing the atomizing air and consequently increasing the spray droplet diameter on the flame properties. The reduction of the droplet size while keeping the fuel and combustion air rates unchanged influences the combustion characteristics as shown by

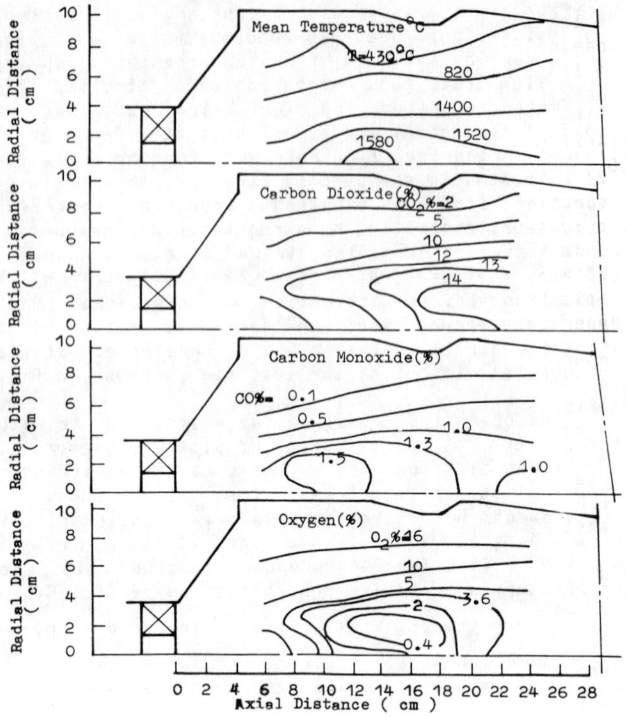

Fig.4 Contours of Mean Temperature and Species Concentration of CO_2, CO and O_2 of Run 1.

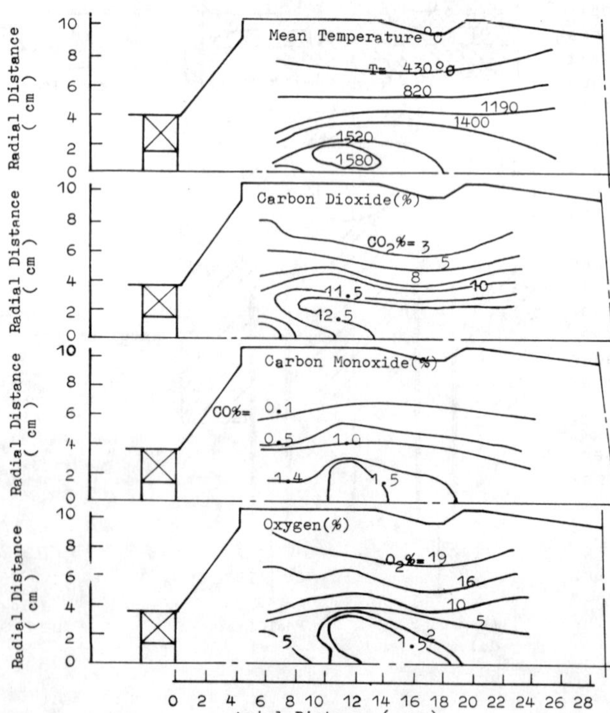

Fig.5 Contours of Mean Temperature and Species Concentration of CO_2, CO and O_2 of Run 2.

the transition of the flow properties in Fig.6 to those of Fig.4 at low combustion air conditions and by the transition of the flow properties from Fig. 7 to those of Fig. 5 at high combustion air conditions. The influence of the reduction in the droplet size is also clarified by the radial distributions of temperature, CO_2, CO and O_2 presented in Figs 8 to 11, respectively. As shown from the contour maps of Figs 4 and 6, the temperature values are increased as the droplet size is decreased generally by about 200°C specially near the centerline. The smaller droplet-size flames of runs 1 and 2 shows higher temperatures and CO_2 values near the centerline as compared to the larger droplet-size flames of runs 3 and 4 as indicated by the radial distributions shown in Figs 8 and 9, respectively at sections 2 to 5. This may be attributed to the shorter time of evaporation and shorter trajectory of small droplets than the large ones. This is explained from the correlation which specifies the decay rate of the droplet diameters given by Agoston et al. [21] as follows:

$$\frac{dD_d}{dt} = \frac{4\lambda_f}{\rho_d C_{p_f} D_d}.(1 + 0.24\, Re_d^{0.25}) \ln[1 + C_{p_f}(T_f - T_s)/L]$$

where ρ_d, D_d and Re_d are the density, diameter and Reynolds number of the droplet, λ_f, C_{pf} and T_f are the thermal conductivity, specific heat and temperature of the gases and T_s and L are the saturation temperature and latent heat of the fuel, respectively

Thus the equation indicates that the small droplets spray of runs 1 and 2 evaporate faster and as a result higher combustion rate is achieved and results in the faster appearence of high values of combustion products This is clear from the comparison between Figs 4 and 6 where the CO_2 values are higher by around 3% . The oxygen values of run 1 are lower than those of run 3, particularly at stations 4 and 5 as shown by the radial profiles of Fig.11 and by the contour maps of Figs 4 and 6. The above observations are also shown from the comparison of the flow properties of Figs 5 and 7 of high and low atomizing air flow rates, respectively. The combustion air flow rates of these two runs is high as compared to the above described two runs (Figs 4 and 6). The comparison of the two figures indicates that the temperature is raised allover by about 270°C, CO_2 is increased by about 2% and O_2 is decreased by 5% in the upper regions as the atomization air flow rate is increased.

The results of the four runs presented in Figs 4 to 7 and also those of the radial profile distributions of Figs 8 to 11 exhibit also the influence of increasing the combustion air flow rate while keeping the atomizing air (and consequently droplet size) and fuel flow rates constant. This influence is shown from comparing Fig. 4 to Fig.5 at high atomizing air flow rate and Fig.6 to Fig.7 at low atomizing air flow rate. The comparison of Figs 4 and 5 indicates clearly that the increase in the combustion air results in transition of the high temperature region towards the swirler exit. As shown from the radial profiles of the temperature in Fig.8

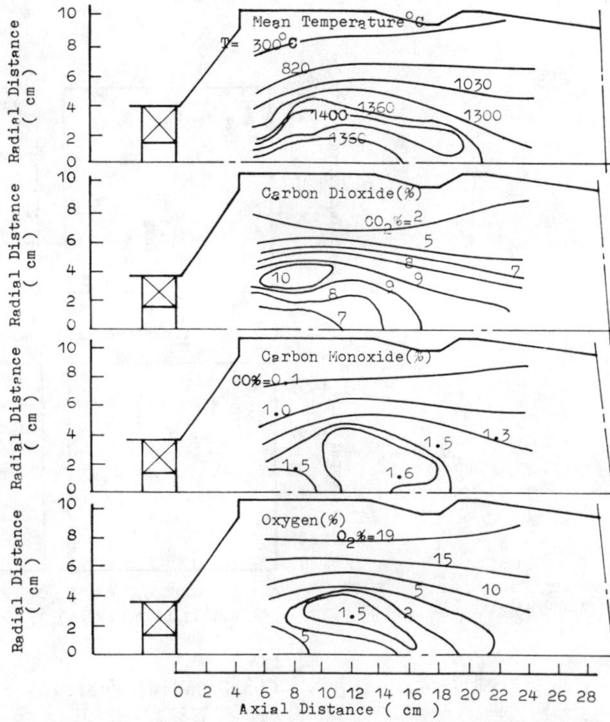

Fig.6 Contours of Mean Temperature and Species Concentration of CO_2, CO and O_2 of Run 3.

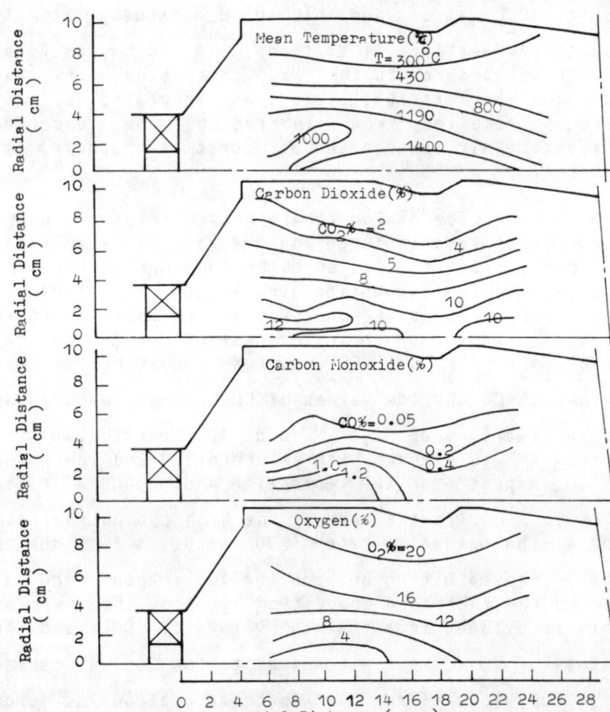

Fig.7 Contours of Mean Temperature and Species Concentration of CO_2, CO and O_2 of Run 4.

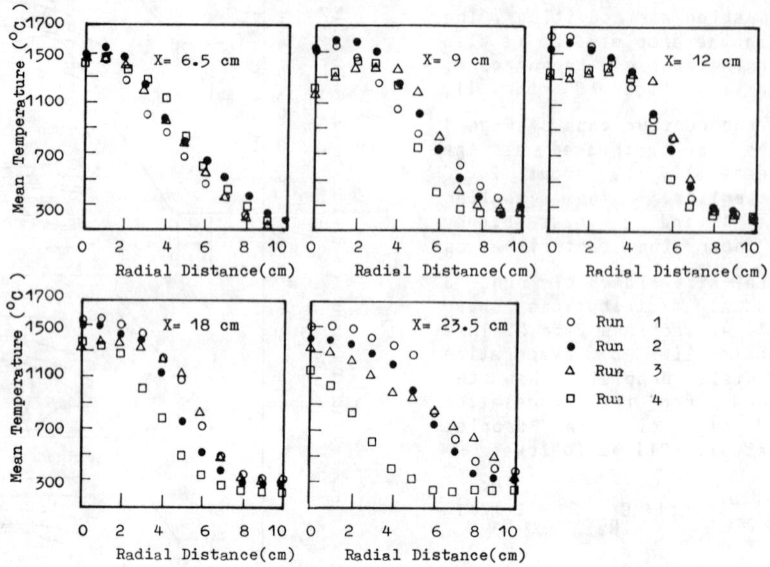

Fig.8 Radial Profiles of Mean Temperature of Runs 1, 2, 3 and 4.

the maximum temperature values appear at station 3 for run1 (low combustion air) while they appear at station 2 for run2 (high combustion air). At the downstream sections the temperature values have decreased by about 150°C corresponding to the increase in the combustion air from 165 to 260 kg/h. It is also shown that the high combustion air case is characterized by lower CO_2 values and higher O_2 values at the downstream sections at stations 3, 4 and 5 , of Figs 9 and 11 as compared to the low combustion air case. This may be attributed to the increased dilution effects resulted from increasing the secondary combustion air through the air ports surrounding the swirler.

In the case of low atomizing air, Figs 6 and 7 the droplet size is large and the rate of evaporation and combustion is expected to be low particularly in the case of low combustion air, Fig.6. As a result of the reduced rate of evaporation and in addition to the lack of combustion air in the case of run 3 the rate of oxidation of CO to CO_2 was low resulting in high values of CO and low values of CO_2. These features are shown clearly from Figs 9 and 10, particularly at stations 2 to 4 where high values of CO and low vaues of CO_2 appear near the centerline and around a radius of 2 to 3 cm. In the case of the high combustion air, run 4, the oxidation rate of CO to CO_2 is fast due to the increased mixing between the fuel vapour and air due to the increased combustion air from the swirler. This is evident from the high values of CO_2 and low values of CO for run 4 compared to low CO_2 values and high CO values of run 3 at upstream sections as shown in Figs 6 and 7. This is in accordance with the increased mean temperatures at the upstream region, especially at X values of less than 12 cm, for run 4 than those of run 3.

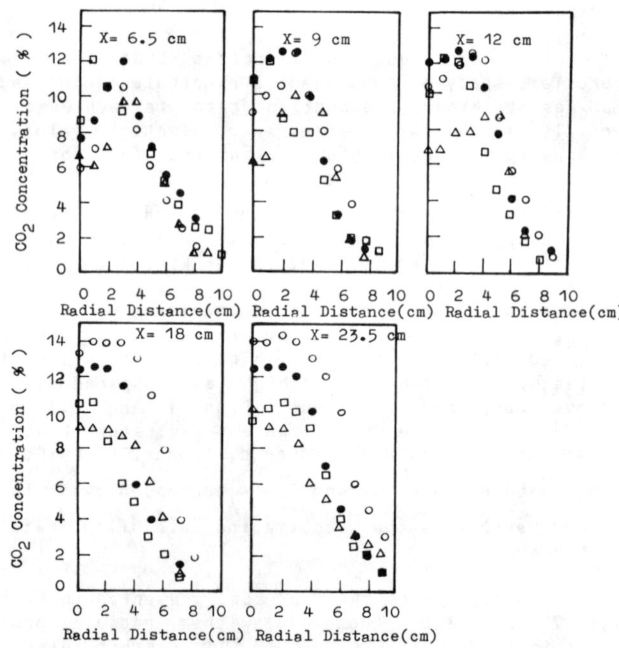

Fig.9 Radial Profiles of Mean Carbon Dioxide Concentration of Runs 1, 2, 3 and 4. (Notation as in Fig.8)

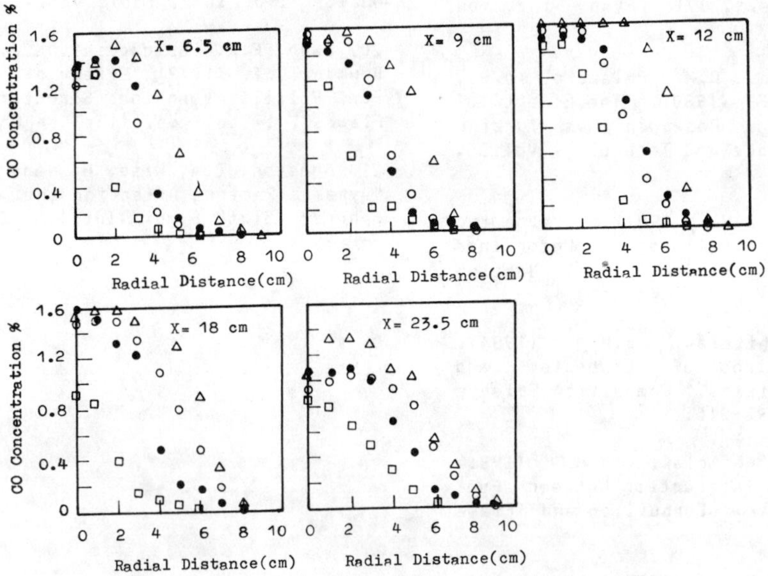

Fig.10 Radial Profiles of Mean Carbon Monoxide Concentration of Runs 1, 2, 3 and 4. (Notation as in Fig.8)

4- CONCLUSIONS

Measurements of the mean temperature and species concentrations of carbon dioxide, carbon monoxide and oxygen were obtained inside a jet-engine combustor. The measurements allowed the investigation of the influence of the combustion air and atomizing air on the combustion characteristics inside the combustor.

The results indicate an annular combustion type in the upstream regions with high temperature and CO_2 values off the centerline. The increase in the atomizing air is shown to enhance the rate of evaporation and to augment the combustion process resulting in high temperature and CO_2 values in the upstream regions. The increase in the combustion air causes the flame to be shorter as the rate of mixing between the fuel vapour and air is improved. In particular, in the case of the low atomizing air, the increase in the combustion air enhances the rate of the oxidation of CO to CO_2.

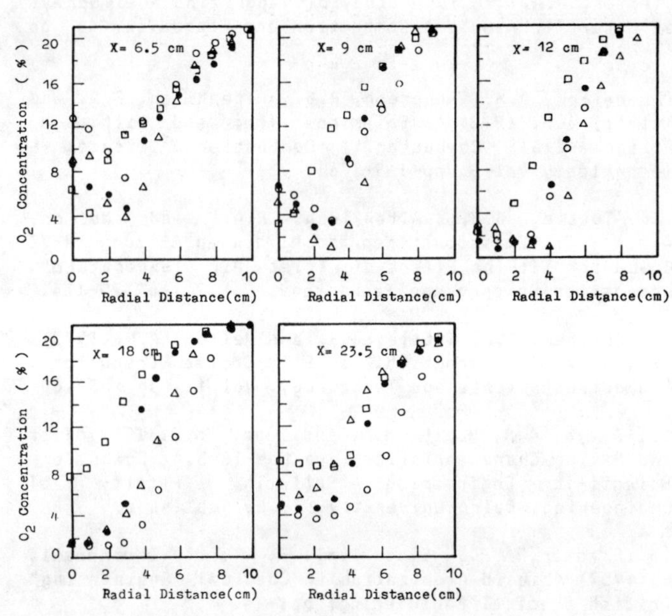

Fig.11 Radial Profiles of Mean Oxygen Concentration of Runs 1, 2, 3 and 4. (Notation as in Fig.8)

REFERENCES

1. Williams, A. (1976), " Fundamentals of OilCombustion.", Progress Energy Combustion Science, Vol. 2, pp 187-179.

2. Law, C.K. (1982), "Recent Advances in Droplet Vaporizationand Combustion.", Progress Energy Combustion Science, Vol. 8, pp 171-201.

3. Chigier, N.A. (1976), "The Atomization and Burning of LiquidFuel Sprays.", Progress Energy Combustion Science, Vol.2, pp 97-114.

4. Sangiovanni, J.J. and Dodge, L.G. (1979), "Observations of Flame Structure in the Combustion of Monodispersed Droplet Streams.", 17th Symp. Int. on Combustion, pp 455-465.

5. Nicholls, J.A., Kauffman, C.W. Pelaccio, D.G., Glass, D.R. and Driscoll, J.F. (1980), "The Effect of Fuel Sprays on Emission from a Research Gas Turbine Combustor.", Combustion Science and Technology Vol.23, pp 203-213.

6. Attya, A.M. and Whitelaw, J.H. (1981), "Velocity, Temperature and Species Concentrations in Unconfined Kerosene Spray Flames." ASME Winter Annual Meeting Nov., Paper No. 81-WA/HT-47.

7. Attya, A.M. and Whitelaw, J.H. (1984), "Measurements and Calculations of Preheated and Unpreheated Confined Spray Flames." Combustion Science and Technology Vol.40, pp 193-215.

8. El-Banhawy, Y.H. and Whitelaw, J.H. (1981) "Experimental Study of the Interaction between Fuel and Surrounding Combustion Air." Combustion and Flame Vol.42, pp 253-275.

9. Styles, A.C. and Chigier, N.A. (1977) "Combustion of Air Blast Atomized Spray Flames." 16th Symp. Int. on Combustion pp 619-630.

10. Noyce, J.R., Sheppard, C.G. and Yamba, F.D. (1981) "Measurements of Mixing and Species Concentrations Within A Gas Turbine Type Combustor." Combustion Science and Technology, Vol.15, pp 209-217.

11. Yule, A., Seng, C., Felton, P., Ungut, A. and Chigier, N.A.(1982)."A Study of Vaporizing Fuel Sprays by Laser Techniques " Combustion and Flame Vol.44, pp 71-84.

12. Mellor, A.M., Anderson, R.D., Altenkirch, R.A. and Tuttle, J.H.(1972)." Emissions from and within an Allison J-33 Combustor." Combustion Science & Technology, Vol.6, pp 169-176.

13. Tuttle, J.H., Altenkirch, R.A. and Mellor, A.M.(1973)."Emissions from and Within an Allison J-33 Combustor II: The Effect of Inlet Air Temperature". Combustion Science and Technology, Vol.7, pp 125-134.

14. Shisher, R.A. Tuttle, J.H. and Mellor, A.M.(1975). "Emissions from and Within a Film-Cooled Combustor". Combustion Science and Technology, Vol.11, pp 153-160

15. Attya, A.M, Habib, M.A. and Taha, M.R.(1989)"Flow and Mixing Characteristics in a Gas-Turbine Combustor" Scientific Engineering Bulletin, Faculty of Engineering, Cairo University, To be Published.

16..Fraster,R P., Eisenklam, P. and Dombrowski, N.(1957) "Liquid Atomization in Chemical Engineering" British Chemical Engineering, pp1-24.

17. Attya, A.M.(1983) "Kerosene Spray Flames". Ph.D. Thesis, University of London.

18. Bilger, R.W.(1977) "Probe Measurements in Turbulent Combustion". AIAA Progress in Astronautics and Aerodynamics". Experimental Diagnotics in Gas Phase Combustion System, Vol.53, pp 333-351.

19. Ramshaw, C. (1968) "Notes on Measurements of Gas Composition and Mixture Ratio in a Rocket Combustion Chamber". J. Inst. Fuel, Vol.41, pp 455-460.

20. Owen, F.K., Spadaccini, L.J., Kennedy. J.B. and Bowman, C.T. (1979) "Effect of Inlet Air Swirl and Fuel Volatility on the Structure of Confined Spray Flames". 17 th Symp. (Int) on Combustion., pp 467-473

21. Agoston, G.A, Wise, H. and Rosser, W.A. (1957) "Dynamic Factors Affecting the Combustion of Liquid Sphers", Sixth Symp. (Int.) on Combustion, pp708-717.